Essentials of Glycobiology

SECOND EDITION

Other Titles from Cold Spring Harbor Laboratory Press

Binding and Kinetics for Molecular Biologists

Epigenetics

Live Cell Imaging: A Laboratory Manual

Proteins and Proteomics: A Laboratory Manual

Essentials of Glycobiology

SECOND EDITION

EDITED BY

Ajit Varki

Richard D. Cummings

Jeffrey D. Esko

Hudson H. Freeze

Pamela Stanley

Carolyn R. Bertozzi

Gerald W. Hart

Marilynn E. Etzler

Full text of this work with associated links is available at http://www.ncbi.nlm.nih.gov/bookshelf/br.fcgi?book=glyco2

CSH PRESS

COLD SPRING HARBOR LABORATORY PRESS

Cold Spring Harbor, New York • www.cshlpress.com

Essentials of Glycobiology, Second Edition

Published by Cold Spring Harbor Laboratory Press, Cold Spring Harbor, New York

Printed in the United States of America

Publisher and Acquisition Editor	John Inglis
Director of Development, Marketing, & Sales	Jan Argentine
Developmental Editor	Kaaren Janssen
Project Manager and Editor	Catriona Simpson
Project Coordinator	Mary Cozza
Permissions Coordinator	Carol Brown
Production Editors	Rena Steuer and Kathleen Bubbeo
Desktop Editor	Lauren Heller
Production Manager	Denise Weiss
Marketing Manager	Ingrid Benirschke
Sales Account Manager	Elizabeth Powers
Cover Designer	Ed Atkeson

Front cover artwork: Influenza A viruses initiate infection by binding to cell-surface sialic acids, via a hemagglutinin molecule (HA). The cover figure background shows a ribbon diagram of the HA structure from the H5N1 influenza strain A/Vietnam/1203/2004 (pdb:2FKO). The top figure in the foreground shows a sialic acid in an α2-3 linkage to a vicinal galactose, modeled into the receptor binding sites of this H5 "bird flu" strain. An equivalent sialoside in an α2-6 linkage is shown for human H3 hemagglutinin (A/Aichi/2/1968, pdb:2HMG) and for human H1 hemagglutinin (A/Puerto Rico/8/1934, pdb:1RVZ). The latter strain is derived from the one that caused the Great Influenza Pandemic of 1918. It can be seen that H1 could potentially also accommodate an α2-3 linked sialic acid, explaining at least in part how it jumped directly from birds into humans. (Images courtesy of James Stevens, Centers for Disease Control and Prevention, Atlanta, Georgia.)

Library of Congress Cataloging-in-Publication Data

Essentials of glycobiology / edited by Ajit Varki ... [et al.]. -- 2nd ed.
p. ; cm.
Includes bibliographical references and index.
ISBN 978-0-87969-770-9 (hardcover : alk. paper)
1. Glycoproteins. 2. Glycosylation. I. Varki, Ajit. II. Title.
[DNLM: 1. Glycoproteins. 2. Glycosylation. 3. Glycosyltransferases. QU 55 E78 2008]

QP552.G59E87 2008
572'.68--dc22

2008007234

10 9 8 7 6 5 4 3 2 1

All Cold Spring Harbor Laboratory Press publications may be ordered directly from Cold Spring Harbor Laboratory Press, 500 Sunnyside Boulevard, Woodbury, New York 11797-2924. Phone: 1-800-843-4388 in Continental U.S. and Canada. All other locations: (516) 422-4100. Fax: (516) 422-4097. E-mail: cshpress@cshl.edu. For a complete catalog of all Cold Spring Harbor Laboratory Press publications, visit our World Wide Web Site http://www.cshl.com/

In memory of

Rosalind H. Kornfeld and Roger W. Jeanloz,

pioneers in the elucidation of glycan structure and function

Contents

STRUCTURE AND BIOSYNTHESIS

ORGANISMAL DIVERSITY

GLYCAN-BINDING PROTEINS

GLYCANS IN PHYSIOLOGY AND DISEASE

METHODS AND APPLICATIONS

Foreword

IT HAS BEEN ALMOST TEN YEARS SINCE THE FIRST EDITION of *Essentials of Glycobiology* was published, and it is not only the science that has changed. The original edition was thought of as a traditional book, to sit on a shelf and be used as a work of reference. However, in late 2002, we at the National Center for Biotechnology Information were approached by the book editors, who expressed an interest in making the book available on the NCBI Bookshelf. This meant that the complete text would be available for searching and browsing via the Internet, therefore extending the use of the book to those who might not have access to the hard copy. The publisher of the book, Cold Spring Harbor Laboratory Press, was on board to try the experiment, and so, in early 2003, *Essentials of Glycobiology* became one of the pioneering textbooks to be distributed electronically.

In the intervening years, hundreds of people every week have continued to look at *Essentials of Glycobiology* online. These readers are geographically dispersed, including from parts of Africa, South America, and Asia. The audience is also educationally dispersed: The book is referenced in Wikipedia, university libraries all over the world list and link to the electronic book, and major Internet search engines index the content. All of which has extended the reach of this enduringly valuable text.

Today's freshmen were only nine years old when that first edition came out. This generation of students has grown up with online information as an integral part of the information resources they use—the first port of call for many homework assignments is now an Internet search rather than a trip to the library. Furthermore, knowledge seekers from all walks of life—for example, high-school students or those seeking health-related information on the web—now have the opportunity to use the book. In recognition of the wider audience and changing ways in which students learn, this new edition of *Essentials of Glycobiology* will be made freely available online simultaneously with the appearance of the printed edition of the book.

This novel approach to textbook publishing is the result of a three-way collaboration between John Inglis and CSHL Press, the NCBI, and the Academic Editors of the book. However, this first is to the enormous credit and vision of Dr. Ajit Varki and the editors and authors of the book, who, having labored long and hard to write the book, now seek to maximize its use by all.

JO MCENTYRE, SECTION CHIEF
DAVID LIPMAN, DIRECTOR
National Center for Biotechnology Information
National Library of Medicine, National Institutes of Health

http://www.ncbi.nlm.nih.gov/bookshelf/br.fcgi?book=glyco2

Foreword from First Edition

THE *ESSENTIALS OF GLYCOBIOLOGY* could not appear at a more opportune time, for the field is in a period of enormous progress and the prospects for future advances are even greater. Glycobiology has its roots in the nineteenth century, when chemists first began to analyze sugars and polysaccharides. Perhaps the first glycoprotein to be studied was the "glycogenous matter" of liver which the famous French physiologist Claude Bernard identified in 1855 as a storage form of glucose. (Interestingly, the evidence that glycogen is a glycoprotein was not obtained until more than 100 years later.) Advances in this area continued at a steady rate during most of this century, but the past 20 years has witnessed an unparalleled explosion of new knowledge that has transformed the field. There are many reasons for this acceleration of progress, including great technical advances in establishing oligosaccharide structures, but by far the most important is the application of recombinant DNA technology to the field. This has resulted in the molecular characterization of the enzymes involved in the assembly, processing, and degradation of oligosaccharides and proteoglycans. It has also allowed the identification of numerous families of plant and animal lectins that recognize carbohydrate structures. The surprising finding to emerge is the vast number of enzymes and proteins that are devoted to glycoprotein and glycolipid synthesis and function. The understanding of the biologic roles of glycans has also increased to a great extent, and we now know that these molecules serve multiple functions, ranging from assisting the folding of nascent proteins to determining the trafficking of lymphocytes and granulocytes in the circulation. The important role of glycans is underscored by the growing list of human diseases that are the result of defects in glycan assembly. The challenge for the future is to further define the biologic functions played by glycans. In this regard, recombinant DNA technology has provided another valuable tool for the glycobiologist: the ability to disrupt genes of interest in mice and other organisms. This presents an unparalleled opportunity for the scientist interested in elucidating the biologic roles of sugars. Without a doubt, the future in this area is the brightest it has ever been.

Essentials of Glycobiology provides an ideal entry into the field. It contains the basic information needed to understand this area along with the most current work at the forefront of the field. The authors are to be commended for assembling a broad, comprehensive, well-organized overview of this burgeoning field. They have also been successful in conveying the excitement in this area of research.

STUART KORNFELD
Washington University School of Medicine
March 1999

Preface

THERE IS A CERTAIN MAGIC IN DOING SOMETHING THAT HAS NEVER BEEN DONE BEFORE. Apart from the excitement and novelty of the endeavor, one knows that even a reasonable effort will be appreciated and that expectations are moderate. So it was with the first edition of this book, which arose from events described in the original preface. In 1999, working with the Cold Spring Harbor Laboratory Press, a group of six researchers put together a reasonable summary of the state of the field of glycobiology, with the assistance of a few outside experts. Given the interests of the writers and the emphasis of the field in the 1990s, the first edition was somewhat vertebrate-centered, focusing mostly on the glycans of so-called "higher" organisms. But glycobiology is relevant to every living species. A remarkable universal finding in nature is that all cells of all species that have ever been studied are coated with a dense and complex array of glycans. Thus the evolution of life has resulted in repeated selection for the expression of glycans at cell surfaces and in extracellular spaces. This broad "dogma" is now expanded by the realization that glycans are also common in intracellular compartments.

In recognition of the wide expression of glycans, the Consortium of Glycobiology Editors decided to expand the book's coverage by increasing the number of editors and by seeking involvement of an even broader range of experts. Of course, because of the obvious relevance to human health, there continues to be an emphasis on vertebrates, as seen in the symbol nomenclature on the inside front cover and in Figure 1.5, which emphasizes the monosaccharides of vertebrate systems. However, as exemplified by Figure 19.1 (inside back cover), we have also addressed the glycans of a variety of other taxa in nature. The addition of more contributors had many positive outcomes, including wider, more accurate, and more up-to-date coverage. However, the added complexity has lengthened the long gestation period for this edition. Nevertheless, the editors welcome the higher level of expectation and increased scrutiny this new edition deserves as an account of an increasingly important aspect of biology.

One truly novel innovation in the development of this edition is the collaboration with the National Library of Medicine/National Center for Biotechnology Information (NLM/NCBI). As indicated by NCBI Director David Lipman in the Foreword, this is a unique and historic effort involving a three-way collaboration among the Editors, the NCBI, and CSHL Press, led by John Inglis. Many of the textbooks on the NCBI Bookshelf were made available some time after they were published in print, as exemplified by the first edition of this book. This is explained by the assumption of editors and publishers that making full text available online might decrease interest in the print copy. However, print and online publication of some books reaches two, only partly overlapping audiences. So it is that this book is being simultaneously released online at NCBI and as a printed volume by the Press. This approach ensures that everyone, from the layperson to the high school student to the graduate student in a developing country, has free access to the knowledge the book contains, even while increasing awareness of the availability of a printed edition that may be more suitable for certain readers' requirements. Added advantages to the NCBI connection will be searchability and the opportunity to add links to the numerous databases hosted at that site. We hope that this functionality will be developed gradually over time, while maintaining the core body of knowledge as a textbook for the advanced undergraduate or the interested graduate student. During the writing of the book it was tempting to insert many links to useful websites, but we quickly learned that, except for those at sites such as NCBI, such links change over time. Given the power of search engines these days, our listings of the full names of relevant sites should allow the reader to find them easily.

Regarding evolution and phylogeny, the classification of species is an ongoing enterprise for which there is still no final answer. Realizing that this nomenclature is in flux, we chose to use the Tree of Life at the University of Arizona, in combination with the phylogeny section of NCBI. To dispel the still prevalent notion that evolution is a purposeful process, we have also avoided the use of the term "design," unless it is in relation to a human effort such as chemical synthesis. The common usage of the terms "higher" and "lower" to indicate species that are closer or more distant from humans has also been limited and, when used, stated in quotes. These incorrect terms arose from the outdated idea of the "great chain of being," in which humans were felt to be at the apex of evolution. More appropriate terms are "recently evolved" and "ancient" species.

A major attempt has also been made to harmonize the artwork style throughout the book and very special thanks are due to Rick Cummings for working with Donna Boyd and the Press to ensure that this happened. Among the other Editors, special thanks are also due to Jeff Esko, who acted as a de facto Deputy Editor at many stages of the process, including a final review of most chapters as well as the production of the Study Guide, which was prepared with the input of trainees who took the Essentials course at UCSD in 2008. The contributions of the many co-authors and consultants were invaluable, and special mention is due to Nathan Sharon, who provided independent reviews of several chapters he had not written. Many individuals at the Press (Kaaren Janssen, Denise Weiss, Mary Cozza, Carol Brown, Rena Steuer, Kathleen Bubbeo, Catriona Simpson, Lauren Heller, and Jan Argentine), and at the NCBI (Jo McEntyre, Laura Dean, Belinda Beck, Marilu Hoeppner, and Jeff Beck) contributed major efforts. We are also deeply grateful to students and postdocs in our labs, who reviewed chapters and provided invaluable comments from the perspective of trainees. One of them, Dave Rabuka, even helped to draw many chemical structures and co-authored two chapters. Many thanks are also due to Melanie Nieze and Jeri Jenkins for coordinating the workflow from my office.

The field of glycobiology has undergone an enormous expansion since the first edition of this book was published, and the Editors were torn between providing information in sufficient depth for the expert or offering just the basics to interested readers from other specialties. We hope both communities of readers will find this a valuable resource for explorations of the fascinating world of glycans and their numerous roles in fields such as chemistry, evolution, biology, medicine, and biotechnology.

AJIT VARKI
La Jolla, California

Preface from First Edition

EMERGING FROM ITS ROOTS in classical carbohydrate chemistry and biochemistry, glycobiology has become a vibrant, expanding, and important extension of modern molecular biology. Over the years, many outstanding monographs and books have documented important advances in this area and summarized critical methods and concepts (see listing following this preface). These volumes continue to serve as excellent resources for those interested in glycobiology. Why then should one publish an additional book on the subject? Most of these prior volumes have been directed at the specialist, assuming a substantial level of technical sophistication and expertise and a working knowledge of the relevant jargon. We present here a book that seeks to fulfill a somewhat different need: to summarize the current state of the art for the expert and yet serve as a resource for the novice wishing to explore the essentials of glycobiology.

This book had its origins from some independent lines of effort. For several years, some of us have been teaching a short elective course in glycobiology for graduate students at the University of California, San Diego. With the recent arrival of additional faculty with expertise in this field, it was decided to present a more comprehensive course on the subject, to be supplemented by a course book that could be then converted into a formal text. Meanwhile, other experts elsewhere in the country had put forward independent proposals to fill the perceived need for a basic textbook in glycobiology. Following a discussion over a beer after a glycobiology conference, we decided to pool all our efforts in this direction.

Since a major goal was to produce a text that would be accessible to students and other trainees, we used the 1998 *UCSD Spring Quarter Graduate Course in Glycobiology* as the basis for creating the text. By recruiting several additional experts as lecturers, we could present a comprehensive course that covered most aspects of the field. Each lecturer was asked to provide handouts for the students that were essentially the first drafts of chapters for the book. In turn, each student was required to provide anonymous critiques of some chapters as a part of the course requirement. This approach ensured not only that the draft chapters were written early on, but also that they underwent in-depth evaluation by bright young minds with an expressed interest in the field. Additional rounds of internal review by the group of six editors served to produce what we hope will be a valuable resource not only for the expert in the field, but also for the novice who wants to learn about glycobiology. We have tried to be as accurate and up to date as possible and to present a balanced point of view on controversial subjects. Given the current breadth of knowledge, it was not possible to do full justice to all aspects of the field, nor to comprehensively reference the extensive literature that exists. The relative emphasis on vertebrate biology bespeaks the greater volume of information currently available in this area of glycobiology.

The Editors are indebted to many others who made this book possible. Besides the students who took the course for credit, several other trainees audited the course and provided very useful feedback. Although the editors wrote the majority of the chapters in the book, the efforts of the other lecturer/authors were crucial in assuring the depth of expertise needed to cover the field effectively. Special thanks are due to John Inglis and Kaaren Janssen at Cold Spring Harbor Laboratory Press for realizing the potential of this book, and for putting up with our many demands and idiosyncrasies. We also thank the Press staff, Jan Argentine, Inez Sialiano, Mary Cozza, Denise Weiss, Dotty Brown, and Danny deBruin, who deserve much credit for keeping us on track and converting our efforts into an attractive product. Last but not least, we acknowledge our families, lab members, and administrative assistants who supported us through all of the hard work needed to create this text. It now remains for the reader to decide if we have achieved our goals in producing this book.

AJIT VARKI
for
The Editors

Contributors

EXECUTIVE EDITOR

Ajit Varki, Distinguished Professor of Medicine and Cellular and Molecular Medicine and Co-Director of the Glycobiology Research and Training Center, University of California, San Diego

EDITORS

Richard D. Cummings, William Patterson Timmie Professor of Biochemistry and Chair of Biochemistry, Emory University School of Medicine, Atlanta, Georgia

Jeffrey D. Esko, Professor of Cellular and Molecular Medicine and Co-Director of the Glycobiology Research and Training Center, University of California, San Diego

Hudson H. Freeze, Professor of Glycobiology, Burnham Institute for Medical Research, La Jolla, California

Pamela Stanley, Horace W. Goldsmith Professor of Cell Biology, Albert Einstein College of Medicine of Yeshiva University, New York

Carolyn R. Bertozzi, T.Z. and Irmgard Chu Distinguished Professor of Chemistry, HHMI Investigator, University of California, Berkeley

Gerald W. Hart, DeLamar Professor and Director of Biological Chemistry, John Hopkins University School of Medicine, Baltimore, Maryland

Marilynn E. Etzler, Professor of Biochemistry, Section of Molecular and Cell Biology, University of California, Davis

CO-AUTHORS

Yoshihiro Akimoto, Kyorin University School of Medicine, Tokyo, Japan

Inka Brockhausen, Queen's University, Ontario, Canada

Karen J. Colley, University of Illinois at Chicago

Paul R. Crocker, University of Dundee, Scotland, United Kingdom

Tamara L. Doering, Washington University School of Medicine, St. Louis, Missouri

Alan D. Elbein, University of Arkansas for Medical Sciences, Little Rock

Michael A.J. Ferguson, University of Dundee, Scotland, United Kingdom

Nathaniel Finney, University of Zurich, Switzerland

Pascal Gagneux, University of California, San Diego

Robert S. Haltiwanger, Stony Brook University, Stony Brook, New York

Vincent Hascall, Cleveland Clinic, Ohio

Bernard Henrissat, AFMB–CNRS–Aix-Marseille Universities, Marseille, France

Reiji Kannagi, Aichi Cancer Center, Nagoya, Japan

Koji Kimata, Aichi Medical University, Aichi, Japan

Taroh Kinoshita, Osaka University, Osaka, Japan

Stuart Kornfeld, Washington University School of Medicine, St. Louis, Missouri

Ulf Lindahl, Uppsala University, Sweden

Robert J. Linhardt, Rensselaer Polytechnic Institute, Troy, New York

Fu-Tong Liu, University of California, Davis

John B. Lowe, Case Western Reserve University, Cleveland, Ohio

Rodger P. McEver, Oklahoma Medical Research Foundation, Oklahoma City

Debra Mohnen, University of Georgia, Athens

Barbara Mulloy, National Institute of Biological Standards and Control, Hertfordshire, United Kingdom

Victor Nizet, University of California, San Diego

Armando J. Parodi, Leloir Institute Foundation, Buenos Aires, Argentina

David Rabuka, University of California, Berkeley

Christian R.H. Raetz, Duke University Medical Center, Durham, North Carolina

James M. Rini, University of Toronto, Ontario, Canada

Ram Sasisekharan, Massachusetts Institute of Technology, Cambridge

Harry Schachter, University of Toronto and the Hospital for Sick Children, Ontario, Canada

Roland Schauer, Christian-Albrechts-University of Kiel, Germany

Ronald L. Schnaar, The Johns Hopkins School of Medicine, Baltimore, Maryland

Peter H. Seeberger, Swiss Federal Institute of Technology (ETH) Zurich, Switzerland

Scott B. Selleck, University of Minnesota, Minneapolis

Nathan Sharon, The Weizmann Institute of Science, Rehovot, Israel

Avadhesha Surolia, National Institute of Immunology, New Delhi, India

Akemi Suzuki, Tokai University, Hiratsuka, Japan

Naoyuki Taniguchi, Osaka University, Osaka, Japan

Michael Tiemeyer, University of Georgia, Athens

Bryan P. Toole, Medical University of South Carolina, Charleston

Salvatore Turco, University of Kentucky College of Medicine, Lexington

Victor D. Vacquier, University of California, San Diego

Christopher M. West, University of Oklahoma Health Sciences Center, Oklahoma City

Acknowledgments

CONSULTANTS

Dean Crick, Colorado State University, Fort Collins

Anthony Day, University of Oxford, United Kingdom

Paul DeAngelis, University of Oklahoma Health Sciences Center, Oklahoma City

John Gallagher, University of Manchester, United Kingdom

James Huntington, University of Cambridge, United Kingdom

Kelley Moremen, University of Georgia, Athens

Elizabeth Neufeld, University of California, Los Angeles

Sara Olson, University of California, San Diego

Paul Weigel, University of Oklahoma Health Sciences Center, Oklahoma City

REVIEWERS

The Editors are deeply appreciative of the comments and suggestions from our trainees, Jennifer Aguilan, Ed Ballister, Lars Bode, Brian Carlson, Miriam Cohen, Phung Gip, Sarah Hubbard, Hung-Hsiang Huang, Christian Kranz, Pascal Lanctot, Roger Lawrence, Nivedita Mitra, Reto Müller, Bobby Ng, Vered Padler-Karavani, Manuela Schuksz, Lance Stein, and Sean Stowell. In addition to those mentioned by name, we would like to thank all the attendees of the 2008 "Essentials of Glycobiology" course at UCSD who offered their invaluable insight and critique.

Books and Monograph Resources

Gottschalk A., ed. 1960. *The chemistry and biology of sialic acids and related substances.* Cambridge University Press, Cambridge.

Stacey M. and Barker S.A. 1960. *Polysaccharides of micro-organisms.* Oxford University Press, London.

Ginsburg V. and Neufeld E., eds. 1966. Complex carbohydrates, part A. *Methods Enzymol.*, vol. 8. Academic Press, San Diego.

Whistler R., ed. 1968–80. *Methods in carbohydrate chemistry*, vols. I–VIII. Academic Press, San Diego.

Hunt S. 1970. *Polysaccharide–protein complexes in invertebrates.* Academic Press, London.

Ginsburg V., ed. 1972. Complex carbohydrates, part B. *Methods in Enzymol*ogy, vol. 28. Academic Press, San Diego.

Gottschalk A., ed. 1972. *Glycoproteins: Their composition, structure and function.* Elsevier, New York.

Rosenberg A. and Schengrund C.-L., eds. 1976. *Biological roles of sialic acids.* Plenum Press, New York.

Ginsburg V., ed. 1978. Complex carbohydrates, part C. *Methods in Enzymol*ogy, vol. 50. Academic Press, San Diego.

Sweeley C.C., ed. 1979. *Cell surface glycolipids.* American Chemical Society, Washington, D.C.

Lennarz W.J., ed. 1980. *The biochemistry of glycoproteins and proteoglycans.* Plenum Press, New York.

Ginsburg V. and Robbins P., eds. 1981. *Biology of carbohydrates*, vol. 1. Wiley, New York.

Ginsburg V., ed. 1982. Complex carbohydrates, part D. *Methods in Enzymology,* vol. 83. Academic Press, San Diego.

Horowitz M. and Pigman W., eds. 1982. *The glycoconjugates.* Academic Press, New York.

Schauer R., ed. 1982. *Sialic acids, chemistry, metabolism, and function.* Springer-Verlag, New York.

Ginsburg V. and Robbins P., eds. 1984. *Biology of carbohydrates*, vol. 2. Wiley, New York.

Ivatt R.J., ed. 1984. *The biology of glycoproteins.* Plenum Press, New York.

Beeley J.G., ed. 1985. *Glycoprotein and proteoglycan techniques.* Elsevier, Amsterdam.

Evered D. and Whelan J., eds. 1986. *Functions of the proteoglycans.* Ciba Foundation Symposium. John Wiley & Sons Ltd, Chichester, United Kingdom

Liener I.E., Sharon N., and Goldstein I.J., eds. 1986. *The lectins: Properties, functions, and applications in biology and medicine.* Academic Press, Orlando, Florida.

Chaplin M.F. and Kennedy J.F., eds. 1987. *Carbohydrate analysis: A practical approach.* IRL Press, Oxford.

Ginsburg V., ed. 1987. Complex carbohydrates, part E. *Methods in Enzymol*ogy, vol. 138. Academic Press, San Diego.

Wight T.N. and Mecham R.P. 1987. *Biology of proteoglycans.* Academic Press, London.

Evered D. and Whelan J. 1988. *The biology of hyaluronan.* Ciba Foundation Symposium. John Wiley & Sons Ltd, Chichester.

Evered D. and Whelan J., eds. 1989. *The biology of hyaluronan.* Ciba Foundation Symposium, vol. 143. Wiley, New York.

Feizi T. 1989. *Carbohydrate recognition in cellular function.* Ciba Foundation Symposium, vol. 145. Wiley, New York.

Ginsburg V., ed. 1989. Complex carbohydrates, part F. *Methods in Enzymology*, vol. 179. Academic Press, San Diego.

Greiling H. and Scott J.E., eds. 1989. *Keratan sulphate: Chemistry, biology, chemical pathology.* The Biochemical Society, London.

Margolis R.U. and Margolis R.K., eds. 1989. *Neurobiology of glycoconjugates.* Plenum Press, New York.
Lane D.G. and Lindahl U., eds. 1990. *Heparin: Chemical and biological properties.* CRC Press, Boca Raton, Florida.
Ginsburg V. and Robbins P., eds. 1991. *Biology of carbohydrates,* vol. 3. Wiley, New York.
Allen H.J. and Kisailus E.C., eds. 1992. *Glycoconjugates: Composition, structure, and function.* Marcel Dekker, Inc., New York.
Fukuda M., ed. 1992. *Cell surface carbohydrates and cell development.* CRC Press, Boca Raton, Florida.
Roth J., Rutishauser U., and Troy F., eds. 1992. *Polysialic acids.* Birkhauser Verlag, Basel, Switzerland.
Fukuda M., ed. 1992. *Glycobiology: A practical approach.* IRL Press, Oxford.
Roberts D.D. and Mecham R.P., eds. 1993. *Cell surface and extracellular glycoconjugates: Structure and function.* Academic Press, San Diego.
Varki A., Guest Ed. 1993. Analysis of glycoconjugates. In *Current Protocols in Molecular Biology* (ed. F. Ausubel et al.), Chapter 17. Green Publishing/Wiley Interscience, New York.
Bock K. and Clausen H., eds. 1994. *Complex carbohydrates in drug research: Structural and functional aspects.* Munksgaard, Copenhagen.
Fukuda M. and Hindsgaul O., eds. 1994. *Molecular glycobiology.* Oxford University Press, New York.
Lee Y.C. and Lee R.T. 1994. Neoglycoconjugates, Part B Biomedical applications. *Methods in Enzymology,* vol. 247. Academic Press, London.
Lennarz W.J. and Hart G.W., eds. 1994. Guide to techniques in glycobiology. *Methods in Enzymology,* vol. 230. Academic Press, San Diego.
Alavi A. and Axford J.S. 1995. *Advances in experimental medicine and biology,* vol. 376, *Glycoimmunology.* Plenum Press, New York.
Montreuil J., Vliegenthart J.F.G., and Schachter H., eds. 1995. *Glycoproteins.* Elsevier, New York.
Rosenberg A., ed. 1995. *Biology of the sialic acids.* Plenum Press, New York.
Verbert A., ed. 1995. *Methods on glycoconjugates: A laboratory manual.* Harwood Academic Publishers, Switzerland.
Montreuil J., Vliegenthart J.F.G., and Schachter H., eds. 1996. *Glycoproteins and disease.* Elsevier, New York.
Gabius H.J. and Gabius S., eds. 1997. *Glycosciences: Status and perspectives.* Chapman and Hall, New York.
Brockhausen I. and Kuhns W. 1997. *Glycoproteins and human disease.* R.G. Landes, Austin, Texas.
Montreuil J., Vliegenthart J.F.G., and Schachter H., eds. 1997. *Glycoproteins II.* Elsevier, New York.
Townsend R.R. and Hotchkiss A.T., eds. 1997. *Techniques in glycobiology.* Marcel Dekker, Inc., New York.
Conrad. H.E., ed. 1998. *Heparin-binding proteins.* Academic Press, San Diego.
Hounsell E.F., ed. 1998. *Methods in molecular biology,* vol. 76, *Glycoanalysis protocols.* Humana Press, Totowa, New Jersy.
Laurent T.C. 1998. *The chemistry, biology and medical applications of hyaluronan and its derivatives.* Portland Press Ltd, London.
Iozzo R., ed. 2000. *Proteoglycans: Structure, biology and molecular interactions.* Marcel Dekker, Inc., New York.
Osborn H. and Khan T. 2000. *Oligosaccharides—Their synthesis and biological roles.* Oxford University Press, United Kingdom.
Crocker P.R., ed. 2001. *Mammalian carbohydrate recognition systems.* Results and Problems in Cell Differentiation 33. Springer-Verlag, Berlin.
Wu A.M. 2001. *The molecular immunology of complex carbohydrates—2.* Kluwer Academic/Plenum Publishers, New York.
Taniguchi N., Honke K., and Fukuda M., eds. 2002. *Handbook of glycosyltransferases and related genes.* Springer-Verlag, Tokyo.
Sharon N. and Lis H., eds. 2003. *Lectins,* 2nd ed. Kluwer Academic Publishers, Dordrecht, The Netherlands
Wong C.H., ed. 2003. *Carbohydrate-based drug discovery.* Wiley-VCH Verlag, Weinheim, Germany.
Wong S.Y.C. and Arsequell G., eds. 2003. *Immunobiology of carbohydrates.* Kluwer Academic/Plenum Publishers, New York.
Dumitriu S. 2005. *Polysaccharides—Structural diversity and functional versatility.* Marcel Dekker, Inc., New York.
Fukuda M., Rutishauser U., and Schnaar R. 2005. *Neuroglycobiology.* Molecular and Cellular Neurobiology Series. Oxford University Press, Oxford.
Yarema K.J., ed. 2005. *Handbook of carbohydrate engineering.* CRC Press, Taylor and Francis Group, Boca Raton, Florida.

Brockhausen I. 2006. *Glycobiology protocols.* Humana Press, Totowa, New Jersey.

Fukuda M., ed. 2006. Glycobiology. *Methods in Enzymology*, vol. 415. Academic Press, San Diego.

Fukuda M., ed. 2006. Glycomics. *Methods in Enzymology*, vol. 416. Academic Press, San Diego.

Fukuda M., ed. 2006. Functional glycomics. *Methods in Enzymology*, vol. 417. Academic Press, San Diego.

Taylor M.E. and Drickamer K. 2003/2006. *Introduction to glycobiology.* Oxford University Press, United Kingdom.

Kamerling J.P., ed. 2007. *Comprehensive glycoscience, from chemistry to systems biology*, vols. 1–4. Elsevier, Oxford.

Lindhorst T.K., ed. 2007. *Essentials of carbohydrate chemistry and biochemistry,* 3rd edition. Wiley-VCH, Weinheim, Germany.

Sansom C. and Markman O. 2007. *Glycobiology.* Scion Publ. Ltd., Bloxham, United Kingdom.

Fraser-Reid B.O., Tatsuta K., and Thiem J., eds. 2008. *Glycoscience—Chemistry and chemical biology I-III,* 2nd edition. Springer-Verlag, Berlin.

Himmel M., ed. 2008. *Biomass recalcitrance: Deconstructing the plant cell wall for bioenergy.* Blackwell Publishing, London.

Abbreviations

ADP	adenosine diphosphate
AIDS	acquired immunodeficiency syndrome
Ara	arabinose
ASGPR	asialoglycoprotein receptor
ATP	adenosine triphosphate
BCR	B-cell receptor
β-HCG	β-human chorionic gonadotrophin
CAZy	carbohydrate-active enzymes database
CDA	congenital dyserythropoietic anemia
CDG	congenital disorder of glycosylation
CDGS	carbohydrate-deficient glycoprotein syndrome
CD-MPR	cation-dependent mannose-6-phosphate receptor
CDP	cytidine diphosphate
CE	capillary electrophoresis
Cer	ceramide
CHO	Chinese hamster ovary
CI-MPR	cation-independent mannose-6-phosphate receptor
CL-43	collectin-43
CMD	congenital muscular dystrophy
CMP	cytidine monophosphate
CMV	cytomegalovirus
CNBr	cyanogen bromide
CNS	central nervous system
CNX	calnexin
CoA	coenzyme A
COG	conserved oligomeric Golgi (complex)
ConA	concanavalin A
CRD	carbohydrate-recognition domain
CRT	calreticulin
CS	chondroitin sulfate
CTD	carboxy-terminal domain
CTLD	C-type lectin domain
Dol	dolichol
DS	dermatan sulfate
DTD	diastrophic dystrophy
ECM	extracellular matrix
EDEM	ER degradation enhancing α-mannosidase-like proteins
EDTA	ethylenediamine tetraacetic acid
EET	enzyme enhancement therapy
EGF	epidermal growth factor
eIF-2	elongation factor 2
Endo H	endo-β-N-acetylglucosaminidase H
EPL	expressed protein ligation
EPO	erythropoietin
EPS	extracellular polysaccharide
ER	endoplasmic reticulum
ERAD	ER-associated degradation
ERGIC	ER-Golgi intermediate compartment
ERT	enzyme replacement therapy
ES	embryonic stem (cells)
ESI-MS	electrospray ionization–mass spectrometry
EtNP	ethanolamine phosphate
FACE	fluorophore-assisted carbohydrate electrophoresis
FBS	fetal bovine serum
FGF	fibroblast growth factor
FGF2	basic fibroblast growth factor
FITC	fluorescein isothiocyanate
Fru	fructose
Fuc	fucose
GAG	glycosaminoglycan
Gal	galactose
GalA	galacturonic acid
GalN	galactosamine
GalNAc	*N*-acetylgalactosamine
GBP	glycan-binding protein
GLC-MS	gas-liquid chromatography/mass spectrometry
GDP	guanosine diphosphate
GH	glycoside hydrolase
GIPL	glycosylinositolphospholipids

Glc	glucose
GlcA	glucuronic acid
GlcN	glucosamine
GlcNAc	*N*-acetylglucosamine
GM-CSF	granulocyte-macrophage–colony-stimulating factor
GMP	guanosine monophosphate
GPI	glycosylphosphatidylinositol
GSL	glycosphingolipid
GT	glycosyltransferase
HA	hyaluronan
HDL	high-density lipoprotein
HexA	hexuronic acid
HexNAc	*N*-acetylhexosamine
HIV	human immunodeficiency virus
HME	hereditary multiple exostosis
HMG	high-mobility-group proteins
HPLC	high-pressure liquid chromatography
HS	heparan sulfate
HSPG	heparan sulfate proteoglycan
Hyl	hydroxylysine
Hyp	hydroxyproline
ICE	interleukin converting enzyme
IdoA	iduronic acid
IFN-γ	interferon γ
IGF	insulin-like growth factor
IgG	immunoglobulin G
IgM	immunoglobulin M
IgSF	immunoglobulin superfamily
IL	interleukin
Kdn	3-deoxy-D-*glycero*-D-*galacto*-2-nonulosonic acid
Kdo	3-deoxy-D-*manno*-octulosonic acid
KS	keratan sulfate
LacdiNAc	GalNAcβ1-4GlcNAc unit
LacNAc	Galβ1-4GlcNAc unit, *N*-acetyllactosamine
LAMPs	lysosome-associated membrane proteins
LCA	*Lens culinaris* agglutinin
LCO	lipochitooligosaccharides
LDL	low-density lipoprotein
Le^a	$Lewis^a$
Le^b	$Lewis^b$
Leg	legionaminic acid
Le^x	$Lewis^x$
LIF	leukemia-inhibitory factor
LLO	lipid-linked oligosaccharide
LOS	lipooligosaccharides
LPA	*Limulus polyphemus* agglutinin
LPG	lipophosphoglycan
LPPG	lipopeptidophosphoglycan
LPS	lipopolysaccharide
MAG	myelin-associated glycoprotein
MALDI-TOF MS	matrix-assisted laser desorption ionization–time of flight mass spectrometry
Man	mannose
ManA	mannuronic acid
ManN	mannosamine
ManNAc	*N*-acetylmannosamine
Man-6-P	mannose-6-phosphate
MBP	mannose-binding protein
MCD	macular corneal dystrophy
MDO	membrane-derived oligosaccharides
MHC	major histocompatibility complex
M6P	mannose-6-phosphate
MPR	mannose-6-phosphate receptor
MPS	mucopolysaccharidosis
MS	mass spectrometry
MSD	multiple sulfatase deficiency
MurNAc	*N*-acetylmuramic acid
MVB	multivesicular body
NAD	nicotinamide adenine dinucleotide, oxidized
NADH	nicotinamide adenine dinucleotide, reduced
NADP	nicotinamide adenine dinucleotide phosphate, oxidized
NADPH	nicotinamide adenine dinucleotide phosphate, reduced
N-CAM	neural cell adhesion molecule
NCL	native chemical legation
Neu	neuraminic acid
Neu5Ac	*N*-acetylneuraminic acid
Neu5Gc	*N*-glycolylneuraminic acid
NF-κB	nuclear factor κB
NGF	nerve growth factor
NK	natural killer (cells)
NMR	nuclear magnetic resonance
OGT	O-GlcNAc transferase
OST	oligosaccharyltransferase
PAMP	pathogen-associated molecular pattern
PAP	3′-phosphoadenosine-5′-phosphate
PAPS	3′-phosphoadenosine-5′-phosphosulfate
PCR	polymerase chain reaction
PDB	Protein Data Bank
PDGF	platelet-derived growth factor
PDMP	D/L-threo-1-phenyl-2-decanoylamino-3-morpholino-1-propanol
PEP	phosphoenolpyruvate
PGM	phosphoglucomutase
PHA	phytohemagglutinin
PI	phosphatidylinositol
PMI	phosphomannose isomerase

PMM	phosphomannomutase
PNA	peanut agglutinin
PNGase	peptide N-glycosidase; N-glycanase
PNH	paroxysmal nocturnal hemoglobinuria
PNP	*para*-nitrophenyl
PNS	peripheral nervous system
PPG	proteophosphoglycan
PRR	pattern-recognition receptor
PSA	polysialic acid
PSGL-1	P-selectin glycoprotein ligand-1
RCA	*Ricinus communis* agglutinin (I or II)
Rha	rhamnose
Rib	ribose
RNAi	RNA interference
RP-HPLC	reverse-phase high-pressure liquid chromatography
SAM	S-adenosylmethionine
SAP	sphingolipid activator protein
SDS-PAGE	sodium dodecyl sulfate-polyacrylamide gel electrophoresis
Sia	sialic acid
Siglec	sialic acid–recognizing immunoglobulin superfamily lectin
siRNA	small interfering RNA
SLe^a	sialyl Lewis a
SLe^x	sialyl Lewis x
SMP	Schwann cell myelin protein
Sn	sialoadhesin
SNA	*Sambucus nigra* agglutinin
SNP	single-nucleotide polymorphism
SPPS	solid-phase peptide synthesis
SSEA	stage-specific embryonic antigen
ST	sialyltransferase
STn	sialyl-Tn antigen (Neu5Acα2-6GalNAc-Ser/Thr)
SV40	simian virus 40
TCR	T-cell receptor
TF	Thomsen-Friedenreich (antigen) (Galβ1-3GalNAc-Ser/Thr)
TGF-β	transforming growth factor β
TGN	*trans*-Golgi network
TH1	T-helper-1 (cells)
TLC	thin-layer chromatography
TLR	Toll-like receptor
Tn	Tn antigen (GalNAc-Ser/Thr)
TNF-α	tumor necrosis factor α
t-PA	tissue plasminogen activator
TSR	thrombospondin repeats
UDP	uridine diphosphate
UGGT	UDP-Glc:glycoprotein glucosyltransferase
UMP	uridine monophosphate
UPR	unfolded protein response
VNTR	variable number of tandem repeat
VSG	variant surface glycoprotein
VSV	vesicular stomatitis virus
WGA	wheat germ agglutinin
Xyl	xylose

CHAPTER 1

Historical Background and Overview

Ajit Varki and Nathan Sharon

THIS CHAPTER PROVIDES HISTORICAL BACKGROUND TO THE FIELD of glycobiology and an overview of this book. General terms found throughout the volume are also considered. The common monosaccharide units of glycoconjugates are mentioned and a uniform symbol nomenclature used for structural depictions throughout the book is presented. The major glycan classes to be discussed in the book are described, and an overview of the general pathways for their biosynthesis is provided. Topological issues relevant to biosynthesis and functions of glycoconjugates are also briefly considered, and the growing role of these molecules in medicine and biotechnology is briefly surveyed.

WHAT IS GLYCOBIOLOGY?

The central paradigm driving research in molecular biology has been that biological information flows from DNA to RNA to protein. The power of this concept lies in its template-based precision, the ability to manipulate one class of molecules based on knowledge of another, and the patterns of sequence homology and relatedness that predict function and reveal evolutionary relationships. A variety of additional roles for RNA have also recently emerged. With the sequencing of the human genome and those of many other commonly studied organisms, even more spectacular gains in understanding the biology of nucleic acids and proteins are anticipated.

However, the tendency is to assume that a conventional molecular biology approach encompassing just these molecules will explain the makeup of cells, tissues, organs, physiological systems, and intact organisms. In fact, making a cell requires two other major classes of molecules: lipids and carbohydrates. These molecules can serve as intermediates in generating energy and as signaling effectors, recognition markers, and structural components. Taken together with the fact that they encompass some of the major posttranslational modifications of proteins themselves, lipids and carbohydrates help to explain how the relatively small number of genes in the typical genome can generate the enormous biological complexities inherent in the development, growth, and functioning of intact organisms.

The biological roles of carbohydrates are particularly important in the assembly of complex multicellular organs and organisms, which requires interactions between cells and the surrounding matrix. All cells and numerous macromolecules in nature carry an array of covalently attached sugars (monosaccharides) or sugar chains (oligosaccharides), which are generically referred to as "glycans" in this book. Sometimes, these glycans can also be freestanding entities. Because many glycans are on the outer surface of cellular and secreted macromolecules, they are in a position to modulate or mediate a wide variety of events in cell–cell, cell–matrix, and cell–molecule interactions critical to the development and function of a complex multicellular organism. They can also act as mediators in the interactions between different organisms (e.g., between host and a parasite or a symbiont). In addition, simple, rapidly turning over, protein-bound glycans are abundant within the nucleus and cytoplasm, where they can serve as regulatory switches. A more complete paradigm of molecular biology must therefore include glycans, often in covalent combination with other macromolecules, that is, glycoconjugates, such as glycoproteins and glycolipids.

The chemistry and metabolism of carbohydrates were prominent matters of interest in the first part of the 20th century. Although these topics engendered much scientific attention, carbohydrates were primarily considered as a source of energy or as structural materials and were believed to lack other biological activities. Furthermore, during the initial phase of the molecular biology revolution of the 1960s and 1970s, studies of glycans lagged far behind those of other major classes of molecules. This was in large part due to their inherent structural complexity, the great difficulty in determining their sequences, and the fact that their biosynthesis could not be directly predicted from a DNA template. The development of many new technologies for exploring the structures and functions of glycans has since opened a new frontier of molecular biology that has been called "glycobiology"—a word first coined in the late 1980s to recognize the coming together of the traditional disciplines of carbohydrate chemistry and biochemistry with a modern understanding of the cell and molecular biology of glycans and, in particular, their conjugates with proteins and lipids.

From a strictly technical point of view, glycobiology can almost be viewed as an anomaly in the history of the biological sciences. If the development of methodologies for glycan analysis had kept pace with those of other macromolecules in the 1960s and 1970s, glycans would have been an integral part of the initial phase of the molecular and cell biology revolution, and there might have been no need to single them out later for study as a distinct discipline. Regardless, the term glycobiology has gained wide acceptance, with a number of textbooks, a major biomedical journal, a growing scientific society, and many research conferences now bearing this name. Defined in the broadest sense, glycobiology is the study of the structure, biosynthesis, biology, and evolution of saccharides (sugar chains or glycans) that are widely distributed in nature, and the proteins that recognize them. (The *Oxford English Dictionary* definition is "the branch of science concerned with the role of sugars in biological systems.") Glycobiology is one of the more rapidly growing fields in the natural sciences, with broad relevance to many areas of basic research, biomedicine, and biotechnology. The field includes the chemistry of carbohydrates, the enzymology of glycan formation and degradation, the recognition of glycans by specific proteins (lectins and glycosaminoglycan-binding proteins), glycan roles in complex biological systems, and their analysis or manipulation by a variety of techniques. Research in glycobiology thus requires a foundation not only in the nomenclature, biosynthesis, structure, chemical synthesis, and functions of glycans, but also in the general disciplines of molecular genetics, protein chemistry, cell biology, developmental biology, physiology, and medicine. This volume provides an overview of the field, with some emphasis on the glycans of animal systems. It is assumed that the reader has a basic background in advanced undergraduate-level chemistry, biochemistry, and cell biology.

HISTORICAL ORIGINS OF GLYCOBIOLOGY

As mentioned above, glycobiology had its early origins in the fields of carbohydrate chemistry and biochemistry, and has only recently emerged as a major aspect of molecular and cellular biology and physiology. Some of the major investigators and important discoveries that influenced the development of the field are presented in Figure 1.1 and Table 1.1. As with any such attempt, a comprehensive list is impossible, but details regarding some of these discoveries can be found in other chapters of this volume. A summary of the general principles gained from this research is presented in Table 1.2.

MONOSACCHARIDES ARE THE BASIC STRUCTURAL UNITS OF GLYCANS

Carbohydrates are defined as polyhydroxyaldehydes, polyhydroxyketones and their simple derivatives, or larger compounds that can be hydrolyzed into such units. A monosaccharide is a carbohydrate that cannot be hydrolyzed into a simpler form. It has a potential carbonyl group at the end of the carbon chain (an aldehyde group) or at an inner carbon (a ketone group). These two types of monosaccharides are therefore named aldoses and ketoses, respectively (for examples, see below and for more details, see Chapter 2). Free monosaccharides can exist in open-chain or ring forms (Figure 1.2). Ring forms of the monosaccharides are the rule in oligosaccharides, which are linear or branched chains of monosaccharides attached to one another via glycosidic linkages (the term "polysaccharide" is typically reserved for large glycans composed of repeating oligosaccharide motifs). The generic term "glycan" is used throughout this book to refer to any form of mono-, oligo-, or polysaccharide, either free or covalently attached to another molecule.

The ring form of a monosaccharide generates a chiral anomeric center at C-1 for aldo sugars or at C-2 for keto sugars (for details, see Chapter 2). A glycosidic linkage involves the attachment of a monosaccharide to another residue, typically via the hydroxyl group of this anomeric center, generating α linkages or β linkages that are defined based on the relationship of the glycosidic oxygen to the anomeric carbon and ring (see Chapter 2). It is important to realize that these two linkage types confer very different structural properties and bio-

FIGURE 1.1. Nobel laureates in fields related to the early history of glycobiology. Listed are the Laureates and their original Nobel citations: Hermann Emil Fischer (Chemistry, 1902), *"in recognition of the extraordinary services he has rendered by his work on sugar and purine syntheses"*; Karl Landsteiner (Physiology or Medicine, 1930), *"for his discovery of human blood groups"*; Walter Norman Haworth (Chemistry, 1937), *"for his investigations on carbohydrates and vitamin C"*; Carl and Gerty Cori (Physiology or Medicine, 1947), *"for their discovery of the course of the catalytic conversion of glycogen"*; Luis F. Leloir (Chemistry, 1970), *"for his discovery of sugar nucleotides and their role in the biosynthesis of carbohydrates"*; and George E. Palade (Physiology or Medicine, 1974) *"for discoveries concerning the structural and functional organization of the cell."* (Reprinted, with permission, ©The Nobel Foundation.)

logical functions upon sequences that are otherwise identical in composition, as classically illustrated by the marked differences between starch and cellulose (both homopolymers of glucose), the former largely α1-4 linked and the latter β1-4 linked throughout. A glycoconjugate is a compound in which one or more monosaccharide or oligosaccharide units (the glycone) are covalently linked to a noncarbohydrate moiety (the aglycone). An oligosaccharide that is not attached to an aglycone possesses the reducing power of the aldehyde or ketone in its terminal monosaccharide component. This end of a sugar chain is therefore often called the reducing terminus or reducing end, terms that tend to be used even when the sugar chain is attached to an aglycone and has thus lost its reducing power. Correspondingly, the outer end of the chain tends to be called the nonreducing end (note the analogy to the 5′ and 3′ ends of nucleotide chains or the amino and carboxyl termini of polypeptides).

GLYCANS CAN CONSTITUTE A MAJOR PORTION OF THE MASS OF A GLYCOCONJUGATE

In naturally occurring glycoconjugates, the portion of the molecule comprising the glycans can vary greatly in its contribution to the overall size, from being very minor in amount to being the dominant component or even almost the exclusive one. In many cases, the glycans comprise a substantial portion of the mass of glycoconjugates (for a typical example, see Figure 1.3). For this reason, the surfaces of all types of cells in nature (which are heavily decorated with different kinds of glycoconjugates) are effectively covered with a dense array of sugars, giving rise to the so-called glycocalyx. This cell-surface structure was first observed many years ago by electron microscopists as an anionic layer external to the cell surface membrane, which could be decorated with polycationic reagents such as cationized ferritin (for a historical example, see Figure 1.4).

TABLE 1.1. *Important discoveries in the history of glycobiology*

Year(s)	Primary scientist(s)	Discoveries	Relevant chapters[a]
1876	J.L.W. Thudichum	glycosphingolipids (cerebrosides), sphingomyelin, and sphingosine	10
1888	H. Stillmark	lectins as hemagglutinins	26, 28
1891	H.E. Fischer	stereoisomeric structure of glucose and other monosaccharides	2
1900	K. Landsteiner	human ABO blood groups as transfusion barriers	5, 13
1909	P.A. Levene	structure of ribose in RNA	1
1916	J. MacLean	isolation of heparin as an anticoagulant	16
1925	P.A. Levene	characterization of chondroitin sulfate and "mucoitin sulfate" (later, hyaluronan)	15, 16
1929	P.A. Levene	structure of 2-deoxyribose in DNA	1
1929	W.N. Haworth	pyranose and furanose ring structures of monosaccharides	2
1934	K. Meyer	hyaluronan and hyaluronidase	15
1934–1938	G. Blix, E. Klenk	sialic acids	14
1936	C.F. Cori, G.T. Cori	glucose-1-phosphate as an intermediate in glycogen biosynthesis	17
1942–1946	G.K. Hirst, F.M. Burnet	hemagglutination of influenza virus and "receptor-destroying enzyme"	14
1942	E. Klenk, G. Blix	gangliosides in brain	10, 14
1946	Z. Dische	colorimetric determination of deoxypentoses and other carbohydrates	2
1948–1950	E. Jorpes, S. Gardell	occurrence of N-sulfates in heparin and identification of heparan sulfate	16
1949	L.F. Leloir	nucleotide sugars and their role in the biosynthesis of glycans	4
1950	Karl Schmid	isolation of α1-acid glycoprotein (orosomucoid), a major serum glycoprotein	
1952	W.T. Morgan, W.M. Watkins	carbohydrate determinants of ABO blood group types	13
1952	E.A. Kabat	relationship of ABO to Lewis blood groups and secretor vs. nonsecretor status	13
1952	A. Gottschalk	sialic acid as the receptor for influenza virus	14
1952	T. Yamakawa	globoside, the major glycosphingolipid of the erythrocyte membrane	10
1956–1963	M.R.J. Salton, J.M. Ghuysen, R.W. Jeanloz, N. Sharon, H.M. Flowers	bacterial peptidoglycan backbone structure major structural polysaccharides in nature (chitin, cellulose, and peptidoglycan) are β1-4-linked throughout	20
1957	P.W. Robbins, F. Lipmann	biosynthesis and characterization of PAPS, the donor for glycan sulfation	4, 16
1957	H. Faillard, E. Klenk	crystallization of *N*-acetylneuraminic acid as product of influenza virus receptor-destroying enzyme (RDE) ("neuraminidase")	14
1957–1963	J. Strominger, J.T. Park, H.R. Perkins, H.J. Rogers	mechanism of peptidoglycan biosynthesis and site of penicillin action	20
1958	H. Muir	"mucopolysaccharides" are covalently attached to proteins via serine	16
1960	D.C. Comb, S. Roseman	structure and enzymatic synthesis of CMP-*N*-acetylneuraminic acid	4, 14
1960–1965	O. Westphal, O. Lüderitz, H. Nikaido, P.W. Robbins	structure of lipopolysaccharides and endotoxin glycans	20
1960–1970	R. Jeanloz, K. Meyer, A. Dorfman	structural studies of glycosaminoglycans	15, 16
1961	S. Roseman, L. Warren	biosynthesis of sialic acid	4, 14
1961–1965	G.E. Palade	ER-Golgi pathway for glycoprotein biosynthesis and secretion	3
1962	A. Neuberger, R. Marshall, I. Yamashina, L.W. Cunningham	GlcNAc-Asn as the first defined carbohydrate-peptide linkage	8
1962	W.M. Watkins, W.Z. Hassid	enzymatic synthesis of lactose from UDP-galactose and glucose	4
1962	J.A. Cifonelli, J. Ludowieg, A. Dorfman	iduronic acid as a constituent of heparin	16

(Continued)

TABLE 1.1. *(Continued)*

Year(s)	Primary scientist(s)	Discoveries	Relevant chapters[a]
1962–1966	L. Roden, U. Lindahl	identification of tetrasaccharide linking glycosaminoglycans to protein core of proteoglycans	16
1962	E.H. Eylar, R.W. Jeanloz	demonstration of the presence of *N*-acetyllactosamine in α1-acid glycoprotein	13
1963	L. Svennerholm	analysis and nomenclature of gangliosides	10
1963	D. Hamerman, J. Sandson	covalent cross-linkage between hyaluronan and inter-α-trypsin inhibitor	15
1963–1964	B. Anderson, K. Meyer, V.P. Bhavanandan, A. Gottschalk	β-elimination of Ser/Thr-O-linked glycans	9
1963–1965	R. Kuhn, H. Wiegandt	structure of GM1 and other brain gangliosides	10
1963–1967	B.L. Horecker, P.W. Robbins, H. Nakaido, M.J. Osborn	lipid-linked intermediates in bacterial lipopolysaccharide and peptidoglycan biosynthesis	20
1964	V. Ginsburg	GDP-fucose and its biosynthesis from GDP-mannose	4
1964	B. Gesner, V. Ginsburg	glycans control the migration of leukocytes to target organs	26
1965	L.W. Cunningham	microheterogeneity of glycoprotein glycans	2, 8, 9
1965–1966	R.O. Brady	glucocerebrosidase is the enzyme deficient in Gaucher's disease	41
1965–1975	J.E. Silbert, U. Lindahl	cell-free biosynthesis of heparin and chondroitin sulfate	16
1965–1975	W. Pigman	tandem repeat amino acid sequences with Ser or Thr as O-glycosylation sites in mucins	9
1966	M. Neutra, C. Leblond	role of Golgi apparatus in protein glycosylation	3
1966–1969	B. Lindberg, S. Hakomori	refinement of methylation analysis for determination of glycan linkages	47
1966–1976	R. Schauer	multiple modifications of sialic acids in nature, their biosynthesis, and degradation	14
1967	L. Rodén, L.-Å. Fransson	demonstration of a copolymeric structure for dermatan sulfate	16
1967	R.D. Marshall	N-glycosylation occurs only at asparagine residues in the sequence motif Asn-X-Ser/Thr	8
1968	J.A. Cifonelli	description of the domain structure of heparan sulfate	16
1968	R.L. Hill, K. Brew	α-lactalbumin as a modifier of galactosyltransferase specificity	5
1969	L. Warren, M.C. Glick, P.W. Robbins	increased size of N-glycans in malignantly transformed cells	8, 44
1969	R.J. Winzler	structures of O-glycans from erythrocyte membranes	9
1969–1974	V.C. Hascall, S.W. Sajdera, H. Muir, D. Heinegård, T. Hardingham	hyaluronan-proteoglycan interactions in cartilage	16, 17
1969	H. Tuppy, P. Meindl	synthesis of 2-deoxy-2,3-didehydro-Neu5Ac as viral sialidase inhibitor	14
1968–1970	E. Neufeld	identification of lysosomal enzyme deficiencies in the mucopolysaccharidoses	41
1969	G. Ashwell, A. Morell	glycans can control the lifetime of glycoproteins in blood circulation	26
1970	K.O. Lloyd, J. Porath, I.J. Goldstein	use of lectins for affinity purification of glycoproteins	45
1971–1973	L.F. Leloir	dolichylphosphosugars are intermediates in protein N-glycosylation	4, 8
1971–1975	P. Kraemer, J.E. Silbert	heparan sulfate as a common constituent of vertebrate cell surfaces	16
1971–1980	B. Toole	hyaluronan in differentiation, morphogenesis, and development	15
1972–1982	S. Hakomori	lacto- and globo-series glycosphingolipids as developmentally regulated and tumor-associated antigens	10, 44
1972	J.F.G. Vliegenthart	high-field proton NMR spectroscopy for structural analysis of glycans	2

TABLE 1.1. *(Continued)*

Year(s)	Primary scientist(s)	Discoveries	Relevant chapters[a]
1973	W.E. van Heyningen	glycosphingolipids are receptors for bacterial toxins	39
1973	J. Montreuil, R.G. Spiro, R. Kornfeld	a common pentasaccharide core structure of all N-glycans	8
1974	C.E. Ballou	structure of yeast mannans and generation of yeast mannan mutants	8, 46
1975	V.I. Teichberg	the first galectin	33
1975	V.T. Marchesi	primary structure of glycophorin, the first known transmembrane glycoprotein	3, 8, 9
1975–1980	A. Kobata	N- and O-glycan structural elucidation using multiple convergent techniques	2, 8, 9
1975–1980	P. Stanley, S. Kornfeld, R.C. Hughes	lectin-resistant cell lines with glycosylation defects	46
1977	W.J. Lennarz	Asn-X-Ser/Thr necessary and sufficient for lipid-mediated N-glycosylation	8
1977	I. Ofek, D. Mirelman, N. Sharon	cell-surface glycans as attachment sites for infectious bacteria	39
1977–1978	S. Kornfeld, P.W. Robbins	biosynthesis and processing of intermediates of N-glycans in protein glycosylation	8
1977	R.L. Hill, R. Barker	first purification of a glycosyltransferase involved in protein glycosylation	5, 8
1978	C. Svanborg	glycosphingolipids as receptors for bacterial adhesion	10, 39
1979–1982	E. Neufeld, S. Kornfeld, K. Von Figura, W. Sly	the mannose-6-phosphate pathway for lysosomal enzyme trafficking	30
1980–1983	F.A. Troy, J. Finne, S. Inoue, Y. Inoue	structure of polysialic acids in bacteria and vertebrates	14
1980	H. Schachter	role of glycosyltransferases in N- and O-glycan branching	5, 8
1980–1982	V.N. Reinhold, A. Dell, A.L. Burlingame	mass spectrometry for structural analysis of glycans	47, 48
1980–1985	S. Hakomori, Y. Nagai	glycosphingolipids as modulators of transmembrane signaling	10
1981–1985	M.J. Ferguson, I. Silman, M. Low	structural definition of glycosylphosphatidylinositol (GPI) anchors	11
1982	U. Lindahl, R.D. Rosenberg	specific sulfated heparin pentasaccharide sequence recognized by antithrombin	16, 35
1982	C. Hirschberg, R. Fleischer	transport of sugar nucleotides into the Golgi apparatus	3, 14
1984	G. Hart	intracellular protein glycosylation by O-GlcNAc	18
1984	J. Jaeken	description of "carbohydrate-deficient glycoprotein syndromes"	42
1985	M. Klagsbrun, D. Gospodarowicz	discovery of heparin–FGF interactions	35
1986	W.J. Whelan	glycogen is a glycoprotein synthesized on a glycogenin primer	17
1986	J.U. Baenziger	structures of sulfated N-glycans of pituitary hormones	13, 28
1986	Y. Inoue, S. Inoue	discovery of 2-keto-3-deoxynononic acid (Kdn) in rainbow trout eggs	14
1986	P.K. Qasba, J. Shaper, N. Shaper	cloning of first animal glycosyltransferase	5
1987	Y-C. Lee	high-performance anion-exchange chromatography of oligosaccharides with pulsed amperometric detection (HPAEC-PAD)	47

This time line of events is deliberately terminated about 20 years ago, on the assumption that it can take a long time to be certain that a particular discovery has had a major impact on the field. Historical details about several of these discoveries can be found in the "Classics" series of the *Journal of Biological Chemistry* (search by author name).

[a] Indicates main chapter(s) of this book in which the relevant topics are covered.

TABLE 1.2. *"Universal" principles of glycobiology*

Occurrence

All cells in nature are covered with a dense and complex array of sugar chains (glycans).

The cell walls of bacteria and Archaea are composed of several classes of glycans and glycoconjugates.

Most secreted proteins of eukaryotes carry large amounts of covalently attached glycans.

In eukaryotes, these cell-surface and secreted glycans are mostly assembled via the ER-Golgi pathway.

The extracellular matrix of eukaryotes is also rich in such secreted glycans.

Cytoplasmic and nuclear glycans are common in eukaryotes.

For topological, evolutionary, and biophysical reasons, there is little similarity between cell-surface/secreted and nuclear/cytoplasmic glycans.

Chemistry and structure

Glycosidic linkages can be in α- or β-linkage forms, which are biologically recognized as completely distinct.

Glycan chains can be linear or branched.

Glycans can be modified by a variety of different substituents, such as acetylation and sulfation.

Complete sequencing of glycans is feasible but usually requires combinatorial or iterative methods.

Modern methods allow in vitro chemoenzymatic synthesis of both simple and complex glycans.

Biosynthesis

The final products of the genome are posttranslationally modified proteins, with glycosylation being the most common and versatile of these modifications.

The primary units of glycans (monosaccharides) can be synthesized within a cell or salvaged from the environment.

Monosaccharides are activated into nucleotide sugars or lipid-linked sugars before they are used as donors for glycan synthesis.

Whereas lipid-linked sugar donors can be flipped across membranes, nucleotide sugars must be transported into the lumen of the ER-Golgi pathway.

Each linkage unit of a glycan or glycoconjugate is assembled by one or more unique glycosyltransferases.

Many glycosyltransferases are members of multigene families with related functions.

Most glycosyltransferases recognize only the underlying glycan of their acceptor, but some are protein or lipid specific.

Many biosynthetic enzymes (glycosyltransferases, glycosidases, sulfotransferases, etc.) are expressed in a tissue-specific, temporally regulated manner.

Diversity

Monosaccharides generate much greater combinatorial diversity than nucleotides or amino acids.

Further diversity arises from covalent modifications of glycans.

Glycosylation introduces a marked diversity in proteins.

Only a limited subset of the potential diversity is found in a given organism or cell type.

There is intrinsic diversity (microheterogeniety) of glycoprotein glycans within a cell type or even a single glycosylation site.

The total expressed glycan repertoire (glycome) of a given cell type or organism is thus much more complex than the genome or proteome.

The glycome of a given cell type or organism is also dynamic, changing in response to intrinsic and extrinsic signals.

Glycome differences in cell type, space, and time generate biological diversity and can help to explain why only a limited number of genes are expressed from the typical genome.

TABLE 1.2. *(Continued)*

Recognition

Glycans are recognized by specific glycan-binding proteins that are intrinsic to an organism.

Glycans are also recognized by many extrinsic glycan-binding proteins of pathogens and symbionts.

Glycan-binding proteins fall in two general categories: those that can usually be grouped by shared evolutionary origins and/or similarity in structural folds (lectins) and those that emerged by convergent evolution from different ancestors (e.g., GAG-binding proteins).

Lectins often show a high degree of specificity for binding to specific glycan structures, but they typically have relatively low affinities for single-site binding.

Thus, biologically relevant lectin recognition often requires multivalency of both the glycan and glycan-binding protein, to generate high avidity of binding.

Genetics

Naturally occurring genetic defects in glycans seem to be relatively rare in intact organisms. However, this apparent rarity may be due to a failure of detection, caused by unpredictable or pleiotropic phenotypes.

Genetic defects in cell-surface/secreted glycans are easily obtained in cultured cells but have somewhat limited biological consequences.

The same mutations typically have major phenotypic consequences in intact multicellular organisms.

Thus, many of the major roles of glycans likely involve cell–cell or extracellular interactions.

Nuclear/cytoplasmic glycans may have more cell-intrinsic roles, e.g., in signaling.

Complete elimination of major glycan classes generally causes early developmental lethality.

Organisms bearing tissue-specific alteration of glycans often survive, but they exhibit both cell-autonomous and distal biological effects.

Biological roles

Biological roles for glycans span the spectrum from nonessential activities to those that are crucial for the development, function, and survival of an organism.

Many theories regarding the biological roles of glycans appear to be correct, but exceptions occur.

Glycans can have different roles in different tissues or at different times in development.

Terminal sequences, unusual glycans, and modifications are more likely to mediate specific biological roles.

However, terminal sequences, unusual glycans, or modifications may also reflect evolutionary interactions with microorganisms and other noxious agents.

Thus, a priori prediction of the functions of a specific glycan or its relative importance to the organism is difficult.

Evolution

Relatively little is known about glycan evolution.

Interspecies and intraspecies variations in glycan structure are relatively common, suggesting rapid evolution.

The dominant mechanism for such evolution is likely the ongoing selection pressure by pathogens that recognize glycans.

However, glycan evolution must also preserve and/or elaborate critical intrinsic functions.

Interplay between pathogen selection pressure and preservation of intrinsic roles could result in the formation of "junk" glycans.

Such "junk" glycans could be the substrate from which new intrinsic functions arise during evolution.

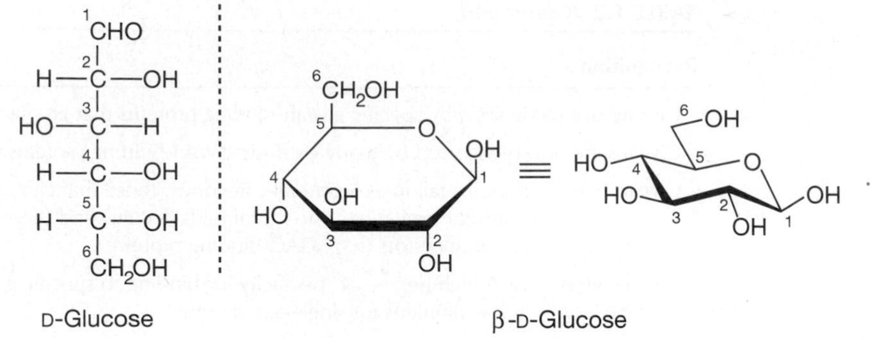

FIGURE 1.2. Open-chain and ring forms of glucose. Changes in the orientation of hydroxyl groups around specific carbon atoms generate new molecules that have a distinct biology and biochemistry (e.g., galactose is the C-4 epimer of glucose).

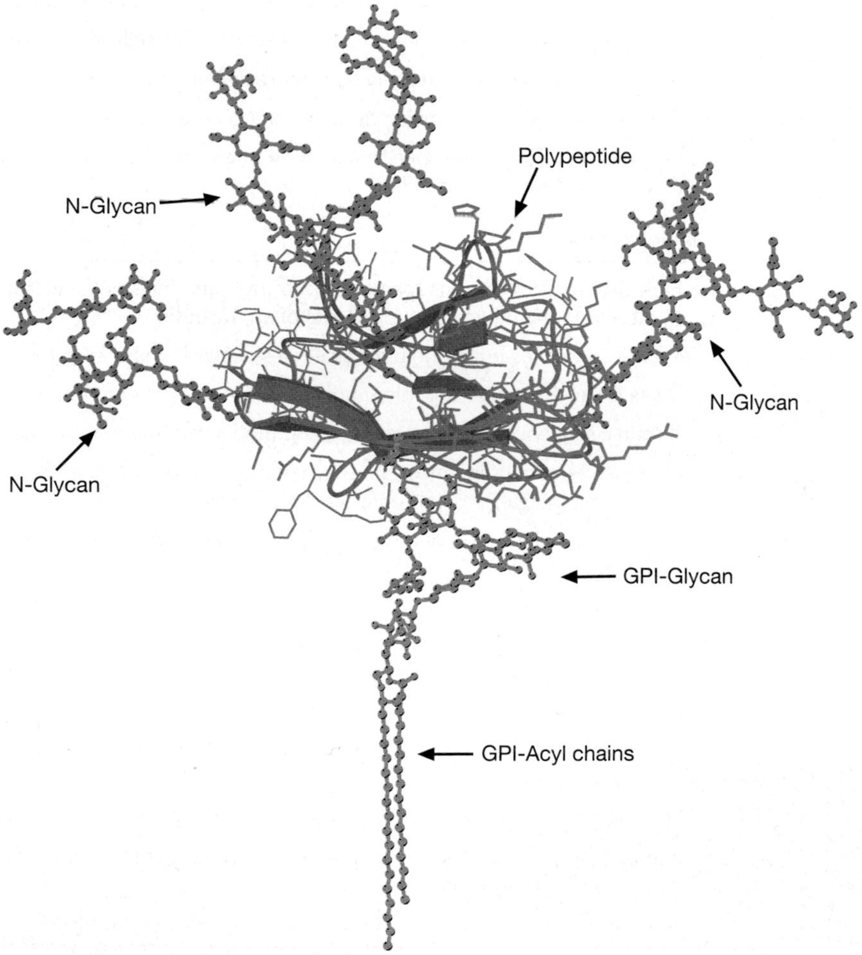

FIGURE 1.3. Schematic representation of the Thy-1 glycoprotein including the three N-glycans (*blue*) and a glycosylphosphatidylinositol (GPI-glycan, *green*) lipid anchor whose acyl chains (*yellow*) would normally be embedded in the membrane bilayer. Note that the polypeptide (*purple*) represents only a relatively small portion of the total mass of the protein. (Original art courtesy of Mark Wormald and Raymond Dwek, Oxford Glycobiology Institute.)

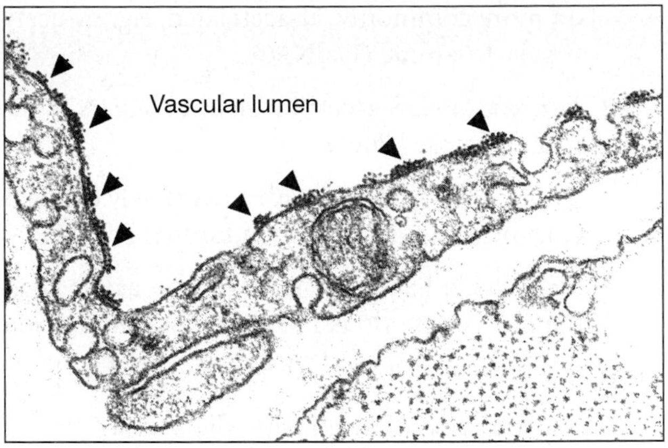

FIGURE 1.4. Historical electron micrograph of endothelial cells from a blood capillary in the diaphragm muscle of a rat, showing the lumenal cell membrane of the cells (facing the blood) decorated with particles of cationized ferritin (*arrowheads*). These particles are binding to acidic residues (sialic-acid-containing glycans and sulfated glycosaminoglycans) contained in the cell-surface glycocalyx. Note that the particles are several layers deep, indicating the remarkable thickness of this layer of glycoconjugates. (Courtesy of George E. Palade.)

MONOSACCHARIDES CAN BE LINKED TOGETHER IN MANY MORE WAYS THAN AMINO ACIDS OR NUCLEOTIDES

Nucleotides and proteins are linear polymers that can each contain only one basic type of linkage between monomers. In contrast, each monosaccharide can theoretically generate either an α or a β linkage to any one of several positions on another monosaccharide in a chain or to another type of molecule. Thus, it has been pointed out that although three different nucleotides or amino acids can only generate six trimers, three different hexoses could produce (depending on which of their forms are considered) anywhere from 1,056 to 27,648 unique trisaccharides. This difference in complexity becomes even greater as the number of monosaccharide units in the glycan increases. For example, a hexasaccharide with six different hexoses could have more than 1 trillion possible combinations. Thus, an almost unimaginable number of possible saccharide units could be theoretically present in biological systems. Fortunately for the student of glycobiology, naturally occurring biological macromolecules are so far known to contain relatively few of the possible monosaccharide units, in a limited number of combinations. However, the great majority of glycans in nature have yet to be discovered and structurally defined.

COMMON MONOSACCHARIDE UNITS OF GLYCOCONJUGATES

Although several hundred distinct monosaccharides are known to occur in nature, only a small number of these are commonly found in animal glycans. They are listed below, along with their standard abbreviations (for details regarding their structures, see Chapter 2).

- *Pentoses*: Five-carbon neutral sugars (e.g., D-xylose [Xyl]).
- *Hexoses*: Six-carbon neutral sugars (e.g., D-glucose [Glc], D-galactose [Gal], and D-mannose [Man]).
- *Hexosamines*: Hexoses with an amino group at the 2-position, which can be either free

or, more commonly, N-acetylated, e.g., *N*-acetyl-D-glucosamine (GlcNAc) and *N*-acetyl-D-galactosamine (GalNAc).

- *Deoxyhexoses*: Six-carbon neutral sugars without the hydroxyl group at the 6-position (e.g., L-fucose [Fuc]).
- *Uronic acids*: Hexoses with a negatively charged carboxylate at the 6-position (e.g., D-glucuronic acid [GlcA] and L-iduronic acid [IdoA]).
- *Sialic acids*: Family of nine-carbon acidic sugars (generic abbreviation is Sia), of which the most common is *N*-acetylneuraminic acid (Neu5Ac, also sometimes called NeuAc or historically, NANA) (for details, see Chapter 14).

For the sake of simplicity, the symbols D- and L- are omitted from the full names of monosaccharides from here on, and only the symbol L- will be used when appropriate (e.g., L-fucose or L-iduronic acid).

This limited set of monosaccharides dominates the glycobiology of more recently evolved (so-called "higher") animals, but several others have been found in "lower" animals (e.g., tyvelose; see Chapters 23 and 24), bacteria (e.g., ketodeoxyoctulosonic acid, rhamnose, L-arabinose, and muramic acid; see Chapter 20), and plants (e.g., arabinose, apiose, and galacturonic acid; see Chapter 22). A variety of modifications of glycans enhance their diversity in nature and often serve to mediate specific biological functions. Thus, the hydroxyl groups of different monosaccharides can be subject to phosphorylation, sulfation, methylation, O-acetylation, or fatty acylation. Although amino groups are commonly N-acetylated, they can be N-sulfated or remain unsubstituted. Carboxyl groups are occasionally subject to lactonization to nearby hydroxyl groups or even lactamization to nearby amino groups.

Details regarding the structural depiction of monosaccharides, linkages, and oligosaccharides are discussed in Chapter 2. Many figures in this volume use a simplified style of depiction of sugar chains (see Figure 1.5). This figure (also reproduced on the inside front cover) uses a monosaccharide symbol set modified from the first edition of *Essentials of*

FIGURE 1.5. Recommended symbols and conventions for drawing glycan structures. (*Top panel*) The monosaccharide symbol set from the first edition of *Essentials of Glycobiology* is modified to avoid using the same shape or color, but with different orientation to represent different sugars. Each monosaccharide class (e.g., hexose) now has the same shape, and isomers are differentiated by color. The same shading/color is used for different monosaccharides of the same stereochemical designation (e.g., Gal, GalNAc, and GalA). To minimize variations, sialic acids and uronic acids are in the same shape, and only the major uronic and sialic acid types are represented. When the type of sialic acid is uncertain, the abbreviation Sia can be used instead. Only common monosaccharides in vertebrate systems are assigned specific symbols. All other monosaccharides are represented by an open hexagon or defined in the figure legend. If there is more than one type of undesignated monosaccharide in a figure, a letter designation can be included to differentiate between them. Unless otherwise indicated, all of these vertebrate monosaccharides are assumed to be in the D configuration (except for fucose and iduronic acid, which are in the L configuration), all glycosidically linked monosaccharides are assumed to be in the pyranose form, and all glycosidic linkages are assumed to originate from the 1-position (except for the sialic acids, which are linked from the 2-position). Anomeric notation and destination linkages can be indicated without spacing/dashes. Although color is useful, these representations will survive black-and-white printing. Modifications of monosaccharides are indicated by lowercase letters, with numbers indicating linkage positions, if known (e.g., 9Ac for the 9-O-acetyl group, 3S for the 3-O-sulfate group, 6P for a 6-O-phosphate group, 8Me for the 8-O-methyl group, 9Acy for the 9-O-acyl group, and 9Lt for the 9-O-lactyl group). Esters and ethers are shown attached to the symbol with a number. For N-substituted groups, it is assumed that only one amino group is on the monosaccharide with an already known position (e.g., NS for an N-sulfate group on glucosamine, assumed to be at the 2-position). (*Middle panel*) Typical branched "biantennary" N-glycan with two types of outer termini, depicted at different levels of structural details. (*Bottom panel*) Some typical glycosaminoglycan (GAG) chains.

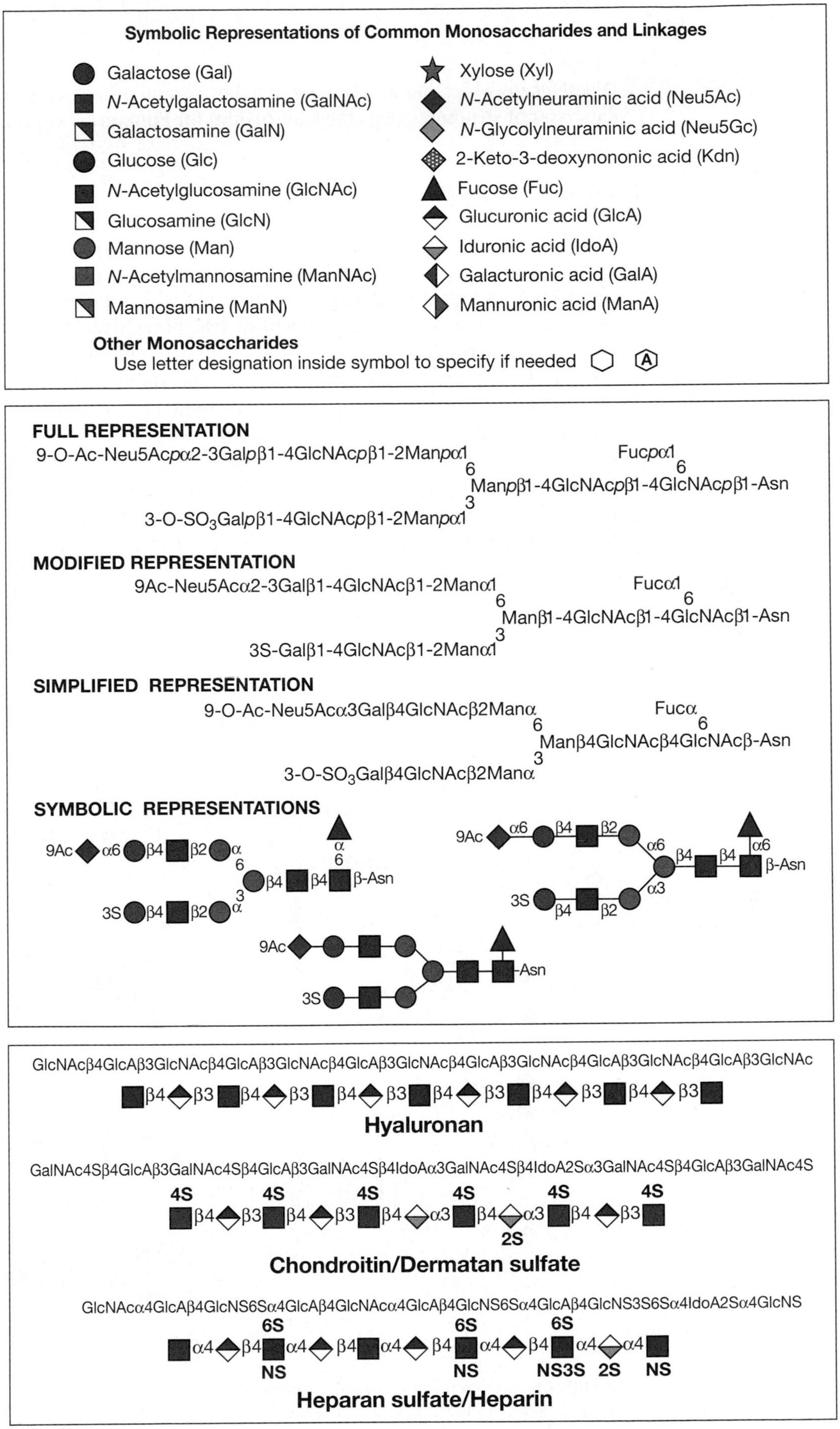
Symbolic Representations of Common Monosaccharides and Linkages
Galactose (Gal)
N-Acetylgalactosamine (GalNAc)
Galactosamine (GalN)
Glucose (Glc)
N-Acetylglucosamine (GlcNAc)
Glucosamine (GlcN)
Mannose (Man)
N-Acetylmannosamine (ManNAc)
Mannosamine (ManN)
Xylose (Xyl)
N-Acetylneuraminic acid (Neu5Ac)
N-Glycolylneuraminic acid (Neu5Gc)
2-Keto-3-deoxynononic acid (Kdn)
Fucose (Fuc)
Glucuronic acid (GlcA)
Iduronic acid (IdoA)
Galacturonic acid (GalA)
Mannuronic acid (ManA)
Other Monosaccharides
Use letter designation inside symbol to specify if needed
A
FULL REPRESENTATION
9-O-Ac-Neu5Acpα2-3Galpβ1-4GlcNAcpβ1-2Manpα1
Fucpα1
6
Manpβ1-4GlcNAcpβ1-4GlcNAcpβ1-Asn
3
3-O-SO3Galpβ1-4GlcNAcpβ1-2Manpα1
MODIFIED REPRESENTATION
9Ac-Neu5Acα2-3Galβ1-4GlcNAcβ1-2Manα1
Fucα1
6
Manβ1-4GlcNAcβ1-4GlcNAcβ1-Asn
3
3S-Galβ1-4GlcNAcβ1-2Manα1
SIMPLIFIED REPRESENTATION
9-O-Ac-Neu5Acα3Galβ4GlcNAcβ2Manα
Fucα
6
Manβ4GlcNAcβ4GlcNAcβ-Asn
3
3-O-SO3Galβ4GlcNAcβ2Manα
SYMBOLIC REPRESENTATIONS
9Ac α6 β4 β2 α 6 β4 β4 β-Asn
3S β4 β2 α 3
α 6
9Ac α6 β4 β2 α6 β4 β4 α6 β-Asn
3S β4 β2 α3
9Ac
3S
-Asn
GlcNAcβ4GlcAβ3GlcNAcβ4GlcAβ3GlcNAcβ4GlcAβ3GlcNAcβ4GlcAβ3GlcNAcβ4GlcAβ3GlcNAcβ4GlcAβ3GlcNAc
β4 β3 β4 β3 β4 β3 β4 β3 β4 β3 β4 β3
Hyaluronan
GalNAc4Sβ4GlcAβ3GalNAc4Sβ4GlcAβ3GalNAc4Sβ4IdoAα3GalNAc4Sβ4IdoA2Sα3GalNAc4Sβ4GlcAβ3GalNAc4S
4S 4S 4S 4S 4S 4S
β4 β3 β4 β3 β4 α3 β4 α3 β4 β3
2S
Chondroitin/Dermatan sulfate
GlcNAcα4GlcAβ4GlcNS6Sα4GlcAβ4GlcNAcα4GlcAβ4GlcNS6Sα4GlcAβ4GlcNS3S6Sα4IdoA2Sα4GlcNS
6S 6S 6S
α4 β4 α4 β4 α4 β4 α4 β4 α4 α4
NS NS NS3S 2S NS
Heparan sulfate/Heparin

FIGURE 1.5. *(See facing page for legend.)*

Glycobiology, which has also been adopted by several other groups interested in presenting databases of structures (e.g., the Consortium for Functional Glycomics).

MAJOR CLASSES OF GLYCOCONJUGATES AND GLYCANS

The common classes of glycans found in or on eukaryotic cells are primarily defined according to the nature of the linkage to the aglycone (protein or lipid) (see Figures 1.6 and 1.7). A glycoprotein is a glycoconjugate in which a protein carries one or more glycans covalently attached to a polypeptide backbone, usually via N or O linkages. An N-glycan (N-linked oligosaccharide, N-(Asn)-linked oligosaccharide) is a sugar chain covalently linked to an asparagine residue of a polypeptide chain, commonly involving a GlcNAc residue and the consensus peptide sequence: Asn-X-Ser/Thr. N-Glycans share a common pentasaccharide core region and can be generally divided into three main classes: oligomannose (or high-mannose) type, complex type, and hybrid type (see Chapter 8). An O-glycan (O-linked oligosaccharide) is frequently linked to the polypeptide via *N*-acetylgalactosamine (GalNAc) to a hydroxyl group of a serine or threonine residue and can be extended into a variety of different structural core classes (see Chapter 9). A mucin is a large glycoprotein that carries many O-glycans that are clustered (closely spaced). Several other types of O-glycans also exist (e.g., attached to proteins via O-linked mannose). A proteoglycan is a glycoconjugate that has one or more glycosamino-

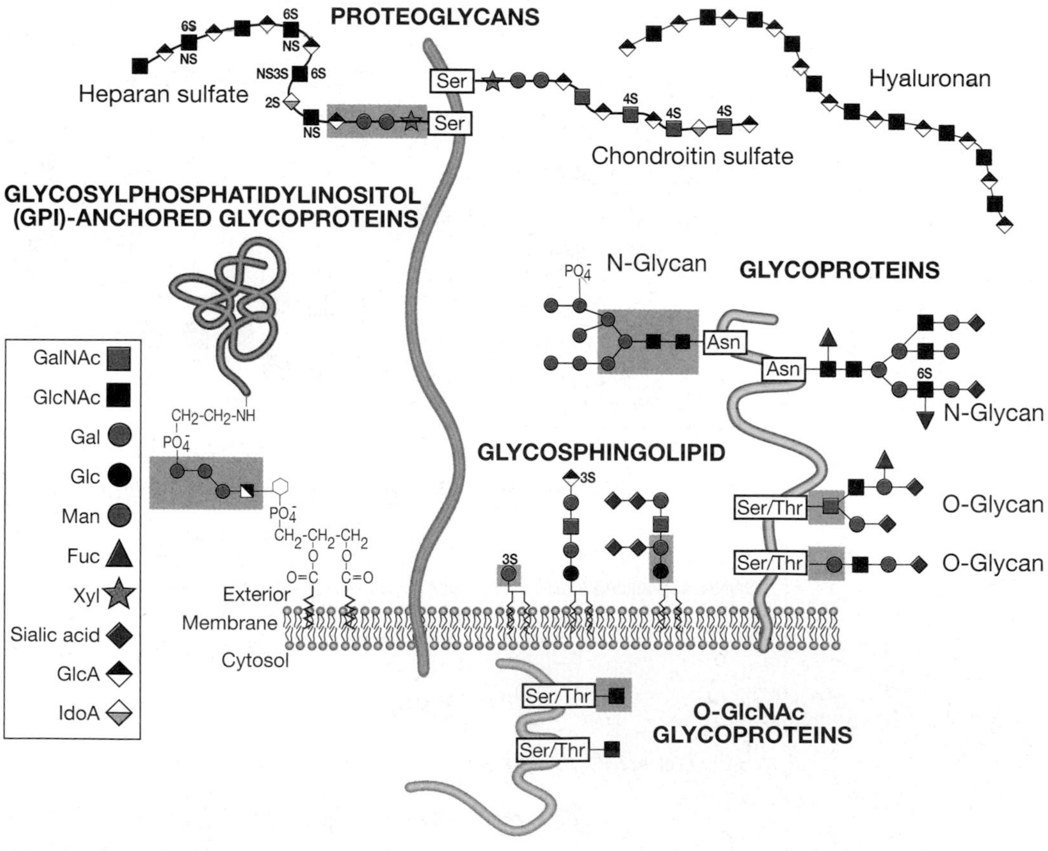

FIGURE 1.6. Common classes of animal glycans. (Modified from Varki A. 1997. *FASEB J.* **11:** 248–255; Fuster M. and Esko J.D. 2005. *Nat. Rev. Can.* **7:** 526–542.)

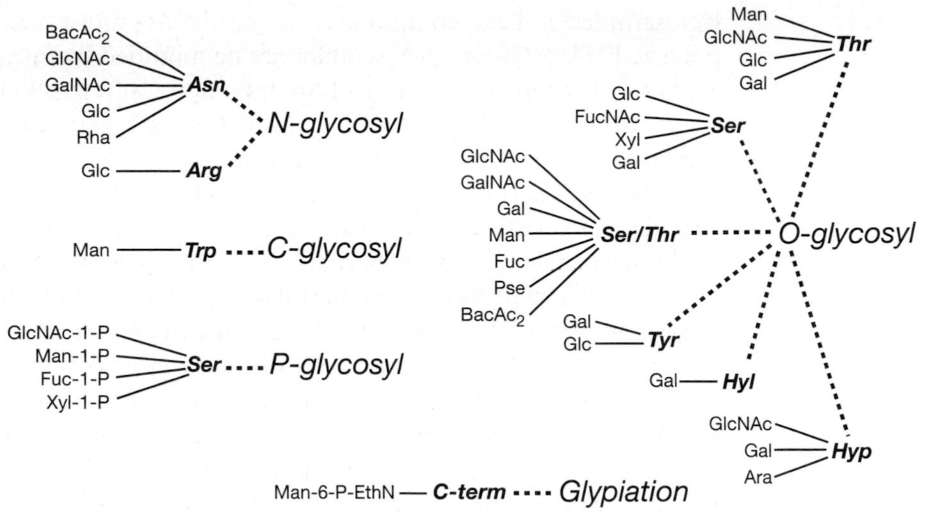

FIGURE 1.7. Glycan-protein linkages reported in nature. (Updated and redrawn, with permission of Oxford University Press, from Spiro R.G. 2002. *Glycobiology*. **12:** 43R–56R.) Diagrammatic representation of the five distinct types of sugar-peptide bonds that have been identified in nature, to date. Monosaccharide abbreviations are as in Figure 1.5. In addition, (Ara) arabinose; (Rha) rhamnose; (FucNAc) *N*-acetylfucosamine (2-acetamido-2,6-dideoxy-D-galactose); (Bac) bacillosamine (2,4-diamino-2,4,6-trideoxy-D-glucose); (Pse) pseudaminic acid (5,7-diacetamido-3,5,7,9-tetradeoxy-L-glycero-L-manno-nonulosonic acid); (Hyl) hydroxylysine; (Hyp) hydroxyproline; (*C-term*) carboxy-terminal amino acid residue. Glypiation is the process by which a glycosylphosphatidylinositol (GPI) anchor is added to a protein. For other details, including anomeric linkages, please see the Spiro (2002) review cited above.

glycan (GAG) chains (see definition below) attached to a "core protein" through a typical core region ending in a xylose residue that is linked to the hydroxyl group of a serine residue. The distinction between a proteoglycan and a glycoprotein is otherwise arbitrary, because some proteoglycan polypeptides can carry both glycosaminoglycan chains and different O- and N-glycans (see Chapter 16). Figure 1.7 provides a listing of known glycan-protein linkages in nature.

A glycosylphosphatidylinositol anchor is a glycan bridge between phosphatidylinositol and an ethanolamine that is in amide linkage to the carboxyl terminus of a protein. This structure typically constitutes the only anchor to the lipid bilayer membrane for such proteins (see Chapter 11). A glycosphingolipid (often called a glycolipid) consists of a glycan usually attached via glucose or galactose to the terminal primary hydroxyl group of the lipid moiety ceramide, which is composed of a long chain base (sphingosine) and a fatty acid (see Chapter 10). Glycolipids can be neutral or anionic. A ganglioside is an anionic glycolipid containing one or more residues of sialic acid. It should be noted that these represent only the most common classes of glycans reported in eukaryotic cells. There are several other less common types found on one or the other side of the cell membrane in animal cells (see Chapters 12 and 17).

Although different glycan classes have unique core regions by which they are distinguished, certain outer structural sequences are often shared among different classes of glycans. For example, N- and O-glycans and glycosphingolipids frequently carry the subterminal disaccharide Galβ1-4GlcNAcβ1- (*N*-acetyllactosamine or LacNAc) or, less commonly, GalNAcβ1-4GlcNAcβ1- (LacdiNAc) units. The LacNAc units can sometimes be repeated, giving extended poly-*N*-acetyllactosamines (sometimes incorrectly called "polylactosamines"). Less commonly, the LacdiNAc motif can also be repeated (termed polyLacdiNAc). Outer LacNAc units can be modified by fucosylation or by branching and are typically capped by sialic acids or, less commonly, by sulfate, Fuc, α-Gal, β-GalNAc, or β-GlcA units (see Chapters 13 and 14). In contrast, glycosaminoglycans are linear copolymers of acidic disaccharide repeating units, each containing a hexosamine (GlcN or GalN) and a hexose (Gal) or hexuronic acid (GlcA or IdoA) (see Chapter 16). The type of disaccharide unit defines the glycosaminoglycan as chondroitin or dermatan sulfate (GalNAcβ1-4GlcA/IdoA), heparin or heparan sulfate (GlcNAcα1-4GlcA/IdoA), or keratan sulfate (Galβ1-4GlcNAc). Keratan sulfate is actually a 6-O-sulfated form of poly-*N*-acetyllactosamine attached to an N- or O-glycan core, rather than to a typical Xyl-Ser-containing proteoglycan linkage region. Another type of glycosaminoglycan, hyaluronan (a polymer of GlcNAcβ1-4GlcA), appears to exist primarily as a free sugar chain unattached to any aglycone (Chapter 15). The glycosaminoglycans (except for hyaluronan) also typically have sulfate esters substituting either amino or hydroyxl groups (i.e., N- or O-sulfate groups). Another anionic polysaccharide that can be extended from LacNAc units is polysialic acid, a homopolymer of sialic acid that is selectively expressed only on a few proteins in vertebrates. Polysialic acids are also found as the capsular polysaccharides of certain pathogenic bacteria (Chapter 14).

GLYCAN STRUCTURES ARE NOT ENCODED DIRECTLY IN THE GENOME

It is important to reemphasize that unlike protein sequences, which are primary gene products, glycan chain structures are not encoded directly in the genome and are secondary gene products. A few percent of known genes in the human genome are dedicated to producing the enzymes and transporters responsible for the biosynthesis and assembly of glycan chains (see Chapter 7), typically as posttranslational modifications of proteins or by glycosylation of core lipids. The glycan chains themselves represent numerous combinatorial possibilities, generated by a variety of competing and sequentially acting glycosidases and glycosyltransferases (see Chapter 5) and the subcompartmentalized "assembly-line" mechanisms of glycan biosynthesis in the Golgi apparatus (see Chapter 3). Thus, even with full knowledge of the expression levels of all relevant gene products, we do not understand enough about the structures and pathways to predict the precise structures of glycans elaborated by a given cell type. Furthermore, small changes in environmental cues can cause dramatic changes in glycans produced by a given cell. It is this variable and dynamic nature of glycosylation that makes it a powerful way to generate biological diversity and complexity. Of course, it also makes glycans more difficult to study than nucleic acids and proteins.

SITE-SPECIFIC STRUCTURAL DIVERSITY IN PROTEIN GLYCOSYLATION

One of the most fascinating and yet frustrating aspects of protein glycosylation is the phenomenon of microheterogeneity. This term indicates that at any given glycan attachment site on a given protein synthesized by a particular cell type, a range of variations can be found in the structures of the attached glycan chain. The extent of this microheterogeneity can vary considerably from one glycosylation site to another, from glycoprotein to glycoprotein, and from cell type to cell type. Thus, a given protein originally encoded by a single gene can exist in numerous "glycoforms," each effectively a distinct molecular species. Mechanistically, microheterogeneity might be explained by the rapidity with which multiple, sequential, partially competitive glycosylation and deglycosylation reactions must take place in the endoplasmic reticulum (ER) and Golgi apparatus, through which a newly synthesized glycoprotein passes (see Chapter 3). An alternate possibility is that each individual cell or cell type is in fact exquisitely specific in the details of the glycosylation that it produces, but that intercellular variations result in the observed microheterogeneity of samples from natural multicellular sources. Whatever the origin of microheterogeneity, it accounts for the anomalous behavior of glycoproteins in various analytical/separation techniques (such as sodium dodecyl sulfate–polyacrylamide gel electrophoresis [SDS-PAGE], in which multiple or diffuse bands are observed) and makes complete structural analysis of a glycoprotein a difficult task. From a functional point of view, the biological significance of microheterogeneity remains unclear. It is possible that this is a type of diversity generator, intended for diversifying endogenous recognition functions and/or for evading microbes and parasites, each of which can bind with high specificity only to certain glycan structures (see Chapters 34 and 39).

CELL BIOLOGY OF GLYCOSYLATION

Most well-characterized pathways for the biosynthesis of major classes of glycans are confined within the ER and Golgi compartments (see Chapter 3). Thus, for example, newly synthesized proteins originating from the ER are either cotranslationally or posttranslationally modified with sugar chains at various stages in their itinerary toward their final destinations. The glycosylation reactions usually use activated forms of monosaccharides (nucleotide sugars; see Chapter 4) as donors for reactions that are catalyzed by glycosyltransferases (for details about their biochemistry, molecular genetics, and cell biology, see Chapters 3, 5, and 7). In almost all cases, these nucleotide donors are synthesized within the cytoplasm or nucleus from monosaccharide precursors of endogenous or exogenous origin (see Chapter 4). To be available to perform the glycosylation reactions, the donors must be actively transported across a membrane bilayer into the lumen of the ER and Golgi compartments. Much effort has gone into understanding the mechanisms of glycosylation within the ER and the Golgi apparatus, and it is clear that a variety of factors determine the final outcome of glycosylation reactions. Some bulky sugar chains are made on the cytoplasmic face of these intracellular organelles and are flipped across their membranes to the other side, but most are synthesized by adding one monosaccharide at a time to the growing glycan chain on the inside of the ER or the Golgi. Regardless, the portion of a glycoconjugate that faces the inside of these compartments will ultimately face the inside of a secretory granule or lysosome and will be topologically unexposed to the cytoplasm. The biosynthetic enzymes (glycosyltransferases, sulfotransferases, etc.) responsible for catalyzing these reactions are well studied (see Chapter 5), and their location has helped to define various functional compartments of the ER-Golgi pathway. A classical model envisioned that these enzymes are physically lined up along this pathway in the precise sequence in which they actually work. This appears to be an oversimplified view, because there is considerable overlap in the distribution of these enzymes, and the actual distribution of a given enzyme seems to depend on the cell type.

All of the topological considerations mentioned above are reversed with regard to nuclear and cytoplasmic glycosylation, because the active sites of the relevant glycosyltransferases face the cytoplasm, which is in direct communication with the interior of the nucleus. Until the mid-1980s, the accepted dogma was that glycoconjugates, such as glycoproteins and glycolipids, occurred exclusively on the outer surface of cells, on the internal (luminal) surface of intracellular organelles, and on secreted molecules. As discussed

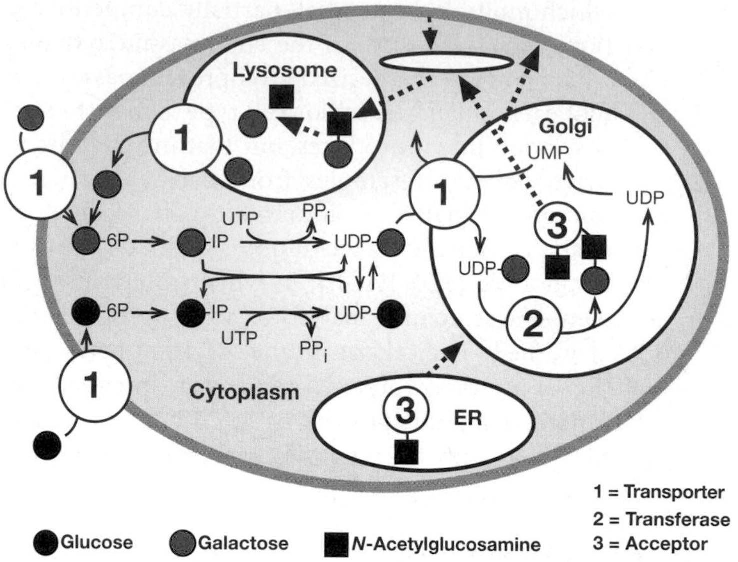

FIGURE 1.8. Biosynthesis, fate, and turnover of galactose, which is a common monosaccharide constituent of animal glycans. Although small amounts of galactose can be taken up from the outside of the cell, most cellular galactose is either synthesized de novo from glucose or recycled from degradation of glycoconjugates in the lysosome. The figure is a simplified view of the generation of the UDP nucleotide sugar UDP-Gal, its equilibrium state with UDP-glucose, and its uptake and use in the Golgi apparatus for synthesis of new glycans. (*Solid lines*) Biochemical pathways; (*dashed lines*) pathways for the trafficking of membranes and glycans.

above, this was consistent with knowledge of the topology of the biosynthesis of the classes of glycans known at the time, which took place within the lumen of the ER-Golgi pathway. Thus, despite some clues to the contrary, the cytoplasm and nucleus were assumed to be devoid of glycosylation capacity. However, it is now clear that certain distinct types of glycoconjugates are synthesized and reside within the cytoplasm and nucleus (see Chapter 17). Indeed, one of them, called O-linked GlcNAc (see Chapter 18), may well be numerically the most common type of glycoconjugate in many cell types. The fact that this major form of glycosylation was missed by so many investigators for so long serves to emphasize the relatively unexplored state of the whole field of glycobiology.

Like all components of living cells, glycans are constantly being degraded and the enzymes that catalyze this process cleave sugar chains either at the outer (nonreducing) terminal end (exoglycosidases) or internally (endoglycosidases) (see Chapters 3 and 41). Some terminal monosaccharide units such as sialic acids are sometimes removed and new units reattached during endosomal recycling, without degradation of the underlying chain. The final complete degradation of most glycans is generally performed by multiple glycosidases in the lysosome. Once broken down, their individual unit monosaccharides are then typically exported from the lysosome into the cytoplasm so that they can be reused (see Figure 1.8). In contrast to the relatively slow turnover of glycans derived from the ER-Golgi pathway, the O-GlcNAc monosaccharide modifications of the nucleus and cytoplasm may be more dynamic and rapidly turned over (see Chapter 18).

TOOLS USED TO STUDY GLYCOSYLATION

Unlike oligonucleotides and proteins, glycans are not commonly expressed in a linear, unbranched fashion. Even when they are found as linear macromolecules (e.g., GAGs), they often contain a variety of substituents, such as sulfate groups. Thus, the complete sequencing of glycans is practically impossible to accomplish by a single method and requires iterative combinations of physical, chemical, and enzymatic approaches that together yield the details of the structure under study (for a discussion of the various forms of low- and high-resolution separation and analysis, including mass spectrometry and NMR, see Chapter 47). Less detailed information on structure may be sufficient to explore the biology of some glycans and can be obtained by simple techniques, such as the use of enzymes (endoglycosidases and exoglycosidases), lectins, and other glycan-binding proteins (see Chapters 45 and 47), chemical modification or cleavage, metabolic radioactive labeling, antibodies, or cloned glycosyltransferases (Chapter 49). Glycosylation can also be perturbed in a variety of ways, for example, by glycosylation inhibitors and primers (Chapter 50) and by genetic manipulation of glycosylation in intact cells and organisms (Chapter 46). The directed in vitro synthesis of glycans by chemical and enzymatic methods has also taken great strides in recent years, providing many new tools for exploring glycobiology (Chapters 49 and 51). The generation of complex glycan libraries by a variety of routes has further enhanced this interface of chemistry and biology (Chapter 49).

GLYCOMICS

Analogous to genomics and proteomics, glycomics represents the systematic methodological elucidation of the "glycome" (the totality of glycan structures) of a given cell type or organism (see Chapter 48). In reality, the glycome is far more complex than the genome or proteome. In addition to the vastly greater structural diversity in glycans, one is faced with the complexities of glycosylation microheterogeneity (see above) and the dynamic changes that occur in the course of development, differentiation, metabolic changes, malignancy, inflammation, or infection. Added diversity arises from intraspecies and interspecies variations in glycosylation. Thus, a given cell type in a given species can manifest a large number of possible glycome states. Glycomic analysis today generally consists of extracting entire cell types, organs, or organisms; releasing all the glycan chains from their linkages; and cataloging them via approaches such as mass spectrometry. In a variation called glycoproteomics, the glycans are analyzed while still attached to protease-generated fragments of glycoproteins. The results obtained represent a spectacular improvement over what was possible a few decades ago, but they still constitute an effort analogous to cutting down all the trees in a forest and cataloging them, without attention to the layout of the forest and the landscape. This type of glycomic analysis needs to be complemented by classical methods such as tissue-section staining or flow cytometry, using lectins or glycan-specific antibodies that aid in understanding the glycome by taking into account the heterogeneity of glycosylation at the level of the different cell types and subcellular domains in the tissue under study. This is even more important because of the common observation that removing cells from their normal milieu and placing them into tissue culture can result in major changes in the glycosylation machinery of the cell. However, such classical approaches suffer from poor quantitation and relative insensitivity to structural details. A combination of the two approaches is now potentially feasible via laser-capture microdissection of specific cell types directly from tissue sections, with the resulting samples being studied by mass spectrometry.

Because most of the genes involved in glycan biosynthetic pathways have been cloned from multiple organisms, it is possible today to obtain an indirect genomic and transcriptomic view of the glycome in a specific cell type (see Chapter 7). However, given the relatively poor correlation between mRNA and protein levels, and the complex assembly line and competitive nature of the cellular Golgi glycosylation pathways, even complete knowledge of the mRNA expression patterns of all relevant genes in a given cell cannot allow accurate prediction of the distribution and structures of glycans in that cell type. In other words, there is as yet no reliable indirect route toward elucidating the glycome, other than by actual structural analysis using an array of methods.

GLYCOSYLATION DEFECTS IN ORGANISMS AND CULTURED CELLS

Many mutant variants of cultured cell lines with altered glycan structures and specific glycan biosynthetic defects have been described, the most common of which are those that are lectin resistant (see Chapter 46). Indeed, with few exceptions, mutants with specific defects at most steps of the major pathways of glycan biosynthesis have been found in cultured animal cells. The use of such cell lines has been of great value in elucidating the details of glycan biosynthetic pathways. Their existence implies that many types of glycans are not crucial to the optimal growth of single cells living in the sheltered and relatively unchanging environment of the culture dish. Rather, most glycan structures must be more important in mediating cell–cell and cell–matrix interactions in intact multicellular organisms and/or interactions between organisms. In keeping with this supposition, genetic defects completely eliminating major glycan classes in intact animals all cause embryonic lethality (see Chapter 42 and Table 6.1). As might be expected, naturally occurring viable animal mutants of this type tend to have disease phenotypes of intermediate severity and show complex phenotypes involving multiple systems. Less severe genetic alterations of outer chain components of glycans tend to give viable organisms with more specific phenotypes (see Chapter 42 and Table 6.1). Overall, there is much to be learned by studying the consequences of natural or induced genetic defects in intact multicellular organisms (see Chapter 42). It is interesting to note that, in the short time since the first edition of this book, we have gone from asking "What is it that glycans do anyway?" to having to explain a large number of complex and sometimes nonviable glycosylation-modified phenotypes in humans, mice, flies, and other organisms.

THE BIOLOGICAL ROLES OF GLYCANS ARE DIVERSE

A major theme of this volume is the exploration and elucidation of the biological roles of glycans. Like any biological system, the optimal approach carefully considers the relationship of structure and biosynthesis to function (see Chapter 6). As might be imagined from their ubiquitous and complex nature, the biological roles of glycans are quite varied. Indeed, asking what these roles are is akin to asking the same question about proteins. Thus, all of the proposed theories regarding glycan function turn out to be partly correct, and exceptions to each can also be found. Not surprisingly for such a diverse group of molecules, the biological roles of glycans span the spectrum from those that are subtle to those that are crucial for the development, growth, function, or survival of an organism (for further discussion, see Chapter 6). The diverse functions ascribed to glycans can be more simply divided into two general categories: (i) structural and modulatory functions (involving the glycans themselves or their modulation of the molecules to which they are attached) and (ii) specific recognition of glycans by glycan-binding proteins. Of course, any given glycan can mediate one or both types of functions. The binding proteins in turn fall into two broad groups: lectins and sulfated GAG-binding proteins (see Chapter 26). Such molecules can be either intrinsic to the organism that synthesized the cognate glycans (e.g., see Chapters 28–33 and 35) or extrinsic (see Chapters 34 and 39 for information concerning microbial proteins that bind to specific glycans on host cells). The atomic details of these glycan-protein interactions have been elucidated in many instances (see Chapter 27). Although there are exceptions to this notion, the following general theme has emerged regarding lectins: Monovalent binding tends to be of relatively low affinity, although there are exceptions to this notion, and such systems typically achieve their specificity and function by achieving high avidity, via interactions of multivalent arrays of glycans with cognate lectin-binding sites (see Chapters 30 and 40).

GLYCOSYLATION CHANGES IN DEVELOPMENT, DIFFERENTIATION, AND MALIGNANCY

Whenever a new tool (e.g., an antibody or lectin) specific for a particular glycan is developed and used to probe its expression in intact organisms, it is common to find exquisitely specific temporal and spatial patterns of expression of that glycan in relation to cellular activation, embryonic development, organogenesis, and differentiation (see Chapter 38). Certain relatively specific changes in expression of glycans are also often found in the course of transformation and progression to malignancy (see Chapter 44), as well as other pathological situations such as inflammation. These spatially and temporally controlled patterns of glycan expression imply the involvement of glycans in many normal and pathological processes, the precise mechanisms of which are understood in only a few cases.

EVOLUTIONARY CONSIDERATIONS IN GLYCOBIOLOGY

Remarkably little is known about the evolution of glycosylation. There are clearly shared and unique features of glycosylation in different kingdoms and taxa. Among animals, there may be a trend toward increasing complexity of N- and O-glycans in more recently evolved ("higher") taxa. Intraspecies and interspecies variations in glycosylation are also relatively common. It has been suggested that the more specific biological roles of glycans are often mediated by uncommon structures, unusual presentations of common structures, or further modifications of the commonly occurring saccharides themselves. Such unusual structures likely result from such unique expression patterns of the relevant glycosyltransferases or other glycan-modifying enzymes. On the other hand, such uncommon glycans can be targets for specific recognition by infectious microorganisms and various toxins. Thus, at least a portion of the diversity in glycan expression in nature must be related to the evolutionary selection pressures generated by interspecies interactions (e.g., of host with pathogen or symbiont). In other words, the two different classes of glycan recognition mentioned above (mediated by intrinsic and extrinsic glycan-binding proteins) are in constant competition with each other, with regard to a particular glycan target. The specialized glycans expressed by parasites and microbes that are of great interest from the biomedical point of view (see Chapters 20, 21, and 40) are themselves presumably subject to evolutionary selection pressures. The evolutionary issues presented above are further considered in Chapter 19, which also discusses the limited information concerning how various glycan biosynthetic pathways appear to have evolved and diverged in different life forms.

GLYCANS IN MEDICINE AND BIOTECHNOLOGY

Numerous natural bioactive molecules are glycoconjugates, and the attached glycans can have dramatic effects on the biosynthesis, stability, action, and turnover of these molecules in intact organisms. For example, heparin, a sulfated glycosaminoglycan, and its derivatives are among the most commonly used drugs in the world. For this and many other reasons, glycobiology and carbohydrate chemistry have become increasingly important in modern biotechnology. Patenting a glycoprotein drug, obtaining an FDA approval for its use, and monitoring its production all require knowledge of the structure of its glycans. Moreover, glycoproteins, which include monoclonal antibodies, enzymes, and hormones, are by now the major products of the biotechnology industry, with sales in the tens of billions of dollars annually, which continues to grow at an increasing rate. In addition, several human disease states are characterized by changes in glycan biosynthesis that can be of diagnostic and/or therapeutic significance. The emerging importance of glycobiology in medicine and biotechnology is further considered in Chapters 43 and 51.

FURTHER READING

Sharon N. 1980. Carbohydrates. *Sci. Am.* **243[5]:** 90–116.

Rademacher T.W., Parekh R.B., and Dwek R.A. 1988. Glycobiology. *Annu. Rev. Biochem.* **57:** 785–838.

Sharon N. and Lis H. 1993. Carbohydrates in cell recognition. *Sci. Am.* **268:** 82–89.

Varki A. 1993. Biological roles of oligosaccharides: All of the theories are correct. *Glycobiology* **3:** 97–130.

Cabezas J.A. 1994. The origins of glycobiology. *Biochem. Edu.* **22:** 3–7.

Drickamer K. and Taylor M.E. 1998. Evolving views of protein glycosylation. *Trends Biochem. Sci.* **23:** 321–324.

Etzler M.E. 1998. Oligosaccharide signaling of plant cells. *J. Cell Biochem.* (suppl.) **30–31:** 123–128.

Gagneux P. and Varki A. 1999. Evolutionary considerations in relating oligosaccharide diversity to biological function. *Glycobiology* **9:** 747–755.

Esko J.D. and Lindahl U. 2001. Molecular diversity of heparan sulfate. *J. Clin. Invest.* **108:** 169–173.

Roseman S. 2001. Reflections on glycobiology. *J. Biol. Chem.* **276:** 41527–41542.

Hakomori S. 2002. The glycosynapse. Inaugural article. *Proc. Natl. Acad. Sci.* **99:** 225–232.

Spiro R.G. 2002. Protein glycosylation: Nature, distribution, enzymatic formation, and disease implications of glycopeptide bonds. *Glycobiology* **12:** 43R–56R.

Haltiwanger R.S. and Lowe J.B. 2004. Role of glycosylation in development. *Annu. Rev. Biochem.* **73:** 491–537.

Sharon N. and Lis H. 2004. History of lectins: From hemagglutinins to biological recognition molecules. *Glycobiology* **14:** 53R–62R.

Toole B.P. 2004. Hyaluronan: From extracellular glue to pericellular cue. *Nat. Rev. Can.* **4:** 528–539.

Drickamer K. and Taylor M.E. 2006. *Introduction to glycobiology,* 2nd ed. Oxford University Press, Oxford.

Freeze H.H. 2006. Genetic defects in the human glycome. *Nat. Rev. Genet.* **7:** 537–551.

Lutteke T., Bohne-Lang A., Loss A., Goetz T., Frank M., and von der Lieth C.W. 2006. GLYCOSCIENCES.de: An Internet portal to support glycomics and glycobiology research. *Glycobiology* **16:** 71R–81R.

Ohtsubo K. and Marth J.D. 2006. Glycosylation in cellular mechanisms of health and disease. *Cell* **126:** 855–867.

Patnaik S.K. and Stanley P. 2006. Lectin-resistant CHO glycosylation mutants. *Methods Enzymol.* **416:** 159–182.

Prescher J.A. and Bertozzi C.R. 2006. Chemical technologies for probing glycans. *Cell* **126:** 851–854.

van Die I. and Cummings R.D. 2006. Glycans modulate immune responses in helminth infections and allergy. *Chem. Immunol. Allergy* **90:** 91–112.

Varki A. 2006. Nothing in glycobiology makes sense, except in the light of evolution. *Cell* **126:** 841–845.

Bishop J.R., Schuksz M., and Esko J.D. 2007. Heparan sulphate proteoglycans fine-tune mammalian physiology. *Nature* **446:** 1030–1037.

Hart G.W., Housley M.P., and Slawson C. 2007. Cycling of O-linked β-*N*-acetylglucosamine on nucleocytoplasmic proteins. *Nature* **446:** 1017–1022.

Kamerling J., Boons G.-J., Lee Y., Suzuki A., Taniguchi N., and Voragen A.G.J. 2007. *Comprehensive glycoscience,* Vols. 1–4. Elsevier Science, London.

CHAPTER 2

Structural Basis of Glycan Diversity

Carolyn R. Bertozzi and David Rabuka

LIKE THE OTHER BIOPOLYMERS, PROTEINS, AND NUCLEIC ACIDS, glycans come in a diversity of structures that underlie a vast array of biological functions. To understand these functions at a molecular level, we must first understand glycan structures at a chemical level. This chapter begins with an introduction to the building blocks—monosaccharides—that are assembled to generate more complex glycans. After a brief summary of nomenclature, we present the salient chemical features of the monosaccharides that define their structural diversity, with an emphasis on stereochemical properties. We then illustrate how monosaccharide diversity, combined with a multiplicity of ways in which they can be linked together, can create the wealth of glycan structures found in nature. An understanding of the structural features that distinguish glycans from other biopolymers will help the reader to appreciate the origin of their biological capabilities.

INTRODUCTION TO GLYCAN TERMINOLOGY

Although we use the term *glycan* in this book, several names for sugar polymers are found in other textbooks and the literature. Early on, sugar-based substances were referred to as *carbohydrates*, a term literally derived from "hydrates of carbon." This name was coined more than 100 years ago to describe naturally occurring substances with the general formula $C_x(H_2O)_n$ that also possess a carbonyl group, either an aldehyde or a ketone. *Monosaccharides* are the simplest of these polyhydroxylated carbonyl compounds (*saccharide* is derived from the Greek *sakchar*, meaning sugar or sweetness).

Monosaccharides are joined together to make oligosaccharides or polysaccharides. Typically, the term *oligosaccharide* refers to any glycan that contains a small number (2–20) of monosaccharide residues connected by glycosidic linkages. The term *polysaccharide* is typically used to denote any linear or branched polymer consisting of monosaccharide residues, such as cellulose (Chapters 13 and 22). Thus, the relationship of monosaccharides to oligosaccharides or polysaccharides is analogous to that of amino acids and proteins, or nucleotides and nucleic acids.

The term *glycoconjugate* is often used to describe a macromolecule that contains monosaccharides covalently linked to proteins or lipids. The prefix *glyco-* and the suffixes *-saccharide* and *-glycan* indicate the presence of carbohydrate constituents (e.g., *glyco*proteins, *glyco*lipids, and proteo*glycans*). Just as is observed with proteins in nature, additional structural diversity can be imparted to glycans by modifying their hydroxy groups with phosphate, sulfate, or acetyl esters.

The designation *complex* is given to a variety of carbohydrates. A carbohydrate may be termed complex if it contains more than one type of monosaccharide building unit. Thus, the glucose-based polymer cellulose is an example of a *simple* carbohydrate, whereas a galactomannan polysaccharide, which possesses both galactose and mannose, is an example of a complex carbohydrate. However, even so-called simple glycans, such as cellulose and starch, often have very complex molecular structures in three dimensions. In the description of N-glycans of glycoproteins (see Chapter 8), the term complex is used more specifically for N-glycans with multiple, extended branches, often containing *N*-acetyllactosamine units. Finally, the term complex carbohydrates includes glycoconjugates, whereas the term carbohydrates per se would not. Additional nomenclature issues are covered in the various sections of this chapter. A more detailed and comprehensive listing of carbohydrate nomenclature rules has been published (see the McNaught reference in Further Reading at the end of the chapter).

MONOSACCHARIDES: BASIC STRUCTURES AND STEREOISOMERISM

The classification of monosaccharide structures began in the late 19th century with the pioneering work of Emil Fischer. All simple monosaccharides have the general empirical formula $C_x(H_2O)_n$, where n is an integer ranging from 3 to 9. As mentioned briefly in Chapter 1, all monosaccharides consist of a chain of chiral hydroxymethylene units, which terminates at one end with a hydroxymethyl group and at the other with either an aldehyde group (aldoses) or an α-hydroxy ketone group (ketoses). Glyceraldehyde is the simplest aldose and dihydroxyacetone is the simplest ketose (Figure 2.1). The structures of glyceraldehyde and dihydroxyacetone are distinct in that glyceraldehyde contains an asymmetric (chiral) carbon atom (Figure 2.1), whereas dihydroxyacetone does not. With the exception of dihydroxyacetone, all monosaccharides have at least one asymmetric carbon atom, the total number being equal to the number of internal (CHOH) groups ($n - 2$ for aldoses and $n - 3$ for ketoses with n carbon atoms). The number of stereoisomers corresponds to 2^k,

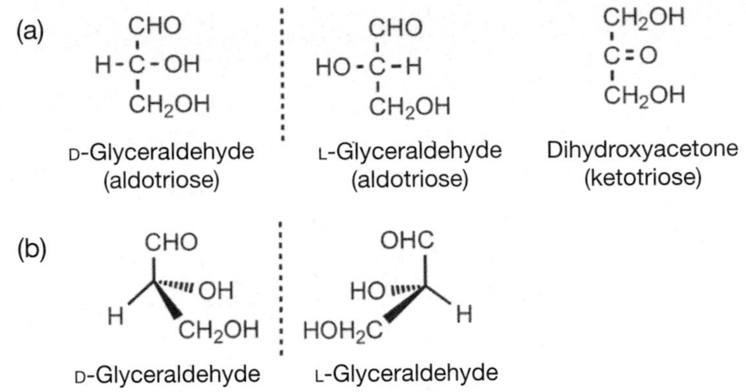

FIGURE 2.1. (*a*) Structures of glyceraldehyde and dihydroxyacetone in Fischer projection; (*b*) D- and L-glyceraldehyde. The chiral nature of the central carbon in glyceraldehyde gives rise to two possible configurations of the molecule, termed D and L.

where k equals the number of asymmetric carbon atoms. For example, an aldohexose with the general formula $C_6H_{12}O_6$ and four asymmetric carbon atoms, that is, four (CHOH) groups, can be described in 16 possible isomeric forms.

The numbering of carbon atoms follows the rules of organic chemistry nomenclature: The aldehyde carbon is referred to as C-1 and the carbonyl group in ketoses is referred to as C-2. The overall configuration (D or L) of each sugar is determined by the absolute configuration of the stereogenic center furthest from the carbonyl group (i.e., with the highest numbered asymmetric carbon atom; this is C-5 in hexoses and C-4 in pentoses). The configuration of a monosaccharide is most easily determined by representing the structure in a Fischer projection. If the OH (or other non-H group) is on the right in the Fischer projection, the overall configuration is D. If the OH (or other non-H group) is on the left, the overall configuration is L (Figure 2.2). This figure also shows D- and L-glucose in the cyclic form (chair conformation) found in solution. Most vertebrate monosaccharides have the D configuration with the exception of fucose and iduronic acid, which are L sugars. The Fischer projections shown in Figure 2.3 illustrate the acyclic structures of all D-aldoses through the aldohexose group.

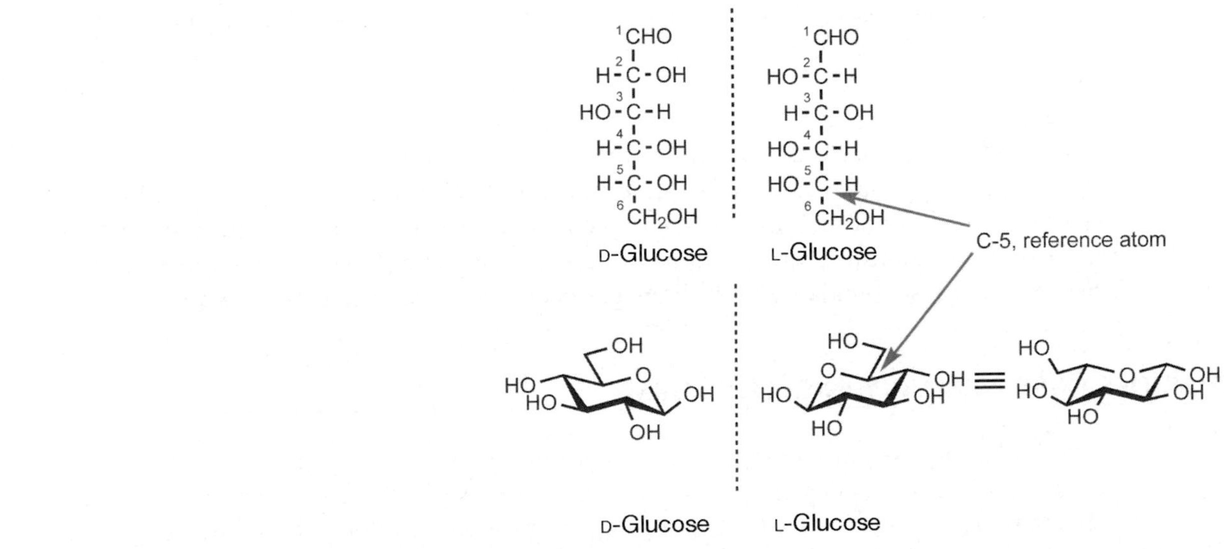

FIGURE 2.2. D- and L-glucopyranose in Fischer projection and chair conformation.

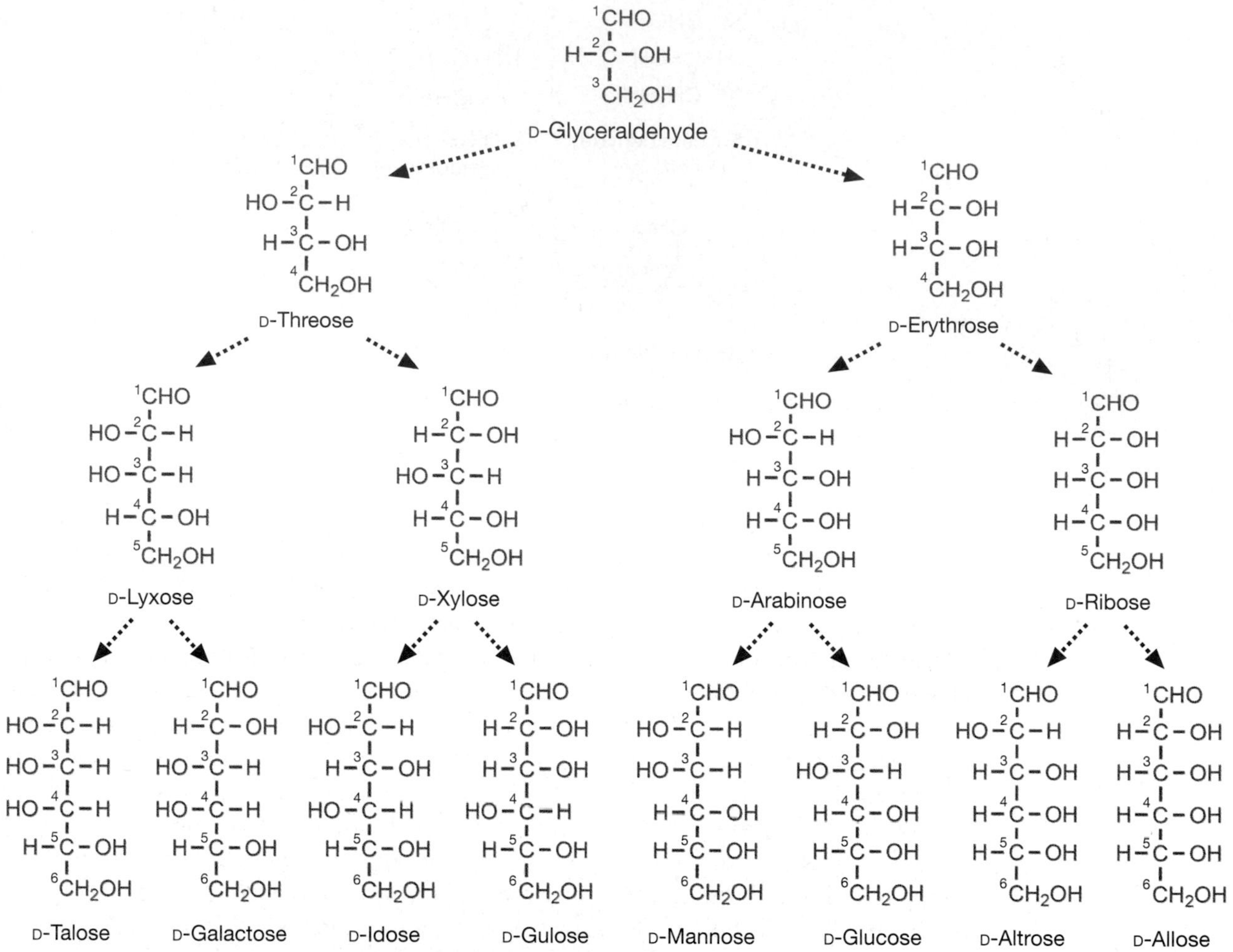

FIGURE 2.3. Fischer projections of the acyclic forms of the D series of aldoses, ranging from triose to hexose.

Any two sugars that differ only in the configuration around a single chiral carbon atom are called *epimers*. For example, D-mannose is the C-2 epimer of D-glucose, whereas D-galactose is the C-4 epimer of D-glucose (Figure 2.4). The names of monosaccharides are frequently abbreviated; most common are three-letter abbreviations for simple monosaccharides (e.g., Gal, Glc, Man, Xyl, Fuc). There are nine common monosaccharides found in vertebrate glycoconjugates (Figure 2.4). Once incorporated into a glycan, these nine monosaccharide building blocks can be further modified to generate additional sugar structures. For example, glucuronic acid can be epimerized at C-5 to generate iduronic acid (IdoA). A great many more monosaccharides exist in glycoconjugates from other species and as intermediates in metabolism. In this volume, we use a symbolic notation for the monosaccharides that are most abundant in vertebrate glycoconjugates (see Chapter 1, Figure 1.5).

MONOSACCHARIDES EXIST PRIMARILY IN CYCLIC FORM

Monosaccharides exist in solution as an equilibrium mixture of acyclic and cyclic forms. The percentage of each form depends on the sugar structure. The cyclic form of a monosaccharide is characterized by a hemiacetal group formed by the reaction of one of the hydroxy groups with the C-1 aldehyde or ketone. The general reaction that cyclizes the monosaccharide is

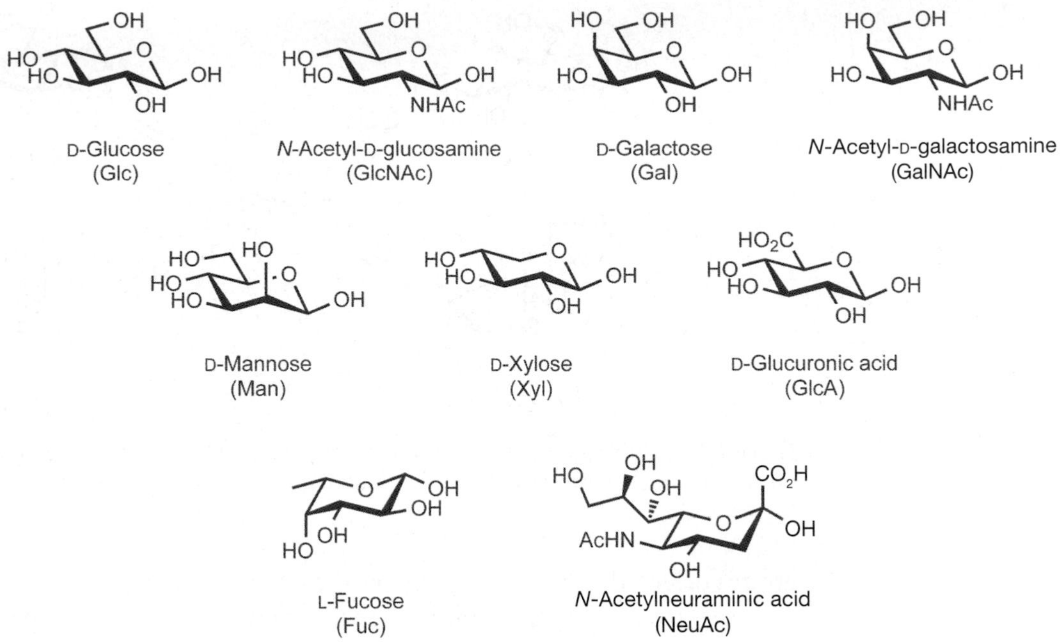

FIGURE 2.4. Common monosaccharides found in vertebrates. *N*-Acetylneuraminic acid is the most common form of sialic acid.

shown in Figure 2.5. For reasons of chemical stability, five- and six-membered rings are most commonly formed from acyclic monosaccharides. Generally, aldohexoses form six-membered rings via a C-1—O—C-5 ring closure; ketohexoses form five-membered rings via a C-2—O—C-5 ring closure; aldohexoses form five-membered rings through a C-1—O—C-4 ring closure or six-membered rings through a C-1—O—C-5 ring closure (Figure 2.6). Because of the structural similarity to the organic compounds furan and pyran, a five-membered cyclic hemiacetal is labeled a *furanose* and a six-membered cyclic hemiacetal is called a *pyranose.*

Monosaccharides can also be represented as Haworth projections in which both five- and six-membered cyclic structures are depicted as planar ring systems, with the hydroxy

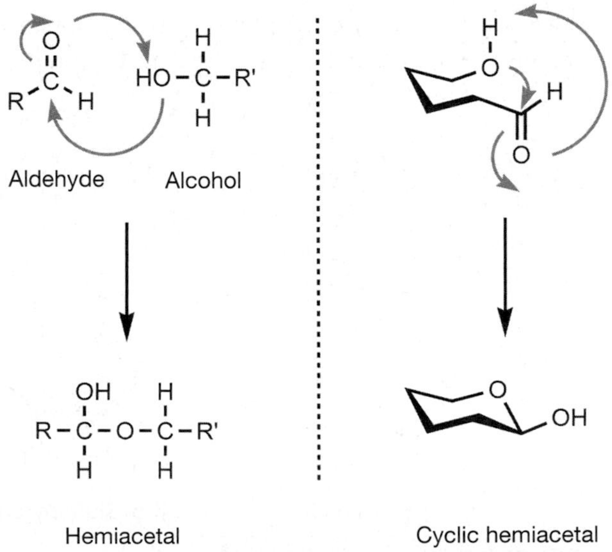

FIGURE 2.5. Formation of hemiacetals.

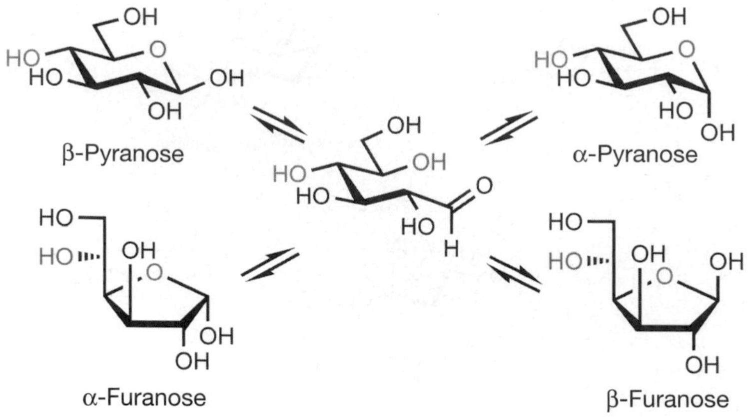

FIGURE 2.6. Cyclization of the acyclic form of D-glucose to form pyranose and furanose structures. The cyclization reaction produces both the α and β anomers (i.e., C-1 epimers).

groups oriented either above or below the plane of the ring (Figure 2.7). Although not truly representative of the three-dimensional structure of a monosaccharide, the Haworth representation has been used since the late 1920s as an easy-to-draw formula that permits a quick evaluation of stereochemistry around the monosaccharide ring. The Haworth representations are preferably drawn with the ring oxygen atom at the top (for furanose) or the top right-hand corner (for pyranose) of the structure; the numbering of the ring carbons increases in a clockwise direction.

For any D sugar, the conversion of a Fischer projection into a Haworth projection proceeds as follows: (1) Any groups (atoms) that are directed to the right in the Fischer structure are given a downward orientation in the Haworth structure, (2) any groups (atoms) that are directed to the left in the Fischer structure are given an upward orientation in the Haworth structure, and (3) the terminal —CH_2OH group is given an upward orientation in the Haworth structure. For an L sugar, (1) and (2) are the same, but the terminal —CH_2OH group is projected downward.

The planar Haworth structures are distorted representations of the actual molecules. The preferred conformation of a pyranose ring is the *chair* conformation, similar to the structure of cyclohexane. The conversion from Haworth projection to chair conformation leaves the downward or upward orientation of ring substituents unaltered. Two chair conformations can be distinguished and designated as $^4C_1$ and $^1C_4$, respectively (Figure 2.8a), and these con-

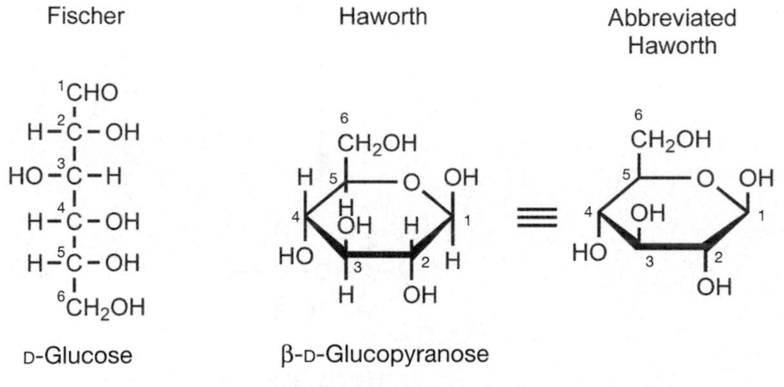

FIGURE 2.7. Conversion from Fischer to Haworth projection formula. Each hydroxy group projected to the right in the Fischer projection points down in the Haworth formula.

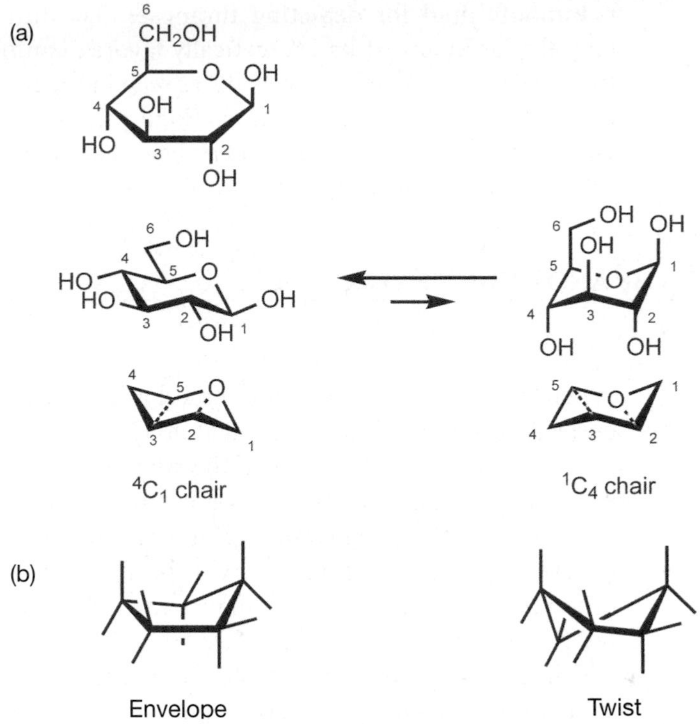

FIGURE 2.8. (*a*) β-D-Glucose in Haworth projection and in its $^4C_1$ and $^1C_4$ chair conformations; (*b*) envelope and twist conformations for a five-membered ring structure.

formers can interconvert by a process called the "ring flip." The first numeral in the chair conformer designation (superscript) indicates the number of the ring carbon atom above the "seat of the chair (C)" and the second numeral (subscript) indicates the number of the ring carbon atom below the plane of the seat (spanned by C-2, C-3, C-5, and the ring O). Chair conformations are designated from structures with the ring oxygen atom in the top right-hand corner of the ring "seat," resulting in the clockwise appearance of the ring numbering. To determine the stereochemistry in the chair form as it corresponds to the Fischer projection, one can locate C-6 and then trace along the carbon skeleton of the sugar, bisecting the C—O and C—H bonds formed from each atom. The OH (or OR) and H groups are found on the right (R) or left (L) sides, just as in the Fischer projection (Figure 2.9).

At present, the more structurally accurate chair representations are preferred to Haworth projections for depicting pyranoses. However, Haworth projections are still

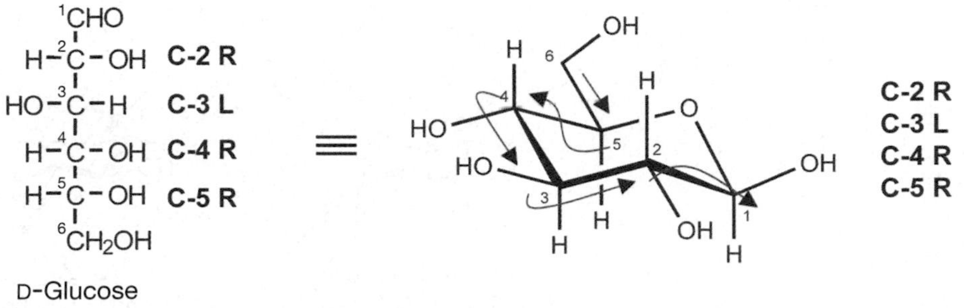

FIGURE 2.9. Conversion from Fischer to chair projection formula; (R) right; (L) left. *Red arrows* illustrate the path to follow along the sugar backbone when correlating the stereochemistry of the Fischer projection with the chair conformation.

commonly used for depicting furanoses. The furanose ring is rather flexible and not entirely flat in any of its energetically favored conformations; for example, it has a slight pucker when viewed from the side, as seen in the representations of the so-called envelope and twist (or *skew*) conformations (Figure 2.8b). Because furanoses can adopt many low-energy conformations, researchers have adopted the Haworth projection as a simple means to avoid this complexity.

CHEMISTRY AT THE ANOMERIC CENTER

Mutarotation

When cyclized into rings, monosaccharides acquire an additional asymmetric center derived from the carbonyl carbon atom (Figure 2.6). The new asymmetric center is termed the *anomeric carbon* (i.e., C-1 in the ring form of glucose). Two stereoisomers are formed by the cyclization reaction because the anomeric hydroxy group can assume two possible orientations. When the configurations (R or S) are the same at the anomeric carbon and the stereogenic center furthest from the anomeric carbon, the monosaccharide is defined as the α *anomer*. When the configurations are different, the monosaccharide is defined as the β *anomer* (Figure 2.10). Unlike the other stereocenters on the monosaccharide ring, which are configurationally stable, the anomeric center can undergo an interconversion of stereoisomers via the process of *mutarotation*. Catalyzed by dilute acid or base, the reaction proceeds by the reverse of the cyclization reaction. The monosaccharide ring opens up and then recloses to form a ring with the other anomeric configuration (Figure 2.6). The term *mutarotation* derives from the rapid change in optical rotation (denoted $[\alpha]_D$) that is observed when an anomerically pure form of a monosaccharide is dissolved in water. For example, β-D-glucopyranose shows an initial rotation of +19°, whereas the α anomer shows an initial rotation of +112°. When either anomer is allowed to undergo the mutarotation reaction, an equilibrium mixture containing both anomers is obtained, producing a rotation of +52.5°.

Oxidation and Reduction

Generally, the acyclic (aldehyde or ketone) form of a monosaccharide is only present in minor amounts in an equilibrium mixture (<0.01%). Nevertheless, the open-chain aldehydes or ketones can participate in chemical reactions that drive the equilibrium and eventually consume the sugar.

Aldoses and ketoses were historically referred to as *reducing sugars* because they responded positively in a chemical test that effected oxidation of their aldehyde and hydroxy-

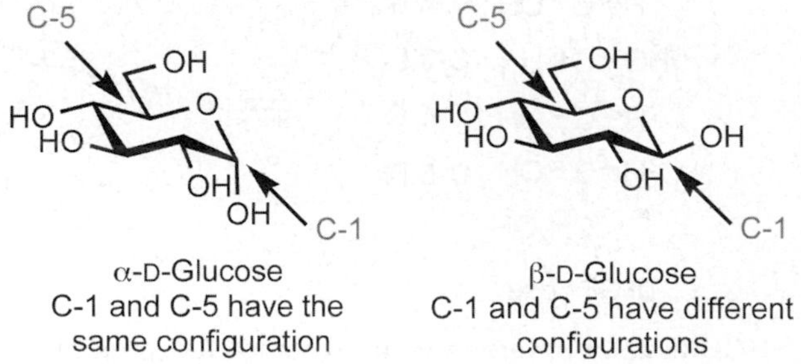

FIGURE 2.10. Determination of configuration at the anomeric center.

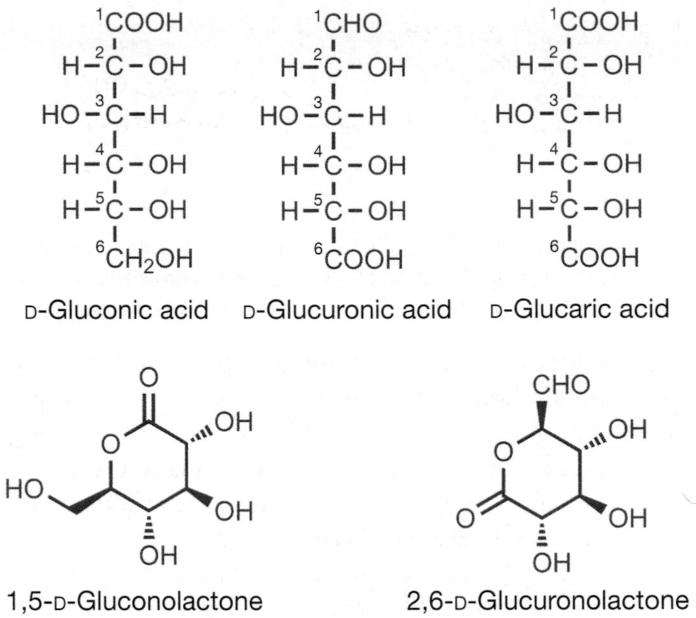

FIGURE 2.11. Oxidized forms of D-glucose.

ketone functionalities, respectively. The carboxylic acid formed by oxidation of the aldehyde in an aldose is referred to as a *glyconic acid* (e.g., gluconic acid is the oxidation product of glucose). It is also possible to oxidize the hydroxy groups of monosaccharides, most notably the terminal OH group (i.e., C-6 of glucose). In this reaction, a *glycuronic acid* is produced, and if both terminal groups are oxidized, the product is a *glycaric acid.* The three acids derived from D-glucose are illustrated in Figure 2.11. These compounds have a tendency to undergo intramolecular cyclization reactions, preferably yielding six-membered lactones. Two examples of lactonization are shown in Figure 2.11. Oxidized forms of monosaccharides can be found in nature. For example, glucuronic acid (GlcA) is an abundant component of many glycosaminoglycans (see Chapter 16).

The carbonyl groups of aldoses and ketoses also can be reduced with sodium borohydride ($NaBH_4$) to form polyhydroxy alcohols, referred to as alditols. This reaction is widely used to introduce a radiolabel at C-1 of the monosaccharide by reduction with NaB^3H_4 (Figure 2.12).

Schiff Base Formation

The aldehyde and ketone groups of monosaccharides can also undergo Schiff base formation with amines or hydrazides, forming imines and hydrazones, respectively (Figure 2.13). This reaction is often used to conjugate the monosaccharide to proteins (via their lysine residues) or to biochemical probes such as biotin hydrazide. It should be noted that the

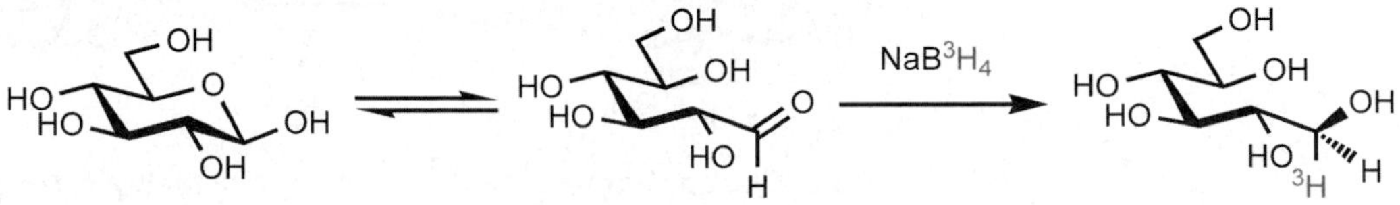

FIGURE 2.12. Conversion of a monosaccharide to a tritium-labeled alditol by reduction with NaB^3H_4.

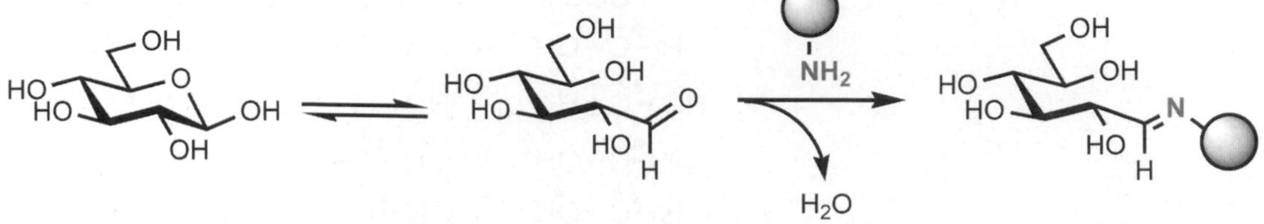

FIGURE 2.13. Conjugation of a monosaccharide to an amino group by formation of an imine. The *filled circle* represents any small molecule or macromolecule containing an amine.

imines formed with amino groups are not stable to water and are typically reduced with sodium cyanoborohydride ($NaCNBH_3$) in a process termed *reductive amination.*

As aldehydes, reducing sugars can also form Schiff bases with amino groups of the lysine residues in proteins. This nonenzymatic process that links glycans to proteins is termed "glycation" and is distinct from "glycosylation," which involves the formation of a glycosidic bond between the sugar and protein. Glycation products can undergo further reactions that lead to the formation of protein cross-links, and these can have pathogenic consequences (i.e., they are immunogenic and change the properties of the protein). Glycation products of glucose accumulate at higher levels in diabetics than in healthy individuals because of elevated blood glucose levels. These modified proteins are thought to underlie some of the pathologies associated with diabetes.

Glycosidic Bond Formation

Two monosaccharide units can be joined together by a *glycosidic bond*—this is the fundamental linkage among the monosaccharide building blocks found in all oligosaccharides. The glycosidic bond is formed between the anomeric carbon of one monosaccharide and a hydroxy group of another. In chemical terms, a hemiacetal group reacts with an alcohol group to form an acetal. Glycosidic bonds can be formed with virtually any hydroxylated compound, including simple alcohols such as methanol or ethanol (Figure 2.14) or hydroxy amino acids such as serine, threonine, and tyrosine. Indeed, glycosidic linkages are formed between sugars and these amino acids within proteins to form glycoproteins (see Chapters 8 and 9). Like the hemiacetal, the acetal or *glycosidic linkage* can exist in two stereoisomeric forms: α and β. But unlike the hemiacetal, the acetal is configurationally stable under most conditions. Thus, once a glycosidic bond is formed, its configuration is maintained indefinitely. Furthermore, no oxidation or reduction can take place at an anomeric center that is involved in a glycosidic bond. Like acetals in general, glycosidic bonds can be hydrolyzed in dilute acid, generating the constituent monosaccharides from oligosaccharides.

In the field of glycan synthesis, there has been considerable effort directed at the development of methods for constructing glycosidic bonds among monosaccharides. An overview of glycan synthesis strategies is provided in Chapter 49.

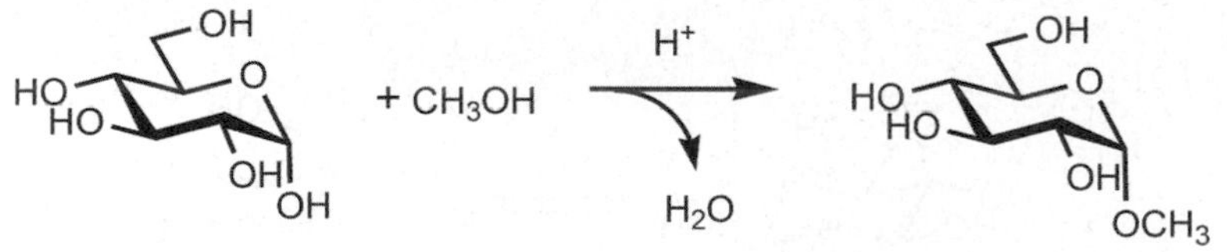

α-D-Glucopyranose

Methyl α-D-Glucopyranoside

FIGURE 2.14. Glycoside formation: Conversion of a hemiacetal into an acetal.

CHEMISTRY OF MONOSACCHARIDE FUNCTIONAL GROUPS

Methylation of Hydroxyl Groups

The hydroxyl groups of both monosaccharides and oligosaccharides can be chemically modified without affecting glycosidic linkages. A common modification is the capping of hydroxyl groups to generate methyl ethers. Methylation is a chemical transformation that is used in the structural analysis of glycans (see Chapter 47). Partially methylated glycans are also known to occur in natural products and a number of methyltransferases have been identified.

Esterification of Hydroxyl Groups

The hydroxyl groups of glycans can be esterified in nature by a variety of different enzymes. Esterification constitutes another element of variation in glycan structure and is sometimes required for interactions with other biomolecules. The most important types of sugar esters in nature are phosphate esters (including diphosphate esters), acyl esters (with acetic acid or fatty acids), and sulfate esters.

Deoxygenation of Hydroxyl Groups

The hydroxyl groups of monosaccharides can be replaced with hydrogen atoms to form deoxysugars. This can be achieved chemically using rather complex procedures, but nature has evolved reductases that perform such reactions in one step. For example, deoxygenation of ribose within a ribonucleotide to form the 2-deoxyribonucleotide is a critical reaction in DNA biosynthesis. Fucose (Fuc), one of the common vertebrate monosaccharides, is deoxygenated at C-6 (Figure 2.4) during its biosynthesis from mannose (see Chapter 4).

Amino Groups

Many monosaccharides have N-acetamido groups, such as GlcNAc, GalNAc, and NeuNAc. In rare cases, the N-acetamido group is de-N-acetylated to form amino groups. These are found in heparan sulfate (see Chapter 16), glycosylphosphatidylinositol (GPI) anchors (see Chapter 11), and many bacterial glycan structures (see Chapter 20). Amino groups can be modified with sulfates, similar to hydroxyl groups, as found in heparan sulfate.

GLYCAN STRUCTURE AND DIVERSITY

The diversity characteristic of glycans on glycoconjugates derives from the many ways in which monosaccharides can be linked together to form higher-order structures. Variety in glycosidic linkages has as its source the many possible isomers that can be formed between two monosaccharides. First, the glycosidic linkage can be formed in two possible stereoisomers at the anomeric carbon of one sugar (α or β). Second, the many hydroxy groups of the other sugar permit several possible regioisomers. For example, two glucose residues can be joined together in numerous ways, as illustrated by maltose (Glcα1-4Glc) and gentiobiose (Glcβ1-6Glc) (Figure 2.15). These isomers have very different three-dimensional structures and biological activities. Finally, a monosaccharide can be involved in more than two glycosidic linkages, thus serving as a branchpoint. The common occurrence of branched sequences (as opposed to the linear sequences that are found in almost all peptides) is unique to glycans and contributes to their structural diversity.

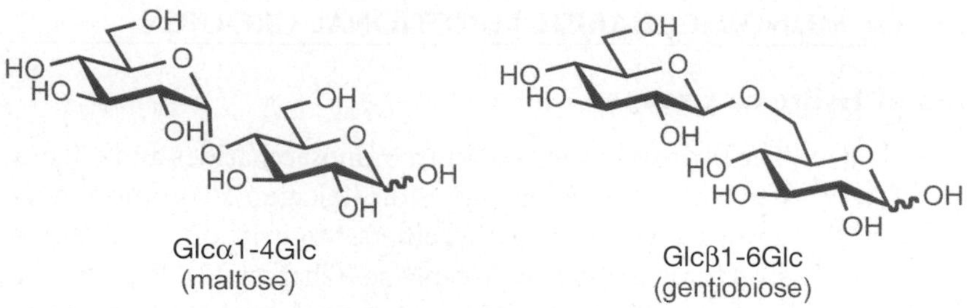

FIGURE 2.15. Two isomeric disaccharides.

The relationship of the glycosidic bond to oligosaccharides is analogous to the relationship of the peptide bond to polypeptides and the phosphodiester bond to polynucleotides. However, amino acids and nucleotides are linked in only one fashion during the formation of polypeptides and nucleic acids, respectively; there is no stereochemical or regiochemical diversity in these biopolymers. The number of monomeric residues contained in an oligosaccharide is designated in the nomenclature—*disaccharide, trisaccharide*, and so on. Just as polypeptides have amino and carboxyl termini and polynucleotides have 5′ and 3′ termini, oligosaccharides have a polarity that is defined by their *reducing* and *nonreducing termini* (Figure 2.16). The reducing end of the oligosaccharide bears a free anomeric center that is not engaged in a glycosidic bond and thus retains the chemical reactivity of the aldehyde. However, it continues to be referred to as the reducing end even when it is engaged in a linkage to another hydroxylic compound, for example, to the hydroxyl of serine or threonine, as in glycoproteins. Structures are commonly written from the nonreducing end on the left toward the reducing end on the right. For some structures, there is no reducing end. For example, the common disaccharides sucrose and trehalose have glycosidic linkages between the anomeric centers of two monosaccharide constituents (Figure 2.17).

The glycosidic linkage is the most flexible part of a disaccharide structure. Whereas the chair conformation of the constituent monosaccharides is relatively rigid, the torsion angles around the glycosidic bond (ϕ, ψ, and ω; Figure 2.18) can vary. Thus, a disaccharide of well-defined primary structure can adopt multiple conformations in solution that differ in the relative orientation of the two monosaccharides. The combination of structural rigidity and flexibility is typical of complex carbohydrates and, more than likely, essential to their biological functions.

Glycans are linked to other biomolecules, such as lipids or amino acids within polypeptides, through glycosidic linkages to form *glycoconjugates* (see Chapters 8–11). Glycans are often referred to as the *glycone* of a glycoconjugate and the noncarbohydrate component is named the *aglycone*. The glycan may be a single monosaccharide or an oligosac-

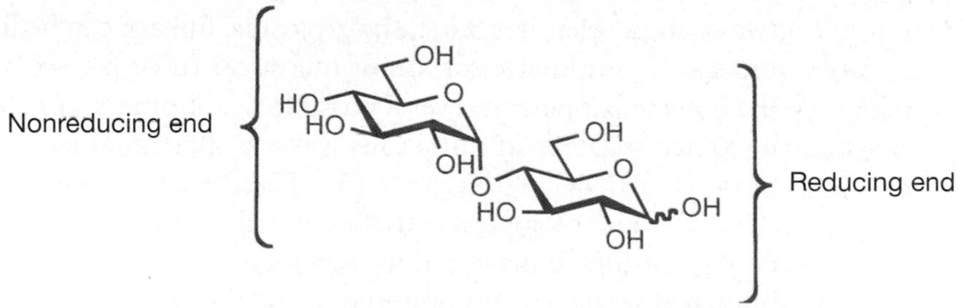

FIGURE 2.16. Reducing and nonreducing ends of a disaccharide.

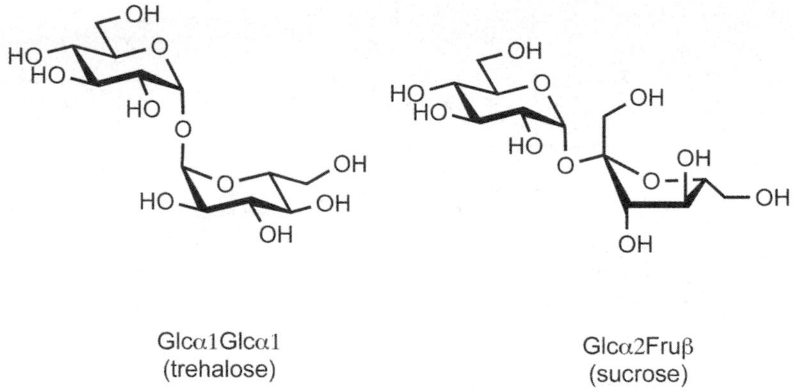

FIGURE 2.17. Nonreducing disaccharides.

charide. The attachment of many glycans to a polypeptide scaffold creates tremendous diversity among glycoproteins. The two general classes of protein-bound glycans are N- and O-linked. The N-linked glycans are bound to the nitrogen atom of asparagine side chains, whereas the O-linked glycans are bound to the oxygen atom of serine or threonine side chains. The structures, biosynthesis, and biological properties of N-linked glycans are discussed in Chapter 8 and those of O-linked glycans are described in Chapter 9. As examples, the structures depicted in Figure 2.19 represent a complex N-linked glycan and an O-linked glycan possessing a branched heptasaccharide structure. Many glycoproteins possess several attached glycans, each of different structure, and both N- and O-linked glycans can occur on the same polypeptide scaffold.

In conclusion, the tremendous diversity of glycoconjugate structures derives from many elements. The multiple monosaccharide building blocks can be linked to various regiochemistries and stereochemistries, and the resulting oligosaccharides can be assembled on protein or lipid scaffolds. Glycoconjugates therefore comprise an "information-rich" system capable of participating in a wide range of biological functions.

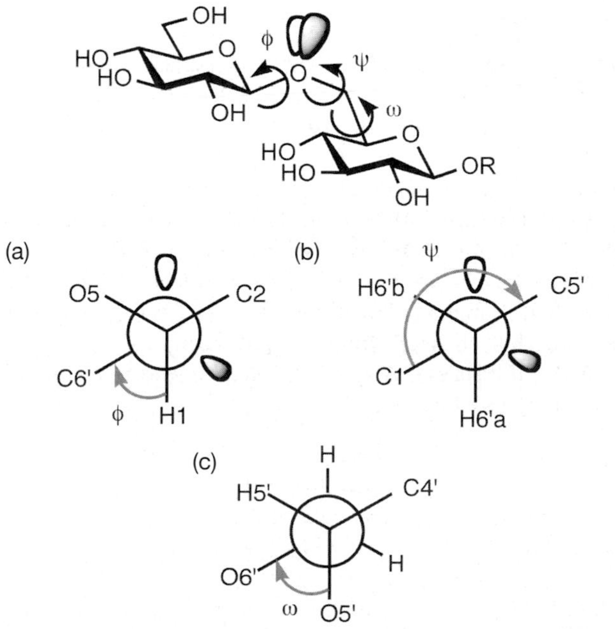

FIGURE 2.18. Torsion angles that define the conformation of the glycosidic linkages φ, ψ, and ω. (*a*) Newman projection along the C1—O1 bond illustrating φ for the 1→6 glycosidic bond. (*b*) Newman projection along the C6′—O1 bond illustrating ψ for a 1→6 linkage. (*c*) Newman projection along the C5′—C6′ bond illustrating ω for a 1→6 linkage. The lobes on the glycosidic oxygen atom represent lone pairs of electrons. The torsion angles depicted are arbitrary and do not necessarily reflect the most stable conformation.

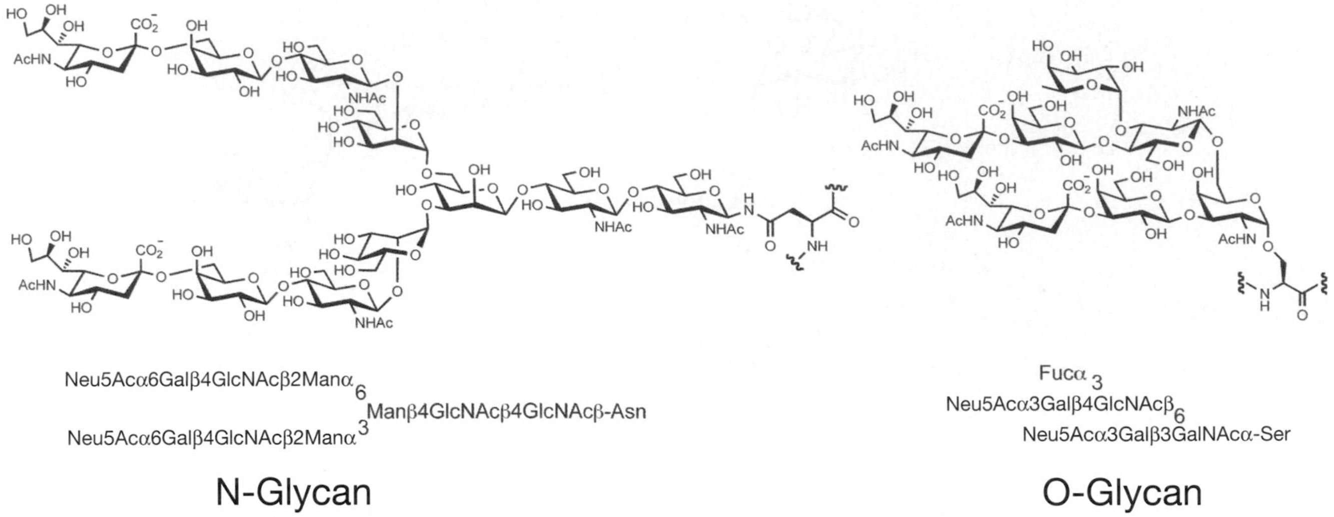

FIGURE 2.19. Examples of typical N- and O-linked glycans.

ONLINE RESOURCES

To assist the scientific community in keeping track of the primary structures and biological functions of glycans, several groups have generated computer-searchable databases, available through several large-scale glycomics initiatives.

The Consortium for Functional Glycomics (CFG) has developed a bioinformatics platform with search options including glycan structures and biological activities. Other major carbohydrate databases include the Kyoto Encyclopedia of Genes and Genomes (KEGG) Glycan database and the KEGG pathways database. A collection of linked carbohydrate databases can be found at the German Cancer Research Center (DKFZ). The Collaborative Glycomics Initiative, EUROCarbDB, can also be accessed.

FURTHER READING

El Khadem H.S. 1988. *Carbohydrate chemistry—Monosaccharides and their oligomers.* Academic Press, San Diego.

Allen H.J. and Kisailus E.C., eds. 1992. *Glycoconjugates: Composition, structure, and function.* Marcel Dekker, New York.

McNaught A.D. 1997. Nomenclature of carbohydrates. *Carbohydr. Res.* **297:** 1–92.

Bill M.R., Revers L., and Wilson I.B.H. 1998. *Protein glycosylation.* Kluwer Academic Publishers, Boston.

Boons G.-J., ed. 1998. *Carbohydrate chemistry.* Blackie Academic & Professional, London.

Stick R.V. 2001. *Carbohydrates: The sweet molecules of life.* Academic Press, New York.

Raman R., Raguram S., Venkataraman G., Paulson J.C., and Sasisekharan R. 2005. Glycomics: An integrated systems approach to structure-function relationships of glycans. *Nat. Methods* **2:** 817–824.

Gary H.G., Cowman M.K., and Hales C.A., eds. 2008. *Carbohydrate chemistry, biology and medical applications,* pp. 1–21. Elsevier, Oxford.

CHAPTER 3

Cellular Organization of Glycosylation

Ajit Varki, Jeffrey D. Esko, and Karen J. Colley

THIS CHAPTER PROVIDES AN OVERVIEW OF GLYCOSYLATION from the perspective of a single cell, taking into account the patterns of expression, topology, and other features of the biosynthetic and degradative enzymes that are common to most cell types. The focus is mainly on eukaryotic cells, for which more information is available than for prokaryotes.

GLYCOSYLATION IS UNIVERSAL IN LIVING ORGANISMS

It is a remarkable fact that every free-living cell and every cell type within multicellular organisms is covered with a dense and complex layer of glycans. Even enveloped viruses that bud from surfaces of infected cells carry with them the glycosylation patterns of the host cell. Additionally, most secreted molecules are glycosylated and the extracellular matrices of multicellular organisms are rich in glycans and glycoconjugates. The matrices secreted by unicellular organisms when they congregate (e.g., bacterial biofilms; see Chapter 20) also contain glycans. The reason for the apparent universality of cell-surface and secreted glycosylation is not clear, but it suggests that evolution has repeatedly selected for glycans as being the most diverse and flexible molecules to position at the interface between cells and the extracellular milieu. For example, the enormous diversity, complexity, and flexibility of glycans may allow host cells to make changes to avoid pathogens, without causing major deleterious effects on cellular functions.

Like membrane proteins, secretory proteins in eukaryotic cells typically pass through an endoplasmic reticulum (ER)-Golgi pathway, the cellular system in which many major glycosylation reactions occur (see below). Perhaps for this reason, most proteins in the blood plasma of animals (with the exception of albumin) are also heavily glycosylated. Glycosylation of secreted proteins may provide solubility, hydrophilicity, and negative charge, thus reducing unwanted nonspecific intermolecular interactions in extracellular spaces and also protecting against proteolysis. Another, not mutually exclusive, hypothesis is that the glycans on secreted molecules act as decoys, binding pathogens that seek to recognize cell-surface glycans to initiate invasion.

In bacteria, Archaea, and fungi, glycans have critical structural roles in forming the cell wall and in resisting large differences in osmolarity between the cytoplasm and the environment. Glycans surrounding bacteria could also have a role in defense against bacteriophages or antibiotics generated by other microorganisms in the environment.

TOPOLOGICAL ISSUES RELEVANT TO GLYCAN BIOSYNTHESIS

The ER-Golgi Pathway of Eukaryotes

Studies stemming from the classic work of George Palade and colleagues have indicated that most cell-surface and secreted molecules in eukaryotic cells originate in the ER. They then make their way via an intermediate compartment through multiple stacks of the Golgi apparatus, finally being distributed to various destinations from the *trans*-Golgi network. Along the way, lipids and proteins are modified by a variety of glycosylation reactions mediated by glycosyltransferases (see Chapter 5). Figure 3.1 superficially depicts some steps in the synthesis of the major glycan classes in the ER-Gogli pathway of animal cells. These pathways are discussed in the following sections and in other chapters of this book. As mentioned earlier, the ER-Golgi pathway is a universal feature of eukaryotic cells and also harbors other glycan-modifying enzymes (see below).

Not all glycans and glycoconjugates assemble within the ER-Golgi pathway. For example, many cytoplasmic and nuclear proteins contain O-GlcNAc, O-Glc, or O-Fuc, and these modifications occur in the cytoplasm (see Chapters 17 and 18). Hyaluronan and chitin assembly occurs at the plasma membrane, with direct extrusion into the extracellular matrix (see Chapter 15). In plant cells, cellulose synthesis also occurs at the plasma membrane (see Chapter 22).

Regardless of their location, most glycosylation reactions use activated forms of monosaccharides (most often nucleotide sugars) as donors for reactions that are catalyzed by enzymes called glycosyltransferases (see Chapter 4 for a listing of these enzymes and details about their biochemistry). A variety of glycan modifications are also found in nature (see Chapter 5). Of these, the most common are generated by sulfotransferases, acetyltransferases, and methyltransferases, which use activated forms of sulfate (3′-phosphoadenosine-5′-phosphosulfate; PAPS), acetate (acetyl-CoA), and methyl groups (*S*-adenosylmethionine; SAM), respectively. Almost all the donors for glycosylation reactions and glycan modifications are synthesized within the cytoplasmic compartment from precursors of endogenous origin. In eukaryotes, most of these donors must be actively transported across

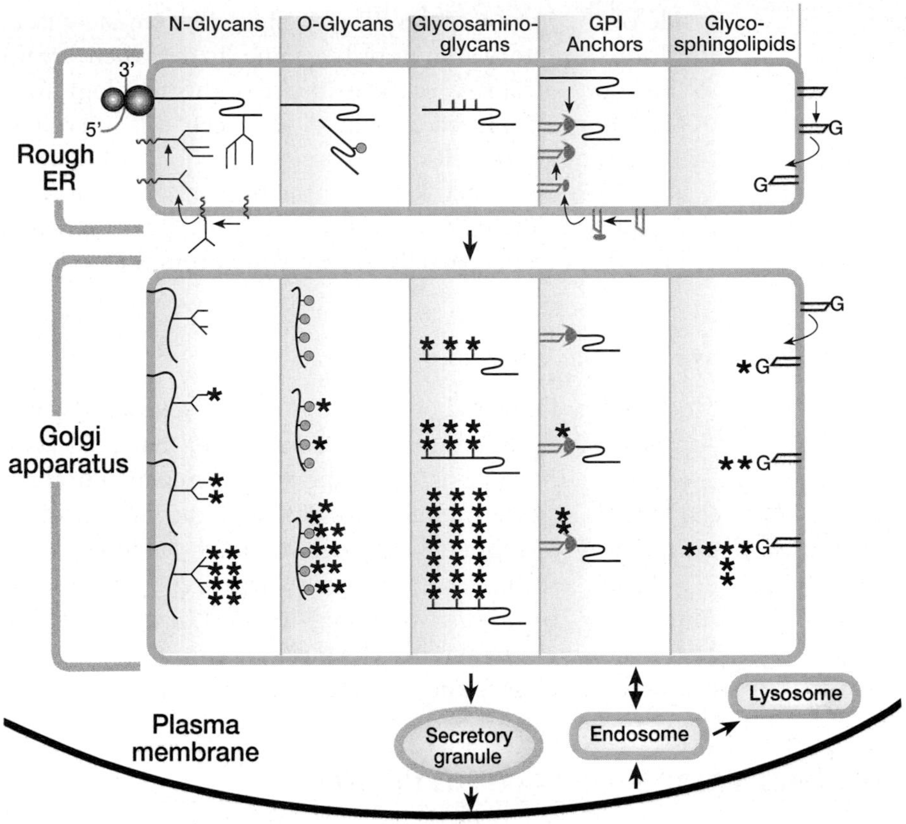

FIGURE 3.1. Initiation and maturation of the major types of eukaryotic glycoconjugates in relation to subcellular trafficking in the ER-Golgi–plasma membrane pathway. This illustration outlines the different mechanisms and topology for initiation, trimming, and elongation of the major glycan classes in animal cells. *Asterisks* represent the addition of outer sugars to glycans in the Golgi apparatus. N-glycans and glycosylphosphatidylinositol (GPI) anchors are initiated by the en-bloc transfer of a large preformed precursor glycan to a newly synthesized glycoprotein. O-glycans and sulfated glycosaminoglycans are initiated by the addition of a single monosaccharide, followed by extension. The most common glycosphingolipids are initiated by the addition of glucose to ceramide on the outer face of the ER-Golgi compartments, and the glycan is then flipped into the lumen to be extended. For a better understanding of the events depicted in this figure, see details in other chapters of this book: N-glycans (Chapter 8); O-glycans (Chapter 9); glycosphingolipids (Chapter 10); GPI anchors (Chapter 11); and sulfated glycosaminoglycans (Chapter 16).

a membrane bilayer in order to become available for reactions within the lumen of the ER-Golgi pathway.

Much effort has gone into understanding the mechanisms of glycosylation and glycan modification within the ER and the Golgi apparatus, and it is clear that a variety of interacting and competing factors determine the final outcome of the reactions. The glycosyltransferases, processing glycosidases, and sulfotransferases are well studied (see Chapter 5), and their location has helped to define various functional compartments of the ER-Golgi pathway. A popular model envisioned these enzymes as being physically lined up along this pathway in the precise sequence in which they actually work. In fact, there is considerable overlap of the enzymes across Golgi stacks and the actual distribution of a given enzyme depends on the cell type.

Some glycan chains are made on the cytoplasmic face of intracellular membranes and flipped across to the other side, but most are added to the growing chain on the *inside* of the ER or the Golgi (see Figure 3.1). Regardless, the portion of a molecule that faces the inside of the lumen of the ER or Golgi will ultimately face the *outside* of the cell or the inside of a secretory granule or lysosome. To date there are no well-documented exceptions to this topological rule. A consequence of this topological asymmetry is that many classes of glycans are optimized to be involved in cell–cell and cell–matrix interactions. Of course, these topological considerations are reversed for nuclear and cytoplasmic glycosylation (see below), because the active sites of the relevant glycosyltransferases for these reactions face the cytoplasm. Perhaps not surprisingly, the types of glycans found on the two sides of the cell membrane are generally quite distinct from each other.

Glycosylation Pathways in Eubacteriae and Archaea

Much less is known about the topology of glycoconjugate assembly in Eubacteriae and Archaea (formerly grouped as prokaryotes). Bacterial cells perform most glycosylation reactions on the inner aspect of the cytoplasmic membrane, using precursors assembled in the cytoplasm (see Chapter 20). These glycan intermediates are then flipped across the cytoplasmic membrane and used to form polymeric (often large) structures in the periplasm (see Chapter 21). In Gram-negative organisms, which have an outer membrane, some of the glycans or glycoconjugates must be transferred between membranes or flipped across the outer membrane. The mechanisms underlying these processes are active areas of research. Even less is known about the topology of glycosylation pathways in Archaea.

GOLGI ENZYMES SHARE SECONDARY STRUCTURE

Despite the lack of sequence homology among different families of glycosyltransferases and sulfotransferases, almost all Golgi enzymes share some features. Early studies of the cell biology and biochemistry of vertebrate glycosyltransferases indicated that some of these activities could be found in soluble form in secretions and body fluids, others were identified as membrane-bound activities within cells, and some exhibited both properties. Cell fractionation studies generally found cell-associated transferase activities in membrane-rich microsomal fractions, which could be liberated in soluble form with the aid of detergents. These observations implied that some transferases probably represent membrane-spanning proteins, whereas others correspond to secreted proteins. However, following the initial molecular cloning efforts that defined the sequences of a β1-4 galactosyltransferase, an α2-6 sialyltransferase, and the blood group A α1-3 *N*-acetylgalactosaminyltransferase, it became clear that Golgi glycosyltransferases share a common secondary structure that could account for all previous findings.

Almost all Golgi glycosyltransferases and sulfotransferases described to date have a single transmembrane domain flanked by a short amino-terminal domain and a longer carboxy-terminal domain. This structure is characteristic of so-called type II transmembrane proteins, whose single amino-terminal membrane-spanning domain functions as a signal-anchor sequence, placing the short amino-terminal segment within the cytoplasm while directing the larger carboxy-terminal domain to the other side of the biological membrane into which the signal anchor has been inserted (Figure 3.2). For plasma-membrane-associated type II proteins, the "other side" is the extracellular surface. For glycosyltransferases, the "other side" is the lumen of the membrane-delimited compartments that constitute the ER-Golgi pathway. These include vesicles that transit from the ER to the *cis* cisterna of the Golgi, the cisternae of the Golgi apparatus itself, the vesiculotubular network of the *trans*-Golgi network, and membrane-delimited structures distal to the *trans*-Golgi network. This arrangement predicts that the larger carboxy-terminal domain

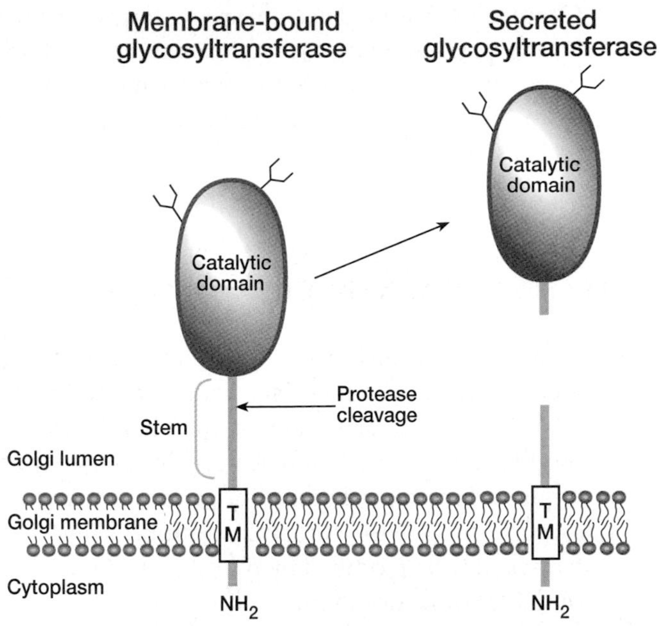

FIGURE 3.2. Typical transmembrane topology and proteolytic processing of Golgi glycosyltransferases. Golgi glycosyltransferases and sulfotransferases generally have a single hydrophobic segment (TM) that functions as a signal-anchor sequence. This segment spans the lipid bilayer of the tubular and vesicular structures of the secretory pathway, including the membrane of the Golgi apparatus. This topology places the catalytic domain of a glycosyltransferase within the lumen of the Golgi apparatus and other membrane-delimited structures of the secretory pathway. The membrane-tethered form of a glycosyltransferase is susceptible to one or more proteolytic cleavage events that transect the enzyme within its "stem" region. Proteolysis can liberate a catalytically active, soluble form of the enzyme that may be released from the cell. With few exceptions, vertebrate glycosyltransferases have one or more potential asparagine-linked glycosylation sites (*forked symbols*). Where examined experimentally, one or more of these sites are used, indicating that most glycosyltransferases are glycoproteins.

contains the catalytic activity of the transferase, and this supposition has extensive experimental support. The intralumenal location of this domain allows it to participate in the synthesis of the growing glycans displayed by glycoproteins and glycolipids during their transit through the secretory pathway.

The type II transmembrane topology predicted by initial sequences of vertebrate glycosyltransferases has been widely confirmed experimentally. The topology may also explain reports of the expression of glycosyltransferases at the surface of mammalian cells. Interestingly, several other types of Golgi enzymes (e.g., processing glycosidases) that have been cloned share a similar topological arrangement. There are a few clear exceptions, such the UDP-GlcNAc:lysosomal enzyme N-acetylglucosamine-1-phosphotransferase (GlcNAc-phosphotransferase) and the GlcNAc-1-phosphodiester α-N-acetylglucosaminidase that are both involved in the synthesis of the Man-6-P targeting signal of newly synthesized lysosomal hydrolases (see Chapter 30). The former is a multisubunit complex, and the latter is a type I membrane-spanning glycoprotein with its amino terminus in the lumen of the Golgi apparatus. One of the sulfotransferases involved in heparan sulfate synthesis (GlcNAc 3-O-sulfotransferase 1) is a resident soluble enzyme in the Golgi. Likewise, the protein O-fucosyltransferase (Notch FucT) appears to be a soluble enzyme in the ER.

All of the above considerations do not apply to the glycosyltransferases involved in nuclear and cytoplasmic glycan synthesis. For example, the soluble GlcNAc transferase responsible for synthesizing the O-linked GlcNAc of nuclear and cytoplasmic proteins (see

Chapter 18) has no detectable homology with the Golgi GlcNAc transferases. Another variation is presented by the hyaluronan and cellulose synthases, which are multipass membrane proteins present in the plasma membrane, extruding their products directly into the extracellular space. Similarly, all of the enzymes that use dolichol-linked precursors (and, in bacteria, bactoprenol-linked precursors) have more complex multi-membrane-spanning structures and contain a motif thought to bind to the isoprenoid chain.

GLYCOSYLTRANSFERASES CAN ALSO BE GLYCOSYLATED

Many Golgi glycosyltransferases have consensus N-glycosylation sequences as well as serine and threonine residues that could be modified by glycosylation processes. Biochemical analyses indicate that many mammalian glycosyltransferases are indeed posttranslationally modified by glycosylation, especially N-glycosylation. Glycosylation is, in some instances, required for proper folding and/or activity, and a few studies indicate that glycosyltransferases are subject to "autoglycosylation." There is also limited evidence that glycosyltransferases may be modified by phosphorylation. The functional relevance of such posttranslational modifications remains unknown.

LOCALIZATION OF GLYCOSYLTRANSFERASES IN GOLGI COMPARTMENTS

Biochemical and ultrastructural studies indicate that glycosyltransferases partially segregate into distinct compartments within the secretory pathway. Generally speaking, enzymes acting early in glycan biosynthetic pathways have been localized to *cis* and *medial* compartments of the Golgi, whereas enzymes acting later in the biosynthetic pathway tend to colocalize in the *trans*-Golgi cisternae and the *trans*-Golgi network. These observations have prompted extensive exploration of the mechanisms whereby glycosyltransferases achieve this compartmental segregation. An effort was made to find Golgi-retention sequences, by analogy with the KDEL tetrapeptide implicated in retention/retrieval of ER-associated proteins. Although some general conclusions arise from these studies, the reader should consider the following caveats:

- Observations made with one enzyme are not necessarily applicable to others.
- The Golgi-retention properties of any given glycosyltransferase may vary depending on the cell type in which localization is examined.
- Variations in the expression level of a glycosyltransferase in an experimental system can have a major influence on retention/localization properties.
- Many studies used chimeric proteins composed of segments of a glycosyltransferase fused to a reporter protein, but conclusions from such experiments have not always been verified using intact glycosyltransferases or chimeras with a different reporter protein.
- In vitro studies using intact Golgi compartments indicate some spatial and functional overlap among enzymes that were previously thought to be segregated on the basis of data from other less sensitive techniques.

Most information relevant to the retention of glycosyltransferases within specific Golgi compartments derives from experiments done with an α2-6 sialyltransferase (ST6Gal-I), a β1-4 galactosyltransferase (GalT-I), and an N-acetylglucosaminyltransferase I (GlcNAcT-I). The former pair of enzymes tends to concentrate in the *trans*-Golgi compartments and the *trans*-Golgi network, whereas GlcNAcT-I localizes mostly to the *medial*-Golgi compartment. With respect to ST6Gal-I, multiple signals and mechanisms may be involved in its Golgi localization. The transmembrane domain and flanking sequences are sufficient to direct heterologous proteins to the Golgi, and it appears that it is the length of the transmembrane segment (provided it is hydrophobic in nature) and not the precise sequence of this domain that is critical. Replacement of this region with a longer unrelated hydrophobic sequence does not compromise the Golgi localization of the intact enzyme, suggesting that other sequences are involved in localization. Recent evidence suggests that the cytoplasmic tail of this protein may mediate an additional localization mechanism and that the enzyme's lumenal sequences are involved in a secondary oligomerization event that stabilizes Golgi retention.

In contrast, an examination of the Golgi-retention determinants of GalT-I points mainly to an important role for the transmembrane domain. The sequences that flank the membrane-spanning domain seem less important in Golgi retention for this enzyme. Retention of GlcNAcT-I is also dictated largely by its transmembrane domain, although its lumenal sequences are also involved in an oligomerization mechanism that has a role in its localization. Considered together, the available experimental observations suggest that the localization of glycosyltransferases within specific regions of the Golgi apparatus is probably not determined by simple primary sequence motifs. Rather, this process is likely determined by several different regions of each enzyme that mediate multiple redundant localization mechanisms.

Golgi localization may also be mediated by retention (in the context of the vesicular transport model) and continuous retrieval to earlier Golgi cisternae (in the context of the cisternal maturation model). Three models have been proposed to account for the localization of glycosyltransferases to specific Golgi subcompartments. In the oligomerization/kin-recognition model, glycosylation enzymes form homooligomers or heterooligomers through interactions between their transmembrane and lumenal sequences after arriving at the proper Golgi compartment. Heterooligomerization or kin recognition among enzymes in the same pathway would also presumably enhance the efficiency of the sequential glycosylation reactions. Some glycosylation enzymes have been found to form homooligomers (e.g., ST6Gal-I, GlcNAcT-1 and GlcNAcT-2), and oligomerization appears to have a role in their stable localization in the correct Golgi compartment. Experimental support for kin recognition comes from the observation that some pairs of glycosyltransferases known to catalyze sequential reactions in the same pathway colocalize to a specific Golgi compartment and are coimmunoprecipitated from cell extracts. However, this type of association has not been demonstrated for most of the enzymes.

A second model depends on partitioning the glycosyltransferases into lipid bilayers of different thicknesses. This bilayer thickness or lipid-partitioning model was proposed on the basis of observations that a cholesterol concentration gradient in the secretory pathway yields lipid bilayers of increasing thickness in the direction of *cis* to *trans* across the Golgi stack and that the transmembrane regions of glycosyltransferases are generally shorter than those of plasma membrane proteins. The model predicts that each glycosyltransferase sorts itself into the proper Golgi location by virtue of the length of its transmembrane segment, which will retain the enzyme once it reaches the proper compartment during the enzyme's transit through the secretory pathway. This model was formulated largely on the basis of experiments involving ST6Gal-I, where the length of the membrane-spanning domain appears to play an important part in Golgi retention. However, the general applicability of this model is not apparent because there is no consistent relationship between the length of the transmembrane segment and retention in a specific Golgi compartment for a variety of glycosyltransferases. This model may therefore help to explain the overall phenomenon of retention of glycosyltransferases in the Golgi apparatus versus the delivery of other membrane proteins to the plasma membrane.

The third mechanism is based largely on the cisternal maturation model of transport through the secretory pathway. In this model, a new Golgi cisterna containing cargo molecules forms at the *cis* face of the stack and it progressively "matures" as Golgi glycosylation enzymes that define each subcompartment are transported into the new cisterna from the old cisterna to modify the cisternal cargo proteins. In this model, the steady-state localization of the Golgi enzymes is maintained by continuous retrograde transport. The cytoplasmic tails of enzymes may mediate interactions with coat proteins that select the enzymes for transport in vesicles or tubules. In support of this mechanism, the cytoplasmic tails of some glycosylation enzymes, such as ST6Gal-I and FucT-I, have been shown to have a role in their Golgi localization.

Taken together, evidence suggests that glycosylation enzymes use multiple mechanisms to maintain their localization in the Golgi. The number of signals and mechanisms used by an enzyme could determine how stable its Golgi localization actually is, whether it is able to move to a later compartment, and whether it can be cleaved and secreted into the extracellular space (see below).

PROTEOLYTIC CLEAVAGE AND SECRETION OF GOLGI GLYCOSYLTRANSFERASES

Many Golgi enzymes are secreted by cells, sometimes in large quantities, and can be found in cell culture supernatants and various body fluids. The nature of the secreted forms of glycosyltransferases first became clear from analysis of amino-terminal peptide sequences of purified mammalian glycosyltransferases that had been isolated as soluble forms without the aid of detergents. These studies showed that the soluble forms were actually derived from their membrane-associated forms by virtue of one or more proteolytic cleavage events that occurred a short distance away from the transmembrane segment within the stem region (Figure 3.2). These proteolytic cleavage events release a catalytically active fragment of the glycosyltransferase from its transmembrane tether and allow the cell to export this fragment to the extracellular space. The existence of catalytically active fragments of glycosyltransferases that are also deficient in various portions of the stem region together with mutational analyses imply that the stem region contributes little to the catalytic function of a glycosyltransferase. Nevertheless, some experimental analyses suggest that peptide sequences within the stem region can contribute to acceptor substrate preference.

The signals within the glycosyltransferase sequence that direct proteolysis are not defined, but it appears that the proteolytic cleavages are relatively specific and are generated by proteases functioning in the *trans* regions of the Golgi apparatus and beyond. The production of these soluble enzymes from cell types such as hepatocytes and endothelium can also be dramatically up-regulated under certain inflammatory conditions. Because these circulating enzymes do not have access to adequate concentrations of donor nucleotide sugars (primarily located inside cells), they should be functionally incapable of performing a transfer reaction in the extracellular spaces. The biological significance of these soluble transferases therefore remains a mystery. Possibilities to consider include a lectin-like activity recognizing their acceptor substrates and/or a role in scavenging small amounts of circulating sugar nucleotides that might otherwise be available to certain microbes, such as gonococci.

TURNOVER AND RECYCLING OF GLYCANS

Like all components of living cells, glycans turn over constantly. Some glycoconjugates, such as transmembrane heparan sulfate proteoglycans, turn over by shedding from the cell surface through limited proteolysis. Most glycoconjugate turnover occurs by endocytosis and subsequent degradation in lysosomes (see Chapter 41). Endoglycosidases can initially cleave glycans internally, producing substrates for exoglycosidases in the lysosome. Once broken down, individual monosaccharides are then typically exported from the lysosome into the cytoplasm, so that they can be reused (see Figure 1.8, Chapter 1). In contrast to the relatively slow turnover of glycans derived from the ER-Golgi pathway, glycans of the nucleus and cytoplasm may be more dynamic and rapidly turned over (see Chapters 17 and 18). Glycans in bacterial cells (especially those in the cell wall) also turn over during cell division when the cell wall undergoes cleavage and remodeling.

NUCLEAR AND CYTOPLASMIC GLYCOSYLATION IS VERY COMMON

Until the mid-1980s, a commonly stated dogma was that glycoconjugates such as glycoproteins and glycolipids occur exclusively on the outer surface of cells, on the internal (lumenal) surface of intracellular organelles, and on secreted molecules. As discussed above, this was in accord with information about the topology of the biosynthesis of the classes of glycans known at the time, which took place within the lumen of the Golgi-ER pathway. Thus, despite some clues to the contrary, the cytoplasm and nucleus (which are topologically semicontinuous because of the existence of nuclear pores) were assumed to be devoid of glycosylation capacity. However, it has now become clear that certain types of glycoconjugates are synthesized and reside within the cytoplasm and nucleus. Indeed, one of them (O-linked GlcNAc; see Chapter 18) may well be numerically the most common type of glycoconjugate in many cells.

GLYCOSYLATION REACTIONS AT THE PLASMA MEMBRANE

Because prokaryotic cells do not have an ER-Golgi pathway, they typically generate their cell-surface glycans at the interface of the cytoplasmic membrane and the cytoplasm or in the periplasm (see Chapter 20). Other glycans assembled at the plasma membrane include hyaluronan in vertebrate cells, chitin in invertebrate cells, and cellulose in plant cells. The topological difficulties of using cytoplasmic hydrophilic sugar nucleotides to manufacture glycans that are found on the opposite side of the cell-surface membrane are obvious. In eukaryotes, the process appears to involve enzyme molecules with multiple passes through the membrane that also act as a pore. In bacteria, membrane flippases may exist to aid in the transfer of intermediates across the various membranes. On the other hand, typical Golgi enzymes have also been claimed to be present at the cell surface in some animal cell types, with their active sites facing the extracellular space. It is not known how they could function at this location or how the nucleotide sugar donors would be delivered to this location. On the other hand, there are examples of remodeling of cell-surface glycans in animal cells: for instance, the Sulf enzymes (endo-6-sulfatases) that modify heparan sulfate glycosaminoglycan (see Chapter 16) and sialidases that may remove cell-surface sialic acids (see Chapter 14).

GLYCOSYLATION IN UNEXPECTED SUBCELLULAR LOCATIONS

There are scattered reports of glycosylation in unexpected subcellular locations, for example, gangliosides in mitochondria and N-glycans in the nucleus. Many of these claims are based on incomplete evidence (see Chapter 17 for some discussion of these issues). One possibility is that there are indeed glycans in these unexpected cellular locations, but that their true structures are actually novel. Conversely, there are instances where the structural evidence is strong, but there is inadequate evidence to be certain about the topology of the claimed structures. Regardless, past experience tells us that the cell biology of glycosylation can hold many surprises, and dogmatic positions about such controversial issues are not warranted.

FURTHER READING

Paulson J.C. and Colley K.J. 1989. Glycosyltransferases. Structure, localization, and control of cell type-specific glycosylation. *J. Biol. Chem.* **264:** 17615–17618.

Bretscher M.S. and Munro S. 1993. Cholesterol and the Golgi apparatus. *Science* **261:** 1280–1281.

Baenziger J.U. 1994. Protein-specific glycosyltransferases: How and why they do it. *FASEB J.* **8:** 1019–1025.

Colley K.J. 1997. Golgi localization of glycosyltransferases: More questions than answers. *Glycobiology* **7:** 1–13.

Glick B.S., Elston T., and Oster G. 1997. A cisternal maturation mechanism can explain the asymmetry of the Golgi stack. *FEBS Lett.* **414:** 177–181.

Varki A. 1998. Factors controlling the glycosylation potential of the Golgi apparatus. *Trends Cell Biol.* **8:** 34–40.

Munro S. 1998. Localization of proteins to the Golgi apparatus. *Trends Cell Biol.* **8:** 11–15.

Opat A.S., van Vliet C., and Gleeson P.A. 2001. Trafficking and localization of resident Golgi glycosylation enzymes. *Biochimie* **83:** 763–773.

Mironov A.A., Beznoussenko G.V., Polishchuk R.S., and Trucco A. 2005. Intra-Golgi transport: A way to a new paradigm? *Biochim. Biophys. Acta* **1744:** 340–350.

Czlapinski J.L. and Bertozzi C.R. 2006. Synthetic glycobiology: Exploits in the Golgi compartment. *Curr. Opin. Chem. Biol.* **10:** 645–651.

CHAPTER 4

Glycosylation Precursors

Hudson H. Freeze and Alan D. Elbein

GLYCOCONJUGATE BIOSYNTHESIS REQUIRES ACTIVATION OF MONOSACCHARIDES to nucleotide sugars. Monosaccharides are imported into the cell, salvaged from degraded glycans, or derived from other sugars within the cell. Although most glycosylation reactions occur in the Golgi, precursor activation and interconversions occur mostly in the cytoplasm. Nucleotide sugar–specific transporters carry the activated donors into the Golgi. This chapter describes how eukaryotic cells accomplish these tasks.

GENERAL PRINCIPLES

Glucose and fructose are the major carbon and energy sources for organisms as diverse as yeast and humans. Both organisms can derive the other monosaccharides needed for glycan biosynthesis from these sources. Not all of these biosynthetic pathways are equally active in all types of cells. However, there are some general principles. Glycan synthesis requires the monosaccharides to be activated to a high-energy donor form. This process requires nucleoside triphosphates (such as UTP or GTP) and a glycosyl-1-P (monosaccharide with a phosphate at the anomeric carbon). Several variations are used, but regardless of the monosaccharide, all must be either activated by a kinase (reaction 1) or generated from a previously synthesized activated nucleotide sugar (reactions 2 and 3):

Reaction 1 Sugar + NTP ⟶ Sugar-P $\xrightarrow{\text{NTP} \to \text{PPi}}$ Sugar-NDP

Reaction 2 Sugar(A)-NDP ⇄ Sugar(B)-NDP

Reaction 3 Sugar(A)-NDP + Sugar(B)-1-P + Sugar(B)-NDP + Sugar(A)-1-P

The most common nucleotide sugar donors in animal cells are shown in Table 4.1. Sialic acids are the only monosaccharides in animals activated as a mononucleotide, CMP-Sia. Iduronic acid does not have a nucleotide sugar parent because it is formed by epimerization of glucuronic acid after its incorporation into glycosaminoglycan (GAG) chains. In some instances, one nucleotide sugar can be formed from another by a nucleotide exchange reaction (reaction 3 above). For example, UDP-Gal is made from UDP-Glc by exchange of Gal-1-P for Glc-1-P.

EXTERNAL SOURCES AND TRANSPORTERS

Three types of sugar transporters carry sugars across the plasma membrane into cells. First are energy-independent facilitated diffusion transporters such as the glucose transporter (GLUT) family of hexose transporters found in yeast and mammalian cells. The genes encoding these proteins are named *SLC2A* (solute carriers 2A). Second are energy-dependent transporters, for example, the sodium-dependent glucose transporters (SGLT; gene names *SLC5A*) in intestinal and kidney epithelial cells. The third type are transporters that couple ATP-dependent phosphorylation with sugar import. These are found in bacteria and are not covered in this chapter.

GLUT family transporters were first described in yeast, where at least 18 genes are now known. Humans have 14 GLUT homologs. All of the yeast GLUT transporters are about the same size (40–55 kD) and have similar structures containing 12 membrane-spanning domains, which is typical of many eukaryotic transporters. The transmembrane domains form a barrel with a small pore for the sugar to pass through. Compared to GLUT1, the other family members have a modest 28–65% amino acid identity. The only "sugar transporter signatures" are a few widely scattered glycine and tryptophan residues and one PET tripeptide sequence. That is, there are no obvious major transporter motifs.

Typically, the GLUTs have K_m values for glucose uptake in the 2–20-mM range. In yeast, many transport glucose, but others are specific for galactose, fructose, or disaccharides. Most mammalian GLUT proteins transport glucose or fructose with variable efficiency, but

TABLE 4.1. *Activated sugar donors*

Sugar	Activated form
Glc	
Gal	
GlcNAc	UDP-sugar
GalNAc	
GlcA	
Xyl	
Man	GDP-sugar
Fuc	
Sia	CMP-Sia

detailed specificity studies for this family have not been done. However, GLUT5 transports fructose, and the GLUT called HMIT is a proton-coupled *myo*-inositol transporter. GLUT2 also efficiently transports glucosamine.

Glucose is transported from the gut by an energy-requiring Na^+-dependent glucose transporter (SGLT-1) and is recovered from the kidney filtrates by a related transporter (SGLT-2). The SGLT-type transporters have K_m values of less than 1 mM.

GLUT1–5 have different distributions in mammalian cells and different K_m values that enable them to respond to the availability of glucose (Table 4.2). Although most of the human GLUT members are located on the cell surface, a portion of GLUT4 is associated with intracellular vesicles and recruited to the cell surface in response to insulin. Following carbohydrate-rich meals, glucose transported by SGLT1 in the intestine is thought to promote the recruitment of GLUT2 to the apical surface for enhanced glucose uptake.

Mannose and Fucose Transport

Two types of mannose transporters are known. One is an energy-dependent transporter analogous to the SGLT variety for glucose. It is located on the brush border of enterocytes and on the surface of kidney tubule epithelial cells. It presumably transports mannose liberated from digested macromolecules and retrieves mannose from the kidney filtrates, much like those used for glucose transport. The second type is a GLUT-like transporter found on the surface of many types of mammalian cells. It is relatively insensitive to glucose inhibition and its K_m is near the concentration of mannose found in the blood of many mammals (50–100 μM). It is likely to be one of the known GLUT family members.

A fucose transporter has also been reported in several types of mammalian cells. Its K_m is approximately 250 μM, which is probably much higher than the fucose concentration in blood. Much of what is taken up can be converted into GDP-Fuc and incorporated into glycoproteins, but its contribution to glycosylation as compared to synthesis from GDP-Man is not known (see below).

Providing exogenous mannose to patients with a genetic deficiency in phosphomannose isomerase remarkably improves their clinical condition (see Chapter 42). However, it is important to stress that, except for these rare inherited diseases, there is no clinically proven

TABLE 4.2. *Monosaccharide transporters*

Protein	Gene	Localization	Primary sugar transported
GLUT1	*SLC2A1*	erythrocytes, brain, ubiquitous	glucose
GLUT2	*SLC2A2*	liver, pancreas, intestine, kidney	glucose (low affinity); fructose; glucosamine
GLUT3	*SLC2A3*	brain	glucose (high affinity)
GLUT4	*SLC2A4*	heart, muscle, fat, brain	glucose (high affinity)
GLUT5	*SLC2A5*	intestine, testes, kidney	fructose; glucose (very low affinity)
GLUT6	*SLC2A6*	brain, spleen, leukocytes	glucose
GLUT7	*SLC2A7*	n.d.	n.d.
GLUT8	*SLC2A8*	testes, brain, blastocyst	glucose
GLUT9	*SLC2A9*	liver, kidney	n.d.
GLUT10	*SLC2A10*	liver, pancreas	glucose
GLUT11	*SLC2A11*	heart, muscle	glucose (low affinity); fructose (long form)
GLUT12	*SLC2A12*	heart, prostate, muscle, small intestine	n.d.
GLUT14	*SLC2A14*	testes-specific	n.d.
HMIT	*SLC2A13*	brain	H^+-*myo*-inositol

(n.d.) Not determined.

benefit to providing either simple or complex sources of monosaccharides to healthy individuals. A number of commercial vendors tout the beneficial effects of complex "phytonutrients" or "glyconutrients" as sources of monosaccharides to improve human health. These plant polysaccharides are not digestible to monosaccharides by humans. Anaerobic bacteria in the colon can convert them to metabolic waste products such as butyric acid or other short-chain fatty acids, but not to monosaccharides available to the host. There is no evidence that these complex polysaccharides provide any bioavailable monosaccharides for humans, and claims to the contrary should be viewed with skepticism.

INTRACELLULAR SOURCES OF MONOSACCHARIDES

Salvage

Monosaccharides can also be salvaged from glycoconjugates degraded within cells (see Chapter 41). Most of the degradation occurs at low pH in the lysosomes. Salvage pathways have received relatively little attention, but their contribution to glycosylation may be quite substantial. For example, 80% of the radiolabeled *N*-acetylglucosamine from glycoproteins degraded in liver lysosomes is converted into UDP-GlcNAc and at least one-third is used to synthesize secreted glycoproteins. Also, fibroblasts endocytose labeled glycans and reuse about 50% of the amino sugars for new glycoprotein synthesis. Efficient salvage is not limited to *N*-acetylglucosamine. Glycosylation-impaired CHO cells require supplements of galactose and *N*-acetylgalactosamine for normal O-glycosylation and glycosaminoglycan biosynthesis (see Chapter 46), but these monosaccharides can be derived in sufficient quantities from glycoproteins in the serum supplement. Multiple studies show that much of the sialic acid from extracellular glycans enters via micropinocytosis of glycoproteins, which are degraded in the lysosome and reused for new glycoprotein synthesis.

Reuse of salvaged sugars requires that the monosaccharides exit the lysosome. Distinct lysosomal carriers are used for neutral hexoses (glucose, mannose, and galactose), N-acetylated amino sugars, and acidic hexoses. The neutral sugar carrier has a K_m value of 50–75 mM, but it also transports fucose and xylose. Although these sugars may slowly diffuse through the lysosomal membrane, their efflux rate (hours) would not be fast enough for efficient use. The *N*-acetylhexosamine carrier (K_m ~ 4 mM) cannot use nonacetylated amino sugars. The sialic acid and glucuronic acid carrier (K_m ~ 300–550 μM) is important because its loss leads to an accumulation of these sugars in the lysosome, secretion into the urine, and a human lysosomal storage disease (see Chapter 41). Most monosaccharides that reach the cytoplasm are activated, as described below. However, the uronic acids cannot be reused and are degraded via the pentose phosphate pathway.

Activation and Interconversion of Monosaccharides

Glycogen

Glycogen is an immense molecule that contains up to 100,000 glucose units, arranged in Glcα1-4Glc repeating disaccharides with periodic α1-6Glc branches. It is synthesized on a cytoplasmic protein called glycogenin (see Chapter 19). Glycogen is the major storage polysaccharide in animal cells, and its synthesis and degradation (glycogenolysis) are highly regulated for energy use. Glycogen is synthesized by the addition of single glucose units from UDP-Glc, and it is degraded by glycogen phosphorylase. The non-ATP-dependent reaction forms glucose-1-P by phosphorolysis of glycogen, which can be used directly to form UDP-Glc or converted to glucose-6-P for further catabolism via glycolysis or direct oxidation via glucose-6-phosphate dehydrogenase.

Glucose

Glucose is the central monosaccharide in carbohydrate metabolism, and it can be converted into all other sugars (Figure 4.1). Glucose is first converted to glucose-6-P by hexokinase. In the glycolytic pathway, glucose-6-P is converted to fructose-6-P by phosphoglucose isomerase or into glucose-1-P by phosphoglucomutase. Reaction of glucose-1-P with UTP forms the high-energy donor UDP-Glc. The UDP-Glc pool is quite large, and it is used to synthesize glycogen and other glucose-containing molecules such as glucosylceramide (see Chapter 10) and dolichol-P-glucose, which is used in the N-linked glycan biosynthetic pathway (see Chapter 8).

Some glucose-6-phosphate is acted on by glucose-6-P dehydrogenase, the entry point for the oxidation via the pentose phosphate pathway, to form 6-phosphogluconate and ribose-5-phosphate. These reactions generate $NADPH^+$, which is needed to maintain proper redox status.

Glucuronic Acid

UDP-GlcA is synthesized directly from UDP-Glc by a two-stage reaction requiring two NAD^+-dependent oxidations at C-6. UDP-GlcA is used primarily for GAG biosynthesis (see Chapters 15 and 16), but some N- and O-linked glycans and glycosphingolipids con-

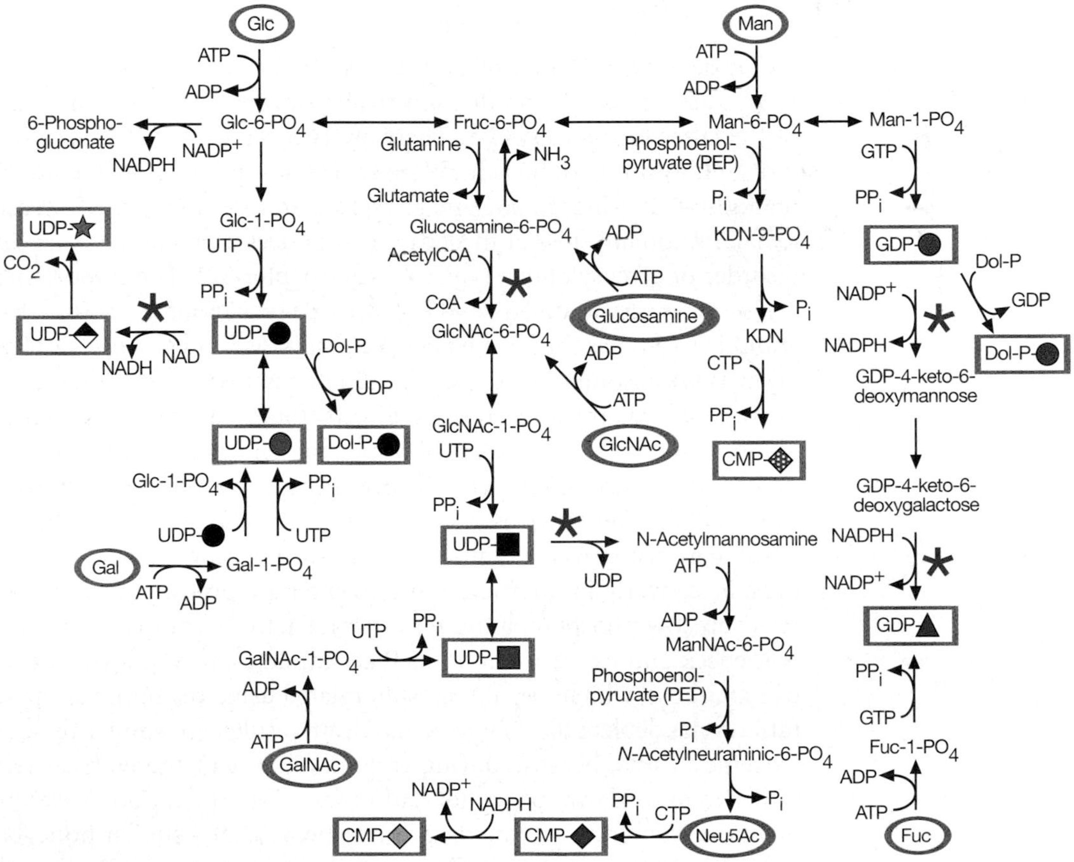

FIGURE 4.1. Biosynthesis and interconversion of monosaccharides. The relative contributions of each pathway under physiological conditions are unknown. (*Rectangles*) Donors; (*ovals*) monosaccharides; (*asterisks*) control points; (6PG) 6-phosphogluconate; (PEP) phosphoenolpyruvate; (KDN) 2-keto-3-deoxy-D-glycero-D-galactononic acid; (Dol) dolichol.

tain glucuronic acid as well. The addition of glucuronic acid to bile acids and xenobiotic compounds increases their solubility, and a large class of glucuronosyl transferases is devoted to such reactions.

Iduronic Acid

Iduronic acid is the C-5 epimer of glucuronic acid, and it is found in the GAGs, dermatan sulfate, heparan sulfate, and heparin. Unlike all other monosaccharides found in glycans, iduronic acid is not directly synthesized from a nucleotide sugar donor. Instead, it is created by epimerization of glucuronic acid after it has been incorporated into the growing GAG chain (see Chapter 16).

Xylose

Decarboxylation of UDP-GlcA gives UDP-Xyl, which is used primarily to initiate GAG synthesis (see Chapter 16). Xylose is also found on proteins that have O-glucose modifications in EGF modules (see Chapter 12). A type II membrane protein performs the decarboxylation reaction using UDP-GlcA transported into the ER or Golgi. In *C. elegans*, the decarboxylase is called SQV-1, and it colocalizes with the UDP-GlcA transporter (see Chapter 23). In *Arabidopsis*, another UDP-GlcA decarboxylase also occurs in the cytoplasm, but no ortholog has been identified in animals.

Mannose

Mannose is used for multiple types of glycans (see Chapters 8–12). GDP-Man is the primary activated donor. Its production requires prior synthesis of mannose-6-P and conversion of the latter to mannose-1-P. There are two ways to produce mannose-6-P. The first is by direct phosphorylation via hexokinase. The second is through conversion of fructose-6-P to mannose-6-P using the enzyme phosphomannose isomerase. In yeast, loss of this enzyme is lethal. In humans, loss of this enzyme produces a potentially fatal disease called congenital disorder of glycosylation (type Ib) (see Chapter 42). The importance of phosphomannose isomerase is easy to understand because free exogenous mannose is not common in the diet, and this enzyme is the key link between mannose and glucose. Both yeast and human phosphomannose isomerase deficiencies can be rescued by providing exogenous mannose.

Although mannose can be used directly for glycan synthesis, in the "honeybee syndrome" mannose can be lethal! This curious phenomenon occurs when honeybees are fed mannose instead of sucrose or glucose. The bees behave normally for several minutes, but then they suddenly die. The reason is that mannose enters the cells and is phosphorylated by abundant hexokinase using ATP. Because mannose-6-P is now the sole energy source for the bees, it must be converted into fructose-6-P to enter glycolysis. The problem is that honeybees have relatively low phosphomannose isomerase activity compared to hexokinase. This creates a bottleneck and excess mannose-6-P accumulates. It is quickly degraded by phosphatase to free mannose, which is again phosphorylated using the diminishing supply of ATP. Multiple futile cycles deplete the ATP pool and death results. For similar reasons, high mannose is teratogenic in rats, because during early development the embryo relies more on glycolysis rather than oxidative phosphorylation for ATP production. Mice totally lacking phosphomannose isomerase activity die in utero because of a similar honeybee effect.

In mammals, mannose-6-P is converted to mannose-1-P using phosphomannomutase. For this conversion, two isozymes are known in humans, and the loss of one of them produces another congenital disorder of glycosylation (type Ia) that results from underglycosylation of proteins (see Chapter 42). Because mannose-6-P and mannose-1-P are both obligate precursors of GDP-Man, failure to make sufficient amounts of either one reduces the formation of GDP-Man. GDP-Man can be used directly for the formation of the lipid-linked oligosaccharide on the cytoplasmic face of the ER. GDP-Man can also transfer mannose to dolichol phosphate to form dolichol-P-mannose in the ER membrane.

Mannose-6-P can also condense with phosphoenolpyruvate to form 3-deoxy-D-*glycero*-D-*galacto*-2-nonulosonic acid (Kdn). This molecule is activated with CTP to produce CMP-Kdn. Kdn is abundant in trout testis and on their sperm, where it is thought to be important for sperm-egg adhesion.

Fucose

GDP-Fuc can be derived from GDP-Man by the sequential action of two enzymes involving three steps. In the first step, the C-4 hydroxyl group of GDP-Man is oxidized to a ketone (GDP-4-keto-6-deoxy-mannose) by the enzyme GDP-Man 4,6-dehydratase along with the reduction of NADP to NADPH. The next two reactions are catalyzed by a single polypeptide that has epimerase and reductase activity and is well conserved from bacteria to mammals. GDP-4-keto-6-deoxy-mannose is epimerized at C-3 and C-5 to form GDP-4-keto-6-deoxygalactose, which is then reduced with NADPH at C-4 to form GDP-Fuc (Figure 4.2). The first dehydration step is feedback inhibited by GDP-Fuc. GDP-Fuc can also be synthesized directly from fucose. The first step uses a kinase to make fucose-1-P, which is then converted to GDP-Fuc. Mutant CHO cells that cannot convert GDP-Man to GDP-Fuc form hypofucosylated proteins, but this can be corrected by providing exogenous fucose in the medium. Also, mice genetically deficient in the GDP-Man to GDP-Fuc conversion can be rescued by providing fucose in their food or drinking water. This finding suggests that some transporters may carry other sugars besides glucose, but the quantitative contribu-

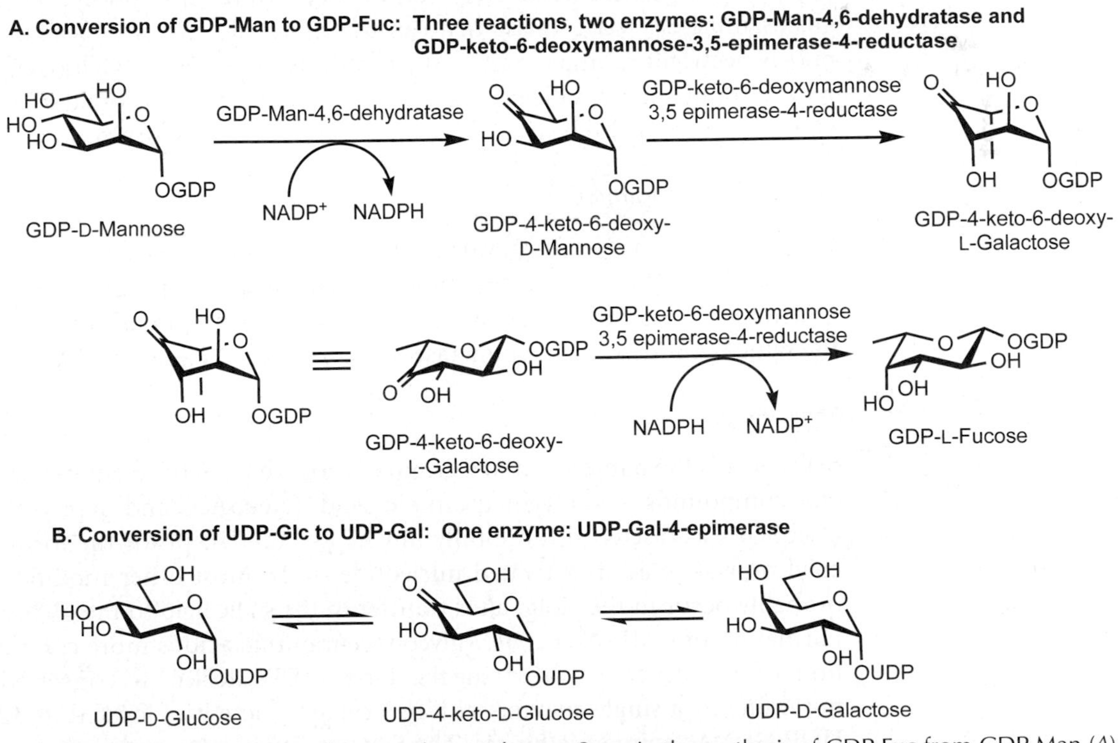

FIGURE 4.2. Conversion of activated sugar donors. Steps in the synthesis of GDP-Fuc from GDP-Man (*A*) and UDP-Gal from UDP-Glc (*B*). Details of the various enzymes are given in the text. GDP-Fuc synthesis by this route is not reversible, whereas the interconversion of UDP-Glc and UDP-Gal is easily reversible.

tion of this pathway is not known. The free fucose concentration in the blood is very low, a few micromolar at most.

Galactose

Activated UDP-Gal can be made in several ways. The first is by direct phosphorylation at C-1 to give galactose-1-P, which reacts with UTP to form UDP-Gal. Alternatively, galactose-1-P can be converted to UDP-Gal via a uridyl transferase exchange reaction with UDP-Glc that displaces glucose-1-P. A deficiency in this activity results in a severe human disease called galactosemia, which leads to mental retardation, liver damage, and eventual death if galactose intake is not controlled (see Chapter 42). Finally, UDP-Gal can be formed from UDP-Glc by the NAD-dependent reaction catalyzed by UDP-Gal 4-epimerase. The enzyme first converts the C-4 hydroxyl group to a keto derivative forming NADH from bound NAD^+. In the next step, the keto group is converted back to a hydroxyl group with opposite orientation and NAD^+ reforms (Figure 4.2). The same enzyme interconverts UDP-GalNAc and UDP-GlcNAc.

Galactose usually occurs as a pyranose (*p*) ring in higher animals, but bacteria and pathogenic eukaryotes such as *Leishmania* and *Aspergillus* incorporate galactofuranose (*f*) into their glycans (see Chapter 20). The donor is formed by conversion of UDP-Gal(*p*)→UDP-Gal(*f*) using a flavin-adenine-dinucleotide-dependent mutase.

N-Acetylglucosamine

Synthesis of UDP-GlcNAc begins with the formation of glucosamine-6-P from fructose-6-P by transamination using glutamine as the $-NH_2$ donor. Glucosamine-6-P is then N-acetylated via an acetyl-CoA-mediated reaction to form *N*-acetylglucosamine-6-P and then isomerized to *N*-acetylglucosamine-1-P via a 1,6-bis-phosphate intermediate. Similar to the other activation reactions, *N*-acetylglucosamine-1-P then reacts with UTP to form UDP-GlcNAc and pyrophosphate. Alternatively, GlcNAc can be directly phosphorylated to form *N*-acetylglucosamine-6-P via a kinase that can use either GlcNAc or *N*-acetylmannosamine. Phospho-*N*-acetylglucosamine mutase then converts this to *N*-acetylglucosamine-1-P. This route may account for the efficient salvage of GlcNAc from lysosomal degradation of glycans. Glucosamine can also be used following sequential phosphorylation and acetylation.

N-Acetylgalactosamine

UDP-GalNAc can arise from two routes. One is the direct reaction of *N*-acetylgalactosamine-1-P with UTP. *N*-Acetylgalactosamine-1-P is formed by a specific kinase that is distinct from galactose-1-kinase. UDP-GalNAc can also be formed by epimerization of UDP-GlcNAc using the same NAD-dependent epimerase that converts UDP-Glc to UDP-Gal.

Sialic Acids

Sialic acid is the name given to a group of more than 50 different variations of the two parent compounds, *N*-acetylneuraminic acid (Neu5Ac) and *N*-glycolylneuraminic acid (Neu5Gc), as discussed more fully in Chapter 14. Except for the formation of the N-glycolyl derivative as an activated nucleotide sugar, most other modifications of sialic acid probably occur in the Golgi after transfer of the sialic acid to the oligosaccharide acceptor. Formation of CMP-*N*-acetyl/*N*-glycolylneuraminic acid is more complicated than formation of the other activated sugars. First, UDP-GlcNAc is converted to *N*-acetylmannosamine by a single enzyme with two catalytic activities located in different domains of UDP-GlcNAc epimerase/kinase. The first activity involves epimerization at C-2 and cleavage of the UDP to yield *N*-acetylmannosamine. In the next reaction, this enzyme acts as a kinase using ATP to form *N*-acetylmannosamine-6-P. Mutations in this enzyme cause two completely distinct metabolic disorders: sialuria and inclusion body myopathy type 2 (see Chapter 42). Knocking out this gene in mice causes early embryonic lethality. In the next step, *N*-acetylmannosamine-6-P is condensed with phosphoenolpyruvate to form *N*-acetylneuraminic acid-9-P. Phosphate is then removed by a phosphatase. Activation with CTP yields CMP-*N*-acetylneuraminic acid. The last step occurs in the nucleus with subsequent export of the activated precursor to the cytoplasm.

There also appear to be alternate pathways for the synthesis of activated sialic acid donors. UDP-GlcNAc epimerase/kinase activity is prominent in relatively few tissues including the liver, salivary gland, and intestinal mucosa, but clearly many other tissues contain sialylated glycoproteins and glycolipids. This finding suggests that other pathways probably exist. Sialic acids can be salvaged from glycoprotein turnover or from plasma and activated by phosphorylation and addition of CMP from CDP. In addition, *N*-acetylglucosamine 2′-epimerase can generate *N*-acetylmannosamine, which can be converted to *N*-acetylmannosamine-6-P by *N*-acetylglucosamine kinase.

Unusual Sugars in Bacteria and Plants

Fucose is the only deoxyhexose found in animal cell glycans. In contrast, bacterial and plant polysaccharides and glycoproteins frequently contain a variety of deoxysugars, deoxyaminosugars, and branched-chain sugars. These unusual sugars often have potent biological properties. For example, the glycan moiety of streptomycin binds to the minor groove of DNA to form stable antibiotic-DNA complexes. Deoxyhexoses are often immunological determinants of lipopolysaccharides or O-antigens of the *Salmonella* species. Five of the eight possible 3,6-dideoxyhexoses have been found in these organisms at the nonreducing end of the Gram-negative cell wall lipopolysaccharide. Other deoxyhexoses, such as a 4,6-dideoxyhexose and a 2,3,6-trideoxyhexose, are also biologically significant but uncommon in nature.

Biosynthesis of both deoxysugars and dideoxysugars begins with the oxidation of C-4, which is catalyzed by a 4,6-dehydratase to produce an NDP-4-keto-6-deoxyhexose. This is similar to the first step of the conversion of GDP-Man to GDP-Fuc. The nucleotide (N) differs for the various sugars, and the individual pathways use different dehydratases. For example, biosynthesis of most 3,6-dideoxyhexoses (except colitose) begins with conversion of CDP-glucose to CDP-4-keto-6-deoxyhexose by NAD^+-dependent CDP-glucose dehydratase. In the biosynthesis of abequose (3,6-dideoxy-D-xylohexose), the product, CDP-6-deoxy-L-threo-D-glycero-hexulose, is then converted in two additional steps to CDP-3,6-dideoxy-D-glycero-D-glycero-4-hexulose by a second dehydratase followed by a reductase.

Amino sugars, such as glucosamine, arise from keto sugars by the addition of an amino group from glutamine (Figure 4.1). In addition, bacteria and plants have many 6-deoxyhexoses with amino groups in the 2, 3, or 4 positions. For example, duanosamine is a 3-amino-6-deoxyhexose that is found in the antibiotic duanomycin. Here, TDP-glucose is dehydrated to 3-keto-6-deoxyglucose and the amino group is added via a transamination reaction probably involving a vitamin B6-dependent reaction.

Plants and bacteria also contain a number of branched-chain sugars. For instance, apiose is a component of the polysaccharide apiogalacturonan of *Lemna minor*, and strepose is a component of the antibiotic streptomycin produced by *Streptomyces griseus*. Apiose (see Figure 4.3A) is synthesized from UDP-GlcA via a 4-keto intermediate that can yield UDP-Xyl or UDP-apiose. Apiose synthesis removes carbon 3 from the chain to give the branched sugar by an unknown mechanism. Streptose (5-deoxy-3-C-formyl-L-lyxose) originates by an intramolecular rearrangement of the hexose of TDP-4-keto-6-deoxy-L-lyxohexose (Figure 4.3B). Although the synthesis of other branched-chain sugars has not been delineated, they probably follow similar reaction pathways.

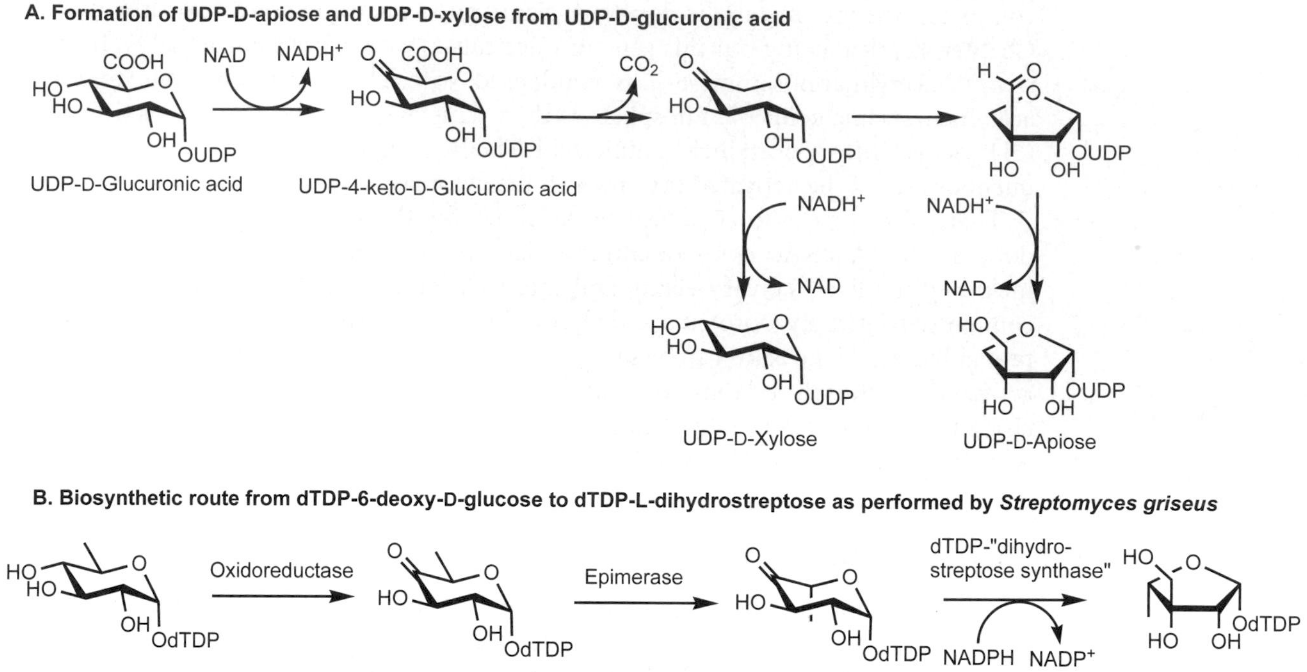

FIGURE 4.3. Steps in the biosynthesis of (*A*) the branched sugar donor UDP-apiose from UDP-GlcA and (*B*) dTDP-dihydrostreptose from dTDP-glucose. *A* shows the conversion of UDP-GlcA into UDP-apiose, which is used in plant polysaccharides such as apiogalacturonan in *Lemna minor*. Note the similarity and overlap to UDP-xylose synthesis. The only difference is that the C-3 is removed by an unknown mechanism, and the reduction of the newly formed aldehyde creates the branched sugar donor. *B* shows the synthesis of the donor for dehydrostreptose that results from a complex intramolecular rearrangement.

NUCLEOTIDE SUGAR TRANSPORTERS

In eukaryotes, the nucleotide sugars synthesized in the cytoplasm or the nucleus are on the "wrong" side of the membrane and must be transported into the ER and Golgi. Negative charge prevents these donors from simply diffusing into these compartments. To overcome this problem, eukaryotic cells have a set of energy-independent nucleotide sugar antiporters that deliver nucleotide sugars into the lumen of these organelles with the simultaneous exiting of nucleoside monophosphates; the latter must first be generated from the nucleoside diphosphates by a nucleoside diphosphatase (Figure 4.4). This transport mechanism was established biochemically in isolated vesicles and genetically in various mutant cell lines. The K_m of the transporters ranges from 1 to 10 μM. Using in vitro systems, the transporters have been shown to increase the concentration of the nucleotide sugars within the Golgi lumen by 10- to 50-fold. This is usually sufficient to reach or exceed the calculated K_m of glycosyltransferases that use these donors.

Most of the antiporters are found in the Golgi, but some are also found in the ER. Thus, they are organelle specific and their location usually corresponds to the location of the downstream glycosyltransferases (Table 4.3 and Figure 4.4). Nucleotide sugar import into the Golgi is not energy dependent or affected by ionophores. However, the import is competitively inhibited by the corresponding nucleoside monophosphates and diphosphates but not by the monosaccharides. ATP and PAPS (3′-phosphoadenosine-5′-phosphosulfate) transporters are also known; the latter is used for carbohydrate and protein sulfation.

Glucuronidation of bile and xenobiotic compounds in the ER is consistent with the presence of the UDP-GlcA transporter in the ER. The presence of a Golgi transporter is consistent with the location of the polymerases that use UDP-GlcA for the formation of

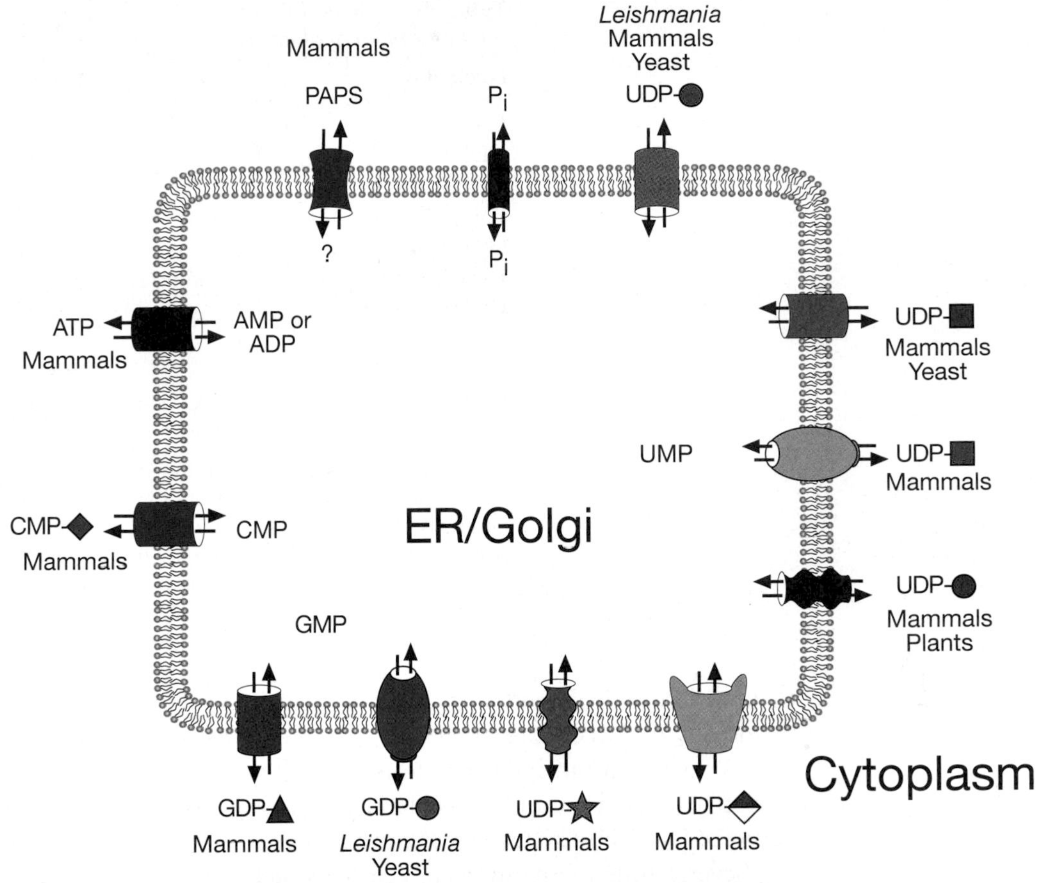

FIGURE 4.4. Transporters for nucleotide sugars, PAPS (3′-phosphoadenosine-5′-phosphosulfate), and ATP are located in the Golgi membranes of mammals, yeast, protozoa, and plants. These proteins are antiporters and return the corresponding nucleoside monophosphate back to the cytoplasm when the nucleotide sugar is delivered into the ER or Golgi compartment. Because most glycosylation reactions produce a nucleoside diphosphate, this requires conversion to the nucleoside monophosphate. For PAPS, the corresponding exiting molecule is unknown and for ATP, it is AMP, ADP, or both. The phosphate (P_i) transporter is hypothetical.

GAG chains. The observation that reglycosylation of misfolded glycoproteins occurs in the ER (see Chapter 36) explains the need for an ER UDP-Glc transporter. Under stressful conditions that activate the unfolded protein response, synthesis of lumenal uridine diphosphatase increases to accommodate increased transport of UDP-Glc needed for reglucosylation of misfolded glycoproteins. The existence of UDP-GlcNAc, UDP-GalNAc, and UDP-Xyl transporters in the ER may mean that some reactions thought to occur exclusively in the Golgi may also occur in the ER. A good example is the synthesis of O-fucosylated proteins such as Notch in the ER versus fucosylation of N- and O-linked chains in the Golgi (see Chapter 12). Other, as-yet-undiscovered, glycosylation reactions may also occur in the ER.

For most glycosylation reactions, the nucleotide sugar donates the sugar, resulting in the formation of nucleoside diphosphate, which must be converted into a monophosphate by the nucleoside diphosphatase in the Golgi lumen. Exchange through the antiporters is electroneutral, because the nucleotide sugar with two negative charges (one on each phosphate) enters the Golgi and a nucleotide with a doubly charged single phosphomonoester exits. The antiporter system has the advantage of coupling the rate of nucleotide sugar use with its import. However, the Golgi can accumulate a pool of unused nucleotide sugars for

TABLE 4.3. *Nucleotide transport in Golgi and endoplasmic reticulum (ER)*

Nucleotide	ER	Golgi
CMP-Sia	–	+++
GDP-Fuc	+	++++
UDP-Gal	–	++++
PAPS	–	++++
GDP-Man	–	++++
UDP-GlcNAc	++	++++
UDP-GalNAc	++	++++
UDP-Xyl	++	++++
ATP	+++	++++
UDP-GlcA	++++	++++
UDP-Glc	++++	+

The relative distribution of the nucleotide transporters in ER and Golgi is indicated by the number of plus signs (+). A minus sign (–) indicates that the transporter is not found in that compartment. See Figure 4.4 for the types of organisms that contain these nucleotide transporters.

glycosylation. Isolated Golgi preparations are capable of glycosylating partially completed endogenous glycoproteins as well as freely diffusible glycoside acceptors. Another advantage of using the antiporter system is that the nucleoside monophosphate is returned to the cytoplasm where it will be available for another round of activation. This creates a highly efficient recycling system for the precursors.

Several mutant mammalian cell lines lack specific nucleotide sugar transporters (e.g., UDP-Gal and CMP-Sia), and as a result, they make incomplete sugar chains (see Chapter 46). However, there is some "leakiness" in such mutants. For instance, loss of the UDP-Gal in the Golgi of mutant MDCK cells decreases the synthesis of keratan sulfate and galactosylated glycoproteins and glycolipids, but leaves heparan and chondroitin sulfate unaffected. This is probably because the galactosyltransferases that synthesize the core region tetrasaccharide common to GAG chains have lower K_m values for their donors (see Chapter 16).

Many putative transporters were identified by homology in the genomes of mammals, *D. melanogaster*, *C. elegans*, plants, and yeast. Like the GLUT transporters discussed above, all are multi-membrane-spanning (type III) proteins, but the level of amino acid identity does not give any clue to the substrate specificity. The UDP-GlcNAc transporters from mammalian cells and yeast are 22% identical, whereas mammalian CMP-Sia, UDP-Gal, and UDP-GlcNAc transporters have 40–50% identity. Clever domain-swapping experiments show that distinct regions are responsible for functional transport, and engineered chimeric transporters can carry both CMP-Sia and UDP-Gal.

Heterologous expression or rescue of transporter-deficient cell lines can be used to analyze the function of the putative transporters. For example, expressing the *C. elegans* gene *SQV-7* in yeast showed that this one protein transports UDP-GlcA, UDP-GalNAc, and UDP-Gal, whereas mutant alleles cannot transport any of these donors. The human gene *SLC35B4* encodes a bifunctional transporter that recognizes UDP-Xyl and UDP-GlcNAc. Also, the GDP-Man transporter of *Leishmania* can also transport GDP-Fuc and GDP-arabinose. Even though multisubstrate transport is somewhat unusual, it illustrates that functional, biochemical analysis is essential; homology is an insufficient criterion to infer functional specificity. Moreover, not all of the potential transport-like genes have been assigned a specific substrate.

Theoretically, glycosylation may be controlled in part by regulating availability of nucleotide sugars within the Golgi, presumably by regulating the transporters. The subcompartmental location (*cis*, *medial*, *trans*) of the transporters in the Golgi is not known nor are their physical relationships to the various glycosyltransferases they service. Clearly, a functional Golgi compartment requires both the nucleotide sugar donor and the acceptor with a colocalized transferase. There have been few studies on how the actual glycosylation reactions occur within the Golgi. Is it more like solution chemistry or like solid-state transfers? Are there really "soluble pools" of nucleotide sugars? Dramatic time-lapse videos of green fluorescent protein (GFP)-tagged glycosyltransferases show that the proteins are highly mobile within the Golgi, but there is also physical evidence for multiglycosyltransferase complexes involved in the biosynthesis of N-linked glycans, glycosphingolipids, and heparan sulfate. Many transporters appear to function as homodimers, and the GDP-Man transporter in *Saccharomyces cerevisiae* (VRG4) oligomerizes in the ER and appears to be transported to the Golgi by an active process. Also, synthesis of galactosylceramide occurs in the ER and a portion of the UDP-Gal transporter binds specifically to galactosylceramide transferase and is retained in the ER to provide donor substrate (see Chapter 10).

CONTROL OF NUCLEOTIDE SUGAR LEVELS

How nucleotide sugar levels are controlled and maintained has not been studied extensively, but this topic is likely very important. Table 4.4 shows several key biosynthetic enzymes that are inhibited by their final products in vitro. A human genetic disorder called sialuria validates the importance of this feedback process. In this condition, massive amounts of sialic acid (5–7 gm/day) are secreted into the urine along with various intermediates in the CMP-Sia biosynthetic pathway. This was found to be due to a defective feedback inhibition of the N-acetylglucosamine-2-epimerase/N-acetylmannosamine kinase, the first step in the pathway (see Chapter 42).

Most of the precursor pools turn over within a matter of a few minutes, and the size of various nucleotide sugar pools have been determined by methods with different reliability. But even with the best methods, it is hard to interpret the measured numbers and translate them into a clear picture, because the relative distribution of the precursors in the cytosol and Golgi is not known. The average cellular concentrations of nucleotide sugar precursors may not be very meaningful, because "cytosol" is an operational definition (100,000*g* supernatant) that may not reflect compartments of cytoplasmic organization.

In whole-animal studies, the GDP-Fuc pool and fucosylated glycans in the intestine can be regulated by the diet and time of weaning. Considering that resident bacteria in the small intestine participate in the induction of fucosylation pathways in the enterocytes, dietary manipulation of glycosylation introduces another level of complexity that has barely been explored. The relationship of amino acid and nucleotide metabolism to nucleotide sugar metabolism is also potentially important, but remains largely unexplored (see Chapter 34).

TABLE 4.4. *Some control points for nucleotide sugar synthesis*

Enzyme	Inhibitor
UDP-Glc dehydrogenase	UDP-Xyl
GDP-Man 4,6-dehydratase	GDP-Fuc
Glutamine:fructose-6-P amidotransferase	UDP-GlcNAc
UDP-GlcNAc epimerase/kinase	CMP-Sia

TABLE 4.5. *Donors for glycan modifications*

Modification	Precursor	Transporter
Phosphate	ATP (?)	yes
Sulfate	PAPS	yes
Methyl	*S*-adenosylmethionine	?
Acetyl	acetyl-CoA	yes
Pyruvate	phosphoenolpyruvate	?
Acyl	acyl-CoA (?)	?
Succinyl	succinyl-CoA (?)	?

DONORS FOR GLYCAN MODIFICATION

Glycans can be modified, imparting additional complexity and biological information. Sulfation, phosphorylation, methylation, pyruvylation, acetylation, and acylation have been found and their donors are listed in Table 4.5. In "lower" eukaryotes and bacteria, pyruvic acid is often found as a 1-carboxyethylidene bridge between two hydroxyl groups on a sugar such as galactose. Because all of these reactions occur in the Golgi, clearly there must be carriers or transporters that deliver and orient activated donors for efficient synthesis. As additional modifications of sugar chains made in the ER-Golgi pathway are uncovered, they will likely require specific transporters to carry the activated donors into the lumen of these compartments.

SYNTHESIS OF CARRIER LIPIDS

Multiple glycosylation pathways in prokaryotes and eukaryotes require lipid carriers to present monosaccharides and oligosaccharides at the proper location. Undecaprenyl-P (bactoprenol) is the glycosyl carrier for O-antigen, peptidoglycan, capsular polysaccharides, teichoic acid, and mannans in bacteria (see Chapter 20). Dolichol-P serves the same function in eukaryotic cells (see Chapter 8). Dolichol-P-mannose provides all of the mannose for glycophospholipid anchors, C-mannosylated proteins, O-mannose-based chains, and four of the mannose residues of the precursor oligosaccharide used for N-glycan biosynthesis. Dolichol-P-glucose provides glucose for the mature N-linked glycan precursor $Glc_3Man_9GlcNAc_2$, which itself is built on dolichol pyrophosphate (dolichol-PP).

The formation of dolichol-P involves elongation of farnesyl pyrophosphate with multiple *cis*-isopentenyl pyrophosphate units. The total number of isoprene units can vary from typically 11 in bacteria (making a C_{55} bactoprenol chain) to 21 in mammals. In eukaryotes, the double bond nearest the pyrophosphate must be reduced for the carrier to be functional, but it is unclear whether the phosphates are removed before or after this step. The evolutionary significance of the different chain lengths and reduction of the double bond is not known. Dolichol is phosphorylated by an ATP-dependent dolichol kinase to generate dolichol-P as needed. Because dolichol, dolichol-P, and dolichol-PP are all generated from a common metabolically stable pool, they must be recycled and interconverted as needed. Dolichol occurs in the ER and Golgi and turns over very slowly.

FURTHER READING

Neufeld E.F. and Ginsburg V. 1965. Carbohydrate metabolism. *Annu. Rev. Biochem.* **34:** 297–312.

Krieger M., Reddy P., Kozarsky K., Kingsley D., Hobbie L., and Penman M. 1989. Analysis of the synthesis, intracellular sorting, and function of glycoproteins using a mammalian cell mutant with reversible glycosylation defects. *Methods Cell Biol.* **32:** 57–84.

Lloyd J.B. 1996. Metabolite efflux and influx across the lysosome membrane. *Subcell. Biochem.* **27:** 361–386.

Hirschberg C.B., Robbins P.W., and Abeijon C. 1998. Transporters of nucleotide sugars, ATP, and nucleotide sulfate in the endoplasmic reticulum and Golgi apparatus. *Annu. Rev. Biochem.* **67:** 49–69.

Varki A. 1998. Factors controlling the glycosylation potential of the Golgi apparatus. *Trends Cell Biol.* **8:** 34–40.

Schenk B., Fernandez F., and Waechter C.J. 2001. The ins(ide) and out(side) of dolichyl phosphate biosynthesis and recycling in the endoplasmic reticulum. *Glycobiology* **11:** 61R–70R.

He X.M. and Liu H.W. 2002. Formation of unusual sugars: Mechanistic studies and biosynthetic applications. *Annu. Rev. Biochem.* **71:** 701–754.

Ishida N. and Kawakita M. 2004. Molecular physiology and pathology of the nucleotide sugar transporter family (SLC35). *Pflugers Arch.* **447:** 768–775.

Tettamanti G. 2004. Ganglioside/glycosphingolipid turnover: New concepts. *Glycoconj. J.* **20:** 301–317.

Uldry M. and Thorens B. 2004. The SLC2 family of facilitated hexose and polyol transporters. *Pflugers Arch.* **447:** 480–489.

Eklund E., Bode L., and Freeze H.H. 2007. Diseases associated with carbohydrates/glycoconjugates. In *Comprehensive Glycoscience*, Vol. 4, Chap. 4.19, pp. 339–372. Elsevier, London.

Sagné C. and Gasnier B. 2008. Molecular physiology and pathophysiology of lysosomal membrane transporters. *J. Inherit. Metab. Dis.* Apr 15 [Epub ahead of print] PMID: 18425435.

CHAPTER 5

Glycosyltransferases and Glycan-processing Enzymes

James M. Rini, Jeffrey D. Esko, and Ajit Varki

THIS CHAPTER COVERS THE GENERAL CHARACTERISTICS OF THE ENZYMES involved in glycan biosynthesis and modification, including aspects of substrate specificity, primary sequence relationships, structures, and enzyme mechanisms.

GENERAL PROPERTIES

The biosynthesis of glycans is primarily determined by the glycosyltransferases that assemble monosaccharide moieties into linear and branched glycan chains. As might be expected from the complex array of glycan structures found in nature, the glycosyltransferases constitute a very large family of enzymes. However, they have in common the ability to catalyze a group-transfer reaction in which the monosaccharide moiety of a simple nucleotide sugar donor substrate (e.g., UDP-Gal, GDP-Fuc, or CMP-Sia; see Chapter 4) is transferred to the acceptor substrate (Figure 5.1). In some instances, the donor substrates are lipids, such as dolichol-phosphate linked to mannose or glucose or a dolichol-oligosaccharide precursor (see Chapter 8). Lipid-linked donor sugars (e.g., undecaprenyl-pyrophosphoryl-*N*-acetylmuramic acid-pentapeptide-*N*-acetylglucosamine) are also used by bacterial glycosyltransferases in the assembly of peptidoglycan, lipopolysaccharide, and capsules (see Chapter 20).

Acceptor + Glycosyl donor —Glycosyltransferase→ Glycosylated acceptor + Nucleotide or isoprenoid-P

Glycosyl donors

CMP-sialic acid
GDP-fucose
GDP-mannose
UDP-glucose
UDP-galactose
UDP-xylose
UDP-*N*-acetylglucosamine
UDP-*N*-acetylgalactosamine
UDP-glucuronic acid
Dolichol-P-glucose
Dolichol-P-mannose
Dolichol-P-P-(glucose$_3$-mannose$_9$-GlcNAc$_2$)
Undecaprenyl-P-P-*N*-acetylmuramic acid-pentapeptide-GlcNAc

Acceptors

Oligosaccharides
Monosaccharides
Proteins
Lipids (ceramides)
DNA

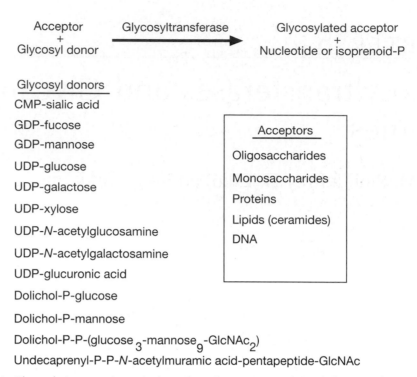

FIGURE 5.1. Glycosylation reactions. A glycosyltransferase uses a glycosyl donor and an acceptor substrate. In animals, glycosyl donors include nucleotide sugars and dolichol-phosphate-linked monosaccharides and oligosaccharides. Bacteria also use undecaprenyl-pyrophosphate (PP)-linked donors. Acceptors are most commonly oligosaccharides, but (in rare cases) they can be monosaccharides. Proteins and ceramides are also acceptors for the glycosyltransferases that initiate glycoprotein, proteoglycan, and glycolipid synthesis. Many other targets, such as drugs and other small molecules, can be glycosylated, but these are not discussed in this chapter. Even DNA can be glycosylated.

The glycosyltransferases that initiate the synthesis of glycoconjugates use acceptor substrates that include oligosaccharides, monosaccharides, polypeptides, lipids, small organic molecules, and even DNA. Only enzymes involved in the biosynthesis of glycoproteins, proteoglycans, and glycolipids are discussed in this chapter. The vast majority of these glycosyltransferases are responsible for elongating glycan chains. Generally speaking, these enzymes act sequentially, so that the product of one enzyme yields a preferred acceptor substrate for the subsequent action of another. The end result is a linear and/or branched polymer composed of monosaccharides linked to one another. In most cases, acceptor recognition does not involve the underlying polypeptide or glycolipid substrate, but several notable exceptions exist that are described below. Despite this finding, specific glycosylation sites of many glycoproteins carry predictable glycan structures, suggesting a complex mechanism for controlling composition.

A few of the enzymes involved in the biosynthesis of glycans are glycosidases that remove monosaccharides to form intermediates that are then acted on by glycosyltransferases. Processing of this type is especially relevant in the formation of N-glycans; in this case, the nascent glycoprotein glycan $Glc_3Man_9GlcNAc_2$-Asn is sequentially trimmed by glucosidases and mannosidases before potential modification by glycosyltransferases that lead to complex and hybrid-type chains (see Chapter 8). In addition, glycans can be modified in many other ways, for example, by sulfotransferases, phosphotransferases, O-acetyltransferases, O-methyltransferases, pyruvyltransferases, and ethanolamine phosphate transferases (Figure 5.2).

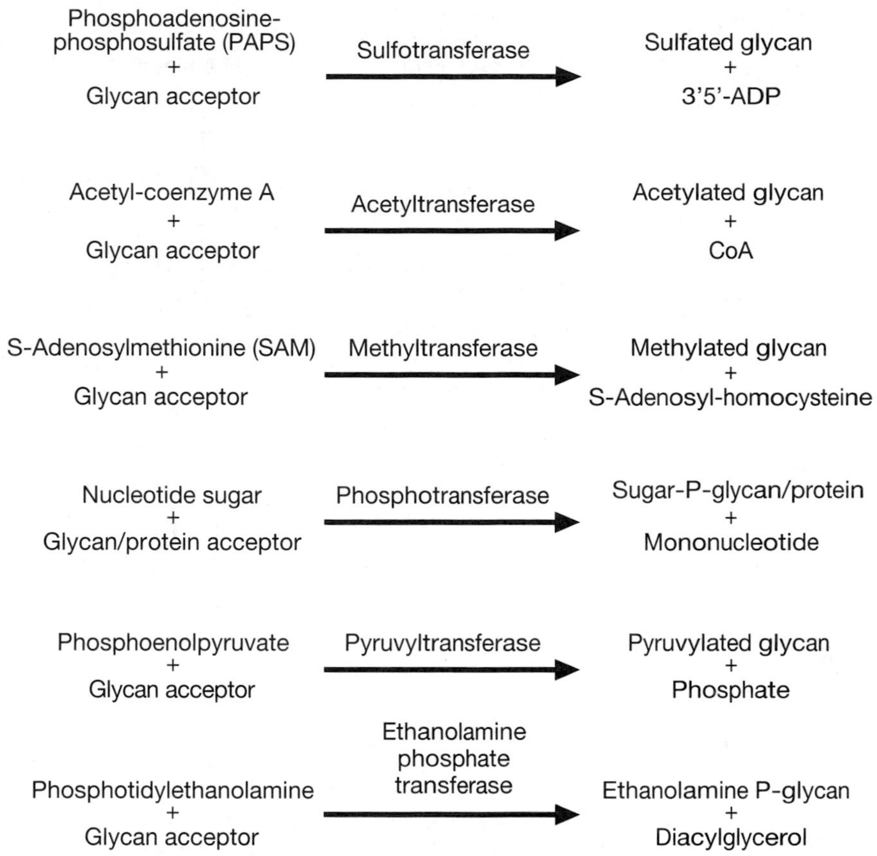

FIGURE 5.2. Glycan-modifying enzymes. A variety of donors are used to modify glycans.

GLYCOSYLTRANSFERASE SPECIFICITY

The specificity of glycosyltransferases, with respect to nucleotide sugar donor and glycan acceptor, led early on to the concept that each glycosidic linkage is the product of a single enzyme. This so-called "one enzyme–one linkage" hypothesis was advanced by Saul Roseman and coworkers. The human B blood group α1-3 galactosyltransferase provides an excellent example of such specificity. This enzyme catalyzes a glycosylation reaction in which galactose is added in α linkage to the C-3 hydroxyl group of a galactose residue on the acceptor substrate (Figure 5.3). However, the enzyme only acts on galactose containing fucose in α1-2 linkage. Prior modification by other monosaccharides, such as α2-6-linked sialic acid, yields a glycan that is not a substrate (Figure 5.3).

We now know that there are also instances in which more than one enzyme can use the same acceptor to make the same linkage. The human fucosyltransferases III–VII, for example, all attach fucose in α1-3 linkage to *N*-acetyllactosamine moieties on glycans (see Chapter 13). Other examples include α2-3 sialyltransferases that act on galactose, β1-4 galactosyltransferases that act on *N*-acetylglucosamine, and the family of GlcNAc 3-O-sulfotransferases that participate in heparin/heparan sulfate synthesis. In some rare cases, a single enzyme can catalyze more than one reaction. Human fucosyltransferase III can attach fucose in either α1-3 or α1-4 linkage, and an enzyme called EXTL2 can attach either *N*-acetylgalactosamine or *N*-acetylglucosamine in α linkage to glucuronic acid. The β1-4 galactosyltransferase involved in *N*-acetyllactosamine formation exhibits an unusual flexibility in specificity. When β1-4 galactosyltransferase binds α-lactalbumin (the complex is called lactose synthase), it switch-

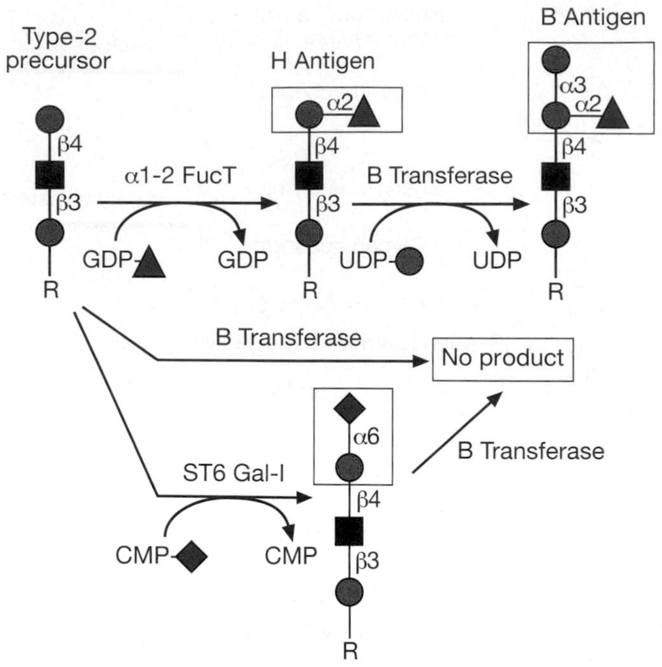

FIGURE 5.3. Strict acceptor substrate specificity of glycosyltransferases as exemplified by the human B blood group α1-3 galactosyltransferase. The B transferase adds galactose in α1-3 linkage to the H antigen (*top*). This enzyme requires the α1-2-linked fucose modification of the H antigen for activity because the B transferase does not add to an unmodified type-2 precursor (*middle*), or precursors modified by sialyl residues (*bottom*) or other monosaccharides (not shown). (For the monosaccharide symbol code, see Figure 1.5, which is reproduced on the inside front cover.)

es its acceptor specificity from *N*-acetylglucosamine to glucose, which enables lactose synthesis during milk formation.

Finally, some glycosyltransferases (such as those that synthesize the backbones of glycosaminoglycans) have two separate active sites, for example, one that catalyzes the attachment of *N*-acetylglucosamine to glucuronic acid and another that attaches glucuronic acid to *N*-acetylglucosamine (see Chapters 15 and 16). However, the examples described above are all exceptions to the generally strict donor, acceptor, and linkage specificity exhibited by most glycosyltransferases, a property that serves to define and limit the number and type of glycan structures observed in a given organism.

In contrast to those glycosyltransferases that elongate glycan chains, initiation of the biosynthesis of glycoproteins and glycolipids requires glycosyltransferases that attach saccharides to either a polypeptide side chain or a sphingolipid base. As might be expected, these enzymes also show a high degree of specificity for their substrates. Among the polypeptide glycosyltransferases, those that initiate the biosynthesis of O-glycans typically transfer a specific monosaccharide to the side chain of serine or threonine (see Chapters 9 and 16). In contrast, N-linked glycans are initiated by the action of oligosaccharyltransferase, an enzyme that transfers the glycan $Glc_3Man_9GlcNAc_2$ to the side chain of asparagine residues in the sequence motif Asn-X-Ser/Thr (where X can be any amino acid except proline). For amino acid–consensus sequences or glycosylation motifs used in the formation of glycopeptide bonds, see Table 5.1. The glycosyltransferases that initiate the synthesis of glycosphingolipids transfer a monosaccharide moiety to what was originally a serine residue in the ceramide lipid precursor of sphingolipids (see Chapter 10). Because different glycolipids have different ceramide moieties, it appears that some glycosyltransferases, such as the sialyltransferases, differentially recognize their substrates based on the nature of the ceramide moiety.

TABLE 5.1. *Amino acid–consensus sequences or glycosylation motifs for the formation of glycopeptide bonds*

Glycopeptide bond	Consensus sequence or peptide motif
GlcNAc-β-Asn	Asn-X-Ser/Thr (X = any amino acid except Pro)
Glc-β-Asn	Asn-X-Ser/Thr
GalNAc-α-Ser/Thr	repeat domains rich in Ser, Thr, Pro, Gly, Ala in no special sequence
GlcNAc-α-Thr	Thr-rich domain near Pro residues
GlcNAc-β-Ser/Thr	Ser/Thr-rich domains near Pro, Val, Ala, Gly
Man-α-Ser/Thr	Ser/Thr-rich domains
Fuc-α-Ser/Thr	EGF modules (Cys-X-X-Gly-Gly-Thr/Ser-Cys)
	TSR modules ($TrpX_5CysX_{2\text{-}3}Ser/ThrCysX_2Gly$)
Glc-β-Ser	EGF modules (Cys-X-Ser-X-Pro-Cys)
Xyl-β-Ser	Ser-Gly (in the vicinity of one or more acidic residues)
Glc/GlcNAc-Thr	Rho: Thr-37; Ras, Rac, and Cdc42: Thr-35
Gal-Thr	Gly-X-Thr (X = Ala, Arg, Pro, Hyp, Ser) (vent worm)
Gal-β-Hyl	collagen repeats (X-Hyl-Gly)
Ara-α-Hyp	repetitive Hyp-rich domains (e.g., Lys-Pro-Hyp-Hyp-Val)
GlcNAc-Hyp	Skp1: Hyp-143
Glc-α-Tyr	glycogenin: Tyr-194
GlcNAc-α-1-P-Ser	Ser-rich domains (e.g., Ala-Ser-Ser-Ala)
Man-α-1-P-Ser	Ser-rich repeat domains
Man-α-C-Trp	Trp-X-X-Trp
Man-6-P-ethanolamine-protein	GPI attached after cleavage of carboxy-terminal peptide

Based on Spiro R.G. 2002. *Glycobiology* **12:** 43R–56R.

(EGF) Epidermal growth factor; (TSR) thrombospondin repeats; (Hyp) hydroxyproline; (Hyl) hydroxylysine; (GPI) glycosylphosphatidylinositol; (Ara) arabinose.

PROTEIN/GLYCOPROTEIN ACCEPTORS AND GLYCOSYLTRANSFERASE ACTION

Among the glycosyltransferases that use proteins/glycoproteins as acceptor substrates, the underlying polypeptide chain of the acceptor is used in different ways to confer specificity on the glycosylation reaction. As mentioned above, all N-glycans are initiated by oligosaccharyltransferase, an enzyme that cotranslationally transfers the N-glycan precursor to asparagine residues in the sequence motif Asn-X-Ser/Thr. In contrast, many enzymes can glycosylate serine and threonine residues leading to O-glycans possessing O-GlcNAc, O-GalNAc, O-fucose, O-mannose, and O-xylose linkages. Among the enzymes responsible for these linkages, specificity for a particular serine or threonine residue is achieved in different ways. For example, O-xylosylation has an absolute requirement for a glycine residue located amino terminally to the serine with nearby acidic residues on one or both sides of the glycosylation site. Some polypeptide O-GalNAc transferases even possess a lectin domain that serves to direct the glycosyltransferase to regions of polypeptide already possessing glycan chains. In this way, regions of polypeptide that have a high degree of carbohydrate substitution, typical of mucin structures, can be synthesized.

A few glycosyltransferases are specific for a particular protein type, and in these cases recognition seems to be dependent on the final folded form of the protein acceptor. The O-fucosyltransferases that act on polypeptides selectively modify epidermal growth factor (EGF) or thrombospondin repeats (TSR), protein domains that must be properly folded and disulfide-bonded to serve as acceptor substrates. The glycoprotein hormone GalNAc transferase provides an interesting example in which modification of an N-glycan is dependent on the presence of the sequence motif Pro-X-Arg/Lys positioned several amino acids amino terminally to the N-glycan being modified. The motif is typically followed closely by additional positively charged

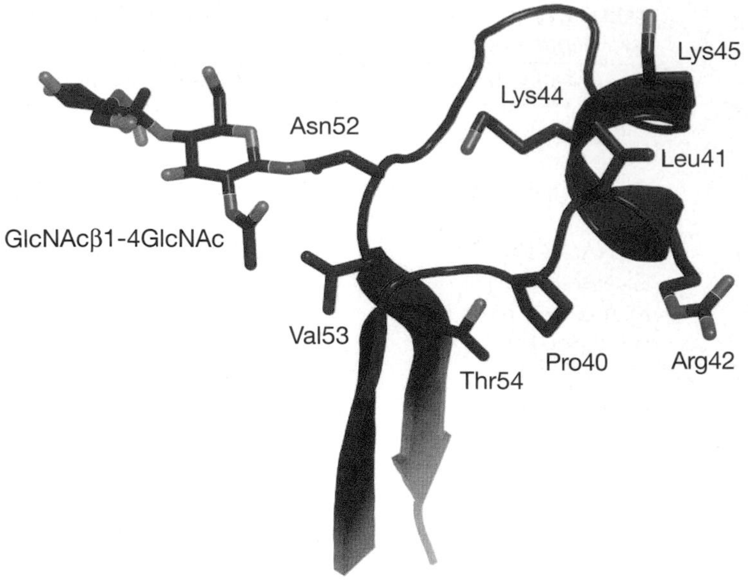

FIGURE 5.4. Recognition site for glycoprotein hormone N-acetylgalactosaminyltransferase on human chorionic gonadotropin. Ribbon diagram of a fragment (residues 34–58) of human chorionic gonadotropin (PDB [protein data bank] ID 1hrp). The Pro40-Leu41-Arg42 tripeptide and residues Lys44 and Lys45 correspond to residues essential for recognition by glycoprotein hormone N-acetylgalactosaminyltransferase. Residues Asn52-Val53-Thr54 correspond to the N-glycosylation sequence motif of the N-glycan (at Asn52) modified by the glycoprotein hormone N-acetylgalactosaminyltransferase. Only the chitobiose core (GlcNAcβ1-4GlcNAc) of the acceptor N-glycan is shown.

residues. The X-ray crystal structure of the acceptor substrate, human chorionic gonadotropin, shows that the Pro-X-Arg/Lys motif is at the beginning of a short surface-exposed helix that also contains the additional positively charged residues (Figure 5.4). The *N*-acetylgalactosamine residues transferred by the glycoprotein hormone GalNAc transferase subsequently undergo a biologically important 4-O-sulfation reaction that, in the case of lutenizing hormone and follicle-stimulating hormone, generates a determinant recognized by specific liver clearance receptors (see Chapter 31). Additional examples are provided by the polysialyltransferases specific for neural cell adhesion molecule (N-CAM) (see Chapter 14) and by EXTL3, an N-acetylglucosaminyltransferase that adds GlcNAc in α1-4 linkage to glucuronic acid in the first committed step to heparan sulfate biosynthesis on proteoglycans (see Chapter 16).

In a variation on this theme, GlcNAc-1-phosphotransferase is able to modify a family of enzymes that are specifically destined for lysosomes (see Chapter 30). In this case, lysine residues appropriately spaced and positioned relative to the N-glycan being modified have been shown to be important determinants of acceptor specificity.

The ER resident glucosyltransferase (GGT) has an acceptor substrate specificity unique among glycosyltransferases. It specifically reglucosylates N-glycans on misfolded glycoproteins, rendering them substrates for the chaperones calnexin and calreticulin (see Chapter 36). Its ability to differentiate between misfolded and properly folded glycoproteins is a critical component of the ER quality control machinery.

GLYCOSYLTRANSFERASE SEQUENCE FAMILIES AND FOLD TYPES

The cloning and sequencing of more than 500 genomes has now shown that glycosyltransferases are a very prevalent enzyme type, representing 1–2% of the genome. More than

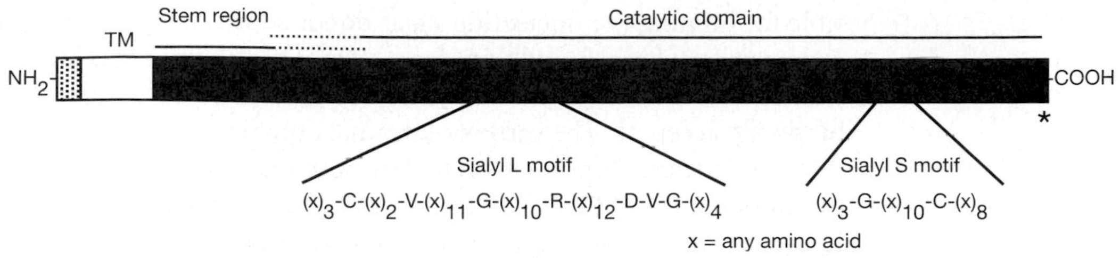

FIGURE 5.5. Sialyl motifs. Domain structure of a typical sialyltransferase, showing the sialyl motifs shared by this family of enzymes. The sialyl L motif of 48–49 amino acids shares significant similarity among members and may be up to 65% identical in amino acid sequence. The sialyl S motif is smaller (~23 amino acids) and diverges more among members of the family, with only two absolutely conserved residues. In both cases, identical residues are indicated and residues showing similarity are denoted by parentheses. (*Asterisk*) Position of a highly conserved sequence H-X_4-E. Additional conserved motifs may exist as well.

30,000 glycosyltransferase sequences are known across all kingdoms, and they comprise approximately 90 glycosyltransferase families defined by primary sequence analysis. These are described in the *c*arbohydrate-*a*ctive en*zy*mes (CAZy) database (see Chapter 7). Although the absence of significant sequence similarity between members of one family and another constitutes the basis for this classification, short sequence motifs common to the members of more than one family have been identified. These sequence elements are typically found among glycosyltransferases with a given donor substrate specificity; the sialyl motifs common to eukaryotic sialyltransferases are a good example (Figure 5.5). Sequence motifs common to galactosyltransferases, fucosyltransferases, and N-acetylglucosaminyltransferases have also been identified. In contrast, the so-called "DXD" motif (Asp–any residue–Asp) is not associated with any particular substrate specificity; rather, the motif is involved in metal ion binding and catalysis, and it is discussed in more detail below.

Despite the large number of sequence families that have been defined, structural analysis has shown that glycosyltransferases possess a limited number of fold types. To date, structures for members of 29 of the 90 families have been determined by X-ray crystallography, and of these all but a few possess what have been termed the GT-A or GT-B folds (Figure 5.6). The catalytic domain of the GT-A-fold enzymes can be viewed as a single domain. The first approximately 120 amino acids show similarity to a structural motif called the Rossmann fold, which is found in proteins that bind nucleotides and are respon-

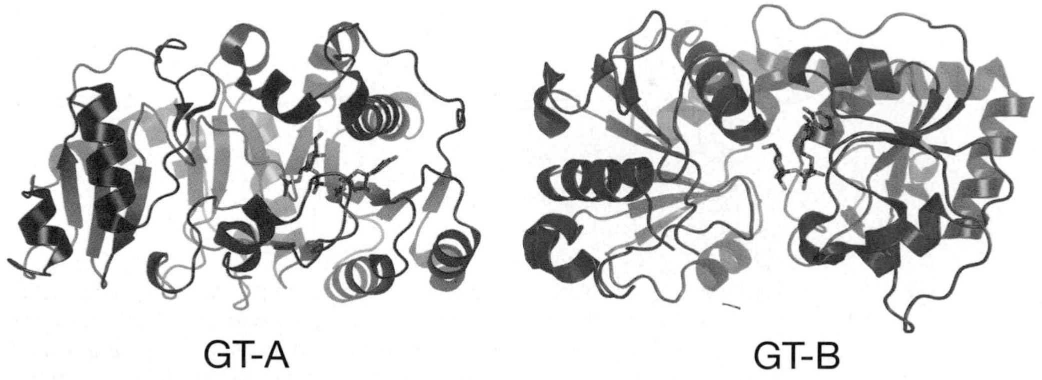

FIGURE 5.6. Ribbon diagrams of representative GT-A and GT-B folds. The GT-A and -B structures correspond to those of rabbit β1-2 N-acetylglucosaminyltransferase I (PDB ID 1foa) and T4 phage β-glucosyltransferase (PDB ID 1j39), respectively. In both cases, bound nucleotide sugar donor substrate is shown in stick representation.

sible for binding the nucleotide sugar donor substrate. With only one exception, the GT-A enzymes have been found to possess a DXD motif and are metal-ion-dependent glycosyltransferases. The GT-B-fold enzymes possess two distinct domains separated by a cleft that binds the acceptor. The carboxy-terminal domain is primarily responsible for binding the nucleotide sugar donor substrate, but both domains possess elements similar to those of the Rossmann fold. Unlike enzymes that contain the GT-A fold, the GT-B glycosyltransferases are metal-ion independent and do not possess a DXD motif.

CATALYTIC MECHANISMS

Glycosyltransferases catalyze their reactions with either inversion or retention of stereochemistry at the anomeric carbon atom of the donor substrate (Figure 5.7). For example, β1-4 galactosyltransferase, an inverting glycosyltransferase, transfers galactose from UDP-α-Gal (naturally occurring nucleotide sugars, except for CMP-Sia and CMP-Kdo [3-deoxy-octulosonic acid], are typically α linked) to generate a β1-4-linked galactose-containing product. Inversion of stereochemistry follows from the fact that the enzyme uses an S_N2 (*s*ubstitution *n*ucleophilic bimolecular) reaction mechanism where an acceptor hydroxyl group attacks the anomeric carbon atom from one side and UDP leaves from the other (Figure 5.7a). Typically, enzymes of this type possess an aspartic acid or glutamic acid

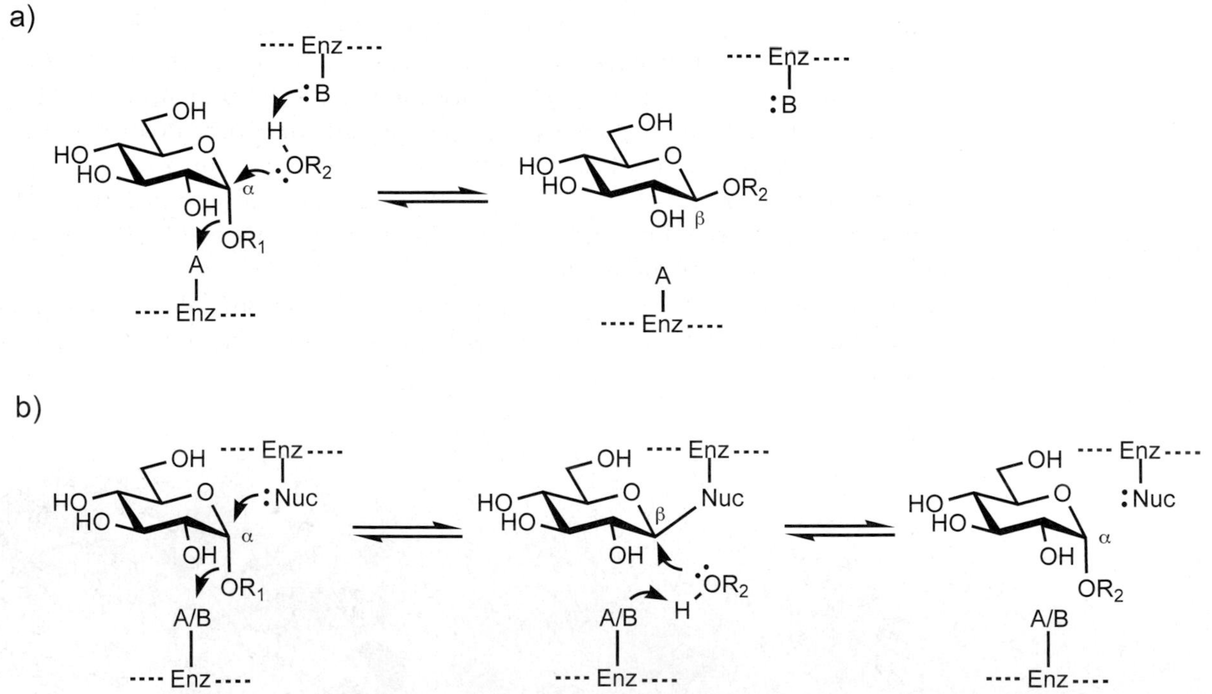

FIGURE 5.7. Schematic representation of (*a*) inverting and (*b*) retaining catalytic mechanisms. (*a*) S_N2-like attack of the acceptor leads to inversion of stereochemistry at C1. For a glycosidase reaction, R2 would correspond to a proton and R1 would be the remainder of the glycan. A and B label general acid and base groups in the catalytic site of the enzyme. For a glycosyltransferase reaction, R2 would correspond to the remainder of the acceptor substrate and R1 would typically be the nucleoside monophosphate or diphosphate moiety of the donor substrate. (*b*) This mechanism has only been established for glycosidases. Two successive S_N2-like reactions separated by a glycosyl-enzyme intermediate lead to retention of the configuration at C1. R1 corresponds to the remainder of the glycan and R2 to a proton.

residue whose side chain serves to partially deprotonate the incoming acceptor hydroxyl group, rendering it a better nucleophile (as shown in Figure 5.8 for the β1-4 galactosyltransferase). In addition, these enzymes promote catalysis by features that help to promote leaving-group departure. In the GT-A enzymes, a metal ion, bound by the DXD motif, is typically positioned to interact with the diphosphate moiety. The positively charged metal ion serves to electrostatically stabilize the additional negative charge that develops on the UDP leaving group during bond breakage (Figure 5.8). In the one GT-A enzyme that is not metal-ion dependent, positively charged side chains stabilize the leaving group, a strategy also used by some of the GT-B-fold enzymes.

Although the mechanism used by inverting glycosyltransferases is not well understood, insight into how they might work is provided by our knowledge of glycosidase mechanism. On the basis of much structural and enzyme kinetic analysis, it is well established that inverting glycosidases proceed via a single S_N2 displacement mechanism (Figure 5.7a), whereas retaining glycosidases use a double displacement mechanism involving a covalent glycosyl-enzyme intermediate (Figure 5.7b). In the double displacement mechanism, an aspartic acid or glutamic acid side chain in the active site makes the first attack and inversion, followed by a second water-mediated attack (on the glycosyl-enzyme intermediate) and inversion, to give overall retention of stereochemistry. In fact, using mechanism-based inhibitors, the glycosyl-enzyme intermediate has been trapped and studied by X-ray crystallography for a number of glycosidases. However, similar attempts to trap and study glycosyltransferase reaction intermediates have not yet been successful. An alternate mechanism, also proposed for glycogen phosphorylase, is that of the so-called S_Ni mechanism. In this case, the incoming nucleophile attacks from the same side as the leaving group, leading to retention of configuration. Further work is required to establish whether glycosyltransferases use this mechanism.

Despite similarities in the mechanisms and transition states used by glycosidases and glycosyltransferases, small-molecule inhibitors of many glycosidases exist (see Chapter 50), whereas there are relatively few glycosyltransferase inhibitors. This probably reflects the

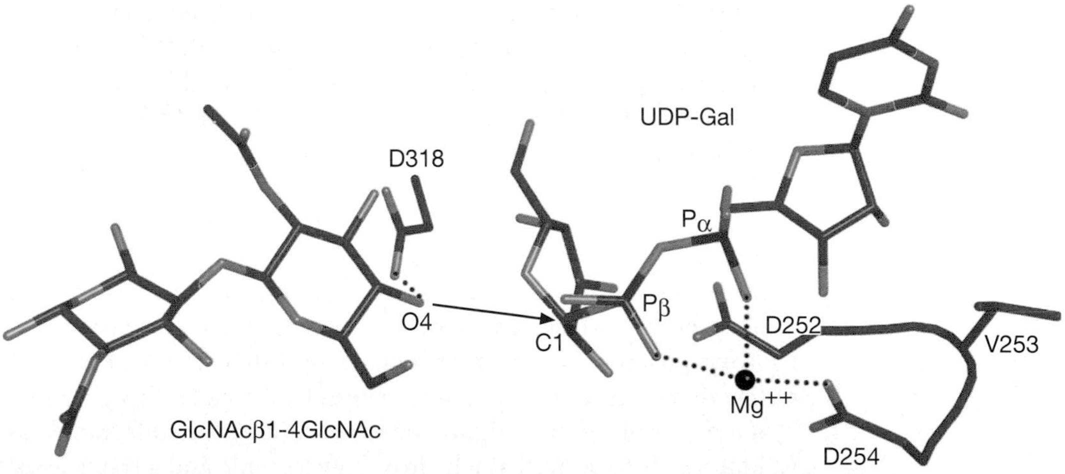

FIGURE 5.8. Catalytic site of bovine β1-4 galactosyltransferase. Composite figure shows selected residues/atoms of the superimposition of the donor complex (PDB ID 1tw1) on the acceptor complex (PDB ID 1tw5). O4 designates the C4 hydroxyl group of the GlcNAcβ1-4GlcNAc acceptor substrate positioned for in-line S_N2 attack (*arrow*) on C1 of the UDP-Gal donor substrate. The base form of D318 serves to partially deprotonate the C4 hydroxyl group rendering it a better nucleophile. The positively charged Mg^{++} ion coordinates the two phosphates of the UDP leaving group, promoting cleavage of the C1-P_β bond by stabilizing the additional negative charge that develops on P_β of the leaving group. D252-V253-D254 corresponds to the DXD motif in bovine β1-4 galactosyltransferase.

fact that many good enzyme inhibitors mimic the transition state, a relatively simple task in the case of the glycosidase reaction where the acceptor substrate is a simple water molecule. Simple high-affinity competitive inhibitors of substrate binding provide another means of inhibiting an enzyme reaction (see Chapter 50).

KINETIC MECHANISMS

In addition to understanding the structure and catalytic mechanisms of glycosyltransferases, there is much interest in understanding the kinetic mechanism. Many glycosyltransferases have been shown to possess a Bi Bi sequential kinetic mechanism in which the donor substrate binds before the acceptor substrate, and the glycosylated acceptor is released before the nucleoside monosphosphate or disphosphate, depending on the reaction. Such kinetics are easily explained by a structural model in which the active site represents a deep pocket, with the nucleotide sugar substrate at the bottom and the acceptor substrate stacked on top. If the acceptor substrate were to bind first, it would sterically preclude donor substrate binding. Necessarily, release of the glycosylated product must precede release of the nucleoside phosphate. Although largely consistent with such a model, the X-ray crystal structures of glycosyltransferase-substrate complexes also shows that substrate-dependent ordering of flexible loops is a feature common to glycosyltransferases. Typically, donor substrate binding orders a loop(s) that in turn facilitates acceptor substrate binding.

SULFATION AND OTHER MODIFICATIONS

Glycans can undergo a number of modifications, depending on the major subclass and species of origin. The sulfotransferases are a large family of enzymes found in both the cytoplasm and the Golgi. Enzymes from the cytoplasm sulfate low-molecular-weight substrates such as steroids and exogenous compounds, whereas those from the Golgi sulfate a large number of cell-surface and secreted glycans. Sulfotransferases have a particularly important role in the production of glycosaminoglycans, molecules involved in embryological development and physiology (see Chapters 16 and 24). They are also involved in the formation of L-selectin ligands, glycans required for the trafficking of lymphocytes across high endothelial venules in lymph nodes (see Chapter 31). All sulfotransferases use 3′-phosphoadenosine-5′-phosphosulfate (PAPS) as the sulfate donor (see Chapter 4). Although the sequence similarity among sulfotransferases can be very low, they all possess conserved sequence motifs responsible for binding the 5′ and 3′ phosphate groups of PAPS. Moreover, structural analysis has shown that all of the sulfotransferases solved to date possess the same basic structure. Mechanistically, the enzyme proceeds via an S_N2-like reaction, with the hydroxyl group of the acceptor making an in-line attack on the sulfate group. In conjunction with structural analysis, mutagenesis has provided insight into the catalytic role played by a number of residues highly conserved among the sulfotransferases. Of these, a histidine residue serves to activate the hydroxyl nucleophile and a lysine assists in stabilizing the PAP leaving group. Interestingly, a conserved serine seems to be involved in modulating the activity of these enzymes to prevent PAPS hydrolysis in the absence of acceptor substrate.

Phosphorylation of sugar residues also occurs (e.g., at C-2 of xylose in proteoglycans), but little is known about the nature of this reaction. Phosphoglycosylation, a process in which a sugar phosphate is transferred from a nucleotide sugar donor directly to a serine residue on a protein (e.g., GlcNAc-P-serine), occurs in *Dictyostelium.* Mannose-6-phosphate on the N-glycans of lysosomal enzymes serves as a targeting signal. However, the formation of this phosphate ester is not mediated by a typical ATP-using kinase, but rather by a two-step reaction involving the donor UDP-GlcNAc (see Chapter 30). Detailed studies of reaction mechanisms have not yet been undertaken.

O-Acetylation occurs in bacteria and in the modification of sialic acids, but little information is available about the chemistry of these reactions. N-Deacetylation of *N*-acetylglucosamine residues occurs during heparin/heparan sulfate formation (see Chapter 16), lipopolysaccharide assembly (see Chapter 20), and GPI-anchor synthesis (see Chapter 11). The bacterial enzyme is zinc dependent, but in-depth studies of the vertebrate N-deacetylases have not been performed. N-Deacetylation of *N*-acetylneuraminic acid (the most common sialic acid) has also been reported (see Chapter 14).

Finally, glycans can be modified in many other ways, some of which are listed in Figure 5.2. These include pyruvylation (e.g., in the formation of *N*-acetylmuramic acid; see Chapter 20), the addition of ethanolamine phosphate (e.g., during GPI-anchor synthesis; see Chapter 11), and alkylation, deoxygenation, and halogenation in microbial glycans. All of these reactions are catalyzed by unique transferases or oxidoreductases and represent areas of active research.

FURTHER READING

Manzella S.M., Hooper L.V., and Baenziger J.U. 1996. Oligosaccharides containing β1,4-linked N-acetylgalactosamine, a paradigm for protein-specific glycosylation. *J. Biol. Chem.* **271:** 12117–12120.

Negishi M., Pedersen L.G., Petrotchenko E., Shevtsov S., Gorokhov A., Kakuta Y., and Pedersen L.C. 2001. Structure and function of sulfotransferases. *Arch. Biochem. Biophys.* **390:** 149–157.

Spiro R.G. 2002. Protein glycosylation: Nature, distribution, enzymatic formation, and disease implications of glycopeptide bonds. *Glycobiology* **12:** 43R–56R.

Coutinho P.M., Deleury E., Davies G.J., and Henrissat B. 2003. An evolving hierarchical family classification for glycosyltransferases. *J. Mol. Biol.* **328:** 307–317.

Negishi M., Dong J., Darden T.A., Pedersen L.G., and Pedersen L.C. 2003. Glucosaminylglycan biosynthesis: What we can learn from the X-ray crystal structures of glycosyltransferases GlcAT1 and EXTL2. *Biochem. Biophys. Res. Commun.* **303:** 393–398.

Ten Hagen K.G., Fritz T.A., and Tabak L.A. 2003. All in the family: The UDP-GalNAc:polypeptide N-acetylgalactosaminyltransferases. *Glycobiology* **13:** 1R–16R.

Haltiwanger R.S. and Lowe J.B. 2004. Role of glycosylation in development. *Annu. Rev. Biochem.* **73:** 491–537.

Lairson L.L. and Withers S.G. 2004. Mechanistic analogies amongst carbohydrate modifying enzymes. *Chem. Commun.* **20:** 2243–2248.

Davies G.J., Gloster T.M., and Henrissat B. 2005. Recent structural insights into the expanding world of carbohydrate-active enzymes. *Curr. Opin. Struct. Biol.* **15:** 637–645.

Jones J., Krag S.S., and Betenbaugh M.J. 2005. Controlling N-linked glycan site occupancy. *Biochim. Biophys. Acta* **1726:** 121–137.

Qasba P.K., Ramakrishnan B., and Boeggeman E. 2005. Substrate-induced conformational changes in glycosyltransferases. *Trends Biochem. Sci.* **30:** 53–62.

Blanchard S. and Thorson J.S. 2006. Enzymatic tools for engineering natural product glycosylation. *Curr. Opin. Chem. Biol.* **10:** 263–271.

Letts J.A., Rose N.L., Fang Y.R., Barry C.H., Borisova S.N., Seto N.O.L., Palcic M.M., and Evans S.V. 2006. Differential recognition of the Type I and II H antigen acceptors by the human ABO(H) blood group A and B glycosyltransferases. *J. Biol. Chem.* **281:** 3625–3632.

Luo Y., Nita-Lazar A., and Haltiwanger R.S. 2006. Two distinct pathways for O-fucosylation of epidermal growth factor-like or thrombospondin type 1 repeats. *J. Biol. Chem.* **281:** 9385–9392.

Narimatsu H. 2006. Human glycogene cloning: Focus on β3-glycosyltransferase and β4-glycosyltransferase families. *Curr. Opin. Struct. Biol.* **16:** 567–575.

Yu H. and Chen X. 2007. Carbohydrate post-glycosylational modifications. *Org. Biomol. Chem.* **5:** 865–872.

CHAPTER 6

Biological Roles of Glycans

Ajit Varki and John B. Lowe

THIS CHAPTER PROVIDES AN OVERVIEW OF THE BIOLOGICAL roles of glycans and attempts to synthesize some general principles for understanding and exploring these roles. For details, see the reviews cited and the other chapters in this book.

GENERAL PRINCIPLES

As with other major classes of macromolecules, the biological roles of glycans span the spectrum from those that appear to be relatively subtle, to those that are crucial for the development, growth, functioning, or survival of the organism that synthesizes them. Many glycans have not yet been assigned a function, because efforts to study them have not been made or a function is not yet evident. Over the years, many theories have been advanced regarding the biological roles of glycans. Although there is evidence to support all of these theories, exceptions to each can also be easily found. This should not be surprising, given the enormous diversity of glycans in nature. Added complexities arise from the fact that glycans are frequently targets for the binding of microbes and microbial toxins, that is, they can be detrimental to the organism that synthesizes them.

The biological roles of glycans can be divided into two broad categories: (1) the structural and modulatory properties of glycans and (2) the specific recognition of glycans by other molecules—most commonly, glycan-binding proteins (GBPs) (Figure 6.1). The GBPs can be subdivided into two major groups: (1) intrinsic GBPs, which recognize glycans from the same organism and (2) extrinsic GBPs, which recognize glycans from a different organism. Intrinsic GBPs typically mediate cell–cell interactions or recognize extracellular molecules, but they can also recognize glycans on the same cell. Extrinsic GBPs consist mostly of pathogenic microbial adhesins, agglutinins, or toxins, but some also mediate symbiotic relationships. As discussed in Chapter 19, these two types of glycan recognition likely act as opposing selective forces driving evolutionary change, at least partly accounting for the enormous diversity of glycan structure found in nature. Further complexity arises from the fact that some microbial pathogens engage in "molecular mimicry," evading immune reactions by decorating themselves with glycans typical of their hosts. Finally, some microbes are themselves targets of their own pathogens (e.g., bacteriophages that invade bacteria), and glycan recognition is a common feature of these interactions as well.

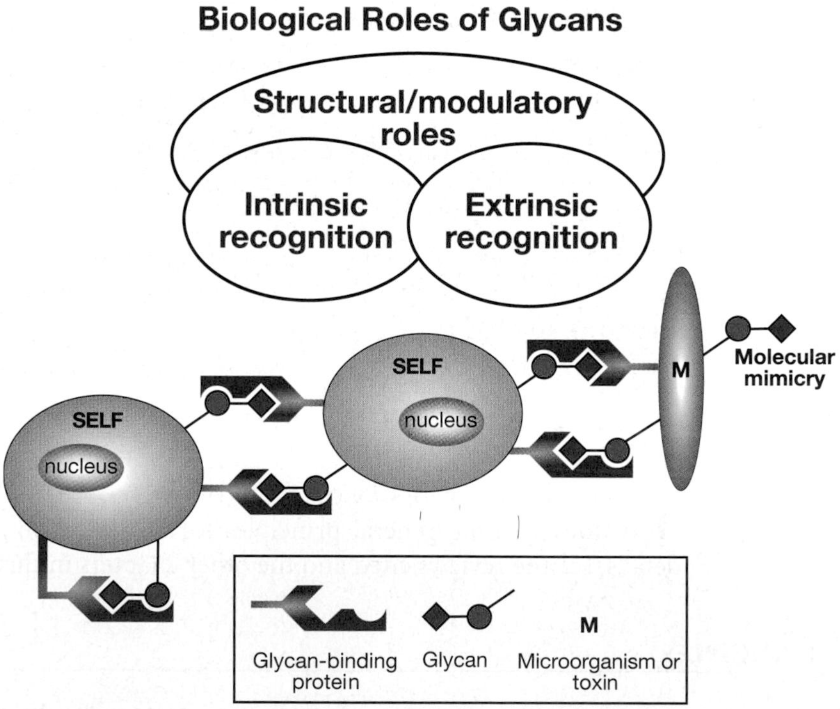

FIGURE 6.1. General classification of the biological roles of glycans. A simplified and broad classification is presented, emphasizing the roles of organism-intrinsic and -extrinsic glycan-binding proteins in recognizing glycans. There is some overlap between the groups (e.g., some structural properties involve specific recognition of glycans). In the lower part of the figure, intrinsic recognition is represented by the binding shown on the left of the central "self" cell and extrinsic recognition is represented by the binding shown to the right of that cell. (Modified and redrawn, with permission of Oxford University Press, from Gagneux P. and Varki A. 1999. *Glycobiology* **9:** 747–755.)

Other general principles emerge after reviewing the extant literature on this subject. The biological consequences of altering glycosylation in various systems seem to be highly variable and unpredictable. A given glycan can have different roles in different tissues or at different times in development (organism-intrinsic functions) or in different environmental contexts (organism-extrinsic functions). As a broad generalization, it can be stated that terminal sequences, unusual structures, and modifications of glycans probably mediate the more specific biological roles within the organism. However, such glycans or modifications are also more likely to be targets for pathogens and toxins. Perhaps as a consequence, intraspecies and interspecies variations in glycosylation are relatively common, and at least some of the diversity of glycans in nature may represent the signatures of past or current host-pathogen interactions (for discussion see Chapter 19). Finally, genetic defects in glycosylation are easily obtained in cultured cells, but often have limited biological consequences. In contrast, the same defects in intact organisms can have major and even catastrophic consequences. This indicates that many major functions of glycans are operative only within an intact organism. Each of these principles is briefly discussed below.

BIOLOGICAL CONSEQUENCES OF ALTERING GLYCOSYLATION ARE VARIABLE

Approaches taken to understand the biological roles of glycans include the prevention of initial glycosylation, prevention of glycan chain elongation, alteration of glycan processing, enzymatic or chemical deglycosylation of completed chains, genetic elimination of glycosylation sites, and the study of naturally occurring genetic variants and mutants in glycosylation (see further discussion below). The consequences of such manipulations range from being essentially undetectable to the complete loss of particular functions or even loss of the entire glycoconjugate bearing the altered glycan. Even within a particular class of molecules, for example cell-surface receptors, the effects of altering glycosylation are variable and unpredictable. Moreover, the same glycosylation change can have markedly different effects in different cell types, or when studied in vivo or in vitro. The answer obtained may depend on the structure of the glycan, the biological context (intrinsic or extrinsic interaction), and the specific biological question being asked. Given all of the above considerations, it is difficult to predict a priori the functions that a given glycan on a given glycoconjugate might mediate and its relative importance to the organism.

STRUCTURAL AND MODULATORY ROLES OF GLYCANS

There is little doubt that glycans have many protective, stabilizing, organizational, and barrier functions. As discussed in Chapter 1, the glycocalyx that covers all eukaryotic cells and the polysaccharide coats of various prokaryotes can represent a substantial physical barrier. Glycans attached to matrix molecules, such as proteoglycans, are important for the maintenance of tissue structure, porosity, and integrity. Such molecules can also contain binding sites for other specific types of glycans that in turn aid the overall organization of the matrix. The external location of glycans on most glycoproteins can provide a general shield, protecting the underlying polypeptide from recognition by proteases or antibodies. Glycans are also involved in the proper folding of newly synthesized polypeptides in the endoplasmic reticulum (ER) and/or in the subsequent maintenance of protein solubility and conformation. Indeed, if some proteins are incorrectly glycosylated, they will fail to fold properly and/or to exit the ER, being consigned instead to degradation in proteasomes. Conversely, there are also examples of glycoproteins whose synthesis, folding, trafficking, sensitivity to proteolysis, or immune recognition seem quite unaffected by altering their glycosylation. Moreover, inhibitors (see Chapter 50) or genetic mutations (see Table 6.1 and Chapter 42) that only affect the later steps of glycan processing often do not interfere with basic structural functions. Although the structural functions of glycans are obviously of great importance to the intact organism, they do not explain why such a diverse range of complex glycan molecules has evolved.

Glycosylation can also modulate the interaction of proteins with one another. Some growth factor receptors seem to acquire their binding abilities in a glycosylation-dependent manner while they are in transit through the Golgi apparatus. This may limit unwanted early interactions of a newly synthesized receptor with a growth factor that is synthesized in the same cell. Glycosylation of a polypeptide can also mediate an on-off or switching effect. For example, when the hormone β-human chorionic gonadotrophin is deglycosylated, it is still able to bind to its receptor with similar affinity, but it fails to stimulate adenylate cyclase. In most instances, the effects of glycosylation are incomplete, that is, glycosylation appears to be "tuning" a primary function of the protein rather than turning it on or off. For example, the activity of some glycosylated growth factors and hormones can be modulated over a wide range by the extent and type of their glycosylation. This becomes particularly evident when recombinant forms of such molecules are produced in biotechnology, bearing different types and extents of glycosylation. A striking example is the role of polysialic acid chains attached to the neural cell adhesion molecule (NCAM). This adhesion receptor normally mediates homophilic binding between neuronal cells. In the embryonic state, or in other states of neural "plasticity," these anionic polysialic chains tend to be very long, thereby interfering with homophilic binding (see Chapter 14). There are also instances wherein protein functions can be tuned by glycans attached to other neighboring structures. For example, the polysialic acids of embryonic NCAM can interfere with the interactions of other unrelated receptor–ligand pairs, simply by physically separating the cells. Also, the tyrosine phosphorylation activity of the epidermal growth factor (EGF) receptor and the insulin receptor can be modulated by endogenous cell-surface gangliosides, possibly by organizing them into membrane microdomains (see Chapter 10). Although the precise mechanisms of these latter effects are uncertain, specificity is implied by the requirement for a defined glycan sequence in the ganglioside. Because most such tuning effects of glycans are partial, their overall importance might be questioned. However, the sum total of several such partial effects can be a dramatic effect on the final biological outcome. Thus, glycosylation appears to be a mechanism for generating important functional diversity from the limited set of basic receptor–ligand interactions that are possible, when using the gene products derived from the typical genome. Of course, as with most other functions of glycans, exceptions to these concepts can be found. There are many receptors whose ligand binding is not acquired in a glycosylation-dependent manner and many peptide ligands whose binding and action are not obviously affected by glycosylation.

Another structural/modulatory function of glycans appears to be to act as a protective storage depot for biologically important molecules. For example, many heparin-binding growth factors (see Chapters 16 and 35) are found attached to the glycosaminoglycan (GAG) chains of the extracellular matrix, adjacent to cells that need to be stimulated, for example, in the basement membrane underlying epithelial and endothelial cells. This prevents diffusion of the factors away from the site (sometimes generating morphogenic gradients), protects them from nonspecific proteolysis, prolongs their active lives, and allows them to be released under specific conditions. Likewise, the GAG chains found in secretory granules seem to bind and protect the protein contents of the granule and modulate their functions. There are several other instances in which glycans act as sinks or depots for biologically important molecules, such as water, ions, and immune regulatory proteins.

TABLE 6.1. *A few examples of mouse and human genetic defects in glycosylation*

Glycan class	Enzyme/*gene* (species)	Location in pathway	Phenotype
N-glycans	phosphomannose isomerase *MPI* (human)	pre-ER assembly of lipid-linked N-glycan precursors	congenital disorder of glycosylation type Ib (CDG-Ib); hepatic fibrosis; hypoglycemia; protein-losing enteropathy; coagulopathy; treatable with oral mannose
N-glycans	GlcNAcT-I *Mgat1* (mouse)	proximal aspect of post-ER elongation	embryonic lethality (E9.5); defective vascularization; defective neural tube formation; situs inversus of the heart
N-glycans	GlcNAcT-V *Mgat5* (mouse)	distal aspect of post-ER elongation	autoimmune kidney disease; hyperactive T-cell receptor signaling; defective maternal nurturing behavior; reduced tumorigenesis of mammary epithelium
O-glycans	core 2 GlcNAcT-I *C2gnt1* (mouse)	core 2 branch of GalNAc-linked O-glycans	defective leukocyte P- and L-selectin ligand activity
O-glycans	polypeptide O-fucosyltransferase I *Pofut1* (mouse)	serine and threonine O-fucosylation	embryonic lethality (E9.5); defects that phenocopy Notch nullizygosity; defective vasculogenesis; neurogenesis; cardiogenesis; somito genesis
Glycosphingolipids	UDP-galactose: ceramide Gal-T *Cgt* (mouse)	galactosylation of ceramide	loss of galactocerebrosides in peripheral nerve myelin; nerve conduction defects; tremor and death from ataxia; loss of sympathetic nervous-system-dependent egress of hematopoietic stem and progenitor cells from bone marrow
Shared outer sequences	β1-4 galactosyltransferase-1 *GalT1* (mouse)	outer-chain galactosylation of N- and O-glycans	defective growth; faulty differentiation of epithelia; endocrine insufficiency
Sialic acids	ST3Gal-I *ST3Gal1* (mouse)	α2-3 sialylation of core 1 O-glycans	enhanced CD8αβ binding avidity to MHC class I molecules; proapoptotic phenotype in $CD8^+$ lymphocytes
Glycophospholipid anchors	GlcNAc phosphatidylinositol synthetase *PIG-A* (mouse, human)	synthesis of the first intermediate in GPI anchor formation	germ-line knockout embryonic lethal in mice; somatic knockout in hematopoietic stem cells removes complement-regulating molecules CD59 and DAF and results in enhanced complement-mediated red cell lysis (called paroxysmal nocturnal hemoglobinuria in humans)
Hyaluronan	hyaluronan synthetase 2 *Has2* (mouse)	synthesis of hyaluronan	embryonic lethality (E9.5–E10); absence of cardiac endothelial transformation to mesenchyme; deficiency in formation of the atrioventricular canal
Sulfated glycosaminoglycans	glucosaminyl *N*-deacetylase/*N*-sulfotransferase-1 *Ndst1* (mouse)	synthesis of heparan sulfate	perinatal lethality due to brain/skull defects and lung surfactant problems
Sulfated glycosaminoglycans	glucosaminyl *N*-deacetylase/*N*-sulfotransferase-2 *Ndst2* (mouse)	synthesis of heparan sulfate	loss of heparin sulfate from mast cells; defective granule formation in mast cells
O-GlcNAc modification	O-GlcNAc transferase *Ogt* (mouse)	GlcNAc addition to serines and threonines on nuclear and cytoplasmic proteins	hemizygosity for this X-linked locus is not compatible even with cellular viability of the targeted embryonic stem cell

GLYCANS AS SPECIFIC LIGANDS FOR CELL–CELL INTERACTIONS (INTRINSIC RECOGNITION)

The first intrinsic glycan receptors to be identified were those that mediate clearance, turnover, and intracellular trafficking of soluble blood-plasma glycoproteins (for examples, see Chapter 31). Most of these receptors specifically recognize certain terminal or subterminal glycans on the soluble glycoprotein. However, even the most elegantly precise examples, such as the role of mannose-6-phosphate (Man-6-P) in the trafficking of lysosomal enzymes to lysosomes (see Chapter 30), feature some exceptions. Thus, Man-6-phosphorylation is not absolutely required for the trafficking of lysosomal enzymes in certain cell types nor is it operative at all in some lower eukaryotes. There are also endocytic receptors, whose functions have yet to be assigned, that recognize specific glycan sequences. Several instances exist wherein free glycans can have hormonal actions that induce specific responses in a highly structure-specific manner. Examples include the interaction of small glycans from bacterial symbionts with plant roots (see Chapter 37) and the bioactive properties of fragments of hyaluronan in mammalian systems (see Chapter 15), both of which can induce biological responses in a size- and structure-dependent manner. Likewise, free heparan or dermatan sulfate fragments released by certain cell types can have major biological effects in complex situations such as wound healing. In many of these instances, the putative receptors for these molecules and their precise mechanisms of action are still being defined.

It is now clear that glycans have many specific biological roles in cell–cell recognition and cell–matrix interactions. One of the best characterized examples concerns the selectin family of adhesion molecules, which recognize glycan structures on their ligands and thereby mediate critical interactions between blood cells and vascular cells in a wide variety of normal and pathological situations (see Chapter 31). As indicated above, GBPs and glycans present on cell surfaces can interact specifically with molecules in the matrix or even with glycans on the same cell surface. In some such instances, the specific biological significance of recognition has yet to be conclusively demonstrated in the intact animal. Also, it is becoming clear that some critical recognition sites are actually combinations of glycans and protein. For example, P-selectin recognizes the generic selectin ligand sialyl Lewis^x with high affinity only in the context of the amino-terminal 13 amino acids of P-selectin glycoprotein ligand-1 (PSGL-1), which include certain required sulfated tyrosine residues (see Chapter 31). More recently, a different form of intrinsic recognition has been described, in which glycan-binding sites of cell-surface receptors are masked by cognate glycans on the same cell surface, making them unavailable for recognition by external ligands (see Chapter 32).

Carbohydrate–carbohydrate interactions may also have a specific role in cell–cell interactions and adhesion. A dramatic example is the species-specific interaction between marine sponges, which is mediated via homotypic binding of the glycans on a large cell-surface glycoprotein. Another example is the compaction of the mouse embryo at the morula stage, which seems to be facilitated by a Lewis^x–Lewis^x interaction. The single-site affinities of such interactions are not very strong and are sometimes difficult to measure. However, if the molecules in question are present in very high copy numbers on the cell surface, a large number of relatively low-affinity interactions can collaborate to produce a high-avidity "Velcro" effect that is sufficient to mediate biologically relevant interactions.

GLYCANS AS SPECIFIC LIGANDS FOR CELL–MICROBE INTERACTIONS (EXTRINSIC RECOGNITION)

As will be discussed in Chapter 34, certain glycans act as specific binding sites for a variety of viruses, bacteria, and parasites, and as recognition targets for many plant and bacterial toxins. In such situations, there is typically excellent recognition specificity for the sequence of the glycan involved. For example, the hemagglutinins of many viruses specifically recognize the type of host sialic acid, its modifications, and its linkage to the underlying sugar chain. Likewise, various toxins bind with great specificity to certain gangliosides but not to related structures (see Chapters 10 and 34). There is little doubt about the importance of structural specificity with respect to these functions of glycans. Indeed, many of the microbial binding proteins involved have been harnessed as specific tools for studying the expression of the cognate sugar chains. However, providing signposts to aid the success of pathogenic microorganisms has little obvious value to the organism that synthesized such glycans. Perhaps to counter such deleterious consequences, some organisms may have also evolved the ability to mask or modify glycans recognized by microorganisms or toxins. Conversely, glycan sequences on soluble glycoconjugates, such as secreted mucins, can also act as decoys for microorganisms and parasites. Thus, a pathogenic organism or toxin seeking to bind to mucosal cell membranes may first encounter the specific glycan ligand attached to a soluble mucin, which can then be washed away, removing the potential danger to the cells underneath. In contrast, instances occur in which symbiosis is mediated by specific glycan recognition, such as some commensal bacteria in the gut lumen of animals and the bacteria involved in forming plant root nodules (see Chapter 37).

MOLECULAR MIMICRY OF HOST GLYCANS BY PATHOGENS

Pathogens that invade multicellular animals sometimes decorate themselves with glycan structures that appear to be identical or nearly identical to those found on their host cell surfaces (see Chapters 39 and 40). These glycans form a thick coating on the surface of the microbe and therefore represent a very successful strategy for evading host immune responses. Perhaps not surprisingly, pathogens appear to have evolved to achieve this state of molecular mimicry by making use of "every possible trick in the book," for example, direct or indirect appropriation of host glycans, convergent evolution toward similar biosynthetic pathways, and even lateral gene transfer. In some instances, the impact of the pathogen is aggravated by autoimmune reactions, resulting from host reactions to these host-like antigens.

THE SAME GLYCAN CAN HAVE DIFFERENT ROLES WITHIN AN ORGANISM

The expression of certain types of glycans on different glycoconjugates in different tissues at different times of development implies that these structures have diverse roles within the same organism. For example, Man-6-P-containing glycans were first found on lysosomal enzymes and are involved in lysosomal trafficking (see Chapter 30). However, Man-6-P-containing glycans are now known to occur on a variety of apparently unrelated proteins, including proliferin, thyroglobulin, the EGF receptor, and the transforming growth factor-β (TGF-β) precursor, and they have certain functional roles in the biology of these proteins or their role remains to be discovered. Likewise, the sialylated fucosylated lactosamines critical for selectin recognition (see Chapter 31) are found in a variety of unrelated cell types in mammals, and the polysialic acid chains that play such an important part in embryonic NCAM function (Chapter 14) are found on fish-egg-jelly coat proteins and on a sodium channel protein. Given that glycans are added posttranslationally, these observations should not be surprising. Once a new glycan or modification has been expressed in an organism, several distinct functions could evolve independently in different tissues and at different times in development. If any of these situations mediated a function valuable to the survival of the organism, the genetic mechanisms responsible for expression of the glycan and its expression pattern would remain conserved in evolution.

INTRASPECIES AND INTERSPECIES VARIATIONS IN GLYCOSYLATION

The underlying core structures of the major classes of glycans tend to be conserved across many species, for example, the core structure of N-glycans is conserved across all eukaryotes and at least some of the Archaea (see Chapter 8). However, as outlined in Chapters 19–25, there can be considerable diversity in outer-chain glycosylation, even among relatively similar species. Such interspecies variations in glycan structure indicate that some glycan sequences do not have fundamental and universal roles in all tissues and cell types in which they are expressed. Of course, such diversity could be involved in generating differences in morphology and function observed between species. Such variations could also reflect differing selection pressures resulting from exposure to different pathogen regimes. Furthermore, significant intraspecies polymorphism in glycan structure can exist without obvious functional value. The potential role of such polymorphisms in the interplay between parasites and host populations is discussed in Chapter 19. Extensive interspecies variability in primary sequence also occurs in conventional genes and proteins, without any obvious consequences to essential functions. For example, some yeast proteins are functional when transfected into mammalian cells and vice versa, despite relatively limited sequence homology.

IMPORTANCE OF TERMINAL SEQUENCES, MODIFICATIONS, AND UNUSUAL STRUCTURES

Given all of the above, it is challenging to predict which glycan structures are likely to mediate the more specific or crucial biological roles within an organism. As mentioned above, terminal sugar sequences, unusual structures, or modifications of the glycans are more likely to be involved in such specific roles. The predictive value of this observation is reduced by the fact that such terminal sequences, unusual glycans, or modifications are also more likely to be involved in interactions with microorganisms and other noxious agents, because the balance between the organism-intrinsic and -extrinsic functions of glycans, discussed above, tends to involve such structures. A further complexity arises from "microheterogeneity" in glycan structure (see discussion in Chapter 19), wherein the same glycosylation site on the same protein in the same species can carry a variety of related glycan structures. The challenge then is to predict and sort out which of these two distinct roles is to be assigned to a given glycan structure in a given cell type in a given organism.

ARE THERE "JUNK" GLYCANS?

Because microorganisms and parasites that bind glycans evolve in parallel with their multicellular hosts, they must adapt their glycan-binding "repertoire" to any change in glycan structure presented by the host. In response, the host population may select for new modifications of the target structure, especially if the latter had meanwhile evolved a vital function elsewhere within the organism. Thus, there would be no choice but to preserve the underlying scaffolding upon which the latest modification was placed, while adding yet another layer of complexity to its glycans. Such cycles of evolutionary interaction between microbes and hosts might explain some of the complex and extended sugar chains found in multicellular organisms, especially in areas of frequent microbial contact, such as on mucosal surfaces and secreted mucins. In this manner, "junk" glycans could accumulate, akin to "junk" DNA. Although such structures may still function as structural scaffolding, they may have no other specific role in that particular cell type, organism, or at that particular time in evolution. They would, of course, provide fodder for future evolutionary selection, either for new organism-intrinsic functions or for population-based selective responses to a new pathogen. Additionally, neutral unselected drift (which is now acknowledged as a major process in evolution) can also explain some of the "junk" glycans.

APPROACHES TO ELUCIDATING SPECIFIC BIOLOGICAL ROLES OF GLYCANS

Some functions of glycans are discovered serendipitously. In other instances, the investigator who has elucidated complete details of the structure and biosynthesis of a specific glycan is left without knowing its functions. It is necessary to design experiments that can differentiate between the trivial and crucial functions mediated by each glycan. Various approaches that can be considered are discussed below, and we emphasize the pros and cons of each. These approaches are also presented in schematic form in Figure 6.2.

Localization of or Interference with Specific Glycans Using Glycan-binding Proteins or Antibodies

Most current approaches to understanding glycan diversity (see Chapter 48) involve the extraction and identification of the entire complement of glycans found in a given organ or tissue, without regard to the fact that individual cell types within that organ or tissue can have widely varying patterns of glycan expression. However, the cell-type-specific localization of glycans can be explored using the numerous highly specific GBPs and antibodies now available (see Chapter 45). Once a specific glycan has been localized in an interesting biological context, it is natural to consider introducing the cognate GBP or antibody into the intact system, hoping that it will interfere with a specific function and generate an interpretable phenotype. A similar approach with antibodies can be very successful when investigating the function of a protein, but with rare exceptions, this strategy is likely to give confusing results with regard to glycan function. Most antibodies against glycans are of the IgM variety and hence tend to have weak affinity and show cross-reactivity between species. Although high-affinity IgG antibodies are preferred, they are hard to obtain, because glycans tend to be T-independent antigens, and often do not generate high-titer immune responses. Likewise, although some plant lectins seem to be very specific for animal glycans, they originate from organisms that typically do not contain the same ligand. Thus, their apparent specificity may not be as reliable when introducing them into complex animal biological systems where unknown cross-reacting glycan structures are potentially present. Finally, both antibodies and GBPs are multivalent, and their cognate ligands (the glycans) tend to be present in multiple copies on multiple glycoconjugates. Thus, introduction of a GBP or antibody into a complex biological system is likely to cause nonspecific aggregation of various molecules and cell types, and the effects seen may have nothing to do with the biological functions of the glycan in question. It would seem more worthwhile to develop recombinant monovalent GBP modules that are derived from the same system being investigated. Providing they are of high enough affinity, the effects of introducing such monovalent GBPs into a complex system as competitors of the native function may yield more interpretable clues.

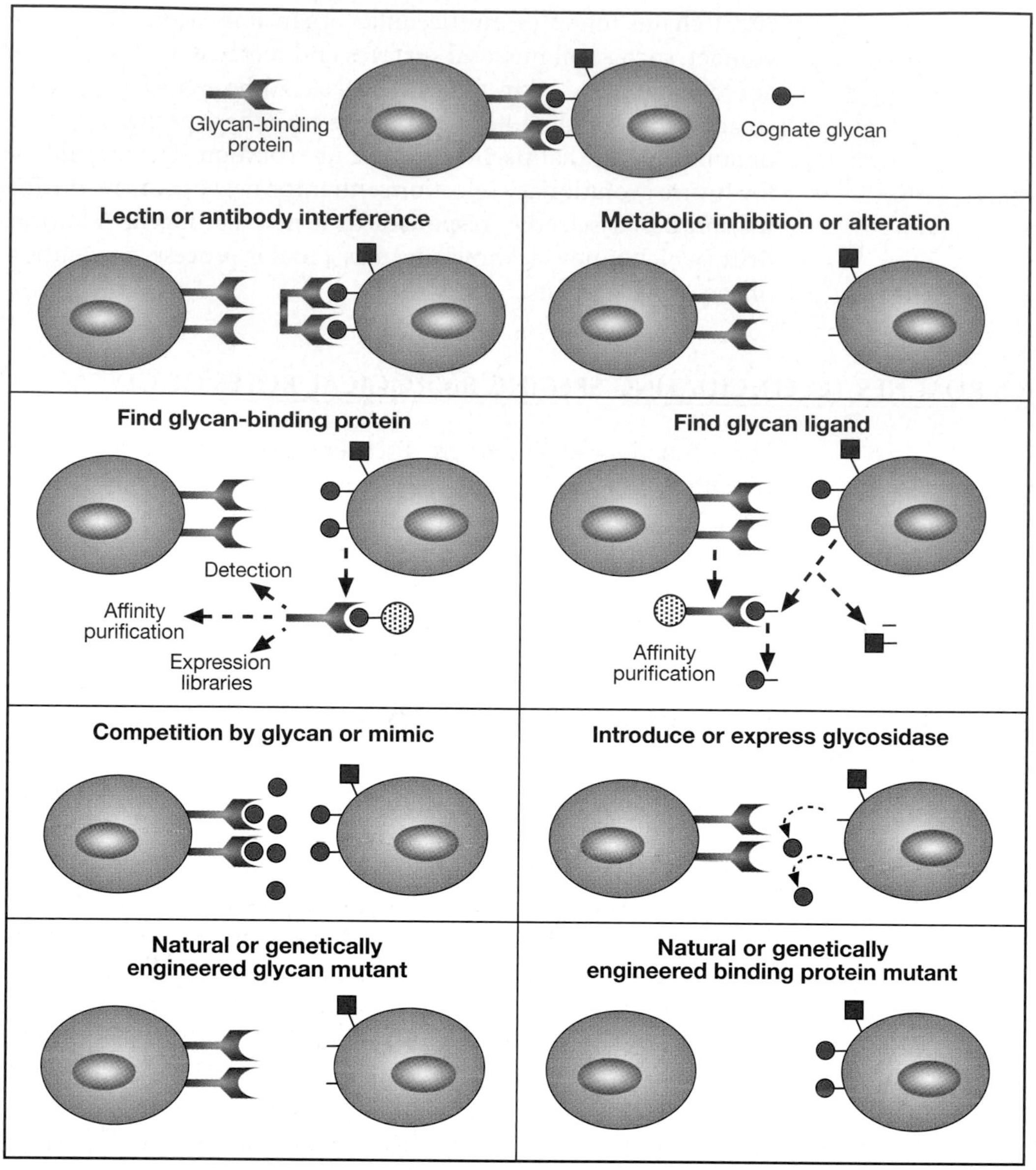

FIGURE 6.2. Approaches for elucidating the biological roles of glycans. The figure assumes that a specific biological role is being mediated by recognition of a certain glycan structure by a specific glycan-binding protein. Clues to this biological role could be obtained by a variety of different approaches (for discussion of each approach, see text).

Metabolic Inhibition or Alteration of Glycosylation

As outlined in Chapter 50, many pharmacological agents can metabolically inhibit or alter glycosylation in intact cells and animals. Although metabolic inhibitors are powerful tools to elucidate biosynthetic pathways, they can sometimes yield confusing results in complex systems. One concern is that the inhibitor may have effects on other unrelated pathways. For example, the inhibitor tunicamycin that blocks N-linked glycosylation can also inhibit UDP-Gal uptake into the Golgi. The second concern is that the inhibitor may cause such massive changes in glycan synthesis that the physical properties of the glycoconjugates and/or membranes are altered, making it difficult to interpret the results. Somewhat more useful results can be obtained by introducing low-molecular-weight primers of terminal glycosylation (see Chapter 50), which can act as alternate substrates for Golgi enzymes, diverting synthesis away from the endogenous glycoproteins. However, this approach can simultaneously generate incomplete glycans on the endogenous glycoconjugates, as well as produce secreted glycan chains, each of which could have its own biological effects.

Finding Natural Glycan Ligands for Specific Receptors

Because specific "carbohydrate-recognition domains" can be identified within a primary amino acid sequence (see Chapters 26 and 27), it is now possible to predict whether a newly cloned protein can bind glycans. If a potential GBP can be produced in sufficient quantities, techniques such as hemagglutination, flow cytometry, surface plasmon resonance, and affinity chromatography (see Chapter 27) can then be used to search for specific ligands. However, the monovalent affinity of the putative GBP for its ligand may not be high. Thus, high densities and/or multivalent arrays may be needed to avoid missing a biologically relevant interaction. The question also arises as to where exactly to look for the biologically relevant ligands in a complex multicellular system. Furthermore, because many glycan structures can be expressed in different tissues at different times in development and growth, a recombinant GBP may detect a cognate structure in a location and at a time that it is not actually of major biological relevance. Careful consideration of the natural occurrence and expression profile of the GBPs should lead to a rational decision as to where to look for its biologically relevant glycan ligands.

Finding Receptors that Recognize Specific Glycans

The converse situation arises when an unusual glycan is found to be expressed in an interesting context and is hypothesized to be a ligand for a specific receptor. It is possible to search for such a receptor by techniques similar to those mentioned above, such as hemagglutination, flow cytometry, and affinity chromatography (see Chapter 27). To facilitate the search, it is necessary to have reasonable quantities of the pure defined glycan in question, as well as a variety of closely related structures that can act as negative controls. Because many biologically relevant lectin-like interactions are of low affinity, it is probably advisable to use a multivalent form of the glycan as the probe. Finally, it may not be obvious where to look for the glycan-binding protein. For example, the receptor that specifically recognizes the unusual sulfated N-glycans of pituitary glycoprotein hormones was eventually found not in the pituitary itself nor in any of the target tissues for these hormones, but in the endothelial cells of the liver, where it serves to regulate the circulating half-life of the hormones (see Chapter 28). Indeed, the most biologically relevant receptor for a particular glycan might even be found in another organism (a pathogen or a symbiont).

Interference by Soluble Glycans or Structural Mimics

The addition of soluble glycans or their structural mimics into the system can cause interference with the interaction between an endogenous GBP and a specific glycan (see Chapter 27). If a sufficient concentration of the specific inhibitor can be achieved, the resulting phenotypic changes can be instructive. When studying in vitro systems, even monosaccharides can be used to advantage in such experiments, as exemplified by the exploration of the Man-6-P receptor pathway (Chapter 30). However, it is often necessary to use competing glycans in somewhat large quantities to block the relatively low-affinity interactions between a GBP and its specific ligand. Effective blockade may also require multivalency of the cognate glycan. Finally, especially when studying complex multicellular systems, the glycans introduced could be cross-recognized by other as-yet-unknown binding proteins, giving a confusing phenotypic readout.

Eliminating Specific Glycan Structures by Glycosidases

A powerful approach to understanding the biological roles of glycans is to use degradative enzymes known to be highly specific for a particular glycan sequence. Many such specific enzymes can be obtained from microbial pathogens. The advantage of this approach is that one is not interfering with the basic biosynthetic cellular machinery, but simply eliminating certain structures selectively after normal synthesis has been completed. Thus, for example, sialidase treatment abolished lymphocyte binding to the high endothelial venules of lymph nodes and provided the first prediction of the nature of endogenous ligands for L-selectin (Chapter 31); injection of endoneuraminidase into the developing retina suggested specific roles for polysialic acids (Chapter 14); and injection of heparanase into developing embryos resulted in a randomization of left–right axis formation (Chapter 16). In all such studies, the purity of the enzyme used is critical and appropriate controls are necessary (including, if possible, a specific inhibitor of the enzyme or a catalytically inactive version of the enzyme). If the enzyme is of bacterial origin, trace amounts of potent contaminants such as endotoxin are also of concern. A genetic approach can be used to avoid problems of contamination by expressing a cDNA for the glycan-modifying enzyme in the intact cell or animal. For example, transgenic expression in mice of an influenza sialic-acid-specific 9-O-acetylesterase gave either early or late abnormalities in development, depending on the promoter used. Unfortunately, many such glycosidases may not function well or at all in the context of an intact animal, which can limit the spectrum of glycan structures that may be probed for function with this approach.

Studying Natural or Genetically Engineered Glycan Mutants

This is intuitively a powerful approach for understanding glycan function. Technically, it is easiest to study glycosylation mutants in cultured cell lines (see Chapter 46). However, although genetic or acquired defects in glycosylation are obtained relatively easily in cultured cell lines, these defects may have limited or not easily discernible biological consequences. This may be because of the lack of other factors or cell types that would be present in the intact organism. For example, the cognate receptor for the glycan may not be present in the same cell type. Of course, such mutants can still be used to analyze basic structural functions of the glycans and their relevance to the physiology of a single cell. Furthermore, one can add back external factors or other cell types thought to interact with the modified glycan. Some mutants can also be reintroduced into intact organisms, for example, to study tumorigenicity or metastatic behavior of malignant cells.

Although much useful information can be gained by such approaches, many of the more specific roles of glycans need to be uncovered by studying mutations in the intact multicellular organism. Genetic defects in glycosylation in intact organisms were initially thought to be relatively uncommon. Looking back on the many glycosylation mutants that have been recently discovered in flies, worms, mice, and humans (see Table 6.1 and Chapter 42), it is clear that glycan changes often affect multiple systems and that the phenotypes are unpredictable and highly variable. In retrospect, the apparent rarity of naturally occurring mutations can now be explained in several ways. In some cases, they appear to have limited biological consequences in the otherwise healthy animal because of redundant pathways. For example, our understanding of the role of GalNAc-initiated O-glycans in animals has been hampered by the remarkable multiplicity of polypeptide GalNAc transferase loci and the consequent inability to easily "delete" O-glycan synthesis using gene "knockout" approaches in mice. In other cases, specific challenges are needed to elucidate the phenotype, and the nature of such challenges is not always initially apparent. Alternatively, many mutations cause lethal aberrations that prevent completion of embryogenesis. This has become apparent, in part, by comparing the genotype-phenotype relationships in naturally occuring human disorders of glycosylation and in experimentally induced glycosylation disorders in mice. In humans, naturally occurring disease-associated mutations in glycosylation pathways almost always correspond to missense mutations that leave some residual enzymatic function intact, whereas deletion of the corresponding enzymatic locus in mice often leads to a lethal phenotype during embryogenesis. Another possibility is that such genetic abnormalities remain undetected because their consequences are pleiotropic. Indeed, it has only been recently discovered that several pediatric developmental disorders are caused by genetic abnormalities in glycan biosynthesis (see Chapter 42). Regardless, the value of constructing glycosylation mutants in intact animals is evident. Indeed, it can now be stated that complete elimination of most of the major glycan classes of vertebrate cells has been genetically accomplished in the mouse, and every instance has lead to embryonic lethality. Given the complex phenotypes and the potential for early developmental lethality, the ability to disrupt glycosylation-related genes in a temporally controlled and cell-type-specific manner can be particularly valuable.

Studying Natural or Genetically Engineered Glycan Receptor Mutants

Eliminating a specific glycan receptor can yield a phenotype that may be very instructive with regard to the functions of the glycan. As with genetic modification of the glycan, the results are more likely to be useful if studied in the intact organism. However, the receptor protein may have other functions unrelated to glycan recognition. Conversely, the glycan in question may have other functions not mediated by the receptor. Thus, for example, the genetic elimination of the CD22/Siglec-2 receptor and the ST6Gal I enzyme that generates its ligand gave complementary, but not identical, phenotypes (see Chapter 32). However, breeding the two mutations into the same mouse indicated that there were indeed epistatic interactions. Similar results were obtained by mating mice deficient in making polysialic acid and in synthesizing the protein carrier of polysialic acids, NCAM.

FURTHER READING

Roseman S. 1970. The synthesis of carbohydrates by multiglycosyltransferase systems and their potential function in intercellular adhesion. *Chem. Phys. Lipids* **5:** 270–297.

Montreuil J. 1980. Primary structure of glycoprotein glycans: Basis for the molecular biology of glycoproteins. *Adv. Carbohydr. Chem. Biochem.* **37:** 157–223.

Berger E.G., Buddecke E., Kamerling J.P., Kobata A., Paulson J.C., and Vliegenthart J.F.G. 1982. Structure, biosynthesis and functions of glycoprotein glycans. *Experientia* **38:** 1129–1162.

Rademacher T.W., Parekh R.B., and Dwek R.A. 1988. Glycobiology. *Annu. Rev. Biochem.* **57:** 785–838.

Hart G.W. 1992. Glycosylation. *Curr. Opin. Cell. Biol.* **4:** 1017–1023.

Kobata A. 1992. Structures and functions of the sugar chains of glycoproteins. *Eur. J. Biochem.* **209:** 483–501.

Lis H. and Sharon N. 1993. Protein glycosylation—Structural and functional aspects. *Eur. J. Biochem.* **218:** 1–27.

Varki A. 1993. Biological roles of oligosaccharides: All of the theories are correct. *Glycobiology* **3:** 97–130.

Stanley P. and Ioffe E. 1995. Glycosyltransferase mutants: Key to new insights in glycobiology. *FASEB J.* **9:** 1436–1444.

Gahmberg C.G. and Tolvanen M. 1996. Why mammalian cell surface proteins are glycoproteins. *Trends Biochem. Sci.* **21:** 308–311.

Hooper L.V., Manzella S.M., and Baenziger J.U. 1996. From legumes to leukocytes: Biological roles for sulfated carbohydrates. *FASEB J.* **10:** 1137–1146.

Salmivirta M., Lidholt K., and Lindahl U. 1996. Heparan sulfate: A piece of information. *FASEB J.* **10:** 1270–1279.

Spillmann D. and Burger M.M. 1996. Carbohydrate–carbohydrate interactions in adhesion. *J. Cell. Biochem.* **61:** 562–568.

Drickamer K. and Taylor M.E. 1998. Evolving views of protein glycosylation. *Trends Biochem. Sci.* **23:** 321–324.

Ferguson M.A.J. 1999. The structure, biosynthesis and functions of glycosylphosphatidylinositol anchors, and the contributions of trypanosome research. *J. Cell. Sci.* **112:** 2799–2809.

Gagneux P. and Varki A. 1999. Evolutionary considerations in relating oligosaccharide diversity to biological function. *Glycobiology* **9:** 747–755.

Angata T. and Varki A. 2002. Chemical diversity in the sialic acids and related α-keto acids: An evolutionary perspective. *Chem. Rev.* **102:** 439–470.

Esko J.D. and Selleck S.B. 2002. Order out of chaos: Assembly of ligand binding sites in heparan sulfate. *Annu. Rev. Biochem.* **71:** 435–471.

Freeze H.H. 2002. Human disorders in N-glycosylation and animal models. *Biochim. Biophys. Acta.* **1573:** 388–393.

Hakomori S.I. 2002. The glycosynapse. *Proc. Natl. Acad. Sci.* **99:** 225–232.

Spiro R.G. 2002. Protein glycosylation: Nature, distribution, enzymatic formation, and disease implications of glycopeptide bonds. *Glycobiology* **12:** 43R–56R.

Lowe J.B and Marth J.D. 2003. A genetic approach to mammalian glycan function. *Annu. Rev. Biochem.* **72:** 643–691.

Wells L. and Hart G.W. 2003. O-GlcNAc turns twenty: Functional implications for post-translational modification of nuclear and cytosolic proteins with a sugar. *FEBS Lett.* **546:** 154–158.

Haltiwanger R.S. and Lowe J.B. 2004. Role of glycosylation in development. *Annu. Rev. Biochem.* **73:** 491–537.

Ohtsubo K. and Marth J.D. 2006. Glycosylation in cellular mechanisms of health and disease. *Cell* **126:** 855–867.

Varki A. 2006. Nothing in glycobiology makes sense, except in the light of evolution. *Cell* **126:** 841–845.

Varki N.M. and Varki A. 2007. Diversity in cell surface sialic acid presentations: Implications for biology and disease. *Lab. Invest.* **87:** 851–857.

Bishop J.R., Schuksz M., and Esko J.D. 2007. Heparan sulphate proteoglycans fine-tune mammalian physiology. *Nature* **446:** 1030–1037.

CHAPTER 7

A Genomic View of Glycobiology

Bernard Henrissat, Avadesha Surolia, and Pamela Stanley

A MULTITUDE OF GLYCOSYLTRANSFERASES (GTs), glycoside hydrolases (GHs; also called glycosidases), nucleotide sugar transporters, and other enzymes are required to synthesize and metabolize glycans. It follows that many genes in the genome are devoted to encoding these activities. In addition, many genes encode glycan-binding proteins (GBPs), which recognize specific glycan structures. This chapter provides a genomic perspective of the genes that code for GTs, GHs, and GBPs.

THE GLYCOME

The glycome comprises all of the glycan structures that may be synthesized by an organism. It is analogous to the genome (all the genes), the transcriptome (all the transcripts), or the proteome (all the proteins) of an organism. Cells of different types will synthesize a subset of the glycome and this will vary depending on their differentiation state and their physiological environment. The human and mouse glycomes have many glycan structures in common, but they also contain glycans that are unique or have divergent functional properties. For example, mice and other rodents have a functional CMP-sialic acid hydroxylase gene that allows them to synthesize CMP-Neu5Gc (cytidine monophospho-*N*-glycolylneuraminic acid), which is transferred to N- and O-glycans (see Chapter 14). Humans have an altered, nonfunctional gene for CMP-sialic acid hydroxylase and therefore do not synthesize CMP-Neu5Gc. A similar situation exists for the gene encoding α1-3 galactosyltransferase, which is functional in the mouse but not in the human genome. The human and fly genomes include orthologous genes that encode GTs that catalyze the same reaction, but they also have GTs that appear to be unique for glycans synthesized only in the human or fly, respectively. The enzyme termed protein O-fucosyltransferase 1 transfers fucose to Notch receptors in mammals and flies and is an example of a conserved GT. On the other hand, flies do not make complex N-glycans with four branches, which are common in mammalian glycoproteins (see Chapter 8). In addition, flies make glycolipids that are not found in humans, but are important in flies for conserved signaling pathways mediated by the epidermal growth factor (EGF) receptor or the Notch receptor (see Chapter 12). These observations predict that a large repertoire of genes is required for glycan assembly and function, some shared across phylogeny and others unique to different organisms.

GENOMICS OF GLYCOSYLATION

The genome encodes all of the enzymes, transporters, and other activities necessary to construct the glycome of an organism. When the first edition of this book appeared in 1999, only a few complete genomes were available, but more than 650 genomes have been published since then. Similarly, the number of GTs with known three-dimensional structures has also grown dramatically, from one structure in 1999 to almost 30 in 2007. The great increases in sequence and structural data have led to the emergence of several databases for comparing genomic and structural information. In the pregenomic era, the glycobiology of mammals, invertebrates, plants, bacteria, and viruses did not overlap extensively, and progress in one domain did not immediately benefit the others. With several genomes now being released each week, a common evolutionary history of GT sequences has emerged. Furthermore, GTs from various origins display the same basic structural folds (see Chapter 5). Thus, the content of genomes from a glycobiological perspective can be analyzed (e.g., by listing candidate GHs or GTs contained in a genome) and compared across genomes to see whether families have expanded or disappeared during evolution. Examination of 650 completely sequenced organisms listed in the CAZy database (*c*arbohydrate-*a*ctive en*zy*mes; www.cazy.org) suggests that about 5% of the genome encodes enzymes that are involved in glycan synthesis, degradation, or recognition. About 1–2% of the genes of an organism encode GTs (Figure 7.1). Overall, the average number of GHs encoded is slightly smaller (~1%), but much more variable between species. For example, several free-living Archaea are apparently completely devoid of GHs, despite encoding a range of GTs. Organisms that have evolved to symbiotic or parasitic lifestyles also tend to encode fewer GHs than do free-living organisms, and they sometimes have very few, if any, GT genes. Although the average number of GTs usually exceeds the number of GHs, the situation is reversed in organisms that specialize in utilizing plants as energy and carbon sources (e.g., saprophytic fungi and soil bacteria) and several bacteria of the intestinal flora. The number of genes that encode GBPs is more difficult to define because many have other functional domains and therefore have not been annotated as GBPs. A conservative estimate would be that mammalian genomes encode 100–200 GBPs.

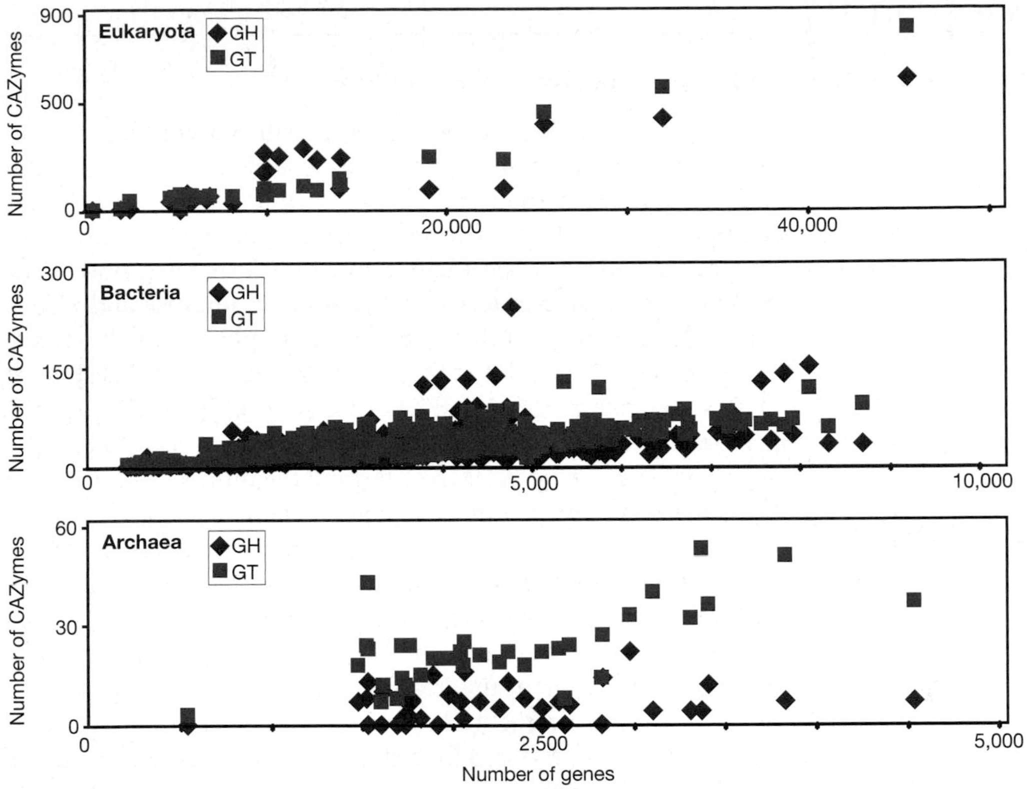

FIGURE 7.1. Plot showing the number of carbohydrate-active enzymes (CAZymes) as a function of the number of genes in the genomes of eukaryotic species (*top*), bacteria (*middle*), and Archaea (*bottom*). (*Pink squares*) Glycosyltransferases (GTs); (*blue diamonds*) glycoside hydrolases (GHs).

As the genomes of new organisms become available, their sequences are routinely searched for genes that show similarity to known GTs, GHs, or GBPs. If a gene identified in a search is not annotated, the sequence can be used to find genes with related sequences that have been annotated, either in the same species or across phylogeny, and this provides a starting point for determining the biochemical activity of the encoded protein. In the absence of annotation (or suggested activity), the sequence of a new gene can be analyzed for functional motifs using computer algorithms. Some examples include algorithms to determine whether there is (1) a signal peptide at the amino terminus that would target the protein to the endoplasmic reticulum (ER) and the secretory pathway, (2) a transmembrane domain that would anchor the protein in a membrane of the secretory pathway, (3) a carboxy-terminal signal peptide for the addition of a glycosylphosphatidylinositol (GPI) anchor (see Chapter 11), or (4) a KDEL carboxy-terminal retrieval sequence that allows proteins to recycle between the Golgi and the ER. Most GTs are type II transmembrane proteins with a short cytoplasmic tail, and many contain a DXD (Asp–any amino acid residue–Asp) sequence. Several GTs with similar transferase activities have common structural motifs (see Chapter 5). Care must be taken in this endeavor. The GTs are among the most versatile category of proteins known in terms of the substrates they may recognize. Thus, it is very difficult to predict the reaction catalyzed by a GT or GH encoded by a gene that shows homology with other GT or GH genes based simply on primary amino acid sequence.

GENE FAMILIES

Glycosyltransferase and Glycoside Hydrolase Families

Carbohydrate-active enzymes can be classified according to several criteria. The simplest classification is based on their substrate specificities and forms the basis of the recommendations of the International Union of Biochemistry and Molecular Biology (IUBMB), which assigns an EC number for a given enzyme. GHs are given the codes EC 3.2.1.x, where x represents substrate specificity. Similarly, GTs are described as EC 2.4.y.z, where y defines whether the sugar transferred is a pentose or a hexose and z describes the acceptor specificity. The advantage of this system is its simplicity, which has led to its widespread use. It should be kept in mind that the founding principle of this system is that the EC number is given only *after* experimental determination of enzymatic specificity. Several hundred or perhaps thousands of different EC numbers are needed to classify GTs, but only about 240 are available from the IUBMB so far, and the number of new EC numbers is growing slowly compared to the growth in sequence data.

The intrinsic problem with a classification system based on substrate (or product) specificity is that it does not appropriately accommodate enzymes that act on several different substrates. This is particularly relevant with carbohydrate-active enzymes, which operate on complex substrates and sometimes display broad, overlapping specificities. In addition, it has become apparent that classification based on substrate specificity fails to reflect the three-dimensional structural features of these enzymes. To circumvent these problems, a novel system was proposed for classifying GHs and GTs, which builds on the fact that there is a direct relationship between amino acid sequence and folding similarities. The basic principle behind this new classification system is simple: Regardless of activity and substrate specificity, sequences that display similarity are grouped in the same family. For GTs, a family classification based on the similarity of amino acid sequences was initiated in 1997 (~500 sequences, 27 families) and updated in 2003. This classification is now continuously updated in the CAZy database, which features more than 30,000 GT sequences in approximately 90 families. CAZy gives access to the various families of GTs, GHs, polysaccharide lyases, and their carbohydrate-binding modules. Each family is annotated with information regarding known enzyme activities and includes catalytic and structural features. This summary is followed by a list of the proteins and open reading frames (ORFs) belonging to the family, with links to sequence and structural information available in public databases. Since information on the repertoire of carbohydrate-active enzymes present in a given organism can provide interesting insights into its carbohydrate metabolism, CAZy features summary pages for more than 650 publicly available genomes.

The earliest feature that appeared from the sequence-based families was that many families are "polyspecific," containing enzymes of different substrate specificity (i.e., with different EC numbers). The existence of polyspecific families indicates that (1) the acquisition of new specificities by GHs and GTs is a common evolutionary event, (2) the substrate specificity of carbohydrate-active enzymes could be engineered for experimental or applied purposes, and (3) the substrate (or product) specificity of these enzymes is governed by fine details of three-dimensional structure, not by a global fold. Human GTs, whose experimentally determined activities have been established, are compiled in several excellent resources (KEGG glycan at http://www.genome.jp/kegg/glycan/ or the Consortium for Functional Glycomics). In contrast, assignments for other organisms are often erroneous, in part because of erroneous assignment of EC numbers based on primary sequence. The challenge of the postgenomic era is therefore to dissect this ever-growing list of ORFs whose encoded proteins are candidate GTs of unknown donor, acceptor, and product specificities.

Glycan-binding Proteins

The information presented by the wide variety of glycans on glycoconjugates is deciphered by an equally versatile number of proteins that recognize glycans. To understand the biology behind protein–glycan interactions, it is imperative to identify all of the GBPs and the glycan ligands to which they bind.

GBPs were identified in the past by systematic biochemical studies that determined their glycan-recognition properties. However, the recent explosion of sequenced genomes makes it possible to identify genes that encode GBPs by sequence homology. For example, the mannose-binding lectins (MBLs) can easily be identified because they exhibit motifs found in C-type lectins (see Chapter 31) as well as a collagen-like domain that promotes their oligomerization and is necessary for host defense through complement activation. A variant allele with changes in both the promoter and structural regions of the human MBL gene (*MBL2*) influences the stability and serum concentration of the protein. Epidemiological studies have suggested that genetically determined variation in MBL serum concentration influences the susceptibility to and the course of different types of infections, autoimmune reactions, and metabolic and cardiovascular diseases. The fact that genetic variations in MBL are frequent indicates a dual role for MBL in host defense and highlights the power of genomics to aid research and the understanding of human disease.

Most of the studies on GBPs have been restricted to mammalian proteins (e.g., C-type lectins, galectins, and siglecs), and their counterparts in plants and other lower organisms have not yet been compiled and explored from a genomics point of view. However, sequence and structural databases dealing with the tertiary and quaternary structure of GBPs and their glycan-binding intricacies have been compiled in considerable detail (http://www.cermav.cnrs.fr/lectines/; http://nscdb.bic.physics.iisc.ernet.in/). A proposed extended classification of GBPs is presented in Chapter 26 (see Table 26.2).

Ligand specificities are the hallmark of in vivo roles of GBPs. Hence, knowledge of ligand specificity of a particular GBP is an absolute necessity. In cases where there is a dearth of information about the glycan specificities of these proteins, scientists have used rational predictive schemes based on the framework and sequence of the existing carbohydrate-recognition domains (CRDs). Legume lectins represent a class of GBPs identified several decades ago that still continues to provide perhaps the best model for protein–glycan recognition. Moreover, discovery of the legume lectin fold (jelly-roll motif) in mammalian lectins such as galectins (see Chapter 33), calnexin, and calreticulin (see Chapter 31) highlights the preeminence of this fold in carbohydrate recognition across phylogeny. Hence, identification of all GBPs in plants and animals is important for progress in glycobiology. In this context, earlier work in the identification of monosaccharide-binding specificities of legume lectins provided the framework for finding their relatives in all forms of life (see Chapter 29). In fact, this approach led to the assignment of glycan specificities for proteins involved in the sorting of vesicular compartments and in glycoprotein folding in the ER and Golgi compartments of mammalian cells (see Chapter 36). Of similar importance is the discovery of new galectins in the galectin-10 family and galectin-like proteins in genome databases using BLAST and PSI-BLAST search algorithms (see Chapter 33). Likewise, siglecs, a family of sialic-acid-binding lectins involved in regulating immune responses (see Chapter 32), exhibit their own signature sequence motifs, and 15 members have been identified in primates to date.

At times, however, predictions are not satisfactory, as the mere presence of a CRD does not necessarily translate into functional glycan-recognizing activity. This is because sequence motifs used to identify CRD domains in proteins often end up in a functionally inactive, lectin-like CRD fold (see Chapter 31). Another issue is that glycan-binding properties are tunable by very subtle reorganization of lectin-like sequences. Thus, at times, novel lectins involved in important biological functions might not be identified in a genome-based screen using known lectin sequences as templates. This suggests that there are new topologies for glycan targets yet to be identified. Hence, both biochemical and in silico approaches need to be pursued simultaneously in identifying new GBPs and the glycans they recognize.

A new biochemical approach is the technique of glycan microarrays to identify binding domains of new GBPs (see Chapter 27). Such microarrays provide a high-throughput means of detecting the interactions of proteins with diverse oligosaccharide sequences of glycoproteins, glycolipids, and polysaccharides. The use of glass slides and microarray printing technology allows the production of glycan microarrays with the potential to examine GBP binding to several thousand unique glycans simultaneously. All types of GBPs may be compared including lectins, antiglycan monoclonal or serum antibodies, and glycan-binding cytokines or chemokines. Binding is assessed by conventional biochemical and biophysical techniques such as fluorescent or spectrometric techniques.

GENOMICS OF GLYCOSYLATION IN VARIOUS ORGANISMS

Viruses

It has long been known that many viruses use host glycans as specific binding receptors for entering the cell. Similarly, many viruses encode lytic enzymes that break down host cell-wall glycans to release viral particles at the end of viral replication. Genome sequencing reveals that many double-stranded DNA viruses, and especially those with a large genome, also take advantage by adding sugars to host glycoproteins through the use of GTs. Although the biological role of viral GTs is often poorly understood, in several cases, a function has been identified. For example, the T4 bacteriophage encodes nucleases that degrade host-cell DNA. To protect its own genome, the phage modifies its DNA by replacing cytosine with 5-hydroxymethylcytosine and then transferring glucose to the 5-hydroxymethylcytosine using a specific UDP-glucose:DNA glucosyltransferase. The baculovirus enzyme ecdysteroid UDP-glucosyltransferase (EGT) disrupts the hormonal balance of the insect host by catalyzing the conjugation of ecdysteroid hormones with glucose or galactose. Expression of the EGT gene allows the virus to block molting and pupation of the infected insect larvae. Serotype conversion in *Shigella flexneri* is mediated by temperate bacteriophages, which encode GTs that mediate O-antigen conversion by the addition of glucose to O-antigen units. Finally, *Acanthamoeba polyphaga* mimivirus, the virus with the largest known genome, encodes no less than 12 putative GTs whose precise targets are unknown, although it is likely that this virus N- and O-glycosylates its own proteins.

Bacteria

Some bacteria such as *Campylobacter* are able to N-glycosylate their proteins, but the most universal role for bacterial glycosylation is probably in the synthesis of cell-wall peptidoglycan and various "decorations" that range from simple glycolipids to lipopolysaccharides to complex exopolysaccharides (see Chapter 20). The GTs involved in peptidoglycan biosynthesis are GT28 (MurG), which adds *N*-acetylglucosamine (GlcNAc) to undecaprenyl diphospho-*N*-acetylmuramic acid (MurNAc), and GT51, which polymerizes undecaprenyl diphospho-MurNAc-GlcNAc. A bacterium such as *Mycobacterium tuberculosis* produces an extremely complex envelope that includes all of the above glycan structures. In bacteria, the role of these glycan structures is to provide a barrier that affords mechanical, chemical, and biological protection to the cell. Some pathogenic or commensal bacteria produce an outer glycan layer, which mimics that of their hosts in order to evade host immune surveillance (see Chapter 39). For instance, *Pasteurella multocida* produces a thick hyaluronan capsule. Other pathogens such as *Escherichia coli* K1 and *Neisseria meningitidis* produce a polyα2-8 sialic acid capsule as a means of seeming like a mammalian glycoprotein, thereby preventing stimulation of an immune response. It is thought that human gut bacteria may also produce capsular polysaccharides similar to those of the gut mucosa.

Archaea

Archaea also devote about 1% of their genes to GTs, with variations in sequence greater than GTs of other organisms. However, on average, Archaea devote only 0.25% of their genes to GHs, and there is almost no correlation with the number of GH genes and the overall number of genes. Surprisingly, one quarter of Archaea genomes sequenced to date appear to be completely devoid of GHs. The most striking example is *Methanosphaera stadtmanae*, whose genome encodes at least 43 GTs, but apparently no GHs. This is not due to sequence divergence, because GHs were readily detected in some Archaea species. These observations suggest that (1) horizontal transfer is likely the determining factor behind archaeal GH repertoires and (2) the Archaea in question do not recycle the glycosidic bonds catalyzed by their own GTs. Although they do not make peptidoglycans like bacteria, Archaea utilize nucleotide-activated oligosaccharides to produce a variety of extracellular polysaccharides such as the heteropolysaccharide "methanochondroitin" made by *Methanosarcina barkeri*, which resembles eukaryotic chondroitin (see Chapter 16). Archaea also make glycophospholipids and one relevant GT is GDP-glucose:glucosyl-3-phosphoglycerate synthase (α-glucosyltransferase) from family GT81 in *Methanococcoides burtonii*. In family GT55, there are several archaeal GDP-mannose:mannosyl-3-phosphoglycerate synthases (α-mannosyltransferases). A number of Archaea have GTs related to bacterial and eukaryotic oligosaccharyltransferases (CAZy family GT66), which is consistent with the fact that Archaea use oligosaccharyldiphospholipids as sugar donors. It has been shown that the archaeon *Methanococcus voltae* uses this strategy to transfer N-glycans to flagellin and S-layer proteins. Like bacteria and eukaryotes, evolution toward an obligate symbiont lifestyle is also accompanied by gene loss in Archaea. For example, the tiny genome of *Nanoarchaeum equitans* appears to encode only three GTs and no GHs.

Eukaryotes

Several eukaryotes have undergone genome reduction and lost most of their GT genes. Thus, *Plasmodium falciparum* and *Encephalitozoon cuniculi* have eight GTs, whereas the minuscule genome (464 genes) of *Guillardia theta* encodes only one GT. Overall, the abundance of GTs in eukaryotes correlates with evolution to multicellularity. Free-living fungi and the unicellular marine green alga *Ostreococcus tauri* have a number of GTs similar to certain bacteria. On the other hand, with their large genomes and complicated body plans, plants and animals encode many more GTs and GHs than bacteria and Archaea.

Plants

The genomes of higher plants encode more GTs than any other organism sequenced to date, with approximately 450 in *Arabidopsis*, more than 560 in rice, and more than 800 in poplar! Higher plants have huge genomes resulting from several rounds of complete genome duplication. The massive number of GTs in plants is due to the expansion of several extremely populated GT families. For example, *Arabidopsis*, rice, and poplar have approximately 120, 200, and 300 family GT2 genes, respectively. Higher plants are characterized by extremely complex cell walls made of various polysaccharides that can be rather simple like cellulose, more complex as in hemicelluloses (e.g., xylans, glucuronoxylans, galactomannans, xyloglucans, or rhamnogalacturonans), or extremely complex like the "hairy" regions of pectin (see Chapter 22). It has been estimated that the biosynthesis of pectin alone requires the action of several dozen different GTs. The composition and properties of cell-wall polysaccharides in the various organs of plants (roots, stem, leaves, fruits, pollen, etc.) can be extremely different. Differential expression in various tissues is probably one of the driving forces behind the accumulation of hundreds of genes encoding GTs in plants. This large spectrum of GTs is accompanied by a similarly numerous and diverse array of GHs, which are involved in the remodeling of the plant cell wall during plant growth. Thus, the *Arabidopsis* genome encodes about 400 GHs and poplar has more than 600.

Vertebrates

Although they do not have such an enormous number of GT genes as plants, vertebrates are characterized by a large diversity of GTs. The human GT genes fall into 43 families, a number similar to that of plants. Families that are present only in vertebrates are GT6, GT12, and most GT29 family members (invertebrates have only one member of a particular sialyltransferase subfamily, whereas vertebrates usually have many different GT29 sialyltransferases belonging to several distinct subfamilies). However, there are no GT families that are unique to humans or primates. The completion of the first animal genomes also revealed a relative paucity in the number of encoded GHs. Thus, the human genome codes for only about 100 GHs, with only a very few able to digest some plant polysaccharides. In fact, the digestion of the great majority of the plant cell-wall polysaccharides of our food is "outsourced" to the multitude of different microorganism species that colonize the human gut. This "microbiome" greatly enlarges our limited genome. For instance, a single species of the gut bacteria *Bacteroides thetaiotaomicron* encodes 2.3 times more GHs than does our own genome!

Invertebrates

One of the initial surprises that came with the completion of the first genomes was that the human genome encodes fewer GTs than the genome of the nematode *Caenorhabditis elegans* (~230 compared to 242, respectively). *Drosophila melanogaster* has only 144 known GT genes. These gross numbers, however, mask important biological differences. The comparative abundance of GTs in *C. elegans* compared to humans is essentially due to three GT families that are more highly represented in the nematode: GT1 glucuronyltransferases (77 members in *C. elegans* vs. 26 in *Homo sapiens*), GT11 fucosyltransferases (26 members in *C. elegans* vs. 11–13 in *H. sapiens*), and GT14 β-xylosyltransferases and β1-6 N-acetylglucosaminyltransferases (20 members in *C. elegans* vs. 11 in *H. sapiens*). For most other GT families, *C. elegans* appears to have the same number or fewer GT genes than humans.

MODULAR GLYCOSYLTRANSFERASES AND GLYCOSYL HYDROLASES

In addition to catalytic specificity, the amino acid sequences of GTs may suggest the existence of one or more additional domains that modulate the activity of the GT. Modular GTs fall into several categories:

A GT domain is appended to another GT domain. The most striking example is with the two-domain mammalian heparan synthases that have evolved for the addition of alternating sugars to form a polysaccharide (see Chapters 15 and 16). The amino-terminal domain, which adds β1-4 glucuronic acid residues, belongs to family GT47, whereas the carboxy-terminal domain, which adds α1-4 *N*-acetylglucosamine residues, belongs to family GT64. Some strains of bacteria have a heparan synthase, which also consists of two catalytic modules, but from families distinct from those used to form the animal heparan synthases, namely, families GT2 and GT45 (Figure 7.2), thereby providing a beautiful example of convergent evolution. A similar example of convergent evolution is found among chondroitin synthases, where the human enzyme is made of GT31 and GT7 catalytic domains, whereas the bacterial equivalent is made of tandem GT2 catalytic domains. Not all modular GTs are designed like heparan and chondroitin synthases. Some remain uncharacterized, such as human LARGE (a putative GT for which the identity of sugar donors and acceptors of the constitutive GT8 and GT49 domains are still unknown).

A GT domain is appended to a GBP domain. The best-known examples are polypeptide α-N-acetylgalactosaminyltransferases (ppGalNAcTs) that transfer GalNAc to serine or threonine residues in proteins (see Chapter 9). In these enzymes, a family GT27 catalytic domain is linked to a GBP domain related to ricin. The GBP domain binds to the GalNAc residue transferred to protein by the GT27 catalytic domain and tethers the enzyme to the substrate. Another example is mouse polypeptide β-xylosyltransferase 2, where a family GT14 catalytic domain is linked to a carboxy-terminal domain of about 100 residues that is thought to act as a GBP.

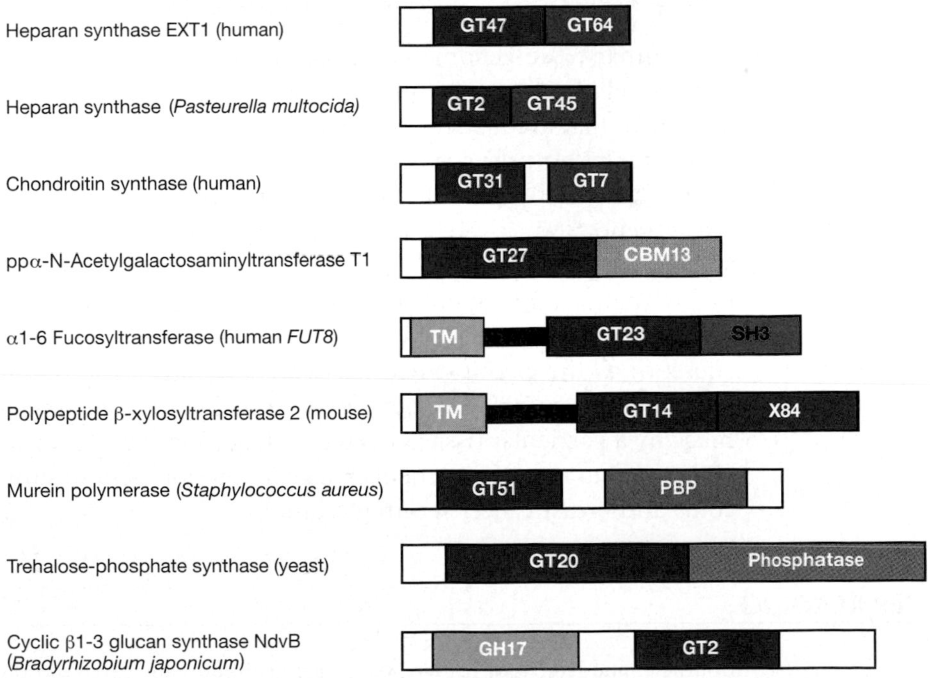

FIGURE 7.2. Schematic examples of modular glycosyltransferases (GTs). (*Red* and *blue boxes*) GT family number. Other modules shown are (CBM13) ricin-like carbohydrate-binding module; (SH3) src homology domain 3; (X84) putative glycan-binding module; (PBP) penicillin-binding protein; (TM) transmembrane domains; (phosphatase) trehalose 6-phosphate phosphatase; (GH17) glycoside hydrolase family 17 module. Boxes without labels represent regions of the polypeptide with unknown function.

A GT domain is appended to a domain with different properties. Prominent examples are the bacterial murein polymerases (which feature a penicillin-binding transpeptidase domain attached to a family GT51 catalytic domain) and yeast trehalose synthase (whose family GT20 catalytic domain produces trehalose-6-phosphate that is converted into trehalose by an appended trehalose phosphate phosphatase domain). Other examples include a GT domain linked to a transglycosidase domain such as *Bradyrhizobium japonicum* cyclic β1-3 glucan synthase NdvB whose family GT2 domain bears an amino-terminal GH17 transglycosidase domain or *Schizosaccharomyces pombe* α-glucan synthase with an amino-terminal GH13 transglycosidase domain followed by a carboxy-terminal GT5 catalytic module. The mammalian α1-6 fucosyltransferase Fut8 has a family GT23 catalytic domain with an SH3-homology domain at the carboxyl terminus. The SH3 domain is a candidate substrate recognition domain for signaling membrane proteins that have acquired a phosphotyrosine.

Modular glycosyl hydrolases. The GHs can also be modular, with the catalytic module appended to one or more other modules, whose role is frequently to bind polysaccharides. Although human GHs are infrequently modular, those of microbes involved in plant cell-wall degradation can sometimes have more than five different modules assembled in a single polypeptide. The most intricate architecture of GHs is probably found in certain environmental bacteria, such as *Clostridium thermocellum*, which elaborate a macromolecular complex called a “cellulosome” in which a large variety of modular plant cell-wall hydrolases are assembled together on a scaffolding protein. This strategy enables the construction of entities comprising dozens of modules targeting the various polysaccharides that make up the plant cell wall.

RELATIONSHIPS OF GENOMICS TO GLYCOMICS

In summary, the genome comprises all of the DNA of an organism, and it therefore includes all the genes that produce the glycome. The glycome comprises all of the glycan structures that are made by an organism. Although within an organism, almost every cell that contains a nucleus and mitochondria has an identical genome, cells typically differ in the portion of the glycome they synthesize because their glycan complement depends on which genes are actively transcribed and which transcripts are translated. Transcription, splicing, translation, and posttranslational processing may vary depending on the state of differentiation and physiological environment of a cell. Therefore, during development and differentiation or under different conditions, the glycan repertoire of a cell will represent a subset of all the glycan structures that an organism is capable of making. To describe this variation, it is common to qualify the term glycome when refering to the glycan structures made by a particular tissue or cell type (e.g., T-cell glycome, liver glycome, and serum glycome) and to note the particular stage of development or growth state (e.g., fetal liver glycome and breast cancer serum glycome).

FURTHER READING

Bourne Y. and Henrissat B. 2001. Glycoside hydrolases and glycosyltransferases: Families and functional modules. *Curr. Opin. Struct. Biol.* **11:** 593–600.

Coutinho P.M., Deleury E., Davies G.J., and Henrissat B. 2003. An evolving hierarchical family classification for glycosyltransferases. *J. Mol. Biol.* **328:** 307–317.

Coutinho P.M., Stam M., Blanc E., and Henrissat B. 2003. Why so many carbohydrate-active enzymes related genes in plants? *Trends Plant Sci.* **8:** 563–565.

Davies G.J., Gloster T.M., and Henrissat B. 2005. Recent structural insights into the expanding world of carbohydrate-active enzymes. *Curr. Opin. Struct. Biol.* **15:** 637–645.

Raman R., Raguram S., Venkataraman G., Paulson J.C., and Sasisekharan R. 2005. Glycomics: An integrated systems approach to structure-function relationships of glycans. *Nat. Methods* **2:** 817–824.

Mitra N., Sinha S., Ramya T.N., and Surolia A. 2006. N-linked oligosaccharides as outfitters for glycoprotein folding, form and function. *Trends Biochem. Sci.* **31:** 156–163. Erratum in *Trends Biochem. Sci.* 2006. **31:** 251.

Gupta G. and Surolia A. 2007. Collectins: Sentinels of innate immunity. *BioEssays* **29:** 452–464.

CHAPTER 8

N-Glycans

Pamela Stanley, Harry Schachter, and Naoyuki Taniguchi

N-GLYCANS ARE COVALENTLY ATTACHED TO PROTEIN AT ASPARAGINE (Asn) residues by an N-glycosidic bond. Five different N-glycan linkages have been reported, of which *N*-acetylglucosamine to asparagine (GlcNAcβ1-Asn) is the most common. This chapter describes only the GlcNAcβ1-Asn N-glycans, their general structure, the steps in their synthesis and processing, and the origins of their structural diversity. Terminal sugars that determine much of the diversity of N-glycans are referred to only briefly because they are described in detail in Chapter 13. Similarly, a discussion of glycosylation-mediated quality control of protein folding is reserved for Chapter 36, and the synthesis of the mannose-6-phosphate (Man-6-P) recognition determinant necessary for targeting lysosomal hydrolases to lysosomes is described in Chapter 30. Congenital disorders of glycosylation that arise from defects in N-glycan synthesis are discussed in Chapter 42.

DISCOVERY AND BACKGROUND

The GlcNAcβ1-Asn linkage was initially discovered by biochemical analyses of abundant glycoproteins found in serum, such as immunoglobulins. Importantly, early experiments on glycoprotein synthesis showed that not all asparagine residues can accept an N-glycan. The minimal amino acid sequence begins with asparagine followed by any amino acid except proline and it ends with serine or threonine (Asn-X-Ser/Thr). Thus, Asn-X-Ser/Thr "sequons" in a protein are candidates for receiving an N-glycan. A fascinating aspect of N-glycans is their complicated biosynthesis. Sensitive and rapid labeling techniques using [2-^{3}H]-mannose revealed how these glycans are initially synthesized on a lipid-like molecule termed dolichol phosphate (Dol-P), followed by "en bloc" transfer of the entire glycan of 14 sugars to protein. This synthetic pathway is conserved in all of the metazoa, in plants, and in yeast. Bacteria use a related pathway to synthesize their cell wall. Other linkages to asparagine have been described, including glucose to asparagine in laminin of both mammals and Archaea, *N*-acetylgalactosamine (GalNAc) to asparagine in Archaea, and rhamnose to asparagine in bacteria. In a sweet corn glycoprotein, arginine is found in N-linkage to glucose.

Understanding these N-glycan pathways is important because N-glycans affect many properties of glycoproteins including their conformation, solubility, antigenicity, and recognition by glycan-binding proteins. N-glycans are used as tags by cell biologists to localize a glycoprotein or to follow its movement through the cell. Defects in N-glycan synthesis lead to a variety of human diseases.

MAJOR STRUCTURAL CLASSES AND NOMENCLATURE

All N-glycans share a common core sugar sequence, Manα1-6(Manα1-3)Manβ1-4GlcNAcβ1-4GlcNAcβ1-Asn-X-Ser/Thr, and are classified into three types: (1) oligomannose, in which only mannose residues are attached to the core; (2) complex, in which "antennae" initiated by N-acetylglucosaminyltransferases (GlcNAcTs) are attached to the core; and (3) hybrid, in which only mannose residues are attached to the Manα1-6 arm of the core and one or two antennae are on the Manα1-3 arm. An example of each N-glycan type is given in Figure 8.1.

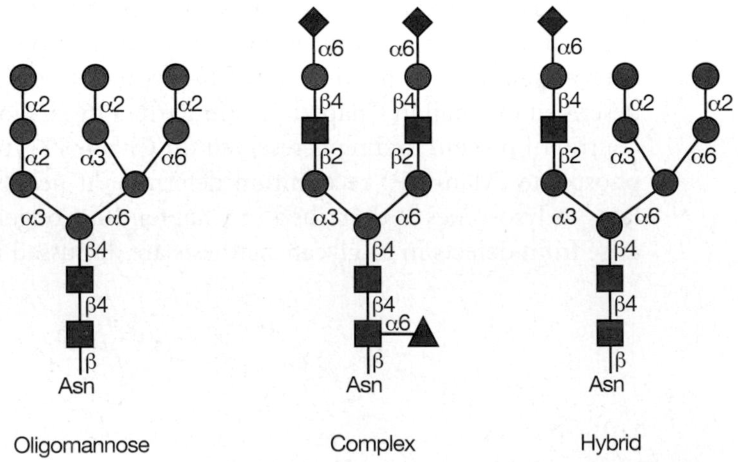

FIGURE 8.1. Types of N-glycans. N-glycans added to protein at Asn-X-Ser/Thr sequons are of three general types in a mature glycoprotein: oligomannose, complex, and hybrid. Each N-glycan contains the common core $Man_3GlcNAc_2Asn$.

PREDICTING SITES OF N-GLYCOSYLATION

N-Glycans occur on many secreted and membrane-bound glycoproteins at Asn-X-Ser/Thr sequons. Analyses of protein sequence databases have revealed that about two thirds of the entries contain this consensus sequence. It is estimated that at least two thirds of those sequons are likely to be N-glycosylated. Occasionally, N-glycans are found at Asn-X-Cys, provided that the cysteine is in the reduced form. Although there have been several published reports of nucleocytoplasmic or cytoplasmic N-glycans, there exists no definitive structural evidence that N-glycans actually occur on cytoplasmic or nuclear proteins nor on the cytoplasmic portions of membrane proteins. The transfer of N-glycans to Asn-X-Ser/Thr sequons occurs on the lumenal side of the endoplasmic reticulum (ER) membrane while the protein moiety is being synthesized on ER-bound ribosomes and is translocating through the translocon in the ER membrane. Membrane glycoproteins remain anchored in the ER membrane with portion(s) exposed to the ER lumen, other portions embedded in the membrane, and yet other regions within the cytoplasm. Only domains that are accessible to the ER lumen will receive an N-glycan. Glycoproteins that lack a transmembrane domain also receive N-glycans cotranslationally, and ultimately they translocate completely into the lumen of the ER.

It is important to note that whereas the presence of the Asn-X-Ser/Thr sequon is necessary for the receipt of an N-glycan, transfer of the N-glycan to this sequon does not always occur, due to conformational or other constraints during glycoprotein folding. Also, the identity of "X" may reduce the efficiency of glycosylation, such as when "X" is acidic (aspartate or glutamate). Thus, when Asn-X-Ser/Thr sequons are present in a deduced amino acid sequence encoded by a cDNA, they are not identified categorically as N-glycan sites, but are referred to as potential N-glycan sites. Proof that an N-glycan is actually present at a potential site requires experimental evidence, as described later in this chapter.

ISOLATION, PURIFICATION, AND ANALYSIS

N-Glycans may be released from asparagine using a bacterial enzyme known as peptide-N-glycosidase F (N-glycanase F, PNGase F). This enzyme will remove oligomannose, hybrid, and complex N-glycans attached to asparagine. However, it will not remove N-glycans with certain modifications of the N-glycan core found so far only in slime molds, plants, insects, and parasites. Another bacterial enzyme termed PNGase A will remove these structures as well as all structures removed by PNGase F. Both enzymes are amidases that release N-glycans attached to the nitrogen of asparagine, thereby converting asparagine to aspartate. Therefore, sites of glycosylation can be deduced by amino acid sequence analysis performed before and after PNGase F treatment. Other bacterial enzymes cleave between the two core *N*-acetylglucosamine residues, leaving one *N*-acetylglucosamine attached to asparagine. These endoglycosidases are more specific in terms of the N-glycan structures they will cleave. Endoglycosidase H will release oligomannose and hybrid N-glycans, but not complex N-glycans. Endoglycosidase F will release simple biantennary N-glycans, but not oligomannose or hybrid N-glycans. N-Glycans may also be obtained free of protein by hydrazinolysis or by exhaustive digestion with a protease that removes all amino acids except for the asparagine. Released N-glycans may be purified by conventional ion-exchange and size-exclusion chromatography, high-pressure liquid chromotography (HPLC) methods, and affinity chromatography on glycan-binding proteins called lectins. Lectins for glycan analysis are usually obtained from plants and are described briefly in Chapter 1 and in detail in Chapters 29 and 45. Purification and analytical methods for obtaining and characterizing N-glycans are described in Chapter 47.

SYNTHESIS OF N-GLYCANS

The biosynthesis of all eukaryotic N-glycans begins on the cytoplasmic face of the ER membrane with the transfer of GlcNAc-P from UDP-GlcNAc to the lipid-like precursor dolichol phosphate (Dol-P) to generate dolichol pyrophosphate *N*-acetylglucosamine (Dol-P-P-GlcNAc). Fourteen sugars are sequentially added to Dol-P before en bloc transfer of the entire glycan to an Asn-X-Ser/Thr sequon in a protein that is being synthesized and translocated through the ER membrane. The protein-bound N-glycan is subsequently remodeled in the ER and Golgi by a complex series of reactions catalyzed by membrane-bound glycosidases and glycosyltransferases. Many of these enzymes are exquisitely sensitive to the physiological and biochemical state of the cell in which the glycoprotein is expressed. Thus, the populations of sugars attached to each glycosylated asparagine in a mature glycoprotein will depend on the cell type in which the glycoprotein is expressed and on the physiological status of the cell, a status that may be regulated during development and differentiation and altered in disease.

Synthesis of the Dolichol-P-P-Glycan Precursor

Dolichol is a polyisoprenol lipid comprised of five-carbon isoprene units linked linearly in a head-to-tail fashion. Dol-P is used in N-glycan synthesis (Figure 8.2). The number of isoprene units in dolichol varies within cells and between cell types and organisms. For example, the most common yeast dolichol has 14 isoprene units, whereas dolichols from other eukaryotes, including mammals, are longer and may have up to 19 isoprene units. The structure of the N-glycan precursor that is synthesized on Dol-P is shown in Figure 8.3. The enzymes that catalyze each step in the biosynthesis have been identified primarily from studies of mutants of the yeast *Saccharomyces cerevisiae.* The gene affected by each yeast mutation is known as an *ALG* gene (for *al*tered in *g*lycosylation) as shown in Figure 8.3.

Synthesis of the N-glycan precursor is initiated on the cytoplasmic face of the ER by the transfer of GlcNAc-P from UDP-GlcNAc to membrane-bound Dol-P, forming GlcNAc-P-P-Dol (Figure 8.3). This step is catalyzed by the enzyme GlcNAc-1-phosphotransferase that transfers GlcNAc-1-P from UDP-GlcNAc. Abolition of N-glycosylation in cells and embryos can be achieved by treatment with tunicamycin, an analog of UDP-GlcNAc that inhibits this enzyme. A second *N*-acetylglucosamine and five mannose residues are subsequently transferred in a stepwise manner from UDP-GlcNAc and GDP-Man, respectively, to generate $Man_5GlcNAc_2$-P-P-Dol on the cytoplasmic side of the ER. Each of the sugar additions is catalyzed by a specific glycosyltransferase. Only GlcNAc-1-phosphotransferase transfers a sugar linked to phosphate (*N*-acetylglucosamine-1-P). All of the other enzymes transfer only the sugar portion of the nucleotide sugar. By a mechanism that is not fully understood, the $Man_5GlcNAc_2$-P-P-Dol precursor translocates across the ER membrane bilayer so that the glycan becomes exposed to the lumen of the ER. This translocation is mediated by a "flippase," which has been genetically linked to the Rtf1 locus in yeast. $Man_5GlcNAc_2$-P-P-Dol is extended by the addition of four mannose residues transferred from Dol-P-Man. Assembly of the Dol-P-P-glycan precursor is completed with the addition of three glucose residues donated by Dol-P-Glc. Dol-P-Man and Dol-P-Glc donors are formed on the cyto-

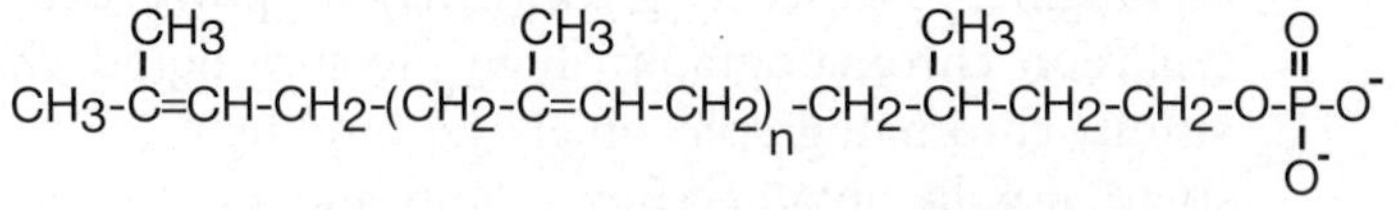

FIGURE 8.2. Dolichol phosphate (Dol-P). N-Glycan synthesis begins by the transfer of GlcNAc-1-P from UDP-GlcNAc to Dol-P to generate dolichol pyrophosphate *N*-acetylglucosamine (Dol-P-P-GlcNAc). This reaction is inhibited by tunicamycin.

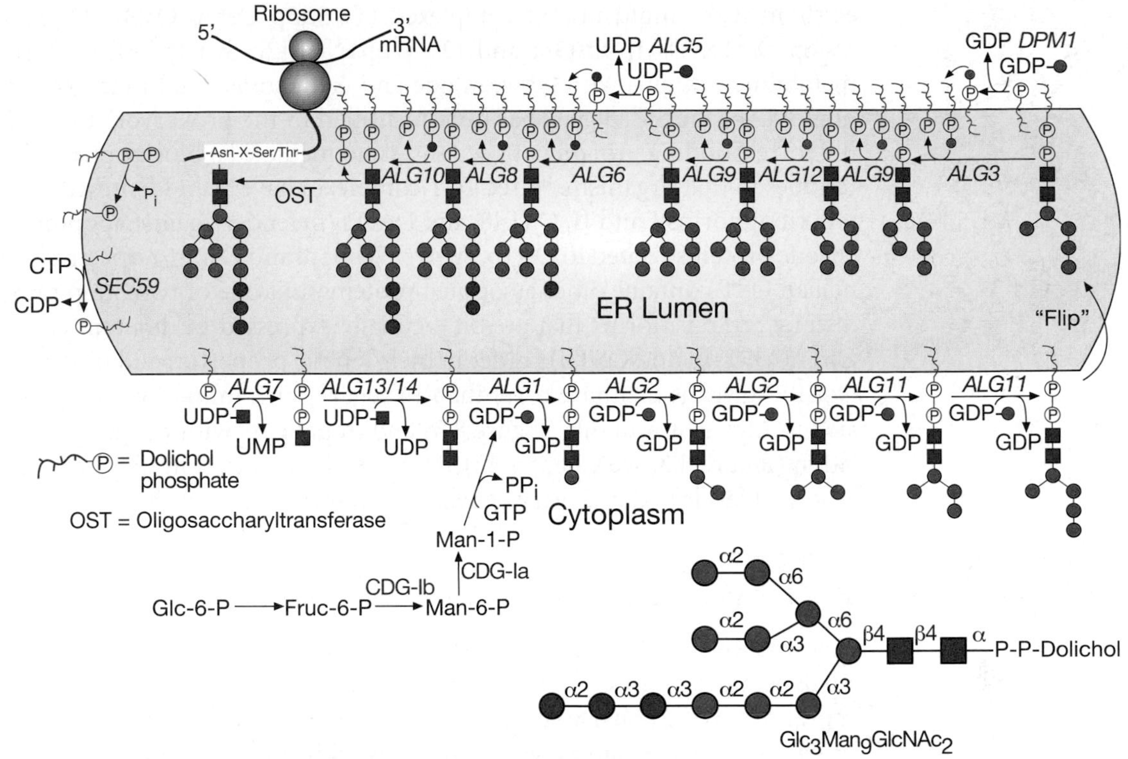

FIGURE 8.3. Synthesis of dolichol-P-P-GlcNAc$_2$Man$_9$Glc$_3$. Dolichol (*red squiggle*) phosphate (Dol-P) located on the cytoplasmic face of the ER membrane receives GlcNAc-1-P from UDP-GlcNAc in the cytoplasm to generate Dol-P-P-GlcNAc. Dol-P-P-GlcNAc is extended to Dol-P-P-GlcNAc$_2$Man$_5$ before being "flipped" across the ER membrane to the lumenal side. On the lumenal face of the ER membrane, four mannose residues are added from Dol-P-Man and three glucose residues from Dol-P-Glc. Dol-P-Man and Dol-P-Glc are also made on the cytoplasmic face of the ER and "flipped" onto the lumenal face. Yeast mutants defective in an *ALG* gene have been used to identify the gene that encodes the enzyme responsible for each transfer (see Chapter 42). Some reactions affected in congenital disorders of glycosylation (CDG) are noted.

plasmic side of the ER from GDP-Man and UDP-Glc by transfer of the respective sugar to Dol-P. Dol-P-Man and Dol-P-Glc must also be flipped across the ER bilayer. Mpdu1 protein is necessary for the bioavailability of Dol-P-Man and Dol-P-Glc in the ER for use by the mannosyltransferases and glucosyltransferases that catalyze the synthesis of the mature N-glycan precursor Glc$_3$Man$_9$GlcNAc$_2$-P-P-Dol (Figure 8.3). This 14-sugar glycan is now ready for transfer to asparagine in receptive Asn-X-Ser/Thr sequons of protein regions that have translocated across the ER membrane.

Transfer of the Dolichol-linked Precursor to Nascent Proteins

A multisubunit protein complex in the ER membrane is responsible for catalyzing the transfer of Glc$_3$Man$_9$GlcNAc$_2$ from Dol-P-P to Asn-X-Ser/Thr in newly synthesized regions of proteins as they emerge from the translocon in the ER membrane. The complex that transfers the 14-sugar glycan is termed oligosaccharyltransferase (OST). The OST complex binds to the membrane-anchored Dol-P-P-oligosaccharide and transfers the glycan to nascent protein by cleavage of the high-energy GlcNAc-P bond, releasing Dol-P-P in the process (Figure 8.3). Yeast OST complexes have been purified and are comprised of nine different membrane-bound subunits denoted by their gene names Ost1p, Wbp1p, Swp1p, Ost2p, Ost3p, Ost6p, Ost4p, Ost5p, and Stt3p. Five subunits (Ost1p, Wbp1p, Swp1p, Ost2p, and Stt3p) are essential for yeast viability. Stt3p appears to be the catalytic subunit. Yeast OST exists in three multisubunit complexes: (1) Stt3p-Ost3p-Ost4p, Ost1p-Ost5p; (2) Stt3p-Ost6p-Ost4p, Ost1p-Ost5p; and (3) Wbp1p-Ost2p-Swp1p. All OST subunits are transmembrane proteins with between one and eight transmembrane domains. Interestingly, a homolog of the STT3 gene has been identified in the prokaryote *Campylobacter jejuni* and the corresponding protein has been shown to mediate en bloc N-glycosylation of asparagine residues in that organism. Three OST complexes have been identified in mammals. All contain ribophorins I and II, OST48, and DAD1 (defender against apoptotic cell death), which encode proteins related to Ost1p, Swp1p, Wbp1p, and Ost2p, respectively. In addition, mammalian OST contains other associated proteins and one of two Stt3p proteins (A or B), two distinct Stt3p isoforms that are differentially expressed in different cell types. Mammalian OST-I, OST-II, and OST-III differ in their kinetic properties and in their abilities to transfer Dol-P-P-glycans that have fewer than 14 sugars. Such immature N-glycan species are generated in Alg yeast mutants (Figure 8.3) and in patients with congenital disorders of glycosylation (Figure 8.3; see Chapter 42). Thus, N-glycan addition may be differentially regulated at several levels and may vary depending on metabolic conditions.

Early Processing Steps: $Glc_3Man_9GlcNAc_2Asn$ to $Man_5GlcNAc_2Asn$

Following the covalent attachment of the 14-sugar oligomannose glycan to Asn-X-Ser/Thr in a polypeptide, a series of processing reactions trims the N-glycan in the ER. The initial steps appear to be conserved among all eukaryotes and are now known to have key roles in regulating glycoprotein folding via interactions with ER chaperones that recognize specific features of the trimmed glycan. Details of the interactions between N-glycans and the chaperones calnexin and calreticulin in glycoprotein folding are presented in Chapter 36. Processing or trimming of $Glc_3Man_9GlcNAc_2Asn$ begins with the sequential removal of glucose residues by α-glucosidases I and II (Figure 8.4). Both glucosidases function in the lumen of the ER, with α-glucosidase I acting specifically on the terminal α1-2Glc and α-glucosidase II sequentially removing the two inner α1-3Glc residues. Removal of glucose residues and the transient re-addition of the innermost glucose during protein folding contribute to the ER retention time of a given glycoprotein. The removal of glucose may be prevented by glucosidase I inhibitors such as castanospermine and deoxynojirimycin (see Chapter 50). In the presence of either of these inhibitors, N-glycans retain the three glucose residues and usually lose one

FIGURE 8.4. *(Facing page)* Processing and maturation of an N-glycan. The mature Dol-P-P-glycan, synthesized as described in Figure 8.3, is transferred to Asn-X-Ser/Thr sequons during protein synthesis as proteins are being translocated into the ER. Following transfer of the 14-sugar $Glc_3Man_9GlcNAc_2$ glycan to protein, glucosidases in the ER remove the three glucose residues, and ER mannosidase removes a mannose residue. These reactions are intimately associated with the folding of the glycoprotein assisted by the lectins calnexin and calreticulin, and they determine whether the glycoprotein continues to the Golgi or is degraded. Another lectin, termed EDEM (ER degradation-enhancing α-mannosidase I–like protein), binds to mannose residues on misfolded glycoproteins and escorts them via retrotranslocation into the cytoplasm for degradation. The removal of the first glucose (and therefore all glucose) can be blocked by castanospermine, leaving $Glc_3Man_9GlcNAc_2Asn$, which may subsequently have terminal mannose residues removed during passage through the Golgi. For most glycoproteins, additional mannose residues are removed in the *cis* compartment of the Golgi until $Man_5GlcNAc_2Asn$ is generated. The mannosidase inhibitor deoxymannojirimycin blocks the removal of these mannose residues, leaving $Man_8GlcNAc_2Asn$, which is not further processed. The action of GlcNAcT-1 on $Man_5GlcNAc_2Asn$ in the *medial*-Golgi initiates the first branch of an N-glycan. This reaction is blocked in the Lec1 CHO mutant in which GlcNAcT-I is inactive, leaving $Man_5GlcNAc_2Asn$, which is not further processed. α-Mannosidase II removes two outer mannose residues in a reaction that is blocked by the inhibitor swainsonine. The action of α-mannosidase II generates the substrate for GlcNAcT-II.

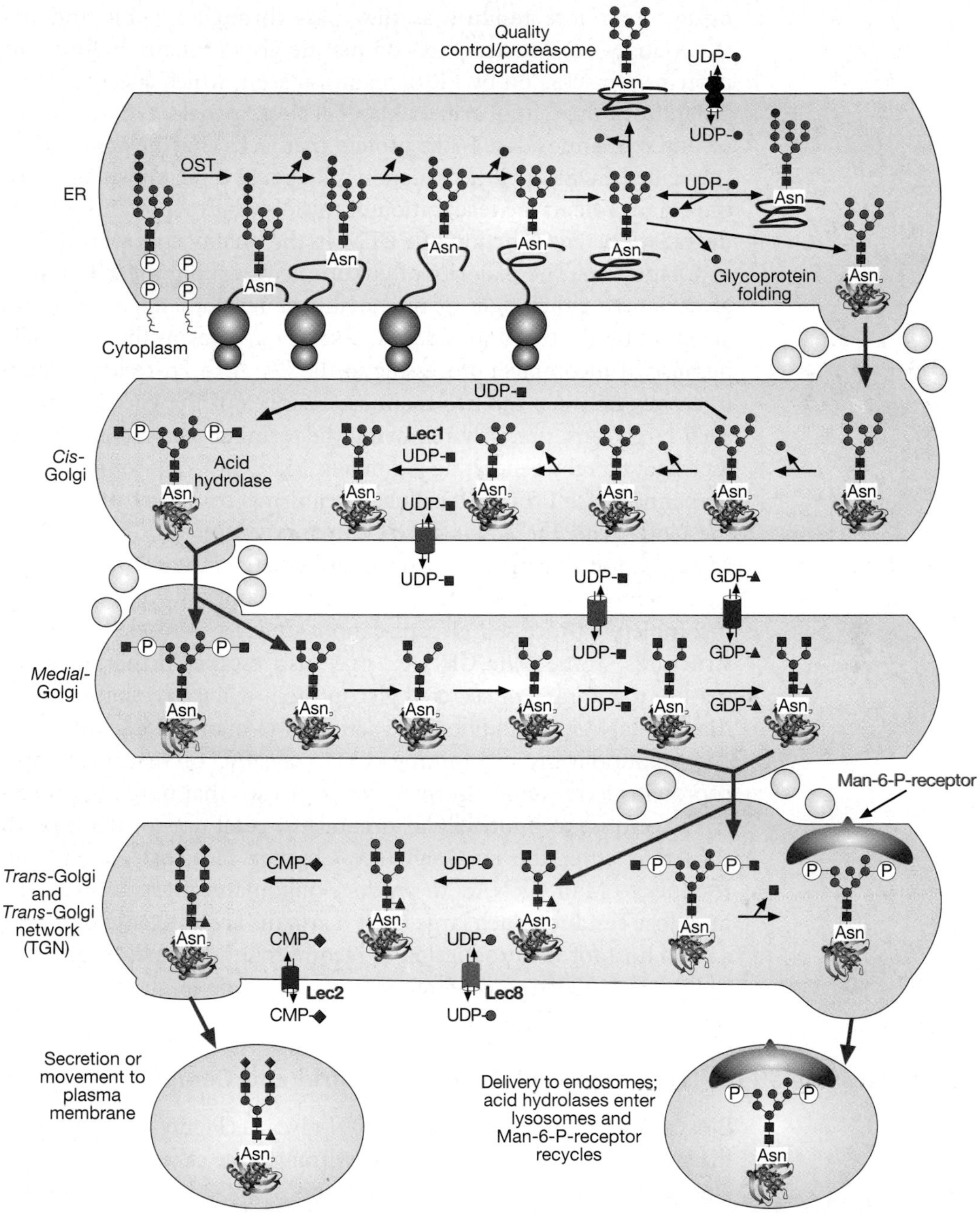

FIGURE 8.4. (*continued*) The resulting biantennary N-glycan is extended by the addition of fucose, galactose, and sialic acid to generate a complex N-glycan with two branches. The addition of galactose does not occur in the Lec8 CHO mutant, which has an inactive UDP-Gal transporter. In Lec8 mutants, complex N-glycans terminate in *N*-acetylglucosamine. The addition of sialic acid does not occur in the Lec2 CHO mutant, which has an inactive CMP-sialic acid transporter. In Lec2 mutants, complex N-glycans terminate with galactose. Complex N-glycans can have many more sugars than shown in this figure, including additional residues attached to the core, additional branches, branches extended with poly-*N*-acetyllactosamine units, and different "capping" structures (see Chapter 13). Also shown is the special case of lysosomal hydrolases that acquire a GlcNAc-1-P at C-6 of mannose residues on oligomannose N-glycans in the *cis*-Golgi. The *N*-acetylglucosamine is removed in the *trans*-Golgi by a glycosidase, thereby exposing Man-6-P residues that are recognized by a Man-6-P receptor and routed to an acidified, prelysosomal compartment, as described in Chapter 30. The inhibitors of N-glycan processing are described in Chapter 50, and the CHO mutants blocked in N-glycan synthesis are described in Chapter 46. (Adapted, with permission of the Annual Review of Biochemistry, from R. Kornfeld and S. Kornfeld. 1985. *Annu. Rev. Biochem.* **54:** 631–634.)

or two mannose residues as they pass through the ER and *medial*-Golgi, resulting in $Glc_3Man_{7-9}GlcNAc_2$ structures on mature glycoproteins. Before exiting the ER, many glycoproteins are acted on by ER α-mannosidase I, which specifically removes the terminal α1-2Man from the central arm of $Man_9GlcNAc_2$ to yield a $Man_8GlcNAc_2$ isomer (Figure 8.4). A second α-mannosidase I–like protein that lacks enzyme activity is also a resident of the ER. It is called EDEM (*ER d*egradation-*e*nhancing α-*m*annosidase I–like protein) and it has an important role in the recognition of misfolded glycoproteins, thereby targeting them for ER degradation. The function of EDEM in the quality control of ER glycoproteins is described in Chapter 36. The majority of glycoproteins exiting the ER en route to the Golgi carry N-glycans with either eight or nine mannose residues, depending on whether they have been acted on by ER α-mannosidase I. Some N-glycans in the *cis*-Golgi retain a glucose residue because of incomplete processing in the ER. A Golgi-resident endo-α-mannosidase cleaves internally between the two mannose residues of the Glcα1-3Manα1-2Manα1-2 moiety of such N-glycans, precisely removing the terminal glucose and the mannnose to which it is attached, thereby generating a different $Man_8GlcNAc_2$ isomer to that produced in the ER by α-mannosidase I. In multicellular organisms, trimming of α1-2Man residues continues in the Golgi with the action of α1-2 mannosidases IA, IB, and 1C in the *cis*-Golgi to give $Man_5GlcNAc_2$ (Figure 8.4), a key intermediate in the pathway to hybrid and complex N-glycans (Figure 8.1). However, all N-glycans are not fully processed to $Man_5GlcNAc_2$, and those incompletely processed glycans cannot undergo remodeling to form hybrid and complex structures. Some $Man_5GlcNAc_2$ may also escape further modification. In these cases, a mature membrane or secreted glycoprotein will carry oligomannose N-glycans of the type $Man_{5-9}GlcNAc_2$. In addition, the action of α-mannosidase I can be blocked by the inhibitor deoxymannojirimycin, resulting in $Man_8GlcNAc_2$ on mature glycoproteins. Most mature glycoproteins have some oligomannose N-glycans that were not processed in the *cis*-Golgi.

In contrast to multicellular organisms, yeast do not truncate the $Man_8GlcNAc_2$ N-glycans that enter the *cis*-Golgi (see Chapter 21). Instead, they add additional mannose residues to $Man_8GlcNAc_2$ to produce oligomannose structures containing many branched mannose residues. Such large yeast mannans are antigenic in humans, and thus yeast is not a good host for the production of recombinant therapeutic glycoproteins unless it is genetically manipulated to generate mammalian-type N-glycans.

Late Processing Steps: From $Man_5GlcNAc_2Asn$ to Hybrid and Complex N-Glycans

Biosynthesis of hybrid and complex N-glycans (Figure 8.1) is initiated in the *medial*-Golgi by the action of an N-acetylglucosaminyltransferase called GlcNAcT-I, which adds an *N*-acetylglucosamine residue to C-2 of the mannose α1-3 in the core of $Man_5GlcNAc_2$ (Figure 8.4). Once this step has occurred, the majority of N-glycans are trimmed by α-mannosidase II, another resident of the *medial*-Golgi, which removes the terminal α1-3Man and α1-6Man residues from $GlcNAcMan_5GlcNAc_2$ to form $GlcNAcMan_3GlcNAc_2$. It is important to note that α-mannosidase II cannot trim the $Man_5GlcNAc_2$ intermediate unless it is first acted on by GlcNAcT-I. Once the two mannose residues are removed, a second *N*-acetylglucosamine is added to C-2 of the mannose α1-6 in the core by the action of GlcNAcT-II to yield the precursor for all biantennary, complex N-glycans. Hybrid N-glycans are formed if the $GlcNAcMan_5GlcNAc_2$ glycan is not acted on by α-mannosidase II, leaving the peripheral α1-3Man and α1-6Man residues intact and unmodified in the mature glycoprotein. Incomplete action of α-mannosidase II can result in $GlcNAcMan_4GlcNAc_2$ hybrids. Another Golgi mannosidase, discovered in mutant mice lacking functional α-mannosidase II, is termed α-mannosidase IIX and also acts on the $GlcNAcMan_5GlcNAc_2$ generated by GlcNAcT-1. Inactivation of both α-mannosidase II and α-mannosidase IIX in the mouse leads to embryos lacking all complex N-glycans. Small oligomannose N-glycans have been found in relatively large amounts in invertebrates and plants. These $Man_{3-4}GlcNAc_2$ N-glycans (called paucimannose N-glycans) are formed from $GlcNAcMan_{3-4}GlcNAc_2$, following the removal of the peripheral *N*-acetylglucosamine residue by a Golgi hexosaminidase that acts after α-mannosidase II.

The complex N-glycan shown in the *medial*-Golgi of Figure 8.4 has two antennae or branches initiated by the addition of two terminal *N*-acetylglucosamine residues. Additional branches can be initiated at C-4 of the core mannose α1-3 (by GlcNAcT-IV) and C-6 of the core mannose α1-6 (by GlcNAcT-V) to yield tri- and tetra-antennary N-glycans (Figure 8.5). Another enzyme, termed GlcNAcT-IX or GlcNAcT-Vb, catalyzes the same reaction as GlcNAcT-V on the C-6 core mannose α1-6, but in contrast to GlcNAcT-V, GlcNAcT-IX/Vb can also transfer *N*-acetylglucosamine to C-6 of the core mannose α1-3. Another branch can be initiated at C-4 of the core mannose α1-3 by GlcNAcT-VI (see Figure 8.5). Highly branched hepta-antennary structures have so far been found in birds and fish, but not yet in mammals, although genes related to GlcNAcT-VI exist in mammalian genomes.

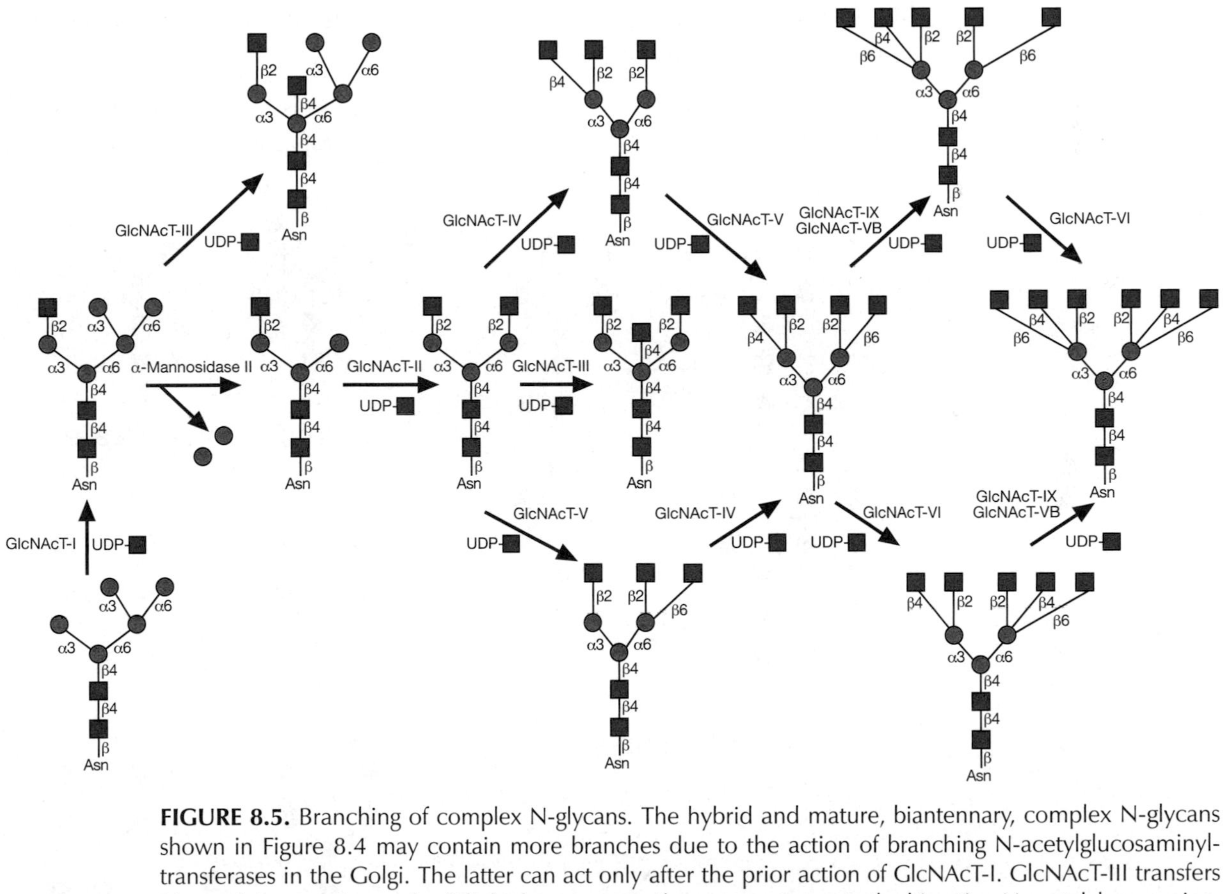

FIGURE 8.5. Branching of complex N-glycans. The hybrid and mature, biantennary, complex N-glycans shown in Figure 8.4 may contain more branches due to the action of branching N-acetylglucosaminyltransferases in the Golgi. The latter can act only after the prior action of GlcNAcT-I. GlcNAcT-III transfers *N*-acetylglucosamine to the β-linked mannose in the core to generate the bisecting *N*-acetylglucosamine. The presence of this residue inhibits the action of α-mannosidase II, thereby generating hybrid structures. A biantennary N-glycan may also accept the bisecting *N*-acetylglucosamine. More highly branched N-glycans are generated by the action of GlcNAcT-IV, GlcNAcT-V, and GlcNAcT-VI and may also carry the bisecting *N*-acetylglucosamine. The most highly branched structures with seven *N*-acetylglucosamine residues (including the bisecting *N*-acetylglucosamine) on the core of N-glycans have been found in a bird glycoprotein. Mammals have the potential for generating similarly complex structures. Each *N*-acetylglucosamine branch may be elongated with galactose, poly-*N*-acetyllactosamine, sialic acid, and fucose, as described in Chapter 13. The bisecting *N*-acetylglucosamine is not further elongated unless the branch initiated by GlcNAcT-II is missing.

Complex and hybrid N-glycans may carry a "bisecting" *N*-acetylglucosamine residue that is attached to the β-mannose of the core by GlcNAcT-III (Figure 8.5). The presence of a bisecting *N*-acetylglucosamine inhibits trimming by α-mannosidase II and also prevents the actions of GlcNAcT-II, GlcNAcT-IV, and GlcNAcT-V in in vitro assays. However, bi-, tri- and tetra-antennary complex N-glycans with a bisecting *N*-acetylglucosamine are synthesized when GlcNAcT-III acts after α-mannosidase II and the initiation of branches by GlcNAcT-II, GlcNAcT-IV, and/or GlcNAcT-V. A bisecting *N*-acetylglucosamine on a biantennary N-glycan is shown in Figure 8.5, and it may be present in all of the more highly branched structures.

Maturation of N-Glycans

Further sugar additions, mostly occuring in the *trans*-Golgi, convert the limited repertoire of hybrid and branched N-glycans into an extensive array of mature, complex N-glycans. For convenience, this part of the biosynthetic process can be divided into three components: (1) sugar additions to the core, (2) elongation of branching *N*-acetylglucosamine residues by sugar additions, and (3) "capping" or "decoration" of elongated branches. These processes are summarized below.

1. In vertebrate N-glycans, the main core modification is the addition of fucose in an α1-6 linkage to the *N*-acetylglucosamine adjacent to asparagine in the core (Figure 8.6A).

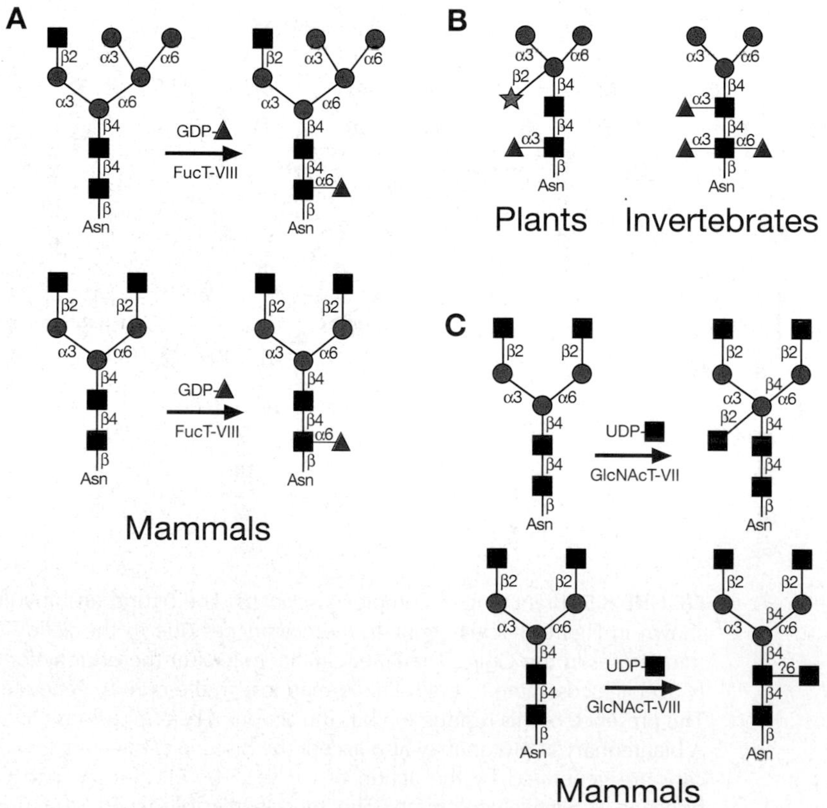

FIGURE 8.6. Modifications of the core of N-glycans. (*A*) A fucose residue may be transferred to the core of N-glycans after GlcNAcT-I has acted. Thus, hybrid and biantennary, complex N-glycans may have a core fucose. (*B*) Plants and invertebrates may have additional modifications to the core, with fucose on either *N*-acetylglucosamine residue (or both for invertebrates) and a xylose attached to the core β-linked mannose residue of plant N-glycans. (*C*) Other additions to the core have been detected in mammalian cells.

Fucosylation in invertebrate glycoproteins also occurs on this *N*-acetylglucosamine, but the fucose can be added in α1-3 and/or α1-6 linkages (Figure 8.6B). Invertebrate glycoproteins may have up to four fucose residues on the two *N*-acetylglucosamines of the N-glycan core (Figure 8.6B). In plants, the fucose is transferred to the asparagine *N*-acetylglucosamine only in α1-3 linkage (Figure 8.6B). The fucosyltransferases involved in transferring fucose to this core *N*-acetylglucosamine require the prior action of GlcNAcT-I (Figure 8.4). Another common modification of the core, notably in plant and helminth glycoproteins, is the addition of xylose in β1-2 linkage to the β-mannose of the core. This xylosyltransferase also requires the prior action of GlcNAcT-I. Xylose has not been detected in vertebrate N-glycans. A few additional core structures have been identified in mammals, but they appear to be very rare. Two examples are given in Figure 8.6C.

2. The majority of complex and hybrid N-glycans (denoted by "R" in Figure 8.7) have elongated branches that are made by the addition of a β-linked galactose residue to the initiating *N*-acetylglucosamine to produce the ubiquitous building block Galβ1-4GlcNAc, referred to as a type-2 *N*-acetyllactosamine or "LacNAc" sequence (Figure 8.7A). Antennae can be further lengthened by the sequential addition of *N*-acetylglucosamine and galactose residues, resulting in tandem repeats of LacNAc $(\text{-3Gal}\beta\text{1-4GlcNAc}\beta\text{1-})_n$, termed poly-*N*-acetyllactosamine or polyLacNAc (Figure 8.7A). In a variation, β-linked galactose is added to the C-3 of the *N*-acetylglucosamine to yield Galβ1-3GlcNAc (Figure 8.7B), referred to as a type-1 *N*-acetyllactosamine (LacNAc) sequence. In some glycoproteins, β-linked *N*-acetylgalactosamine is added to *N*-acetylglucosamine instead of β-linked galactose, yielding antennae with a GalNAcβ1-4GlcNAc ("LacdiNAc") extension (Figure 8.7C). In contrast to poly-*N*-acetyllactosamine, which is a relatively common structure, tandem repeats of LacdiNAc and type-1 sequences are uncommon, although poly-*N*-acetyllactosamine sequences are sometimes terminated with a type-1 unit. The structures and biosynthesis of polylactosamines are discussed further in Chapter 13.

3. The most important "capping" or "decorating" reactions involve the addition of sialic acid, fucose, galactose, *N*-acetylgalactosamine, and sulfate to the branches described in the preceding paragraph. Capping sugars are most commonly α-linked and therefore protrude away from the β-linked ribbon-like poly-*N*-acetyllactosamine branches, thus facilitating the presentation of terminal sugars to lectins and antibodies. Many of these

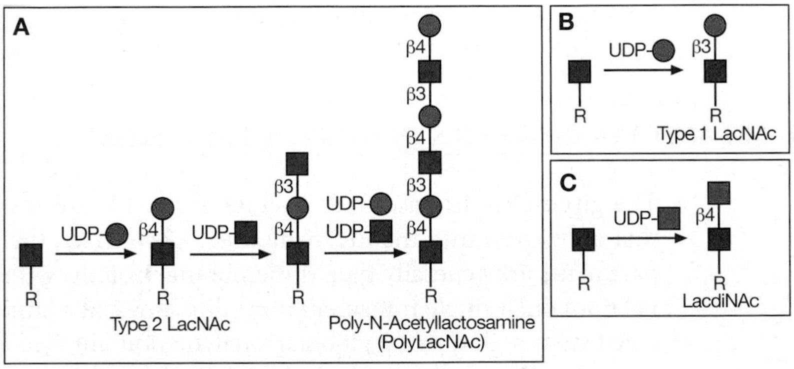

FIGURE 8.7. Elongation of branch *N*-acetylglucosamine residues of N-glycans. (*A*) A single *N*-acetyllactosamine unit is generated when galactose is transferred to a branch *N*-acetylglucosamine on an N-glycan (R). Further elongation to form poly-*N*-acetyllactosamine occurs by sequential addition of galactose and *N*-acetylglucosamine, as shown. This structure is composed of type-2 poly-*N*-acetyllactosamine units. (*B*) Type-1 *N*-acetyllactosamine units can also be present in poly-*N*-acetyllactosamine. (*C*) Transfer of *N*-acetylgalactosamine to *N*-acetylglucosamine generates LacdiNAc.

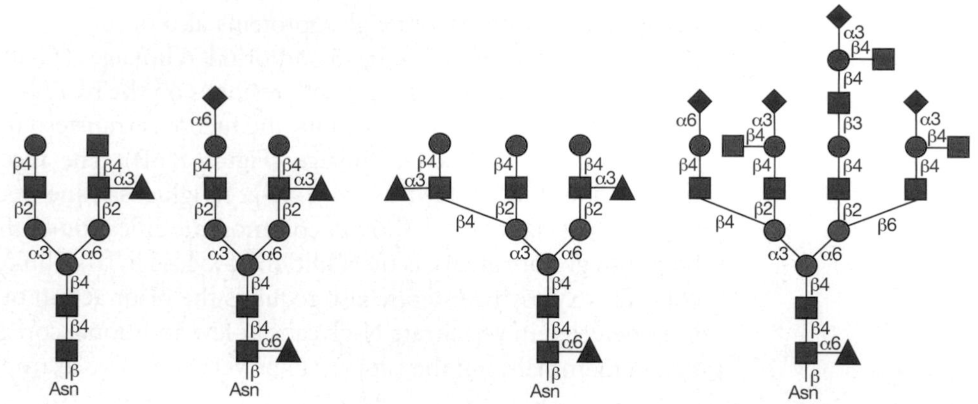

FIGURE 8.8. Typical complex N-glycan structures found on mature glycoproteins.

structures are shared by N- and O-glycans and by glycolipids, and for this reason their detailed description is presented in Chapter 13.

The various processes described above potentially yield a myriad of complex N-glycans that differ in branch number, composition, length, capping arrangements, and core modifications. Some examples to illustrate this diversity are shown in Figure 8.8. Many more examples may be found throughout this book.

THE SPECIAL CASE OF THE PHOSPHORYLATED N-GLYCANS ON LYSOSOMAL HYDROLASES

Hydrolase enzymes reside in lysosomes where they degrade proteins, lipids, and glycans. Many of these enzymes are targeted to lysosomes by a specialized trafficking pathway that requires the generation of phosphorylated N-glycans. The phosphorylation step occurs in the *cis*-Golgi and involves the transfer of GlcNAc-1-P to C-6 of mannose residues of oligomannose N-glycans on lysosomal hydrolases (Figure 8.4). A glycosidase in the *trans*-Golgi removes the *N*-acetylglucosamine to generate Man-6-P residues. Such residues are recognized by lectin receptors (termed Man-6-P receptors) that transport the lysosomal hydrolase to an acidified compartment where it is released from the receptor and ultimately ends up in a lysosome. The details of the synthesis of the Man-6-P recognition marker on lysosomal hydrolases and the trafficking pathway are presented in Chapter 30.

TRANSFERASES AND TRANSPORTERS IN N-GLYCAN SYNTHESIS

The glycosyltransferases that operate in the ER are mainly multitransmembrane proteins that are woven into the ER membrane. In contrast, the glycosyltransferases in Golgi compartments are generally type II membrane proteins with a small cytoplasmic amino-terminal domain, a single transmembrane domain, and a large lumenal domain that has an elongated stem region and a globular catalytic domain (see Chapter 5). The stem region can be cleaved, releasing the catalytic domain into the lumen of the Golgi and allowing its secretion. Thus, extracellular soluble forms of many glycosyltransferases exist in tissues and sera. At least one soluble form has a unique function. Soluble GlcNAcT-V is an angiogenic factor in tumors. However, extracellular soluble glycosyltransferases are not expected to function as transferases because nucleotide sugars are not known to be present extracellularly. Nucleotide sugars are synthesized in the cytoplasm, except for CMP-sialic acid, which is synthesized in the nucleus (see Chapter 4). They are subsequently concentrated in the appropriate compartment following transport across the membrane by specialized nucleotide sugar transporters. Specific genes encode the nucleotide sugar transporters, which translocate CMP-sialic acid, UDP-Gal, UDP-GlcNAc, GDP-Fuc, and other nucleotide sugars. A few of these transporters can transport more than one nucleotide sugar. Each transporter is a multitransmembrane protein and usually contains ten membrane-spanning domains.

N-LINKED GLYCOPROTEINS COMPRISE MANY GLYCOFORMS

As mentioned above, glycoproteins are heterogeneous with respect to their content of N-glycans and often have a range of different N-glycans on a particular Asn-X-Ser/Thr N-glycosylation sequon. For example, 58 different complex N-glycan structures have been identified at one N-glycan site in mouse zona pellucida glycoprotein 3. Furthermore, if there is more than one Asn-X-Ser/Thr sequon per molecule, different molecules in a population may have different subsets of N-glycans on different sequons. This is referred to as site-specific glycan heterogeneity or microheterogeneity. A homogeneous glycoprotein component of a population is called a glycoform. It appears that the protein sequence or conformation can cause N-glycan diversity, presumably by affecting substrate availability for Golgi glycosidases or glycosyltransferases. Other factors affecting N-glycan heterogeneity include nucleotide sugar metabolism, transport rates of the glycoprotein through the lumen of the ER and Golgi, and the proximity of an N-glycan attachment sequon to a transmembrane domain. Also, localization of glycosyltransferases within subcompartments of the Golgi can determine which enzymes encounter N-glycan substrates first. With respect to the last point, it is important to note that processing enzymes often compete for the same substrate and that most glycosyltransferases and glycosidases require the prior actions of other glycosyltransferases and glycosidases before they can carry out their reactions.

FUNCTIONS OF N-GLYCANS

Determining the functions of N-glycans may be accomplished by the use of inhibitors of N-linked glycosylation such as tunicamycin; inhibitors of N-glycan processing such as castanospermine, deoxynojirimycin, and swainsonine; or the generation of mutants in a gene that codes for a glycosylation activity in model organisms such as yeast, cultured mammalian cells, *Drosophila*, *C. elegans*, zebrafish, and mouse. The various chemical inhibitors of N-glycan synthesis are discussed in Chapter 50. Many yeast mutants in the synthesis and initial processing of N-glycans are identified in Figure 8.3, and three of the many mutants of cultured cells with altered glycosylation are identified in Figure 8.4 and described in detail in Chapter 46. Mutant cells or organisms with an altered N-glycosylation ability shed enormous insights not only into the biological functions of N-glycans, but also into their contributions to the biochemical properties of a glycoprotein in terms of structure, susceptibility to proteases, and antigenicity. In addition, mutant cells and organisms allow the glycosylation pathways that operate in vivo to be defined. A cell or organism with a loss-of-function mutation often accumulates the biosynthetic intermediate that is the substrate of the activity lost by the mutant. Gain-of-function mutations reveal alternative pathways or glycosylation reactions that may occur. Thus, the study of cells and organisms mutated in a specific gene that affects N-glycosylation has been a major source of information regarding the functions of N-glycans. N-glycan functions have also been determined from the features of human diseases that arise from a defect in N-glycan synthesis. An important group of such diseases is called congenital disorders of glycosylation (CDG), and these diseases are described in Chapter 42.

Mouse mutants in particular have provided enormous insights into the functions of individual sugars present in N-glycans as well as the functions of whole classes of N-glycans. Thus, deletion of the *Mgat1* gene that encodes GlcNAcT-I prevents the synthesis of complex and hybrid N-glycans, and $Man_5GlcNAc_2$ is found at all complex and hybrid N-glycan sites. Whereas the absence of GlcNAcT-I does not affect the viability or growth of Lec1 cultured cells (Figure 8.4), elimination of GlcNAcT-I in the mouse results in death during embryonic development (see Chapter 25). The complex N-glycans that fail to be synthesized in mice lacking GlcNAcT-I, GlcNAcT-V, GlcNAcT-IVb, and FucT-VIII are important in retaining growth factor and cytokine receptors at the cell surface, probably through interactions with glycan-binding proteins such as galectins or cytokines such as transforming growth factor β. Cell-surface receptors and a glucose transporter lacking branches of a complex N-glycan have a shorter residence time on the cell surface and their signaling is attenuated. Deletion of genes encoding sialyltransferases, fucosyltransferases, or branching N-acetylglucosaminyltranferases other than GlcNAcT-I has generally produced viable mice with defects in immunity or neuronal cell migration, emphysema of the lung, or inflammation. N-glycans may carry the sugar determinants recognized by selectins that mediate cell–cell interactions important for leukocyte extravasation from the blood stream and regulate lymphocyte homing to lymph nodes (see Chapter 31). N-Glycans are known to become more branched when cells become cancerous, and this change facilitates cancer progression (see Chapter 44). Tumors formed in mice lacking GlcNAcT-V or GlcNAcT-III (Figure 8.5) may be retarded in their progression. Thus, certain glycosyltransferases may be appropriate targets for the design of cancer therapeutics.

FURTHER READING

Waechter C.J. and Lennarz W.J. 1976. The role of polyprenol-linked sugars in glycoprotein synthesis. *Annu. Rev. Biochem.* **45:** 95–112.

Snider M.D., Sultzman L.A., and Robbins P.W. 1980. Transmembrane location of oligosaccharide-lipid synthesis in microsomal vesicles. *Cell* **21:** 385–392.

Kornfeld R. and Kornfeld S. 1985. Assembly of asparagine-linked oligosaccharides. *Annu. Rev. Biochem.* **54:** 631–664.

Herscovics A. 1999a. Importance of glycosidases in mammalian glycoprotein biosynthesis. *Biochim. Biophys. Acta* **1473:** 96–107.

Herscovics A. 1999b. Processing glycosidases of *Saccharomyces cerevisiae. Biochim. Biophys. Acta* **1426:** 275–285.

Berninsone P.M. and Hirschberg C.B. 2000. Nucleotide sugar transporters of the Golgi apparatus. *Curr. Opin. Struct. Biol.* **10:** 542–547.

Schachter H. 2000. The joys of HexNAc. The synthesis and function of N- and O-glycan branches. *Glycoconj. J.* **17:** 465–483.

Schenk B., Fernandez F., and Waechter C.J. 2001. The ins(ide) and outs(ide) of dolichyl phosphate biosynthesis and recycling in the endoplasmic reticulum. *Glycobiology* **11:** 61R–70R.

Lowe J.B. and Marth J.D. 2003. A genetic approach to mammalian glycan function. *Annu. Rev. Biochem.* **72:** 643–691.

Freeze H.H. and Aebi M. 2005. Altered glycan structures: The molecular basis of congenital disorders of glycosylation. *Curr. Opin. Struct. Biol.* **15:** 490–498.

Kelleher D.J. and Gilmore R. 2006. An evolving view of the eukaryotic oligosaccharyltransferase. *Glycobiology* **16:** 47R–62R.

Lau K.S., Partridge E.A., Grigorian A., Silvescu C.I., Reinhold V.N., Demetriou M., and Dennis J.W. 2007. Complex N-glycan number and degree of branching cooperate to regulate cell proliferation and differentiation. *Cell* **129:** 123–134.

CHAPTER 9

O-GalNAc Glycans

Inka Brockhausen, Harry Schachter, and Pamela Stanley

O-GLYCOSYLATION IS A COMMON COVALENT MODIFICATION OF SERINE AND THREONINE residues of mammalian glycoproteins. This chapter describes the structures, biosynthesis, and functions of glycoproteins that are often termed mucins. In mucins, O-glycans are covalently α-linked via an *N*-acetylgalactosamine (GalNAc) moiety to the -OH of serine or threonine by an O-glycosidic bond, and the structures are named mucin O-glycans or O-GalNAc glycans. The focus of this chapter is on mucins and mucin-like glycoproteins that are heavily O-glycosylated, although glycoproteins that carry only one or a few O-GalNAc glycans are also briefly discussed. There are also several types of nonmucin O-glycans, including α-linked O-fucose, β-linked O-xylose, α-linked O-mannose, β-linked O-GlcNAc (*N*-acetylglucosamine), α- or β-linked O-galactose, and α- or β-linked O-glucose glycans (discussed in Chapters 12 and 16–18). In this chapter, however, the term O-glycan refers to mucin O-glycans, unless otherwise specified. Mucin glycoproteins are ubiquitous in mucous secretions on cell surfaces and in body fluids.

DISCOVERY AND BACKGROUND

In 1865, E. Eichwald, a Russian physician working in Germany, provided the first chemical evidence that mucins are proteins conjugated to carbohydrate. He also showed that mucins are widely distributed in the animal body. In 1877, F. Hoppe-Seyler discovered that mucins produced by the epithelial cells of mucous membranes and salivary glands contain an acidic carbohydrate, identified many years later to be sialic acid. Mucins are present at many epithe-

TABLE 9.1. *Structures of O-glycan cores and antigenic epitopes found in mucins*

O-Glycan	Structure
Core	
Tn antigen	GalNAcαSer/Thr
Sialyl-Tn antigen	Siaα2-6GalNAcαSer/Thr
Core 1 or T antigen	Galβ1-3GalNAcαSer/Thr
Core 2	GlcNAcβ1-6(Galβ1-3)GalNAcαSer/Thr
Core 3	GlcNAcβ1-3GalNAcαSer/Thr
Core 4	GlcNAcβ1-6(GlcNAcβ1-3)GalNAcαSer/Thr
Core 5	GalNAcα1-3GalNAcαSer/Thr
Core 6	GlcNAcβ1-6GalNAcαSer/Thr
Core 7	GalNAcα1-6GalNAcαSer/Thr
Core 8	Galα1-3GalNAcαSer/Thr
Epitope	
Blood groups O, H	Fucα1-2Gal-
Blood group A	GalNAcα1-3(Fucα1-2)Gal-
Blood group B	Galα1-3(Fucα1-2)Gal-
Linear B	Galα1-3Gal-
Blood group i	Galβ1-4GlcNAcβ1-3Gal-
Blood group I	Galβ1-4GlcNAcβ1-6(Galβ1-4GlcNAcβ1-3)Gal-
Blood group Sd(a), Cad	GalNAcβ1-4(Siaα2-3)Gal-
Blood group Lewis^a	Galβ1-3(Fucα1-4)GlcNAc-
Blood group Lewis^x	Galβ1-4(Fucα1-3)GlcNAc-
Blood group sialyl-Lewis^x	Siaα2-3Galβ1-4(Fucα1-3)GlcNAc-
Blood group Lewis^y	Fucα1-2Galβ1-4(Fucα1-3)GlcNAc-

lial surfaces of the body, including the gastrointestinal, genitourinary, and respiratory tracts, where they shield the epithelial surfaces against physical and chemical damage and protect against infection by pathogens. Mucin O-glycans begin with an α-linked *N*-acetylgalactosamine residue linked to serine or threonine. The *N*-acetylgalactosamine may be extended with sugars including galactose, *N*-acetylglucosamine, fucose, or sialic acid, but not mannose, glucose, or xylose residues. There are four common O-GalNAc glycan core structures, designated cores 1 through 4 and an additional four designated cores 5 though 8 (Table 9.1). Mucin O-glycans can be branched, and many sugars or groups of sugars on mucin O-glycans are antigenic (Table 9.1). Important modifications of mucin O-glycans include O-acetylation of sialic acid and O-sulfation of galactose and *N*-acetylglucosamine. Thus, mucin O-glycans are often very heterogeneous, with hundreds of different chains being present in some mucins. Numerous chemical, enzymatic, and spectroscopic methods are applied to analyze the sugar composition and the linkages among sugars of mucin O-glycans.

MUCIN GLYCOPROTEINS

Mucins are heavily O-glycosylated glycoproteins found in mucous secretions and as transmembrane glycoproteins of the cell surface with the glycan exposed to the external environment. The mucins in mucous secretions can be large and polymeric (gel-forming mucins) or smaller and monomeric (soluble mucins). Many epithelial cells produce mucin, but gel-forming mucins are produced primarily in the goblet or mucous cells of the tracheobronchial, gastrointestinal, and genitourinary tracts. In goblet cells, mucins are stored intracellularly in mucin granules from which they can be quickly secreted upon external stimuli.

The hallmark of mucins is the presence of repeated peptide stretches called "variable number of tandem repeat" (VNTR) regions that are rich in serine or threonine O-glycan acceptor sites and have an abundance of clustered mucin O-glycans that may comprise 80% of the molecule by weight. The tandem repeats are usually rich in proline residues that appear to facilitate O-GalNAc glycosylation. Mucins may have hundreds of O-GalNAc glycans attached to serine or threonine residues in the VNTR regions. The clustering of O-GalNAc glycans causes mucin glycoproteins to adopt an extended "bottle brush" conformation (Figure 9.1).

About 20 different mucin genes have been cloned, and they are expressed in a tissue-specific fashion. For example, different mucin genes are expressed in different regions of the gastrointestinal tract, suggesting that they serve specific functions. The first mucin gene to be cloned encoded a transmembrane mucin termed MUC1. Mucins that span the plasma membrane are known to be involved in signal transduction, to mediate cell–cell adhesion, or to have an antiadhesive function. Cell-surface mucins contain an extracellular domain with a central VNTR region that carries O-GalNAc glycan chains, a single transmembrane domain, and a small cytoplasmic tail at the carboxyl terminus.

The mucin genes, their transcripts, the resulting mucin proteins, and the attached O-GalNAc glycans all exhibit extreme variability. Different mucins vary in the number and composition of the peptide repeats in their VNTR regions. Within the same mucin, the repeats usually vary in their amino acid sequences. It has not yet been possible to determine exactly which of the many different O-GalNAc glycan structures isolated from a purified mucin preparation are attached to specific serine or threonine residues in the tandem repeats. However, in some less densely glycosylated glycoproteins, specific O-GalNAc glycosylation sites have been identified. The structures of O-glycans at these defined sites are usually heterogeneous and any mucin includes a range of glycoforms. One of the exceptions is the Antarctic fish antifreeze glycoprotein, which carries only Galβ1-3GalNAc- at threonine residues of alanine-alanine-threonine repeat sequences.

Some mucins contain a hydrophobic transmembrane domain that serves to insert the molecule into the cell membrane. Secreted mucins have cysteine-rich regions and cystine knots that are responsible for their polymerization and the formation of extremely large molecules of several million daltons. Mucins may also have protein domains similar to motifs present in other glycoproteins. For example, D domains found in mucins are also found in von Willebrand factor, a glycoprotein important in blood clotting. The D domain may regulate mucin polymerization (Figure 9.1).

Mucin-like glycoproteins such as GlyCAM-1, CD34, and PSGL-1 are less densely O-glycosylated membrane-associated glycoproteins that mediate cell–cell adhesion. Most glycoproteins with O-GalNAc glycans carry several O-glycans, although some, such as interleukin-2 and erythropoietin, carry a single O-GalNAc glycan. The expression of mucin genes is regulated by a large number of cytokines and growth factors, differentiation factors, and bacterial products.

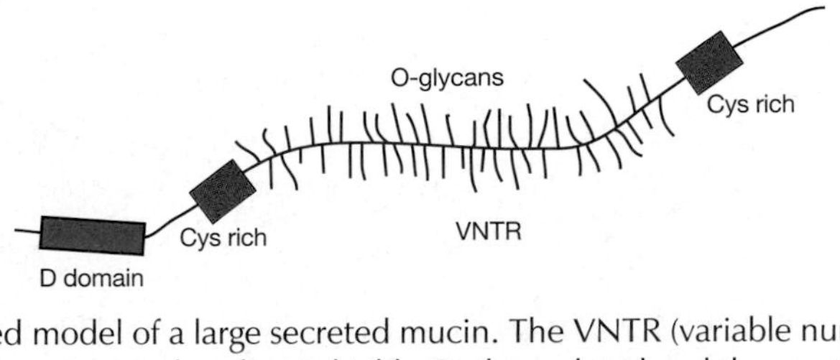

FIGURE 9.1. A simplified model of a large secreted mucin. The VNTR (variable number of tandem repeat) region rich in serine, threonine, and proline is highly O-glycosylated and the peptide assumes an extended "bottle brush" conformation. Hundreds of O-GalNAc glycans with many different structures may be attached to serine or threonine residues in the VNTR domains. The cysteine-rich regions at the ends of the molecules are involved in disulfide bond formation to form large polymers of several million daltons. D domains have similarity to von Willebrand factor and are also involved in polymerization.

The purification of gel-forming mucins is achieved by density-gradient centrifugation. Purified mucins have distinct viscoelastic properties that contribute to the high viscosity of mucous secretions. They are hydrophilic and contain charges that attract water and salts. Bacteria, viruses, and other microbes are trapped by mucins and sometimes adhere to specific O-GalNAc glycans that serve as receptors (see Tables 34.1 and 34.2). The ciliary action of the mucous membrane epithelium aids in the removal of microbes and small particles.

Mucins hydrate and protect the underlying epithelial cells, but they have also been shown to have roles in fertilization, blastocyst implantation, and the immune response. Cell-surface MUC1 and mucin-like glycoproteins have roles in cell adhesion (see Chapter 31). A number of diseases are associated with abnormal mucin gene expression and abnormal mucin carbohydrate structures and properties. These include cancer, inflammatory bowel disease, lung disease, and cystic fibrosis. The expression of underglycosylated MUC1 is often increased in individuals with cancer (see Chapters 43 and 44).

O-GalNAc GLYCAN STRUCTURES

The simplest mucin O-glycan is a single *N*-acetylgalactosamine residue linked to serine or threonine. Named the Tn antigen, this glycan is often antigenic. The most common O-GalNAc glycan is Galβ1-3GalNAc-, and it is found in many glycoproteins and mucins (Table 9.1, Figure 9.2). It is termed a core 1 O-GalNAc glycan because it forms the core of many longer, more complex structures. It is antigenic and is also named the T antigen. Both Tn and T antigens may be modified by sialic acid to form sialylated-Tn or -T antigens, respectively.

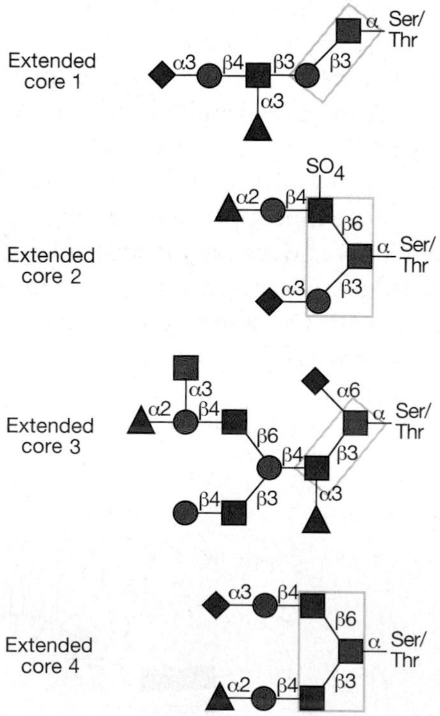

FIGURE 9.2. Complex O-GalNAc glycans with different core structures. Representative examples of complex O-GalNAc glycans with extended core 1, 2, or 4 structures from human respiratory mucins and an O-GalNAc glycan with an extended core 3 structure from human colonic mucins. All four core structures (in *boxes*) can be extended, branched, and terminated by fucose in various linkages, sialic acid in α2-3 linkage, or blood group antigenic determinants (Table 9.1). Core structures 1 and 3 may also carry sialic acid α2-6-linked to the core *N*-acetylgalactosamine.

Another common core structure contains a branching *N*-acetylglucosamine attached to core 1 and is termed core 2 (Table 9.1, Figure 9.2). Core 2 O-GalNAc glycans are found in both glycoproteins and mucins from a variety of cells and tissues. Linear core 3 and branched core 4 O-GalNAc glycans (Table 9.1, Figure 9.2) have been found only in secreted mucins of certain mucin-secreting tissues, such as bronchi, colon, and salivary glands. Core structures 5–8 have an extremely restricted occurrence. Mucins with core 5 have been reported in human meconium and intestinal adenocarcinoma tissue, whereas core 6 structures are found in human intestinal mucin and ovarian cyst mucin. Core 8 has been reported in human respiratory mucin. Bovine submaxillary mucin was shown to contain O-GalNAc glycans with relatively short chains, including those containing a core 7 structure (Table 9.1).

All of the core structures can be sialylated. However, only cores 1–4 and core 6 have been shown to occur as extended, complex O-glycans that carry antigens such as the ABO and Lewis blood group determinants, the linear i antigen (Galβ1-4GlcNAcβ1-3Gal-), and the GlcNAcβ1-6 branched I antigens (Table 9.1). Both type 1 (based on the Galβ1-3GlcNAc sequence) and type 2 (based on the Galβ1-4GlcNAc sequence) extensions can be repeated several times in poly-*N*-acetyllactosamine units, which provide a scaffold for the attachment of additional sugars or functional groups. The terminal structures of O-GalNAc glycans may contain fucose, galactose, *N*-acetylglucosamine, and sialic acid in α-linkages, *N*-acetylgalactosamine in both α- and β-linkages, and sulfate. Many of these terminal sugar structures are antigenic or represent recognition sites for lectins. In particular, the sialylated and sulfated Lewis antigens are ligands for selectins (see Chapter 31). Poly-*N*-acetyllactosamine units and terminal structures may also be found on N-glycans and glycolipids (see Chapter 13).

ISOLATION, PURIFICATION, AND ANALYSIS OF MUCIN O-GLYCANS

All O-GalNAc glycans may be released from the peptide by an alkali treatment called β-elimination. The *N*-acetylgalactosamine residue attached to serine or threonine is reduced to *N*-acetylgalactosaminitol by borohydride as it is released. Conversion to *N*-acetylgalactosaminitol prevents the rapid alkali-catalyzed degradation of the released O-GalNAc glycan by a "peeling" reaction. Glycoproteins that contain both N- and O-glycans will retain their N-glycans and specifically lose O-glycans when subjected to β-elimination. However, harsh alkali treatment will also disrupt peptide bonds and degrade the mucin. Unsubstituted *N*-acetylgalactosamine residues may also be released from serine or threonine by a specific N-acetylgalactosaminidase, and unsubstituted Galβ1-3GalNAc (core 1) can be released with an enzyme termed O-glycanase. For core 1 O-GalNAc glycans substituted with sialic acids, the sialic acid must first be removed by treatment with sialidase or mild acid before O-glycanase treatment will be successful. No enzymes are known to release larger O-glycans from the peptide, in contrast to the availability of several enzymes that release N-glycans from asparagine (see Chapter 8).

Released O-GalNAc glycans may be separated into different species by chromatographic methods such as gel filtration, anion-exchange chromatography, and high-pressure liquid chromatography. Chemical derivatization of purified reduced O-glycans helps in the analysis of sugar composition and the determination of linkages between sugars by gas chromatography and mass spectrometry (see Chapter 47). With improvements in the sensitivity of mass spectrometry, it is now possible to analyze small amounts and identify structures on the basis of their mass in complex mixtures. Structural analysis by nuclear magnetic resonance (NMR) spectroscopy is more precise, but requires more material and highly purified compounds. The NMR spectra of hundreds of O-glycans have been published and are used as references to identify known and new O-glycan structures. Other tools to analyze the structures of O-GalNAc glycans include specific exoglycosidases that remove terminal sugars, antibodies to *N*-acetylgalactosamine (anti-Tn) and core 1 (anti-T), and lectins such as peanut agglutinin (which binds to core 1) and *Helix pomatia* agglutinin (which binds to terminal *N*-acetylgalactosamine) (see Chapter 45). O-GalNAc glycans in a specific cell line or tissue may be predicted based on the presence of active glycosyltransferases required for their synthesis.

BIOSYNTHESIS OF O-GALNAC GLYCANS

Polypeptide-N-Acetylgalactosaminyltransferases

The first step of mucin O-glycosylation is the transfer of *N*-acetylgalactosamine from UDP-GalNAc to serine or threonine residues, which is catalyzed by a polypeptide-N-acetylgalactosaminyltransferase (ppGalNAcT). There are at least 21 polypeptide-N-acetylgalactosaminetransferases (ppGalNAcT-1 to -21) that differ in their amino acid sequences and are encoded by different genes. Homologs of ppGalNAcT genes are expressed in all eukaryotic organisms and a high degree of sequence identity exists between mouse and human ppGalNAcTs.

Immunohistochemistry studies have demonstrated that some ppGalNAcTs localize to the *cis*-Golgi in submaxillary glands (see Chapter 3). However, more recent studies suggest that in human cervical cancer cells, several enzymes of this family are located throughout the Golgi. The subcellular localization of ppGalNAcTs and other glycosyltransferases involved in O-glycosylation has a critical role in determining the range of O-glycans synthesized by a cell. Thus, a ppGalNAcT localized to a late compartment of the Golgi will act just before the secretion of a glycoprotein and is likely to produce a number of "incomplete" shorter O-GalNAc glycans, a factor that will contribute to the glycan heterogeneity of a mucin.

ppGalNAcT expression levels vary considerably between cell types and mammalian tissues. All ppGalNAcTs bind UDP-GalNAc (the donor of *N*-acetylgalactosamine), but they often differ in the protein substrates to which they transfer *N*-acetylgalactosamine. Such differences allow ppGalNAcTs to be distinguished. Many ppGalNAcTs do not efficiently transfer *N*-acetylgalactosamine to serine, but only to threonine in peptide acceptors used in in vitro assays. Nevertheless, the VNTR regions of mucins appear to be fully O-glycosylated. Many ppGalNAcTs appear to have a hierarchical relationship with one another, such that one enzyme cannot attach an *N*-acetylgalactosamine until an adjacent serine or threonine is glycosylated by a different ppGalNAcT. Thus, coexpression in the same cell of ppGalNAcTs with complementary, partly overlapping acceptor substrate specificities probably ensures efficient O-GalNAc glycosylation.

Although defined amino acid sequons that accept *N*-acetylgalactosamine have not been identified, certain amino acids are preferred in the substrate. Proline residues near the site of *N*-acetylgalactosamine addition are usually favorable to mucin O-glycosylation, whereas charged amino acids may interfere with ppGalNAcT activity. It is possible that the role of proline is to expose serine or threonine residues in a β-turn conformation, leading to more efficient O-glycosylation. Databases are available that estimate the likelihood of O-glycosylation at specific sites based on known sequences around O-glycosylation sites of several hundreds of glycoproteins (e.g., OGLYCBASE and NetOGlyc).

The domain structure of ppGalNAcTs is that of a type II membrane protein, typical of all Golgi glycosyltransferases (see Chapter 3). In addition, a distinct lectin-like domain has been discovered at the carboxyl terminus of several ppGalNAcTs. The lectin domain is used for binding to *N*-acetylgalactosamine residues that have already been added to the glycoprotein. In vitro studies using O-glycosylated peptides as acceptor substrates have shown that further addition of *N*-acetylgalactosamine is strongly influenced by the presence of existing *N*-acetylgalactosamine and larger O-glycans in the acceptor substrate, suggesting a molecular mechanism for the hierarchical relationships of ppGalNAcTs.

Purification of ppGalNAcTs has been achieved using mucin peptide and nucleotide sugar derivatives as affinity reagents. The recent elucidation of the crystal structure of ppGalNAcT-1 showed that the catalytic site of the enzyme is spatially separated from the lectin-binding site, and that there is a nucleotide sugar- and Mn^{++}-binding groove containing an aspartate-X-histidine motif. The unique structure of ppGalNAcT-1 suggests that this transferase can accommodate a number of different glycosylated acceptor substrates.

Synthesis of O-GalNAc Glycan Core Structures

The transfer of the first sugar from UDP-GalNAc directly to serine or threonine in a protein initiates the biosynthesis of all O-GalNAc glycans. Subsequently, with the addition of the next sugar, different mucin O-glycan core structures are synthesized (Figure 9.2). In contrast to the initial reactions of N-glycosylation and O-mannosylation, no lipid-linked intermediates are involved in O-GalNAc glycan biosynthesis, and no glycosidases appear to be involved in the processing of O-GalNAc glycans within the Golgi.

The O-glycans of mucins produced in a specific tissue predict the enzyme activities that are present in that tissue. The glycosyltransferases that are involved solely in the assembly of mucin O-GalNAc glycans are listed in Table 9.2. The many other enzymes that may be involved also contribute to the synthesis of N-glycans and glycolipids. The first sugar (*N*-acetylgalactosamine) added to the protein creates the Tn antigen (GalNAc-Ser/Thr), which is uncommon in normal mucins, but is often found in mucins derived from tumors. This suggests that the extension of O-GalNAc glycans beyond the first sugar is blocked in some cancer cells. Another common cancer-associated structure found in mucins is the sialyl-Tn antigen, which contains a sialic acid residue linked to C-6 of *N*-acetylgalactosamine (Table 9.1). O-Acetylation of the sialic acid residue masks the recognition of sialyl-Tn by anti-sialyl-Tn antibodies. No other sugars are known to be added to the sialyl-Tn antigen.

There are eight O-GalNAc glycan core structures (Table 9.1), most of which may be further substituted by other sugars. *N*-Acetylgalactosamine is converted to core 1 (Galβ1-3GalNAc-) by a core 1 β1-3 galactosyltransferase termed T synthase or C1GalT-1 (Figure 9.3). This activity is present in most cell types. However, to be exported from the endoplasmic reticulum in vertebrates, T synthase requires a specific molecular chaperone called Cosmc. Cosmc is an ER protein that appears to bind specifically to T synthase and ensures its full activity in the Golgi. The result is high expression of Tn and sialyl-Tn antigens. Lack of core 1 synthesis can be due to either defective T synthase or the absence of functional Cosmc chap-

TABLE 9.2. *Glycosyltransferases specific for mucin O-GalNAc glycans*

Enzyme	Short form
Polypeptide *N*-acetylgalactosaminyltransferase	ppGalNAcT-1 to -24
Core 1 β1-3 galactosyltransferase	C1GalT-1 or T synthase
Core 2 β1-6 *N*-acetylglucosaminyltransferase	C2GnT-1, C2GnT-3
Core 3 β1-3 *N*-acetylglucosaminyltransferase	C3GnT-1
Core 2/4 β1-6 *N*-acetylglucosaminyltransferase	C2GnT-2
Elongation β1-3 *N*-acetylglucosaminyltransferase	elongation β3GnT-1 to -8
Core 1 α2-3 sialyltransferase	ST3Gal I, ST3Gal IV
α2-6 sialyltransferase	ST6GalNAc I, II, III or IV
Core 1 3-O-sulfotransferase	Gal3ST4
Secretor gene α1-2 fucosyltransferase	FucT-I, FucT-II

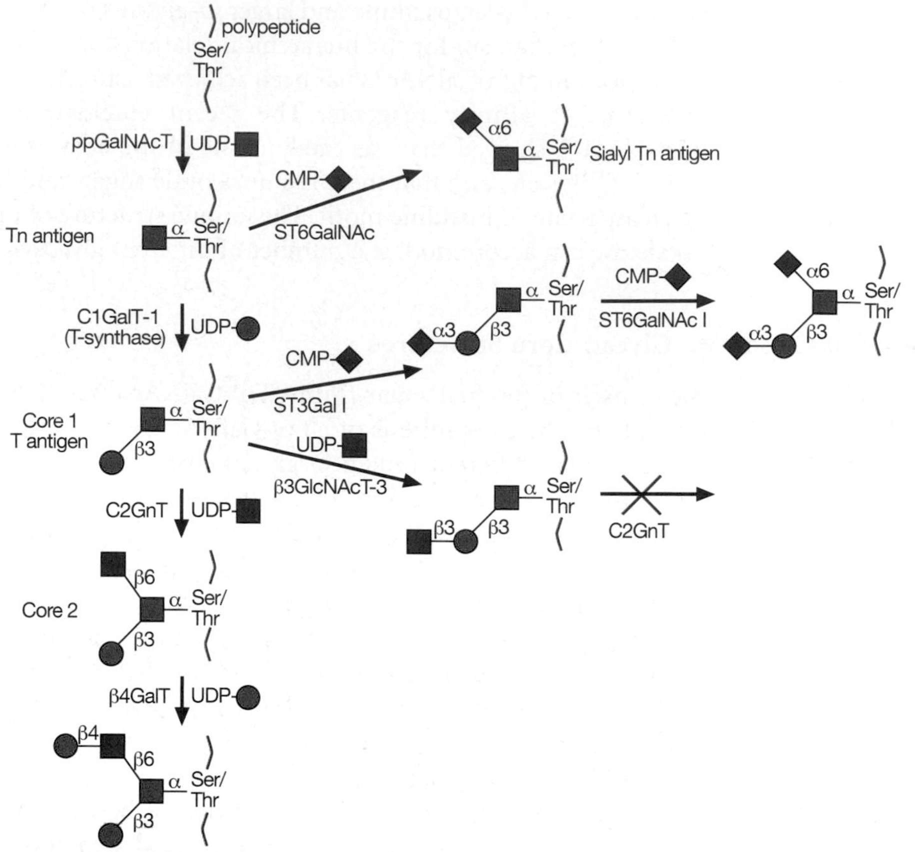

FIGURE 9.3. Biosynthesis of core 1 and 2 O-GalNAc glycans. Shown is the biosynthesis of some extended core 1 and core 2 O-GalNAc glycans. The linkage of *N*-acetylgalactosamine to serine or threonine to form the Tn antigen, catalyzed by polypeptide-N-acetylgalactosaminyltransferases (ppGalNAcTs), is the basis for all core structures. Core 1 β1-3 galactosyltransferase (C1GalT-1) synthesizes core 1 (T antigen). Core 1 may be substituted with sialic acid or *N*-acetylglucosamine (as shown) or by fucose or sulfate (not shown). Substitution of core 1 by a β1-3-linked *N*-acetylglucosamine prevents the synthesis of core 2 by core 2 β1-6 *N*-acetylglucosaminyltransferase (C2GnT), an enzyme that is highly specific for unmodified core 1 as a substrate. *Red cross* marks blocked pathway. In breast cancer cells, C2GnT and α2-3 sialyltransferase (ST3Gal I) compete for the common core 1 substrate. If sialyltransferase activity is high, chains will be mainly short with sialylated core 1 structures; if C2GnT activity is high, chains will be complex and large. The Galβ1-3 residue of the core 1 O-GalNAc glycan and both *N*-acetylglucosamine and galactose branches of the core 2 O-GalNAc glycan may be extended and carry terminal blood group and Lewis antigens (see Table 9.1).

erone. Immunoglobulin A nephropathy in humans is associated with low expression of Cosmc, and a mutation in the Cosmc gene gives rise to a condition called the Tn syndrome.

In many serum glycoproteins and mucins, the T antigen is substituted by sialic acid at C-3 of galactose and at C-6 of *N*-acetylgalactosamine (Figure 9.3). These substitutions add a negative charge to the O-GalNAc glycan. They also prevent other modifications of core 1. The cell surfaces of many leukemia and tumor cells contain large numbers of sialylated core 1 O-GalNAc glycans. On rare occasions, core 1 remains unsubstituted, leaving the T antigen exposed, for example, in cancer and inflammatory bowel disease. In these cases, it is likely that there is an abnormality in either the sialylation of core 1 or further extension and branching of core 1.

Core 2 mucin O-glycans are branched core 1 structures that are produced in many tissues, including the intestinal mucosa. The synthesis of core 2 O-GalNAc glycans is regulated during activation of lymphocytes, cytokine stimulation, and embryonic development.

Leukemia and cancer cells and other diseased tissues have abnormal amounts of core 2 O-GalNAc glycans. The synthesis of core 2 O-GalNAc glycans has been correlated with tumor progression. Because of their branched nature, core 2 O-glycans can block the exposure of mucin peptide epitopes. The enzyme responsible for core 2 synthesis is core 2 β1-6 *N*-acetylglucosaminyltransferase or C2GnT (Figure 9.3). At least three genes encode this subfamily (C2GnT-1 to -3) of a larger family of β1-6 *N*-acetylglucosaminyltransferases that catalyze the synthesis of GlcNAcβ1-6-linked branches. There are two major types of core 2 β1-6 *N*-acetylglucosaminyltransferases. The L type (leukocyte type, C2GnT-1 and -3) synthesizes only the core 2 structure, whereas the M type (mucin type, C2GnT-2) is also involved in the synthesis of core 4 and other GlcNAcβ1-6-linked branches. The L enzyme is active in many tissues and cell types, but the M enzyme is found only in mucin-secreting cell types. The expression and activity of both the L and M enzymes are altered in certain tumors.

The synthesis of core 3 O-GalNAc glycans appears to be restricted mostly to mucous epithelia from the gastrointestinal and respiratory tracts and the salivary glands. The enzyme responsible is core 3 β1-3 *N*-acetylglucosaminetransferase (C3GnT; Figure 9.4). Although in vitro assays of this enzyme suggest that it is relatively inefficient, the enzyme must act efficiently in vivo in colonic goblet or mucous cells because secreted colonic mucins mainly carry core 3 and few core 1, core 2, or core 4 O-GalNAc glycans. The activity of C3GnT is especially low in colonic tumors and virtually absent from tumor cells in culture. The synthesis of core 4 by the M-type β1-6 *N*-acetylglucosaminetransferase (C2GnT-2; see above) requires the prior synthesis of a core 3 O-GalNAc glycan (Figure 9.4). The glycosyl-

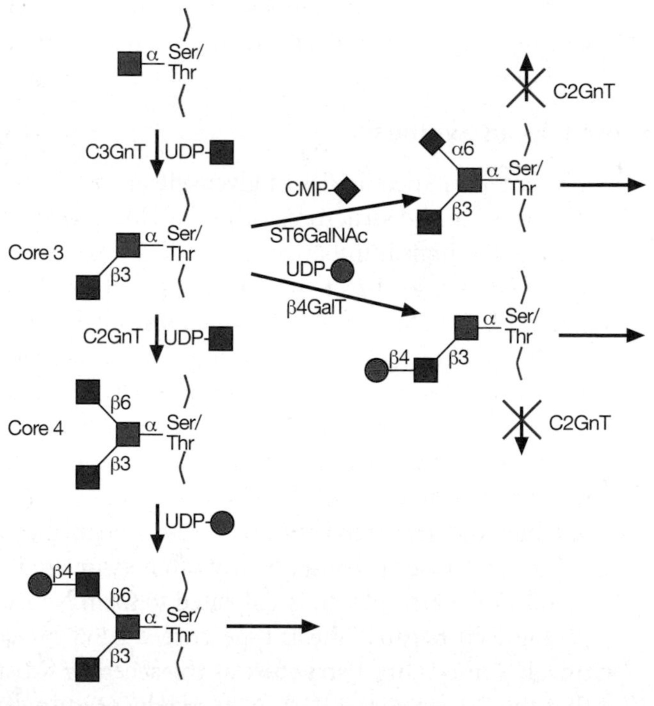

FIGURE 9.4. Biosynthesis of core 3 and core 4 O-GalNAc glycans. *N*-Acetylgalactosamine is substituted by a β1-3 *N*-acetylglucosamine residue to form core 3. This reaction is catalyzed by core 3 β1-3 *N*-acetylglucosaminyltransferase (C3GnT), which is active in colonic mucosa. Core 3 may be substituted with sialic acid in α2-6 linkage to *N*-acetylgalactosamine or with galactose in β1-4 linkage to *N*-acetylglucosamine. These substitutions prevent the synthesis of core 4 by core 2/4 β1-6 *N*-acetylglucosaminyltransferase (C2GnT-2). *Red crosses* mark blocked pathways. Unmodified core 3 can be branched by C2GnT-2 to form core 4. Both core 3 and core 4 O-GalNAc glycans can be extended to complex chains carrying determinants similar to those of core 1 and core 2 O-GalNAc glycans.

transferase synthesizing core 5 must exist in colonic tissues and in colonic adenocarcinoma because these tissues produce mucin with core 5. The enzymes synthesizing cores 5 to 8 remain to be characterized.

Synthesis of Complex O-GalNAc Glycans

The elongation of the galactose residue of core 1 and core 2 O-GalNAc glycans is catalyzed by a β1-3 *N*-acetylglucosaminyltransferase (Table 9.2) that is specific for O-GalNAc glycans. These O-glycans can also be extended by N-acetylglucosaminyltransferases and galactosyltransferases to form repeated GlcNAcβ1-3Galβ1-4 (poly-*N*-acetyllactosamine) sequences that represent the little i antigen. Less common elongation reactions are the formation of GalNAcβ1-4GlcNAc- (LacdiNAc) and Galβ1-3GlcNAc- sequences. Linear poly-*N*-acetyllactosamine units can be branched by members of the β1-6 *N*-acetylglucosaminyltransferase family, resulting in the large I antigen. Some of these branching reactions are common to other O- and N-glycans and glycolipids (see Chapter 13).

The ABO and other glycan-based blood groups as well as sialic acids, fucose, and sulfate are common terminal structures in mucins as in other glycoconjugates. The families of glycosyltransferases that catalyze the addition of these terminal structures are described in more detail in Chapters 13 and 14. In contrast to N-glycans, O-GalNAc glycans do not have NeuAcα2-6Gal linkages, although the NeuAcα2-6GalNAc moiety is common, for example, in the sialyl-Tn antigen and in elongated core 1 and 3 structures. Thus, in most mammalian mucin-producing cells, α2-6 sialyltransferases act on *N*-acetylgalactosamine, and α2-3 sialyltransferases act on galactose. Some of the sialyltransferases and sulfotransferases prefer O-GalNAc glycans as their substrate (Table 9.2), but many of these enzymes have an overlapping specificity and also act on N-glycan structures as acceptor substrates.

Control of O-GalNAc Glycan Synthesis

The acceptor specificities of glycosyltransferases and sulfotransferases are the main factors determining the structures of O-GalNAc glycans found in mucins, and these specificities restrict the high number of theoretically possible O-glycans to "only" a few hundred. The specificities also direct the pathways that are feasible. For example, the sialylated core 1 structure NeuAcα2-6(Galβ1-3)GalNAc- can be synthesized only by adding a sialic acid residue to core 1, but not by adding a galactose residue to the sialyl-Tn antigen, because the sialic acid prevents the action of core 1 β1-3 galactosyltransferase. The disialylated core 1 structure NeuAcα2-3Galβ1-3(NeuAcα2-6)GalNAc- is synthesized by the addition of the α2-3-linked sialic acid first to core 1, followed by the addition of the α2-6-linked sialic acid to *N*-acetylgalactosamine. These enzymes can therefore recognize at least two sugar residues and are often (but not always) blocked by the presence of sialic acid. Expression of sialyltransferases is consequently often associated with shorter O-GalNAc glycans. When a core 1 O-GalNAc glycan is sialylated to form NeuAcα2-3Galβ1-3GalNAc-, core 2 cannot be synthesized because the L-type core 2 β1-6 *N*-acetylglucosaminyltransferase requires an unsubstituted core 1 structure as the acceptor substrate. Detailed specificity studies suggest that the L-type core 2 β1-6 *N*-acetylglucosaminyltransferase recognizes unsubstituted C-4 and C-6 hydroxyl groups of both the galactose and *N*-acetylgalactosamine residues of a core 1 O-GalNAc glycan.

Another important control factor is that the relative activities of the glycosyltransferases determine the relative amounts of O-GalNAc glycans in mucins. Because of partially overlapping localization in the *cis*- and *medial*-Golgi compartments of breast cancer cells, core 2 β1-6 *N*-acetylglucosaminyltransferase (C2GnT-1) competes with α2-3 sialyltransferase for the common core 1 substrate (Figure 9.3). Therefore, the relative activities of these two enzymes determine the nature of the O-GalNAc glycans synthesized and the antigenic properties of the cell surface. Depending on which activity predominates, the cell-surface mucin MUC1 may be highly sialylated and its O-glycans relatively small or its O-glycans may be large and more complex due to the presence of extended, branched core 2 O-GalNAc glycans. In MUC1 from breast cancer cells, the presence of these sialylated small O-glycans leads to the exposure of underlying mucin peptide epitopes.

In vitro studies have shown that the activities of transferases are controlled by many factors such as metal ions and membrane components. Only two activities involved in O-glycan biosynthesis have been found to be regulated by the presence of specific binding proteins. The first is β1-4 galactosyltransferase 1 (β4GalT-1), which can bind to α-lactalbumin in mammary gland to change its preferred acceptor substrate from *N*-acetylglucosamine to glucose, thereby forming lactose. The second enzyme is core 1 β1-3 galactosyltransferase (C1GalT-1), which requires the coexpression of the molecular chaperone Cosmc.

The first step of O-GalNAc glycosylation is clearly regulated by the amino acid sequence of the acceptor substrate. Although individual ppGalNAcTs may prefer certain sites in the acceptor substrate, a combination of enzymes expressed in the same cell compartment ensures efficient mucin O-glycosylation. Several enzymes of the O-glycosylation pathway recognize their substrates in the context of the underlying peptide. The sequence of the peptide moiety of the acceptor substrate may direct the synthesis of cores 1, 2, and 3 as well as the extension of *N*-acetylglucosamine-terminating core structures by galactose (Figures 9.3 and 9.4).

It appears that glycosyltransferases are arranged in a diffuse "assembly line" within the Golgi compartments. This intracellular localization of enzymes allows them to act on mucin substrates in a specific sequence. Thus, early-acting enzymes synthesize the substrates for the next reactions. Because several enzymes can often use the same substrate, and there is extensive overlap of both localization and specificities, large numbers of different O-glycan structures are synthesized, leading to the heterogeneity seen in mucins. The concentrations of nucleotide sugar and sulfate-donor substrates within the Golgi and the rate of substrate transport throughout the Golgi are additional important control factors (see Chapter 4). The sensitive and complex balance of these various factors appears to be altered upon cell differentiation, cytokine stimulation, and in disease states.

FUNCTIONS OF O-GALNAC GLYCANS

O-GalNAc glycosylation is probably an essential process because all mammalian cell types studied to date express ppGalNAcTs. However, when the ppGalNAcT-1 gene was deleted in mice the animals appeared to be unaffected, possibly due to the fact that another ppGalNAcT replaces the function of ppGalNAcT-1. In the secreted mucins of the respiratory, gastrointestinal, and genitourinary tracts, as well as those of the eyes, the O-GalNAc glycans of mucous glycoproteins are essential for their ability to hydrate and protect the underlying epithelium. Mucins also trap bacteria via specific receptor sites within the O-glycans of the mucin. Some sugar residues or their modifications can serve as "decoys," thus masking underlying antigens or receptors. For example, O-acetyl groups on the sialic acid residue of the sialyl-Tn antigen prevent recognition by anti-sialyl-Tn antibodies. Gut bacteria often actively remove this decoy. Bacteria can cleave sulfate with sulfatases or terminal sugars with glycosidases. Because the O-glycans are hydrophilic and usually negatively charged, they promote binding of water and salts and are major contributors to the viscosity and adhesiveness of mucus, which forms a physical barrier between lumen and epithelium. The removal of microbes and particles trapped in mucus is an important physiological process.

However, in diseases such as cystic fibrosis, the abnormally high viscosity of the mucus leads to obstruction and life-threatening tissue malfunction.

O-GalNAc glycans, especially in the highly glycosylated mucins, have a significant effect on the conformation of the attached protein. Depending on the size and bulkiness of O-glycans, underlying peptide epitopes can be variably recognized by antibodies. O-glycosylation of mucins provides almost complete protection from protease degradation, and it is possible that the sparse O-glycosylation of some secreted glycoproteins, such as the single O-GalNAc glycan on interleukin-2, has a similar protective role.

Cell lines defective in specific O-glycosylation pathways can be used as models to study the role of O-glycosylation (see Chapter 46). For example, due to a defect in UDP-Glc-4-epimerase that reversibly converts UDP-GlcNAc to UDP-GalNAc, the ldlD cell line does not produce GalNAc-Ser/Thr linkages unless *N*-acetylgalactosamine is added to the cell medium. The addition of galactose as well as *N*-acetylgalactosamine results in complete O-GalNAc glycans. This model has been used to demonstrate that O-glycans of cell-surface receptors may regulate receptor stability and expression levels.

O-GalNAc glycans change during lymphocyte activation and are abnormal in leukemic cells where an increase in core 2 and a decrease in core 1 O-glycans are often seen. The ligands for selectin-mediated interactions between endothelial cells and leukocytes are commonly based upon sialyl Lewis^x epitopes attached to core 2 O-GalNAc glycans. This type of selectin–glycan interaction is important for the attachment of leukocytes to the capillary endothelium during homing of lymphocytes or the extravasation of leukocytes during the inflammatory response (see Chapter 31). Removal of core 2 O-GalNAc glycans by eliminating the C2GnT-1 gene in mice results in a severe deficiency in the immune system, particularly in the selectin-binding capability of leukocytes (see Chapter 50).

Cancer cells often express sialyl Lewis^x epitopes and may thus use the selectin-binding properties of their cell surface as a mechanism to invade tissues. The critical role of O-glycans in this process has been studied in cell lines treated with the O-glycan extension inhibitor benzyl-α-GalNAc (see Chapter 50). This compound acts as a competitive substrate for the synthesis of core 1, core 2, core 3, and core 4 O-glycans in cells and thus causes a reduction in the synthesis of complex O-GalNAc glycans. As a consequence, mucins carry a higher number of unmodified *N*-acetylgalactosamine residues and shorter O-GalNAc glycans. Inhibitor-treated cancer cells lose the ability to bind to E-selectin and endothelial cells in vitro.

Reproductive tissues produce mucins and O-glycosylated glycoproteins that may have important roles in fertilization. Specific terminal O-glycan structures have been shown to form the ligands for sperm-egg interactions in several species (see Chapter 25).

The changes of O-glycans commonly observed in diseases can be due to the actions of cytokines or growth factors that affect cell growth, differentiation, and cell death and alter the expression of glycosyltransferase genes. Although these glycosylation changes may be considered a "side effect" of a pathological condition, they can also significantly contribute to the ultimate pathology and the course of disease. In cancer, the biosynthesis of O-GalNAc glycans is often abnormal due to either decreased or increased expression and activities of specific glycosyltransferases. An altered cell-surface glycocalyx may then affect the biology and survival of the cancer cell.

FURTHER READING

Tabak L.A. 1995. In defense of the oral cavity: Structure, biosynthesis, and function of salivary mucins. *Annu. Rev. Physiol.* **57:** 547–564.

van Klinken B.J.W., Dekker J., Büller H.A., and Einerhand A.W.C. 1995. Mucin gene structure and expression: Protection vs. adhesion. *Am. J. Physiol.* **269:** G613–G627.

Hansen J.E., Lund O., Nielsen J.O., and Brunak S. 1996. O-GLYCBASE: A revised database of O-glycosylated proteins. *Nucleic Acids Res.* **24:** 248–252.

Brockhausen I. 1999. Pathways of O-glycan biosynthesis in cancer cells. *Biochim. Biophys. Acta* **1473:** 67–95.

Fukuda M. 2002. Roles of mucin-type O-glycans in cell adhesion. *Biochim. Biophys. Acta* **1573:** 394–405.

Brockhausen I. 2003. Sulphotransferases acting on mucin-type oligosaccharides. *Biochem. Soc. Trans.* **31:** 318–325.

Lowe J.B. and Marth J. 2003. A genetic approach to mammalian glycan function. *Annu. Rev. Biochem.* **72:** 643–691.

ten Hagen K.G., Fritz T.A., and Tabak L.A. 2003. All in the family: The UDP-GalNAc:polypeptide *N*-acetylgalactosaminyltransferases. *Glycobiology* **13:** 1R–16R.

Hollingsworth M.A. and Swanson B.J. 2004. Mucins in cancer: Protection and control of the cell surface. *Nat. Rev. Cancer* **4:** 45–60.

Robbe C., Capon C., Coddeville B., and Michalski J.C. 2004. Structural diversity and specific distribution of O-glycans in normal human mucins along the intestinal tract. *Biochem. J.* **384:** 307–316.

Wandall H.H., Irazoqui F., Tarp M.A., Bennett E.P., Mandel U., Takeuchi H., Kato K., Irimura T., Suryanarayanan G., Hollingsworth M.A., and Clausen H. 2007. The lectin domains of polypeptide GalNAc-transferases exhibit carbohydrate-binding specificity for GalNAc: Lectin binding to GalNAc-glycopeptide substrates is required for high density GalNAc-O-glycosylation. *Glycobiology* **17:** 374–387.

Hattrup C.L. and Gendler S.J. 2008. Structure and function of the cell surface (tethered) mucins. *Annu. Rev. Physiol.* **70:** 431–457.

CHAPTER 10

Glycosphingolipids

Ronald L. Schnaar, Akemi Suzuki, and Pamela Stanley

GLYCOSPHINGOLIPIDS (GSLs) ARE A TYPE OF GLYCOLIPID. They are found in the cell membranes of organisms from bacteria to man, and are the major glycans of the vertebrate brain, where more than 80% of glycoconjugates are in the form of glycolipids. The emphasis of this chapter is on vertebrate glycosphingolipids. Information on glycolipids of fungi, plants, and invertebrates is covered elsewhere in this volume; see Chapters 19 and 21–24. Glycosylphosphatidylinositols (GPIs), a distinct family of glycolipids that covalently attach to proteins and serve as membrane anchors, are discussed separately in Chapter 11. The lipopolysaccharide of Gram-negative bacteria, sometimes called a saccharolipid, is discussed in Chapter 20. This chapter describes characteristic features of glycosphingolipids, pathways for their biosynthesis, and insights into their biological roles in membrane structure, host–pathogen interactions, cell–cell recognition, and modulation of membrane protein function. Some mention is also made of glycoglycerolipids.

DISCOVERY AND BACKGROUND

The first glycolipid to be structurally characterized was galactosylceramide (GalCer). Among the simplest of glycolipids, it is also one of the most abundant molecules in the brain. It consists of a single galactose residue in glycosidic linkage to the C-1 hydroxyl group of a lipid moiety called ceramide (Figure 10.1). When its structure was deduced in 1884, the lipid moiety was particularly novel, consisting of a long-chain amino alcohol in amide linkage to a fatty acid (Figure 10.1). The structure was so difficult to determine that the amino alcohol was dubbed "sphingosine" after the enigmatic Egyptian Sphinx, "in commemoration of the many enigmas which it presented to the inquirer." We now know that nearly all

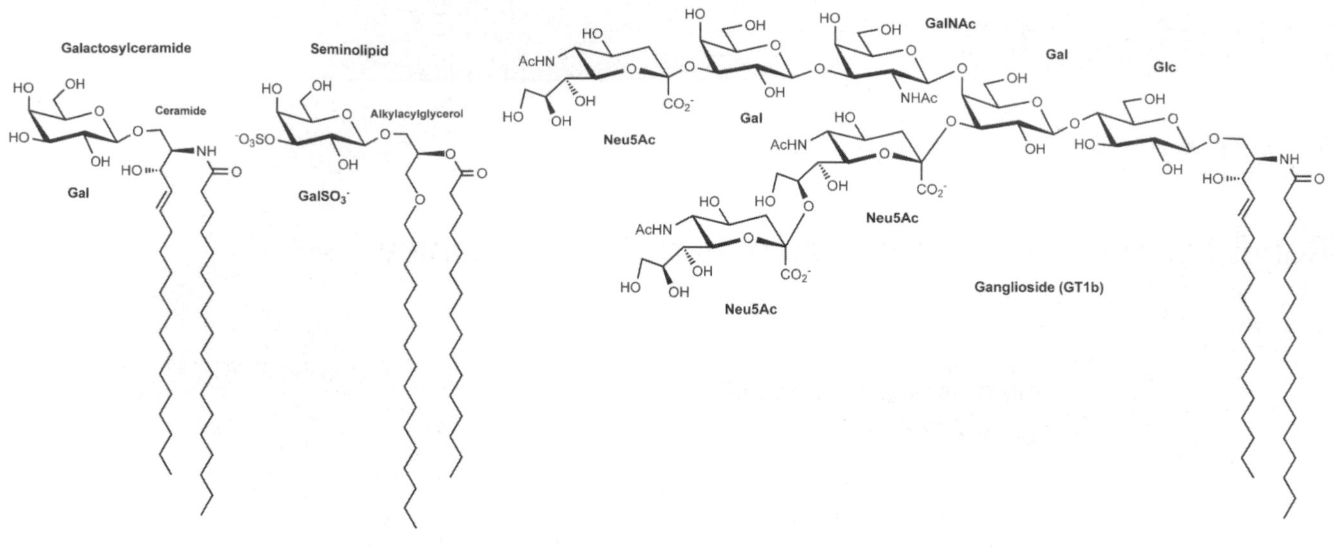

FIGURE 10.1. Structures of representative glycosphingolipids and glycoglycerolipids. Glycosphingolipids, such as GalCer, are built on a ceramide lipid moiety that consists of a long-chain amino alcohol (sphingosine) in amide linkage to a fatty acid. In comparison, glycoglycerolipids, such as seminolipid, are built on a diacyl or acylalkylglycerol lipid moiety. Most animal glycolipids are glycosphingolipids, which have a large and diverse family of glycans attached to ceramide. Shown is one example of a complex sialylated glycosphingolipid, GT1b ($IV^3NeuAcII^3[NeuAc]_2Gg_4Cer$).

glycolipids in vertebrates are glycosphingolipids, which, in turn, are part of the larger family of sphingolipids (lipids having ceramide as their core structure) that includes the major membrane phospholipid, sphingomyelin. Other glycosphingolipids were later identified because they accumulate to pathological levels in tissues of patients suffering from lysosomal storage diseases, genetic disorders in which the enzymes that degrade glycans are faulty or missing (see Chapter 41). For example, a sialic acid–containing glycosphingolipid was first isolated from the brain of a victim of Tay-Sachs disease, in which it accumulates, and was named a "ganglioside" (according to the location of accumulation in a neuronal cluster or "ganglion" in the brain). Likewise, glucosylceramide (GlcCer) was first isolated from the spleen of a Gaucher's disease patient. As purification, separation, and analytical techniques improved, glycosphingolipids were found in all vertebrate tissues. Hundreds of unique glycosphingolipid structures have been described based on glycan structural variations alone—and there are many more when lipid variations are taken into account.

Another class of animal glycolipids are the glycoglycerolipids, distinguished from glycosphingolipids by their lipid moieties, having glycans linked to the C-3 hydroxyl of diacylglycerol (Figure 10.1). These are very minor constituents of most animal tissues, other than the testes. They are more widely distributed in microbes and plants. Glycosphingolipids are by far the most abundant and diverse class of glycolipids in animals and have also been discovered in fungi, plants, and invertebrates (see Chapters 19 and 21–24). This large and diverse family of glycans has important roles in physiology and pathology that are becoming increasingly amenable to exploration.

MAJOR CLASSES AND NOMENCLATURE

As mentioned above, the ceramide component of glycosphingolipids consists of a long-chain amino alcohol, sphingosine, in amide linkage to a fatty acid. Ceramide structures vary in length, hydroxylation, and saturation of both the sphingosine and fatty acid moieties, resulting in lipid structural diversity that impacts the presentation of the attached glycan at membrane surfaces.

Although ceramide variations add diversity to glycosphingolipid structures, major structural and functional classifications have traditionally been based on the glycans. The first sugars linked to ceramide in higher animals are typically β-linked galactose (GalCer) or glucose (GlcCer). GalCer and its analog sulfatide, with sulfate at the C-3 hydroxyl of galactose, are the major glycans in the brain, where they have essential roles in the structure and function of myelin, the insulator that allows for rapid nerve conduction. Interestingly, a related sulfogalactoglycerolipid, seminolipid (Figure 10.1), is abundant only in the male reproductive tract, where it is essential for spermatogenesis. Sialylated GalCer (NeuAcα2-3GalβCer; GM4; see Figure 10.2) is also found in myelin. These galactolipids are seldom extended with larger saccharide chains; rather, most other members of the large and diverse family of glycosphingolipids in higher animals are built on GlcCer.

GlcCer is abundant in certain tissues. In skin, GlcCer and its derivatives, acylGlcCer and ceramide produced by the hydrolysis of GlcCer, have important functions in the formation of the water barrier (see below). In more complex vertebrate glycosphingolipids, the glucose moiety is typically substituted with β-linked galactose on the C-4 hydroxyl of glucose, to give lactosylceramide (Galβ1-4GlcβCer). Further extensions of the glycan generate a series of neutral "core" structures that form the basis for the nomenclature of glycosphingolipids (Table 10.1). Thus, ganglio-series glycosphingolipids are based on the neutral core structure Galβ1-3GalNAcβ1-4Galβ1-4GlcβCer, whereas neolacto-series glycosphingolipids are based on the core structure Galβ1-4GlcNAcβ1-3Galβ1-4GlcβCer, lacto-series on Galβ1-3GlcNAcβ1-3Galβ1-4GlcβCer, globo-series on Galα1-4Galβ1-4GlcβCer, and isoglobo-series on Galα1-3Galβ1-4GlcβCer (key differences among the core structures are underlined). These glycosphingolipid subfamilies are expressed in tissue-specific patterns. In mammals, for example, ganglio-series glycosphingolipids, although broadly distributed, predominate in the brain, whereas neolacto-series glycolipids are common on certain hematopoietic cells including leukocytes. In contrast, lacto-series glycolipids are prominent in secretory organs and globo-series are the most abundant in erythrocytes. This diversity evidently reflects important differences in glycosphingolipid functions.

Glycosphingolipids are further subclassified as neutral (no charged sugars or ionic groups), sialylated (having one or more sialic acid residues), or sulfated. Traditionally, all sialylated glycosphingolipids are known as "gangliosides," regardless of whether they are based on the ganglio-series neutral core structure mentioned above. In the official nomenclature, saccharide and other substituents that extend or branch from the neutral core structures in Table 10.1 are indicated by a roman numeral, designating which of the neutral core sugars

TABLE 10.1. *Names and abbreviations of major core structures of vertebrate glycosphingolipids*

Subfamily series	Structure	Abbreviation
Lacto	GlcNAcβ1-3Galβ1-4GlcβCer	Lc_3Cer
	Galβ1-3GlcNAcβ1-3Galβ1-4GlcβCer	Lc_4Cer
Neolacto	Galβ1-4GlcNAcβ1-3Galβ1-4GlcβCer	nLc_4Cer
	Galβ1-4GlcNAcβ1-3Galβ1-4GlcNAcβ1-3Galβ1-4GlcβCer	nLc_6Cer
Ganglio	GalNAcβ1-4Galβ1-4GlcβCer	Gg_3Cer
	Galβ1-3GalNAcβ1-4Galβ1-4GlcβCer	Gg_4Cer
Globo	Galα1-4Galβ1-4GlcβCer	Gb_3Cer
	GalNAcβ1-3Galα1-4Galβ1-4GlcβCer	Gb_4Cer

Key structural signatures defining each family are underlined.

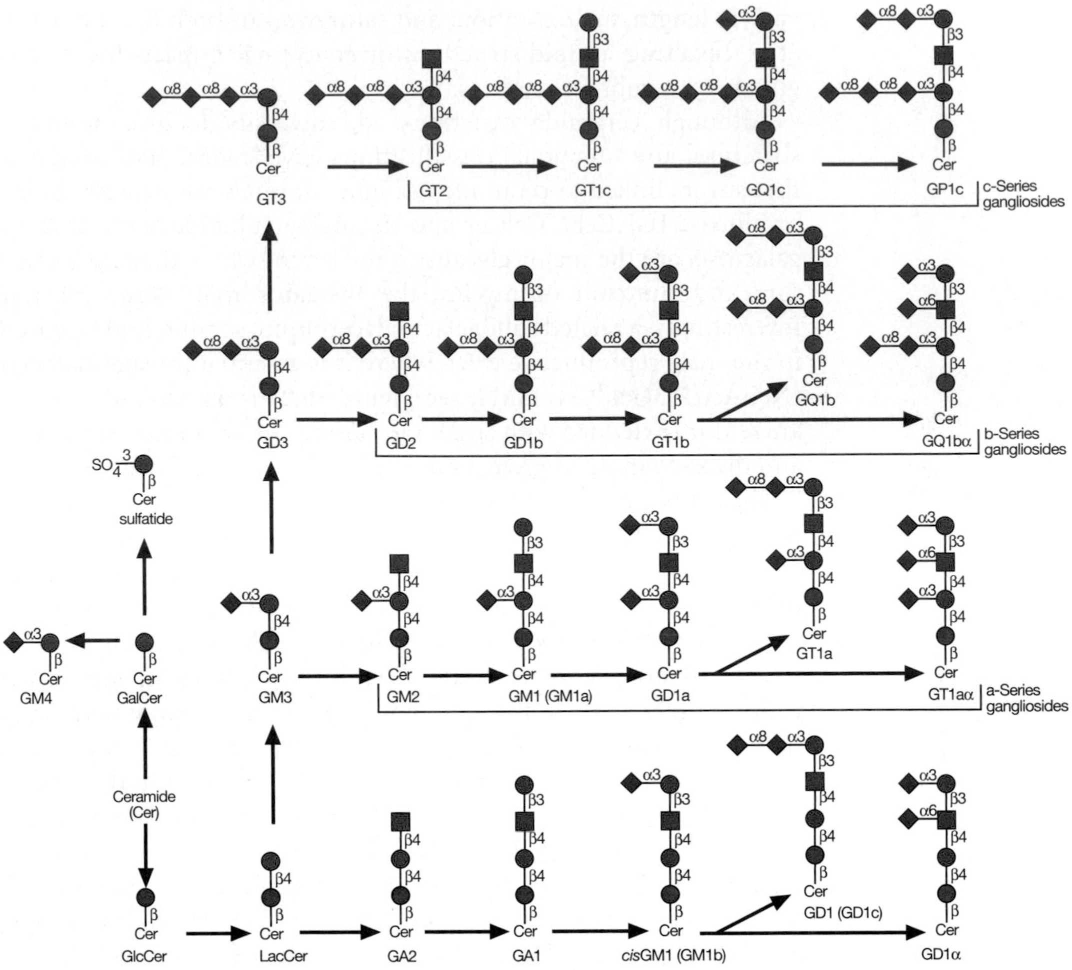

FIGURE 10.2. Biosynthetic pathway for brain glycosphingolipids. Glycosphingolipids are synthesized by the stepwise addition of sugars first to ceramide, then to the growing glycan. Shown as examples are brain glycosphingolipids. Ceramide (Cer) is the acceptor for UDP-Gal:ceramide β-galactosyltransferase or UDP-Glc:ceramide β-glucosyltransferase in the major pathways to glycosphingolipid biosynthesis in oligodendrocytes and nerve cells, respectively. GalCer is the acceptor for GalCer sulfotransferase, which adds a sulfate group to the C-3 of galactose to form sulfatide. Extension of GlcCer to the major brain gangliosides occurs by the action of UDP-Gal:GlcCer β1-4 galactosyltransferase to make lactosylceramide (LacCer), then CMP-NeuAc:lactosylceramide α2-3 sialyltransferase to make the simple ganglioside GM3. GM3 is a branch point and acts as the acceptor for UDP-GalNAc:GM3/GD3 β1-4 N-acetylgalactosaminyltransferase to generate a-series gangliosides and for CMP-NeuAc:GM3 α2-8 sialyltransferase to generate GD3 and the b-series gangliosides. Similarly, the action of an α2-8 sialyltransferase on GD3 gives rise to GT3 and the c-series gangliosides. Enzymes for subsequent elongation are common to the a-, b-, and c-series gangliosides. In mammals and birds, the major gangliosides in brain are GM1, GD1a, GD1b, and GT1b.

(counting the sugar closest to the ceramide as "I") carries the substituent, and a superscript designating which hydroxyl on that sugar is modified. Thus, the abundant ganglioside Galβ1-3GalNAcβ1-4(NeuAcα2-3)Galβ1-4GlcβCer is shortened to II^{3}NeuAc-Gg$_4$Cer, meaning that the galactose closest to Cer (II) carries sialic acid at C-3 (II^3), and Gg$_4$ designates the ganglio-series tetrasaccharide core. This nomenclature is clearly too complex for daily use, so the most common glycosphingolipids are usually referred to by unofficial names. In the widely used ganglioside nomenclature of Svennerholm, for example, the structure above is designated simply as "GM1" (Figure 10.2). In this nomenclature, G refers to ganglioside series, the second letter refers to the number of sialic acid residues (mono, di, tri, etc.), and the number (1, 2, 3, etc.) refers to the order of migration of the ganglioside on thin-layer chromatography (e.g., GM3 > GM2 > GM1).

Invertebrates (e.g., insects and mollusks) express glycosphingolipids based on the core structure Manβ1-4GlcβCer (the "arthro" core structure), and inositolphosphate ceramides are major sphingolipids in fungi and plants (see Chapters 19 and 21–24).

ISOLATION, PURIFICATION, AND ANALYSIS

Organic solvents are used to solubilize glycolipids from tissues and cells, where they are found primarily in the external leaflet of plasma membranes. Extraction procedures have been developed, often using defined chloroform-methanol-water mixtures added in a specific solvent sequence to optimize precipitation and removal of proteins and nucleic acids, while maximizing solubilization of glycosphingolipids (along with other lipids). Because glycosphingolipids aggregate with one another and other lipids in aqueous solution, organic solvents are used throughout subsequent purification steps, which typically involve solvent partition, ion-exchange chromatography, and silicic acid chromatography. Base treatment, a traditional approach to removing phospholipids, is no longer recommended because it can damage labile aspects of the molecule, such as acyl groups. High-pressure liquid chromatography (HPLC) on silicic acid–based columns is often the final step in separating glycolipids to obtain homogeneous molecular species. HPLC on hydrophobic columns can also separate glycolipids with identical glycans into different species based on their different ceramide structures.

Because of their amphipathic nature, glycolipids are well suited for thin-layer chromatography (TLC) analysis, which is useful for monitoring their purification, qualitative and quantitative determination of their expression in normal and diseased tissues, partial structural analysis, and detecting biological activities including immunoreactivity and binding activity toward toxins, viruses, bacteria, and eukaryotic cells. After separation by TLC, submicrogram quantities of glycolipids may be chemically detected with orcinol reagent for hexoses and with resorcinol-HCl reagent for sialic acid. For detection of proteins or receptors that bind to glycolipids, glycolipids are immobilized following separation and the plate is subsequently overlaid with the binding protein or organism to be tested (e.g., antibodies, lectins, toxins, viruses, or bacteria). After extensive washing, glycolipid species that bind can be identified by detection of the bound material at a precise position on the TLC plate.

Complete structural analyses of glycolipids requires a combination of techniques to determine the composition, sequence, linkage positions, and anomeric configurations of the glycan moiety and the fatty acid and long-chain base of the ceramide moiety. Glycan composition is determined by hydrolysis and analysis of the released monosaccharides. Mass spectroscopy of underivatized glycosphingolipids or of their permethylated derivatives is a powerful tool for sequence determination and ceramide identification and can sometimes be done directly from the TLC plate. Linkage determination is most rigorously performed by methylation analysis, and anomeric configurations can be obtained by nuclear magnetic resonance (NMR) spectroscopy (given sufficient quantities). Information on glycan sequence, linkage, and anomeric configuration can also be obtained by combining enzymatic hydrolysis using specific glycosidases with TLC. Release of the intact glycans from most glycosphingolipids is accomplished enzymatically using ceramide glycanases, and the released oligosaccharides are then subject to analysis using the methods described in Chapter 47.

BIOSYNTHESIS, TRAFFICKING, AND DEGRADATION

Glycosphingolipid biosynthesis occurs in a stepwise fashion, with an individual sugar added first to ceramide and then subsequent sugars transferred by glycosyltransferases from nucleotide sugar donors. Ceramide is synthesized on the cytoplasmic face of the endoplasmic reticulum (ER); it subsequently equilibrates to the lumenal face and trafficks to the Golgi compartment. GlcCer is synthesized on the cytoplasmic face of the ER and early Golgi apparatus; it then flips into the Golgi lumen, where it is typically elongated by a series of glycosyltransferases. In contrast, GalCer is synthesized on the lumenal face of the ER and then trafficks through the Golgi, where it may be sulfated to form sulfatide. In both cases, the final orientation of glycosphingolipids during biosynthesis is consistent with their nearly exclusive appearance on the outer leaflet of the plasma membrane, facing the extracellular milieu. Although ceramide resides on intracellular organelles such as mitochondria, glycosphingolipids beyond GlcCer are not known to exist on membranes facing the cytoplasm.

The biosynthesis of glycosphingolipids in the brain provides an example of how competing biosynthetic pathways can lead to glycan structural diversity (Figure 10.2). In the brain, stepwise biosynthesis of GalCer and sulfatide occurs in oligodendrocytes, the cells that elaborate myelin. Gangliosides, in contrast, are synthesized by all cells, with concentrations of the different forms varying according to cell type. Expression patterns of glycosphingolipids are determined by the expression and intracellular distribution of the enzymes required for their biosynthesis. In some cases, multiple glycosyltransferases compete for the same glycosphingolipid precursor (see Figure 10.2). For example, the ganglioside GM3 may be acted on by N-acetylgalactosaminyltransferase, thereby forming GM2, the simplest of the "a-series" gangliosides, or by sialyltransferase, thereby forming GD3, the simplest of the "b-series" gangliosides. Each branch is a committed pathway, because sialyltransferases cannot directly convert a-series gangliosides (beyond GM3) to their corresponding b-series gangliosides. Due to this branch exclusivity, competition between two enzymes at a key branch point determines the relative expression levels of the final glycosphingolipid products. The transfer of *N*-acetylgalactosamine to a-, b-, and c-series gangliosides, transforming GM3 into GM2, GD3 into GD2, or GT3 into GT2, is catalyzed by the same N-acetylgalactosaminyltransferase. Likewise, the transfer of galactose to GM2 to form GM1, to GD2 to form GD1b, or to GT2 to form GT1c is accomplished by a single galactosyltransferase. The levels of nucleotide sugar donors used by glycosyltransferases in the Golgi lumen (including UDP-Gal, UDP-Glc, UDP-GlcNAc, UDP-GalNAc, and CMP-NeuAc) ultimately affect the final structure of glycans and are regulated by synthetic enzymes in the cytoplasm or nucleus and by the activity of nucleotide sugar transporters in the Golgi membrane (see Chapter 4). The sialic acids of human gangliosides are exclusively in the form of *N*-acetylneuraminic acid (NeuAc) and its O-acetylated derivatives, but those of other mammals, even the closely related chimpanzee, contain both *N*-acetylneuraminic acid and *N*-glycolylneuraminic acid (NeuGc). This is due to a specific mutation in humans of the enzyme that hydroxylates CMP-NeuAc to form CMP-NeuGc. Even in animals with predominantly *N*-glycolylneuraminic acid in the gangliosides of non-neural tissues, brain gangliosides have *N*-acetylneuraminic acid nearly exclusively. Sialic acids on gangliosides may be further modified by substitutents such as O-acetyl groups or by removal of the N-acyl group (see Chapter 14).

An additional level of regulation may occur via stable association of different glycosphingolipid glycosyltransferases into functional "multiglycosyltransferase" complexes. The multiple enzymes are then thought to act concertedly on the growing glycosphingolipid without releasing intermediate structures, ensuring progression to the preferred end product.

Although the enzymes that catalyze the initial steps in glycosphingolipid biosynthesis are specific and used only for glycosphingolipid biosynthesis, outer sugars, such as the outermost sialic acid, fucose, or glucuronic acid residues, are sometimes added by glycosyltransferases that also act on glycoproteins, resulting in terminal structures being shared by glycosphingolipid and glycoprotein glycans (see Chapter 13). One typical example is the blood group ABO antigen system. The α1-3 N-acetylgalactosaminyltransferase encoded by the blood group A gene produces the blood group A determinant GalNAcα1-3(Fucα1-2)Galβ on glycoproteins and glycolipids. Correspondingly, the α1-3 galactosyltransferase encoded by the allelic blood group B gene transfers galactose to both glycoproteins and glycolipids.

The major brain ganglioside structures are highly conserved among individual humans and even among different mammalian species. In contrast, individual differences in expressed blood cell glycolipids are well known in humans, as in the case of the ABO, Lewis, and P blood group antigens (see Chapter 13). These differences result from mutations of glycosyltransferases responsible for synthesizing these antigens. For example, the α1-3 N-acetylgalactosaminyltransferase encoded by the A gene and the α1-3 N-galactosyltransferase encoded by the B gene of the ABO locus differ by only four amino acids. The blood group O precursor (Fucα1-2Galβ1-) results when either the A or B gene at the ABO locus is mutated and produces no active enzyme. Species differences among glycolipids also occur, one example being the expression of Forssman antigen, GalNAcα1-3GalNAcβ1-3Galα1-4Galβ1-4GlcβCer. This molecule is a good immunogen in Forssman antigen-negative species, such as rabbit, rat, and human, which have a mutated α1-3 N-acetylgalactosaminyltransferase that cannot transfer *N*-acetylgalactosamine to the precursor Gb4Cer. In contrast, guinea pig, mouse, sheep, and goat are Forssman antigen positive.

The breakdown of glycosphingolipids occurs stepwise by the action of lysosomal hydrolases. Glycosphingolipids on the outer surface of the plasma membrane are internalized, along with other membrane components, in invaginated vesicles that then fuse with endosomes, resulting in the glycosphingolipid glycan facing the endosome lumen. Glycosphingolipid-enriched areas of the endosomal membrane may then invaginate once again to form multivesicular bodies within the endosome. When endosomes fuse with primary lysosomes, glycosphingolipids become exposed to lysosomal hydrolases.

As glycosphingolipids are successively cleaved to smaller structures, the remaining "core" monosaccharides become inaccessible to the water-soluble lysosomal hydrolases and require assistance from activator proteins that are referred to as "liftases." These include GM2-activator protein and four structurally related saposins, all of which are derived from a single polypeptide saposin precursor by proteolytic cleavage. Saposins are thought to bind to their glycolipid substrate, disrupt its interaction with the local membrane environment, and facilitate access of the glycans to hydrolytic enzymes. In certain lysosomal storage diseases (see Chapter 41), mutations in these activator proteins result in pathological accumulation of glycolipids, even though there is an abundance of the hydrolases responsible for degradation, thus demonstrating the essential role of activator proteins in glycosphingolipid catabolism in vivo. Glycosphingolipids are eventually broken down to their individual components, which are then available for reuse.

Ceramide, the hydrophobic portion of glycosphingolipids, is composed of a sphingoid (long-chain) base in amide linkage to a fatty acid. Sphingoid bases are of three general chemical types: sphingosine, sphinganine, and phytosphingosine (Figure 10.3). The 18-carbon form of sphingosine (d18:1, where d = dihydroxy and one double bond is located between C-4 and C-5) is the most common sphingoid base in ceramide of mammals. Its stereochemistry (D-*erythro* configuration; 2S,3R) is important to its biological functions.

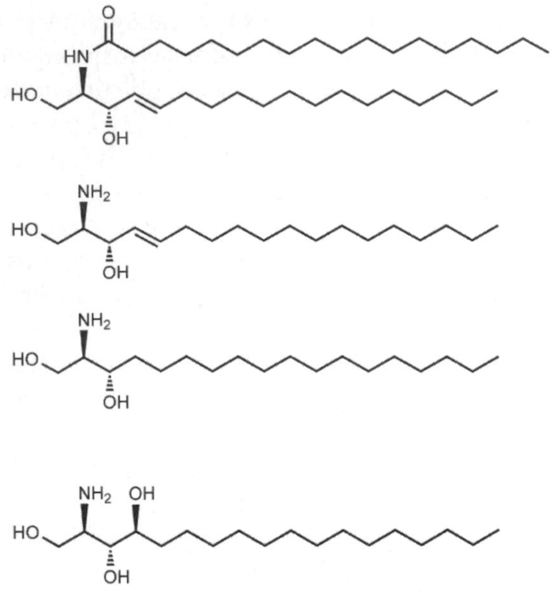

FIGURE 10.3. Structures of a ceramide and three sphingoid bases. A common mammalian ceramide with an 18-carbon sphingosine (d18:1) in amide linkage to a C16:0 fatty acid is shown, below which three sphingoid bases, sphingosine, sphinganine, and phytosphingosine, are compared. Sphingoid base and fatty acid amide length, stereochemistry, and structure can affect the presentation of glycan head groups at the cell surface, thereby modulating glycosphingolipid functions.

Ceramide biosynthesis starts when the enzyme serine palmitoyltransferase condenses serine and palmityl-CoA to form 3-ketosphinganine, which is then reduced to sphinganine. Enzymatic N-acylation using fatty acid CoA provides dihydroceramide, which is then desaturated to give ceramide. Saturated (dihydroceramide-containing) gangliosides are minor brain components that increase with age, and ceramides containing phytosphingosine (4-hydoxysphinganine) are prominent in the glycosphingolipids of plants and fungi (including yeasts). Although rare in mammals, phytosphingosine-containing glycosphingolipids are found in membranes of cells of the small intestine and kidney.

The fatty acid components of ceramides vary widely. Although C16:0 and C18:0 fatty acids are common, fatty acids in ceramides range from C14 to C30 (or greater), may be unsaturated, and may contain α-hydroxy groups (common in GalCer and sulfatide of myelin and glycolipids of small intestine and kidney). The biological significance of ceramide structural variations is unknown, but they are thought to modulate membrane structure and functions of glycosphingolipids.

BIOLOGICAL AND PATHOLOGICAL ROLES

Glycosphingolipid-enriched Membrane Microdomains

Glycosphingolipids found in the plasma membrane of all cells in "higher" animals comprise from less than 5% (erythrocytes) to more than 20% (myelin) of the membrane lipids. Glycosphingolipids are not uniformly distributed in the membrane, but cluster in "lipid rafts," small lateral microdomains of self-associating membrane molecules. Although the precise structure and makeup of lipid rafts is a matter of debate, their outer leaflets are believed to be enriched in sphingolipids, including glycosphingolipids and sphingomyelin (the phosphocholine derivative of ceramide). The self-association of sphingolipids is driven by the unique biophysical properties afforded by their long unsaturated carbon chains (Figure 10.3). Besides sphingolipids, lipid rafts are enriched in cholesterol and selected proteins, including GPI-anchored proteins and selected transmembrane signaling proteins such as receptor tyrosine kinases. From the cytosplasmic side, acylated proteins such as the Src family of protein tyrosine kinases and Gα subunits of heterotrimeric G proteins are known to associate with these rafts.

Lipid rafts are apparently small (10–50 nm diameter), each containing perhaps hundreds of lipid molecules along with a few protein molecules. It has been argued that external clustering of lipid rafts into larger structures might bring signaling molecules (such as kinases and their substrates) together to initiate intracellular signaling. Thus, glycosphingolipids may act as intermediaries in the flow of information from the outside to the inside of cells. This idea is supported by the observation that antibody-induced glycosphingolipid clustering activates lipid-raft-associated signaling and has led to the concept of plasma membrane "glycosignaling domains" or "glycosynapse." Interactions between glycans and glycan-binding proteins, as well as glycan–glycan interactions, are also very much influenced by the density of glycans in terms of binding affinity. Thus, ten glycans clustered in a very limited area such as a lipid raft or liposome greatly increase binding affinity of cognate proteins compared to a single molecule of a glycan (see Chapter 27). This multivalency adds unique functional properties to plasma membrane glycolipids. Indeed, several growth factor receptors including the epidermal growth factor (EGF) receptor, insulin receptor, and the nerve growth factor receptor are localized in membrane microdomains, and evidence indicates that their signaling functions are significantly modulated by glycolipids.

Physiological Functions of Glycosphingolipids

Glycosphingolipids are primarily expressed in the outer leaflet of the limiting plasma membrane of cells, with their glycans facing the external milieu. Their functions fall into two major categories: mediating cell–cell interactions via binding to complementary molecules on apposing plasma membranes (*trans* recognition) and modulating activities of proteins in the same plasma membrane (*cis* regulation). Biochemical, cell biological, and, more recently, genetic experiments and human diseases are establishing these diverse functions of glycolipids (see Chapters 46 and 50).

At the single-cell level, glycosphingolipids are not essential for life. Using specific chemical inhibitors and genetic ablation of biosynthetic genes, cells without glycosphingolipids survive, proliferate, and even differentiate. However, glycosphingolipids are required for development at the whole-animal level. Mice engineered to lack the gene for GlcCer synthesis fail to develop, with arrest occurring just past the gastrula stage due to extensive apoptosis in the embryo. These and other recent observations lead to a basic principle: Glycosphingolipids mediate and modulate intercellular coordination in multicellular organisms. Sometimes this occurs in quite subtle ways, as exemplified by the role of GalCer and sulfatide in myelination.

GalCer and its 3-O-sulfated derivative, sulfatide, are predominant glycans in the brain, where they constitute more than 50% of the total glycoconjugate. In the brain, they are expressed by oligodendrocytes, which elaborate myelin, the multilayered membrane insulation that ensheathes nerve axons. GalCer and sulfatide constitute more than 20% of myelin lipids and were widely believed to be essential to myelin structure. This turned out to be true, but in a much more subtle way than anticipated. Mice engineered to lack the enzyme responsible for GalCer synthesis (UDP-Gal:ceramide β-galactosyltransferase) do not make any GalCer or sulfatide. However, they do myelinate axons, and the myelin appears grossly normal. Nevertheless, the mice show all the signs of failed myelination, including tremor, ataxia, slow nerve conduction, and early death. Ultrastructural studies revealed the reason. In both normal and mutant mice, myelination occurs in short stretches along axons, with intermittent gaps called "nodes of Ranvier," where concentrated ion channels pass nerve impulses along to the next gap. At the edge of the node, myelin membranes normally curve downward and attach to the axon to seal off the node. In animals lacking GalCer and sulfatide, these myelin "end feet" fail to attach to the axon, instead turning upward, away from the axon. The result is a faulty node of Ranvier, with ion channels and adhesion molecules in disarray. Without the proper structure at the node of Ranvier, rapid nerve conduction is disrupted. A similar phenotype is shared by mice lacking the enzyme that adds the sulfate group to GalCer to make sulfatide. The conclusion is that sulfatide is essential for myelin-axon interactions, and its absence results in severe neurological deficits.

As mentioned earlier, development is blocked before birth in mice engineered to lack GlcCer. This may be due to a need for GlcCer itself or for any of the many glycosphingolipids biosynthetically downstream from GlcCer, making conclusions about GlcCer function difficult to elucidate from mutant embryonic mice. However, one key function of GlcCer has been learned from studies on its catabolism in postnatal animals. Ceramide is a key component of the outer layer of the skin and is responsible for the epidermal permeability barrier—a key defense against dehydration. Infants with severe Gaucher's disease, in which β-glucocerebrosidase activity is nearly absent and GlcCer is not catabolized, are prone to dehydration due to high skin permeability. The relationship among GlcCer, ceramide, and skin permeability was confirmed in mice engineered with the same mutation in β-glucocerebrosidase as found in these infants. The mice unable to catabolize GlcCer died within days of birth by dehydration through the skin. This established the role of GlcCer as the obligate precursor to the ceramide required to build the outermost protective layer (stratum corneum) of the skin. GlcCer is synthesized, transported to the stratum corneum, and then enzymatically hydrolyzed, resulting in ceramide deposition.

More complex glycosphingolipids function both in cell–cell recognition and in the regulation of signal transduction. As with sulfatide, these functions are sometimes subtle, as exemplified by the effects of blocking ganglioside biosynthesis on nervous system physiology. Given the complexity of complex ganglioside biosynthesis (Figure 10.2), it was surprising to discover that major alterations in ganglioside expression resulted in only modest phenotypic changes in mice. When the N-acetylgalactosaminyltransferase responsible for ganglioside elongation (GM2/GD2 synthase) was inactivated in mice, none of the major complex gangliosides (GM1, GD1a, GD1b, or GT1b) was expressed. Instead, a comparable concentration of the simple gangliosides GM3 and GD3 were the major gangliosides found in the adult brain. Nevertheless, the resulting mice were grossly normal. As they aged, however, mice without the normal spectrum of brain gangliosides displayed signs of axon degeneration and demyelination, hallmarks of a problem in myelin-axon cell–cell communication. By the time these mice were 1 year old, they were severely impaired, dragging their hindlimbs and walking in short labored movements. These deficits may arise from altered interactions of GM2/GD2 with a well-characterized protein on the myelin membrane, myelin-associated glycoprotein (MAG), a member of the Siglec family of sialic-acid-dependent carbohydrate-binding proteins (see Chapter 32). MAG is expressed on the innermost myelin wrap, directly across from the axon surface. Mice engineered to lack MAG have some of the same phenotypic changes as mice lacking GM2/GD2 synthase, and biochemical and cell biological studies demonstrated that the major brain gangliosides GD1a and GT1b are excellent ligands for MAG. The results support the conclusion that MAG on the innermost myelin membrane binds to GD1a and GT1b on the axon cell surface to stabilize myelin-axon interactions. Genetic disruption of MAG or its target gangliosides results in similar long-term destabilization of axons and myelin.

A second *trans* recognition role for glycosphingolipids may be in the interaction of leukocytes with the blood vessel wall during the process of inflammation, the body's protection against bacterial infection. As discussed in Chapter 31, the first step in inflammation is the binding of white blood cells (leukocytes) to the endothelial cells lining the blood vessel near sites of infection (activated endothelium). This cell–cell interaction is initiated when glycan-binding proteins of the selectin family, expressed on the activated endothelium, bind to complementary glycans on the surface of passing leukocytes. One of the selectins, E-selectin, binds to as-yet-undetermined targets on human leukocytes. The receptor(s) are resistant to protease treatment, indicating that they may be glycosphingolipids. A candidate class of glycosphingolipids, myeloglycans, has been identified in leukocytes. The candidate glycosphingolipids have long sugar chains consisting of a neutral core with Galβ1-4GlcNAcβ1-3 repeats, substituted with a terminal sialic acid and fucose residues on one or more of the *N*-acetylglucosamine residues. Although the data are intriguing, the role of these minor glycosphingolipids in inflammation has yet to be established.

NKT cells, which carry both T- and NK-cell receptors, are involved in the suppression of autoimmune reactions, cancer metastasis, and the graft rejection response. The MHC class I molecule (CD1d) of dendritic cells presents glycolipid antigens via T-cell receptor recognition to activate NKT cells. The activation of NKT cells was originally demonstrated using Galα-ceramide, the isoglobo-series glycolipid iGb_3Cer (Galα1-3Galβ1-4GlcβCer), which has been proposed as an endogenous NKT activator.

In addition to their action as *trans* recognition molecules, glycosphingolipids also interact laterally with proteins in the same membrane to modulate their activities. Notable among these *cis* regulatory interactions are those between gangliosides and members of the receptor tyrosine-kinase family. Gangliosides regulate the activity of the EGF receptor, platelet-derived growth factor receptor, fibroblast growth factor receptor, TrkA neutrotrophin receptor, and insulin receptor. Ganglioside GM3, for example, down-regulates the response of the insulin receptor to insulin. Mice engineered to lack the enzyme responsible for the biosynthesis of GM3 (CMP-NeuAc:lactosylceramide sialyltransferase) lack all GM3. They do synthesize gangliosides, but with a Gg_4Cer core (Table 10.1) bearing sialic acids on the terminal galactose and/or *N*-acetylgalactosamine residues, but not on the internal galactose. The resulting mice appear grossly normal, but display a change in insulin sensitivity. These mice, which do not express ganglioside GM3, have increased insulin receptor phosphorylation, enhanced glucose tolerance, enhanced insulin sensitivity, and are less susceptible to induced insulin resistance. These data implicate GM3 in the regulation of insulin responsiveness and support other data demonstrating the modulation of various receptors by gangliosides residing in the same membrane (*cis* regulation).

Glycosphingolipids in Human Pathology

Glycosphingolipid storage diseases are rare genetic disorders that lead to the accumulation of glycosphingolipids in lysosomes. They typically result from mutations in glycosidases, and less frequently, to mutations in activator proteins (see Chapter 41). The symptoms depend on the tissues in which the unhydrolyzed glycosphingolipid accumulates and on the extent of loss of enzyme activity. The most common glycosphingolipid storage disease is Gaucher's disease, which is caused by mutations in the enzyme β-glucocerebrosidase, resulting in the accumulation of GlcCer in the liver and spleen (and other tissues in more severe cases). Enzyme replacement therapy has been successful in treating Gaucher's disease, and drugs to block GlcCer synthesis ("substrate reduction therapy") are in clinical use (see Chapter 50). In enzyme replacement therapy, Gaucher's patients used to receive β-glucocerebrosidase purified from human placenta and modified by sialidase, β-galactosidase, and β-hexosaminidase treatment to expose the Manβ1-6(Manβ1-3)Manα core structure of N-glycans (see Chapter 8) attached to the enzyme. Such modified enzyme molecules are efficiently taken up by the same phagocytotic cells in which GlcCer accumulates, via their mannose-binding lectin. Recombinant β-glucocerebrosidase made in the Lec1 CHO mutant and thus carrying only oligomannose N-glycans (see Chapters 8 and 46) is the latest treatment for Gaucher's disease. Another example, Tay-Sachs disease, is caused by mutations in a hexosaminidase and results in the buildup of GM2, culminating in irreversible fatal deterioration of brain function. Glycosphingolipid storage and related diseases are considered more extensively in Chapter 41.

Antiglycosphingolipid antibodies are involved in certain autoimmune diseases. Although their role in disease progression has been controversial, some forms of Guillain-Barré syndrome, the most common form of paralytic disease worldwide, clearly involve autoantibodies against gangliosides. One form of Guillain-Barré syndrome occurs subsequent to infection with particular strains of the common diarrheal bacterial agent *Campylobacter jejuni* (see Chapter 34). These bacteria produce near-exact replicas of brain ganglioside glycans (such as GD1a) attached to their lipopolysaccharide cores. Following infection and immune clearance of the bacteria, the antiglycan antibodies produced to fight the bacteria go on to attack the patient's own nerves, causing paralysis. In some patients with multiple myeloma (a malignancy of antibody-producing plasma cells), the tumor cells secrete monoclonal antibodies against glycolipids, such as the rare sulfoglucuronyl epitope of nervous system glycosphingolipids termed HNK-1 (IV^3GlcA[3-sulfate]-nLc_4Cer). These patients suffer severe peripheral neuropathy.

Several bacterial toxins take advantage of glycosphingolipids to gain access to cells (see Chapter 34). Cholera toxin and the structurally related *Escherichia coli* heat-labile enterotoxins are produced in the intestinal tract of infected individuals, bind to intestinal epithelial cell surfaces, and insert their toxic polypeptide "payload" through the cell membrane, where it disrupts ion fluxes, causing severe diarrhea. The toxins behave as docking modules with gangliosides acting at the site of attachment. Five identical polypeptide B subunits in a ring each bind to ganglioside GM1 (Galβ1-3GalNAcβ1-4[NeuAcα2-3]Galβ1-4GlcβCer) on the cell surface, and a sixth A subunit (the "payload") is then inserted through the membrane. A similar mechanism is used by *Shiga* toxin (also called verotoxin), which binds to the glycolipid Gb_3Cer (globotriaosylceramide, Galα1-4Galβ1-4GlcβCer) via five subunits in a ring, each with three glycosphingolipid-binding sites. In contrast, tetanus and related botulinum toxins are multidomain single polypeptides. One domain binds b-series gangliosides on nerve cells, whereas the other domains translocate the toxin into cells and disrupt proteins essential for synaptic transmission. Custom-designed multivalent sugars are being evaluated as high-affinity blockers of certain bacterial toxins. In addition to soluble toxins, certain intact bacteria also bind to specific glycosphingolipids via bacterial surface proteins called adhesins. This adherence is essential for successful colonization and symbiosis. Microbial adhesins are addressed in more detail in Chapter 34.

Malignant transformation in cancer progression is often associated with changes in the glycan structures of glycoproteins and glycolipids. The changes result mainly from altered levels of glycosyltransferase activities involved in glycolipid biosynthesis. The increase of GD3 or GM2 in melanoma, and of sialyl-Lewis^a (NeuAcα2-3Galβ1-3[Fucα1-4]GlcNAcβ1-3Galβ1-4GlcβCer) in gastrointestinal cancers are typical examples (see Chapter 44). Certain cancers also produce and shed large amounts of gangliosides that have an immunosuppressive effect.

FURTHER READING

Hakomori S. 2002. The glycosynapse. *Proc. Natl. Acad. Sci.* **99:** 225–232.

Kolter T., Proia R.L., and Sandhoff K. 2002. Combinatorial ganglioside biosynthesis. *J. Biol. Chem.* **277:** 25859–25862.

Hakomori S. 2003. Structure, organization, and function of glycosphingolipids in membrane. *Curr. Opin. Hematol.* **10:** 16–24.

Proia R.L. 2003. Glycosphingolipid functions: Insights from engineered mouse models. *Philos. Trans. R. Soc. Lond. B Biol. Sci.* **358:** 879–883.

Degroote S., Wolthoorn J., and van Meer G. 2004. The cell biology of glycosphingolipids. *Semin. Cell Dev. Biol.* **15:** 375–387.

Furukawa K., Tokuda N., Okuda T., Tajima O., and Furukawa K. 2004. Glycosphingolipids in engineered mice: Insights into function. *Semin. Cell Dev. Biol.* **15:** 389–396.

Schnaar R.L. 2004. Glycolipid-mediated cell–cell recognition in inflammation and nerve regeneration. *Arch. Biochem. Biophys.* **426**: 163–172.

Hancock J.F. 2006. Lipid rafts: Contentious only from simplistic standpoints. *Nature Rev. Mol. Cell Biol.* **7:** 456–462.

Sonnino S., Mauri L, Chigorno V., and Prinetti A. 2007. Gangliosides as components of lipid membrane domains. *Glycobiology* **17:** 1R–13R.

CHAPTER 11

Glycosylphosphatidylinositol Anchors

Michael A.J. Ferguson, Taroh Kinoshita, and Gerald W. Hart

PLASMA MEMBRANE PROTEINS ARE EITHER PERIPHERAL PROTEINS or integral membrane proteins. The latter include proteins that span the lipid bilayer once or several times and those that are covalently attached to lipids. Proteins attached to glycosylphosphatidylinositol (GPI) via their carboxyl termini are found in the outer leaflet of the lipid bilayer and face the extracellular environment. The GPI membrane anchor may be conveniently thought of as an alternative to the single transmembrane domain of type-1 integral membrane proteins. This chapter reviews the discovery, distribution, structure, biosynthesis, properties, and suggested functions of GPI membrane anchors and related molecules.

BACKGROUND AND DISCOVERY

The first data suggesting the existence of protein-phospholipid anchors appeared in 1963 with the finding that crude bacterial phospholipase C (PLC) selectively releases alkaline phosphatase from mammalian cells. Phosphatidylinositol-protein anchors were first postulated in the mid-1970s based on the ability of highly purified bacterial phosphatidylinositol-specific PLC enzymes to release certain proteins, such as alkaline phosphatase and 5′-nucleotidase, from mammalian plasma membranes. By 1985, these predictions were confirmed by compositional and structural data from studies on *Torpedo* acetylcholinesterase, human and bovine erythrocyte acetylcholinesterase, rat Thy-1, and the sleeping sickness parasite *Trypanosoma brucei* variant surface glycoprotein (VSG). The first complete GPI structures, which were those for *T. brucei* VSG and rat Thy-1, were solved in 1988 (see Chapter 1, Figure 1.3).

DIVERSITY OF PROTEINS WITH GPI ANCHORS

To date, hundreds of GPI-anchored proteins have been identified in many eukaryotes, ranging from protozoa and fungi to humans (Table 11.1). The range of GPI-anchored

TABLE 11.1. *Examples of GPI-anchored proteins*

GPI-anchored Protein	Function
Protozoa	
Trypanosoma brucei variant surface glycoprotein (VSG)	protective coat
Leishmania major promastigote surface protease (PSP)	bound complement degradation
Trypanosoma cruzi GPI-anchored mucins	host cell invasion
Plasmodium falciparum merozoite surface protein 1 (MSP-1)	erythrocyte invasion
Toxoplasma gondii surface antigen 1 (SAG-1)	host cell invasion
Entamoeba histolytica GPI proteophosphoglycans	virulence factor
Yeast, fungi, slime mold	
Saccharomyces cerevisiae α-agglutinin	adhesion molecule
Saccharomyces cerevisiae GAS1p	cell-wall biogenesis
Aspergillus fumigatus GEL1p	cell-wall biogenesis
Candida albicans HWP1	adhesion molecule
Dictyostelium discoideum prespore antigen (PsA)	adhesion molecule
Dictyostelium discoideum contact-site A (CsA)	adhesion molecule
Plants	
Pyrus arabinogalactan proteins (AGP)	cell-wall biogenesis
Arabidopsis thaliana metallo and aspartyl proteases	pollen tube development
Arabidopsis thaliana β1-3glucanase	cell-wall biogenesis
Animals (nonmammalian)	
Schistosoma mansoni gp200	praziquantel drug target
Caenorhabditis elegans odorant response abnormal 2 (ODR-2)	olfaction
Torpedo marmorata acetylcholinesterase (AchE)	hydrolase
Mammals	
Erythrocyte CD59 and decay acceleration factor (DAF)	complement regulation
Alkaline phosphatase	cell-surface hydrolase
5′-Nucleotidase	cell-surface hydrolase
Renal dipeptidase	cell-surface hydrolase
Trehalase	cell-surface hydrolase
[a]Neural cell adhesion molecule 120 (NCAM-120)	adhesion molecule
Neural cell adhesion molecule TAG-1	adhesion molecule
[a]CD58	adhesion molecule
[a]FcγIII receptor	Fc receptor
Ciliary neurotrophic factor receptor (CNTFR) α subunit	neural receptor
Glial-cell-derived neurotrophic factor receptor (GDNFR) α subunit	neural receptor
CD14	LPS receptor
Prion protein (PrP)	unknown
Glypican family of GPI-anchored proteoglycans	extracellular matrix component

[a]Examples for which differential mRNA splicing can result in transmembrane and GPI-anchored forms of the same protein.

proteins suggests that (1) GPI anchors are ubiquitous among eukaryotes and are particularly abundant in protozoa; (2) GPI-anchored proteins are functionally diverse and include hydrolytic enzymes, adhesion molecules, complement regulatory proteins, receptors, protozoan coat proteins, and prion proteins; and (3) in mammals, alternative mRNA splicing may lead to the expression of transmembrane and/or soluble and GPI-anchored forms of the same gene product. These variants may be developmentally regulated. For example, neural cell adhesion molecule (NCAM) exists in GPI-anchored and soluble forms when expressed in muscle and in GPI-anchored and transmembrane forms when expressed in brain.

STRUCTURE OF GPI ANCHORS

Virtually all protein-linked GPI anchors share a common core structure (Figure 11.1). The structural arrangements of GPI anchors are unique among protein-carbohydrate associations in that the reducing terminus of the GPI oligosaccharide is not attached to the protein. Rather, the reducing terminal glucosamine residue is in α1-6 glycosidic linkage to the D-*myo*-inositol head group of a phosphatidylinositol (PI) moiety. A distal, nonreducing mannose residue is attached to the protein via an ethanolamine phosphate (EtNP) bridge between the C-6 hydroxyl group of mannose and the α-carboxyl group of the carboxy-terminal amino acid. GPIs are one of the rare instances in nature where glucosamine is found without either an acetyl group (as in most glycoconjugates) or a sulfate moiety (as in proteoglycans) modifying the amine group at the C-2 (see Chapter 16). The substructure Manα1-4GlcNα1-6*myo*-inositol-1-*P*-lipid is a universal hallmark of GPI anchors and related structures.

The structures of GPI anchors are quite diverse, depending on both the protein to which they are attached and the organism in which they are synthesized (Figure 11.2). With one known exception, protein-linked GPI anchor cores contain a minimum of three mannosyl residues in the sequence EtNP-6Manαl-2Manα1-6Manα1-4GlcNα1-6(PI) (Figure 11.1). Superimposed on this highly conserved core are additional EtNP substituents and a wide variety of linear and branched glycosyl substituents (Figure 11.2). The functional significance of these substituents are largely unknown.

There is considerable variation in the PI moiety. Indeed, GPI is a rather loose term because, strictly speaking, PI refers specifically to D-*myo*-inositol-1-*P*-3(*sn*-1,2-diacylglycerol) (i.e., diacyl-PI), whereas many GPIs contain other types of inositolphospholipids, specifically, *lyso*-acyl-PI (e.g., *Trypanosoma cruzi* Tc85), alkylacyl-PI (e.g., *Leishmania* promastigote surface protease [PSP]), alkenylacyl-PI (e.g., bovine eAChE), and inositolphosphoceramide (e.g., *Dictyostelium* prespore antigen [PsA]) (see Figure 11.2). Another variation, termed inositol acylation, is characterized by the presence of an ester-linked fatty acid attached to the C-2 hydroxyl of the inositol residue (e.g., human eAChE). The presence of this modification makes the anchor inherently resistant to the action of bacterial PI-specific PLC. The available lipid structural data suggest that (1) inositolphosphoceramide-based protein-linked GPIs are only found in lower eukaryotes, such as *Saccharomyces cerevisiae, Aspergillus niger, Dictyostelium discoideum,* and *T. cruzi*; (2) the lipid structures of GPIs generally do not reflect those of the general cellular PI or inositolphosphoceramide pool; and (3) the lipid structures of some (e.g., trypanosome) GPI-anchored proteins are under developmental control.

The factors that control the synthesis of a mature GPI anchor found on a given protein appear to be similar to those for other posttranslational modifications such as N- and O-glycosylation. Thus, primary control is at the cellular level, whereby the levels of specific

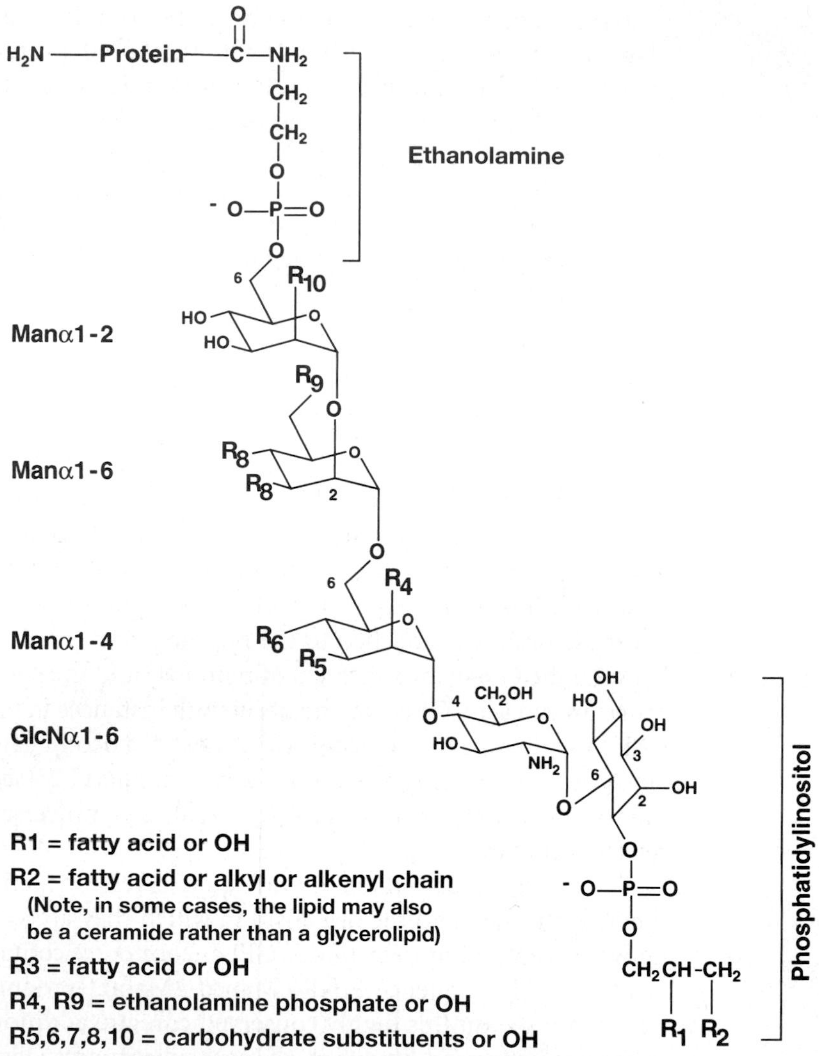

FIGURE 11.1. General structure of GPI anchors. All characterized GPI anchors share a common core consisting of ethanolamine-PO_4-6Manα1-2Manα1-6Manα1-4GlcNα1-6*myo*-inositol-1-PO_4-lipid. Heterogeneity in GPI anchors is derived from various substitutions of this core structure and is represented as R groups. R1 may be a long-chain fatty acyl chain or OH and R2 may be a long-chain fatty acyl, alkyl, or alkenyl chain. In some cases the lipid is a ceramide rather than a glycerolipid. R3 is is most often a palmitate group attached to C-2 of the inositol ring and, when present, renders anchors resistant to PI-PLC. R4 and R9 can be OH or additional ethanolamine phosphate groups that are not attached to protein. R5, R6, R7, R8 and R10 can be OH or monosaccharide or oligosaccharide side-chain attachment points (see Figure 11.2). (Adapted, with permission, from Cole R.N. and Hart G.W. 1997. In *Glycoproteins II* [ed. J. Montreuil et al.], pp. 69–88. Elsevier, Amsterdam, © Elsevier.)

biosynthetic and processing enzymes dictate the final repertoire of structures. Secondary control is at the level of the tertiary/quaternary structure of the protein bearing the GPI anchor, which affects accessibility to processing enzymes. Examples of primary control include (1) differences in GPI glycan side chains in human versus bovine membrane dipeptidase and brain versus thymocyte rat Thy-1 and (2) differences in carbohydrate side chains and lipid structure when *T. brucei* VSG is expressed in bloodstream and insect life-cycle stages of the parasite. An example of secondary control is the difference in VSG glycan side chains when VSGs with different carboxy-terminal sequences are expressed in the same trypanosome.

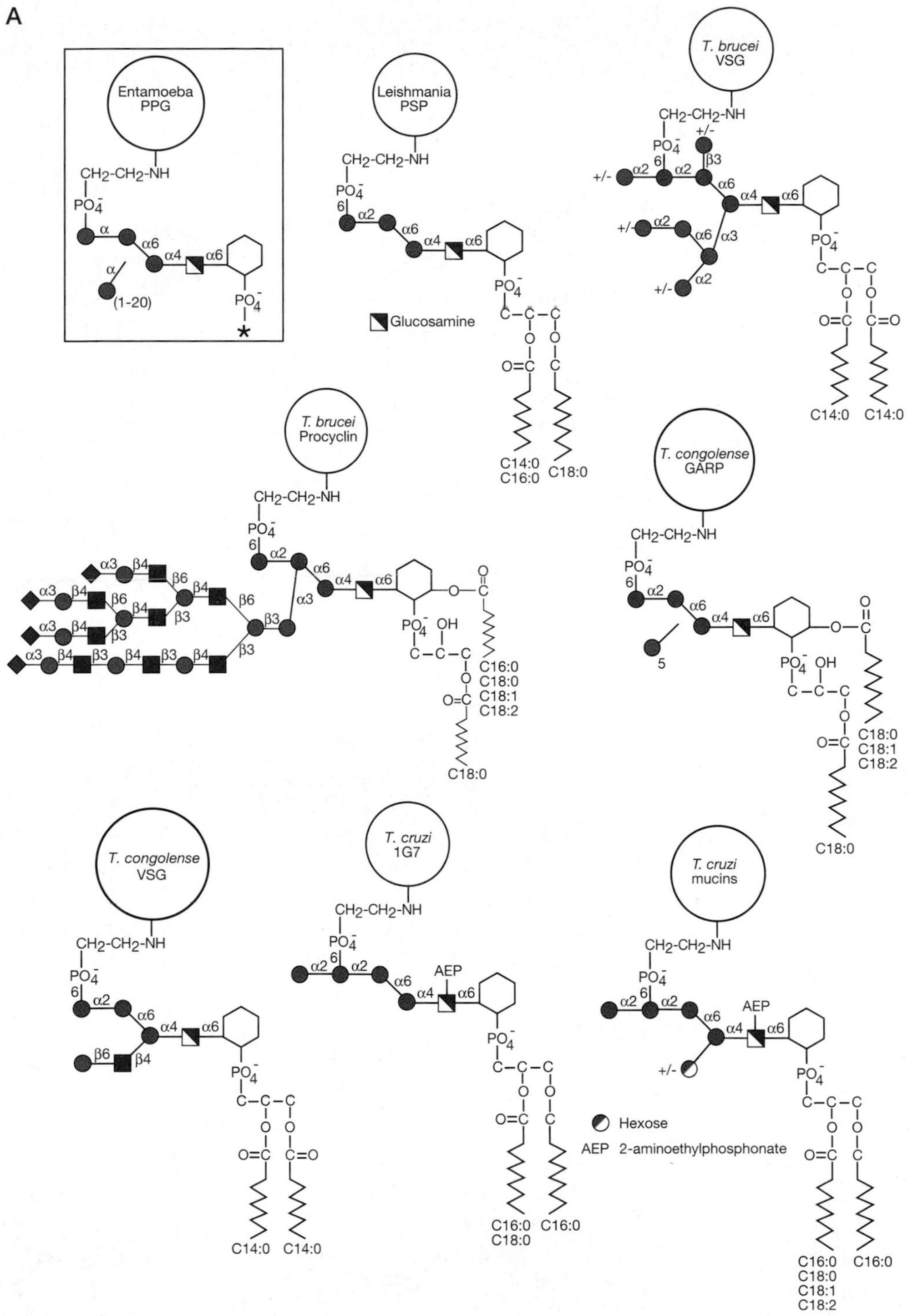

FIGURE 11.2A. *(See following page for part B.)*

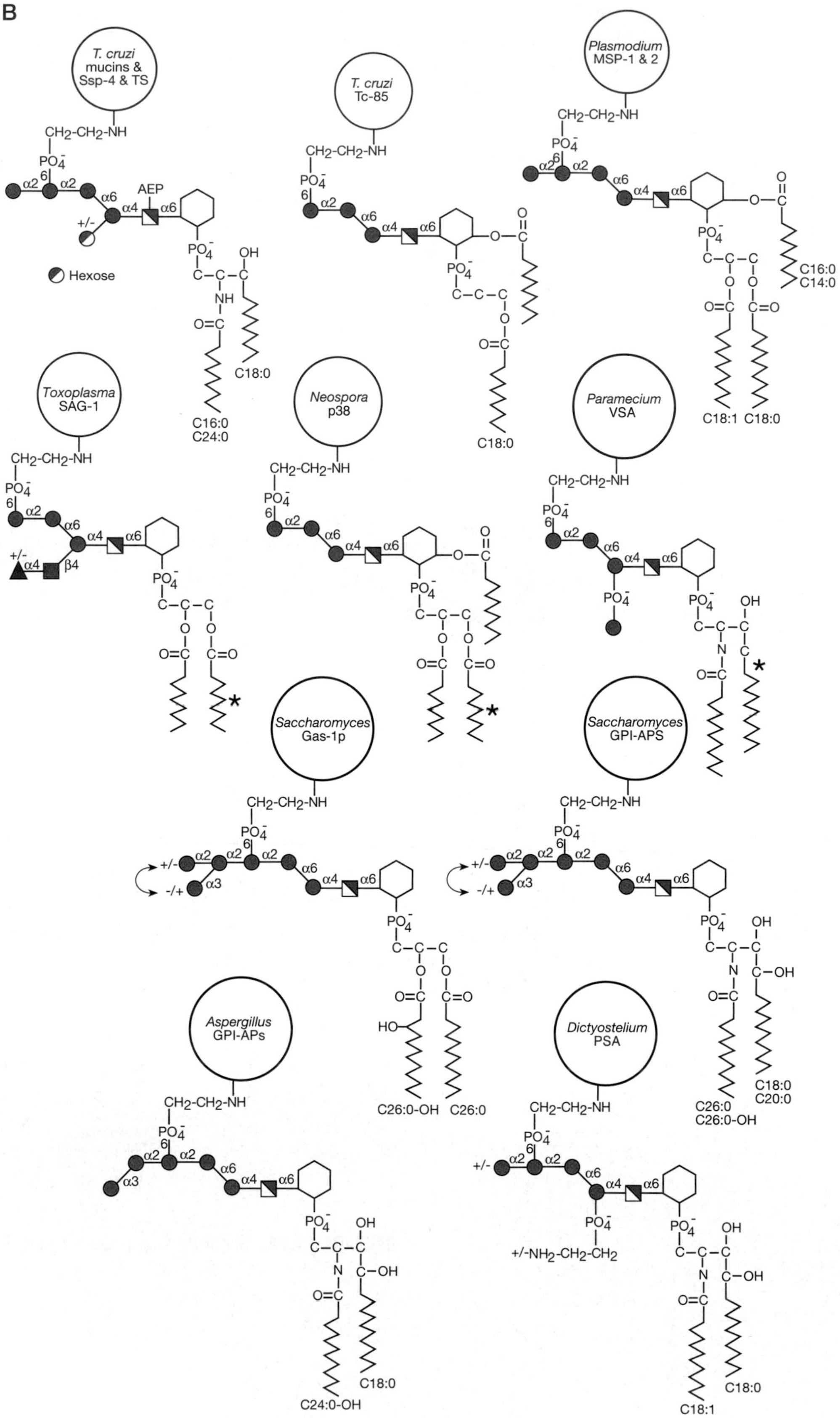

FIGURE 11.2B. *(See facing page for part C.)*

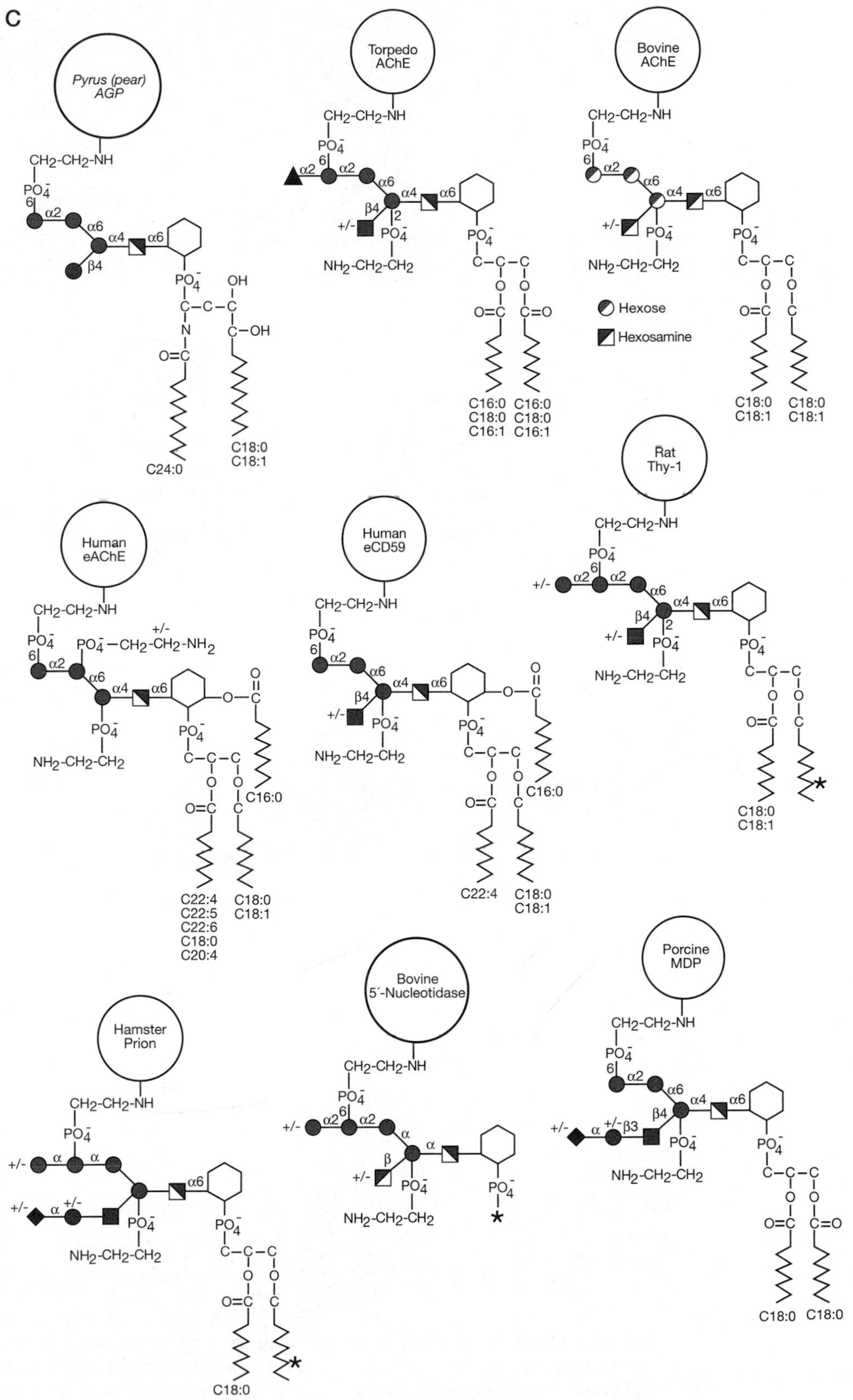

FIGURE 11.2C. *(See following page for part D and legend.)*

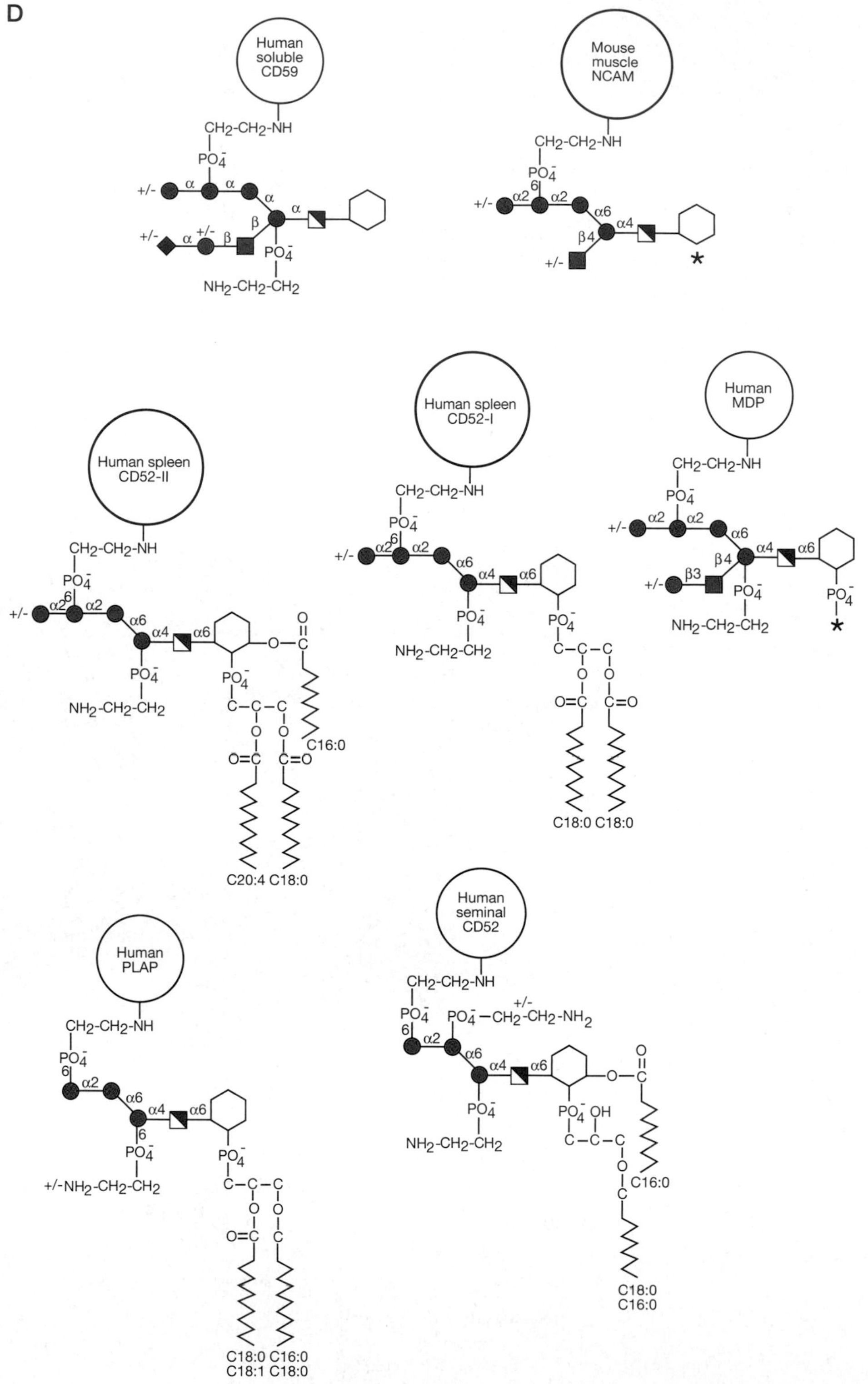

FIGURE 11.2D. Complex and varied structures of GPI anchors. With the exception of *Entamoeba* proteophosphoglycan (PPG) (*A, box*), all known protein-linked GPI anchors have the same minimal core structure embellished with species- and tissue-specific side-chain and lipid variations. (±) Glycosyl heterogeneity; (*curved arrows* for the *Saccharomyces* examples) the two terminal αMan residues are mutually exclusive; (*) unknown lipid type or chain length. Chain lengths and degrees of unsaturation of the lipids are indicated. (For the monosaccharide symbol code, see Figure 1.5, which is also reproduced on the inside front cover.)

Non-Protein-Linked GPI Structures

In mammalian cells, some free GPIs (GPI-anchor biosynthetic intermediates) are found at the cell surface, but their functional significance is unknown. On the other hand, several protozoa (particularly trypanosomatids) express high numbers (>10^7 copies per cell) of free GPIs on their cell surface as metabolic end products. These include the so-called glycoinositol phospholipids (GIPLs) and lipophosphoglycans (LPGs) of the *Leishmania*. Some protozoan (type-1) GIPLs conform to the Manα1-6Manα1-4GlcNα1-6PI sequence common to protein-linked GPIs, whereas others contain a (type-2) Manα1-3Manα1-4GlcNα1-6PI motif, and still others are hybrid structures containing the branched motif (Manα1-6)Manα1-3Manα1-4GlcNα1-6PI.

THE CHEMISTRY OF GPI ANCHORS

GPI anchors are complex molecules that include amide, glycosidic, phosphodiester, and hydroxyester linkages between their various components. The challenge of their total organic synthesis was first met by groups from several countries, including Japan, the United States, Germany, and the United Kingdom. Recently, a solid-phase synthesis of a GPI (without lipid) has been described. The synthesis of many analogs of the GlcN-PI substructure has been instrumental in probing the comparative enzymology of GPI biosynthesis in lower and higher eukaryotes (see below).

The GPI-anchor structure lends itself to selective cleavage by several chemical and enzyme reagents. These were originally used to help determine GPI structures and are now applied to confirm the presence of a GPI anchor and/or obtain partial structural information from native or [^{3}H]mannose-, [^{3}H]glucosamine-, [^{3}H]inositol-, or [^{3}H]fatty-acid-radiolabeled GPI-anchored proteins or GPI biosynthetic intermediates. Some of these reactions and their applications are illustrated in Figure 11.3. A key reaction, from an analytical perspective, is nitrous acid deamination of the glucosamine residue. This gentle (room temperature, pH 4.0) reaction is dependent on the free amino group of the glucosamine residue and gives a highly selective cleavage of the glucosamine-inositol glycosidic bond. The reaction liberates the PI moiety, which can be isolated by solvent partition and analyzed by mass spectrometry, and generates a free reducing terminus on the GPI glycan in the form of 2,5-anhydromannose. This reducing sugar can be reduced to [1-^{3}H]2,5-anhydromannitol (AHM) by sodium borotritide reduction, thus introducing a radiolabel, or it may be attached to a fluorophore such as 2-aminobenzamide (2-AB) by reductive amination. Once the GPI glycan is radioactively or fluorescently labeled and dephosphorylated with aqueous hydrogen fluoride, the glycan can often be conveniently sequenced using exoglycosidases (Figure 11.3).

GPI BIOSYNTHESIS AND TRAFFICKING

The biosynthesis of GPI anchors occurs in three stages: (1) preassembly of a GPI precursor in the ER membrane, (2) attachment of the GPI to newly synthesized protein in the lumen of the ER with concomitant cleavage of a carboxy-terminal GPI-addition signal peptide, and (3) lipid remodeling and/or carbohydrate side-chain modifications in the ER and the Golgi.

Analysis of GPI precursor biosynthesis was first made possible by the development of a cell-free system in *T. brucei*. Each trypanosome has 1 x 10^7 molecules of GPI-linked VSG on its surface. Therefore, enzymes and intermediates in the GPI-biosynthetic pathway are relatively abundant in microsomal membrane preparations produced from this organism. The sequence of events underlying GPI biosynthesis has been studied in *T. brucei*, *T. cruzi*, *Toxoplasma gondii*, *Plasmodium falciparum*, *Leishmania major*, *Paramecium*, *S. cerevisiae*,

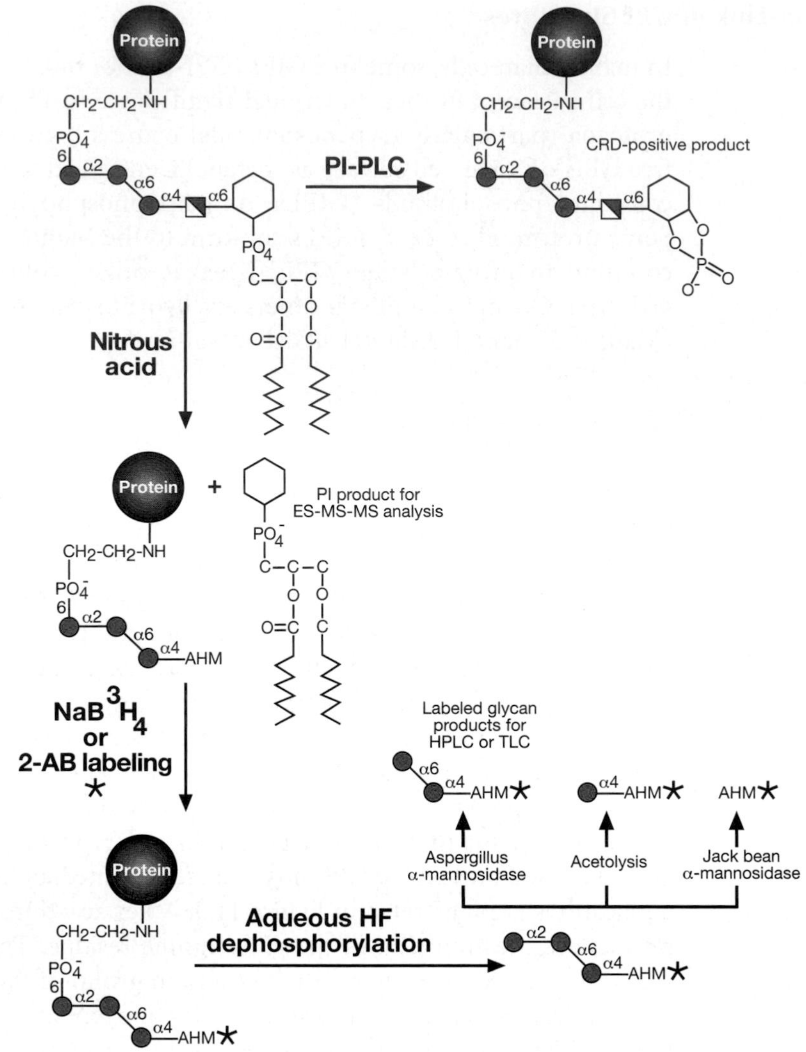

FIGURE 11.3. Chemical and enzymatic reactions of GPI anchors. Nitrous acid deamination of the GlcN residue generates the sugar 2,5-anhydromannose (AHM) that can be radiolabeled by sodium borotritide reduction or fluorescently labeled by reductive amination with 2-aminobenzamide (2-AB). Following dephosphorylation with cold aqueous hydrogen fluoride (HF), the labeled glycans can be conveniently sequenced with exoglycosidases. (CRD) Cross-reacting determinant.

Cryptococcus neoformans, and mammalian cells. The emphasis on eukaryotic microbes reflects the adundance of GPI-anchored proteins in these organisms and the potential of GPI inhibition for chemotherapeutic intervention. This notion has been genetically validated in the bloodstream form of *T. brucei*, in yeast, and in *Candida albicans.*

The essential events in GPI precursor biosynthesis are, like the core structure, highly conserved. There are, however, variations on the theme, and *T. brucei* and mammalian cell GPI pathways are used here to represent these differences (Figure 11.4). In all cases, GPI biosynthesis involves the transfer of *N*-acetylglucosamine from UDP-GlcNAc to phosphatidylinositol PI, to give GlcNAc-PI via an ER-membrane-bound multiprotein complex (Table 11.2, Figure 11.5). This step occurs on the cytoplasmic face of the ER, as does the second step of the pathway, the de-N-acetylation of GlcNAc-PI to GlcN-PI. Notable differences between the *T. brucei* and mammalian GPI-biosynthetic pathways occur from GlcN-PI onward, including (1) the timing and reversibility of inositol acylation, (2) substrate channeling

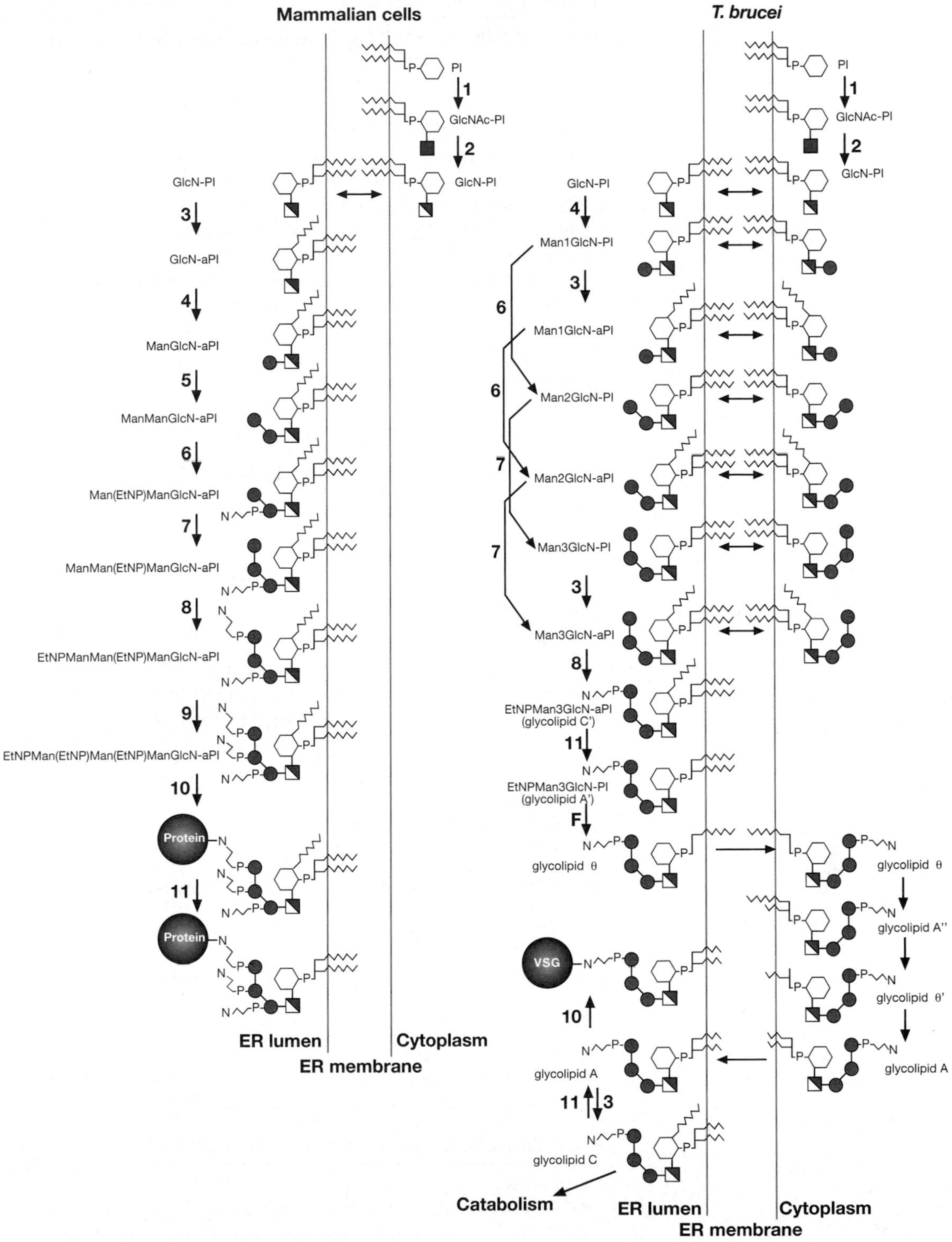

FIGURE 11.4. GPI-biosynthetic pathways of mammalian cells and *Trypanosoma brucei*. These examples show that, despite the highly conserved core structure of GPI anchors, some diversity in GPI-anchor biosynthesis exists. In particular, mammalian GPI intermediates are inositol acylated at the level of GlcN-PI and are not deacylated until after transfer of the mature GPI to protein, whereas *T. brucei* GPI intermediates are acylated at the level of Man1GlcN-PI and undergo rounds of deacylation and reacylation throughout the pathway. The protein components of the pathway that catalyze Steps 1–11 in mammalian cells are described in Table 11.2 and Figure 11.5.

TABLE 11.2 *Components of the mammalian GPI-biosynthetic machinery*

Step 1. UDP-GlcNAc:PI α1-6 *N*-acetylglucosaminyltransferase (GlcNAc-T)

Components

PIG-A (GPI3), 484 AA, 1 TM, catalytic subunit
PIG-H (GPI15), 188 AA, 2 TM, associated with PIG-A
PIG-C (GPI2), 292 AA, 8 TM
PIG-Q (GPI1), 581 AA, 6 TM, associated with PIG-C
PIG-P (GPI19), 134 AA, 2 TM
DPM-2 (n/a), 84 AA, 2 TM, subunit of Dol-P-Man synthase
PIG-Y (ERI1), 71 AA, 2 TM, associated with PIG-A
Notes: PIG-A/PIG-H complex is associated with the PIG-C/PIG-Q complex. Active site is on cytoplasmic face of the ER.

Step 2. GlcNAc-PI de-N-acetylase

Components

PIG-L (GPI12), 252 AA, 1 TM
Note: Active site of PIG-L is on cytoplasmic face of ER.

Step 3. Inositol acyltransferase (inositol acyl-T)

Components

PIG-W (GWT1), 504 AA, 13 TM
Note: Active site of PIG-W is on the lumenal face of ER.

Step 4. Mannosyltransferase-1 (MT-1)

Components

PIG-M (GPI14), 423 AA, 8 TM, catalytic subunit
PIG-X (PBN1), 252 AA, 1 TM, associated with PIG-M
Note: Active site of PIG-M is on the lumenal face of ER.

Step 5. Mannosyltransferase-2 (MT-2)

Components

PIG-V (GPI18), 493 AA, 8 TM
Note: Active site of PIG-V is on the lumenal face of ER.

Step 6. Ethanolamine phosphate transferase-1 (EtNPT-1)

Components

PIG-N (MCD4), 931 AA, 15 TM
Note: Active site of PIG-N is on the lumenal face of ER.

Step 7. Mannosyltransferase-3 (MT-3)

Components

PIG-B (GPI10), 554 AA, 9 TM
Note: Active site of PIG-B is on the lumenal face of ER.

Step 8. Ethanolamine phosphate transferase-2 (EtNP-T2)

Components

PIG-O (GPI13), 880 AA, 10 TM, catalytic subunit
PIG-F (GPI11), 219 AA, 6 TM, associated with PIG-O
Note: Active site of PIG-O is on the lumenal face of ER.

Step 9. Ethanolamine phosphate transferase-3 (EtNP-T3)

Components

GPI7 (PIG-G), 983 AA, multiple TM, catalytic subunit
PIG-F (GPI11), 219 AA, 6 TM, associated with GPI7
Note: Active site of GPI7 is on the lumenal face of ER.

TABLE 11.2 *(Continued)*

Step 10. GPI transamidase

Components

GAA1 (GAA1), 621 AA, 7 TM
PIG-K (GPI8), 367 AA, 1 TM, catalytic subunit
PIG-T (GPI16), 557 AA, 1 TM, disulphide bonded to PIG-K
PIG-S (GPI17), 555 AA, 2 TM
PIG-U (GAB1), 435 AA, 8 TM
Note: Active site of PIG-K is on the lumenal face of ER.

Step 11. Inositol deacylase

Components

PGAP1 (BST1) 922 AA, 6 TM
Note: Active site of PGAP-1 is on the lumenal face of ER.

The step numbers refer to Figure 11.4. The names in brackets are the yeast homologs. The size of the predicted protein in amino acids (AA) and the number of predicted transmembrane (TM) domains are given.

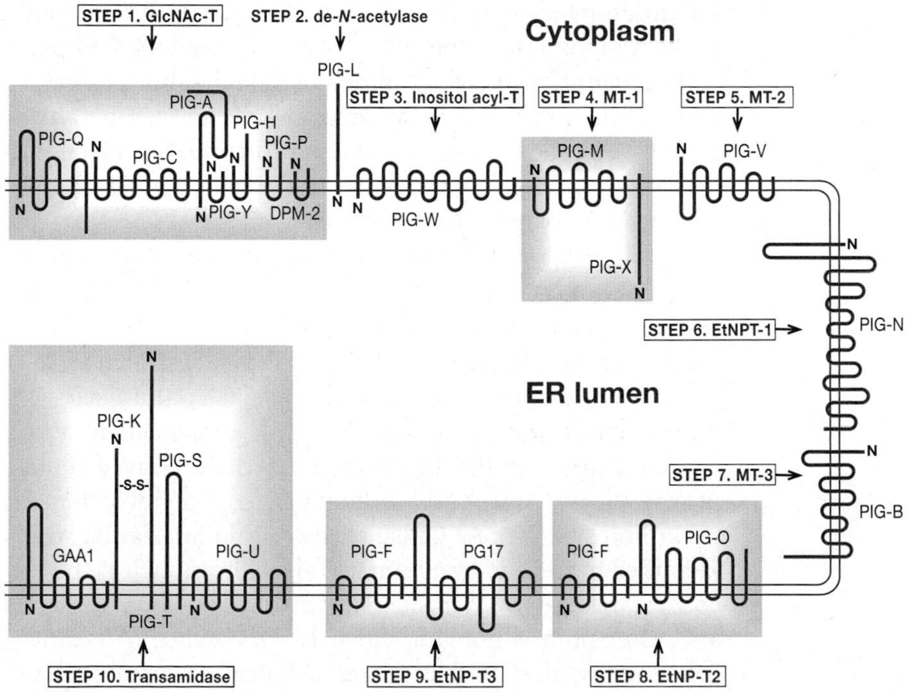

FIGURE 11.5. Predicted topologies of the components of GPI biosynthesis in mammalian cells. Components within boxes belong to multisubunit complexes. The step numbers refer to those in Figure 11.4 and Table 11.2. The topologies of the "multispanning" membrane proteins are mostly those predicted by bioinformatics methods. (N) Amino terminus.

between de-N-acetylase and the first mannosyltransferase (MT-1) in *T. brucei*, (3) the addition of ethanolamine phosphate groups to all three mannose residues in mammalian cells, and (4) fatty-acid remodeling of *T. brucei* GPI anchors before attachment to proteins.

Inositol acylation (the transfer of fatty acid to the C-2 hydroxyl group of the D-*myo*-inositol residue) of GlcN-PI strictly follows the action of the first mannosyltransferase in *T. brucei*, whereas these steps are temporally reversed in mammalian cells. This difference was exploited in the discovery of the first generation of specific substrates and inhibitors of the *T. brucei* GPI-biosynthetic pathway in vitro. In the mammalian pathway, inositol acylation and inositol deacylation are discrete steps that occur only at the beginning and end of the pathway, respectively, whereas in *T. brucei* these reactions occur on multiple GPI intermediates. Inositol acylation is inhibited by phenylmethylsulfonylfluoride in *T. brucei*, but not in mammalian cells. Furthermore, in some mammalian cells such as human erythroblasts, the inositol-linked fatty acid is never removed and the mature GPI protein retains three fatty chains (see Figure 11.2). Fatty-acid remodeling in *T. brucei* occurs at the end of the pathway, but before transfer to VSG protein, and involves exchanging the *sn*-2 fatty acids (a mixture of C18–C22 species) and the *sn*-1 fatty acid (C18:0) exclusively for C14:0 myristate. In contrast, lipid remodeling in mammalian cells is more complex. Many protein-linked GPIs contain *sn*-1-alkyl-2-acyl-PI with two saturated fatty chains, whereas major cellular PI is predominantly *sn*-1-stearoyl-2-arachidonoyl-PI (i.e., with C18:0 and C20:4 fatty acids and few, if any, alkyl or alkenyl species). Two processes are involved in these structural changes. First, remodeling from the diacyl PI to the 1-alkyl-2-acyl form having unsaturated fatty acid at the *sn*2 position occurs in GlcN-aPI. The reaction that mediates this remodeling is yet to be determined. Second, fatty-acid remodeling occurs after GPI is transferred to proteins and the inositol-linked acyl chain is removed. This is accomplished by exchanging the unsaturated *sn*2 fatty acid with saturated fatty acid, mainly stearate (C18:0). Lipid remodeling of GPIs in yeast also involves two processes. The first is fatty-acid remodeling that exchanges the unsaturated *sn*2 fatty acid with C26:0 chain. The second process involves the exchange of diacylglycerol for ceramide on many, but not all, GPI proteins.

The identification of GPI pathway genes has been principally by expression cloning using GPI-deficient mutants of mammalian cells and temperature-sensitive yeast mutants. More recently, epitope tagging/pull-down/proteomic approaches have been used to identify GPI pathway-associated components. The known mammalian and yeast components and their respective topologies in the ER membrane are shown in Table 11.2 and Figure 11.5.

The transfer of the preassembled GPI precursor to protein occurs via a multisubunit transamidase complex with a cysteine-protease-like catalytic subunit. The reaction involves two complex sustrates: the GPI precursor and the carboxyl terminus of partially folded nascent protein (Figure 11.6). The carboxy-terminal GPI-addition signal peptide (GPIsp) has three domains: (1) three relatively small amino acids (Ala, Asn, Asp, Cys, Gly, or Ser) located at ω, $\omega + 1$, $\omega + 2$, where ω is the amino acid attached to the GPI anchor and where $\omega + 1$ and $\omega + 2$ are the first two residues of the cleaved peptide; (2) a relatively polar domain of typically five to ten residues; and (3) a hydrophobic domain of typically 15–20 hydrophobic amino acids. These GPIsp sequences do not have a strict consensus, but they are easily identified by eye and by automated algorithms, such as the big-PI predictor and DGPI, available via the ExPASy Web site. The final hydrophobic stretch of amino acids often resembles a transmembrane domain, but it is the absence of positively charged and polar residues immediately after it that makes a GPIsp easy to spot. Of course, like N-glycosylation sequons, a GPIsp will only be functional if the protein is translocated into the ER. Essentially all GPI-anchored proteins have an amino-terminal signal peptide as well. There are some differences between the transamidases of different species and their fine specificity for GPIsp sequences. For example, the yeast transamidase will not use cysteine as the ω amino

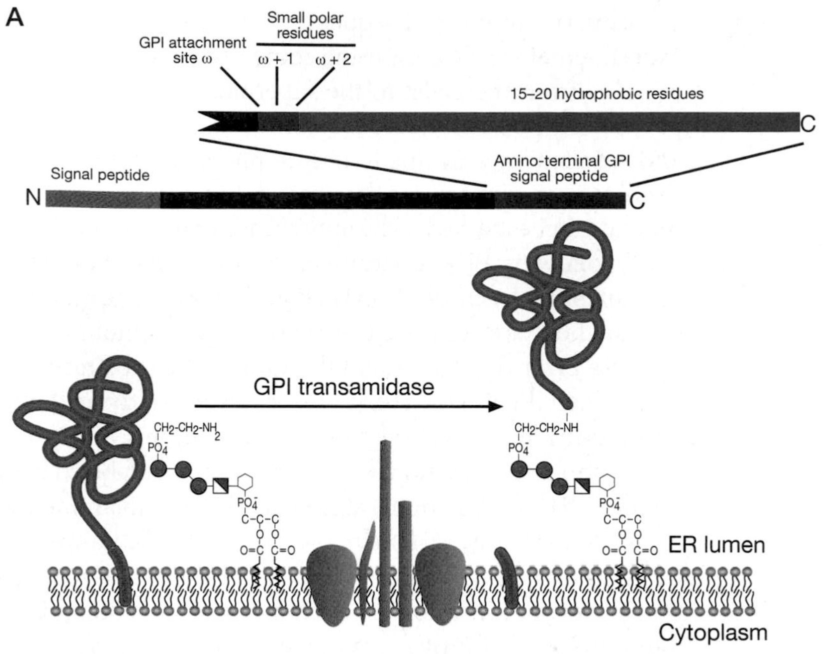

B

Protein	GPI signal sequence	
Acetylcholinesterase *(Torpedo)*	NQFLPKLLNATA**C**	DGELSSSGTSSSKGIIFYVLFSILYLIFY
Alkaline phosphatasee *(placenta)*	TACDLAPPAGTT**D**	AAHPGRSVVPALLPLLAGTLLLLETATAP
Decay accelerating factor	HETTPNKGSGTT**S**	GTTRLLSGHTCFTLTGLLGTLVTMGLLT
PARP *(T. Brucei)*	EPEPEPEPEPEP**G**	AATLKSVALPFAIAAAALVAAF
Prion protein *(hamster)*	QKESQAYYDGRR**S**	SAVLFSSPPVILLISFLIFLMVG
Thy-1 *(rat)*	KTINVIRDKLVK**C**	GGISLLVQNTSWLLLLLLSLSFLQATDFISI
Variant surface glycoprotein *(T. Brucei)*	ESNCKWENNACK**D**	SSILVTKKFALTVVSAAFVALLF

Boldfaced amino acid is the site of attachment of the GPI. Sequence to the right of the space is cleaved from the protein by the transpeptidase upon anchor addition.

FIGURE 11.6. (*A*) Features of GPI-anchored proteins and their processing by GPI transamidase. GPI-anchored proteins have an amino-terminal signal peptide and a carboxy-terminal GPI-addition signal peptide (*top*) that is removed and directly replaced by a GPI precursor (*bottom*). (*B*) Examples of carboxyl-terminal sequences signaling the addition of GPI anchors.

acid, and protozoan and metazoan GPIsp are not always functionally interchangeable.

Apart from the final glycan structures (Figure 11.2), almost nothing is known about the genes and enzymes involved in adding carbohydrate side chains to the conserved GPI core. The only exception is the mannosyltransferase (encoded by *SMP3*) that adds the fourth αMan residue to yeast and mammalian GPI anchors. This occurs during GPI precursor assembly, whereas the other modifications (e.g., the addition of galactose to *T. brucei* VSG) occur after transfer of the GPI to protein. Interestingly, some of these steps occur in the ER and others in the Golgi.

IDENTIFICATION OF GPI-ANCHORED PROTEINS

In the postgenome era, GPI anchoring is most often inferred by the identification of an amino-terminal signal peptide and a carboxy-terminal GPI signal peptide from the predicted primary amino acid sequence of a given gene (Figure 11.6). Such predictions require experimental verification using some of the reactions displayed in Figure 11.3. The shift of proteins from the pellet to the supernatant after treatment of whole cells with PI-PLC is a particularly popular method. Variations on this theme include solubilization in Triton X-114 and subsequent warming to induce phase separation. GPI-anchored proteins partition into the detergent-rich lower phase before, but not after, PI-PLC treatment. An additional criterion that can be applied is the appearance of an epitope known as the "cross-reacting determinant" following PI-PLC cleavage. Unfortunately, these approaches are limited because many GPI anchors are inositol acylated and therefore resistant to PI-PLC. On the other hand, all GPI anchors are sensitive to serum GPI-phospholipase D (GPI-PLD). However, GPI-PLD requires prior detergent solubilization of the substrate and is generally inactive on cell surfaces. The GPI-PLD reaction also leaves one fatty acid attached to the protein (the inositol acyl group) and, depending on the protein, this may prevent complete Triton X-114 phase separation after digestion. GPI anchors may be labeled biosynthetically with [^{3}H]*myo*-inositol and [^{3}H]ethanolamine, although [^{3}H]ethanolamine incorporation is not unique to GPI proteins. Similarly, [^{3}H]mannose and [^{3}H]glucosamine labeling, in the presence of tunicamycin to inhibit N-glycosylation, can be useful. [^{3}H]Fatty-acid labeling can also be helpful, but incorporation into GPI anchors must be distinguished from N-myristoylation and S-palmitoylation of proteins by careful product analysis. Certain pore-forming bacterial toxins such as aerolysin have been shown to bind to GPI anchors, and these may be used to probe one- and two-dimensional gel western blots.

MEMBRANE PROPERTIES OF GPI-ANCHORED PROTEINS

GPI-anchored proteins with two fatty acid chains (i.e., those containing diacylglycerol, alkylacylglycerol, alkenylacylglycerol, or ceramide) provide a stable association with the lipid bilayer. It follows that inositol-acylated GPI proteins with three fatty acid chains should be more stably associated and, presumably, dimeric GPI-anchored proteins benefit from effectively four or six fatty acid chains per molecule. On the other hand, a GPI-related structure with only a single C24:0 alkyl chain (e.g., the lipophosphoglycan of *Leishmania*) has a half-life of only minutes at the cell surface and is secreted intact into the medium. The thermodynamics of bilayer interactions also depend on the nature of the fatty acid chains (length and degree of saturation). In this regard, the saturated nature of several (but not all) mammalian GPI anchors (Figure 11.2) is thought to explain why GPI proteins associate with "lipid rafts," which are lipid-dependent membrane microdomains. The concept of lipid rafts has gained popularity in recent years and was originally devised to explain the sorting of certain cell-surface glycosphingolipids to the apical surface of Madin-Darby canine kidney (MDCK) cells, a process that suggested physical association of particular lipid contents in the *trans*-Golgi to allow packaging and vectorial delivery. The preferential delivery of GPI-anchored proteins to the apical membrane of polarized epithelial cells further suggested that they might coassociate with glycosphingolipid-rich domains in the Golgi. Because it was known that most GPI-anchored proteins are insoluble in cold neutral detergents, pulse-chase studies that recorded the acquisition of detergent insolubility by GPI-anchored proteins as they entered the Golgi (the site of glycosphingolipid synthesis) provided considerable support for the formation of glycosphingolipid- and GPI-protein-containing rafts in the Golgi. The lipid raft hypothesis has received further support from a variety of studies, many of which use cold neutral detergent extraction to generate "detergent-resistant membranes" (DRMs). Extraction with Triton X-100 at 0–4°C followed by flotation in a sucrose density gradient is typically used to isolate DRMs. These fractions are relatively rich in sphingolipids, glycosphingolipids, cholesterol, GPI-anchored proteins, and certain nonreceptor tyrosine kinases and relatively poor in phospholipids and transmembrane proteins. The tyrosine kinases found in these complexes such as p56lck and p59lyn are acylated with at least two saturated fatty acids. The cosequestration of GPI-anchored proteins and nonreceptor tyrosine kinases provides a possible explanation for the perplexing, but well-characterized, ability of GPI-anchored proteins to transduce signals across the plasma membrane.

There are many examples of transmembrane signaling via the cross-linking of GPI-anchored proteins with antibody and clustering with a second antibody on various cells, particularly leukocytes. Cellular responses include rises in intracellular Ca^{++}, tyrosine phosphorylation, proliferation, cytokine induction, and oxidative burst. These antibody-induced signaling events are clearly dependent on the presence of a GPI anchor and might be due to the coalescence of small lipid rafts. Lipid-raft-resident GPI-binding transmembrane proteins have also been postulated to account for the missing link between the outer and inner leaflets of the plasma membrane bilayer. One candidate for this in leukocytes is the β2-integrin complement receptor type 3. Despite the plethora of GPI-protein cross-linking/signal-transduction examples, it should be noted that there are no receptor/ligand pairs of established physiological relevance that signal in a GPI-dependent way. Thus, GPI-anchored proteins known to be involved in transmembrane signaling, such as the glial-cell-line-derived neurotrophic factor receptor-α (GDNFR-α), need to be associated with transmembrane β coreceptors to transmit their signals. Similarly, GPI-anchored CD14 (the LPS/LPS-binding protein receptor) functions equally well with a GPI anchor or with a spliced transmembrane domain, and the signal-transducing partner for CD14 has been identified as the transmembrane Toll-like receptor-2.

GPI ANCHORS AS TOOLS IN CELL BIOLOGY

The replacement of carboxy-terminal transmembrane domains by GPI-addition signal peptides allows their expression on the plasma membrane of transfected mammalian cells in GPI-anchored form. This can be a useful way to produce soluble forms of membrane proteins. For example, the T-cell receptor could not be expressed in a soluble form by simply deleting the transmembrane domain, but it could be expressed in GPI-anchored form and then rendered soluble by the action of bacterial PI-PLC. In addition, purified GPI-anchored proteins can be used to coat hydrophobic surface plasmon resonance chips, thus providing a convenient way of orienting and presenting proteins for binding studies. There are several examples of the exchange of GPI-anchored protein from one cell surface to another. The precise mechanism of this exchange is unknown, but it is clear that purified GPI-anchored proteins will spontaneously insert into lipid bilayers. The key difference between GPI-anchored and transmembrane proteins, in this regard, is the lack of any cytoplasmic domain in GPI-anchored proteins. The physiological significance of GPI-protein exchange between membranes is still uncertain, particularly because all mammals express potent GPI-PLD activity in serum that can remove the lipid (phosphatidic acid) component of the anchor and, therefore, prevent GPI-protein reinsertion. However, the transfer property of GPI-anchored proteins has been exploited experimentally to "paint" exogenous proteins onto cell surfaces.

BIOLOGICAL FUNCTIONS OF GPI ANCHORS

GPI anchors are essential for life in some, but not all, eukaryotic microbes. In the yeast *S. cerevisiae*, and probably most fungi, the presence of a GPI anchor is used to target certain mannoproteins for covalent incorporation into the β-glucan cell wall. The cross-linking occurs via a transglycosylation reaction, whereby a mannose residue within the GPI-anchor core is transferred to the β-glucan polymer. Defects in cell-wall biosynthesis are known to be detrimental to yeast and this may be why GPI biosynthesis is essential to this organism. Gene knockout studies have shown that GPI biosynthesis is also essential for the bloodstream form of *T. brucei*, even in tissue culture. This may be due to nutritional stress because this parasite uses an essential GPI-anchored transferrin receptor. On the other hand, surprisingly, GPI biosynthesis and/or transfer to protein are not essential for the insect-dwelling forms of *T. brucei* or *Leishmania*. The availability of GPI-deficient mammalian cell lines demonstrates that GPI-anchored proteins are not essential at a cellular level. However, mouse knockouts and tissue-specific conditional knockouts of the *PIG-A* gene (the catalytic subunit of the UDP-GlcNAc:PI α1-6 *N*-acetylglucosaminyltransferase) clearly show that GPI-anchored proteins are essential for early embryo and tissue development, respectively. In the plant *Arabidopsis*, GPI biosynthesis is required for cell-wall synthesis, morphogenesis, and pollen tube development. It was thought, originally, that GPI anchors might impart certain properties to their attached proteins, such as a high degree of lateral mobility in the bilayer and the ability to be shed in soluble form from the cell surface through the action of cellular or serum phospholipases. However, fluorescence photobleaching measurements have indicated that, whereas GPI-anchored proteins have the potential for high lateral mobility, in biological membranes mobility is principally modulated by interactions with other surface components. With respect to shedding via phospholipase action, it would appear that in living cells there are few examples of this phenomenon and that this is the exception rather than the rule. In lower eukaryotes, GPI anchors may be useful for assembling particularly dense cell-surface protein coats, such as the VSG coat of *T. brucei*. In this case, each parasite expresses 5 million VSG dimers on the cell surface to protect it against complement-mediated lysis. If each VSG monomer had, instead of a GPI anchor, a single transmembrane domain, there would be little room for other integral membrane proteins such as hexose and nucleoside transporters. Generally, GPI-anchored proteins do recycle through intracellular compartments but, compared with typical transmembrane proteins, they reside in higher proportion on the cell surface and have longer half-lives. GPI-anchored proteins are often enriched in membrane microvesicles that are shed from cell surfaces in response to physical or chemical stress.

GPI ANCHORS AND DISEASE

Paroxysmal nocturnal hemoglobinuria (PNH) is a human disease in which patients suffer from hemolytic anemia. The condition arises from improper expression of several GPI-anchored proteins that protect their blood cells from lysis by the complement system (e.g., decay accelerating factor and CD59). The defect in PNH cells is a somatic mutation in the X-linked *PIGA* gene. The mutation appears to occur in a bone marrow stem cell. Unlike other enzymes in the pathway, which are encoded by autosomal genes, PNH caused by *PIGA* mutations is thought to arise at a higher frequency because of X inactivation. In a PNH heterozygote, X inactivation of the one active allele of *PIGA* results in the complete loss of a functional UDP-GlcNAc:PI α1-6 *N*-acetylglucosaminyltransferase.

As mentioned above, GPI biosynthesis and transfer to protein are essential for yeast, probably for pathogenic fungi, and for the African sleeping sickness parasite *T. brucei*. Several key surface molecules of the apicomplexan parasites *Plasmodium* (malaria), *Toxoplasma*, and *Cryptosporidium* are GPI anchored, and it is thought that the GPI pathway is likely to be essential to these pathogens. Thus, pathogen-specific GPI pathway inhibitors are being actively sought as potential drugs. In addition, there is evidence that some parasite GPI anchors have a direct role in modulating the host immune response to infection. This was first proposed for the malaria parasite and similar results have been reported for the Chagas' disease parasite *T. cruzi.* These GPIs have toxin-like proinflammatory activities that may be responsible, in part, for the bursts of TNFα and fever in malaria and for controlling the acute infection in Chagas' disease, thereby extending the host's survival and the parasite's chances of transmission.

Like other glycoconjugates, GPI-anchored proteins can be exploited by pathogens. For example, the GPI anchors themselves are receptors for hemolytic pore-forming toxins such as aerolysin from *Aeromonas hydrophilia*, which causes gastroenteritis, deep wound infections, and septicemia in humans. In addition, the GPI-anchored protein CD55/DAF is the principal cell-surface ligand for enterovirus and several echoviruses. Finally, the endogenous prion protein is GPI anchored and it is thought that the conformational changes that it undergoes to become the aberrant spongiform-encephalopathy-causing form (e.g., in sheep scrapie, mad cow disease, and human Creutzfeldt-Jakob disease) may be associated with a clathrin-independent endocytic pathway followed by GPI-anchored prion protein in neurons.

FURTHER READING

Ferguson M.A. and Williams A.F. 1988. Cell-surface anchoring of proteins via glycosyl-phosphatidylinositol structures. *Annu. Rev. Biochem.* **57:** 285–320.

Ferguson M.A., Homans S.W., Dwek R.A., and Rademacher T.W. 1988. Glycosyl-phosphatidylinositol moiety that anchors *Trypanosoma brucei* variant surface glycoprotein to the membrane. *Science* **239:** 753–759.

Homans S.W., Ferguson M.A.J., Dwek R.A., Rademacher T.W., Anand R., and Williams A.F. 1988. Complete structure of the glycosyl phosphatidylinositol membrane anchor of rat brain Thy-1 glycoprotein. *Nature* **333:** 269–272.

Simons K. and Ikonen E. 1997. Functional rafts in cell membranes. *Nature* **387:** 569–572.

Kinoshita T. and Inoue N. 2000. Dissecting and manipulating the pathway for glycosylphosphatidylinositol-anchor biosynthesis. *Curr. Opin. Chem. Biol.* **4:** 632–638.

Simons K. and Toomre D. 2000. Lipid rafts and signal transduction. *Nat. Rev. Mol. Cell. Biol.* **1:** 31–39.

Chatterjee S. and Mayor S. 2001. The GPI-anchor and protein sorting. *Cell. Mol. Life Sci.* **58:** 1969–1987.

Guha-Niyogi A., Sullivan D.R., and Turco S.J. 2001. Glycoconjugate structures of parasitic protozoa. *Glycobiology* **11:** 45R–59R.

de Macedo C.S., Shams-Eldin H., Smith T.K., Schwarz R.T., and Azzouz N. 2003. Inhibitors of glycosylphosphatidylinositol anchor biosynthesis. *Biochimie* **85:** 465–472.

Eisenhaber B., Kubina W., Maurer-Stroh S., Neuberger G., Schneider G., Wildpaner M., and Eisenhaber F. 2003. Prediction of lipid posttranslational modifications and localization signals from protein sequences: Big-Pi, NMT and PTS1. *Nucleic Acids Res.* **31:** 3631–3634.

Maeda Y., Ashida H., and Kinoshita T. 2006. CHO glycosylation mutants: GPI anchor. *Methods Enzymol.* **416:** 182–205.

Rege T.A. and Hagood J.S. 2006. Thy-1, a versatile modulator of signaling affecting celluar adhesion, proliferation, survival, and cytokine/growth factor responses. *Biochim. Biophys. Acta* **1763:** 991–999.

Gowda D.C. 2007. TLR-mediated cell signaling by malaria GPIs. *Trends Parasitol.* **23:** 596–604.

Orlean P. and Menon A.K. 2007. Thematic review series: Lipid posttranslational modifactions. GPI anchoring of protein in yeast in mammalian cells, or: How we learned to stop worrying and love glycophospholipids. *J. Lipid Res.* **48:** 993–1011.

Young S.G., Davies B.S., Fong L.G., Gin P., Weinstein M.M., Bensadoun A., and Beigneux A.P. 2007. GPI-HBP1. An endothelial cell molecule important for the lipolytic processing of chylomicrons. *Curr. Opin. Lipidol.* **18:** 389–396.

Paulick M.G. and Bertozzi C.R. 2008. The glycosylphosphatidylinositol anchor: A complex membrane-anchoring structure for proteins. *Biochemistry* **47:** 6991–7000.

CHAPTER 12

Other Classes of ER/Golgi-derived Glycans

Hudson H. Freeze and Robert S. Haltiwanger

THIS CHAPTER FOCUSES ON THE LESS COMMON TYPES of glycoprotein linkages made in the endoplasmic reticulum (ER)/Golgi pathway, the systems in which they occur, the specific proteins that are modified, the glycosyltransferases involved, and the functions of the glycans.

FINDING NOVEL TYPES OF GLYCOSYLATION

The linkage between the first sugar of a glycan and an aglycone, the noncarbohydrate moiety, defines the types of glycosylation. Figure 1.7 shows 32 known types of linkages, including the common GlcNAc-N-Asn, GalNAc-O-Ser/Thr, and Xyl-O-Ser present in glycoproteins and proteoglycans. The first indication of the existence of novel, nonclassical types of glycosylation came from the analysis of human urine, which revealed unusual amino acid glycosides such as Glcβ1-3Fucα-Thr. This discovery garnered little interest at the time, in part because fucose was regarded exclusively as a terminal sugar. However, finding O-fucose directly linked to various clotting proteins and signaling receptors, such as Notch, broadened this perspective. Monoclonal antibodies that detected glycans on specific proteins such as α-dystroglycan provided tools to identify other novel structures. In other

TABLE 12.1. *Additional types of glycosylation in the ER/Golgi*

Modification	Proteins	Location of Glycan
O-α-Fuc		
Fucα	urokinase, t-PA, factor XII, factor VII, Cripto	EGF-like repeat
Siaα2-3/6Galβ1-4GlcNAcβ1-3Fucα	factor IX, Notch, Delta, Serrate, Jagged (all contain both mono and tetrasaccharides)	EGF-like repeat
Glcβ1-3Fuc-α	thrombospondin 1, properdin, F-spondin, ADAMTS13, ADAMTS-like 1	TSR
O-β-Glc		
Xylα1-3Xylα1-3Glcβ	factor VII, factor IX, protein Z, Notch	EGF-like repeat
O-β-Gal		
Glcα1-2Galβ-O-Hyl/Hyp	collagens, surfactant protein, complement factor Clq, mannan-binding proteins	collagen repeat
Man-α	α-dystroglycan	mucin-like domain
Glcβ-Asn	laminin	?
C-mannosylation	RNase 2, thrombospondin1, properdin,	WXXW motif

(t-PA) Tissue plasminogen activator; (TSR) thrombospondin type-1 repeat.

cases, mass spectrometry of proteins indicated unusual modifications, such as mannose linked to protein as a C-glycoside. Table 12.1 describes many of the unusual types of linkages made in the ER/Golgi pathway. Chapters 17 and 18 describe linkages made in the nucleus and cytoplasm.

O-LINKED MODIFICATIONS IN EGF-LIKE REPEATS

Epidermal growth factor (EGF)-like repeats are small protein motifs (~40 amino acids) that are defined by six conserved cysteine residues, which form three disulfide bonds (Figure 12.1). They are found in hundreds of cell-surface and secreted proteins in meta-

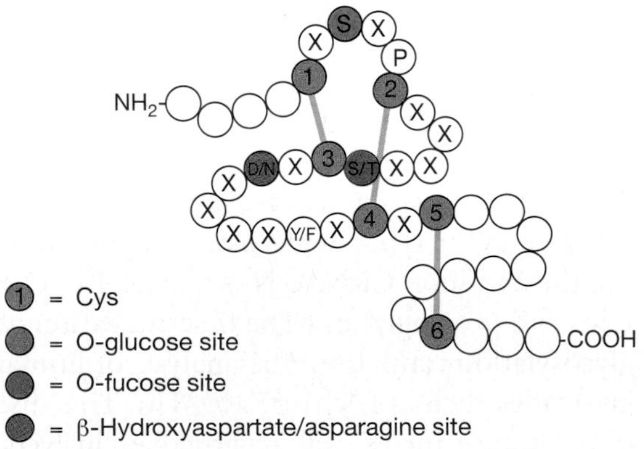

FIGURE 12.1. Epidermal growth factor (EGF)-like repeats can be modified by either O-fucose or O-glucose. A schematic representation of an EGF-like repeat is shown. (*Yellow*) Conserved cysteine residues; (*gray lines* between the cysteines) disulfide-bonding pattern. The modification sites for O-glucose (*green*), O-fucose (*blue*), and β-hydroxyaspartate/asparagine (*orange*) are shown in the context of the putative consensus sequences for each. (S) Serine; (T) threonine; (P) proline; (D) aspartic acid; (N) asparagine; (Y) tyrosine; (F) phenylalanine; (X) any amino acid. (Modified, with permission, from Haltiwanger R.S. 2004. In *Encyclopedia of biological chemistry*, Vol. 2, pp. 277–282, ©Elsevier.)

zoans, and some are modified with unusual O-linked glycans. These modified proteins include several involved in blood clot formation and dissolution, the Notch family of receptors and ligands (Delta and Serrate/Jagged), and the Cripto/FRL/Criptic (CFC) family of proteins. The glycan modifications are important because they can regulate signal transduction involved in early development and the growth of some cancers.

O-α-Fucose

The O-fucose modification site immediately precedes the third conserved cysteine of the EGF-like repeat (Figure 12.1). A consensus motif for fucosylation has been identified: $C^2X_{4-5}(S/T)C^3$, where C^2 and C^3 are the second and third conserved cysteines of the EGF-like repeat. Fucose can be elongated to a tetrasaccharide (Siaα2-3/6Galβ1-4GlcNAcβ1-3 Fuc-O-Ser/Thr) in certain contexts (e.g., EGF-1 from human clotting factor IX and EGF-26 from mouse Notch1) (Figure 12.2), but not in others (e.g., EGF-1 from human clotting factor VII and EGF-3 from mouse Notch1).

Protein O-fucosyltransferase 1 (POFUT1) transfers fucose from GDP-Fuc to a properly folded EGF-like repeat containing the appropriate consensus sequence (Figure 12.2). O-Fucose can be elongated by a β1-3 *N*-acetylglucosaminyltransferase (β3GnT) that is specific for fucose residues in O-linkage to properly folded EGF-like repeats (Figure 12.2). The gene for the O-fucose-specific β3GnT was originally identified in *Drosophila* as a modifier of Notch activity called *fringe* (see below). There are three mammalian homologs: *Manic fringe*, *Lunatic fringe*, and *Radical fringe*. Each of the Fringe proteins catalyzes the transfer of *N*-acetylglucosamine from UDP-GlcNAc to fucose on an EGF-like repeat. The *N*-acetylglucosamine residue can be further elongated with a β1-4-linked galactose by β4GalT-1 and capped with an α2-3- or α2-6-linked sialic acid, probably using the same sialyltransferases as those that act on N- or O-glycans. Elongation past the disaccharide does not appear to occur in *Drosophila*.

The O-fucose glycans can modulate signal transduction pathways by altering the interaction of EGF-like repeats with other proteins. The best example of this is Notch receptor signaling. Notch is a cell-surface receptor involved in numerous stages of development. It was originally identified in *Drosophila*, but homologs have been found in all metazoans, with four Notch receptors in mammals. Notch activation is controlled at numerous levels, and deregulation of Notch signaling results in a number of human diseases, including several types of cancer and a variety of developmental disorders. Two classes of ligands bind to and activate Notch: Delta and Serrate. Mammals have three Delta homologs (Delta-like 1, 3, and 4) and

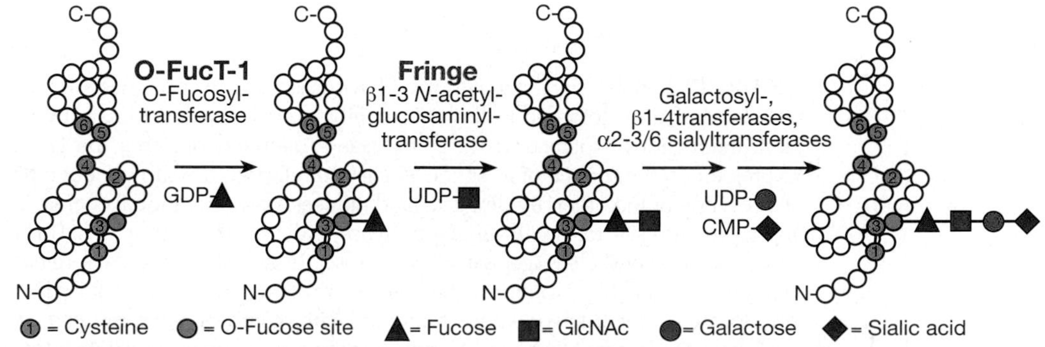

FIGURE 12.2. O-Fucose glycosylation pathway. The sequential modification of an EGF-like repeat with an O-fucose tetrasaccharide is shown. (Modified, with permission of MacMillan Publishers Ltd., from Moloney D.J., Panin V.M., Johnston S.H., et al. 2000. *Nature* **406:** 369–375.)

two Serrate homologs (Jagged-1 and -2). These ligands are single-pass transmembrane glycoproteins, so they must be expressed on the surface of an adjacent cell to bind to and activate Notch (Figure 12.3). The extracellular domain of Notch contains up to 36 tandem EGF-like repeats (Figure 12.4), many of which contain consensus sites for O-fucosylation.

RNA interference (RNAi) of POFUT1 expression in *Drosophila* results in strong Notch-like phenotypes, suggesting that O-fucose modification of Notch is essential for proper Notch function. POFUT1 null mice have an embryonic-lethal phenotype that is even more severe than the elimination of any single Notch receptor, suggesting that all Notch receptors need O-fucose for full function. POFUT1 is localized to the ER, where it only modifies properly folded EGF-like repeats, suggesting that O-fucosylation may have some role in

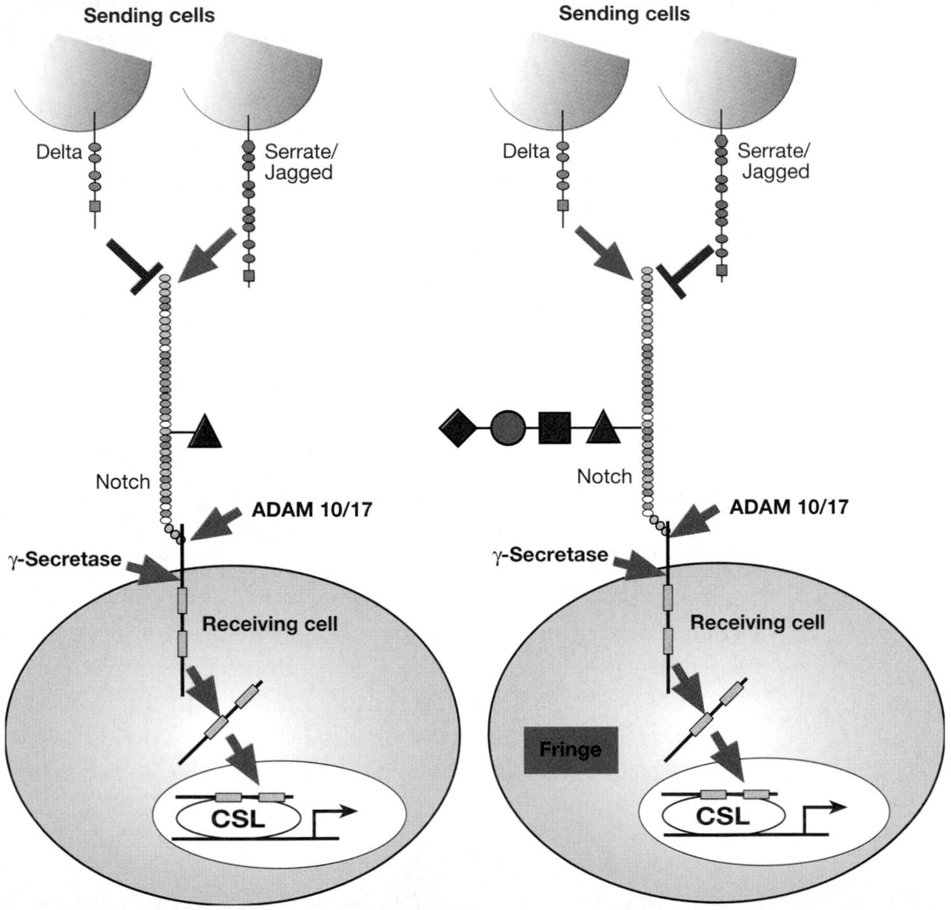

FIGURE 12.3. Notch signaling pathway. Notch exists on the cell surface as a heterodimer where the extracellular domain is tethered to the transmembrane/intracellular domain by noncovalent, calcium-dependent interactions. Notch is activated by ligand (members of either the Delta or Serrate/Jagged families) expressed on the surface of a "sending cell." Both ligands are shown, but typically the sending cell would express one or the other. Binding of ligand induces a conformational change in the Notch receptor, exposing a protease site for a cell-surface protease (either ADAM 10 or 17). This extracellular cleavage is followed by a second cleavage, catalyzed by γ-secretase, which results in release of the Notch intracellular domain as a soluble protein in the cytoplasm of the "receiving cell." The Notch intracellular domain translocates to the nucleus where it interacts with members of the CSL family of transcriptional regulators and activates transcription of a number of downstream gene products. (*Left panel*) Notch expressed in cells without Fringe; (*right panel*) with Fringe. The presence of Fringe results in elongation of O-fucose to a tetrasaccharide and alters the ability of Notch to respond to ligands on the sending cell. (Modified, with permission of the *Annual Review of Biochemistry,* from Haltiwanger R.S. and Lowe J.B. 2004. *Annu. Rev. Biochem.* **73:** 491–537, ©Annual Reviews.)

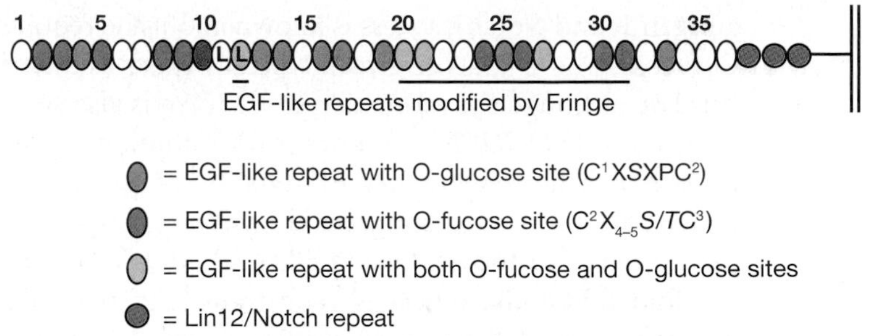

FIGURE 12.4. Extracellular domain of Notch showing the numerous sites for O-fucose and O-glucose modification that are evolutionarily conserved in *Drosophila* Notch, mouse Notch1 and 2, and human Notch1 and 2. EGF-like repeats with an O-fucose on mouse Notch1 that can be modified by one of the Fringe proteins are indicated by the bar. (L) EGF-like repeats responsible for ligand binding. (Adapted, with permission of the American Society for Biochemistry and Molecular Biology, from Shao L., Moloney D.J., and Haltiwanger R.S. 2003. *J. Biol. Chem.* **278:** 7775–7782.)

quality control. The *Drosophila* homolog of POFUT1 also has chaperone activity, suggesting a role in the proper folding of Notch. Mutation of specific, highly conserved O-fucose glycosylation sites in mouse Notch1 (within EGF-like repeats 12, 26, and 27; Figure 12.4) significantly alters ligand-mediated Notch activation in cell-based Notch signaling assays. Clearly, the O-fucose modification is important for Notch function.

The O-fucose glycans also modulate Notch activity during development. The expression of O-fucose-specific β3GnT Fringe enzymes is dynamic, increasing or decreasing depending on the developmental stage or tissue. Mutations in *Fringe* show phenotypes in several, but not all, contexts where Notch functions. Elimination of *Fringe* in *Drosophila* causes defects in wing, eye, and leg development, whereas ablation of *Lunatic fringe* in mice causes a severe defect in somite formation. The Fringe proteins mediate their effects on Notch activity by extending O-fucose on the Notch protein (Figure 12.2). A major question is how changing the O-fucose glycan structure alters Notch activation. The most likely answer is that Fringe-mediated elongation of O-fucose changes the affinity between Notch and its ligands. Fringe may directly alter interactions between Notch and ligands by modifying an O-fucose residue on EGF-like repeat 12, known to be part of the ligand-binding sites (Figure 12.4). It could also work indirectly by causing a change in the conformation of the extracellular domain of Notch. Whatever the correct explanation, it must account for the finding that Fringe has opposite effects on Notch activation depending on the ligand. Fringe inhibits Notch activation by Serrate/Jagged ligands, but it potentiates Notch activation by Delta ligands (Figure 12.3). The finding that Fringe is a glycosyltransferase that modifies O-fucose residues on Notch provides one of the clearest examples known of how a signal transduction pathway can be regulated by altering the glycosylation of the receptor.

O-Fucose has been implicated in another signaling pathway involving the binding of urokinase-type plasminogen activator to its receptor. As with Notch, the lack of O-fucose on an EGF-like repeat alters activation of the receptor. There are few details about the precise mechanism, but this serves to strengthen the argument that the O-fucosylated glycans fine-tune interactions between EGF-like repeats and their ligands.

O-β-Glucose

The O-glucose modification site occurs between the first and second conserved cysteines of EGF-like repeats at the putative consensus sequence C^1XSXPC^2 (see Figure 12.1). Only a few proteins are known to have the O-glucose modification (factor VII, factor IX, protein Z, fetal antigen 1, and Notch), so less is known about the requirements for this modification than for O-fucose. The O-glucose glycan typically exists as the trisaccharide Xylα1-3Xylα1-3Glcβ-O-Ser/Thr, although the monosaccharide form is also seen. The gene encoding the O-glucosyltransferase (*POGLUT*) is known, and mutations in the *Drosophila* homolog of *POGLUT* cause Notch-like phenotypes. Elimination of specific O-glucose sites on mouse Notch1 alters Notch activation, suggesting that they have an important role in Notch biology, possibly similar to that of O-fucose. Like POFUT1, POGLUT is localized to the ER and requires a properly folded EGF-like repeat as a substrate. Enzymatic activities for the two α1-3 xylosyltransferases have been detected in extracts of cells, but their genes remain unidentified.

Some EGF-like repeats also contain a consensus sequence C^3X*D*/*N*XXXXY/FXC^4XC^5 (Figure 12.1) that allows β-hydroxylation of aspartate or asparagine residues via aspartyl(asparaginyl) β-hydroxylase. This modification may also fine-tune Notch signaling, because the Notch ligand Jagged-1 can be hydroxylated and mice lacking this hydroxylase have subtle developmental abnormalities similar to mice deficient in the Notch ligand, Jagged-2.

MODIFICATIONS OF THROMBOSPONDIN TYPE-1 REPEATS

O-Fucose is also found in thrombospondin type-1 repeats (TSRs). Similar to EGF-like repeats, TSRs are small protein motifs (50–60 amino acids) with six conserved cysteines that form three disulfide bonds (Figure 12.5). Like EGF repeats, TSRs are found in many cell-surface and secreted proteins in metazoans, and they appear to function in protein–protein interactions. The O-fucose site occurs between the first and second conserved cysteine residues (again denoted by the superscripts) at the putative consensus sequence $C^1X_{2/3}$(*S*/*T*)C^2X_2G. Several dozen proteins contain these consensus sequences in the context of a TSR (see Table 12.1). The O-fucose on TSRs can be elongated to form a disaccharide with the structure Glcβ1-3Fucα-O-Ser/Thr. Interestingly, amino acid glycosides with this structure were isolated from human urine nearly 30 years ago. The enzyme that fucosylates TSRs in vitro is called POFUT2. It is expressed in many cells and tissues and is distinct from POFUT1. The gene encoding the enzyme that adds β1-3 glucose (β3GlcT) to O-fucose on TSRs is known, and mutations in the human homolog cause a genetic disorder known as Peter's plus syndrome, but the molecular details are unknown. TSRs are also

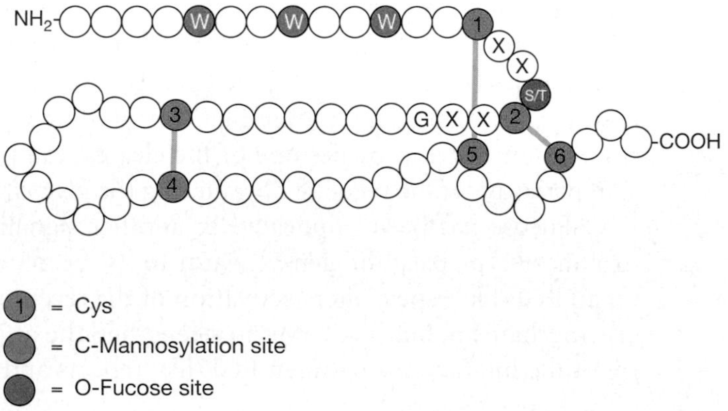

FIGURE 12.5. Modifications of thrombospondin type-1 repeats (TSRs). These repeats have six conserved cysteines (*yellow*), three disulfide bonds (*gray lines*), and consensus sites for O-fucose (*blue*) and C-mannose (*green*) addition. (W) Tryptophan; (S) serine; (T) threonine; (G) glycine; (X) any amino acid. (Modified, with permission, from Haltiwanger R.S. 2004. In *Encyclopedia of biological chemistry* [ed. W.J. Lennarz and M.D. Lane], Vol. 2, pp. 277–282, ©Elsevier.)

often modified with C-mannose (Figure 12.5; see below), although the relationship between the two modifications is not yet clear. A number of biologically interesting proteins are known or predicted to contain this modification, including thrombospondin-1 and -2, all ADAMTS (*a disintegrin and metalloproteinase with thrombospondin motifs*; ADAMTS 1–20), properdin, and F-spondin (Table 12.1). O-Fucosylation of TSRs in ADAMTS 13 and ADAMTS-like 1 is needed for their secretion, suggesting that O-fucosylation of TSRs may be involved in quality control or folding, similar to the predicted role of O-fucose on EGF-like repeats. Like POFUT1, POFUT2 is also an ER-localized enzyme that only fucosylates properly folded TSRs. In thrombospondin-1, the region of the TSR modified with O-fucose is involved in interactions between TSRs and several cell-surface receptors. Thus, the presence of the fucose in the midst of this region could influence interactions between TSRs and cells, much as O-fucose on EGF-like repeats alters binding events. TSRs allow proteins like the thrombospondins to bind to cells and/or to other components in the extracellular matrix. Cell binding is an essential step in the biological roles of thrombospondins, including the anti-angiogenic function of these proteins.

O-LINKED MANNOSE

α-Linked O-mannose was identified in yeast in the 1950s and will be discussed in more detail in Chapter 21. Yeasts elongate this mannose with a battery of mannosyltransferases to form structures that resemble extensions on their N-linked chains. Higher organisms adapted and modified this basic pathway. Mannose was first identified as the linkage sugar in a rat brain proteoglycan in 1979. Galβ1-4[Fucα1-3]GlcNAcβ1-3Man, GlcNAcβ1-3Man, and mannose by itself were found. O-Mannose-based glycans account for one third of all O-linked glycans in the brain. Some of the O-mannose-based structures can be quite complex as shown in Figure 12.6. For instance, some contain a second *N*-acetylglucosamine residue as a branch on the linkage mannose residue. In addition, a small portion carry the HNK-1 antigenic epitope, an unusual 3-O-sulfated glucuronic acid found at the nonreducing end of the chain. This epitope is implicated in neuronal cell adhesion. Of special interest is the protein α-dystroglycan, in which two thirds of the O-linked chains are Siaα2-3Galβ1-4GlcNAcβ1-2ManSer/Thr. This protein is part of the dystrophin-glycoprotein complex,

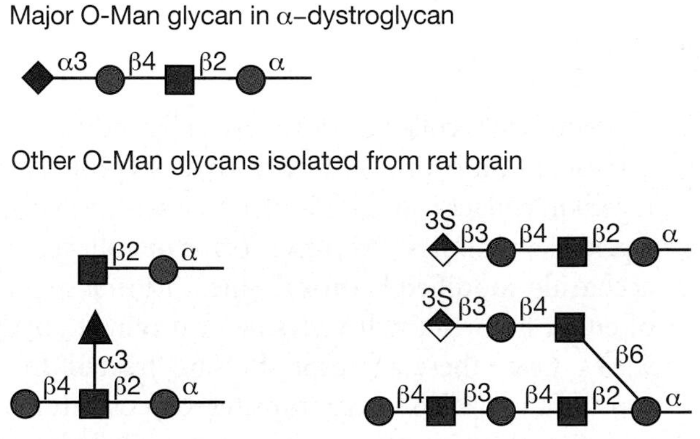

FIGURE 12.6. O-Mannose-linked glycans are complex. The most abundant structure on α-dystroglycan is shown here. This sialylated modification is thought to be involved in sialic-acid-dependent binding to laminin-2. Analysis of O-mannose-based glycans from rat brain glycopeptides reveals a complex array of structures including some containing Lewis[x], glucuronic acid-3-sulfate (HNK-1 antigen), and at least two branches.

which bridges the cytoskeleton and the extracellular matrix. α-Dystroglycan binds to β-dystroglycan in the cell membrane and to laminin in the extracellular matrix. The O-mannose glycans are found in a mucin-like domain of α-dystroglycan, and glycans are critical for binding to laminin in the extracellular matrix. As described in Chapter 42, an entire set of muscular dystrophies, which can result in severe neurological abnormalities, are caused by mutations in genes that contribute to the synthesis of O-mannose glycans.

The initiating O-mannosyltransferase and *N*-acetylglucosaminyltransferases are unique to this pathway. The mannose residue is added by a heterodimer composed of protein O-mannosyltransferase (POMT) 1 and 2. The products of these closely related genes are both required to mannosylate a glycopeptide acceptor derived from α-dystroglycan. Defects in these genes cause Walker–Warburg syndrome. The *N*-acetylglucosaminyltransferase (POMGNT1) adds β1-2-linked *N*-acetylglucosamine to O-mannose-containing glycoproteins and glycopeptides. Defects in this activity cause muscle–eye–brain disease. It is unknown whether remaining monosaccharides use pathway-specific glycosyltransferases or those with more general specificity. Three additional genes called *fukutin* (*FCMD*), *fukutin-related protein* (*FKRP*), and *Large* appear to enable glycosylation of dystroglycan, but they have not been shown carry out glycosylation reactions. Mutations in these genes cause other less severe forms of muscular dystrophy and alter glycosylation of α-dystroglycan based on the binding of a glycan-recognizing monoclonal antibody.

Studies in mice and *Drosophila* further confirm the importance of this class of glycans. Mice lacking *Pomt1* die in early embryogenesis. Growth arrest starts about embryonic day 7.5 and embryos die within a day or two. Their demise is very similar to that of mice lacking the entire α-dystroglycan gene. Both *Pomt1*- and α-dystroglycan-deficient mice have defects in the formation of Reichert's membrane, which is a thick basement membrane in the parietal wall of the yolk sac that surrounds the early embryo. Incomplete glycosylation is thought to interfere with its binding to laminin. Disruption of the *fukutin* gene in mice causes embryonic lethality around day 6.5–7.5. In *Drosophila*, mutations in the *pomt* homologs, *rotated abdomen* (*rt*) and *twisted* (*tw*), cause a clockwise helical rotation of the body. Similarly, POMT2 (*dPOMT2*) RNAi knockdown causes a twisted abdomen phenotype. Viable null animals show defects in embryonic muscle development; the entire larva body is rotated. This most likely results from insufficient torque in muscle architecture needed to align the segmented histoblasts during development. In adults, this in turn generates a helical staggering of the abdominal cuticle.

O-LINKED GLYCANS IN COLLAGENS

Proteins with collagen domains are modified by a disaccharide, Glcα1-2Galβ-, which is assembled on hydroxylysine or hydroxyproline residues. This modification was first recognized on collagen in the 1960s, and it has been observed in a few other proteins that have collagen-like modules. Primitive forms of collagen in sponges and sea anemones also have disaccharide-modified hydroxylysine. The first step in the pathway is the obligate hydroxylation of either lysine to hydroxylysine or proline to hydroxyproline using the appropriate hydroxylases. Once these acceptor sites are created, the two-step glycosylation rapidly follows on most of the available acceptor sites. Glycosylation is thought to proceed until the protein folds and assembles into the well-known triple helix. Some studies suggest that the extent of glycosylation controls or influences the rate of triple-helix formation and, in turn, the size of the fibrils. Other studies suggest that excessive glycosylation of hydroxylysines makes them poor substrates for extracellular enzymatic deamination, the first step in cross-linking collagen. Mice lacking one of the lysyl hydroxylases (LH3) fail to glycosylate collagen IV properly, which causes embryonic lethality and deposition of misfolded collagen in the ER. There is little information on the regulation of either of the pathway-specific galactosyl- or glucosyltransferases, except that they are more abundant in embryonic tissues than in adults.

Several other proteins with conserved collagen-like domains that form triple-helical structures also carry the disaccharide. These include pulmonary surfactant proteins, complement factor C1q, and rat and human mannan-binding proteins made by the liver. The mannan-binding protein found in the serum assembles into large aggregates by interchain disulfide bonds, which depend on the creation of hydroxylysine and presumably on glycosylation. Triple-helix formation of collagen can also prevent the addition of galactose to the glucosylated form.

NOVEL N-LINKED GLYCANS

GalNAcβ-Asn and Glcβ-Asn occur in bacteria. N-glycosylation is an ancient protein modification that occurs in Archaea, where it exploits the same consensus sequence as an acceptor for a preformed lipid-linked oligosaccharide. Glcβ-Asn was first identified in halobacteria as the major N-glycosylation pathway. Only one animal cell glycoprotein, laminin, has been shown to have this modification based on recognition by a specific antibody. The antigen on laminin was only exposed after treatment with acid sufficient to hydrolyze other sugars, suggesting that it may be the linkage monosaccharide that defines a more complex biosynthetic pathway.

C-MANNOSYLATION

Another novel type of glycosylation is called C-mannosylation. The C-1 atom of a single mannose residue is added in an α-linkage to the C-2 atom of the indole moiety of tryptophan 7 in RNase 2 (Figure 12.7). Note that this is not a typical glycosidic linkage. It is a C—C bond rather than a C—O or C—N bond. This modification was first detected in human urine. Structural analysis, biosynthetic studies in mammalian cell lines, and a specific antibody against the modification show that C-mannosylation is widespread, but it appears to be absent in bacteria and yeast. Site-directed mutagenesis shows that the critical glycosylation sequence in RNase 2 is W-X-X-W-, with the first tryptophan being modified. Synthetic peptides with this motif can be C-mannosylated in vitro, and a database search shows that more than 300 mammalian cell proteins have this sequence. Several complement proteins

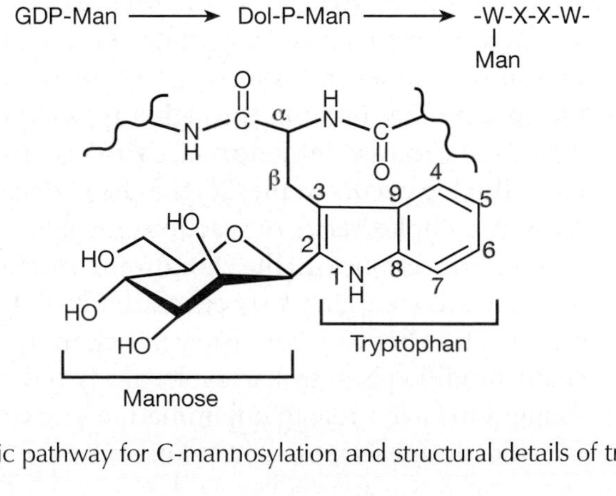

FIGURE 12.7. Biosynthetic pathway for C-mannosylation and structural details of tryptophan-7 in RNase 2.

contain TSRs with C-mannose (see above). For instance, in properdin (which is a positive regulator of complement) 14 of 17 consensus sites are fully modified. Dolichol-P-mannose produced from GDP-Man (see Chapter 4) is the biosynthetic donor for C-mannosylation in the ER. The modification appears to either compete with or aid in protein folding because already folded proteins are poor acceptor substrates in vitro, and expression of some proteins in C-mannosylation-disabled cells leads to their accumulation in the ER. The search continues for specific functions of this modification.

FREE GLYCANS SECRETED BY THE GOLGI APPARATUS

Diverse plant polysaccharides such as pectins (see Chapter 22) are also made as free chains. They are assembled in the various stacks of the Golgi, but many details remain to be worked out. Compared to animals, plant polysaccharides are more complex because they use additional monosaccharides and modifications such as O-methylation. Few enzymes have been purified or relevant genes identified.

Mammals make a series of free oligosaccharides and secrete them into the milk. Humans have a relatively high concentration of complex, sialylated oligosaccharides, which appear to protect infants against enteric pathogens. Human milk contains more than 130 different free glycans with some containing up to 15 monosaccharides. Other mammals synthesize species-specific arrays of glycans that may have similar functions. Surprisingly, there is essentially no information on the hormone-regulated biosynthesis of these abundant glycans. β1-4 galactosyltransferase makes lactose (Galβ1-4Glc), but it must bind lactalbumin to use glucose efficiently rather than *N*-acetylglucosamine as an acceptor. Even though this process has been demonstrated to occur in intact Golgi, it is unknown whether the other milk oligosaccharides result from sequential elongation of lactose.

CROSS-LINKING OF MACROMOLECULES VIA GLYCANS

Glycans can cross-link other molecules. In bacteria, the pentapeptide of peptidoglycan is used to cross-link the individual glycan chains (see Chapter 20), making the wall into one giant covalent macromolecule. In plants, pectins such as rhamnogalacturonan I (RG-I), can be nonenzymatically cross-linked into dimers with borate by formation of 1:2 borate-diol esters involving the *cis*-diol groups of apiose.

An example of a different type of cross-linking exists in mammals and involves the biosynthesis of inter-α-trypsin inhibitor (IαI), the major urinary proteinase inhibitor. The complex is composed of one or two heavy chains that are linked via the α-carboxyl group on the carboxy-terminal aspartate residue to the C-6 hydroxyl groups of *N*-acetylgalactosamine residues in a short (12–18 disaccharides) chondroitin sulfate chain attached to the core protein bikunin through a typical proteoglycan linkage region (see Chapter 16). All of the protease inhibitory activity occurs in the bikunin molecule (Figure 12.8). The cross-linking occurs in the Golgi of hepatocytes, but other cells can also carry out the reactions. A complex series of reactions involving proteolytic cleavage of the heavy chains and their esterification with the chondroitin sulfate chains occurs in the Golgi. When these proteins are secreted, they may encounter hyaluronan, which serves as an acceptor in a trans-esterification reaction releasing the chondroitin sulfate–bikunin molecule. These heavy-chain-modified hyaluronan molecules can then assemble into cable-like structures that are thought to have a role in inflammation (see Chapter 15). Details of the specificity for trans-esterification reactions are not known.

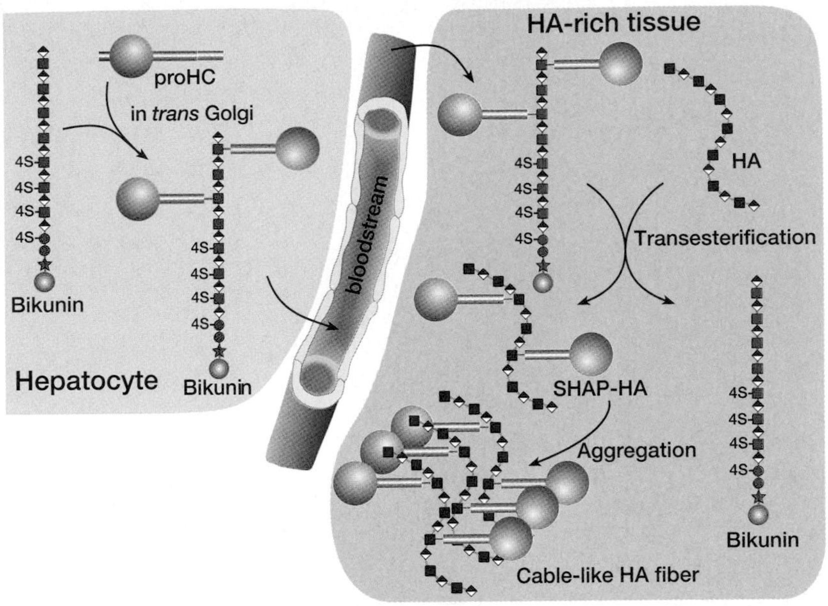

FIGURE 12.8. Model for the structure and function of a chrondroitin sulfate (CS) glycan that bridges several proteins. The inter-α-trypsin inhibitor (IαI) family of molecules are synthesized in the Golgi of hepatocytes. The inhibitor contains three protein chains, HC1 (heavy chain 1), HC2, and serine protease inhibitor bikunin, all held together via a chondroitin sulfate chain. The CS chain is attached to Ser-10 of bikunin via a standard glycosaminoglycan attachment (see Chapter 16), whereas the HCs are linked through ester bonds between carboxylate groups of their carboxy-terminal aspartic acid residues and the C-6 hydroxyls of internal *N*-acetylgalactosamine units in the CS chain. The CS-bound HCs are then transferred to C-6 hydroxyls of *N*-acetylglucosamine residues in HA via a transesterification involving the carboxy-terminal aspartic acids. This leads to the construction of extracellular matrices by the aggregation of SHAP (serum-derived hyaluronan-associated protein)-HA molecules into a "cable-like structure" containing the complexes and the interaction of the matrices with inflammatory cells. (Redrawn, with permission of the American Society for Biochemistry and Molecular Biology, from Zhuo L, Hascall V.C., and Kimata K. 2004. *J. Biol. Chem.* **279:** 38079–38082.)

FURTHER READING

Furmanek A. and Hofsteenge J. 2000. Protein C-mannosylation: Facts and questions. *Acta Biochim. Pol.* **47:** 781–789.

Haltiwanger R.S. and Lowe J.B. 2004. Role of glycosylation in development. *Annu. Rev. Biochem.* **73:** 491–537.

Muntoni F. 2004. Journey into muscular dystrophies caused by abnormal glycosylation. *Acta Myol.* **23:** 79–84.

Myllyharju J. and Kivirikko K.I. 2004. Collagens, modifying enzymes and their mutations in humans, flies and worms. *Trends Genet.* **20:** 33–43.

O'Neill M.A., Ishii T., Albersheim P., and Darvill A.G. 2004. Rhamnogalacturonan II: Structure and function of a borate cross-linked cell wall pectic polysaccharide. *Annu. Rev. Plant Biol.* **55:** 109–139.

Zhuo L., Hascall V.C., and Kimata K. 2004. Inter-α-trypsin inhibitor, a covalent protein-glycosaminoglycan-protein complex. *J. Biol. Chem.* **279:** 38079–38082.

Newburg D.S., Ruiz-Palacios G.M., and Morrow A.L. 2005. Human milk glycans protect infants against enteric pathogens. *Annu. Rev. Nutr.* **25:** 37–58.

Martin P.T. 2007. Congenital muscular dystrophies involving the O-mannose pathway. *Curr. Mol. Med.* **7:** 417–425.

Rampal R., Luther K.B., and Haltiwanger R.S. 2007. Notch signaling in normal and disease states: Possible therapies related to glycosylation. *Curr. Mol. Med.* **7:** 427–445.

Stanley P. 2007. Regulation of notch signaling by glycosylation. *Curr. Opin. Struct. Biol.* **17:** 530–535.

CHAPTER 13

Structures Common to Different Glycans

Pamela Stanley and Richard D. Cummings

THIS CHAPTER DESCRIBES TERMINAL GLYCAN STRUCTURES that are common to different classes of glycans. They are the variable portions of N-glycans, O-glycans, and glycolipids that are attached to the core sugars characteristic of each glycan class. The core structures of N-glycans, O-glycans, and glycolipids (Figure 13.1) are the product of biosynthetic pathways discussed in Chapters 8, 9, and 10. More complicated cores can result from tissue- or cell-type-specific pathways and lead to further structural diversification of glycan chains. The glycan units formed by subultimate and terminal sugars at "outer" positions of a glycan often determine the function(s) or recognition properties of a glycoconjugate.

REGULATED GLYCOSYLATION OF CORE GLYCAN STRUCTURES

In contrast to core glycan synthesis, which is constitutive in most cell types, the addition of branching and terminal sugars is often regulated in a tissue- or cell lineage–specific man-

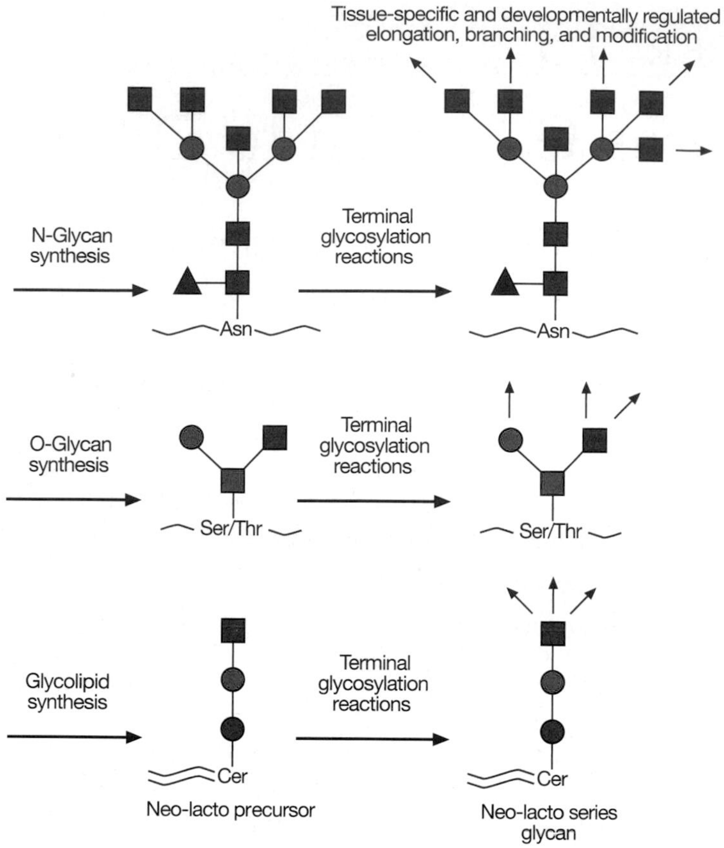

FIGURE 13.1. N-Glycan synthesis leads to structures (a biantennary N-glycan in this example; see Chapter 8) that are subsequently modified by branching *N*-acetylglucosamine (GlcNAc) residues and other sugars (*arrows*) in glycosylation reactions that may be tissue-specific, developmentally regulated, or even protein-specific. The GlcNAc linked to the core mannose is termed the bisecting GlcNAc, and it is not modified. O-Glycan synthesis leads to core structures (a core-1 O-glycan in this example; see Chapter 9) that are modified subsequently by many of the same enzymes that act on N-glycans. Similarly glycolipid core structures (neo-lactosylceramide in this example; see Chapter 10) are also modified by many of the same enzymes that act on N- and O-glycans.

ner (Figure 13.1). Many of these reactions are regulated during embryogenesis and in the postnatal period as part of the normal developmental program (see Chapter 38). Changes in terminal glycan structure are also often associated with malignant transformation in cancer (see Chapter 44). As discussed in Chapters 5 and 38, tissue- and/or lineage-specific regulation of outer-chain biosynthesis is largely a function of the regulated expression of the relevant glycosyltransferases. Biological consequences that correlate with a number of changes in outer-glycan units are discussed throughout this volume. However, the majority of regulated terminal glycosylations observed do not yet have a defined function.

Type-2 Glycan Units

The more mature glycans shown in Figure 13.1 contain terminal *N*-acetylglucosamine (GlcNAc) residues. The subsequent addition of galactose in β1-4 linkage generates a type-2 unit, which is composed of the disaccharide Galβ1-4GlcNAc, also called *N*-acetyllactosamine (LacNAc) (Figure 13.2). The terminal galactose so generated can be modified by the addition of a GlcNAc residue, which in turn can receive a galactose in β1-4 linkage, thus

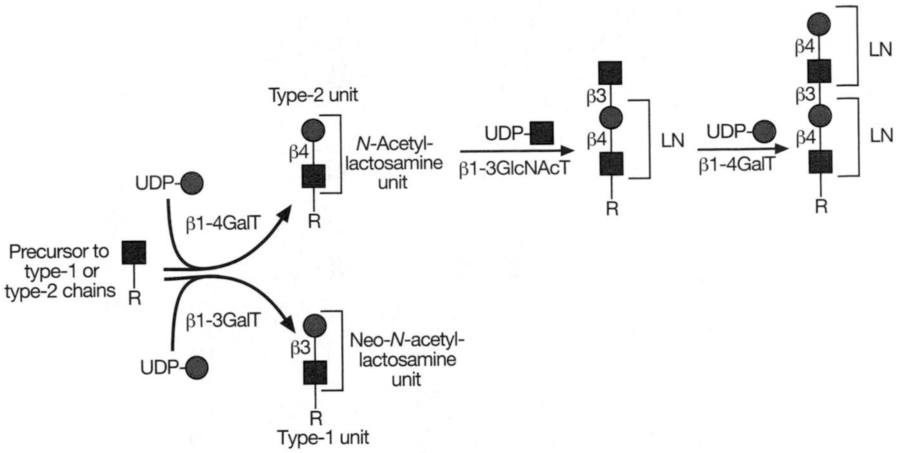

FIGURE 13.2. Terminal GlcNAc residues are usually galactosylated. Modification by β1-4Gal residues (*top*) occurs in all mammalian tissues. This reaction is catalyzed by β1-4 galactosyltransferase (β1-4GalT) and yields the Galβ1-4GlcNAc (*N*-acetyllactosamine) unit termed type-2. Transfer of β1-3Gal residues (*bottom*) is restricted to certain tissues. This reaction is catalyzed by β1-3 galactosyltransferase (β1-3GalT) and yields the Galβ1-3GlcNAc (neo-*N*-acetyllactosamine) unit termed type-1. R indicates N- and O-glycans or glycolipids. Type-2 and -1 units can be further modified by subsequent glycosylation reactions. Poly-*N*-acetyllactosamine chain initiation is also shown. LN designates *N*-acetyllactosamine unit.

forming two LacNAc units. These reactions may continue to form poly-*N*-acetyllactosamine $[\text{Gal}\beta1\text{-}4\text{GlcNAc}]_n$. Poly-*N*-acetyllactosamine chains are found in glycans from most cell types. An alternative to type-2 LacNAc units is a chain composed of so-called LacdiNAc glycan units generated by the action of a β1-4 *N*-acetylgalactosaminyltransferase, to form $[\text{GalNAc}\beta1\text{-}4\text{GlcNAc}]_n$. Efforts are being made to define the biological role(s) for type-2 chains by the generation and analysis of mice deficient in β1-4 galactosyltransferase genes, of which there are six that can generate type-2 chains.

Type-1 Glycan Units

GlcNAc residues in N-glycans, O-glycans, and glycolipids may instead be modified by galactose in β1-3 linkage (Figure 13.2) to obtain a type-1 unit composed of a disaccharide termed neo-*N*-acetyllactosamine. In humans, expression of type-1 units is mostly restricted to epithelia of the gastrointestinal or reproductive tracts. Type-1 units may be modified by glycosyltransferases that transfer sugars to either terminal galactose or subterminal GlcNAc, generating sialylated structures or blood group determinants. The synthesis of type-1 units and expression of the corresponding β1-3 galactosyltransferases are regulated in a tissue-specific manner, but biological roles for this disaccharide remain to be defined.

Poly-*N*-acetyllactosamines

Glycoproteins and glycolipids frequently bear N-glycans, O-glycans, and glycolipids with poly-*N*-acetyllactosamine chains. As noted above, poly-*N*-acetyllactosamine biosynthesis is directed by the alternating actions of β1-4 galactosyltransferases and β1-3 *N*-acetylglucosaminyltransferases (Figure 13.2). Some glycoproteins and glycolipids are preferentially modified to carry poly-*N*-acetyllactosamine. This implies that the glycosyltransferases responsible for poly-*N*-acetyllactosamine biosynthesis can discriminate between glycoproteins or glycolipids that appear to present the same terminal GlcNAc and galactose glycan

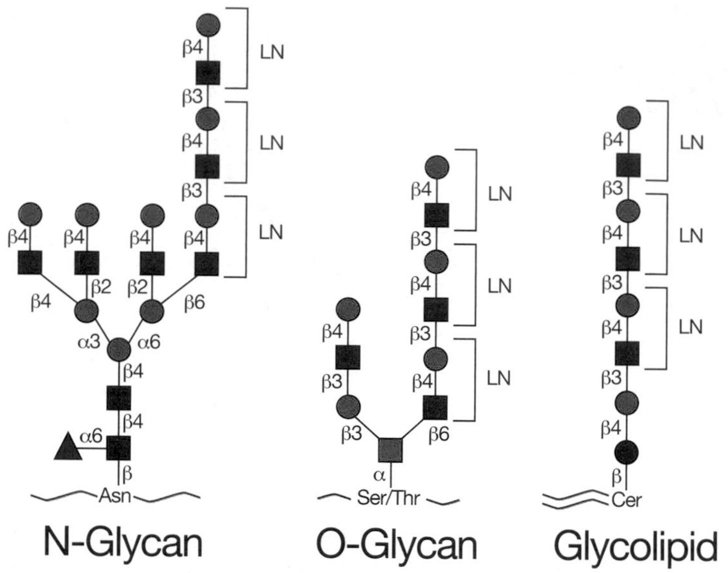

FIGURE 13.3. Poly-*N*-acetyllactosamine chains are found on N-glycans, O-glycans, and glycolipids. LN designates *N*-acetyllactosamine unit.

acceptors. For example, poly-*N*-acetyllactosamine extensions preferentially occur on multiantennary N-glycans and particularly on the β1-6GlcNAc branch, the synthesis of which is under the control of GlcNAc transferase V (GlcNAcT-V) (Figure 13.3; see Chapter 8). Similarly, poly-*N*-acetyllactosamine extensions on O-glycans associated with mucin glycoproteins often preferentially occur on the β1-6GlcNAc transferred by a core 2 β1-6GlcNAcT (Figure 13.3). Poly-*N*-acetyllactosamine chain length is also under regulatory control, with N-glycans generally having longer chains than O-glycans. Poly-*N*-acetyllactosamine chains serve as acceptors for subsequent glycosylations, including fucosylation and sialylation. The linear nature of these chains together with their hydrophilic character predicts that they have an extended linear conformation. Thus, poly-*N*-acetyllactosamine chains may serve as scaffolds for the presentation of specific terminal glycans, whose functions require them to be presented at a certain distance from the plasma membrane. Poly-*N*-acetyllactosamine chains are also recognized with high affinity by mammalian galectins (see Chapter 33) in a manner that implies recognition of internal sugars as well as terminal galactose.

β1-6GlcNAc-branched Poly-*N*-acetyllactosamines

Poly-*N*-acetyllactosamine chains may be branched by the addition of GlcNAc in β1-6 linkage to internal galactose residues through the action of β1-6 *N*-acetylglucosaminyltransferases (β1-6GlcNAcTs) (Figure 13.4). Branched and nonbranched poly-*N*-acetyllactosamine chains correspond to the "I" and "i" antigens, respectively. These antigens were discovered by the analysis of a cold-dependent agglutinating antibody (cold agglutinin) in a patient with acquired hemolytic anemia (see Chapter 43). Cold agglutinin antibodies react with red blood cells of most blood donors. Nonreactive donors are classified as having the i blood group, whereas reactive donors are assigned to the I blood group. Thus, β1-6-branched poly-*N*-acetyllactosamine units correspond to the I blood group antigen and linear poly-*N*-acetyllactosamine chains that lack branching correspond to the i blood group antigen. Distinct β1-6GlcNAcTs yield different types of β1-6-branched structures (Figure 13.4). Glycans reactive to I and i blood group antibodies are expressed by many human cells. The i antigen is abundantly expressed on the surface of embryonic red cells

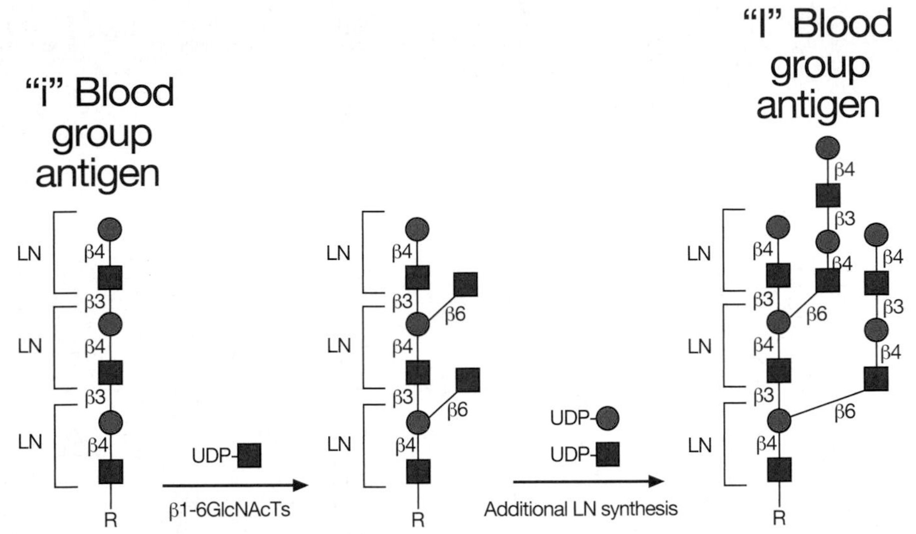

FIGURE 13.4. i and I antigen synthesis. Linear poly-*N*-acetyllactosamine chains (i antigen) synthesized on N- and O-glycans or glycolipids (R) may be modified by β1-6 *N*-acetylglucosaminyltransferases (β1-6GlcNAcTs). These enzymes transfer *N*-acetylglucosamine in β1-6 linkage to internal galactose residues. The newly added β1-6 *N*-acetylglucosamine branch may serve as substrate for subsequent poly-*N*-acetyllactosamine biosynthesis (see Figure 13.3). LN designates *N*-acetyllactosamine unit.

and on red cells during times of altered erythropoiesis. Such cells are relatively deficient in the expression of the I antigen. However, during the first 18 months of life, I antigen reactivity on red cells reaches adult levels, and i antigen reactivity declines to very low levels. This developmental regulation is presumed to be due to regulated expression of I β1-6GlcNAcT genes. Rare individuals never express the I antigen on red cells and maintain embryonic levels of red cell i antigen expression as adults. Individuals with this i phenotype may be homozygous for inactive alleles at β1-6GlcNAcT loci. There is no obvious pathophysiology associated as yet with the absence of the I blood group.

THE A, B, AND H BLOOD GROUPS

In humans, poly-*N*-acetyllactosamine chains and their β1-6-branched variants are subject to tissue-specific glycosylations that form the ABO blood group antigens. The ABO system was discovered early in the twentieth century by Karl Landsteiner and colleagues. They showed that humans could be divided into different classes according to the presence or absence of serum factors that would agglutinate red cells isolated from other humans. We now know that these serum factors are antibodies and that the corresponding antigens are glycan epitopes, whose structures are determined by the inheritance of genes that encode glycosyltransferases with different activities.

The A, B, and H blood group antigens are glycans presented on the type-1 or type-2 *N*-acetyllactosamines described above (Figure 13.5), on O-GalNAc glycans (type-3; Figure 13.6), or on glycolipids (type-4; Figure 13.7). The A, B, and H antigens are formed by the sequential action of glycosyltransferases encoded by three genetic loci (the *ABO*, *H*, and *Secretor* [*Se*] loci) (Figure 13.8). The pathway of ABO blood group antigen synthesis begins with modification of type-1 or type-2 *N*-acetyllactosamine by α1-2 fucosyltransferases (α1-2FucTs). The transfer of fucose in α1-2 linkage to the galactose in type-1 or type-2 *N*-acetyllactosamine forms the blood group H determinant. The human genome encodes two different α1-2FucTs, encoded by the *H* and *Se* blood group loci, respectively. The *H* α1-

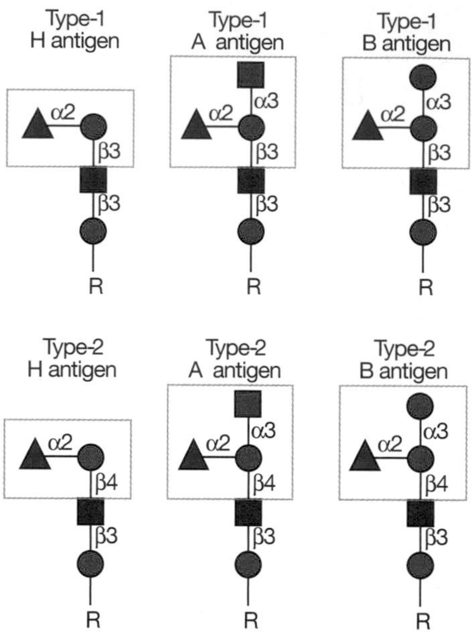

FIGURE 13.5. Type-1 and -2 H, A, and B antigens that form the O (H), A, and B blood group determinants on N- and O-glycans.

2FucT is expressed in red cell precursors, and it transfers fucose to type-2 and type-4 glycan units to form the H antigen on red cells (Figure 13.5 and Figure 13.7). The *Se* α1-2FucT is expressed in epithelial cells, and it uses type-1 and type-3 *N*-acetyllactosamines to form the H antigen (Figure 13.5 and Figure 13.6) in epithelia lining the lumen of the gastrointestinal, respiratory, and reproductive tracts and in salivary glands.

A or B blood group determinants are subsequently formed from type-1, -2, -3, or -4 H determinants by glycosyltransferases encoded by the *ABO* blood group locus. The blood group A glycan epitope is formed by the α1-3GalNAcT encoded by the *A* allele of the *ABO* locus (Figure 13.8). The blood group *B* allele of the *ABO* locus encodes the α1-3GalT that forms the blood group B glycan determinant (Figure 13.8). *O* alleles at the *ABO* locus encode a functionally inactive A/B glycosyltransferase. Individuals who synthesize exclusively A determinants are blood group A and have the genotype *AA* or *AO*, whereas blood group B individuals have the genotype *BB* or *BO*. Individuals who express both A and B determinants are blood group AB and have the genotype *AB*. Blood group O individuals do not express either A or B determinants (their H antigen is unmodified). They are homozygous for the inactive *O* allele and their genotype is *OO*.

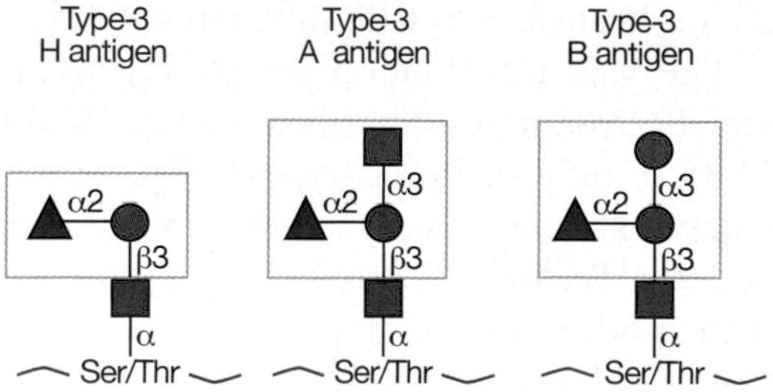

FIGURE 13.6. Type-3 H, A, and B antigens that form the O (H), A, and B blood group determinants on N- and O-glycans.

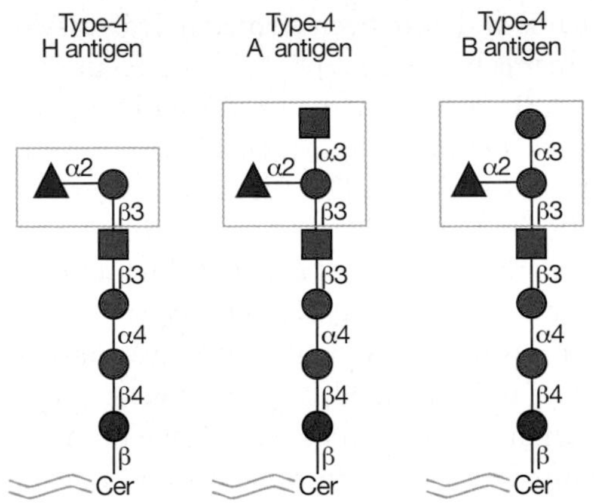

FIGURE 13.7. Type-4 H, A, and B antigens that form the O (H), A, and B blood group determinants on glycolipids.

The ABO antigens are expressed on membrane glycoproteins and glycolipids on the surface of red cells and other cells in many tissues, including the vascular endothelium and a variety of epithelia. Some tissues also synthesize soluble, secreted forms of these molecules as glycans on secreted glycoproteins, glycolipids, and free glycans. As discussed below, the ability to secrete soluble molecules carrying ABH blood group antigens is a genetically determined trait that is a function of alleles at the *Se* locus.

On each human red blood cell, approximately 1–2 million ABH determinants (~80% of the total) are attached to the anion transport protein (also known as band 3). About half a million ABH determinants are carried by the red cell glucose transport protein (band 4.5). Both of these integral membrane proteins carry ABH antigens on a single branched N-glycan with poly-*N*-acetyllactosamines whose terminal branches may carry several ABH determinants. Small numbers of ABH antigens are expressed by other red cell glycoproteins. Each red cell also expresses approximately half a million glycolipids with ABH determinants. Many of these glycolipids have A, B, and H determinants on poly-*N*-acetyllactosamine

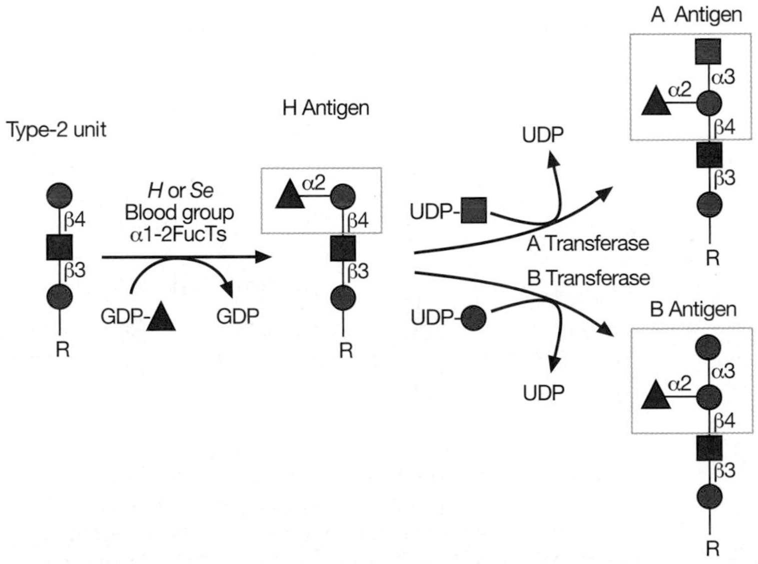

FIGURE 13.8. Synthesis of H (O), A, and B blood group determinants. For details, see text.

chains and have been termed polyglycosylceramides or macroglycolipids. A, B, and H determinants based on type-4 chains (Figure 13.7) are also present in human red cell glycolipids.

ABH determinants expressed by the epidermis are primarily constructed from type-2 units (Figure 13.5). Mucins derived from the gastric mucosa and from ovarian cyst fluid express A, B, and H antigens on type-3 units (Figure 13.6). As noted above, the epithelial cells lining the digestive, respiratory, urinary, and reproductive tracts express type-1 units, as do the epithelia of some salivary and other exocrine glands. These tissues synthesize soluble forms of the ABH determinants, which are therefore largely carried on type-1 units. Expression of the A, B, and H determinants in such secretory tissues is a function of the α1-2FucT encoded by the *Se* gene, since the *H* gene is not expressed in these tissues. Humans with an inactive *Se* gene do not express soluble forms of the A, B, or H determinants in saliva or in other tissues, even though they express the soluble precursors. The term nonsecretor is used to describe these individuals and refers to the fact that soluble blood group A, B, or H antigens cannot be detected in their saliva.

Serology used to characterize red cells for transfusion has identified variants of the A and B blood group determinants that typically yield weak reactivity with typing reagents. For example, the lectin *Dolichos biflorus* agglutinates red cells from most blood group A individuals (termed A1 individuals), but it does not agglutinate red cells from individuals of the A2 subgroup. The molecular structures of A1 and A2 subgroup antigens are distinct (Figure 13.9). These structural differences reflect the different catalytic activities of the A transferases encoded by the A1 and A2 alleles.

The heritable red cell antigenic polymorphisms determined by the *ABO* locus have important medical implications. Early in the postnatal period, the immune system generates IgM antibodies against ABO antigen(s) when they are *absent* from an individual's red cells. The antibodies likely reflect an immune response to glycan antigens presented by bacteria and fungi that express glycans similar or identical to the A and B blood group determinants. For instance, type-O individuals do not make A or B determinants and they exhibit relatively high titers of circulating IgM antibodies (termed isoagglutinins) against A or B blood group determinants. Similarly, blood group B individuals exhibit circulating IgM anti-A isoagglutinins, but they do not make isoagglutinins against the blood group B determinant, a "self" antigen. Sera taken from blood group A individuals contain anti-B but not anti-A antibodies. Finally, those individuals with the AB blood group do not make

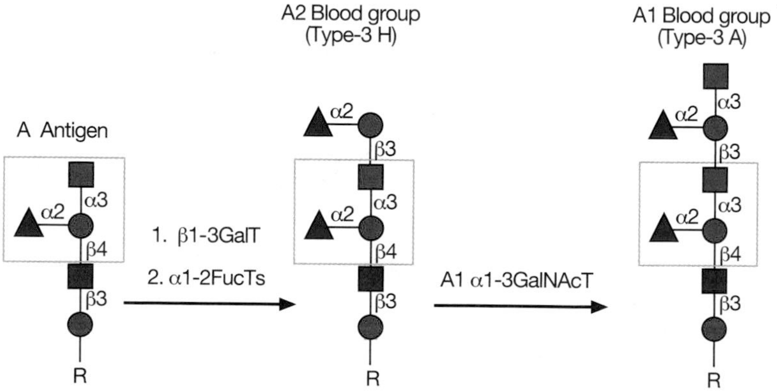

FIGURE 13.9. A1 and A2 blood group antigens. The type-2 A glycan is modified by a β1-3GalT and the α1-2FucT from the *H* locus to form a type-3 H determinant and the A2 blood group. The type-3 H glycan is a substrate for the A1 transferase to form the repeated A determinant unit proposed to be responsible for the strong serological reactivity of the A1 phenotype. The A2 transferase is unable to complete this last reaction efficiently. R represents glycoprotein or glycolipid. A-reactive epitopes are enclosed in boxes.

either anti-A or anti-B IgM isoagglutinins. Anti-H antibodies are not made in most people because a substantial fraction of H structures are converted to A or B determinants.

IgM isoagglutinins efficiently trigger the complement cascade and they circulate in human plasma at titers sufficient to cause complement-dependent lysis of transfused erythrocytes that display the corresponding blood group antigens. Such rapid red cell lysis is associated with an immediate, or acute, transfusion reaction, which can lead to hypotension, shock, acute renal failure, and death from circulatory collapse. This problem is avoided by ensuring that the ABO type of transfused red cells is compatible with the recipient's ABO type. Thus, an A recipient may receive red cells from another A person or from an O person, but not from a person of type B or AB. Blood banks perform typing and cross-matching assays. First, units of red cell products typed for the A and B antigens are chosen to match the patient's ABO type. To ensure that these are truly "compatible," the patient's serum is cross-matched by mixing with a small aliquot of each prospective red cell unit. Red cells of compatible units do not agglutinate (form a clump), whereas incompatibility is indicated by agglutinated red cells formed by antibodies in the patient's serum. Blood typing is used to ensure compatibility not only for red blood cell transfusions, but also for transfusion with plasma. Similar ABO compatibility concerns are important in heart, kidney, liver, and bone marrow transplantation procedures. The "type and cross" procedures have virtually eliminated ABO blood group transfusion reactions in the developed world. In the rare instances where a transfusion reaction occurs, the cause is usually human error.

Cross-matching procedures helped to identify a rare ABO blood group phenotype termed the Bombay phenotype, so named because the first identified individual lived in that city. Affected persons have red cells deficient in H, A, and B antigens, whereas their sera contain IgM antibodies that react with red cells from virtually all donors, including O red cells (H antigen-positive, A and B antigen-negative). Bombay people have inactive *H* and *Se* genes and are therefore incapable of synthesizing A, B, or H determinants in any tissue because of the absence of both α1-2FucT enzymes. They exhibit robust titers of anti-H, anti-A, and anti-B IgM antibodies and are therefore incompatible with red cells of all donors except those of the same Bombay (H-deficient) blood type. A related phenotype, termed para-Bombay, occurs in people with an inactive *H* gene but at least one functional *Se* allele (secretor-positive).

The fact that Bombay individuals appear healthy implies that developmental or physiological functions for the A, B, and H antigens, if they ever existed, are no longer relevant. It has been proposed that polymorphisms at the *ABO* locus may provide a selective advantage for protection from certain infectious agents, and recent genetic evidence tends to support this contention. Such selective pressures might therefore have maintained population polymorphisms at the *ABO* locus. In fact, a variety of associations have been made between the ABO blood group phenotype and relative risk for a spectrum of diseases.

One possible association concerns the well-known correlation between ABO status and plasma levels of von Willebrand factor (VWF), which is a glycoprotein involved in hemostasis. In the mouse, the circulating half-life of VWF (and thus its level in plasma) varies according to the nature of the alleles at a locus that encodes a β1-4GalNAcT, which modifies VWF glycans. A low serum VWF level is associated with an allele that directs expression of the β1-4GalNAcT to vascular endothelial cells, a major site of VWF synthesis. Glycoforms of VWF that carry the β1-4GalNAc modification are rapidly removed from plasma by the asialoglycoprotein receptor in liver. In contrast, mice with normal VWF levels do not express the β1-4GalNAcT in vascular endothelium, so their VWF is cleared less rapidly from the circulation. In humans, a similar mechanism could also account for the generally lower VWF levels in people with the O blood group, since VWF carries A and B blood group determinants in humans, and these structures may be recognized with lower affinities by the asialoglycoprotein receptor.

A second possible association is the proposed role for gastrointestinal ABO status and Lewis blood group antigens in the pathogenesis of the ulcerogenic spirochete *Helicobacter pylori.* The A blood group is associated with a modest increase in the risk of stomach cancer, whereas blood group O is associated with a slight increase in the risk of developing peptic ulcers. Although both of these disorders are now clearly associated with infection by *H. pylori*, it remains to be determined if there is a mechanistic relationship between ABO status and the consequences of *H. pylori* infection. A third explanation is that enveloped viruses carry the ABH glycans of their hosts and are thus susceptible to lysis upon first encountering the body fluids of another individual with an ABO-incompatible type. Finally, differences in susceptibility to severe complications of malaria appear to be affected by ABO blood groups. Several or all of these mechanisms may explain why the ABH system has remained active for more than 50 million years of primate evolution.

LEWIS BLOOD GROUPS

The Lewis blood group antigens are a related set of glycans that carry α1-3/α1-4 fucose residues (Figure 13.10). The term Lewis derives from a family who suffered from a red blood cell incompatibility that helped lead to the discovery of this blood group. The Lewis[a]

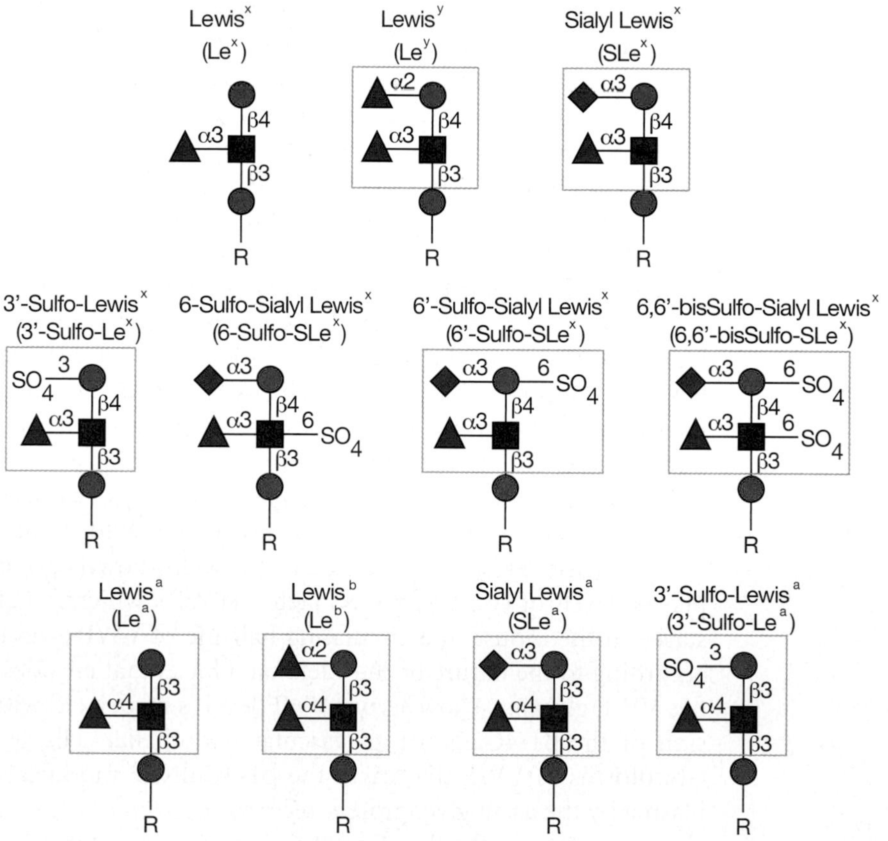

FIGURE 13.10. Type-1 and -2 Lewis determinants. Type-1 and -2 units differ in the linkage of the outermost galactose (β1-3 or β1-4, respectively) and in the linkage of fucose to the internal GlcNAc (α1-4 or α1-3, respectively). R represents N- or O-glycan or glycolipid. Lewis[x] determinants may contain sulfate at C-6 of galactose and GlcNAc. The 6-O-sulfated GlcNAc is necessary for lymphocyte homing to peripheral lymph nodes.

antigen (Le[a]) is synthesized by an α1-3/α1-4FucT encoded by the *Lewis* (*Le*) blood group locus (Figure 13.11). The Lewis[b] antigen (Le[b]) is synthesized by the concerted actions of the Le[a] α1-3/α1-4FucT and the α1-2FucT encoded by the *Se* gene (Figure 13.11). The nature of the alleles at the *Le* and *Se* loci determines the complement of Lewis glycans that are synthesized by an individual. Secretor-positive individuals convert type-1 units to type-1 H determinants that may be acted on by the Lewis α1-3/α1-4FucT to form the Le[b] determinant (Figure 13.11). Nonsecretors who do not synthesize type-1 H determinants in secretory epithelia express the Le[a] determinant through the action of the Lewis α1-3/α1-4FucT (Figure 13.11). Individuals with an inactive *Le* locus are termed Lewis-negative. Lewis-negative secretors express type-I H determinants that cannot be converted to Le[a] or Le[b] determinants (Figure 13.11). Lewis-negative nonsecretors express type-1 units that are devoid of fucose.

Expression of Le[a] and Le[b] glycans and the Lewis α1-3/α1-4FucT is largely restricted to the same epithelia that express the *Se* α1-2FucT. Thus, soluble forms of these antigens are released into secretions and body fluids. Le[a] and Le[b] antigens are also detectable on red

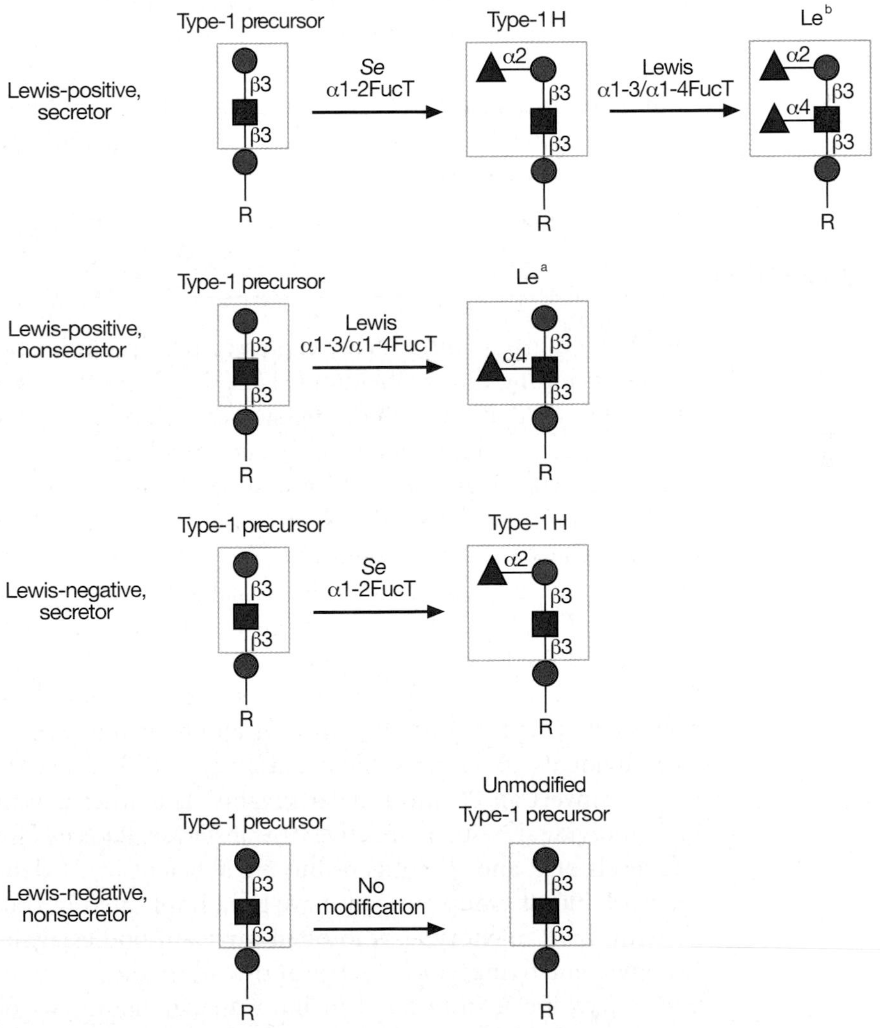

FIGURE 13.11. Lewis blood groups. The type-1 Lewis blood group determinants on glycoproteins and glycolipids (R) are characterized by the presence or absence of the *Secretor* locus α1-2FucT and the *Lewis* locus α1-3/α1-4FucT.

cells. However, the precursors of red cells do not synthesize these determinants. Instead, Lewis antigens are acquired by the red cell membrane through passive adsorption of Lewis-positive glycolipids that circulate in plasma as lipoprotein complexes and aqueous dispersions. Antibodies against the Le^a antigens have been implicated in occasional transfusion reactions.

Other members of the Lewis blood group family include the $Lewis^x$ (Le^x) and $Lewis^y$ (Le^y) determinants and forms of the Le^a and Le^x determinants that are sialylated and/or sulfated (Figure 13.10). These structures are formed through the actions of one or more α1-3FucTs that are distinct from the Lewis α1-3/α1-4FucT. Some members of the Lewis blood group antigen family have important functions in selectin-dependent leukocyte and tumor cell adhesion processes. Most strongly implicated are the sialylated and/or sulfated determinants represented by sialyl Le^x and its sulfated variants (Figure 13.10), which function as selectin ligands on glycoproteins and glycolipids of leukocytes and probably also on tumor cells (Chapters 31 and 44).

The Lewis blood group antigens have also been proposed to function in the pathogenesis of *H. pylori*, the causative agent in chronic active gastritis and which is associated with hypertrophic gastropathy, duodenal ulcer, gastric adenocarcinoma, and gastrointestinal lymphoma (see Chapter 34). This organism adheres to the gastric mucosa that express Lewis blood group antigens. *H. pylori* itself also expresses Lewis antigens in a case of molecular mimicry. Antibodies generated against the bacterial Lewis antigens have been found in humans and may function to enhance colonization by *H. pylori*. However, the physiological relationship between *H. pylori* pathobiology and Lewis or ABO blood group status remains to be established.

P BLOOD GROUPS

The P blood group antigens are expressed on glycolipids of the red cell membrane and in tissues, including the urothelium (Figure 13.12). Little is known about the genes or glycosyltransferases that direct the expression of P blood group antigens. The synthesis of P antigens involves two different pathways, each of which begins with lactosylceramide as a common precursor (Figure 13.12). In one pathway, P antigen biosynthesis is initiated by an α1-4GalT (P^k transferase) that synthesizes the P^k antigen. The P^k antigen may be modified by a β1-3GalNAcT (P transferase) to form the P antigen. In the second pathway, P1-antigen biosynthesis involves three sequential glycosylation reactions, again starting with lactosylceramide. The first two reactions lead to paragloboside synthesis. Paragloboside is a substrate of the P1 transferase, which forms the P1 antigen. Expression of these antigens is polymorphic in humans. The most common P blood group is P1. These individuals express the enzymes of both pathways and their red cells express both P and P1 antigens. P1 individuals also express small amounts of P^k because the P transferase does not completely convert all P^k into P determinants. The other common blood group phenotype is P2 in individuals with an inactive P1 transferase. Red cells from P2 individuals express normal levels of P and P^k antigens but are deficient in P1 determinants. Antibodies directed against P blood group antigens have been implicated in transfusion reactions in individuals with the P phenotype, who often carry antibodies against P, P1, and P^k determinants. Complement-fixing, cold-reactive anti-P antibodies known as "Donath–Landsteiner" antibodies have been implicated in intravascular hemolysis observed in a syndrome called paroxysmal cold hemoglobinuria (see Chapter 43).

Physiological functions for the P blood group antigens are not known. However, a role in the pathogenesis of urinary tract infections is suggested because various uropathogenic

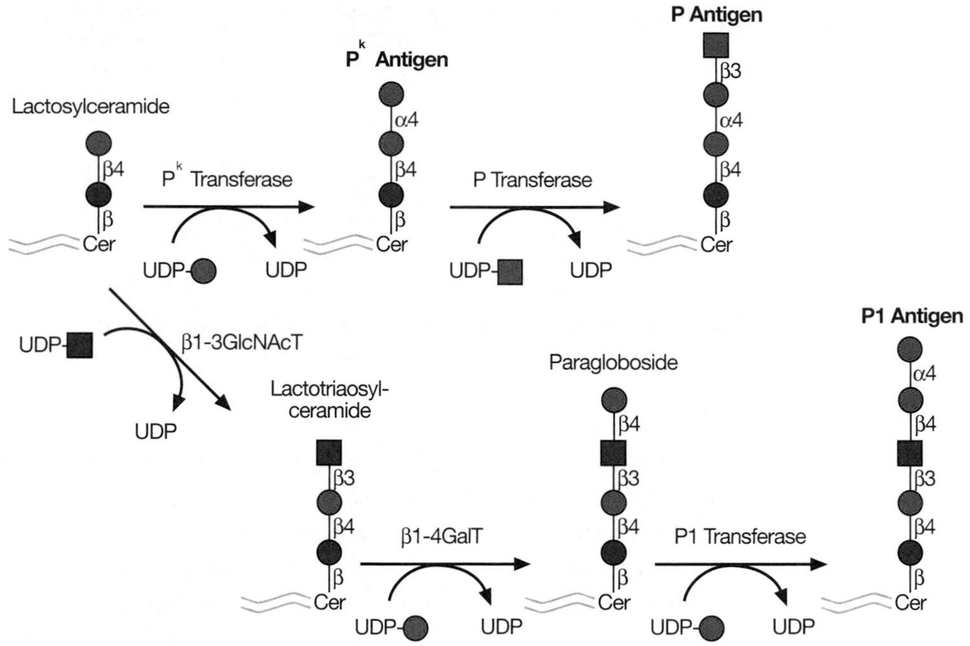

FIGURE 13.12. Biosynthesis of antigens of the P blood group system: P^k, P, and P1.

strains of *Escherichia coli* express adhesins that bind to the Galα1-4Gal moiety of the P^k and P1 antigens (see Chapter 34). The P1 determinant is expressed on the urothelium and may facilitate infection by mediating attachment of bacteria to the lining of the urinary tract. In fact, P1 individuals have a higher relative risk for urinary tract infections and pyelonephritis. In addition, the adhesion of a pyelonephritic strain of *E. coli* to renal tissue is mediated by a bacterial adhesin specific for the Galα1-4Gal structure, and deficiency of the adhesin severely attenuates the pyelonephritic activity of the organism.

The P blood group antigens may also have a role as receptors for the human parvovirus B19. This virus causes erythema infectiosum and leads to congenital anemia and hydrops fetalis following infection in utero. It is also associated with transient aplastic crisis in patients with hemolytic anemia and with cases of pure red cell aplasia and chronic anemia in immunocompromised individuals. Parvovirus B19 replication is restricted to erythroid progenitor cells. An adhesive interaction between the virion and P antigen–active glycolipids is involved in viral infection of erythroid progenitors. Individuals with the P blood group phenotype are apparently resistant to parvovirus B19 infection.

THE GALα1-3GAL TERMINUS

The Galα1-3Gal epitope is synthesized on type-2 units on glycolipids and glycoproteins by a specific α1-3GalT (Figure 13.13). This epitope and the α1-3GalT are expressed by New World primates and many nonprimate mammals, but they are absent from the cells and tissues of Old World primates including *Homo sapiens*. The molecular basis for the clade-specific absence of this enzyme is inactivation of the gene encoding the α1-3GalT in primate taxa. A physiological function for the Galα1-3Gal epitope has not been identified.

Species that do not express the Galα1-3Gal epitope, including humans, carry naturally occurring anti-Galα1-3Gal antibodies in their serum, likely because of immunization

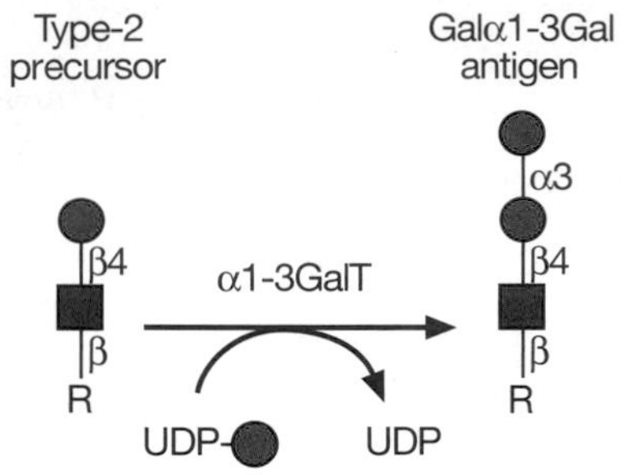

FIGURE 13.13. Structure and synthesis of the Galα1-3Gal antigen. The α1-3GalT uses unsubstituted type-2 units on glycoproteins or glycolipids (R) to form the Galα1-3Gal terminal epitope.

through exposure to microbial antigens similar or identical to the Galα1-3Gal epitope. Anti-Galα1-3Gal antibodies present a major barrier to the use of porcine and other non-primate organs for xenotransplantation in humans, because they bind to Galα1-3Gal epitopes on the vascular endothelium of the xenotransplants and cause hyperacute graft rejection through complement-mediated endothelial cell cytotoxicity. Efforts are in progress to overcome this barrier by using animal organ donors that have been genetically modified (see Chapter 38). Approaches include transgenic expression of enzymes such as the H α1-2FucT that diminishes Galα1-3Gal expression by diverting type-2 units toward H antigen synthesis. The creation of animal organ donors with an inactive α1-3GalT locus represents a definitive solution to the problem. Unfortunately, however, pig tissues lacking Galα1-3Gal elicit a graft rejection reaction to other pig antigens. Anti-Galα1-3Gal antibodies have also been shown to significantly diminish the infective efficiency of recombinant retroviruses because packaging cell lines used to propagate these viruses express the Galα1-3Gal epitope. As with the ABO antigens, this may reflect a natural mechanism that acts to restrict the interspecies spread of retroviral genomes. The problem has been solved through the generation of packaging cell lines that are deficient in α1-3GalT. Recombinant glycoproteins for therapeutic use in humans must also be prepared in cells that do not express α1-3GalT.

THE FORSSMAN ANTIGEN

The Forssman antigen (also known as globopentosylceramide) is a glycolipid that contains terminal *N*-acetylgalactosamine (GalNAc) in α1-3 linkage to the terminal GlcNAc of globoside (Figure 13.14). The Forssman antigen is expressed during embryonic and adult stages in many mammals. Humans carry anti-Forssman antibodies in their serum, suggesting that we do not synthesize the Forssman antigen. Although such antibodies are not consistently present, they may contribute to the pathogenesis of Guillain–Barré syndrome by binding to glycolipid components of peripheral nerve myelin. Similarly, there is evidence that small amounts of Forssman antigen may be found on human gastrointestinal epithelium, in various human cultured cells, and in pulmonary and gastointestinal tract carcinomas. These conflicting observations may reflect different specificities of the anti-Forssman monoclonal antibodies and differences in epitope reactivities with respect to detection methods. The function of the Forssman antigen is not known. However, anti-Forssman antibodies can disrupt tight junction formation, apical-basal polarization, and cell adhesion, suggesting that this molecule may participate in cell–cell adhesion and communication processes.

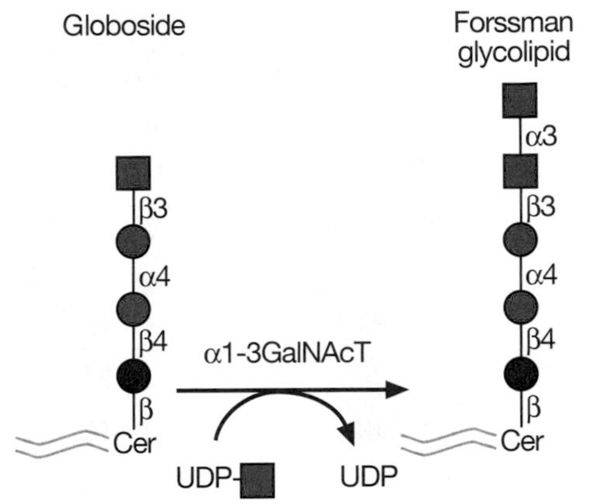

FIGURE 13.14. Forssman antigen biosynthesis. The glycolipid globoside serves as the substrate for the Forssman α1-3GalNAcT that forms globopentosylceramide, also termed the Forssman glycolipid.

SULFATED TERMINAL β-LINKED *N*-ACETYLGALACTOSAMINE ON PITUITARY GLYCOPROTEIN HORMONES

Glycans with sulfated terminal β-linked GalNAc are found on the pituitary glycoprotein hormones lutropin (LH) and thyrotropin (TSH) but not on follicle-stimulating hormone (FSH), although it is made in the same cells. These heterodimeric glycoproteins contain a common α subunit and a unique β subunit. Each subunit carries biantennary N-glycans. The biantennary N-glycans on TSH and LH contain an unusual 4-O-sulfated GalNAc attached to one or both GlcNAc residues (Figure 13.15). This contrasts with the N-glycans on FSH and on most N-glycans, in which GlcNAc residues are substituted with β1-4 galactose, which is often extended by α2-3 or α2-6 sialic acid (see Chapter 8). A free α subunit common to LH, TSH, and FSH is present in pituitary cells and it also carries this determinant, as do other glycoproteins synthesized by the pituitary and elsewhere (e.g., on the O-glycans of proopiomelanocortin). Synthesis of the sulfated GalNAc determinant is controlled by a β1-4GalNAcT that acts on GlcNAc residues of biantennary N-glycans (Figure 13.15). The terminal β1-4GalNAc is then sulfated by a sulfotransferase also expressed in pituitary cells. In some tissues, including the pituitary, the β1-4GalNAc is subsequently modified by an α2-6 sialic acid. GlcNAc residues extended with a β1-4GalNAc may also be modified by an α1-3 fucose residue. Both β1-4GalNAcT and β1-4GalT enzymes are expressed in pituitary cells, and the N-glycans on LH and TSH carry the unusual β1-4GalNAc. However, N-glycans on the related glycoprotein FSH carry the more common β1-4Gal residue. This protein-specific glycosylation is a consequence of interactions between the β1-4GalNAcT and a specific peptide motif present on the combined αβ subunits of LH and TSH. This interaction causes an increase in the catalytic efficiency of the β1-4GalNAcT that modifies biantennary N-glycans on LH and TSH at the expense of the competing β1-4GalT, which does not interact with the peptide motif. In contrast, the peptide motif recognized by the β1-4GalNAcT is not present in the β subunit of FSH, and the recognition motif on the α subunit of FSH is not accessible to the enzyme. Consequently, the biantennary N-glycans on FSH are modified exclusively by the competing β1-4GalT.

These differential glycosylation events have profound consequences for the ovulatory cycle in vertebrates. Circulating LH levels rise and fall in a highly pulsatile manner. This assures maximal stimulation of the ovarian LH receptor at the preovulatory surge, since sustained high LH levels would lead to LH receptor desensitization. The rise and fall in LH levels is due in part to pulsatile release of the hormone by the pituitary. However, it is also

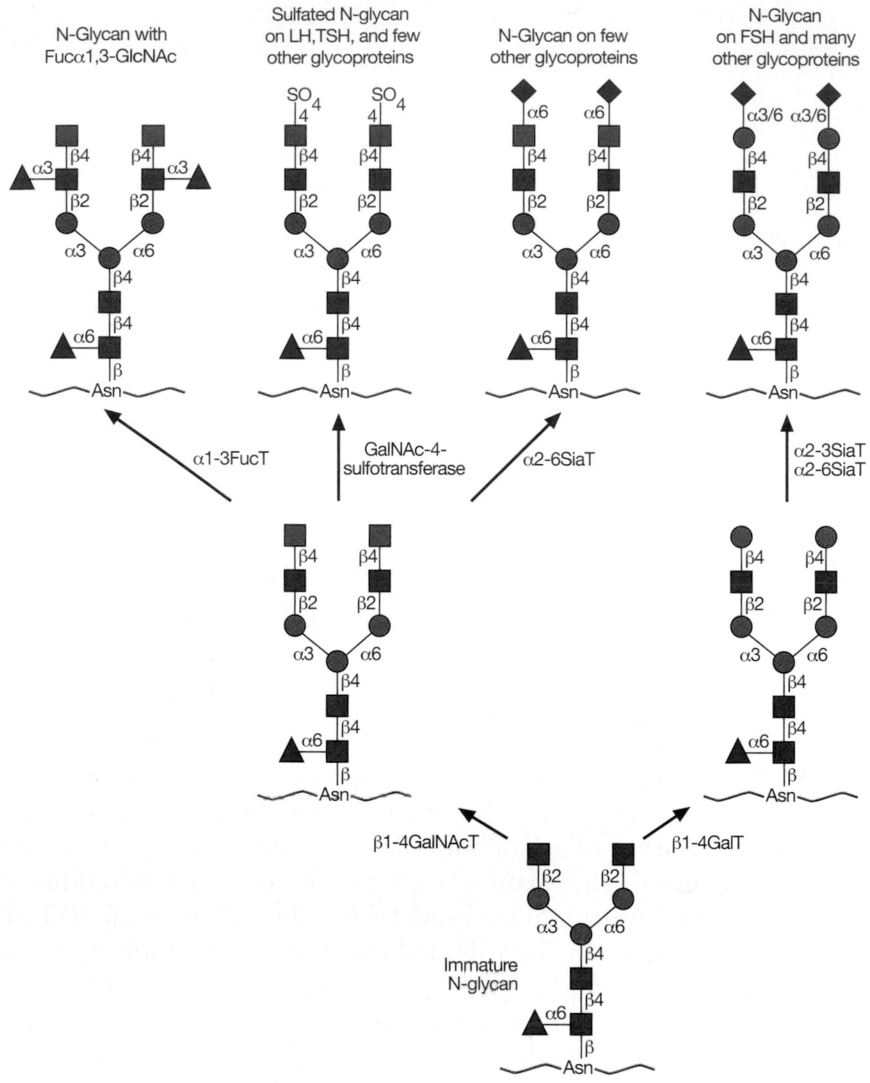

FIGURE 13.15. Structure and synthesis of N-glycans bearing terminal *N*-acetylgalactosamine (GalNAc), including those with sulfated-GalNAc found on the pituitary hormones lutropin (LH) and thyrotropin (TSH).

the consequence of rapid clearance of LH from the circulation mediated by a receptor that binds the terminal sulfated-GalNAcβ1-4GlcNAc determinant. This receptor is expressed in the liver by hepatic endothelial cells and by Kupffer cells. LH binding is followed by internalization and lysosomal degradation. The liver receptor is identical in protein sequence to that of the macrophage mannose receptor. However, in liver, the receptor recognizes sulfated-GalNAc via an R-type lectin domain, whereas in macrophages, the receptor recognizes mannose via an L-type lectin domain (see Chapter 28). An additional molecular basis for this difference is that in liver, the receptor is a dimer, whereas in macrophages, the receptor is a monomer. The receptor is now called the Man/GalNAc-SO_4 receptor.

Similar GalNAcβ1-4GlcNAc (LacdiNAc) termini occur on N-glycans from other vertebrate sources, such as bovine milk, rat prolactin, and kidney epithelial cells, as well as in invertebrates such as snails. These residues generally do not become sulfated as in pituitary hormones, but they are frequently α2-6-sialylated in vertebrates. It is not yet clear how the GalNAcTs responsible for producing LacdiNAc are related to the enzyme that acts specifically on pituitary hormones.

TERMINAL β-LINKED *N*-ACETYLGALACTOSAMINE

The addition of GalNAc to glycans terminating with α2-3 sialic acid may occur on glycoproteins and glycolipids (Figure 13.16). On glycoproteins, this structure forms the human Sd^a blood group. In mouse, it was first described on mouse cytotoxic T lymphocytes (CTLs) and was termed the CT antigen. On glycolipids, this terminus forms the ganglioside GM2. The human Sd^a antigen was first identified in N-glycans of Tamm–Horsfall glycoprotein from human urine. The Sd^a determinant is formed by the addition of β1-4GalNAc to the Gal of α2-3-sialylated type-2 units (Figure 13.16). A β1-4GalNAcT capable of catalyzing this reaction exists in human kidney and urine, intestine, colon, and blood plasma. Both human and mouse β1-4GalNAcTs transfer GalNAc to N- and O-glycans present on glycoproteins but not to the glycolipid GM3 (Siaα2-3Galβ1-4Glc-Cer), even though both can efficiently use 3′-sialyllactose (Siaα2-3Galβ1-4Glc) as a substrate in vitro. In the mouse, the Sd^a antigen(s) is recognized by IgM monoclonal antibodies termed CT1 and CT2, which were isolated for their ability to block lysis of cellular targets by a murine CTL clone. The function of the Sd^a determinant in humans or mice is unknown. Rare humans lack the determinant and form naturally occurring antibodies against it, but they exhibit no apparent pathophysiology. As noted earlier, rare strains of mice with a dominantly inherited form of von Willebrand's disease express the β1-4GalNAcT aberrantly in vascular endothelium. Since the blood-clotting protein VWF is also expressed by endothelial cells, the presence of the β1-4GalNAcT in this unusual location generates VWF carrying the Sd^a determinant. This VWF glycoform is rapidly cleared from the circulation, accounting for the low levels of VWF circulating in such strains of mice. Rapid clearance is mediated by the asialoglycoprotein receptor in liver, which exhibits affinity for terminal GalNAc.

The glycolipid form of the Sd^a determinant, termed GM2, is widely expressed in the central and peripheral nervous systems and in the adrenal gland, suggesting potential biological roles in these organs. In fact, mice homozygous for a null mutation at the GM2 synthase locus exhibit modest conduction defects in the peripheral nervous system and male sterility (see Chapter 38).

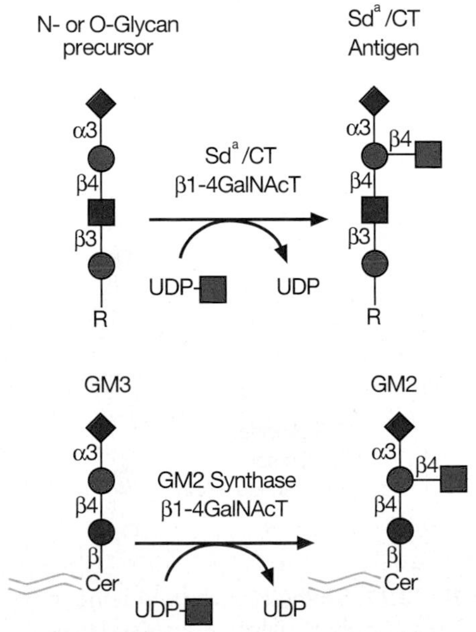

FIGURE 13.16. Synthesis of the human Sd^a or mouse CT antigen and the glycolipid GM2.

α2-3-SIALYLATED GLYCANS

Sialic acid in α2-3 linkage is found on N-glycans, O-glycans, and glycolipids. A family of at least six different α2-3 sialyltransferases (ST3Gal-I to ST3Gal-VI) are responsible for its synthesis (Figure 13.17). ST3Gal-III and ST3Gal-IV are expressed in most cells of adult mammals. In contrast, ST3Gal-I transcripts in humans and mice are abundant in the spleen, liver, bone marrow, thymus, and salivary glands and are less abundant in other tissues. ST3Gal-II is also more restricted in expression pattern with transcripts most abundant in the brain, where α2-3-sialylated glycolipids are abundant (Figure 13.17). ST3Gal-V is expressed in the brain, skeletal muscle, testes, and liver. In vertebrates, α2-3 sialic acid residues are found on

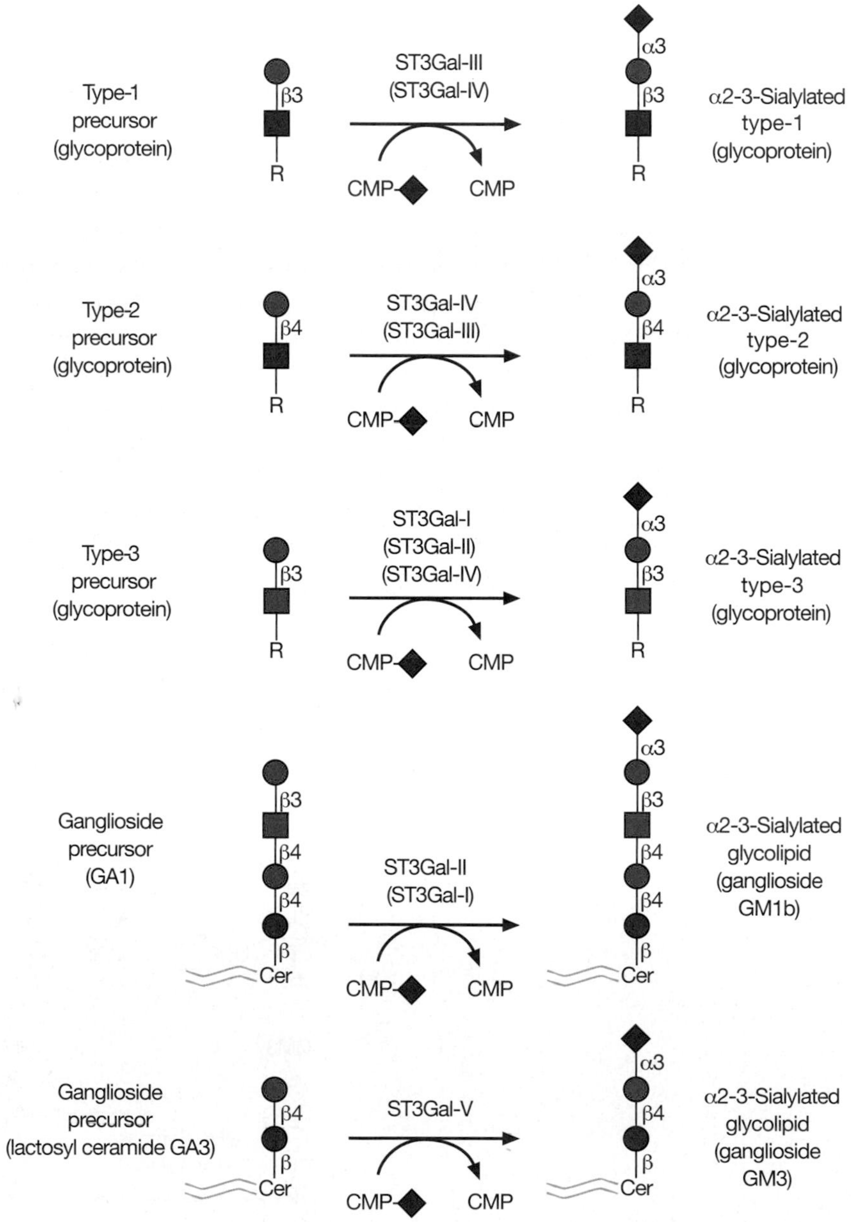

FIGURE 13.17. Synthesis of glycoproteins and glycolipids bearing terminal α2-3 sialic acid transferred by the ST3Gal family of sialyltransferases. Enzymes in parentheses contribute at relatively low levels in vitro to the reactions indicated.

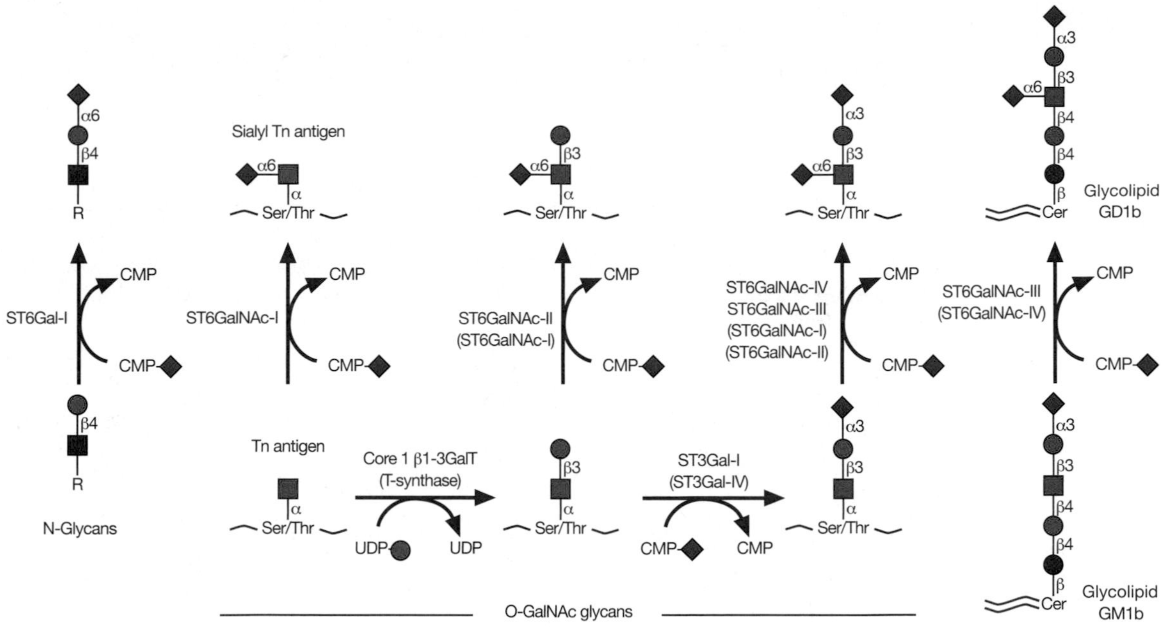

FIGURE 13.18. Synthesis of α2-6 sialic acid on O-glycans and glycolipids (see Chapters 9 and 10) by the ST6GalNAc family of sialyltransferases. Enzymes in parentheses contribute at relatively low levels in vitro to the reactions indicated.

terminal galactose (Figure 13.17). Glycans terminating with galactose are substrates for the transfer of α1-3 fucose (Figure 13.10 and Chapter 31), β1-4GalNAc in limited circumstances (Figure 13.15), α2-6 sialic acid (Figure 13.18), α2-8 sialic acid (Figure 13.19 and Figure 13.20), and SO_4 (Figure 13.10). In contrast, the α2-3-sialylated glycans presented in Figure 13.17 are generally not substrates for other enzymes, including α1-2FucTs, α1-3GalT, GlcNAcTs, or GalNAcTs. The latter enzymes compete with terminal α2-3 sialyltransferases. Although most α2-3 sialic acid on glycoproteins is found on complex N-glycans (see Chapter 8) and O-GalNAc glycans (see Chapter 9), it also occurs on O-fucose and O-man-

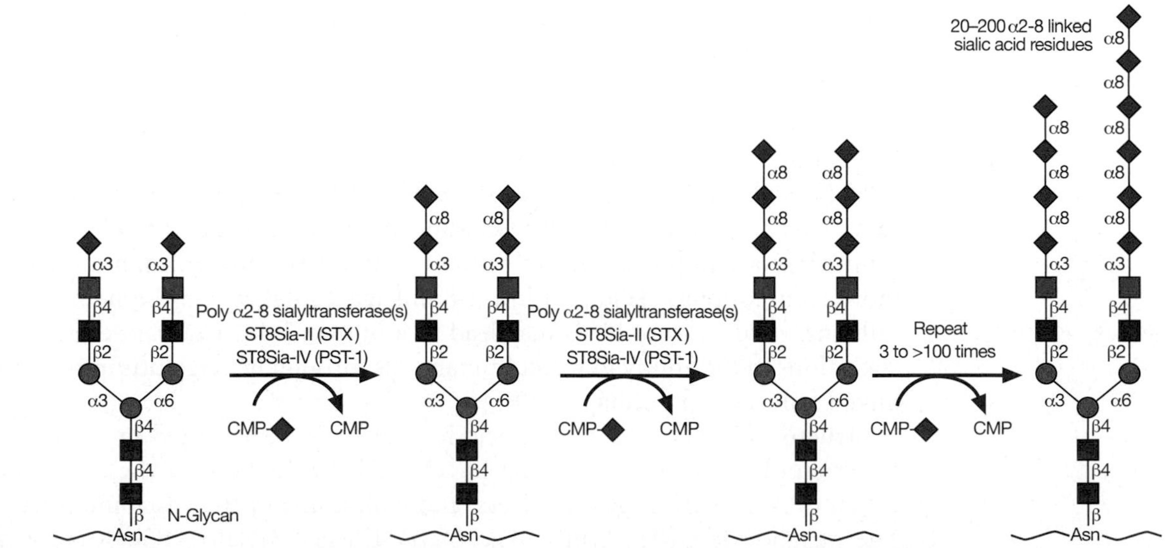

FIGURE 13.19. Structure and synthesis of polysialic acid on N-glycans.

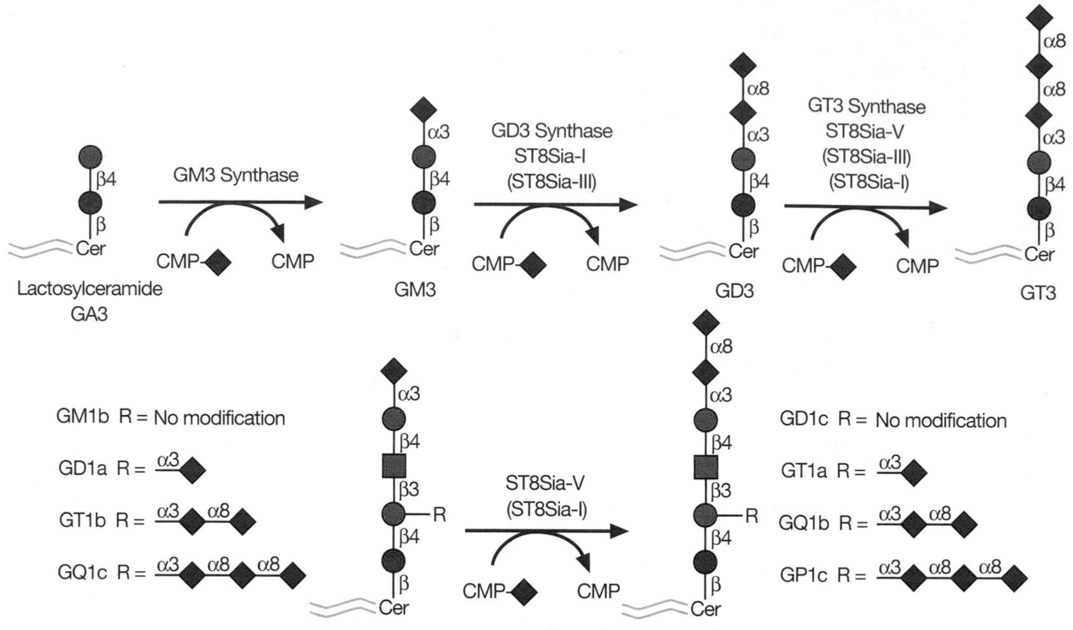

FIGURE 13.20. Structure and synthesis of polysialic acid on glycolipids. Enzymes in parentheses contribute at low levels in vitro to the reactions indicated.

nose glycans that are found on a limited subset of glycoproteins (see Chapter 12). As discussed in Chapter 31, the ligands recognized by the selectin family of leukocyte adhesion molecules include α2-3-sialylated glycans, modified additionally by α1-3 fucose or by α1-3 fucose and sulfate. Glycans bearing α2-3 sialic acid may contribute to the circulating half-life of plasma glycoproteins by virtue of "masking" terminal galactose residues that would contribute to the removal of glycoproteins from serum by the asialoglycoprotein receptor (see Chapter 31). However, mice deficient in this receptor do not have increased plasma levels of desialylated glycoproteins or lipoproteins in their circulation. In fact, recent studies demonstrate that the asialoglycoprotein receptor can also recognize glycans terminating in α2-6 sialic acid attached to GalNAcβ1-4 or Galβ1-4 (Figure 13.15)

In other studies, the role of ST3Gal-I in the production of the Siaα2-3Galβ1-3GalNAcα-Ser/Thr glycan is important for the viability of peripheral CD8$^+$ T cells. Mice lacking ST3Gal-I exhibit decreased cytotoxic T cell responses with an increase in the apoptotic death of naïve CD8$^+$ T cells. The timing, specificity, and location of this enhanced apoptosis indicate an endogenous lectin, yet to be recognized, that binds to Galβ1-3GalNAcα-Ser/Thr glycans (see Chapter 33).

Sialic acid in α2-3 linkage is recognized by the hemagglutinin in the envelope of influenza viruses from birds and pigs. Binding to sialic acid and subsequent release by neuraminidase are important for infection by influenza virus. Human influenza viruses bind more commonly to sialic acid in α2-6 linkage. Mutations in the hemagglutinin genes of influenza viruses from birds may lead to a human influenza pandemic partly because of alterations in the ability to infect human cells through hemagglutinin recognition of α2-6-linked sialic acid (see Chapter 39).

Structures bearing α2-3 sialic acid have also been implicated in bacterial pathogenesis. For example, glycans terminating in α2-3 sialic acid support the adhesion of *H. pylori*, which causes gastritis, gastric ulcers, and lymphoma of the gastrointestinal tract mucosa. The ganglioside GM1 (Galβ1-3GlcNAcβ1-4[Siaα2-3]Galβ1-4GlcβCer) is a receptor for cholera toxin produced by *Vibrio cholerae* and the heat-labile enterotoxin (LT-1) produced by enterotoxigenic *E. coli* (see Chapter 39). Glycan-based inhibitors are currently under evaluation in humans for their ability to diminish the symptoms and progression of cholera.

α2-6-SIALYLATED GLYCANS

Sialic acid in α2-6 linkage is found on N-glycans, O-glycans, and glycolipids. A family of at least six different α2-6 sialyltransferases (Figure 13.18) (including ST6Gal-I, ST6Gal-II, and ST6GalNAc-I to ST6GalNAc-IV) catalyzes its transfer. In vertebrates, α2-6 sialic acid is found on terminal galactose, on terminal or subterminal GlcNAc, or, in the case of reactions catalyzed by ST6GalNAc-III, on an internal GalNAc. α2-6 sialic acid appears less ubiquitously on glycans compared to α2-3 sialic acid. Glycans with terminal α2-6 sialic acid are generally not modified further, except possibly by some members of the poly α2-8 sialyltransferase family, as discussed below. The products of ST6Gal-I are typically found on N-glycans, although in vitro assays indicate that β1-4 galactose on O-GalNAc glycans can be modified by this enzyme. ST6Gal-I is expressed at a relatively high level in hepatocytes and lymphocytes and is responsible for α2-6 sialylation of serum glycoproteins and glycoproteins of the antigen receptor complex in lymphocytes. In contrast, the α2-6-sialylated glycans from ST6GalNAc-I and ST6GalNAc-II are restricted to O-glycans. ST6GalNAc-III is responsible for transferring α2-6 sialic acid to the GalNAcα-Ser/Thr core of O-glycans, whereas ST6GalNAc-IV appears to use glycolipids as preferred acceptors.

As noted above (and in Chapter 39), α2-6 sialic acid serves as a receptor for many strains of influenza virus infectious for humans, and glycoproteins bearing α2-6 sialic acid are recognized and cleared from the circulation by asialoglycoprotein receptor.

Mice lacking the enzyme ST6Gal-I exhibit an immunodeficiency characterized by a diminished antibody response to T-lymphocyte-dependent and -independent antigens, reduced B-lymphocyte proliferation in response to cross-linking of the B cell surface glycoproteins CD40 and surface IgM, reductions in the expression of B cell surface IgM and CD22, and an approximately 65% reduction in serum IgM levels. B lymphocyte antigen receptor–dependent signal transduction processes are also attenuated in these mice. These observations disclosed an important role for ST6Gal-I in the immune system and indicated defects in B cell receptor (BCR) signaling (see Chapter 32). The extracellular domain of CD22 on B lymphocytes specifically recognizes Siaα2-6Galβ1-4GlcNAc-. In the absence of α2-6 sialic acid, CD22 exhibits increased clustering with the BCR and thus BCR signaling is dysregulated. Interestingly, mice lacking both ST6Gal-I and CD22 are not immunodeficient, suggesting that α2-6 sialic acid is a functional ligand for CD22.

α2-8-SIALYLATED GLYCANS

Glycans modified by α2-8 sialic acid are developmentally regulated in the vertebrate central nervous system. α2-8-Sialylated glycans may also be expressed in nonneuronal tissues during embryogenesis and in the adult vertebrate, and these glycans have been observed on transformed cells. At least five different α2-8 sialyltransferases direct the synthesis of α2-8-sialylated glycans (Figure 13.19 and Figure 13.20). Two of these enzymes, termed ST8Sia-II (also called STX) and ST8Sia-IV (also called PST-1 or PST) catalyze the synthesis of linear polymers of 200 or more α2-8 sialic acid residues to give polysialic acid (PSA) (Figure 13.19). N-glycans with terminal α2-3 sialic acid (and possibly α2-6 sialic acid) serve as the substrate for attachment of the initial α2-8 sialic acid. This "initiation" reaction is followed by an "elongation" reaction in which each α2-8 sialic acid serves as the attachment site for the next α2-8 sialic acid. Both activities are performed by the ST8Sia-II and ST8Sia-IV enzymes.

PSA has been most extensively studied as a component of the neural cell-adhesion molecule (NCAM), a member of the immunoglobulin superfamily (see Chapter 14). The α subunit of the voltage-gated sodium channel may also carry PSA. Both ST8Sia-II and ST8Sia-IV are autocatalytic and synthesize PSA on their own N-glycans, although polysialylation is not a prerequisite for their polysialyltransferase activity. Autocatalysis may yield membrane-associated, poly α2-8-sialylated forms of the enzymes, which may be why some cultured cell lines that do not express NCAM or the sodium channel express surface PSA when transfected with ST8Sia-II or with ST8Sia-IV.

Regulated expression of PSA during development is directed in part by tissue-specific regulation of the expression of the ST8Sia-II and ST8Sia-IV genes (see Chapter 38). ST8Sia-II is prominently expressed in fetal brain, but it declines markedly during the postnatal period. In contrast, ST8Sia-IV continues to be expressed postnatally in the brain and in nonneuronal tissues such as the heart, lung, and spleen. Posttranscriptional mechanisms may also regulate PSA expression. PSA negatively modulates the homotypic adhesive properties of NCAM on opposing cells. The embryonic form of NCAM is extensively modified by PSA and is less able to participate in homotypic adhesive interactions than is the adult form of NCAM, which is not highly modified by PSA. PSA can also diminish interactions promoted by other adhesion molecules, including L1-dependent attachment to laminin or collagen. Since PSA is highly negatively charged, highly hydrated, and contributes up to one third of the molecular mass of NCAM, it is believed that PSA negatively modulates cell-adhesion processes by physical interference with apposition of the plasma membranes of adjacent cells. Mice lacking ST8Sia-IV exhibit reduced PSA in certain brain regions and have altered neuronal responses in the hippocampal CA1 region. Mice lacking ST8Sia-II have a distinct neuronal phenotype because of misguided migration of a subset of hippocampal neurons and ectopic synapses. When both ST8Sia-II and ST8Sia-IV are inactivated, mice lack PSA in the brain, have severe neuronal and other problems, and die precociously. However, this phenotype is reversed by also removing NCAM in a triple-gene-knockout strain. This shows that the loss of PSA from NCAM is the basis of the neuronal defects in the ST8Sia-II and ST8Sia-IV mutant mice and that the presence of NCAM lacking PSA causes the severe defects in double-knockout mice (see Chapter 38).

Certain glycolipids also carry α2-8 sialic acid linkages, which are constructed by three α2-8 sialyltransferases termed ST8Sia-I (also known as GD3 synthase), ST8Sia-III, and ST8Sia-V (Figure 13.20). They generate single or oligomeric α2-8 sialic acid but not the polymeric PSA synthesized by ST8Sia-II or ST8Sia-IV. The three enzymes are generally thought to act primarily on glycolipid substrates, but in vitro studies suggest that ST8Sia-III can also utilize N-glycans. The nature of products formed by these enzymes in vivo is not clear, but their in vitro substrate specificities are presented in Figure 13.20. These α2-8 sialyltransferases are expressed in the brain, where each exhibits a distinct developmentally regulated expression pattern. ST8Sia-I is also expressed in kidney and thymus. In vitro experiments imply that certain α2-8-sialylated glycolipids may participate in signal transduction processes in neuronal cell types. Inactivation of the ST8Sia-I gene in the mouse causes alterations in sensory neuron responses to pain.

SULFATED GLYCANS: L-SELECTIN LIGANDS, HNK-1, AND KERATAN SULFATE

In principle, any free hydroxyl group on a monosaccharide has the potential to be modified by sulfation. However, in vertebrates, glycan sulfation is generally restricted to galactose, GlcNAc, glucuronic acid, and GalNAc at internal or terminal positions. The internally sulfated glycans represented by heparin, heparan sulfate, and chondroitin sulfate proteoglycans are discussed in Chapter 16. This chapter discusses the sulfated glycans that contribute to L-selectin ligands, the HNK-1 epitope, keratan sulfate, and some pituitary glycoprotein hormones as mentioned above.

Sulfated glycans on the specialized endothelium that lines the postcapillary venules in peripheral lymph nodes and other secondary lymphoid organs mediate lymphocyte homing to peripheral lymph nodes. L-selectin on lymphocytes binds to the high endothelial venules (HEV) through recognition of L-selectin ligands present on O-GalNAc glycans of HEV glycoproteins. Sulfated forms of α2-3-sialylated, α1-3-fucosylated glycans represented by the sialyl Lewis^{x} determinant (Figure 13.10) provide an essential contribution to L-selectin recognition of these glycoproteins. Sulfation occurs at C-6 of the galactose and the GlcNAc, both of which contribute to L-selectin ligand activity. Mice lacking these two sulfotransferases exhibit almost no homing of lymphocytes to HEV. The biosynthesis of sulfated L-selectin ligands and the enzymes that participate in this process are discussed in Chapter 31.

The HNK-1 antigen is a terminally sulfated glycan that was first described on human natural killer cells, and it is also called CD57. The HNK-1 epitope is expressed by a variety of cell types in the vertebrate nervous system, where its cell type–specific expression patterns change during neural development (see Chapter 38). The HNK-1 determinant is composed of a 3-O-sulfated glucuronic acid (GlcA) attached in α1-3 linkage to a terminal galactose (Figure 13.21). This structure is synthesized by specific glucuronosyltransferases that act on terminal (poly)-*N*-acetyllactosamine units of N-glycans. Glucuronylation is followed by 3-O-sulfation of the GlcA by one or more specific sulfotransferases. The HNK-1 epitope has also been described on O-glycans of glycoproteins, on proteoglycans, and on glycolipids. Two different glucuronic acid transferases participate in HNK-1 GlcA addition: GlcAT-P and GlcAT-S. They have very different activities for glycoprotein or glycolipid substrates in vitro and thus may generate functionally different HNK-1 epitopes in vivo. The HNK-1 epitope is present on a variety of neuronal cell glycoproteins, including NCAM, contactin, myelin-associated glycoprotein, telencephalin, L1, and P0 (the major glycoprotein of peripheral nerve myelin). There is evidence that HNK-1 can function as a ligand for laminin, L-selectin, P-selectin, and a cerebellar adhesion protein termed amphoterin. HNK-1 has also been shown to mediate homotypic adhesive interactions involving P0. HNK-1-dependent adhesive interactions have been implicated in cell migration processes involving cell–cell and cell–matrix interactions and are proposed to participate in reinnervation of muscles by motor neurons.

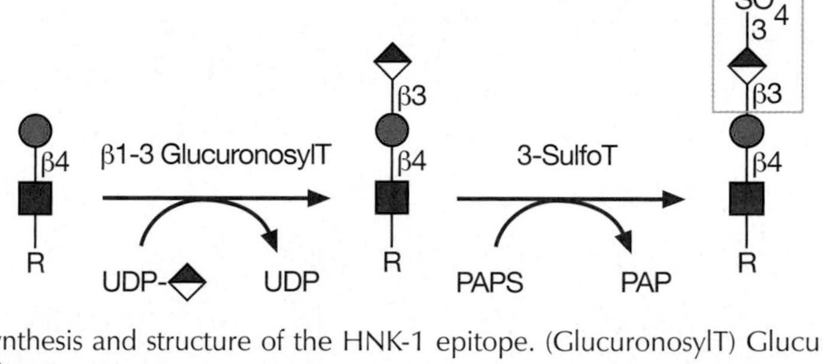

FIGURE 13.21. Synthesis and structure of the HNK-1 epitope. (GlucuronosylT) Glucuronosyltransferase; (SulfoT) sulfotransferase.

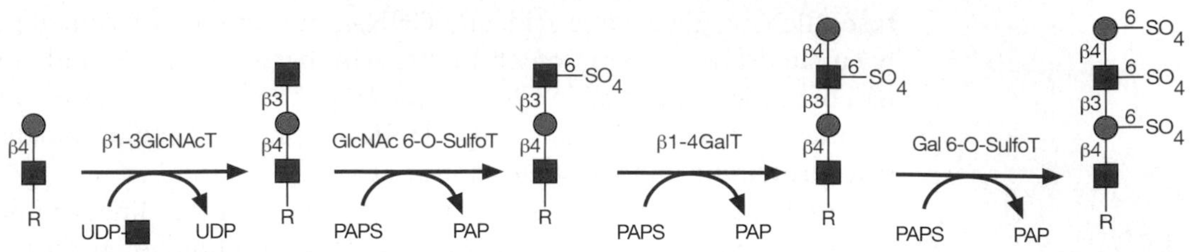

FIGURE 13.22. Synthesis of keratan sulfate. Different 6-O-sulfotransferases (SulfoT) transfer sulfate to the C-6 of galactose and GlcNAc in poly-*N*-acetyllactosamine chains. Keratan sulfate is attached to proteins (R) via an N-glycan (KS type I) or an O-glycan (KS type II).

Keratan sulfate represents a prominent posttranslational modification of several extracellular matrix proteins (see Chapter 16). Keratan sulfate is a poly-*N*-acetyllactosamine chain attached to proteins via a typical N-glycan (KS type I) or a typical O-glycan (KS type II). It is believed that keratan sulfate is synthesized in a stepwise manner during poly-*N*-acetyllactosamine synthesis, in coordination with a pair of 6-O-sulfotransferase activities (Figure 13.22). A keratan-specific *N*-acetylglucosamine 6-O-sulfotransferase activity (CHST6) transfers sulfate to terminal GlcNAc residues and will not effectively modify internal GlcNAc nor GalNAc in chondroitin sulfate. Mutations in this gene have been associated with macular corneal dystrophy in humans. A second, distinct 6-O-sulfotransferase (CHST5) also catalyzes 6-O-sulfation of GlcNAc in keratan sulfate.

The galactose 6-O-sulfotransferase CHST3 efficiently sulfates galactose within keratan sulfate poly-*N*-acetyllactosamine units, but it is apparently more efficient when the galactose is adjacent to a 6-O-sulfated GlcNAc. These observations imply that sulfation of GlcNAc residues can only occur when terminal GlcNAc exists during synthesis and elongation of poly-*N*-acetyllactosamine, whereas sulfation of galactose may occur at any time during poly-*N*-acetyllactosamine synthesis.

Corneal keratan sulfate (KS I) apparently maintains the proper spatial organization of the type I collagen fibrils in the cornea that promote transparency of this structure. Alterations in the degree of sulfation of keratan sulfate are associated with corneal opacity in a disease known as macular corneal dystrophy. Alterations in the amount of sulfation of brain keratan sulfate have been found in the brains of Alzheimer's disease patients.

FURTHER READING

Blumenfeld O.O. and Patnaik S.K. 2004. Allelic genes of blood group antigens: A source of human mutations and cSNPs documented in the Blood Group Antigen Gene Mutation Database. *Hum. Mutat.* **23:** 8–16. http://www.ncbi.nlm.nih.gov/gv/mhc/xslcgi.cgi?cmd=bgmut/home

Yamamoto F. 2004. ABO blood group system—ABH oligosaccharide antigens, anti-A and anti-B, A and B glycosyltransferases, and ABO genes. *Immunohematol.* **20:** 3–22.

CHAPTER 14

Sialic Acids

Ajit Varki and Roland Schauer

THIS CHAPTER DESCRIBES THE SIALIC ACID FAMILY OF MONOSACCHARIDES, with respect to their biosynthesis, structural diversity, and linkage to the underlying glycan chain. Also mentioned are the general principles behind different methods for their study. The biological and pathophysiological roles of sialic acids are briefly considered, particularly the functional significance of lectins that recognize sialic acids.

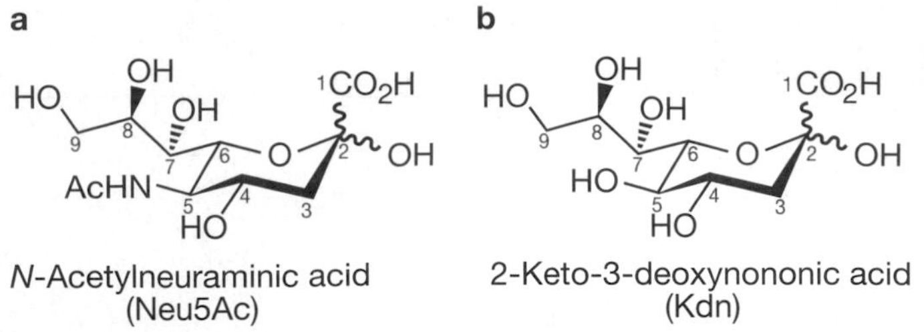

FIGURE 14.1. Two common "primary" sialic acids. Shown are 5-acetamido-2-keto-3,5-dideoxy-D-glycero-D-galactonononic acid (*N*-acetylneuraminic acid, Neu5Ac; *a*) and 2-keto-3-deoxy-D-glycero-D-galactonononic acid (2-keto-3-deoxynononic acid, Kdn; *b*). The only difference is the substitution at the C-5 position. All other sialic acids are apparently metabolically derived from these two, with the exception of some bacterial molecules such as legionaminic acid (not shown; see text for discussion). Neu5Ac is more common than Kdn in most vertebrate cell types. The bonds of the anomeric center (C-2) are drawn to indicate mutarotation between α- and β-anomeric forms in solution. Glycosidically bound sialic acids in naturally occurring glycans are in the α form, and free sialic acids in solution are mainly in the β form.

HISTORICAL BACKGROUND

About 70 years ago, Gunnar Blix, Ernst Klenk, and other investigators discovered sialic acid as a major product released by mild acid hydrolysis of brain glycolipids or salivary mucins. The structure, chemistry, and biosynthesis of the compound that they obtained (*N*-acetylneuraminic acid or Neu5Ac, a 9-carbon, acidic α-keto sugar; see Figure 14.1) were elucidated in the 1950s and 1960s by multiple groups. Sialic acid had already been shown to be the cellular receptor for influenza viruses by George Hirst and Frank Macfarlane Burnet in the 1940s. Erwin Chargaff's group then discovered that the "receptor-destroying enzyme" (RDE, a term coined by Burnet) of influenza viruses acts as a sialidase, releasing sialic acids from macromolecules, and Karl Meyer's group found a similar activity in bacteria. Alfred Gottschalk suggested the name "neuraminidase" for this activity in 1957. The pathways for biosynthesis of Neu5Ac were then worked out, largely by the groups led by Saul Roseman and Leonard Warren. From the earliest days, it was apparent that Neu5Ac was the most common member of a large family of related molecules derived from neuraminic acid. Partly because of its discovery in salivary mucins (Greek: *sialos*), this family was christened the "sialic acids." By the 1980s, more than 30 types of sialic acid had been described. The discovery of 2-keto-3-deoxynononic acid (Kdn; also called 3-deoxy-non-2-ulosonic acid, a desamino form of neuraminic acid; see Figure 14.1) further expanded the family of sialic acids, which now contains more than 50 members.

DIVERSITY IN STRUCTURE AND LINKAGE

Sialic acids (Sias) are typically found to be terminating branches of N-glycans, O-glycans, and glycosphingolipids (gangliosides) (and occasionally capping side chains of GPI anchors) (see Chapter 1, Figure 1.6). Their remarkable potential for biologically significant diversity justifies a separate chapter devoted to this one type of monosaccharide. The first level of diversity results from the different α linkages (Figure 14.2) that may be formed between the C-2 of Sias and underlying sugars by specific sialyltransferases, using CMP-Sias as high-energy donors (see also Chapters 5 and 13). The most common linkages are to the C-3 or C-6 positions of galactose residues or to the C-6 position of *N*-acetylgalactosamine residues. Sialic acids can also occupy internal positions within glycans, the most

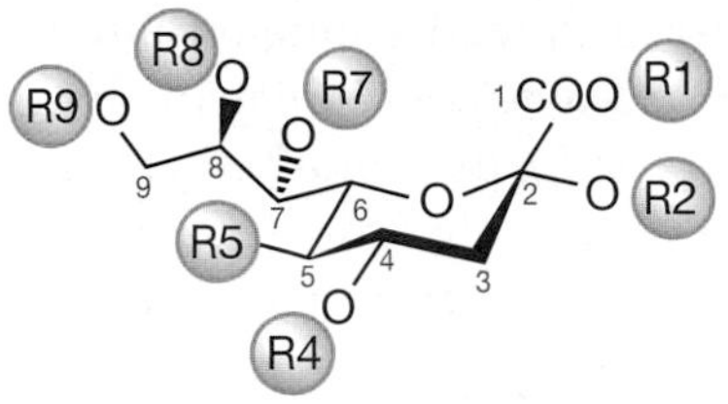

FIGURE 14.2. Diversity in the sialic acids. The nine-carbon backbone common to all known Sias is shown, in the α configuration. The following variations can occur at the carbon positions indicated:
R1 = H (on dissociation at physiological pH, gives the negative charge of Sia); can form lactones with hydroxyl groups on the same molecule or on other glycans; can form lactams with a free amino group at C-5; tauryl group.
R2 = H; alpha linkage to Gal(3/4/6), GalNAc(6), GlcNAc(4/6), Sia (8/9), or 5-O-Neu5Gc; oxygen linked to C-7 in 2,7-anhydro molecule; anomeric hydroxyl eliminated in Neu2en5Ac (double bond to C-3).
R4 = H; -acetyl; anhydro to C-8; Fuc; Gal.
R5 = Amino; N-acetyl; N-glycolyl; hydroxyl; N-acetimidoyl; N-glycolyl-O-acetyl; N-glycolyl-O-methyl; N-glycolyl-O-2-Neu5Gc.
R7 = H; -acetyl; anhydro to C-2; substituted by amino and N-acetyl in Leg.
R8 = H; -acetyl; anhydro to C-4; -methyl; -sulfate; Sia; Glc.
R9 = -H; -acetyl; -lactyl; -phosphate; -sulfate; Sia; OH substituted by H in Leg.

common being when one Sia residue is attached to another, often at C-8 position (see the section on oligosialic and polysialic acids below). In addition, internal Sias can occur in the repeating units of some bacterial polysaccharides and echinodermal oligosaccharides. In echinoderms, other monosaccharides (e.g., fucose and galactose) can be linked to C-4 of glycosidically bound Sia residues (see Figure 14.2).

The second level of diversity arises from a variety of natural modifications (Figure 14.2). As mentioned above, C-5 position can have an N-acetyl group (giving Neu5Ac) or a hydroxyl group (as in Kdn). The 5-N-acetyl group can also be hydroxylated, giving *N*-glycolylneuraminic acid (Neu5Gc). Less commonly, the 5-amino group is not N-acylated, giving neuraminic acid (Neu). These four "core" Sia molecules (Neu5Ac, Neu5Gc, Kdn, and Neu) can carry one or more additional substitutions at the hydroxyl groups on C-4, C-7, C-8, and C-9 (O-acetyl, O-methyl, O-sulfate, O-lactyl, or phosphate groups). The carboxylate group at the C-1 is typically ionized at physiological pH, but can also be condensed into a lactone with hydroxyl groups of adjacent saccharides or into a lactam with a free amino group at C-5. Combinations of different glycosidic linkages with the multitude of possible modifications generate hundreds of ways in which Sias can present themselves. Unsaturated and anhydro forms of free Sias also exist; 2-deoxy-2,3-didehydro-Neu5Ac (Neu2en5Ac) is the most common. This pronounced chemical diversity of Sias contributes to the enormous variety of glycan structures on cell surfaces and the distinctive makeup of different cell types. This, in turn, can determine and/or modify recognition by antibodies and by a variety of Sia-binding lectins of intrinsic or extrinsic origin (see below). Despite this complexity, it may be sufficient in some biological studies to simply know that a sialic acid residue is present at the terminal position, and just label it with the generic abbreviation "Sia."

THE EXTENDED 2-KETO-3-DEOXYNONONIC ACID FAMILY

The number and diversity of 2-keto-3-deoxynononic acids that have been identified in eukaryotic and prokaryotic cells are increasing. However, these molecules are not identical with regard to their configuration. Legionaminic acid (Leg) from the lipopolysaccharide (LPS) of *Legionella pneumophila* has recently been determined to be a member of the Sia family. This 5,7-diamino-3,5,7,9-tetradeoxy-non-2-ulosonic acid is a true Sia, because it has the same D-glycero-D-galacto configuration found in Neu5Ac and Kdn (see Figure 14.1). The amino groups are substituted in the native LPS, yielding 5-acetimidoylamino-7-acetamido-Leg, and 8-O-acetylation can also occur. The superficially similar nine-carbon pseudaminic acid (Pse) found in the LPS of *Pseudomonas* species is actually in the L-glycero-L-manno configuration and therefore isomeric to Sia. There also exist other epilegionaminic acids in bacteria that do not fit this rule. However, all of these keto acids use phosphoenolpyruvate (PEP) for initial biosynthesis, and catalysis proceeds through a mechanism similar to Neu5Ac synthase. Furthermore, Pse synthase is evolutionarily homologous to Kdn, Leg, and Neu5Ac synthases. Finally, in all cases studied so far, the high-energy donor form is a CMP glycoside. It is therefore possible to redefine the Sia family as 2-keto-3-deoxynononic acids of various configurations, all of which are the products of an evolutionarily related synthase family. However, according to the presently accepted IUPAC carbohydrate nomenclature, only a nonulosonic acid with the D-glycero-D-galacto configuration should be defined as a Sia.

Also structurally related to Sias are eight- and seven-carbon 2-keto-3-deoxyoctonic acids and heptonic acids. Kdo belongs to the former group, although because of its different configuration it cannot be considered a true eight-carbon analog of the Sia Kdn. The biosynthetic pathways of Kdn and Kdo are also similar and they appear to share common ancestral genetic origins.

NOMENCLATURE AND ABBREVIATIONS

The complete chemical names of Sias are too cumbersome for routine use. A uniform and simple nomenclature system is being increasingly used, in which the abbreviation Neu denotes the core structure neuraminic acid, and Kdn denotes the core structure 2-keto-3-deoxynononic acid. Various substitutions are then designated by letter codes (Ac = acetyl, Gc = glycolyl, Me = methyl, Lt = lactyl, S = sulfate), and these are listed along with numbers indicating their location relative to the carbon positions. For example, *N*-glycolylneuraminic acid is Neu5Gc, 9-*O*-acetyl-8-*O*-methyl-*N*-acetylneuraminic acid is Neu5,9$Ac_2$8Me, and 7,8,9-tri-*O*-acetyl-*N*-glycolylneuraminic acid is Neu5Gc7,8,9Ac_3. If one is uncertain of the type of the Sia present at a particular location, the generic abbreviation Sia should be used. If other partial information is available, this can be incorporated, for example, a Sia of otherwise unknown type with an acetyl substitution at the C9 position could be written as Sia9Ac.

OLIGOSIALIC AND POLYSIALIC ACIDS

Polysialic acid (polySia) is an extended homopolymer of Sia found on only a few animal glycoproteins (e.g., the N-glycans of the neural cell adhesion molecule [NCAM] and O-glycans of fish egg glycoproteins), as well as in the capsular polysaccharides of certain pathogenic bacteria (e.g., colominic acid in K1 *Escherichia coli*) (Figure 14.3). The expression of polySia on NCAM decreases markedly during postnatal development and apparently plays a part in maintaining developmental plasticity by interfering with both homotypic and heterotypic interactions involving neuronal cells. In keeping with this role, increases in polySia expression are correlated with "neural plasticity," that is, neurite

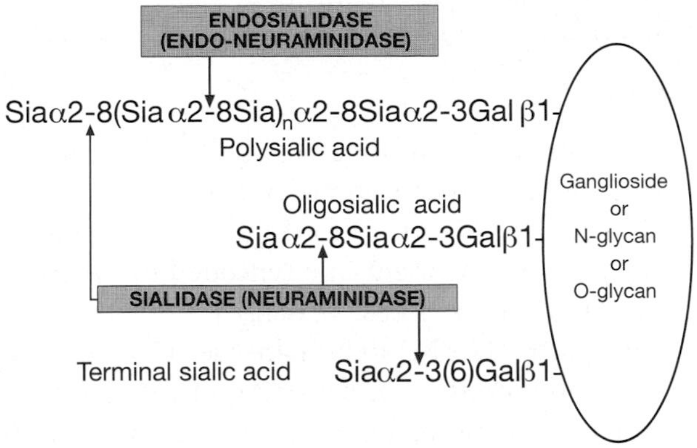

FIGURE 14.3. Terminal sialic, oligosialic, and polysialic acids, and the enzymes that can degrade them. (*Arrows*) Typical cleavage points for the action of the enzymes. Bacteria can also express some forms of sialic acids, but the linkage to the underlying core region is not always the same; in some instances, it is unknown (e.g., colominic acid, a bacterial polysialic acid that can also be cleaved by endosialidase).

sprouting and other situations involving neuronal damage repair or axonal migration, as well as the regulation of circadian rhythms. PolySia is often "primed" on an initiating α2-3-linked sialic acid residue. PolySia structures based on Neu5Gc, Neu5Ac, Kdn, or Leg have been reported. The linkages between the Sia units in a polySia chain can vary; the most common is an α2-8 linkage. Such a polySia polymer can also be O-acetylated at the C-7 or C-9. A bacteriophage that attacks polySia-expressing bacteria produces a highly specific endosialidase that is also a powerful tool for studying polySia biology. Shorter oligosialic acids (Figure 14.3) consisting of two to three Sia units can terminate the N-glycans of glycoconjugates, particularly in the brain or in milk, but much less is known about their significance. The biosynthesis and enzymology of oligosialic and polysialic acids are discussed briefly in Chapters 5 and 13.

TISSUE- AND MOLECULE-SPECIFIC EXPRESSION OF LINKAGES AND MODIFICATIONS

Certain linkages and modifications of Sias typically show tissue-specific and developmentally regulated expression. Some linkages and modifications are even molecule-specific, that is, they are found only on certain types of glycoconjugates in a given cell type. Even within a particular glycoconjugate group, a modification such as O-acetylation may be restricted to certain Sia residues at particular positions on a glycan. Such findings indicate the occurrence of specific enzymatic mechanisms for the generation and regulation of Sias (see below); they also suggest specific roles for these linkages and modifications. On the other hand, available evidence indicates substantial species-specific variations in the cell- and tissue-type distribution of different Sia linkages and modifications. Thus, at least some of this regulated expression may be unrelated to intrinsic functions of Sias. Rather, it may be the signature of the evolutionary history of a species in relation to the Sia-binding preferences of its pathogens and/or symbionts. Effectively, each species expresses a distinct "sialome," a term defined as the total array of sialic acid types and linkages expressed by a particular organelle, cell, tissue, organ, or organism. Of course, unlike the genome, which is the same in every cell type of an organism and undergoes very few changes during the lifetime of the organism, the sialome differs among cell types and varies markedly with regard to time, space, and environmental cues.

METABOLISM

Synthesis of Sialic Acids

Neu5Ac and Kdn appear to be the metabolic precursors for all known animal Sias (see Figure 14.4). In vertebrate systems, they are derived by condensation of ManNAc-6-P (for Neu5Ac) or Man-6-P (for Kdn) with phosphoenolpyruvate. The ManNAc-6-P is produced by a bifunctional enzyme (encoded by *GNE*) that converts UDP-GlcNAc to ManNAc-6-P and UDP in two steps. Missense mutations in this gene give rise to hereditary inclusion body myopathy (HIBM) in humans (see Chapter 42), and inactivation causes embryonic lethality in the mouse. Condensation of the sugar phosphates with phosphoenolpyruvate yields the corresponding Sia-9-phosphates, which must be dephosphorylated by a specific phosphatase (encoded by *NANP*), giving free Sias in the cytoplasm. In contrast, Neu5Ac biosynthesis in prokaryotes involves condensation of ManNAc with phosphoenolpyruvate, giving nonphosphorylated Neu5Ac (Figure 14.4). Notably, various synthetic unnatural mannosamine derivatives can be utilized by the Sia biosynthetic machinery, allowing manipulation of the chemical structures of cell-surface sialic acids (see Chapters 49 and 50).

Activation to Form CMP–Sialic Acids

Free Sia derived from biosynthesis (or recycled/recovered from the lysosome; see below) can be used for glycan biosynthesis only after activation into the nucleotide donor CMP-

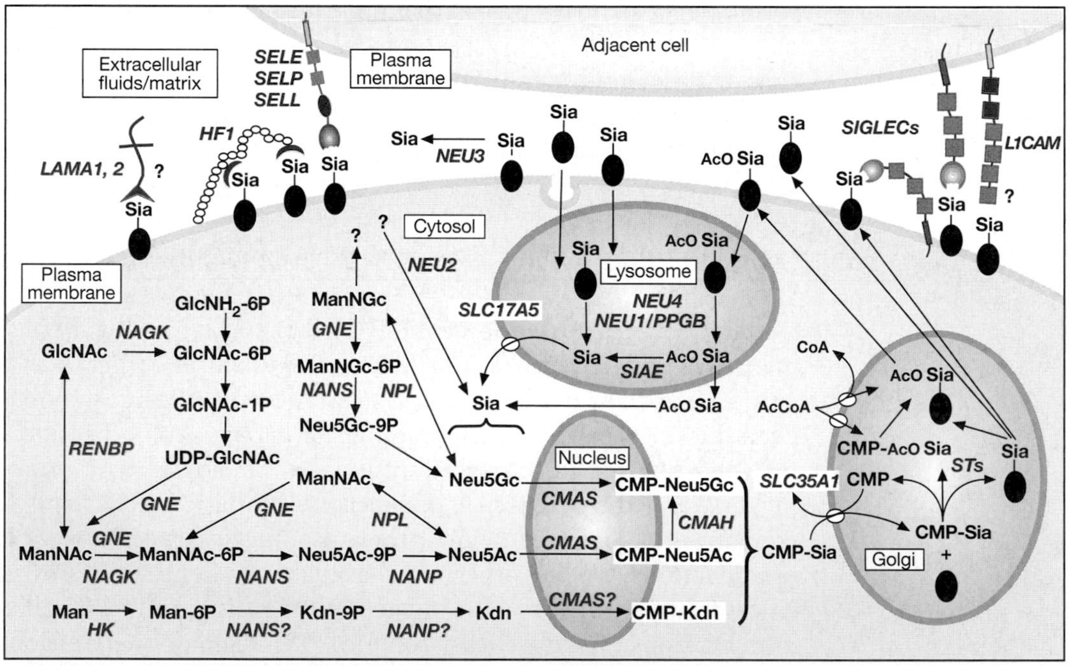

FIGURE 14.4. Genes and pathways involved in the biology of animal sialic acids. The general pathways for biosynthesis, activation, transfer, and recycling of the three common core sialic acids are shown in the context of two cells, one including the relevant organelles such as the Golgi apparatus, the nucleus, and the lysosome. Details on each gene product can be found by searching the gene name given for each reaction at the website of the Human Gene Nomenclature Committee, and related links. Pathways for Sia modification other than CMP-Neu5Gc production and O-acetylation/de-O-acetylation reactions are not shown (see text for discussion). (*Question marks*) Unknown or hypothetical pathways. (Modified and redrawn, with permission, from Altheide T.K. et al. 2006. *J. Biol. Chem.* **281:** 25689–25702, ©American Society for Biochemistry and Molecular Biology.)

Sia, a reaction catalyzed by CMP-Sia synthases (encoded by *CMAS*) using CTP as a donor. For reasons that are unclear, in all eukaryotic cells studied so far, this particular reaction takes place within the nucleus. The CMP-Sia products then return to the cytoplasm, where they are delivered into the lumen of Golgi compartments by the action of a specific antiporter (balanced by the export of CMP), which allows the generation of a higher concentration of CMP-Sias within the Golgi lumen than would be possible with passive transport (see Figure 14.4 and Chapter 4). These topological issues do not apply in prokaryotes, where CMP-Sias are synthesized in the cytoplasm and directly used in the coordinated assembly of cell-surface glycans, before their delivery to the surface. In eukaryotes, the levels of cytoplasmic free CMP-Sia can also cause feedback inhibition of UDP-GlcNAc 2′epimerase (encoded by *GNE*), the rate-limiting enzyme in the endogenous synthesis of the Sia precursor ManNAc. A genetic disease called "sialuria" arises from the failure of feedback regulation of this enzyme, which results in overproduction and excretion of sialic acids.

Transfer of Sialic Acids to Glycans

The transfer of Sias from CMP-Sias onto newly synthesized glycoconjugates passing through eukaryotic Golgi compartments is catalyzed by a family of linkage-specific sialyltransferases (STs), most of which have been cloned and characterized from multiple species. As with most other glycosyltransferases, STs are type II membrane proteins with complex signals dictating Golgi localization. Shared amino acid sequence motifs (called sialylmotifs) were found in the first STs cloned and were then used to clone new family members (see also Chapters 5 and 7). These evolutionarily conserved regions seem to represent substrate-binding sites, especially for CMP-Sia recognition. In striking contrast, prokaryotic STs do not have sialylmotifs, and in several instances they do not even show homology to one another. This suggests that prokaryotic STs have arisen independently on more than one occasion.

Regarding substrate specificity, several eukaryotic STs exhibit distinct preferences for glycolipids, glycoproteins, or poly/oligosaccharides, the structure of the acceptor glycan, the nature of the accepting terminal monosaccharide, or the type of Sia linkage formed. Interestingly, the specificity of prokaryotic STs is less pronounced. Modified Sias, such as Neu5Gc or O-acetylated species, are also transferred after activation to the CMP form. Some mammalian STs transfer both Neu5Ac and Kdn, but others transfer only one or the other. A "*trans*-sialidase" activity is present in some pathogenic trypanosome species and some bacteria, which directly transfers Sia from one glycosidic linkage to another (on galactose), without using CMP-activated Sia (see below and Chapter 40). Although *trans*-sialidases are specific with regard to the glycosidic linkage they generate (α2-3), they are rather promiscuous with regard to the nature of donor or acceptor substrates.

Modification of Sialic Acids

The remarkable chemical diversity of Sias is generated by multiple enzymatic mechanisms. The synthesis of Neu5Gc occurs by conversion of CMP-Neu5Ac to CMP-Neu5Gc in the cytoplasm (Figure 14.4). The CMP-Neu5Ac hydroxylase (encoded by the *CMAH* gene) responsible for this reaction is a cytoplasmic, iron-dependent enzyme that uses molecular oxygen and the common electron transport chain of cytochrome *b5* and *b5* reductase. Alternative pathways for generation of Neu5Gc are being explored, because Neu5Gc has been found in low quantities even in species such as humans that lack the hydroxylase (but see discussion below). Once a Neu5Ac residue has been converted into Neu5Gc, there is no known way to reverse the reaction, perhaps accounting for the accumulation of Neu5Gc in cells that do express it (for the cellular pathways involving Neu5Gc, see Figure 14.4). In contrast to this cytoplasmic conversion reaction, the addition of O-acetyl esters and other hydroxyl group modifications seem to occur mostly in the lumen of the Golgi or in Golgi-related organelles, either onto the CMP-Sia precursor or after the transfer of Sias to glycoconjugates. Regarding O-acetyltransferases, there is evidence for distinct enzymatic activities catalyzing O-acetylation of specific positions on Sias (e.g., C-4 vs. C-9), as well as specificity for O-acetylation of Sias on different linkages on different classes of glycoconjugates (e.g., gangliosides vs. N-glycans). Side-chain (C-7/8/9) O-acetyl groups appear to be initially added to C-7, followed by nonenzymatic migration to C-9 under physiological conditions, perhaps assisted by a "migrase" enzyme. The purification and cloning of these labile eukaryotic O-acetyltransferases has proven to be an intractable problem. O-Acetyltransferase genes from a few microorganisms were recently identified, but they show no homology to any eukaryotic gene. In mammalian systems, a protein complex in Golgi membranes is thought to be involved.

Other substitutions of the hydroxyl groups arise from use of the appropriate donors (e.g., *S*-adenosylmethionine for methylated Sias or 3′-phosphoadenosine 5′-phosphosulfate for sulfated molecules). With 9-O-lactyl groups, even the donor is still unknown. Appropriate enzymes should also exist to permit the turnover of each of these substitutions. Notably, with the exception of Neu5Gc, the other modified Sias studied so far do not appear to be effective substrates for reactivation by most CMP-Sia synthases. Thus, O-acetyl esters need to be removed at some point in the life cycle of the parent molecule, either for terminal degradation or as part of an acetylation/ deacetylation cycle (see below).

The de-N-acetylated form of Neu5Ac (neuraminic acid, Neu) is unstable in the free state and thus had been assumed not to exist in nature. However, the glycosidically bound form of Neu is stable, and there is evidence that small amounts do exist in nature and that these molecules can be re-N-acetylated. The search is underway for enzymes that presumably remove and add back the N-acetyl group. In some instances, such a free amino group can react with the carboxylate at C-2, giving an intramolecular lactam ring. Various dehydrated or unsaturated Sias also occur in nature, including 2,7-anhydro Sias released following cleavage of bound Sias by certain unusual sialidases; 4,8-anhydro compounds formed during release or deacetylation of 4-O-acetylated Sias; and the 2-deoxy-2,3-didehydro Sias resulting from mild alkali-catalyzed breakdown of CMP-Sias or as products from sialidase reactions. Although many of these substances have been detected in free form in biological fluids, their biological significance is not known. Interestingly, the 2,3-didehydro forms are inhibitors of microbial sialidases and led to the development of potent anti-influenza drugs (see below and Chapter 50).

Release of Sialic Acids

Sialic acids attached to a glycoconjugate must eventually be removed at some point in the life cycle of the molecule (Figure 14.4). In eukaryotic systems, this occurs by the action of specific sialidases (encoded by NEUs; the term "sialidase" is now preferred over the older term "neuraminidase," which is now used only in reference to viral enzymes, for historical reasons). Glycoconjugates are desialylated in endosomal/lysosomal compartments during recycling of cell-surface molecules and can sometimes return to the Golgi to undergo resialylation. In addition to endosomal/lysosomal sialidases, mammalian cells also have cell-surface (plasma membrane) and cytoplasmic sialidases. Cell-surface sialidases were originally thought to be involved in the abrupt shedding of cell-surface Sias that occurs upon activation of certain cell types (e.g., leukocytes). However, direct evidence for this is still lacking, and the plasma membrane sialidase appears to be specific for gangliosides, with claims for its involvement in signalling processes, apoptosis, and cell–cell contacts. The functions of cytoplasmic sialidases also remain quite obscure, because there is as yet no convincing evidence for glycosidically bound Sias in the cytoplasm nor on the cytoplasm-facing leaflet of cellular membranes. Recently, another sialidase was reported in human mitochondria. These enzymes are claimed to be involved in many cell biological processes, such as differentiation and cancer cell metastasis.

Many microorganisms also express sialidases, several of which have been cloned and characterized. Whereas the viral sialidases represent two distinct families, the bacterial, fungal, and invertebrate enzymes are evolutionarily related to mammalian families (in this instance, horizontal gene transfer between animals and pathogens seems possible). Most sialidases share a set of common "Asp boxes" (Ser-X-Asp-X-Gly-X-Thr-Tyr) that are probably involved in the maintenance of the enzyme protein conformation, together with a number of other highly conserved amino acids. The three-dimensional structures of several viral and bacterial sialidases have been elucidated, some in a complex with their substrates or with transition-state analogs. Interestingly, some have additional lectin domains that recognize underlying sugar chains and appear to direct the action of the enzyme.

Most sialidases exhibit substrate specificity regarding Sia linkage or the presence of substituents. Generally, α2-3 linkages are hydrolyzed more easily than α2-6 bonds, with the hydrolysis rate of α2-8-Sia being intermediate. A known exception is the enzyme from *Arthrobacter ureafaciens,* which acts best on α2-6 bonds. O-Methylation and O-acetylation of Sias can hinder (or even prevent, in the case of 4-O-acetyl groups) hydrolysis of the glycosidic bond by sialidases. These properties are both biologically and practically significant. The only known β-sialidase is CMP-Sia hydrolase, a poorly studied enzyme of unknown function, localized in the plasma membrane of some cell types.

A different type of sialidase is the "*trans*-sialidase" expressed by certain pathogenic protozoa (e.g., trypanosomes). These novel enzymes remove Sias from mammalian cell surfaces and transfer the sugar directly onto the parasite's own cell-surface acceptors, apparently providing protection from the host immune system (see Chapter 40). Microbial sialidases and *trans*-sialidases are powerful virulence factors that may assist invasion, unmask potential binding sites, and, in addition, provide nutrients for some bacteria. Viral neuraminidases ("receptor-destroying enzymes") are thought to assist viral entry by cleaving interfering sialic acids on inappropriate targets. They also facilitate release and spreading of newly formed viruses. Neuraminidase inhibitors are already in practical use as antiviral drugs (see Chapters 50 and 51).

Recycling of Sialic Acids

Once a Sia is released into the lysosome of a vertebrate cell, it is delivered back to the cytoplasm by a specific exporter called "Sialin" (see Chapter 4). This allows Sias to be either efficiently reutilized or degraded (Figure 14.4). Genetic defects in Sialin cause Salla disease and infantile sialic acid storage disease, resulting in accumulation of Sia in lysosomes and excretion of excess Sia in the urine. Some microorganisms can also directly scavenge Sias from the extracellular space, using high-efficiency transporters. In contrast, there is no evidence for plasma membrane Sia transporters in eukaryotic cells. However, free Sias can be relatively efficiently taken up into mammalian cells via fluid-phase macropinocytosis, eventually arriving in the lysosomes, from which they are exported into the cytoplasm by Sialin. Sialic acids that are glycosidically bound to soluble extracellular glycoproteins can be similarly transported to the lysosomes, where lysosomal sialidases can release them for delivery to the cytoplasm and eventual utilization by the cellular CMP-Sia synthase. The extent to which various eukaryotic cell types rely on such exogenous sources of Sias and/or on internal recycling is unknown. As discussed above, O-acetylated Sias probably need to be de-O-acetylated by specific 9-O-acetylesterases before they can be reutilized by cells. The acylmannosamines derived from the degradative activity of lyases (see next section) may also be reused for Sia synthesis. At the whole-body level, free Sias in the bloodstream (derived from cellular sources or digestive processes in the intestine) are rapidly excreted in the urine.

Degradation of Sialic Acids

If Sias are not reused in eukaryotic cells, degradation can occur, catalyzed by cytoplasmic Sia-specific pyruvate lyases (encoded by *NPL*) that cleave the molecule into *N*-acetylmannosamine and pyruvate. Similar pyruvate lyases exist in various microorganisms, and some can therefore use Sias as a food source. Current data suggest that there are at least two Sia-9-O-acetylesterases in mammalian systems. One is a cytoplasmic activity that may facilitate "recycling" of O-acetylated Sias that are exported from lysosomes into the cytoplasm. The other is a glycoprotein that traverses the ER-Golgi pathway and is targeted to lysosomal and endosomal compartments. However, this enzyme has a relatively high K_m value for its substrate, and unlike classic lysosomal enzymes, it has a neutral pH optimum. At present, it is not possible to reconcile these properties with a specific role for this enzyme in the lysosomal turnover of O-acetylated Sias. Enzymes with Sia-specific 9-O-acetylesterase activity have also been reported from bacterial and viral sources. The esterases from influenza C virus and coronaviruses are better characterized and act as receptor-destroying activities that are incorporated into the hemagglutinin molecule of the virus. Notably, all of these O-acetylesterases are specific for esters at C-9 and are incapable of releasing O-acetyl esters from C-7. However, 7-O-acetyl groups can migrate to C-9 under physiological conditions and thus become substrates for these enzymes. Esterases specific for 4-O-acetyl groups are present in horse liver and some coronaviruses. The mechanisms for removal and turnover of other Sia modifications (including the Gc group of Neu5Gc) remain unknown.

METHODS FOR STUDYING SIALIC ACIDS

Linkage-specific sialidases, esterases, Sia lyases, and/or lectins can all help to define some aspects of the Sias on a given glycan of a glycoconjugate. Monoclonal antibodies, lectins, and combinations of mild periodate oxidation with saponification have also been used to identify Sias and/or O-acetyl groups histochemically on tissue sections. A recombinant soluble form of the 9-O-acetyl-specific hemagglutinin of influenza C virus can probe for such molecules on thin-layer chromatograms, microwells, cells, and tissues. In some instances, information derived from such simple analyses is sufficient to reach biologically relevant conclusions. Various mass spectrometric (MS) and nuclear magnetic resonance (NMR) methods allow Sias to be more precisely characterized while they are still attached to the underlying glycan. The most accurate analysis of Sias from biological sources requires complete release and purification, with their modifications intact. However, some methods used to release, purify, or characterize the glycans can result in loss of labile Sia modifications (see below). Released and purified Sias can be analyzed by spectrophotometry, thin-layer chromatography (TLC), gas-liquid chromatography, MS, or NMR spectroscopy. Derivatization with 1,2-diamino-4,5-methylenedioxybenzene dihydrochloride (DMB) followed by high-pressure liquid chromatography analysis with fluorescent detection has proven to be particularly sensitive, specific, and applicable to most Sias. The adaptation of this technique to on-line electrospray mass spectrometry has been a powerful enhancement. Several techniques have also been developed for the detailed analysis of substitutions on metabolically labeled Sias.

For technical reasons, studies of sialoglycoconjugates continue to miss the extent of naturally occurring Sia structural complexity. Some Sia linkages may be partially or completely resistant to certain sialidases. Some substitutions are particularly labile (e.g., O-acetylation) and/or can alter the behavior of Sias during release, purification, and analysis. In addition, substitutions can slow down or even completely prevent release of Sias by commonly used sialidases or by acid hydrolysis. On the other hand, when stronger acidic conditions are used, destruction of some substitutions and of Sias themselves occurs. Furthermore, many methods used in structural analysis of intact glycans (e.g., alkaline conditions) cause the destruction of Sia modifications. Additionally, the presence of sialidases or esterases in crude cell extracts can alter the natural spectrum of sialoglycoconjugates. Because Sia modifications can affect size, shape, hydrophilicity, net charge, and biological properties of a glycoconjugate, a careful analysis for their presence is worthwhile in situations in which Sias are thought to have biological roles. With regard to side-chain O-acetylation, chemical and enzymatic improvements now allow near-quantitative release and purification of such molecules, without loss or migration of the ester groups. With rarer molecules such as O-lactylated, O-methylated, or sulfated Sias, much less is known about their susceptibility to sialidases or their optimal release with acid, and other methods for their direct detection are not available. It is evident that much needs to be done to improve methods for the detection, release, and purification of Sias from biological sources.

GENERAL FUNCTIONS OF SIALIC ACIDS

The high expression of Sias on outer cell membranes (e.g., more than 10 million molecules per human erythrocyte) on the interior of lysosomal membranes and on secreted glycoproteins (such as blood proteins and mucins) suggests that they have roles in the stabilization of molecules and membranes, as well as in modulating interactions with the environment. Some functions arise from the relatively strong electronegative charge of Sias, for example, binding and transport of ions and drugs, stabilizing the conformation of proteins including enzymes, and enhancing the viscosity of mucins. Sias can also protect molecules and cells from attack by proteases or glycosidases, extending their lifetime and function. Furthermore, Sias can regulate the affinity of receptors and are reported to modulate processes involved in transmembrane signaling, fertilization, growth, and differentiation. In one system, apoptosis was reported to be inhibited by Sia O-acetylation. A recently described general property of Sias seems to be their free-radical scavenging antioxidative effect, which could be particularly significant on endothelia of blood vessels.

Another prominent role of Sias is dualistic; they act either as masks or recognition sites. In the first case, they mask antigenic sites, receptors, and, most importantly, penultimate galactose residues. After Sia loss, molecules and cells can be bound, for example, by macrophages and hepatocytes, via Gal-recognizing receptors, and can even be taken up and degraded. This phenomenon has been most extensively studied with serum glycoproteins and blood cells. On the other hand, Sias themselves can serve as ligands for a variety of microbial and animal lectins, as is discussed in the following section.

Chemical modification of Sias can strongly influence all of these properties, in particular ligand functions. For example, 9-O-acetylation or N-acetyl-hydroxylation of Neu5Ac can create new receptor functions or decrease the affinity of binding.

SIALIC-ACID-RECOGNIZING LECTINS

Sias can be critical components of glycan ligands recognized by specific lectins. Table 14.1 lists examples of Sia-binding lectins from a variety of animal, plant, and microbial origins (see also Chapters 29, 31, and 32). Some of these lectins were first discovered in viruses because of their ability to agglutinate red blood cells in vitro and by the observation that this hemagglutination capacity was lost upon sialidase treatment of target cells. Others were discovered during investigations of cell–cell interactions when it was noted that binding was sensitive to sialidase treatments. In recent times, Sia-binding lectins have been found purely by virtue of their sequence homology. The three-dimensional structures of some of these molecules have been elucidated, sometimes in a complex with a sialylated oligosaccharide. In most examples studied, the negatively charged carboxylate group at C-1 of the Sia has proven critical for recognition. The role of divalent cations and the underlying oligosaccharide can range from being absolutely essential to being unimportant. The linkage of the Sia is recognized specifically by most of the lectins, sometimes in the context of the underlying sugar chain (for some examples, see Figure 14.5). This selectivity in recognition provides a "biological readout" for some of the complex pathways of Golgi glycosylation that terminate in sialylation. The structural diversity in the Sias men-

TABLE 14.1. *Examples of naturally occurring sialic-acid-binding lectins*

Vertebrate
C-type: Selectins (see Chapter 31)
I-type: Siglecs (see Chapter 32)
Unclassified: Complement factor H

Arthropod
Crab lectins: Limulin (American horseshoe crab, *Limulus polyphemus*)
Lobster and prawn lectins: L-agglutinin (lobster, *Homarus americanus*)
Scorpion lectins: Whip scorpion lectin (*Mastigoproctus giganteus*)
Insect lectins: Allo A-II (beetle lectin, *Allomyrina dichotoma*)

Mollusk
Slug lectins: *Limax flavus* agglutinin (LFA) (*Limax flavus*)
Mussel and oyster lectins: Pacific oyster lectin (*Crassostrea gigas*)
Snail lectins: Achatinin-H (*Achatina fulica*)

Protozoa
Parasite lectins: Merozoite erythrocyte-binding antigens (EBAs) (*Plasmodium falciparum*)

Plant
SN agglutinin (SNA) (elderberry bark lectin, *Sambucus nigra*), PS agglutinin (*Polyporus squamosus*), MA agglutinin (MAH) (*Maackia amurensis*), wheat-germ agglutinin (*Triticum vulgaris*)

Bacteria
Bacterial adhesins: S-adhesin (*Escherichia coli* K99), SabA and SabB (*Helicobacter pylori*)
Bacterial toxins: Cholera toxin (*Vibrio cholerae*), tetanus toxin (*Clostridium tetani*), botulinum toxin (*Clostridium botulinum*), pertussis toxin (*Bordetella pertussis*)
Mycoplasma lectins: *Mycoplasma pneumoniae* hemagglutinin

Viruses
Hemagglutinins: Influenza A and B viruses, primate polyomaviruses, rotaviruses
Hemagglutinin neuraminidases: Newcastle disease virus, Sendai virus, fowl plague virus
Hemagglutinin esterases: Influenza C viruses, human and bovine coronaviruses

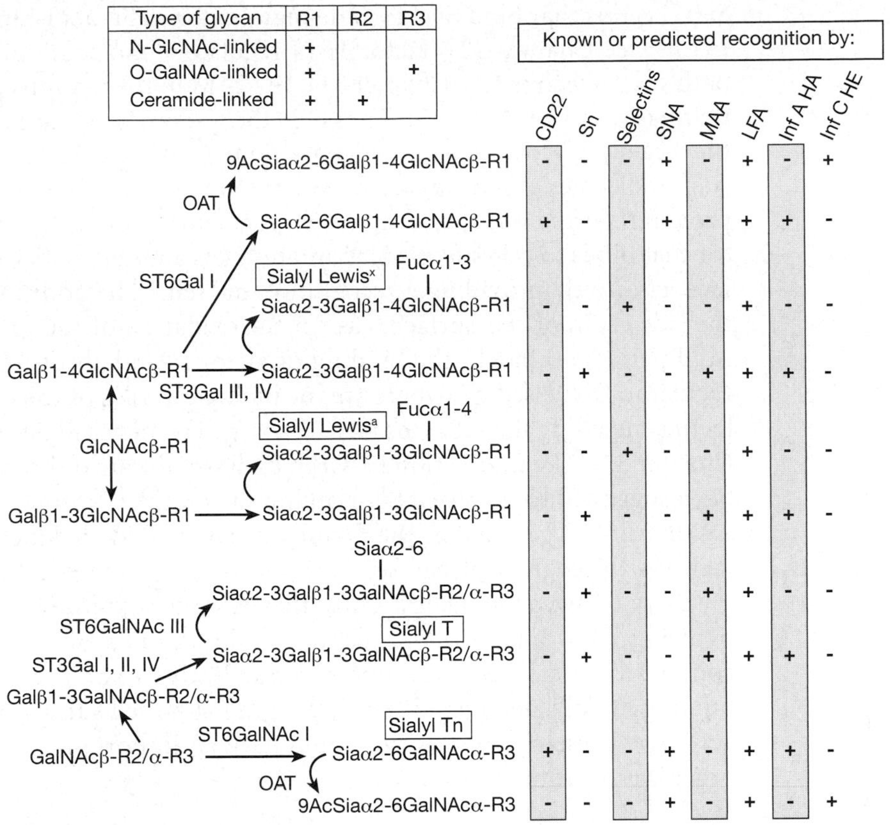

FIGURE 14.5. Examples of terminal glycan sequences recognized by some sialic-acid-binding proteins. *N*-acetylglucosamine (GlcNAc) or *N*-acetylgalactosamine (GalNAc) residues on glycoproteins and/or glycosphingolipids can be extended by several biosynthetic pathways, some examples of which are shown. The sialylated sequences recognized by various binding proteins are based on published literature and/or reasonable predictions based on known specificities. The sequences shown are the minimal structural motifs necessary for binding, and relative differences in binding strength are not shown. Natural high-affinity ligands may be more complex. Recognition can be affected by modifications of sialic acid other than O-acetylation or by sulfation of adjacent monosaccharides (not shown). (ST) Sialyltransferase; (OAT) O-acetyltransferase; (CD22) Siglec-2; (Sn) Sialoadhesin/Siglec-1; (SNA) *Sambucus nigra* agglutinin; (MAA) *Maackia amurensis* agglutinin; (LFA) *Limax flavus* agglutinin; (Inf A HA) influenza A hemagglutinin; (Inf C HA) influenza C hemagglutinin. With Inf A HA, the relative preference for α2-3 and α2-6 Sia linkages can vary with the viral strain. Note that "MAA" is typically a mixure of at least two lectins, *Maackia amurensis* hemagglutinin (MAH) and *Maackia amurensis* leukogglutinin (MAL), each with somewhat different preferences for the glycan underlying the α2-3-linked Sia. The latter (MAL) can also recognize 3-O-sulfated LacNAc termini.

tioned above also affects lectin recognition. The role of various linkages and substitutions is highly variable, ranging from being completely unimportant to being crucial for recognition. Various combinations of treatments with sialidases, 9-O-acetylesterases, and mild periodate oxidation can be used to explore lectin specificities.

Intrinsic Lectins in Vertebrates

Elimination of Sia production in mice causes embryonic lethality, suggesting that there are critical endogenous functions for Sias in development. Nevertheless, so far relatively few examples of Sia-specific lectins are intrinsic to an organism that synthesizes its own

Sias. Lectins that bind Sias include Siglecs (for *si*alic-acid-binding *i*mmuno*g*lobulin-like *lec*tins; see Chapter 32), factor H (a regulatory molecule of the alternate complement pathway), selectins (see Chapter 31), L1-CAM in the nervous system, a uterine agglutinin that has yet to be cloned, and possibly the G-domain of some laminins, which recognize the heavily glycosylated mucin-type domain of α-dystroglycan. The relative rarity of such molecules could be due to ascertainment bias. The first mammalian Sia-binding protein reported was the complement regulatory molecule factor H, a soluble serum factor that binds to cell-surface Sias and restricts alternative pathway activation on that surface, effectively providing a recognition of "self." The addition of a 9-O-acetyl group to the side chain of cell-surface Sias (or the oxidation of the unsubstituted side chain with mild periodate) blocks the binding of factor H and abrogates its function as a negative regulator. Discussed elsewhere are the biological roles of the other vertebrate Sia-binding lectins including the selectins (see Chapter 31) and the Siglec subset of I-type lectins (see Chapter 32). The interaction between α-dystroglycan and certain laminins in muscle has been suggested to involve a Sia-binding site on the G-domain of the latter and sialylated O-Man-linked glycans on the former. Analysis of such functions is complicated by the fact that the cognate glycan sequences for some of these lectins are commonly found on a variety of glycoconjugates. Thus, these lectins sometimes function by specifically recognizing a few high-affinity ligands within a milieu of low-affinity inhibitors. Further complexity arises because some of these lectins (e.g., the Siglecs) can be occupied by binding to sialylated ligands present on the same cell surface as the lectin itself (*cis* interactions). These *cis* interactions could have an important role in receptor functions and organization at the cell surface.

Extrinsic Lectins on Pathogens and Toxins

Sia-specific lectins extrinsic to the organisms that synthesize Sias are widespread in nature and include numerous viral hemagglutinins, bacterial adhesions, and toxins (see Table 14.1 for a very limited listing). This should not be surprising, given the location of Sias at the outermost reaches of the cell surface, where pathogens make first contact with target cells. A large number of microbial-host interactions are dependent on recognition of specific sialylated ligands (see Table 14.1 and Chapter 9). Examples of medical relevance include the recognition of airway epithelial Sias by influenza viruses, binding of *Helicobacter pylori* (the cause of peptic ulcer disease) to gastric mucins and glycosphingolipids via at least two different Sia-dependent mechanisms, interaction of cholera and tetanus toxins with target gangliosides on mammalian cells, and binding of the merozoite stage of the malarial parasite *Plasmodium falciparum* to erythrocyte sialoglycophorins. The interactions of some of these lectins with Sias can be abolished by substitutions such as O-acetyl and N-glycolyl groups that are found on mammalian mucosal surfaces. Thus, it has been suggested that such modifications serve a specific protective purpose in this location. Indeed, it is possible that many of the complexities of Sia diversification are the outcome of the ongoing evolutionary "arms race" between animals and microbial pathogens (see Chapter 19). In this regard, expression of O-acetyl and N-glycolyl groups on cell surfaces can also limit the action of bacterial sialidases and block the binding of some pathogenic viruses. Alternately, such modifications can facilitate binding of viruses that have adapted to them. With regard to the unsaturated Sias found in free form in biological fluids, it is possible that they provide protection by virtue of their powerful inhibition of microbial sialidases. Of course, the evolutionary persistence of modified Sias in some cell types suggest that these glycans have critical structural roles and/or are required for recognition by endogenous lectins.

Lectins in Organisms without Sialic Acids

Many Sia-binding lectins are found in organisms that do not themselves seem to express Sias (see Table 14.1 for examples). One explanation is that their primary function is defense against exogenous sialylated pathogens. In keeping with this, limulin in the hemolymph of the horseshoe crab can trigger foreign cell hemolysis. Sia-binding lectins may also protect plants from being eaten by mammals, for example, elderberry shrubs. Of course, some of these Sia-binding properties might be serendipitous, with the real lectin ligands being other similar anionic glycans, such as 3-deoxy-octulosonic acid (Kdo, Pse, or Leg) found in prokaryotes and in some plants.

PRACTICAL USES OF SIALIC-ACID-BINDING LECTINS

Regardless of the nature of their natural ligands, some Sia-binding lectins have proven to be powerful tools for studying the biology of Sias (see Chapter 45). For example, wheat-germ agglutinin and *Limax flavus* agglutinin have been used as general tools to detect sialylated glycoconjugates, and combinations of *Sambucus nigra, Polyporus squamosus,* and *Maackia amurensis* agglutinins can distinguish among different types of Sia linkages on terminal *N*-acetyllactosamines. Caution is needed in the case of *Maackia amurensis,* because this seed has multiple lectins with differing specificity (see legend of Figure 14.5). Recombinant soluble forms of the Siglecs can also be used for this purpose. A recombinant soluble form of the influenza C hemagglutinin–esterase can specifically probe for 9-*O*-acetylated Sias, which can also be detected by the Achatinin H lectin from the snail *Achatina fulica.* Of course, in all situations in which a lectin is used as a detection tool, the absence of binding does not necessarily imply the absence of the expected glycan structure, and false positive results are possible as well.

EVOLUTIONARY HISTORY OF SIALIC ACIDS

Early studies suggested species specificity in the occurrence of different types of Sias. However, with improvements in detection and analysis techniques, it is evident that most Sia types are widely expressed and simply occur at differing levels of detectability. As a group, Sias became prominent late in evolution, primarily in animals of deuterostome lineage (see Chapter 25), which comprises the vertebrates and some "higher" invertebrates (such as echinoderms) that emerged at the Cambrian expansion (~530 million years ago). Indeed, with rare exceptions (some that remain controversial), Sias are not generally found in plants or in most prokaryotes or invertebrates. However, there have been a few credible reports of Sias in mollusks, such as octopus and squid, and insects such as *Drosophila.* Also, genes structurally related to those involved in vertebrate Sia metabolism have been reported in insects and plants, and even in Archaea. With improved analysis techniques, Sias are now often found in membrane macromolecules of microorganisms. Overall, it appears that Sias may be a more ancient Precambrian invention, but they were then either eliminated or used only sparingly in many lineages—finally "flowering" into prominence only in deuterostome lineage. In this regard, genetic evidence also suggests that the original invention of Sias may have derived from homologous gene products that synthesize keto-deoxy-octulosonic acid (Kdo). Meanwhile, certain strains of bacteria can contain large amounts of Sias or other 2-keto-3-deoxynononic acids in their capsular polysaccharides and/or lipooligosaccharides. Some of these bacteria are pathogenic and cell-surface sialic acids protect them from complement activation and/or antibody production. Thus, although definitive proof has not been obtained, the possibility of gene transfer from host eukaryotes exists. However, it does seem that many of the bacterial enzymes involved in synthesizing and metabolizing Sias have evolved independently, possibly being "reinvented" from the Kdo pathway. Interestingly, there is wide variation in Sia expression and complexity within deuterostome lineage, with the sialome of echinoderms appearing very complex and that of humans being more simple. However, expression of Neu5Gc and O-acetylated Sias is highly conserved in deuterostomes, although exceptions exist, such as the lack of Neu5Gc in man, chicken, and some other birds.

LOSS OF *N*-GLYCOLYLNEURAMINIC ACID PRODUCTION IN HUMANS

The common mammalian Sia Neu5Gc was once thought to be an onco-fetal antigen in humans, being apparently absent from normal adult human tissues but expressed in fetal samples and certain human tumors and tumor cell lines. Indeed, upon human intravenous exposure to horse antiserum (still sometimes used in situations such as snake bite), the resulting "serum sickness" (Hanganutziu–Deicher or "HD") antibodies are prominently directed against Neu5Gc. Spontaneously appearing HD antibodies were also reported in patients with cancer and certain infectious diseases, as well as in chickens with Marek's disease, a malignant herpesvirus infection. In humans, the explanation is homozygosity for an inactivating exon deletion in the *CMAH* gene that occurred after our last common ancestor with the African great apes. Meanwhile, using sensitive techniques, traces of Neu5Gc have been found in normal human tissues. This, as well as the higher level of reexpression of Neu5Gc reported in malignant tissues, seems to represent incorporation from dietary sources such as red meats and milk products. However, an alternate pathway for Neu5Gc synthesis in tumor cells has not been conclusively ruled out. With the discovery that most or all healthy humans have some levels of circulating anti-Neu5Gc antibodies, the possibility has been raised that this might account for the high frequency of atherosclerosis and epithelial cancers in humans, diseases that seem uncommon in the great apes and have been correlated with red meat consumption in humans.

A potentially related observation is the suppression of CMAH/Neu5Gc expression in the brains of all animals studied, including those that have high levels expressed in other tissues. Because the loss of Neu5Gc in human lineage may have predated the appearance of the genus *Homo*, it is possible that the complete elimination of Neu5Gc may have somehow facilitated human brain evolution; however, no testable hypothesis has been advanced. Additional consequences for the evolution of humans may relate to the ancestral condition of many CD33-related Siglecs (see Chapter 32), which selectively bind to Neu5Gc. Thus, loss of Neu5Gc during human evolution would have caused a temporary loss of ligands for these inhibitory molecules of the innate immune system, a situation that has apparently been corrected by multiple human-specific changes in this family of receptors, leading to better binding to Neu5Ac. Other possible consequences include human resistance to veterinary microbial pathogens such as *Escherichia coli* K99, and the successful emergence of Neu5Ac-preferring pathogens such as the human-specific malarial parasite *P. falciparum*.

SIALIC ACIDS IN DEVELOPMENT AND MALIGNANCY

Cultured cell lines that are grossly deficient in sialylated glycans show generally normal growth patterns. Thus, more critical biological roles of Sias may only be evident in multicellular or intact vertebrate systems. Indeed, as already mentioned, Sias are critically required for early mammalian development. However, apart from the function of polySias in allowing "neural plasticity," the exact roles of Sias during development remain uncertain. The role of Neu5Ac expression in the larvae of the insects *Drosophila* and the cicada *Philaenus spumarius* is also unknown. Several examples of Sia regulation have been reported in living animals. Certain classes of T lymphocytes have O-acetylated Sias, whereas others do not. The expression of polysialylation and O-acetylation in neural gangliosides varies with developmental stage and location, and differences in O-acetylation of brain gangliosides have been reported between cold- and warm-blooded species, and between awake and hibernating animals. Developmental regulation of Neu5Gc expression and O-acetylation expression in the gut mucosa may occur in response to microbial colonization and has been suggested to have a role in protecting against certain microorganisms. Similarly, although adult bovine submandibular glands produce large amounts of highly O-acetylated mucins, this Sia modification is scarcely expressed in the corresponding fetal tissue. The type and linkages of endothelial, plasma protein, and erythrocyte Sias can undergo marked changes in responses to inflammatory stimuli. Interesting abnormalities have also been reported in transgenic mice expressing influenza C 9-O-acetylesterase and following genetic inactivation of various sialyltransferases in the intact mouse. A variety of sialyltransferase-null mice have been produced that show interesting and specific phenotypes, ranging from altered Siglec-2/CD22 function (ST6Gal-I null) to defects in T-cell maturation (ST3Gal-I null) and changes in brain development (ST8Sia-II and ST8Sia-IV null).

In addition to the accumulation of Neu5Gc, several other specific changes in Sias occur in malignancy (see Chapter 44). In general, the total amount of Sia increases and switches occur in linkages, with α2-6 linkages becoming particularly prominent. O-Acetylation at C-9 can either disappear (as occurs in colon carcinomas) or become prominent (as in 9-*O*-acetyl-G_{D3}, which is much increased in melanomas and basal cell carcinomas). With the exception of the role of Sia in selectin ligands (see Chapter 31), the precise mechanisms by which these Sia changes enhance tumorigenesis and/or invasive behavior remain uncertain. Increased sialylation may also enhance the masking effect of Sia on antigenic sites of tumor cells, which become more like "self" and therefore more invasive. Regardless of the mechanisms involved, certain sialylated molecules are specific markers for some cancers and potential ligands for targeted therapies (see Chapter 44).

SIALIC ACIDS IN PATHOLOGY AND PHARMACOLOGY

Because Sias are involved in so many cellular functions, disturbances of their biosynthesis or degradation can lead to medical problems. Because of their exposed position, Sias are vulnerable to the action of microbial esterases, sialidases, and lyases. The actions of these enzymes can affect the amount of ligand present, masking of antigenic sites, stabilization of membranes, and immunological and other functions of Sias. In this regard, microbial lectins, sialidases, and *trans*-sialidases are potent virulence factors. Many bacterial toxins (e.g., cholera, tetanus, and pertussis toxins) and species of virus (e.g., influenza viruses) bind to sialylated glycoconjugates (see Chapter 34). Bacteria may also create new binding sites by sialidase-mediated unmasking of penultimate galactose residues. *Trans*-sialidases of some pathogenic trypanosome species make these parasites fitter for survival in the vector or host, and they strongly disturb the host's immune system by compromising the cytokine network and influencing signaling processes.

Changes in Sias have also been found to be involved in degenerative diseases such as artherosclerosis and diabetes as well as neurological disorders such as Alzheimer's disease and alcoholism. Mucins also have to be properly and highly sialylated in order to exert their physiological functions as lubricants and in innate immunity. Selectins recognize sialyl

Lewis^x glycans that generally contain a terminal Sia, and Sias are thus involved in rolling and extravasation of leukocytes during inflammation (see Chapter 31).

Several human genetic Sia disorders are known: for example, hereditary inclusion body myopathy (HIBM) (caused by missense mutations of the UDP-GlcNAc-2-epimerase/ManNAc kinase [*GNE*] gene), sialuria (a defect of GNE feedback inhibition by CMP-Neu5Ac), Salla disease (a defect in the lysosomal Sia transporter Sialin), and galactosialidosis (galactosidase-peptidase-sialidase complex deficiency). Many of the congenital disorders of glycosylation (CDGs) may also lead to altered sialylation (see Chapter 42), but less is known regarding the molecular basis of phenotypic consequences.

On the basis of these diverse pathophysiological roles of Sias, many efforts have been undertaken to create appropriate pharmacologically active agents. Best known are the competitive inhibitors of sialidases, derived from the natural sialidase inhibitor Neu2en5Ac (2-deoxy-2,3-didehydro-Neu5Ac), which hinder budding and spreading of influenza A and B viruses (see Chapter 50). Inhibitors of bacterial sialidases and trypanosomal *trans*-sialidases are urgently needed. Although sialyltransferase inhibitors could be useful in cancer, potent agents are not yet available. Many attempts have been made to generate agents derived from sialyl Lewis^x (sLe^x) to compete with the selectins and to affect inflammatory processes, reperfusion injury, or tumor cell metastasis. Antiadhesive sialylated molecules could also be potentially useful for the treatment of bacterial and viral infections; for example, corresponding sialylated glycodendrimers can inhibit binding of the influenza virus hemagglutinin. Sialylated milk oligosaccharides are claimed to fulfill this task in a natural way in the intestine and stomach, perhaps reducing *H. pylori* infection. Sulfated polysialic acid has been found to suppress human immunodeficiency virus (HIV). Strategies for attacking cancer cells include vaccination with sialoglycoconjugates (e.g., gangliosides) in the case of melanoma or vaccination with polysialic acid modified with unnatural Sia in other cancers.

Manipulation and control of sialylation levels are also important in biotechnology (see Chapter 51). Engineering of erythropoietin to a hypersialylated form gives better pharmacokinetic properties, especially a longer lifetime in the bloodstream. The same strategy is presently being investigated with other naturally carbohydrate-free peptide hormones such as insulin or cytokines by adding N-glycosylation sites. These procedures require sophisticated chemical, recombinant, and other methods in order to achieve proper and possibly variable sialylation with Neu5Ac (Neu5Gc should be avoided because of its antigenicity in man). The production of recombinant glycoproteins of pharmaceutical value in large quantities is most promising in yeast, insect, and plant cells that do not express Sia, but these are now being engineered by transfection of the appropriate enzymes so they can make "humanized" glycoproteins with complex sialylated N-glycans (see Chapter 51).

FURTHER READING

Blix G., Gottschalk A., and Klenk E. 1957. Proposed nomenclature in the field of neuraminic and sialic acids. *Nature* **179:** 1088.

Roseman S. 1970. The synthesis of carbohydrates by multiglycosyltransferase systems and their potential function in intercellular adhesion. *Chem. Phys. Lipids* **5:** 270–297.

Schauer R. 1982. Chemistry, metabolism, and biological functions of sialic acids. *Adv. Carbohydr. Chem. Biochem.* **40:** 131–234.

Schauer R. 1985. Sialic acids and their roles as biological masks. *Trends Biochem. Sci.* **10:** 357–360.

Faillard H. 1989. The early history of sialic acids. *Trends Biochem. Sci.* **14:** 237–241.

Troy F.A. 1992. Polysialylation: From bacteria to brains. *Glycobiology* **2:** 5–23.

Varki A. 1992. Diversity in the sialic acids. *Glycobiology* **2:** 25–40.

Reuter G. and Schauer R. 1994. Determination of sialic acids. *Methods Enzymol.* **230:** 168–199.

Taylor G. 1996. Sialidases: Structures, biological significance and therapeutic potential. *Curr. Opin. Struct. Biol.* **6:** 830–837.

Kelm S. and Schauer R. 1997. Sialic acids in molecular and cellular interactions. *Int. Rev. Cytol.* **175:** 137–240.

Schauer R. and Kamerling J.P. 1998. Chemistry, biochemistry and biology of sialic acids. *Cell. Mol. Life Sci.* **54:** 1330–1349.

Angata T. and Varki A. 2002. Chemical diversity in the sialic acids and related α-keto acids: An evolutionary perspective. *Chem. Rev.* **102:** 439–470.

Knirel Y.A., Shashkov A.S., Tsvetkov Y.E., Jansson P.-E., and Zähringer U. 2003. 5,7-Diamino-3,5,7-tetradeoxynon-2-ulosonic acids in bacterial glycopolymers: Chemistry and biochemistry. *Adv. Carbohydr. Chem. Biochem.* **58:** 371–417.

Miyagi T., Wada T., Yamaguchi K., and Hata K. 2004. Sialidase and malignancy: A minireview. *Glycoconj. J.* **20:** 189–198.

Schauer R. 2004. Sialic acids: Fascinating sugars in higher animals and man. *Zoology* **107:** 49–64.

Vimr E.R., Kalivoda K.A., Deszo E.L., and Steenbergen S.M. 2004. Diversity of microbial sialic acid metabolism. *Microbiol. Mol. Biol. Rev.* **68:** 132–153.

Harduin-Lepers A., Mollicone R., Delannoy P., and Oriol R. 2005. The animal sialyltransferases and sialyltransferase-related genes: A phylogenetic approach. *Glycobiology* **15:** 805–817.

Varki A. and Angata T. 2006. Siglecs—The major subfamily of I-type lectins. *Glycobiology* **16:** 1R–27R.

Varki A. 2007. Glycan-based interactions involving vertebrate sialic acid-recognizing proteins. *Nature* **446:** 1023–1029.

CHAPTER 15

Hyaluronan

Vince Hascall and Jeffrey D. Esko

THIS CHAPTER DESCRIBES THE STRUCTURE AND METABOLISM of the nonsulfated glycosaminoglycan hyaluronan and how its chemical attributes contribute to its highly diverse and versatile biological functions. Chapter 16 describes sulfated glycosaminoglycans.

HISTORICAL AND EVOLUTIONARY PERSPECTIVES

Sulfated glycosaminoglycans were first isolated in the late 1800s, and the isolation of hyaluronic acid (now called hyaluronan) followed in the early 1930s. In their classic paper, Karl Meyer and John Palmer named the "polysaccharide acid of high molecular weight" that they purified from bovine vitreous humor *hyaluronic acid* (from hyaloid [meaning vitreous] and uronic acid) and they showed that it contained "uronic acid (and) an amino sugar." It took almost 20 years to determine the actual structure of the disaccharide that forms the repeating disaccharide motif of hyaluronan (Figure 15.1). In contrast to the other classes of glycosaminoglycans, hyaluronan is not further modified by sulfation or by epimerization of the glucuronic acid moiety to iduronic acid. Thus, the chemical structure shown in Figure 15.1 is faithfully reproduced by any cell that synthesizes hyaluronan, including animal cells and bacteria.

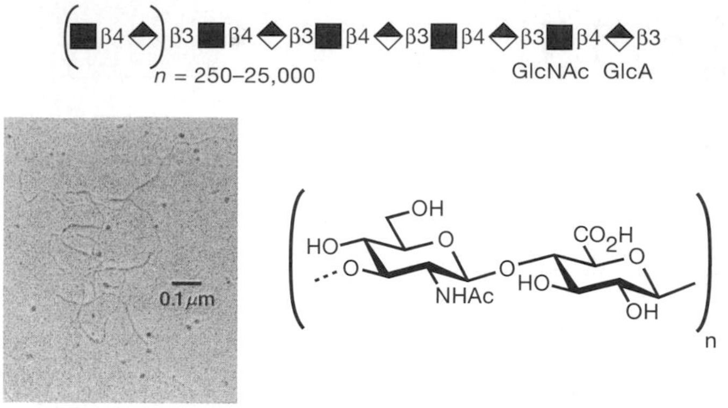

FIGURE 15.1. Hyaluronan consists of repeating disaccharides composed of *N*-acetylglucosamine (GlcNAc) and glucuronic acid (GlcA). It is the largest polysaccharide found in vertebrates, and it forms hydrated matrices. (Electron micrograph provided by Drs. Richard Mayne and Randolph Brewton, University of Alabama at Birmingham.)

At first glance, the simplicity of hyaluronan might suggest that it arose early in evolution relative to other glycosaminoglycans (see Chapter 16). However, this is not the case, because *Drosophila melanogaster* and *Caenorhabditis elegans* do not contain the necessary synthases for its assembly. Instead, it appears that hyaluronan arose during the evolution of the notocord shortly before or concurrent with the advent of cartilage and appendicular skeletons. The simplicity of hyaluronan's structure may be the key to its success, because numerous hyaluronan-binding proteins have evolved (often referred to as hyaladherins) and several enzymes responsible for its synthesis and degradation are present. Virtually all cells from vertebrate species can produce hyaluronan, and its expression correlates with tissue expansion and cell motility. As discussed below, hyaluronan has essential roles in development, tissue organization, cell proliferation, signaling reactions across the plasma membrane, and microbial virulence.

STRUCTURE AND BIOPHYSICAL PROPERTIES

As described in the next section, hyaluronan is the only glycosaminoglycan synthesized in the cytoplasm at the plasma membrane, with the growing polymer being extruded into the extracellular environment. This allows hyaluronan to have an indefinite and very large degree of polymerization, typically in the range of 10^4 disaccharides (~3.7×10^6 D as the sodium salt) and with an end-to-end length of approximately 10 μm (~1 nm/disaccharide). Thus, a single molecule of hyaluronan could stretch about halfway around the circumference of a typical mammalian cell. The carboxyl groups on the glucuronic acid residues (pK_a 4–5) are predominantly negatively charged at physiological pH and ionic strength, making hyaluronan polyanionic. The anionic nature of hyaluronan together with spatial restrictions around the glycosidic bonds confer a relatively stiff, random coil structure to individual hyaluronan molecules in most biological settings. Hyaluronan chains occupy a large hydrodynamic volume such that individual molecules of high molecular weight in a 3–5 mg/ml physiological solution occupy essentially all of the solvent. This arrangement creates a size-selective barrier in which small molecules can diffuse freely, whereas larger molecules are partially or completely excluded. Such a solution would have a swelling pressure and exhibit high viscosity with viscoelastic properties, conditions found in the vitreous humor of the human eye and in joints. Hyaluronan in synovial fluids of articular joints is essential for distributing load during joint motion and for protecting the cartilaginous surfaces. Thus, in both eye and joint tissues, the physical properties of hyaluronan relate directly to tissue function.

BIOSYNTHESIS

Hyaluronan synthesis is catalyzed by hyaluronan synthases (HAS), each of which contains dual catalytic activities required for the transfer of *N*-acetylglucosamine and glucuronic acid units from the corresponding nucleotide sugars (Figure 15.2). The first bona fide HAS gene (*spHas*) was cloned from *Streptococcus*, and the protein expressed in *Escherichia coli* was shown to synthesize high-molecular-weight hyaluronan from the UDP-sugar substrates. The gene shows homology with a *Xenopus* gene, *DG42* (now known as *xlHAS1*; see Chapter 25). The homology was instrumental in the subsequent identification of the three members of the mammalian *Has* gene family, *Has1–3*. These genes code for homologous proteins predicted to contain five to six membrane-spanning segments and a central cytoplasmic domain.

As described in Chapter 16, cells synthesize sulfated glycosaminoglycans (heparan sulfate, chondroitin sulfate, and keratan sulfate) on core proteins of proteoglycans as they transit through the Golgi, and elongation of the chains occurs at their nonreducing ends. In contrast, hyaluronan synthesis normally occurs at the inner surface of the plasma membrane in eukaryotic cells and at the cytoplasmic membrane of bacteria that produce hyaluronan capsules. The synthases use the substrates UDP-GlcA and UDP-GlcNAc and extrude the growing polymer through the membrane (Figure 15.2). According to the model, the reducing end of the growing chain would have a UDP moiety that is displaced when the next sugar is added.

Hyaluronan biosynthesis in bacteria involves the expression of multiple enzymes, usually as an operon. For example, in *Streptococcus, hasC* encodes an enzyme that makes UDP-Glc from UTP and glucose-1-P; *hasB* encodes the dehydrogenase that converts UDP-Glc to UDP-GlcA; *hasD* generates UDP-GlcNAc from glucosamine-1-P, acetyl CoA, and UTP; and *hasA* (*spHas*) encodes the hyaluronan synthase. The *Streptococcus hasA* gene encodes a bifunctional protein that contains both transferase activities and uses a nucleotide sugar as the acceptor. Thus, *spHas* assembles the polysaccharide from the reducing end. The synthase spans the membrane multiple times, presumably forming a pore for hyaluronan extrusion during capsule formation. In contrast, *Pasteurella* synthesizes hyaluronan from

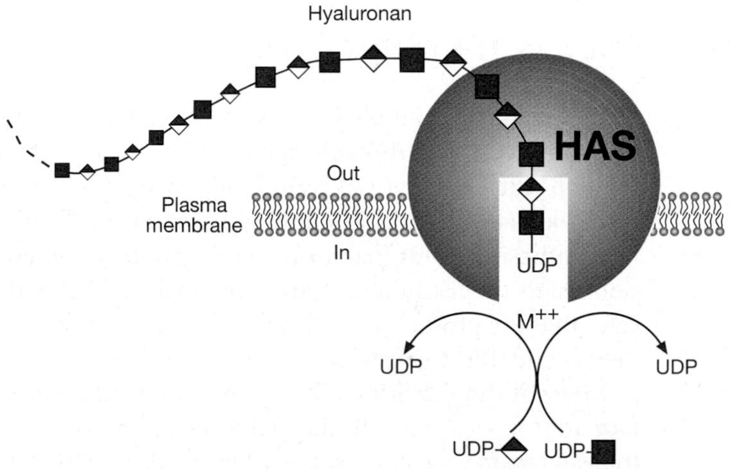

FIGURE 15.2. Hyaluronan biosynthesis by hyaluronan synthase (Has) occurs by addition of sugars (*N*-acetylglucosamine and glucuronic acid) to the reducing end of the polymer. M^{++} refers to a metal ion cofactor.

the nonreducing end by an enzyme that is unrelated to *hasA* and the mammalian *Has* gene family. In this case, the enzyme has two separable domains with independent glycosyltransferase activities—one for UDP-GlcNAc and the other for UDP-GlcA.

THE HYALURONIDASES AND HYALURONAN TURNOVER

Animal cells express a set of catabolic enzymes that degrade hyaluronan. The human hyaluronidase gene (*HYAL*) family is complex, with two sets of three contiguous genes located on two chromosomes, a pattern that suggests two ancient gene duplications followed by a block duplication. In humans, the cluster on chromosome 3p21.3 (*HYAL1, 2,* and *3*) appears to have major roles in somatic tissues. *HYAL4* in the cluster on chromosome 7q31.3 codes for a protein that appears to have chondroitinase, but not hyaluronidase, activity; *PHYAL1* is a pseudogene; and *SPAM1* (sperm adhesion molecule1, PH-20) is restricted to testes. The role of SPAM1 in fertilization is discussed below.

The turnover of hyaluronan in most tissues is rapid (e.g., a half-life of approximately 1 day in epidermal tissues), but its residence time in some tissues can be quite long and dependent on location (e.g., in cartilage). It has been estimated that an adult human contains approximately 15 g of hyaluronan and that about one-third turns over daily. Turnover appears to occur by receptor-mediated endocytosis and lysosomal degradation either locally or after transport by lymph to lymph nodes or by blood to liver. The endothelial cells of the lymph node and liver sinusoids remove hyaluronan via specific receptors such as LYVE-1 (a homolog of CD44) and HARE (hyaluronan receptor for endocytosis). HARE appears to be the major clearance receptor for hyaluronan delivered systemically by lymph and blood. The current understanding of this catabolic process is that hyaluronidases at the cell surface and in the lysosome cooperate to degrade the chains. Large hyaluronan molecules in the extracellular space interact with cell-surface receptors that internalize fragments produced by a membrane-associated, GPI-anchored hyaluronidase, most likely Hyal2. The fragments are transported into a unique vesicular compartment and eventually enter a pathway to lysosomes for complete degradation to monosaccharides, probably involving Hyal1 and the two exoglycosidases β-glucuronidase and β-*N*-acetylglucosaminidase. The importance of this process is demonstrated by the fact that *Hyal2*-null mice are embryonic lethal and by the identification of a lysosomal storage disorder in a person with a mutation in *HYAL1*.

HYALURONAN FUNCTION IN THE EXTRACELLULAR MATRIX

Hyaluronan has multiple roles in early development, tissue organization, and cell proliferation. The *Has2*-null mouse exhibits an embryonic lethal phenotype at the time of heart formation, whereas *Has1*- and *Has3*-null mice as well as *Has1/3* compound mutant mice show no obvious phenotype. Interestingly, explanted cells from the *Has2*-null embryonic heart do not synthesize hyaluronan or undergo endothelial-mesenchymal transformation and migration unless small amounts of hyaluronan are added to the culture medium. This finding indicates that the production of hyaluronan at key points may be essential for many tissue morphogenetic transformations—in this case, formation of the tricuspid and mitral valves.

Many of the activities of hyaluronan depend on binding proteins present on the cell surface and/or secreted into the extracellular matrix. A class of proteins that bind selectively to hyaluronan was first discovered in cartilage. This class is now referred to as the link module family of hyaladherins (Figure 15.3). Proteoglycans were efficiently extracted from this tissue with denaturing solvents and were shown to reaggregate when restored to renatur-

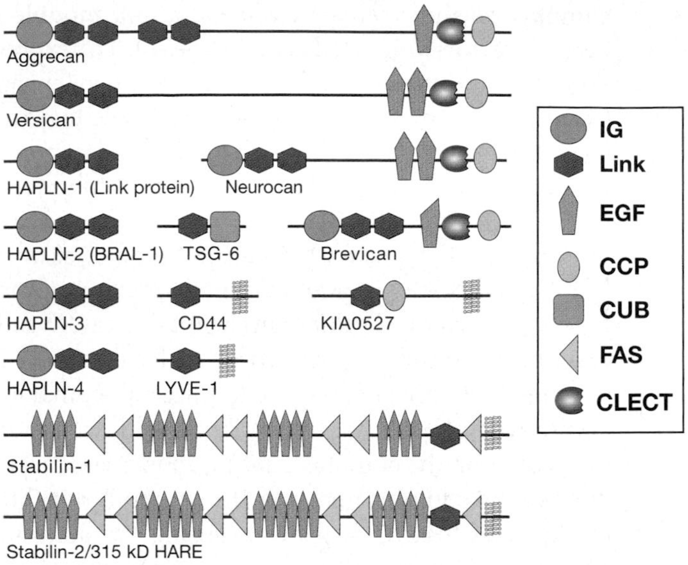

FIGURE 15.3. Modular organization of the link module superfamily of hyaluronan-binding proteins. These proteins contain one or more link modules that bind to hyaluronan. Like many extracellular matrix proteins, the link module superfamily members contain various subdomains, depicted by the following annotated symbols: (IG) immunoglobulin-like domain; (EGF) epidermal growth factor–like domain; (CLECT) C-type lectin domain; (CCP) complement control protein module; (Link) hyaluronan-binding module; (CUB) domain found in some complement proteins, peptidases, and bone morphogenetic protein; (FAS) domain found in fasciclin I family of proteins. (See the SMART database at EMBL for additional information on these domains: http://smart.embl-heidelberg.de/.) (Redrawn, with permission, from Blundell C.D. 2004. In *Chemistry and biology of hyaluronan* [ed. H.G. Garg and C.A. Hales], pp.189–204, © Elsevier.)

ing conditions. An essential protein, referred to as the link protein, was shown to be necessary for stabilizing the proteoglycan aggregates, and subsequently, the structure of the aggregate was defined (see Chapter 16, Figure 16.1). The link protein contains two repeats of a motif, the link module, that interact specifically with hyaluronan and form the backbone on which the proteoglycan aggregate assembles. The proteoglycan, now named aggrecan, also contains a globular domain, the G1 domain, with two homologous link modules that interact with hyaluronan. An additional domain in the link protein cooperatively interacts with a homologous domain in G1, which locks the proteoglycan on the hyaluronan chain. Absence of the link molecule fails to anchor the proteoglycan, and null mice deficient in link protein show defects in cartilage development and delayed bone formation (short limbs and craniofacial anomalies). Most mutant mice die shortly after birth as a result of respiratory failure, and the few survivors develop progressive skeletal deformities.

Interestingly, there are four proteoglycan genes with homologous G1 domains interspersed in the genome (versican, neurocan, brevican, and aggrecan), and each contains a contiguous homologous link molecule (Figure 15.3). Versican is a major component of many soft tissues and is especially important in vascular biology. Neurocan and brevican are expressed predominantly in brain tissue. Versican and aggrecan are anchored to hyaluronan in tissues by similar mechanisms, and it is likely that neurocan and brevican are organized similarly. Thus, hyaluronan acts as a scaffold on which to build proteoglycan structures adapted to diverse tissue functions.

An impressive example of the requirement for a hyaluronan-based matrix occurs during the process of cumulus oophorus expansion in the mammalian preovulatory follicle. At the beginning of this process, the oocyte is surrounded by about 1000 cumulus cells tightly compacted and in gap-junction contact with the oocyte. In response to hormonal stimuli, the cumulus cells up-regulate Has2 and a link module family hyaladherin encoded by tumor necrosis factor-stimulated gene 6 (*TSG-6*). The expression of these proteins initiates production of hyaluronan and its organization into an expanding matrix around the cumulus cells. Concurrently, the follicle becomes permeable to serum, which introduces an unusual molecule called inter-α-trypsin inhibitor (IαI), composed of the trypsin inhibitor bikunin and two heavy chains all covalently bound to a chondroitin sulfate chain (see Chapter 12). In a complex process, TSG-6 catalyzes the transfer of heavy chains from chondroitin sulfate onto the newly synthesized hyaluronan. In the absence of either TSG-6 or ITI, the matrix does not form, and the phenotype of mice null for either of these molecules is female infertility. At the time of ovulation, hyaluronan synthesis ceases, and ovulation of the expanded cumulus cell–oocyte complex occurs. Prior to fertilization, individual sperm undergo capacitation enabling them to penetrate and fertilize an ovum. During this process, SPAM1/PH20, a GPI-anchored hyaluronidase, redistributes and accumulates in the sperm head. SPAM1 binds hyaluronan in the cumulus, causing an increase in Ca^{++} flux and sperm motility. It also helps dissolve the cumulus matrix as the sperm moves through the hyaluronan vestment. A soluble form of SPAM1 is secreted during the acrosome reaction. The release of acrosomal hyaluronidase and proteases renders the sperm capable of fusing with the egg and eventually destroys the entire matrix to allow the fertilized oocyte to implant and develop.

HYALURONAN-BINDING PROTEINS: THE LINK PROTEIN MOTIF

Many of the hyaluronan-binding proteins described above have in common a protein motif called the link module, first described in cartilage link protein (Figure 15.4). The link proteins belong to a subfamily called the hyaluronan and proteoglycan link proteins (HAPLN) and are expressed in many tissues. Four cell-surface receptors have extracellular

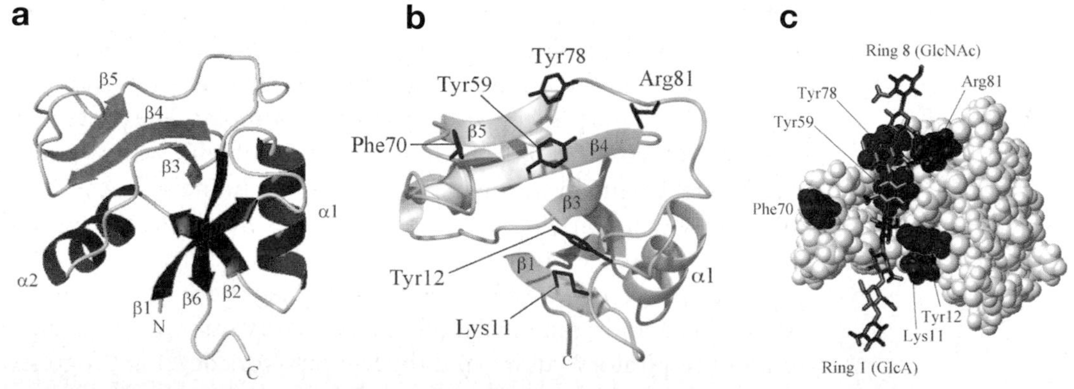

FIGURE 15.4. Structure of the link module. (*a*) TSG-6 contains a prototypical link fold defined by two α helices (α1 and α2) and two triple-stranded antiparallel β sheets (β1,2,6 and β3–5). (Redrawn, with permission, from Blundell C.D. et al. 2003. *J. Biol. Chem.* **278:** 49261–49270, ©American Society for Biochemistry and Molecular Biology.) (*b*) The hyaluronan-bound conformation of the protein, with a view showing key amino acids and the position of hyaluronan. (Redrawn, with permission, from Blundell C.D. et al. 2004. In *Chemistry and biology of hyaluronan* [eds. H.G. Garg and C.A. Hales], pp. 189–204, ©Elsevier.) (*c*) A model of a link protein/hyaluronan complex. Binding of the hyaluronan to the protein is mediated through ionic interactions between positively charged amino acid residues (*green*) and the carboxylate groups of the uronic acids, and by a combination of aromatic ring stacking and H-bond interactions between the polysaccharide and aromatic amino acids (*red*). In addition, hydrophobic pockets (on either side of Tyr59) can accommodate the methyl groups of two *N*-acetylglucosamine side chains, which is likely to be a major determinant in specificity of link modules for hyaluronan. (Redrawn, with permission, from Blundell C.D. et al. 2005. *J. Biol. Chem.* **280:** 18189–18201, ©American Society for Biochemistry and Molecular Biology.)

domains with link module motifs: CD44, LYVE-1 (lymphatic vessel endothelial hyaluronan receptor), HARE/STABILIN-2 (hepatic hyaluronan clearance receptor), and STABILIN-1, which are expressed on discontinuous endothelial cells and some activated macrophages. The other members of the superfamily are secreted and include the chondroitin-sulfate-containing proteoglycans that comprise the aggrecan superfamily (aggrecan, versican, brevican, and neurocan) and TSG-6.

The three-dimensional structure of the module in TSG-6 has been determined by nuclear magnetic resonance and defines a consensus fold of the two α-helices and two triple-stranded antiparallel β-sheets. The link module consists of about 100 amino acids and contains four cysteines disulfide-bonded in the pattern Cys1-Cys4 and Cys2-Cys3. This fold has so far only been found in vertebrates, consistent with the fact that hyaluronan is a relatively recent evolutionary invention. The link module fold is related to that found in the C-type lectins, but it lacks the Ca^{++} binding motif (see Chapter 31). In the case of TSG-6, the interaction of hyaluronan with the protein involves (1) ionic interactions between positively charged amino acid residues and the carboxyl groups of the uronic acids and (2) hydrophobic interactions between the acetamido side chains of two *N*-acetylglucosamine residues and hydrophobic pockets on either side of adjacent tyrosines (Figure 15.4). Many of these features are conserved in other members of the superfamily. Different subgroups of the link module superfamily differ in size and length of hyaluronan recognized (e.g., hexasaccharides to decasaccharides).

Some hyaluronan-binding proteins do not contain a link module (RHAMM, ITI, SPACR, SPACRCAN, CD38, CDC37, HABP1/P-32, and IHABP4), and most of these are unrelated to one another by primary sequence. Some of these proteins contain clusters of basic amino acids, referred to as BX_7B motifs (where B is either lysine or arginine and X can be any amino acid other than acidic residues), but the actual docking site of the chain with this motif has not been established. Thus, the presence of this motif should not be taken as proof that the protein interacts with hyaluronan.

HYALURONAN AND CELL SIGNALING

Hyaluronan expression has long been implicated in enhanced cell adhesion and locomotion because it is expressed abundantly during morphogenesis and in both physiological and pathological invasive processes. A search for cell-surface receptors revealed two major hyaluronan-binding proteins, CD44 and RHAMM (receptor for hyaluronan-mediated motility). CD44 is a transmembrane receptor expressed by many cell types and it varies markedly in glycosylation, oligomerization, and protein sequence because of differential mRNA splicing. CD44H (the isoform expressed by hematopoietic cells) binds to hyaluronan, and the interaction can mediate leukocyte rolling and extravasation in some tissues. Changes in CD44 expression are associated with a wide variety of tumors and the metastatic spread of cancer, although as with other tumor-associated factors, a strict correlation does not exist. Many cells also express the receptor RHAMM, which is involved in cell motility and cell transformation. The RHAMM pathway is thought to induce focal adhesions to signal the cytoskeletal changes required for elevated cell motility seen in tumor progression, invasion, and metastasis. Like CD44, RHAMM splice variants exist, some of which may be intracellular.

CD44 contains a cytoplasmic domain, a transmembrane segment, and an ectodomain with a single link module that can bind hyaluronan. When hyaluronan binds to CD44, the cytoplasmic tail interacts with regulatory and adaptor molecules, such as SRC kinases, RHO (ras homolog) GTPases, VAV2 (a human proto-oncogene), GAB1 (a GRB2-associated binding protein), and ankyrin and ezrin (which regulate cytoskeletal assembly/disassembly and cell migration). Hyaluronan binding to RHAMM also transduces signals that influence growth and motility, for example, by activating SRC, FAK (focal adhesion kinase), ERK (extracellular mitogen-regulated protein kinase), and PKC (protein tyrosine kinase C) (see Chapter 37).

Interaction of hyaluronan with CD44 can also regulate ERBB-family (epithelial growth factor receptor) signaling, activating the PI3K (phophoinositide-3-kinase)–PKB/AKT (protein kinase B) signaling pathway and phosphorylation of FAK and BAD (BCL2-antagonist of cell death), which promote cell survival. RHAMM can interact with and activate ERK1, which can also phosphorylate BAD. Thus, both CD44 and RHAMM interactions with hyaluronan can influence cell survival. These pathways are relevant to tumor cell survival and invasion; their inhibition by hyaluronan oligomers and soluble hyaluronan-binding proteins suggests novel therapeutic approaches for treating cancer (see Chapter 44).

HYALURONAN CAPSULES IN BACTERIA

Some pathogenic bacteria (e.g., certain strains of *Streptococcus* and *Pasteurella*) produce hyaluronan and deposit it as an extracellular capsule (Figure 15.5; also see Chapter 20). Capsular hyaluronan, like other capsular polysaccharides, increases virulence by helping to shield the microbe from host defenses. For example, the capsule blocks phagocytosis and protects against complement. Because bacterial hyaluronan is identical in structure to host hyaluronan, the capsule can also prevent the formation of protective antibodies. Thus, the formation of hyaluronan capsules by bacteria is a form of molecular mimicry. The capsule also can aid in bacterial adhesion to host tissue, facilitating colonization (see Chapter 34). Finally, the production of hyaluronan by invading bacteria can also induce a number of signaling events through hyaluronan-binding proteins that modulate the host physiology, for example, cytokine production (see Chapter 39).

In addition to bacteria, an algal virus (*Chlorella*) also encodes a hyaluronan synthase. The functional significance of viral hyaluronan production is unknown, but could be related to prevention of secondary viral infection, increase in host capacity to produce virus, or viral burst size. The origin of viral HAS is unknown, but based on sequence homology it most likely arose from a vertebrate.

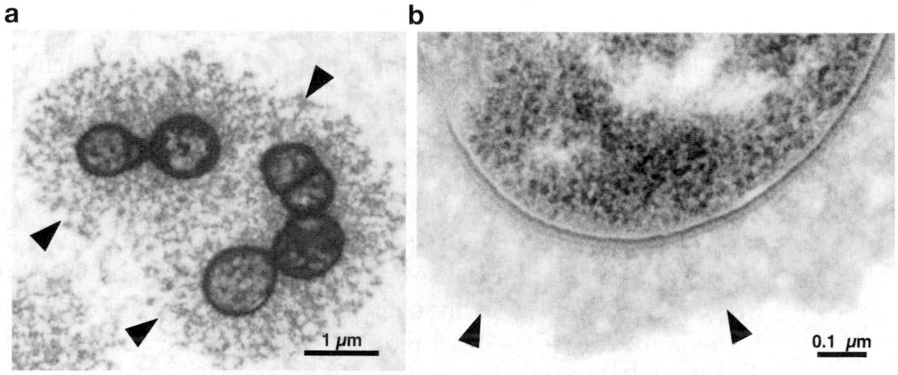

FIGURE 15.5. Hyaluronan capsule. Cross-sectioned *Streptococcus zooepidemicus* cells surrounded by a hyaluronan capsule and observed at differing magnifications. (*a*) 1 μm. (Reprinted, with permission, from Chong B.F., Blank L.M., McLaughlin R., and Nielsen L.K. 2005. *Appl. Microbiol. Biotechnol.* **66:** 341–351, ©Springer Verlag.) (*b*) 0.1 μm. (Reprinted from Goh L.-T. 1998. "Effect of culture conditions on rates of intrinsic hyaluronan production by *Streptococcus equi* subsp. *zooepidemicus*." Ph.D. thesis. University of Queensland, Australia.)

HYALURONAN AS A THERAPEUTIC AGENT

Hyaluronan has been used therapeutically for a number of years. In some countries, patients with osteoarthritis are successfully treated by direct injection of high-molecular-weight hyaluronan into the synovial space of an affected joint. The mechanism of action is complex and probably involves both the viscoelastic properties of the polymer as well as effects on synovial cells in the joint capsule. Hyaluronan suppresses cartilage degeneration, acts as a lubricant (thereby protecting the surface of articular cartilage), and reduces pain perception. It can also suppress prostaglandin E2 and IL-1 production, which in turn can affect proliferation of synovial cells.

The application of hyaluronan in ophthalmology is widespread. During surgery for lens replacement due to cataracts, a high potential for injury of fragile intraocular tissues exists, especially for the endothelial layer of the cornea. High-molecular-weight hyaluronan is injected to maintain operative space and structure and to protect the endothelial layer from physical damage. Hyaluronan has also been approved for cosmetic use (e.g., by subdermal injection to fill wrinkles or pockets under the skin).

Low-molecular-weight hyaluronan oligosaccharides ($\sim 10^3$–10^4 D) also have potent biological activities by altering selective signaling pathways. In cancer cells, hyaluronan oligosaccharides induce apoptosis and inhibit tumor growth in vivo. Short hyaluronan chains may prove useful for preventing cancer metastasis by boosting certain immune responses or altering new blood vessel growth. Recently, recombinant forms of the *Pasteurella* synthase (pmHas) have been engineered to produce hyaluronan oligosaccharides of defined size. This strategy has great promise for exploring the relationship of hyaluronan size to function, which may in turn yield new therapeutic agents with selective activities.

FURTHER READING

Day A.J. and Prestwich G.D. 2002. Hyaluronan-binding proteins: Tying up the giant. *J. Biol. Chem.* **277:** 4585–4588.

DeAngelis P.L. 2002. Microbial glycosaminoglycan glycosyltransferases. *Glycobiology* **12:** 9R–16R.

Itano N. and Kimata K. 2002. Mammalian hyaluronan synthases. *IUBMB Life* **54:** 195–199.

Tammi M.I., Day A.J., and Turley E.A. 2002. Hyaluronan and homeostasis: A balancing act. *J. Biol. Chem.* **277:** 4581–4584.

Turley E.A., Noble P.W., and Bourguignon L.Y.W. 2002. Signaling properties of hyaluronan receptors. *J. Biol. Chem.* **277:** 4589–4592.

Weigel P.H. 2002. Functional characteristics and catalytic mechanisms of the bacterial hyaluronan synthases. *IUBMB Life* **54:** 201–211.

Hascall V.C., Majors A.K., De La Motte C.A., Evanko S.P., Wang A., Drazba J.A., Strong S.A., and Wight T.N. 2004. Intracellular hyaluronan: A new frontier for inflammation? *Biochim. Biophys. Acta* **1673:** 3–12.

Toole B.P. 2004. Hyaluronan: From extracellular glue to pericellular cue. *Nat. Rev. Cancer* **4:** 528–539.

Zhuo L., Hascall V.C., and Kimata K. 2004. Inter-α-trypsin inhibitor, a covalent protein-glycosaminoglycan-protein complex. *J. Biol. Chem.* **279:** 38079–38082.

Day A.J. and de la Motte C.A. 2005. Hyaluronan cross-linking: A protective mechanism in inflammation? *Trends Immunol.* **26:** 637–643.

Taylor K.R. and Gallo R.L. 2006. Glycosaminoglycans and their proteoglycans: Host-associated molecular patterns for initiation and modulation of inflammation. *FASEB J.* **20:** 9–22.

Volpi N. 2006. Therapeutic applications of glycosaminoglycans. *Curr. Med. Chem.* **13:** 1799–1810.

Jiang D., Liang J., and Noble P.W. 2007. Hyaluronan in tissue injury and repair. *Annu. Rev. Cell Dev. Biol.* **23:** 435–461.

Weigel P.H. and DeAngelis P.L. 2007. Hyaluronan synthases: A decade-plus of novel glycosyltransferases. *J. Biol. Chem.* **282:** 36777–36781.

CHAPTER 16

Proteoglycans and Sulfated Glycosaminoglycans

Jeffrey D. Esko, Koji Kimata, and Ulf Lindahl

THIS CHAPTER FOCUSES ON THE STRUCTURE, BIOSYNTHESIS, and general biology of proteoglycans. Topics include a description of the major families of proteoglycans, their characteristic polysaccharide chains (glycosaminoglycans), biosynthetic pathways, and general concepts about proteoglycan function. Proteoglycans, like other glycoconjugates, are enormously diverse and have many essential roles in biology.

HISTORICAL PERSPECTIVE

The study of proteoglycans dates back to the beginning of the 20th century with investigations of "chondromucoid" from cartilage and anticoagulant preparations from liver (heparin). From 1930 to 1960, great strides were made in analyzing the chemistry of the polysaccharides of these preparations (also known as "mucopolysaccharides"), yielding the structure of hyaluronan (see Chapter 15), dermatan sulfate (DS), keratan sulfate (KS), different isomeric forms of chondroitin sulfate (CS), heparin, and heparan sulfate (HS).

Together, these polysaccharides came to be known as glycosaminoglycans (sometimes abbreviated as GAGs) to indicate the presence of amino sugars and other sugars in a polymeric form. Subsequent studies provided insights into the linkage of the chains to core proteins, and these structural studies paved the way for the biosynthetic studies that followed.

The 1970s marked a turning point in the field, when improved isolation and chromatographic procedures were developed to purify and analyze tissue proteoglycans and glycosaminoglycans. Density-gradient ultracentrifugation allowed separation of the large aggregating proteoglycans from cartilage, revealing a complex of proteoglycan, hyaluronan, and link protein (Figure 16.1). Also during this period, it was realized that the production of proteoglycans was a general property of animal cells and that proteoglycans and glycosaminoglycans were present on the cell surface, inside the cell, and in the extracellular matrix (ECM). This observation led to a rapid expansion of the field and the eventual appreciation of proteoglycan function in cell adhesion and signaling, as well as a host of other biological activities (see Chapter 35). Today, studies with somatic cell mutants (see Chapter 46) as well as experiments using gene knockout and silencing techniques in a variety of model organisms, including nematode worms (*Caenorhabditis elegans)*, fruit flies (*Drosophila melanogaster*), African clawed frogs (*Xenopus laevis*), zebrafish (*Danio rerio*), and mice (*Mus musculus*), are aimed at extending our understanding of the role of proteoglycans in development and physiology (see Chapters 23–25). In turn, human diseases associated with aberrant biosynthesis or degradation of proteoglycans have been identified (see Chapters 38–44), and some are classified as congenital disorders of glycosylation. A variety of analytical techniques have been developed, including mass spectroscopic methods and glycan array applications, that provide new tools for understanding proteoglycan structure and function.

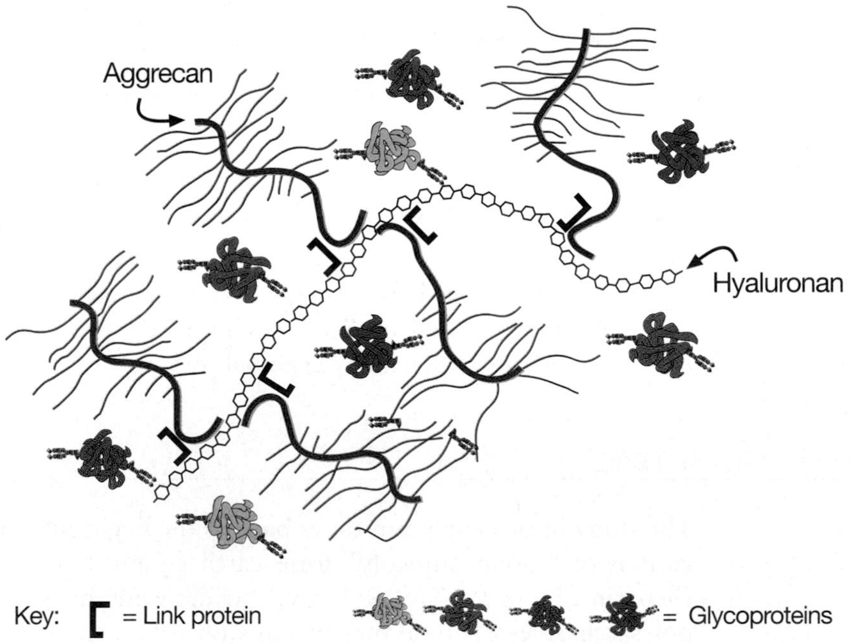

FIGURE 16.1. The large cartilage CS proteoglycan (aggrecan) forms an aggregate with hyaluronan and link protein (see Chapter 15). (Redrawn, with permission of Springer Science and Business Media, from Rodén L. 1980. In *The biochemistry of glycoproteins and proteoglycans* [ed. W.J. Lennarz], p. 291. Plenum Press, New York.)

PROTEOGLYCANS AND GLYCOSAMINOGLYCAN DIVERSITY

Proteoglycans consist of a core protein and one or more covalently attached glycosaminoglycan chains (Figure 16.2). Glycosaminoglycans are linear polysaccharides, whose disaccharide building blocks consist of an amino sugar (*N*-acetylglucosamine, glucosamine that is variously N-substituted, or *N*-acetylgalactosamine) and a uronic acid (glucuronic acid or iduronic acid) or galactose. Figure 16.3 depicts characteristic features of the major types of glycosaminoglycans found in vertebrates. Hyaluronan is shown for comparison: It does not occur covalently linked to proteoglycans, but instead interacts noncovalently with proteoglycans via hyaluronan-binding motifs (see Chapter 15). Generally, invertebrates produce the same types of glycosaminoglycans as vertebrates, except that hyaluronan is not present and the chondroitin chains tend to be nonsulfated. Most proteoglycans also contain O- and N-glycans typically found on glycoproteins (see Chapters 8 and 9). The glycosaminoglycan chains are much larger than these other types of glycans (e.g., a 20-kD glycosaminoglycan chain contains ~80 sugar residues, whereas a typical biantennary N-glycan contains 10–12 residues). Therefore, the properties of the glycosaminoglycans tend to dominate the chemical properties of proteoglycans (although N- or O-glycans on proteoglycans may have distinct biological properties as described for glycoproteins).

Virtually all mammalian cells produce proteoglycans and secrete them into the ECM, insert them into the plasma membrane, or store them in secretory granules. The ECM

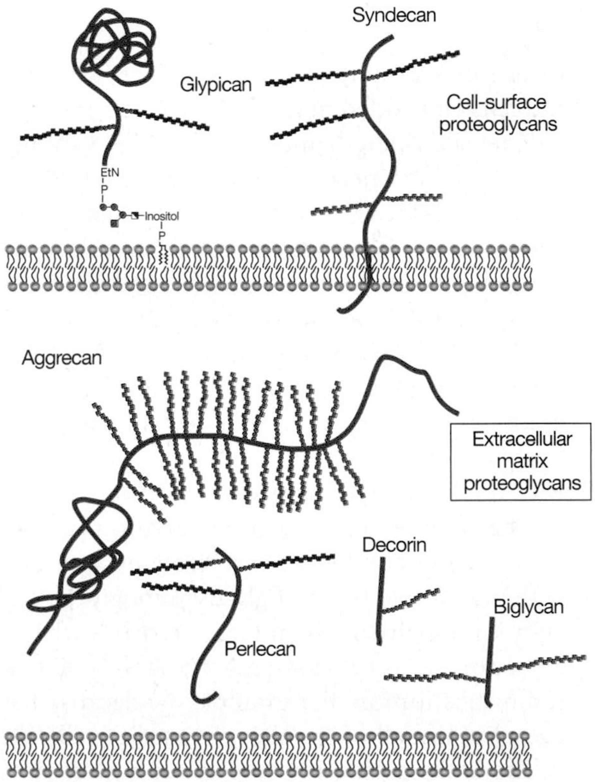

FIGURE 16.2. Proteoglycans consist of a protein core (*brown*) and one or more covalently attached glycosaminoglycan chains ([*blue*] HS; [*yellow*] CS/DS). Membrane proteoglycans either span the plasma membrane (type I membrane proteins) or are linked by a GPI anchor. ECM proteoglycans are usually secreted, but some proteoglycans can be proteolytically cleaved and shed from the cell surface (not shown).

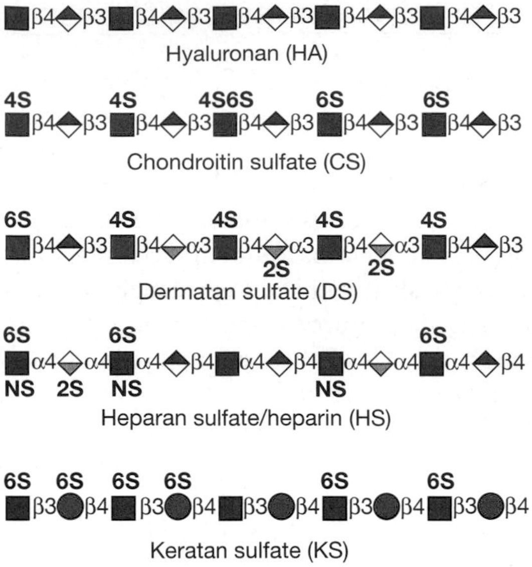

FIGURE 16.3. Glycosaminoglycans consist of repeating disaccharide units composed of an N-acetylated or N-sulfated hexosamine and either a uronic acid (glucuronic acid or iduronic acid) or galactose. Hyaluronan lacks sulfate groups, but the rest of the glycosaminoglycans contain sulfates at various positions. DS is distinguished from CS by the presence of iduronic acid. Keratan sulfates lack uronic acids and instead consist of sulfated galactose and *N*-acetylglucosamine residues.

determines the physical characteristics of tissues and many of the biological properties of the cells embedded in it. The major components of the ECM are fibrous proteins that provide tensile strength and elasticity (e.g., various collagens and elastins), adhesive glycoproteins (e.g., fibronectin, laminin, and tenascin), and proteoglycans that interact with other ECM components to provide a hydrated gel that resists compressive forces. Together, the various ECM components provide an extracellular environment that regulates cell proliferation and differentiation. Cells synthesize a diverse group of membrane proteoglycans. These typically have a type I orientation with a single membrane-spanning domain or a glycosylphosphatidylinositol (GPI) anchor (Figure 16.2). Additionally, some cells concentrate proteoglycans along with other secretory products in secretory granules. Secretory granule proteoglycans are thought to help sequester and regulate the availability of positively charged components, such as proteases and bioactive amines, through interaction with the negatively charged glycosaminoglycan chains.

The tremendous structural variation of proteoglycans is due to a number of factors. First, a large number of core proteins have been identified, and these can be substituted with one or two types of glycosaminoglycan chains. Some proteoglycans contain only one glycosaminoglycan chain (e.g., decorin), whereas others have more than 100 chains (e.g., aggrecan). Another source of variability lies in the stoichiometry of glycosaminoglycan chain substitution. For example, syndecan-1 has five attachment sites for glycosaminoglycans, but not all of the sites are used equally. Other proteoglycans can be "part time," that is, they may exist with or without a glycosaminoglycan chain or with only a truncated oligosaccharide. A given proteoglycan present in different cell types often exhibits differences in the number of glycosaminoglycan chains, their lengths, and the arrangement of sulfated residues along the chains. Thus, a preparation of syndecan-1 represents a diverse population of molecules, each potentially representing a unique structural entity. These characteristics, typical of all proteoglycans, create diversity that may facilitate the formation of binding sites of variable density and affinity for different ligands.

The major classes of proteoglycans can be defined by distribution, homologies, and function. The following groupings provide an overview of the classes of proteoglycans currently known.

Interstitial Proteoglycans and the Aggrecan Family

A large number of proteoglycans are present in the ECM, and their distribution depends on the nature of the ECM. The interstitial proteoglycans represent a diverse class of molecules, differing in size and glycosaminoglycan composition. The small leucine-rich proteoglycans (SLRPs) contain leucine-rich repeats flanked by cysteines in their central domain. At least nine members of this family are known and they carry CS, DS, or KS chains (Table 16.1 and Table 16.2). These proteoglycans help to stabilize and organize collagen fibers, for example, in tendons. In the cornea, KS proteoglycans maintain the register of collagen fibers required for transparency. Decorin also can bind transforming growth factor β (TGF-β), serving as a sink to keep the growth factor sequestered in the matrix surrounding most cells.

The aggrecan family of proteoglycans consists of aggrecan, versican, brevican, and neurocan. In all four members, the protein moiety contains an amino-terminal domain capable of binding hyaluronan, a central region that contains covalently bound CS chains, and a carboxy-terminal domain containing a C-type lectin domain (see Chapter 31). Aggrecan is the best-studied member of this family, because it represents the major proteoglycan in cartilage. It contains as many as 100 CS chains and, in humans, it contains KS chains as well. Versican, which is produced predominantly by connective tissue cells, undergoes alternative splicing events that generate a family of proteins of differing complexity that may have a role in neural crest cell and axonal migration. Neurocan is expressed in the late

TABLE 16.1. *Examples of chondroitin sulfate proteoglycans*

Proteoglycan	Core protein (kD)	Number of chondroitin sulfate chains	Tissue distribution
Aggrecan family			
Aggrecan	208–220	~100	secreted; cartilage
Versican/PG-M	265	12–15	secreted; connective tissue cells; aorta; brain
Neurocan	145	1–2	secreted; brain
Brevican	96	0–4	secreted; brain
SLRPs			
Decorin	36	1	secreted; connective tissue cells
Biglycan	38	1–2	secreted; connective tissue cells
Other examples			
Leprecan	82	1–2	secreted; basement membranes
Type IX collagen, α2 chain	68	1	secreted; cartilage; vitreous humor
Phosphacan	175	2–5	membrane bound; brain
Thrombomodulin	58	1	membrane bound; endothelial cells
CD44	37	1–4	membrane bound; lymphocytes
NG2	251	2–3	membrane bound; neural cells
Invariant chain	31	1	membrane bound; antigen-processing cells
Serglycin	10–19	10–15	intracellular granules; myeloid cells

(SLRP) Small leucine-rich proteoglycans.

TABLE 16.2. *Examples of keratan sulfate proteoglycans*

Proteoglycan	Type	Core protein (kD)	Tissue distribution
SLRPs			
Lumican	KS I	37	secreted; broad
Keratocan	KS I	37	secreted; broad, but sulfated only in cornea
Fibromodulin	KS I	59	secreted; broad
Mimecan	KS I	25	secreted; broad, but sulfated only in cornea
Other examples			
SV2	KS I	80	membrane bound; synaptic vesicles
Claustrin	KS II	105	membrane bound; CNS
Aggrecan (human)	KS II	200	secreted; cartilage

KS I is found on an N-glycan core structure, whereas KS II (skeletal KS) is found on an O-glycan core-2 structure.

embryonic central nervous system (CNS) and can inhibit neurite outgrowth. Brevican is expressed in the terminally differentiated CNS, particularly in perineuronal nets.

The interstitial proteoglycans and aggrecan family of proteoglycans appear to be unique to vertebrates. Other proteoglycans are expressed in *C. elegans* and *D. melanogaster*, suggesting that the core proteins have undergone enormous diversification during evolution.

Secretory Granule Proteoglycans

Serglycin is the major proteoglycan present in cytoplasmic secretory granules in endothelial, endocrine, and hematopoietic cells. Depending on the species, it has a variable number of glycosaminoglycan attachment sites that can carry CS or heparin chains. Heparin is a highly sulfated form of HS (discussed below) and is made exclusively on serglycin present in connective-tissue-type mast cells. Other granular proteoglycans may exist as well, such as chromogranin A, but the extent of glycosaminoglycan substitution appears to be substoichiometric, making them part-time proteoglycans.

Basement Membrane Proteoglycans

The basement membrane is an organized layer of the ECM that lies flush against epithelial cells and consists largely of laminin, nidogen, collagens, and proteoglycans. Basement membranes contain at least four types of proteoglycans depending on tissue type: perlecan, agrin, and collagen type XVIII (Table 16.3), which carry HS chains (although perlecan has been shown to carry CS in cartilage), and leprecan, which carries CS chains (Table 16.1). Perlecan has a mass of 400 kD and consists of multiple domains that have numerous functions. It has a role in embryogenesis and tissue morphogenesis and a particularly important role in cartilage development. Agrin acts in neuromuscular junctions (where it aggregates acetylcholine receptors) and in renal tubules (where it has an important role in determining the filtration properties of the glomerulus).

Membrane-bound Proteoglycans

The membrane proteoglycans are diverse (Tables 16.1–16.3). The syndecan family consists of four members, each with a short hydrophobic domain that spans the membrane, link-

TABLE 16.3. *Examples of heparan sulfate proteoglycans*

Proteoglycan	Core protein (kD)	Number of glycosaminoglycan chains	Tissue distribution
Perlecan	400	1–3 HS	secreted; basement membranes; cartilage
Agrin	200	1–3 HS	secreted; neuromuscular junctions
Collagen type XVIII	147	2–3 HS	secreted; basement membranes
Syndecans 1–4	31–45	1–3 CS 1–2 HS	membrane bound epithelial cells and fibroblasts
Betaglycan	110	1 HS 1 CS	membrane bound fibroblasts
Glypicans 1–6	~60	1–3 HS	membrane bound; epithelial cells and fibroblasts
Serglycin	10–19	10–15 heparin/CS	intracellular granules; mast cells

ing the larger extracellular domain containing the glycosaminoglycan attachment sites to a smaller intracellular cytoplasmic domain. Syndecan-1 and syndecan-3 carry CS chains on the membrane proximal regions and HS chains at the more distal sites further away from the membrane. In contrast, syndecan-2 and syndecan-4 carry only HS chains. These are expressed in a tissue-specific manner and facilitate cellular interactions with a wide range of extracellular ligands, such as growth factors and matrix molecules. Because of their membrane-spanning properties, the syndecans can transmit signals from the extracellular environment to the intracellular cytoskeleton via their cytoplasmic tails. For example, binding of a ligand to the HS chain can induce oligomerization of syndecans at the cell surface, which leads to recruitment of factors at their cytoplasmic tails, such as kinases (e.g., c-Src), PDZ-domain proteins, or cytoskeletal proteins. The recruitment of cytoplasmic proteins in turn triggers a signal that affects actin assembly. Proteolytic cleavage of the syndecans occurs by matrix metalloproteases, resulting in shedding of the ectodomains bearing the glycosaminoglycan chains. These ectodomains can have potent biological activity as well, for example, by binding the same ligands as cell-surface proteoglycans (see Chapter 35). *C. elegans* and *D. melanogaster* express only one syndecan.

Each member of the glypican family of cell-surface proteoglycans has a GPI anchor attached at the carboxyl terminus, which embeds these proteoglycans in the outer leaflet of the plasma membrane. Thus, the glypicans do not have a cytoplasmic tail like the syndecans. The amino-terminal portion of the protein has multiple cysteine residues and a globular shape that distinguishes the glypicans from the syndecan ectodomains, which tend to be extended structures (Figure 16.2). Glypicans carry only HS chains, which can bind a wide array of factors essential for development and morphogenesis. Six glypican family members exist in mammals, two in *D. melanogaster* and one in *C. elegans.* Glypican-3 (GPC3) is the best-studied member of the family in vertebrates. Humans lacking functional GPC3 exhibit Simpson-Golabi-Behmel syndrome, characterized as an overgrowth disorder. The overgrowth phenotype suggests that GPC3 normally functions to inhibit cell proliferation, but the mechanism by which this occurs is unknown.

In addition to these two gene families, a number of other membrane proteoglycans are expressed on the surface of many different cell types. The CS proteoglycan NG2 is a surface marker expressed on stem cell populations, cartilage chondroblasts, myoblasts, endothelial cells of the brain, and glial progenitors. CD44, a transmembrane cell-surface receptor present on leukocytes and other cells, has a role in processes as diverse as immune cell trafficking and function, axon guidance, and organ development. Only certain splice forms of CD44 carry a glycosaminoglycan chain, and, like the aggrecan family, it can bind hyaluronan (see Chapter 15). Phosphacan is expressed as three different splice variants in the CNS, and, depending on the isoform, it can carry KS or CS chains. One splice variant is present in the ECM, whereas two other forms represent short- and full-length versions of a protein-tyrosine-phosphatase type of transmembrane receptor.

Undoubtedly, new proteoglycans will be discovered as different tissues and model organisms are studied. Genomic analysis has shown that many of the same HS proteoglycans are present in vertebrates and invertebrates, but the CS-containing proteoglycans appear to vary. The application of proteomic technologies to *C. elegans* has revealed a new family of chondroitin-containing proteoglycans that do not have orthologs in vertebrates or other invertebrates. Thus, we can expect new species of proteoglycans to emerge in the future as genomic and proteomic methods are applied across phylogeny.

KERATAN SULFATE, A SULFATED POLY-*N*-ACETYLLACTOSAMINE

KS consists of a sulfated poly-*N*-acetyllactosamine chain. The poly-*N*-acetyllactosamine structure is identical to that found on conventional glycoproteins and mucins (see Chapters 8 and 9). There are two types of KS, distinguished by the nature of their linkage to protein. KS I, originally described in cornea, is found on an N-glycan core structure linked to protein through an asparagine residue. KS II (skeletal KS) is found on an O-glycan core 2 structure and is thus linked through *N*-acetylgalactosamine to serine or threonine. Examples of KS proteoglycans are shown in Figure 16.4 and Table 16.2.

As mentioned above, KS proteoglycans maintain the even spacing of type I collagen fibrils in the cornea, allowing the passage of light without scattering. Defects in sulfation (macular corneal dystrophy) or chain formation (keratoconus) cause distortions in fibril

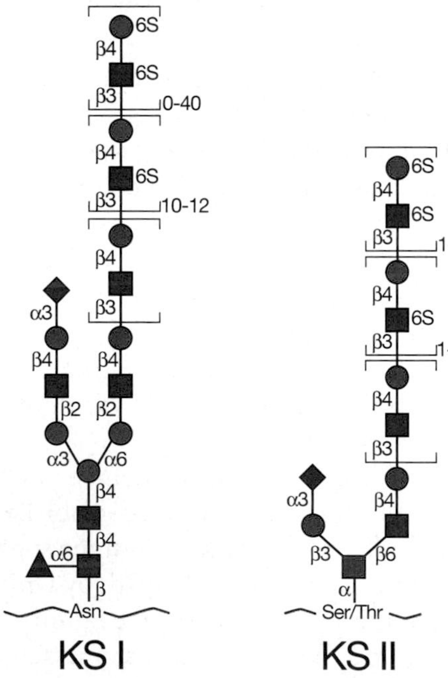

FIGURE 16.4. Keratan sulfates contain a sulfated poly-*N*-acetyllactosamine chain linked to either Asn or Ser/Thr residues. The actual order of the various sulfated and nonsulfated disaccharides occurs somewhat randomly along the chain. Not shown are sialic acids that may be present at the termini of the chains and fucose residues attached to *N*-acetylglucosamine units.

TABLE 16.4. *Keratanases*

Enzyme	Specificity
Endo-β-galactosidase (*Flavobacterium*)	Galβ1-4GlcNAc (no sulfate tolerated)
Keratanase I (*Pseudomonas* species) (endo β-galactosidase)	Galβ1-4GlcNAc6S
Keratanase II (Gram-negative organisms) (endo β-glucosaminidase)	GlcNAcβ1-3Gal±6S

organization and corneal opacity. In cartilage, the function of KS II is unclear. In humans and cows, the large CS proteoglycan found in cartilage (aggrecan) contains a segment of 4–23 hexapeptide repeats (E-E/L-P-F-P-S) where the KS chains are located, but aggrecan in rats and other rodents lacks this motif and does not contain KS.

The poly-*N*-acetyllactosamine of KS I can be quite long (~50 disaccharides, 20–25 kD) and contains a mixture of nonsulfated, monosulfated (Gal-GlcNAc6S), and disulfated (Gal6S-GlcNAc6S) disaccharide units. The biosynthesis of poly-*N*-acetyllactosamine and the underlying linkage structure is covered in Chapters 8 and 9. At least two classes of sulfotransferases, one or more GlcNAc 6-O-sulfotransferases, and one Gal 6-O-sulfotransferase catalyze the sulfation reactions. These enzymes, like other sulfotransferases, use activated sulfate (PAPS [3′-phosphoadenyl-5′-phosphosulfate]) as a high-energy donor (see Chapter 4). GlcNAc 6-O sulfation occurs only on the nonreducing terminal *N*-acetylglucosamine residue, whereas sulfation of galactose residues takes place on nonreducing terminal and internal galactose residues, with a preference for galactose units adjacent to a sulfated *N*-acetylglucosamine. Sulfation of a nonreducing terminal galactose residue blocks further elongation of the chain, providing a potential mechanism for controlling chain length. Only one galactose sulfotransferase has been identified, whereas multiple sulfotransferases catalyze the sulfation of *N*-acetylglucosamine residues. Mutations in the corneal GlcNAc 6-O-sulfotransferase result in macular corneal dystrophy. The relationship of enzymes involved in KS I and KS II sulfation is unclear. The chains can be fucosylated and sialylated as well (see Chapter 13).

Bacterial keratanases degrade KS at characteristic positions (Table 16.4). In animals, KS is degraded in lysosomes by the sequential action of exoglycosidases (β-galactosidase and β-hexosaminidase) after removal of the sulfate groups on the terminal residue by sulfatases (see Chapter 41).

HEPARAN SULFATE AND CHONDROITIN SULFATE ARE LINKED BY XYLOSE TO SERINE

Two classes of glycosaminoglycan chains, CS/DS and HS/heparin, are linked to serine residues in core proteins by way of xylose (Figure 16.5). Xylosyltransferase initiates the process using UDP-xylose as donor. Two isoforms of the enzyme are known in vertebrates (XT-1 and XT-2), but only one isozyme exists in *C. elegans* and *D. melanogaster*. A glycine residue invariably lies to the carboxy-terminal side of the serine attachment site, but a perfect consensus sequence for xylosylation does not exist. At least two acidic amino acid residues are usually present, and they can be located on one or both sides of the serine, usually within a few residues. Several proteoglycans contain clustered glycosaminoglycan attachment sites, raising the possibility that xylosyltransferase could act in a processive manner. Xylosylation is an incomplete process in some proteoglycans, which may explain why proteoglycans with multiple potential attachment sites contain different numbers of chains in different cells. Variation in the degree of glycosaminogly-

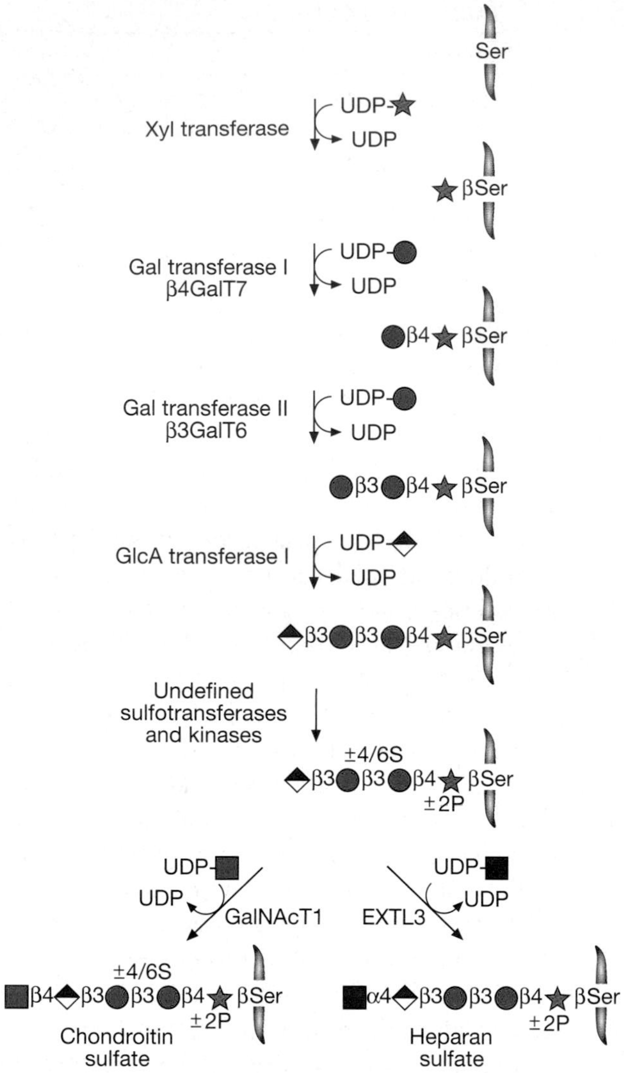

FIGURE 16.5. The biosynthesis of CS (*left chain*) and HS (*right chain*) is initiated by the formation of a linkage region tetrasaccharide. Addition of the first hexosamine commits the intermediate to either CS (GalNAc) or HS (GlcNAc).

can substitution also might result from low levels of UDP-xylose, low activity of the xylosyltransferases, or competing reactions such as serine phosphorylation, acylation, or other forms of glycosylation.

After xylose addition, a linkage tetrasaccharide assembles by the transfer of two galactose residues catalyzed by unique members of the β1-4 galactosyl-, β1-3 galactosyl-, and β1-3 glucuronosyltransferase families of enzymes (Figure 16.5). This intermediate can undergo phosphorylation at the C-2 position of xylose and sulfation of the galactose residues. In general, phosphorylation and sulfation occur substoichiometrically. The lack of chain specificity for phosphorylation would seem to exclude it as a signal for controlling composition. Phosphorylation may be transient, suggesting a role in processing or sorting. Galactose sulfation is found only in CS, but its role in chain initiation, polymerization, or turnover remains unclear.

The linkage tetrasaccharide lies at a bifurcation in the biosynthetic pathway. Two types of reactions occur: addition of β1-4GalNAc (initiation of CS) or addition of α1-

4GlcNAc (initiation of HS) (Figure 16.5). Genetic evidence from studies of *C. elegans* suggests that *N*-acetylgalactosamine addition is mediated by the same enzyme that is involved in chain polymerization (SQV5), but biochemical evidence suggests that more than one enzyme may exist in vertebrates. In heparin/HS formation, the addition of the first *N*-acetylglucosamine residue is catalyzed by an enzyme called EXTL3, which differs from the transferases involved in heparan polymerization (called EXT1 and EXT2). These enzymes are important control points because they ultimately regulate the type of glycosaminoglycan chain that will assemble. Control of the addition of β4GalNAc or α4GlcNAc appears to be manifested at the level of enzyme recognition of the polypeptide substrate. In HS formation, EXTL3 recognizes the linear amino acid sequence immediately adjacent to the attachment site in the core protein. Attachment sites for HS formation usually contain a cluster of acidic residues within seven to nine residues of the serine. Several HS proteoglycans contain multiple contiguous Ser-Gly attachment sites. In the example of syndecan-2 shown below, the underlined sequence indicates the sites of glycosaminoglycan attachment and the boldface letters refer to the clustered acidic residues. Additionally, these sites often contain hydrophobic amino acids in close proximity (e.g., valine, tyrosine, and alanine in syndecan-2). Distant effects of polypeptide structure also can act as a negative factor, for example, by preventing the action of β1-4 *N*-acetylgalactosaminyltransferase.

-SSI**EE**ASGVYPI**DDDD**YSSASGSGA**DED**IESPVLTTS-

CHONDROITIN SULFATE/DERMATAN SULFATE BIOSYNTHESIS

Vertebrate CS consists of repeating sulfate-substituted GalNAc-GlcA disaccharide units polymerized into long chains (see Figure 16.3). In contrast, invertebrates such as *C. elegans* and *D. melanogaster* make either nonsulfated or undersulfated chains, respectively. The assembly process for the backbone appears to be highly conserved, based on the presence of homologous genes for all of the reactions (see Chapters 23 and 24). As described above, the assembly process is initiated by the transfer of *N*-acetylgalactosamine to the linkage tetrasaccharide (Figure 16.5). In both vertebrates and invertebrates, the polymerization step is catalyzed by one or more bifunctional enzymes (chondroitin synthases) that have both β1-3 glucuronosyltransferase and β1-4 *N*-acetylgalactosaminyltransferase activities. Vertebrates also express homologs that can transfer individual sugars to the chain, but genetic data demonstrating the functionality of these isoforms have not yet been obtained. Chondroitin polymerization also requires the action of the chondroitin polymerizing factor, a protein that lacks independent activity but collaborates with the polymerases to enhance the formation of polymers.

Sulfation of chondroitin in vertebrates is a complex process, with multiple sulfotransferases involved in 4-O sulfation and 6-O sulfation of *N*-acetylgalactosamine residues (Figure 16.6). Additional enzymes exist for epimerization of glucuronic acid to iduronic acid in DS, sulfation at the C-2 position of the uronic acids, and other patterns of sulfation found in unusual species of chondroitin (Table 16.5). The location of sulfate groups is easily assessed using bacterial chondroitinases (ABC, B, and ACII) that cleave the chains into disaccharides. Many chains are hybrid structures containing more than one type of chondroitin disaccharide unit. For example, DS contains one or more iduronic acid–containing disaccharide units (CS B) as well as glucuronic acid–containing disaccharides (CS A and C). Animal cells also degrade CS in lysosomes using a series of exolytic activities (see Chapter 41).

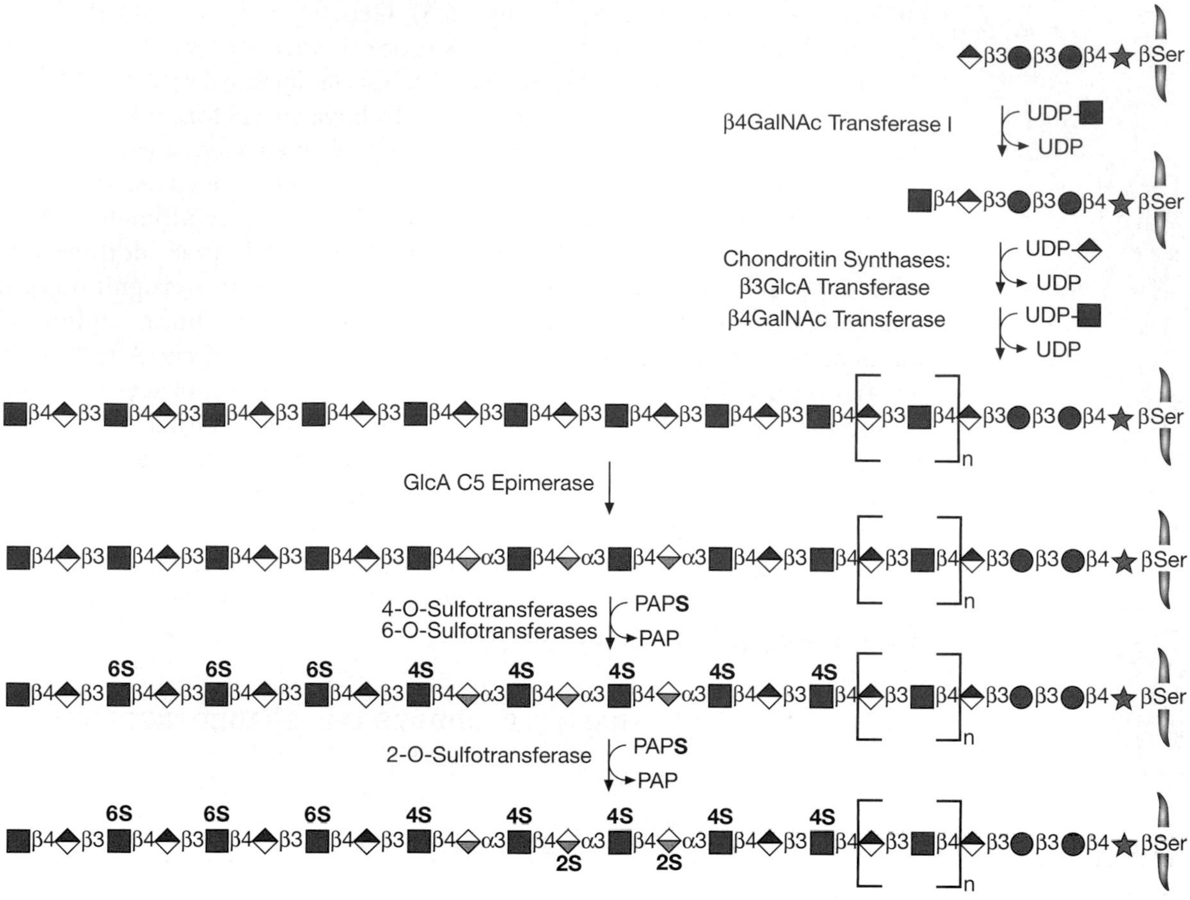

FIGURE 16.6. Biosynthesis of CS/DS involves the polymerization of *N*-acetylgalactosamine and glucuronic acid units and a series of modification reactions including sulfation and epimerization of glucuronic acid to iduronic acid. Chain polymerization and modification are thought to occur simultaneously. (PAPS) 3′-phosphoadenyl-5′-phosphosulfate, the high-energy donor of sulfate groups.

TABLE 16.5. *Types of chondroitin sulfates*

Chondroitin sulfate type	Disaccharide repeat	Source
A	GlcAβ1-3GalNAc4S	cartilage and other tissues
B	IdoAα1-3GalNAc4S	skin; tendon
C	GlcAβ1-3GalNAc6S	cartilage and other tissues
D	GlcA2Sβ1-3GalNAc6S	shark cartilage; brain
E	GlcAβ1-3GalNAc4,6diS	squid; secretory granules

This list is not meant to be exhaustive because many types of chondroitins exist with unusual modifications. For example, DS disaccharide can also contain sulfate at the C-2 position of iduronic acid and sulfate at C-6 instead of C-4, and 2-O-sulfated and 3-O-sulfated glucuronic has been described in some cartilage CS.

HEPARIN AND HEPARAN SULFATE BIOSYNTHESIS

Heparin and HS assemble as copolymers of GlcNAcα1-4GlcAβ1-4 (see Figure 16.3), which then undergoes extensive modification reactions. Heparin is produced exclusively by mast cells, whereas HS is made by virtually all types of cells. Heparin also differs from HS in the degree of modification of sugar residues, as described below. As the chains polymerize, they undergo a series of modification reactions catalyzed by at least four families of sulfotransferases and one epimerase (Figure 16.7). GlcNAc N-deacetylase/N-sulfotransferases act on a subset of *N*-acetylglucosamine residues to generate N-sulfated glucosamine units, some of which occur in clusters along the chain. Generally, the enzyme deacetylates *N*-acetylglucosamine and rapidly adds sulfate to the free amino group to form $GlcNSO_3$. A small number of glucosamine residues with free amino groups are present, which may arise from

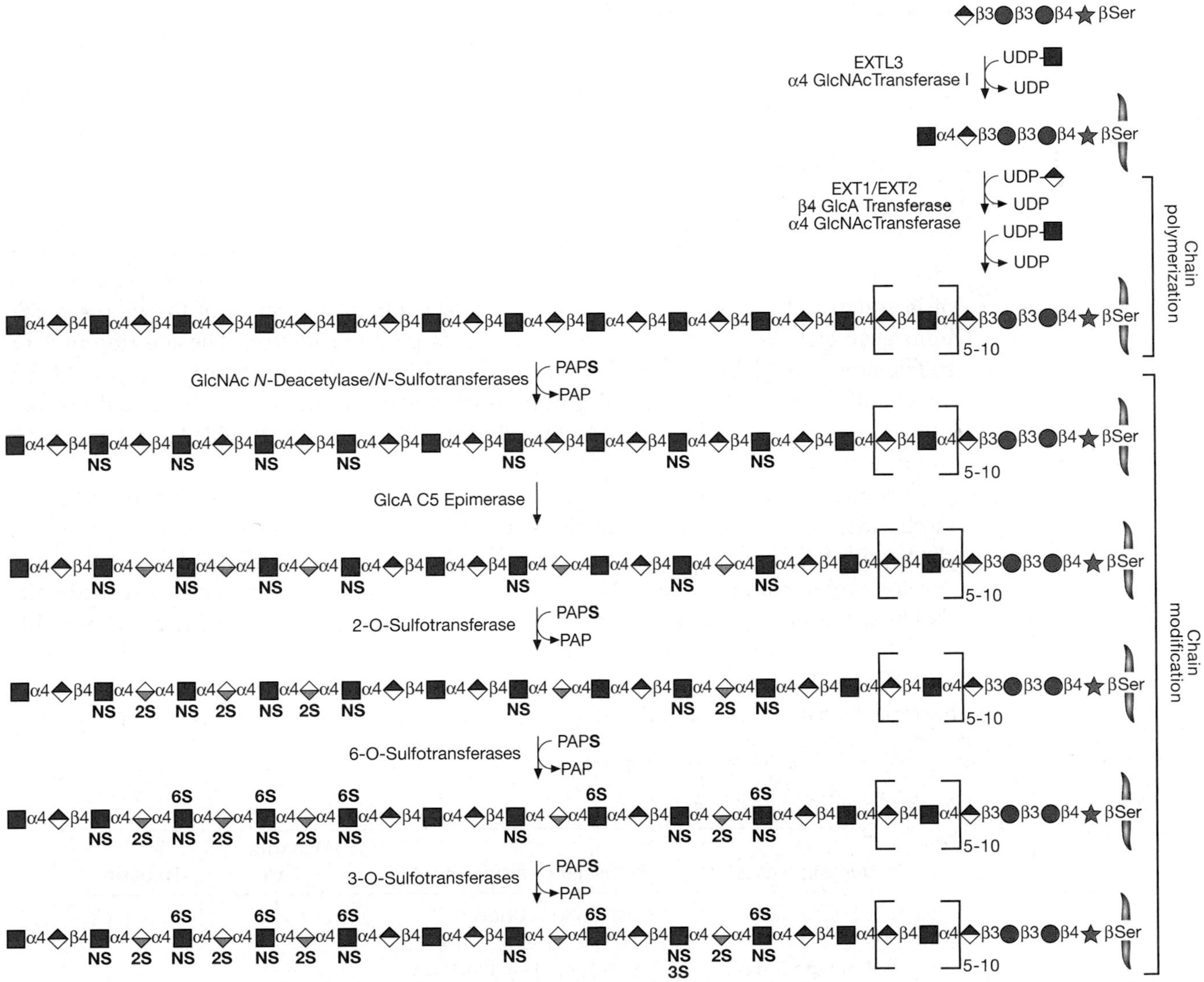

FIGURE 16.7. Heparan sulfate biosynthesis involves copolymerization of *N*-acetylglucosamine and glucuronic acid residues. A series of modification reactions including sulfation and epimerization of glucuronic acid to iduronic acid occurs; chain polymerization and modification are thought to occur simultaneously. (PAPS) 3′-phosphoadenyl-5′-phosphosulfate, the high-energy donor of sulfate groups.

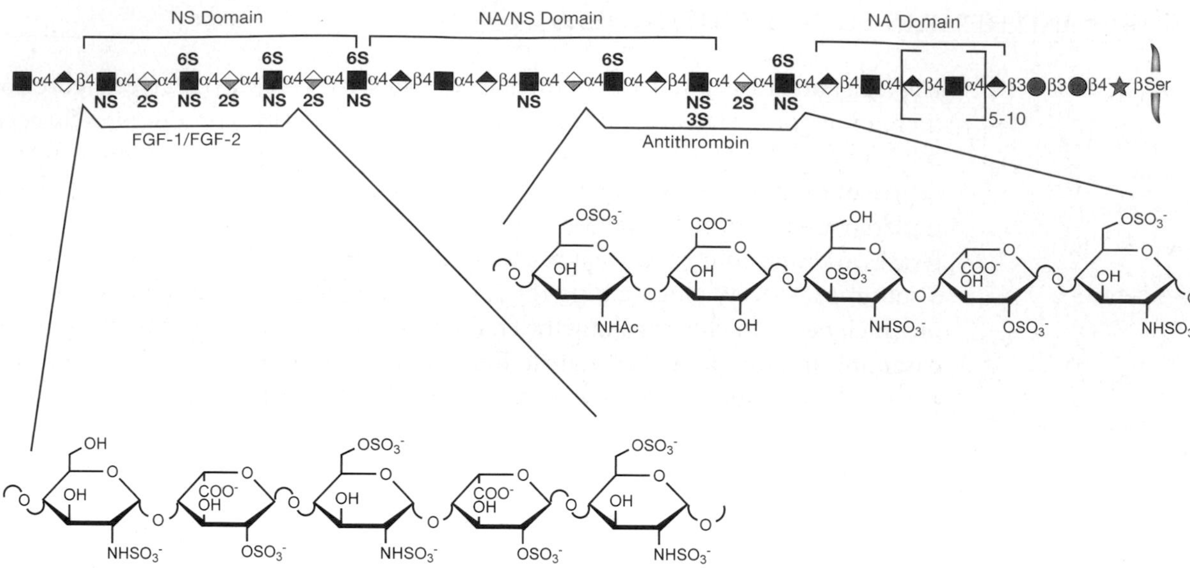

FIGURE 16.8. The heparan sulfate chain consists of different domains that vary in the extent of modification by sulfation and epimerization. (NS) N-sulfated domains; (NA/NS) domains of variable sulfation; (NA) N-acetylated domains. Modified domains make up binding sites for various ligands, for example, fibroblast growth factors (FGFs) and antithrombin.

incomplete N-sulfation. An epimerase, different from the one involved in DS synthesis, then acts on glucuronic acid residues immediately adjacent to and toward the reducing side of N-sulfated glucosamine units, followed by 2-O sulfation of some of the iduronic acid units generated. Some glucuronic units also undergo 2-O sulfation. The addition of 2-O-sulfate groups to glucuronic acid or iduronic acid blocks the epimerization reaction. Next, a 6-O-sulfotransferase adds sulfate groups to selected glucosamine residues. Finally, certain subsequences of sulfated sugar residues and uronic acid epimers provide targets for a 3-O-sulfotransferase.

In contrast to chondroitin chains, which tend to have long tracts of fully modified disaccharides, the modification reactions in HS biosynthesis occur in clusters along the chain, with regions devoid of sulfate separating the modified tracts. This arrangement gives rise to segments referred to as N-acetylated (NA), N-sulfated (NS), and mixed domains (NA/NS) (Figure 16.8). In general, the reactions proceed in the order indicated, but they often fail to go to completion, resulting in tremendous chemical heterogeneity within the modified regions. The disaccharide composition of the chains can be readily assessed using bacterial heparin lyases (Table 16.6) or chemical degradation methods, which are more

TABLE 16.6. *Heparin lyases*

Heparin lyases[a]	Preferred cleavage site	Heparan sulfate	Heparin
I (Heparinase)	GlcNS6Sα1-4IdoA2S	<5%	+
II	GlcNAc/S±6Sα1-4IdoA±2S	+	+
III (Heparitinase)	GlcNAc/Sα1-4GlcA/IdoA	+	<5%

The percentages indicate the extent of cleavage of the substrate by the enzyme. + indicates that the chain is extensively cleaved.

[a]The commercial names of these enzymes depend on their source. Some companies designate heparin lyase III as heparitinase I, and heparin lyase I (heparinase) as heparitinase III.

useful for differentiating glucuronic acid/iduronic acid, but direct sequencing of the chains has proven difficult because of their heterogeneity.

The specific arrangement of sulfated residues and uronic acid epimers in heparin and HS gives rise to binding sequences for ligands (Figure 16.8). The two examples shown in Figure 16.8 demonstrate minimal sequences that can interact with fibroblast growth factors (FGFs) or antithrombin. More modified sequences can interact as well, and the binding of most FGFs is actually more sensitive to overall sulfation than to the specific position of the sulfate groups. Binding of glycosaminoglycans to proteins is described in greater detail in Chapters 15 and 35. A major question remains regarding how the enzymes and pathway of HS/heparin biosynthesis are regulated to achieve tissue-specific expression of ligand-binding sequences. During the last decade, all of the enzymes involved in HS synthesis have been purified and molecularly cloned from mammals. Several important features have emerged from these studies, which may shed light on how different binding sequences arise.

- Several of the enzymes appear to have dual catalytic activities. Thus, a single protein bearing two catalytic domains catalyzes N-deacetylation of *N*-acetylglucosamine residues and subsequent N-sulfation (NDSTs). The same is true of the copolymerase (EXT1/EXT2), which transfers *N*-acetylglucosamine and glucuronic acid from the corresponding UDP sugars to the growing polymer. In contrast, epimerase and 2-O-, 3-O-, and 6-O-sulfotransferase activities appear to be unique properties of independent enzymes.
- In several cases, multiple isozymes exist that can catalyze either a single or a pair of reactions. Thus, four N-deacetylase/N-sulfotransferases, three 6-O-sulfotransferases, and seven 3-O-sulfotransferases have been identified. Their tissue distribution varies and differences exist in substrate preference, which may cause differences in the pattern of sulfation. However, some overlap in expression and in substrate use occurs as well.
- The polymerization and polymer modification reactions probably colocalize in the same stacks of the Golgi complex. Thus, the enzymes may form a supramolecular complex that coordinates these reactions. The composition of these complexes may play a part in regulating the fine structure of the chains.
- In general, the composition of HS on a given proteoglycan varies more between cell types than that of HS on different core proteins expressed in the same cell. This observation suggests that each cell type may express a unique array of enzymes and potential regulatory factors. However, these general compositional differences may mask underlying differences in the arrangement of the various disaccharide units, which might confer differences in ligand binding.
- Recombinant enzymes are increasingly used to define substrate specificities. Chemoenzymatic methods are being used to generate saccharide products of predetermined structure, and glycan arrays are being developed to probe ligand-binding affinities and specificities.

THE DIFFERENCES BETWEEN HEPARIN AND HEPARAN SULFATE

Considerable confusion exists regarding the definition of heparin and HS; the major differences are summarized in Table 16.7. Heparin derived from porcine and bovine entrails is prepared commercially by selective precipitation and is sold by pharmaceutical companies as an anticoagulant due to its capacity to bind to antithrombin. Low-

TABLE 16.7. *Major differences between heparin and heparan sulfate*

Characteristics	Heparan sulfate	Heparin
Soluble in 2 M potassium acetate (pH 5.7, 4°C)	yes	no
Size	10–70 kD	7–20 kD
Sulfate/hexosamine ratio	0.8–1.8	1.8–2.6
$GlcNSO_3$	40–60%	≥80%
Iduronic Acid	30–50%	≥70%
Binding to antithrombin	0–0.3%	~30%
Site of synthesis	virtually all cells	connective-tissue-type mast cells

molecular-weight heparins (LMWH) are derived from commercial unfractionated heparin (UFH) by chemical and enzymatic cleavage, depending on the brand. The active sequence is a pentasaccharide shown in Figure 16.8, which is now sold as a purely synthetic anticoagulant (Arixtra). Selectively desulfated forms of heparin are also commercially available, some of which lack anticoagulant activity, but still retain other potentially useful properties (e.g., inhibition of inflammation and cell proliferation, and antimetastatic activity).

Heparin is made solely as serglycin proteoglycan by connective-tissue-type mast cells, whereas HS is made by virtually all cells. HS can also contain anticoagulant activity, but typical preparations are much less active than heparin. During biosynthesis, heparin undergoes more extensive sulfation and uronic acid epimerization, such that more than 80% of the *N*-acetylglucosamine residues are N-deacetylated and N-sulfated and more than 70% of the uronic acid is converted to iduronic acid. Another way to distinguish heparin from HS is by susceptibility to bacterial (*Flavobacterium*) heparin lyases (Table 16.6).

PROTEOGLYCAN PROCESSING AND TURNOVER

As described above, cells secrete matrix proteoglycans directly into the extracellular environment (e.g., the basement membrane proteoglycans, SLRPs, serglycin, and members of the aggrecan family). However, others are shed from the cell surface through proteolytic cleavage of the core protein (e.g., the syndecans). Cells also internalize a large fraction of cell-surface proteoglycans by endocytosis (Figure 16.9). These internalized proteoglycans first encounter proteases that cleave the core protein and heparanase that cleaves the HS chains at a limited number of sites, depending on sequence. These smaller fragments eventually appear in the lysosome and undergo complete degradation by way of a series of exoglycosidases and sulfatases (see Chapter 41). The main purpose of intracellular heparanase may be to increase the number of target sites for exolytic degradative enzymes. CS and DS proteoglycans follow a similar endocytic route, but endoglycosidases that degrade the chains before the lysosome have not been described.

Cells secrete heparanase as well (Figure 16.10). Extracellular heparanase can cleave HS chains at restricted sites, resulting in release of growth factors or chemokines immobilized on HS proteoglycans at cell surfaces or in the ECM. In particular, invading cells secrete heparanase, thought to help degrade the ECM. Thus, heparanase may act with matrix metalloproteases to remodel the ECM.

Recently, a family of plasma membrane endosulfatases has been described that can remove sulfate groups from internal 6-O-sulfated glucosamine residues (Sulfs; see Fig.

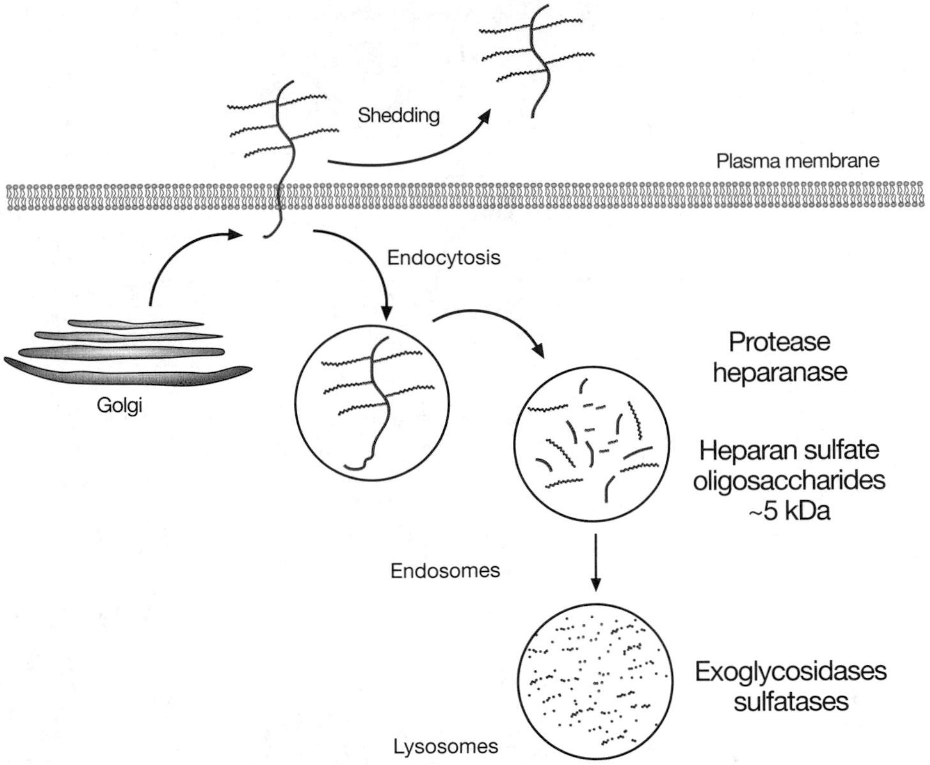

FIGURE 16.9. Heparan sulfate proteoglycans turn over by proteolytic shedding from the cell surface and endocytosis, as well as step-wise degradation inside lysosomes. (*Brown*) Core protein; (*magenta*) polysaccharide chains. (Adapted, with permission, from Yanagishita M. and Hascall V. 1992. *J. Biol. Chem.* **267:** 9451–9454, ©The American Society for Biochemistry & Molecular Biology.)

16.10). This postassembly processing of the chains at the cell surface results in altered response of cells to growth factors and morphogens. The discovery of this reaction and the action of extracellular heparanase suggests that other processing enzymes may exist that modify the structure and function of HS after secretion from the cell.

PROTEOGLYCANS HAVE DIVERSE FUNCTIONS

Several generalizations can be made regarding proteoglycan function.

- A common property of the interstitial proteoglycans containing CS and DS chains is their capacity to bind water and form hydrated matrices. Thus, these molecules fill the space between cells. In cartilage, the aggregates of proteoglycans and hyaluronan provide a stable matrix capable of absorbing high compressive loads by water desorption and resorption. Interstitial proteoglycans can interact with collagen, thus aiding in the structural organization of most tissues.
- Proteoglycans help to organize basement membranes, thus providing a scaffold for epithelial cell migration, proliferation, and differentiation. They can regulate the permeability properties of specialized basement membranes.
- Proteoglycans in secretory vesicles have a role in packaging granular contents, maintaining proteases in an active state, and regulating various biological activities after secretion, such as coagulation, host defense, and wound repair.

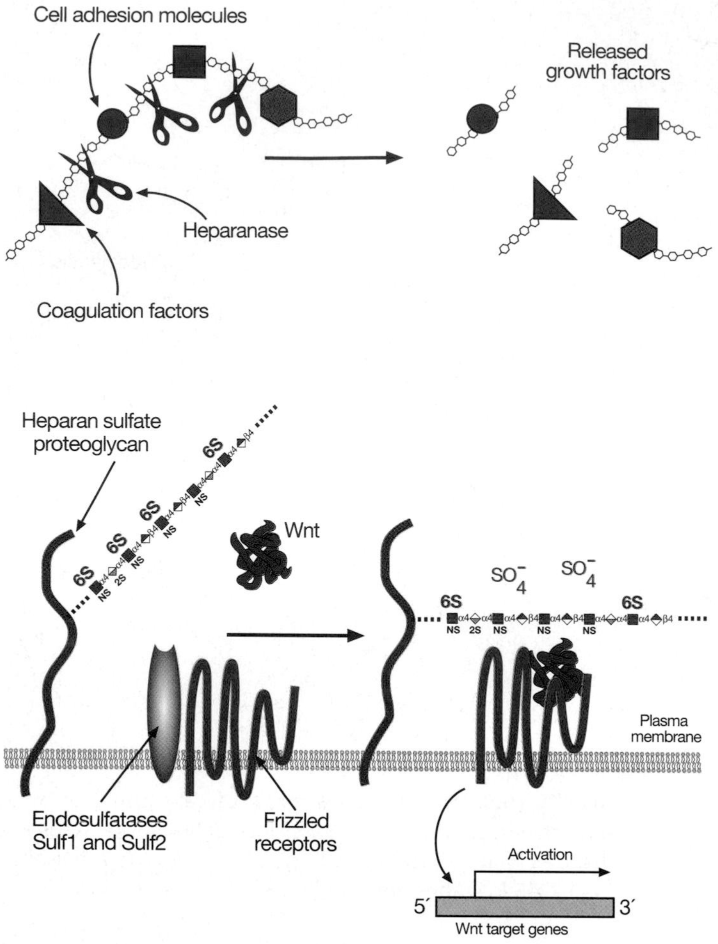

FIGURE 16.10. After secretion, cell-surface heparan sulfate proteoglycans can be processed further. Heparanase cleaves the HS chain, liberating oligosaccharides and bound ligands. The Sulfs remove sulfate groups from subsets of 6-O-sulfated N-sulfated glucosamine units and can modulate growth factor signaling.

- Proteoglycans in the ECM can bind cytokines, chemokines, growth factors, and morphogens, protecting them against proteolysis. These interactions provide a depot of regulatory factors that can be liberated by selective degradation of the matrix. They also facilitate the formation of morphogen gradients essential for cell specification during development.
- Proteoglycans can act as receptors for proteases and protease inhibitors regulating their spatial distribution and activity.
- Membrane proteoglycans can act as coreceptors for various tyrosine-kinase-type growth factor receptors, lowering the threshold or changing the duration of signaling reactions.
- Membrane proteoglycans can act as endocytic receptors for clearance of bound ligands.
- Membrane proteoglycans cooperate with integrins and other cell adhesion receptors to facilitate cell attachment, cell–cell interactions, and cell motility. Proteoglycans in the ECM can regulate cell migration as well.

To a large extent, the biological functions of proteoglycans depend on the interaction of the glycosaminoglycan chains with different protein ligands. Table 16.8 lists examples of

TABLE 16.8. *Examples of proteins that bind to sulfated glycosaminoglycans*

Cell/matrix interactions	Coagulation/ fibrinolysis	Lipolysis	Inflammation	Growth factors and morphogens
laminin, fibronectin, vitronectin, thrombospondin, tenascin, various collagens, amyloid proteins	antithrombin, heparin cofactor II, tissue factor pathway inhibitor, thrombin, protein C inhibitor, tPA and PAI-1	lipoprotein lipase, hepatic lipase, apoE, apoB apoA-V	cytokines (IL-2, IL-7 IL-8), chemokines (e.g., MIP-1β, SDF-1, etc.), TNF-α, L and P selectins, superoxide dismutase, microbial adhesins	FGFs and FGF receptors, HGF; scatter factor, VEGF, TGF-β, BMPs, Hedgehogs, Wnts

proteins known to interact with glycosaminoglycans (also see Chapter 35). Most of the proteins bind to HS or DS as opposed to CS, which may reflect the greater chemical diversity and capacity of these glycosaminoglycans to interact with proteins through the varied arrangements of sulfate groups and glucuronic acid/iduronic acid residues. Indeed, unusual CS species containing additional sulfate groups bind growth factors and matrix proteins more avidly. Proteins that bind to the sulfated GAG chains appear to have evolved by convergent evolution (i.e., they do not contain a specific fold present in all glycosaminoglycan binding proteins, in contrast to other groups of glycan binding proteins) (see Chapter 26).

These interactions have profound physiological effects. For example, injection of heparin into the bloodstream results in rapid anticoagulation because of binding and activation of antithrombin, release of lipoprotein lipase, transient blockade of P and L selectins, displacement of growth factors and chemokines, and undoubtedly a number of other activities. In some cases, the interaction depends on a very specific sequence of modified sugars in the glycosaminoglycan chain. The best-studied example is antithrombin-heparin, which depends on a specific pentasaccharide sequence (Figure 16.8). However, in other cases, the glycosaminoglycan sequences that interact with proteins, such as growth factors and their receptors, show preference for certain modifications or spatial arrangements of sulfated sugars rather than a specific sequence. As more refined methods become available for isolation of bioactive oligosaccharides and their sequence analysis, unusual arrangements of sulfated sequences may be discovered to confer selective ligand-binding and biological activity. This is an active area of research.

FURTHER READING

Bernfield M., Gotte M., Park P.W., Reizes O., Fitzgerald M.L., Lincecum J., and Zako M. 1999. Functions of cell surface heparan sulfate proteoglycans. *Annu. Rev. Biochem.* **68:** 729–777.

Lindahl U. 1999. What else can "heparin" do? *Haemostasis* **29:** 38–47.

Yamaguchi Y. 2000. Lecticans: Organizers of the brain extracellular matrix. *Cell Mol. Life Sci.* **57:** 276–289.

Gallagher J.T. 2001. Heparan sulfate: Growth control with a restricted sequence menu. *J. Clin. Invest.* **108:** 357–361.

Esko J.D. and Selleck S.B. 2002. Order out of chaos: Assembly of ligand binding sites in heparan sulfate. *Annu. Rev. Biochem.* **71:** 435–471.

Funderburgh J.L. 2002. Keratan sulfate biosynthesis. *IUBMB Life* **54:** 187–194.

Song H.H. and Filmus J. 2002. The role of glypicans in mammalian development. *Biochim. Biophys. Acta* **1573:** 241–246.

Trowbridge J.M. and Gallo R.L. 2002. Dermatan sulfate: New functions from an old glycosaminoglycan. *Glycobiology* **12:** 117R–125R.

Vlodavsky I., Goldshmidt O., Zcharia E., Atzmon R., Rangini-Guatta Z., Elkin M., Peretz T., and Friedmann Y. 2002. Mammalian heparanase: Involvement in cancer metastasis, angiogenesis and normal development. *Semin. Cancer Biol.* **12:** 121–129.

Habuchi H., Habuchi O., and Kimata K. 2004. Sulfation pattern in glycosaminoglycan: Does it have a code? *Glycoconj. J.* **21:** 47–52.

Kolset S.O., Prydz K., and Pejler G. 2004. Intracellular proteoglycans. *Biochem. J.* **379:** 217–227.

Iozzo R.V. 2005. Basement membrane proteoglycans: From cellar to ceiling. *Nat. Rev. Mol. Cell Biol.* **6:** 646–656.

Volpi N., ed. 2006. Chondroitin sulfate: Structure, role and pharmacological activity. *Adv. Pharmacol.* **53:** 1–568.

Bishop J.R., Schuksz M., and Esko J.D. 2007. Heparan sulphate proteoglycans fine-tune mammalian physiology. *Nature* **446:** 1030–1037.

Gorsi B. and Stringer S.E. 2007. Tinkering with heparan sulfate sulfation to steer development. *Trends Cell Biol.* **17:** 173–177.

Lamanna W.C., Kalus I., Padva M., Baldwin R.J., Merry C.L., and Dierks T. 2007. The heparanome—The enigma of encoding and decoding heparan sulfate sulfation. *J. Biotechnol.* **129:** 290–307.

Vreys V. and David G. 2007. Mammalian heparanase: What is the message? *J. Cell. Mol.* **11:** 427–452.

CHAPTER 17

Nucleocytoplasmic Glycosylation

Gerald W. Hart and Christopher M. West

THIS CHAPTER REVIEWS THE EVIDENCE FOR COMPLEX glycoconjugates in the nucleoplasmic and cytoplasmic compartments of the cell. The modification of nuclear and cytoplasmic proteins by a single β-linked *N*-acetylglucosamine (O-β-GlcNAc) is described in Chapter 18. Here the focus is on resident glycoproteins of the cytoplasm/nucleus with glycans of length two units or more. Current views regarding the putative functions of nuclear lectins and the possible roles of glycans as nuclear localization signals are also presented.

HISTORICAL BACKGROUND

Most general biochemistry and cell biology texts still persist in the view that glycosylation is largely restricted to the addition of complex glycans to proteins or lipids localized on the cell surface or within lumenal compartments of subcellular organelles, such as the endoplasmic reticulum (ER), Golgi, endosomes, or lysosomes. In fact, a large body of data documents the presence of both simple and complex glycoconjugates within the nucleus and cytoplasm. Many of these studies have used plant lectins, anticarbohydrate antibodies, or natural polysaccharide (e.g., hyaluronan)-binding proteins as tools to detect glycosylated molecules within the nucleoplasmic or cytoplasmic compartments. Even though there have been more than 50 papers that have used the binding of plant lectins to detect complex glycans within the nucleus of cytoplasm, these binding studies have largely been ignored for several reasons: (1) Most glycosyltransferases involved in complex glycan biosynthesis are type II membrane proteins with their active sites within the lumen of the ER or Golgi. Thus, the "enzymatic machinery" is on the "wrong" side of the membrane for the biosynthesis of complex nuclear or cytoplasmic glycoconjugates. (2) Most of the lectin-binding studies that make claims of cytoplasmic or nuclear glycosylation present no supporting structural data to establish firmly the identity of these glycans. (3) Under some circumstances, plant lectins, such as concanavalin A, can bind to molecules by nonspecific hydrophobic interactions. Also, it is possible that the competing saccharide ligand can alter the conformation of such lectins to induce a loss of this hydrophobic binding. Therefore, although lectins are important tools in glycobiology, conclusions based exclusively on their use, particularly those not compatible with known pathways and established concepts of biochemistry and cell biology, must be followed up by rigorous structural analyses. Nevertheless, these many lectin-binding studies indicate that *N*-acetylglucosamine-, mannose-, and fucose-containing glycans are abundant within the nucleus and cytoplasm, suggesting that further, more rigorous analyses are warranted.

ARE THERE GOLGI-TYPE GLYCANS IN THE NUCLEOCYTOPLASMIC COMPARTMENT?

There are several possible explanations for the potential origins of complex nucleocytoplasmic glycans suggested by cytological studies, as summarized in Figure 17.1. Most simply, nonconventional soluble glycosyltransferases may reside in the cytoplasm or nucleus and directly modify proteins there (Figure 17.1a). Nucleotide sugar precursors are available, because the enzymes that synthesize them reside within these compartments. As dis-

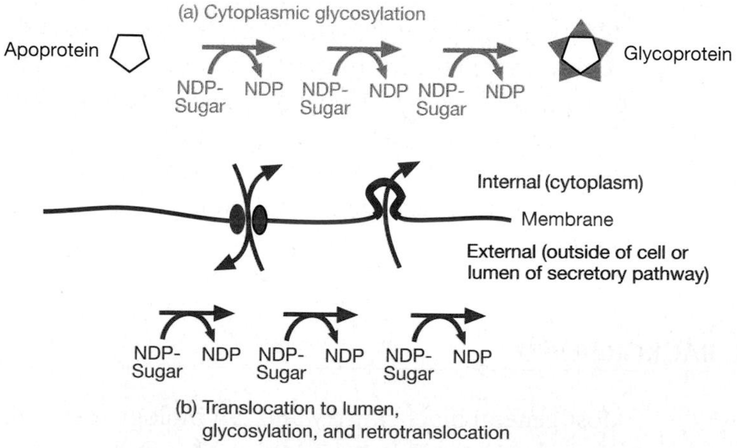

FIGURE 17.1. Potential mechanisms for the glycosylation of cytoplasmic proteins. (*a*) In the simplest model, the glycosyltransferases are present in the cytoplasm and transfer sugars from nucleotide sugar donors to the acceptor protein in the cytoplasm. This is the mechanism of glycosylation of Skp1 in *Dictyostelium* (see Figure 17.4 below). (*b*) A more complex scenario would involve the translocation of the protein into, for example, the secretory pathway where glycosyltransferases are normally present, modification by sugars transferred from nucleotide sugars, and subsequent return to the cytoplasm. Retrotranslocation from the rough endoplasmic reticulum to the cytoplasm might be mediated by the Sec61/63 complex. Alternatively, secreted glycoproteins might traverse the plasma membrane via a poorly characterized pathway invoked to explain movement of growth factors and/or their receptors to the nucleus.

cussed below, this is the mechanism by which glycogen synthesis is primed and by which Skp1 is modified by multiple sugar types in the cytoplasm of *Dictyostelium*. Glycosylated proteins might then be transferred to the nucleus through nuclear pores via conventional nuclear transport pathways.

A second potential source of cytoplasmic complex glycans is the secretory pathway itself. An emerging area of cell biology has identified numerous proteins that return to the cytoplasm after entry into the ER or after secretion (Figure 17.1b). The Sec61/63 translocator that transports nascent proteins into the ER is also thought to support the transfer of misfolded nascent proteins back into the cytoplasm where they are subsequently degraded by the 26S proteasome. In addition, there is evidence that endocytosed proteins, such as cholera toxin, can gain access to the cytoplasm via this pathway. Therefore, retrotranslocation is also a potential source of cytoplasmic and ultimately nuclear glycoproteins. Furthermore, there is a growing body of literature that supports the concept that cytokines, growth factors, and even their transmembrane receptors, many of which are glycosylated, can translocate to the nucleus to exert direct effects on transcription. These novel translocation pathway(s) are obscure but may involve lipid rafts, endocytosis, and conventional nuclear localization signals (Figure 17.1b). Although the existence of these transport pathways is controversial, they might provide a source of cytoplasmic and nuclear glycoproteins with conventional glycan modifications. Finally, occasional reports of conventional O-glycosylation of cytoplasmic proteins in pathological tissues raises the possibility that compartmental breakdowns result in the exposure of latent cytoplasmic proteins to Golgi glycosyltransferases.

Finally, the biosynthesis of several classes of glycans is initiated by membrane-associated glycosyltransferases whose catalytic domains are situated at the cytoplasm-facing surface. This includes early steps in the synthesis of glucosylceramides and GPI anchors, as well as early assembly of dolichyl-linked N-glycosylation precursors that ultimately "flip" to the other side of the membrane and are transferred to proteins within the lumen of the ER in the secretory pathway (see Chapter 8). Other membrane-associated glycosyltransferases whose catalytic domains face the cytoplasm include those that polymerize hyaluronan (see Chapter 15), cellulose, chitin, and lipopolysaccharides, and these products are subsequently translocated across the plasma membrane. The glycan products of these membrane-associated glycosyltransferases normally exit the cytoplasmic space, but if this did not occur, they could serve as a source of novel, cytoplasmic, nonprotein-associated glycans. Alternatively, homologs of these enzymes might be capable of modifying protein acceptors.

Cytoplasmic N-Linked Glycans?

The secretory pathway is home to the biosynthesis of several classes of glycans, including the N- and O-linked types and the more rare glycans linked via C-mannosyl tryptophan. Numerous reports offer partial evidence to support the existence of N-linked glycoproteins within the cytoplasm. The α subunit of the sodium pump (Na^+, K^+-ATPase) from dog kidney, which is a transmembrane protein, was reported to contain N-linked glycans with terminal *N*-acetylglucosamine residues in its cytoplasmic domain. This conclusion was based on the enzymatic attachment of radioactive galactose to *N*-acetylglucosamine residues by galactosyltransferase labeling of permeabilized right-side-out membrane vesicles. N-Glycanase sensitivity of the radiolabeled products suggested that the acceptors are N-linked glycans (see Chapter 8). In such vesicles, extracellular domains would be protected from enzymatic probes by being inside sealed membranes. This provocative claim has been largely ignored because of the lack of any site-mapping and structural analysis of the putative cytoplasmic glycans, even though the primary sequence of the protein has been known for years.

Several studies have also strongly indicated that the so-called high-mobility-group proteins (HMGs), which are important structural components of chromatin, are bona fide "classical type" glycoproteins. Highly purified preparations of HMGs 14 and 17 were found to contain *N*-acetylglucosamine, mannose, galactose, glucose, fucose, and possibly xylose. A fucose-specific lectin, *Ulex europeus* agglutinin I, bound to both HMGs and isolated nucleosomes. Alkali resistance of the saccharides on the HMGs is consistent with attachment of the glycan via an N-linkage. In addition, these HMGs could be metabolically radiolabeled with tritiated fucose, galactose, mannose, or N-unsubstituted glucosamine. Later studies suggested that the glycans on HMGs 14 and 17 were required for their binding to the nuclear matrix, where the HMGs have a role in modulating chromatin structure at active sites of gene transcription. Even though these HMG glycosylation studies are widely cited as evidence for nuclear N-glycans, they also suffer from a lack of definitive structural data, such as physical proof of the linkage to protein and physical definition of the structure of the putative N-glycan(s). In fact, more recent studies have not been able to verify the presence of complex glycans of any type on the HMGs, but did find that the HMGs are modified by O-β-GlcNAc (see Chapter 18).

Sialic acid–containing glycoproteins have been reported on the cytoplasmic face of the nuclear envelope. The sialic acid–specific lectin *Sambucus nigra* agglutinin (SNA) was shown to bind to several proteins, previously shown to also contain O-GlcNAc. Two of these proteins were identified as major nucleoporins: p62 and p180. Prior sialidase treatment blocked binding of SNA to these nuclear pore proteins. On the basis of N-glycanase sensitivity, the sialic acids on p180 appeared to be on N-linked glycans, and those on p62 appeared to be on O-linked glycans. The authors also demonstrated that SNA blocked nuclear protein import in neuroblastoma cells, suggesting that the sialic acids might have functional importance on the nuclear pore proteins. Again, the significance of these provocative findings must await systematic structural proof of the nature of the putative sialylated glycans on these nucleoporins.

Weighing in against the possibility of resident N-glycoproteins in the cytoplasm is the existence of catabolic pathways that rapidly degrade them. Unfolded proteins that are translocated from the ER are deglycosylated by a cytoplasmic N-glycanase, polyubiquitinated, and degraded by the 26S proteasome. Alternatively, these proteins can be recognized directly by a ubiquitin ligase that is selective for high mannose N-glycans, also resulting in polyubiquitination and degradation. These turnover pathways suggest that cytoplasmic N-linked glycoproteins are short-lived and therefore unlikely to function in the cytoplasm.

A heat-shock-like nuclear chaperone protein with *N*-acetylglucosamine-binding activity (CBP70) has also been suggested to be N-glycosylated. A subpopulation of prion protein may be N-glycosylated and found in the nucleus, possibly in association with the *N*-acetylglucosamine-binding lectin. It is possible that the compartmentalization of N-glycoproteins within the nucleus or mitochondrion protects them from degradation by the above-mentioned cytoplasmic pathways.

Mitochondrial Glycosylation?

On the basis of lectin-binding studies, mitochondria have been suggested to contain complex glycoconjugates. Most mitochondrial proteins are synthesized within the cytoplasm under the direction of nuclear genes, suggesting that these glycoconjugates might originate in the cytoplasm. Recently, two glycoproteins located on the inner mitochondrial membrane have been identified, and they appear to be conventionally N-glycosylated in the rER based on pulse-chase labeling studies and susceptibility to N-glycanase. These glycoproteins might transit to mitochondria via a novel vesicle transport pathway or by retro-translocation to the cytoplasm followed by conventional uptake into mitochondria. High-resolution electron microscopy shows that some mitochondrial membranes are in contact with portions of the ER, offering another option for direct membrane transfer into the mitochondria. In plants, a galactolipid is synthesized in mitochondrial membranes, which would provide an alternative explanation for mitochondrial lectin labeling.

Nucleocytoplasmic N-Glycan Glycosyltransferases?

In support of the existence of novel nucleocytoplasmic glycosyltransferases, there are several reports in the literature of glycosyltransferase activities in highly purified preparations of rat liver nuclei judged to be >99% pure by marker enzyme analysis. These studies document the transfer of *N*-acetylglucosamine from UDP-GlcNAc to endogenous acceptors and show that at least 80% of the activity is blocked by low concentrations of the antibiotic tunicamycin, suggesting the involvement of N-linked biosynthetic intermediates, such as *N*-acetylglucosaminyl-pyrophosphoryldolichol (GlcNAc-PP-dolichol). Later studies demonstrated the direct transfer of chitobiose (GlcNAcβ1-4GlcNAc) from chitobiosyl-dolichol to endogenous nuclear acceptors by these nuclear preparations, suggesting a novel pathway of N-glycosylation. The products of these in vitro reactions were found to be N-linked chitobiosyl moieties, based on their sensitivity to peptide N-glycosidase F (N-glycanase) and hydrazinolysis, but also on their insensitivity to alkali-induced β-elimination (see Chapter 47). Similar studies have documented the presence of nuclear mannosyltransferases. These studies are provocative, but they must also be interpreted with caution. The ER, which is the widely accepted site of N-glycosylation (see Chapter 8), is functionally contiguous with the outer nuclear envelope, and cell biologists have detected infoldings of this membrane into the nuclear interior in some cells. Even a minor contamination of nuclear envelope with ER could lead to misinterpretation of these findings. In addition, it is very difficult to purify nuclei such that other cellular components do not adhere nonspecifically to the otherwise "pure" nuclei during their preparation. Given these potential problems, widespread acceptance of the existence of these nuclear glycosyltransferases must await independent confirmation by more direct criteria.

Recently, an unusual N-linked glycan has been detected crystallographically on a capsid protein of a virus, PBCV-1. This capsid protein (VP54) is assembled, posttranslationally modified, and incorporated into viral structures within the cytoplasm of *Chlorella* algae. The PBCV-1 genome encodes multiple sequences predicted to encode cytoplasmically localized glycosyltransferases that lack targeting sequences for the secretory pathway. N-glycosylation of VP54 is unusual in that the asparagine attachment site is not associated with the canonical N-glycosylation sequon used for N-glycosylation in the secretory pathway. This example raises the possibility that a variant form of N-glycosylation has evolved within the cytoplasm, but, because PBCV-1 eventually exits the host after cell lysis, the product glycoprotein appears to ultimately function outside of the cell.

CYTOPLASMIC O-GLYCOSYLATION

The hydroxyl side chains of threonine, serine, tyrosine, hydroxyproline, and hydroxylysine constitute potential sites of attachment of so-called O-linked glycans. Threonine and serine are the predominant linkage residues in the secretory pathway, but interestingly, the best-characterized examples of cytoplasmic O-glycosylation (other than O-GlcNAc; see Chapter 18) occur on tyrosine and hydroxyproline. Unlike secretory pathway glycosylation, cytoplasmic glycosylation pathways appear to target only single proteins. The following sections review evidence for these and other kinds of cytoplasmic O-glycosylation.

Glycogenin

Glycogen is a large homopolysaccharide of glucose. It serves as a short-term storage form of glucose and can be broken down to provide glucose to liver, muscle, and blood during stress. Several reports have unequivocally established that glycogen is covalently attached at its reducing terminus to a polypeptide and thus is a glycoprotein. In 1975, liver extracts were shown to synthesize from UDP-glucose a glycogen-like product that was precipitated by trichloroacetic acid, indicating that it had a protein component. In vitro, maltotetraose [(Glcα1-4)$_4$] is the smallest saccharide that primes glycogen elongation. Therefore, it was proposed that protein, not carbohydrate, was the original primer for glycogen synthesis. In 1985, the glycogen protein primer was named glycogenin. Glycogenin was then purified in a complex with glycogen synthase. Glycogenin has 322 amino acids and the first glucosyl residue is attached through a glycosidic linkage to the hydroxyl group of tyrosine-194 of the protein. In muscle, each molecule of glycogenin is covalently attached to one molecule of glycogen (i.e., a ratio of 1, where the molecular weight of glycogen is 10^7, approximately the maximum mass of a glycogen molecule in muscle). In liver, the mass ratio of glycogenin to glycogen is about 0.0025. This small ratio reflects both the enormous size of liver glycogen (the molecular weight can be as much as 5×10^9) and the existence of large numbers of free glycogen chains in liver glycogen granules.

Glycogenin is autocatalytic, attaching glucose to itself from the donor (UDP-Glc) and is classified as a retaining hexosyltransferase (see Chapter 5). Glycogenin catalyzes two chemically distinct reactions: formation of the Glcα1-Tyr linkage and the subsequent Glcα1-4Glc linkages. The first reaction was established using recombinant glycogenin from an *Escherichia coli* strain unable to synthesize UDP-Glc. A mutant glycogenin lacking Tyr-194 can glucosylate a catalytically inactive glycogenin, showing interpolypeptide transfer. Glycogenin occurs as a dimer in solution and in crystals, and elongation of the glucan chain is essentially unaffected by dilution. Therefore, glycogenin appears to extend the glucan chain on the other subunit of a functional dimer, as depicted in Figure 17.2. This self-glucosylation yields a malto-octaose (eight glucose units), and both *p*-nitrophenyl α-glucoside and α-malto-oligosaccharides are competitive substrates. However, the distances between UDP-Glc and Tyr-194 (of either subunit) are too great in the known crystal structures of glycogenin for this mechanism to explain initial priming, which war-

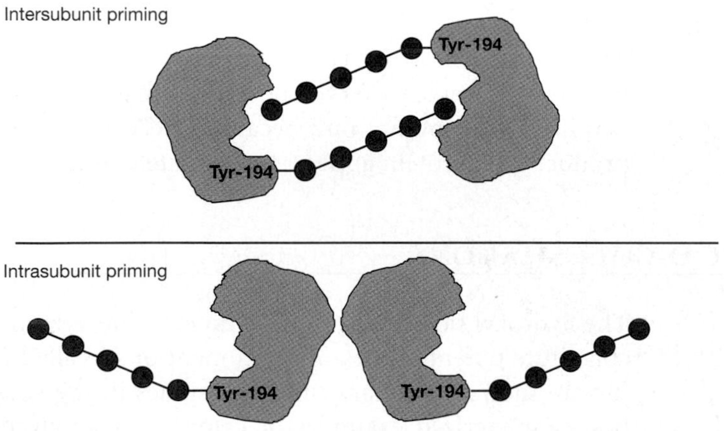

FIGURE 17.2. Opposing models for the self-glucosylation of glycogenin. Structural data support the hypothesis of intersubunit elongation rather than intrasubunit elongation. (*Top*) Active site of one subunit of the glycogenin dimer adding α1-4Glc residues to the adjacent subunit. (*Bottom*) Model in which each subunit adds glucose to itself.

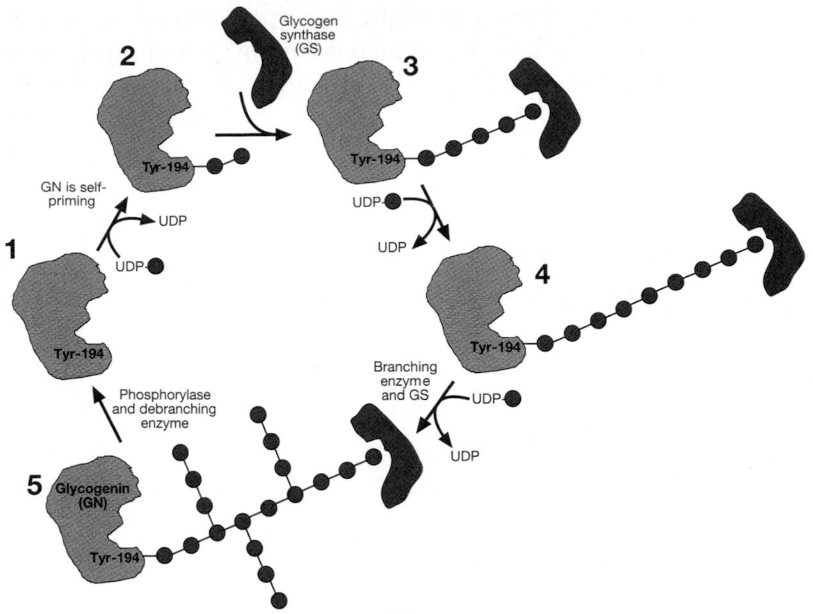

FIGURE 17.3. Model for glycogen biosynthesis via glycogenin (GN), glycogen synthase (GS), and branching enzyme, as proposed by Carl Smythe and Philip Cohen in the late 1980s. (*Step 1*) GN is self-priming, attaching glucose to Tyr-194. (*Step 2*) GN elongates the glucose polymer to contain up to eight monosaccharide units. (*Step 3*) GS binds. (*Step 4*) GS elongates the polysaccharide (glycogen) further using the primer produced by GN. (*Step 5*) Branching enzyme attaches α6 branches to the α4 cores to produce highly branched glycogen polymers.

rants further analysis. To date, glycogenin is the only known example of the glycosylation of tyrosine residues in animals.

Glycogenin-like proteins are found in a wide range of plants, animals, and free-living lower eukaryotes. *E. coli* also appears to have a glycogenin-like molecule, but it is incapable of glucosylating recombinant mammalian glycogenin. Based on its primary sequence, glycogenin has been classified in the CAZy database (see Chapter 7) as a member of the family GT-8 glycosyltransfereases, which includes bacterial lipopolysaccharide glucosyl, galactosyl transferases, and galactinol synthases, all of which are apparently cytoplasmic like glycogenin. Remarkably, a novel Glcβ1-Arg linkage has been described in a plant protein associated with starch synthesis, but this protein has no apparent sequence similarity to glycogenin.

Biochemically, glycogenin may have a key role in the regulation of glucose homeostasis. Glycogenin has a K_m for UDP-Glc that is about 1000 times lower than the K_m of glycogen synthase for the sugar nucleotide. Production of the glycogenin primer, especially in muscle cells, can be the rate-limiting step in glycogen formation. Thus, levels of glycogenin could override the better understood, hormonally controlled mechanisms that involve protein phosphorylation/dephosphorylation and regulate glycogen elongation. Figure 17.3 illustrates a model of glycogen synthesis by glycogenin, glycogen synthase, and branching enzyme.

Hydroxyproline-linked Glycans

A small fucosylated protein, Skp1, is found in the cytoplasm of a free-living soil amoeba, the cellular slime mold *Dictyostelium*. Skp1 has an established role in yeast, plants, and animals as an adaptor in certain E3-ubiquitin ligase complexes that are active in the cytoplasm and nucleus. $E3^{SCF}$-ubiquitin ligases promote the polyubiquitination and eventual proteasomal degradation of cell-cycle regulatory proteins, transcriptional factors, and other signal-transduction proteins. Mass spectrometric structural analyses of Skp1 peptides established that it is glycosylated at Pro-143 after prior hydroxylation. The glycan is a pentasaccharide consisting of a core trisaccharide equivalent to the type 1 blood group H structure: Fucα1-2Galβ1-3GlcNAc- (Figure 17.4, top). This glycan is further modified by an α1-3-linked galactose residue on the fucose and another α-linked galactose on the α-Gal or fucose residue. In these studies, the great majority of Pro-143 is hydroxylated, and nearly all of the hydroxylated form is glycosylated. The linkage to protein, localization of this type of structure in the cytoplasm, and the structure itself are all novel, which suggests a very atypical glycosylation pathway in this organism and perhaps in other eukaryotes. Because

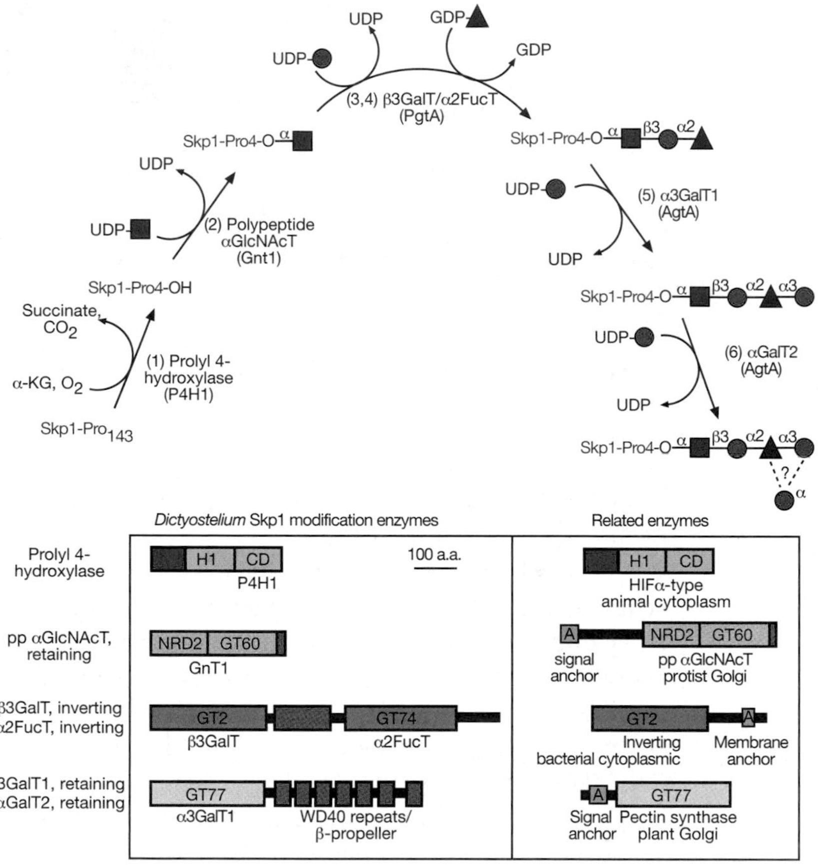

FIGURE 17.4. Mechanism of glycosylation of Skp1 in the cytoplasm of *Dictyostelium*. (*Top*) Skp1 Pro-143 is sequentially modified by a soluble prolyl 4-hydroxylase and five soluble, sugar nucleotide–dependent glycosyltransferase activities. (αGlcNAcT) αN-acetylglucosamine transferase; (β3GalT) β1-3galactosyltransferase; (α2FucT) α1-2fucosyltransferase; (α3GalT1) α1-3galactosyltransferase-1; (αGalT2) αgalactosyltransferase-2; (α-KG) α-ketoglutarate. (*Bottom*) Skp1 is hydroxylated by an apparent ortholog of the cytoplasmic animal HIFα-type prolyl 4-hydroxylase. Addition of the first sugar is catalyzed by a hydroxyPro polypeptide αGlcNAcT, whose catalytic domain is distantly related to the corresponding region of microbial Golgi Thr/Ser polypeptide αGlcNAcT and animal Golgi Thr/Ser polypeptide αGalNAcT. However, the Skp1 αGlcNAcT lacks the amino-terminal signal anchor sequence characteristic of Golgi type 2 membrane proteins. Addition of the subsequent two sugars is catalyzed by an unusual processive diglycosyltransferase with separate domains mediating each of the transfer reactions. The first glycosyltransferase domain is homologous to a large family of glycolipid glycosyltransferases (GT2) that are usually anchored to the cytoplasmic surfaces of membranes from both prokaryotes and eukaryotes, often via a carboxy-terminal region. The first αGal and probably also the second αGal are attached by another two-domain glycosyltransferase consisting of a catalytic domain and a β-propeller-like protein–protein interaction domain. Its catalytic domain has sequence homologs in a large number of putative plant pectin synthases.

Pro-143 is conserved in the Skp1 of plants, invertebrates, and lower eukaryotes, and sequences related to those of known Skp1 modification enzymes are present in the genomes of other protists, similar glycosylation may occur in other organisms.

Extracts of *Dictyostelium* can hydroxylate and glycosylate recombinant Skp1 in vitro. The Skp1 prolyl 4-hydroxylase is the *Dictyostelium* ortholog of animal cytoplasmic prolyl 4-hydroxylases that regulate the O_2 and $E3^{VHL}$ ubiquitin-ligase-dependent degradation of the hypoxia-inducible factor-1α transcriptional factor subunit (HIFα) (Figure 17.4, bottom). The biochemical assays suggest that the sugars are added sequentially by soluble UDP-sugar-dependent enzymes that reside within the cytoplasm. The proteins that catalyze the first four glycosyltransferase reactions have been purified to homogeneity. Cloning of the gene for the *N*-acetylglucosaminyltransferase, which adds the first sugar, showed it to be related in sequence to the Thr/Ser polypeptide α-*N*-acetylgalactosaminyl- and α-*N*-acetylglucosaminyltransferases that initiate mucin-type O-glycosylation in the Golgi of animals and protists, and, therefore, it is expected to form an anomeric linkage opposite to that of O-β-*N*-acetylglucosaminyltransferase (see Chapter 18). However, unlike all other known polypeptide α-*N*-acetylgalactosaminyl- and α-*N*-acetylglucosaminyltransferases, the Skp1 *N*-acetylglucosaminyltransferase does not possess an amino-terminal signal anchor, consistent with its biochemical fractionation as a soluble cytoplasmic protein (Figure 17.4, bottom). These findings suggest that the mechanism of initiation of Skp1 glycosylation is similar to that of mucin-type domains of animal secretory proteins, except that a different *N*-acetylhexosamine (*N*-acetylglucosamine vs. *N*-acetylgalactosamine) is attached to a distinct hydroxyamino acid (hydroxyproline, not threonine or serine) in the cytoplasm instead of the Golgi lumen. Cloning of the gene for the fucosyltransferase, which adds the third sugar, led to the discovery that both the second (β-Gal) and third (α-L-Fuc) sugars are added by separate domains of the same protein (Figure 17.4, bottom). This protein is also predicted to be soluble. The β1-3-galactosyltransferase catalytic domain is most similar to bacterial lipopolysaccharide and capsular glycosyltransferases, which are also cytoplasmic, suggesting that cytoplasmic glycosylation in *Dictyostelium* has its evolutionary origins in bacterial glycolipid synthesis. The β-galactosyltransferase catalytic domain belongs to the inverting GT family 2 (CAZy database), which is diagnostic for glycosyltransferases with their catalytic domains exposed to the cytoplasm. However, in both bacteria and eukaryotes, these enzyme family members are usually membrane-associated and their products, including lipopolysaccharides, dolichol phosphate sugars, and polysaccharides such as hyaluronan, cellulose, and chitin, are translocated to the lumen or cell exterior. Bifunctional diglycosyltransferases with a similar architecture are used for cell-surface glycosaminoglycan and polysaccharide synthesis in certain bacteria and in the eukaryotic Golgi. The fourth sugar is added by another two-domain glycosyltransferase in which the α1-3-galactosyltransferase domain is fused to a β-propeller-like domain of unknown function. The catalytic domain is related to a large number of plant Golgi proteins implicated in pectin biosynthesis.

In contrast to glycogenin, *Dictyostelium* Skp1 possesses a hetero-oligomeric glycan that is the product of multiple enzymes that act sequentially on the acceptor protein substrate by a mechanism most similar to that of O-linked glycosylation in the eukaryotic Golgi. Like glycogenin, Skp1 is the only known protein substrate for its glycan modification. Both glycogenin and Skp1 have novel sugar–amino acid linkages. Mutagenesis of Pro-143 in *Dictyostelium* Skp1 or inhibition of Pro-143 hydroxylation reduces the steady-state level of Skp1 in the nucleus, suggesting an involvement of glycosylation in the subcellular compartmentalization of this protein. Disruption of the Skp1 prolyl 4-hydroxylase gene interferes with an O_2-dependent step in *Dictyostelium* development. This reflects a function for the early part of the modification pathway, possibly involving addition of α-GlcNAc, because disruption of the subsequently acting diglycosyltransferase gene has only a minimal effect on development.

Phosphoglucosylation and Cytoplasmic O-Linked Mannose

α-Glc-1-P can be transferred from UDP-Glc to O-linked mannosyl residues situated predominantly on a single 62-kD protein called parafusin. Note that this form of cytoplasmic O-mannose appears to be distinct from the O-mannosylation that is common in the lumen of the ER, Golgi, and cell surface of many eukaryotes, particularly fungi (see Chapter 12). Parafusin responds to a secretagogue by reversibly dissociating from cellular membranes at the same time that Glc-P is released. Considerable biochemical evidence for the occurrence of Glcα1-P and Man-P linkages, which are likely to be conjoined in a single phosphodiester moiety, has been obtained from in vitro studies of the modification of parafusin in extracts from organisms as diverse as *Paramecium*, birds, and mammals. Parafusin is a member of the phosphoglucomutase (PGM) protein family, named for the enzyme that catalyzes the interconversion of Glc-1-P and Glc-6-P. Although parafusin itself appears not to exhibit PGM activity, there is evidence that the active PGM protein from yeast and rat liver is similarly modified in extracts. Latency studies established that both the novel glucose phosphotransferase that acts on parafusin and a Glc-1-phosphate phosphodiesterase that cleaves Glc-P from parafusin have their active sites within the cytoplasm. Cochromatography studies of hydrolyzed parafusin from cells metabolically labeled with radioactive sugar precursors suggest that PGM-like proteins are similarly modified in vivo. The modification of PGM-like proteins is highly responsive to calcium levels and appears to regulate the association of parafusin with the membrane. Unfortunately, nothing is known about the mannosyltransferases that modify the PGM-like proteins, and little is known about the underlying mannosyl glycan structures modified by Glc-1-P. Identification of the enzyme, genes, and structural confirmation of the sugar linkages (Chapter 47) is needed to confirm this modification. Nonetheless, these findings suggest that both the O-mannosylation and dynamic Glc-1-phosphoryation of PGM-like proteins may be very important regulatory forms of cytoplasmic glycosylation.

Nucleocytoplasmic Glycosaminoglycans?

As early as 1964, Yamashina and colleagues proposed the presence of glycosaminoglycans (GAGs) in purified nuclei. From 1971 to 1983, a series of studies addressed nuclear heparan sulfates in developing sea urchin embryos. Heparin stimulated messenger RNA (mRNA) synthesis in pregastrula, but not in postgastrula embryos. Other glycosaminoglycans such as chondroitin sulfates (CSs) and hyaluronans were without effect. Pulse-chase studies suggested that the nuclear glycosaminoglycans appeared first in the cytoplasm and were transported to the nucleus. β-Xylosides, which are primers of CS and heparan sulfate (HS) biosynthesis (see Chapter 16), blocked development of sea urchins at the blastula stage. Addition of postgastrula (but not pregastrula) proteoglycans to the cultures prevented this block in development. Microinjection of proteoglycans into embryos blocked development at the stage from which they were isolated. Given recent advances in the tools to study proteoglycans, these potentially exciting observations warrant a critical reexamination.

The first structural data proposing nuclear GAGs, which were not easily dismissed on the basis of potential contamination with extracellular GAGs, were published in 1989. Rat hepatocytes were radiolabeled with $^{35}SO_4$, and nuclear fractions were found to contain 11% of the total cellular HS, but this fraction was greatly enriched in a relatively rare structure: GlcA2SGlcNS6S disaccharides. The existence of this unique disaccharide makes it hard to explain the nuclear GAGs as resulting from a contamination of the nuclear fraction by cell-surface HS molecules, where such structures are not present. Later pulse-chase studies suggested that cell-surface HS is taken into the nucleus and subsequently modified to form this unique species. The uptake of HS into the nucleus was not affected by chloroquine. In support of the uptake model, when unlabeled HS proteoglycan that contains no GlcA2S was incubated with cells, about 10% ended up within the nucleus and resulted in free HS chains containing the unusual disaccharide. Given the ability of heparan sulfates to influence gene transcription in vitro, these studies could eventually prove to be highly significant. How nuclei would take up and modify such large and negatively charged molecules remains unknown. Other studies using different nuclear isolation methods on rat ovarian granulosa cells have found dermatan sulfates, but not HSs, associated with the nucleus. No unique structures were found in this study.

Using another approach, immunocytochemical studies probing for a GPI-anchored HS proteoglycan, glypican, have provided strong evidence for the nuclear accumulation of this proteoglycan, in addition to its traditionally accepted presence at the cell surface. Although the mechanism of nuclear compartmentalization is not known, there is evidence that the glypican polypeptide harbors a nuclear localization sequence that is functional when grafted onto a neutral carrier expressed in the cytoplasm. In a separate study, overexpression of a CS proteoglycan core protein in cultured cells resulted in accumulation of the core polypeptide in the cytoplasm and nucleus in addition to the secretory pathway, indicating, as has been suggested for other proteins, the possibility of dual compartmentalization. However, although these studies provide an explanation for how proteoglycan core proteins might accumulate in the nucleus, it is difficult to imagine how they have been modified by the copolymerases residing in the Golgi lumen.

Cytochemical studies using naturally occurring proteins with high-affinity binding to hyaluronan (HA) have been used to document accumulation of HA in the nuclei of cultured cells. Intracellular hyaluronan might represent a pool of polysaccharide that failed to translocate following biosynthesis, because the catalytic domain of HA synthases are oriented at the cytoplasmic surface of the plasma membrane (see Chapter 15). Some of the HA-binding proteins occur naturally within the nucleus, raising the possibility of a physiological role for nuclear hyaluronan via interaction with these HA-binding proteins. However, it is not certain that HA is the natural ligand for these binding proteins. An alternative possibility is that they interact with nuclear proteins that have been cluster-modified by O-β-GlcNAc (see Chapter 18).

Alpha Toxins of *Clostridium* Species

Small cytoplasmic G proteins (GTP-binding proteins) of the Rho family are involved in regulating the cytoskeleton. As shown in Figure 17.5, Cdc42 is involved in the formation of filopodia, Rac regulates membrane ruffling, and Rho regulates focal adhesions and stress fibers. Certain toxins from anaerobic bacteria were found to possess glycosyltransferase activities that attach a glycosyl moiety to a threonine residue in the GTP-binding site of these proteins (e.g., on Rho, glycosylation occurs at Thr-37; Figure 17.6). These secreted toxins exhibit the remarkable ability of translocating across the surface membrane into the cytoplasm of mammalian target cells. The enterotoxins from *Clostridium difficile* (ToxA) and *Clostridium sordellii* are α-glucosyltransferases that use UDP-Glc as a donor. In contrast, a similar toxin from *Clostridium novyi* is a UDP-GlcNAc-dependent O-α-*N*-acetylglucosaminyltransferase. The *C. novyi* toxin has no primary sequence relationship to the endogenous inverting animal O-β-GlcNAc (see Chapter 18) or Skp1 α-*N*-acetylglucosaminyltransferases that have been cloned. These glycosyltransferase toxins might have mammalian homologs and should prove valuable as tools to study the functions of these small G proteins; they also represent a unique example of abnormal cytoplasmic protein glycosylation.

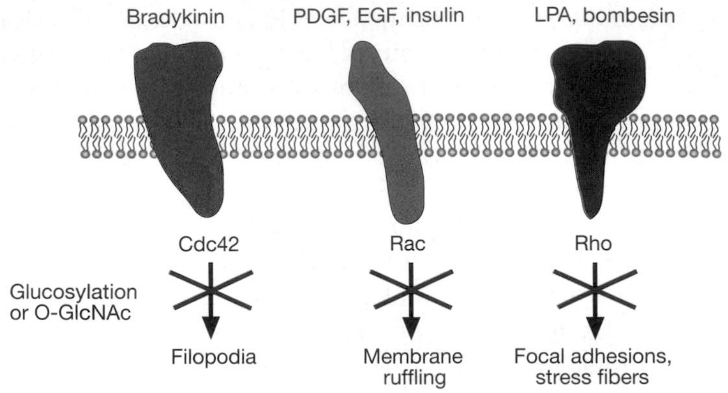

FIGURE 17.5. Rho family proteins regulate receptor-mediated actin cytoskeleton perturbations and are targets for bacterial toxins. Note how the different Rho family members are involved in regulating different cytoskeletal structures. The figure depicts transmembrane receptors mediating cytoskeletal structures via different Rho family members. (X) Blockage by glycosylation of small G-protein- and cytoskeleton-mediated assembly of the structures indicated.

Other Possible Examples of Cytoplasmic O-Glycosylation

A potential example of a novel O-linked glycan has been reported on a plant nuclear-pore-associated protein. This protein is recognized by the lectin wheat-germ agglutinin and can be labeled using tritiated UDP-Gal and β1-4 galactosyltransferase, which is specific for nonreducing terminal *N*-acetylglucosamine. The glycan, with an approximate size of five sugars, can be released by mild alkaline degradation consistent with β-elimination from a threonine or serine residue. Confirmation by mass spectrometry is needed to establish more firmly the existence of this modification. In another study, purified mammalian cytokeratin was found to bind lectins that recognize α1,3-linked *N*-acetylgalactosamine, a finding that was supported by compositional studies. Lectin binding was sensitive to pretreatment with *N*-acetylgalactosaminidase, but this does not rule out the possibility of a contaminant protein until direct structural evidence for glycosylation is obtained. Additional potential examples are oxygen-regulated protein 150, a putative cytoplasmic glycoprotein whose nuclear uptake is inhibited by a plant lectin thought to be internalized by epithelial cells, and a cytoplasmic isoform of carbonic anhydrase II secreted from male coagulating gland epithelial

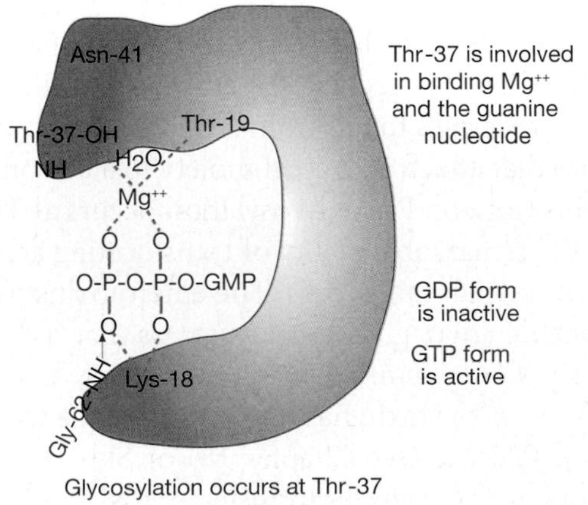

FIGURE 17.6. Glycosylation of Rho proteins by toxins occurs at Thr-37 and this modification has a key role in GTP/GDP nucleotide exchange by these G proteins. The cartoon depicts the GTP-binding site of the Rho protein. Note the essential role of Thr-37 in the active site.

cells by ectocytosis. A final example is the VP54 PBCV-1 capsid protein described above. The X-ray-diffraction analysis demonstrates that in addition to four sites for unconventional N-glycosylation, there are two additional sites of serine glycosylation that appear to be generated on the cytoplasmic precursor of the mature capsid protein.

NUCLEAR AND CYTOPLASMIC LECTINS

Many studies using labeled neoglycoproteins suggest the existence of multiple nuclear lectins. For example, bovine serum albumin (BSA) derivatized with L-rhamnose, D-*N*-acetylglucosamine, D-glucose, lactose, Man-6-P, and L-fucose all bind to nuclei at threefold higher affinity than underivatized BSA. BSA-lactose binds to cryostat sections of nuclei. However, little is known about the molecular nature of these putative carbohydrate-binding proteins except for galectins-1 and -3 (see below) and a recently identified heat-shock chaperone protein CBP70 (described above). CBP70 has GlcNAc-binding activity that might recognize O-β-GlcNAc–modified proteins.

The galectin family of lectins in animals and the discoidin family of lectins in *Dictyostelium* are high-abundance cytoplasmic proteins that have carbohydrate-binding specificity generally directed toward β-linked galactose and *N*-acetylgalactosamine (see Chapter 33). Large pools of these proteins are soluble in the cytoplasm, but cell biological studies have shown that these proteins also reside at the cell surface and in the pericellular matrix. These proteins lack typical amino-terminal signal peptides and exit the cytoplasm by a poorly characterized posttranslational pathway. Biochemical and genetic studies have uncovered primarily extracellular functions for these lectins (Chapter 33), suggesting that the cytoplasmic pools are simply precursors of functional extracellular pools. Nevertheless, CBP35 (now called galectin-3) and possibly galectin-1 are present in the nucleus as part of heterogeneous ribonucleoprotein complexes (HnRnP), where they appear to be required for normal mRNA splicing in extracts. Interestingly, a neoglycoprotein comprising the blood group A tetrasaccharide attached to BSA was found to inhibit specifically mRNA splicing in vitro. In addition, a cytoplasmic function in regulating apoptosis via Bcl-2 binding has been implicated for galectin-3. These intracellular activities may be mediated by protein–protein-binding activities ascribed to these proteins, because roles for their glycan-binding activities have not been established. These findings certainly leave open the possibility that galectins have important untested functions via cytoplasmic glycoconjugates.

Like galectins, members of the annexin family of proteins reside in the cytoplasm, at the cell surface, and in the extracellular matrix. Individual isoforms exhibit glycan-binding activity directed toward various GAGs, GPI anchors, or the bisecting *N*-acetylglucosamine of N-glycans. These known ligands generally reside outside of the cell and have not been specifically implicated in intracellular annexin functions, but they could be receptors for secreted annexins.

Numerous other soluble cytoplasmic proteins have been assigned sugar-binding activity that point to the potential significance of cytoplasmic glycoconjugates. A Golgi-associated actin-binding protein in *Dictyostelium* (comitin) has a mannose-binding activity, although the significance of this activity in vivo has not been tested. In plants, cytoplasmic mannose- and *N*-acetylglucosamine-binding lectins are induced by various kinds of biotic and abiotic effectors. In animals, numerous proteins have an affinity for glycogen, which exists primarily in the cytoplasm. One of these, laforin, is a protein phosphatase with a discrete glycogen-binding domain related to bacterial sugar-binding proteins. This domain is required for localization near glycogen, and evidence exists that this activity is specifically mutated in a congenital form of juvenile epilepsy.

GLYCANS AS NUCLEAR LOCALIZATION SIGNALS

Molecules larger than approximately 40 kD cannot diffuse freely through nuclear pores and must be specifically and actively transported into and out of the nucleus. Generally, nuclear localization sequences (NLSs) consist of one or two stretches of positive amino acids, as are found, for example, on SV40 large T antigen and nucleoplasmin. However, during the past several years, evidence has accumulated that sugars may also serve as nuclear localization signals. The neoglycoproteins BSA-glucose, BSA-fucose, and BSA-mannose are rapidly transported into the nucleus of permeabilized or microinjected living HeLa cells, whereas BSA itself is not. Sugar-mediated nuclear transport appears distinct from the basic peptide-mediated NLS pathway. Like the classical pathway, sugar-mediated nuclear transport requires energy and is blocked by the lectin, wheat-germ agglutinin. However, unlike the basic peptide system, the sugar-mediated pathway does not require cytosolic factors and is not blocked by sulfhydryl-reactive chemicals. Additional evidence shows that BSA substituted with GlcNAcβ1-4GlcNAc is rapidly localized to purified nuclei in vitro by a pathway distinct from the classically defined NLS systems. Validation of these fascinating results awaits characterization of the components involved and the identification of natural counterparts to the neoglycoproteins. A possible candidate is the Skp1 glycoprotein from *Dictyostelium,* although a distinct sugar specificity would be required in this species. Interference with Skp1 glycosylation inhibits nuclear accumulation in vivo, although other evidence suggests that the role of the glycan is indirect.

FURTHER READING

Marchase R.B. and Hiller A.M. 1986. Glucose phosphotransferase and intracellular trafficking. *Mol. Cell Biochem.* **72:** 101–107.

Hart G.W., Haltiwanger R.S., Holt G.D., and Kelly W.G. 1989. Glycosylation in the nucleus and cytoplasm. *Annu. Rev. Biochem.* **58:** 841–874.

Alonso M.D., Lomako J., Lomako W.M., and Whelan W.J. 1995. A new look at the biogenesis of glycogen. *FASEB J.* **9:** 1126–1137.

Busch C. and Aktories K. 2000. Microbial toxins and the glycosylation of Rho family GTPases. *Curr. Opin. Struct. Biol.* **10:** 528–535.

Lee J.Y. and Spicer A.P. 2000. Hyaluronan: A multifunctional, megadalton, stealth molecule. *Curr. Opin. Cell Biol.* **12:** 581–586.

Graves M.V., Bernadt C.T., Cerny R., and Van Etten J.L. 2001. Molecular and genetic evidence for a virus-encoded glycosyltransferase involved in protein glycosylation. *Virology* **285:** 332–345.

Tsai B., Ye Y., and Rapaport T.A. 2002. Retro-translocation of proteins from the endoplasmic reticulum into the cytosol. *Nat. Rev. Mol. Cell Biol.* **3:** 246–255.

Monsigny M., Rondanino C., Duverger E., Fajac I., and Roche A.C. 2004. Glyco-dependent nuclear import of glycoproteins, glycoplexes and glycosylated plasmids. *Biochim. Biophys. Acta* **1673:** 94–103.

Wang J.L., Gray R.M., Haudek K.C., and Patterson R.J. 2004. Nucleocytoplasmic lectins. *Biochim. Biophys. Acta* **1673:** 75–93.

West C.M., van der Wel H., Sassi S., and Gaucher E.A. 2004. Cytoplasmic glycosylation of protein-hydroxyproline and its relationship to other glycosylation pathways. *Biochim. Biophys. Acta* **1673:** 29–44.

CHAPTER 18

The O-GlcNAc Modification

Gerald W. Hart and Yoshihiro Akimoto

THIS CHAPTER PRESENTS AN OVERVIEW OF THE DYNAMIC modification of serine or threonine hydroxyl moieties on nuclear and cytoplasmic proteins by β-linked *N*-acetylglucosamine, termed O-β-GlcNAc or simply O-GlcNAc.

HISTORICAL BACKGROUND

O-GlcNAc is distinct from all other common forms of protein glycosylation in several major respects (Figure 18.1). (1) It occurs exclusively within the nuclear and cytoplasmic compartments of the cell. (2) The GlcNAc is generally not elongated or modified to form more complex structures. (3) It is attached and removed multiple times in the life of a polypeptide, often cycling rapidly and at different rates at different sites on a polypeptide. In most ways, O-GlcNAcylation is more similar to protein phosphorylation than it is to "classical" protein glycosylation. Research on O-GlcNAc is particularly relevant to chronic human diseases including diabetes, cardiovascular disease, neurodegenerative disorders, and cancer. A long-held dogma, still promulgated in many cell biology and biochemistry textbooks today, is that protein glycosylation does not occur within the cytoplasmic and nuclear compartments of the cell. This view is apparently true for typical N-glycans and

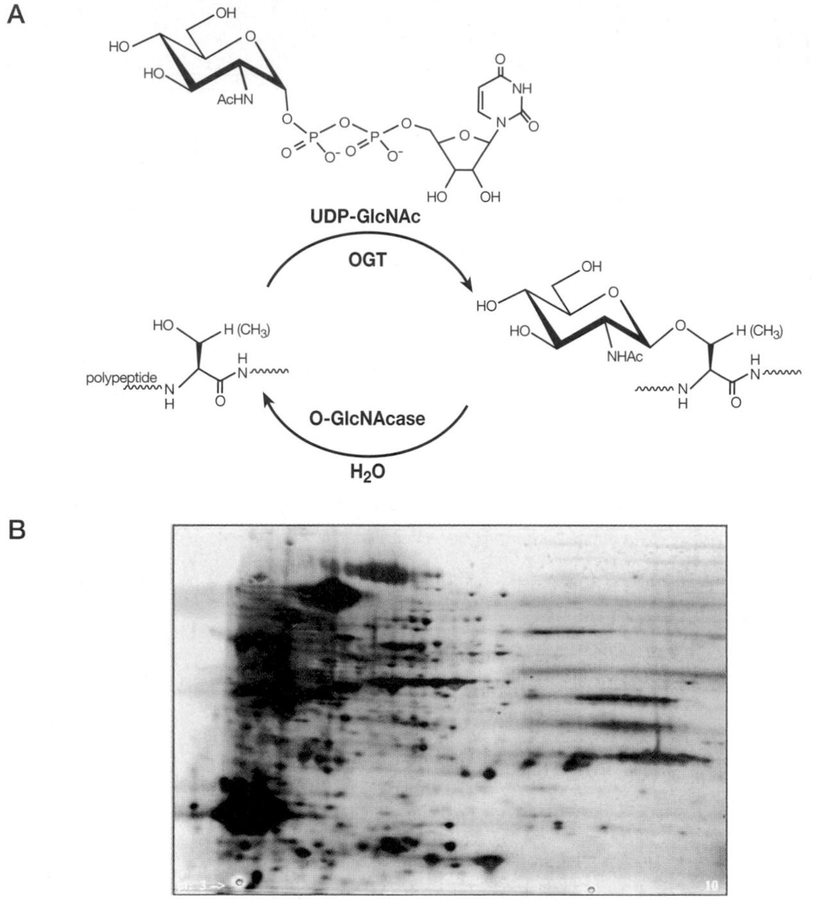

FIGURE 18.1. (*A*) O-Linked β-*N*-acetylglucosamine (O-GlcNAc) is a dynamic protein modification abundant within the nucleus and cytoplasm. The O-GlcNAc transferase (OGT) attaches O-GlcNAc to proteins at specific serine or threonine residues. O-GlcNAcase hydrolyzes O-GlcNAc from proteins. The two enzymes often occur within the same complex and are therefore highly regulated in controlling O-GlcNAc cycling. (*B*) Two-dimensional (2D) gel analysis of immunoaffinity-isolated O-GlcNAc–modified proteins from nucleocytoplasmic extracts of HeLa cells. The 2D gel analysis reveals that many proteins are modified by O-GlcNAc.

O-glycans that are attached to protein by glycosyltransferases with their active sites within the lumenal compartments of the ER and Golgi (see Chapters 3 and 8–12). However, it is now clear that certain specialized types of protein glycosylation do, indeed, occur within the nucleus and the cytoplasm (see Chapter 17). In fact, during the past two decades, it has become clear that O-GlcNAc is one of the most abundant posttranslational modifications within the nucleocytoplasmic compartments of all metazoans including plants, animals, and the viruses that infect them (Figures 18.1 and 18.2 and Table 18.1).

O-GlcNAc was discovered in 1983, when purified bovine milk galactosyltransferase and its radiolabeled donor substrate (UDP-[^{3}H]galactose) were used to probe for GlcNAc-terminating glycoconjugates in living murine thymocytes, splenic B- and T-lymphocytes, and macrophages. Galactosyltransferase is a Golgi glycosyltransferase that attaches galactose in a β1-4 linkage to almost any terminal *N*-acetylglucosamine residue. Contrary to expectations, product analyses, including release from protein by alkali-induced β-elimination and resistance to cleavage by peptide-N-glycosidase F (PNGase F; see Chapters 8 and 47), showed that most of the galactosylated glycans existed within the cell as single O-

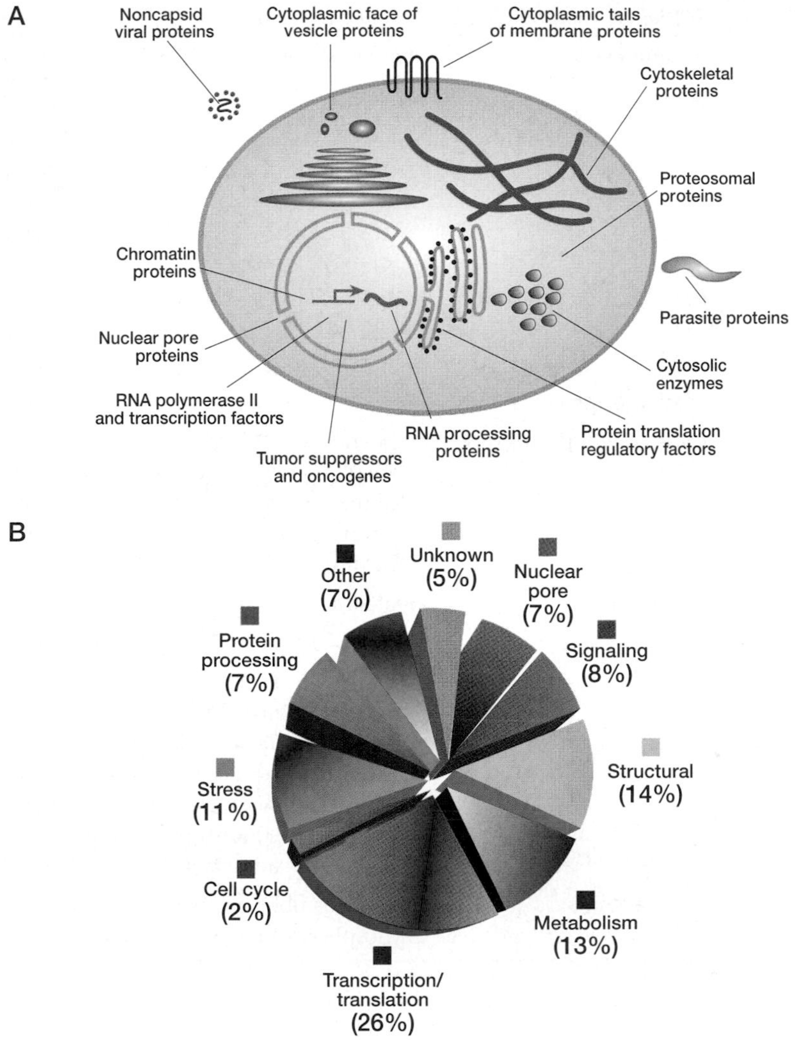

FIGURE 18.2. (*A*) O-GlcNAcylated proteins occur in many different cellular compartments. Found mostly within the nucleus, they are also in the cytoplasm and within viruses that infect eukaryotic cells. The subcellular distribution of O-GlcNAc is similar to that of Ser(Thr)-O-phosphorylation. (Redrawn, with permission, from Comer F.I. and Hart G.W. 2000. *J. Biol. Chem.* **275:** 29179–29182.) (*B*) O-GlcNAcylated proteins belong to many different functional classes. Pie chart illustrates the functional distribution of O-GlcNAcylated proteins identified thus far. (Redrawn, with permission, from Love D.C. and Hanover J.A. 2005. *Science STKE* **312:** re13.)

linked GlcNAc moieties. Studies of the subcellular localization of O-GlcNAc in rat liver established that it is most abundant within chromatin, most concentrated on the nuclear pores of the nuclear envelope, and also present within the cytoplasm of the cell. There is so far no evidence for O-GlcNAc on the regions of polypeptides that reside on the outside of cells or within lumenal compartments. Galactosyltransferase in combination with donor substrates of either radiolabeled UDP-galactose or chemically modified UDP-galactose, to which fluorescent or biotin tags can be attached, still represents one of the best methods for detecting and studying O-GlcNAc. More recent studies have used metabolic incorporation of *N*-azidoacetylglucosamine in living cells, combined with chemical tagging, affinity isolation, and mass spectrometry, to identify larger numbers of O-GlcNAc–modified proteins. Unlike many other methods, these covalent probes for O-GlcNAc permit direct chemical and enzymatic analyses of the products formed.

TABLE 18.1. *Some O-GlcNAcylated proteins*

Protein class	Some examples
Nuclear pore proteins	p62, Nup54, Nup155, Nup180, Nup153, Nup214, Nup358
Transcription factors	Sp1, c-fos, c-jun, CTF, HNF1, v-ErbA, pancreas-specific TF, serum response factor, c-Myc, p53, estrogen receptors, β-catenin, Nf-κB, elf-1, pax-6, enhancer factor D, human C1, Oct1, plakoglobin, YY1, PDX-1, CREB, Rb, p107, RNA polymerase II, ATF-2, HCF-1, SRC-1, TLE-4, CCR4-NOT4
RNA-binding proteins	HnRNP-G, Ewing sarcoma RNA-binding protein, EF4A1, EF-1α, RNA-binding motif protein 14, 15 ribosomal proteins
Phosphatases, kinases, adapter proteins	nuclear tyrosine phosphatase p65, casein kinase II, insulin receptor substrates 1 and 2, GSK3β, p85 and p110 PI3-kinase subunits
Cytoskeletal proteins	cytokeratins 8, 13, and 18, neurofilaments H, M, and L, band 4.1, talin, vinculin, ankyrin, synapsin 1, myosin, E-cadherin, cofilin, tau, microtubule-associated proteins 1B, 2 and 4, dynein, α-tubulin, AP3, AP180, β-amyloid precursor protein, β-synuclein, piccolo, spectrin β-chain, WNK-1, PDZ-GEF, synaptopodin
Chaperones	Heat-shock protein 27, α-crystallin, HSC70, HSP70, HSP90
Metabolic enzymes	endothelial nitric oxide synthase (eNOS), enolase, glyceraldehyde-3-phosphate dehydrogenase, phosphoglycerate kinase, pyruvate kinase, UDP-glucose pyrophosphorylase, glycogen synthase
Other regulatory proteins	eukaryotic peptide chain initiation factor 2, p67, O-GlcNAc transferase, CRMP-2, ubiquitin carboxyhydrolase (UCH), GLUT1, annexin 1, nucleophosmin, many proteasome subunits, Q04323 UCH homolog, Sec24, Sec23, Ran, peptidyl prolylisomerase, Rho GDP-dissociation inhibitor, γ-aminobutyric acid (GABA) receptor interacting protein
Viral proteins	adenovirus fiber protein, SV40 larger T-antigen, cytomegalovirus basic phosphoprotein, NS26 rotaviral protein, baculoviral tegument protein

O-GlcNAc IS A HIGHLY DYNAMIC MODIFICATION

Unlike the relatively stable nature of mature N- and O-glycans on glycoproteins, O-GlcNAc cycles rapidly on and off most proteins because of the combined action of O-GlcNAc transferase (sometimes abbreviated to OGT) and O-GlcNAcase (Figure 18.3). Early studies showed that mitogen or antigen activation of lymphocytes rapidly decreased O-GlcNAcylation of many cytoplasmic proteins, but concomitantly increased O-GlcNAcylation of many nuclear proteins. Likewise, neutrophils were shown to rapidly modulate O-GlcNAcylation of several proteins in response to chemotactic agents. Pulse-chase analyses have revealed that O-GlcNAc residues on small heat-shock protein in lens (α-crystallin) and on intermediate filament proteins (cytokeratins) turn over more rapidly than the polypeptide chains to which they are attached. Similar studies on several proteins have confirmed that O-GlcNAc cycles, sometimes very rapidly, depending on the protein as well as the site on the protein to which it is attached. Recent proteomic studies that have quantified O-GlcNAc cycling in response to external signals (e.g., insulin, high glucose, or cellular stress) and/or inhibitors of O-GlcNAcase have established that the rates of O-GlcNAc cycling are similar to those often seen for protein phosphorylation, with half-lives of 1 minute or even less. It is also now clear that O-GlcNAc levels change during the cell cycle and that O-GlcNAc is involved in cell-cycle progression. However, on some proteins (e.g., nuclear pore proteins) O-GlcNAc cycles slowly or only in response to external signals that remain to be identified.

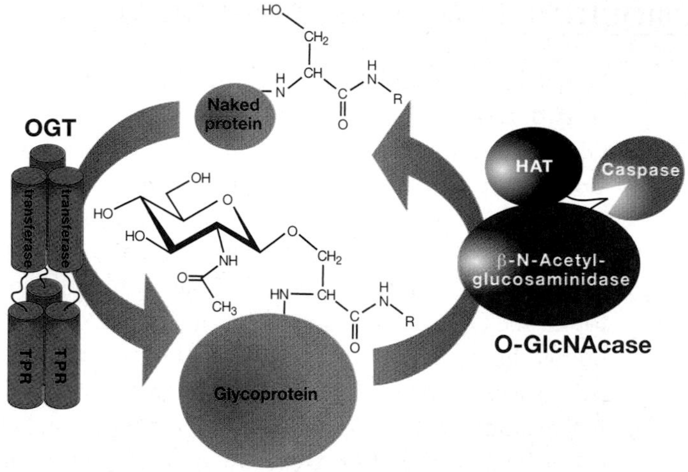

FIGURE 18.3. O-GlcNAc cycling is regulated by the concerted action of O-GlcNAc transferase (OGT) and O-GlcNAcase. OGT has two distinct domains: the transferase domain, which contains the catalytic site and is related to glycogen phosphorylase, and the tetratricopeptide repeat (TPR) domain, which contains protein:protein interaction sites. O-GlcNAcase also has two functional domains, which are separated by a caspase-3 cleavage site. O-GlcNAcase has an active histone acetyltransferase (HAT) domain at its carboxyl terminus and an N-acetylglucosaminidase domain at its amino terminus. (Redrawn, with permission, from Slawson C., Housley M.P., and Hart G.W. 2006. *J. Cell. Biochem.* **97**: 71–83.)

Why Did O-GlcNAc Remain Undetected for So Long?

As shown in Figures 18.1 and 18.2, O-GlcNAc is found on hundreds or even thousands of nucleocytoplasmic proteins. In fact, many of the most heavily studied proteins in biology are O-GlcNAcylated (Table 18.1). Yet most researchers devoted to studying the structure and functions of these same proteins remain surprisingly unaware of the presence of O-GlcNAc. So why did O-GlcNAc remain undetected until 1983, and why is it still largely unseen and unconsidered by investigators studying transcription, signaling, and the cytoskeleton? First, unlike charged modifcations (e.g., phosphate), addition and removal of O-GlcNAc generally does not affect the migration of polypeptides on SDS-PAGE (sodium dodecyl sulfate–polyacrylamide gel electrophoresis) or even on high-resolution 2D gels. A small shift in protein migration might be seen if the O-GlcNAc residues are highly clustered or if the protein is extensively O-GlcNAcylated at multiple sites (e.g., p62 nuclear pore protein or Sp1 transcription factor). Second, all cells contain high levels of hydrolases, including abundant lysosomal hexosaminidases and nucleocytoplasmic β-*N*-acetylglucosaminidases, that rapidly remove O-GlcNAc from intracellular proteins when the cell is damaged or lysed. Thus, O-GlcNAc is often lost during the isolation of a protein. Third, O-GlcNAc is particularly difficult to detect by physical techniques such as mass spectrometry, because it often occurs at substoichiometric amounts at different sites on a protein and readily falls off the polypeptide during the ionization process in a mass spectrometer. In both electrospray mass spectrometry and in matrix-assisted laser desorption time-of-flight (MALDI-TOF) mass spectrometry analyses of a mixture of unmodified and O-GlcNAc–modified peptides, not only is the O-GlcNAc lost during ionization, but the signal from the O-GlcNAc–modified peptides that remain is nearly, or even entirely, suppressed by the presence of the unmodified peptide. However, the recent development of monoclonal antibodies that detect O-GlcNAc and the invention of more sophisticated mass spectrometric methods, when combined with new potent inhibitors of β-*N*-acetylglucosaminidases, have substantially improved our ability to detect O-GlcNAc on proteins.

O-GlcNAc IS UBIQUITOUS AND ESSENTIAL IN METAZOANS

To date, nucleocytoplasmic O-β-GlcNAc has been found in all multicellular organisms investigated, ranging from filamentous fungi, worms, insects, and plants to humans. However, O-GlcNAc and the enzymes that control its cycling (see below) do not appear to exist in yeast, such as *Saccharomyces cerevisiae* or *Schizosaccharomyces pombe.* O-α-GlcNAc appears to be a common modification of cell-surface and extracellular proteins in protozoa, where it is attached by enzymes structurally related to the lumenal protein:O-GalNAc transferases (ppGalNAc-Ts) in mammals (see Chapter 9). However, it remains unclear whether nucleocytoplasmic O-β-GlcNAc occurs within either free-living or parasitic protozoa. For example, the known genes that encode the O-GlcNAc transferase and O-GlcNAcase are apparently absent from the genome of the parasitic flagellated protozoan *Trypanosoma brucei* (see Chapter 40). It now appears that earlier studies reporting O-GlcNAc in protozoa were, in fact, based mostly on the presence of O-α-GlcNAc on extracellular proteins, which appears to serve the same functions as mucin-type α-GalNAc O-glycans in higher organisms, perhaps because many protozoa lack the C-4 epimerase required to convert UDP-GlcNAc to UDP-GalNAc (see Chapter 4). Genetic and subsequent biochemical studies in *Arabidopsis* showed that the genes *SPY* and *SECRET AGENT* regulate growth hormone (gibberellic acid) signaling, and later studies established that both of these genes encode O-β-GlcNAc transferases. Mutations in either *SPY* or *SECRET AGENT* cause severe growth phenotypes, but they are not lethal. However, simultaneous mutation of both genes is lethal. Studies in rice, potatoes, and other plants have also indicated that O-GlcNAcylation is important for growth regulation.

Unlike plants, mammals and insects appear to have only a single gene encoding the catalytic subunit of the O-GlcNAc transferase (*OGT*). Using *Cre-LoxP* conditional gene disruption in mice, *OGT* was shown to be required for the viability of embryonic stem cells. Tissue-targeted disruption in mice and disruption of *OGT* expression in cell culture have established that O-GlcNAcylation is essential for viability at the single-cell level in mammalian cells. Disruption of *OGT* in the worm *Caenorhabditis elegans* causes defective carbohydrate metabolism and abnormalities in dauer formation, a metabolically inactive larval form of the worm that is long lived and stress resistant, analogous to spores in bacteria. O-GlcNAc–modified proteins have also been found on many viruses that infect metazoans (e.g., adenovirus, SV40, cytomegalovirus, rotovirus, baculovirus, plum pox, HIV, and others). In viruses, instead of being localized on the outside capsid of the virus where "classical" N- or O-glycans are found (see Chapter 39), O-GlcNAc is found deep within the viruses, on tegument and other regulatory proteins, close to the nucleic acid components. Thus far, at least one pathogenic bacterium (*Clostridium novyi*) has been found to exploit the O-GlcNAc modification. The α toxin of *C. novyi* is an O-GlcNAc transferase (no homology to the mammalian enzyme) that kills cells by attaching an O-GlcNAc residue to GTP-binding sites on the small G proteins such as Rho, Rac, and Ran that regulate the cytoskeleton. Some of these G proteins are normally O-GlcNAcylated by an endogenous O-GlcNAc transferase, but the sites of attachment of O-GlcNAc are not known.

O-GlcNAc Is Found on a Wide Range of Proteins

Studies from several laboratories have described O-GlcNAcylated proteins from virtually all cellular compartments representing nearly all functional classes of protein (Figure 18.2). In short, O-GlcNAc occurs everywhere that Ser(Thr)-O-phosphorylation has been found. O-GlcNAc is particularly abundant within the nucleus, where it occurs on the transcriptional regulatory machinery including RNA polymerase II catalytic subunit carboxy-terminal domain (CTD) and myriad basal and specialized RNA polymerase II transcription factors. About one quarter of O-GlcNAcylated proteins are transcription or translation reg-

TABLE 18.2. *Some sites of O-GlcNAc attachment*

PV(S/T) motif (underlined)	
Human erythrocyte 65-kD protein	...D**S**PVSQPSLVGSK...
Human serum response factor	...YLAPV**S**ASVSP**S**AV**S**A...
Rhesus monkey α-B-crystallin	...EEEKPAV**T**AAPK...
Bovine lens α-A-crystallin	...DIPV**S**REEK...
Rat neurofilament (NF-L)	...YSAPV**SSS**LSVR...
Rat neurofilament (NF-M)	...GSPS**T**VSSSYK...
	...QPSV**T**ISSK...
HCMV (UL32) BPP	...PPSVPV**S**GSAPGR...
HCMV (UL32) BPP	...PPSVPV**S**GSAPGR...
No PV(S/T) motif	
Human erythrocyte band 4.1	...AQTIT**S**ETPSSTT...
Human serum response factor	...VTNLPGTT**S**TIQTAPSTSTT...
	...TQT**SSS**GTVTLPATIM...
Calf thymus RNA polymerase II	...(S/T)P(**S/T**)SP...TP**T**SPN...SP**T**SP**T**...
Chicken talin	...MAXQNLVDPAX**T**Q...
	...NQL**T**NDYGQLAQQ...
Rat nuclear pore p62	...MAGGPADT **S** DPL...
Rat neurofilament (NF-L)	...YVE**T**PRVHI **S** SVR...
HCMV (UL32) BPP	...STTPTYPAVTTVYPPS **S** TAK...

Boldface indicates the site of the modification.

ulatory factors; nuclear pore proteins are among the most heavily O-GlcNAcylated proteins. O-GlcNAcylation also appears to be particularly abundant on proteins involved in signaling, stress responses, and energy metabolism. It also occurs on many cytoskeletal regulatory proteins, such as those regulating actin assembly (e.g., vinculin, talin, vimentin, and ankyrin) and tubulin assembly (e.g., microtubule-associated proteins [MAPs], dynein, and tau). Even α-tubulin itself is dynamically modified by O-GlcNAc, but stoichiometry appears to be low. Intermediate filaments, such as cytokeratins and neurofilaments in brain, are also heavily modified by O-GlcNAc. To date, more than 600 O-GlcNAcylated proteins have been identified, but this large number probably still represents only a portion of the most abundant proteins that are dynamically modified within the cell. Table 18.1 shows a partial list of some of the known O-GlcNAcylated proteins.

Individual sites modified by O-GlcNAc have been identified on several proteins (Table 18.2). About half of the mapped sites have a proline and valine residue on either side of the modified hydroxy amino acid. The "PVS" motif is similar to that recognized by proline-directed kinases, such as glycogen synthase kinase-3β (GSK3β). The other O-GlcNAc sites seemingly have little in common at the primary sequence level, but they have sequences similar to those also recognized by different kinases. Many of the identified O-GlcNAc sites have high "PEST" scores, a sequence motif that is associated with rapid degradation of a protein, and a few studies suggest that O-GlcNAcylation within the PEST sequence prevents rapid degradation of the protein.

O-GLCNAC HAS A COMPLEX DYNAMIC INTERPLAY WITH O-PHOSPHATE

O-GlcNAc and O-phosphate exhibit a complex interplay on signaling, transcriptional, and cytoskeletal regulatory proteins within the cell (Figure 18.4). In fact, a major enzyme that removes O-phosphate, protein phosphatase 1 catalytic subunit (PP1c), is in a dynamic complex with OGT, indicating that, in many cases, the same enzyme complex both removes

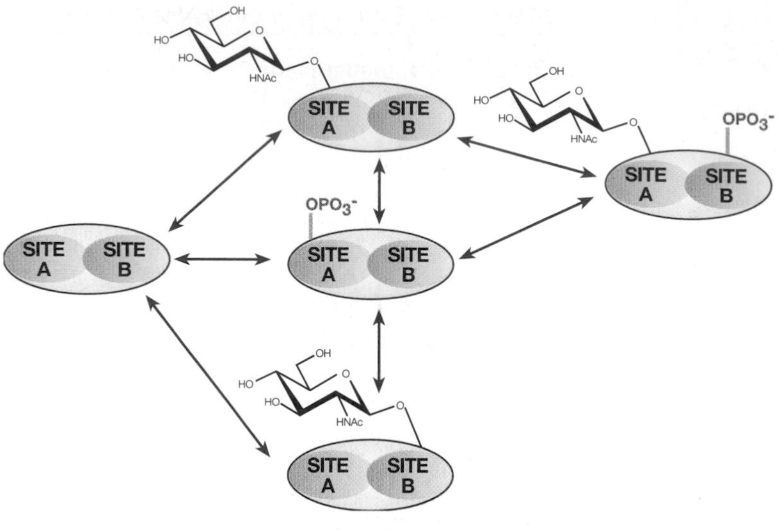

FIGURE 18.4. O-GlcNAcylation exhibits a complex dynamic interplay with O-phosphorylation. The example shown depicts the complexity created by two potential sites on a protein. On some proteins, O-GlcNAc and O-phosphate compete for the same site. On others, their respective sites are close to each other and they are still mutually exclusive. On yet other proteins, O-GlcNAcylation and O-phosphorylation occur independently of each other. (Redrawn, with permission, from Comer F.I. and Hart G.W. 2000. *J. Biol. Chem.* **275:** 29179–29182.)

O-phosphate and concomitantly attaches O-GlcNAc (Figure 18.5). The microtubule regulatory protein tau is abundantly phosphorylated in the early stages of brain development, but throughout adult life it is apparently extensively O-GlcNAcylated and not significantly phosphorylated. However, in the brains of patients with Alzheimer's disease, tau becomes less O-GlcNAcylated and more extensively phosphorylated, leading to the formation of intraneuronal "tangles" composed of paired helical filamentous (PHF)-tau, which are characteristic of this disease. The CTD repeat domain of RNA polymerase II is alternately extensively O-GlcNAcylated (with as many as three O-GlcNAc residues per repeat) and phosphorylated. At some time during the transcription cycle, O-GlcNAc on the CTD is removed and replaced with O-phosphate, initiating the elongation phase of transcription. In vitro studies have shown that synthetic peptides with up to ten CTD repeats (comprising 70 amino acids) cannot be phosphorylated by CTD kinases if they contain even a single O-GlcNAc moiety. Likewise, these CTD peptides cannot be O-GlcNAcylated if even one of the repeats contains a single O-phosphate residue, despite the many potential modification sites available on the same or other repeat sequences.

O-GlcNAc occurs reciprocally on several transcription factors important to cancer. For example, Thr-58, located within the transactivation domain of the c-*myc* proto-oncogene, which is the major site mutated in human lymphomas, is alternatively phosphorylated by GSK3β kinase and O-GlcNAcylated by OGT. Phosphorylation at Thr-58 is the culmination of a hierarchical signaling cascade penultimately involving a "priming" phosphorylation reaction characteristic of many GSK3β substrates. First, Ser-62 on c-Myc is phosphorylated by the growth factor kinase ERK, which creates a recognition site that allows Thr-58 of c-Myc to be phosphorylated by GSK3β. Studies with site- and modification-specific antibodies have shown that in nongrowing cells, Thr-58 of c-Myc is modified by O-GlcNAc. However, after applying a growth stimulus, O-GlcNAc is very rapidly removed and replaced by O-phosphate at this site.

In a similar fashion, O-GlcNAcylation of Ser-16 on estrogen receptor β causes it to become less transcriptionally active, but more long lived in the cell. In contrast, O-phos-

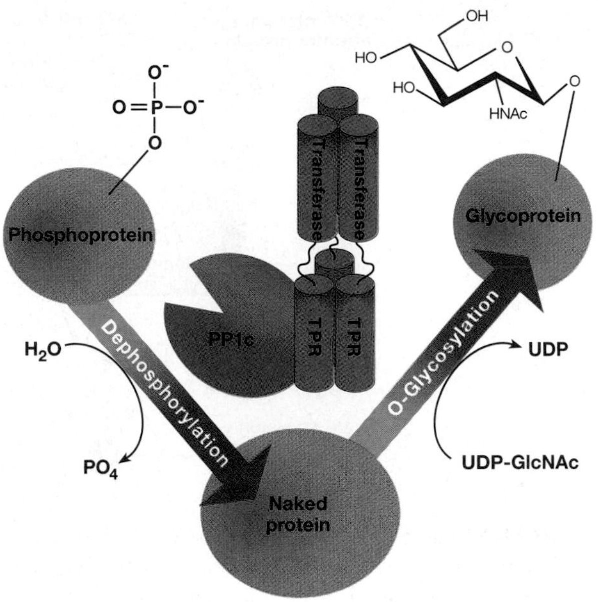

FIGURE 18.5. Protein phosphatase 1 catalytic subunit (PP1c), a key enzyme that removes O-phosphate residues from many important regulatory proteins, exists in a complex with O-GlcNAc transferase, which attaches O-GlcNAc. This suggests that the same enzyme complex both removes phosphate and attaches O-GlcNAc, perhaps without detaching from the substrate. (TPR) Tetratricopeptide repeat. (Redrawn, with permission, from Slawson C., Housley M.P., and Hart G.W. 2006. *J. Cell. Biochem.* **97:** 71–83.)

phorylation of Ser-16 causes estrogen receptor β to become more transcriptionally active, but degraded much more rapidly.

Hyper-O-GlcNAcylation, as occurs in diabetes, also blocks phosphate-dependent activation of some proteins. For example, the critical AKT kinase regulatory site on endothelial nitric oxide synthase (eNOS) is occupied by an O-GlcNAc moiety in individuals with diabetes, preventing activation of the enzyme. This O-GlcNAc–mediated block in eNOS activation is associated with the vascular complications and erectile dysfunction seen in patients with diabetes.

On some proteins (e.g., casein kinase II), O-GlcNAc and O-phosphate occur at separate but adjacent sites, yet they still appear to be reciprocal with each other. However, on other proteins, the relationship between O-GlcNAc and O-phosphate dynamics remains unclear. For example, on cytokeratins, O-GlcNAcylation and O-phosphorylation appear to be independently regulated, but may occur mutually exclusively on completely different subsets of the same polypeptides.

Thus, it is now clear that the dynamic interplay between O-GlcNAcylation and O-phosphorylation is widespread and quite complex (Figure 18.4), requiring us to modify current dogma with respect to cellular signaling. We can no longer view the dynamic regulation of signaling, transcription, and other cellular processes simply as binary "on" or "off" systems controlled by the addition and removal of O-phosphate, but rather we must view these processes as an "analog" system in which the dynamic interplay between these two (and perhaps other) modifications, often at multiple sites on a protein, creates an enormous range of highly dynamic structural and functional molecular possibilities.

ENZYMES CONTROLLING O-GLCNAC CYCLING

Even though O-GlcNAc cycles like O-phosphate and it often even competes for the same site, O-GlcNAc regulation differs remarkably from the regulation of protein phosphoryla-

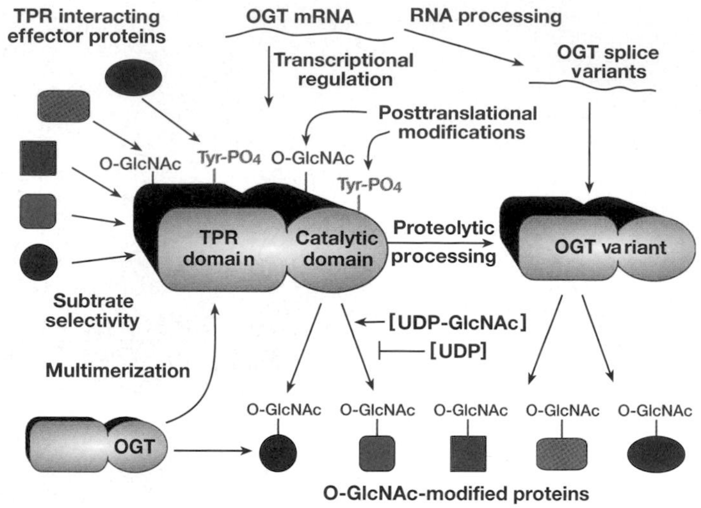

FIGURE 18.6. O-GlcNAc transferase (OGT) is regulated by multiple complex mechanisms, including transcriptional regulation of its expression, differential mRNA splicing, proteolytic processing, posttranslational modification, and multimerization with itself and other proteins. The target specificity of OGT is regulated by a large number of proteins that bind to the tetratricopeptide repeats (TPRs). However, the most important regulator of OGT is the level of its donor substrate UDP-GlcNAc. (Redrawn, with permission, from Comer F.I. and Hart G.W. 2000. *J. Biol. Chem.* **275:** 29179–29182.)

tion by specific kinases. Mammalian genomes encode hundreds of site-specific kinase catalytic subunits that are each independently regulated to act like "microswitches" controlling the signaling circuits of the cell. In contrast, mammalian genomes, thus far, have been found to encode only a single, very highly conserved, catalytic subunit for each of the enzymes OGT and O-GlcNAcase (Figure 18.3). Current dogma dictates that different stimuli cause specific phosphate cycling to change on only a few proteins within a signaling cascade or functional complex. However, in the case of O-GlcNAc, different stimuli cause rapid changes in O-GlcNAc cycling concomitantly on many proteins, presumably even in different pathways and complexes. Thus, the dynamics of O-GlcNAcylation appear to act at a "global" level, whereas the dynamics of O-phosphorylation appear to act at a local level. However, although both OGT and O-GlcNAcase each have only a single catalytic subunit, they actually exist within the cell as a multitude of different holoenzymes in which their catalytic subunits are noncovalently bound to a myriad of accessory proteins that appear to control their targeting specificities (Figure 18.6).

O-GlcNAc Transferase

O-GlcNAc transferase (OGT; uridine diphospho-*N*-acetylglucosamine:polypeptide β-*N*-acetylglucosaminyltransferase) catalyzes the addition of *N*-acetylglucosamine from UDP-GlcNAc to specific serine or threonine residues to form a β-glycosidic linkage. The enzyme was first identified and purified from rat liver and subsequently cloned from rat, human, *C. elegans*, and other organisms. *OGT* resides on the X chromosome (Xq13 in humans) near the centromere and is among the most highly conserved proteins from worms to man. As described above, *OGT* is required for single-cell viability in mammals and plants, but not in *C. elegans*. In rat liver, OGT appears to be a heterotrimer of two 110-kD subunits and one 78-kD subunit. The 78-kD subunit is most abundant in kidney, muscle, and liver, but in most tissues such as the brain and pancreas the enzyme contains only the 110-kD subunits. The functional significance of these variants is unclear.

Also, a unique mitochondrial form of OGT has been described. OGT is present in all cells, but is most abundant by far in the glucose-sensing cells (α and β cells) of the pancreas and also in the brain. OGT has two distinct domains separated by a putative nuclear localization sequence. The amino terminus of each OGT subunit contains up to 11.5 tetratricopeptide repeats (TPRs) that serve as protein:protein interaction domains and represent the sites to which most of the OGT targeting/regulatory proteins bind. The crystal structure of human TPRs of OGT show that the repeats occur as stacked α-helical domains, forming a "tube-like" structure with a remarkable structural similarity to the TPR region of the nuclear transport protein importin-α, suggesting that targeting of OGT occurs in a manner similar to importin-α's mechanism of transporting proteins across the nuclear envelope. The TPRs of OGT are also required for the multimerization of OGT subunits. The carboxy-terminal domain of OGT appears to have evolved from the glycogen phosphorylase superfamily of enzymes, and it contains the UDP-GlcNAc-binding and catalytic sites of the enzyme.

The regulation of OGT is quite complex and still not well understood (Figure 18.6). OGT is itself O-GlcNAcylated and also tyrosine phosphorylated. Tyrosine phosphorylation appears to activate the enzyme, but the role of O-GlcNAc on OGT is not yet clear. Purified or recombinant OGT modifies small synthetic peptides based on known sites from O-GlcNAcylated proteins (Table 18.2), but OGT appears to require accessory proteins bound to the TPRs to modify full-length protein substrates efficiently. Yeast two-hybrid analyses have identified many of these OGT targeting proteins. The best characterized is OIP106, which is involved in targeting OGT to RNA polymerase II–containing transcriptional machinery. However, the most important factor regulating overall OGT activity within the cell is the level of its donor substrate UDP-GlcNAc. The concentrations of cellular UDP-GlcNAc are exquisitely sensitive to carbohydrate, fatty acid, energy, and nitrogen metabolic fluxes (Figure 18.7), making it an ideal metabolic sensor. In cells, 2–5% of all glucose use occurs through the hexosamine biosynthetic pathway, which culminates in the biosynthesis of UDP-GlcNAc (see Chapter 4). Upon transfer of GlcNAc to proteins, UDP is released, which is a potent feedback inhibitor of OGT. Under conditions during which UDP is rapidly removed (as occurs within the cell), OGT activity is dependent on the UDP-GlcNAc level over a remarkable range of concentrations (from the low nM range to well over 50 mM). The concentration of UDP-GlcNAc in a cell typically ranges from approximately 0.1 to 1 mM, second only to ATP for high-energy small molecules. Thus, rapid changes in UDP-GlcNAc concentrations may serve as a nutrient sensor by directly affecting the extent of O-GlcNAcylation of regulatory proteins.

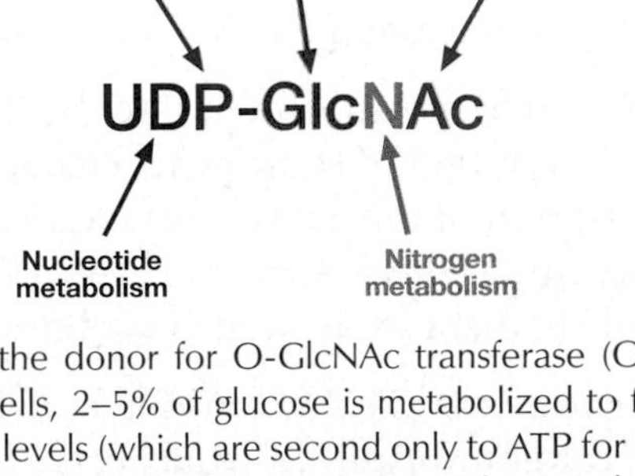

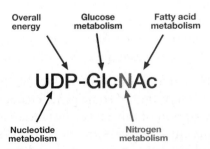

FIGURE 18.7. UDP-GlcNAc is the donor for O-GlcNAc transferase (OGT) and an ideal sensor of the metabolic status of the cell. In cells, 2–5% of glucose is metabolized to form UDP-GlcNAc. Unlike other sugar nucleotides, UDP-GlcNAc levels (which are second only to ATP for high-energy cellular compounds) are exquisitely sensitive to fluxes through the metabolic pathways shown. In turn, the extent of O-GlcNAcylation of many proteins is highly responsive to UDP-GlcNAc levels. (Adapted, with permission of Birkhauser Verlag AG, from Cooke S.F. and Bliss T.V. 2003. *Cell Mol. Life Sci.* **60:** 1–7.)

O-GlcNAcase

Nucleocytoplasmic β-*N*-acetylglucosaminidase (O-GlcNAcase) was first identified as a neutral cytosolic hexosaminidase (referred to as "hexosaminidase C") and often compared to the better-studied lysosomal hexosaminidases A and B. O-GlcNAcase was purified from rat kidney and bovine brain, and the human gene was cloned using peptide sequence information. The O-GlcNAcase gene was found to be identical to *MGEA5*, a putative hyaluronidase genetically identified because of its association with meningiomas. The O-GlcNAcase gene maps to the late-onset Alzheimer's disease locus in humans (10q24.1) and lies adjacent to the gene for insulin-degrading enzyme. Sequence analyses suggest that O-GlcNAcase is a bifunctional enzyme (Figure 18.3). The amino terminus of O-GlcNAcase encodes the glycosidase domain, whereas the carboxyl terminus shows homology to the *GCN5* HAT family, which specifically acetylates histones to activate gene expression. O-GlcNAcase has both HAT and O-GlcNAcase activities in vitro. During the latter stages of apoptosis (regulated cell death), the executioner caspase (caspase-3) cleaves O-GlcNAcase to separate the HAT and O-GlcNAcase domains. Like OGT, O-GlcNAcase also interacts in a dynamic fashion with a very large number of cellular proteins, but the location of interaction sites on the enzyme have not been mapped. O-GlcNAcase and OGT are often within the same functional complex, particularly in transcription complexes, where both enzymes are found with transcriptional regulators, such as the transcriptional repressor mSin3A. Thus, both enzymes appear to be tightly regulated to prevent futile cycling of O-GlcNAc.

BIOLOGICAL FUNCTIONS OF O-GLCNAC

Like O-phosphorylation, the specific functions of O-GlcNAc depend on the protein and sites to which the glycan is attached. However, on the basis of many studies, some generalities are emerging. As discussed above, one of the major functions of O-GlcNAc is to prevent O-phosphorylation and, by doing so, to modulate signaling and transcription in response to cellular nutrients or stress.

O-GlcNAc Regulates Transcription and Translation

O-GlcNAc directly regulates the activities of a variety of transcription factors including Sp1, estrogen receptors, STAT5, NF-κB, p53, YY1, Elf-1, c-Myc, Rb, PDX-1, CREB, forkhead, and others. As described above, RNA polymerase II and the basal transcription factors are also extensively modified by O-GlcNAc. O-GlcNAc modification has been reported to suppress or enhance transcription, depending on the promoter and other associated proteins. Some examples are outlined below.

1. Hyperglycosylation of Sp1 occurs in diabetic states and this modification enhances the transcription of some genes characteristic of this disease, while suppressing the transcription of others. For instance, Sp1 with high O-GlcNAc drives the transcription of plasminogen activator and extracellular matrix proteins thought to have an important role in diabetes-associated cardiovascular disease.

2. Increased O-GlcNAc contributes to impaired calcium cycling in the heart and to diabetic cardiomyopathy by reducing the transcription of a sarcoplasmic reticulum (SR) calcium ATPase (SERCA2a). This protein has an important role in removing cytoplasmic Ca^{++} and is the main mechanism for restoring SR Ca^{++} load during the contraction cycle.

3. The transcription factor STAT5 requires O-GlcNAc to bind to the coactivator of transcription (CBP), which is an essential interaction for STAT5-mediated gene transcriptions. Other members of the STAT family are also O-GlcNAcylated, but the function of O-GlcNAc on these proteins is unknown.
4. Increased O-GlcNAcylation of PDX-1, a pancreatic β-cell transcription factor that controls insulin transcription, increases its affinity for DNA and results in increased proinsulin transcription.
5. O-GlcNAcylation prevents YY1 (a zinc-finger, DNA-binding transcription factor that regulates the expression of a wide variety of cellular and viral genes and is essential for the development of mammalian embryos) from binding the retinoblastoma Rb protein. Upon dissociation from Rb, the O-GlcNAcylated YY1 is free to bind DNA and activate transcription.
6. The Rb tumor suppressor protein is itself O-GlcNAcylated during the stages of the cell cycle when it is not O-phosphorylated. It is the O-GlcNAcylated form of Rb that binds to E2F transcription factors.
7. Elf-1 (a member of the Ets transcription factor family) is involved in the transcriptional regulation of several hematopoietic genes. Following O-GlcNAcylation and phosphorylation by PKC-θ, the cytoplasm-located Elf-1 translocates to the nucleus where it binds to the promoter of the T-cell receptor (TCR) ζ gene and promotes its transcription.
8. O-GlcNAcylation of CREB (cyclic-AMP-responsive element-binding protein), a transcription factor essential for long-term memory, disrupts the interaction between CREB and $TAF_{II}130$, a component of the basal transcription machinery, thereby repressing the transcriptional activity of CREB.
9. O-GlcNAcylation of Sp1 and estrogen receptors increases the steady-state levels of these transcription factors by preventing their degradation.

Thus, it appears that the glycan is important in regulating the complex assembly and protein interactions required to control both the specificity and activity of the transcriptional machinery. At least 15 well-characterized ribosomal proteins and several associated translational factors are O-GlcNAcylated. It has been proposed that the activity of eukaryotic initiation factor 2 (eIF2) is regulated by its binding to p67. The p67 protein is O-GlcNAcylated, and its interactions with eIF2 are controlled by the O-GlcNAc modification. Thus, although little work has been done with respect to O-GlcNAcylation and the regulation of protein translation, it is likely that O-GlcNAc is also involved in these processes.

O-GlcNAc Regulates Protein Trafficking and Turnover

In neurons, O-GlcNAcylation of synapsin proteins appears to prevent release of synaptic vesicles from the cytoskeleton, thus regulating the release of neurotransmitters at the synapse. O-GlcNAcylation also appears to regulate the trafficking of β-catenin and E-cadherin to the cell surface of epithelial cells. When these proteins are O-GlcNAcylated, they remain bound to the cytoskeleton and are not transported to the cell surface. Because E-cadherin is an important cell adhesion molecule, blockage of its transport to the cell surface may be important to epithelial tumor cell metastasis. Likewise, there is growing evidence that increased O-GlcNAcylation of GLUT4 vesicle proteins such as Munc18 and others has a role in the inhibition of glucose transport in diabetes. GLUT4 is the insulin-sensitive glucose transporter that, when stimulated by insulin in muscle and adipose tissues, is rapidly transported to the plasma membrane to increase glucose uptake.

As mentioned above, several studies have shown that O-GlcNAcylation of proteins, either directly or indirectly, slows their proteolytic degradation. The 26S proteasome is a complex structure that degrades many ubiquitin-tagged regulatory proteins within the cell. Recent proteomic analyses of the 26S proteasome have demonstrated that 5 of 19 and 9 of 14 proteins of the catalytic and regulatory cores, respectively, are modified by O-GlcNAc. Increased O-GlcNAcylation of the Rpt2 ATPase, a component of the 19S cap of the proteasome, blocks its ATPase activity, reducing proteasome-catalyzed degradation. It has been suggested that O-GlcNAcylation of the proteasome allows the cell to respond to metabolic needs by controlling the availability of amino acids and altering the half-lives of key regulatory proteins.

O-GlcNAc Is Involved in Neurodegenerative Disease

The *OGT* gene has been mapped to Xq13, a locus linked to several neurological diseases including Parkinson's dystonia. Similarly, as noted above, the gene encoding O-GlcNAcase is located at 10q24.1, a locus linked to late-onset Alzheimer's disease (AD). Brain metabolic processes, including glycolysis, and overall levels of glucose are known to decline with age, particularly in AD neurons. As discussed above, one normal function of O-GlcNAc is to "cap" phosphorylation sites to shut them off. Most of the proteins involved in the pathology of AD are modified by O-GlcNAc. Tau, a microtubule-associated protein that has multiple microtubule-binding repeats, is required for polymerization and stability of microtubules in neurons. There are numerous kinases known to phosphorylate tau including glycogen synthase kinase 3 (GSK-3), CKII, and extracellular regulated protein kinase (ERK). Hyperphosphorylated tau aggregates to form neurofibrillary tangles (PHF-tau) in AD neurons. Normally, in adult brain, tau protein is extensively modified by O-GlcNAc at more than 12 sites. At least one identified O-GlcNAc site in tau is in its microtubule-binding domain. This site (Ser-262) accounts for 70% of PHF-tau-forming activity in vitro. It is postulated that O-GlcNAc modification of tau may affect its microtubule-binding ability, compete with phosphorylation, or even affect its nucleocytoplasmic localization. Recent studies on human brain tau have established that O-GlcNAcylation negatively regulates its O-phosphorylation in a site-specific manner both in vitro and in vivo. Using an animal model of starved mice, low glucose uptake/metabolism mimicked the hypo-O-GlcNAcylation and hyper-O-phosphorylation of tau seen in human AD brain. Quantitative comparisons between normal human brain tissue and that of AD patients indeed showed that global O-GlcNAcylation is reduced in AD brains.

The β-amyloid precursor protein (APP) is O-GlcNAc modified in its cytoplasmic tail; however, the sites of modification are not known. APP is also phosphorylated in its cytoplasmic tail by cyclin-dependent kinase 1, which is known to affect its proteolytic processing. Upon abnormal proteolysis, APP gives rise to the toxic β1-42 peptide fragment, which forms the amyloid plaques present in AD brains. The cytoplasmic tail also contains a sequence controlling endocytosis and a G-protein-binding site, suggesting that O-GlcNAcylation of this region of APP may affect signaling events, trafficking, and metabolism of APP.

Altered O-GlcNAcylation has been reported for other proteins involved in neurodegenerative disease. Neurofilaments appear to be hypo-O-GlcNAcylated in neurons from patients with Lou Gehrig's disease (ALS; amyotrophic lateral sclerosis). Clathrin-assembly proteins AP-3 and AP-180 are both modified by O-GlcNAc, and these modifications decline in AD, suggesting that reduced O-GlcNAc is associated with the loss of synaptic vesicle recycling. Altogether, current data point to potentially significant roles of the O-

GlcNAc modification in normal neuronal function and in the molecular mechanisms underlying the pathology of neurodegenerative disease.

Elevated O-GlcNAc Underlies Diabetes and Glucose Toxicity

Perhaps the best understood function of O-GlcNAc is its role in the regulation of insulin signaling and as a mediator of glucose toxicity. Increased O-GlcNAc in adipocytes or muscle blocks insulin signaling at several points (Figure 18.8), and overexpression of *OGT* in muscle or adipose tissue in transgenic mice causes overt diabetes. Many of the toxic effects of glucose require its conversion to glucosamine, which concomitantly elevates UDP-GlcNAc and increases O-GlcNAcylation. Current models suggest that abnormal increases in O-GlcNAcylation, due to hyperglycemia, hyperlipidemia, and/or hyperinsulinemia, disturb the normal dynamic balance between O-GlcNAcylation and O-phosphorylation that controls signaling, transcription, and other cellular functions, leading to the toxicity associated with diabetes.

The first studies to link glucosamine metabolism directly with the toxicity of glucose in diabetes showed that glucosamine is many times more potent than glucose in inducing insulin resistance in cultured adipocytes. These studies also demonstrated that the ability of glucose to induce insulin resistance could be blocked by deoxynorleucine (DON), a drug that inhibits glucose:fructose amidotransferase (GFAT, the enzyme that converts fructose-6-P to glucosamine-6-P; Figure 18.8), and that this blockage could be bypassed by adding glu-

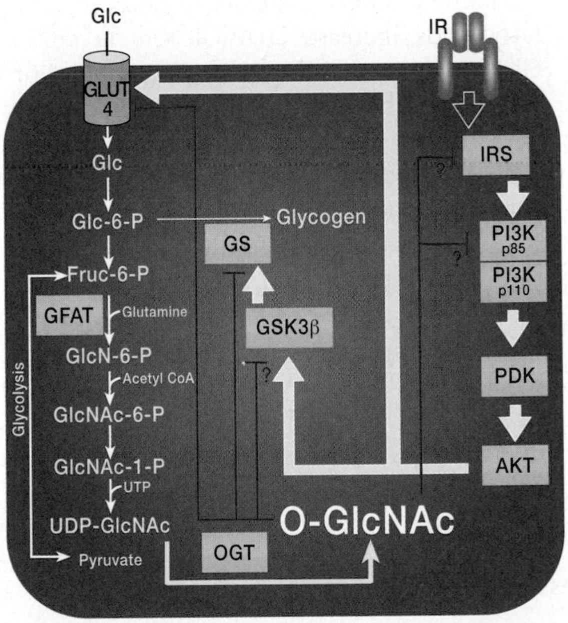

FIGURE 18.8. Elevating O-GlcNAc blocks insulin signaling at many points. Glucose flux via glucose transporters (e.g., GLUT4 in insulin-sensitive cells) through the hexosamine biosynthetic pathway (which accounts for 2–5% of total glucose utilization) leads to the production of UDP-GlcNAc, the donor for O-GlcNAcylation. O-GlcNAc cycles on cellular proteins. Elevation of O-GlcNAc through increased activity of O-GlcNAc transferase or decreased activity of O-GlcNAcase blocks insulin signaling at several early steps and also blocks insulin-stimulated glucose uptake and prevents insulin from activating glycogen synthase. (IR) Insulin receptor; (IRS) insulin receptor substrate; (PI3K) phosphatidylinositol 3-kinase; (PDK) phosphatidylinositol-dependent kinase; (AKT) AKT kinase; (GS) glycogen synthase; (GSK3β) glycogen synthase kinase-3β; (GFAT) glucose:fructose-6-phosphate amidotransferase; (OGT) O-GlcNAc transferase. (Redrawn, with permission, from Slawson C., Housley M.P., and Hart G.W. 2006. *J. Cell. Biochem.* **97:** 71–83.)

cosamine to the culture media (see Chapter 4). Insulin resistance is the hallmark of type II diabetes. Later studies found that increased glucosamine availability induced severe skeletal muscle insulin resistance in normal rats and also showed that glucosamine metabolism has a role in fat-induced insulin resistance in vivo. Elevated fatty acids, like hyperglycemia, induce insulin resistance by increasing production of glucosamine metabolites. O-GlcNAcylation of glycogen synthase prevents its activation by insulin, blocking the storage of glucose. A recent genetic study showed a strong association between a single nucleotide polymorphism in O-GlcNAcase and a population of individuals with type II diabetes. In rat skeletal muscle, hyperinsulinemia, as seen in type II diabetes, dramatically increases the levels of O-GlcNAc–modified proteins. Infusion of glucosamine also increases the levels of O-GlcNAc–modified proteins in muscle, even in the absence of elevated insulin.

Many studies have shown that hyperglycemia and/or hyperinsulinemia globally increases the O-GlcNAcylation of proteins in different cell types. Several studies have shown that elevated glucose increases the O-GlcNAcylation of proteins in pancreatic β-cells (insulin-secreting cells), and as described above, O-GlcNAcylation of transcription factors may regulate insulin gene transcription. A single dose of a chemically reactive analog of GlcNAc (streptozotocin, STZ) given to animals selectively kills the β-cells of their pancreas, resulting in type I diabetes. The highly selective nature of STZ suggests that β-cells are particularly sensitive to altered glucosamine metabolism. Biochemical and histological studies of rat aorta and cultured rat aortic smooth muscle show that hyperglycemia qualitatively and quantitatively alters the expression of many O-GlcNAcylated proteins, suggesting a role of O-GlcNAc in glucose toxicity to vascular tissues. Transgenic mice that overexpress the glucose transporter protein (GLUT1) in muscle have increased UDP-GlcNAc levels, are insulin resistant, and have increased O-GlcNAcylation of GLUT4-vesicle-associated proteins. Free fatty acids increase UDP-GlcNAc in cultured human myotubes and increase the O-GlcNAc-dependent DNA-binding activity of the Sp1 transcription factor. Studies of cultured myotubes grown in high glucose and/or insulin showed enhanced O-GlcNAcylation of numerous proteins. Proteomic analyses of these myotubes show that hyper-O-GlcNAcylated proteins include HSP70, α-tubulin, and Sp1. The roles of O-GlcNAc in diabetic cardiomyopathy have been examined in a rat model. Myocytes exposed to high glucose display altered O-GlcNAcylation of several transcription factors involved in cardiomyocyte function, concomitant with altered calcium signaling. In contrast, when the cardiomyocytes are transfected with adenovirus encoding O-GlcNAcase, cardiomyocyte functions improve dramatically. Thus, it is now clear that both hyperglycemia and insulin regulate the O-GlcNAcylation state of myriad cellular proteins, which in turn modulates signaling and transcription. Although much remains to be learned, it is increasingly clear that dysregulation of O-GlcNAcylation underlies the molecular basis of glucose toxicity and insulin resistance, two major hallmarks of diabetes.

O-GlcNAc and Stress Survival

In every mammalian cell type examined to date, one of the earliest responses to cellular stress is a rapid and global increase in O-GlcNAcylation on a multitude of proteins. Levels of O-GlcNAc increase very rapidly in response to every form of stress (heat shock, ethanol, UV, hypoxia, reductive, oxidative, and osmotic stress). Moreover, the stress-induced increase in O-GlcNAcylation of many proteins is dynamic, returning to normal after 24–48 hours. This increase in O-GlcNAc could be a result of increased glucose flux into the cells that occurs in response to stress, or increased activity of OGT, decreased activity of O-GlcNAcase, or all three combined.

Artificial modulation of the levels of O-GlcNAc results in altered tolerance to lethal levels of cellular stress. Decreasing OGT and O-GlcNAc levels results in cells that are less tolerant, whereas increasing the levels results in cells that are more tolerant. Short-term elevation in O-GlcNAcylation appears to be an important survival mechanism. The overexpression of heat-shock proteins (chaperones) is also known to protect cells from stress. The induction of heat-shock proteins (HSP70 and HSP40) occurs faster, and these proteins turn over more slowly in the presence of increased O-GlcNAc. Furthermore, reduced levels of HSP70 and HSP40 are observed in cell lines in which OGT is reduced. It is clear that O-GlcNAc mediates stress tolerance, in part, by altering the levels of heat-shock proteins. However, O-GlcNAc may also protect cells in other ways, for example, by (1) stabilizing protein structure; (2) preventing protein–protein aggregation, possibly directly or through HSP70's O-GlcNAc-binding activity; (3) modulating signal transduction pathways implicated previously in glucose protection, such as JNK activation; and finally (4) altering the activity or localization of heat-shock proteins, many of which are known to be modified by O-GlcNAc. Regardless of the underlying mechanisms, it is clear that increased O-GlcNAcylation on many proteins is one of the most rapid cellular responses to stress from a wide variety of unrelated sources. Elucidating how increasing O-GlcNAcylation helps a cell to survive stressful conditions should improve our understanding of the roles of this ubiquitous posttranslational modification.

FURTHER READING

Torres C.-R. and Hart G.W. 1984. Topography and polypeptide distribution of terminal *N*-acetylglucosamine residues on the surfaces of intact lymphocytes. Evidence for O-linked GlcNAc. *J. Biol. Chem.* **259:** 3308–3317.

Hart G.W. 1997. Dynamic O-linked glycosylation of nuclear and cytoskeletal proteins. *Annu. Rev. Biochem.* **66:** 315–335.

Comer F.I. and Hart G.W. 2000. O-Glycosylation of nuclear and cytosolic proteins. Dynamic interplay between O-GlcNAc and O-phosphate. *J. Biol. Chem.* **275:** 29179–29182.

Wells L., Vosseller K., and Hart G.W. 2001. Glycosylation of nucleocytoplasmic proteins: Signal transduction and O-GlcNAc. *Science* **291:** 2376–2378.

Vocadlo D.J., Hang H.C., Kim E.-J., Hanover J.A., and Bertozzi C.R. 2003. A chemical approach for identifying O-GlcNAc-modified proteins in cells. *Proc. Natl. Acad. Sci.* **100:** 9116–9121.

Wells L., Vosseller K., and Hart G.W. 2003. A role for *N*-acetylglucosamine as a nutrient sensor and mediator of insulin resistance. *Cell Mol. Life Sci.* **60:** 222–228.

Whelan S.A. and Hart G.W. 2003. Proteomic approaches to analyze the dynamic relationships between nucleocytoplasmic protein glycosylation and phosphorylation. *Circ. Res.* **93:** 1047–1058.

Khidekel N., Ficarro S.B., Peters E.C., and Hsieh-Wilson L.C. 2004. Exploring the O-GlcNAc proteome: Direct identification of O-GlcNAc-modified proteins from the brain. *Proc. Natl. Acad. Sci.* **101:** 13132–13137.

Liu F., Iqbal K., Grundke-Iqbal I., Hart G.W., and Gong C.-X. 2004. O-GlcNAcylation regulates phosphorylation of tau: A mechanism involved in Alzheimer's disease. *Proc. Natl. Acad. Sci.* **101:** 10804–10809.

Zachara N.E. and Hart G.W. 2004. O-GlcNAc a sensor of cellular state: The role of nucleocytoplasmic glycosylation in modulating cellular function in response to nutrition and stress. *Biochim. Biophys. Acta* **1673:** 13–28.

Love D.C. and Hanover J.A. 2005. The hexosamine signaling pathway: Deciphering the "O-GlcNAc code." *Science STKE* **312:** re13.

Slawson C., Housley M.P., and Hart G.W. 2006. O-GlcNAc cycling: How a single sugar post-translational modification is changing the way we think about signaling networks. *J. Cell. Biochem.* **97:** 71–83.

Hart G.W., Housley M.P., and Slawson C. 2007. Cycling of O-linked β-*N*-acetylglucosamine on nucleocytoplasmic proteins. *Nature* **446:** 1017–1022.

CHAPTER 19

Evolution of Glycan Diversity

Ajit Varki, Hudson H. Freeze, and Pascal Gagneux

THIS CHAPTER PROVIDES A BRIEF COMPARATIVE OVERVIEW of the patterns of glycosylation in various taxa of living organisms and discusses the complexity and diversity of these glycans from an evolutionary perspective. Because much of the currently available information concerns vertebrates, this chapter focuses on comparisons between the glycans of vertebrates and those of other taxa. The evolutionary processes that likely determine the generation of glycan diversity are briefly considered, including intrinsic host glycan-binding protein functions and interactions of hosts with extrinsic pathogens or symbionts.

RELATIVELY LITTLE IS KNOWN ABOUT GLYCAN DIVERSITY IN NATURE

The genetic code is essentially the same in all known living organisms, and several core functions such as gene transcription and energy generation tend to be conserved across various taxa. Although glycans are also found in all organisms, considerable diversity of their structure and expression exists in nature, both within and between evolutionary lineages. Partly because of the inherent difficulties in studying their structures, relatively little is known about the details of this glycan diversity, and there are few comprehensive data sets on this subject. For many taxa, essentially no information is available on their glycan profiles. Sufficient data are available, however, to indicate that there is no universal "glycan structure code" akin to the genetic code. Indeed, the glycans expressed by most free-living Eubacteria and Archaea (formerly grouped as prokaryotes) (see Chapter 20) have relatively little in common structurally with those of eukaryotes (an exception occurs when bacterial pathogens mimic eukaryotic host structures). On the other hand, most major glycan classes identified in animal cells seem to be represented in some related form among other eukaryotes, and sometimes in Archaea. Figure 19.1 outlines the major branches of the eukaryotic tree of life, with an emphasis on the protostome-deuterostome split and the phylogeny of "model" organisms, for which whole-

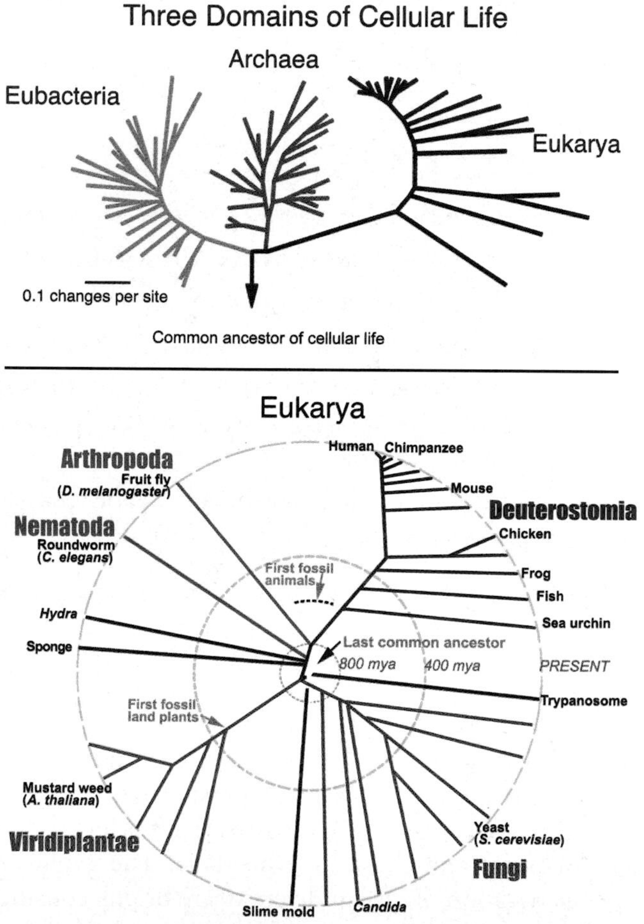

FIGURE 19.1. Phylogenetic trees of the three domains of cellular life (*upper panel*) and of the multicellular Eukarya (*lower panel*). The universal tree of life (*upper panel*) is inferred from maximum likelihood analysis of 1620 homologous nucleotide positions of small-subunit ribosomal RNA sequences from each organism. (The tree is redrawn, with permission, from Barns S.M. et al. 1996. *Proc. Natl. Acad. Sci.* **93:** 9188–9193, ©National Academy of Sciences, U.S.A. The eukaryotic phylogeny is redrawn and modified, with permission, from Pollard T.D. et al. 2007. *Cell Biology*, 2nd Edition. Saunders, New York, ©Elsevier.) Common eukaryotic "model" organisms are indicated. Except for the sponge, all indicated species have had their genomes sequenced. (*Gray dotted rings*) Approximate time before present (mya = millions of years ago). Major groups are indicated by different colors and refer to specific chapters (see text for discussion). The unicellular alveolates (e.g., trypanosomes) and slime mold diverged more than 1 billion years ago. Thus, their branching points are not shown.

genome sequence data have been generated. In contrast, far fewer organisms have been the subject of in-depth glycan structural analyses. The high levels of diversity encountered in the best-studied vertebrate species are a predictor of similar diversity in other groups of organisms. The existing information on the distribution of glycan types points to complicated patterns. On the one hand, glycan patterns can form "trends" and characterize entire phylogenetic lineages where one encounters further biochemical variation with subsets unique to certain sublineages. On the other hand, many glycans show rather discontinuous distribution across the tree of life and distantly related organisms can express surprisingly similar glycans.

EVOLUTIONARY VARIATIONS IN GLYCANS

O-Glycans

Homologs of the UDP-GalNAc:polypeptide *N*-acetylgalactosaminyltransferases (ppGalNAcTs) that initiate synthesis of the most common O-glycan class in vertebrates have been found throughout the animal kingdom (see Chapter 9). Multiple isoforms of ppGalNAcTs exist in most species. The common core-1 Galβ1-3GalNAcα1-O-Ser/Thr structure of vertebrates is present in insects, where it also forms part of a mucin-like protective layer in the gut. In contrast, plants do not appear to have O-linked GalNAc. Instead, they express arabinose O-linked to hydroxyproline and galactose O-linked to serine and threonine (see Chapter 22). Far less is known about bacterial O-glycosylation, although it is clear that novel O-glycans can be found within bacterial "S" layers, for example, a Galβ1-O-Tyr core (see Chapter 20).

Glycosphingolipids

Glucosylceramide is found in both plants and animals (see Chapter 10). However, the commonest core structure of vertebrate glycosphingolipids (Galβ1-4Glc-Cer) is varied in other organisms, for example, Manβ1-4Glc-Cer and GlcNAcβ1-4Glc-Cer in certain invertebrates. Other variations are inositol-1-O-phosphorylceramide, for example, mannosyldiinositolphosphorylceramide, which is the most abundant sphingolipid of yeast, and GlcNAcα1-4GlcAα1-2-*myo*-inositol-1-O-phosphorylceramide, which is found in tobacco leaves. Galactosylceramide and its derivatives seem to be limited to the nervous system of the deuterostome lineage of "higher" animals (see Figure 19.1). In contrast, all protostome nerves contain mainly glucocerebrosides. An evolutionary trend is suggested: A transition from gluco- to galactocerebrosides corresponds with changes in the nervous system from loosely structured to highly structured myelin. With regard to the complex gangliosides of the deuterostome nervous system, some general trends are seen in comparing reptiles to fish to mammals: an increase in sialic acid content, a decrease in the complexity of ganglioside composition, and a decrease in "alkali-labile" molecules (bearing O-acetylated sialic acids). A general rule has also been suggested: the lower the environmental temperature, the more polar the composition of brain gangliosides. Thus, poikilothermic (cold-blooded) animals tend to express many polysialylated gangliosides in the brain.

N-Glycans

Perhaps the broadest base of evolutionary information concerns asparagine–N-linked glycans (see Chapter 8). All plants and animals studied to date seem to share the same early stages of the classic N-glycan processing pathway (see Chapter 8), including the generation and transfer of $Glc_3Man_9GlcNAc_2$ from a dolichol-linked precursor to asparagine residues on newly synthesized proteins. Such an extreme degree of conservation is understandable, given the critical role of this glycan structure in modulating the folding and maturation of newly synthesized glycoproteins in the endoplasmic reticulum (ER) (see Chapter 36). However, some parasitic protists can transfer truncated forms of otherwise similar lipid-linked oligosaccharides, sometimes even just the core $GlcNAc_2$ sequence. The intracellular localization of these structures also makes them less likely to be involved in rapid evolutionary arms races due to exploitation by parasites and pathogens (see below). The trimming and extension steps that occur thereafter along the vertebrate N-glycan processing pathway are recapitulated to varying extents in other eukaryotic taxa (Figure 19.2). Yeasts and vegetative slime molds do not appear to complete the trimming of mannose residues, and are thus unable to generate typical "complex-type" N-glycans. Yeast often further specialize their high-mannose glycans by extending them into large mannans. In contrast, developing slime molds trim down the high-mannose forms to some extent but then do not extend them. In insects, mannose trimming appears to be generally completed, as in mammals, down to a $Man_3GlcNAc_2$ structure. The subsequent addition of GlcNAc residues is frequently followed by removal of these residues by a very active β-hexosaminidase (see Chapter 24). Thus, the final structure found in insects often has only the three core mannose residues (Figure 19.2). Prior to the removal of the GlcNAc residues in insect cells, an α1-3-linked fucose unit is often added to the core GlcNAc residue (frequently in addition to the α1-6-linked fucose typically found on the core GlcNAc of vertebrate N-glycans). Plants follow a pathway similar to that in vertebrates in the initial stages, but then often add a bisecting β1-2-linked xylose residue on the β-linked mannose residue (see Chapter 22 and Figure 19.2). The latter structure is also present in some invertebrates, but it appears to be immunogenic in vertebrates. In keeping with the above findings, the early-processing α-mannosidases of the N-glycan pathway have a wide evolutionary distribution. In contrast, the endo-α-D-mannosidase processing enzyme that provides an "alternate deglucosylating pathway" for N-glycans (see Chapter 8) appears to be limited to members of the

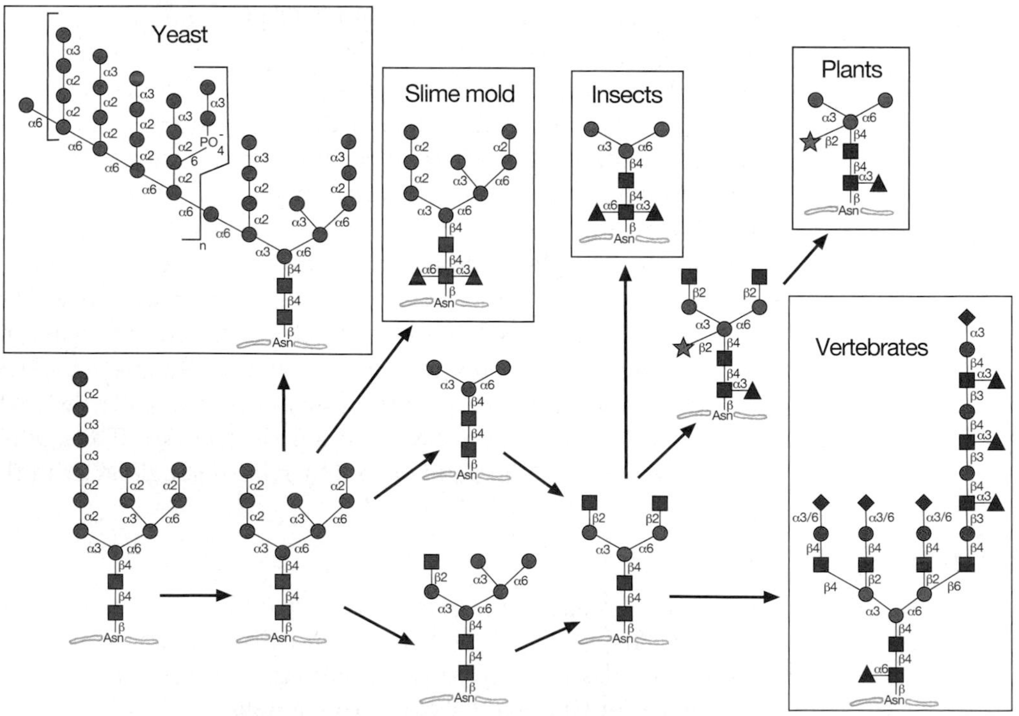

FIGURE 19.2. Dominant pathways of N-glycan processing among different taxa. See text for discussion.

chordate phylum, with the exception of the *Mollusca*, where it was detected in three distinct classes. The absence of this enzyme in other invertebrates examined, as well as in yeasts, various protozoa, and higher plants, suggests that the need for an alternate deglucosylation route paralleled the development of complex N-glycans in higher animals.

Overall, it is clear that the N-glycan pathway is evolutionarily ancient and found throughout the eukaryotes. However, limited information exists regarding the presence and distribution of most N-glycan pathway enzymes and genes in most animals and plants. Thus, there is still insufficient information to paint a clear picture of exactly how it has been diversified and specialized during evolution. It was once thought that only eukaryotes express N-glycans. However, it is now clear that Archaea express GalNAc-Asn and Glc-Asn linkages, and recent data indicate that a few bacteria such as *Campylobacter jejuni* express a very similar N-glycan pathway, but with a novel linkage unit, involving bacillosamine (2,4-diacetamido-2,4,6-trideoxy-D-glucose or $BacAc_2$) linked to asparagine.

Shared Outer Chains of Glycans

Outer terminal sequences are often shared among N- and O-glycans and glycosphingolipids (Chapter 13). The common outer-chain Galβ1-4GlcNAcβ1- (*N*-acetyllactosamine or "LacNAc") structure of vertebrates (see Chapter 13) is also found in plants. Some plants even add outer-chain Fucα1-3 residues to the GlcNAc residues of LacNAc units, generating $Lewis^x$-like structures identical to those found in animal cells. In some taxa, such as mollusks, an outer GalNAcβ1-4GlcNAcβ1- structure (the so-called LacDiNAc or LDN unit) tends to dominate, in place of the typical LacNAc structure more commonly seen in vertebrates. The SO_4-4-GalNAcβ1-4GlcNAcβ1- terminal units of pituitary glycoprotein hormones (see Chapter 13) have been conserved throughout vertebrate evolution, suggesting that they are critical for biological activity. Controversy exists as to whether or not further extensions and terminations of N-glycan antennae typical of vertebrates occur in insect or plant cells (see discussion regarding sialic acids below). The genetic elimination of the common vertebrate terminal sequence Galα1-3Galβ1-4GlcNAcβ1- in Old World primates and variations in sialic acids are discussed below.

Sialic Acids

Sialic acids are prominently expressed at the outer termini of N-glycans, O-glycans, and glycosphingolipids of the deuterostome lineage of animals (see Figure 19.1 and Chapter 14). It was once thought that sialic acids were an evolutionary innovation unique to this lineage that originated during the Cambrian Expansion, and that all other reports of sialic acids in a few scattered taxa reflected lateral gene transfer and/or convergent evolution (i.e., independent evolution of sialic acid synthesis in these taxa). However, although such lateral transfer mechanisms exist and may explain the presence of sialic acids in some bacterial taxa, sialic acids are also reported in some fungi and mollusks. Together with evidence for a limited set of genes for sialic acid production and addition in some protostomes (e.g., in insects such as *Drosophila*), the situation is indicative of an earlier evolutionary origin for sialic acids. *Caenorhabditis elegans*, the free-living nematode, does not contain genes for synthesizing or metabolizing sialic acid. In addition, prior claims of the presence of sialic acids in plants are probably due to environmental contamination and/or incorrect identification of the chemically related sugar Kdo (3-deoxy-octulosonic acid). However, recent studies have found that the sialic acid biosynthetic genes of some insect and bacterial species share homology with those of vertebrates. Overall, it appears likely that sialic acids were an ancient invention derived from genes of the related pathway for Kdo synthesis. In this scenario, sialic acids were differentially exploited during evolution, becoming prominent only in the deuterostome lineage, while being abandoned or substantially reduced in complexity and/or biological importance in other animal and fungal taxa. Meanwhile, a variety of bacteria synthesize other sialic-acid-like molecules using a very similar biosynthetic pathway (see Chapter 14).

It is curious that the most sialic acid diversity tends to be found in invertebrate deuterostomes, such as echinoderms (sea urchins and starfish), and the simplest profiles are found in humans. Likewise, whereas complex substituted polysialic acids are found in echinoderms and fish eggs, simpler polysialic acids are found in humans. Thus, sialic acids are not subject to incremental sophistication along recently evolved lineages. Rather, they seem to have evolved in many possible directions, disappearing altogether, or complicating or simplifying their structures. Although there is a tendency for some types of sialic acids to be dominant in certain mammalian species (e.g., *N*-glycolylneuraminic acid [Neu5Gc] in pigs and 4-O-acetylated sialic acids in horses), careful investigation reveals the presence of lower quantities of such sialic acids in most other species. Polysialosyl groups and sialic acid O-acetylation in gangliosides seem to be particularly enriched in poikilothermic (cold-blooded) animals. An interesting finding is that humans are "knockout" primates for the enzyme CMP–Neu5Ac hydroxylase (CMAH), because they contain a mutated and inactive *CMAH* gene. Thus, unlike the closely related great apes, humans are deficient in expression of Neu5Gc acid (see Chapter 14). As chickens also make an immune reaction against Neu5Gc, it remains to be determined whether birds represent another lineage that lost this sialic acid.

Glycosaminoglycans

Structures thought to be typical of "higher" animal heparan and chondroitin sulfate chains have been found in many invertebrates, including insects (see Chapter 24) and mollusks. The most widely distributed and evolutionarily ancient class appears to be chondroitin chains, which are not always sulfated (e.g., in *C. elegans*) (see Chapter 23). The more highly sulfated and epimerized forms of heparin and dermatan sulfate tend to be found primarily in "higher" animal species of the deuterostome lineage. The same is true of hyaluronan. Echinoderms such as the sea cucumber make typical chondroitin chains, but some glucuronic acids have branches containing fucose sulfate. Simpler multicellular animals such as sponges can have novel glycosaminoglycans that include uronic acids, but they do not have the typical repeat units of chondroitin sulfate and heparan sulfate. Plants do not have typical animal glycosaminoglycans. Instead, they have acidic pectin polysaccharides, characterized by the presence of galacturonic acid and its methyl ester derivative (see Chapter 22). Bacteria have completely distinct polysaccharides (Chapter 20), although certain pathogenic strains can mimic mammalian glycosaminoglycan chains (see below).

Glycosylphosphatidylinositol Anchors

Glycosylphosphatidylinositol (GPI)-anchored proteins and lipids (see Chapter 11) that share the "core" motif Manα1-4GlcNα1-6-*myo*-inositol-1-P lipid are distributed ubiquitously in eukaryotes. In some species (e.g., yeasts and slime molds), the lipid tail can be a ceramide instead of a phosphatidylinositol (see Chapter 21). GPI-anchored lipids and proteins can constitute the major components of the highly variable outer membranes of some parasitic protozoans, such as *Leishmania* and *Trypanosoma* (see Chapter 40). GPI anchors are generally thought to be absent in prokaryotes. However, at least one archaeal organism has been reported to have a GPI-anchored protein.

Nuclear and Cytoplasmic Glycans

The O-β-GlcNAc modification commonly found on cytoplasmic and nuclear proteins (see Chapter 18) is widely expressed in "higher" animals and in plants. Conserved homologs of the O-GlcNAc transferase that is responsible for synthesizing this structure have been found in many eukaryotic taxa. No clear homolog is evident in the yeast genome. Although the structure has been claimed in *Dictyostelium*, this is actually a distinct α-linked O-GlcNAc. There is currently no evidence that bacteria or Archaea can generate this modification.

VIRUSES ACQUIRE GLYCOSYLATION FROM THEIR HOSTS

Viruses often carry minimalist genomes that typically do not direct glycosylation of their own glycoproteins but instead utilize host-cell machinery. Thus, the glycosylation of viruses reflects that of host cells from which they emerge. However, there are some exceptions to this rule, with some viruses and especially bacteriophages containing genes encoding unusual glycosyltransferases. For example, the chlorella virus generates a glycoprotein termed PBCV-1 that is modified by a "Fringe-type" glycosyltransferase encoded in the viral genome. Some of the phage viral glycosyltransferases can modify their surface antigens to change the serotype of their host bacteria or glycosylate their own DNA to block it from degradation by restriction enzymes. Baculoviruses also encode their own glycosyltransferases to glycosylate host ecdysteroids, allowing them to block molting of the insect host.

Utilization of host glycosylation machinery is particularly prominent in the case of enveloped viruses. Most viral envelope glycoproteins are glycosylated (mostly with N-glycans) during the passage of these proteins through the host Golgi apparatus. This glycosylation is typically quite extensive and appears to protect the virus from host immune reactions directed against the underlying viral polypeptide. In this regard, it has been suggested that the relatively common occurrence of the heterozygous state for congenital disorders of glycosylation in humans (see Chapter 42) may reflect selection for heterozygous individuals whose genomes interfere with viral replication by preventing complete glycosylation of proteins of invading viruses. In other instances, host lectins may be "hijacked" by the glycans on viral surface glycoproteins, aiding attachment and/or entry into target host cells.

VAST DIVERSITY IN BACTERIAL AND ARCHAEAL GLYCOSYLATION

Despite the enormous potential for structural diversity built into monosaccharides, a rather limited subset of all possible monosaccharides and their possible linkages and modifications are found in eukaryotic cells. Why one encounters only such a limited subset of the possible glycan structures is one of the puzzling questions of glycobiology. On the other hand, this limited subset has allowed extensive elucidation of the structure of eukaryotic glycans. In contrast, Bacteriae and Archaea have had several billion additional years to experiment with glycan variation. These organisms also have short generation times and are capable of exchanging genetic material across vast phylogenetic distances, via plasmid-mediated horizontal gene flow. They also inhabit a much wider range of ecological niches with innumerable physicochemical and biological conditions, ranging from the deep litho- and hydrosphere to the stratosphere. Thus, it should not come as a surprise that bacteria and Archaea express a much greater diversity in glycosylation, both in terms of range of monosaccharides that they utilize or synthesize and with regard to their types of linkages and modifications. Some discussion of such structures is presented in Chapter 20. However, much of the work to date has focused on the glycans of pathogens, and it is safe to say that we have barely scratched the surface of this diversity.

TABLE 19.1. *Examples of molecular mimicry of animal glycans by pathogenic bacteria*

Organism	Animal glycans synthesized
E. coli K1, *Meningococcus* group B	polysialic acid
E. coli K5	heparosan (heparan sulfate backbone)
Group A *Streptococcus*	hyaluronan
Group B *Streptococcus*	sialylated *N*-acetyllactosamines
Campylobacter jejuni	sialylated ganglioside-like glycans

MOLECULAR MIMICRY OF HOST GLYCANS BY PATHOGENS

It is evident that great differences exist between the pathways generating the glycan structures of bacteria and those of vertebrates. Despite this, occasional microbial surface structures are found to be strikingly similar to those of mammalian cells. Interestingly, most examples of this type of "molecular mimicry" occur in pathogenic microorganisms, presumably adapting them for better survival in the host by avoiding, reducing, or manipulating host immunity. A few examples are listed in Table 19.1. The initial hope of scientists trying to clone vertebrate glycosyltransferases was that most of the responsible microbial genes arose from lateral gene transfer and that these would provide a backdoor approach to isolating the corresponding ones from eukaryotes. However, in most instances where full genetic information has become available, the evidence points toward convergent evolution rather than gene transfer as the dominant mechanism. For example, the genes involved in synthesizing sialic acids in bacteria seem to have been mainly derived from the preexisting bacterial pathways for the biosynthesis and transfer of Kdo, a bacterial sugar with a structural resemblance to sialic acids. Meanwhile, bacterial sialyltransferases bear little resemblance to those of eukaryotes, and the vast sequence differences between different bacterial sialyltransferases indicate that these have even been reinvented on several separate occasions. On the other hand, lateral gene transfer appears to have been quite common among the bacteria and Archaea themselves, facilitating rapid phylogenetic dissemination of such enzymatic "inventions."

INTERSPECIES AND INTRASPECIES DIFFERENCES IN GLYCOSYLATION

Why do closely related species differ with regard to the presence or absence of certain glycans? Does the same glycoprotein have the same type of glycosylation in different but related species? Relatively little data are available concerning these issues, but examples of both extreme conservation and extreme diversification can be found. A reasonable explanation is that conservation of glycan structure is only required when there are very specific functions for the glycans in question. In other instances, considerable drift in the details of glycan structure might be tolerated, as long as the underlying protein is able to carry out its primary functions (changes with no consequences for survival or reproduction, i.e., selectively neutral).

Even in the absence of important functions within an organism, glycans can have important roles in the mediation of interactions with symbionts and pathogens. The evolution of diversity and microheterogeneity (across tissues and cell types) in glycosylation could well be of value to the organisms in evading pathogens that use glycans as signposts for attachment and entry. Glycans can also have important roles in attracting the important symbiont microbial communities needed for gastrointestinal functions and in accommodating or restricting these to particular areas of the host.

It is also clear that there can be significant variation in glycosylation among members of the same species, particularly with regard to terminal glycan sequences. The classic example is that of the ABH(O) blood group system (see Chapter 13), a glycan-defined polymorphism found in all human populations, which has also persisted for tens of millions of years of primate evolution and has even been independently rederived in some instances. Somewhat surprisingly, despite its great clinical importance for blood transfusion, this polymorphism appears to cause no major differences to the intrinsic biology of individuals of the species (see Chapter 13). Like other blood groups, the ABO polymorphism is accompanied by the production of antibodies against the other variants. It has been suggested that these antibodies are protective, by causing complement-mediated lysis of enveloped viruses generated within other individuals who can express the target structure for the antibody. Thus, an enveloped virus generated in a B blood group individual might bear this structure and be susceptible to lysis upon contact with an A or O blood group individual, who would express anti-B antibodies. Recent experimental evidence is supportive of such a mechanism. However, this mechanism alone should strongly favor O individuals, as these form antibodies against the A and B variants and should lead to higher frequencies of O type than are observed.

Another possible explanation for interspecies diversity is the selection exerted by pathogens that recognize glycans as targets for attachment and entry into cells. This mechanism is likely operative in generating the diversity of sialic acid types and linkages (see above). However, it should result in selection of ABO subtypes and result in approximately even frequencies of each phenotype, not what is observed. Recent analyses have tried to combine the two mechanisms: the antibody-mediated protection from intracellular viruses and possible frequency-dependent protection from glycan-exploiting extracellular pathogens, such as Noroviruses and *Plasmodium falciparum* malaria. Modeling approaches have successfully generated observed frequencies of ABO by incorporating these two simultaneous selection pressures. It is fair to say that the evolutionary persistence of the ABO system needs further explanation.

Another unexplained phenomenon is the genetic inactivation in Old World primates of the ability to synthesize the otherwise very common terminal Galα1-3Galβ1-4GlcNAc-R structure. This glycan variation system is also associated with spontaneously appearing and persistently circulating antibodies against the missing glycan determinant, thus forming a kind of interspecies "blood group." It has been proposed that this glycan difference is protective for the primate lineage which lost "α Gal" and has a high-titer circulating antibody, as it is now better protected against infection by viruses emanating from other mammals.

Regardless of the precise underlying purposes of these types of polymorphic systems, such intra- and interspecies diversity might also provide for "herd immunity," a phenomenon whereby one glycan-variant-resistant individual can effectively protect other susceptible individuals by limiting the spread of a pathogen through the population. It is also important to emphasize that these proposed protective functions of glycan diversity are only apparent at the level of populations and not the individual. This complicates their study in model organisms, where the focus is classically on the individual.

Future studies will have to test precisely how much of interspecies and intraspecies glycan variation is directly driven by such host–pathogen interactions. Despite a lack of comprehensive studies of such phenomena, it is becoming clear that glycan variation forms an important determinant of host susceptibility and must be considered when trying to understand disease, especially epidemics or zoonotics involving different host species and their interactions, for example, influenza A (see Chapter 14).

"MODEL" ORGANISMS FOR STUDYING GLYCAN DIVERSITY

Details of glycan expression patterns in various popular "model" organisms can be found in other chapters in this volume. In recent years, there has been increasing definition of the structures of bacterial inner cell wall peptidoglycans and the outer membranes that are composed of lipooligosaccharides and lipopolysaccharides, particularly those of *Escherichia coli* (see Chapter 20). Chapter 21 provides some details about various genera and species of fungi and protists, including *Saccharomyces cerevisiae* and *Dictyostelium discoideum.* Pathogenic protists such as trypanosomes and leishmanial parasites express very high densities of surface GPI anchors and are discussed in Chapters 11 and 40. Chapter 23 presents an overview of the roundworm *C. elegans*, its development, glycan structures, and expression of glycosyltransferases. The functional insights derived from this organism cover all classes of glycans. For some details about glycobiology of *Drosophila*, see Chapter 24. Like *C. elegans*, this workhorse of genetics has made important functional contributions to the field during the last 5 years. It has been especially important in understanding how O-fucosylation modifies Notch signaling (Chapter 12) and how heparan sulfate proteoglycans determine morphogen and growth factor gradients. Chapter 22 discusses the glycans of plants, including those of the model organism, *Arabidopsis.* Various aspects of sea urchin glycobiology, including the acrosome reaction, egg/sperm interactions, and the role of proteoglycans and lectins, are covered in Chapter 25, which also discusses aspects of *Xenopus* glycobiology, including the synthesis of chitin oligosaccharides, the role of proteoglycans in determining left–right asymmetry, and lectins that help in fertilization and the innate immune system. The same chapter also discusses aspects of zebrafish glycobiology, including glycoproteins, proteoglycans, and lectins.

The recent discovery that rodents are the closest evolutionary cousins to primates has provided added justification for the use of rats and mice as model organisms to understand the mechanisms of human disease (see Chapter 25). Last but not least, Nobel Laureate Sydney Brenner has suggested that we now have enough information about humans to consider ourselves to be the "next model organism." Indeed, there is an increasing tendency to focus tractable questions about glycans and their biology directly on humans and on naturally occurring human mutants. Most recently, there has also been interest in studying the great apes (our closest evolutionary cousins) and an independent realization of recent hominid evolution, particularly with regard to several differences in sialic acid biology (see Chapter 14).

WHY DO WIDELY EXPRESSED GLYCOSYLTRANSFERASES SOMETIMES HAVE LIMITED INTRINSIC FUNCTIONS?

Prior to the generation of glycosyltransferase-deficient mice, it was popular to suggest that every single glycan on every single cell type must have a critical intrinsic host function. Analysis of available gene disruption data indicates that this is not the case. For example, the ST6Gal-I α2-6 sialyltransferase is the main enzyme that produces Siaα2-6Galβ1-4GlcNAcβ1- termini on vertebrate glycans. Although this sequence serves as a specific ligand for the B-cell regulatory molecule CD22 (Siglec-2; see Chapter 32), it is also found on many other cell types, as well as on many soluble secreted glycoproteins. Furthermore, the ST6Gal-I mRNA varies markedly among cell types, and its transcription is regulated by several cell-type-specific promoters, which are in turn modulated by hormones and cytokines. Despite all these data suggesting very diverse and complex roles for this enzyme and its products, the prominent functional consequences of eliminating its expression in mice so far seem to be restricted to the B cell, with decreased signaling and proliferative responses and impaired antibody production (see Chapter 32). Few other obvious abnormalities have yet been found in organ structure and physiology, morphology, or behavior. If the specific intrinsic functions of the ST6Gal-I glycan product are in fact restricted to B cells, why does the organism express it in so many other locations? Even more puzzling, why up-regulate its expression so markedly in the liver and endothelium during a so-called "acute phase" inflammatory response? Could it be that scattered expression of this structure in other locations represents a "smoke-screen" effect, restricting intraorganismal spread of an invading pathogen? Could it be that heavily glycosylated nonnucleated cells like erythrocytes act as a "sink" to divert viral pathogens that need nucleated cells for replication? The answers to these questions must take into account the evolutionary selection pressures (both intrinsic and extrinsic recognition phenomena such as host–pathogen interactions and innate immune contributions) on glycosyltransferase products. Many of these effects may also not be apparent in inbred genetically modified mice living in hygienic vivaria but may rather require population studies of animals in a natural, pathogen-rich environment. It is also possible that other gene products are masking the phenotypes in these model systems, by compensating for the genetic loss. Furthermore, it is likely we have not looked hard enough at such genetically modified mice nor applied the right environmental pressures to elicit phenotypes.

EVOLUTIONARY FORCES DRIVING GLYCAN DIVERSIFICATION IN NATURE

There is too little information available today to allow a comprehensive exposition of the evolution of even the major classes of glycans. On the basis of the available data, it is reasonable to suggest that glycan diversification in complex multicellular organisms has been driven by evolutionary selection pressures of both intrinsic and extrinsic origin relative to the organism under study (see Chapter 6). It is reasonable to postulate that glycans are particularly susceptible to the "Red Queen" effect, in which host glycans must keep on changing in order to stay ahead of the pathogens, which have extremely rapid evolutionary rates because of short generation times, high mutation rates, and much horizontal gene transfer. Given the rapid evolution of extrinsic pathogens and their frequent use of glycans as targets for host recognition, it seems likely that a significant portion of the overall diversity in vertebrate cell-surface glycan structure reflects such pathogen-mediated selection processes. Meanwhile, even one critical intrinsic role of a glycan would disallow its elimination as a mechanism to evade pathogens. Thus, the glycan expression patterns of a given organism may represent a compromise between evading pathogens and preserving intrinsic functions.

More gene disruption studies in intact animals would be helpful to differentiate between these intrinsic and extrinsic glycan functions. More systematic comparative glycobiology could also contribute, by making predictions about intrinsic glycan function; that is, the consistent (conserved) expression of the same structure in the same cell type across several taxa would imply a critical intrinsic role. Such work might also help define the rate of glycan diversification during evolution, better define the relative roles of the intrinsic and extrinsic selective forces, and eventually lead to a better understanding of the functional significance of glycan diversification during evolution. The possibility that glycan diversification might even drive the process of speciation (via reproductive isolation) also needs to be considered.

FURTHER READING

Warren L. 1963. The distribution of sialic acids in nature. *Comp. Biochem. Physiol.* **10:** 153–171.

Kishimoto Y. 1986. Phylogenetic development of myelin glycosphingolipids. *Chem. Phys. Lipids* **42:** 117–128.

Galili U. 1993. Evolution and pathophysiology of the human natural anti-α-galactosyl IgG (anti-Gal) antibody. *Springer Semin. Immunopathol.* **15:** 155–171.

Kappel T., Hilbig R., and Rahmann H. 1993. Variability in brain ganglioside content and composition of endothermic mammals, heterothermic hibernators and ectothermic fishes. *Neurochem. Int.* **22:** 555–566.

Martinko J.M., Vincek V., Klein D., and Klein J. 1993. Primate ABO glycosyltransferases: Evidence for trans-species evolution. *Immunogenetics* **37:** 274–278.

Dairaku K. and Spiro R.G. 1997. Phylogenetic survey of endomannosidase indicates late evolutionary appearance of this N-linked oligosaccharide processing enzyme. *Glycobiology* **7:** 579–586.

Drickamer K. and Taylor M.E. 1998. Evolving views of protein glycosylation. *Trends Biochem. Sci.* **23:** 321–324.

Gagneux P. and Varki A. 1999. Evolutionary considerations in relating oligosaccharide diversity to biological function. *Glycobiology* **9:** 747–755.

Freeze H.H. 2001. The pathology of N-glycosylation—Stay the middle, avoid the risks. *Glycobiology* **11:** 37G–38G.

Angata T. and Varki A. 2002. Chemical diversity in the sialic acids and related α-keto acids: An evolutionary perspective. *Chem. Rev.* **102:** 439–470.

Varki A. 2006. Nothing in glycobiology makes sense, except in the light of evolution. *Cell* **126:** 841–845.

Bishop J.R. and Gagneux P. 2007. Evolution of carbohydrate antigens—Microbial forces shaping host glycomes? *Glycobiology* **17:** 23R–34R.

CHAPTER 20

Eubacteria and Archaea

Jeffrey D. Esko, Tamara L. Doering, and Christian R.H. Raetz

THIS CHAPTER DESCRIBES THE STRUCTURE AND ASSEMBLY of the glycans present in Eubacteria (bacteria) and Archaea. Bacterial glycans include peptidoglycan, periplasmic glucans, lipopolysaccharide, and extracellular polysaccharides that make up capsules and biofilms. In Archaea, the cell wall has a surface layer (S layer), which is composed of glycoproteins and pseudomurein. This chapter also includes a description of the recently discovered glycoproteins found in both kingdoms.

BACKGROUND

Eubacteria and Archaea (often grouped together as prokaryotes) produce a variety of glycoconjugates and polysaccharides of enormous structural diversity and complexity. These glycans include many unusual sugars not found in vertebrates, such as Kdo (3-deoxy-D-*manno*-octulosonic acid), heptoses, and variously modified hexoses, which have important roles in the biology, and sometimes the pathogenicity, of bacterial cells.

Eubacteria were historically divided into two major groups based on whether or not they retain crystal violet dye during a Gram staining procedure (Gram-positive and Gram-negative organisms). Later studies showed that this difference in staining depends on the nature of the cell wall. In Gram-negative bacteria, such as *Escherichia coli*, the cell wall consists of inner and outer membranes separated by a space termed the periplasm (Figure 20.1). Peptidoglycan constitutes the major structural component of the periplasm, and it consists of a polysaccharide covalently cross-linked by short peptides. The periplasmic space of Gram-negative bacteria may also contain β-glucans. The outer membrane is an asymmetric lipid bilayer, because it contains mostly lipopolysaccharide (LPS) in the outer leaflet. Mucoid (slimy) strains also contain a polysaccharide capsule that surrounds the whole cell, which may have a role in virulence. Gram-positive bacteria lack the outer membrane (Figure 20.2), but they have a much thicker peptidoglycan layer, which is modified with additional specialized polymers known as teichoic acids.

The polysaccharide components of the cell envelope, which surrounds the cell and encloses the cytoplasm, have important structural and functional roles in the life of bacteria. Capsular polysaccharides and LPS provide defense against bacteriophages and contain

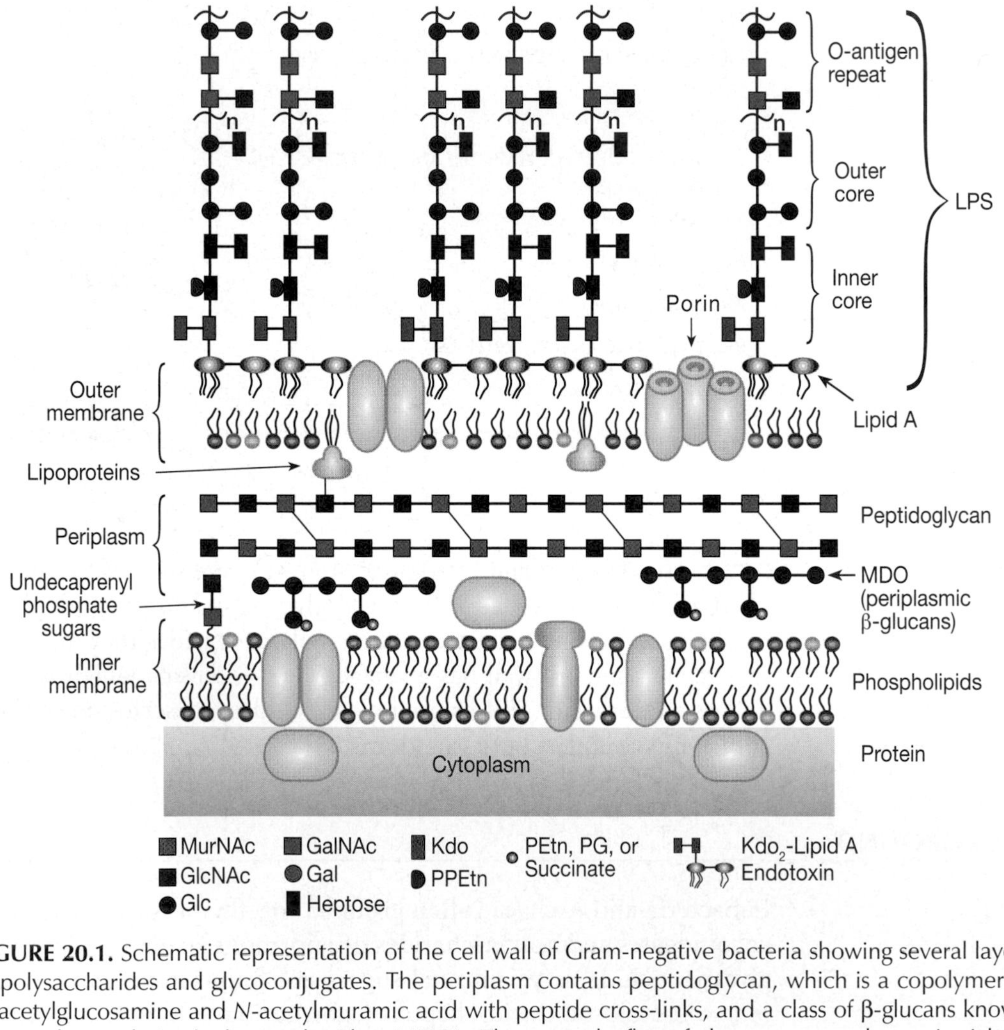

FIGURE 20.1. Schematic representation of the cell wall of Gram-negative bacteria showing several layers of polysaccharides and glycoconjugates. The periplasm contains peptidoglycan, which is a copolymer of *N*-acetylglucosamine and *N*-acetylmuramic acid with peptide cross-links, and a class of β-glucans known as membrane-derived oligosaccharides (MDO). The outer leaflet of the outer membrane is rich in lipopolysaccharide (LPS). In mucoid strains, a capsular polysaccharide covers the entire cell (not shown). (*Red* lipids) Phosphatidylethanolamine; (*yellow* lipids) phosphatidylglycerol; (Kdo) 3-deoxy-D-*manno*-octulosonic acid; (heptose) L-glycero-D-manno-heptose; (n) variable number of O-antigen repeats; (PPEtn) pyrophosphoethanolamine.

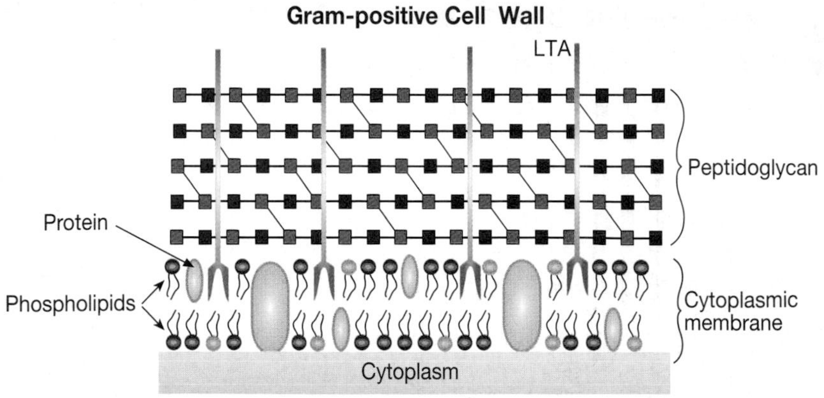

FIGURE 20.2. Cell wall from Gram-positive bacteria. Gram-positive bacteria lack the outer membrane and associated lipopolysaccharide (LPS) that is present in Gram-negative organisms (see Fig. 20.1). In Gram-positive bacteria, the peptidoglycan layer is thicker and contains teichoic acids.

the major antigenic determinants that distinguish various serotypes of bacteria. In the context of mammalian infection, these structures represent the first line of defense against complement, and they can have profoundly detrimental effects on the host. For example, LPS contains lipid A, also known as endotoxin, a potent stimulator of innate immunity that contributes to secondary complications of infections including septic shock, multiple organ failure, and mortality. Thus, beyond fundamental interest in understanding these unique structures and how they are built, considerable interest exists in developing agents to block the deleterious effects that these compounds exert during pathogenesis. Inhibitors of peptidoglycan biosynthesis include many aminoglycoside antibiotics, such as penicillins, cephalosporins, and vancomycin.

Archaea have rigid cell walls with diverse structures. They contain many unusual lipids with repeating isoprenyl groups linked to glycerol and an S layer of glycoproteins in a lattice-like arrangement attached to the membrane. They lack the peptidoglycan found in almost all bacteria and instead contain a pseudomurein layer, which is similar to the peptidoglycan structure. In general, Archaea structures have been studied in less detail than the corresponding bacterial structures.

Bacteria and Archaea produce numerous glycan-binding proteins. These proteins include adhesins that facilitate bacterial colonization, exotoxins that bind to host membrane glycans, and single-sugar-binding proteins involved in metabolism. Bacterial glycan-binding proteins are discussed in detail in Chapter 34. Contrary to prior misconceptions, Eubacteria and Archaea also produce glycoproteins that contain a number of different linkages. As discussed below, some of the pathways appear to be highly conserved between Eubacteria, Archaea, and Eukaryota. However, many differences also exist, which has allowed extensive use of bacteria as expression system for making glycan-free recombinant glycoproteins (see Chapter 51).

PEPTIDOGLYCAN PROVIDES A RIGID BUT DYNAMIC FRAMEWORK

Function, Structure, and Arrangement

Peptidoglycan (also known as murein) makes up about 10% of the dry weight of the cell wall in Gram-negative bacteria and as much as 20–25% of the dry weight in Gram-positive bacteria. It consists of parallel strands of polysaccharide composed of *N*-acetylglucosamine and *N*-acetylmuramic acid (MurNAc) in β1-4-linkage, which are thought to surround the bacterium (Figure 20.1). The average chain length of peptidoglycan in *E. coli* is 25–35 disaccha-

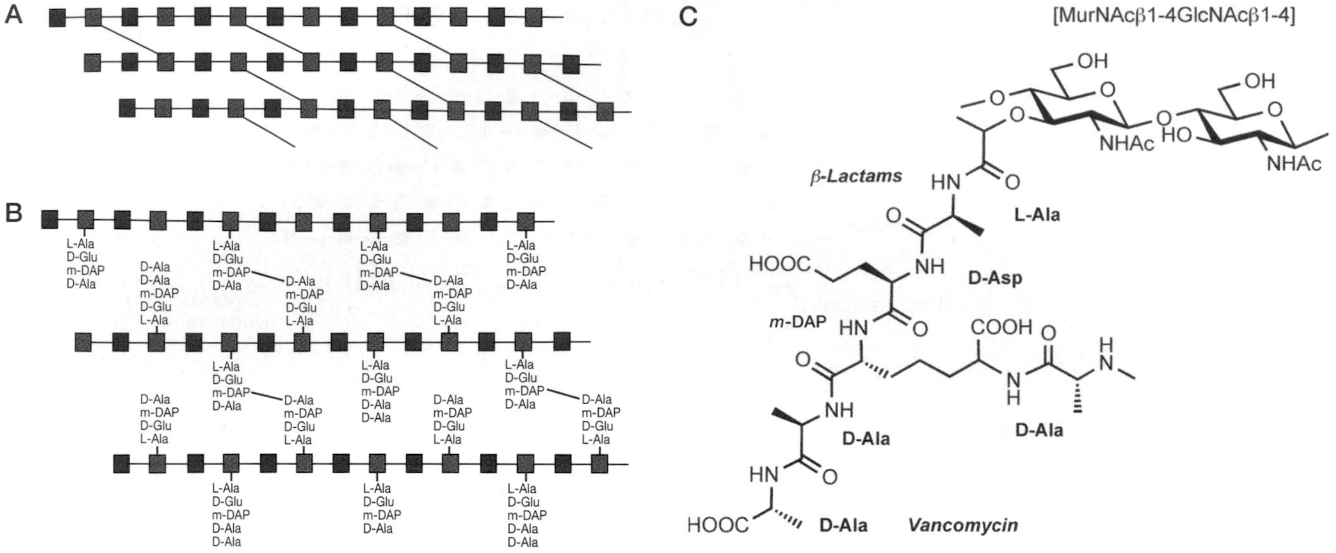

FIGURE 20.3. Peptidoglycan. (*A*) Schematic representation of peptidoglycan strands composed of alternating residues of *N*-acetylmuramic acid (MurNAc; *orange*) and *N*-acetylglucosamine (GlcNAc; *blue*) showing periodic cross-links. (*B*) Pentapeptides consisting of L-Ala, D-Glu, *meso*-diaminopimelic acid (*m*-DAP), and D-Ala form cross-links between the peptidoglycan strands. (*C*) Structure of the muropeptide subunit showing a cross-link. β-Lactam antibiotics block the assembly of the pentapeptide. Vancomycin blocks the cross-linking reaction, which requires removal of the terminal D-Ala residue.

ride units. Short peptides attached to the MurNAc units extend in all directions from the chain and form cross-links at frequent intervals with adjacent strands (Figure 20.3A). Although the amino acid composition varies in different bacteria, the short peptides are typically composed of L-Ala, unusual D-amino acids (D-Glu and D-Ala), and the dibasic amino acid, *meso*-diaminopimelic acid (*m*-DAP), which form the cross-links (Figure 20.3B). The structure of one disaccharide-peptide subunit is shown in Figure 20.3C.

The structures of the peptidoglycan are very similar in Gram-positive and Gram-negative bacteria, but the thickness varies from 10–20 layers in Gram-positive bacteria to only 1–3 layers in Gram-negative bacteria (Figure 20.2). In Gram-negative bacteria, the peptidoglycan is covalently attached to lipoproteins in the outer membrane. In Gram-positive bacteria, which lack the outer membrane, the peptidoglycan is cross-linked to teichoic acids and teichuronic acids. The cross-linked structure of peptidoglycan confers mechanical strength and shape on the cell and provides a barrier to withstand internal osmotic pressure. Nevertheless, it has sufficient plasticity to allow cell growth and division. This plasticity results from the highly dynamic nature of the polymer, which undergoes about 50% turnover every generation catalyzed by autolysins (glycosidases, peptidases, and amidases). This fast turnover (a generation is about 30 minutes) also makes peptidoglycan an attractive target for antibiotics that interfere with new synthesis. These compounds cause a rapid loss of cell wall integrity, followed by osmotic swelling and cell lysis. The actions of specific antibiotics are described below.

Assembly Scheme

The steps of peptidoglycan biosynthesis in Gram-negative bacteria are outlined below and in Figure 20.4.

1. Biosynthesis starts with the formation of UDP-MurNAc through the condensation of phosphoenolpyruvate (PEP) with UDP-GlcNAc and subsequent reduction. Sequential

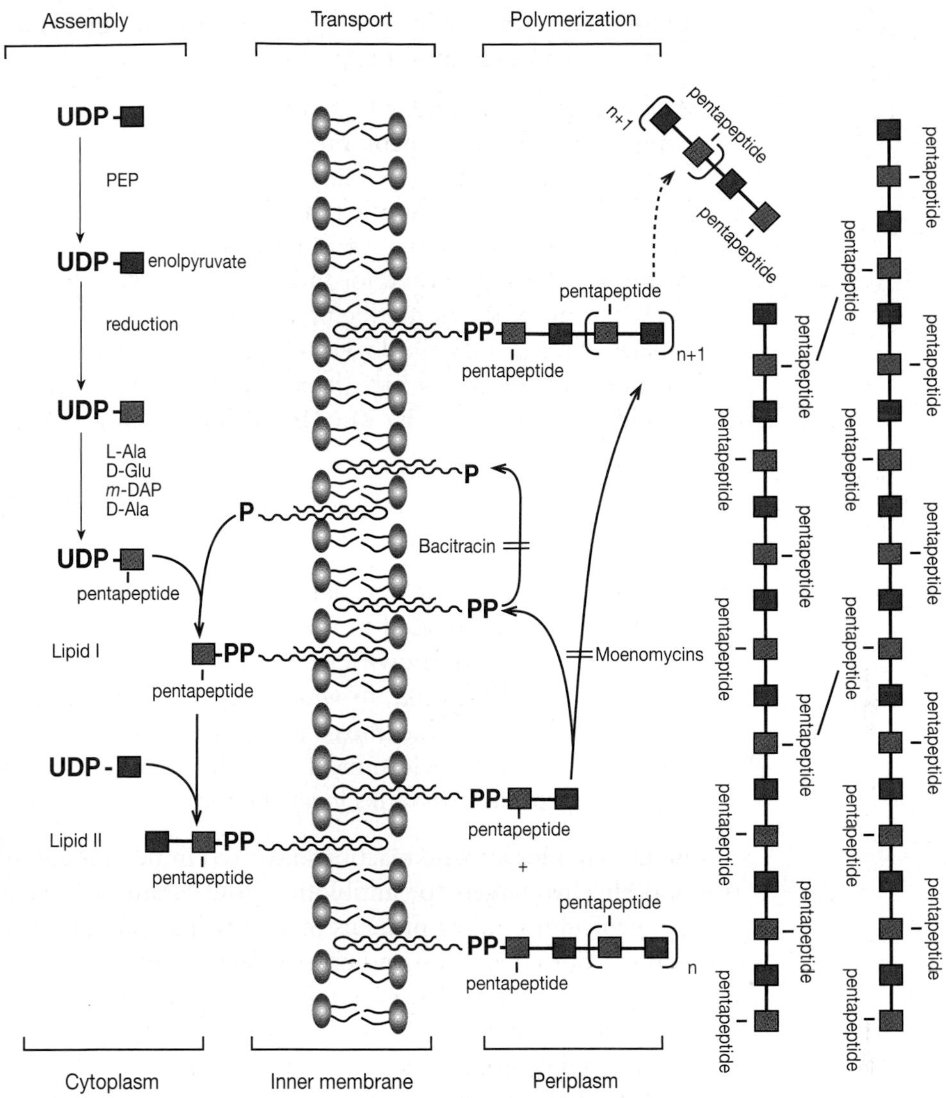

FIGURE 20.4. Synthesis of peptidoglycan occurs in three phases: assembly of precursor in the cytoplasm, transport across the inner membrane, and polymerization. The lipid-linked muropeptide (lipid I) is generated in the cytoplasm from amino acids and UDP-MurNAc (MurNAc is depicted by *orange squares*). Transfer of *N*-acetylglucosamine (*blue squares*) from UDP-GlcNAc completes formation of the precursor lipid II. Translocation across the inner membrane occurs, and subsequently, the chain polymerizes while attached to the lipid carrier. The unit is then transferred to existing peptidoglycan. (PEP) Phosphoenolpyruvate; (*m*-DAP) *meso*-diaminopimelic acid.

addition of L-Ala, D-Glu, *m*-DAP, and D-Ala results in the formation of UDP-MurNAc-pentapeptide. The addition of each amino acid requires a specific ATP-dependent amino acid ligase and the final two amino acids (D-Ala-D-Ala) added as a dipeptide unit. Cytoplasmic enzymes catalyze all of these reactions.

2. A membrane translocase transfers the MurNAc-pentapeptide to undecaprenyl (C55) phosphate (also known as bactoprenol phosphate) on the inner face of the inner membrane. This lipid resembles the eukaryotic dolichol carrier used in N-glycan synthesis (see Chapter 8). The final product, called lipid I, contains a pyrophosphate linkage.

3. A transferase on the same face of the inner membrane then transfers *N*-acetylglucosamine from UDP-GlcNAc to the undecaprenyl-PP-MurNAc-pentapeptide. This

lipid-linked disaccharide pentapeptide is called muropeptide or lipid II, and it represents the basic subunit for peptidoglycan assembly.

4. The undecaprenol lipid acts in a poorly understood way to move muropeptide subunits across the inner membrane. Shape-determining genes have been identified that affect cell wall synthesis, possibly by regulating this flipping reaction (e.g., *rodA*). Once reoriented to the periplasmic face of the plasma membrane, the muropeptide is transferred en bloc to existing peptidoglycan in a transglycosylation reaction. Two mechanisms have been proposed for this reaction: growth from the reducing end (where the 4-OH group of the nonreducing *N*-acetylglucosamine residue attacks the MurNAc-P linkage of a nascent peptidoglycan chain displacing undecaprenyl-PP), or growth from the nonreducing end (where the nonreducing end *N*-acetylglucosamine of the nascent peptidoglycan attacks the MurNAc-P linkage in a subunit, again with liberation of undecaprenyl-PP).

5. The undecaprenyl-PP undergoes cleavage of one phosphate group, which makes it available for another round of transfer.

6. The mechanism controlling chain length is unknown. The final release of a new peptidoglycan chain is coupled to the formation of 1,6 anhydroMurNAc on the reducing end of the chain (not shown). Release is followed by the formation of interchain cross-links by transpeptidation, in which cleavage of the terminal D-Ala residue results in transfer of the liberated carboxyl group of the new terminal D-Ala residue to the amino group of an *m*-DAP acid unit of a neighboring strand. Thus, the final structure contains tetrapeptide cross-links located on average every other subunit (Figure 20.3).

The unique biosynthetic reactions involved in peptidoglycan synthesis represent attractive and effective targets for antibiotics. Both Gram-positive and Gram-negative bacteria contain penicillin-binding proteins that participate in the transglycosylation and transpeptidation reactions. Penicillin and other β-lactam antibiotics bind these proteins and inhibit peptidoglycan synthesis. The antibiotic vancomycin inhibits a different step in synthesis by binding to the D-Ala-D-Ala dipeptide in the muropeptide and blocking further polymerization. Interestingly, some resistant bacteria contain an altered ligase that generates D-Ala-D-lactate instead of D-Ala-D-Ala and thereby resist the action of vancomycin. Bacitracin blocks the dephosphorylation and recycling of bactoprenol pyrophosphate (Figure 20.4).

Lysozyme cleaves peptidoglycan between the GlcNAc-MurNAc disaccharides. In the laboratory, lysozyme is used to prepare subcellular fractions from bacteria. For example, treating Gram-negative bacteria with lysozyme renders the cells osmotically sensitive and they can be easily disrupted for biochemical studies. In the context of a bacterial infection, host complement perforates the outer membrane, allowing lysozyme secreted by leukocytes to penetrate and disrupt the peptidoglycan layer. Thus, lysozyme plays an integral part in innate immunity.

TEICHOIC ACIDS

As mentioned above, Gram-positive bacteria lack an outer membrane and have a thicker peptidoglycan layer than Gram-negative bacteria (Figure 20.2). In Gram-positive organisms, the polysaccharide backbone of peptidoglycan typically contains 100 disaccharides and the peptide/peptide cross-bridge between strands varies (Figure 20.5). On average, every tenth unit contains a teichoic acid. In *Streptococcus pyogenes*, the teichoic acid consists of polyglycerophosphate (Figure 20.5), whereas *Bacillus* strains produce teichoic acids

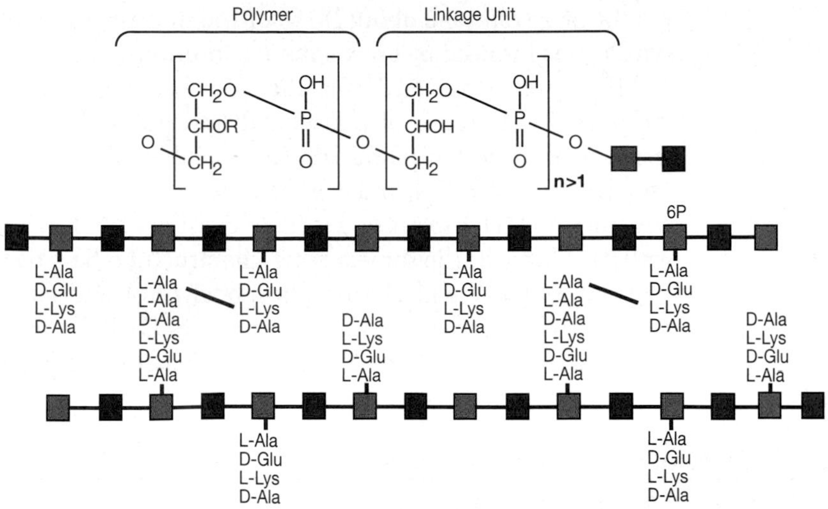

FIGURE 20.5. Structure of *Streptococcus pyogenes* peptidoglycan with teichoic acid. The strands of peptidoglycan are cross-linked by a pentapeptide with an amino acid composition different from that of *E. coli*. The teichoic acid consists of a polymer of glycerol phosphate linked to peptidoglycan. Lipoteichoic acids and teichuronic acids are not shown. (*Orange squares*) *N*-acetylmuramic acid (MurNAc); (*blue squares*) *N*-acetylglucosamine (GlcNAc); (*green square*) *N*-acetylmannosamine (ManNAc); (R) monosaccharides or D-alanine.

composed of polyribitolphosphate or polyglycerophosphate. All teichoic acids bound to peptidoglycan appear to be linked by a conserved unit consisting of ManNAc-GlcNAc-1-P linked to C-6 of a MurNAc residue. The teichoic acid polymer composed of polyribitolphosphate or polyglycerophosphate forms on the linkage unit while it is still attached to core disaccharide precursor, and the entire unit (e.g., [glycerophosphate]n-ManNAc-GlcNAc-1-P-undecaprenol) is thought to be transferred en bloc to the growing peptidoglycan. The glycerol and ribitol subunits also undergo modification by the addition of monosaccharides and D-Ala residues. Lipoteichoic acids contain a reducing terminal phosphatidic acid and are not linked to peptidoglycan.

The teichoic acids impart a high negative charge to the cell wall, which may have a role in selective uptake of charged molecules or as a barrier to uptake of antibiotics. Not all Gram-positive organisms contain teichoic acids, but those that lack them contain other types of polyanionic cell wall constituents, such as succinylated lipomannan. Under conditions of phosphate limitation, some bacteria produce teichuronic acids containing glucuronic acid–glucose or galacturonic acid–glucose copolymers instead of polyribitolphosphate and polyglycerophosphate polymers, demonstrating the necessity of charged constituents in the cell wall. Although the precise function of teichoic acids is unknown, mutant cells that do not synthesize these compounds are unable to grow.

MEMBRANE-DERIVED OLIGOSACCHARIDES OF THE PERIPLASMIC SPACE

Bacteria encounter extreme differences in osmolarity in the environment (like a bicycle tire; i.e., up to 6 atmospheres of turgor pressure!) and have evolved both physical and chemical mechanisms to resist disruption. Peptidoglycan, as discussed in the previous section, provides a structural barrier to osmotic swelling. In Gram-negative bacteria, a chemical mechanism also exists to protect the inner membrane. An osmotic buffer provided by highly charged β-glucans, termed membrane-derived oligosaccharides (MDOs), is created in the periplasmic space. MDO compounds constitute approximately 1–5% of the dry weight of *E. coli* (and about 0.1% of the dry weight of Gram-positive bacteria), and their synthesis is induced by low osmotic conditions.

MDOs were discovered in studies of phospholipid turnover in *E. coli*, which showed that the polar head groups of phosphatidylglycerol and phosphatidylethanolamine were transferred to low-molecular-weight, water-soluble oligosaccharides. Other organisms including *Pseudomonas*, *Rhizobia*, and *Agrobacteria* also make these compounds. Although the precise structure of MDO varies, it generally consists of 6–12 glucose units, mostly in β1-2 linkage with β1-6 branches; in some cases, cyclic structures have been found. In addition, the oligosaccharides contain ethanolamine, phosphoglycerol, and succinyl groups, which confer the high net negative charge. MDO assembly requires UDP-Glc as a donor, and undecaprenyl-PP-Glc as both a primer and a carrier for transport across the inner membrane.

LIPOPOLYSACCHARIDE AND ENDOTOXIN

Function

The outer membrane of Gram-negative bacteria consists of a lipid bilayer, but unlike other cell membranes composed of a bilayer of phospholipids, the outer leaflet contains mostly LPS (see Figure 20.1). Each bacterial cell contains about 10^6 molecules of LPS (vs. 10^7 total phospholipids). Divalent cations bound to the phosphate groups stabilize the outer membrane structure; chelators that bind divalent cations (e.g., EDTA) make it permeable, even to large proteins such as lysozyme.

LPS was first discovered more than a century ago as a heat-stable toxin associated with bacteria, in contrast to heat-labile exotoxins (see Chapter 39). The description of this complex molecule is based largely on studies of *E. coli*, but the structures of LPS from *Salmonella*, *Pseudomonas*, and *Rhizobium* are now known. LPS consists of lipid A (a type of glycolipid, recently termed a saccharolipid, and defined as a glycoconjugate containing fatty acyl chains covalently attached directly to a sugar backbone), an inner core region, and outer O-antigen oligosaccharides (Figure 20.1). Lipid A anchors LPS in the outer membrane and serves as the scaffold for assembly of the inner core region and the outer O-antigen oligosaccharides. Chemical synthesis of *E. coli* LPS in 1985 confirmed that lipid A corresponds to the heat-stable endotoxin.

Lipid A (endotoxin) has powerful biological effects in mammals, causing fever, septic shock, and other deleterious physiological effects (see Chapter 39). When released into the circulation, LPS binds to CD14 and Toll-like receptor 4 (TLR4) on monocytes and macrophages, which triggers secretion of proinflammatory mediators. At low levels, lipid A serves as an adjuvant causing polyclonal expansion of B cells, but at high levels, it causes morbidity and mortality. Because of its clinical significance, substantial effort has been invested in studying its biosynthesis and in developing approaches to target the enzymes involved in its formation.

Structural Components and Assembly

The glycan structure of lipid A consists of two glucosamine residues in β1-6 linkage. In *E. coli*, the reducing terminal sugar contains phosphate at C-1 and two units of β-hydroxymyristic acid, one in ester linkage at C-3 and one in amide linkage at C-2 (Figure 20.6). The stereochemistry at these chiral centers is identical to that found in the glycerol backbone of glycerophospholipids, and the acyl amide is reminiscent of the one found in mammalian sphingolipids (see Chapter 10). The second glucosamine residue also contains two β-hydroxymyristic acids, in ester linkage at C-3′ and amide linkage at C-2′, with additional

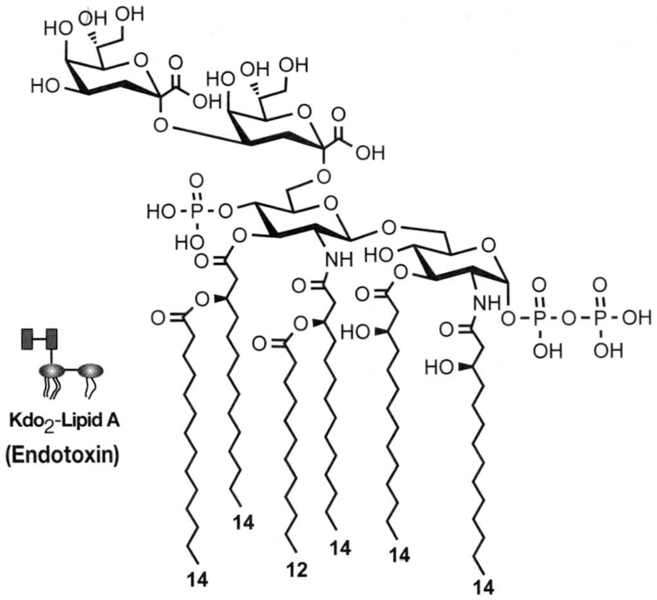

FIGURE 20.6. Structure of Kdo_2–lipid A in *E. coli*. The chemical structure of the schematic cartoon (*left*) is depicted (see Fig. 20.1). (Adapted, with permission, from Raetz C.R.H., Reynolds C.M., Trent M.S., et al. 2007. *Annu. Rev. Biochem.* **76:** 295–329, ©Annual Reviews.)

lauroyl groups on the β-hydroxyls (which resemble waxes). All lipid-A molecules contain 1–4 units of an unusual sugar, Kdo (Figure 20.6). The Kdo moieties, together with the phosphate groups at C-1 and C-4′ on the lipid-A disaccharide, are the binding sites for the divalent cations that stabilize the outer membrane. The extent of phosphorylation and the extent and type of acylation vary considerably among Gram-negative bacteria and determine the endotoxicity of a particular LPS.

The outermost portion of LPS consists of the O-antigen, which contains 1–8 sugars, repeated up to 50 times and capped by an additional 0–50 residues. A broad range of sugars is present in these antigens, including free and amidated uronic acids, amino sugars, methylated and deoxygenated derivatives, acetylated sugars, and others that contain covalently bound amino acids and phosphate. This diversity gives rise to hundreds of bacterial serotypes, distinguished by their reactivity with human antisera. For *E. coli* alone, more than 170 serotypes have been defined. Although no strict correlation exists between serotypes and disease, some infections are more typical of certain serotypes. For example, bladder infections are typically associated with *E. coli* O157H7, which bears a specific type of O-antigen coupled with flagellar antigen H7 (see Chapter 39). The O-antigen apparently provides a hydrophilic barrier that protects against hydrophobic antibiotics (natural fungal and bacterial metabolites), bile acids (in enterobacteria), and complement. Some bacteria that colonize mucosal surfaces, such as *Neisseria*, express a truncated and nonrepeating O-antigen glycan; in these cells, the lipid-A-based structure is called lipooligosaccharide (LOS) instead of LPS.

LPS biosynthesis begins with the formation of lipid A. UDP-GlcNAc is acylated at C-3, followed by N-deacetylation, N-acylation, and cleavage of the pyrophosphate linkage to form 2,3-diacylglucosamine-1-P (Figure 20.7). Two of these molecules condense to form the tetra-acyl disaccharide core, which is then phosphorylated on C-4′ of the nonreducing sugar and modified with Kdo. Interestingly, the two β-hydroxyl-linked fatty acids are added only after the Kdo units. The formation of Kdo_2–lipid A is essential for survival of *E. coli*, but this may not be true for all Gram-negative bacteria. This assembly process occurs in the inner membrane, and newly made molecules are "flipped" across the membrane to face the periplasm by a trans-

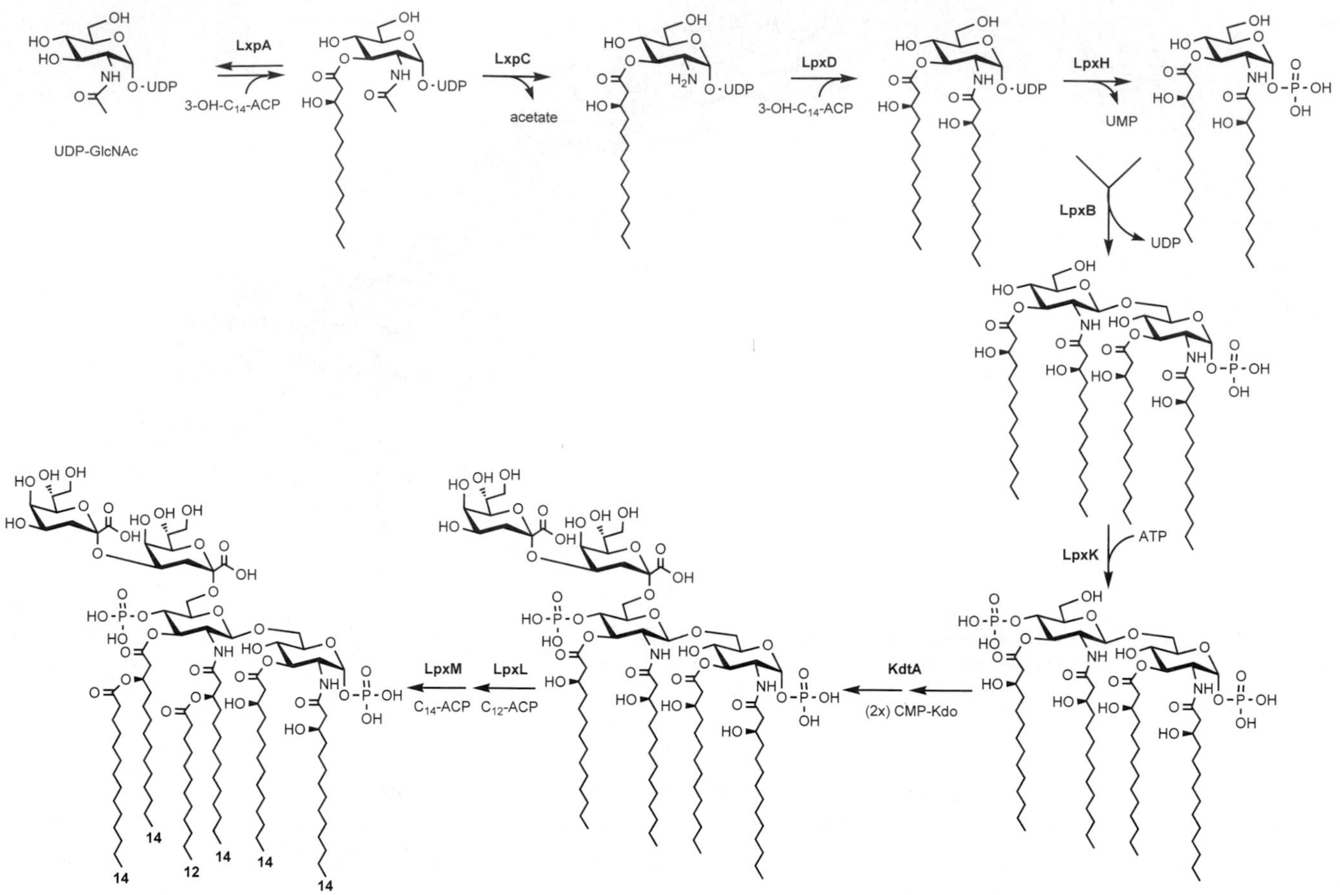

FIGURE 20.7. Assembly of Kdo$_2$–lipid A in *E. coli*. LpxA, LpxC, and LpxD are cytoplasmic enzymes, whereas LpxH and LpxB are peripheral membrane proteins. The remaining enzymes are integral inner-membrane proteins, with their active sites facing the cytoplasm. The numbers below the fatty acid chains indicate the predominant fatty acid chain lengths found in *E. coli* lipid A. (Adapted, with permission, from Raetz C.R.H., Reynolds C.M., Trent M.S., et al. 2007. *Annu. Rev. Biochem.* **76:** 295–329, ©Annual Reviews.)

porter related to ATP-binding cassette (ABC) transporters found in animal cells. Little is known about how lipid A translocates to the outer leaflet of the outer membrane.

Although there are mutants of *E. coli* and *Salmonella* that make only a Kdo-bearing lipid A, in most cases, this structure is further modified by the addition of heptoses, hexoses, and phosphate groups to form the core region. This process requires numerous transferases.

The assembly of O-antigen occurs independently of lipid A and the inner core. The chain is built on undecaprenyl-P by membrane-bound enzymes facing the cytosol, which transfer sugars from nucleotide sugar donors to the reducing end. The completed subunits are then flipped across the inner membrane, presumably by a system analogous to the one required for lipid-A translocation, but details about the actual mechanism are unknown. Once outer-chain synthesis is completed, the chain is transferred en bloc to the core region of LPS and the entire complex is transferred to the outer membrane.

Although much has been discovered about the structure of bacterial glycans and the pathways for their assembly, little information is available about the topology of assembly. Glycan synthetic precursors arise in the cytoplasm from water-soluble nucleotides and various sugars, but the assembly process for membrane glycans takes place at the interface of the membrane and the cytoplasm, and peptidoglycan biogenesis takes place in the periplasmic space. How biosynthetic intermediates traverse the barriers between these compartments provides a fascinating topic for further study. The enzymes of LPS biosynthesis and the translocation proteins are good candidates for the development of new antimicrobial agents. The idea of targeting the formation of essential intermediates, such as lipid A or the unusual sugars unique to bacteria (e.g., Kdo and heptoses), is especially appealing. Development of effective drugs is of paramount importance because bacteria resistant to all known antibiotics are becoming more prevalent.

CAPSULAR POLYSACCHARIDES AND BIOFILMS

In addition to the glycoconjugates described above, bacteria produce polysaccharide capsules, which form the outermost layer of the cell and impart a mucoid appearance to bacterial colonies (see Chapter 39). These are termed K-antigens, to distinguish them from the O-antigens of LPS and the F-antigens of fimbriae and flagella. Capsule polysaccharides exhibit extraordinary diversity in structure, but the presence of certain common sugars forms the basis for their classification. Group Ia capsules contain hexuronic acids and neutral sugars, group Ib capsules contain hexuronic acids and N-acetylated hexosamines, and group II capsules contain hexuronic acids, Kdo, or sialic acids in combination with neutral or amino sugars, such as rhamnose and *N*-acetylglucosamine.

Function

Capsules have multiple functions. They are highly hydrated and may help to deter desiccation. Because they form the outer surface of the cell, they mediate adhesion to inert surfaces or living tissues, which is important for colonization of host organisms and the formation of biofilms. During infection, capsules have a role in virulence, allowing bacteria to evade host defenses including complement-mediated lysis, phagocytosis, and cell-mediated immune mechanisms (see Chapter 39). Some capsules are poorly immunogenic because of the presence of structures identical to those found in the host. For example, the hyaluronan in the capsule of *Streptococcus* A is identical in composition to hyaluronan generated by mammalian cells (see Chapter 15). Molecular mimicry like this also occurs in bacteria with K1 capsules (polysialic acid is found in the brain), in bacteria with K5 capsules (*N*-acetylheparosan is the backbone of heparan sulfate), and in group IIIB *Streptococcus* (Neu5Acα2-3Galβ1-4Glc resembles ganglioside GM_3) (Table 20.1).

Biofilms consist of communities of bacteria adhering to a moist surface—for example, on the surface of ponds, on teeth, in the gastrointestinal tract, or even on the surfaces of rocks. Microbial mats are specialized biofilms composed mainly of photosynthetic bacteria. Both biofilms and microbial mats generate an extracellular slime layer composed of capsular polysaccharides. Some films are beneficial—for example, those used to treat

TABLE 20.1. *Examples of capsular polysaccharides*

Capsule type	Structure
K1 (polysialic acid)	◆α8◆α8◆α8◆α8◆α8◆
K5 (*N*-acetylheparosan)	■α4⬘β4■α4⬘β4■α4⬘
Group B *Streptococcus* (hyaluronan)	■β4⬘β3■β4⬘β3■β4⬘

wastewater. Others can pose a serious health threat, such as biofilms that form on indwelling catheters. The glycans in biofilms protect bacteria from the effects of antibiotics by acting as a physical barrier, and thus biofilms can facilitate the maintenance and spread of bacterial infection. In the environment, biofilms entrap particulate materials such as clay, organic materials, and precipitated minerals, which then facilitates the attachment of other organisms. Thus, biofilms often support a complex ecosystem.

Assembly of Capsules

Group I capsule polysaccharide assembly takes place by polymerization of oligosaccharide repeat units linked to undecaprenol-PP. Bacitracin, which blocks the recycling of undecaprenyl-P (see Figure 20.4), inhibits the formation of these capsules. However, other capsular polysaccharides do not require this carrier and are therefore resistant to bacitracin. Some of these capsules contain phosphatidic acid–Kdo conjugates or lipid A at the reducing termini of the oligosaccharides, suggesting that different types of primers may exist. Sugars may be added to capsule oligosaccharides in a simple processive way, as in the assembly of K1 capsules composed of α2-8-linked polysialic acid, in which biosynthesis occurs from the reducing end. Hyaluronan capsules in *Streptococcus* A strains are constructed in a similar way by an enzyme with both *N*-acetylglucosamine and glucuronic acid transferase activities (see Chapter 15). In contrast, *Pasteurella* synthesizes hyaluronan from the nonreducing end by an enzyme that is unrelated to the synthases found in *Streptococcus* or vertebrate cells. The *Pasteurella* enzyme has two separable domains with independent glycosyltransferase activities, one for UDP-GlcNAc and the other for UDP-GlcA. The formation of other copolymeric capsules also may involve dual-function enzymes, but additional research is needed to elucidate these assembly processes.

The assembly of capsules involves genes that are clustered in the bacterial chromosome in three contiguous regions. This arrangement allows a simple mechanism for changing capsule types by merely swapping different serotype cassettes. In fact, the serotype locus can be transferred on plasmids between compatible bacteria, resulting in altered capsule composition. Genes in region II, known as the serotype region, encode enzymes involved in nucleotide sugar formation and capsule-specific transferases. Genes in regions I and III encode serotype-independent transport activities required for movement of the polysaccharides across the inner membrane and periplasm, but genes required for transport across the outer membrane have not yet been identified. In *Streptococcus*, the hyaluronan synthase spans the membrane multiple times, presumably forming a pore for hyaluronan extrusion during capsule formation.

MYCOBACTERIA

Mycobacteria are classified as Gram-positive bacteria, but they have features of both Gram-positive and Gram-negative organisms. Pathogenic mycobacteria include the organisms that cause tuberculosis and leprosy and are intracellular parasites, replicating within modified phagosomes of macrophages. In this location, they prevent formation of phagolysosomes and inhibit the macrophage responses to infection such as apoptosis and secretion of inflammatory cytokines. These activities depend on cell wall constituents, such as plasma membrane lipoarabinomannans, which consist of phosphatidylinositol mannosides covalently linked to arabinogalactans that extend out from the cell surface (Figure 20.8). Lipoarabinomannans are also associated with another layer in the cell wall composed of mycolic acids, a type of fatty acid. Mycobacteria are surrounded by a mycolic acid–arabinogalactan–peptidoglycan complex and a polysaccharide-rich capsule of arabinomannan and mannan (Figure 20.8). Because of

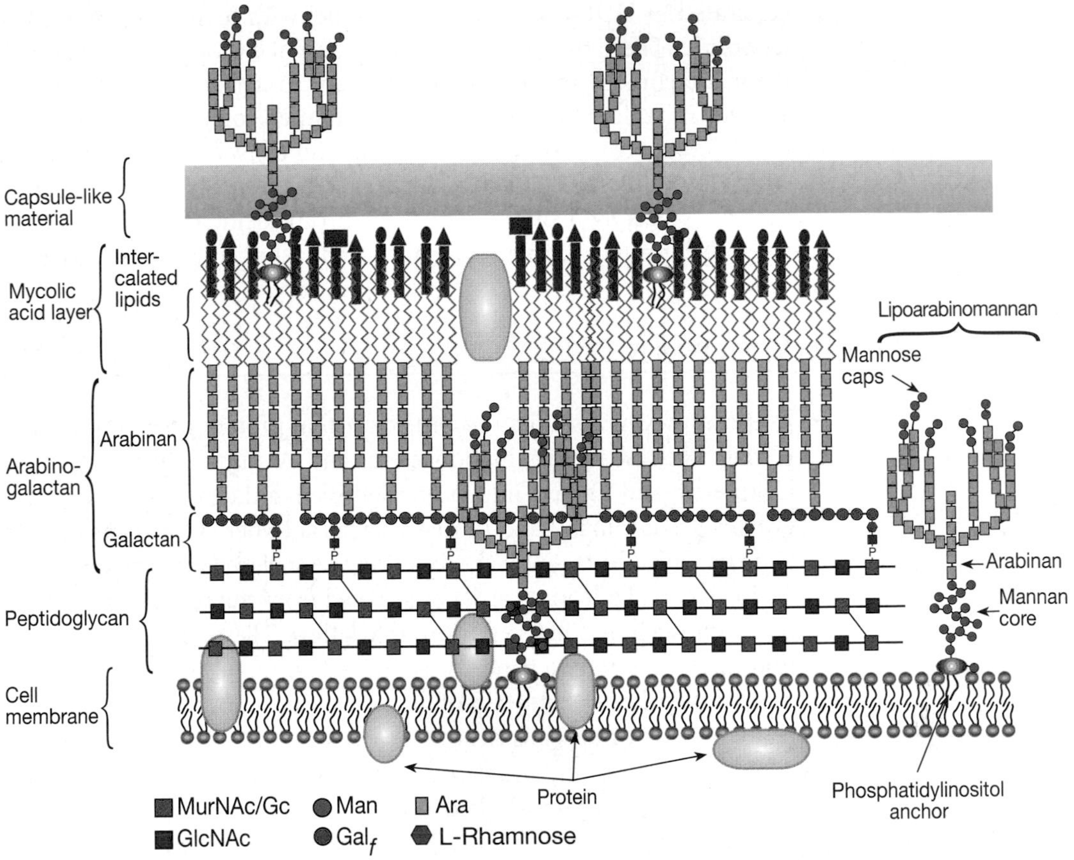

FIGURE 20.8. Structure of the cell wall of mycobacteria. (Derived from Brennan P.J. and Crick D.C. 2007. *Curr. Top. Med. Chem.* **7:** 475–488.)

their unique structures, assembly of these complex glycans represents an excellent target for the development of inhibitors that could prove useful for treating human disease.

ARCHAEA

Archaea constitute the third domain of life (in addition to Eukaryota and Eubacteria), and they are single-cell bacteria-like organisms. They can inhabit extreme environments, typically characterized by high temperature or pressure (e.g., deep sea thermal vents) or extreme salinity, alkalinity, or acidity. These versatile microbes reside in the digestive tracts of ruminants, termites, and marine life (where they produce methane) and in the soil. They can live under anoxic conditions in mud, at the bottom of the ocean, and even in petroleum deposits.

Like Eubacteria, Archaea contain a cell wall composed of various polysaccharides and glycoconjugates. Archaea lack peptidoglycan, but they still form rigid cell boundaries that confer resistance to high internal osmotic pressure. To do this, they elaborate protein or glycoprotein coats or reinforce their cytoplasmic membranes. The S-layer glycoproteins are the best characterized glycoproteins of Archaea. Organisms of the genus *Methanopyrus* form a compound called pseudomurein, which is similar in structure to peptidoglycan (murein) but completely different in its fine detail. Its polysaccharide portion is formed of β1-3-linked *N*-acetylglucosamine and L-talosaminuronic acid, and its peptide branches are composed of L-amino acids. Pseudomurein is also formed by unusual mechanisms. The disaccharide is generated in UDP-linked form in parallel with formation of a UDP-linked branched pentapeptide. These two precursors are then linked together and transferred to undecaprenyl-P for attachment to the growing glycan. Archaeal cell wall polymers other than pseudomurein include methanochondroitin, named for its similarities to vertebrate chondroitin (see Chapter 16), and the glutaminylglycan of *Natrococcus*. The latter is a novel structure in which two different oligosaccharides, consisting of either *N*-acetylgalactosamine and glucose or *N*-acetylglucosamine and glucuronic acid, are linked to polyglutamic acid. Most S-layer glycoproteins contain N-linked and O-linked glycans (described below).

PROKARYOTIC GLYCOPROTEINS

In the last few years, it has become clear that protein glycosylation is not limited to eukaryotic cells but also occurs in Eubacteria and Archaea. Diverse N- and O-linked glycans (see Chapters 8 and 9) have now been identified and characterized in archaeal species (where N-glycans predominate) and in many bacteria (where O-glycans are more common). The proteins modified in this way are found in the surface layer, the outer and inner membranes, the periplasm, and pili and flagella or secreted from the cell. The glycans also undergo other modifications such as sulfation and methylation. These glycoproteins have important biological roles including maintenance of shape, adhesion, antigenicity, and biosynthetic processes. In many cases, the contribution of glycosylation to normal protein function has not been investigated, but in others, it is clear that glycosylation is required. Examples include effects on protein conformation and protease susceptibility in bacteria and flagellar assembly in the Archaea.

Structurally, protein-linked glycans in Eubacteria and Archaea are incredibly diverse, and relatively few have been fully characterized in detail. Sugars O-linked to serine, threonine, or tyrosine include galactose, glucose, mannose, rhamnose, *N*-acetylglucosamine, *N*-acetylgalactosamine, and other compounds, and much of the same repertoire of sugars is found in N-glycans. Although many of the structures formed are unique to Eubacteria and Archaea, such as a glycosylated polyglutamyl polymer found in *Natronococcus occultus*, bacteria also make glycoconjugates closely related to those found in mammals. For example, *Chlamydia trachomatis* makes high-mannose N-glycans quite similar to those of eukaryotes (see Chapter 8). The assembly of such "eukaryotic" glycans by pathogenic bacteria may be a form of molecular mimicry, which aids in evading the host immune response.

Study of the biosynthetic pathways of prokaryotic glycoproteins is just beginning, but genetic information suggests that these systems range in complexity. Several species of *Campylobacter* contain gene clusters encoding a "general" glycosylation system that modifies various proteins, and in other cases, the genes whose products modify a specific protein are located adjacent to the gene encoding the protein itself. In *Campylobacter jejuni*, a cluster of 12 genes called the *pgl* locus (for protein glycosylation) is responsible for the synthesis of a heptasaccharide GalNAcα1-4GalNAcα1-4(Glcβ1-3)GalNAcα1-4GalNAcα1-4GalNAcα1-3Bac (where Bac is 2,4-diacetamido-2,4,6-trideoxy-D-Glc) on bactoprenol and the transfer of the chain en bloc to asparagine residues present in the same consensus sequence found in eukaryotic glycoproteins (N-X-S/T). The topology of protein glycosylation in bacteria and Archaea presents additional challenges because the intracellular compartments that serve to organize most protein glycosylation in eukaryotes are absent. Glycans assembled on lipid-linked precursors are presumably translocated across a membrane to the site of protein glycosylation, in this case, the external surface of the cell. Oligosaccharyl transferases homologous to the eukaryotic enzymes catalyze the en bloc transfer of the lipid-linked intermediate, but details of the complex and reaction chemistry are still under study. One distinguishing feature of bacterial systems is the apparent independence of *N*-glycosylation from the protein translocation machinery: Bacterial oligosaccharyltransferase can glycosylate folded bacterial proteins with high efficiency, whereas glycosylation generally occurs cotranslationally in eukaryotes. Additional research will be required to unravel these fascinating systems.

FURTHER READING

van Heijenoort J. 2001. Formation of the glycan chains in the synthesis of bacterial peptidoglycan. *Glycobiology* **11:** 25R–36R.

Wacker M., Linton D., Hitchen P.G., Nita-Lazar M., Haslam S.M., North S.J., Panico M., Morris H.R., Dell A., Wren B.W., and Aebi M. 2002. N-linked glycosylation in *Campylobacter jejuni* and its functional transfer into *E. coli*. *Science* **298:** 1790–1793.

Eichler J. and Adams M.W. 2005. Posttranslational protein modification in Archaea. *Microbiol. Mol. Biol. Rev.* **69:** 393–425.

Ostash B. and Walker S. 2005. Bacterial transglycosylase inhibitors. *Curr. Opin. Chem. Biol.* **9:** 459–466.

Schaffer C. and Messner P. 2005. The structure of secondary cell wall polymers: How Gram-positive bacteria stick their cell walls together. *Microbiology* **151:** 643–651.

Bhavsar A.P. and Brown E.D. 2006. Cell wall assembly in *Bacillus subtilis*: How spirals and spaces challenge paradigms. *Mol. Microbiol.* **60:** 1077–1090.

Weerapana E. and Imperiali B. 2006. Asparagine-linked protein glycosylation: From eukaryotic to prokaryotic systems. *Glycobiology* **16:** 91R–101R.

Berg S., Kaur D., Jackson M., and Brennan P.J. 2007. The glycosyltransferases of *Mycobacterium tuberculosis*—Roles in the synthesis of arabinogalactan, lipoarabinomannan, and other glycoconjugates. *Glycobiology* **17:** 35R–56R.

Brennan P.J. and Crick D.C. 2007. The cell-wall core of *Mycobacterium tuberculosis* in the context of drug discovery. *Curr. Top. Med. Chem.* **7:** 475–488.

Raetz C.R.H., Reynolds C.M., Trent M.S., and Bishop R.E. 2007. Lipid A modification systems in Gram-negative bacteria. *Annu. Rev. Biochem.* **76:** 295–329.

Weidenmaier C. and Peschel A. 2008. Teichoic acids and related cell-wall glycopolymers in Gram-positive physiology and host interactions. *Nat. Rev. Microbiol.* **6:** 276–287.

CHAPTER 21

Fungi

Richard D. Cummings and Tamara L. Doering

FUNGI ARE FASCINATING ORGANISMS THAT HAVE BEEN INSTRUMENTAL in defining the fundamental processes of glycosylation. This chapter describes the glycan structures of these diverse organisms and their synthesis, offers some insights into glycobiology revealed through studying fungal systems, and delineates the relationships of several important glycoconjugates to fungal biology and pathogenesis.

FUNGAL DIVERSITY

More than 70,000 species of fungi have been described, and it is likely that many more exist. The known species include yeast, molds, rusts, smuts, puffballs, and mushrooms. Study of these eukaryotes has been motivated by their unique and fascinating biology, their many useful products (including wine, cheese, and antibiotics), their utility as experimental systems for basic biology and protein expression, and their importance as animal and plant pathogens. Fungi are eukaryotic heterotrophs and absorb food from their environment. They are nonmotile and have life cycles that incorporate both sexual and asexual reproduction. They typically have elongated filaments or hyphae, which have cell walls that comprise complex polysaccharides including mannans, galactans, glucans, and chitin. There are four major phyla of fungi; each is extremely diverse, and fungi are assigned to a phylum on the basis of their mechanism for producing asexual spores. These phyla are the Chytridiomycota (primitive aquatic fungi), Zygomycota (e.g., black bread mold, *Rhizopus nigricans*), Ascomycota (sac fungi, e.g., *Saccharomyces, Candida, Aspergillus, Neurospora*, and morel mushrooms), and Basidiomycota (e.g., mushrooms, rot fungi, and puffballs). Some fungi are extremely beneficial to humans (e.g., the yeast *Saccharomyces*, which is used in fermentation) and some are not (e.g., pathogenic yeasts such as species of *Candida* and *Crypotococcus*).

YEAST AS A MODEL SYSTEM FOR BIOCHEMISTRY AND GLYCOBIOLOGY

More than 100 years ago, Louis Pasteur discovered that fermentation requires a viable organism, and since then yeast have been used as a model system to study cellular metabolism. In fact, Pasteur coined the word "ferment" during his work on alcohol production by yeast. *Saccharomyces cerevisiae*, or baker's yeast, has been a wonderful resource for biologists and glycobiologists, especially because many of the fundamental enzymes in aerobic and anerobic metabolism (terms also invented by Pasteur) are shared between yeast and animals. Breakthroughs in enzymology occurred following the discovery by the Buchner brothers in 1897 that extracts of yeast could make ethanol and carbon dioxide from glucose, just like intact cells. Mannose is a major component of the yeast cell wall; it was discovered by Emil Fischer in 1888 and the mannose-rich glycans in yeast, historically called yeast gum, have been known since the 1890s. It is also noteworthy that the great chemist Sir Walter Norman Haworth discovered that the cell wall of yeast was composed of D-mannose. Haworth won the Nobel Prize in 1937 for his work on determining the chemical structures of carbohydrates (and vitamin C). The nucleotide sugar UDP-glucose was discovered by Luis Leloir in 1950 using yeast extracts, and he went on to discover GDP-mannose and other nucleotide sugars. Leloir won the Nobel Prize in chemistry in 1970 for this discovery of sugar nucleotides and their functions in carbohydrate synthesis. Studies with yeast were also important in defining the dolichol-linked sugars and their roles as intermediates in protein N-glycosylation and in elucidating numerous other fundamental biochemical pathways including the synthesis of sterols from acetate and fatty acid oxidation.

Yeast cells have a nucleus, a nuclear envelope associated with the endoplasmic reticulum (ER), a Golgi apparatus, mitochondria, and peroxisomes. These simple eukaryotes provide both a convenient source of enzymes for study and a powerful genetic system. Yeast secretory mutants (*Sec* mutants) helped to define the components of the protein secretory pathway, by which proteins travel from the ER to the cell surface and the exterior of the cell, becoming glycosylated en route. Yeast cells with various mutations have also been instrumental in elucidating mechanisms and pathways of protein N-glycosylation and steps in the biosynthesis of glycosylphosphatidylinositol (GPI)-anchored glycoproteins. At the same time, studies in yeast have certain limitations. These cells do not synthesize typical complex N-glycans, mucins or mucin-type O-glycans, O-linked *N*-acetylglucosamine (O-GlcNAc), or glycosaminoglycans of the types found in vertebrates. In addition, although yeast are valuable for studying sphingosine and sphingolipid metabolism, they do not synthesize complex glycosphingolipids or gangliosides like those found in mammals. However, they do make some very unusual glycolipids, including short, mannose-containing compounds. Finally, most fungi do not synthesize sialic acid, have relatively low amounts of fucose, and lack long-chain glycolipids (apart from those glycolipids participating in the synthesis of GPI anchors or glycoproteins in the ER).

GLYCAN STRUCTURES OF YEAST

All fungi have cell walls, which are critical to maintaining cell shape and integrity in environments that range from the surface of grapes to human tissues. Cell walls are highly cross-

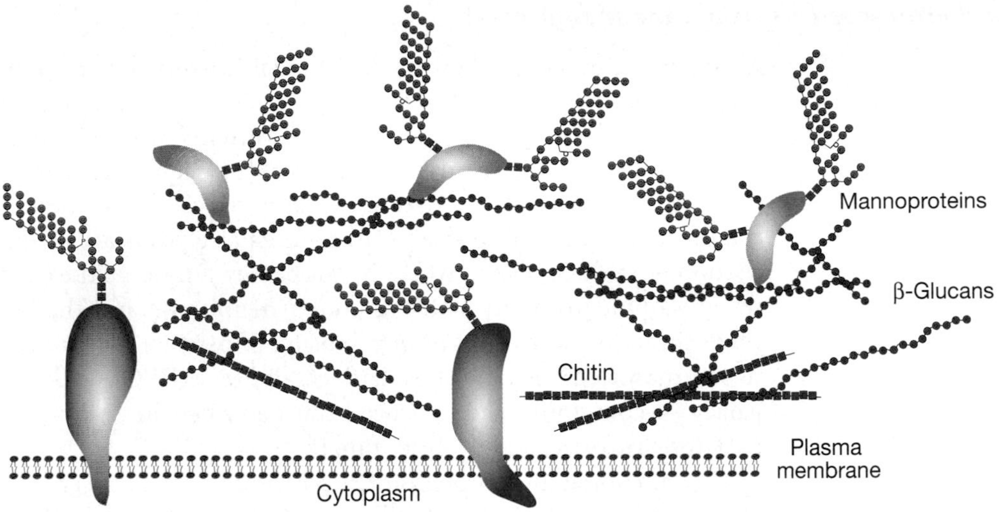

FIGURE 21.1. Illustration of the cell wall of fungi, showing the presence of glycoproteins and mannoproteins in the layer of the wall and an inner layer of different polysaccharides. The presence of different types of glucans and chitin varies between different fungal species.

linked structures, which adapt to growth conditions in a dynamic and flexible way (Fig. 21.1). The cell-wall polysaccharides that have been described so far are composed of polymers of mannose, glucose, galactose, *N*-acetylglucosamine, and/or rhamnose, and these include mannans, glucans, chitin, galactomannans, glucomannans, rhamnomannans, and phosphomannans. In mannans, the mannose residues of the polymeric backbone are α-linked (usually α1-6), whereas in glucans, the glucose residues are β-linked (mostly β1-3, although some are β1-6). The interconnected polymers may have other sugars attached to these backbone structures and additional modifications that are specific to each organism. Fungal cell walls also contain covalently and noncovalently linked glycoproteins that bear N- and O-glycans of myriad structures; some of these glycoproteins begin as GPI-anchored proteins.

Chitin is a polymer of β1-4-linked GlcNAc, which occurs in chains that typically exceed 1000 residues. These chains self-associate to form microfibrils and are deposited primarily at the bud neck of yeast or at septa in filamentous fungi. Chitin synthesis is highly regulated by the coordinated action of multiple chitin synthases, so that deposition occurs at the specific sites and times required for normal cell growth and division. In *S. cerevisiae* three chitin synthases (Chs1p, Chs2p, and Chs3p) have been described, along with multiple additional gene products that participate in the regulation and localization of chitin synthesis. Chitin may also be deacetylated to form another polymer, chitosan. In *S. cerevisiae* this process occurs during spore formation, although in other fungi chitosan is required for cell wall functions during vegetative growth. Chitin is also the primary component of exoskeletons of arthropods (see Chapter 24) and in the cuticle of nematodes (see Chapter 23).

Yeast express a relatively simple array of glycolipids, although *Candida albicans* is notable for its large lipid-linked mannans. Many fungi make short-chain glycolipids, commonly containing *myo*-inositol phosphate linkers to mannose that may be modified by galactofuranose (as in *Histoplasma capsulatum*) or an additional mannose residue. Some longer galactose- and mannose-containing glycolipids are found in *Aspergillus niger*. Short-chain glycosylceramides such as Glc-Cer and Gal-Cer are also found in the fungi *Schizophyllum commune* and *Aspergillus fumigatus*, respectively. Examples of fungal species are highlighted below to illustrate the many general glycan features that yeast share while emphasizing the glycoconjugates that are unique to each organism. The last two sections of this chapter provide notable fungal examples of glycoconjugate biosynthesis.

Saccharomyces cerevisiae, the Model Yeast

S. cerevisiae, by far the best-studied fungal organism, is a round budding yeast approximately 5–10 μm in diameter. This single-celled eukaryote contains an amazing assortment of different glycoconjugates, including proteins with N- and O-glycan modifications, polypeptides bearing mannan polysaccharides (Figures 21.2 and 21.3a), GPI-anchored glycoproteins (Figure 21.4), glycolipids, and cell-wall polymers. The O-glycans have core mannose residues linked to Ser/Thr and the N-glycans are all linked to Asn residues in the sequon -Asn-X-Ser/Thr-, where X can be any amino acid except proline. *S. cerevisiae* has few glycolipid components in the outer membrane and the best characterized are small phytoceramide derivatives of *myo*-inositol phosphate that contain a single residue of mannose (mannose inositol-phosphate-ceramide [MIPC] and $M(IP)_2C$). Interestingly, no galactose-containing glycans have been described in *S. cerevisiae*, although glycans with galactose are expressed by other fungi.

GPI-anchored glycoproteins of *S. cerevisiae* have a conserved trimannose core linked to glucosamine-inositol-phosphatide, substituted with an additional α1-2-linked mannose and ethanolamine phosphodiesters (Figure 21.4) (see Chapter 11). The biosynthesis of these structures is complex, involving more than 20 genes, and it proceeds by the formation of a lipid precursor whose synthesis involves many of the same steps as those found in the synthesis of mammalian GPI anchors. The GPI precursor is incorporated near the carboxyl terminus of nascent glycoproteins in the ER by a GPI transamidase, which is directed to the site of addition by a signal sequence that is concomitantly cleaved. In an interesting divergence from higher eukaryotes, the GPI anchor is partly removed from some yeast glycoproteins, and the remaining glycosyl portion of the GPI structure, with the attached protein, becomes covalently linked to cell-wall β1-6 glucans.

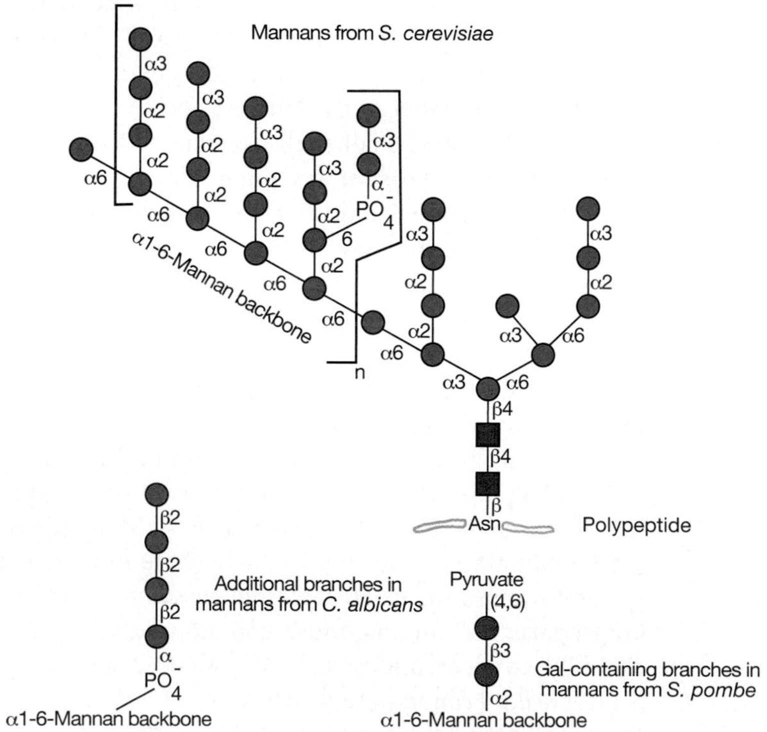

FIGURE 21.2. Structures of selected yeast mannans. Note that a single pyruvate is (R)4,6 acetyl-(ketal)-linked to the terminal galactose residue in the pyruvylated structure.

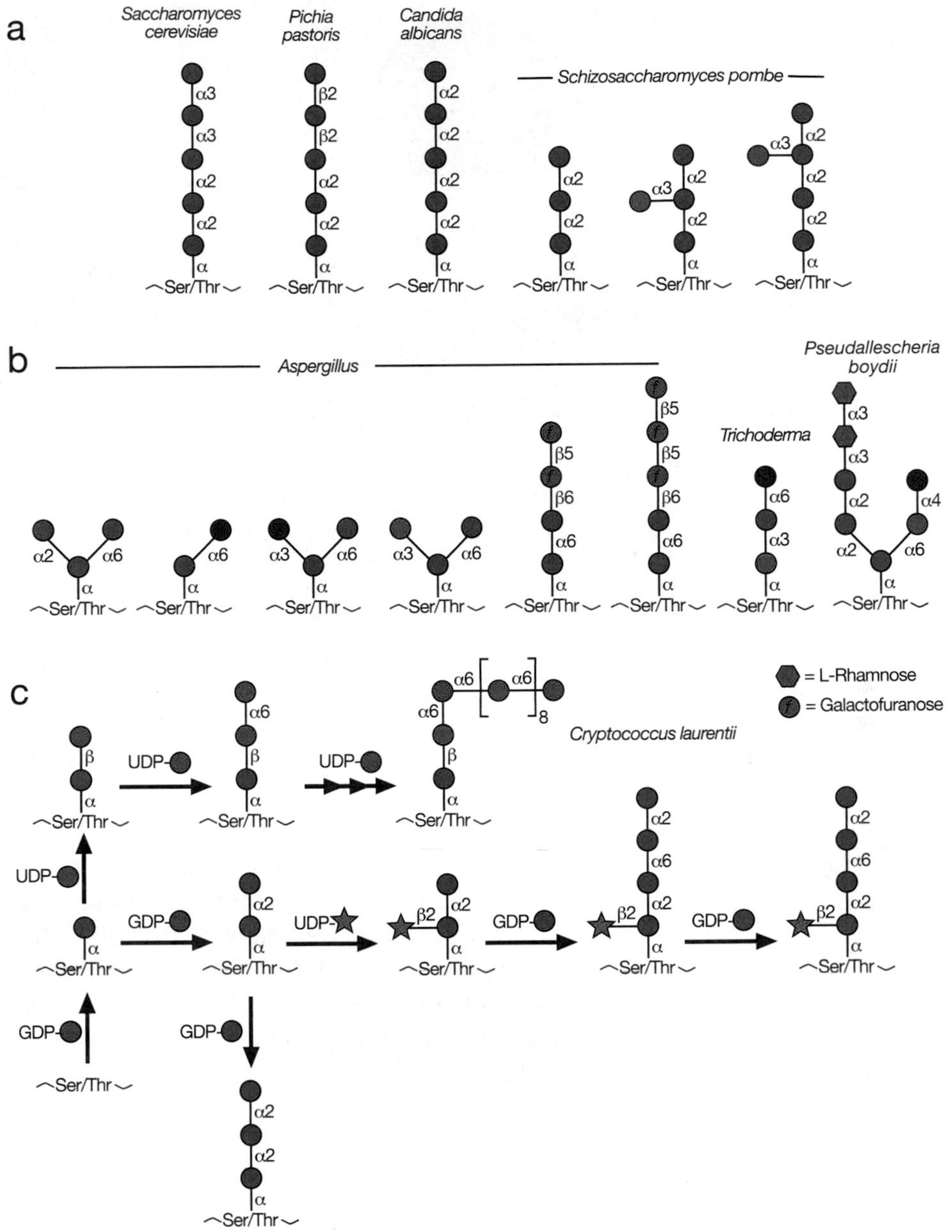

FIGURE 21.3. Structures of selected O-linked glycans in fungi: (a) yeast, (b) *Aspergillus*, and (c) *Cryptococuus*.

The cell wall of *S. cerevisiae* is a formidable structure and may account for up to one third of the dry weight of the cell. The major components of this cell wall are membrane-associated glycoproteins (mannoproteins) and β-linked glucans, along with some chitin that is primarily in the bud scars. The wall is a layered structure, with the rigid β-glucans proximal to the membrane surrounded by linked mannoproteins and additional β-glucans to form an elastic network. Cell-wall mannoproteins have the conserved N-glycan core structure linked to Asn residues, but this structure is further elaborated with an extensive repeating α1-6-linked mannose chain. The repeating α1-6-linked mannose backbone is usually branched by short chains of α1-2- and α1-3-linked mannose structures; some of these side chains may be in phosphodiester linkage (Figure 21.2). The mannans are highly heterogeneous in length and branching. In addition to large N-glycans

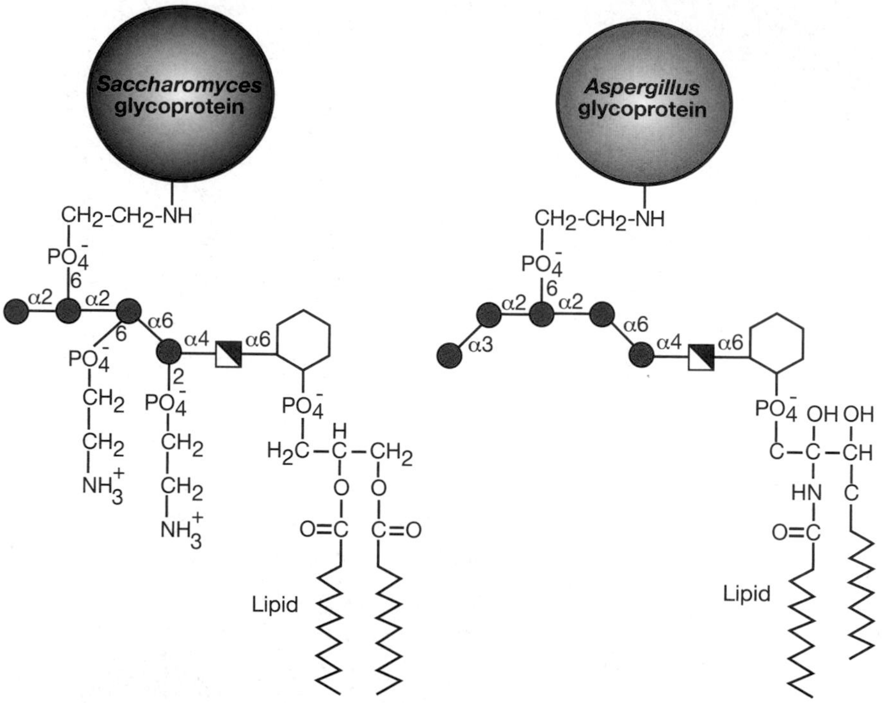

FIGURE 21.4. Structures of two yeast glycosylphosphatidylinositol (GPI)-anchors. *Hexagon* indicates *myo*-D-inositol.

rich in mannose, yeast cell walls contain proteins bearing Ser/Thr-linked O-mannose glycans, although these structures are of modest size compared to the mannans on N-glycans (Figure 21.3a). As mentioned above, many of the cell-wall mannoproteins in yeast also originate on GPI anchors (Figure 21.4). A wide assortment of mutants in mannan biosynthesis (termed *mnn* mutants) have been identified in *S. cerevisiae*, and a significant percentage of the genome (up to 20%) may be involved in elaborating the cell wall and its varied components.

Schizosaccharomyces pombe, a Fission Yeast

S. pombe is a rod-shaped fission yeast, which is 3–4 μm in diameter and 7–14 μm in length. Rather than budding, this organism grows by elongation and fission to give equal-sized daughter cells. Like *S. cerevisiae*, it has a relatively small genome of approximately 14 million base pairs. *S. pombe* has been useful as a genetically manipulatable model organism for studying the cell cycle. Because it has well-defined organelle structures compared to other yeast, it is a popular choice for studies of intracellular structure. *S. pombe* also synthesizes mannoproteins and mannans, some containing caps of α1-2-linked galactose residues that may also be pyruvylated. The galactose residues are important in lectin recognition in nonsexual flocculation (clumping) of *S. pombe*, as evidenced by inhibition of this process by free galactose. In contrast, flocculation in *S. cerevisiae* is mannose-dependent and inhibited by free mannose. The newly synthesized N-glycans in both *S. cerevisiae* and *S. pombe* have nine mannose residues ($Man_9GlcNAc_2Asn$; Figure 21.5). In *S. cerevisiae*, these structures are trimmed in the ER to form $Man_8GlcNAc_2Asn$ (as discussed below), but this does not occur in *S. pombe*. In contrast to the model yeast, some *S. pombe* O-linked mannose structures contain galactose (Figure 21.3a).

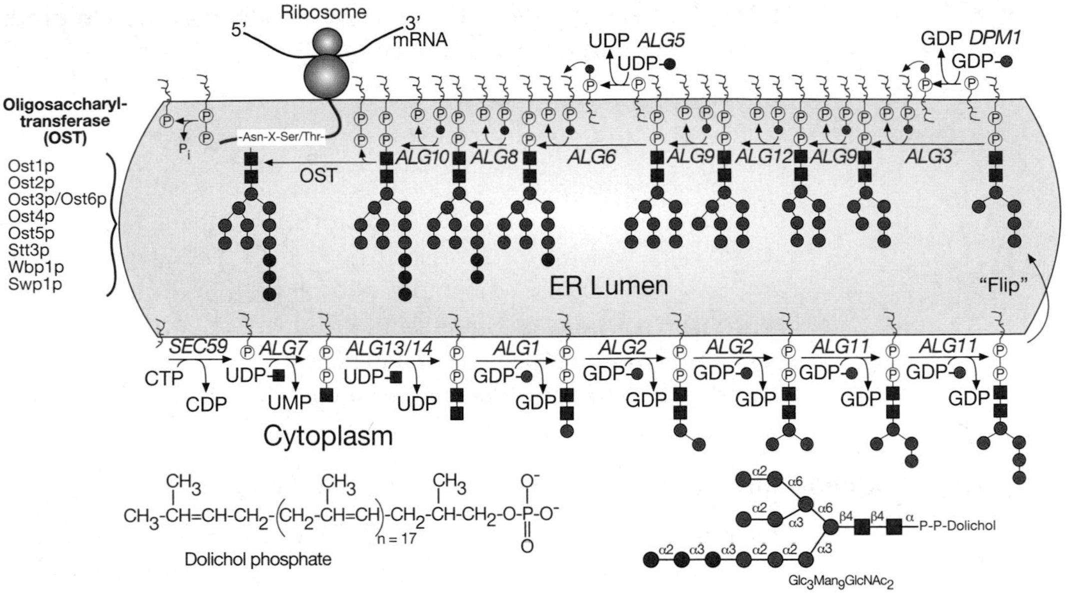

FIGURE 21.5. Biosynthesis of N-glycans and addition of N-glycans to -Asn-X-Ser/Thr- residues in newly synthesized glycoproteins in the yeast endoplasmic reticulum (ER). Individual steps in the biosynthesis pathway from dolichol phosphate and the structure of dolichol are shown. Steps in the biosynthesis identified by mutants in yeast are shown as genes designated as *ALG* or *SEC*. The structure of the intermediate $Glc_3Man_9GlcNAc_2$-P-P-Dol, which is common to yeast and mammals, is shown.

Pichia pastoris, a Popular Expression System

P. pastoris is a methylotrophic, nonpathogenic organism that was discovered in 1969 in a screen for yeast capable of utilizing methanol. Methanol is oxidized to formaldehyde and hydrogen peroxide by alcohol oxidase (AOX) in the peroxisome. The formaldehyde exits the peroxisome and is oxidized to formate and carbon dioxide in the cytoplasm for the purpose of energy production. Any remaining formaldehyde is assimilated into glyceraldehyde-3-phosphate and dihydroxyacetone by condensation with xylulose-5-monophosphate in a reaction catalyzed by the peroxisomal enzyme diydroxyacetone synthase. *P. pastoris* has become popular as a model system for making recombinant proteins because it is easy to manipulate genetically and can be grown to very high densities. The promoter for AOX is methanol inducible, and transcripts driven by this promoter may comprise up to 5% of the total poly(A)$^+$ RNA in induced cells.

P. pastoris has several advantages over *Escherichia coli* as an expression system in that it does not produce inclusion bodies and it promotes the correct folding of eukaryotic proteins. It also has certain advantages over yeast. First, although its basic pathway of N-glycosylation is similar to that of *S. cerevisiae* and yields glycoproteins with oligomannose-type N-glycans, these structures in *P. pastoris* have only 5–15 mannose residues (typically $Man_9GlcNAc_2Asn$ and $Man_8GlcNAc_2Asn$) compared with the 50–150 mannose residues found in glycoproteins from *S. cerevisiae* (see Figure 21.2). Hyperglycosylation in *S. cerevisiae* can interfere with protein folding, requiring use of *mnn* mutants to limit hyperglycosylation and avoid this problem. Also, *P. pastoris* does not add outer α1-3-linked mannose residues to its N-glycans. These structures are highly antigenic to humans, making proteins expressed in *S. cerevisiae* unsuitable for human pharmaceutical use. *P. pastoris* synthesizes O-glycans with an O-linked mannose core attached to Ser/Thr residues; most of these are short α1-2-linked mannose structures (see Figure

21.3a). Recently, *P. pastoris* has been genetically modified to produce therapeutic glycoproteins with humanized glycosylation.

Kluyveromyces lactis, a Yeast of Industrial Interest

K. lactis metabolizes lactose to lactic acid and, along with *A. niger* and *E. coli*, it is grown to produce rennet for making cheese and other products. *K. lactis* is also a rich source of β-galactosidase, which hydrolyzes lactose. *K. lactis* synthesizes mannans similar to those in *S. cerevisiae*, but they lack mannose phosphate modifications, and some side chains are capped with a residue of *N*-acetylglucosamine. A *K. lactis* mutant that lacks these *N*-acetylglucosamine residues because of a deficiency of the Golgi UDP-GlcNAc nucleotide sugar transporter has been productively exploited in studies of heterologous transporters.

Cryptococcus albidus and *C. laurentii*, Two Environmental Fungi

Cryptococcus is an encapsulated basidiomycete found in all environments. There are nearly 40 known species of *Cryptococcus*, including the pathogen *C. neoformans* (which is discussed below). Two common species of *Cryptococcus* that are not pathogenic are *C. albidus* and *C. laurentii*. The *C. albidus* capsule is composed of polysaccharides with a backbone of α1-3-linked mannose residues, some of which are O-acetylated and/or substituted with β1-2-linked xylose or glucuronic acid residues, similar to the glucuronoxylomannans in *C. neoformans*. *C. laurentii* is an encapsulated yeast that is prevalent in the tundra and it has the unusual property of producing toxins that kill the yeast *Candida albicans*. The O-glycans of *C. laurentii* have been well studied and are unusual in that they contain mannose, xylose, and galactose (Figure 21.3c) and they are synthesized by a unique set of mannosyl-, xylosyl-, and galactosyltransferases that are not homologous to human enzymes.

Pseudallescheria boydii, an Emerging Pathogen

P. boydii is found in soils and polluted water and is an emerging pathogen in immunocompromised individuals. The mycelial form of *P. boydii* synthesizes a novel peptidorhamnomannan that contains both N- and O-glycans. It is characterized by immunodominant α1-3-linked rhamnopyranose caps on the glycans, as shown in Figure 21.3b.

Candida albicans, an Important Pathogenic Yeast

C. albicans is a normal commensal organism that under various circumstances causes illness ranging from irritations of mucosal surfaces to life-threatening systemic infections. The *Candida* cell wall consists of mannans similar to those in *S. cerevisiae*, but they are termed phosphopeptidomannans. It also produces short β1-2-linked mannose glycans that are highly antigenic (Figure 21.2). These unusual β1-2-linked mannose glycans are also expressed on phospholipomannan antigens (PLMs). PLMs contain phytoceramide derivatives of *myo*-inositol phosphate to which are linked mannose and long polysaccharides of β1-2-linked mannose. The cell wall also contains β1-3- and β1-6-linked glucans and chitin. The O-glycans of *C. albicans* are short mannose-containing chains that have α1-2-linked mannose (Figure 21.3a) but lack the α1-3-linked mannose caps found in *S. cerevisiae*. As in *S. cerevisiae*, deficiencies of O-mannose addition generated through genetic deletions are lethal, indicating that O-mannosylation is essential in this yeast. *C. albicans* mannans are also important in its interactions with host cells, including macrophages and dendritic cells. In particular, these structures are recognized by the mannose receptor and by dectin-2. These are C-type lectins expressed by immune cells and they are important in both innate and adaptive immune responses (see Chapter 31). PLMs may be shed by *C. albicans*, and through interactions with Toll receptors (TLR-2), they can induce NF-κB activation and cytokine responses such as tumor necrosis factor-α (TNF-α) secretion. Galectin-3, a ubiquitous member of the galectin family of lectins that is highly expressed in macrophages, also appears to recognize *C. albicans* expressing β1-2-linked mannose residues, resulting in opsonization of the yeast.

Dictyonema glabratum Adopts a Symbiotic Lifestyle

Basidiomycetes are fungi that produce spores formed from a pedestal-like structure called the basidium. Examples of basidiomycetes include fungi that have gills or pores, such as common mushrooms and bracket fungi as well as the *Cryptococcus* species discussed above. *D. glabratum* illustrates another fungal lifestyle as this organism lives in symbiosis with cyanobacteria *Scytonema* sp. forming a lichen. This lichen synthesizes many glycans that are unusual compared to those of other lichen and fungi, including β-glucans, mannans, and xylans. For example, the β-glucans of most lichens are linear, but in *D. glabratum* they are branched with β1-3 and β1-6 linkages. The mannans of *D. glabratum* have an α1-3-linked backbone, rather than the typical α1-6 linkages found in other lichens, along with branches at the 2 and 4 positions. Finally, the xylans of this organism are linear β1-4-linked polymers of xylose, more typical of those found in higher plants and algae than in fungi. *D. glabratum* also synthesizes some unusual short glycolipids, including glycosyldiacylglycerolipids, which are similar to plant glycolipids and contain mono-, di-, and linear trisaccharides of α1,6-linked galactopyranose. This fungus thus serves as one more example of the extensive and surprising diversity seen in the fungal kingdom.

PROTEIN GLYCOSYLATION IN *S. CEREVISIAE*

The yeast *S. cerevisiae* has been an excellent model system for the study of certain glycan synthetic pathways. Many successful studies have used various approaches to select for glycosylation-deficient mutants. For example, mannan biosynthesis was partly defined by selecting mutants for altered antibody binding. Other important mutants that have proven valuable in studying N-glycosylation pathways were generated by [^{3}H]mannose suicide selection; only cells that incorporate unusually low levels of [^{3}H]mannose survive growth in the presence of this compound. The survivors included yeast mutants that were defective in N-glycosylation, termed the *alg* mutants for their deficiency in asparagine-linked glycosylation. The *alg* mutants were sorted into complementation groups and found to be defective in adding glycans to polypeptide precursors, thus explaining the reduced mannose incorporation. At nonpermissive temperatures, these mutants accumulated specific lipid-linked oligosaccharides immediately upstream of the defective biosynthetic reactions. Analysis of the accumulated intermediates allowed correlation of the mutants with steps in the assembly of the dolichol-P-P-oligosaccharide precursor (Figure 21.5) and identification of the enzymes involved. Mammalian cells have homologs to each of these yeast enzymes and they act to synthesize the conserved lipid-linked core glycan donor, $Glc_3Man_9GlcNAc_2$-P-P-Dol, which is transferred to nascent polypeptides in the ER (see Chapter 8).

Following core N-glycosylation, the $Glc_3Man_9GlcNAc_2$Asn-R is processed in mammals and yeast, with removal of glucose residues by α-glucosidases I and II to generate $Man_9GlcNAc_2$Asn-R. In mammals and *S. cerevisiae*, the $Man_9GlcNAc_2$Asn-R is further trimmed to $Man_8GlcNAc_2$ Asn-R by an ER-mannosidase. *S. pombe* lacks this enzyme and stops processing at $Man_9GlcNAc_2$Asn-R. The $Man_8GlcNAc_2$Asn-R in mammals and the

$Man_9GlcNAc_2Asn$-R in *S. pombe* are substrates for the UDP-Glc:glycoprotein glucosyltransferase that generates $Glc_1Man_8GlcNAc_2Asn$ and $Glc_1Man_9GlcNAc_2Asn$ in mammals and *S. pombe*, respectively. This reglucosylation is part of the quality control system in protein folding in the ER (see Chapter 36). The monoglucosylated structure is a ligand for the chaperone lectins calnexin and calreticulin in mammalian cells. Although *S. cerevisiae* lacks this glucosyltransferase, both *S. cerevisiae* and *S. pombe* express calnexin, but lack a calreticulin homolog. The newly synthesized *S. cerevisiae* glycoproteins may acquire monoglucosylated, oligomannose N-glycans during partial deglucosylation by α-glucosidase II, which may be sufficient to allow them to interact with the quality-control pathway. However, the quality-control system in *S. cerevisiae* may also be weak compared to that of *S. pombe* and mammalian cells, because the survival of *S. cerevisiae* is not affected by mutations causing loss of the calnexin homolog.

Yeast protein O-linked mannose is added by ER protein mannosyltransferases (PMTs), which use Dol-P-Man as a mannose donor. There are several PMTs in *S. cerevisiae* and each may have a different substrate specificity and glycoprotein preference. Mammals also have protein *O*-mannosyltransferases (POMTs) that use Man-P-Dol as a donor to generate Manα-Ser/Thr-modified glycoproteins in the ER (see Chapter 12). The Man-P-Dol for the yeast PMTs is synthesized as shown in Figure 21.5 (top right) and is used for both N- and O-glycosylation pathways. Subsequent additions of mannose residues to growing chains occur in the Golgi apparatus, where GDP-Man serves as the donor for reactions catalyzed by Mn^{++}-dependent mannosyltransferases. This is interesting in topological terms, because GDP-Man is used to generate both the biosynthetic $Man_5GlcNAc_2$-P-P-Dol precursor, which is then "flipped" into the ER, and the Dol-P-Man donor of additional mannosylation reactions (Figure 21.5). Thus, GDP-Man does not appear to be transported into the ER, although it is transported into the Golgi apparatus.

Mannan extensions are generated in the Golgi apparatus by mannosyltransferases that utilize GDP-Man as the donor, with one or more specific glycosyltransferases to catalyze synthesis of each linkage and branch. Phosphomannose addition occurs by transfer of Man-1-P from GDP-Man donors in the Golgi apparatus. Many mutants in mannan biosynthesis (*mnn*) have been identified that lead to truncated mannans.

CAPSULE BIOSYNTHESIS IN THE PATHOGEN *CRYPTOCOCCUS NEOFORMANS*

C. neoformans is a common soil-dwelling, encapsulated fungus. It latently infects healthy people but causes severe disease in immunocompromised individuals and is an AIDS-defining illness. *C. neoformans* is unique among pathogenic fungi in having an extensive polysaccharide capsule that is required for its virulence (Figure 21.6). Two major capsular polysaccharides named for their monosaccharide components are glucuronoxylomannan (GXM) and a galactoxylomannan (GalXM). GXM is an extended α1-3 mannan substitut-

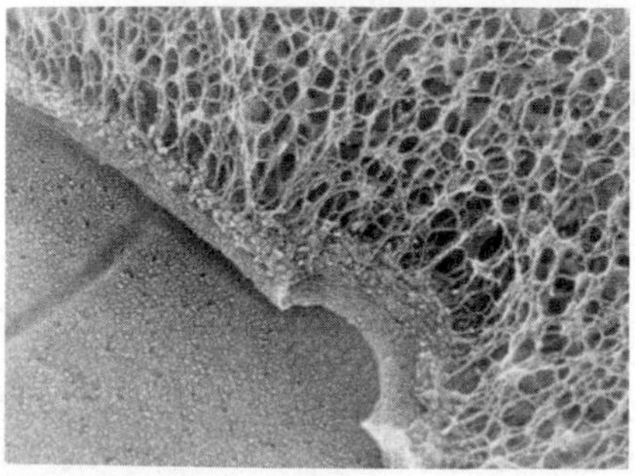

FIGURE 21.6. A quick-freeze deep-etch image of the edge of a *C. neoformans* cell. The polysaccharide capsule (open meshwork at right) is linked to the cell wall (central structure dividing the image from upper left to lower right) via α1-3 glucan. The region at lower left is the cell membrane, and the arc in the cell wall represents formation of a new bud. (Image by John Heuser and Tamara Doering.) (Reprinted, with permission of Blackwell Publishing, from Reese A.J., Yoneda A., Berger J.A., et al. 2007. *Mol. Microbiol.* **63:** 1385–1398.)

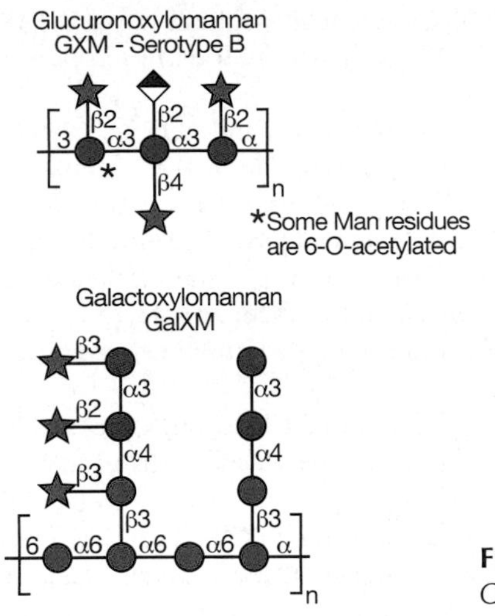

FIGURE 21.7. Structures of capsular polysaccharides in *Cryptococcus neoformans*.

ed with β1-2Xyl, β1-4Xyl, and β1-2GlcA, and a subset of the mannose residues are 6-O-acetylated. Several serotypes of *C. neoformans* differ in the xylose modifications of the repeating unit; serotype B is shown in Figure 21.7. The second polymer, GalXM, is based on α1-6 galactan, with side chains of galactose, mannose, and xylose (Figure 21.7). In GalXM, most of the galactose is galactopyranose, although galactofuranose has been reported in this structure. The cryptococcal galactofuranose donor is generated by the enzyme UDP-galactopyranose mutase. Molecularly, GXM is substantially larger than GalXM (1.7–7.4 x 10^6 daltons for GXM vs. 10^5 daltons for GalXM).

The capsule is a dynamic structure that can change in thickness and composition depending on the environment and growth conditions. Under standard in vitro conditions, the capsule is approximately 1–2 µm in diameter, but it can be much thicker, especially in the context of mammalian infection. Association of the capsule with the cell surface relies on a cell-wall component, α1-3 glucan. Although α1-3 glucan is not present in the cell walls of *S. cerevisiae* or *C. albicans*, it is common in other fungi.

Cell biological studies indicate that cryptococcal capsule components are generated intracellularly. Investigations of capsule biosynthesis have taken both molecular and biochemical approaches. For example, mutants in capsule structure have been identified through genetic screening approaches based on colony morphology and antibody reactivity, allowing identification of genes required for capsule synthesis. Genome analysis has also identified candidate biosynthetic enzymes. Finally, direct biochemical studies of cryptococcal enzymes have been applied to this question. Together, these approaches have begun to elucidate the biosynthetic pathways required to generate this elaborate and important structure.

FURTHER READING

Spencer J.F. and Gorin P.A. 1973. Mannose-containing polysaccharides of yeasts. *Biotechnol. Bioeng.* **15:** 1–12.

Ballou C.E. 1974. Some aspects of the structure, immunochemistry, and genetic control of yeast mannans. *Adv. Enzymol. Relat. Areas Mol. Biol.* **40:** 239–270.

Ballou C.E. and Raschke W.C. 1974. Polymorphism of the somatic antigen of yeast. *Science* **184:** 127–134.

Ballou C.E., Lipke P.N., and Raschke W.C. 1974. Structure and immunochemistry of the cell wall mannans from *Saccharomyces chevalieri*, *Saccharomyces italicus*, *Saccharomyces diastaticus*, and *Saccharomyces carlsbergensis*. *J. Bacteriol.* **117:** 461–467.

Huffaker T.C. and Robbins P.W. 1983. Yeast mutants deficient in protein glycosylation. *Proc. Natl. Acad. Sci.* **80:** 7466–7470.

Herscovics A. and Orlean P. 1993. Glycoprotein biosynthesis in yeast. *FASEB J.* **7:** 540–550.

Bretthauer R.K. and Castellino F.J. 1999. Glycosylation of *Pichia pastoris*-derived proteins. *Biotechnol. Appl. Biochem.* **30:** 193–200.

Burda P., Jakob C.A., Beinhauer J., Hegemann J.H., and Aebi M. 1999. Ordered assembly of the asymmetrically branched lipid-linked oligosaccharide in the endoplasmic reticulum is ensured by the substrate specificity of the individual glycosyltransferases. *Glycobiology* **9:** 617–625.

Dickson R.C. and Lester R.L. 1999. Yeast sphingolipids. *Biochim. Biophys. Acta* **1426:** 347–357.

Gemmill T.R. and Trimble R.B. 1999. Overview of N- and O-linked oligosaccharide structures found in various yeast species. *Biochim. Biophys. Acta* **1426:** 227–237.

Herscovics A. 1999. Processing glycosidases of *Saccharomyces cerevisiae. Biochim. Biophys. Acta* **1426:** 275–285.

Parodi A.J. 1999. Reglucosylation of glycoproteins and quality control of glycoprotein folding in the endoplasmic reticulum of yeast cells. *Biochim. Biophys. Acta* **1426:** 287–295.

Calderone R., Suzuki S., Cannon R., Cho T., Boyd D., Calera J., Chibana H., Herman D., Holmes A., Jeng H.W., Kaminishi H., Matsumoto T., Mikami T., O'Sullivan J.M., Sudoh M., Suzuki M., Nakashima Y., Tanaka T., Tompkins G.R., and Watanabe T. 2000. *Candida albicans*: Adherence, signaling and virulence. *Med. Mycol.* (suppl. 1) **38:** 125–137.

Bencurova M., Rendic D., Fabini G., Kopecky E.M., Altmann F., and Wilson I.B. 2003. Expression of eukaryotic glycosyltransferases in the yeast *Pichia pastoris. Biochimie* **85:** 413–422.

Bose I., Reese A.J., Ory J.J., Janbon G., and Doering T.L. 2003. A yeast under cover: The capsule of *Cryptococcus neoformans. Eukaryot. Cell* **2:** 655–663.

Hamilton J.F., Bobrowicz P., Bobrowicz B., Davidson R.C., Li H., Mitchell T., Nett J.H., Rausch S., Stadheim T.A., Wischnewski H., Wildt S., and Gerngross T.U. 2003. Production of complex human glycoproteins in yeast. *Science* **301:** 1244–1246.

Poulain D. and Jouault T. 2004. *Candida albicans* cell wall glycans, host receptors and responses: Elements for a decisive crosstalk. *Curr. Opin. Microbiol.* **7:** 342–349.

Daly R. and Hearn M.T. 2005. Expression of heterologous proteins in *Pichia pastoris*: A useful experimental tool in protein engineering and production. *J. Mol. Recognit.* **18:** 119–138.

Wildt S. and Gerngross T.U. 2005. The humanization of N-glycosylation pathways in yeast. *Nat. Rev. Microbiol.* **3:** 119–128.

Kelleher D.J. and Gilmore R. 2006. An evolving view of the eukaryotic oligosaccharyltransferase. *Glycobiology* **16:** 47R–62R.

Klutts J.S., Yoneda A., Reilly M.C., Bose I., and Doering T.L. 2006. Glycosyltransferases and their products: Cryptococcal variations on fungal themes. *FEMS Yeast Res.* **6:** 499–512.

Taylor J.W. and Berbee M.L. 2006. Dating divergences in the Fungal Tree of Life: Review and new analyses. *Mycologia* **98:** 838–849.

Klis F.M., Ram A.F., and De Groot P.W. 2007. A molecular and genomic view of the fungal cell wall. *The Mycota VIII.* In *Biology of the fungal cell,* 2nd ed. (ed. R.J. Howard and N.A.R. Gow), pp. 97–120. Springer-Verlag, Berlin.

Schutzbach J., Ankel H., and Brockhausen I. 2007. Synthesis of cell envelope glycoproteins of *Cryptococcus laurentii. Carbohydr. Res.* **342:** 881–893.

CHAPTER 22

Viridiplantae

Marilynn E. Etzler and Debra Mohnen

GREEN PLANTS CONSTITUTE ABOUT HALF OF THE LIVING MATTER on Earth and have a diversity ranging from simple green algae to flowering plants. Plants synthesize many of the same glycans that are found in animals and also produce a wide variety of unique glycans that have important roles during the life cycle of the plant. This chapter summarizes our current knowledge of plant glycobiology and shows some unique functions of glycans in these organisms.

BACKGROUND

Plants contain many glycans, especially in their cell walls. In fact, some of the earliest glycobiology studies were conducted on cellulose and other glycans of plant cell walls as well as on seed lectins, particularly those from leguminous plants. In the past several decades, however, most glycobiological research has focused on animal systems because of an increased interest in biomedicine and the availability of funding in that area. This biomedical emphasis is readily demonstrated by the high proportion of chapters of this book that are devoted to animal systems.

It is important to realize, however, that plant glycobiology has produced a number of findings of extreme interest to the biomedical community. The concept of oligosaccharide signaling in development originated in studies with plants (see Chapter 37), and a wide variety of plant lectins of different glycan specificities are currently used in biomedical research (see Chapter 26). Unique plant glycans provide major sources of food and fiber. Some plant species have easily manipulatable systems for genetic and molecular biology studies, including the expression of large amounts of glycoproteins.

Limited and unstable resources of oil have sparked a renewed interest in the use of plant-wall carbohydrates as sources of biomass for biofuel production. This may usher in a "golden age" for plant glycobiology, with rapid increases in our understanding of the function and synthesis of what appears to be the most structurally complex reservoir of carbohydrate structures in nature.

UNIQUE FEATURES OF THE PLANT SECRETORY PATHWAY

Most plants are not mobile and must grow in the environment in which their seeds are sown. They have evolved special defenses against predators such as insects, fungi, and other pathogens. Defense responses often involve the production of phenols and other toxic reagents, which are stored in the plant vacuoles. The vacuole is usually the largest organelle of mature plant cells and often accounts for most of the volume of the cell. This organelle is the plant equivalent of the animal lysosome and is a major component of the plant secretory system. In the seeds of many plant species, the vacuoles condense into compact organelles called protein bodies, which are packed with glycoproteins that are used as food upon germination.

Although glycoproteins enter the vacuole by way of the secretory system, plant cells do not use mannose-6-phosphate for targeting of proteins as is the case in mammalian cells (see Chapter 30), and the phosphotransferase employed by animal cells to direct proteins to lysosomes has not been found in plants. Several signals have been identified for vacuolar protein targeting, but none are glycan based.

The vacuole is also the site of synthesis and storage of a variety of polysaccharides, including fructans. These polysaccharides serve as reserve storage materials for energy production for the plants and may play a role in helping plants survive harsh environmental conditions and seed desiccation.

UNIQUE FEATURES OF PLANT N-GLYCANS

Most of the proteins that pass through the plant secretory system contain N-linked glycans of four main types: oligomannose, complex, hybrid, and paucimannose (Figure 22.1). No significant difference has been found between plants and animals in the initial stages of N-glycan synthesis, including the transfer of the dolichol oligosaccharide precursor and the control of protein folding in the endoplasmic reticulum (ER) (see Chapter 8). However, plants show some unique modifications of N-glycans upon passage through the Golgi.

As in vertebrates, the glycoproteins enter the plant Golgi with oligomannose-type N-glycans, and many of these glycans are trimmed in the *cis*-Golgi. As mentioned previously, in contrast to animal cells, no phosphorylation of mannose occurs in the plant *cis*-Golgi. After passage to the *medial*-Golgi and the addition of *N*-acetylglucosamine to the distal mannose of the core by N-acetylglucosamine transferase I (GnT-I) as in vertebrates, two specific plant modifications occur (Figure 22.2). The first modification is the addition of xylose in β1-2 linkage to the core β-mannose. This modification occurs in the *medial*-Golgi and is unique to plants. The second modification, which occurs in the *trans*-Golgi, is the addition of fucose in α1-3-linkage to the asparagine-linked *N*-acetylglucosamine residue. This core fucosylation has also been found in invertebrates. Both the xylotransferase and fucosyltransferase that catalyze the above reactions have been isolated, and substrate specificity

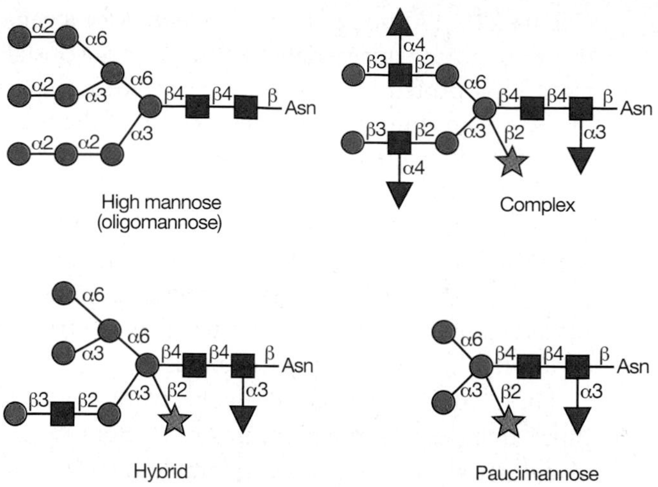

FIGURE 22.1. Types of N-glycans found in plants.

studies have shown that they require at least one terminal *N*-acetylglucosamine residue. Each reaction is independent because neither enzyme requires the addition of the other sugar for activity. Following these additions, the glycan is trimmed by α-mannosidase II and a second *N*-acetylglucosamine is added to the trimmed chain by GnT-II. As in animal cells, some plant glycans do not undergo further mannose trimming and proceed through the Golgi as hybrid-type glycans (Figure 22.2).

The complex and hybrid-type glycans are further processed as they pass through the *trans*-Golgi. As in animals, these modifications include the addition of galactose and fucose

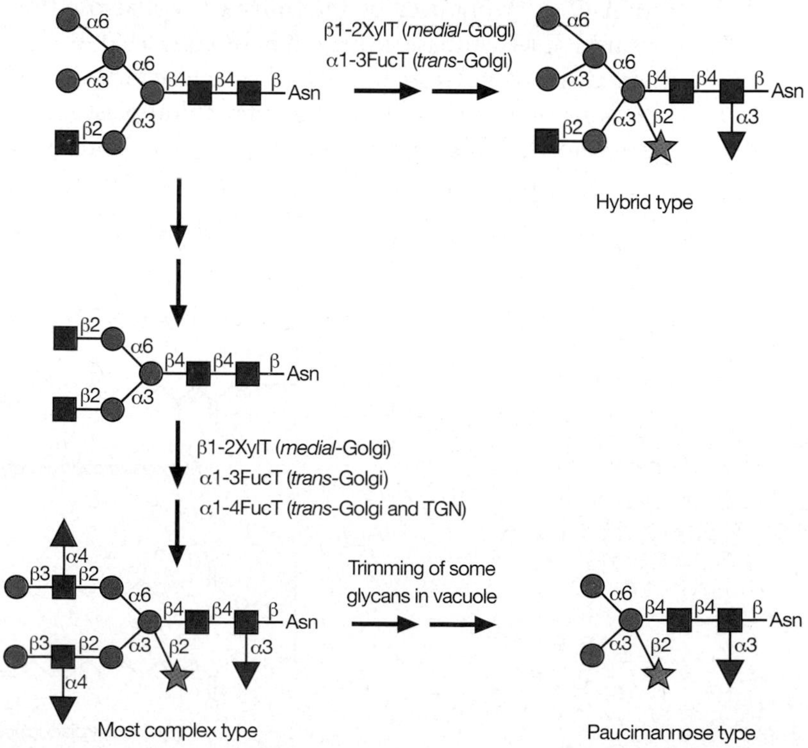

FIGURE 22.2. Processing of N-glycans in the plant secretory system. Only those events that are unique to plants are shown in detail.

(Figure 22.2). Although there is evidence for a gene resembling a sialyltransferase in plants, there are no conclusive data for the presence of sialic acid in plants. After leaving the Golgi, plant glycoproteins are either secreted from the cell or transported to the vacuoles. Many of the glycans found in the vacuoles are of the paucimannose type, suggesting that they are trimmed by vacuolar glycosidases.

Many of the glycosyltransferases that catalyze the processing of glycans in the plant Golgi have specificities similar to those of their animal homologs. cDNAs show homologies with the animal genes. The cDNAs of the unique core β1-2 xylosyltransferase and α1-3 fucosyltransferase have been sequenced. Core α1-3 fucosyltransferase is related to the Lewis fucosyltransferase family, whereas the β1-2 xylosyltransferase is unique and unrelated to other known glycosyltransferases.

These core modifications make plant N-glycans highly immunogenic and may also cause allergies to some plant products. Because plants are being used for mass production of animal glycoproteins, considerable attention is focused on preventing the addition of these residues to glycans to prevent an immune response.

PLANT GLYCOLIPIDS

The most abundant glycolipids in plants are the glycoglycerol lipids. The two main members of this class are mono- and digalactosyldiacylglycerol (Figure 22.3), which are present in all plants. Some species of plants also contain tri- and tetragalactosyldiacylglycerol. The synthesis of all of these galactolipids begins with the synthesis of diacylglycerol, which can occur in either the ER membrane or the chloroplast membrane. The former pathway produces mostly C16 fatty acids at the *sn*2 position and C18 fatty acids at the *sn*3 position, whereas the latter pathway produces C18 fatty acids at both positions. Each of these fatty acids is then desaturated to 16:3 or 18:3 acyl groups. Monogalactosyldiacylglycerol (MGDG) is synthesized by the transfer of galactose from UDP-galactose to diacylglycerol by an MGDG synthase. It may then be converted to digalactosyldiacylglycerol (DGDG) by the transfer of galactose from UDP-galactose to MGDG by a DGDG synthase. Both of these reactions occur primarily in the outer chloroplast membrane and the products are transported to the inner membrane and the thylakoid membranes of the chloroplasts. Small

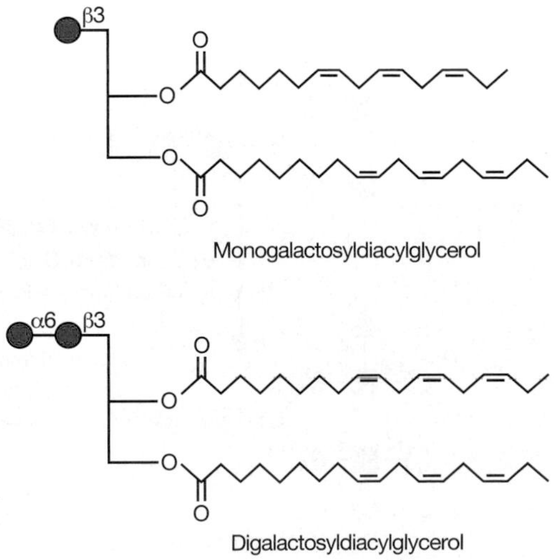

FIGURE 22.3. Most abundant plant galactolipids.

amounts of both galactolipids can also be found in the plasma membranes of the cells. The mechanism of galactolipid exchange among the membranes is not yet understood.

The thylakoid membranes of chloroplasts are the site of photosynthesis, and it has long been suggested that the abundance of galactolipids in the thylakoids may play an important role in photosynthesis. Crystallization of the photosystem I complex of cyanobacteria and studies of a MGDG synthase mutant in *Arabidopsis* support this idea. Another glycerolipid synthesized from diacylglycerol is the sulfolipid, sulfoquinovosyldiacylglycerol. It is abundant in the thylakoid membrane, which suggests that it may also play a role in photosynthesis.

In addition to the above glycolipids, plant membranes also contain glycosphingolipids (see Chapter 10), which are found as components of the plasma membrane. Compared to animals, few plant glycosphingolipids have been characterized. Although such sugars as galactose and fucose are found in these glycans, no gangliosides have been identified.

Some plant glycoproteins have glycosylphosphatidylinositol-type (GPI) anchors (see Chapter 11). Only a few plant GPI anchors have been studied, and these contain a phosphoceramide core characteristic of yeast and *Dictyostelium*. A complete structure of the glycan of the GPI anchor of a pea arabinogalactan protein has been determined as -Manα1-2Manα1-6Manα1-4GlcNH2-inositol. A unique feature of this structure is that more than 50% of the anchors are further substituted at the mannose attached to the glucosamine unit (see Chapter 11) with β1-4 galactose.

PLANTS CONTAIN A LARGE VARIETY OF O-GLYCANS

The most abundant O-glycans in plants are the hydroxyproline-rich glycoproteins (HRGPs) found in the cell wall. These proteins have a high content of proline that is converted to hydroxyproline by prolyl hydroxylases in the ER. Hydroxyproline residues are O-glycosylated as they pass through the ER and Golgi apparatus. The degree and type of hydroxyproline glycosylation depend on the primary structure of each protein backbone and are dictated by the arrangement of hydroxyprolines in the sequence. The glycosylation of hydroxyproline is found in plants and *Dictyostelium* (see Chapter 17) and is initiated by the addition of an arabinose or galactose residue in plants. Contiguous hydroxyproline residues are arabinosylated, whereas clustered noncontiguous hydroxyproline residues are galactosylated. Plant proteins with these modifications are usually divided into three subclasses as discussed below. These proteins can also be O-glycosylated through serine residues and occasionally through threonine. Another feature of these cell wall proteins is that they can form intra- and intermolecular cross-linkages between tyrosines, a reaction that is catalyzed by specific peroxidases in the cell wall.

The first subclass of HRGPs are the extensins, which are characterized by a $Ser(Hyp)_4$ repeat and contain about 50–60% (w/w) glycan, most of which consists of short chains of one to four arabinose residues bound to hydroxyproline. They also have a small number of single galactose residues α-linked to serine residues. The second HRGP subclass, termed the proline/hydroxyproline-rich glycoproteins, is primarily distinguished from the extensins by amino acid sequence. Not as much is known of their glycosylation patterns, but the extent of glycosylation is known to range from 3% to 70% (w/w). Both of these two subclasses of HRGPs are thought to play a structural role in the cell wall, and their expression is developmentally regulated. Wounding and fungal attack of plant tissues have been found to induce the synthesis of some of these proteins.

The third HRGP subclass is the arabinogalactans. They have a neutral to acidic peptide backbone and a glycan content of more than 90% (w/w) containing both galactosyl-*O*-serine and galactosyl-*O*-hydroxyproline linkages of 30–150 residues. Most of these chains consist of a β1-3 galactose backbone with extensive branches of β1-6-linked galactose residues. These chains are substituted with β1-3-linked and terminal arabinose and other residues. The arabinogalactans are secreted into the cell wall and can also be anchored to the plasma membrane. Proposed roles for these glycans include participation in signaling, development, cell expansion, cell proliferation, and somatic embryogenesis.

Another interesting group of hydroxyproline-rich, O-glycosylated glycoproteins are the lectins of the Solanaceous family of plants. This group includes the potato, tomato, and thorn-apple lectins, which contain two domains: a nonglycosylated carbohydrate-binding domain that is rich in cysteine and glycine and a highly glycosylated hydroxyproline-rich domain. The glycan chains in this latter domain are similar to those of the extensins and most contain three or four residues of arabinose β-linked to hydroxyproline and single galactose residues α-linked to serine. These lectins are found in the vacuoles of Solanaceous plants and they are thought to play a role in defense by binding to chitin present on the surface of pathogens.

In addition to unique types of O-glycans, plants also contain O-GalNAc in α-linkage to serine and threonine residues (see Chapter 9). O-GlcNAc occurs in cytoplasmic and nuclear proteins at serine and threonine residues where phosphorylation occurs (see Chapter 17).

THE GLYCAN-RICH CELL WALL HELPS CONTROL GROWTH OF PLANT CELLS

The cell wall is the extracellular matrix of the plant cell. It must be strong enough to support the plant and withstand the internal turgor pressure of the cell. Yet, it must also be able to extend during cell growth and participate in interactions with the environment. The cell wall is composed of a network of strong rod-like molecules tethered together by cross-linked glycans and embedded in a highly organized matrix of acidic polysaccharides, some of which partially resemble the proteoglycans of animal cells.

Plant cells use two types of cell walls to perform these functions, termed the primary and secondary walls. The primary wall is the first wall laid down in dividing and growing plant cells, and it is the terminal wall in many cells in the soft parts of the plant (e.g., surrounding palisade cells in leaves). The primary wall contains 80–90% polysaccharide and 10–20% protein. Cellulose, hemicellulose, and pectin are the main polysaccharide components in the primary wall (Figure 22.4). Secondary walls, which surround specialized cells that serve a structural role such as fiber cells and xylem cells in vascular bundles, generally have less pectin, more cellulose, and a different class of hemicellulose (β1-4 xylans) and are often rigidified by lignification.

Instead of the collagen fibrils used by animal cells to provide the tensile strength of the extracellular matrix, plant cells synthesize cellulose, a polysaccharide that consists of long, parallel, linear β1-4 glucan chains hydrogen-bonded into crystalline microfibrils containing about 36 polysaccharide chains. Cellulose is the most abundant biopolymer in nature. The synthesis of these chains occurs on the plasma membrane of the cell where cellulose synthase is assembled into rosette structures containing six subunits. Each of these subunits is thought to be composed of six cellulose synthases, each of which adds glucose onto the growing glucan chains using cytoplasmic UDP-glucose generated from sucrose by the reverse action of sucrose synthase. The cellulose fibers that are produced are parallel with one another (Figure 22.4) and are in alignment with the cytoskeleton underlying the membrane.

The cellulose fibers are coated and cross-linked with one another by glycans called hemicelluloses. Xyloglucan is the major hemicellulose in the type I primary cell walls of most higher plants (Figure 22.5). It is a polymer consisting of repetitive segments of four residues of a β1-4 glucan backbone substituted on the first three positions with α1-6 xylose. The xyloses at positions 2 and 3 can have galactose attached in a β1-2 linkage and a

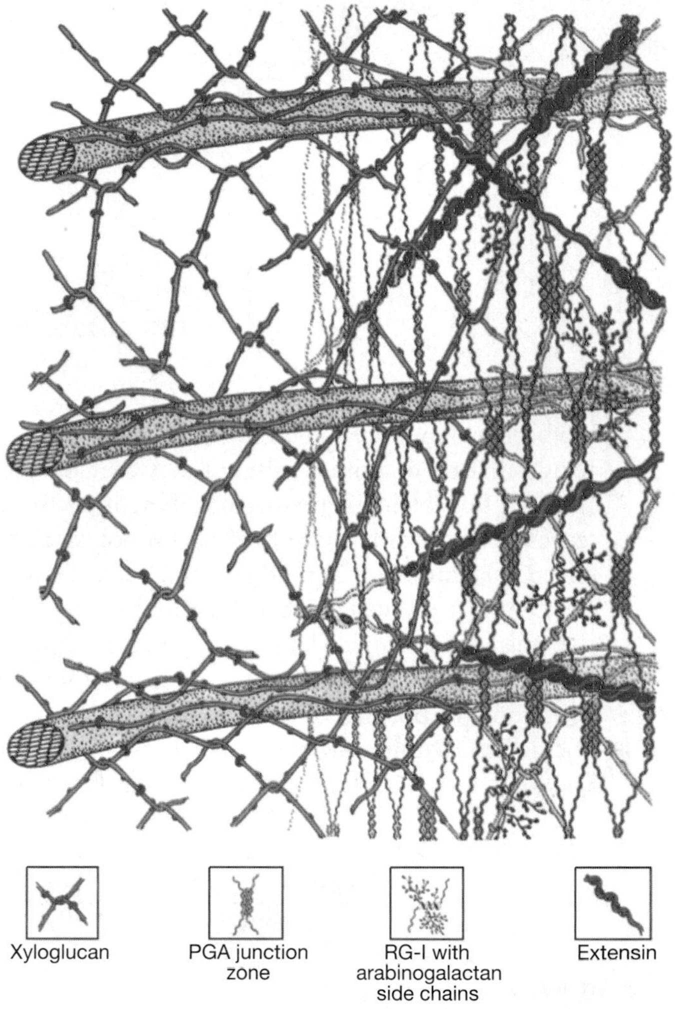

FIGURE 22.4. Model of the primary cell wall (type I) found in most flowering plants (except grasses). Cellulose microfibrils are embedded in a hemicellulose (e.g., xyloglucan) and pectin matrix. The model depicts partial hydrogen bonding of xyloglucan to cellulose along with bond breakage in xyloglucan that may accompany cell expansion. Pectin (homogalacturonan [HGA, PGA]), rhamnogalacturonan I (RG-I), and rhamnogalacturonan II (RG-II, not shown) form a structurally complex hydrated and adhesive matrix. The orientation of the rod-like extensin molecules is not known. (Reprinted, with permission, from Carpita N.C. and Gibeaut D.M. 1993. *Plant J.* **3:** 1–30.)

fucose is usually found in an α1-2 linkage to the galactose at C-2. Other, less abundant polysaccharides, such as the glucomannans and galactomannans, are also thought to help orient the cellulose fibrils. In grasses and other commelinoid monocots with type II walls, glucuronoarabinoxylan is the major hemicellulose, whereas during expansion in grasses β1-3/β1-4 mixed-linkage glucans are prevalent.

As cells increase in volume, there is a loosening of the hydrogen bonds that bind hemicelluloses to cellulose, and an enzymatic cleavage of some of the xyloglucan chains allows the internal osmotic pressure of the cell to push apart the cellulose microfibrils. This process is accompanied by a laying down of layers of new cellulose microfibrils with their associated hemicellulose polymers. In the cell plate of dividing cells or at times during growth, callose (a β1-3 glucan polymer) is found among the cellulose fibrils.

The cellulose-xyloglucan network is embedded in a matrix of complex acidic polysaccharides called pectins (Figure 22.6). Pectins are the most structurally complex plant-wall poly-

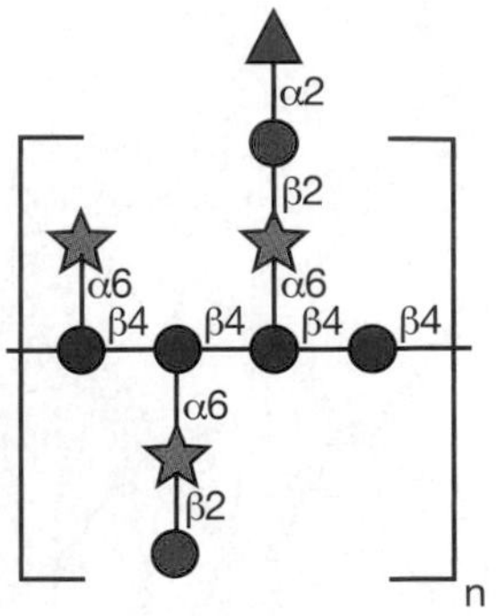

FIGURE 22.5. Repeating subunit found in xyloglucan.

saccharides. They are somewhat comparable to the glycosaminoglycans found in the extracellular matrix of animal cells, but they are not sulfated and it is not clear which, if any, are attached to protein. A number of different pectins have been isolated. Homogalacturonans represent approximately 65% of pectin polysaccharides and they are linear chains of α1-4-linked galacturonic acid residues, which may be methyl-esterified on some of their carboxyl groups and acetylated at the C-2 or C-3 hydroxyls. Pectin methyl esterases in the cell wall can alter the extent of methylation of the pectin, allowing it to associate with other pectin molecules by calcium links between the acidic groups. The extent of methylation can therefore determine the porosity of the pectin matrix in the wall and the extent of cell-wall stiffening. The pectin rhamnogalacturonan-I (RG-I) has a repeating backbone of GalAα1-2Rhaα1-4. The galacturonic acid in the backbone may be O-acetylated at C-2 or C-3. Approximately 20–

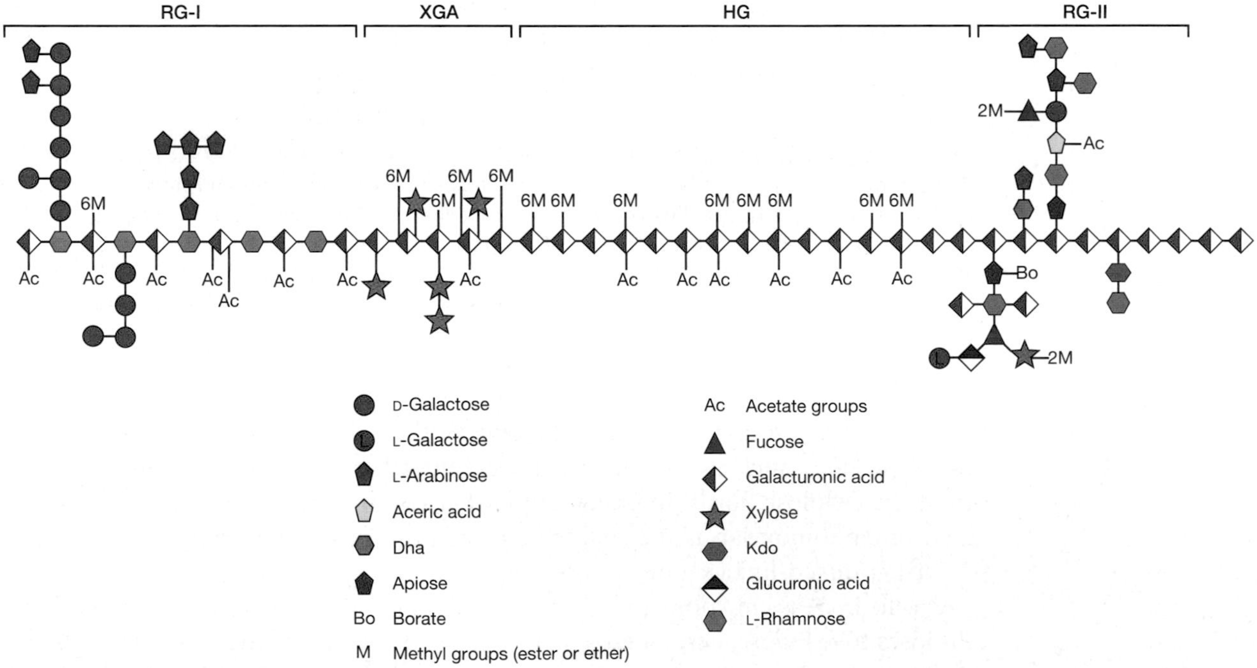

FIGURE 22.6. Schematic structure of pectin showing the three main pectic polysaccharides: rhamnogalacturonan I (RG-I, *left*) and rhamnogalacturonan II (RG-II, *right*), at each side of a homogalacturonan (HG) chain. A region of substituted galacturonan, known as xylogalacturonan (XGA), is also shown between the HG and the RG-1. Note that the representative pectin structure shown is not quantitatively accurate. HG should be increased 12.5-fold and RG-I increased 2.5-fold to approximate the amounts of these polysaccharides relative to each other in plant walls. (Dha) 3-Deoxy-D-lyxo-heptulosaric acid. (Modified from http://www.uk.plbio.kvl.dk/plbio/cellwall.htm.)

80% of the rhamnose residues are substituted at C-4 with linear or branched side chains of arabinose, galactose, or arabinogalactans. Another pectin, rhamnogalacturonan-II (RG-II), has a backbone of α1-4-linked galacturonic acid residues and is substituted with four highly complex and conserved side chains that contain 12 different sugars in more than 20 different linkages. A nonasaccharide and an octasaccharide chain are attached to C-2 of the backbone galacturonic acid residues, and two different disaccharide chains are attached to C-3 of the backbone making a highly branched structure. RG-II exists most often in plant cell walls as a borate diester formed between the apiose residue in two RG-II octasaccharide side chains of two distinct RG-II molecules. Because RG-II is embedded in a homogalacturonan chain, these RG-II dimers affect wall porosity and function. Mutations in RG-II structure that cause reduced RG-II dimer formation have dramatic effects on plant growth including dwarfism (see the next section).

Depending on the cell type and the stage of cell growth, there are many variations in wall fine structure and in the interactions between cellulose, hemicellulose, and pectin. It is expected that ongoing intensive studies will elucidate the role of wall glycans in cell expansion and tissue morphology. A number of cell wall proteins such as the HRGP classes of proteins discussed in the previous section are also immersed in the pectin matrix. The cell wall is thus emerging as a highly complex intricate network (Figure 22.4). It contains a wide variety of glycans that can interact with one another and with proteins to help adapt the wall to growth and mediate the interaction of the cell with its environment. The structure of this matrix is altered by the cell during cell division and development and in response to environmental stimuli by differential synthesis and modifications of the matrix components or by addition of new ones. Recent success in identifying genes that encode cell-wall polysaccharide biosynthetic enzymes, including many of those required for xyloglucan and cellulose synthesis and some of those required for pectin synthesis, is providing a framework for an increased understanding of how the plant wall is synthesized. Progress in this area will impact our ability to provide superior plant wall–based products. The complexity of the cell wall, including the crystalline nature of cellulose microfibrils and the interactions of hemicellulose, pectin, and lignin with cellulose, also makes the wall recalcitrant to deconstruction to the simple sugars needed for bioethanol production. An increased understanding of wall synthesis will also improve our ability to generate plants that can serve as better sources of biomass for bioethanol production.

PLANT MUTANTS ARE PROVIDING CLUES TO GLYCAN FUNCTION

In recent years, the production and characterization of plant mutants have provided clues to the function of plant glycans. *Arabidopsis thaliana* is a good model plant because of its relatively small genome, short life cycle, and ease of growth. A variety of banks of mutants are available for screening for glycosylation defects. An early mutant identified by this process was the *cgl* mutant, which was found by screening leaf extracts of chemically produced mutants with antisera against complex glycans. This mutant lacked GnT-I activity. It produced no complex glycans, but accumulated $Man_5GlcNAc_2$. This mutant had no apparent effect on the development and morphology of the plants. More recent studies on a mutant of a subunit of the oligosaccharyltransferase complex, *dgll*, demonstrated that reduced N-linked glycosylation leads to a variety of effects in plants, ranging from reduced cell elongation to embryonic lethality. These results show that N-glycosylation is essential for normal plant growth and development.

Another mutant *Arabidopsis* line, *mgd1*, was identified by screening transfer-inserted mutant lines for plants with defects in chloroplast biogenesis. The plants of this line were chlorotic (deficient in chlorophyl production) and had a change in chloroplast ultrastructure. They were found to contain only about 50% of normal levels of MGDG, thus helping to support the proposed role for this glycolipid in photosynthesis.

The *mur1* mutant of *Arabidopsis* was identified by making hydrolysates of cell walls of chemically produced mutants and screening the alditol acetate derivatives by gas-liquid chromatography (see Chapter 47). These plants were found to be deficient in an isoform of GDP-mannose-4,6-dehydrase (see Chapter 4) and therefore they were deficient in fucose. As a result, reduced amounts of RG-II dimer were formed because L-galactose substituted for fucose in RG-II, and dwarf plants with fragile cell walls were created. These observations provide evidence for a crucial role for RG-II in plant growth.

GLYCANS AND GLYCOCONJUGATES SERVE AS SIGNALS TO REGULATE PLANT DEVELOPMENT AND INTERACTIONS WITH THE ENVIRONMENT

Glycan signals help regulate plant development, defense, and other interactions of plants with the environment. Early studies of plant–pathogen interactions showed that glycans arising from the degradation of the cell wall of either the plant or the pathogen can act as elicitors to trigger defense responses to the pathogen. These responses consist of the activation of a number of plant genes that lead to the cross-linking of cell wall polymers, the production of additional cell wall components (such as HRGPs), the production of glycanases (such as β1-3 glucanases and chitinases) that can degrade the pathogen cell wall, and the production of phytoalexins that kill the invading pathogen. The end result is that necrotic spots appear at the site of infection and limit the spread of the pathogen. The structures of some of the glycans that are released upon degradation of plant and pathogen cell walls and that can act as elicitors are shown in Figure 22.7. These glycans are quite small, containing 4–20 sugar residues. As mentioned below, a few plant receptors that recognize some of these signals have been identified, but little is known about the signaling mechanism.

Xyloglucan fragments from plant cell walls act as signals to trigger the expansion of cell walls and inhibit auxin-induced growth in stems. The most potent fragment is a nonasaccharide. Partial hydrolysates of plant cell walls also affect plant growth and organogenesis. Oligosaccharides released from the pectic polysaccharides (including oligogalacturonides

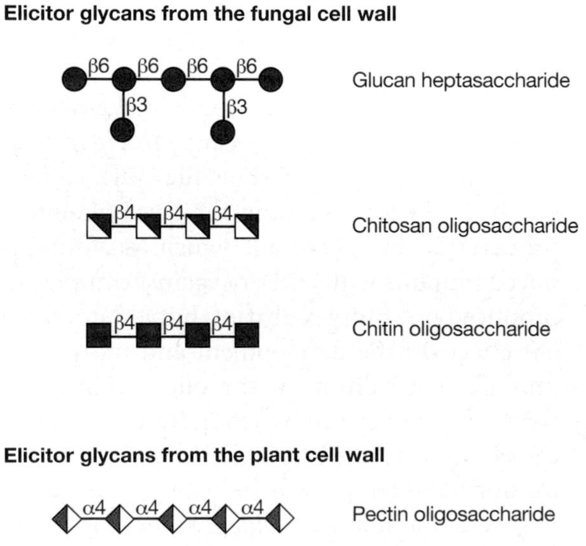

FIGURE 22.7. Glycans from fungal and plant cell walls that elicit plant defense responses.

released from homogalacturonan) can alter plant organogenesis in vitro and stimulate plant defense responses. Changes in the concentrations of some plant hormones have been found to stimulate the production of the glycosidases involved in cell-wall degradation and these hormones may regulate this process.

Glycan signals also initiate the nitrogen-fixing *Rhizobium*–legume symbiosis. In response to flavonoids produced by the legume roots, *Rhizobium* bacteria are stimulated to produce increased amounts of a lipochitooligosaccharide called a Nod factor. The Nod factor consists of a short chitin backbone with a long-chain fatty acid at the nonreducing end. The Nod factor triggers the deformation of the root hairs and the differentiation of some of the cells in the root cortex to form a new organ called the nodule. The *Rhizobia* nestle in the curl of the root hairs and cause the plant to synthesize an infection thread, which they follow to take up residence in the nodule. Picomolar amounts of isolated Nod factor added to the roots can cause root hair deformation and nodule formation. Recognition between particular strains of *Rhizobia* and certain species of legumes is specific. This specificity is due to the fact that the different *Rhizobia* strains make different derivatives of Nod factor that vary in the substituents attached at different positions of the chitin backbone. A generic Nod factor structure is shown in Figure 37.4 (see Chapter 37 on glycan signals).

GLYCAN-BINDING PROTEINS IN PLANTS

Plant lectins were first discovered in the late 1800s and since then they have been found in every plant species studied. They are particularly abundant in the seeds of leguminous plants and can account for about 10% of the nitrogen in the total seed extract. The lectins were initially identified on the basis of their carbohydrate-binding activities by hemagglutination or precipitin techniques. More recently, they have been tentatively identified by screening for DNA homology.

In general, lectins within the same family of plants are very similar in structure, but they differ from the structures of lectins from other plant families. Chapters 26–29 discuss various types of plant lectins, but little is known about their natural function. Their abundance in seeds and in the bark of some trees led to hypotheses that they may be involved in pathogen recognition, and some lectins are clearly toxic to insects. An initial idea that lectins recognize glycan signals produced by *Rhizobia* for symbiosis is incorrect. In fact, none of the conventional lectins from plants have been shown to recognize any of the glycan signals discussed above. Searches in plants for ligands for these conventional lectins have so far been unsuccessful. Studies of the distribution and localization of lectins throughout the life cycles of plants show that some species contain more than one lectin encoded by separate genes with different spatial and temporal expression. The question of biological function for these highly characterized lectins remains open at this time.

Some of the conventional legume lectins have been found to have a separate hydrophobic binding site in addition to their carbohydrate site. This hydrophobic site binds to adenine and derivatives of the plant hormone cytokinin. A recent finding has shown that the seed lectin from the legume *Dolichos biflorus* is also a lipoxygenase. Putative receptors for glycans that play signaling roles in plants have recently been identified. Although these proteins are lectins by definition, they do not resemble the conventional lectins identified from these plants. Receptors for chitin oligosaccharides that act as elicitors have been identified in the plasma membranes of several plants, and one of them was recently isolated and cloned. A novel lectin was isolated from legume roots and found to bind with Nod factors produced by rhizobial symbionts of that plant; this lectin was also found to have nucleotide phosphorylase activity and was thus called an LNP (lectin nucleotide phosphorylase). In both cases, suppression of expression of the proteins inhibited the signaling response suggesting that these proteins indeed play roles in signal recognition (see Chapter 37).

FURTHER READING

Carpita N.C. and Gibeaut D.M. 1993. Structural models of primary cell walls in flowering plants: Consistency of molecular structure with the physical properties of the walls during growth. *Plant J.* **3:** 1–30.

Lerouge P., Cabanes-Macheteau M., Rayon C., Fischette-Laine A.C., Gomord V., and Faye L. 1998. N-Glycoprotein biosynthesis in plants: Recent developments and future trends. *Plant Mol. Biol.* **38:** 31–48.

Sommer-Knudsen J., Bacic A., and Clarke A. 1998. Hydroxyproline-rich plant glycoproteins. *Phytochemistry* **4:** 483–497.

Gaspar Y., Johnson K.L., McKenna J.A., Bacic A., and Schultz C.I. 2001. The complex structures of arabinogalactan-proteins and the journey towards understanding function. *Plant Mol. Biol.* **47:** 161–176.

Kieliszewski M.J. 2001. The latest hype on Hyp-*O*-glycosylation codes. *Phytochemistry* **57:** 319–323.

Ridley B.L., O'Neill M.A., and Mohnen D. 2001. Pectins: Structure, biosynthesis, and oligogalacturonide-related signaling. *Phytochemistry* **57:** 929–967.

Dormann P. and Benning C. 2002. Galactolipids rule in seed plants. *Trends Plant Sci.* **7:** 112–118.

Reiter W.D. 2002. Biosynthesis and properties of the plant cell wall. *Curr. Opin. Plant Biol.* **5:** 536–542.

Wilson I.B.H. 2002. Glycosylation of proteins in plants and invertebrates. *Curr. Opin. Struct. Biol.* **12:** 569–577.

Ritsema T. and Smeekens S. 2003. Fructans: Beneficial for plants and humans. *Curr. Opin. Plant Biol.* **6:** 223–230.

Kelly A.A. and Dormann P. 2004. Green light for galactolipid trafficking. *Curr. Opin. Plant Biol.* **7:** 262–269.

Lerouxel O., Cavalier D.M., Liepman A.H., and Keegstra K. 2006. Biosynthesis of plant cell wall polysaccharides—A complex process. *Curr. Opin. Plant Biol.* **9:** 621–630.

Himmel M.E., Ding S.Y., Johnson D.K., Adney W.S., Nimlos M.R., Brady J.W., and Foust T.D. 2007. Biomass recalcitrance: Engineering plants and enzymes for biofuels production. *Science* **315:** 804–807.

Somerville C. 2007. Biofuels. *Curr. Biol.* **17:** R115–R119.

CHAPTER 23

Nematoda

Richard D. Cummings and Jeffrey D. Esko

THIS CHAPTER FOCUSES ON THE NEMATODE (roundworm) *Caenorhabditis elegans* as an example of the phylum Nematoda. *C. elegans* provides a powerful genetic system for studying glycans during embryological development and in primitive organ systems.

C. ELEGANS AS A MODEL ORGANISM

C. elegans is one of about 10,000 known species of nematodes. These organisms are protostomes, which are distinguished from deuterostomes by the order of formation of the mouth and anus (see Chapter 25) and by the exact specification of cell fates during embryological development. Biologists have known about nematodes since ancient times (as long ago as the era of Hippocrates) because of the association of some species with parasitic disease. However, *C. elegans* is free-living and normally resides in the soil.

The pioneering molecular biologist Sydney Brenner introduced *C. elegans* as a model organism for studying neurobiology and developmental biology. In 1974, Brenner wrote "The ease of handling of the nematode coupled with its small genome size suggests that it is feasible to look for mutants in all of the genes to try to discover how they participate in the development and functioning of a simple multicellular organism." *C. elegans* is an ideal model organism because the adult hermaphrodite contains exactly 959 cells and the position of each cell in the adult is derived from a strict cell lineage, which was precisely defined in the 1980s by John Sulston and colleagues. In studies by Craig Mello and Andrew Fire, it was the first organism in which small interfering RNAs were used successfully for gene silencing. It was also the first animal to be studied with a reporter protein—namely, green fluorescent protein (GFP) from the jellyfish *Aequorea victoria*. It was the organism in which apoptosis was discovered by H. Robert Horvitz, and it was the first metazoan to have its genome sequenced. Nobel Prizes in Medicine were awarded in 2002 to Brenner, Horvitz, and Sulston and in 2006 to Fire and Mello for their work on *C. elegans*.

DEVELOPMENTAL BIOLOGY OF *C. ELEGANS*

C. elegans is transparent, and individual cells can be easily visualized in the living organism through all stages of development. Basically, the worm is a tube within a tube (Figure 23.1). A cuticle composed of a collagenous, multilayered, protective exoskeleton surrounds the worm. The "mouth" at the anterior end connects to a tubular intestinal system, which is composed of a muscular pharynx and an intestine. The gonad occupies most of the body cavity. In the hermaphrodite, the gonad is bilobed, with each lobe connecting via an oviduct and spermatheca to a shared midventral vulva and uterus. The worm exists as two sexes, hermaphrodite or male (no female organisms exist). Eggs pass through the spermatheca where fertilization takes place by stored sperm, and the eggs begin to develop inside the mother. During sexual reproduction, males fertilize hermaphrodites. The male sperm is also stored in the spermatheca and is preferentially used during fertilization.

Gastrulation begins prior to egg laying, and at this stage, the embryo contains about 30 cells (Figure 23.2). Proliferation results in an embryo of 558 relatively undifferentiated cells. Following this, organogenesis/morphogenesis begins, terminal differentiation occurs, and the embryo hatches. The animal normally passes through four larval stages, termed L1, L2, L3, and L4 (Figure 23.2). The end of each larval stage is marked by molting, when the cuticle is shed. In L1 larvae, the nervous system, the reproductive system, and the digestive tract begin to develop, and this is completed by the L4 stage. Mature adults develop about 45–50 hours after hatching, and they contain 959 somatic cells, including 302 neurons and

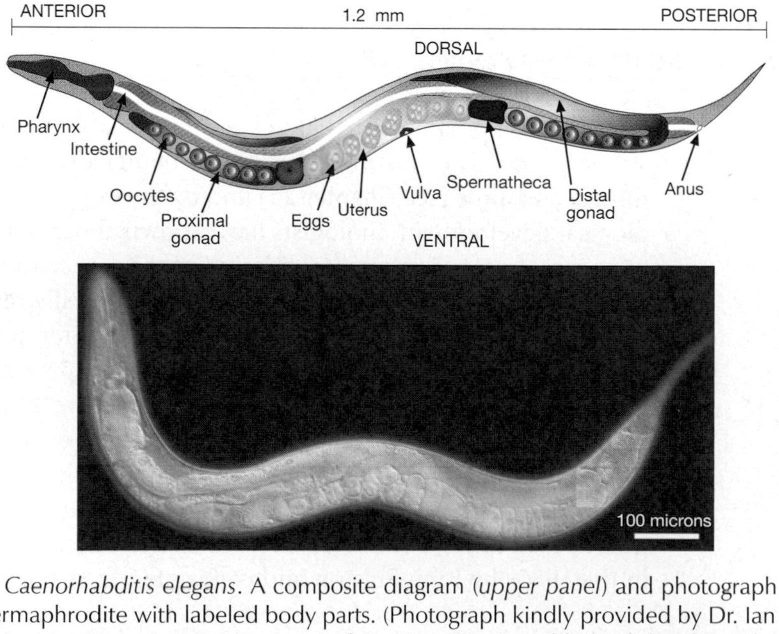

FIGURE 23.1. *Caenorhabditis elegans*. A composite diagram (*upper panel*) and photograph (*lower panel*) of the adult hermaphrodite with labeled body parts. (Photograph kindly provided by Dr. Ian D. Chin-Sang at Queen's University, Kingston, Ontario [www.chin-sang.ca]). For additional details on the biology of *C. elegans*, see www.wormatlas.org.

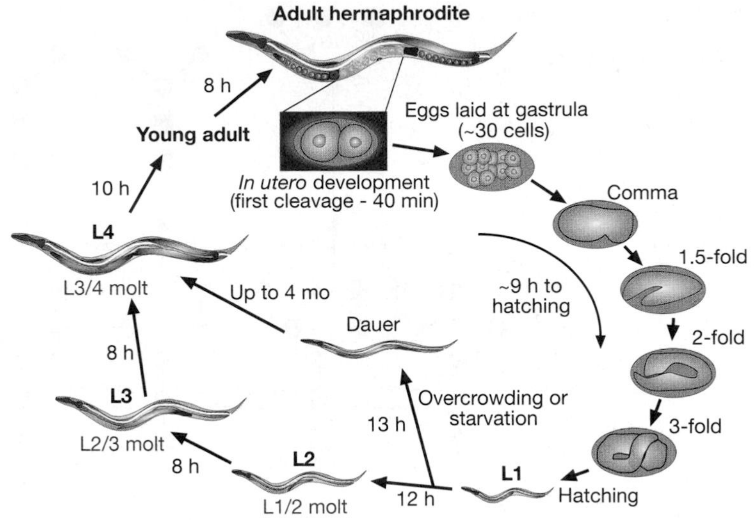

FIGURE 23.2. Life cycle of *C. elegans*. For additional details on the biology of *C. elegans,* see www.wormatlas.org.

95 body-wall muscle cells. At this time, the hermaphrodite can lay its first eggs, thus completing the 3.5-day life cycle. The adult hermaphrodite produces oocytes for about 4 days, resulting in approximately 300 progeny; afterward, the animal lives for another 10–15 days. Overcrowding or starvation results in the formation of dauer larvae, which are easily distinguished from other developmental stages by morphology and behavior.

GLYCANS IN *C. ELEGANS*

The *C. elegans* genome encodes homologs of many of the enzymes for glycoconjugate biosynthesis that are found in higher animals and humans, including enzymes for the synthesis of O-GalNAc (mucin-type) glycans, N-glycans, glycosaminoglycans (GAGs), glycosphingolipids (GSLs), and glycosylphosphatidylinositol (GPI) anchors. However, there are also many notable differences between the types of glycans made by *C. elegans* and those in higher animals. For example, the worm lacks sialic acids and any enzymes associated with sialic acid biosynthesis or utilization. Moreover, besides lacking sialic acids, worm glycolipids have different core structures. The N-glycans are generally truncated and lack the extensions and modifications found in higher animals.

The N-glycans of *C. elegans* are synthesized by the common eukaryotic pathway involving generation of the 14-sugar precursor $Glc_3Man_9GlcNAc_2$-P-P-Dol and transfer of the glycan to asparagine residues within the sequon -Asn-X-Ser/Thr- of nascent polypeptides in the endoplasmic reticulum (see Chapter 8). These glycans can be trimmed to $Man_9GlcNAc_2$-Asn-R and further to $Man_5GlcNAc_2$-Asn by α-glucosidases I and II and α-mannosidases I and II. The $Man_5GlcNAc_2$-Asn can be modified by the addition of *N*-acetylglucosamine to generate $GlcNAc_1Man_5GlcNAc_2$-Asn, which is then further processed in a unique way in *C. elegans* (Figure 23.3). Subsequent trimming and modification by β-hexosaminidase, α-mannosidases, and α-fucosyltransferases generate truncated N-glycans with three or fewer mannose residues (so-called paucimannose) and a core α1-3 fucose residue, in addition to the common vertebrate modification of α1-6 fucose residues. Although truncated N-glycans are common in *C. elegans*, the organism may generate smaller amounts of more complex branched structures. In addition, many of the hybrid and complex N-glycans in *C. elegans* contain phosphorylcholine linked to outer-chain and

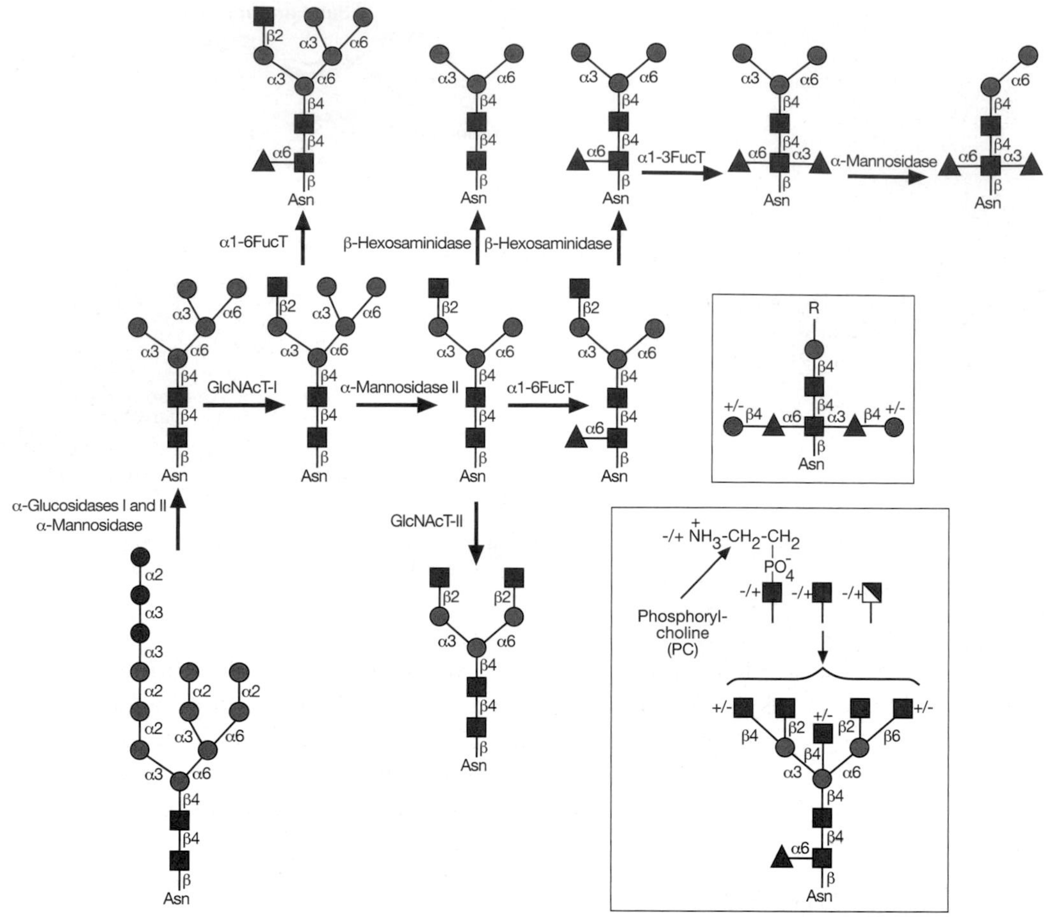

FIGURE 23.3. Biosynthesis of paucimannose and core fucosylated N-glycans in *C. elegans*.

core chitobiose *N*-acetylglucosamine residues, a modification that appears to be a common feature of nematode glycoproteins. Another modification is the addition of galactose to core fucose residues and to the outer antennae in a minor population of N-glycans (Figure 23.3).

The O-glycans of *C. elegans* contain core-1 structures common to vertebrates (see Chapter 9), but they also include unusual GlcAβ1-3GalNAc linkages and extended core-1 O-glycans containing β-glucose and glucuronic acid residues, α1-2 fucose residues, and 2-O-methylated fucose (Figure 23.4). Interestingly, *C. elegans* contains multiple genes for the UDP-GalNAc:polypeptide α-*N*-acetylgalactosaminyltransferases (ppGalNAcTs) that modify Ser/Thr residues in mucin O-glycans, which is similar to the situation in vertebrates. *C. elegans* also contains a gene encoding β1-3 galactosyltransferase (T-synthase), which generates core 1 of O-glycan.

C. elegans synthesizes O-GlcNAc on Ser/Thr residues of cytoplasmic and nuclear proteins, using a highly conserved O-GlcNAc transferase (OGT); the worm also has the O-GlcNAcase that removes O-GlcNAc (see Chapter 18). Interestingly, deletion of OGT, although fatal to vertebrate cells, is not lethal to *C. elegans* but is accompanied by a phenotype that resembles human insulin resistance. A related phenotype is also produced by deletion of the O-GlcNAcase.

Glycosphingolipids in *C. elegans* have a core consisting of GlcNAcβ1-3Manβ1-4Glcβ1-Cer, which is based on the arthroseries Manβ1-4Glcβ1-Cer core instead of the common Galβ1-4Glcβ1-Cer core found in vertebrates (see Chapter 10). In addition, *C. elegans* has genes encoding the enzymes for synthesis of GPI-anchored glycoproteins, but the GPI anchor structures are not yet defined (see Chapter 11).

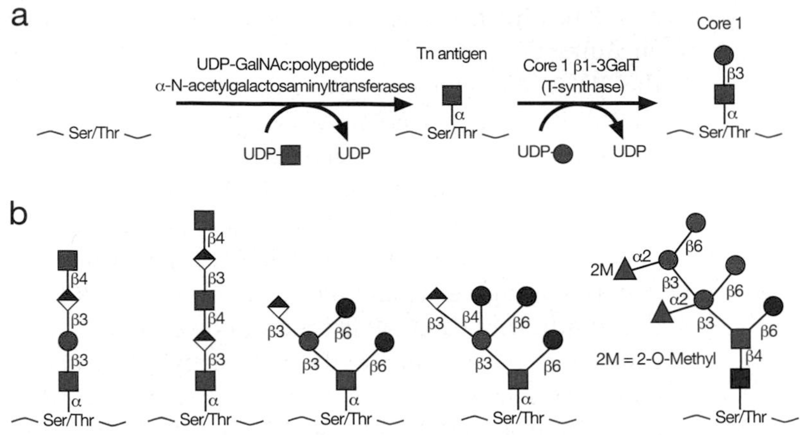

FIGURE 23.4. Biosynthesis of core-1 O-glycan in *C. elegans* (*a*) and some O-glycans identified in adult worms (*b*).

C. elegans has all of the enzymes for synthesizing the GAGs chondroitin and heparan sulfate (HS) but not those for keratan sulfate or dermatan sulfate. However, *C. elegans* only generates unsulfated chondroitin chains because it lacks the sulfotransferases and epimerase present in vertebrates (see Chapter 16). In contrast, *C. elegans* does have the sulfotransferases and epimerase for generating HS, and the overall structure is similar to the chains elaborated by vertebrates. *C. elegans* does not make hyaluronan (see Chapter 15).

C. elegans also synthesizes glycoproteins with O-fucose on epidermal growth factor (EGF)-like repeats (Fucα1-Ser/Thr) with a precise consensus sequence (see Chapter 12). This may be further modified to generate elongated structures R-GlcNAcβ1-3Fucα1-Ser/Thr, although the structures of such glycans in *C. elegans* are not defined and there is no evidence as yet for the necessary Fringe-related β1-3GlcNAcT. POFUT2, which transfers O-fucose to thrombospondin type-1 repeats (TSRs), is also present in *C. elegans.*

GLYCOSYLTRANSFERASE GENES IN *C. ELEGANS*

On the basis of the structures of its N- and O-glycans, glycolipids, and GAGs, *C. elegans* is predicted to express a wide assortment of enzymes involved in glycoconjugate metabolism. Indeed, the genome of *C. elegans* appears to encode about 300 carbohydrate-active enzymes (CAZy database http://www.cazy.org/), including glycosyltransferases, glycosidases, polysaccharide lyases, and carbohydrate esterases (see Chapter 7). The glycosyltransferase family contains more than 240 genes. To date, most of these putative enzymes are not characterized, and very little is known overall about their expression or function in *C. elegans.* Some information has arisen from phenotypes generated by mutagenesis, as discussed below. Together, these studies have identified some interesting differences and similarities between *C. elegans* and vertebrates.

C. elegans contains a few extremely large classes of glycosyltransferases and lacks others. Some large classes include the putative ppGalNAcTs (11 enzymes), α-fucosyltransferases (5 α1-3FucTs, 26 α1-2FucTs, and 1 α1-6FucT), and β-*N*-acetylglucosaminyltransferases (9 enzymes). In contrast, *C. elegans* completely lacks any sialyltransferases or enzymes involved in sialic acid synthesis and turnover, which are numerous in most vertebrates (see Chapters 13 and 14).

Relatively few of these enzymes predicted to be in *C. elegans* have been shown to be functional, and in many cases, their acceptor specificity has not been characterized. Eleven of the ppGalNAcTs have been prepared as recombinant proteins, but only five of them are active toward mammalian peptide acceptors. Additionally, *C. elegans* has a single gene encoding OGT, as in humans. Humans and other vertebrates have a single gene encoding the β1-2 *N*-acetylglucosaminyltransferase (GlcNAcT-1) that catalyzes formation of GlcNAcβ1-2Man$_5$GlcNAc$_2$-Asn (see Figure 23.3), whereas *C. elegans* contains three genes encoding this GlcNAcT-1 activity. This product is then acted upon by a single α-mannosidase II to generate GlcNAcβ1-2Man$_3$GlcNAc$_2$-Asn. Interestingly, *C. elegans* has an unusual β-hexosaminidase, which cleaves GlcNAcβ1-2Man$_3$GlcNAc$_2$-Asn to generate the paucimannose structure Man$_3$GlcNAc$_2$-Asn, a reaction not found in vertebrates. *C. elegans* has a single gene that has high homology with the human β1-4 galactosyltransferase that synthesizes Galβ1-4GlcNAc-R; however, the recombinant *C. elegans* β1-4 galactosyltransferase is actually a β1-4 *N*-acetylgalactosaminyltransferase and generates the sequence GalNAcβ1-4GlcNAc-R. *C. elegans* can synthesize the canonical core tetrasaccharide of GAGs, GlcAβ1-3Galβ1-3Galβ1-4Xylβ1-Ser, and can extend this to generate chondroitin and HS. The glycolipids in *C. elegans* have the unusual arthroseries core structure, and the organism lacks a gene encoding the β1-4 galactosyltransferase that would synthesize Galβ1-4Glcβ1-Cer.

Expression of the known glycosyltransferases in *C. elegans* has not been systematically mapped at the cellular level during development or in the adult organism. Many of the studies are based on promoter analyses using GFP as a reporter. The promoter region for a gene of interest (usually 0.5–1.5 kb upstream of the gene and sometimes including the upstream elements and the first few exons of the gene) is ligated to the cDNA encoding the GFP in an appropriate vector. Transgenic animals are then produced by direct injection of vectors into the hermaphrodite gonad. Newly developing animals take up this DNA and become transgenic. Thus, one can observe promoter utilization in the early stages of embryogenesis (see Figure 23.2). This approach has led to some surprising findings. For example, the T-synthase, which is a common β1-3 galactosyltransferase for generating the core 1 of O-glycans, is expressed in all cells in *C. elegans* from early embryogenesis. In addition, the galactosyltransferase that adds the second galactose residue to the core tetrasaccharide of GAGs (Figure 23.5) is expressed in all cells of early and late embryos. Similarly, the two protein O-fucosyltransferases I and II (POFUT1 and POFUT2) are widely expressed. In contrast, expression of the six individual core-2 *N*-acetylglucosaminyltransferases with sequences related to core-2 and core-1 enzymes in vertebrates (see Chapters 9 and 13) occur in selective tissues. One gene, *gly-15*, is expressed only in two gland cells. Similarly, among the 26 α1-2FucTs that are predicted in *C. elegans*, one of them (*CE2FT-1*) is expressed in a single cell in embryos and exclusively in 20 intestinal cells of larval stages L1–L4 and adult worms. Thus, in large gene families, individual members may be expressed in a localized fashion and have unique activities toward certain substrates, whereas single gene families appear to be expressed in all cells.

GLYCAN-BINDING PROTEINS IN *C. ELEGANS*

Although the *C. elegans* genome encodes a number of predicted glycan-binding proteins (GBPs), only a few of them have been characterized biochemically or explored by genetic manipulation. The first GBP found in *C. elegans* was a galectin, which was isolated by affinity chromatography and sequenced in 1992. This was surprising because until this observation, galectins were thought to be expressed only in vertebrates (see Chapter 33). Amazingly, the *C. elegans* genome encodes 28 putative galectins, nearly twice as many as in humans. Only two of these proteins have been studied in detail: a tandem-repeat 32-kD

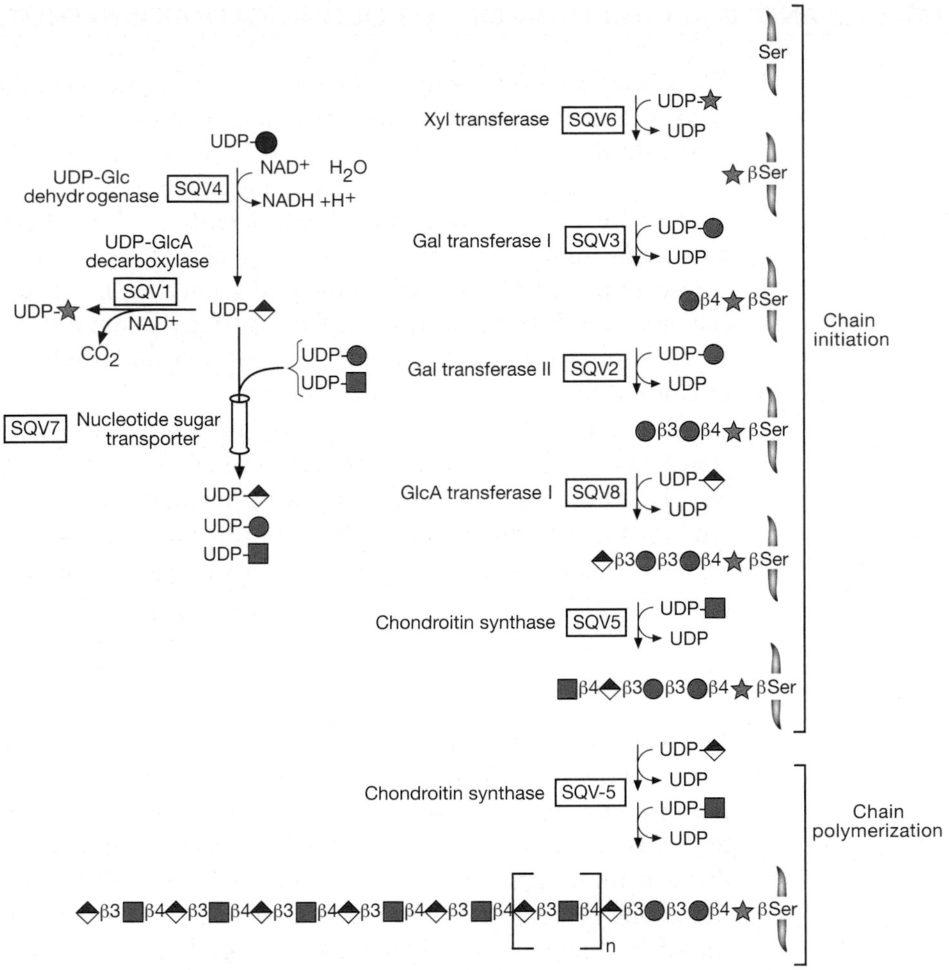

FIGURE 23.5. Biosynthesis of chondroitin in *C. elegans*. Mutations in individual steps were identified as *sqv* mutants, as described in the text.

galectin and a prototypical 16-kD galectin. Both galectins can bind to galactose-containing ligands, but the endogenous ligands in *C. elegans* are unknown. Interestingly, only a few of the glycans identified to date from *C. elegans* contain galactose.

A total of 183 C-type lectin domains (CTLD) identified in the *C. elegans* genome are contained within 135 proteins (some proteins have multiple CTLDs), but only 19 of these CTLDs have sequences predicting carbohydrate recognition. The functional roles of the CTLDs have not yet been studied in detail in *C. elegans.* In contrast to the CTLD-containing proteins in vertebrates (see Chapter 31), most of the proteins with CTLDs in *C. elegans* have signal sequences and no transmembrane domains, indicating that they are secreted proteins.

The *C. elegans* genome encodes many predicted lectins involved in quality control of glycoprotein biosynthesis and glycoprotein targeting to organelles; these include L-type lectins related to those in vertebrates (such as calnexin, calreticulin, ERGIC-53, and VIP36) as well as a gene encoding a mannose-6-phosphate receptor (see Chapters 29, 30, and 36). In addition, *C. elegans* contains members of the R-type lectin group within the ppGalNAcT family, as found in humans and vertebrates (see Chapter 28). Not surprisingly, *C. elegans* lacks I-type lectins, which in vertebrates are involved in sialic acid biology (see Chapter 32).

MUTATIONS AND RNAi INHIBITION OF GLYCOCONJUGATE BIOSYNTHESIS IN *C. ELEGANS*

The advantages of studying *C. elegans* as a model system are the ease of manipulation of its genome and its surprising similarity to mammals—about 40% of the human genes associated with disease have homologs in the *C. elegans* genome. The hermaphrodite, which is self-fertilizing, can maintain homozygous mutations and there is no need for mating. A large number of mutant and RNA interference (RNAi) phenotypes have been described, and many of the mutant strains are available from the *C. elegans* Gene Knockout Consortium (www.celeganskoconsortium.omrf.org/). Gene silencing by RNAi can be achieved simply by growing *C. elegans* on bacteria containing the appropriate knockdown RNAi vector. Unfortunately, direct gene targeting and modification by homologous recombination cannot yet be done in *C. elegans.*

These approaches have yielded important information about the functions of glycosyltransferases and their glycan products. Several dozen genes involved in glycosylation pathways have been shown to be developmentally important in *C. elegans* or important in resistance or susceptibility to pathogens in the innate immunity of the worm. Only some of the highlights of this work are described here. More details are available in the literature cited at the end of the chapter.

Proteoglycans and Glycosaminoglycans

The first studies of glycosyltransferases and glycans in *C. elegans* came out of the analysis of mutations affecting vulva formation (see Figure 23.1). During egg laying, fertilized eggs must pass through the vulva, which is a simple tubular structure that links the gonads with the external cuticle. During postembryonic development, vulva morphogenesis arises through the invagination of a single layer of epithelial cells. Using mutagenesis, several mutations that perturb invagination of the vulva were identified (designated *sqv* or squashed vulva). In the original screen, 25 mutations were identified in eight genes named *sqv1* through *sqv8.* All of the mutations produced a similar phenotype: that is, partial collapse of vulval invagination, elongation of the central vulval cells, hermaphrodite sterility associated with maternal-effect lethality, and cytokinesis defects in the early embryo. All eight *sqv* genes show homology with vertebrate enzymes that are involved in the biosynthesis of GAGs (Figure 23.5). *sqv1, sqv4,* and *sqv7* encode proteins that have roles in nucleotide sugar metabolism and transport. The SQV7 nucleotide transporter was the first example of a carrier that could import more than one nucleotide sugar into the Golgi (see Chapter 4). SQV4 and SQV1 proteins represent sequential enzymes involved in the formation of UDP-glucuronic acid and UDP-xylose, respectively, demonstrating that the *sqv* mutations most likely affect GAG synthesis. Biochemical analysis of *sqv6, sqv3, sqv2,* and *sqv8* demonstrated that they encode worm orthologs of the vertebrate transferases required for the assembly of the linkage region tetrasaccharide common to HS and chondroitin. Finally, characterization of *sqv5* showed that it encodes the chondroitin synthase. Thus, the various phenotypes (failed invagination of the epithelial layer that forms the vulva, maternal-effect lethality, and cytokinesis defects) result from defective chondroitin formation.

The requirement for chondroitin assembly in seemingly disparate systems may result from biophysical changes in the lumen of the vulva or between the eggshell and the embryo. One idea is that the high negative charge imparted by the glucuronic acids in chondroitin attracts counterions that raise the local osmolarity, causing a swelling pressure. Another possibility is that the chondroitin acts as a physical scaffold bound to the cell membrane or eggshell. Interestingly, the *sqv* screen did not detect mutations affecting genes that encode proteoglycan core proteins on which the chondroitin chains assemble. Inspection of the *C.*

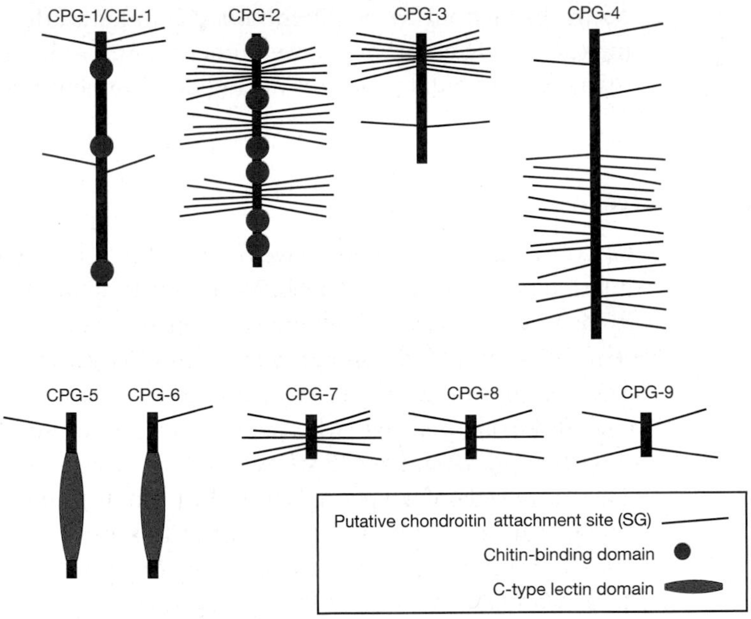

FIGURE 23.6. Chondroitin proteoglycans (CPGs) of *C. elegans*.

elegans genome also did not reveal any homologs of vertebrate chondroitin sulfate proteoglycans. Proteomic analysis subsequently led to the identification of nine novel chondroitin proteoglycan (CPG) core proteins that contain chondroitin chains (Figure 23.6). Two of these (CPG-1 and CPG-2) contain chitin-binding domains that presumably allow the proteoglycans to interact with chitin in the eggshell, thus positioning the proteoglycans between the eggshell and the plasma membrane of the embryo, where they could serve as spacers or osmotic regulators. Silencing *cpg-1* and *cpg-2* expression by RNAi recapitulates the cytokinesis defect observed in *sqv* mutants, suggesting that these are the relevant proteoglycans. The proteoglycans involved in epithelial invagination have not yet been determined.

HS biosynthesis in *C. elegans* follows the same pattern observed in vertebrate systems (see Chapter 16). Mutations in the pathway for HS biosynthesis are lethal in *C. elegans.* Two of the key genes involved in this pathway are *rib-1* and *rib-2,* homologs of the vertebrate genes *Ext2* and *Ext1*, respectively, which catalyze the polymerization of the backbone of HS chains (GlcAβ1-4GlcNAcα1-4) (see Chapter 16). Mutants in *rib-2,* the worm homolog of *Ext1,* have defects in development and egg laying. The worm genome also contains a single gene for glucuronic acid C-5 epimerase (*hse-5*) and four genes for sulfotransferase activities (GlcNAc N-deacetylase/N-sulfotransferase [Ndst], *hst-1*; uronyl 2-O-sulfotransferase, *hst-2*; 3-O-sulfotransferase, *hst-3*; and 6-O-sulfotransferase, *hst-6*), all of which are homologs of vertebrate genes involved in HS synthesis (see Chapter 16). In contrast, vertebrates contain four Ndsts, three 6-O-sulfotransferases, and seven 3-O-sulfotransferases. Although mutations in the epimerase (*hse-5*) and two of the sulfotransferases (*hst-6* and *hst-2*) do not affect viability, they cause defects in specific cell migration and axonal outgrowth. Consistent with this finding, inactivation of the cell surface HS proteoglycan syndecan affects neural migration and axonal guidance. *C. elegans* also produces two GPI-anchored HS proteoglycans. LON-2, a member of the glypican family, negatively regulates a bone morphogenetic protein-like signaling pathway that controls body length in *C. elegans.* Worms also contain a homolog of the vertebrate basement membrane proteoglycan perlecan (encoded by *unc-52*). At least three major classes of UNC-52 isoforms are produced through alternative splicing, and distinct spatial and temporal expression patterns occur throughout development. *unc-52* mutants affect myofilament assembly in body-wall muscle during embryonic development. Thus, as in vertebrates, HS proteoglycans mediate many fundamental processes during development and in the adult animal.

N-Glycans and O-Glycans on Glycoproteins

Glycoproteins in *C. elegans* have both N- and O-glycans, as discussed above. In vertebrates, interference of the early steps in N-glycosylation or O-glycosylation causes embryonic lethality or results in severe developmental phenotypes. As in vertebrates, interference of the later steps in N- and O-glycan biosynthesis in *C. elegans* does not cause developmental problems. The genome of *C. elegans* contains three genes (*gly-12*, *gly-13*, and *gly-14*) that encode GlcNAcT-I-like enzymes, whereas mammals have a single GlcNAcT-I gene. Promoter analyses show that *gly-12* and *gly-13* are expressed in all cells beginning in embryogenesis, whereas *gly-14* is expressed only in gut cells from L1 to adults. Unexpectedly, deletion of any of these genes singly or in combination does not affect *C. elegans* development. The triple-knockout worms also do not generate paucimannose-type N-glycans or fucosylated versions, essentially generating $Man_5GlcNAc_2$-Asn as the major structure. These studies suggest that lack of processing of N-glycans by GlcNAcT-I does not cause an observable developmental phenotype. This is in marked contrast to mammals, which die at midgestation in the absence of GlcNAcT-I.

C. elegans contains three α-mannosidase activities: one α-mannosidase II/IIx-like activity involved in N-glycan processing and encoded by the *F58H1.1* gene, one lysosomal enzyme, and a third cytoplasmic enzyme. A mutant harboring a large deletion in the *F58H1.1* gene generates largely $Man_5GlcNAc_2$-Asn, $GlcNAc_1Man_5GlcNAc_2$-Asn-R, and fucosylated and phosphorylcholine-modified $Man_5GlcNAc_2$-Asn, but lacks the paucimannose-type structures. In addition, the mutant lacks the core α1-3 fucose antigen associated with antibodies to horseradish peroxidase. Interestingly, promoter analyses show that the *F58H1.1* gene is expressed in most cells of the organism, but there is no obvious developmental phenotype.

Two of the largest gene families in *C. elegans* are the ppGalNAcT and the α-fucosyltransferase families. To date, however, there have been no developmental phenotypes associated with the loss of any members of these families, suggesting that some of these enzyme activities may be redundant.

The biosynthesis of all fucosylated ligands requires the precursor GDP-fucose and its transport into the Golgi apparatus by nucleotide sugar transporters (see Chapter 4). Interestingly, the human disease leukocyte-deficiency type II (LAD II) is caused by a defect in the transport of GDP-fucose and the subsequent loss of fucosylated ligands important in leukocyte trafficking and recognition by selectins (see Chapter 31). A search of the *C. elegans* genome for putative nucleotide sugar transporters led to the identification of several candidates. One of these *C. elegans* genes complemented the transport and fucosylation defect in LAD II fibroblasts, leading to the identification of the defect in these patients.

C. elegans is also an interesting model system to study infection and innate immunity. The organism may be colonized by different bacterial pathogens, including *Pseudomonas aeruginosa*, *Yersinia pestis*, and *Yersinia pseudotuberculosis*. The two *Yersinia* species generate a sticky biofilm (an exopolysaccharide matrix encasing a community of bacteria) on the exterior of the worm's head that impairs viability. *P. aeruginosa*, in contrast, colonizes the intestinal tissues. Another bacterium, *Microbacterium nematophilum*, sticks to the anus of the animals and induces an irritation in the underlying hypodermal tissue. *Bacillus thuringiensis* infection leads to destruction of the intestine, which is discussed in more detail below in regard to glycolipids. Mutations in the worm have been found that affect colonization by these bacteria.

An especially interesting set of mutations are the *srf* mutants (altered surface antigenicity mutants). Some of the *srf* mutants were identified by altered antibody or lectin binding to the cuticle, indicating that loss of cuticle components exposed new antigens. *srf-3* mutants are resistant to infection by *M. nematophilum*. *srf-3* encodes a nucleotide sugar transporter that can transport both UDP-galactose and UDP-*N*-acetylglucosamine, suggesting that altered sugar composition of the cuticle resulting from mutations in this transporter confer resistance to *M. nematophilum*. Interestingly, there are 18 putative nucleotide sugar transporters in the genome of *C. elegans*, which is a considerably larger number than the known nucleotide sugars (UDP-galactose, UDP-glucose, UDP-*N*-acetylglucosamine, UDP-*N*-acetylgalactosamine, UDP-xylose, GDP-mannose, and GDP-fucose), suggesting possible functional overlap in these transporters. *srf-3* mutants are deficient in glycosylation, in particular, they lack O-linked glycoconjugates containing glucuronic acid and galactose and they also have reduced levels of N-glycans and fucose.

C. elegans, like other metazoa, generates O-fucose-containing glycoproteins, which in animals are usually linked to serine/threonine residues within cysteine-rich domains such as the EGF-like repeats and TSRs (see Chapter 12). In vertebrates, the Fucα1-Ser/Thr may be further modified to generate elongated structures R-GlcNAcβ1-3Fucα1-Ser/Thr, although it is not yet clear whether *C. elegans* generates extended structures. O-fucose in glycoproteins is essential for signaling pathways in development and Notch signaling. The *C. elegans* genome contains two genes that have homology with the protein O-fucosyltransferases POFUT1 and POFUT2 in humans and *Drosophila*.

Glycolipids

The arthroseries glycolipids in *C. elegans* have a unique core structure composed of GlcNAcβ1-3Manβ1-4Glcβ1-Cer (Figure 23.7). Some of these glycolipids also contain phosphorylcholine modifications on *N*-acetylglucosamine residues. Recent studies on the resistance of *C. elegans* to bacterial toxins have led to interesting insights into the structures and functions of glycolipids. *Bacillus thuringiensis* (Bt) toxins are used in both transgenic and organic farming because of their ability to kill insect pests. Bt is toxic to *C. elegans*, but following mutagenesis, Bt-resistant strains were identified and classified as *bre-1* through *bre-5*. None of the strains demonstrated altered development, but they were highly resistant to Bt. Unexpectedly, the Bt-resistant mutants had abnormal biosynthesis of glycolipids, which turned out to be the ligands for Bt in the intestinal epithelium. The *C. elegans* glycolipids are shown in Figure 23.7, along with the genetic steps associated with different *bre* mutants. Mutations in *bre* genes caused truncated glycolipids. However, *bre-1* encodes a GDP-mannose-4,6-dehydratase that is required for the synthesis of GDP-fucose. *bre-1* mutants are deficient in fucosylated glycans, which raises questions about the roles of fucosylated glycans in *C. elegans* development. It is likely that the *bre-1* mutations have some residual activity, allowing some synthesis of GDP-fucose for utilization by the POFUT enzymes and other fucosyltransferases. Interestingly, these studies also showed that *C. elegans* lacks a salvage pathway for fucose, since feeding the *bre-1* mutants with fucose did not restore synthesis of fucosylated glycolipids.

GLYCOBIOLOGY OF OTHER NEMATODES

Studies in the nonparasitic nematode *C. elegans* have been incredibly rewarding because of the ease of genetic manipulation and culture. Much less is known about parasitic nematodes, which cause tremendous death and suffering in animals and people throughout the

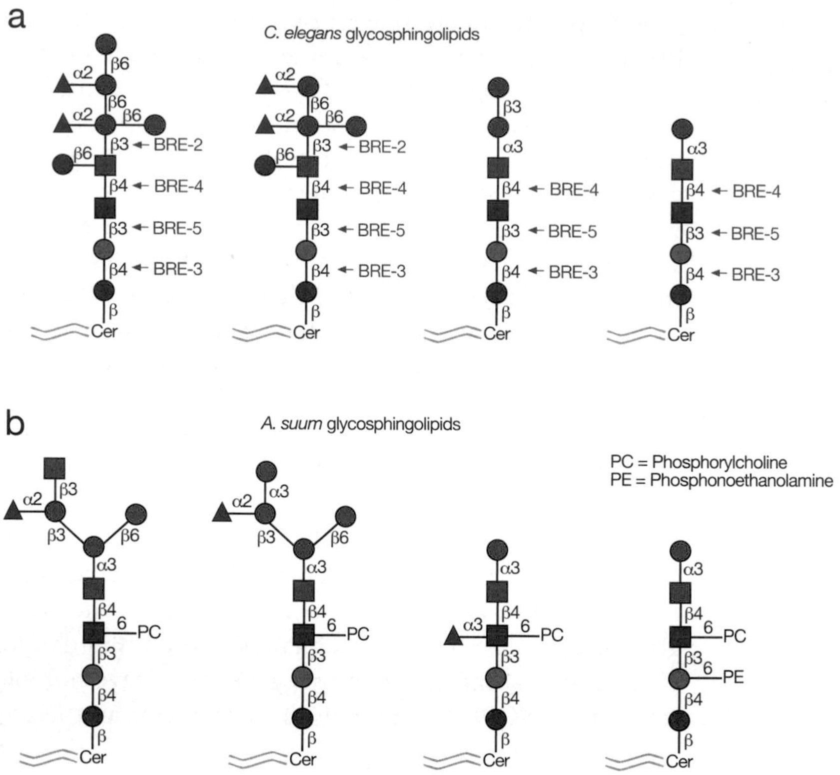

FIGURE 23.7. Examples of nematode glycolipids. (*a*) Structures of glycolipids from *C. elegans*. Mutations that result in truncated glycolipids have been identified as *bre* mutants, as described in the text. (*b*) Structures of glycolipids from *Ascaris suum*.

world. It might be expected that *C. elegans* and other nematodes share much in common in terms of glycoconjugate structures and biosynthesis, but each nematode has differences in glycans compared to *C. elegans.* These may pertain to their virulence and parasitic requirements.

Some of the major parasitic nematodes that have been studied in terms of glycoconjugates include *Ascaris suum, Trichinella spiralis, Dictyocaulus viviparous, Haemonchus contortus, Onchocerca volvulus, Necator americanus, Toxocara canis,* and *T. cati* (see Chapter 40). *A. suum* is a parasitic intestinal nematode of pigs. Like *C. elegans,* the *A. suum* N-glycans are paucimannose-rich and contain phosphorylcholine and core fucose residues. In contrast, the N-glycans of the parasitic nematode of deer *Parelaphostrongylus tenuis* are extensively terminally modified with galactose and carry the terminal structure Galα1-3Galβ1-4GlcNAc-R. The cattle parasite *D. viviparous* has N-glycans with Lewis antigens (see Chapter 13), including Lewis^{x}. The sheep parasite *H. contortus* synthesizes N-glycans containing the fucosylated LacdiNAc antigen GalNAcβ1-4(Fucα1-3)GlcNAc-R (see Chapter 13). *T. spiralis* synthesizes N-glycans that have the unusual and antigenic sugar tyvelose (3,6-dideoxy-D-arabinohexose). Nematodes also make unusual glycolipids, and it is likely that each type of nematode synthesizes different glycolipid structures. One of the best-studied nematodes in terms of glycolipids is *A. suum.* Many of these glycolipids, which also contain the arthroseries, have galactose and fucose modifications, in addition to phosphorylcholine and phosphonoethanolamine. Virtually nothing is known about the genetics regulating glycosylation in these parasitic nematodes, as most studies have focused on the analysis of N- and O-glycans and some glycolipids.

FURTHER READING

Brenner S. 1974. The genetics of *Caenorhabditis elegans*. *Genetics* **77:** 71–94.

Drickamer K. and Dodd R.B. 1999. C-Type lectin-like domains in *Caenorhabditis elegans*: Predictions from the complete genome sequence. *Glycobiology* **9:** 1357–1369.

Oriol R., Mollicone R., Cailleau A., Balanzino L., and Breton C. 1999. Divergent evolution of fucosyltransferase genes from vertebrates, invertebrates, and bacteria. *Glycobiology* **9:** 323–334.

Dodd R.B. and Drickamer K. 2001. Lectin-like proteins in model organisms: Implications for evolution of carbohydrate-binding activity. *Glycobiology* **11:** 71R–79R.

Hirabayashi J., Arata Y., and Kasai K. 2001. Glycome project: Concept, strategy and preliminary application to *Caenorhabditis elegans*. *Proteomics* **1:** 295–303.

Haslam S.M. and Dell A. 2003. Hallmarks of *Caenorhabditis elegans* N-glycosylation: Complexity and controversy. *Biochimie* **85:** 25–32.

Schachter H. 2004. Protein glycosylation lessons from *Caenorhabditis elegans*. *Curr. Opin. Struct. Biol.* **14:** 607–616.

Olson S.K., Bishop J.R., Yates J.R., Oegema K., and Esko J.D. 2006. Identification of novel chondroitin proteoglycans in *C. elegans*: Embryonic cell division depends on CPG-1 and CPG-2. *J. Cell Biol.* **173:** 985–994.

Shi H., Tan J., and Schachter H. 2006. N-glycans are involved in the response of *Caenorhabditis elegans* to bacterial pathogens. *Methods Enzymol.* **417:** 359–389.

Antoshechkin I. and Sternberg P.W. 2007. The versatile worm: Genetic and genomic resources for *Caenorhabditis elegans* research. *Nat. Rev. Genet.* **8:** 518–532.

CHAPTER 24

Arthropoda

Michael Tiemeyer, Scott B. Selleck, and Jeffrey D. Esko

THIS CHAPTER DESCRIBES GLYCOSYLATION IN *DROSOPHILA MELANOGASTER* as the best-studied example of an arthropod. The major glycan classes are similar to those described in vertebrates, with interesting and often subtle differences. The powerful genetic systems available for studying gene function in *D. melanogaster* have proved to be effective means for understanding glycan function in early development and have provided some of the first examples of how glycans affect growth factor signaling reactions and morphogen gradients in vivo.

HISTORICAL PERSPECTIVE

The phylum Arthropoda includes Chelicerata (e.g., spiders and horseshoe crabs), Myriapoda (e.g., centipedes and millipedes), Hexapoda (e.g., insects), and Crustacea (e.g., shrimp, barnacles, and lobsters). Arthropods are among the most successful species on Earth and are found in all types of environments. One of their characteristic features is an exoskeleton composed of chitin, which provides support and physical protection. The best-studied example is the fruit fly, *D. melanogaster*. In 1910, T.H. Morgan published the first paper about the genetics of *D. melanogaster*, which showed that white eye color was a sex-linked trait. Since then, this organism has been the predominant model system for genetic analysis. Its advantages include an easily studied developmental program, a relatively complex neural system, and the ability to discern literally thousands of different phenotypes in morphology, development, and behavior. The *D. melanogaster* genome has been sequenced, providing all of the additional advantages of homology screening for the identification of orthologs between species. Unwary *Drosophila* geneticists in pursuit of genes that regulate development ran head-on into glycobiology. Their studies have revealed intimate interrelationships between glycan expression and genes that ultimately control morphogenesis by establishing boundaries and regulating cell communication and cell division via signal transduction pathways. These emerging areas hold great promise for detecting new functional roles for glycoconjugates and unraveling their mechanisms of action.

Drosophila biologists have aimed toward functional analysis of proteins, resulting in a relative paucity of data describing the organism's glycoconjugates. During the past few years, the availability of sensitive analytic techniques has begun to fill this void by permitting structural analysis on small amounts of material. Many of the exciting functional implications drawn from studying mutants are based on homology with genes that code for mammalian enzymes and lectins. However, few of the putative *Drosophila* glycosyltransferase gene products have actually been shown to generate the structures predicted by the vertebrate homolog. With the new analytical techniques in hand, it should now become more common to link interesting phenotypes with altered glycan expression.

INSECT GLYCOPROTEINS

N-Linked Glycan Synthesis and Diversity

It has been a long-accepted and overgeneralized rule that oligosaccharides linked through asparagine to the glycoproteins of arthropods are exclusively of the high-mannose or paucimannose type (see Chapter 8). The question of whether insects have the capacity to process N-linked glycans into hybrid or complex type structures has remained unanswered. The completion and annotation of the *D. melanogaster* genome sequence predicted the existence of the necessary enzymatic machinery to generate complex glycans. Moreover, increases in the sensitivity of key analytic techniques, primarily those based on mass spectrometry, allowed the detection of very minor glycans. In addition, the commercial and experimental demand for eukaryotic expression systems that might deliver reasonable amounts of glycosylated protein led to the characterization of the glycosylation capacity of cells derived from the moth *Spodoptera frugiperda* (Sf9 cells) and from *D. melanogaster* (S2 cells). It is now clear that high-mannose and paucimannose glycans account for greater than 90% of the total N-linked glycan complexity in *Drosophila* throughout its life cycle. However, hybrid and complex glycans, including sialylated structures, are present as minor components (Figure 24.1).

Arthropod N-linked glycans are both similar to and distinct from those of other animals. The similarities and differences start at the chitobiose core, where *Drosophila* adds fucose in both α1-3 and α1-6 linkages to the reducing terminal *N*-acetylglucosamine, whereas vertebrates restrict this linkage to α1-6. The Fucα1-3GlcNAc structure is frequently found on plant glycoproteins, such as horseradish peroxidase (HRP), and on some glycoproteins of *Caenorhabditis elegans*, but it has never been conclusively demonstrated in mammals. In fact, Fucα1-3GlcNAc is immunogenic in humans and rabbits, resulting in the production of antibodies against the so-called HRP epitope. Anti-HRP antibodies demonstrate that the Fucα1-3GlcNAc epitope is restricted primarily to neural tissue in a broad range of arthropods. *Drosophila* does not extend its core fucose residues with additional capping monosaccharides nor modify its N-linked glycans by O-methylation, both of which occur extensively in *C. elegans* (see Chapter 23). The recent demonstration in *Drosophila* embryos of fucosylated, sialylated, hybrid, biantennary complex, and triantennary complex glycans brings the total diversity of the arthropod glycan pool into closer parallel with that of mammals (see Chapter 25).

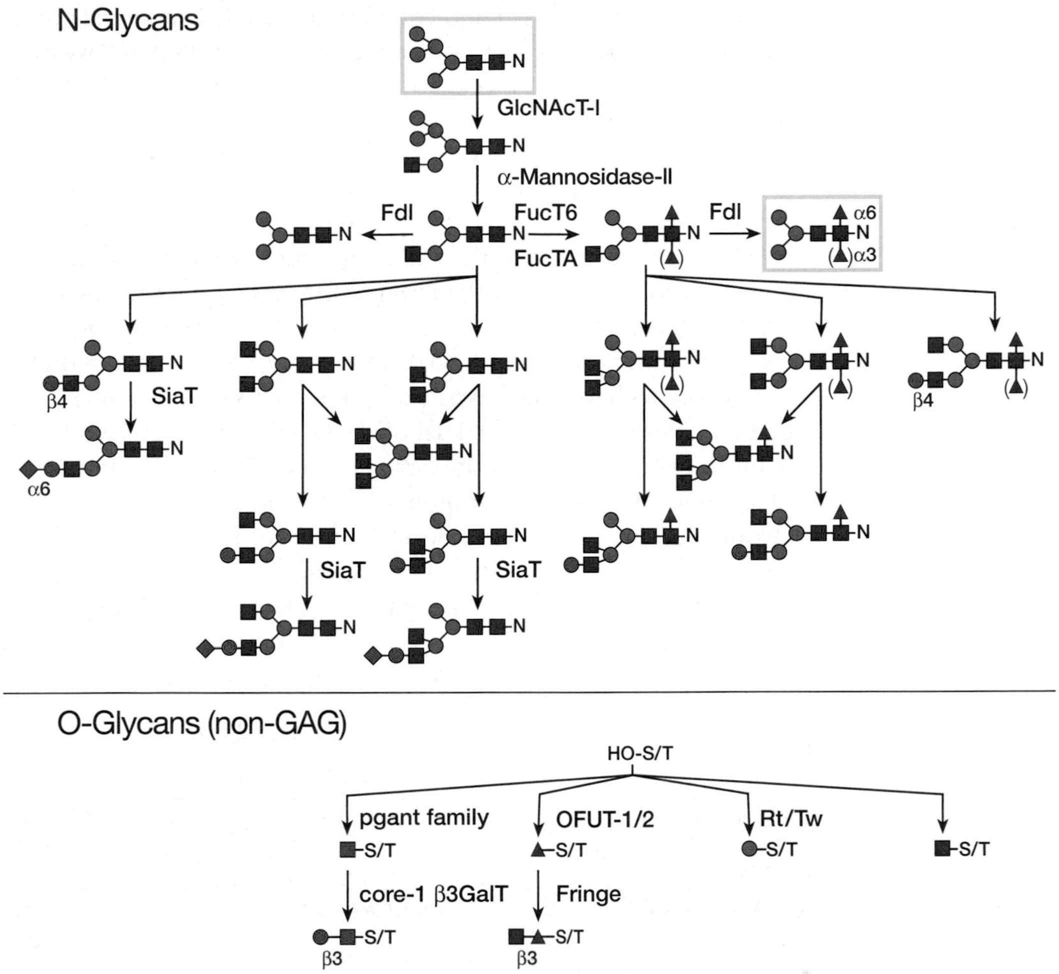

FIGURE 24.1. N-Linked and O-linked glycan diversity. For N-linked glycans, processing after ER mannosidase trimming to the $Man_5GlcNAc_2$ structure is shown. (*Gray boxes*) The predominant N-linked glycans, $Man_5GlcNAc_2$ and $Man_3GlcNAc_2Fuc$, found in embryonic or adult stages of *Drosophila*. N-linked glycan complexity is limited by the Fdl hexosaminidase and expanded by expression of uncharacterized branching GlcNAcT activities, undefined GalT activities, and a single SiaT activity. For O-linked glycans, *Drosophila* was instrumental in defining a role for simple modifications with mono- or disaccharides on serine/threonine residues. The identification of a large family of ppGalNAcTs (*pgants*) in *Drosophila* indicates that arthropods will become increasingly important for understanding the function of mucin-like glycans.

Hybrid and complex structures comprise less than 10% of the total glycan pool in arthropods, in contrast to their abundance in mammalian tissues. The relative paucity of complex glycans in *Drosophila* has been attributed at least partially to the presence of a hexosaminidase that is active in the secretory pathway of many arthropod cells. Encoded by the *fused lobes* (*fdl*) gene, the enzyme is capable of efficiently removing *N*-acetylglucosamine residues that are added by *N*-acetylglucosaminyl transferase I (GlcNAcT-I) to the nonreducing terminal Manα1-3 arm of the $Man_5GlcNAc_2$ core glycan. GlcNAcT-I catalyzes the first committed step toward the generation of a hybrid or complex glycan (see Chapter 8). Therefore, removal of this *N*-acetylglucosamine residue effectively blocks subsequent extension reactions, generating the observed predominance of high-mannose or paucimannose glycans on glycoproteins. It is likely that down-regulation of this enzyme is a key permissive requirement for the elaboration of complex glycans in arthropods.

The presence of hybrid and complex glycans in *Drosophila* predicts the existence of specific enzymes, each of which acts on acceptor substrates that have escaped Fdl-mediated trimming. For instance, α1-6 fucosyltransferase, GlcNAcT-I, -II, -III, -IV, galactosyltransferase (GalT), and sialyltransferase (SiaT) activities are required to generate structures more complex than paucimannose-type glycans. Of these enzymes, only GlcNAcT-I and SiaT are reasonably well characterized. Genome annotations predict the existence of the other enzymes, although their activity and expression have not been demonstrated. In fact, as recombinant proteins assayed in vitro, the candidates for GalT-like enzymes preferentially transfer *N*-acetylgalactosamine rather than galactose to GlcNAc-terminated acceptors. Thus, additional enzyme characterization studies, preferably performed in whole animals, will be necessary to completely delineate the N-linked synthesis and processing pathway in arthropods.

Phenotypes have been described for mutations in key N-linked glycan processing and synthetic enzymes (Table 24.1). Loss of the Golgi-trimming enzyme, α1-2 mannosidase (Golgi mannosidase I, *mas1*), was the first mutation identified to affect a glycan-processing step in *Drosophila.* The *mas1* mutation has little or no impact on the processing of high-mannose glycans, leading to the identification of an alternate mannosidase activity that effectively bypasses loss of mannosidase I. However, the embryonic peripheral nervous system, the wing, and the adult eye exhibit mild alterations in the *Drosophila mas1* mutant, suggesting that the bypass is not complete. A similar, partial bypass was unmasked in mammals following targeted disruption of mouse α1-6 mannosidase (Golgi mannosidase II) (see Chapter 8).

TABLE 24.1. *Mutations that affect the synthesis or function of* Drosophila *glycans*

Mutant	Affected protein	Affected glycan	Phenotype
N-linked or O-linked pathways			
mas1	α-mannosidase I	$Man_8GlcNAc_2$	peripheral nervous system, wing, and eye morphology
fused lobes (fdl)	glycoprotein glycan processing β-hexosaminidase	complex glycans	larval and adult brain lobe morphology
mgat1	GlcNAcT-I	complex and difucosylated glycans	reduced viability and locomotor activity
tollo/toll-8	Toll-like receptor 8	difucosylated glycans (HRP-epitope)	altered glycosylation in embryonic CNS
neurally altered carbohydrate (nac)	unknown	difucosylated glycans (HRP-epitope)	altered glycosylation in larval, pupal, adult CNS
dβ4GalNAcTA	β4GalNAcT (homology with vertebrate β4GalT)	GalNAcβ1-4GlcNAc on unknown glycoprotein glycans	locomotor behavioral deficits
pgant35a	polypeptide GalNAcT	O-GalNAc linked to serine/threonine	pupal lethality
rotated abdomen (rt)	protein-*O*-mannosyltransferase I	O-mannose linked to serine/threonine	abdominal morphology
twisted (tw)	protein-*O*-mannosyltransferase II	O-mannose linked to serine/threonine	abdominal morphology
Ofut-1	*O*-fucosyltransferase I	O-fucose linked to serine/threonine	*notch*-like defects in cellular differentiation
fringe (fng)	fucose-specific β3GlcNAcT	*N*-acetylglucosamine addition to O-linked fucose	pattern formation defects resulting from altered Notch activation

TABLE 24.1. *(Continued)*

Gene	Protein/activity	Affected glycan	Signaling deficits
GAG/PG pathways			
sugarless (*sgl*)	UDP glucose dehydrogenase	chondroitin and heparan sulfates	Wg, Hh, Dpp signaling
slalom (*sll*)	PAPS transporter	all sulfated glycans	Wg, Hh signaling
fringe connection (*frc*)	UDP-sugar transporter	glycosaminoglycans and O-linked glycans	Wg, Hh, FGF, Notch signaling
tout-velu (*ttv*)	heparan sulfate polymerase (Ext1)	heparan sulfate	Wg, Hh, Dpp signaling
brother-of tout-velu (*botv*)	*N*-acetylglucosamine transferase (ExtL3)	heparan sulfate	Wg, Hh, Dpp signaling
sister-of-tout-velu (*sotv*)	heparan sulfate polymerase (Ext2)	heparan sulfate	Wg, Hh, Dpp signaling
sulfateless (*sfl*)	*N*-deacetylase *N*-sulfotransferase	sulfation of heparan sulfate	Wg, Hh, Dpp, FGF signaling
D-Hs2st	Hs2st	heparan sulfate	FGF, tracheal development
D-Hs6st	Hs6st	heparan sulfate	FGF, tracheal development
D-Hs3st-B	Hs3st	heparan sulfate?	*notch*-like defects
dally (*dly*)	glypican-related proteoglycan		Dpp, Wg signaling
dally-like protein (*dlp*)	glypican-related proteoglycan		Wg, Hh, D-LAR signaling
syndecan (*sdc*)	syndecan ortholog		Slit-Robo, D-LAR signaling
terribly reduced optic lobes (*trol*)	perlecan ortholog		Hh, FGF signaling
Chitin pathways			
cystic/mummy (*cyst/mmy*)	UDP-*N*-acetylglucosamine diphosphorylase	UDP-GlcNAc and chitin	tracheal morphology and axon guidance defects
krotzkopf verkhert (*kkv*)	chitin synthase	chitin	tracheal morphology
serpentine (*serp*) and *vermiform* (*verm*)	multidomain proteins possessing chitin *N*-deacetylase domains	chitin	tracheal morphology
Glycosphingolipid pathways			
egghead (*egh*)	GlcCer-specific β4ManT	mactosylceramide (Manβ1-4Glcβ1-Cer)	*notch*-like defects in cellular differentiation
brainiac (*brn*)	mactosylceramide-specific β3GlcNAcT	arthrotriosylceramide (GlcNAcβ1-3Manβ1-4Glcβ-Cer)	*notch*-like defects in cellular differentiation
Lectin function			
furrowed (*fw*)	furrowed, a C-type lectin	unknown binding preference	bristle and eye morphology
gliolectin (*glec*)	gliolectin, an unclassified carbohydrate-binding protein	binding preference for βGlcNAc-terminated glycosphingolipids	axon pathfinding

Mutations in genes responsible for subsequent processing steps have begun to reveal the importance of glycan complexity in *Drosophila.* In addition to increased levels of complex glycans, reduction in the activity of the Fdl hexosaminidase results in altered brain structure. Brain lobes that are normally separated in wild-type adults are instead fused together through a continuous stalk at the midline in *fdl* mutants, hence, the original name of the mutation, *fused lobe* (Figure 24.2). In the wild-type adult, the separated lobes form portions of the mushroom body, a brain structure whose function has been implicated in *Drosophila* learning and memory. In an interesting convergence, a null mutation in *mgat1* (GlcNAcT-I) generates an apparently identical fused-lobe phenotype in *Drosophila* adults, although the glycan expression profile shifts in the opposite direction (from more complex in *fdl* mutants to high mannose and paucimannose in *mgat1* mutants) relative to *fdl.* It is still unclear how loss of complex glycans in one case (*mgat1*) or enrichment of complex glycans in another case (*fdl*) generates the same neural phenotype. GlcNAcT-I loss-of-function mutant adults also exhibit reduced locomotor activity and decreased life span.

A terminal synthetic step in the production of N-linked glycans in mammals is the addition of sialic acid residues as caps at the nonreducing ends of antennae. In vitro assays of recombinant enzyme demonstrate that the single SiaT identified in *Drosophila* exhibits a twofold preference for transferring sialic acid to a LacdiNAc (GalNAcβ1-4GlcNAc) over a type II LacNAc (Galβ1-4GlcNAc) acceptor. However, the only identified sialylated N-linked glycans in *Drosophila* possess subterminal LacNAc instead of LacdiNAc. It remains to be determined whether N-linked LacdiNAc acceptors are present in *Drosophila*, perhaps at levels below current detection thresholds. Alternatively, the acceptor specificity of endogenous SiaT may differ from that of recombinant enzyme. The expression pattern of the enzyme is consistent with the observed low abundance of sialylated glycans. SiaT is detected in a very small number of neurons beginning in late

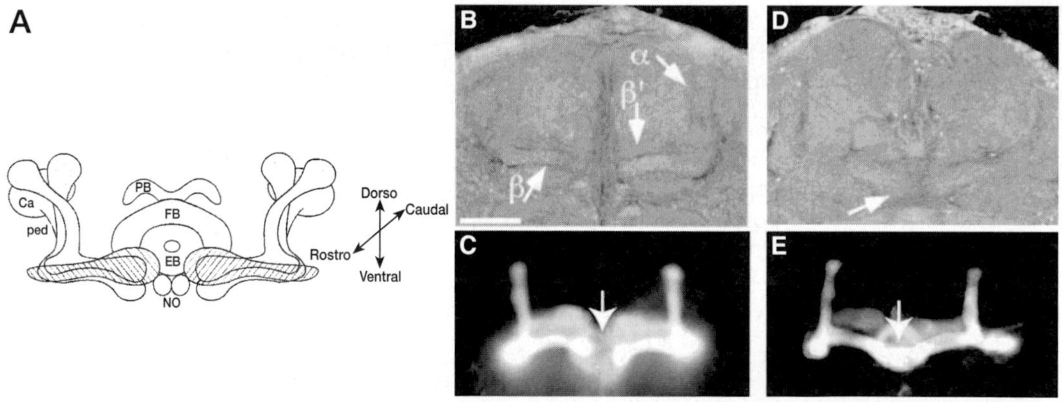

FIGURE 24.2. Mutations in enzymes that process complex N-linked glycans alter adult brain morphology in *D. melanogaster*. (*A*) The major lobes of the adult brain are shown in cross section. (*Hatched areas*) γ lobes; (ped) peduncle; (Ca) calyx; (EB) ellipsoid body; (FB) fan-shaped body; (NO) nodulli; (PB) protocerebral bridge. (*B,C*) Symmetrical sets of α and β lobes (*B*) are distinctly separated at the midline (*C, arrow*) in the wild-type adult. (*D*) A mutation in the *fused lobes* (*fdl*) gene, which encodes for a hexosaminidase that removes the *N*-acetylglucosamine added by GlcNAcT1, yields increased expression of complex glycans and midline fusion of the β lobes (*arrow*). (*E*) A loss-of-function mutation in GlcNAcT1 decreases complex N-linked glycan expression, but it also fuses the β lobes (*arrow*). Despite driving glycosylation in opposite directions, these two mutations yield convergent brain phenotypes. (*A,B,D*: Reprinted, with permission of Wiley-Liss, Inc., from Boquet I. et al. 2000. *J. Neurobiol.* **42:** 33–48. *C,E*: Reprinted, with permission of the American Society for Biochemistry and Molecular Biology, from Sarkar M. et al. 2006. *J. Biol. Chem.* **281:** 12776–12785.)

embryonic stages. The extremely restricted expression of the SiaT enzyme reinforces proposals that minor glycans may be restricted to small subsets of cells during limited developmental or adult stages, where they may fulfill very specific cellular functions (see Chapter 6).

O-Linked Glycan Synthesis and Diversity

Insects add glycans in O-linkage to serine and threonine residues on secreted, cell-surface, and intracellular proteins. Structural complexity ranges from single monosaccharides (*N*-acetylgalactosamine, *N*-acetylglucosamine, mannose, glucose, or fucose) to extensively modified glycosaminoglycan chains (see below). Between these two extremes, the core-1 structure (Galβ1-3GalNAcβ-Thr/Ser), which is found extensively on vertebrate mucin-like proteins (see Chapter 9), has also been described in insects (Figure 24.1). Cultured *Drosophila* cell lines secrete mucin-like proteins bearing glycans that are recognized by peanut agglutinin and removed by *O*-glycanase. As much as 40% of the total mass of these proteins is contributed by carbohydrate, although the glycan composition is dependent on culture conditions. In *Drosophila* tissues, peanut agglutinin and mucin-specific antibodies reveal developmentally regulated and spatially restricted expression of Galβ1-3GalNAc on currently unidentified proteins. Extended or branched O-linked glycans (core 2, etc.) are yet to be detected in insect tissues (see Chapter 9). The core-1 disaccharide is also present in other insects and insect cell lines.

One functional role for Galβ1-3GalNAc appears to be in defending against microbial attack. A 19-amino-acid peptide, produced upon trauma or pathogen challenge to *Drosophila*, requires the core-1 disaccharide for full activity. Insects also use mucins as protective barriers in the gut. A mucin-like protein, called insect intestinal mucin (IIM), is associated with the chitin-containing peritrophic membrane that coats the epithelial cells lining the insect gut. Invading baculovirus produces an enhancin, which possesses mucin-degrading activity. Enhancin-mediated degradation of IIM boosts viral infectivity by facilitating the penetration of the mucinous barrier of the lepidopteran gut epithelium.

As in vertebrates, it has been demonstrated that *Drosophila* possesses a large family of polypeptide *N*-acetylgalactosamine transferases (ppGalNAcT encoded by *pgant* genes), the initiating enzymes for the synthesis of O-linked mucin cores (see Chapter 9). Currently, there are 12 identified *pgant* genes that are likely to encode functional enzymes and multiple isoforms. Mutational analysis is just beginning for the ppGalNAcT family in *Drosophila*, but it has already been shown that loss of one particular family member (*pgant35A*) is lethal. Many of the ppGalNAcTs demonstrate their own signature (sometimes partially overlapping) and dynamic expression patterns. The generation of single and multiple ppGalNAcT mutants and the subsequent use of these mutants in targeted genetic screens should lead to a better understanding of the unique tissue requirements and polypeptide acceptor specificities that drive O-linked glycan function across species.

Other types of O-linked glycan modifications are also present in insects. The discovery that O-linked *N*-acetylglucosamine (O-GlcNAc) decorates protein components of the *Drosophila* polytene chromosome added weight to initial demonstrations of the existence of nucleocytoplasmic glycosylation in animal cells (see Chapters 17 and 18). Unfortunately, remarkably little work on the function of O-GlcNAc has subsequently been undertaken in arthropods. O-Linked mannose, a modification that contributes to the pathophysiology of some human muscular dystrophies (see Chapter 42), is also predicted to be present in *Drosophila* based on sequence similarity of two putative transferase genes. Mutations in either of these two genes, *rotated abdomen* (*rt*) or *twisted* (*tw*), cause clockwise helical rotations of abdominal morphology in adult flies. The two genes interact such that the abdominal rotation observed with reductions in both genes is more severe than that of either single mutant. Both the *rt* and *tw* genes encode transmembrane proteins that closely resemble the yeast protein O-mannosyltransferases, PMT1 and PMT2, and the vertebrate protein O-mannosyltransferases, POMT-1 and POMT-2. Using transient expression in insect cell lines, it was demonstrated that Rt and Tw proteins must be expressed together in the same cell to achieve transfer of mannose onto α-dystroglycan, a potential physiologic acceptor for O-mannosylation. The prevalence of O-linked mannose on proteins expressed in *Drosophila* tissues is not currently known.

Addition of a simple glycan, O-linked fucose, modulates complex developmental signals. The *notch* gene was first identified and characterized in *Drosophila* as encoding a large, multimodular receptor protein involved in cell-fate determination. Members of the Notch family of receptor proteins are widely distributed from *C. elegans* to humans. One characteristic of Notch is the presence of epidermal growth factor (EGF)-like modules. The EGF-like domain modules in the human Notch homolog have been shown to contain two unusual types of O-linked glycosylation, O-glucose and O-fucose, which have been implicated in Notch's interactions with its ligands (see Chapter 12). The extracellular ligands for Notch (Delta and Serrate/Jagged) also contain EGF-like domains that are similarly O-glycosylated. As in mouse and humans, *Drosophila* possesses protein O-glucosyltransferase activity and two protein O-fucosyltransferases, OFUT-1 and OFUT-2. In vitro enzyme activity studies have demonstrated that in both *Drosophila* and vertebrates, OFUT-1 transfers fucose to EGF-like domains and OFUT-2 transfers fucose to thrombospondin receptor repeats on a different set of proteins. Loss of OFUT-1 in *Drosophila* yields *notch* mutant phenotypes (Figure 24.3), indicating that fucosylation of Notch is essential for ligand-induced activation.

Fucosylated Notch is a substrate for an *N*-acetylglucosaminyltransferase encoded by the *fringe* gene. Elongation of O-linked fucose with *N*-acetylglucosamine yields Notch protein that is activated by Delta but not by Serrate/Jagged. Therefore, O-linked glycosylation functions as a switch that activates the Notch receptor or alters its ligand preference. Differential Notch activation by cell-specific expression of OFUT-1 and Fringe generates distinct cell fate choices that lead to pattern formation in the embryo. The ability of the *Drosophila*

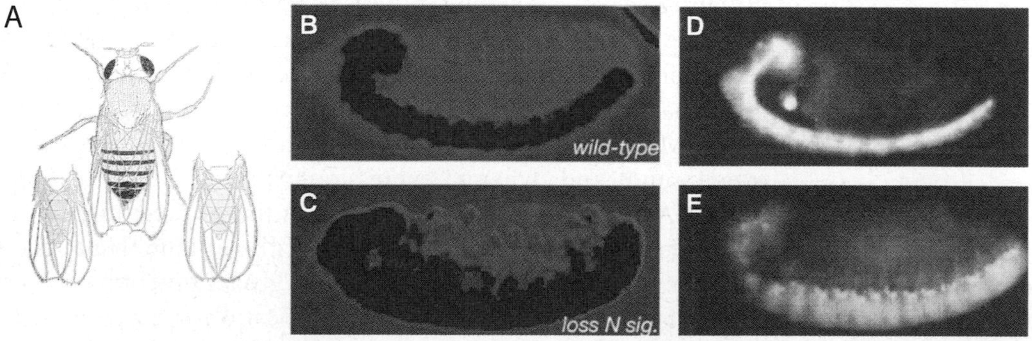

FIGURE 24.3. Cell fate choices dependent on Notch require appropriate glycan expression. (*A*) *notch* mutations were originally identified based on aberrant wing morphology. Changes in cell fate generate wings that are notched at their margins. Wing notches arise from insufficient numbers of nonneural cells in the developing wing. In the embryo, loss of Notch signaling expands neural tissue at the expense of nonneural ectodermal cell fates. (*B,C*) Staining with a neuron-specific antibody reveals increased neural cell numbers in a *notch* mutant (*C*) compared to wild type (*B*). (*D,E*) Loss of the *egghead* or *brainiac* genes, which are essential for GSL synthesis, results in neurogenic phenotypes similar to loss of Notch signaling. (*B,C*: Reprinted, with permission of the Company of Biologists, from Lai E.C. 2004. *Development* **131:** 965–973. *D,E*: Reprinted, with permission of the Company of Biologists, from Goode S. et al. 1992. *Development* **116:** 177–192.)

OFUT-1 and Fringe proteins to rescue or modify Notch protein or *notch* mutant phenotypes is well documented in cultured cells and in whole embryos, although the demonstration of O-fucose or the GlcNAc-extended disaccharide on Notch protein extracted from *Drosophila* tissue has not yet been achieved. In mice and humans, but apparently not in *Drosophila*, the disaccharide can be extended by the addition of galactose and then capped with sialic acid.

INSECT PROTEOGLYCANS AND GLYCOSAMINOGLYCANS

Structure

The greatest amount of structural data on arthropod glycosaminoglycans (GAGs) derives again from studies of *Drosophila.* Mutations affecting a number of genes involved in GAG biosynthesis have been identified in genetic screens for patterning mutants, providing the means for determining the relationship between specific biosynthetic enzymes, GAG structures, and their function in discrete developmental events. Many of the heparan sulfate (HS)-bearing proteoglycan core proteins known from vertebrate systems are also represented in *Drosophila,* including the glypicans, syndecan, and perlecan. Of these, the glypicans are the best studied, and mutations affecting the glypican ortholog "division abnormally delayed" (Dally) were identified in a screen for cell division patterning mutants. Analysis of *dally* mutants provided some of the first evidence that these molecules are critical for patterning during development. Biochemical analysis of Dally synthesized in *Drosophila* S2 cells showed it to be a true HS proteoglycan, with GAG chains sensitive to both heparinase digestion and low pH nitrous acid cleavage, the latter demonstrating the presence of N-sulfated glucosamine residues characteristic of vertebrate HS and heparin (see Chapter 16).

Much of the structural analysis of GAGs from *Drosophila* has employed a disaccharide profiling method based on digestion of the polymer chains with cocktails of either heparan or chondroitin sulfate (CS) lyases, followed by chemical tagging with a fluorophore (see Chapter 16). Disaccharide profiling of GAGs from *Drosophila* demonstrated the presence of HS and CS remarkably similar in structure to that of GAGs found in vertebrates (see Chapter 16). The principal disaccharide species of HS-derived units include N-, 2-O, and 6-O sulfated forms, and mono-, di-, and trisulfated disaccharides. CS detected from whole embryos or larvae was largely unsulfated or 4-O-sulfated, but 6-O-sulfated disaccharides have been detected as well. The covalent attachment of GAG chains to proteins via the canonical tetrasaccharide linker (GlcA-Gal-Gal-Xyl) has also been demonstrated for *Drosophila.* Not surprisingly, CS and HS have been documented in many other arthropod species, establishing the conservation of these macromolecules throughout this phylum. Hyaluronan, a structurally simple glycosaminoglycan synthesized by a plasma-membrane-associated synthase in vertebrates (see Chapter 15), is not found in *Drosophila.* Expression of a mouse hyaluronan synthase gene in *Drosophila* produces massive accumulation of extracellular hyaluronan that disrupts normal development.

Mutations affecting GAG biosynthesis and modification have provided important insights concerning proteoglycan biosynthesis and the modulation of structures in response to perturbations. For example, loss of the single *N*-deacetylase/*N*-sulfotransferase (*Ndst*) gene in *Drosophila (sulfateless, sfl)* does not reduce polymer synthesis but results in essentially an unsulfated GAG, *N*-acetylheparosan, without N, 2-O, or 6-O modifications, which results in defective patterning decisions orchestrated by several growth factors. This established that loss of NDST activity essentially abolished O-linked sulfation, as in vertebrates. Both *Ext1*- and *Ext2*-related genes, *ttv* and *sotv*, are required for HS polymerization in *Drosophila.* Mutations affecting the single *D-Hs2st* or *D-Hs6st* genes have particularly interesting effects on HS assembly. Loss of the uronyl 2-O sulfotransferase eliminates 2-O sulfate groups as expected, but results in compensatory increases in 6-O sulfation. The converse is also true; loss of 6-O sulfotransferase activity produces a polymer with no 6-O sulfate groups but elevations of 2-O sulfate and N-sulfate sufficient to retain the overall sulfation state of the polymer. Compensation between 2-O and 6-O modifications has been observed in mouse embryonic fibroblasts derived from *Hs2st* mutants and indicates that this is a conserved and likely important mechanism for retaining the activity of HS proteoglycans in vivo. The signaling deficits found in *Drosophila Hs2st* and *Hs6st* mutants support this view; fibroblast growth factor (FGF)-directed patterning is only modestly affected in single mutants, but it is severely compromised in the double-mutant animals.

Genetic studies have demonstrated that HS and CS biosynthesis are linked; mutations that reduce, but do not eliminate, Ext activity required for HS polymerization increase the net amount of chondroitin polymer formed, similar to observations made in vertebrate systems. These in vivo data support the idea of a "gagosome," a complex of biosynthetic and modification enzymes that resides in the Golgi and coordinates GAG production (see Chapter 16). There may well be some protein components shared between complexes that generate and modify different GAGs, so that loss of one component can have effects on more than one type of GAG polymer.

Growth Factor Signaling and Morphogen Gradients

Genetic studies in *Drosophila* demonstrated that GAGs are critical for signaling mediated by a number of growth factors during development. Mutations affecting Dally disrupted patterning of many tissues and affected cell-cycle progression in the nervous system. Subsequent studies confirmed that Dally affected signaling mediated by Wnt and bone morphogenetic protein (BMP)-related growth factors, Wingless (Wg) and Decapentaplegic (Dpp; a BMP4 ortholog). Mutations compromising HS biosynthesis have profound effects on patterning decisions orchestrated by numerous growth factors, including Wg, Dpp, FGF, and Hedgehog. Loss of HS production or modification compromises signaling mediated by these secreted protein factors to varying degrees (see Table 24.1). Genes required for HS production have the most severe effects on multiple pathways, whereas those affecting specific modifications have more limited phenotypes, perhaps only compromising a subset of growth factor signaling processes. These genetic studies also show that proteoglycans serve a modulatory role in signaling, not as "essential" components required for any signaling to occur. For example, although HS proteoglycans are critical for normal Dpp signaling in the wing disc, mutations compromising HS biosynthesis have no detectable effect on Dpp signaling in the early *Drosophila* embryo. Embryos lacking enzymes critical for HS synthesis can still respond to Wg if its expression is sufficiently elevated with Wg-encoding transgenes.

Two molecular mechanisms contribute to defective signaling when HS biosynthesis is compromised: (1) reduced cell responsiveness to the growth factor with the HS proteoglycan normally serving a coreceptor role, affecting the assembly of signaling complexes, and (2) lower levels of growth factor in the matrix. The best direct evidence in vivo for HS proteoglycans contributing to increased cell responsiveness comes from studies of Dally, where patches of cells expressing Dally at high levels produce markedly elevated levels of Dpp (BMP4) signaling in a cell-autonomous manner. Loss of HS synthesis can also affect signaling by reducing the levels of growth factor in the matrix. The ability of HS proteoglycans to alter the distribution and levels of growth factors in the matrix is a critical aspect of their function because many of these secreted proteins are morphogens.

The term "morphogen" applies to secreted protein factors that show graded distributions across tissues and distinct cell responses to different levels of that signaling molecule. Morphogens provide an essential mechanism for generating cell diversity during tissue assembly. Morphogen activity has been demonstrated in vivo for Wg, Dpp, and Hh in *Drosophila*; these proteins are representatives of three families of signaling molecules essential for normal development in virtually all multicellular organisms. Controlling the generation and maintenance of morphogen gradients is therefore of critical importance during development. Genetic studies in *Drosophila* have established that HS proteoglycans are important regulators of morphogen gradients. Loss of HS biosynthesis makes many morphogen gradients more "shallow" with lowered levels of morphogen at a distance from the producing cells. Mutations affecting individual HS proteoglycans can also produce local increases in morphogen levels, as observed in *dlp* mutants, where Wg immunoreactivity and signaling near the Wg-producing cells is elevated (Figure 24.4).

There is good evidence that HS proteoglycans are integrated into regulatory circuits that are designed to maintain the integrity of morphogen gradients and the patterning

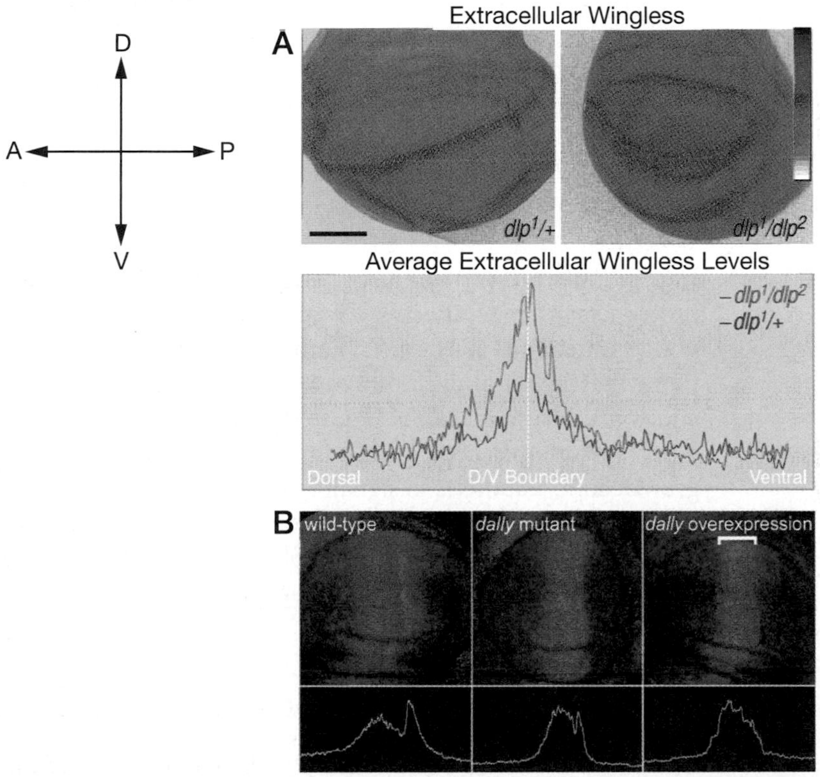

FIGURE 24.4. Proteoglycan deficiencies affect morphogen gradients and signaling. (*A*) Effects of Dally-like (dlp) on Wingless (Wg) morphogen gradient in the developing wing of *Drosophila*. Loss of *dlp* (dlp^1/dlp^2) produces an increase in the levels of Wg in the extracellular matrix compared to a heterozygous control ($dlp^1/+$), detected here with anti-Wg antibody. Wg is produced by a band of four to five cells at the center of the Wg stripe, located at the dorsoventral boundary of the developing wing disc. Wing discs from *dlp* homozygous animals show an increase in Wg across a large segment of the wing epithelium, evident in both the confocal images of the wing disc (intensity of staining is color-coded according to the inset lookup scale) and the intensity tracing represented below the photograph for both control and *dlp* mutant discs. Bar, 50 μm. (*B*) Influence of Dally expression on Dpp (a BMP4 ortholog). Either reductions (*dally* mutant) or increases in Dally expression limit the gradient of Dpp signaling, compared to wild-type control discs, measured here by detection of phosphorylated Mad, a direct target of Dpp activity. (*A*: Reprinted, with permission of Elsevier, from Kirkpatrick C.A. et al. 2004. *Dev. Cell* **7:** 513–523. *B*: Reprinted, with permission of the Company of Biologists, from Fujise M. et al. 2003. *Development* **130:** 1515–1522.)

events directed by these gradients. For example, both expression of a Dpp receptor (Thickvein) and that of the Dally coreceptor for Dpp are negatively regulated by Dpp signaling. This provides the potential for correcting for inappropriate (by either genetic or environmental mechanisms) reductions in Dpp signaling by increasing receptor and coreceptor levels. Computer modeling of morphogen gradients suggests that HS proteoglycans serve to increase the robustness of the system; that is, they can compensate for changes that would otherwise disrupt the normal patterning of the tissue orchestrated by the morphogen system.

Genetic studies in *Drosophila* have also revealed that individual proteoglycans can have very different influences on signaling. For example, mutations compromising the function of Dally or its related glypican family member, Dally-like, have opposite effects on Wg signaling in the same set of cells. *dally* mutants have reduced Wg signaling and loss of sensory bristles at the wing margin, whereas *dlp* mutants have increased Wg signaling and ectopic bristle formation. Likewise, the *dally* mutation has very modest effects on Wg or Hh signaling in the embryo, but loss of *dlp* produces Hh signaling deficits as severe as *hh* null mutants. A *Drosophila* ortholog of perlecan encoded by *terribly reduced optic lobes (trol)* results in cell division arrest of larval neuroblasts, and both biochemical and genetic studies suggest that Trol-mediated regulation of FGF and Hh signaling is responsible for these phenotypes. Syndecan and Dally-like have distinct functions in neuromuscular synapse assembly and apparently opposing effects on the protein tyrosine phosphatase, LAR.

Axon guidance is another developmental process in which proteoglycans are now known to serve critical roles. Syndecan (*sdc*) mutants show disrupted axon guidance in *Drosophila* embryos, and several findings suggest that Slit-Robo signaling is compromised. Interestingly, Slit protein distribution is altered in *sdc* mutant embryos, suggesting that HS proteoglycans may affect axon guidance in a manner consistent with their role in morphogen-mediated patterning, namely, by controlling the levels of key signaling molecules in the matrix. *dlp* and *sdc* mutants also have axon-guidance abnormalities in the *Drosophila* adult visual system. *dlp* is required in both retinal neurons and in the brain for distinct aspects of photoreceptor axon targeting, indicating that it may serve both in controlling growth cone responses to signaling molecules and in the distribution of guidance cues provided in the brain for incoming photoreceptor axons. In addition, there is evidence that Sdc and Dlp have unique activities in photoreceptor axon guidance, indicating that these molecules are not simply delivery systems for HS chains.

CHITIN

Chitin, a polymer of $(\text{GlcNAc}\beta4)_n$, is the second most abundant biopolymer on Earth (cellulose is the most abundant). Chitin is a major component of the rigid, cuticular exoskeleton of all Arthropoda, and therefore these animals expend considerable resources toward its assembly. Details of chitin biosynthesis are described in Chapter 21. Chitin fibrils also form a more subtle, protective layer at the apical surface of gut epithelial cells (the peritrophic membrane) and in the lumen of the forming tracheal system. Recently, elegant genetic screens have demonstrated that mutations in genes necessary for chitin polymerization, modification, and disassembly independently affect tracheal diameter or length (see Table 24.1). Specific structural characteristics of the chitin produced during defined developmental stages influence the morphogenesis of epithelial tube structures. The larger potential for chitin or for small chitin-based oligosaccharides to modulate cell signaling and morphogenesis remains unexplored in *Drosophila* (see Chapters 25 and 37).

INSECT GLYCOSPHINGOLIPIDS

Insects were once classified as "animals without gangliosides" in reference to the class of sialylated glycosphingolipids (GSLs) found broadly distributed in animal families other than the arthropods (see Chapter 10). Implicit in this label was a sense of surprise that an animal could function without this family of glycolipids. But arthropods possess their own parallel family of GSLs, designated as the arthroseries. Although it is still true that sialylated GSLs (gangliosides) have not been found in *Drosophila*, the arthroseries glycolipids, like vertebrate GSLs, come in acidic as well as neutral varieties (Figure 24.5). The acidic residue is supplied as glucuronic acid rather than sialic acid, but like many of the ganglioside sialic acids (see Chapter 10), the glucuronic acid is linked to a terminal galactose residue. Arthroseries glycolipids with more than one glucuronic acid residue or with a branching glucuronic acid residue have not yet been described.

Neutral and acidic arthroseries glycolipids share core structures. The core of the arthroseries is built from a glucosylceramide, which is frequently heterogeneous with respect to the length of its sphingosine base and its N-acyl chain (see Chapter 10). The arthroseries distinguishes itself from other classes of animal GSLs with the addition of the next monosaccharide residue. Vertebrates generally build their GSLs on a lactosylceramide core (Galβ1-4Glcβ-ceramide). In contrast, the arthropods add a mannose residue to Glcβ-Cer to generate a core (Manβ1-4Glcβ-ceramide) that has been called "mactosylceramide." Addition of the next monosaccharide, an *N*-acetylglucosamine in β1-3 linkage to mannose,

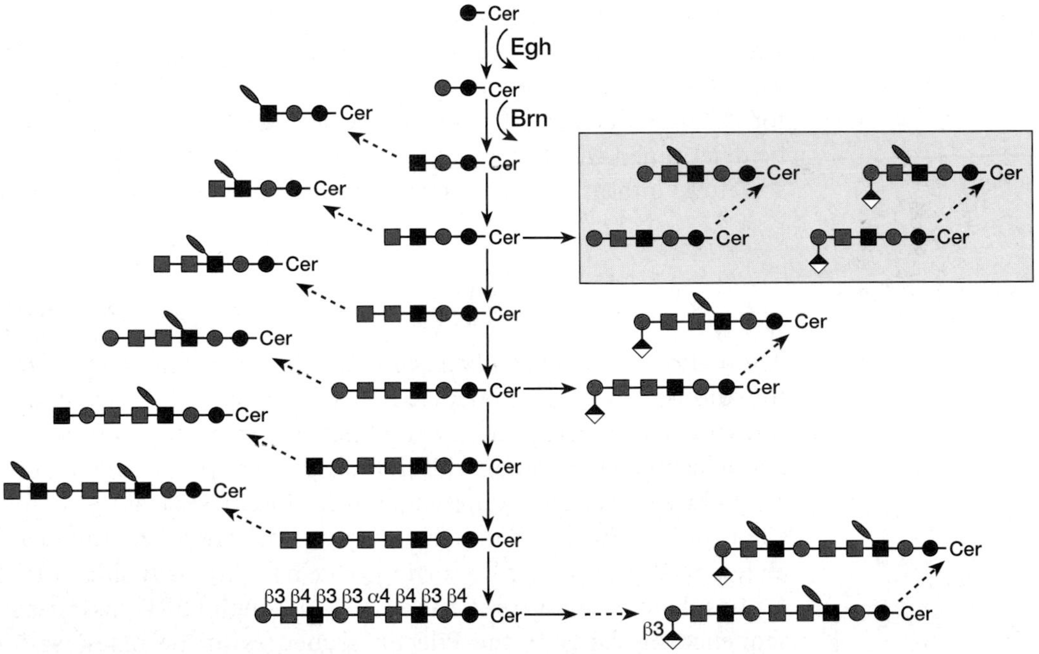

FIGURE 24.5. Glycosphingolipid glycan diversity. The arthroseries of glycosphingolipids are built by extension from a mactosylceramide core (Manβ1-4Glcβ-ceramide). Neutral arthroseries glycans are frequently modified by the addition of ethanolamine (*orange ovals*) on *N*-acetylglucosamine residues, forming a zwitterionic structure. The relationship between glycan extension and modification with ethanolamine is unknown. Acidic charge is imparted by the addition of glucuronic acid to nonreducing terminal galactose residues on neutral or zwitterionic cores. Extension of the GalNAc-terminated arthrotetraose by addition of galactose (*shaded box*) has been demonstrated in different developmental stages of two dipterans, the larvae of the greenbottle fly (*Lucilia caesar*) and the pupae of the blowfly (*Calliphora vicina*), but not in embryos of *Drosophila*.

produces a substrate for a postsynthetic modification not found in vertebrate tissues, ethanolamine (EthNP) in diester linkage to C-6 of *N*-acetylglucosamine (EthNP-6GlcNAcβ1-3Manβ1-4Glcβ-Cer). Therefore, the neutral core of most arthroseries glycolipids is better described as zwitterionic rather than neutral. Arthroseries cores with one or more *N*-acetylglucosamine residues are found with 0, 1, or 2 EthNP additions. Although the number and the position of the EthNP groups impart additional structural diversity, the functional significance of any particular feature of the arthroseries glycans is currently unknown.

Mutations that affect the first steps in *Drosophila* GSL synthesis were originally identified as modulators of cell fate. Biochemical analysis demonstrated that genes called *egghead* (*egh*) and *brainiac* (*brn*), respectively, encode the mannosyltransferase and *N*-acetylglucosaminyltransferase enzymes that add the second and third monosaccharide residues to the arthroseries core lipid, Glcβ-Cer (Figure 24.3). Phenotypes associated with loss of maternal and zygotic *egh* and *brn* are identical. Overproliferation of neural cells occurs at the expense of epithelial lineages. These *egh/brn* phenotypes closely resemble mutations in the *notch* gene.

Signaling through Notch and other pathways requires appropriate membrane glycolipid composition. Perturbation of GSL synthesis alters the partition of signaling receptors between modified lipid rafts and other signaling or nonsignaling membrane domains. Mutations that act early in the assembly of glycolipid glycans, such as *egh* and *brn*, will impact glycan heterogeneity and membrane dynamics more severely than mutations that act later and thus affect a smaller portion of the total GSL pool. Interestingly, the mammalian B4GalT that adds galactose to Glcβ-Cer to generate lactosylceramide (see Chapter 10) rescues the *egh* phenotype when expressed in mutant embryos. This rescue is dependent on Brn activity, demonstrating that GalT-generated lactosylceramide cores are able to act as acceptor substrates for subsequent elongation of glycosphingolipid oligosaccharides in *Drosophila*. Therefore, the chemical identity of the second residue in the arthroseries core is not essential for GSL function or for the formation of functional membrane microdomains. It remains to be determined whether arthroseries glycans elongated beyond the triaosylceramide possess structural information that imparts specific function in any particular context.

INSECT LECTINS

Annotation of the *Drosophila* genome indicates the presence of representatives for each of the known classes of animal lectins (see Chapter 26). Intracellular carbohydrate-binding proteins that in vertebrates are associated with protein folding and quality control in the endoplasmic reticulum or with trafficking through the early compartments of the secretory pathway (calnexin, calreticulin, VIP, ERGIC-53) are also found in *Drosophila* (see Chapter 36). Further downstream, lysosomal targeting is mediated by LERP (lysosomal enzyme receptor protein), a sorting protein functionally similar to the vertebrate mannose-6-phosphate receptor (see Chapter 30). Although LERP possesses domain architecture and sequence similarity to the P-lectin sequences of the mannose-6-phosphate receptor, its binding does not appear to be glycan-dependent, suggesting evolutionary divergence in the mechanisms that control the biogenesis of lysosomes. For galectin (see Chapter 33), C-type (see Chapter 31), and I-type lectin families (see Chapter 32), whose influences are primarily exerted at the cell surface, the binding specificities and functional character of insect family members are a relatively new area of investigation.

The genome sequence of *D. melanogaster* predicts the presence of at least four galectins, three of the tandem repeat type and one of the monomer type (see Chapter 33). The mosquito, *Anopheles gambiae*, also expresses a galectin of the monomer type. Developmental functions for the only well-characterized *Drosophila* galectin are suggested by its embryonic expression pattern, which demonstrates elevated mRNA in the presumptive mesoderm, in the early extended germ band, and in the late embryonic central nervous system. Expression in larval hemocytes of *Drosophila* and in the salivary gland of *Anopholes* is consistent with a role for galectins in mediating innate immune responses to pathogen.

C-type lectins (see Chapter 31) have also been implicated in insect defense responses, especially in the larvae and pupae of *Periplaneta* (cockroach), *Sarcophaga* (flesh fly), and *Drosophila.* C-type lectin domains (CTLDs) have been identified in as many as 32 *Drosophila* gene sequences, although not all of these predicted proteins are likely to encode functional lectins. In fact, binding specificity has been investigated for only one of these putative CTLDs, a Fuc/Man-binding protein that may serve as a receptor for difucosylated N-linked cores such as the HRP-epitope. Another putative C-type lectin, called furrowed (Fw), influences cell fate decisions in sensory organs. In addition to its CTLD, Fw also has a series of ten complement-binding domains, reminiscent of the architecture of the vertebrate selectin family of C-type lectins. The Fw lectin domain is well matched in the position of critical cysteine residues and hydrophobic clusters when compared to the selectins. However, the expressed protein has not yet been shown to bind any carbohydrate structure and it is important to remember that relatively small changes in the primary structures of lectins can substantially alter or even eliminate their carbohydrate-binding properties.

The vertebrate I-type lectins are distinguished by the specific organization of their amino-terminal immunoglobulin-type domains and by their high degree of specificity for binding sialylated glycoconjugates (see Chapter 32). Until recently, the relative lack of evidence for the existence of sialylated glycans in insects has limited the enthusiasm for pursuing this family of lectins in insects. However, proteins that bear amino-terminal immunoglobulin repeats have crucial roles in many aspects of *Drosophila* embryonic development, especially in the nervous system where the only *Drosophila* sialyltransferase that has been characterized is expressed. Regardless of whether a subset of these proteins can bind to the extremely low-abundance sialylated glycans of insects, the importance of their adhesive and signaling functions may now warrant efforts toward detecting possible lectin activities.

The current classification scheme for animal lectins reflects the long history of studying carbohydrate-binding proteins in vertebrates (see Chapter 26). At least two lectin activities, first identified in *Drosophila*, challenge the comprehensiveness of current animal lectin designations. One of these was identified in a search for secreted proteins that stimulate the proliferation and motility of imaginal disc cells. The imaginal disc growth factor (IDGF) family members are structurally related to chitinases but lack amino acid residues essential for catalytic activity. It has been suggested that the IDGF family has evolved away from hydrolysis while maintaining a glycan-binding activity that facilitates mitogenic and trophic support. Another currently unclassified lectin, called "gliolectin," was identified in a screen of *Drosophila* cDNAs for proteins that mediate cell adhesion to immobilized glycans. Gliolectin is expressed in a subset of embryonic glial cells found at the midline of the developing central nervous system. Mutants that lack gliolectin exhibit defects in axonal pathfinding consistent with a role for glycan-mediated cell adhesion in facilitating the transmission of signals between cells.

FURTHER READING

Lander A.D. and Selleck S.B. 2000. The elusive functions of proteoglycans: In vivo veritas. *J. Cell Biol.* **148:** 227–232.

Seppo A. and Tiemeyer M. 2000. Function and structure of *Drosophila* glycans. *Glycobiology* **10:** 751–760.

Altmann F., Fabini G., Ahorn H., and Wilson I.B. 2001. Genetic model organisms in the study of N-glycans. *Biochimie* **83:** 703–712.

Cadigan K.M. 2002. Regulating morphogen gradients in the *Drosophila* wing. *Semin. Cell Dev. Biol.* **13:** 83–90.

Haines N. and Irvine K.D. 2003. Glycosylation regulates Notch signalling. *Nat. Rev. Mol. Cell Biol.* **4:** 786–797.

Ten Hagen K.G., Fritz T.A., and Tabak L.A. 2003. All in the family: The UDP-GalNAc:polypeptide N-acetylgalactosaminyltransferases. *Glycobiology* **13:** 1R–16R.

Haltiwanger R.S. and Lowe J.B. 2004. Role of glycosylation in development. *Annu. Rev. Biochem.* **73:** 491–537.

Lin X. 2004. Functions of heparan sulfate proteoglycans in cell signaling during development. *Development* **131:** 6009–6021.

Aoki K., Perlman M., Lim J.-M., Cantu R., Wells L., and Tiemeyer M. 2007. Dynamic developmental elaboration of N-linked glycan complexity in the *Drosophila melanogaster* embryo. *J. Biol. Chem.* **282:** 9127–9142.

CHAPTER 25

Deuterostomes

Hudson H. Freeze, Victor D. Vacquier, and Jeffrey D. Esko

THIS CHAPTER DISCUSSES UNIQUE ASPECTS OF GLYCANS in model organisms that belong to the deuterostome lineage, with particular emphasis on sea urchins, frogs, zebrafish, and mice. These organisms provide models for studying the function of glycans in development and physiology.

EVOLUTIONARY BACKGROUND

Animal evolution split into two major lineages approximately 600 million years ago: the deuterostomes and the protostomes. The superphylum Deuterostomia contains the major phyla of Echinodermata (sea urchins and starfish), Hemichordata (acorn worms), Urochordata (ascidians), Cephalochordata (amphioxus), and Vertebrata (fish, amphibia, reptiles, birds, and mammals). In deuterostomes, cell divisions of the zygote occur by radial cleavage, and cell fates are not precisely determined. Another characteristic feature is that the first opening to form in the blastula becomes the anus and the second opening becomes the mouth. The other superphylum, the Protostomia, contains the major phyla of Porifera (corals and sponges), Cnidara (anemones and hydra), Annelida (segmented worms),

Mollusca (clams, oysters, snails, and slugs), and Arthropoda (insects, spiders, and crustacea). In protostomes, unlike deuterostomes, cell division during early development occurs in a highly organized manner and cell fate is precisely determined. Examples of protostome model organisms include the nematode *Caenorhabditis elegans* (see Chapter 23) and the arthropod *Drosophila melanogaster* (see Chapter 24).

Much effort has been devoted to studying the deuterostomes because of their relatedness to humans and because they provide excellent model systems for studying vertebrate biology. In particular, sea urchins, frogs, zebrafish, and mice have received the most attention. Each of these systems provides certain advantages for studying the function of glycans (e.g., in fertilization, early development, or adult physiology). In some cases, these models have revealed aspects of glycobiology that were later confirmed in more evolutionarily advanced organisms, and they often serve as models for studying human disease. Each of these models is briefly described below with reference to other sections of the book for additional details.

SEA URCHINS

Sea urchin glycans that are involved in fertilization have been studied extensively. In general, most of what we know about the biochemistry of fertilization was first discovered in this organism (Figure 25.1), and subsequently the information was applied to mammalian sperm–egg interaction. One of the advantages of studying fertilization in sea urchins is that eggs and sperm are easy to obtain in large quantities. Because fertilization occurs outside the adult body, it is easy to manipulate sperm–egg interaction.

Egg Glycans and Fertilization

Sea urchin eggs are covered by a hydrated jelly coat. About 80% of the weight of egg jelly is a high-molecular-weight linear fucose sulfate polymer (FSP), which has a molecular mass of more than 10^6 daltons. The FSP binds to receptor proteins on the sperm, triggering the opening of two pharmacologically distinct calcium channels that induce the exocytosis of the sperm's acrosome vesicle (the "acrosome reaction"). Additionally, exiting H^+ raises the internal pH of the sperm from 7.50 to 7.75, activating the polymerization of actin to form a finger-like acrosomal process, which fuses with the egg, uniting the two cells into one cell. The ionic mechanisms that trigger the acrosome reaction are conserved in mammals, but the nature of the sperm surface receptors varies. Blocking the acrosome reaction in mammals has long been a target for nonhormonal contraception.

FSP is a species-selective inducer of the sperm acrosomal reaction. FSP has no associated protein with only one exception (*Strongylocentrotus droebachiensis*) and consists of linear α1-3-linked fucose polymers. The individual fucose residues are sulfated differently, so that most FSPs are made of tri- or tetrasaccharide repeats. The number of fucose residues per repeat, the linkage, and sulfation patterns all help ensure species selectivity for inducing the acrosome reaction (Figure 25.2). For instance, the α1-3-linked polymer of *S. franciscanus* is sulfated only at C-2, whereas *S. pallidus* has an α1-3 tetrafucosyl repeat, with site-specific C-2 and C-4 sulfate groups. Other variations are also known (e.g., the egg jelly of *Echinometra lacunter* contains a sulfated galactan).

Glycan-binding proteins are important for sea urchin fertilization. FSP binds to at least one sperm surface receptor (REJ1; receptor for egg jelly-1) that is located on the sperm plasma membrane over the acrosome vesicle. REJ1 is a 1450-amino-acid protein with two C-type lectin domains in its amino-terminal region (see Chapter 31).

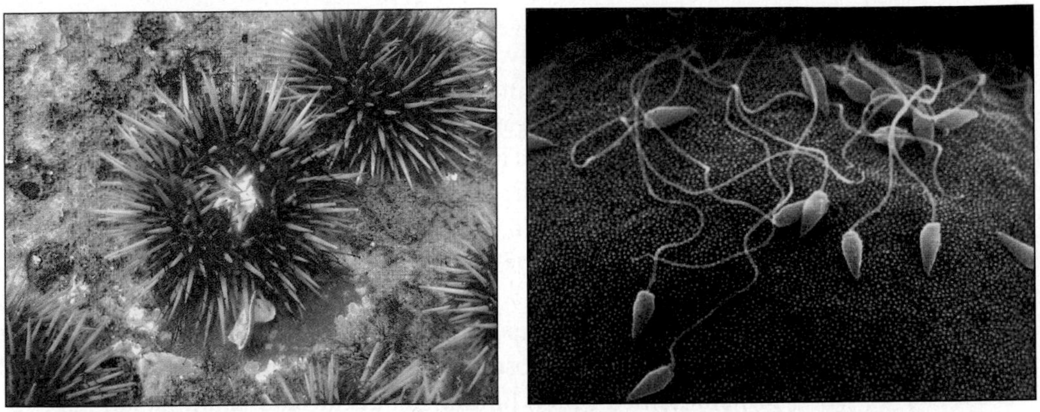

FIGURE 25.1. (*Left*) The purple sea urchin *Strongylocentrotus purpuratus*. (Reprinted, with permission, courtesy of Charles Hollahan, Santa Barbara Marine Biologicals.) (*Right*) Sperm binding to a sea urchin egg. (Reprinted, with permission, courtesy of M. Tegner and D. Epel.) Sea urchins have been used extensively in studies of fertilization. Many of the key discoveries regarding glycans and their role in determining species-specific sperm–egg binding and blocks to polyspermy were made in sea urchins.

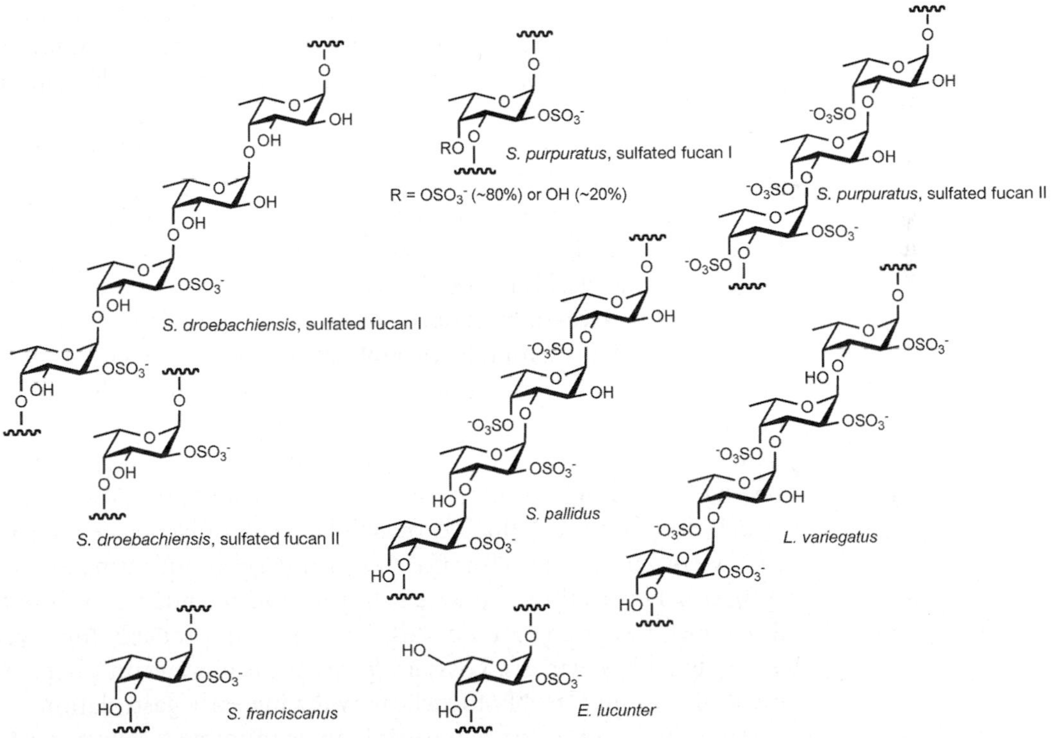

FIGURE 25.2. Fucose sulfate polymers. Structures of sulfated α-fucans and a sulfated α-galactan from sea urchin egg jelly. Shown are eight fully characterized structures of sulfated polysaccharides from the egg jellies of seven species of sea urchins. The specific pattern of sulfation, the position of the glycosidic linkage, and the constituent monosaccharide vary among sulfated polysaccharides from different species. (Redrawn, with permission, from Vilela-Silva A.C., Castro M.O., Valente A.P., et al. 2002. *J. Biol. Chem.* **277:** 379–387, © American Society for Biochemistry and Molecular Biology.)

About 20% of the egg jelly mass is a large glycoprotein containing a unique form of polysialic acid (sialoglycan), which can be purified from crude sea urchin egg jelly by β-elimination. Sialoglycan potentiates the acrosome reaction induced by FSP, possibly by regulating sperm pH. Treating sialoglycan with neuraminidase or mild periodate oxidation completely destroys its biological activity. Studies on two closely related sea urchin species show that the polysialic acid is attached to protein backbones of very different sizes, suggesting that the protein components are encoded by different genes. The sialoglycan has a novel structure—$[\text{Neu5Gc}\alpha\text{2-5-}O\text{-glycolylNeu5Gc}]_n$, but its receptor on the sperm membrane is unknown.

When sea urchin sperm undergo the acrosome reaction, a protein named bindin is released from the acrosomal vesicle that cements the sperm to a large egg surface glycoprotein named EBR1. Bindin can agglutinate mammalian red blood cells much like plant lectins (see Chapter 29), and glycopeptide fragments of unfertilized eggs can block bindin-induced red cell agglutination. Bindin is thought to recognize glycans on EBR1. EBR1 itself has lectin-like domains, but its ligands are not known.

Mammalian fertilization shares many features with the sea urchin system. All mammalian eggs are enclosed within an extracellular matrix called the zona pellucida (ZP). The ZP is made of three heavily glycosylated mucin-like proteins, termed ZP1, ZP2, and ZP3. O-linked glycans of ZP3 are recognized by sperm membrane proteins and bind the sperm loosely to the egg in a lectin-type interaction. Not only are the glycans of ZP3 involved in binding sperm to the ZP, but they also induce the acrosome reaction. After the acrosome reaction, binding of ZP2 to the sperm's inner acrosomal membrane enhances the binding of the sperm to the ZP in another lectin-like interaction. Two clusters of residues on ZP3 contain O-linked glycans that show species selectivity in sperm–egg binding and induction of the acrosome reaction. The mammalian sperm protein strongly implicated in binding ZP3 is called zonadhesin, and it has several domains that could be involved in lectin-like binding to ZP3 glycans.

Lectins

Sea urchin eggs contain other lectins of unknown function. A 12-kD galactose-binding lectin called SUEL can be purified from eggs and is present in all developmental stages of sea urchin embryos. It is stored in secretory granules in the cortex region of the embryonic cells and is released by exocytosis. Much of SUEL is embedded in the extracellular hyaline layer gel that surrounds the embryo, where it may have a role in cell–cell adhesion or in binding microbes attempting to colonize the embryo. The surface of the unfertilized sea urchin egg also contains a 350-kD heavily O-glycosylated protein that has lectin activity for sialic acid. This lectin also binds sperm gangliosides in detergent-insoluble lipid rafts, suggesting that it might be involved in signal transduction in sperm.

Echinonectin is a 230-kD extracellular matrix glycoprotein also found in the sea urchin embryo hyaline layer (a clear Ca^{++}-requiring gel about 3 μm in diameter that surrounds the embryo). This lectin recognizes galactose and fucoidan and shows differential affinity for different types of embryonic cells. Echinonectin probably functions in cell adhesion during early embryogenesis. Loss of affinity for echinonectin might have a role in the movement of cells into the blastocoel cavity during early gastrulation.

Adult sea urchins have a powerful innate immune system to prevent microbial infection. Sea urchins, like mammals, have a true body cavity filled with phagocytes and other types of defense cells. The body cavity fluid contains lectins that bind to invading microorganisms, causing the foreign cells to aggregate and be more easily bound to and ingested by the phagocytes. Thus, lectins are used in sea urchins and all other invertebrates (whether aquatic or terrestrial) to facilitate capture and destruction of any foreign cell or particle.

Adult sea urchins have hundreds of three-jawed, pincer-like structures called pedicellaria, which are located over their entire outer surface where they are used for defense. Some species have very large pedicellaria associated with a venom gland that can cause painful stings to humans handling these animals. Lectins have been purified from pedicellaria of the species *Toxopneustes pileolus* and *T. gratilla.* SUL-1 and -2 are galactose-recognizing lectins from *T. pileolus*, and TGL-1 is a calcium-dependent, heparin-binding lectin from *T. gratilla.* One lectin, purified from *T. pileolus*, binds galactose and D-fucose (but not the more commonly seen L-fucose); it will also block histamine release from rat mast cells.

FROGS

The African clawed frog, *Xenopus laevis*, is a well-established model organism for studying fertilization and embryonic development (Figure 25.3). Adults are easy to maintain and can be induced to lay eggs up to three times a year by injection of human chorionic gonadotropin. The very large eggs make microinjection experiments easy and allow them to be used as an expression system. The embryos also develop quickly. However, the generation time is long (1–2 years) and the genome is tetraploid, making genetic studies difficult. Nevertheless, interesting insights into glycan signaling have recently emerged from studies of chitin formation in *Xenopus* and a new class of lectins has been discovered.

DG42, Hyaluronan, and Chitin Oligosaccharide Synthesis

The *DG42* gene has homology with fungal chitin synthase and a bacterial *Rhizobium NodC* gene, which makes short chitin oligomers $(GlcNAc\beta1\text{-}4GlcNAc)_n$. These products and some of their modified derivatives are required for *Rhizobium* to invade plant root hairs and produce a nodule for nitrogen fixation (see Chapter 37). *DG42* is expressed in *Xenopus* endoderm cells briefly during mid-late gastrulation. Transgenic expression of *DG42* results in chitin oligomers that are digestible by chitinase, which is taken as evidence that vertebrates, at least vertebrate embryos, can synthesize short chitin oligosaccharides. However, *DG42* also has considerable homology with the *hasA* gene from

FIGURE 25.3. *Xenopus laevis*. (*Left*) An adult specimen of the African clawed frog. (Image courtesy of Bruce Blumberg, University of California, Irvine.) (*Right*) Embryonic development from the neurula stage until just prior to hatching, about 24 hours postfertilization. (Reprinted, courtesy of Thierry Brassac, Exploratorium at www.exploratorium.edu.)

Streptococcus, which is responsible for hyaluronan synthesis (GlcAβ1-3GlcNAcβ1-4)$_n$ (see Chapter 15). Expression of *DG42* in several cell lines and in yeast (which does not make hyaluronan) showed that the expressed gene increased the synthesis of hyaluronan in those cells. In fact, all mammalian hyaluronan synthase genes isolated and expressed to date have considerable homology with *DG42*. Thus, by homology and initial biochemical characterization, there is evidence that *DG42* can assemble both chitin and hyaluronan. One idea is that chitin oligosaccharides could act as a primer for the assembly of hyaluronan, but supporting data are lacking.

A New Family of Lectins in *Xenopus*

X. laevis oocyte cortical granules contain a lectin called XL35, which is released at fertilization and binds to mucin-type O-glycans in the egg jelly. This complex forms a layer on the fertilization envelope and blocks polyspermy. Other species such as lamprey, trout, and sea squirts have related proteins, as do humans (HL-1 and HL-2) and mice (interlectin-1 and interlectin-2). Other homologous sequences exist as well. XL35 binds to melibiose (a disaccharide consisting of one galactose and one glucose moiety in an α1-6 linkage) and requires calcium, but it does not have the typical sequence motif for C-type lectins (see Chapter 31). Thus, XL35 and related proteins appear to represent a new family of lectins. Although the specific glycan-binding properties of family members differ, they have high degrees of amino acid sequence homology with a common fibrinogen-like motif that may involve glycan binding. Several studies on this new family of lectins suggest that they are stored in specialized vesicles and released by pathogen infections. Some of these proteins bind to bacterial pathogen glycans, suggesting that they may be used for pathogen surveillance by the innate immune system.

ZEBRAFISH

Zebrafish (*Danio rerio*) provide an excellent genetic model organism for understanding the function of genes in early vertebrate development (Figure 25.4). Females lay several hundred eggs weekly and the eggs undergo external fertilization. Embryos develop rapidly (in hours), and their translucence allows visualization of early development. Zebrafish are relatively easy to maintain compared to other vertebrate species. Furthermore, they can be mutagenized and outbred to produce mutants, or studied using morpholinos to silence genes transiently during early development (gene "knockdowns").

When the first edition of this book was published in 1999, there was only brief mention of zebrafish and only a handful of papers had been published on glycans in this species. Since then, all of the major glycan classes present in mammals have been described in zebrafish and many of them have been manipulated genetically, yielding fresh insights into glycan function (Table 25.1). Because zebrafish live in a nonterrestrial environment and because widespread gene duplications and diversification occurred during teleost evolution, one can predict that novel glycans, pathways of assembly, and biological activities may exist.

Zebrafish Lectins

In addition to the major families of glycans, various orthologs of mammalian lectins have been described in zebrafish. These include all three major galectin types: proto, chimera, and tandem-repeat (see Chapter 33). Zebrafish galectins exhibit distinct patterns of temporal expression during embryo development, suggesting that they have

FIGURE 25.4. Zebrafish (*Danio rerio*). (Reprinted, with permission, from Sprague J., Bayraktaroglu L., Clements D., et al. 2006. The Zebrafish Information Network: The zebrafish model organism database. *Nucleic Acids Res.* **34:** D581–D585.)

important roles in different tissues. With regard to Siglecs, the only unambiguous ortholog to mammalian molecules corresponds to Siglec-4 (see Chapter 32). Like its mammalian counterparts, zebrafish Siglec-4 is only present in the nervous system, but its function has not yet been established. Zebrafish also express both hyaluronan and hyaluronan-binding proteins, such as members of the aggrecan family (see Chapter 16). Dermacan is distinct from any other identified lecticans and has a restricted pattern of expression. Silencing of dermacan in zebrafish embryos results in defective craniofacial dermal bone development.

TABLE 25.1. *Examples of zebrafish glycans under study*

Glycan	Affected system	Phenotype
N-linked glycoproteins	egg chorion	fertilization
Heparan sulfate: knypek (Glypican4/6)	embryonic development	convergent extension
Chitin oligosaccharide	embryonic development	gastrulation; trunk and tail defects
Syndecan-2; Hs3st	embryonic development	left–right patterning
Heparan sulfate: HsC5epi; b4gt	embryonic development	dorsoventral patterning
Heparan sulfate: Ext2, ExtL3, agrin	nervous	axon guidance
Chondroitin sulfate	nervous	axon growth
HNK-1	nervous	axon growth
Heparan sulfate	musculoskeletal	muscle differentiation
Heparan sulfate Ext2, ExtL3	musculoskeletal	limb development
Galα1-3Gal	digestive	microbiota alters glycan pattern in gut
Heparan sulfate: syndecan-2; Hs6st	cardiovascular	vascular branching and embryonic angiogenesis
Jekyll; UDP-Glc dehydrogenase	cardiovascular	cardiac valve formation
Chitin oligosaccharides	cardiovascular	angiogenesis

Novel Zebrafish Genes

Zebrafish also allow development of sophisticated screening methods for the identification of novel mutants that affect specific physiological systems. For example, high-throughput screenings have identified groups of genes required for fin development and regeneration, which serve as models for cartilage and bone development in mammals. Mutants altered in vascular development and hemostasis have been identified as well. As the candidate genes are identified and cloned, a number of these will undoubtedly encode glycan-synthesizing enzymes or lectins. The ability to identify novel genes in the system and manipulate glycan expression via silencing methods should make this model organism a focus of future studies in vertebrate glycobiology.

MICE (*MUS MUSCULUS*)

In recent years, mice (*Mus musculus*) have become the most studied deuterostome because they are mammals, they reproduce quickly, they have manageable genetics, and their genes can be functionally deleted (gene knockouts). Another advantage is that highly inbred mouse strains exist and spontaneous mutations occur that have generated specific morphological or pathophysiological phenotypes. Examples include brachymorphic mice, which are partially deficient in the synthesis of PAPS, the universal sulfate donor (see Chapter 4), and have shortened limbs because of defective chondroitin sulfate synthesis during endochondral ossification (see Chapter 16). Another example is a mutation that causes a muscular dystrophy that was eventually traced to $Large^{myd}$, a putative glycosyltransferase that when mutated in humans causes muscular dystrophy because of abnormal synthesis of O-mannose-linked glycan on α-dystroglycan (see Chapters 12 and 42). The majority of information about mouse glycobiology comes from the use of homologous recombination methods to knock out a specific gene systemically, in specific tissues or at selected times in development. In a few cases, human pathological mutations that represent hypomorphic alleles have been engineered into their conserved murine homologs in an attempt to generate models for disease. Table 6.1 lists a sampling of defects based on the phenotype and the organs or systems most affected by the mutations. Not surprisingly, many of these mutations impact multiple systems.

Mutations That Affect Entire Pathways

Because homologous recombination has been used to inactivate numerous genes involved in glycosylation, only a few mutants are mentioned here. Mutant lines have been created in early genes involved in the assembly of all the major subclasses of vertebrate glycans. In most cases, these mutations prevent initiation of the biosynthetic pathways and show embryonic lethality. These include the single O-GlcNAc transferase (see Chapter 18), which is lethal even at the single-cell stage. Loss of the entire N-linked pathway caused by ablation of phosphomannomutase 2 (which generates mannose-1-P and all mannose-containing glycans) and loss of GlcNAc-1-P transferase (which initiates the lipid-linked oligosaccharide precursor) are lethal within a few days of fertilization. Loss of heparan sulfate biosynthesis due to deletion of *Ext1* or *Ext2* is also lethal at the E6–7 stage because of a failure to differentiate mesoderm. Systemic deletion of the N-glycan-processing enzyme MgatI is lethal around mid-development (E9–10) showing that early development through gastrulation can proceed in the complete absence of hybrid or complex glycans. Chondroitin sulfate and hyaluronan synthesis are essential as well, although developmental arrest occurs during limb generation and organogenesis. Because there are more than 20 partially redundant polypeptide *N*-acetylgalactosaminyl transferases in mouse, knocking out this pathway completely has not been possible. However, deletion of the single copy core-1 galactosyltransferase (T-synthase, which generates Galβ1-3GalNAcα-O-Ser/Thr) is lethal by stage E14 because of deranged and distorted angiogenesis. Knocking out the O-fucosyltransferase that modifies epidermal growth factor (EGF) modules in Notch is lethal and causes developmental defects similar to those seen in Notch1 knockout mice. Absence of glycosylphosphatidylinositol anchors or O-mannose-based glycans (α-dystroglycan) also results in embryonic lethality.

Mutations That Affect Specific Extensions or Modifications

Mutations in genes that encode glycosyltransferases used to extend the basic glycan chains usually have less severe phenotypes than those that delete entire pathways. Chapters 8–18 describe the effects of mutations on selected classes of glycans or specific modifications found on multiple types of glycans. For example, Chapters 13 and 14 discuss the consequences of gene knockouts that affect the immune system, including leukocyte rolling, lymphocyte homing, and T-cell homeostasis. Knocking out sequential biosynthetic steps or multiple, seemingly redundant, genes frequently reveals another layer of glycan-dependent functions, novel structures, and unrecognized biosynthetic pathways. Many mouse mutants serve as models of human genetic disorders (see Chapters 41 and 42).

Tissue-specific Mutations

To study the function of glycans in specific tissues, tissue-specific lesions in a gene can be created by crossing strains bearing conditional null alleles to strains that selectively delete the gene only in those tissues of interest. The most common method is to flank a portion of the gene with *loxP* recombination sites and then mate these mice to a strain that produces bacteriophage Cre-recombinase under the control of a tissue-specific or inducible promoter (Figure 25.5). When activated, the Cre-recombinase recognizes the *loxP* sites and excises the DNA segment contained between them. These techniques can be combined with an environmental stress, dietary modification, or inflammatory challenge to delete genes in both a tissue-specific and temporally regulated fashion.

Sometimes, altering a gene can have a highly specific effect on a given tissue without affecting other tissues because of the unforeseen presence of related isozymes. α-ManII was thought to have the dominant role in N-glycan processing (see Chapter 8), but α-ManII knockouts survive and have a dyserythropoietic anemia. These mice produce many unusual glycans from the abnormally high levels of incomplete precursors that accumulate and drive the production of other atypical products. They eventually develop a progressive autoimmune nephropathy and accumulate immune complexes in the kidney tubules, probably in response to the abnormal glycan antigens. Their erythrocytes lack complex N-glycans, but other tissues continue to produce them using a different isozyme called α-ManIIx. α-ManIIx-null mice are mostly normal, except that male mice are sterile because spermatocytes fail to make a glycan required for their binding to Sertoli cells in the testes, causing nonfunctional immature spermatocytes to be released into the epididymis. Knock out of both α-ManII and α-ManIIx prevents synthesis of all complex glycans, and most embryos die between E15 and E18 or shortly after birth from respiratory failure.

Strain Differences and Genetic Modifiers

Determining the effect of a gene knockout is usually done in highly inbred strains that are genetically identical except for the gene in question. Sometimes, the phenotype of a

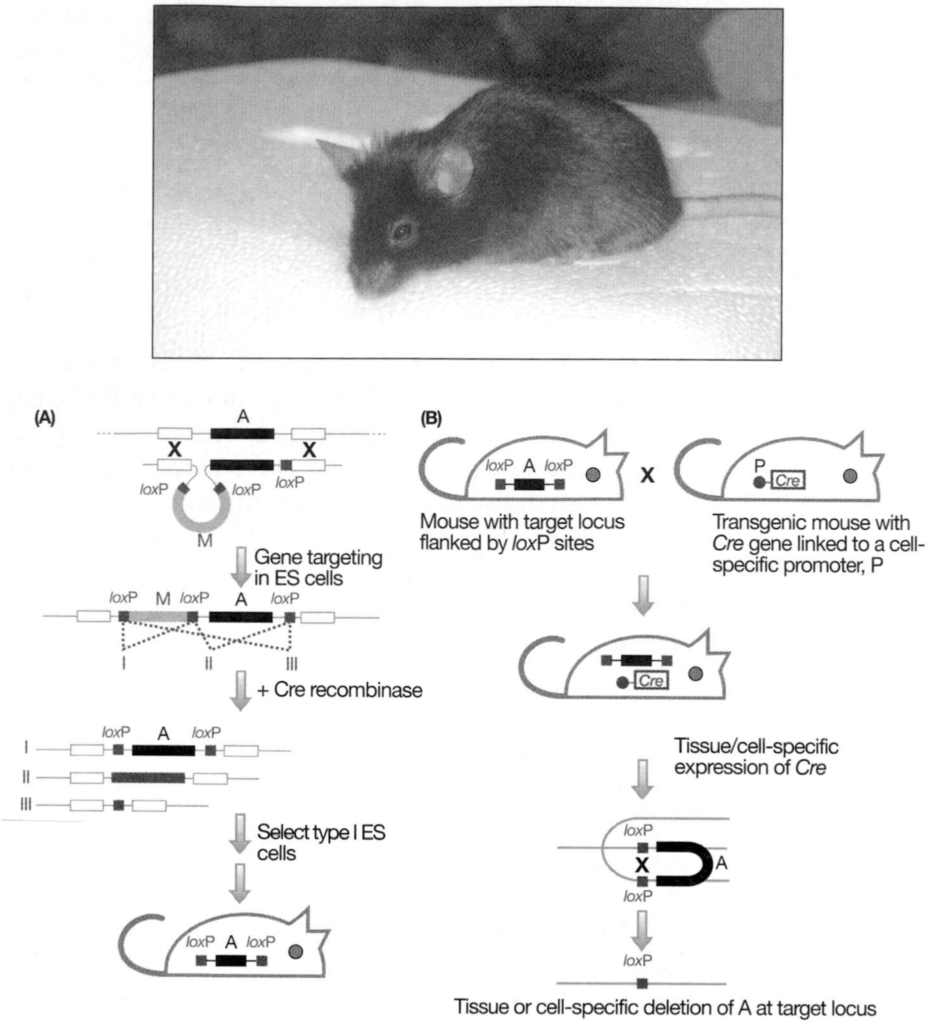

FIGURE 25.5. *Mus musculus*. (*Top*) An adult mouse. (*Bottom*) Cre-loxP targeting for making conditional knockouts.

mutation changes according to the genetic background (strain) in which the mutation is expressed. For instance, inactivating the N-glycan-processing gene *Mgat2*, which encodes GlcNAcT-II, is uniformly lethal after birth in one background, but placing it in a different background generates rare survivors with severe gastrointestinal, hematological, and osteogenic abnormalities. Interestingly, these mutants phenocopy the human disorder, CDG-IIa, in which patients lack the same enzyme (see Chapter 42). As another example, lines deficient in global fucosylation due to mutations in the enzyme that converts GDP-mannose to GDP-fucose (see Chapter 4) die as embryos, but when bred into a different background, they survive until birth and can then be maintained on a fucose-supplemented diet. A third example is the formation of bony outgrowths (exostoses) in the growth plates of endochondral bones. In one background, heterozygous mutations in *Ext1* or *Ext2*, which encode the copolymerase involved in heparan sulfate biosynthesis, cause infrequent exostoses, whereas in another background, the incidence of tumors increases severalfold. These experiments suggest the existence of modifiers that could alter the expression levels of enzymes, assembly of precursors, or salvage of sugars from the diet or glycan turnover.

Environmentally Driven Phenotypes

In some mutants, phenotypes are only evident under environmental challenge. For example, deletion of both enzymes responsible for polysialic acid synthesis produces "no-fear" mice that tend to be more aggressive and ignore normally stressful or anxiety-producing situations. Deletion of syndecan-1, a major heparan sulfate core protein (see Chapter 16), shows no obvious changes under normal laboratory conditions, but the mice are more resistant to bacterial lung infections. Apparently, bacteria exploit the shed syndecan in the lung to enhance their virulence and modulate host defenses.

Mice deficient in the synthesis of ganglioside GM3 appear to be fairly normal except for enhanced insulin sensitivity. These mice shunt glycosphingolipid synthesis toward more complex gangliosides that substitute for GM3 loss (see Chapter 10). On the other hand, mice that synthesize only ganglioside GM3, due to inactivation of both *Galgt1* and *Siat8a*, exhibit tonic-clonic (epileptic) seizures, and 90% die in response to sharp sounds. These examples show the importance of having the correct balance of related glycans. They also show how phenotypes depend to a large extent on other genetic factors, diet, and environmental cues. Thus, many mutations that appear silent may in fact elicit strong phenotypes in the natural environment, rather than the controlled laboratory setting.

FURTHER READING

Kramer K.L. and Yost H.J. 2003. Heparan sulfate core proteins in cell–cell signaling. *Annu. Rev. Genet.* **37:** 461–484.

Lowe J.B. and Marth J.D. 2003. A genetic approach to mammalian glycan function. *Annu. Rev. Biochem.* **72:** 643–691.

Biermann C.H., Marks J.A., Vilela-Silva A.C., Castro M.O., and Mourao P.A. 2004. Carbohydrate-based species recognition in sea urchin fertilization: Another avenue for speciation? *Evol. Dev.* **6:** 353–361.

Lee J.K., Baum L.G., Moremen K., and Pierce M. 2004. The X-lectins: A new family with homology to the *Xenopus laevis* oocyte lectin XL-35. *Glycoconj. J.* **21:** 443–450.

Neill A.T. and Vacquier V.D. 2004. Ligands and receptors mediating signal transduction in sea urchin spermatozoa. *Reproduction* **127:** 141–149.

Guerardel Y., Chang L.Y., Maes E., Huang C.J., and Khoo K.H. 2006. Glycomic survey mapping of zebrafish identifies unique sialylation pattern. *Glycobiology* **16:** 244–257.

Ohtsubo K. and Marth J.D. 2006. Glycosylation in cellular mechanisms of health and disease. *Cell* **126:** 855–867.

Bishop J.R., Schuksz M., and Esko J.D. 2007. Heparan sulphate proteoglycans fine-tune mammalian physiology. *Nature* **446:** 1030–1037.

Mourao P.A. 2007. A carbohydrate-based mechanism of species recognition in sea urchin fertilization. *Braz. J. Med. Biol. Res.* **40:** 5–17.

CHAPTER 26

Discovery and Classification of Glycan-Binding Proteins

Ajit Varki, Marilynn E. Etzler, Richard D. Cummings, and Jeffrey D. Esko

THIS CHAPTER PROVIDES AN OVERVIEW REGARDING naturally occurring glycan-binding proteins (GBPs), with an emphasis on their discovery and current classification schemes. Some general principles regarding the structure and function of GBPs are also considered, as well as aspects of their glycan-binding properties. Further information regarding most of the major classes of GBPs can be found in Chapters 28–35. For details regarding the analysis of glycan–protein interactions and the physical principles involved, see Chapter 27.

GLYCAN RECOGNITION BY PROTEINS: A KEY TO SPECIFICITY IN GLYCOBIOLOGY

Glycans can mediate a wide variety of biological roles by virtue of their mass, shape, charge, or other physical properties. However, many of their more specific biological roles are mediated via recognition by GBPs. Nature appears to have taken full advantage of the vast diversity of glycans expressed in organisms by evolving protein modules to recognize discrete glycans that mediate specific physiological or pathological processes. Indeed, there are no living organisms in which GBPs have not been found.

TWO MAJOR CLASSES OF GLYCAN-BINDING PROTEINS—LECTINS AND GLYCOSAMINOGLYCAN-BINDING PROTEINS

Excluding glycan-specific antibodies, it is possible to classify GBPs broadly into two major groups—lectins and glycosaminoglycan-binding proteins (see Table 26.1). Most lectins are members of families with defined "carbohydrate-recognition domains" (CRDs) that apparently evolved from shared ancestral genes, often retaining specific features of primary amino acid sequence or three-dimensional structure. Thus, new family members can be identified by searching protein sequence or structural databases. Despite this ability to predict new GBPs, the structures of glycans recognized by members of a single lectin family can be quite diverse. Single-site-binding affinities in many lectins appear to be low (with K_d values in the micromolar range), although some lectins recognize glycans with much higher affinity (with K_d values in the nanomolar range). For those lectins with low affinity, multivalent interactions between multiple CRDs and multiple glycans are often required to produce the high-avidity binding interactions that are relevant in vivo. Lectins tend to recognize specific terminal aspects of glycan chains by fitting them into shallow, but relatively well-defined, binding pockets. In contrast, protein interactions with sulfated glycosaminoglycans seem to involve surface clusters of positively charged amino acids that line up against internal regions of extended anionic glycosaminoglycan chains (see Chapter 35). Thus, despite the fact that the glycosaminoglycan structural motifs recognized by a given molecule can be quite specific, most glycosaminoglycan-binding proteins do not

TABLE 26.1. *Comparison of the two major classes of glycan-binding proteins*

	Lectins[a]	Glycosaminoglycan-binding proteins[b]
Shared evolutionary origins	yes (within each group)	no
Shared structural features	yes (within each group)	no
Defining AA residues involved in binding	often typical for each group	patch of basic amino acid residues
Type of glycans recognized	N-glycans, O-glycans, glycosphingolipids (a few also recognize sulfated glycosaminoglycans)	different types of sulfated glycosaminoglycans
Location of cognate residues within glycans	typically in sequences at outer ends of glycan chains	typically in sequences internal to an extended sulfated glycosaminoglycan chain
Specificity for glycans recognized	stereospecificity high for specific glycan structures	often recognize a range of related sulfated glycosaminoglycan structures
Single-site binding affinity	often low; high avidity generated by multivalency	often moderate to high
Valency of binding sites	multivalency common (either within native structure or by clustering)	often monovalent
Subgroups	C-type lectins, galectins, P-type lectins, I-type lectins, L-type lectins, R-type lectins etc.	heparan sulfate–binding proteins, chondroitin sulfate–binding proteins, dermatan sulfate–binding proteins
Types of glycans recognized within each group	can be similar (e.g., galectins) or variable (e.g., C-type lectins)	classification itself is based on type of glycosaminoglycan chain recognized

Modified from Varki A. and Angata T. 2006. *Glycobiology* **16:** 1R–27R.

[a]There are other animal proteins that recognize glycans in a lectin-like manner and do not appear to fall into one of the well-recognized classes (e.g., various cytokines).

[b]Hyaluronan (HA)-binding proteins (hyaloadherins) fall in between these two classes. On the one hand, some (but not all) of the hyaloadherins have shared evolutionary origins. On the other hand, recognition involves internal regions of HA, which is a nonsulfated glycosaminoglycan.

seem to be evolutionarily related to each other. Rather, their defining feature (the ability to recognize sulfated glycosaminoglycans) seems to have emerged by convergent evolution. Hyaluronan-binding proteins (hyaladherins) are exceptions, having a characteristic evolutionarily conserved fold, which docks with short segments of the hyaluronan chain (see Chapter 15). Thus, although hyaluronan is structurally a nonsulfated glycosaminoglycan, hyaladherins may be better classified with the lectins, rather than with other glycosaminoglycan-binding proteins.

DISCOVERY AND HISTORY OF LECTINS

Lectins were first discovered more than 100 years ago in plants (see below and Chapters 28 and 29), but they are now known to be present throughout nature. Lectins are also prevalent in the microbial world, wherein they tend to be called by other names, such as hemagglutinins and adhesins (see below and Chapter 34).

Plant Lectins

In 1888, Stillmark found that extracts of castor bean seeds contained a protein that could agglutinate animal red blood cells. Soon thereafter, a number of other plant seeds were found to contain such "agglutinins," but interest in them began to wane with the advent and development of the field of immunochemistry. It was not until the Second World War and the resulting interest in blood typing for blood transfusion that some lectins were found to be specific for various ABO blood types (see Chapter 13) and others were found to have specificities for different glycans. These agglutinins were thus renamed "lectins," a term derived from the Latin word "legere," meaning "to select." Lectins have since been found in almost every plant species studied and are particularly abundant in the seeds of leguminous plants. These "L-type" lectins have been intensively studied and are discussed in further detail in Chapter 29. Plant lectins that fit into the "R-type" category are described in Chapter 28. Other categories of plant lectins have been found and are included in Table 26.2.

Viral Lectins (Hemagglutinins)

The influenza virus hemagglutinin was the first GBP isolated from a microorganism (~1950), and it is now one of the most thoroughly studied of all lectins. Wiley and associates crystallized the viral hemagglutinin, determined its structure in 1981, and later solved the structure of cocrystals prepared with sialyllactose. Since then, the crystal structures for several other viral hemagglutinins have been determined as well. Like animal cell lectins, most viral lectins bind to terminal sugar residues, but some can bind to internal sequences found in linear or branched glycans. The specificity of these interactions can be highly selective. For example, the human influenza viruses bind primarily to cells containing Siaα2-6Gal linkages, whereas other animal and bird influenza viruses preferentially bind to Siaα2-3Gal termini (see book cover figure). Influenza C, in contrast, binds preferentially to glycoproteins containing terminal 9-O-acetylated sialic acids. Many other viruses (e.g., reovirus, rotavirus, Sendai, and polyomavirus) also appear to use sialic acids in specific linkages for infection. Other viruses display glycosaminoglycan-binding proteins that can bind to heparan sulfate proteoglycans, often with high specificity for certain sulfated sequences (e.g., the gD protein of herpes simplex virus).

TABLE 26.2. *A general classification of lectins and lectin-like proteins*

Category I—Defined lectin families with structural and/or evolutionary sequence similarities[a]

β-prism lectins—Jacalin-related (B-type?)
C-type lectins (e.g., calcium-dependent lectins such as selectins, collectins, etc.)
eel fucolectins (E-type?)
ficolins—fibrinogen/collagen-domain-containing lectins (F-type?)
garlic and snowdrop lectins and related proteins (G-type?)
galectins (formerly S-type lectins)
hyaluronan-binding proteins or hyaladherins (H-type?)
I-type lectins—immunoglobulin superfamily members, including the Siglec family
amoeba lectins—Jacob and related chitin-binding proteins (J-type?)
L-type lectins, plant legume seed lectins, ERGIC-53 in ER-Golgi pathway, calnexin family
M-type lectins—α-mannosidase-related lectins (e.g., EDEM)
N-type lectin nucleotide phosphohydrolases (LNPs) with glycan-binding and apyrase domains
P-type (i.e., mannose-6-phosphate receptors)
R-type (e.g., ricin, other plant lectins, GalNAc-SO_4 receptors)
tachylectins from horseshoe crab *Tachypleus tridentatus* (T-type?)
haevin-domain lectins (e.g., wheat germ agglutinin, haevin, etc.) (W-type?)
Xenopus egg lectins/eglectins (X-type?)

Category II—Lectin-like proteins without established evolutionary classification

some annexins
pentraxins with pentavalent domain structure
some laminin G domains, recognizing glycans on α-dystroglycan
CD11b/CD18 (β3-integrin, CR3) recognizes fungal glucans and exposed GlcNAc residues on glycoproteins
complement factor H, recognizes cell-surface sialic acids as "self"
tumor necrosis factor-α, binds to oligomannose N-glycans
interleukins IL-1α, IL-1β, IL-4, IL-6, and IL-7 bind various glycans
amphoterin binds carboxylated N-glycans

[a]Category I includes all lectin families with generally agreed-upon names (e.g., C-type and R-type). In addition, the question marks (e.g., F-type? and X-type?) indicate suggested names for other families. Final acceptance of the latter terms will require a consensus among those scientists who study each respective family.

Bacterial Lectins (Adhesins and Toxins)

Many bacterial lectins have been described, and they fall into two classes: (1) lectins (adhesins) that reside on the bacterial surface and facilitate bacterial adhesion and colonization and (2) secreted bacterial toxins. Many bacterial lectins bind to membrane glycolipids, whereas only a few bind to glycoproteins. In some cases, binding specificity can explain the tissue tropism of the organism; for example, urinary tract infection by specific serotypes of *Escherichia coli* depends on binding mannose or P blood group structures. Not all of these interactions are pathogenic, and some glycan–protein interactions between bacteria and host tissues play important roles in symbiosis. For instance, the columnar epithelium that lines the large intestine expresses Galα1-4Galβ1-4Glcβ-ceramide, whereas the cells lining the small intestine may not. Thus, *Bacterioides, Clostridium, E. coli,* and *Lactobacillus* species that recognize Galα4Gal only colonize the large intestine under normal conditions. The granulocytotropic bacterium *Anaplasma phagocytophilum* infects leukocytes, particularly neutrophils, by binding to the cell surface mucin P-selectin glycoprotein ligand-1 (PSGL-1) and recognizing both the peptide of PSGL-1 and the associated sialyl Lewis^x antigen.

Many bacterial surface lectins are present in the form of long, hairy appendages known as fimbriae or pili that extend away from the cell. The presence of multiple glycan-binding subunits in the fimbriae allows multivalent interactions, thus increasing the avidity of bacterial host-cell binding. Several examples of heparan sulfate–binding adhesins are also now known.

Many heat-sensitive secreted bacterial toxins bind to glycans. The best-studied example is cholera toxin from *Vibrio cholerae*, which consists of A and B subunits, in the ratio AB_5. The B subunits bind to multiple GM_1 ganglioside receptors through glycan-recognition domains located on the base of the subunits, facilitating delivery of the toxic A subunit into the cytoplasm (see Chapter 34). The structures of related toxins from *Shigella dysenteria*, *Bordetella pertussis*, and *E. coli* have also been solved.

Animal Lectins

The first suggestion that lectins existed in animals was inferred from the reaggregation of dissociated marine sponge cells, a form of species-specific recognition. However, at least some of these phenomena appear to be mediated by glycan–glycan interactions. Agglutinating proteins, similar to the hemagglutinins of viruses, were described in the body fluids of various crustaceans and arachnids. The first suggestion that there might be lectins endogenous to vertebrates came from the studies of Victor Ginsburg and colleagues in the 1960s, in which blood leukocytes from rats were treated with bacterial glycosidases and then injected back into the circulation. The treatments resulted in changes in the homing of cells to different internal organs. Given the nature of the tools available at that time, it was difficult to ascertain whether these phenomena were mediated by an endogenous lectin. In retrospect, these studies were probably demonstrating the generation of ligands for a hepatic galactose receptor (due to sialidase action) or, possibly, loss of ligands for the selectins.

The first direct evidence for a mammalian lectin arose serendipitously during work by Gilbert Ashwell and colleagues, who were studying mechanisms that controlled the turnover of glycoproteins in blood circulation. Looking for an improved way to introduce a radioactive tracer into purified glycoproteins before reinjection, they attempted to transfer the tritium label from tritiated borohydride into the sugar chains of these proteins, which they knew contained the terminal sequence Sia-Gal-GlcNAc. This was successfully done either following enzymatic oxidation of the 6-position of galactose residues (which required first removing the outer sialic acid residues) or following mild periodate oxidation of sialic acid side chains (which left the rest of the sialic acid molecule intact). Surprisingly, there was a dramatic difference between the circulating half-lives of these two preparations, which terminated in either sialic acid or galactose residues. The molecules that retained labeled sialic acids remained in the circulation for many days, whereas those that had lost them (and now terminated in galactose residues) disappeared within minutes. The importance of these terminal β-linked galactose residues was confirmed by in vitro resialylation or by β-galactosidase treatment, both of which partially restored stability in circulation. The site of accumulation of the desialylated glycoproteins was found to be primarily the liver. This led to the discovery of the "asialoglycoprotein receptor," a hepatocyte membrane protein complex that specifically recognizes terminal β-linked galactose or *N*-acetylgalactosamine on circulating glycoproteins or cells. This lectin activity was markedly inhibited by plasma from birds and reptiles. This, in turn, led to the discovery of a chicken hepatic lectin, which preferentially recognizes terminal *N*-acetylglucosamine. Thus, chickens were found to have a *N*-acetylglucosamine-specific hepatic receptor instead of a galactose-specific one, permitting the circulation of asialoglycoproteins, but not glycoproteins that also have galactose removed. Affinity columns of immobilized asialoglycoproteins were then used to purify the mammalian hepatic asialoglycoprotein receptor.

The same approach was then applied by investigators working in other systems, uncovering a variety of galactose-binding lectins in cell types ranging from the slime mold *Dictyostelium discoideum* to mammalian tissues such as the heart. Most of these lectins eventually proved to be quite different from the original hepatic asialoglycoprotein receptor, being water-soluble and of relatively low molecular weight; they are now known as the galectins (see Chapter 33). Robert Hill and colleagues reported yet another type of hepatic uptake system that seemed to involve recognition of fucose moieties. In the early 1970s, Elizabeth Neufeld and colleagues reported a carbohydrate-dependent system that mediates the uptake of lysosomal enzymes by fibroblasts. In 1977, William Sly and colleagues demonstrated specific blockade of this uptake by the monosaccharide Man-6-P. Further work by the groups of Stuart Kornfeld, Bill Jourdian, Kurt von Figura, and others led to the discovery of Man-6-P receptors, which recognize phosphorylated high-mannose-type glycans that are selectively expressed on lysosomal enzymes (see Chapter 30). Encouraged by the prospect of using this uptake system to correct lysosomal enzyme deficiency diseases in humans, other investigators infused labeled lysosomal enzymes into intact animals and followed their fate. As it turned out, mature lysosomal enzymes were not rich in Man-6-P (the phosphate esters are mostly removed after initial targeting to lysosomes; see Chapter 30). Instead, rapid clearance mostly involved recognition of terminal mannose and *N*-acetylglucosamine residues. This led to the discovery of the macrophage mannose receptor.

Thus, by the beginning of the 1980s, the concept of vertebrate lectins that recognize specific endogenous ligands had become firmly established. Several circulating soluble lectins were also discovered in the blood plasma of various species, with varied glycan-binding specificities. Initially, it had been thought that sialic acids, while serving as ligands for exogenous microbial pathogens, generally acted as "masks" within vertebrates, preventing binding by endogenous lectins that recognized underlying saccharides. The discovery of some arachnid and crustacean lectins that could recognize sialic acids in vitro did not change this impression because these organisms did not themselves express endogenous sialic acids. The first indication that sialic acids might serve as endogenous ligands within vertebrates came from the discovery that binding of the complement regulatory H protein to "self" cell surfaces was dependent on sialic acid residues. Steve Rosen and colleagues then showed that sialidase treatment of rat lymph node sections abolished binding of lymphocytes to high endothelial venules. This was the first demonstration of the involvement of glycans in recognition by what turned out to be the selectin family of vascular receptors (see Chapter 31).

In 1972, Tim Hardingham and Helen Muir showed that hyaluronan could aggregate cartilage proteoglycans. Studies by Vincent Hascall and Richard Heinegard in 1974 documented that there is specific binding between hyaluronan, the amino-terminal globular part of the proteoglycan, and a link protein (see Chapter 15). By the late 1980s, it was realized that the primary amino acid sequence of a protein could be used to predict glycan-recognition properties of proteins (see below). This led to the recognition of the hyaluronan-binding properties of CD44 and the correct prediction that selectins would recognize carbohydrates. More recently, a specific clearance system was also discovered that recognized the sulfated *N*-acetylgalactosamine residues on pituitary glycoprotein hormones (see Chapter 28). In addition, the discovery of sialic acid–dependent binding by the B-cell molecule CD22 and the cloning of the macrophage sialic acid–dependent receptor sialoadhesin led to the identification of another new family of lectins (the Siglecs), which belong to the immunoglobulin superfamily (I-type lectins; see Chapter 32). Most recently, several lectins have been discovered within the endoplasmic reticulum (ER)-Golgi pathway itself, where glycan biosynthesis occurs (see Chapter 36).

Sulfated Glycosaminoglycan-binding Proteins

As mentioned above, a large group of GBPs that defy classification based on sequence data or general structure are those that recognize sulfated glycosaminoglycans such as heparin, chondroitin sulfate, and dermatan sulfate (see Chapter 35). The best-studied example is the interaction of heparin with antithrombin. Heparin was actually discovered in 1916 by a medical student named Jay McLean, but it was not until 1939 that heparin was shown to be an anticoagulant in the presence of heparin cofactor, which was then identified as antithrombin in the 1950s. The interaction of antithrombin and heparin depends on specific positively charged amino acid residues provided by two opposing α-helices in antithrombin, which bind to a specific pentasaccharide sequence of unusual structure within a heparin chain. Most other sulfated glycosaminoglycan-binding proteins have clusters of positively charged residues arrayed along the surface of the protein. Many of these proteins can also interact with sulfated glycosaminoglycans in a looser fashion—that is, they do not always show a high level of specificity like that demonstrated by antithrombin. However, in most cases, the actual sequence of sulfated sugars required for biological function is unknown, leaving open the possibility that specific sequences mediate the formation of higher-order complexes, for example, between a ligand and its receptor. In many cases, the sulfated glycosaminoglycan acts as a template for homo-oligomerization or for approximation of two proteins.

CLASSIFICATION OF LECTINS BASED ON SEQUENCE AND STRUCTURAL HOMOLOGY

The first classifications came from the plant lectin field and were based on the glycan sequences to which they bound best (e.g., β-galactoside-binding lectins). Only with the advent of molecular cloning did a more consistent classification emerge based on amino acid sequence homology and evolutionary relatedness. The first such classification was proposed by Kurt Drickamer using some highly conserved amino acid sequence motifs in the CRDs of two groups of lectins. One group required calcium for recognition and the members were therefore called C-type lectins; the other group required free thiols for stability and the members were termed S-type lectins (later renamed the galectins, as not all of them were thiol-dependent). Meanwhile, the two lectins that recognized Man-6-P were sequenced and found to be homologous but distinct from all the others, justifying their designation as P-type lectins. Although some classes of lectins, such as the P-type and galectins, appear to recognize a single class of sugars (Man-6-P and β-galactosides, respectively), others like the C-type lectins recognize a variety of molecules that share only a lectin protein module in common. Figure 26.1 shows generic structures of several of these classes of lectins. With the availability of sequence data and crystal structures, it also became possible to classify the plant lectins into several distinct groups (see Table 26.2). Interestingly, it turns out that several of these groups have structural or sequence similarity with animal lectins, revealing how evolutionarily ancient these CRDs are.

A major breakthrough occurred when the independent cloning of three homologous vascular adhesin receptors revealed a common amino-terminal C-type lectin motif; these three molecules eventually turned out to be the selectins (see Chapter 31). This was the first time that glycan recognition had been predicted on the basis of the primary amino acid sequence of a cloned protein, validating the concept of classification based on sequence homology. The cloning of a variety of circulating soluble lectins also led to the recognition of a subset of C-type lectins designated as the "collectin family." In addition, two Ca^{++}-binding lectins (calnexin and calcireticulin) are unrelated to the C-type lectins (not all Ca^{++}-requiring lectins are C-type lectins), and they specifically recognize glucose residues on newly synthesized glycoproteins (see Chapter 36).

Studies in the 1990s also revealed that immunoglobulin superfamily members can recognize carbohydrates, leading to a new group of I-type lectins (see Chapter 32). A subgroup of these molecules, which specifically recognizes sialic acids, has been designated the "Siglecs"

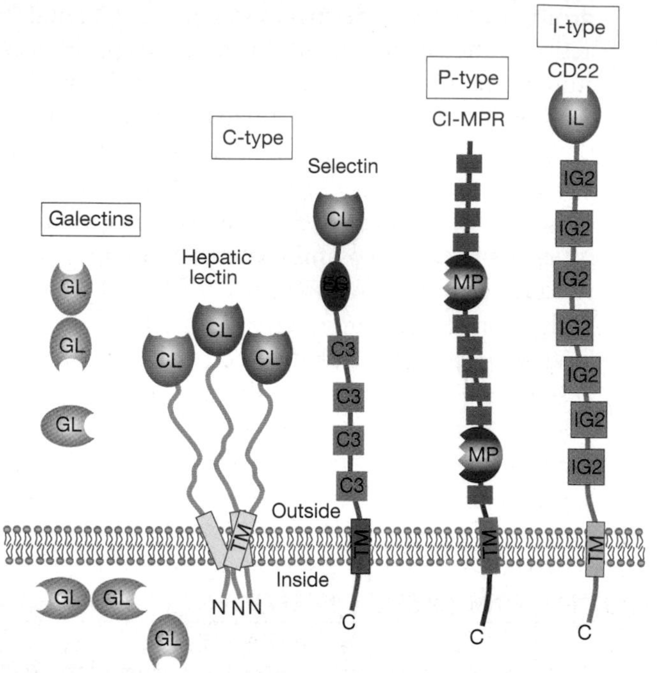

FIGURE 26.1. Schematic examples of major types of animal lectins, based on protein structure. Examples of some of the major families are shown. The emphasis is on the extracellular domain structure and topology. The following are the defined carbohydrate-binding domains (CRDs) shown: (CL) C-type lectin CRD; (GL) S-type lectin CRD; (MP) P-type lectin CRD; (IL) I-type lectin CRD. Other domains are (EG) EGF-like domain; (IG2) immunoglobulin C2-set domain; (TM) transmembrane region; and (C3) complement regulatory repeat. The number of domains underlying the CRD can vary among family members.

(for sialic acid–recognizing immunoglobulin superfamily lectins) (see Chapter 32). These and other general groupings are based primarily on sequence homologies and probable evolutionary relatedness and include the majority of known animal lectins. However, many others do not show any obvious sequence homologies or evolutionary relationships. For example, a class of evolutionarily very ancient circulating soluble lectins called the pentraxins are recognizable not so much by primary sequence homologies, but by a consistent pentameric structural organization and a role in the primary host immune response.

At the present time, there is no single universally accepted classification of lectins. Table 26.2 summarizes currently known lectins into two categories and makes some suggestions for naming of some of the groups in category I, which are characterized by sequence homologies or evolutionary relationships. As with all nomenclature issues, it will be up to investigators within the relevant fields to communicate with each other and to decide whether these suggestions are acceptable.

As mentioned earlier, glycosaminoglycan-binding proteins cannot be classified based on shared evolutionary origins, as they seem to have mostly evolved by convergent evolution. Table 16.8 in Chapter 16 presents a limited listing of examples of these types of GBPs. For further details, see Chapter 35.

NATURE OF GLYCAN–PROTEIN INTERACTIONS

Several crystal structures of GBPs with their cognate ligands have been determined, allowing an understanding of these interactions at the level of atomic resolution (see Chapter 27 for details). These can be divided into two general groups: those involving sulfated glycosaminoglycan chains (mostly mediated by ordered arrays of surface charge contacts; see Chapter 35) and those involving other classes of glycans (e.g., N- and O-glycans). Several principles have emerged about the latter group of glycan–protein interactions. First, the binding sites for low-molecular-weight glycans are of relatively low affinity (K_d values in the high micromolar to low millimolar range) and comprise shallow indentations on the surface of the proteins. Second, selectivity is mostly achieved via a combination of hydrogen bonds (involving the hydroxyl groups of the sugars) and by van der Waals packing of the hydrophobic face of monosaccharide rings against aromatic amino acid side chains. Third, further selectivity and enhanced affinity can be achieved by additional contacts between the glycan and the protein, sometimes involving bridging water molecules or divalent cations. Finally, the actual region of contact between the saccharide and the polypeptide typically involves only one to three monosaccharide residues, although there are many exceptions to this observation. As a consequence of all of the above, these lectin-binding sites tend to be of relatively low affinity, although they can exhibit high specificity. The ability of such low-affinity sites to mediate biologically relevant interactions in the intact system thus usually requires multivalency. Glycosaminoglycan-binding proteins differ from other GBPs in that binding usually involves five or more sugar residues. In some cases, the affinity can be quite high (K_d values in nanomolar to micromolar range), but the specificity in terms of linear sequence of sulfated residues is often relaxed.

MANY GLYCAN-BINDING PROTEINS ARE FUNCTIONALLY MULTIVALENT

Until the 1990s, all of the plant and animal lectins discovered were found to be naturally multivalent, either because of their defined multisubunit structure or by virtue of having multiple carbohydrate-binding sites within a single polypeptide. Indeed, increased avidity generated by multivalent binding of lower-affinity single sites appears to be a common mechanism for optimizing lectin function in nature, and a traditional definition for a lectin was "a multivalent carbohydrate-binding protein that is not an antibody." Notable exceptions to this general rule are the selectins, which have only a single CRD site within their extracellular polypeptide domains (see Chapter 31). The same situation applies to the Siglecs (see Chapter 32) and many glycosaminoglycan-binding proteins (see Chapter 35). However, in these instances, these molecules can become functionally multimeric by noncovalent association and clustering on cell surfaces. Since many lectins may function as signaling molecules, their multivalency may promote cross-linking of relevant cell-surface receptors and may be required for signaling. Whether biologically significant binding by any plant or animal lectin can arise from a strictly monovalent interaction remains a difficult problem to address. It is also of note that a single lectin can carry multiple binding sites for multiple ligands; for example, the macrophage mannose receptor is now known to bind not only to mannans, but also via a distinct CRD to the 4-O-sulfated GalNAc residues of pituitary glycoprotein hormones.

DEFINING NATURAL LIGANDS FOR GLYCAN-BINDING PROTEINS

Despite the fact that they were discovered first, remarkably little is known about the natural intrinsic ligands for plant lectins. Indeed, to date the only definitive example is that of the recognition of *Rhizobium* Nod factors by LNP (lectin-nucleotide phosphohydrolase) lectins (see Chapter 37). On the other hand, plant lectins recognize many animal or prokaryotic glycans with a high degree of specificity (see Chapters 28, 29, 31, and 45), and some (but not all) have toxic properties. This suggests that some plant lectins were probably evolved for protection against other species and/or for mutually beneficial (symbiotic) reasons. Details about the natural ligands for animal lectins can be found in other chapters in this volume.

Despite their specificity, monosaccharides or small oligosaccharide units tend to be weak inhibitors of lectin interactions. The natural ligands for most lectins are typically complex glycoconjugates that carry clustered arrays of the cognate carbohydrate or unique glycan structures, thus cooperating with clustered lectin-binding sites to generate high-avidity binding, which is further enhanced by mass transport effects (high local concentrations of ligands). In some instances (e.g., the selectins), the nature of this clustering is not easily defined, and cooperation with other aspects of the underlying polypeptide may be necessary to generate optimal binding. For example, optimal binding of P- and L-selectin to the ligand PSGL-1 requires a precise ordered combination of an amino-terminal peptide sequence bearing an O-glycan with a terminal sialyl Lewis^x motif adjacent to a peptide rich in acidic residues and sulfated tyrosines (see Chapter 31). Partly for this reason, it is common to see the names of the underlying polypeptide backbone used to define the nature of a ligand, for example, "PSGL-1 is the ligand for P-selectin." However, it should be recognized that unless it is correctly glycosylated or otherwise modified (e.g., sulfated), the PSGL-1 polypeptide is not itself the ligand. Typically, these polypeptides are simply carriers of the true ligands for lectins, which are made up of combinations of glycan units.

Recombinant lectins that are often used to identify potential biological ligands are usually multimeric in structure or are presented in multivalent clustered arrays in soluble complexes or on solid supports (see Chapters 45 and 48). Thus, although a variety of molecules may be found to bind to a given recombinant lectin in a glycosylation-dependent manner, only a few of these "ligands" may actually be involved in mediating biologically significant interactions. The challenge then is to tell the difference between what *can* bind to a recombinant lectin in an in vitro experiment and what actually *does* bind to the native lectin in a biologically relevant manner in vivo. Indeed, the term "ligand" should probably be reserved for the latter type of biologically relevant structures. It should also be kept in mind that the natural ligands of some animal lectins may be present primarily on foreign invaders. For example, the circulating soluble mannan-binding protein may serve to bind and opsonize (mark for phagocytosis) microorganisms that bear high densities of mannose, such as yeasts and other fungi. All of these issues are dealt with in more detail in Chapters 27–35.

A more complex variation is the concept of a "clustered saccharide patch" (CSP), which is generated by a peptide carrying multiple closely spaced glycans. Free glycans in solution or at single peptide attachment sites have significant freedom of motion in an aqueous environment. Thus, lectins that recognize linear glycan chains do so by capturing and "freezing" one of the many possible solution conformations of the glycan. However, glycans packed closely together (e.g., O-glycans in mucins or clustered glycosphingolipids in cell-surface rafts) have less mobility. Such clustering of common glycans could present uncommon CSPs, generated by multiple glycans closely spaced enough to restrict their motion and together they could form a unique epitope, perhaps aided or stabilized by peptide interactions with adjacent glycans. Releasing the glycans from the polypeptide backbone would result in complete loss of recognition by the lectin. There are examples of antibodies and enzymes whose specificities cannot be explained by single glycan chain recognition and which likely recognize such CSPs. Selectin recognition of heterogeneous carcinoma mucins and some features of plant lectin-recognition phenomena are also probably explained by CSPs. However, this concept has been more difficult to visualize and define at the scale of atomic resolution.

BIOSYNTHESIS, TRAFFICKING, AND REGULATION OF GLYCAN-BINDING PROTEINS

From a functional point of view, it is worth considering GBPs in two physical classes: soluble and membrane-bound. Cell membrane–bound GBPs are more likely to be involved in endocytosis, cell adhesion, or cell signaling and to stay confined to the cell type of their original synthesis. On the other hand, soluble GBPs are capable of diffusing locally in tissues and/or entering the blood circulation. Although useful in functional terms, this type of physical classification is confounded by two issues: First, some GBPs that start out as membrane-bound proteins can be proteolytically shed into the extracellular fluid. Second, soluble GBPs can become attached to cell surfaces or matrices via their glycan-binding sites. Figure 26.2 indicates some examples of these relationships and the nature of the potential interactions with natural ligands, which, in turn, can also be soluble or membrane-bound.

All membrane-bound and many soluble lectins are synthesized on ER-bound ribosomes and delivered to their eventual destinations via the ER-Golgi pathway. Thus, the GBPs themselves are often glycoproteins. However, a significant subset of soluble GBPs (galectins, heparin-binding growth factors, and some cytokines) are synthesized on free ribosomes in the cytoplasm and then delivered directly to the exterior of the cell by an as yet poorly understood mechanism. This makes functional sense, since several of these lectins can recognize biosynthetic intermediates that occur in the Golgi-ER pathway (e.g., galactosides and high-mannose oligosaccharides). The circumvention of the conventional pathway of secretion may allow these molecules to avoid unwanted premature interactions with potential ligands that are synthesized within the same cell.

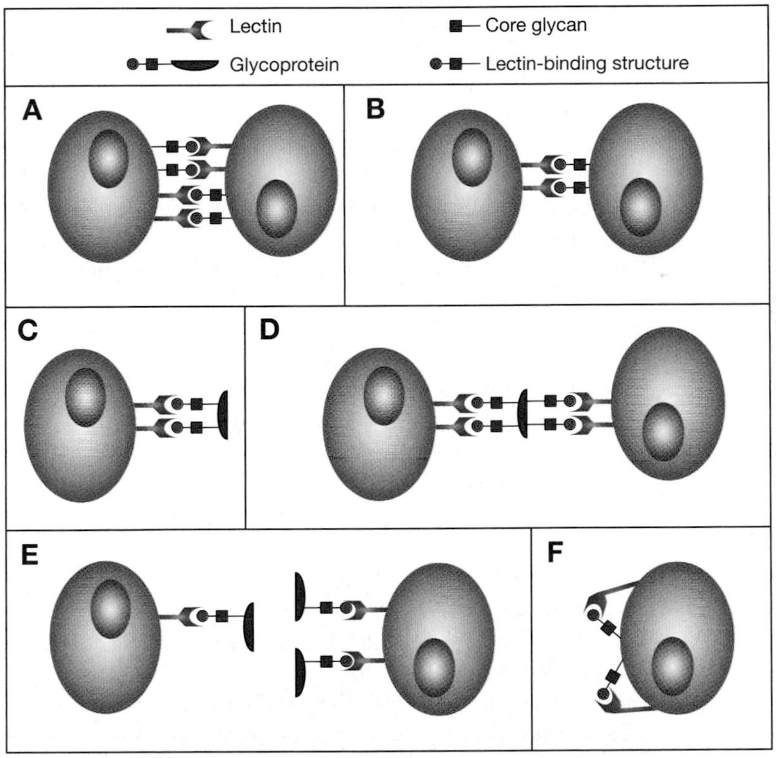

FIGURE 26.2. Possible mechanisms of regulation of an animal lectin by cognate ligands. Potential ligands can be on the cell surface and/or on soluble glycoproteins (including sugar chains attached to the lectin itself). As discussed in the text, the "ligand" depicted in this cartoon can either be a simple linear or terminal glycan sequence or a very complex motif that includes more than one monosaccharide and/or additional components. Direct cell–cell interactions could occur among lectin-positive cells (*A*) or between lectin-positive cells and other cell types bearing cognate ligands (*B*). Soluble glycoprotein ligands could interact directly with lectin-positive cells (*C*), bridge between two such cells (*D*), or inhibit cell–cell interactions involving the lectin (*E*). Expression of ligands on lectin-positive cells could inactivate the lectin function by either inter- or intramolecular interactions (*F*).

Some GBP genes are expressed constitutively, whereas others are induced by gene activation under specific biological circumstances. E-selectin is a good example of a lectin that is transcriptionally regulated and expressed by endothelial cells only after activation by inflammatory mediators (see Chapter 31). Many lectins are regulated posttranslationally or after secretion. For example, some members of the galectin family of GBPs are sensitive to the redox state of the environment and can remain active only in the reducing environment of the cytoplasm. Upon entering the oxidizing environment of the extracellular space, they must therefore bind to ligands or become progressively inactivated. Another form of regulation occurs when the lectin binds to cognate sugar chains present on the same molecule or the same cell surface and hence becomes functionally inactive (e.g., the Siglecs, where sialic acid–bearing ligands from the same cell surface can restrict the availability of the lectin-binding sites to external ligands) (see Chapter 32). Some membrane-bound lectins are also internalized upon binding to ligands, with delivery to internal acidic compartments (endosomes). The cargo can be released, allowing some of the receptors to recycle back to their original location, for example, the mannose 6-phosphate receptor. In some cases, the lectin is then transferred to the lysosome, where it undergoes degradation.

FURTHER READING

Stillmark H. 1888. "Uber Ricin, Ein giftiges Ferment aus den Samen von *Ricinus communis* L. und einigen anderen Euphoribiaceen." Inaugural dissertation. University of Dorpat, Dorpat (now Tartu), Estonia.

Goldstein I.J., Hughes R.C., Monsigny M., Osawa T., and Sharon N. 1980. What should be called a lectin? *Nature* **285:** 66.

Ashwell G. and Harford J. 1982. Carbohydrate-specific receptors of the liver. *Annu. Rev. Biochem.* **51:** 531–554.

Drickamer K. 1988. Two distinct classes of carbohydrate-recognition domains in animal lectins. *J. Biol. Chem.* **263:** 9557–9560.

Kornfeld S. 1992. Structure and function of the mannose 6-phosphate/insulinlike growth factor II receptors. *Annu. Rev. Biochem.* **61:** 307–330.

Drickamer K. and Taylor M.E. 1993. Biology of animal lectins. *Annu. Rev. Cell Biol.* **9:** 237–264.

Varki A. 1994. Selectin ligands. *Proc. Natl. Acad. Sci.* **91:** 7390–7397.

Powell L.D. and Varki A. 1995. I-type lectins. *J. Biol. Chem.* **270:** 14243–14246.

McEver R.P. and Cummings R.D. 1997. Role of PSGL-1 binding to selectins in leukocyte recruitment. *J. Clin. Invest.* **100:** 485–492.

Etzler M.E. 1998. Oligosaccharide signaling of plant cells. *J. Cell Biochem. Suppl.* **30–31:** 123–128.

Cebo C., Vergoten G., and Zanetta J.P. 2002. Lectin activities of cytokines: Functions and putative carbohydrate-recognition domains. *Biochim. Biophys. Acta* **1572:** 422–434.

Esko J.D. and Selleck S.B. 2002. Order out of chaos: Assembly of ligand binding sites in heparan sulfate. *Annu. Rev. Biochem.* **71:** 435–471.

Helenius A. and Aebi M. 2004. Roles of N-linked glycans in the endoplasmic reticulum. *Annu. Rev. Biochem.* **73:** 1019–1049.

Varki A. and Angata T. 2006. Siglecs—The major subfamily of I-type lectins. *Glycobiology* **16:** 1R–27R.

Rabinovich G.A., Toscano M.A., Jackson S.S., and Vasta G.R. 2007. Functions of cell surface galectin-glycoprotein lattices. *Curr. Opin. Struct. Biol.* **17:** 513–520.

Sharon N. 2007. Lectins: Carbohydrate-specific reagents and biological recognition molecules. *J. Biol. Chem.* **282:** 2753–2764.

Varki A. 2007. Glycan-based interactions involving vertebrate sialic-acid-recognizing proteins. *Nature* **446:** 1023–1029.

CHAPTER 27

Principles of Glycan Recognition

Richard D. Cummings and Jeffrey D. Esko

GLYCANS INTERACT WITH MANY TYPES OF PROTEINS, as exemplified by their binding to enzymes, antibodies, and glycan-binding proteins (GBPs). Binding of glycans to proteins represents the major way in which the information contained in glycan structures is recognized, deciphered, and put into biological action. This chapter describes the structural and thermodynamic principles of glycan–GBP interactions and the methods used to measure association constants.

GLYCAN–PROTEIN RECOGNITION

The general classes of GBPs are described in Chapter 26. GBPs differ in the types of glycans they recognize and in their binding affinity. A common question is: How does a GBP (such as a bacterial adhesin, toxin, plant lectin, viral hemagglutinin, antibody, or animal GBP) bind to only a very limited number of glycans (or even a single glycan) among the thousands that are produced by a cell? A fundamental question at the molecular level is: What forces allow a specific glycan to leave the aqueous phase to enter a protein-combining site? Understanding the molecular basis of the interaction between glycans and proteins is not merely an academic exercise in chemistry and physics, it is essential for the development of a real molecular appreciation of how glycoconjugates affect fundamental biological responses. Furthermore, the information that emerges from these kinds of studies provides a springboard for designing specifically tailored compounds that inhibit a glycan-binding site. For example, understanding the principles of glycan–protein interactions led to the development of high-affinity inhibitors of viral neuraminidase, which is involved in the removal of sialic acids from host glycoprotein receptors, and these inhibitors have proven utility in reducing the duration and spread of influenza infection (see Chapters 39, 50, and 51).

HISTORICAL BACKGROUND

Many of the early studies of glycan–protein interactions focused on the recognition of glycans by enzymes, including the endoglycosidase lysozyme (which can degrade bacterial cell walls) and certain enzymes involved in intermediary metabolism (such as glycogen and starch synthases and phosphorylases). Following their discovery of the antibacterial action of lysozyme and penicillin, Sir Alexander Fleming, Ernst Chain, and Sir Howard Florey received the Nobel Prize in 1945. Lysozyme was subsequently shown to be a highly specific endoglycosidase capable of cleaving α1-4-linkages in bacterial peptidoglycan (see Chapter 20). Glycogen synthase was found to generate α1-4-glucosyl residues in glycogen, whereas other branching and debranching enzymes recognized α1-6-branched glucose residues. Studies on the metabolism of glycogen by these enzymes led to the award of the Nobel Prize in 1970 to Luis F. Leloir for his discovery of sugar nucleotides and their role in the biosynthesis of glycans.

The concept of glycans being specifically recognized by proteins dates back to Emil Fischer, who used the phrase "lock and key" to refer to enzymes that recognize specific glycan substrates. Lysozyme was the first "carbohydrate-binding protein" to be crystallized, and determination of its three-dimensional structure led to the realization that specific interactions occur between sugars and proteins. The structure of lysozyme was solved in a complex with a tetrasaccharide in an elegant series of studies in the late 1960s. Lysozyme is an ellipsoid protein, which has a long cleft that runs for most of the length of one surface of the molecule. This cleft is astonishingly large, considering that lysozyme has only 129 amino acids, and it is capable of accommodating a hexasaccharide and cleaving it into di- and tetrasaccharide products. Other GBPs whose three-dimensional structures are of historical significance are concanavalin A (ConA; crystal structure reported in 1972) and influenza virus hemagglutinin (crystal structure reported in 1981). Much of the initial work important to our understanding of glycan–protein interactions was gathered in studies on the combining sites of plant lectins and antibodies toward specific blood group antigens. From these and many recent studies, it is now clear that the specific recognition of sugars by proteins occurs by multiple mechanisms.

BIOPHYSICAL METHODS TO DETERMINE GLYCAN–PROTEIN INTERACTIONS

Two of the most widely used biophysical approaches for examining protein–glycan interactions at the molecular level are X-ray crystallography and nuclear magnetic resonance (NMR). Several hundred three-dimensional structures of GBPs are listed in various structural protein data banks (e.g., Research Collaboratory for Structural Bioinformatics [RCSB], Protein Data Bank [PDB], and the National Center for Biotechnology Information [NCBI]). Additionally, several thousand three-dimensional structures of mono- and oligosaccharides are stored in the Cambridge Structural Database.

A resolution of at least 2–2.5 Å is required to identify accurately the positions and mode of binding of glycans. However, such resolution is often difficult to obtain. In some cases, a crystal structure can be obtained for a GBP independent of ligand, and its structure and potential binding site can be predicted using information from the three-dimensional structure of a homologous protein. Since small molecules often cocrystallize with a GBP better than large molecules, a lot of our knowledge about glycan–GBP interactions at the atomic level is based on cocrystals of GBPs with unnatural ligands. Thus, a great challenge exists in attempting to understand glycan–GBP interactions in the context of natural glycans present as a glycoprotein, glycolipid, or proteoglycan.

In NMR, the proton–proton distances in small molecules (generally <2000 daltons) can be obtained following assignment of the proton resonances through multidimensional techniques, such as measuring the nuclear Overhauser effect (NOE). This information coupled with computational methods that employ computer modeling allows the prediction of the free-state glycan conformation in solution. NMR studies of free glycans have shown that monosaccharides are relatively rigid molecules in solution, but an oligosaccharide actually has a higher degree of flexibility because of the general freedom of rotation about glycosidic bonds (see Chapter 2). The main approaches for using NMR to define the conformation of bound glycans are NOE spectroscopy (NOESY), transfer NOE (trNOESY; transferred rotating-frame Overhauser effect spectroscopy), and heteronuclear single-quantum correlation (HSQC). Computer-assisted modeling strategies and information from glycan solution conformations and protein three-dimensional structures can be combined with NMR to provide even more information about the molecular details of the interactions between glycans and proteins. Although these approaches are highly informative, they are limited by the degree to which small glycans structurally mimic the larger macromolecule to which they are usually attached.

ATOMIC DETAILS OF PROTEIN–GLYCAN INTERACTIONS

Several hundred structures of glycan–protein complexes have been solved by X-ray crystallography and NMR spectroscopy. In most cases, the glycan-binding sites typically accommodate one to four sugar residues, although in some complexes, recognition extends over larger numbers of residues and may include aglycone components, such as the peptide or lipids to which the glycans are attached. The complex of cholera toxin, a pentameric protein, bound to the ganglioside GM1 pentasaccharide illustrates this general pattern (Figure 27.1). Cholera toxin is a soluble protein generated by *Vibrio cholerae* that has an AB_5 subunit structure, where the single A subunit (~22 kD) is an enzyme (ADP-ribosyltransferase) and the five identical B subunits (~55 kD) each bind GM1 (see Chapter 34). Only three of the five monosaccharides of each glycolipid ligand actually make contact with each B subunit. The terminal β-linked galactose residue is almost completely buried in the protein-combining site, making numerous hydrogen-bonding interactions with protein functional groups, some via bound water molecules. In addition, an aromatic amino acid (Trp-88) stacks against the hydrophobic "underside" of one of the sugar residues, in this case, the terminal β-linked galactose residue. Hydrophobic interactions are very common in glycan–protein complexes and can involve aromatic residues as well as alkyl side chains of amino acids in the combining site. The penultimate β-linked GalNAc residue makes only minor contacts with the protein, whereas the Galβ1-4Glc reducing-end disaccharide makes no contact at all. About 350 Å of the solvent-accessible surface of the pentasaccharide is buried in the B subunit. About 43% of this area is due to interactions with the sialic acid, 39% is due to galactose, and 17% is due to *N*-acetylgalactosamine.

Knowing the three-dimensional structure of a glycan-protein complex can reveal much about the specificity of binding, changes in conformation that take place on binding, and the contribution of specific amino acids to the interaction. However, one would also like to

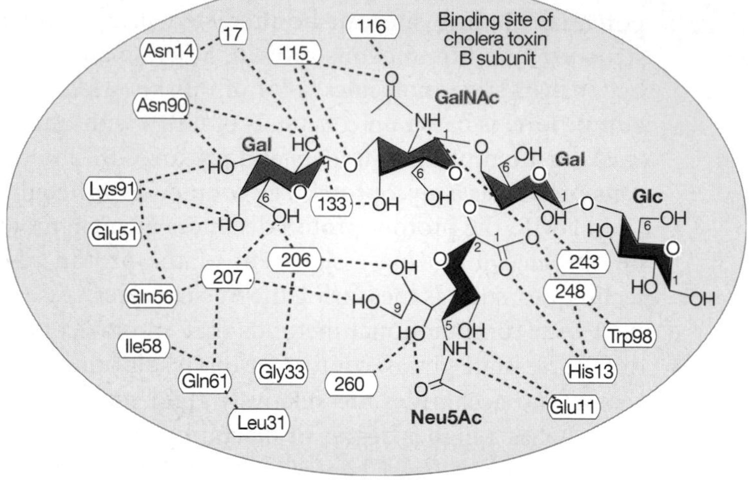

FIGURE 27.1. Simplified structure of the GM1 pentasaccharide bound to the B subunit of cholera toxin. Identification of the amino acid residues in cholera toxin that contact the glycan directly or through hydrogen bonds with water. (Redrawn, with permission, from Merritt E.A., Sarfaty S., van den Akker F., et al. 1994. *Protein Sci.* **3:** 166–175.)

determine the affinity of the interaction. Since the forces involved in the binding of a glycan to a protein are the same as for the binding of any ligand to its receptor (hydrogen bonding, electrostatic or charge interactions, van der Waals interactions, and dipole attraction), it is tempting to try to calculate their contribution to overall binding energy, which can then be related to an affinity constant (K_a). Unfortunately, calculating the free energy of association is difficult for several reasons, including problems in defining the conformation of the unbound versus the bound glycan, changes in bound water within the glycan and the binding site (as discussed below), and conformational changes in the GBP upon binding.

The "Water Problem"

As shown in Figure 27.2, the overall process of binding typically involves the union of a hydrated polyhydroxylated glycan and a hydrated protein-combining site. If a surface on the glycan is complementary to the protein-combining site, water can be displaced and

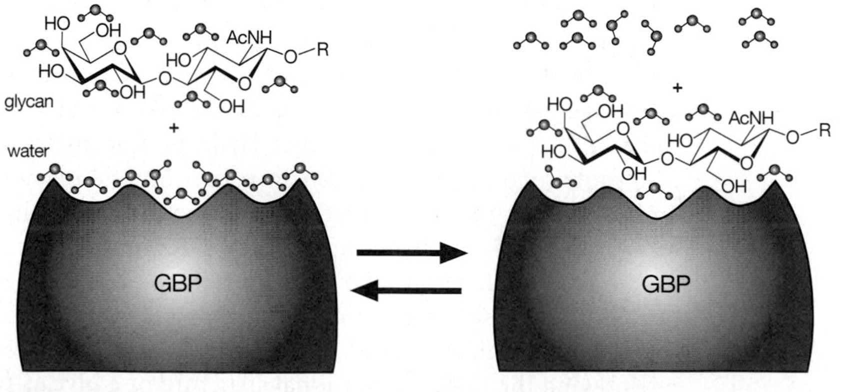

FIGURE 27.2. Schematic diagram of the binding of a glycan to a GBP in water, which results in the displacement of water.

binding occurs. When the complex finally forms, it presents a new surface to the surrounding medium, which will also be hydrated. Solvation/desolvation energies are very large because of entropy from the disordering of water molecules, but their contribution to binding cannot be reliably determined with existing models. Furthermore, glycans may undergo a conformational change upon binding, changing their internal energy and solvation. Thus, although one can estimate the energetic contributions of van der Waals and hydrogen bonding interactions in the combining site, errors in the estimation of the attendant solvation energy changes render the overall calculations of binding energy problematic.

Valency

Another complicating issue concerns valency, which greatly increases the association of GBPs with their glycan ligands under biological conditions. The oligosaccharide of ganglioside GM1 is unusual in that it binds strongly to the B subunits of cholera toxin with high affinity because of the specific and multiple interactions in the combining site (K_d of ~40 nM). However, the affinity of most single glycan–protein interactions is generally low (mM–μM K_d values). In nature, many GBPs are oligomeric or may be membrane-associated proteins, which allows aggregation of the GBP in the plane of the membrane. Many of the glycan ligands for GBPs are also multivalent. In the case of cholera toxin, five B subunits present in the holotoxin interact with five molecules of GM1 normally present in the cell membrane. The interaction of multiple subunits with a multivalent display of GM1 raises the affinity of the interaction by several orders of magnitude (K_d of ~40 pM). The term *avidity* is used to refer to the strength of multivalent ligand binding.

The density of binding sites on the ligand can also affect the affinity of binding. Ligands for some GBPs may be glycoproteins that carry one or more multiantennary N-linked chains. Mucins present potentially hundreds of glycans, and their proximity can affect conformation and presentation of the ligands. Some polysaccharides, such as glycosaminoglycans, have multiple binding sites located along a single chain (see Chapter 16). In addition, nonglycan components, such as tyrosine sulfate, lipid, or peptide determinants, may also cooperate with glycans to provide relatively high affinity and specific interactions. A good example of this is the interaction of PSGL-1 (P-selectin glycoprotein ligand-1) with P- and L-selectin (see Chapter 31), where the combination of sialyl Lewis^x on a core-2 O-glycan coupled with multiple adjacent tyrosine sulfate and amino acid residues increases the binding affinity manyfold, from low affinity (K_d ~ 0.1 mM) to a physiologically relevant level (K_d ~ 0.1–0.3 μM).

THERMODYNAMICS OF BINDING

The interaction of glycans with GBPs can also be described thermodynamically and kinetically. The binding of a lectin (L) to a glycan (G) is governed by Equation 1. The affinity constant, K, is defined as an association constant (or K_a) by Equation 2 and is equal to k_1/k_2. Like any equilibrium constant, K is related to the standard free-energy change of the binding reaction at pH 7 (ΔG_o) in kcal per mole, as shown by Equation 4, where R is the gas constant (0.00198 kcal/mol-degree) and T is the absolute temperature (298°K). The affinity constant K is related to the thermodynamic parameters ΔG, ΔH, and ΔS (see Equation 4), which represent the changes in free energy, enthalpy, and entropy of binding, respectively.

$$\text{Lectin (L)} + \text{Glycan (G)} \underset{K_2}{\overset{K_1}{\rightleftarrows}} \text{LG} \quad (1)$$
$$K_a = [LG]/[L][G] = k_1/k_2 \quad (2)$$
$$K_d = [L][G]/[LG] = 1/K_a = k_2/k_1 \quad (3)$$
$$\Delta G_o = RT \ln K = \Delta H - T\Delta S \quad (4)$$

The rate constant k_1 is expressed in units of $M^{-1}sec^{-1}$ or $M^{-1}min^{-1}$, whereas k_2 is expressed in units of sec^{-1} or min^{-1}. Although it is important to define K_a, k_1, k_2, ΔG, ΔH, and ΔS for each binding phenomenon under consideration, investigators often discuss data in terms of the $K_d = 1/K_a$ (see Equation 3), because the units are in concentration (millimolar, micromolar, nanomolar, etc.).

Binding of a monovalent GBP to a monovalent ligand is easily defined by the equilibrium kinetics described in the above equations. However, with multivalent ligands or GBPs, multiple affinities occur and a more complex binding equilibrium (more accurately described by a set of equilibrium constants) must be used. Typically, for multivalent ligands and GBPs, the values reported for affinity are apparent affinity constants and usually measure the avidity.

There are many different ways to study binding of glycans to proteins, and each approach has its advantages and disadvantages in terms of thermodynamic rigor, amounts of protein and glycan needed, and the speed of analysis. Below is a discussion of some of the major ways in which the binding between a glycan and protein can be studied.

TECHNIQUES TO STUDY GLYCAN–PROTEIN INTERACTIONS

Much of the available information about glycan–protein interactions derives from studies of relatively small glycan ligands interacting with a protein. In examining these interactions, two broad categories of techniques have been applied: (1) kinetic and near-equilibrium methods, such as equilibrium dialysis and titration calorimetry; and (2) nonequilibrium methods such as glycan microarray screening, hapten inhibition, ELISA-based approaches, and agglutination.

Kinetics and Near-Equilibrium Methods

Equilibrium Dialysis for Measuring K_d Values and Interaction Valency. The simple concept of equilibrium dialysis involves placing a solution of a GBP (e.g., a lectin or an antibody) in a dialysis chamber that is permeable to a glycan or other small hapten. In glycobiology, the term hapten, which was coined by Karl Landsteiner in his study of immunity to antigens, is often used to denote a small glycan that competitively binds to a lectin and competes for its binding to a more complex ligand. Thus, it can be used as a surrogate ligand. The chamber is then placed in a known volume of buffer that contains the glycan in the concentration range of the expected K_d. At equilibrium, the concentration of bound plus free glycan inside the bag [In] and free glycan outside the bag [Out] will depend on the concentration and affinity of the GBP inside the bag. From this information, both the K_a and the valence n can be determined from the relationship shown in Equation 5, where r is the molar ratio of glycan bound to GBP; c is the concentration of unbound glycan [Out]. The amount of glycan bound is determined by simply subtracting [Out] from [In].

$$r/c = K_a n - K_a r \quad (5)$$

A plot of *r/c* versus *r* for different hapten concentrations is known as a Scatchard plot and will approximate to a straight line with a slope of $-K_a$. The valence of binding (number of binding sites per mol) is defined by the *r* intercept at an infinite hapten concentration. If such an analysis were done with cholera toxin, for example, one would obtain five binding sites per mol of AB_5 complex, or one mol per mol of B subunit.

As in any technique for determining binding constants, a number of important assumptions are made and their validity must be considered. These include demonstrating that the protein and its hapten are stable and active over the course of the experiment, the hapten is freely diffusible, the complex is at equilibrium, and structurally unrelated haptens, not expected to bind, show no apparent binding in the experimental setup. The following are several advantages to equilibrium dialysis: (1) The ease of the approach and sophisticated equipment is not needed; (2) if the affinity is high, then relatively small amounts of protein are needed (≤1 mmole); (3) if the affinity is high, only small amounts of hapten may be required; (4) if the protein and haptens are very stable, they may be recovered and reused; (5) radioactive haptens may be used; and (6) reliable equilibrium measurements can be made. Some drawbacks of the approach are that (1) it can only provide K_a; (2) if the affinity of the GBP or antibody for the hapten is low, then relatively large amounts of both may be required; and (3) many different measurements must be made and this may require many days or weeks to complete.

A variation of this technique is illustrated by the equilibrium gel-filtration method developed by Hummel and Dreyer. In the Hummel–Dreyer method, a GBP is applied to a gel-filtration column that has been preequilibrated with a glycan of interest that is easily detectable (e.g., by radioactive or fluorescent tagging). As the protein binds to the ligand, a complex is formed that emerges from the column as a "peak" above the baseline of ligand alone, followed by a "trough" (where the concentration of ligand is decreased below the baseline) that extends to the included or salt volume of the column. The amount of complex formed is easily determined by the known specific activity of the ligand. Because the amount of complex formed is directly proportional to the amount of protein (or ligand) applied, it is easy to calculate a binding curve from several different Hummel–Dreyer column profiles at different concentrations of either protein or ligand. This binding curve allows the calculation of the equilibrium constant of the interaction. The advantages and drawbacks of this technique are generally the same as for equilibrium dialysis, except that Hummel–Dreyer analyses are often quicker to perform and can be used with ligands of many different sizes. Such an approach has been invaluable in defining the equilibrium binding of selectins to their ligands (see Chapter 31).

Affinity Chromatography to Assess the Specificity of Binding. In many cases, affinity chromatography is simply used to identify interacting partners. In this technique, a GBP is immobilized to an affinity support, such as Affi-Gel™, CNBr-activated Sepharose, Ultralink™, or some other activated support. If a glycan binds tightly to an immobilized GBP, buffer containing a known hapten may be added to force dissociation of the complex. For example, oligomannose-type and hybrid-type N-glycans will bind avidly to an agarose column containing the plant lectin concanavalin A (ConA-agarose) and 10–100 mM α-methyl mannoside is required to elute the bound material efficiently. In contrast, many highly branched complex-type N-glycans will not bind. Biantennary complex-type N-glycans bind to ConA-agarose, but they do not bind as tightly as high-mannose-type N-glycans and their elution can be effected by using 10 mM α-methyl glucoside. In this manner, one can assess the binding specificity of a GBP. In practice, this approach is rather crude, and although it gives valuable practical information about the capacity of an immobilized lectin to bind specific glycans, it does not provide quantitative affinity measurements. A variant of this method is to immobilize the glycan ligand through covalent linkage or by capturing a biotinylated glycan on a streptavidin-linked surface and then measuring GBP binding.

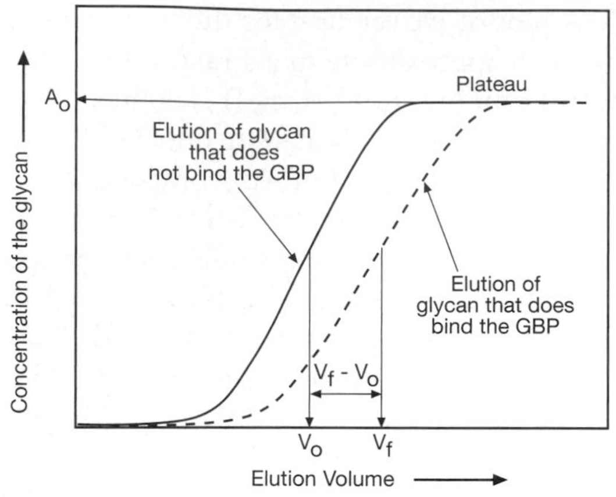

FIGURE 27.3. Example of frontal affinity chromatography, where different concentrations of a glycan are applied to a column of immobilized GBP. The profile depicts the elution of one glycan that binds the GBP and the elution of another glycan that does not bind the GBP. The V_o (void volume) and V_f are determined as shown from the elution volume. High V_f values indicate higher affinity.

A more sophisticated version of this approach, termed frontal affinity chromatography, does provide quantitative measurements of the equilibrium binding constants. In this technique, a solution containing a glycan of known concentration is continuously applied to a column of immobilized GBP, and the elution front of the glycan from the column is monitored. Eventually, enough ligand is added through continuous addition that its concentration in the eluant equals that in the starting material. If the glycan has no affinity for the GBP, it will elute in the void volume V_o; if, however, the glycan interacts with the GBP, it will elute after the V_o and at a volume V_f (Figure 27.3).

The advantages of frontal affinity chromatography are similar to those discussed for equilibrium dialysis: (1) The approach is easy and inexpensive; (2) if the affinity is high, then relatively small amounts of protein are needed (≤1 mmole), and only a single column is required; (3) correspondingly, small amounts of glycan may be used if the K_d is in the range of 10 nM to 10 mM; (4) if the glycans are stable, they may be recovered and reused; (5) radioactive glycans may be used; and (6) reliable equilibrium measurements can be made. There are several drawbacks to this approach, including (1) only the K_d can be derived, not k_{on} or k_{off}; (2) the conjugation of the GBP to the matrix must be stable and the protein must retain reasonable activity for many different column runs; (3) many different column runs must be made with a single glycan; and (4) if the K_d is high (>1 mM), this approach is not really feasible. Overall, frontal affinity chromatography is quite useful and has now been automated.

Titration Calorimetry to Measure K_d *and Binding Enthalpy.* This is one of the most rigorous means of defining the equilibrium binding constant between a glycan and a GBP. The binding of a glycan to the GBP is measured as a change in enthalpy through isothermal titration microcalorimetry using a commercial microcalorimeter. In this technique, a solution containing a glycan of interest is added in increments into a solution containing a fixed concentration of GBP. The glycan is added at many intervals and the heat evolved from binding is measured relative to a reference cell. Over the course of the experiment, the concentration of glycan is increased in the mixing cell over a glycan-to-GBP molar ratio of 0–10. The change of heat capacity of binding is determined and the data are replotted as kcal/mole of injectant versus the molar ratio (Figure 27.4). These data are then analyzed by replotting data to obtain the K_d. The heat change is directly related to the enthalpy of reaction ΔH. From knowledge of the K_d and ΔH, and using Equation 4, it is possible to define the binding entropy ΔS.

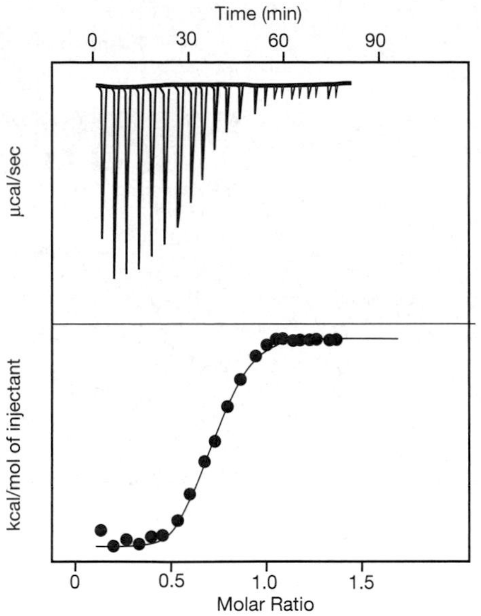

FIGURE 27.4. Example of titration microcalorimetry. Increasing amounts of a glycan are injected to a fixed amount of GBP in a cell, and the heat produced upon binding is measured as µcal/sec. The total kcal/mole of injected glycan relative to the molar ratio is plotted. These data can be used to define directly the thermodynamic parameters of binding and calculate the K_d of interaction between the glycan and the GBP.

The major advantage of this approach is that it can provide all major thermodynamic information about the binding of a glycan to a GBP, and thus, it is highly superior to equilibrium dialysis and affinity chromatography. The disadvantages of this approach are that (1) relatively large amounts of protein may be required to conduct multiple experiments (>10 mg); (2) relatively large amounts of glycans may be required; and (3) because of the above-mentioned problem, it is not typical for such analyses to use a wide range of different glycans. Nevertheless, this approach is rigorous and if the titration cell dimensions could be decreased in the future, then lower amounts of materials would be required.

Surface Plasmon Resonance to Measure the Kinetics of Binding and the K_d. Surface plasmon resonance (SPR) is a technique for measuring the association and dissociation kinetics of ligands (analytes) with a receptor. In SPR, the association of the analyte and receptor with one or the other immobilized on a sensor chip induces a change in the refractive index of the layer in contact with a gold film (Figure 27.5). This is measured as a change in the refractive index at the surface layer and is recorded as the SPR signal or resonance units (RU). Such measurements are often conducted in an instrument made by Biacore Life Sciences called the BIAcore™. Binding is measured in real time (Figure 27.5), and information about the association and dissociation kinetics can thus be obtained, which in turn can be used to obtain K_a and K_d from Equations 2 and 3.

A variety of chemistries are available for the coupling of the ligand or receptor to the surface of the chip, including reaction with amines, thiols, aldehydes, and noncovalent biotin capture. In some approaches, a glycoprotein ligand for a GBP is immobilized and the binding of the GBP is measured directly. It is also possible to degrade the immobilized glycoprotein ligand on the chip sequentially by passing over solutions containing exoglycosidases and reexamining at each step the binding to different GBPs, thereby obtaining structural information about the ligand. The immobilized ligand is usually quite stable and can be used repeatedly for hundreds of runs during a period of months.

The advantages of this approach are that (1) affinities in the range from millimolar to picomolar can be measured; (2) complete measurements of k_1 and k_2 are routine (see Equations 2

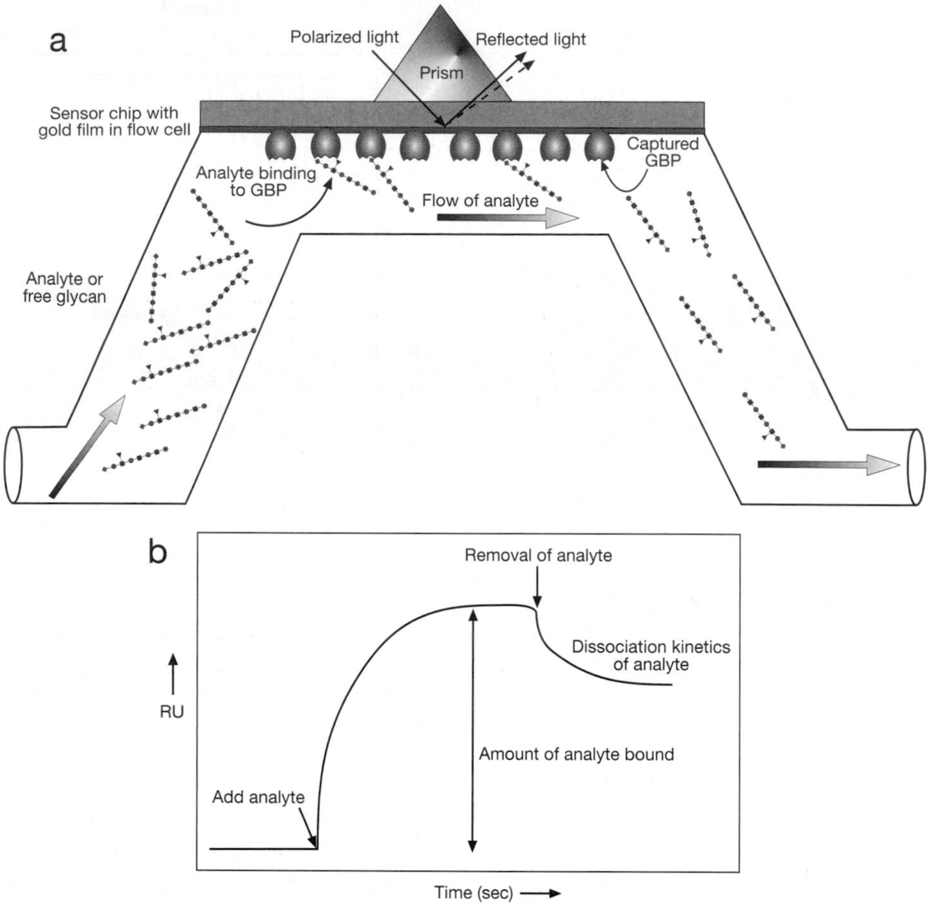

FIGURE 27.5. Example of surface plasmon resonance (SPR). (*a*) In SPR, the reflected light is measured and is altered in response to binding of the analyte in the flow cell to the immobilized GBP. (*b*) An example of a sensorgram showing the binding of the analyte to the ligand and the kinetics of binding and dissociation. RU indicates resonance units.

and 3) and calculations of K_d are straightforward; (3) for immobilization of a molecule using amine coupling, only 1–5 μg is normally sufficient; (4) typically, the concentration range of analyte is 0.1–100 x K_d and the typical volumes needed are in the range of 50–150 μl; and (5) measurements are extremely rapid and complete experimental results can be obtained within a few days. The drawbacks of this approach are that (1) analytes must have sufficient mass to cause a significant change in SPR upon binding (thus, the glycan is usually immobilized instead of the protein), (2) coupling of free glycans to the chip surface is inefficient and thus neoglycoproteins or some other type of large conjugate must be immobilized, and (3) there may some nonhomogeneity in conditions on the BIAcore™ because of mass transport effects, which could affect the dissociation rate and thus provide an inaccurate K_d measurement.

Fluorescence Polarization for Measuring K_d. A relatively new technique for measuring the binding constant of glycans to GBPs is fluorescence polarization. This approach is based on the reduced rotational motion of a relatively small glycan when it is bound to a relatively large protein compared to the free glycan's rotation in solution. The technique utilizes a glycan containing a fluorophore and a filter to select molecules oriented close to the plane of incident polarized light. Light absorbed by the fluorophore is emitted as fluorescence, but the angle of the emission relative to the incident light is depolarized by rotation of the molecule in solution. In practice, a fluorescently labeled glycan is incubated with increasing concentrations of a GBP and the fluorescence depolarization is measured. In the absence of the GBP, the fluorescently labeled glycan tumbles randomly and the degree of polarization is low. However, if the fluorescently labeled glycan binds to the GBP, its rate of rotation is diminished and the polarization remains high. By this approach, one can measure directly the K_d of the interaction as a function of the GBP concentration.

The advantages of this technique are that (1) the technique is a homogeneous assay and provides direct measurements of the K_d in solution without derivatization of the GBP, (2) it is relatively simple and can provide rapid measurements of many compounds using microtiter plate-based approaches, (3) it utilizes relatively small amounts of glycan, (4) the concentrations of all the molecules are known, (5) it avoids complications of multivalent interactions because the glycans are monovalent and free in solution, and (6) it is amenable to inhibition by competitive agents and can be used to determine relative potency of compounds as inhibitors of GBPs. In the latter approach, a single fluorescently labeled glycan is mixed with the GBP, increasing concentrations of inhibitor glycans (which are not fluorescently labeled) are added, and the inhibition of binding is measured. Because the interactions are simple single-site competition, it is possible to use the concentration that causes 50% inhibition (IC_{50}) to derive the K_d for the inhibitor. Some of the disadvantages are that (1) the technique is limited to small molecules (≤2000 daltons), (2) it requires fluorescence derivatization of the glycan (the fluorophore may alter the properties of the glycan), and (3) preparation of the glycan and chemical derivatization may be tedious and require large amounts of glycans. However, once the fluorescently labeled glycans are generated, there is usually enough for thousands of assays.

Nonequilibrium Methods

ELISA to Measure Specificity and Relative Binding Affinity of Ligands. The conventional enzyme-linked immunosorbent assay (ELISA) has been adapted for studying glycans and GBPs in a variety of formats. Of course, many glycans are antigens and antibodies to them can be analyzed in the conventional ELISA format. Some of the earliest approaches used biotinylated bacterial polysaccharides captured on streptavidin-coated microtiter plates to measure interactions of antibodies to the polysaccharides. In most types of ELISAs used in glycobiology, either an antibody or a GBP of interest is immobilized and the binding of a glycan to the protein is measured, or the reagents are reversed. In either approach, the glycans are conjugated in some way, such as to biotin or to another protein with an attached reporter group (e.g., a fluorescent moiety or an enzyme such as peroxidase).

Competition ELISA-type assays have also recently been developed to probe the binding site of a GBP or an antibody. In this approach, a glycan is coupled to a carrier protein (the target) and its binding to an immobilized GBP is detected directly. Competitive glycans are added to the wells and their competition for the GBP is measured as a function of concentration to obtain an IC_{50}. The major advantages of this approach are that (1) it is relatively easy, (2) it has high-throughput capability and can be used in an automated fashion by robotic handling, (3) it can provide relative K_d if the GBP concentration is varied appropriately over a large range and binding is saturable, and (4) it has the capacity to define the relatively binding activity of a panel of glycoconjugates. The major disadvantages are that (1) it does not provide direct information about affinity constants or other thermodynamic parameters, (2) it can require relatively high amounts of GBP and glycans if used as a general screening array, and (3) it usually requires chemical derivatization of glycans or GBPs.

Glycan Microarrays to Measure Specificity. Glycan microarrays are an extension of both ELISA-type formats and modern DNA and protein microarray technology. In the microarray, glycans are captured, usually covalently, through reaction with *N*-hydroxysuccinimide

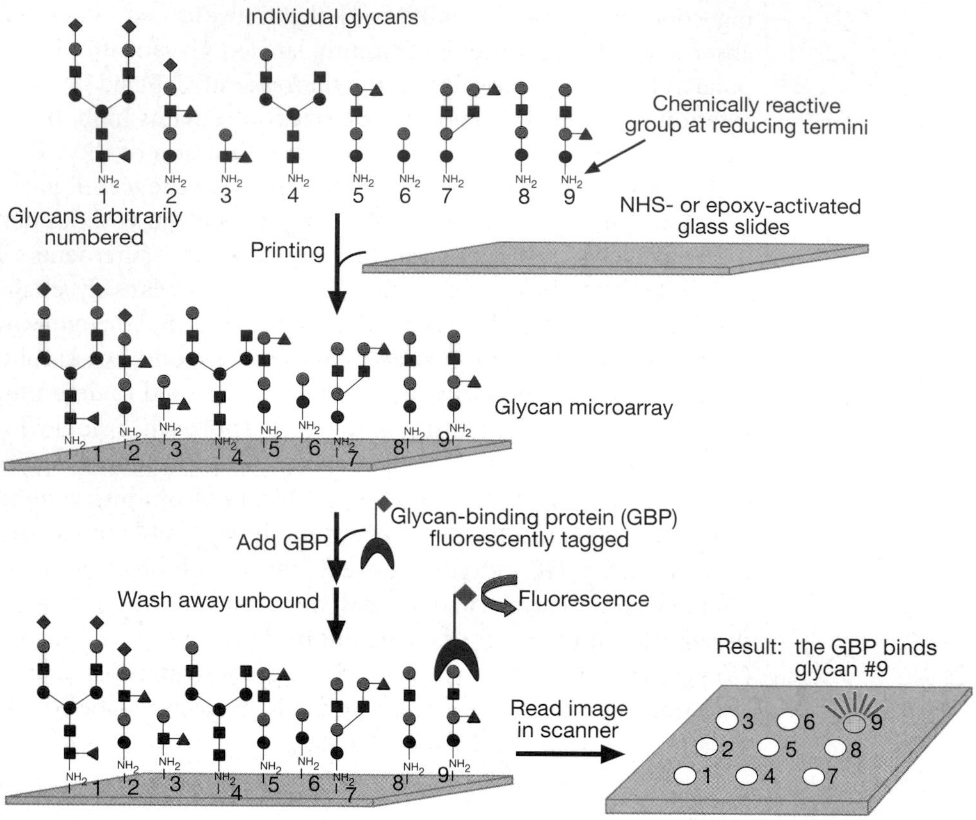

FIGURE 27.6. Preparation of covalent glycan microarrays printed on *N*-hydroxysuccinimide (NHS)- or epoxide-activated glass slides. In this example, the glycans have a free amine at the reducing end and are coupled to the glass slide. After washing, the slide is "interrogated" with a GBP. After washing away the unbound GBP, binding is detected by fluorescence. The GBP may be directly fluorescently labeled or detected using fluorescently labeled secondary reagents. The slides are read in a fluorescence scanner.

(NHS)- or epoxide-containing surfaces on a glass slide. Glycans are printed, much like DNA is printed for DNA microarrays, using contact printers or piezoelectric (noncontact) printing (Figure 27.6). Usually, a few nanoliters of solutions containing glycans in concentrations of 1–100 μM are deposited by a robotic printer on the glass surface in 5–15 μm-diameter spots. Slides are incubated for several hours to allow the chemical reactions to covalently fix the samples on the slides. In most cases, glycans are prepared to contain reactive primary amines at their reducing termini, although other chemical coupling methods are available. These microarrays are then overlaid with a buffer containing the GBP and incubated for several hours to allow maximal binding to occur. The slides are washed to remove unbound GBP and then analyzed. Analyses involve fluorescence detection, which means that either the GBP has to be directly fluorescently labeled or fluorescently tagged antibody to the GBP must be used.

The chief feature of the successful microarrays are the variety of glycans they contain and the clustered and high-density presentation of glycans that promotes binding of even relatively low affinity GBPs. Thus, the density of the ligand should be taken into account when interpreting the results. A publicly available microarray and data on many different GBPs analyzed on the microarray are available from the Consortium for Functional Glycomics. In a typical successful analysis, several glycans may be bound by the GBP and appear as intensely fluorescent spots against the background. For example, in Figure 27.7, the binding of the fluorescently labeled influenza virus (strain A/Oklahoma/323/03) to the

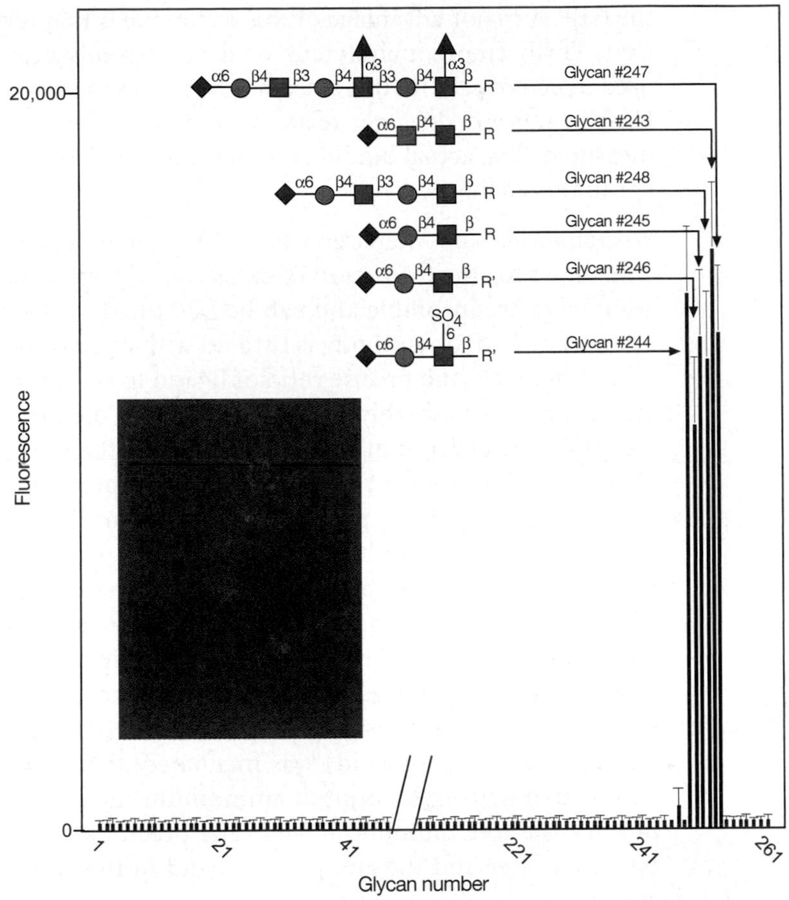

FIGURE 27.7. Binding of a GBP to a covalent glycan microarray. In this example, the fluorescently labeled influenza virus (strain A/Oklahoma/323/03) was labeled with Alexa-488 and applied to the glass slide. The bound virus was detected by fluorescence using a fluorescence scanner. (*Inset*) Type of data obtained for a subarray of the whole slide, where six major fluorescent spots are seen, which represent the six glycans bound well by the GBP. The data are shown as a histogram, where the fluorescent units are plotted versus the glycan number on the microarray. Data are averaged from several replicated patterns of binding in multiple subarrays on the same slide. The structures of the sialylated *N*-acetyllactosamines recognized by this GBP are indicated. (Based on Kumari K., Gulati S., Smith D.F., et al. 2007. *Virology J.* **4:** 42, and available at the website of the Consortium for Functional Glycomics.)

microarray is shown. Six major glycans, each containing terminal α2-6-linked sialic acid, are clearly bound above background, and hundreds of other glycans on the microarray did not bind appreciably. The data are visually imaged on a scanner and then graphically represented. If desired, the GBP can then be tested for its binding to the identified candidates by other methods to define the K_d, such as titration microcalorimetry or fluorescence polarization, as discussed above. The use of microarrays in characterizing GBPs is a central component of functional glycomics (see Chapter 48).

Agglutination. In this approach, one measures the ability of a soluble glycan to block the ability of a multivalent GBP to agglutinate cells expressing glycans recognized by the GBP. The concentration of the soluble glycan that provides 50% inhibition of agglutination is taken as the inhibitory concentration (IC_{50}). Such approaches have been used for many years in studies on lectin agglutination of cells and were useful in elucidating the nature of the human blood group substances. If a sufficiently large panel of soluble glycans is used, then the relative efficacies of each of these can be measured to help define the specificity of the GBP. A major advantage of this technique is that it does not require tagging of the glycans. Furthermore, polystyrene or dextran beads modified with discreet glycans can be used in lieu of cells. In this case, the glycans on the agglutinating particle are better defined. Usually, the IC_{50} does not relate directly to the binding affinity, since inhibition is being measured. The actual binding affinity must be defined by other techniques described earlier in this chapter.

Precipitation. The interaction of a multivalent GBP or antibody with a multivalent ligand allows for the formation of cross-linked complexes in solution. In many cases, these complexes are insoluble and can be identified as precipitates. In this technique, a fixed amount of GBP or antibody is titrated with a glycoprotein or a glycan bound by the protein of interest. At a precise ratio of ligand to receptor, a precipitate is formed. Such precipitation may be highly specific and reflect the affinity constant of the ligand for the receptor. The amount of protein or ligand in the precipitate can be measured directly by chemical means, using assays for glycans or proteins. The technique of precipitation is still useful for studying potentially multivalent ligands, and it has been used recently to demonstrate that each branch of terminally galactosylated complex-type di-, tri-, and tetra-antennary N-glycans is independently recognized by galactose-binding lectins. Another precipitation approach takes advantage of the fact that a complex between a GBP and a glycan can be "salted out" or precipitated by ammonium sulfate. A variation of this approach was used in early studies on the characterization of the hepatocyte Gal/GalNAc receptor (asialoglycoprotein receptor), in which the ligand (in this case ^{125}I-labeled asialoorosomucoid) was incubated with a preparation of receptor. The sample was treated with an amount of ammonium sulfate capable of precipitating the complex, but not the free unbound ligand. The precipitated complex was captured as a precipitate on a filter and the amount of ligand in the complex was directly determined by γ-counting.

Electrophoresis. In this approach, a glycoprotein (or ligand) is mixed with a GBP or antibody and the mixture is electrophoretically separated in polyacrylamide. For glycosaminoglycans, this technique is termed affinity coelectrophoresis (ACE) (see Chapter 16). This method is particularly useful in defining the apparent K_d of the interaction and allows for identification of subpopulations of glycosaminoglycans that differentially interact with the GBP. In another method, termed crossed affinity immunoelectrophoresis, a second step of electrophoresis is conducted in the second perpendicular dimension across an agarose gel that contains precipitating monospecific antibody to the glycoprotein or ligand. The gel is then stained with Coomassie brilliant blue and an immunoelectrophoretogram is obtained. Glycoproteins not interacting with the GBP or antibody have faster mobility than the complex. The amount of glycoprotein or ligand is determined by the area under the curves obtained in the second dimensional analysis. This method is useful for studying glycoforms of proteins and has been particularly valuable in analyzing glycoforms of α1-acid glycoprotein (an acute-phase glycoprotein) in serum and changes in its α1-3-fucosylation.

Expression of cDNAs for Ligands and Receptors. A very indirect approach to studying glycan–protein interactions is to express the cDNA encoding either a glycosyltransferase or a GBP in an animal or bacterial cell. The adhesion of the modified cell to a GBP or antibody is then measured and taken to reflect the binding of the GBP or antibody to the new glycans (neoglycans) on the cell surface. For example, this approach has been highly valuable in studying selectin ligands and helped to lead to the identification of sialyl Lewis X and sialyl Lewis A as important recognition determinants for selectins and the expression cloning of the cDNA encoding the PSGL-1 (see Chapter 31). The expression of selectins or I-type lectins on the cell surface of transfected cells has been helpful in evaluating the specific role of GBPs in cell adhesion under physiological flow conditions (see Chapters 31 and 32).

MODELING GLYCAN–PROTEIN INTERACTIONS

Computational methods have been applied to studies of glycan conformation. Glycans can assume numerous conformations in solution because of flexibility about glycosidic bonds and internal fluctuations in the rings (see Chapter 2). These conformations are dynamic and change rapidly on a nanosecond–microsecond timescale. Thus, the static structures depicted in diagrams represent average structures, which oftentimes may not be representative of the predominant structure that exists in solution. Molecular dynamics simulations can predict these conformational states and determine their relative probability. The relationships between sequence and conformational behavior are highly variable and not all parameters can be fully taken into account (e.g., the role of water, pH, and electrostatic charge). The picture that emerges from these studies is one of highly dynamic molecules, in which torsional rotations about glycosidic bonds are accompanied by internal fluctuations, allowing the molecule to assume stable conformations.

Molecular modeling also can be used to examine glycan conformations when bound to a GBP. These studies usually take into account other experimental data obtained by NMR experiments, X-ray crystallography, and cross-linking studies. Molecular modeling can also be used to predict the relative binding energies for closely related ligands, thus aiding the selection and development of potential antagonists. As computer capacity continues to improve and new algorithms are created to take into account all atoms in the system, modeling will most likely become more commonplace.

GLYCAN–GLYCAN INTERACTIONS

In addition to GBP–glycan interactions, a growing body of evidence suggests that glycan–glycan interactions also may be of importance, especially in the context of cell–cell contacts. Complementary binding between glycans is often considered nonspecific, but biologically relevant high affinities might be achieved by the multivalent interactions made possible by the high density of glycoconjugates on the plasma membrane. One of the best studied biological systems is the sponge, where species-specific cell–cell recognition can occur through multivalent interactions between high-molecular-mass (~200 kD) glycans. The composition of these glycans from different individuals of the same species does not vary, which allows species-specific recognition between cells. In contrast, the composition of glycans on cells from different species varies and prevents their aggregation. Other examples of glycan–glycan interactions include Ca^{++}-dependent cell–cell interactions mediated by glycosphingolipids. For example, interactions between glycosphingolipids expressing the Lewis^x blood group determinant are thought to facilitate compaction of mouse embryonic cells in the 8–32 cell stage and formation of synapses across myelin sheets in the nervous system. Sequestration of glycosphingolipids in microdomains (lipid rafts) of the plasma membrane might provide points of multivalent contact between cells. The molecular forces active between glycans are not well-characterized but presumably involve the same interactions that facilitate protein–glycan interactions, such as van der Waals contacts, hydrogen bonding, and hydrophobic interactions described in previous sections. The requirement for Ca^{++} ions in many of these systems may indicate their role in the clustering of glycans or in conformational changes in the sugar chains to provide optimal interactions. Further studies are needed to understand the importance and frequency of glycan–glycan interactions in various biological processes.

FURTHER READING

Rini J.M. 1995. Lectin structure. *Annu. Rev. Biophys. Biomol. Struct.* **24:** 551–577.

Weis W.I. and Drickamer K. 1996. Structural basis of lectin–carbohydrate recognition. *Annu. Rev. Biochem.* **65:** 441–473.

von der Lieth C., Siebert H., Kozar T., Burchert M., Frank M., Gilleron M., Kaltner H., Kayser G., Tajkhorshid E., Bovin N.V., Vliegenthart J.F. and Gabius H. 1998. Lectin ligands: New insights into their conformations and their dynamic behavior and the discovery of conformer selection by lectins. *Acta Anat.* **161:** 91–109.

Woods R.J. 1998. Computational carbohydrate chemistry: What theoretical methods can tell us. *Glycoconj. J.* **15:** 209–216.

Fred Brewer C. 2002. Binding and cross-linking properties of galectins. *Biochim. Biophys. Acta* **1572:** 255–262.

Duverger E., Frison N., Roche A.C., and Monsigny M. 2003. Carbohydrate–lectin interactions assessed by surface plasmon resonance. *Biochimie* **85:** 167–179.

Hirabayashi J. 2003. Oligosaccharide microarrays for glycomics. *Trends Biotechnol.* **21:** 141–143.

Sorme P., Kahl-Knutson B., Wellmar U., Nilsson U.J., and Leffler H. 2003. Fluorescence polarization to study galectin-ligand interactions. *Methods Enzymol* **362:** 504–512.

Bucior I. and Burger M.M. 2004. Carbohydrate–carbohydrate interactions in cell recognition. *Curr. Opin. Struct. Biol.* **14:** 631–637.

Dam T.K. and Brewer C.F. 2004. Multivalent protein–carbohydrate interactions: Isothermal titration microcalorimetry studies. *Methods Enzymol.* **379:** 107–128.

Feizi T. and Chai W. 2004. Oligosaccharide microarrays to decipher the glyco code. *Nat. Rev. Mol. Cell. Biol.* **5:** 582–588.

Neumann D., Lehr C.M., Lenhof H.P., and Kohlbacher O. 2004. Computational modeling of the sugar–lectin interaction. *Adv. Drug Deliv. Rev.* **56:** 437–457.

Ratner D.M., Adams E.W., Su J., O'Keefe B.R., Mrksich M., and Seeberger P.H. 2004. Probing protein–carbohydrate interactions with microarrays of synthetic oligosaccharides. *Chembiochem.* **5:** 379–382.

Homans S.W. 2005. Probing the binding entropy of ligand–protein interactions by NMR. *Chembiochem.* **6:** 1585–1591.

Nakamura-Tsuruta S., Uchiyama N., and Hirabayashi J. 2006. High-throughput analysis of lectin–oligosaccharide interactions by automated frontal affinity chromatography. *Methods Enzymol.* **415:** 311–325.

Paulson J.C., Blixt O., and Collins B.E. 2006. Sweet spots in functional glycomics. *Nat. Chem. Biol.* **2:** 238–248.

Smith D.F. and Cummings R.D. 2008. Deciphering lectin ligands through glycan arrays. In *Animal lectins: A functional view* (ed. G.R. Vasta and H. Ahmed). CRC Press, Boca Raton, Florida. (In press.)

CHAPTER 28

R-type Lectins

Richard D. Cummings and Marilynn E. Etzler

THE R-TYPE LECTINS ARE MEMBERS OF A SUPERFAMILY of proteins, all of which contain a carbohydrate-recognition domain (CRD) that is structurally similar to the CRD in ricin. Ricin was the first lectin discovered and it is the prototypical lectin in this category. R-type lectins are present in plants, animals, and bacteria, and the plant lectins often contain a separate subunit that is a potent toxin. The structure–function relationships of this group of proteins are discussed in this chapter.

BACKGROUND

In 1888, Peter Hermann Stillmark reported that seed extracts of the poisonous plant *Ricinus communis* (castor bean) contain a toxin that can agglutinate erythrocytes. He named this agglutinin "ricin." Other agglutinins were soon discovered in the seed extracts of other species of poisonous plants, and by the middle of the 20th century, these proteins had become important members of the general class of glycan-binding proteins known as "lectins." As structural studies on ricin proceeded, it was found that the CRD of this lectin is closely related in sequence and three-dimensional structure to the CRDs of a number of other plant lectins. Some of these structural similarities were also noted in a variety of ani-

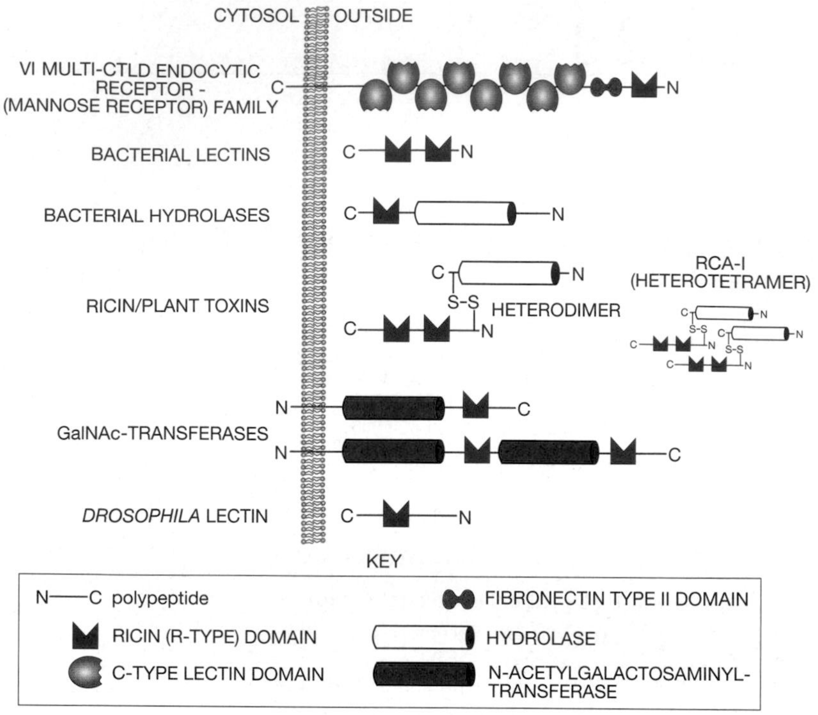

FIGURE 28.1. The R-type lectin superfamily. Different groups within the family are indicated with the domain structures shown.

mal and bacterial glycan-binding proteins discussed in this chapter, causing all proteins containing this ricin-type CRD to be classified as R-type lectins (Figure 28.1).

RICINUS COMMUNIS LECTINS

General Properties

Two different lectins have been purified from *R. communis* seeds, and in the original nomenclature they were termed RCA-I and RCA-II. RCA-I is an agglutinin but a very weak toxin. RCA-II is commonly called ricin, and it is both an agglutinin and a very potent toxin. The designation RCA-II has now been dropped, but the original name for the agglutinin RCA-I has been retained. The molecular mass of RCA-I is approximately 120 kD and that of ricin is approximately 60 kD.

Ricin is synthesized as a single prepropolypeptide with a 35-amino-acid amino-terminal signal sequence, followed by the A-chain domain, a 12-amino-acid linker region, and the B chain. Proteolysis results in cleavage between the A and B chains. Mature ricin contains four intrachain disulfide bonds and a single disulfide bond links the A chain to the B chain. The A chain contains the catalytic activity responsible for the toxicity and the B chain has the glycan-binding activity. The mature A chain has 267 amino acids and the B chain has 262 amino acids. Each B subunit is a product of gene duplications and has two CRDs, each of which is composed of an ancient 40-amino-acid galactose-binding polypeptide region. RCA-I is a tetramer with two ricin heterodimer-like proteins that are noncovalently associated. Each heterodimer in RCA-I contains an A-chain disulfide linked to the galactose-binding B chain. The sequences of the A chain of ricin and the A chain of RCA-I differ in 18 of 267 residues and are 93% identical, whereas the B chains differ in 41 of 262 residues and are 84% identical. All of the subunits are N-glycosylated and usually express oligomannose-type N-glycans. Because ricin is a multimeric glycoprotein, recombinant forms produced in *Escherichia coli* (such as ricin B chain) are usually poorly active, and eukaryotic expression systems must be used. The genome of *R. communis* also encodes several other proteins that have high homology with ricin, and some of these lectins have been designated ricin-A, -B, -C, -D, and -E.

The B chain of ricin has two binding sites for sugars, which are about 35 Å apart. Ricin binds to β-linked galactose and *N*-acetylgalactosamine, whereas RCA-I prefers β-linked galactose. Indeed, these lectins are often separated from one another by differential elution from galactose-based affinity resins in which ricin is eluted first with *N*-acetylgalactosamine and then RCA-I is eluted with galactose. The affinities of these lectins for monosaccharides are quite low (K_d in the range 10^{-3} to 10^{-4} M). In contrast, the binding to cells is of much higher affinity (K_d in the range 10^{-7} to 10^{-8} M), owing to both increased avidity from the multivalency and enhanced binding to glycans terminating in the sequence Galβ1-4GlcNAc-R. Such sequences represent higher-affinity determinants for lectin binding. In general, ricin and RCA-I both preferentially bind to glycans containing nonreducing terminal Galβ1-4GlcNAc-R or GalNAcβ1-4GlcNAc-R, although they will bind weakly to Galβ1-3GlcNAc-R. Neither ricin nor RCA-I binds well to glycoconjugates containing nonreducing terminal α-linked Gal residues. RCA-I is commonly used for glycan isolation and characterization because it is safer than ricin and has higher avidity because of its tetrameric nature. Ricin is often used as a toxin in cell selection for glycosylation mutants, and the A chain of ricin is often used in chimeric proteins as a toxin for specific-cell killing.

R-type Domain

The R-type domain contained in these proteins is the CRD; it is also termed a carbohydrate-binding module (CBM) and has been placed in the CBM13 family in the CAZy database (carbohydrate-active enzymes database) (see Chapter 7). The R-type domain is an ancient type of protein fold that is found in many glycosyltransferases as well as in bacterial and fungal hydrolases. Interestingly, the R-type CRD is the only one conserved between animal and bacterial lectins.

The crystal structure of ricin (Figure 28.2) shows that the A chain has eight α-helices and eight β-strands and it is the catalytic subunit, as discussed above. The B chain, which contains R-type lectin domains, has two tandem CRDs that are about 35 Å apart and have a shape resembling a barbell, with one binding domain at each end. Each R-type domain has a three-lobed organization that is a β-trefoil structure (from the Latin *trifolium* meaning "three-leaved plant"). The β-trefoil structure probably arose evolutionarily through gene fusion events linking a 42-amino-acid peptide subdomain that has galactose-binding activity. The three lobes are termed α, β, and γ and are arranged around a threefold axis. Conceivably, each lobe could be an independent binding site, but in most R-type lectins only one or two of these lobes retain the conserved amino acids required for sugar binding. Sugar binding is relatively shallow in these loops and arises from aromatic amino acid stacking against the Gal/GalNAc residues and from hydrogen bonding between amino acids and hydroxyl groups of the sugar ligands. A characteristic feature of these loops in the R-type domain in ricin is the presence of $(QxW)_3$ repeats (where x is any amino acid), which are found in many, but not all, R-type family members.

Toxicity of Ricin

Because of its incredible potency as a toxin and its ease of production, ricin has been in the news over the years as a possible lethal agent. The toxicity of ingested ricin is severe and symptoms are initiated after a 2–24-hour latent period following a fatal ingestion. The

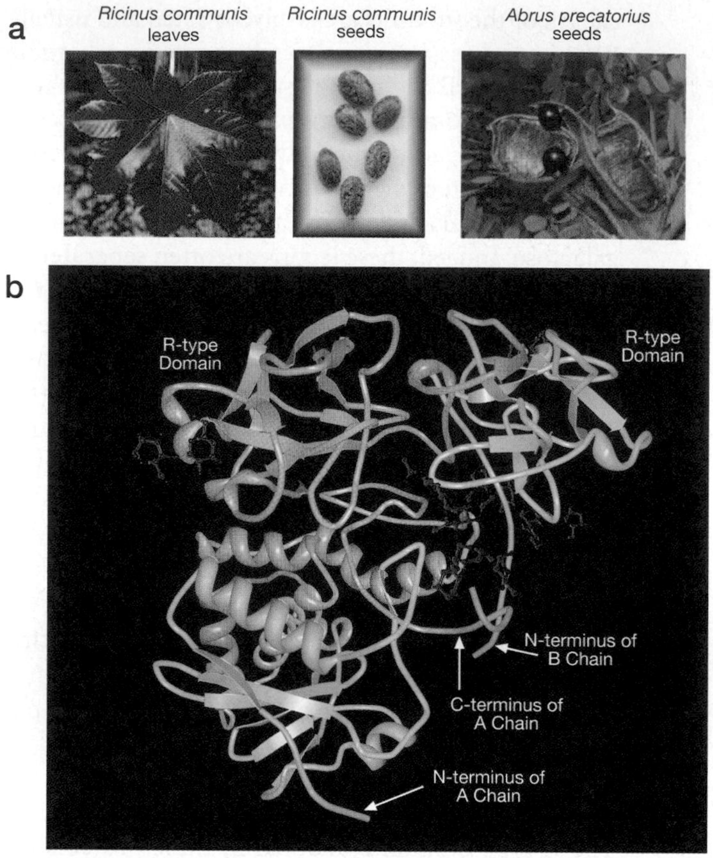

FIGURE 28.2. Ricin and abrin. (*a*) Photographs showing the leaves and seeds of *Ricinus communis* and the seeds of *Abrus precatorius*. *R. communis* is commonly called the castor-oil plant and the seeds are castor beans. The *A. precatorius* seeds are sometimes called rosary pea, coral bead plant, or jequirity bean. (*b*) The crystal structure of ricin refined to 2.5 Å. (*b*, Rutenber E. et al. 1991. *Proteins* **10:** 240–250. Image from PDB file 2aai.)

lethal dose (LD_{50}) of ricin may be as low as 3–5 μg/kg per kilogram body weight, depending on the mode of exposure. Ricin is a type II ribosome-inactivating protein (RIP-II). Although one might predict that RCA-I also would be highly toxic, it has weak activity compared to ricin because it lacks a separate A chain. In contrast, type I ribosome-inactivating proteins (RIP-I), which lack a B subunit, are not lectins and are much less toxic than ricin because of poor ability to enter cells. Examples of RIP-I proteins are saporin, gelonin, and momordin-I.

Ricin binds to cells via interactions with β-linked Gal/GalNAc-containing glycans on the cell surface and is imported into endosomes (Figure 28.3). From there, the protein moves by retrograde trafficking to the endoplasmic reticulum (ER) via the *trans*-Golgi network and Golgi apparatus. In the ER, the A and B chains separate after reduction of the disulfide bond, perhaps through the action of the ER enzyme, protein disulfide isomerase. Some fraction of the free A chain, which may be partly denatured in the ER, may then escape from the ER by retrotranslocation through the Sec61 translocon and thereby enter the cytoplasm. This retrotranslocation may be through the quality control system in the ER known as ER-associated degradation (ERAD), which is involved in removing misfolded proteins from the ER (see Chapter 36). Some of the A chain may be ubiquitinated and degraded in the proteasome, but the A chain may not be an efficient substrate for ubiquitination because it has few lysine residues.

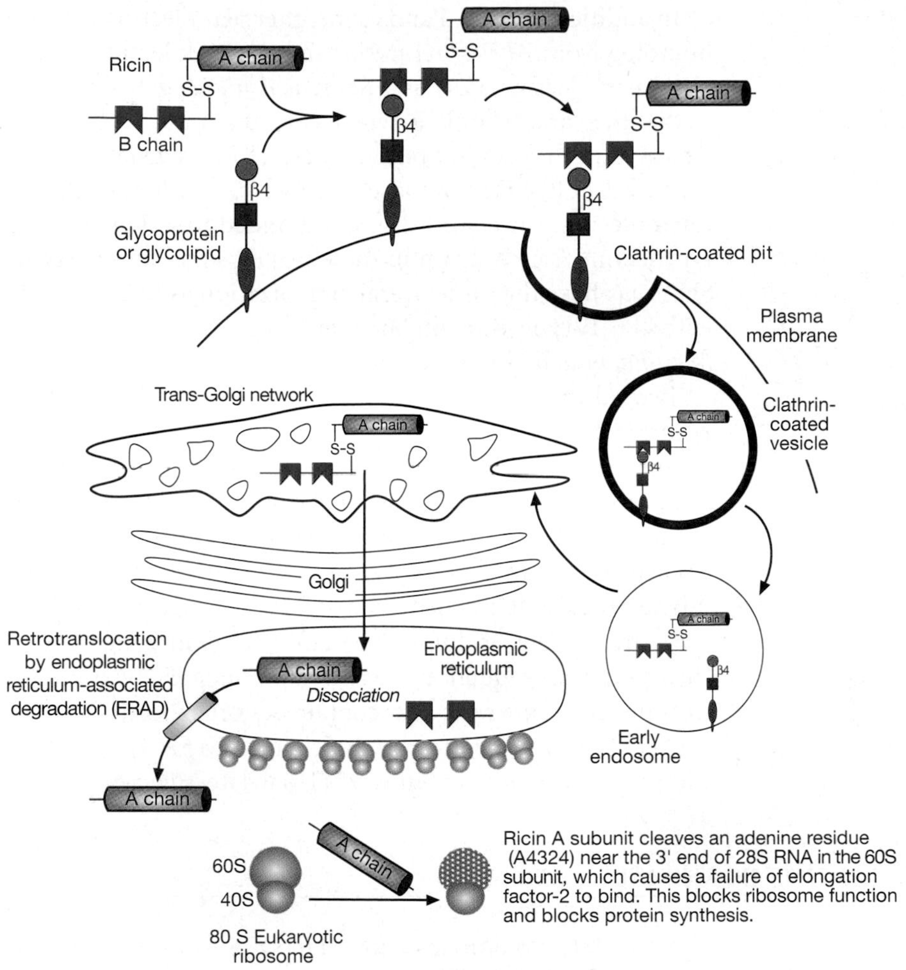

FIGURE 28.3. Pathway of ricin uptake by cells and the mechanism by which the toxic activity of the A chain in the cytoplasm results in cell death.

In the cytoplasm, just a few molecules of ricin A chain appear to be sufficient to kill cells. The A chain is an enzyme, RNA N-glycosidase, that cleaves one adenine residue (A4324) in the exposed GAGA tetraloop of the 28S RNA in the eukaryotic 60S ribosomal subunit. This deletion results in a loss of binding of elongation factor-2 and the inability of the ribosome to promote protein synthesis.

OTHER R-TYPE PLANT LECTINS

The toxins abrin, modeccin from *Adenia digitata*, *Viscum album* agglutinin (VAA or mistletoe lectin), and volkensin also have R-type domains, belong to the RIP-II class, and kill cells in a manner similar to ricin. There are other R-type plant lectins in the RIP-II class that are not toxic, and these include several proteins from the genus *Sambucus* (elderberry), such as nigrin-b, sieboldin-b, ebulin-f, and ebulin-r. All of the B subunits of these proteins appear to bind Gal/GalNAc, but they may have some differences in affinity and may recognize different Gal/GalNAc-containing glycoconjugates. Cell lines selected for resistance to killing by modeccin are not resistant to abrin and ricin, and vice versa. The glycan-binding specificity of these lectins should be explored more fully in the future using glycan microarrays and related screening approaches.

In addition to RCA-I and ricin, other plant lectins with R-type domains include the ricin homolog from *Abrus precatorius* and the bark lectins from the elderberry plant, *Sambucus sieboldiana* lectin (SSA) and *Sambucus nigra* agglutinin (SNA). SSA and SNA are unusual in that they are the only R-type lectins that bind well to α2-6-linked sialic acid–containing ligands and they do not bind to α2-3-linked sialylated ligands. SSA and SNA are heterotetramers (~140 kD) composed of two heterodimers each containing an A chain (which resembles ricin A chain) disulfide-bonded to a B chain (which binds glycans and is an R-type lectin). The A chain in these proteins has very weak RIP-II activity in vitro. SSA and SNA may have the same overall organization as RCA-I (see Figure 28.1). Other plant lectins with the R-type domain include lectins from the seeds of *Trichosanthes dioica* and *T. anguina*, which also appear to bind β-linked galactose.

Two other plant lectins that can bind to sialylated glycans are from the leguminous tree *Maackia amurensis*. Although these lectins are discussed here, they do not show a classical R-type domain but instead have an L-type lectin domain (see Chapter 29). *M. amurensis* leukoagglutinin (MAL) (also referred to commercially as MAA-II) recognizes Siaα2-3Galβ1-4GlcNAcβ-Man-R and does not bind to isomers containing sialic acid in α2-6-linkage. MAL can also bind to sulfated (rather than sialylated) glycans with the sequence sulfate-3Galβ1-4GlcNAcβ-Man-R. Interestingly, the specific sialylated sequences bound by MAL are lacking on human erythrocytes, and hence MAL is a very poor hemagglutinin. However, these sequences are abundant on leukocytes and so the lectin is a strong leukoagglutinin. *M. amurensis* also contains another lectin termed *M. amurensis* hemagglutinin (MAH) (also referred to commercially as MAA-I), which shows less specificity than MAL and probably generally binds α2-3-sialylated glycans.

R-TYPE LECTINS IN ANIMALS

The R-type lectin domain is found in several animal lectins, including the mannose receptor (MR) family, discussed in detail below, and in some invertebrate lectins. EW29 is a galactose-binding lectin from the annelid (earthworm) *Lumbricus terrestris*. EW29 has two homologous domains, an amino-terminal domain and a carboxy-terminal domain, both of which are R-type domains with the three α, β, and γ lobes and both contain the triple-repeated QxW motif. The R-type domain is also found in pierisin-1, which is a cytotoxic protein from the cabbage butterfly *Pieris rapae*, and in the homologous protein pierisin-2, from *Pieris brassicae*. Pierisin-1 is a 98-kD protein (comprising 850 amino acids) that exhibits extreme toxicity to animal cells in culture. It has an amino-terminal domain with ADP-ribosylation activity, as found in some bacterial toxins such as cholera and pertussis toxins, and a carboxy-terminal region with four R-type domains and QxW repeats. Through this R-type domain, pierisin-1 binds to glycosphingolipids, including the major ligands globotriaosylceramide (Gb3) and globotetraosylceramide (Gb4). It is interesting that both the catalytic domain and the lectin domains are in the same polypeptide. The mechanism by which the catalytic domain enters the cytoplasm is not well understood, but it may involve proteolysis to separate the domains. Tandem R-type motifs are found in some other R-type family members, including ricin, but the presence of four such motifs is unique, thus far, to pierisin-1.

Some proteins with the R-type lectin domain are also enzymes and these are found in both animals and microbes. For example, *Limulus* horseshoe crab coagulation factor G has a central R-type lectin domain, which is flanked at the amino terminus by a xylanase Z-like domain and at the carboxyl terminus by a glucanase-like domain. This protein also has a subunit that is a serine protease. Some examples of microbial proteins with R-type domains include *Streptomyces lividans* xylanase 10A, which is discussed in more detail below, and the mosquitocidal toxin (MTX) from *Bacillus sphaericus*. MTX is a 97-kD ADP-ribosyltransferase, which is activated by proteolytic cleavage releasing the active 27-kD enzyme and a 70-kD carboxy-terminal fragment that contains the R-type domain.

The Mannose Receptor Family

There are four known members of the MR family in humans, all of which contain an R-type lectin domain, and on the basis of a survey of the human genome, no other family members are predicted. The MR family includes the MR, the phospholipase A2 (PLA2) receptor, DEC-205/MR6-gp200, and Endo180/urokinase plasminogen activator receptor-associated protein (Figure 28.4). All of these proteins are large type I transmembrane glycoproteins and they contain a single fibronectin type II domain, multiple C-type lectin domains (CTLDs), and an amino-terminal cysteine-rich domain. Some of the CTLDs in the MR and Endo180 bind glycans, but other CTLDs in these proteins do not bind glycans and have other functions. The CTLDs in DEC-205 have not been shown to bind glycans. Each protein is a recycling plasma membrane receptor with a cytoplasmic domain that mediates clathrin-dependent endocytosis and uptake of extracellular glycan-containing ligands (see Chapter 31). All of the MR family members except DEC-205 recycle back to the cell surface from early endosomes, but DEC-205 recycles from late endosomes. However, each receptor has evolved to have distinct functions and distributions. These receptors are unusual among animal lectins in that they can bind ligands in either a "*cis*" or "*trans*" fashion, which means they can bind to cell-surface glycoconjugates on the same cell or to those on other cells and to soluble ligands.

The Mannose Receptor

The MR (CD206) has important roles in the innate and adaptive immune systems. It is expressed at high levels on hepatic endothelial cells and Kupffer cells as well as on many other endothelial and epithelial cells, macrophages, and immature dendritic cells. The MR is part of the innate immune system and it facilitates the phagocytosis of mannose-rich

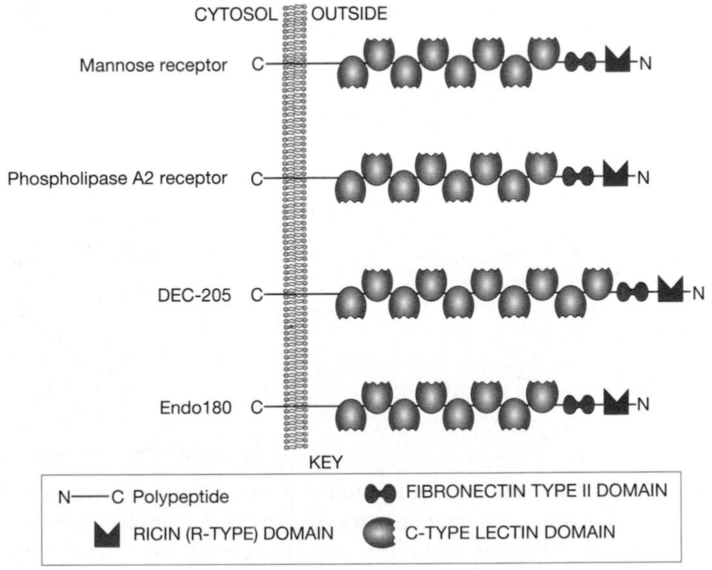

FIGURE 28.4. Members of the mannose receptor family of R-type lectins. These proteins are also in the C-type lectin family and represent a unique group of lectins with more than one type of lectin domain. However, although the mannose receptor and Endo180 can bind to sugar ligands, it is not clear whether any of the R- or C-type lectin domains in either phospholipase A2 receptor or DEC-205 can bind to sugars.

pathogens. It also assists leukocytes in responding appropriately to antigens by promoting trafficking to the germinal center and is also involved in antigen presentation. The receptor was originally discovered in the 1980s in rabbit alveolar macrophages as a large membrane protein (175 kD) that bound mannose-containing ligands as well as pituitary hormones such as lutropin and thyrotropin, which have 4-O-sulfated *N*-acetylgalactosamine residues on N-glycans. However, although the MR is a C-type lectin with multiple C-type lectin domains (see Chapter 31), these domains were not required for binding to the sulfated hormones. A 139-residue amino-terminal domain in the mature protein was originally recognized as a cysteine-rich domain, but its importance in binding glycans was not realized until the late 1990s when the sulfated pituitary hormones were found to be cleared from the circulation by binding to the cysteine-rich region. When the cysteine-rich domain was crystallized in a complex with 4-O-sulfated GalNAc, this domain was found to have a β-trefoil structure similar to that reported for ricin B, soybean trypsin inhibitor, fibroblast growth factors, and interleukin-1 (Figure 28.5). Interestingly, the MR and other members of the MR family are among the few mammalian glycan-binding proteins that have two separate lectin motifs (C-type and R-type) in the same molecule. This group is also unusual in that it is the only known lectin group in mammals with more than two C-type lectin domains in the same molecule. Only CTLDs 4 and 5 of the MR have been shown to bind glycans in a Ca^{++}-dependent manner and to bind mannose, *N*-acetylglucosamine, and fucose.

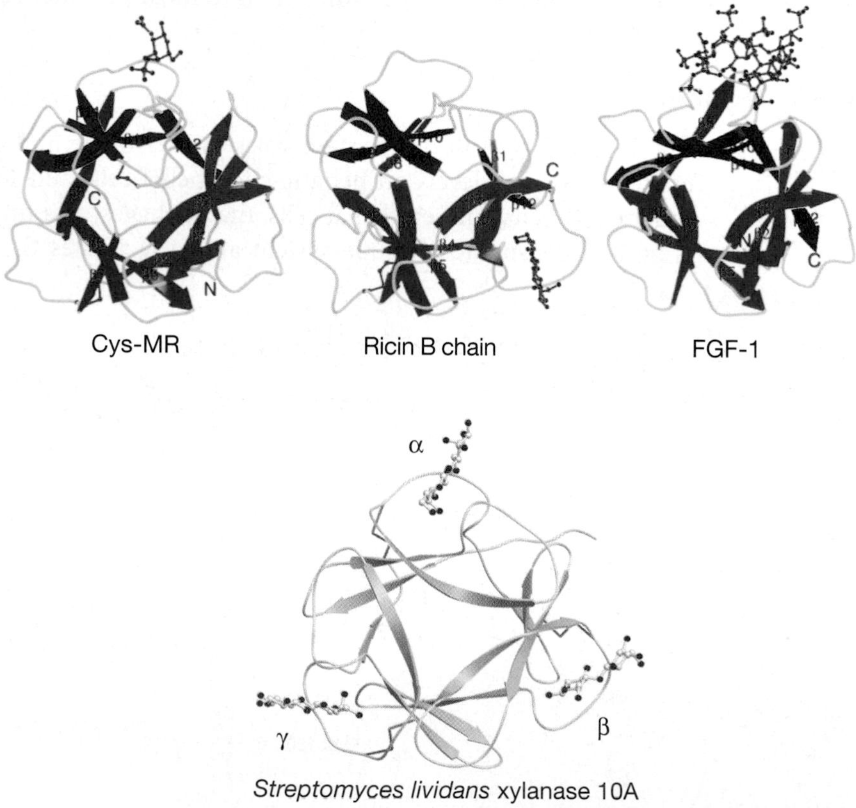

FIGURE 28.5. Structures of the β-trefoil R-type domains in different proteins. (*Top*) Cysteine-rich R-type domain of the mannose receptor (MR) in complex with 4-O-sulfated GalNAc; ricin B chain in complex with galactose; and acidic fibroblast growth factor (FGF-1) in complex with sulfated heparan decasaccharide. (Redrawn, with permission, from Liu Y. et al. 2000. *J. Exp. Med.* **191:** 1105–1116, ©2000, The Rockefeller University Press.) (*Bottom*) *Streptomyces lividans* endo-β1-4xylanase 10A in complex with lactose. (Redrawn, with permission, from Notenboom V. et al. 2002. *Biochemistry* **41:** 4246–4254, ©2002 American Chemical Society.)

The cysteine-rich R-type domain of the MR binds other sulfated glycans and also N-glycans on pituitary glycoprotein hormones containing 4-SO_4-GalNAcβ1-4GlcNAcβ1-2Manα1-R. Other ligands for the MR include chondroitin-4-sulfate proteoglycans on leukocytes that also contain 4-SO_4-GalNAcβ1-R residues and perhaps sulfated glycans, such as those containing 3-O-sulfated galactose, 3-O-sulfated Le^x, and 3-O-sulfated Le^a. The cysteine-rich R-type domain in the MR binds with relatively low affinity to sulfated monosaccharides (K_d in the range of 10^{-3}–10^{-4} M), but oligomeric forms of the protein probably display much higher avidity for glycoproteins with multiple sulfated glycans. Interestingly, although the cysteine-rich R-type domain binds sulfated glycans, the C-type lectin domains bind unsulfated glycans and may also play a role in glycoprotein homeostasis and clearance.

The MR is also unusual in that it is the only member of the MR family that can function both in clathrin-dependent endocytosis and in the phagocytosis of nonopsinized microbes and large ligands. The MR can bind many different microorganisms, including *Candida albicans*, *Pneumocystis carinii*, *Leishmania donovani*, *Mycobacterium tuberculosis*, and *Klebsiella pneumoniae*. The MR may distinguish self from nonself glycans via its multivalent attachment to the mannose-rich glycoconjugates on these organisms, which are presented in unique clusters. The MR also functions in adaptive immunity through its ability to deliver antigens to major histocompatibility (MHC) class II compartments and through its cleavage and release as a soluble protein into blood. Interestingly, it is speculated that the membrane-bound MR may bind antigens through its C-type lectin domains and then, following proteolytic cleavage, the soluble MR bound to its cargo may move to germinal centers where it may bind via its R-type domain to macrophages and dendritic cells expressing ligands such as sialoadhesion or CD45 that may contain sulfated glycans. The function of the R-type domain in the MR may be unique because DEC-205 does not bind lutropin and sequence comparisons predict that neither the phospholipase A2 (PLA2) receptor nor Endo180 will bind sulfated glycans.

The Phospholipase A2 Receptor

The PLA2 receptor was discovered as a receptor for phospholipase A2 neurotoxins in snake venoms and was referred to as the M-type PLA2 receptor to distinguish it from the neuronal or N-type PLA2 receptor. The cDNA encoding this 180-kD glycoprotein has homology with the MR (with 29% identity). It can occur as both a long form that is a type I transmembrane glycoprotein with a domain structure like the MR and a shorter form that is secreted. The membrane-bound form can function as an endocytic receptor to internalize phospholipase A2 ligands. In animal cells, secreted phospholipase A2s represent a large family of enzymes that are important in the degradation of phospholipids and the release of arachidonic acid, which is the precursor for prostaglandins, leukotrienes, thromboxanes, and prostacyclins. The murine PLA2 receptor binds to sPLA2-X enzyme, but the human PLA2 receptor may bind to several different phospholipase A2 isozymes. Although early studies suggested that the PLA2 receptor might be involved in clearance of phospholipase A2s, murine knockouts for the PLA2 receptor had the unusual phenotype of resistance to endotoxic shock. This suggested that the PLA2 receptor might be important in regulating production of proinflammatory cytokines by soluble phospholipase A2s. Thus, the PLA2 receptor might function in signal transduction mediated by phospholipase A2 binding. Some of the C-type lectin domains in the PLA2 receptor function by binding phospholipase A2 ligands, rather than binding directly to glycan ligands. The fibronectin type II domain binds to collagen, which is a feature shared by this domain in other MR family members, except for possibly DEC-205.

DEC-205

DEC-205 is a 205-kD member of the MR family that is expressed by dermal dendritic cells and, at a lower level, by epidermal Langerhans cells. It is now designated as CD205. It is also expressed on some epithelial cells, on bone marrow stroma, and by endothelial cells. It was identified originally in the 1980s by a monoclonal antibody NLDC-145, which recognized a surface antigen on Langerhans cells. Subsequent cloning of the gene revealed its relationship to the MR and its unique characteristic of having ten tandem C-type lectin domains, rather than the eight found in other MR family members. DEC-205 is important in the recognition and internalization of antigens for presentation to T cells. Upon endocytosis, DEC-205 internalizes to late endosomes/lysosomes and recycles to the surface. Expression of DEC-205 is enhanced in both types of cells upon cell maturation induced by inflammatory stimuli. None of the ten CTLDs in DEC-205 have the conserved amino acids known to be important in carbohydrate binding, and thus far there is no evidence that these domains bind glycans. There is also no evidence that the cysteine-rich R-type domain at the amino terminus binds to glycan ligands.

Endo180

The protein now known as Endo180 was discovered independently by several groups. It was found to be part of a trimolecular cell-surface complex with urokinase plasminogen activator (uPA) and its receptor (uPAR) and was termed the UPAR-associated protein or UPARAP. It was also discovered as a novel antigen on macrophages and human fibroblasts. Like the MR, Endo180 is expressed on macrophages, but Endo180 is also expressed on fibroblasts and chondrocytes, some endothelial cells, and tissues undergoing ossification. The cysteine-rich R-type domain at the amino terminus of Endo180 has been shown to be unable to bind sulfated glycans. The function of this domain is unknown. Endo180 is predicted to be important in remodeling of the extracellular matrix. It binds to the carboxy-terminal region of type I collagen, and collagens type II, IV, and V, through its FN II domain, and binds to uPA and uPAR, but whether glycan recognition is important for these interactions is not known. Endo180 (CD280) binds in a Ca^{++}-dependent manner to *N*-acetylglucosamine, mannose, and fucose through CTLD2, and this is the only domain that contains all of the conserved amino acids for binding both Ca^{++} and sugar. Endo180 is an endocytic receptor for these soluble ligands and may be important in matrix turnover. In addition, cells expressing Endo180 demonstrate enhanced adhesion to matrixes, suggesting that Endo180 may also be important in cell adhesion.

UDP-GalNAc:Polypeptide α-*N*-Acetylgalactosaminyltransferases

Mucin-type O-glycans have the common core structure of GalNAcα1-Ser/Thr (Figure 28.6), which may be further modified by addition of galactose or *N*-acetylglucosamine residues (see Chapters 9 and 13). The biosynthesis of the core structure is dependent on a family of UDP-GalNAc:polypeptide α-*N*-acetylgalactosaminyltransferases (ppGalNAcTs) that function in the Golgi apparatus. Such enzymes are found in all animals. More than a dozen such enzymes have been characterized, and animal genomes encode many potential ppGalNAcTs, including 24 potential ppGalNAcTs in humans and 14 in *Drosophila* (see Chapter 9). These ppGalNAcTs may be separated into two general classes: those that can transfer *N*-acetylgalactosamine from UDP-GalNAc to unmodified polypeptide acceptors, and those that prefer acceptor glycopeptides containing GalNAc-Ser/Thr (i.e., a set predicted to have a CRD). Cloning of the genes encoding these enzymes revealed that they are large proteins and have

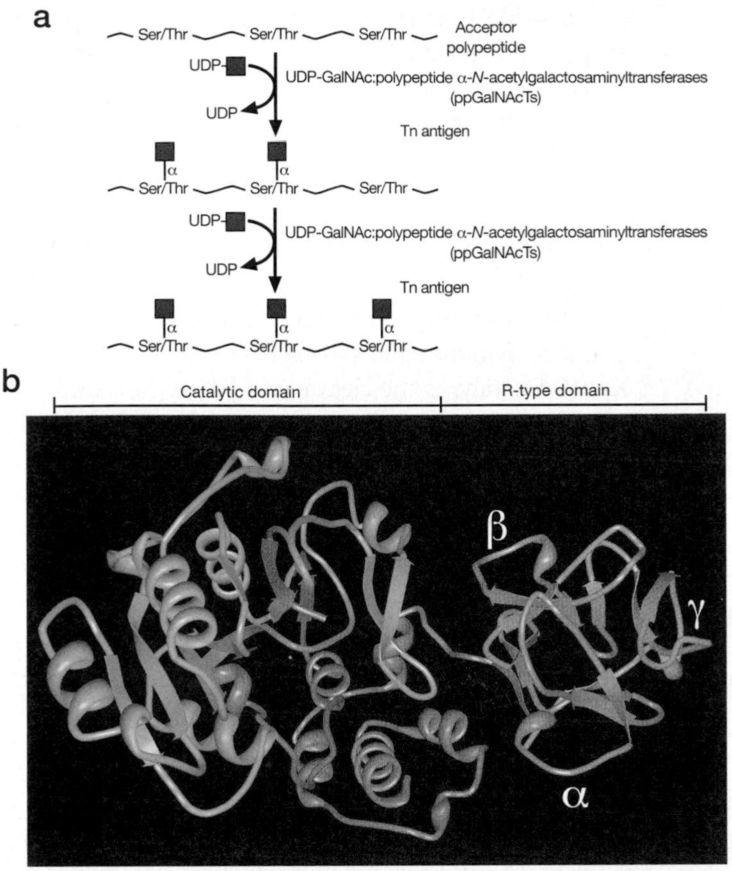

FIGURE 28.6. Structure and function of UDP-GalNAc:polypeptide α-*N*-acetylgalactosaminyltransferases (ppGalNAcTs). (*a*) The activity of the ppGalNAcT is shown using an acceptor peptide with Ser and Thr residues and UDP-GalNAc as the donor. Some of the ppGalNAcTs may also prefer to act on the product of this reaction and use peptides with attached *N*-acetylgalactosamine as the acceptor. (*b*) The crystal structure of murine ppGalNAcT1 (mT1). The catalytic domain is shown on the *left* and the R-type lectin domain on the *right*. The β-trefoil structure and the three-lobe repeating α, β, and γ loops are indicated. The carbohydrate chain attached to the R-type domain is partly visible in the crystal. (*b*, Reprinted, with permission, from Fritz T.A. et al. 2004. *Proc. Natl. Acad. Sci.* **101:** 15307–15312, ©National Academy of Sciences, U.S.A. PDB 1xh.b.)

a unique multidomain structure. The enzymes are all type II transmembrane proteins and have a carboxy-terminal R-type domain of about 130 amino acid residues and an amino-terminal catalytic domain. The R-type domain has the QxW repeat, and recent evidence shows that these domains in the ppGalNAcTs function as lectins during the catalytic process.

The first crystal structure for this family was determined for the murine ppGalNAcT, termed mT1. This enzyme does not require the R-type domain for catalysis with either glycopeptide (GalNAc-peptide) or peptide acceptors. In contrast, in human hT2, the lectin domain is important in catalysis with GalNAc-peptide acceptors, but not peptide acceptors, and the lectin domains of both hT2 and hT4 can directly bind *N*-acetylgalactosamine. In mT1, the R-type domain has the β-trefoil structure and the three-lobe repeating α, β, and γ loops. Mutations in the repeats can abolish binding to GalNAc-peptide acceptors. The ppGalNAcTs are the only known glycosyltransferases that have a lectin and catalytic domain conjoined. The available models suggest that each of the ppGalNAcTs has a somewhat different peptide or glycopeptide acceptor specificity, allowing the assortment of enzymes to be highly efficient at adding *N*-acetylgalactosamine to a tremendous variety of polypeptide substrates, including long mucin polypeptides of thousands of amino acids, and to specific single serine or threonine residues on membrane and secreted glycoproteins. The presence of the two domains in these enzymes and the multiple family members may promote an efficient processive activity and association of these enzymes with acceptor proteins.

MICROBIAL R-TYPE PROTEINS: *STREPTOMYCES LIVIDANS* ENDO-β1-4XYLANASE 10A

A feature of many microbial glycosidases is the presence of both a catalytic domain and a CBM. *S. lividans* endo-β1-4xylanase 10A (Xyn10A) is a good example of such an enzyme. Xyn10A catalyzes the cleavage of β1-4xylans and can bind to xylan and a variety of small soluble sugars, including galactose, lactose, and xylo- and arabino-oligosaccharides. The catalytic domain is at the amino terminus and the carboxyl terminus has an R-type β-trefoil motif. As mentioned above, the R-type domain CBM represents the CBM13 family in the CAZy database. In Xyn10A, all of the original β-trefoil sugar-binding motifs are retained, along with the conserved disulfide bridges, and evidence suggests that each of the three potential sugar-binding sites in the β-trefoil structure interact with sugars and each site may span up to four xylose residues. The binding to monosaccharides is very weak (K_d in the range of 10^{-2} to 10^{-3} M), but multivalent binding to polysaccharides can be of very high affinity.

FURTHER READING

Stirpe F., Barbieri L., Battelli M.G., Soria M., and Lappi D.A. 1992. Ribosome-inactivating proteins from plants: Present status and future prospects. *Biotechnology* **10:** 405–412.

Lord J.M., Roberts L.M., and Robertus J.D. 1994. Ricin: Structure, mode of action, and some current applications. *FASEB J.* **8:** 201–208.

Clausen H. and Bennett E.P. 1996. A family of UDP-GalNAc: Polypeptide *N*-acetylgalactosaminyl-transferases control the initiation of mucin-type O-linked glycosylation. *Glycobiology* **6:** 635–646.

Hazes B. 1996. The $(QxW)_3$ domain: A flexible lectin scaffold. *Protein Sci.* **5:** 1490–1501.

Gabius H.J. 1997. Animal lectins. *Eur. J. Biochem.* **243:** 543–576.

Amado M., Almeida R., Schwientek T., and Clausen H. 1999. Identification and characterization of large galactosyltransferase gene families: Galactosyltransferases for all functions. *Biochim. Biophys. Acta* **1473:** 35–53.

Dodd R.B. and Drickamer K. 2001. Lectin-like proteins in model organisms: Implications for evolution of carbohydrate-binding activity. *Glycobiology* **11:** 71R–79R.

Rudiger H. and Gabius H.J. 2001. Plant lectins: Occurrence, biochemistry, functions and applications. *Glycoconj. J.* **18:** 589–613.

Woodworth A. and Baenziger J.U. 2001. The Man/GalNAc-4-SO4-receptor has multiple specificities and functions. *Results Probl. Cell Differ.* **33:** 123–138.

East L., Rushton S., Taylor M.E., and Isacke C.M. 2002. Characterization of sugar binding by the mannose receptor family member, Endo180. *J. Biol. Chem.* **277:** 50469–50475.

Ten Hagen K.G., Fritz T.A., and Tabak L.A. 2003. All in the family: The UDP-GalNAc:polypeptide N-acetylgalactosaminyltransferases. *Glycobiology* **13:** 1R–16R.

Sharon N. and Lis H. 2004. History of lectins: From hemagglutinins to biological recognition molecules. *Glycobiology* **14:** 53R–62R.

Llorca O. 2008. Extended and bent conformations of the mannose receptor family. *Cell. Mol. Life Sci.* **65:** 1302–1310.

CHAPTER 29

L-type Lectins

Marilynn E. Etzler, Avadhesha Surolia, and Richard D. Cummings

THE L-TYPE LECTINS WERE FIRST DISCOVERED IN THE SEEDS of leguminous plants, and they were found to have structural motifs that are now known to be present in a variety of glycan-binding proteins from other eukaryotic organisms. The structures of many of these lectins have been thoroughly characterized, and many L-type lectins are employed in a wide range of biomedical and analytical procedures. This chapter discusses the structure–function relationships of these lectins and the various biological roles they have in different organisms.

HISTORICAL BACKGROUND

The L-type lectins have a rich history that goes back to the end of the 19th century when it was found that extracts from the seeds of leguminous plants could agglutinate red blood cells. These agglutinins (later named lectins) were found to be soluble proteins that are very abundant in the seeds of leguminous plants, and differences in hemagglutination specificity were found among agglutinins from different species of legumes. A considerable amount of work on these proteins was done in the early part of the 20th century, including the crystallization of concanavalin A (ConA; the hemagglutinin from jack beans) and the finding that the hemagglutinating properties of these proteins are due to their ability to bind glycans on the cell surface.

The abundance of these proteins in the soluble extracts of legume seeds enabled a large number of these lectins to be isolated and characterized. The seed lectins were found to have considerable amino acid sequence homology, and the variety of carbohydrate-binding specificities found among these lectins enabled them to be employed as useful tools in a wide variety of analytical and biomedical procedures.

The crystal structures of a number of legume seed lectins were obtained, and the locations of the carbohydrate-binding sites were determined. Structural similarities were soon noticed among these lectins and some other lectins, including the galectins (see Chapter 33) and a variety of other lectins discussed in this chapter. For this reason, the term "L-lectins" has recently been designated as a classification for all lectins with this legume seed lectin-like structure.

COMMON FEATURES OF L-TYPE LECTINS

The L-type lectins are distinguished from other lectins primarily on the basis of tertiary structure. In general, either the entire lectin monomer or the carbohydrate-recognition domains (CRDs) of the more complex lectins are composed of antiparallel β-sheets connected by short loops and β-bends, and they are usually devoid of any α-helical structure. These sheets form a dome-like structure related to the "jelly-roll fold," and it is often called a "lectin fold." The carbohydrate-binding site is generally localized toward the apex of this dome. The tertiary structure of the monomer of ConA, the lectin from the seeds of the legume *Canavalia ensiformis*, is shown in Figure 29.1. The crystal structures of at least 20 other legume L-type lectin monomers have been determined by high-resolution X-ray crystallography and are almost superimposable on this structure. Thus, it is not surprising that the primary amino acid sequences of these legume lectins show remarkable homology with one another as well as with the sequences of many other legume seed lectins sequenced but not yet crystallized. Fewer, but significant, homologies in primary structure have been found between legume L-type lectins and some L-type lectins from far distant sources, such as ERGIC-53 and VIP36. Yet, in other L-type lectins, no homology is found with the seed lectins, although they contain similar lectin folds. For example, a comparison of the tertiary structure of the legume soybean lectin with the structure of human galectin-3 shows that both proteins contain the typical L-type lectin fold, but no amino acid sequence homology exists between these two lectins (Figure 29.2).

It is clear from comparisons of sequences of L-type lectins from legumes and their relationship to the phylogeny of the various species within the Leguminoseae family of plants that the variety found among these lectins most probably arose from divergent evolution. It remains an open possibility that the tertiary structures of some of the other members of the L-type lectin family arose by convergent evolution. It must also be noted that a protein cannot be firmly placed in the L-type lectin category simply on the basis of its tertiary structure. The protein also must be found to have the glycan-binding activity that would classify it as a lectin.

All soluble L-type lectins found to date are multimeric proteins, although all do not have the same quaternary structure. Thus, these lectins are multivalent with more than one glycan-binding site per lectin molecule. The same multivalent principle applies to the membrane bound L-type lectins because the presence of two or more molecules on a membrane surface essentially presents a multivalent situation. In addition to increasing the avidity of

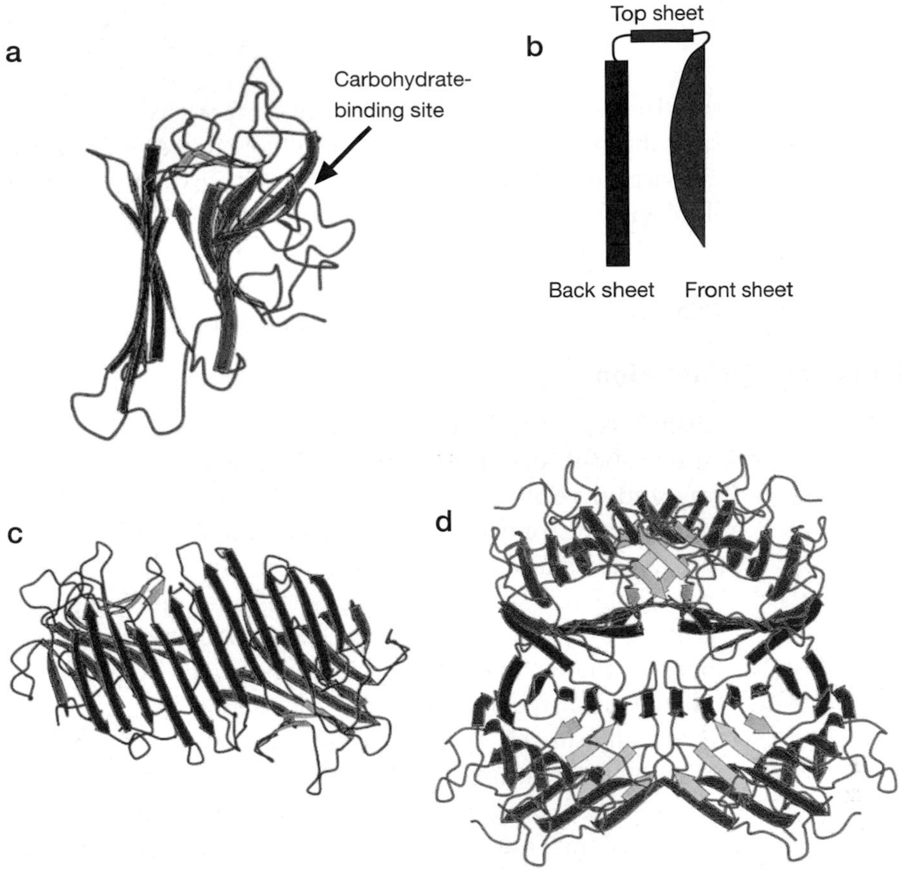

FIGURE 29.1. Structure of concanavalin A (ConA), a legume seed lectin. (*a*) The tertiary structure of the monomer is best described as a "jelly-roll fold." (*b*) This fold consists of a flat six-stranded antiparallel "back" β-sheet (*red*), a curved seven-stranded "front" β-sheet (*green*), and a five-stranded "top" sheet (*pink*) linked by loops of various lengths. (*c*) Dimerization of ConA involves antiparallel side-by-side alignment of the flat six-stranded "back" sheets, giving rise to the formation of a contiguous 12-stranded sheet. (*d*) The tetramerization of ConA occurs by a back-to-back association of two dimers. (Adapted, with permission of Elsevier, from Srinivas V.R., Reddy G.P., Ahmad N., et al. 2001. *Biochim. Biophys. Acta* **1527:** 102–111.)

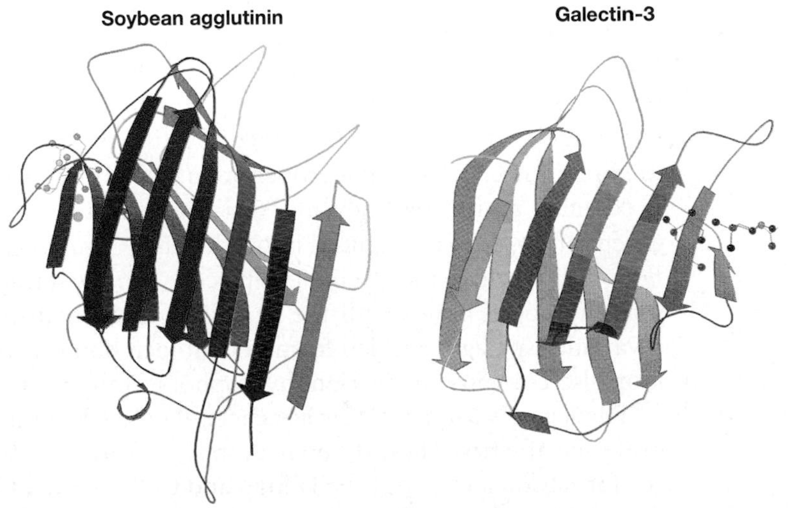

FIGURE 29.2. Comparison of the subunit structures of soybean agglutinin (*left*) complexed with a pentasaccharide containing Galβ1-4GlcNAc-R (Olsen L.R., Dessen A., Gupta D., et al. 1997. *Biochemistry* **36:** 15073–15080) and human galectin-3 (*right*) complexed with lactose (Seetharaman J., Kanigsberg A., Sallby R., et al. 1998. *J. Biol. Chem.* **273:** 13047–13052). Both lectins display a related β-barrel configuration.

the lectins for branched and/or cell-surface glycans, it is becoming increasingly apparent that this multivalence can have great biological significance. Binding of the lectins to the cell surface can lead to aggregation of specific glycan receptors, which can promote a variety of biological responses such as mitogenesis and various signal transduction processes.

PLANT L-TYPE LECTINS

Distribution and Localization

Plant L-type lectins are primarily found in the seeds of leguminous plants where they constitute about 10% of the total soluble protein of the seed extracts. They are synthesized during seed development several weeks after flowering and transported to the vacuole where they become condensed into specialized vesicles called protein bodies. They are stable during desiccation of the seeds and can remain in that state indefinitely until the seeds germinate. They represent one of several classes of proteins stored in high concentrations in the seeds and are often called storage proteins. During seed germination, the storage bodies become the vacuoles of the cotyledons, which appear as the first leafy appendages of the plant. During the first week of development, this cotyledon provides food for the plant and eventually shrivels up and disappears. L-type lectins have also been found in the bark of some leguminous trees, and very low amounts of these lectins are also found in other vegetative tissues of legumes. In some cases, these latter lectins have been found to be encoded by separate but very similar genes. About 100 of the seed legume L-type lectins have been characterized, and they are the most extensively studied proteins of this class.

Structure

A common feature of the legume L-type lectins is their oligomeric structure. The structures of the monomers consist of three antiparallel β-sheets: a flat six-stranded "back" sheet, a concave seven-stranded "front" sheet, and a short "top" sheet that keeps the two major sheets together (Figure 29.1a,b). All of these lectins require Ca^{++} and a transition metal ion (usually Mn^{++}) for their carbohydrate-binding activity. The glycan-binding and metal-binding sites are localized in close proximity to each other at the top of the "front" sheet.

The glycan-binding site is composed of four loops: A, B, C, and D (Figure 29.3, top). These loops contain four invariant amino acids that are essential for carbohydrate binding (Figure 29.3, bottom). Loop A contains an invariant aspartate, which forms hydrogen bonds between its side chain and the glycan ligand. This amino acid is linked to its preceding amino acid (usually alanine) by a rare *cis*-peptide bond, which is stabilized by the metal ions and is necessary for the proper orientation of the aspartate in the combining site. Loop B contains an invariant glycine, which also forms hydrogen bonds with the ligand. An exception to this case is found in two lectins (ConA and the closely related *Dioclea grandiflora* lectin) where the glycine is replaced with an arginine. Both the glycine and arginine form hydrogen bonds with the ligand via their main-chain amides. Loop C contains an invariant asparagine, which forms a hydrogen bond with the ligand via its side chain. This loop also contains an invariant hydrophobic amino acid.

The legume L-type lectins are generally classified into groups based on their carbohydrate specificities. These differences in specificities are brought about by variability in the conformation and size of the D loop and to some extent by the C loop. Although the main specificity regions of the legume lectins are determined by the loops, it is now clear that there are sites other than these that contribute to lectin specificity. There also exist several additional modes of refining these specificities, such as interaction with water, posttranslational modifications, and state of oligomerization.

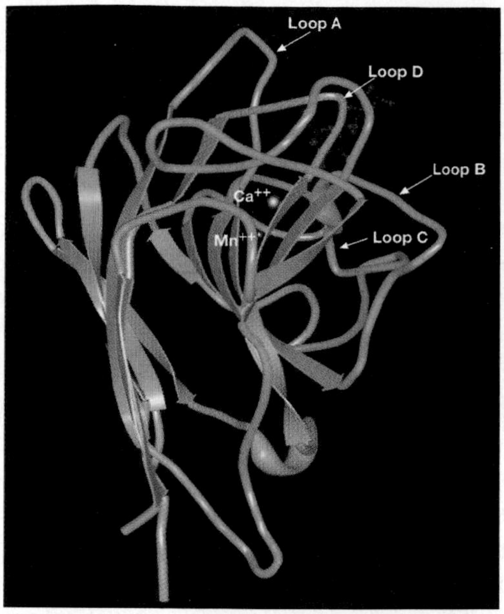

Lectin	Loop A	Loop B	Loop C	Loop D	Specificity
EcorL	GPPYT–RPLPADGLVF	AQ-GYGYLG	VEFDTFSN----PWDP	GLSGATG----AQRDAAETHDVYSW	GalNAc
DBL	APSK---ASFADGIAP	RR-NGGYLG	VEFDTLSNS---GWDP	GFSATTGLSDG----YIETHDVLSW	GalNAc
E-PHA	VPNN---EGPADGLAF	KD-KGGLLG	VEFDTLYNV---HWDP	GFTATTGITKG----NVETNDILSW	Complex
L-PHA	VPNN---AGPADGLAF	KD-KGGFLC	VEFDTLYNK---DWDP	GFSATTGINKG----NVETNDVLSW	Complex
SBA	APDT---KRLADGLAF	QT-HAGYLG	VEFDTFRN----SWDP	GFSAATGLDIP-----GESHDVLSW	GalNAc
PNA	KD--IKDYDPADGIIF	GSIGGGTLG	VEFDTYSNS–-EYNDP	GFSASGSL------GGRQIHLIRSW	Gal
LTA	IR--ELKYTPTDGLVF	GS-TGGFLG	VEFDSYHN----IWDP	GFSATTGN------PEREKHDIYSW	Fuc
UEA-I	SANP---KAATDGLTF	RRA-GGYFG	VEFDTI-GSPVNFDDP	GFSGGTYI------GRQATHEVLNW	Fuc
UEA-II	EPDE--KIDGVDGLAF	GS-SAGMFG	VEFDSYPGKTYNPWDP	GFSGGVGN------AAKFDHDVLSW	GlcNAc
LAAI	PPIQSRKADGVDGLAF	GS-SAGMFG	VEFDTYFGKAYNPWDP	GFSAGVGN------AAKFNHDILSW	GlcNAc
LSL	RPNSDS-QVVADGFTF	RG-DGGLLG	VEFDTFHNQ---PWDP	GLSASTATYY-------SAHEVYSW	Man/Glc
Con A	SPDS----HPADGIAF	GS-TGRLLG	VELDTYPNT--DIGDP	GLSASTGL-------YKETNTILSW	Man/Glc
DiocL	SPDH----EPADGITF	GS-GGRLLG	VELDSYPNT--DIGDP	GLSATTGL-------YKETNTILSW	Man/Glc
LenL	SPNG---YNVADGFTF	QT-GGGYLG	VEFDTFYNA---AWDP	GFSATTGAEF-------AAQEVHSW	Man/Glc
PSL	APNS---YNVADGFTF	QT-GGGYLG	VEFDTFYNA---AWDP	GPSATTGAEY-------AAHEVLSW	Man/Glc
FAVIN	APNG---YNVADGFTF	QT-GGGYLG	VEFDTFYNA---AWDP	GFSATTGAEY-------ATHEVLSW	Man/Glc
	** *	*	** **	* *	

FIGURE 29.3 (*Top*) Three-dimensional structure of a legume lectin (PNA) monomer showing the four loops involved in sugar binding: loops A, B, C, and D. The bound sugar (lactose) is shown as a *ball-and-stick* model. Calcium and manganese ions are required for ligand binding. (*Bottom*) Sequence alignment of loops A–D in legume lectins. The size of binding-site loop D and monosaccharide specificity show an explicit correlation. Monosaccharide specificity and number of gaps are indicated at the *right*. Key residues are highlighted in *blue* and highly conserved residues have been indicated with an *asterisk*. (Adapted, with persmission of Elsevier, from Vijayan M. and Chandra N. 1999. *Curr. Opin. Struct. Biol.* **9:** 707–714; and the PBD deposited structure 1V6I.)

Within these oligomeric proteins, the back β-sheet is involved in formation of the oligomers, which are mostly dimeric or tetrameric in nature. A variety of quaternary structures are found among the lectins. The dimeric and tetrameric structures of ConA are shown in Figure 29.1c,d). Although some of the other lectins occur as dimeric and tetrameric structures, several other different orientations of the β-sheets have been found to account for the variability in dimeric and tetrameric structures of other lectins in this class. It is of interest that some legume lectins have been found to have a hydrophobic binding site that binds adenine and adenine-derived plant hormones with micromolar affinity; this is two to three orders of magnitude higher than their affinity for monosaccharides. Three of these lectins (the soybean agglutinin, phytohemagglutinin L [PHAL], and *Dolichos biflorus* lectin) have been crystallized and found to have a unique tetrameric structure, in that the dimer–dimer interface creates a channel running through the center of the tetramer. Two identical adenine-binding sites are found at opposite ends of this channel.

Another common feature of the legume lectins is that they are secretory proteins and undergo cotranslational signal peptide removal, which accompanies their entry into the secretory system. All but the peanut agglutinin are N-glycosylated; the N-glycans undergo the normal posttranslational modifications that occur as they transit the Golgi apparatus. The lectins vary from one another as to whether the mature proteins contain oligomannose-type, complex-type, or a mixture of both types of N-glycans. The lectins may also undergo a variety of proteolytic modifications as they transit through the secretory system. Some of the lectins are cleaved to generate a β-chain, corresponding to the amino terminus and an α-chain corresponding to the carboxyl terminus. For example, the pea lectin and favin (the lectin from *Vicia faba*) are tetrameric glycoproteins that contain two types of subunits, α and β, which have molecular weights of about 5,000 and 21,000, respectively. These two lectins are each synthesized as single polypeptide precursors that contain the sequences of both chains in the following orientation: β-chain–α-chain. The chains associate to form dimers; they are then proteolytically processed in the protein bodies to form tetramers containing two separate α- and β-chains. Other lectins may undergo carboxy-terminal trimming of only some of their subunits. For example, the soybean agglutinin, phytohemagglutinin E (PHAE), and *Dolichos biflorus* lectins are tetramers of equimolar mixtures of intact and trimmed subunits. The most intriguing proteolytic modification occurs in the case of ConA: A small segment is removed from the interior of the protein and the original amino terminus is ligated with the original carboxyl terminus. This forms what is termed a circularly permuted protein. The glycosylated segment of the protein is removed during this transpeptidation process; thus, the mature ConA is not a glycoprotein. This finding provided an explanation for the fact that the protein sequence of isolated ConA aligns with other seed lectins, whereas the alignment of the DNA encoding the protein with other lectin genes suggested that the gene is circularly permuted.

Function

Despite the extensive amount of information available on the plant L-type lectins, the biological role of these proteins remains a mystery. Many hypotheses have been proposed over the years. One hypothesis is that the lectins simply serve as other storage proteins for the plant. However, it is a puzzle as to why proteins with such exquisite carbohydrate specificity would evolve merely to feed the plant. Another hypothesis was that the L-type lectins are involved in the recognition of glycan signals produced by *Rhizobia* during the initiation of the nitrogen-fixing *Rhizobium*-legume symbiosis. Recently, it has been shown that this hypothesis is not correct. A different type of lectin (an LNP) was found that may have this function (see Chapters 22 and 37). In fact, none of the L-type lectins has been shown to recognize any of the plant glycan signals identified to date (see Chapters 22 and 37); searches in plants for ligands for these lectins have been unsuccessful so far.

The most attractive hypothesis to date for plant L-type lectin function is that these proteins may have a role in plant defense, which may occur in one or more ways. Some of these lectins have been found to be toxic to insects, thus raising the possibility that they may serve as a deterrent to these pests and other pathogens. They also may have a role as pattern-recognition receptors within the plant innate immune system. Interestingly, some animal lectins, as discussed elsewhere, have been found to function in innate immunity. A recent study has shown that the *Dolichos biflorus* seed lectin is also a lipoxygenase. It will be of interest to see how many other L-type lectins have this activity, which is necessary to initiate the wound-induced defense pathway in plants.

L-TYPE LECTINS IN PROTEIN QUALITY CONTROL AND SORTING

Calnexin/Calreticulin

Calnexin (CNX) and calreticulin (CRT) are homologous protein chaperones that mediate quality control of proteins in the endoplasmic reticulum (ER) (see Chapter 36). CNX is membrane-bound and is perhaps closely associated with the protein-translocating channel that is involved in importing nascent proteins into the ER. CRT is a soluble ER luminal component. As discussed in Chapter 36, these two proteins bind to monoglucosylated, high-mannose-type glycans and prevent their exit from the ER until they are properly folded and assembled into correct quaternary structures. During the binding and dissociation from CRT or CNX, if the glycoprotein folds correctly, then glucose removal by glucosidase-II allows its passage out of the ER. In the event that a glycoprotein misfolds or aggregates, it is reglucosylated by UDP-Glc:glycoprotein glucosyltransferase (UGGT); this enzyme only recognizes misfolded or aggregated glycoproteins. Following reglucosylation, the monoglucosylated protein binds again to CRT or CNX. Thus, there is a cycle of glucose removal and addition by the alternating actions of glucosidase-II and UGGT and interactions with CNX/CRT.

Both CRT and CNX are Ca^{++}-binding proteins, and their carbohydrate-binding activity is sensitive to changes in Ca^{++} concentration. CNX is a type I membrane protein with its carboxy-terminal end in the cytoplasm. The lumenal portion of the protein is divided into three domains: a Ca^{++}-binding domain (which is adjacent to the transmembrane domain), a proline-rich long hairpin loop called the P domain, and the amino-terminal L-type lectin domain. CRT has a similar structure, but it is missing the cytoplasmic and transmembrane regions; it is retained in the ER through its KDEL-retrieval signal at the carboxyl terminus (Figure 29.4).

ERGIC-53 and VIP36

ERGIC-53 and VIP36 are type I membrane proteins that have been found to participate in vesicular protein transport in the secretory system (see Chapter 36). ERGIC-53 has been found in the ER-Golgi intermediate compartment (ERGIC) and its cytoplasmic carboxyl terminus contains the dilysine KKFF retention/retrieval motif. The dilysine part of this motif is recognized by the COPI coatomer complex; this binding enables the coated vesicles to be recycled from the ERGIC back to the ER. The diphenylalanine helps to direct the COPII-coated vesicles to ER export sites by binding to the COPII coatomer. The location of VIP36 is uncertain; overexpressed protein has been found in both the ER and ERGIC.

Both ERGIC-53 and VIP36 bind to oligomannose-type glycans and require Ca^{++} for their carbohydrate-binding activity. These two proteins were the first animal lectins that were found to share some sequence and structural homology with the legume seed lectins. Although the overall sequence identity of these proteins to the seed lectins is only about 19–24%, those amino acids important for metal and carbohydrate binding in the seed lectins are conserved, including the invariant aspartate, glycine, and asparagine. The invariant aspartate also participates in a *cis*-peptide bond with its preceding amino acid; this is similar to the case of the legume seed lectins discussed in the above section. The crystal structure of the CRD of ERGIC-53 has been determined and confirms the structural similarity of these lectins.

OTHER L-TYPE LECTINS

A variety of other proteins have been described that have carbohydrate-binding domains with tertiary structures similar to the lectin fold and may be considered as members of the

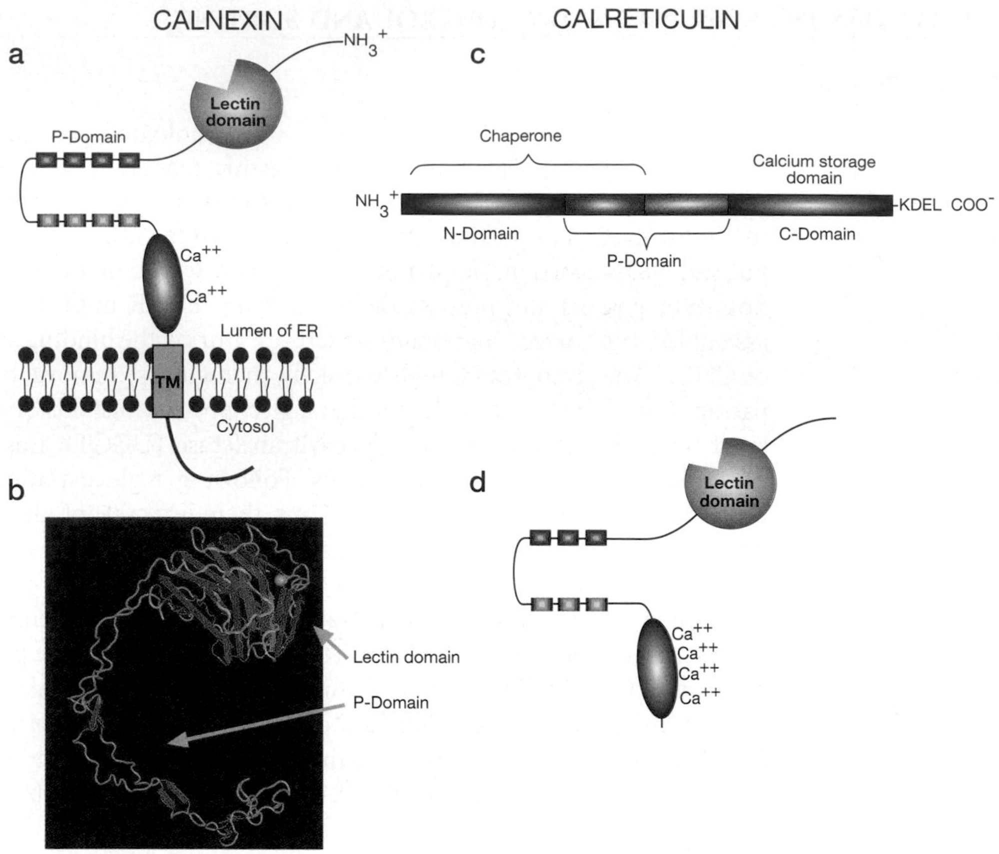

FIGURE 29.4. Schematic representation of calnexin showing the lectin domain, the P domain (containing the proline repeats), and the calcium-binding domain (*a*). Structure of calnexin based on crystallographic data (*b*). (Adapted, with permission of Elsevier, from Schrag J.D., Bergeron J.J.M., Li Y., et al. 2001. *Mol. Cell* **8:** 633–644.) Domain organization of calreticulin (*c*) and its proposed tertiary organization (*d*).

L-type lectin family. Members of the galectin family of lectins fit into this category and are the subject of a separate chapter in this volume (see Chapter 33). Other carbohydrate-binding proteins that may fit into this category are briefly discussed below.

Pentraxins and Related Proteins

The pentraxins are a superfamily of plasma proteins that are involved in innate immunity in invertebrates and vertebrates. They contain L-type lectin folds and require Ca^{++} ions for ligand binding. Their name is based on the pentameric arrangement of their subunits. The short pentraxins, C-reactive protein (which binds phosphocholine) and the serum amyloid P component (which binds carbohydrate), are acute-phase proteins in humans and mice, respectively. This family also contains long pentraxins that have an unrelated long amino-terminal domain coupled to the pentraxin domain. PTX3 is one of these long pentraxins; in addition to its role in innate immunity, it may help in the assembly of a hyaluronan-rich extracellular matrix.

Laminin G domain-like (LG) modules are made of 180–200 amino acid residues and were first identified in proteins like laminins. These modules were predicted to have a fold similar to the folds of pentraxins. Some LG modules share binding properties for cellular receptors and carbohydrate ligands, indicating that the LG fold may have evolved from the L-type lectin fold for participation in related functions.

vp4 Sialic Acid–binding Domain

vp4 is a monomeric sialic acid–binding domain with an L-type lectin fold. This domain is required for infectivity of most animal rotaviruses. vp4 forms the viral spikes and mediates the recognition and attachment of the virus to the surface of animal cells. Unlike other L-type lectins, which bind to neutral glycans, vp4 binds to a charged glycan (sialic acid) through electrostatic interactions. Trypsin cleavage of vp4 gives rise to two fragments: vp8, which binds sialic acid, and vp5, which contains a hydrophobic region that permeabilizes the membrane for entry.

OTHER PROTEINS WITH JELLY-ROLL MOTIFS

Clostridium Neurotoxins, the Second-last Domain

Tetanus and botulism are caused by the toxic effect of neurotoxins produced by *Clostridium tetanus* and *Clostridium botulinum*, respectively. Botulinum neurotoxins (BoNTs) block the release of acetylcholine at the neuromuscular junction; in contrast, tetanus neurotoxin blocks the release of neurotransmitters such as glycine and γ-aminobutyric acid in the inhibitory interneurons of the spinal cord. Their entry into cells is mediated by binding to polysialogangliosides and other acidic lipids on the presynaptic membrane. These neurotoxins usually contain an activating domain (A) and a binding domain (B). The crystal structures of BoNT/A and BoNT/B reveal features of their receptor binding and activation. In the case of BoNT/B, the amino-terminal end of the binding domain consists of two seven-stranded antiparallel β-sheets that form a 14-stranded β-barrel in a jelly-roll motif. The structure of the binding domain is very similar to that of the C-fragment of tetanus toxin and the binding domain of BoNT/A. The study of binding of these neurotoxins to the cell-surface gangliosides is a necessary step toward designing effective inhibitors to combat their adverse effects.

Exotoxin A, Amino-terminal Domain

Pseudomonas aeruginosa exotoxin A is extremely toxic to eukaryotic cells because of its ability to inhibit protein synthesis. This effect is brought about by the ADP ribosylation of a specific posttranslationally modified histidine of eEF-2 (diphthamide). The crystal structure of the exotoxin reveals a tertiary fold, consisting of three distinct structural domains. These domains are individually responsible for amino-terminal receptor binding (domain I), transmembrane targeting (domain II), and ADP ribosyltransferase (domain III) activities. Domain Ia displays a complex 13-stranded jelly-roll structure.

Vibrio cholerae Sialidase, Amino-terminal and Insertion Domains

Bacterial sialidases have been implicated in the pathogenesis of a number of diseases. *Vibrio cholerae* neuraminidase aids in the pathogenesis of cholera by removing sialic acid from larger gangliosides to expose GM1, the receptor for cholera toxin. The crystal structure of the neuraminidase reveals the presence of a three-domain protein with a six-bladed β-propeller neuraminidase domain flanked by two lectin domains. One of these domains is at the amino terminus and the other is between the second and third blade of the propeller.

Leech Intramolecular *trans*-Sialidase, Amino-terminal Domain

The intracellular *trans*-sialidase from the leech *Macrobdella decora* catalyzes an intramolecular *trans*-sialosyl reaction; this is specific for the cleavage of the terminal Neu5Acα2-3Gal linkage in sialoglycoconjugates and releases 2,7-anhydro-Neu5Ac instead of Neu5Ac. This enzyme displays a multidomain architecture with a lectin-like domain II and an irregular β-stranded domain III, which is built around a canonical catalytic domain C. Domain II has a cluster of two histidine and two tyrosine residues on the curved surface of the same side as the active site crater. This domain II may be involved in carbohydrate recognition through sugar ring and aromatic side-chain interactions, as observed in many lectins.

FURTHER READING

Goldstein I.J. and Poretz R.D. 1986. Isolation, physicochemical characterization, and carbohydrate-binding specificity of lectins. In *The lectins: Properties, functions, and applications in biology and medicine* (eds. I.E. Liener, N. Sharon, and I.J. Goldstein), pp. 33–247. Academic Press, Orlando, Florida.

Fiedler K. and Simons K. 1994. A putative novel class of animal lectins in the secretory pathway homologous to leguminous lectins. *Cell* **77:** 625–626.

Taylor G. 1996. Sialidases: Structures, biological significance and therapeutic potential. *Curr. Opin. Struct. Biol.* **6:** 830–837.

Sharma V. and Surolia A. 1997. Analyses of carbohydrate recognition by legume lectins: Size of the combining site loops and their primary specificity. *J. Mol. Biol.* **267:** 433–445.

Lis H. and Sharon N. 1998. Lectins: Carbohydrate-specific proteins that mediate cellular recognition. *Chem. Rev.* **98:** 637–674.

Loris R., Bouckaert J., Hamelryck T., and Wynn L. 1998. Legume lectin structure. *Biochim. Biophys. Acta* **1383:** 9–36.

Hamelryck T.W., Loris R., Bouckaert J., Dao-Thi M.-H., Strecker G., Imberty A., Fernandez E., Wyns L., and Etzler M.E. 1999. Carbohydrate binding, quaternary structure and a novel hydrophobic binding site in two legume lectin oligomers from *Dolichos biflorus. J. Mol. Biol.* **286:** 1161–1177.

Vijayan M. and Chandra N. 1999. Lectins. *Curr. Opin. Struct. Biol.* **9:** 707–714.

Rudenko G., Hohenester E., and Muller Y.A. 2001. LG/LNS domains: Multiple functions—One business end? *Trends Biochem. Sci.* **26:** 363–368.

Schrag J.D., Procopio D.O., Cygler M., Thomas D.Y., and Bergeron J.J.M. 2003. Lectin control of protein folding and sorting in the secretory pathway. *Trends Biochem. Sci.* **28:** 49–57.

Bottazzi B., Garlanda C., Salvatori G., Jeannin P., Manfredi A., and Mantovani A. 2006. Pentraxins as a key component of innate immunity. *Curr. Opin. Immunol.* **18:** 10–15.

Roopashree S., Singh S.A., Gowda L.R., and Rao A.A. 2006. Dual-function protein in plant defence: Seed lectin from *Dolichos biflorus* (horse gram) exhibits lipoxygenase activity. *Biochem. J.* **395:** 629–639.

Dam T.K., Gerken T.A., Cavada B.S., Nascimento K.S., Moura T.R., and Brewer C.F. 2007. Binding studies of α-GalNAc-specific lectins to the α-GalNAc (Tn-antigen) form of porcine submaxillary mucin and its smaller fragments. *J. Biol. Chem.* **282:** 28256–28263.

CHAPTER 30

P-type Lectins

Ajit Varki and Stuart Kornfeld

LYSOSOMES ARE INTRACELLULAR MEMBRANE-BOUND organelles that perform the final degradation of many cellular macromolecules. This is achieved by the action of a number of lysosomal enzymes (originally called "acid hydrolases" because of the low internal pH characteristic of lysosomes). These enzymes are synthesized in the endoplasmic reticulum (ER) on membrane-bound ribosomes and traverse the ER-Golgi pathway along with other newly synthesized proteins. At the terminal Golgi compartment (the *trans*-Golgi network or TGN), they are segregated from all other glycoproteins and selectively delivered to lysosomes. In most "higher" animal cells, this specialized trafficking is achieved primarily by a specific glycan marker that is recognized by certain receptors. This chapter describes the discovery and characterization of this glycan-mediated biological system, which relies on recognition of glycans containing mannose-6-phosphate (M6P) by "P-type" lectins. This was the first clear-cut example of a biological role for glycans on mammalian glycoproteins and the first demonstrated link between glycoprotein biosynthesis and human disease. The interesting history of its discovery is therefore described in some detail.

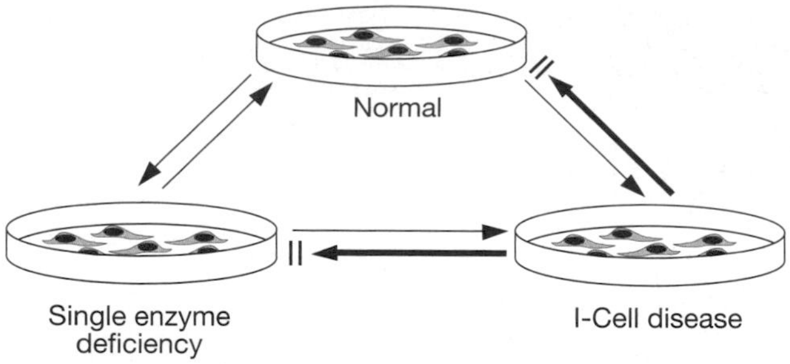

FIGURE 30.1. Historical background regarding "cross-correction" of lysosomal enzyme deficiencies in cultured cells. Small amounts of "high-uptake" lysosomal enzymes secreted by normal fibroblasts (*thin arrows*) were found to be taken up by fibroblasts from a patient with a genetic lack of a single lysosomal enzyme, correcting that deficiency. In contrast, I-cell disease fibroblasts secrete large amounts of multiple lysosomal enzymes of the "low-uptake" variety (*thick arrows*). The latter forms cannot correct lysosomal enzyme deficiencies in other cells. However, I-cells retain the capability to accept "high-uptake" enzymes secreted by other cells (*thin arrows*). Uptake was blocked by the addition of mannose-6-phosphate (M6P) to the media. These data implied the existence of a "common recognition marker" that was missing on I-cells, and the existence of cell-surface receptors for the "high-uptake" forms. Subsequent studies showed that the recognition marker required mannose-6-phosphate on the N-glycans of the enzymes, and that the uptake was mediated by mannose-6-phophate receptors (MPRs).

HISTORICAL BACKGROUND

I-Cell Disease and the "Common Recognition Marker" of Lysosomal Enzymes

Early studies of human genetic "storage disorders" by Elizabeth Neufeld and colleagues indicated a failure of intracellular degradation of cellular components that therefore accumulated in lysosomes (see Chapter 41). Soluble "corrective factors" from normal cells could reverse these defects when added to the culture media. These factors were found to be lysosomal enzymes, and they were deficient in patients with different diseases. They were secreted in small quantities by normal cells in culture (or by cells from patients with a different "complementary" defect) (Figure 30.1). The enzymes existed in two forms: a "high-uptake" form, recognized by saturable, high-affinity receptors that could correct deficient cells, and an inactive "low-uptake" form that could not correct the defect. Meanwhile, fibroblasts from patients with a rare human genetic disease exhibiting prominent inclusion bodies in cultured cells (therefore termed "I-cell" disease) were found to be deficient in not one, but almost all, lysosomal enzymes. In I-cells, all of the lysosomal enzymes are actually made, but they are almost completely secreted into the medium instead of being retained in the lysosomes. Although I-cells can incorporate the high-uptake enzymes secreted by normal cells, enzyme molecules secreted by I-cells are not taken up by other cells (Figure 30.1). Neufeld therefore proposed that I-cell disease results from a failure to add a "common recognition marker" to all lysosomal enzymes. This marker was assumed to be responsible for the retention of lysosomal enzymes in normal cells and also for internal trafficking of the enzymes to lysosomes. Because the high-uptake property was destroyed by periodate treatment, it was predicted that the recognition marker was a glycan.

Discovery of the M6P Recognition Marker

"High" uptake of lysosomal enzymes was next found by William Sly to be specifically blocked by the sugar mannose-6-phosphate (M6P) and its stereoisomer fructose-1-phosphate. Although millimolar concentrations were required for this inhibition, similar concentrations of other sugar phosphates did not inhibit. Treatment of lysosomal enzymes with alkaline phosphatase also abolished high-uptake activity. By this time, the general pathway for N-glycan processing had been defined (see Chapter 8). Because oligomannosyl N-glycans (then called "high-mannose-type" N-glycans) are rich in mannose residues, these were predicted to be phosphorylated on lysosomal enzymes. Indeed, blockade of N-glycosylation by tunicamycin treatment caused excessive secretion of lysosomal enzymes and failure of trafficking to lysosomes. The M6P moiety was also detected on oligomannosyl N-glycans released by endo-β-*N*-acetylglucosaminidase H (endo H) from high-uptake forms of lysosomal enzymes. Surprisingly, the groups of Stuart Kornfeld and Kurt von Figura then found that most of the M6P moieties were "blocked" by α-linked *N*-acetylglucosamine residues attached to the phosphate residue, creating a phosphodiester. The *N*-acetylglucosamine could be removed by mild acid, generating the bioactive M6P phosphomonoester.

Enzymatic Mechanism for Generation of the M6P Recognition Marker

Oligomannosyl N-glycans released from lysosomal enzymes were found to carry one or two phosphate residues at the C-6 of various mannose residues. Comparison of glycans with phosphodiesters and phosphomonoesters predicted that the metabolic precursor was a phosphodiester and that phosphorylation was mediated not by an ATP-dependent kinase, but by a UDP-GlcNAc-dependent GlcNAc-1-phosphotransferase. This was proven using a double-labeled donor substrate and Golgi extracts:

Reactants: Uridine-P-^{32}P-[6-^{3}H]GlcNAc + Manα1-(N-glycan)-lysosomal enzyme
Products: Uridine-P + [6-^{3}H]GlcNAcα1-^{32}P-6-Manα1-(N-glycan)-lysosomal enzyme

Another Golgi enzyme was shown to remove the outer *N*-acetylglucosamine residue and "uncover" the phosphomonoester, generating the M6P moiety. Pulse-chase studies confirmed the order of events, indicating that more than one glycan on a given lysosomal enzyme could be phosphorylated and that removal of some mannose residues on the N-glycan by processing Golgi mannosidases was also required (Figure 30.2). In most cell types, the phosphate is eventually lost from the M6P, presumably after exposure to acid phosphatase in lysosomes. Thus, the overall biochemical pathway is as follows:

Uridine-P-P-GlcNAc + Manα1-(N-glycan)-Lysosomal enzyme
↓ *"Phosphotransferase"*
Uridine-P + GlcNAcα1-6-Manα1-(N-glycan)-Lysosomal enzyme
↓ *"Uncovering Enzyme"*
GlcNAc + P-6-Manα1-(N-glycan)-Lysosomal enzyme
↓ *Lysosomal Phosphatase*
P + Manα1-(N-glycan)-Lysosomal enzyme

However, during passage through the Golgi, the phosphate residues partially block the action of processing mannosidases, maintaining the N-glycans in an oligomannosyl form (Figure 30.2). Because M6P is lacking on lysosomal enzymes in I-cell disease (see below), these glycans are likely to be processed in the Golgi, thus explaining why the secreted enzymes in these patients carry more sialylated, complex, N-linked glycans.

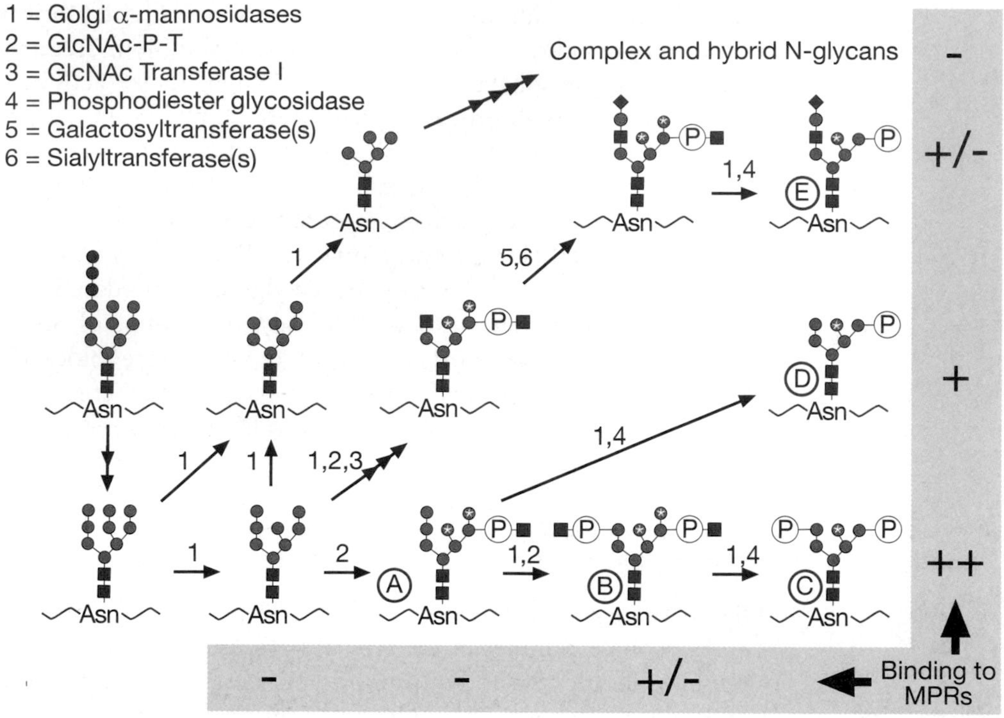

FIGURE 30.2. Pathways for biosynthesis of N-glycans bearing the mannose-6-phosphate (M6P) recognition marker. Following early N-glycan processing (see Chapter 8 for details), a single GlcNAc phosphodiester is added to the N-glycans of lysosomal enzymes on one of three mannose (Man) residues on the arm with the α1-6 Man linked to the core mannose (*structure A*). A second phosphodiester can then be added to the other side of the N-glycan (*structure B*) or onto other mannose residues (alternate sites for phosphorylation are denoted by *asterisks*). Removal of the outer *N*-acetylglucosamine residues and further processing of mannose residues give structures *C* and *D*. Further mannose removal is restricted by the phosphomonoesters. Thus, *C* and *D* represent only two of several possible structures bearing one or two phosphomonoesters. Glycans that are not phosphorylated become typical complex or hybrid N-glycans. Some hybrid N-glycans with M6P are also found (*structure E*). Binding studies of these N-glycans with purified mannose-6-phosphate receptors (MPRs) have shown the relative affinities indicated in the figure ([++] strong; [+] moderate; [+/–] weak; [–] no binding).

The genes encoding both enzymes involved in these reactions have been cloned and characterized. The UDP-*N*-acetylglucosamine:lysosomal-enzyme *N*-acetylglucosamine-1-phosphotransferase (GlcNAc-P-T) is a 540-kD complex composed of two disulfide-linked 166-kD α subunits, two disulfide-linked 51-kD β subunits, and two identical noncovalently associated 56-kD γ subunits. The α and β subunits are encoded by a single gene whose product undergoes proteolysis to give rise to the two subunits, and the γ subunit is encoded by a separate gene. The α subunit was identified as the catalytic subunit by photoaffinity labeling. The second enzyme (α-N-acetylglucosaminyl-1-phosphodiester glycosidase) is a 272-kD complex of four identical 68-kD subunits, arranged as two disulfide-linked homodimers. Unlike other Golgi enzymes, this is a type-I membrane-spanning glycoprotein with its amino terminus in the lumen of the Golgi.

Enzymatic Basis for I-Cell Disease and Pseudo-Hurler Polydystrophy

Analysis of fibroblasts from patients with I-cell disease (also called mucolipidosis-II; ML-II) revealed a deficiency in GlcNAc-P-T enzyme activity. A milder variant called pseudo-Hurler

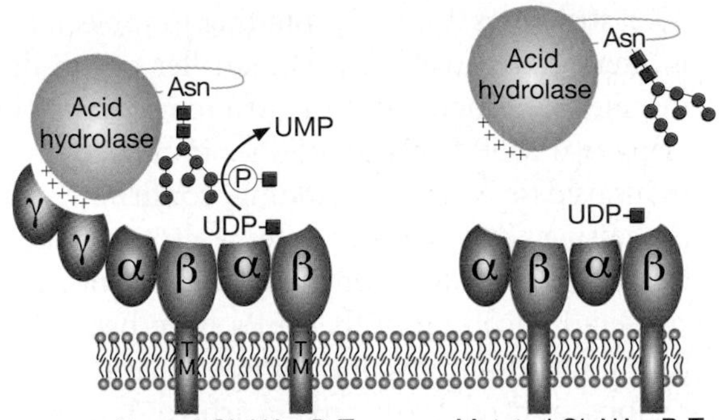

FIGURE 30.3. Selective recognition of lysosomal enzymes by GlcNAc-P-T. GlcNAc-P-T is a multisubunit enzyme ($\alpha_2\beta_2\gamma_2$). In normal cells, it has three independent binding sites, one for the N-glycan-M6P substrate, one for the UDP-GlcNAc donor, and one for the selective recognition of lysosomal enzymes. The γ subunit of GlcNAc-P-T facilitates the recognition of the protein element on lysosomal enzymes (++++) that is a critical determinant of the selective phosphorylation of these substrates; this subunit is lost in some variant cases of pseudo-Hurler polydystrophy (mutant GlcNAc-P-T). (TM) Transmembrane domain.

polydystrophy (mucolipidosis-III; ML-III) showed a less severe deficiency of enzyme activity. Metabolic radiolabeling of fibroblasts corroborated the failure to phosphorylate mannose residues in these diseases, and asymptomatic obligate heterozygotes showed a partial deficiency, with slightly elevated levels of serum lysosomal enzymes. Mutations of various types in the two GlcNAc-P-T genes have since been detected in all examined patients with ML-II and -III, indicating that deficiency of this enzyme is the primary genetic disorder.

Variants of I-Cell Disease and Pseudo-Hurler Polydystrophy

The simplest in vitro substrate for GlcNAc-P-T is α-methyl mannoside. However, it is a poorer substrate than an oligomannosyl N-glycan, which, in turn, is much poorer than a *native* lysosomal enzyme. Typical glycoproteins bearing oligomannosyl N-glycans that are not lysosomal enzymes are also poor substrates. Thus, the GlcNAc-P-T enzyme must specifically recognize oligomannosyl N-glycans on lysosomal enzymes in preference to those on other glycoproteins, via a second protein–protein recognition site (Figure 30.3). In most cases of ML-II and -III, assays with α-methyl mannoside and lysosomal acceptors both give concomitant reductions in activity. However, rare cases of ML-III show *normal* activity with the α-methyl mannoside acceptor, but *markedly decreased* activity with lysosomal enzyme acceptors. The GlcNAc-P-T enzyme in these patients is present in normal catalytic amounts, but because of a mutation in the γ subunit, it is impaired in the recognition of lysosomal enzymes as appropriate acceptors for phosphorylation. This finding provided a genetic basis for specific recognition of lysosomal enzymes by GlcNAc-P-T (see below).

COMMON FEATURES OF P-TYPE LECTINS (M6P RECEPTORS)

The first candidate (~275-kD) receptor for the M6P recognition marker was isolated by affinity chromatography and was found to bind M6P in the absence of cations. Certain cells deficient in this receptor still showed M6P-inhibitable binding of lysosomal enzymes, leading to the discovery of a second M6P receptor (MPR) of approximately 45 kD, which required divalent cations for optimal binding. The larger cation-independent M6P receptor (CI-MPR) binds with highest affinity in a 1:1 stoichiometry to glycans carrying two M6P residues (Figure 30.2, structure C) and poorly to molecules bearing GlcNAc-P-Man phosphodiesters (Figure

30.2, structures A and B). Binding to molecules carrying one M6P (Figure 30.2, structure D) is intermediate in affinity. The smaller, cation-dependent MPR (CD-MPR) has only one binding site for a single M6P. In vitro removal of "blocking" GlcNAc residues from molecules carrying two M6P-GlcNAc residues improved binding to both receptors. Treatment with an α-mannosidase enhanced binding, confirming that removal of outer mannose residues by the processing Golgi mannosidases is also a requirement for recognition. These findings with isolated N-glycans were confirmed and extended by studying their direct uptake into cells.

Genes encoding both MPRs have been cloned and extensively characterized. Both are type I membrane glycoproteins with large extracytoplasmic domains, single transmembrane regions, and relatively small carboxy-terminal cytoplasmic domains. The CI-MPR has 15 unique, contiguous repetitive units of approximately 145 amino acids with partial identity to one another. The CD-MPR has a single extracellular domain, showing homology to the repeating domains of the CI-MPR. Together with conservation of certain intron–exon boundaries, this homology suggests that the two genes evolved from a common ancestor. On the basis of their sequence relationships and unique binding properties to M6P, the two MPRs have been formally classified as P-type lectins. Structural homologs of the MPRs are present in yeast and *Drosophila*, but these lack M6P-binding ability.

The CD-MPR exists mainly as a dimer, with each monomer binding one M6P residue. However, monomeric and tetrameric forms of the CD-MPR exist, and the equilibrium between forms is affected by temperature, pH, and the presence of ligands. The CI-MPR also seems to be a dimer in the membrane, although it readily dissociates upon solubilization. Somewhat surprisingly, this much larger molecule binds only two residues of M6P, using just 2 of its 15 repeating units (a third repeat binds M6P very weakly). Mutagenesis studies have identified specific residues of these receptors involved in M6P binding, and the crystal structure of the single extracytoplasmic domain of the CD-MPR has been obtained in a complex with M6P. This domain crystallized as a dimer, with each monomer folded into a nine-stranded flattened β-barrel that has a striking resemblance to the protein folds in avidin (Figure 30.4). The distance between the two ligand-binding sites of the dimer provides a good explanation for the differences in binding affinity shown by the CD-MPR toward various lysosomal enzymes. The crystal structure of the amino-terminal 432

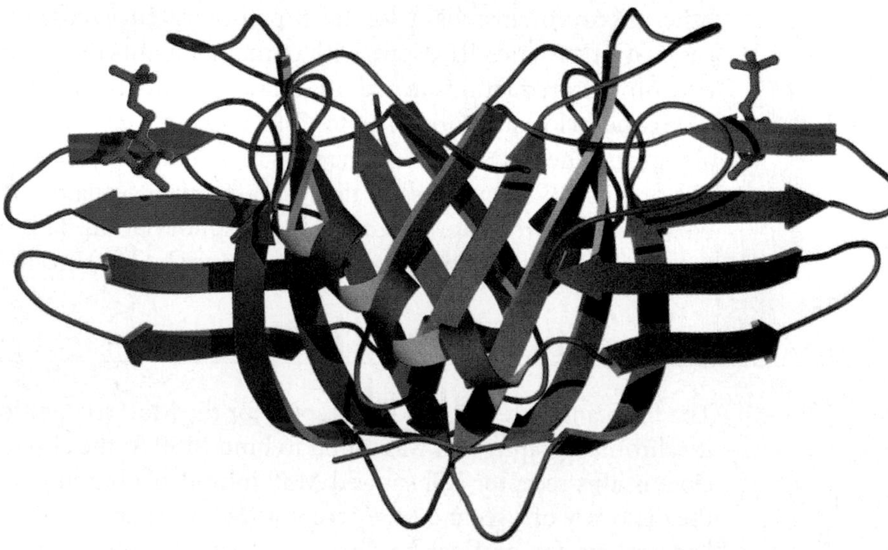

FIGURE 30.4. Ribbon diagram of the bovine cation-dependent M6P receptor (CD-MPR). Shown are the two monomers (*magenta and cyan ribbons*) of the dimer as well as the ligand M6P (*gold ball-and-stick model*). (Modified, with permission, from Roberts et al. 1998. *Cell* **93:** 639–648, ©Cell Press.)

residues of the CI-MPR, encompassing domains 1–3 (domain 3 is one of the M6P-binding domains), has also been solved. Each domain exhibits a topology similar to that of the CD-MPR, and the three domains assemble into a compact structure that provides insight into the arrangement of the entire extracellular region of the CI-MPR. The proposed model does not position the two M6P-binding domains (3 and 9) sufficiently close to bind a single, diphosphorylated N-glycan. This suggests that the high-affinity binding of this N-glycan is due to the spanning of binding sites located on different CI-MPR dimers. An interesting possibility is that the receptor is dynamic, with the spacing between the two M6P-binding sites being flexible, to enhance interactions with lysosomal enzymes containing phosphorylated glycans at various positions on their protein backbones.

GENERATION OF THE "M6P RECOGNITION MARKER" FOR THE MPRs

The M6P recognition marker actually encompasses a family of M6P-bearing N-glycans with varying degrees of affinity for the MPRs, based on the position of phosphate groups and the structure of the underlying N-glycan (Figure 30.2). The number and distribution of such N-glycans on different lysosomal enzymes could further affect binding to the two receptors. Thus, whereas both MPRs have a preference for enzymes containing glycans with two M6P residues, it appears that two appropriately spaced N-glycans with a single M6P each can provide a high-affinity ligand. A cohort of newly synthesized lysosomal enzymes therefore presents a spectrum of affinities for the MPRs. Taken together with factors such as the number, compartmental localization and availability of MPRs, differences in properties of the two receptors, and the concentration of cations, there is clearly much flexibility in this trafficking mechanism. Indeed, different cell types target different M6P-containing proteins to their lysosomes at different rates, with varying proportions being secreted.

Recognition of Lysosomal Enzymes by the GlcNAc Phosphotransferase

The oligomannosyl N-glycans of lysosomal enzymes are identical to those of many other glycoproteins passing through the ER-Golgi pathway. Thus, specific recognition of the former by GlcNAc-P-T is crucial to achieve selective trafficking. This recognition is not explained by any similarities in the primary polypeptide sequences of lysosomal enzymes. Indeed, denatured lysosomal enzymes lose their specialized GlcNAc-P-T acceptor activity, indicating that features of secondary or tertiary structure are critical for recognition by GlcNAc-P-T. Two complementary approaches have been used to define elements of this recognition marker. In loss-of-function studies, various amino acids of the lysosomal enzyme have been replaced with alanine, with the effect on phosphorylation determined. In gain-of-function experiments, residues of the lysosomal protease cathepsin D have been substituted into the homologous secretory protease glycopepsinogen. These studies revealed that selected lysine residues have a critical role in the interaction with GlcNAc-P-T. In fact, as few as two lysines in the correct orientation to each other and to an N-glycan can serve as minimal elements of the recognition domain. However, additional amino acid residues function to enhance the interaction with GlcNAc-P-T. In some instances (e.g., cathepsin D), the enzyme may contain a very extended determinant, or perhaps, more than one recognition domain.

Uncovering the M6P Recognition Marker

The enzyme GlcNAc phosphodiester glycosidase that catalyzes the exposure of the M6P recognition marker is found primarily in the *trans*-Golgi network (TGN), and it cycles between this compartment and the plasma membrane. Thus, uncovering the recognition marker appears to be a late event in the Golgi apparatus, occurring just before loading the enzymes onto the MPRs. Specific cytoplasmic residues have been identified in the glycosidase that determine Golgi retention, facilitate exit from the Golgi, and mediate internalization into clathrin-coated vesicles at the plasma membrane.

NATURAL AND INDUCED GENETIC DEFECTS IN THE MPRs

Targeted disruption of the CD-MPR gene in mice is associated with normal or only slightly elevated levels of lysosomal enzymes in the circulation and an otherwise normal phenotype. However, thymocytes or primary cultured fibroblasts from such mice show an increase in the amount of phosphorylated lysosomal enzymes secreted into the medium. Thus, there must be mechanisms that compensate for the deficiency in vivo. Intravenous injection of inhibitors of other glycan-specific receptors capable of mediating endocytosis (e.g., the mannose receptor of macrophages and the asialoglycoprotein receptor of hepatocytes; see Chapter 31) give rise to a marked increase in lysosomal enzymes in the serum of the deficient mice. Thus, such receptors are likely part of the compensatory mechanisms in vivo.

Other studies have shown that mouse CI-MPR is part of the naturally occurring *Tme* locus, a maternally imprinted region of chromosome 17 (i.e., expressed only from the maternal chromosome). Mice inheriting a deletion of the *Tme* locus from their mother die at day 15 of gestation. Genetic disruption of the CI-MPR gene showed that the lethality is due to lack of this receptor. Maternal inheritance of a null allele or homozygosity for the inactive allele is generally lethal at birth, and mutants are about 30% larger in size. The phenotype is probably due to an excess of insulin-like growth factor II (IGF-II), which is another ligand for the CI-MPR (see below), because the introduction of an IGF-II null allele rescued the mutant mice. CI-MPR mutant mice also have organ and skeletal abnormalities.

Cell lines lacking either or both MPRs were obtained by mating CD-MPR–deficient mice with mice heterozygous for a CI-MPR–deleted allele. Like fibroblasts that lack only CD-MPR, fibroblasts that lack only CI-MPR have a partial impairment in sorting. Fibroblasts from embryos that lack both receptors show a massive missorting of multiple lysosomal enzymes and accumulated undigested material in their endocytotic/lysosomal compartments. Thus, both receptors are required for efficient intracellular targeting of lysosomal enzymes. Comparison of lysosomal enzymes secreted by the different cell types indicates that the two receptors may interact preferentially with different subgroups of enzymes. Thus, the structural heterogeneity of the M6P recognition marker within a single lysosomal enzyme and between different enzymes is one explanation for the evolution of two MPRs with complementary binding properties: that is, to provide an efficient but varied targeting of lysosomal proteins in different cell types or tissues. In the final analysis, what initially appeared to be a precise "digital" "lock-and-key" mechanism turns out to be a far more complex and flexible "analog" system.

SUBCELLULAR TRAFFICKING OF THE MPRs

At steady state, MPRs are most concentrated in the TGN and late endosomes, but they cycle constitutively between these organelles, early (sorting) endosomes, recycling endosomes, and the plasma membrane (Figure 30.5). MPRs avoid delivery to lysosomes, where they would be degraded. This trafficking is directed by a number of short amino acid sorting sig-

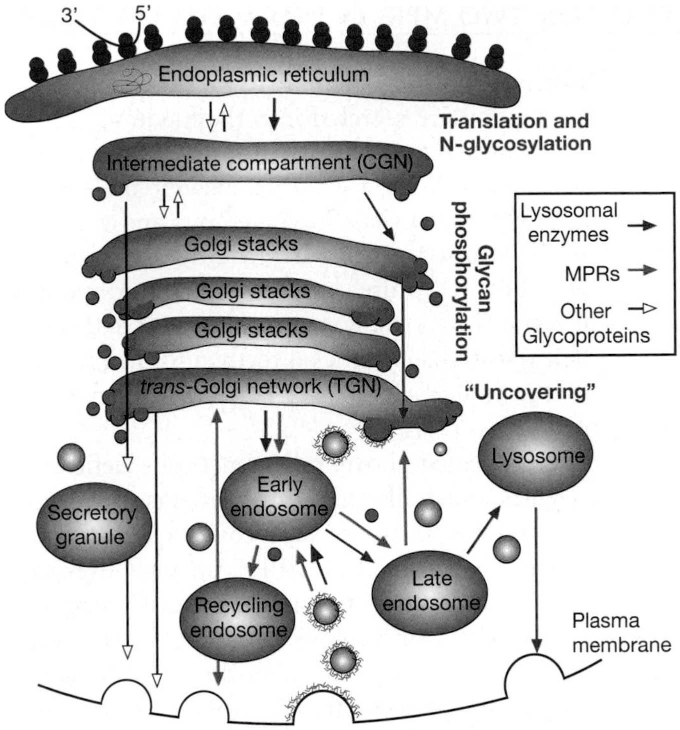

FIGURE 30.5. Subcellular trafficking pathways followed by glycoproteins, lysosomal enzymes, and MPRs. Newly synthesized glycoproteins originating from the rough ER pass through the Golgi stacks and are then sorted to various destinations as indicated. Along this route, lysosomal enzymes are recognized by GlcNAc-P-T and phosphorylated in the intermediate compartment (*cis*-Golgi network; CGN) and then acted upon by "uncovering" enzyme (phosphodiester glycosidase) in the *trans*-Golgi network (TGN). Beyond the TGN, trafficking of lysosomal enzymes is primarily mediated by the MPRs through early endosomes to late endosomes, in which their lysosomal enzyme cargo is released. Smaller amounts of lysosomal enzymes can escape capture by MPRs and be secreted into the extracellular fluid. Lysosomal enzymes can also reenter the cell by binding to cell-surface MPRs and subsequent endocytosis. Once in the endocytic pathway, internalized lysosomal enzymes can intermingle with those following the biosynthetic route, as depicted. (*Open arrowheads*) Pathways general to many nonlysosomal glycoproteins; (*red arrows*) specific itineraries of lysosomal enzymes; (*green arrows*) pathways of MPRs. Additional pathways for MPRs include recycling from the early endosome to the cell surface and back to the TGN, and from the recycling endosome and the late endosome, following release of their lysosomal enzyme cargo.

nals in the cytoplasmic tails of the receptors. The TGN is the site where newly synthesized lysosomal enzymes bind to MPRs that are then collected into clathrin-coated pits and budded into clathrin-coated vesicles for delivery to the early endosome. This process involves interaction of the MPRs with two types of coat proteins: the GGAs (Golgi-localized, γ-ear-containing, ADP-ribosylation factor binding) and AP1 (adapter protein 1). In addition to binding MPRs, the coat proteins recruit clathrin for the assembly of clathrin-coated vesicles. Following delivery to early endosomes, lysosomal enzymes are released from MPRs as the endosomes mature to late endosomes and the pH decreases. Late endosomes then undergo dynamic fusion/fission with lysosomes, allowing selective transfer of lysosomal enzymes to the lysosomes and leaving the MPRs behind in subdomains of the late endosomes. These MPRs may then either return to the TGN or move to the plasma membrane, where internalization via clathrin-coated pits occurs, mediated by the coat protein AP2. There are several pathways for the MPRs to be returned to the TGN from the various endosomal compartments, although the relative importance of the different pathways is unclear at this time.

RELATIVE ROLES OF THE TWO MPRs IN INTRACELLULAR TRAFFICKING

In many cell types, a minority of newly synthesized lysosomal enzyme molecules escape sorting and are secreted into the medium even though they carry M6P residues (Figure 30.5). Such secreted molecules may be recaptured by the same cell or by adjacent cells expressing cell-surface MPRs. Enzyme molecules that bind to such cell-surface MPRs are endocytosed via clathrin-coated pits and vesicles, eventually reaching the same late endosomal compartments where newly synthesized molecules arrive from the Golgi. This "secretion-recapture" pathway is a minor one in most cells, but has potential importance in some situations. For example, some activated macrophages secrete a large portion of their lysosomal enzymes directly into the medium. It is possible that under inflammatory situations, it is useful for such secreted enzymes to be returned to lysosomes via the MPR pathway.

As indicated above, cells genetically deficient in both receptors secrete most of their enzymes, much like cells from patients with I-cell disease. Under normal conditions, only the CI-MPR is responsible for lysosomal enzyme endocytosis from the cell surface. However, when the CD-MPR is strongly overexpressed, it is capable of mediating uptake from the plasma membrane. These differences may be due to the narrow pH optimum of the CD-MPR for binding ligand and/or its variable oligomeric state. Taken together, the results indicate that although the CI-MPR is a major determinant of trafficking in the biosynthetic pathway, the CD-MPR also contributes significantly. Curiously, when the CD-MPR is overexpressed in cells containing the CI-MPR, *increased secretion* of lysosomal enzymes can result. Thus, the CD-MPR may modulate the pathway in the direction of either retention or secretion, perhaps based on factors such as its oligomeric state, expression level, subcompartmental pH, divalent cation availability, amounts of CI-MPR present, and/or differences in precise affinities for multivalent ligands (Table 30.1). In the final analysis, different combinations of amounts and locations of the two receptors, together

TABLE 30.1. *Comparison of the two mammalian M6P receptors (P-type lectins)*

Feature	Cation-independent (large) receptor (CI-MPR)	Cation-dependent (small) receptor (CD-MPR)
Topology	type I membrane glycoprotein	type I membrane glycoprotein
Subunit mass in SDS-PAGE	250–300 kD	~45 kD
Core polypeptide mass	275 kD	28 kD
Optimal pH for binding	6.0–7.0	6.0–6.5
Cation dependence for binding	no	$Mn^{++} > Mg^{++} = Ca^{++}$
Domain structure	15 homologous repeating single units of ~145 amino acids each	155-amino-acid unit homologous to repeating units of CI-MPR
Native oligomeric state	dimer?	dimer or tetramer
Stoichiometry of M6P binding	two per monomer	one per monomer
K_d for N-glycan with two M6P units	2×10^{-9} M	2×10^{-7} M
Role in biosynthetic pathway	yes	yes
Role in endocytotic pathway	yes	no (except at high density)
Binding of other ligands:		
Methylphosphomannose	yes	no
IGF-II	yes (no in chicken/*Xenopus*)	no
Retinoic acid	yes	no
Urokinase-type plasminogen activator receptor (uPAR)	yes	no
Plasminogen	yes	no

with the spectrum of M6P recognition markers on various enzymes, could explain the highly variable physiology of lysosomal enzyme trafficking in different cell types.

IMPLICATIONS FOR ENZYME REPLACEMENT THERAPY

As described in Chapter 41, there are many genetic disorders in glycan degradation that result from decreased activity of a given lysosomal enzyme. Some of these enzymes that are targeted to lysosomes via the M6P pathway have been prepared in large quantities as recombinant soluble proteins and used in enzyme replacement therapy. To date the benefits have been variable but less than optimal. There are a number of potential reasons for this. First, some of the preparations do not contain the physiologic complement of the phosphomannosyl recognition marker. It is reasonable to suggest that the M6P-MPR system might be used to improve the efficacy of enzyme replacement in these patients. This has been shown to be the case in mouse and dog model systems. However, even with fully phosphorylated enzymes there may be obstacles that are difficult to overcome. For example, some cell types in the body may not express adequate levels of the CI-MPR on their surfaces to endocytose sufficient enzyme to restore normal lysosomal function. Also, the organ that is most seriously affected in many of these diseases (the brain) is inaccessible because of the blood–brain barrier. This is an area where further studies are needed, and replacement therapy using the MPR system might end up being more effective in selected circumstances.

EVOLUTIONARY ORIGINS OF THE M6P RECOGNITION SYSTEM

Although the MPR pathway has a major role in vertebrate lysosomal enzyme trafficking, its contribution in invertebrate systems is not prominent. Lysosomal enzymes are targeted in organisms such as *Saccharomyces*, *Trypanosoma*, and *Dictyostelium*, without the aid of identifiable MPRs. The slime mold *Dictyostelium discoideum* produces a novel methylphosphomannose structure on some of its lysosomal enzymes that can be recognized in vitro by the mammalian CI-MPR (not the CD-MPR). However, despite the presence of a GlcNAc-P-T that recognizes α1-2-linked mannose residues, the enzyme does not specifically recognize lysosomal enzymes, and no receptor for M6P has been found in this organism. The protozoan *Acanthamoeba* produces a GlcNAc-P-T that specifically recognizes lysosomal enzymes. However, this organism lacks a gene that could encode an "uncovering" enzyme, and so would not be expected to form M6P available to an MPR. Although some additional "lower" organisms do show evidence for an "uncovering" enzyme, no MPR activity has yet been found. The evolutionary divergence point at which the complete MPR system became established remains to be defined.

THE CATION-INDEPENDENT RECEPTOR BINDS MANY OTHER LIGANDS

Although originally discovered as a receptor for lysosomal enzyme trafficking, the CI-MPR turns out to be a remarkably multifunctional molecule. IGF-II was previously known to bind two receptors, one identical to the IGF-I receptor and another independent receptor. Molecular cloning of the latter receptor revealed the surprising fact that it is identical to the CI-MPR. Importantly, IGF-II does not carry M6P residues. Many studies have since explored potential interactions between these disparate ligands for the CI-MPR. Although the two ligands bind to distinct sites on the receptor, there are conflicting reports regarding synergistic or antagonistic interactions between the two activities. It has also been suggested that the redistribution of the CI-MPR upon IGF-II stimulation could explain some of the known metabolic effects of this hormone on protein degradation, by altering the trafficking of lysosomal enzymes. However, CI-MPRs of the chicken and *Xenopus* do not bind IGF-II, although their cells can respond to IGF-II. This makes it less likely that the overlap in binding specificity is of vital importance to animal cells in general. Rather, it appears that the CI-MPR acts primarily as a general "sink" for excess IGF-II in the extracellular fluid, carrying it to the lysosome for degradation, and reducing the amount available to bind to the IGF-I receptor. A number of reports indicate that binding of IGF-II to the CI-MPR regulates motility and growth in some cell types. It has also been found that the CI-MPR binds retinoic acid with high affinity at a site that is distinct from those for M6P and IGF-II. This binding of retinoic acid seems to enhance the primary functions of the CI-MPR receptor, and the biological consequence appears to be the suppression of cell proliferation and/or induction of apoptosis. The significance of this unexpected observation is still being explored. Other interesting ligands include urokinase-type plasminogen activator receptor (uPAR) and plasminogen.

There are also some unexplained changes in CI-MPR expression in relation to malignancy. Loss of heterozygosity at the CI-MPR locus occurs in dysplastic liver lesions and in hepatocellular carcinomas associated with the high-risk factors of hepatitis virus infection and liver cirrhosis. Mutations in the remaining allele were detected in about 50% of these tumors, which also seem to frequently develop from clonal expansions of phenotypically normal, CI-MPR-mutated hepatocytes. Thus, the CI-MPR fulfills many of the classic criteria to be classified as a liver "tumor-suppressor" gene.

SIGNIFICANCE OF M6P ON NONLYSOSOMAL PROTEINS

Interestingly, M6P-containing N-glycans have been found on a variety of nonlysosomal proteins. Some are hydrolytic enzymes that seem to have taken on a predominantly secretory route, for example, uteroferrin and DNase I. In the first case, the failure of removal of the blocking *N*-acetylglucosamine residues may be the cause for secretion. With DNase I, the native level of phosphorylation simply appears to be low. M6P has been found on the transforming growth factor β (TGF-β) precursor and the phosphate is lost from the mature form. It appears that M6P may serve to target the precursor to CI-MPR for activation. Other nonlysosomal proteins reported to carry M6P include proliferin, CREG (cellular repressor of E1A-stimulated genes), LIF (leukemia inhibitory factor), and thyroglobulin. In the last case, M6P-containing N-glycans are suggested as a mechanism to target the protein for degradation and release of thyroid hormones. One should not necessarily assume that M6P-containing N-glycans on all of these proteins are involved in intracellular trafficking. Just as phosphorylation of serine residues has diverse biological roles, M6P might be used for more than one purpose in a complex multicellular organism. Further investigation of each situation is therefore needed, with an open mind to all of the possibilities.

Herpes simplex virus and varicella zoster virus (VZV) glycoproteins have also been shown to carry N-glycans with M6P. In these cases, M6P is on complex N-glycans, suggesting that it originates from a distinct biosynthetic pathway. Regardless of its mode of synthesis, interaction of cell-free VZV with CI-MPR at the cell surface is required for viral entry into endosomes. Interestingly, intracellular CI-MPR can also divert newly synthesized enveloped VZV to late endosomes where the virions are inactivated before exocytosis. This is suggested as the mechanism by which this successful parasite limits immediate excessive spread and avoids killing the host. Biopsies of VZV-infected human skin showed that CI-MPR expression is lost in maturing superficial epidermal cells, preventing diversion of VZV to endosomes and allowing constitutive secretion of infectious VZV. These data implicate CI-MPR in the complex biology of VZV infection.

OTHER PATHWAYS FOR TRAFFICKING LYSOSOMAL ENZYMES

Although the M6P recognition marker has a crucial role in trafficking newly synthesized lysosomal enzymes to vertebrate lysosomes, alternate mechanisms evidently exist in some cell types. Even in I-cell disease, some cells and tissues (e.g., liver and circulating granulocytes) have essentially normal levels of enzymes. B-lymphoblast lines derived from these patients also do not show the complete phenotype of enzyme deficiency seen in fibroblasts. One interpretation is that the M6P pathway for trafficking of lysosomal enzymes is a specialized form of targeting, superimposed on some other basic mechanisms that evolutionarily more primitive organisms use, some of which remain undefined. In this regard, two lysosomal enzymes, acid phosphatase and β-glucocerebrosidase, are not at all affected in their distribution in I-cell disease fibroblasts. With acid phosphatase, the enzyme is synthesized initially as a membrane-bound protein, and once in the lysosome, it is proteolytically cleaved to generate the mature soluble form. Glucocerebrosidase is also membrane associated, does not show phosphorylation of its glycans, and is targeted to lysosomes independently of the MPR pathway. Likewise, integral membrane proteins of the lysosome such as the LAMP/lgp proteins do not require the M6P recognition marker pathway for trafficking to lysosomes. Rather, they seem to use motifs in their cytoplasmic tails similar to those that target MPRs to clathrin-coated vesicles.

FURTHER READING

Neufeld E.F. 1974. The biochemical basis for mucopolysaccharidoses and mucolipidoses. *Prog. Med. Genet.* **10:** 81–101.

Kornfeld S. 1986. Trafficking of lysosomal enzymes in normal and disease states. *J. Clin. Invest.* **77:** 1–6.

Von Figura K. and Hasilik A. 1986. Lysosomal enzymes and their receptors. *Annu. Rev. Biochem.* **55:** 167–193.

Kornfeld S. and Mellman I. 1989. The biogenesis of lysosomes. *Annu. Rev. Cell Biol.* **5:** 483–525.

Munier-Lehmann H., Mauxion F., and Hoflack B. 1996. Function of the two mannose 6-phosphate receptors in lysosomal enzyme transport. *Biochem. Soc. Trans.* **24:** 133–136.

Dahms N.M. and Hancock M.K. 2002. P-type lectins. *Biochim. Biophys. Acta* **1572:** 317–340.

Bonifacino J.S. and Traub L.M. 2003. Signals for sorting of transmembrane proteins to endosomes and lysosomes. *Annu. Rev. Biochem.* **72:** 395–447.

Ghosh P., Dahms N.M., and Kornfeld S. 2003. Mannose 6-phosphate receptors: New twists in the tale. *Nat. Rev. Mol. Cell Biol.* **4:** 202–213.

Ghosh P. and Kornfeld S. 2004. The GGA proteins: Key players in protein sorting at the *trans*-Golgi network. *Eur. J. Cell. Biol.* **83:** 257–262.

Gelfman C.M., Vogel P., Issa T.M., Turner C.A. Lee W.S., Kornfeld S., and Rice D.S. 2007. Mice lacking α/β subunits of GlcNAc-1-phosphotransferase exhibit growth retardation, retinal degeneration, and secretory cell lesions. *Invest. Ophthalmol. Vis. Sci.* **48:** 5221–5228.

Braulke T., Pohl S., and Storch S. 2008. Molecular analysis of the GlcNAc-1-phosphotransferase. *J. Inherit. Metab. Dis. Biol.* (in press).

CHAPTER 31

C-type Lectins

Richard D. Cummings and Rodger P. McEver

C-TYPE LECTINS ARE Ca^{++}-DEPENDENT GLYCAN-BINDING PROTEINS that share primary and secondary structural homology in their carbohydrate-recognition domains (CRDs). These proteins have a C-type lectin fold, which is a fold with highly variable protein sequence that is also present in many proteins that do not bind carbohydrates (C-type lectin domain [CTLD]-containing proteins). C-type lectins and proteins with CTLDs are found in all organisms. The large family of C-type lectins includes collectins, selectins, endocytic receptors, and proteoglycans. Some of these proteins are secreted and others are transmembrane proteins. They often oligomerize into homodimers, homotrimers, and higher-ordered oligomers, which increases their avidity for multivalent ligands. Although they share structural homology, C-type lectins usually differ significantly in the types of glycans that they recognize with high affinity. These proteins function as adhesion and signaling receptors in many immune functions such as inflammation and immunity to tumor and virally infected cells.

HISTORICAL BACKGROUND AND DISCOVERY OF C-TYPE LECTINS

The first C-type lectin identified in animals was the hepatic asialoglycoprotein receptor (ASGPR), also termed the hepatic galactose/*N*-acetylglucosamine receptor, as introduced in Chapter 26. It was discovered during studies on the structure and function of the sialylated serum glycoprotein, ceruloplasmin. Most serum glycoproteins contain terminal sialic acid residues, and it was observed that desialylated glycoproteins were rapidly removed from the blood. A key experimental result was that the clearance of enzymatically desialylated and radiolabeled ceruloplasmin depended on exposure of the penultimate galactose residue in ceroluplasmin, and removal or modification of the galactose caused retention in the circulation. The desialylated glycoproteins removed from the circulation were sequestered in the liver, principally in lysosomes.

A Ca^{++}-dependent receptor specific for asialoglycoproteins was identified in hepatocyte plasma membrane fractions, and the ASGPR was purified by affinity chromatography on asialo-orosomucoid–Sepharose. The ASGPR was found to consist of a major subunit with a molecular mass of approximately 48 kD and a minor subunit with a mass of about 40 kD. The purified rabbit hepatocyte receptor agglutinated desialylated human and rabbit erythrocytes and also induced mitogenesis in desialylated peripheral lymphocytes. This was the first demonstration that an animal lectin could have such profound effects on cellular metabolism. Binding of rabbit hepatic lectin to cells was inhibited by both *N*-acetylgalactosamine and galactose, with the former being more potent than the latter, and the lectin bound best to glycoproteins containing either nonreducing terminal *N*-acetylgalactosamine or galactose. At the time of the discovery, it was not known that circulating glycoproteins had terminal *N*-acetylgalactosamine residues, but now we know that many pituitary glycoprotein hormones and parasite-derived glycoproteins contain terminal β-linked *N*-acetylgalactosamine residues and may be recognized by this hepatic receptor. A similar lectin in rats, which contained an unusual heterotrimeric structure composed of two subunits, was termed rat hepatic lectin R2/3. A related lectin identified in chicken hepatocytes recognized glycoproteins containing terminal *N*-acetylglucosamine residues. Thus, the avian liver rapidly cleared only glycoproteins that lacked both terminal sialic acid and galactose residues. Interestingly, circulating glycoproteins in birds appear to be constitutively desialylated compared to their mammalian counterparts.

DEFINITION OF C-TYPE LECTINS AND STRUCTURAL MOTIFS

The C-type lectin fold has been found in more than 1000 proteins, and it represents a ligand-binding motif that is not necessarily restricted to binding sugars. The CTLD broadly denotes proteins with a C-type lectin domain, regardless of their ability to bind sugars. In animals, C-type lectins are the major representatives of CTLD-containing proteins. In metazoans, most proteins with a CTLD are not lectins. Proteins use the C-type lectin fold to bind other proteins, lipids, inorganic molecules (e.g., Ca_2CO_3), or even ice (e.g., the antifreeze glycoproteins). Glycan binding by the C-type lectins is always Ca^{++}-dependent because of specific amino acid residues that coordinate Ca^{++} and bind the hydroxyl groups of sugars. The C-type lectin fold is a rigid scaffold that can accommodate a remarkable number of sequence variations. An extreme example is seen in the major tropism determinant (Mtd), which is a receptor-binding protein of *Bordetella* bacteriophage. This C-type lectin fold may occur in at least 10^{13} different sequences, a diversity that rivals the immunoglobulin fold in its conservation of structure using millions of different primary amino acid sequences. Thus, the C-type lectin fold is an evolutionarily ancient structure that is adaptable for many uses.

The C-type lectin fold is unique. It is a compact domain of 110–130 amino acid residues with a double-looped, two-stranded antiparallel β-sheet formed by the amino- and carboxy-terminal residues connected by two α-helices and a three-stranded antiparallel β-sheet (Figure 31.1). The CRD has two highly conserved disulfide bonds and up to

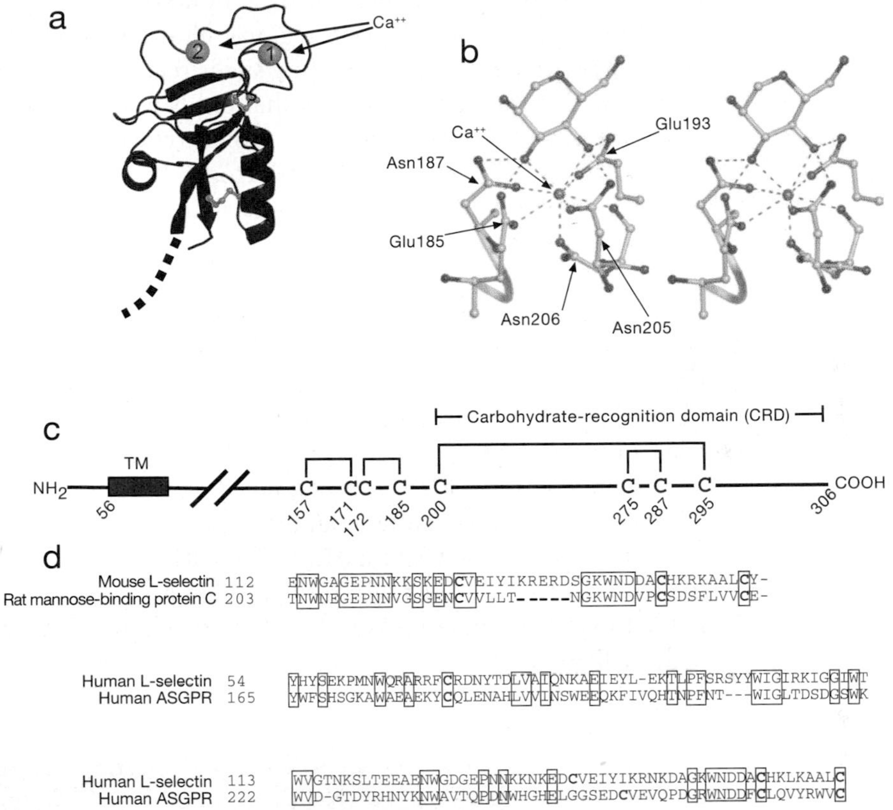

FIGURE 31.1. Structure of C-type lectins. (*a*) Ribbon diagram of the carbohydrate-recognition domain (CRD) of rat mannose-binding protein A (MBP-A). (*Light green spheres*) Ca^{++}-binding sites, where 1 is the auxiliary binding site and 2 is the principal binding site. (*Purple bars*) Disulfide. The long-loop region that binds Ca^{++} ions is shown at the top of the CRD. A single disulfide bond helps to form this loop, and a second disulfide bond at the bottom of the CRD helps to form the whole loop domain. (Redrawn, with permission of the American Society for Biochemistry and Molecular Biology, from Feinberg H., Park-Snyder S., Kolatkar A.R., et al. 2000. *J. Biol. Chem.* **275:** 21539–21548.) (*b*) Stereoview of the complex between rat MBP-A and the terminal mannose residue in the N-glycan Man_6-$GlcNAc_2$-Asn. (*Orange*) Coordination bonds. Hydrogen bonds where sugar hydroxyl groups act as acceptor are *red* and those where they act as donor are *blue*. The interaction is through a ternary complex formed between the terminal mannose of the glycan, the Ca^{++} ion in binding site 2, and the protein. The complex is stabilized by a network of coordination and hydrogen bonds involving the 3- and 4-hydroxyl O atoms in the mannose, two coordination bonds with the Ca^{++} ion, and four hydrogen bonds with the carbonyl side chains that form the Ca^{++}-binding site. (Redrawn from PDB image 2msb and by permission of Macmillan Publishers Ltd. from Weis W.I., Drickamer K., and Hendrickson W.A. 1992. *Nature* **360:** 127–134.) (*c*) The disulfide bonding in the CRD of the asialoglycoprotein receptor (ASGPR). The eight Cys residues in the extracellular domain are shown, with four disulfide bonds, including two in the CRD. The transmembrane domain (TM) is indicated. (Modified and with permission from the American Society for Biochemistry and Molecular Biology, from Yuk M.H. and Lodish H.F. 1995. *J. Biol. Chem.* **270:** 20169–20176.) (*d*) Primary sequence comparisons between different C-type lectins. Residues are numbered from the amino terminus. Cys residues are in *bold red* and homologous residues are *boxed*.

four sites for binding Ca^{++}, with site occupancy depending on the lectin. Amino acid residues with carbonyl side chains are often coordinated to Ca^{++} in the CRD, and these residues directly bind to sugars when Ca^{++} is bound in site 2. A ternary complex may be formed between a sugar in a glycan, the Ca^{++} ion in site 2, and amino acids within the CRD. Changes in amino acids within the CRD may alter sugar specificity. Key conserved residues that bind sugars include the "EPN" and "WND" motifs within the CRD of C-type lectins (see mouse L-selectin and rat mannose-binding protein C in Figure 31.1). The presence of such motifs in the CRD, along with Ca^{++} binding in site 2 and other secondary structures (including hydrogen bond donors and acceptors flanking a conserved Pro residue in the double-loop region), allows predictions as to whether a CRD binds sugar. However, because the binding site is relatively shallow with few contacts to sugars, it is difficult to predict the glycan that binds to a particular C-type lectin. Nevertheless, sequence determinants in the CRD provide clues to the "monosaccharide" specificity of C-type lectins (e.g., mannose vs. galactose). In several C-type lectins, such as P-selectin and the ASGPR, Ca^{++} binding induces structural changes in the CRD that stabilize the double loop region. Loss of Ca^{++} can lead to destabilization of these loops and loss of ligand binding, even when Ca^{++} is not directly involved in complexing the ligand, as seen in the macrophage mannose receptor. This destabilization is also important in pH-induced changes that lead to loss of ligand-binding affinity, because of the pH-induced loss of Ca^{++}. In CTLD-containing proteins such as human tetranectin, which is not known to bind sugars, Ca^{++} is also important for interactions with several kringle-domain-containing proteins.

C-type lectins exist as both oligomers and monomers. Many C-type lectins occur as trimers, including the trimeric rat mannose-binding protein-A (MBP-A) in complex with α-methyl mannoside (Figure 31.2). The CRD of trimeric lectins is angled to the side of the stalk domain through which the protein associates to form the trimer. The CRDs are at the top of the trimer and can function to enhance multivalent interactions with carbohydrate ligands.

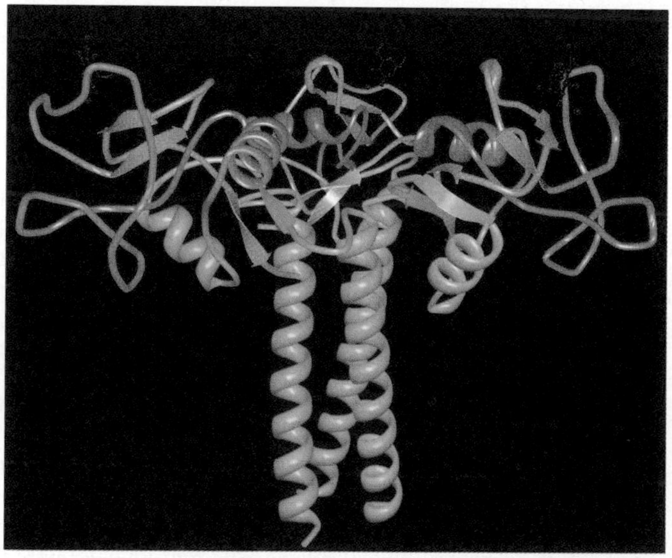

FIGURE 31.2. Crystal structure of trimeric rat mannose-binding protein-A complexed with α-methylmannoside. (Created from PDB deposited structure 1kwu and with permission of the American Society for Biochemistry and Biology from Ng K.K., Kolatkar A.R., Park-Snyder S., et al. 2002. *J. Biol. Chem.* **277:** 16088–16095.)

DIFFERENT SUBFAMILIES OF C-TYPE LECTINS

There are at least 17 groups of proteins with CTLDs, which are distinguished by their domain architecture. Well over 100 different proteins encoded in the human genome contain the CTLD (Figure 31.3). Most of these groups have a single CTLD, but the macrophage mannose receptor (group VI) has eight of these domains. Groups VII (REG), IX (tetranectin), XI (attractin), XIII (DGCR2; DiGeorge syndrome critical region gene 2), XV (BIMLEC), and XVII (CBCP) have no known glycan ligands. From a functional perspective, we know most about collectins, endocytic receptors, myeloid lectins, and selectins, and these groups are discussed in detail below.

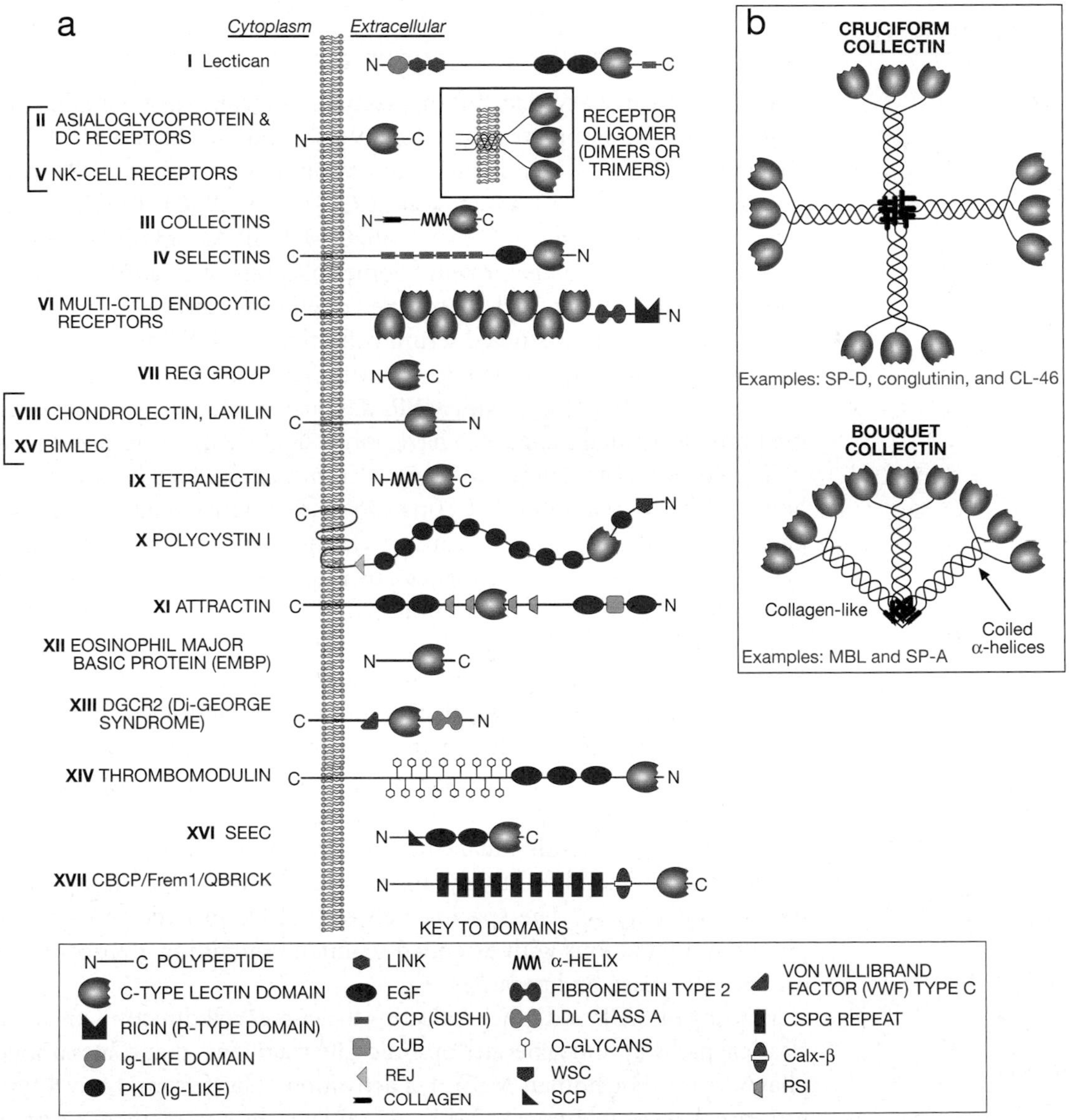

FIGURE 31.3. Different groups of C-type lectins and their domain structures. Seventeen groups are shown, defined by their phylogenetic relationships and domain structures. Some of the groups are soluble proteins and others are transmembrane proteins. The group III collectins form oligomeric structures, shown as cruciform and bouquet structures in the box. Each of the domains is named as indicated in the key. (DC) Dendritic cell; (NK cell) natural killer cell; (CTLD) C-type lectin domain; (MBP) mannose-binding protein; (SP-A and SP-D) surfactants; (CCP) complement control protein.

CTLD-containing proteins are found in all metazoans and many nonmetazoans. This latter group includes bacterial toxins (e.g., pertussis toxin), outer-membrane adhesion proteins (e.g., invasin from *Yersinia pseudotuberculosis*), and viral proteins (e.g., envelope protein in Epstein–Barr virus). Interestingly, the viral proteins have more similarity to mammalian CTLD-containing proteins than the bacterial proteins.

There are at least 135 proteins with CTLDs within the genome of *Caenorhabditis elegans*, which has approximately 19,000 protein-coding sequences. Within this group, 183 C-type CTLDs have been found, because some proteins have multiple CTLDs. Interestingly, in response to pathogens, *C. elegans* expresses at least 10 genes that encode proteins with a CTLD out of the 68 pathogen up-regulated genes described to date.

THE COLLECTINS

The collectins are C-type lectins that contain a collagen-like domain and usually assemble in large oligomeric complexes containing 9–27 subunits (Figure 31.3). To date, nine different collectins have been identified: mannose-binding protein (MBP), conglutinin, surfactant proteins SP-A and SP-D, and collectins CL-43, CL-46, CL-P1, CL-L1, and CL-K1. The collectins MBP, conglutinin, CL-43, CL-46, CL-K1, SP-A, and SP-D are soluble, whereas CL-L1 and CL-P1 are membrane proteins. Some collectins, such as MBP and SP-A, organize into a "bouquet," and others, such as bovine conglutinin and SP-D, organize into a "cruciform" shape. One of the best-studied serum collectins is MBP. Bovine CL-43 is structurally one of the simplest collectins, consisting of only three polypeptides, each of which contains a terminal CTLD. Rats have two serum MBPs designated A and C, sometimes called mannan-binding proteins. Humans appear to have only a single MBP corresponding to the rat MBP-A.

The collectins contribute to "innate" immunity and act before the induction of an antibody-mediated response. Collectins stimulate in vitro phagocytosis by recognizing surface glycans on pathogens, they promote chemotaxis, and they stimulate the production of cytokines and reactive oxygen species by immune cells. Lung surfactant lipids have the ability to suppress a number of immune cell functions such as proliferation, and this suppression of the immune response is further augmented by SP-A. Although SP-A and SP-D were originally found in the lung, they are also expressed in the intestine.

MBP forms a trimeric helical structure via interactions of the collagenous tails that are stabilized by disulfide bonds in the cysteine-rich amino-terminal region. These trimers aggregate to generate three or six trimers in a "bouquet" organization. Each CRD in the trimer is separated by approximately 53 Å, which is critical to the function of the lectin. This is because each individual CRD has a relatively low affinity and low specificity for glycan ligands and can bind to glycans rich in *N*-acetylglucosamine, *N*-acetylmannosamine, fucose, and glucose. The spacing between CRDs provides regulation and enhances the potential interactions with extended mannan-containing glycoconjugates, especially those on bacteria, yeast, and parasites.

Binding of MBP and other collectins to a target cell directly activates complement via the classical pathway and generates opsonic C3b fragments that coat pathogens and lead to their phagocytosis. For human MBP, this activation results from a novel type of C1s-like serine protease that complexes with MBP and initiates the complement cascade in vivo. Some individuals with MBP deficiency syndrome have mutations in the Gly-X-Y repeat encoded within exon-1 of the MBP gene (*MBL2*). Mutations within exon-1, which are highly variable among human populations, inhibit assembly of the MBP subunit, leading to increased risk of microbial infections. Furthermore, several polymorphisms within the promoter region of MBL2 are associated with MBP deficiency and enhanced susceptibility to infections.

THE ENDOCYTIC RECEPTORS

Many C-type lectins function in receptor-mediated endocytosis to deliver bound, soluble ligands to lysosomes (Figure 31.4). The ASGPR and most endocytic receptors are type II transmembrane proteins, whereas the macrophage mannose receptor is type 1. The cytoplasmic domains of endocytic C-type lectins, such as the ASGPR in hepatocytes, DC-SIGN (dendritic cell-specific intercellular adhesion molecule 3-grabbing nonintegrin) in myeloid cells, and P-selectin in platelets and endothelial cells, have internalization motifs that include triads of acidic amino acid and dileucine- or tyrosine-based motifs. Myeloid cells, including dendritic cells and macrophages, express a large number of C-type lectins. Endocytosis of ligands by C-type lectins in dendritic cells and macrophages can lead to

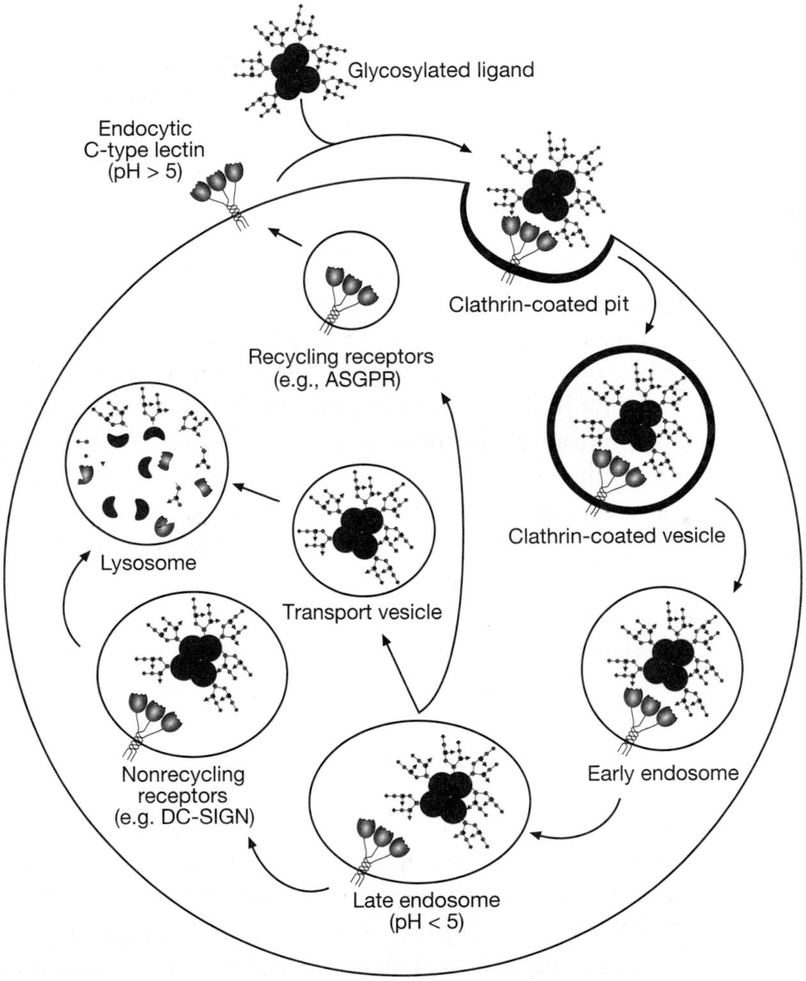

FIGURE 31.4. C-type lectins that function as endocytic receptors. Ligands are internalized by clathrin-dependent pathways and delivered to early and then late endosomes. Receptors may be recycled or degraded, depending on the receptor and the type of ligand it endocytoses. At a pH of less than 5, C-type lectins typically lose Ca^{++}, which shifts the equilibrium to promote ligand dissociation. The cytoplasmic domains of these receptors dictate their fates in the endocytic pathway. The tyrosine-based motif in the cytoplasmic domains of the asialoglycoprotein receptor (ASGPR) and the mannose receptor promotes ligand delivery to early endosomes and receptor recycling to the cell surface. The triacidic motif (EEE or DDD) in the cytoplasmic domain of DEC-205 diverts it to late endosomes/lysosomes. DC-SIGN has a tyrosine-based, coated pit sequence-uptake motif similar to that in the macrophage mannose receptor, along with a dileucine motif essential for internalization and a triacidic cluster, which is key to targeting to endosomes/lysosomes.

receptor accumulation and degradation in phagolysosomes or to recycling of the receptor to the cell surface. The pathway taken is dependent on the bound ligand. For example, dectin-1 is degraded when it internalizes zymosan, but it is recycled when it endocytoses soluble ligands. Stimulation of C-type lectins such as dectin-1 in myeloid cells activates mitogen-activated protein kinase (MAPK) and NF-κB and enhances transcription of genes important in innate immune responses. Internalization of antigens via the C-type lectins in dendritic cells induces production of reactive oxygen species and other responses.

The mammalian ASGPR is composed of major and minor subunits in different ratios that are encoded by two genes. The human ASGPR is a heterotetramer (H1 and H2 in a 3:1 ratio). The H2 transcript, however, exhibits splice variants H2a, H2b, and H2c, and the encoded proteins may contribute to different receptor structures and activities. The rat ASGPR is composed of three subunits, designated rat hepatic lectins (RHL) 1, 2, and 3. RHL1 is derived from a single gene, whereas RHL2 and RHL3 are derived from the same gene and have the same primary sequence, but they differ in glycosylation. In contrast, the chicken hepatic lectin is probably a homotrimer and contains a single subunit. These hepatic lectins occur in oligomeric forms that promote high affinity for specific glycoconjugate ligands. Clustering of the CRDs may determine both the specificity and affinity of the lectins for ligands because each individual CRD can act independently to bind sugar. Studies have shown that the precise orientation of the trimers within the ASGPR can dictate its affinity to specific glycans that have multivalent presentations. Tri- and tetra-antennary N-glycans with appropriate branching and presentation of nonreducing terminal galactose or *N*-acetylgalactosamine residues bind to the rat ASGPR with greater than 100,000 times higher affinity (~nM range) than ligands with a single terminal Gal/GalNAc residue. Interestingly, the rat hepatic ASGPR can also bind sialylated ligands (Siaα2-6Galβ1-4GlcNAc-R). The hepatic lectins have a single C-type CRD, whereas the macrophage mannose receptor has eight CRDs in a single polypeptide. The adjacent CRDs may help it to bind specific multivalent, mannose-containing glycans. The macrophage mannose receptor internalizes lysosomal enzymes containing high-mannose-type N-glycans and facilitates phagocytosis of several pathogens such as yeast, *Pneumocystis carinii*, and *Leishmania*.

Although the ASGPR was predicted to contribute to homeostasis and maintenance of serum glycoprotein levels, mice lacking subunit 2 have no obvious phenotypic abnormalities and do not accumulate endogenous glycoproteins in serum. However, the mutant mice have defective hepatic uptake and clearance of injected asialoglycoproteins. Thus, ASGPR may be required to regulate serum glycoprotein levels in periods of induced stress, which is known to elevate serum glycoprotein levels, and/or to be involved in specific interactions with glycoconjugates from pathogenic organisms. In the latter case, these hepatic receptors may function as part of the innate immune system, perhaps in a manner like the C-type lectins in myeloid cell function. For example, the human ASGPR promotes endocytosis-dependent entry of hepatitis B virus particles into liver cells. It is interesting that autoimmune hepatitis is associated with autoantibodies to the human hepatic ASGPR, and there is evidence that hepatitis virus invasion can augment this response.

Domains of C-type lectins other than the canonical CRD may also have receptor activity. For example, the cysteine-rich domain of the "mannose" receptor binds to glycans containing R-GalNAc-4-SO_4 on pituitary glycoprotein hormones and thus acts to clear these hormones from the circulation.

THE MYELOID C-TYPE LECTINS

Myeloid cells have many C-type lectins, which belong mainly to groups II, V, and VI (see Figure 31.3). In addition, myeloid cells express many members of the galectin and Siglec families of lectins. C-type lectins in myeloid cells include in group II, DC-SIGN (in humans, but there is no murine homolog), SIGN-R1 (in mice, but there is no human homolog), macrophage C-type lectin (MCL), dectin-2, langerin, and the macrophage galactose-binding lectin (MGL); in group V, dectin-1, myeloid-DAP12-associating lectin (MDL-1), and dendritic-cell-associated lectin-1 (DCAL-1); and in group VI, macrophage mannose receptor and DEC-205.

Dendritic cells (DCs) are important antigen-presenting cells. They internalize antigens by fluid-phase pinocytosis and via specific endocytic receptors, and they process antigens for presentation to $CD8^{+}$ cytotoxic T cells. DCs also contribute to the balance between tolerance and the induction of immunity and help to distinguish harmless "self-antigens" from pathogens (Figure 31.5). When DCs are activated, they migrate to lymph nodes where they interact with T cells. The types of antigens that DCs encounter lead to their differentiation into different subtypes, which function to instruct the differentiation of T cells. Toll-like receptors (TLRs) and C-type lectins (for myeloid cells, these have been termed C-type lectin receptors or CLRs) act to help DCs discriminate between pathogens and self-antigens. TLRs are pattern-recognition receptors (PRRs) that interact with "pathogen-associated molecular patterns" (PAMPs). TLRs, unlike CLRs, cannot directly promote phagocytosis of bound ligands. Ligation of TLRs can lead to DC maturation in a pathogen-specific manner. In contrast, DC interaction with pathogens through CLRs, in the absence of TLR activation, leads to internalization and processing of antigens and can result in DCs remaining immature. Interactions of T cells with antigens presented by immature DCs may

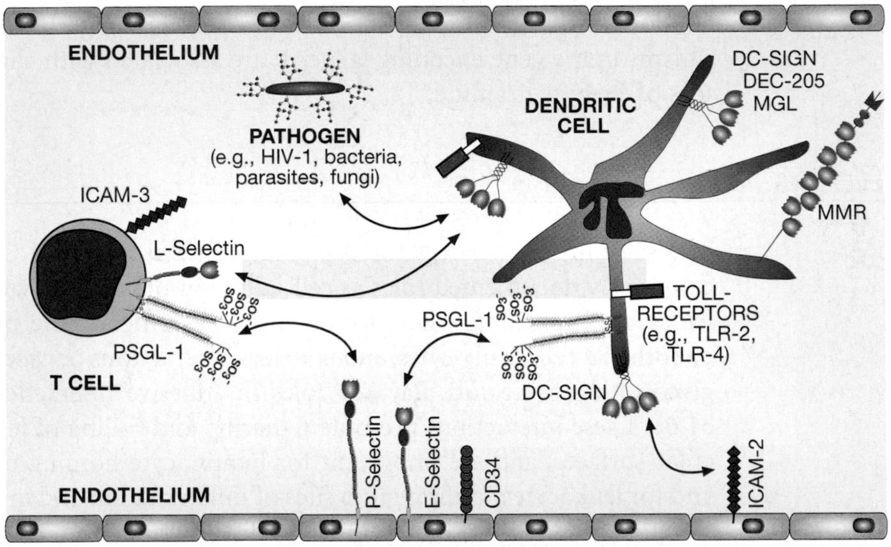

FIGURE 31.5. C-type lectins function in innate immune responses and have a dual function in pathogen recognition and cell adhesion. Lectins expressed on dendritic cells (DCs) and macrophages interact with free pathogen-derived glycans or directly with pathogens, which can lead to pathogen adhesion and possibly internalization if the pathogen is single-celled. Activation of C-type lectins such as DC-SIGN and dectin-1 potentiates cytokine production by DCs. Interactions of pathogens with toll-like receptors (TLRs, such as TLR-2 and TLR-4) can directly activate DCs and macrophages, leading to release of cytokines such as interferon-γ and IL-12, and up-regulation of factors that stimulate T cells. C-type lectins on DCs and macrophages can synergize or antagonize TLR signals, depending on the glycan and pathogen. For example, mycobacterium-derived mannosylated lipoarabinomannans bind to DCs via the macrophage mannose receptor (MMR) and DC-SIGN, which inhibit TLR-mediated IL-12 production. Thus, simultaneous binding of ligands to TLRs and C-type lectins might produce a tolerogenic Th2 response rather than a protective Th1 response. Lectins can bind to specific ligands such as ICAM-2, CD34, and PSGL-1 to promote endothelial and T-cell interactions.

lead to tolerogenic, rather than pathogenic, responses. However, if both TLRs and CLRs are activated and "cross-talk" ensues, DCs mature in different ways. For example, mycobacteria interact with DCs via TLR-2 and TLR-4, resulting in strong T-helper 1 (Th1) responses by the activated DCs. However, some virulent strains of mycobacteria secrete glycosylated factors (e.g., ManLAM) that are bound by CLRs, but not by TLRs, which lead to down-regulation of TLR activation and limitation of DC maturation.

Dectin-1 is a type of natural killer (NK) cell C-type lectin (group V; see Figure 31.3) that binds ligands independently of Ca^{++}. Dectin-1, which is also expressed on human neutrophils and macrophages, is a major PRR that recognizes glycan antigens such as β-glucans. When dectin-1 and TLR-2 are coligated by PAMPs, they coordinate the secretion of proinflammatory cytokines (interleukin-12 and tumor necrosis factor-α [TNF-α]) and the production of reactive oxygen species. Dectin-1 is degraded intracellularly upon internalization of large-sized β-glucans, but it can recycle upon internalizing small-sized β-glucans. Dectin-1 carries an immunoreceptor tyrosine-based activation motif (ITAM) in the cytoplasmic domain, which is involved in cell signaling through Syk family tyrosine kinases. DC-SIGN in human cells has been of special interest because it binds to the oligomannose-type N-glycans in the envelope glycoprotein of HIV-1. This protein also recognizes fucosylated ligands expressed by parasites such as the helminth *Schistosoma mansoni.*

Langerhans cells are a special type of DC and represent immature DCs found in the epidermis and mucosal tissues. These cells probably do not commonly coexpress the macrophage mannose receptor, but they do express many other C-type lectins, including langerin, which is a C-type lectin that largely recognizes glycans rich in mannose. Internalization of ligands by langerin leads to accumulation in Birbeck granules, which are subdomains of endosomes specifically found in Langerhans cells. Interestingly, polymorphisms in the gene encoding langerin are associated with changes in ligand specificity and loss of Birbeck granules.

THE SELECTINS

The selectins are perhaps the best-characterized family of C-type lectins because of their extensively documented roles as cell-adhesion molecules that mediate the earliest stages of leukocyte trafficking. The selectins are type-1 membrane proteins that are expressed on endothelial cells, leukocytes, and platelets. Interactions between the selectins and cell-surface glycoconjugate ligands play key roles in adhesive interactions among these cells (Figure 31.6). These interactions promote tethering and rolling of leukocytes and platelets on vascular surfaces, and are important for lymphocyte homing to secondary lymphoid organs and for leukocyte recruitment to sites of inflammation and injury. Rolling is a form of adhesion that requires rapid formation and dissociation of bonds between selectins and their ligands. Rolling adhesion enables leukocytes to encounter endothelium-bound chemokines. Signaling through chemokine receptors cooperates with signaling through selectin ligands to activate leukocyte integrins, which bind to immunoglobulin superfamily ligands on endothelial cells to slow rolling velocities and arrest leukocytes on vascular surfaces. The arrested leukocytes then emigrate across the vascular wall into the underlying tissues.

The three selectins are L-selectin, which is expressed on all leukocytes; E-selectin, which is expressed by cytokine-activated endothelial cells; and P-selectin, which is expressed constitutively in α-granules of platelets, in Weibel–Palade bodies of endothelial cells, and on the surfaces of activated platelets and endothelial cells (Figure 31.7). Each selectin has a C-type CRD at the amino terminus followed by a consensus epidermal growth factor (EGF)-like domain and a number of short consensus repeats composed of sushi domains (also called complement control protein [CCP] modules). The proteins have a single transmembrane

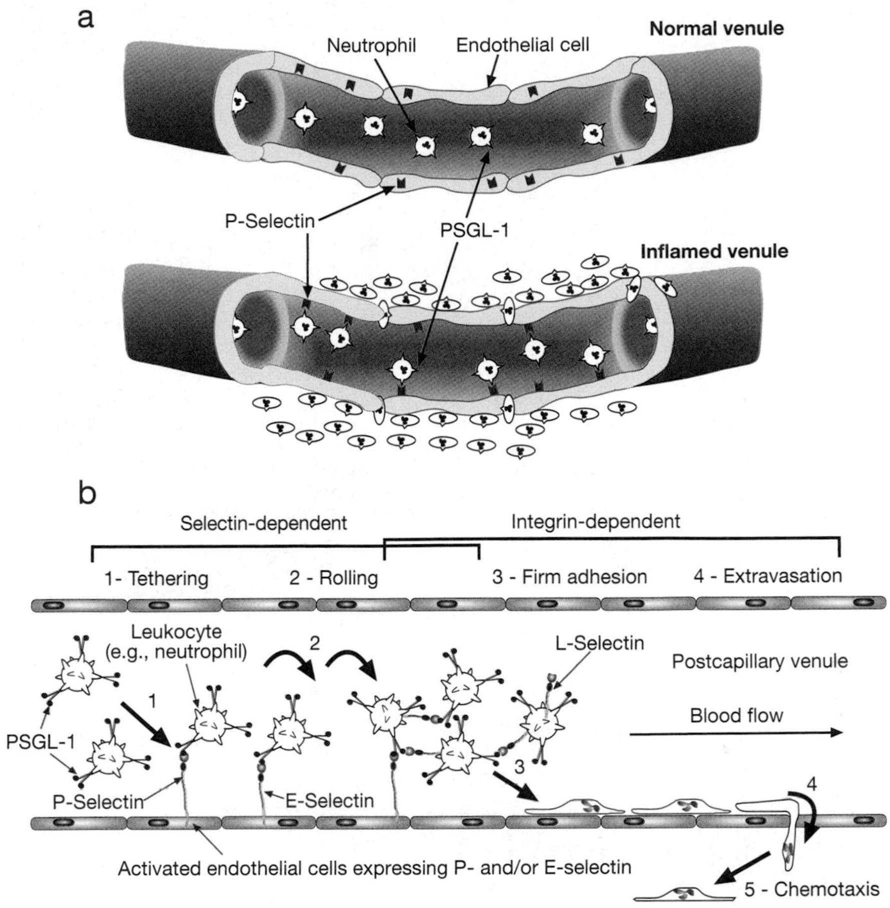

FIGURE 31.6. Tethering of circulating leukocytes to activated endothelium via interactions between selectins and their ligands. (*a*) In normal venules, leukocytes flow without adhesive interactions with the endothelium, but in inflamed vessels, selectins and integrin ligands are expressed on endothelial surfaces. This leads to tethering, rolling, and arrest of circulating leukocytes and their eventual extravasation from the circulation to the surrounding tissue. For example, P-selectin is normally expressed in Weibel–Palade bodies of endothelial cells, but within minutes after endothelial cell activation by thrombin, histamine, hypoxia, or injury, these bodies fuse with the plasma membrane, promoting the expression of P-selectin on the endothelial cell surface. Similarly, P-selectin stored in the α-granules of platelets becomes expressed on the surfaces of platelets within minutes after platelet activation. Leukocytes tether to and roll on activated endothelial cells and platelets. (*b*) The example shows neutrophils constitutively expressing PSGL-1 and L-selectin on their microvilli. These cells interact by random contact with membrane P-selectin on endothelial cells through PSGL-1. E-Selectin may also participate in these interactions by binding to PSGL-1 or other ligands on these cells. These interactions enable leukocytes to tether to and roll along the endothelium. Leukocyte–leukocyte interactions involving L-selectin and PSGL-1 are also depicted. Adherent cells become activated by regionally presented chemokines or lipid autacoids. The activated leukocytes express integrins (e.g., LFA-1, CD11a/CD18 and Mac-1, CD11b/CD18) that interact with immunoglobulin-like counterreceptors on endothelial cells (ICAM-1, ICAM-2) to strengthen the adhesion and promote the transmigration of cells from the circulation into the underlying tissues.

domain (TM) and a cytoplasmic domain and are relatively rigid, extended molecules. The C-type CRD of each selectin has modest affinity for the sialylated, fucosylated structure known as the sialyl Lewis^x antigen NeuAcα2-3Galβ1-4(Fucα1-3)GlcNAcβ1-R (SLe^x). In addition, P- and L-selectin, but not E-selectin, bind to some forms of heparin/heparan sulfate. However, each of the selectins binds with higher affinity to specific macromolecular ligands. Many of the known ligands are mucins containing sialylated, fucosylated O-glycans. The major ligand for P-selectin, termed P-selectin glycoprotein ligand-1 (PSGL-1), has sul-

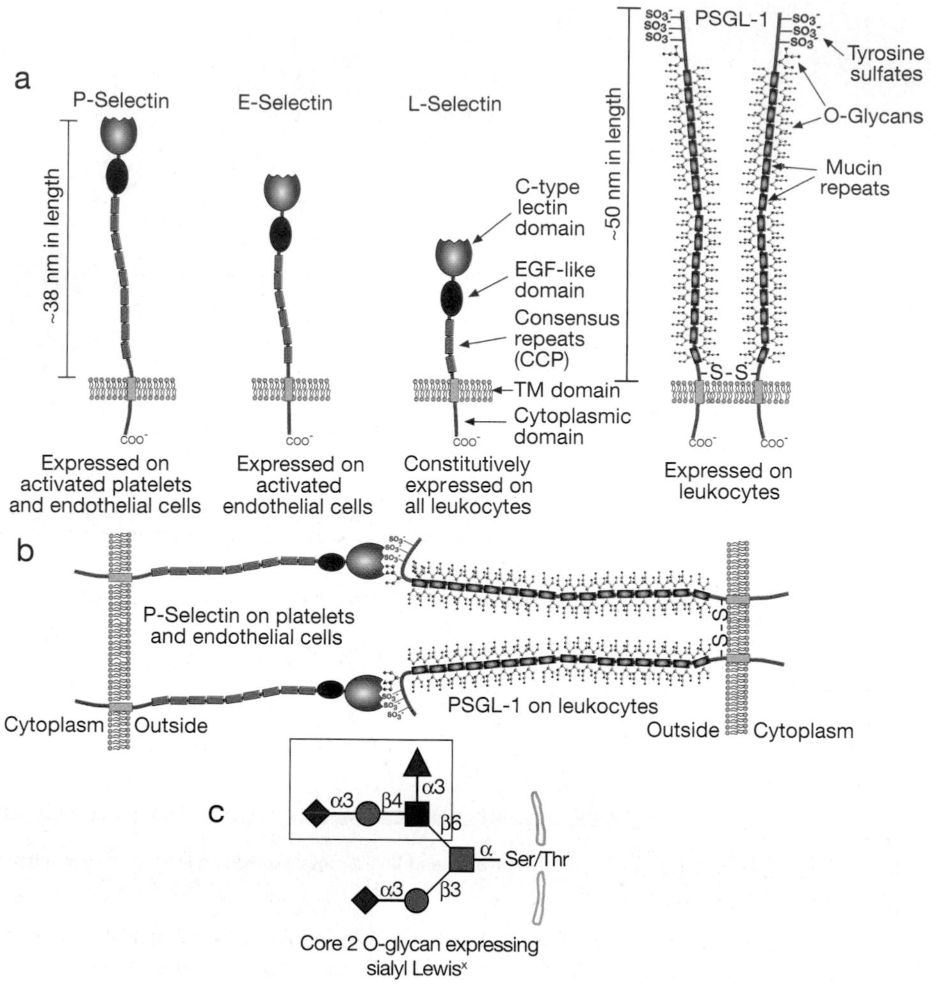

FIGURE 31.7. Overall domain structures of P-selectin, E-selectin, L-selectin, and PSGL-1. (*a*) The domain organization of the selectins and their expression patterns are indicated. P-Selectin also forms homodimers in the membrane. The predicted disulfide-bonded dimeric form of PSGL-1 on leukocytes is shown. PSGL-1 has three potential sites for N-glycosylation and multiple sites for O-glycosylation. The sulfated tyrosine residues are indicated. (The key for the domain structures is shown in Figure 31.3.) (*b*) The interaction between P-selectin and PSGL-1 through their amino-terminal domains is indicated. L-Selectin also binds to the same region of PSGL-1, although with different kinetics and affinity. (*c*) The major fucose-containing core-2 O-glycan identified in PSGL-1 that is required for binding to selectins. The sialyl Lewis^x determinant is *boxed*.

fated tyrosine residues adjacent to a core-2-based O-glycan expressing SLe^x. Ligands for L-selectin that occur within specialized endothelia termed high endothelial venules (HEV) contain 6-sulfo-SLe^x antigens on mucin-type O-glycans and on N-glycans.

P-Selectin

P-Selectin (CD62P) was discovered as an antigen expressed on the surface of activated platelets. It is constitutively expressed in megakaryocytes, where it is packaged into the membranes of α-granules of circulating platelets. It is also expressed in the Weibel–Palade bodies of vascular endothelial cells. Within minutes following activation of either platelets or endothelial cells by proinflammatory secretagogues such as histamine, thrombin, or complement components, P-selectin is expressed on the cell surface because of fusion of the intracellular storage membranes with the plasma membrane. Sequences within the cytoplasmic domain of P-selectin mediate its sorting to secretory granules as well as its rapid endocytosis from the plasma membrane and movement from endosomes to lysosomes, where it is degraded. Splice variants of human P-selectin transcripts yield forms of P-selectin that lack a transmembrane domain, thus contributing to low-level soluble forms of P-selectin in the circulation. Leukocyte adhesion also stimulates proteolytic cleavage of the ectodomain of P-selectin from the plasma membrane, releasing it into the circulation. The inflammatory mediators TNF-α, interleukin 1-β (IL1-β), and lipopolysaccharide (LPS) augment transcription of mRNA for P-selectin in endothelial cells in mice but not in humans.

P-Selectin contributes to leukocyte recruitment in both acute and chronic inflammation. Mice that lack P-selectin exhibit defective rolling on endothelial cells of postcapillary venules, diminished recruitment of neutrophils or monocytes into tissues following injection of inflammatory mediators, and impaired recruitment of T cells into skin or other tissues following challenge with specific antigens. Mobilization of P-selectin to the surfaces of activated endothelial cells is important for all these responses. In addition, P-selectin expressed on the surfaces of activated platelets contributes to inflammation as well as to hemostasis and thrombosis. Activated platelets adhere through P-selectin to neutrophils, monocytes, NK cells, and some subsets of T lymphocytes. This adhesion augments the recruitment of leukocytes and platelets to sites of vascular injury. Expression of P-selectin on both platelets and endothelial cells contributes to experimental atherosclerosis in mice. Platelet-expressed P-selectin also stimulates monocytes to synthesize tissue factor, a key cofactor of blood coagulation that facilitates fibrin deposition during clot formation.

As discussed below, the major leukocyte counterreceptor for P-selectin is PSGL-1. P-Selectin binds weakly to some forms of heparin/heparan sulfate and to some glycoproteins that bear the SLe^x determinant. The physiological significance of this binding is unclear, but the levels of heparin that are clinically prescribed might be able to block P-selectin functions. In addition, P-selectin can interact with mucins containing highly clustered glycans bearing SLe^x antigens, which might be important in metastasis of tumors bearing such ligands (see Chapters 43 and 44). Heparin might be therapeutically useful in blocking cancer metastases through blocking adhesion of tumor cells to P-selectin.

PSGL-1

PSGL-1 (CD162) is a homodimeric, disulfide-bonded mucin with subunits with a molecular mass of approximately 120 kD (351 amino acids). It is the major physiological ligand on leukocytes for P- and L-selectin and is also an important ligand for E-selectin (Figure 31.7). The precursor to the PSGL-1 monomer has 402 amino acids, including an 18-amino-acid signal sequence. Maturation of the protein occurs following cleavage at residues 38–41 by a paired basic amino-acid-converting enzyme in leukocytes. There are 16 decapeptide repeating units with the consensus sequence spanning residues 118–277 in the ectodomain of the long form of the protein, which is the major form expressed in humans. Murine PSGL-1 has 397 amino acids with recognizable sequence similarity to the human sequence, but the mouse protein has only 10 decameric repeats with the consensus sequence -E-T-S-Q/K-P-A-P-T/M-E-A-, which is different from human PSGL-1. The highest homology between human and murine PSGL-1 occurs in the transmembrane and cytoplasmic domains.

PSGL-1 is primarily expressed in hematopoietic cells (including some hematopoietic stem cells), all neutrophils, monocytes, eosinophils, and basophils and certain subsets of T cells. PSGL-1 is also expressed in some activated endothelial cells, most notably the inflamed microvessels of the ileum in a spontaneous model of chronic ileitis in mice. When it undergoes the appropriate posttranslational modifications (see below), PSGL-1 interacts with each of the three selectins to support leukocyte rolling under flow. Engagement of PSGL-1 during rolling also transduces signals into leukocytes that activate leukocyte integrins to slow rolling velocities. Signaling through PSGL-1 also cooperates with signaling through chemokine receptors to elicit other effector responses in leukocytes.

PSGL-1 is heavily glycosylated; each subunit of human PSGL-1 has three potential sites for N-glycosylation and 70 serine and threonine residues in the extracellular domain that are potential sites for O-glycosylation. In addition, human PSGL-1 has three amino-terminal Tyr residues and murine PSGL-1 has two tyrosine residues that are potentially sulfated. Like most mucins, PSGL-1 has an extended structure (Figure 31.7).

Most of the O-glycans of native PSGL-1 purified from human HL60 cells are core-2 structures, of which only a subset are fucosylated. Optimal binding of P-selectin to PSGL-1 requires that the latter presents SLe^x on a specific, amino-terminal core-2 O-glycan and sulfate esters on specific amino-terminal tyrosine residues. Tyrosine sulfate residues are generated by two tyrosylprotein sulfotransferases (TPST-1 and -2), which transfer sulfate from the sulfate donor (phosphoadenosine phosphosulfate) to exposed tyrosine residues. Thus, the key binding domain of PSGL-1 resides in its amino-terminal region. There are many lines of evidence that support this conclusion. Antibodies to this region block binding of PSGL-1 to P- and L-selectin (but not to E-selectin, which interacts with more than one site on PSGL-1 and does not interact with sulfated tyrosines). Treatment of neutrophils with a cobra venom metalloproteinase that removes the first ten amino acids from the amino terminus of PSGL-1 abrogates its binding to P- and L-selectin. Site-directed mutagenesis of a specific O-glycosylated amino-terminal threonine and/or the three tyrosine residues in the amino terminus of PSGL-1 prevents binding to P- and L-selectin. Finally, synthetic glycosulfopeptides with the structure shown in Figure 31.8 bind to P-selectin with the same high affinity (~300 nM) as that of native PSGL-1.

The data point to a model in which the combination of tyrosine sulfate residues and oligosaccharides on the protein are required for high-affinity binding to P-selectin (Figure 31.7). The cocrystal structure of a PSGL-1-derived glycosulfopeptide with P-selectin confirms the complex and tight association between these binding partners. The interactions between the glycosulfopeptide and P-selectin result from a combination of hydrophobic and electrostatic contacts. These include contacts of at least two of the three tyrosine sulfate residues as well as other PSGL-1 amino acids with multiple residues within the P-selectin CRD. In addition, two hydroxyl groups in the fucose residue of SLe^x ligate the lectin domain-bound Ca^{++}, and there are additional binding interactions with the hydroxyl groups of galactose and the –COOH group of Neu5Ac.

E-Selectin

E-Selectin (CD62E) was discovered as a leukocyte adhesion molecule expressed by activated vascular endothelial cells. In most tissues (the bone marrow and skin may be exceptions), endothelial cells do not constitutively express E-selectin. Cytokine-dependent transcriptional processes lead to an inducible expression of E-selectin on the surface of the endothelium. Inducible transcription of the E-selectin locus by TNF-α, IL1-β, and LPS is mediated at least in part through NF-κB-dependent events. In vitro, cytokine-dependent regulation of the E-selectin locus yields E-selectin expression beginning about 2 hours after cytokine treatment, with maximal expression at about 4 hours. E-Selectin expression then declines to basal levels within 12–24 hours in vitro, but E-selectin may be expressed chronically at sites of inflammation in vivo. Decline of E-selectin expression is associated with decreased transcription of the E-selectin locus, degradation of E-selectin transcripts, and internalization and turnover of E-selectin protein. Acute and chronic inflammatory conditions associated

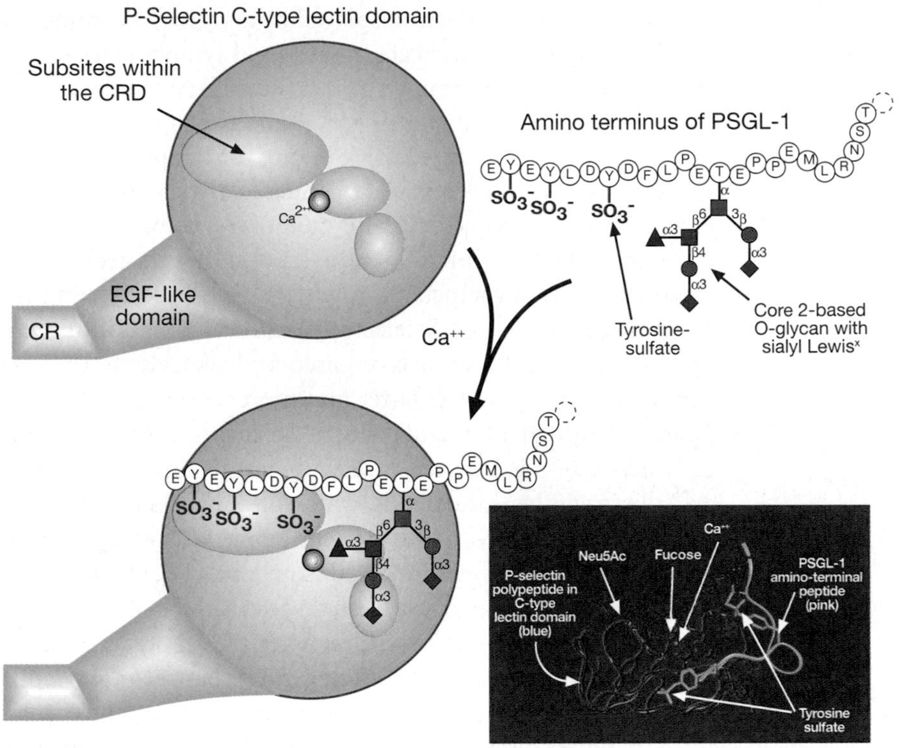

FIGURE 31.8. The molecular interactions between P-selectin and the amino terminus of PSGL-1. The C-type carbohydrate-recognition domain (CRD) at the amino terminus of P-selectin is a relatively large binding site that interacts with tyrosine sulfate residues, peptide regions, and the fucose and sialic acid residues in the sialyl Lewis^x core-2 O-glycan of PSGL-1. These are depicted as subsites within the domain. At the bottom right is the crystal structure of the glycosulfopeptide of PSGL-1 interacting with P-selectin and the coordinated Ca^{++} in the CRD. (Image derived from studies by Somers W.S., Tang J., Shaw G.D., et al. 2000. *Cell* **103:** 467–479.)

with E-selectin expression include sepsis, rheumatoid arthritis, and organ transplantation. E-Selectin cooperates with P- and L-selectin to recruit leukocytes to sites of inflammation.

Physiological ligands for E-selectin contain the SLe^x antigen and occur on neutrophils, monocytes, eosinophils, memory/effector T cells, and NK cells. Each of these cell types is found in acute and chronic inflammatory sites in association with the expression of E-selectin, which implicates E-selectin in the recruitment of these cells to such inflammatory sites. PSGL-1 is one of the physiological ligands for E-selectin, but E-selectin can also interact with several other glycoproteins that express the SLe^x antigen on either N- or O-glycans, including the E-selectin ligand-1, CD44, L-selectin (in humans), and possibly long-chain glycosphingolipids expressing the SLe^x antigen. These interactions may have a potential role in cancer metastasis (see Chapter 44).

L-Selectin

L-Selectin (CD62L) is expressed on the microvilli of most leukocytes, including all myeloid cells, naïve T and B cells, and some memory/effector T cells. L-Selectin was discovered through efforts to define molecules that facilitate recirculation of lymphoid cells from the intravascular compartment to the secondary lymphoid organs, including lymph nodes and Peyer's patches, from which the lymphoid cells then return to the circulation through the lymphatic system. This recirculation process provides lymphocytes with the opportunity to encounter foreign antigens displayed by antigen-presenting cells within secondary lymphoid organs. Early studies indicated that blood lymphocytes enter lymph nodes in postcapillary venules within these organs. These postcapillary venules are lined with a specialized endothelium (HEV). Cells in the HEV are cuboid in shape and their surfaces are decorated with certain glycoproteins required for adhesion of lymphocytes to the HEV, leading to subsequent transmigration. L-Selectin also cooperates with P- and E-selectins to promote leukocyte recruitment of myeloid cells and memory/effector T cells to sites of infection or injury during acute or chronic inflammation. L-Selectin is proteolytically shed from the surface of activated leukocytes by a metalloproteinase TNF-α-converting enzyme (TACE).

Rolling mediated by L-selectin exhibits a counterintuitive "shear threshold" requirement (e.g., a minimum flow rate is required for leukocytes to roll). As the flow rate drops below this threshold, leukocytes roll faster and more unstably and then detach. Flow-enhanced rolling operates through a force-dependent mechanism. As the flow rate increases, the force applied to adhesive bonds between L-selectin and its ligands increases. At threshold levels, the force actually strengthens the bonds, which prolongs their lifetimes. These are called "catch bonds." As the flow rate increases further, the applied force begins to weaken the bonds, which shortens their lifetimes. These are called "slip bonds." Catch bonds are regulated by force-dependent straightening of the angle between the lectin and EGF domains of L-selectin, which affects how ligand dissociates from the binding interface on the lectin domain. Transitions between catch and slip bonds are also seen for interactions between P-selectin and PSGL-1 and probably also occur between E-selectin and its ligands. However, the E-selectin transitions occur at lower forces with less dramatic effects at physiologically relevant shear forces in the circulation.

Lymphocytes home to secondary lymphoid organs where they may encounter antigens. Lymphocytes can return to the blood through the thoracic duct. This process is termed lymphocyte recirculation or homing. Entry to the lymph occurs in these organs through the HEV, where cuboid cells express ligands for lymphocyte L-selectin. These ligands are called peripheral node addressins, a group of mucins expressed on the HEV of lymph nodes. They include the mucins CD34, Sgp200, GlyCAM-1, MAdCAM-1, endoglycan, endomucin, and the podocalyxin-like protein (PCLP).

A unique feature of the L-selectin ligands on HEV is the requirement for sulfated glycans, such as 6-sulfo-SLe^x on both core-2 O-glycans and on extended core-1 O-glycans. The 6-sulfo-SLe^x determinant is associated with the MECA-79 epitope on O-glycans (Figure 31.9),

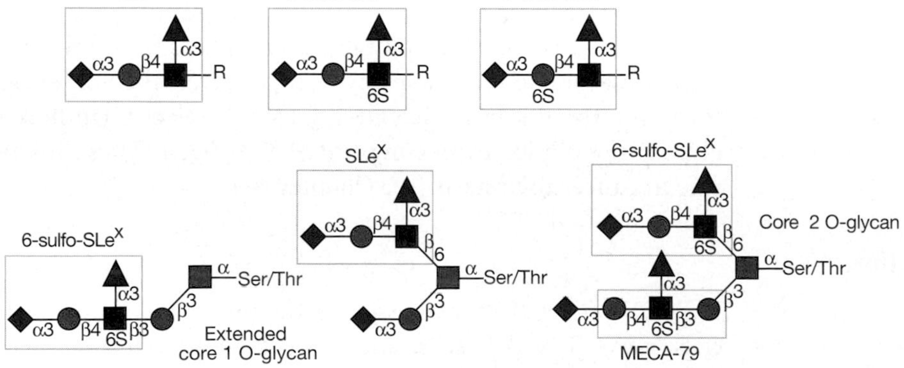

FIGURE 31.9. Structures of fucosylated and sialylated O-glycans that have been shown to mediate selectin-dependent cell adhesion. In addition, the structures at the top may be found on N-glycans and may mediate interactions with L-selectin. The biosynthesis of these glycans is described in Chapter 13. (SLe^x) Sialyl Lewis^x; (R) N- or O-glycans.

an antibody that binds to 6-sulfo-*N*-acetyllactosamine on extended core-1 O-glycans. The biosynthesis of the 6-sulfo-SLe^x determinant depends on two key α1-3 fucosyltransferases, FucT-VII and FucT-IV, along with at least four different sulfotransferases that may form the 6-sulfo-SLe^x determinant. Two of these sulfotransferases, GlcNAc6ST-1 and GlcNAc6ST-2, are both expressed in HEV and appear to be most important. Mice lacking FucT-VII or both FucT-VII and FucT-IV have dramatically reduced homing of lymphocytes to lymph nodes. Mice lacking both GlcNAc6ST-1 and GlcNAc6ST-2 do not express 6-sulfo-SLe^x or the MECA-79 epitope and exhibit markedly diminished lymphocyte homing to lymph nodes. A unique β1-3GlcNAcT generates the extended core-1 O-glycan; mice lacking both this β1-3GlcNAcT and the β1-6GlcNAcT branching enzyme for core-2 O-glycan biosynthesis do not express the MECA-79 antigen, but they have residual lymphocyte rolling on HEV and only a minimal decrease in lymphocyte numbers in peripheral and mesenteric lymph nodes. The residual L-selectin-dependent lymphocyte homing appears to result from 6-sulfo-SLe^x on N-glycans, suggesting that both N- and O-glycans on HEV glycoproteins contribute to L-selectin-dependent lymphocyte recirculation through lymph nodes.

L-Selectin also plays a role in adhesion of neutrophils, eosinophils, and monocytes to nonlymphoid vascular endothelium (see Figure 31.6). The major ligand for L-selectin in these inflammatory settings is PSGL-1 (see Figure 31.7), which is expressed on adherent leukocytes and may also be deposited on inflamed endothelial cells as fragments left behind by previously rolling leukocytes. Unlike the HEV ligands, PSGL-1 interacts with L-selectin through cooperative binding of its amino-terminal core-2 O-glycan capped with SLe^x and its amino-terminal sulfated tyrosines, analogous to but not identical to its interactions with P-selectin. In addition, an alternate ligand on endothelial cells is heparan sulfate. Mice lacking L-selectin have defects in neutrophil recruitment in the context of inflammation as well as defects in homing of naïve lymphocytes to secondary lymphoid organs.

THE NATURAL KILLER LYMPHOCYTE PROTEINS WITH C-TYPE LECTIN DOMAINS

The immunoglobulin-like receptors and NK cell receptors containing a CTLD are two families of major histocompatibility complex (MHC) class-I-specific receptors found on NK cells. NK receptors with the CTLD are largely in group V and are represented by the Ly49 family of receptors in mice and the NK complex (NKC) in humans (see Figure 31.3). The human NKC proteins include CD69, CD94, activation-induced C-type lectin (AICL), and mast-cell-function-associated antigen (MAFA). The murine Ly49 family includes the Ly49 group of about a dozen proteins, including CD94, NKG2-A, -C, -D, and -E.

NK receptors with a CTLD are either activating or inhibitory to immune responses. For the most part, these proteins function in targeting lymphocytes to cells that lack MHC class I antigens. The Ly49 and NKC proteins are disulfide-bonded, type II, homodimeric transmembrane proteins. Most of these proteins express immunoreceptor tyrosine-based inhibitory motif (ITIM) motifs in their cytoplasmic domain. Dectin-1, which is not considered to be part of the Ly49 or NKC complex, is exceptional in that it contains an activating ITAM-like motif. In human NK cells, CD94 forms MHC class-I-specific, disulfide-linked, heterodimers with NKG2 family members, which also contain a CTLD. Ligand binding to the CTLDs of these NK receptors can trigger or inhibit target cell lysis by NK cells or can activate various hematopoietic cells.

Most of the NK receptors with CTLDs are not known to bind carbohydrate ligands and lack the conserved Ca^{++}-binding residues and amino residues in most C-type lectins that bind glycans. In humans, it is known that Ca^{++} is not required for CD94/NKG2 interactions with HLA-E. The CTLD in the NK receptor has a somewhat different structure from typical

C-type lectins and may have evolved to promote interactions with MHC class I glycoproteins. However, some recent studies suggest that murine CD69 may bind some sulfated glycans; in addition, another protein in the NK family, the osteoclast inhibitory lectin (OCIL) from humans and mice, also binds some sulfated polysaccharides in a Ca^{++}-independent fashion. Thus, the possibility that the NK receptors with CTLDs bind to some types of glycan ligands should be explored. As discussed above, dectin-1 lacks the canonical residues for Ca^{++} binding and monosaccharide binding but nevertheless binds β-glucans. Thus, the CTLDs in the NK receptors may have evolved to recognize glycans by novel mechanisms.

The B-cell differentiation antigen (CD72) interacts with the B-cell receptor (BCR) and modulates BCR-mediated signals. CD72 is a type II membrane proteins with a CTLD that lacks the residues associated with Ca^{++} and monosaccharide binding. It has been speculated that CD72 might recognize BCR or other glycoprotein ligands by corecognizing glycan and protein determinants. The low-affinity IgE receptor, CD23, also contains a CTLD that interacts with CD21 and other ligands, but it may bind these ligands in a manner that depends on their correct glycosylation, perhaps a combination of glycan and peptide determinants.

PROTEOGLYCANS WITH C-TYPE LECTIN DOMAINS

The CTLD has also been identified in several proteoglycans (or lecticans) that lack transmembrane domains and occur in the extracellular matrix (group I; see Figure 31.3). These include aggregan, brevican, versican, and neurocan. Like the selectins, each of these core proteins contains a CTLD, an EGF-like domain, and a CCP domain, but their order is different and they are located in the carboxyl terminus of the protein. A large region containing attachment sites for chondroitin sulfate and keratan sulfate is proximal to the lectin domain. A more complete discussion of the proteoglycans is provided in Chapter 16. The exact functions of the CTLD in these proteins are still unknown. The CTLD of rat aggrecan is important for Ca^{++}-dependent binding to the fibronectin type II repeats 3–5 of rat tenascin-R. The CTLDs in other lecticans are probably also responsible for protein–protein interactions with other receptors, including tenascin-R, tenascin-C, and other tenascins, which are extracellular matrix glycoproteins highly expressed in the nervous system. Interestingly, the protein–protein interactions between rat fibronectin and the CTLD of rat aggregan bear resemblance to the protein–protein interactions seen in P-selectin binding to PSGL-1. Tenascin-R is one of the main carriers of the unusual glycan antigen HNK-1, which was named after its identification on human NK cells. Recent studies show that brevican, a lectican found in the nervous system, binds to the HNK-1-containing glycosphingolipids. Thus, the CTLD in lecticans may represent a versatile structural feature that can be used for both protein–protein and protein–glycan interactions.

OTHER TYPES OF C-TYPE LECTINS

A number of proteins with C-type CRDs have been identified in the pancreas and kidney, but the importance of the CRD and glycan binding to the function of these proteins is unclear. Autosomal dominant polycystic kidney disease (ADPKD) is a commonly occurring hereditary disease that accounts for about 10% of end-stage renal disease. PKD1, one of two recently isolated ADPKD gene products, has been implicated in cell–cell and cell–matrix interactions. The *PKD1* gene encodes a novel protein named polycystin that has multiple cell-recognition domains, including a single CTLD at its amino terminus. The function of *PKD1* and the effect of mutations on its possible activity are unclear.

Interestingly, some C-type lectins can be extremely small; for example, HIP (hepatic intestinal pancreatic protein) and PSP (pancreatic stone protein), which are essentially isolated CTLDs that are preceded by a signal sequence.

Lower vertebrates, invertebrates, and some viruses have now been found to contain the CTLD, and some of these proteins have been shown to bind sugars. For example, the galactose-specific lectin from *Crotalus atrox* binds a variety of galactose-containing glycolipids in a Ca^{++}-dependent fashion. A number of related venom proteins inhibit platelet function and/or the coagulation cascade. Alboaggregin A from *Trimeresurus albolabris* (the white-lipped pit viper) contains four subunits, and subunit 1 contains a single CTLD at its amino terminus. The protein binds to the platelet GPIb/IX receptor and stimulates platelet agglutination, but the potential role of glycan recognition in this process is presently unknown.

FURTHER READING

Drickamer K. 1999. C-type lectin-like domains. *Curr. Opin. Struct. Biol.* **9:** 585–590.

Ravetch J.V. and Lanier L.L. 2000. Immune inhibitory receptors. *Science* **290:** 84–89.

McEver R.P. 2002. Selectins: Lectins that initiate cell adhesion under flow. *Curr. Opin. Cell Biol.* **14:** 581–586.

Cambi A. and Figdor C.G. 2003. Dual function of C-type lectin-like receptors in the immune system. *Curr. Opin. Cell Biol.* **15:** 539–546.

McGreal E.P., Martinez-Pomares L., and Gordon S. 2004. Divergent roles for C-type lectins expressed by cells of the innate immune system. *Mol. Immunol.* **41:** 1109–1121.

Cambi A., Koopman M., and Figdor C.G. 2005. How C-type lectins detect pathogens. *Cell. Microbiol.* **7:** 481–488.

Iovanna J.L. and Dagorn J.C. 2005. The multifunctional family of secreted proteins containing a C-type lectin-like domain linked to a short N-terminal peptide. *Biochim. Biophys. Acta* **1723:** 8–18.

McMahon S.A., Miller J.L., Lawton J.A., Kerkow D.E., Hodes A., Marti-Renom M.A., Doulatov S., Narayanan E., Sali A., Miller J.F., and Ghosh P. 2005. The C-type lectin fold as an evolutionary solution for massive sequence variation. *Nat. Struct. Mol. Biol.* **12:** 886–892.

Zelensky A.N. and Gready J.E. 2005. The C-type lectin-like domain superfamily. *FEBS J.* **272:** 6179–6217.

Zhu C. and McEver R.P. 2005. Catch bonds: Physical models and biological functions. *Mol. Cell. Biomech.* **2:** 91–104.

Brown G.D. 2006. Dectin-1: A signalling non-TLR pattern-recognition receptor. *Nat. Rev. Immunol.* **6:** 33–43.

Chen M. and Geng J.G. 2006. P-Selectin mediates adhesion of leukocytes, platelets, and cancer cells in inflammation, thrombosis, and cancer growth and metastasis. *Arch. Immunol. Ther. Exp.* **54:** 75–84.

Kishore U., Greenhough T.J., Waters P., Shrive A.K., Ghai R., Kamran M.F., Bernal A.L., Reid K.B., Madan T., and Chakraborty T. 2006. Surfactant proteins SP-A and SP-D: Structure, function and receptors. *Mol. Immunol.* **43:** 1293–1315.

Ludwig I.S., Geijtenbeek T.B.H., and van Kooyk Y. 2006. Two way communication between neutrophils and dendritic cells. *Curr. Opin. Pharmacol.* **6:** 408–413.

Pyz E., Marshall A.S.J., Gordon S., and Brown G.D. 2006. C-type lectin-like receptors on myeloid cells. *Ann. Med.* **38:** 242–251.

Robinson M.J., Sancho D., Slack E.C., Leibund Gut-Landmann S., and Reis e Sousa C. 2006. Myeloid C-type lectins in innate immunity. *Nat. Immunol.* **7:** 1258–1265.

Sperandio M. 2006. Selectins and glycosyltransferases in leukocyte rolling in vivo. *FEBS J.* **273:** 4377–4389.

Uchimura K. and Rosen S.D. 2006. Sulfated L-selectin ligands as a therapeutic target in chronic inflammation. *Trends Immunol.* **27:** 559–565.

Zhou T., Chen Y., Hao L., and Zhang Y. 2006. DC-SIGN and immunoregulation. *Cell. Mol. Immunol.* **3:** 279–283.

Gupta G. and Surolia A. 2007. Collectins: Sentinels of innate immunity. *BioEssays* **29:** 452–464.

Trinchieri G. and Sher A. 2007. Cooperation of Toll-like receptor signals in innate immune defence. *Nat. Rev. Immunol.* **7:** 179–190.

CHAPTER 32

I-type Lectins

Ajit Varki and Paul R. Crocker

I-TYPE LECTINS ARE GLYCAN-BINDING PROTEINS that belong to the immunoglobulin superfamily (IgSF), excluding antibodies and T-cell receptors. Bioinformatics analyses of mammalian genomes predict more than 500 proteins of the IgSF, other than antibodies and T-cell receptors. Thus, there is considerable potential for assignment to the I-type lectin family. In fact, the Siglec family of sialic acid–binding lectins is the only well-characterized group of I-type lectins, both structurally and functionally. These proteins are thus the major focus of this chapter. Details of their discovery, characterization, binding properties, and biology are provided, along with discussions of their functional implications in mammalian biology.

HISTORICAL BACKGROUND AND OVERVIEW

Members of the IgSF must contain at least one immunoglobulin (Ig)-like fold, but often contain other structural features such as fibronectin type III repeats. The Ig fold was first discovered in antibodies and is made up of antiparallel β-strands organized into a β-sandwich containing 100–120 amino acids and usually stabilized by an intersheet disulfide bond. Three types or "sets" of Ig domains have been defined on the basis of similarities in sequence and structure to the domains of antibodies: the V-set variable-like domain, the C1- and C2-set constant-like domains, and the I-set domain that combines features of both V- and C-set domains.

Prior to the 1990s, it was thought that antibodies were the only IgSF members capable of recognizing glycans. The first direct evidence for non-antibody IgSF glycan-binding proteins was from independent studies on sialoadhesin (Sn), a sialic acid (Sia)-dependent binding receptor on certain mouse macrophage subtypes, and on CD22, a molecule already previously cloned as a B-cell marker. A variety of techniques were used to show that Sn functions as a lectin, including loss of binding following sialidase treatment of ligands, inhibition assays with sialylated compounds, and Sia-dependent binding of the purified receptor to glycoproteins and to red blood cells derivatized to carry Sias in α2-3 linkages. In the case of recombinant CD22, loss of cell adhesive interactions caused by sialidase treatment led to the discovery that it was a Sia-binding lectin, with a high degree of specificity for α2-6-linked Sias. The cloning of Sn then showed that it was an IgSF member that had homology with CD22 and with two other previously cloned proteins, CD33 and myelin-associated glycoprotein (MAG). Demonstration of Sia recognition by CD33 and MAG resulted in the definition of a new family of Sia-binding molecules, which were initially termed the "sialoadhesins." Meanwhile, evidence for glycan binding by additional IgSF members had emerged, and a suggestion was made to classify all of these molecules as "I-type" lectins. However, it became clear that the Sia-binding molecules were a distinct group sharing both sequence homology and Ig domain organization and that they were not all involved in adhesion. The term Siglec (sialic acid–binding, immunoglobulin-like lectin) was therefore proposed in 1998. Subsequently, most of the CD33-related Siglecs (CD33rSiglecs) were discovered as a direct result of the large-scale genomic sequencing projects, which allowed in silico identification of novel Siglec-related genes and cDNAs.

I-TYPE LECTINS OTHER THAN SIGLECS

Several IgSF members other than Siglecs have been claimed to bind glycans, but in many cases, the evidence is indirect. The best evidence is probably for L1 cell-adhesion molecule (L1-CAM) in the nervous system. However, further studies have indicated that this molecule does not bind Sias through an IgSF domain and it therefore does not strictly qualify as an I-type lectin. The neural cell-adhesion molecule (NCAM) has been claimed to recognize and bind oligomannose-type glycans on adjacent glycoproteins in the nervous system. Similar findings have recently been reported for another IgSF molecule called basigin. The cell-adhesion molecule ICAM-1 has been shown to bind hyaluronan and possibly certain mucin-type glycoproteins. Hemolin is an IgSF plasma protein from lepidopteran insects that binds lipopolysaccharide (LPS) from Gram-negative bacteria and lipoteichoic acid from Gram-positive bacteria. Hemolin appears to have two binding sites for LPS, one that interacts with the phosphate groups of lipid A and another that interacts with the O-specific glycan antigen and the outer-core glycans of LPS. There is indirect and less convincing evidence for interactions of other IgSF molecules with glycans, such as P0 with HNK-1, CD83 with Sias, PILR with Sias on CD99, and CD2 with Lewis^{x}. Further studies are needed to ascertain whether these are indeed I-type lectins. The rest of this chapter will be devoted to the Siglecs, which are the best-characterized I-type lectins.

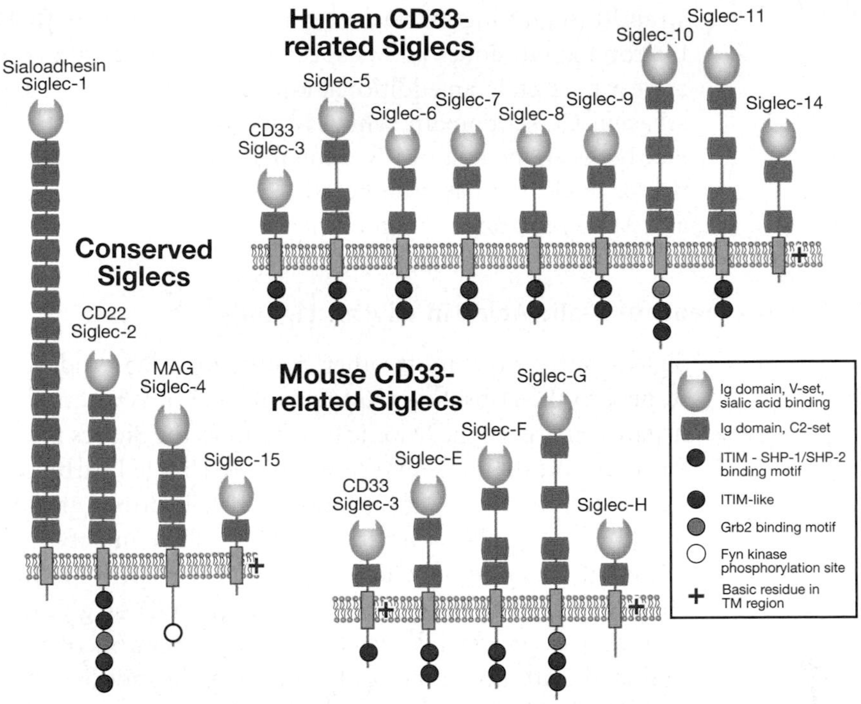

FIGURE 32.1. Domain structures of the known Siglecs in humans and mice. There are two subgroups of Siglecs: One group contains sialoadhesin (Siglec-1), CD22 (Siglec-2), MAG (Siglec-4), and Siglec-15, and the other group contains CD33-related Siglecs. In humans, Siglec-12 has lost its arginine residue required for sialic acid binding and Siglec-13 is deleted. The *plus sign* indicates the presence of a charged residue in the transmembrane domain, which has been shown to interact with the immunoreceptor tyrosine-based activatory motif (ITAM)-containing adaptor proteins DAP12 and DAP10. (ITIM) Immunoreceptor tyrosine-based inhibitory motif.

TWO MAJOR SUBFAMILIES OF SIGLECS

The Siglecs can be divided into two major subgroups based primarily on sequence similarity (Figure 32.1) and on conservation between different mammalian species. The first group comprises Sn (Siglec-1), CD22 (Siglec-2), MAG (Siglec-4), and Siglec-15 for which there are clear-cut orthologs in all mammalian species examined and which share only about 25–30% sequence identity among each other. The second group comprises the CD33rSiglecs, which share about 50–80% sequence similarity but appear to be evolving rapidly and undergoing shuffling of Ig-domain-encoding exons, making it difficult to define orthologs even between rodents and primates (see details below).

COMMON FEATURES OF SIGLECS

The Amino-terminal V-set Sialic Acid–binding Domain

All Siglecs are type-1 membrane proteins that contain a Sia-binding, amino-terminal V-set domain and varying numbers of C2-set Ig domains that act as spacers, projecting the Sia-binding site away from the plasma membrane. The V-set domain and the adjacent C2-set domain contain a small number of invariant amino acid residues, including an "essential" arginine on the F β-strand that is required for sialic acid binding and an unusual organization of cysteine residues. Instead of the typical intersheet disulfide bond between the B and F β-strands, Siglecs display an intrasheet disulfide bond between the B and E β-strands, permitting increased separation between the β-sheets. The resulting exposure of hydrophobic residues allows specific interactions with constituents of Sia. All Siglecs also appear to contain an additional unusual disulfide bond between the V-set domain and the adjacent C2-set domain, which would be expected to promote tight packing at the interface between the first two Ig domains. Although the significance of this for ligand recognition is unclear, it has been noted that the Sia-binding activity of some Siglecs (e.g., CD22 and MAG) appears to require the adjacent C2-set domain, probably for correct folding.

Absolute Requirement for Sialic Acids in Glycan Ligands

Siglecs differ from most other mammalian Sia-binding lectins (such as selectins) with respect to their absolute requirement for Sia. Whereas the selectins use Sias as carriers of negative charge to make ionic interactions, the Siglecs make more extensive molecular contacts, exploiting not only the negatively charged carboxylate group, but also the glycerol side chain, the N-acyl group, and the C-4-hydroxyl group. In contrast to selectins, substitution of Sia in oligosaccharides with a sulfate moiety results in loss of binding to Siglecs (although addition of sulfate esters to other parts of a the underlying glycan can enhance affinity). In addition, treatment of target cells or glycoconjugates with broad specificity sialidases is an effective way to destroy glycan recognition by Siglecs. Similar to many lectins, the affinity of Siglecs for sialylated ligands is low, with binding constants typically in the high micromolar to low millimolar range, as revealed in surface plasmon resonance, equilibrium dialysis, nuclear magnetic resonance (NMR), or thermal calorimetry measurements (see Chapter 27). Multimerization of Siglecs is likely to occur naturally on plasma membranes, leading to high avidity binding to clustered glycan ligands.

Masking and Unmasking

The cell-surface glycocalyx of most mammalian cells is richly decorated in glycoconjugates that contain Sias. The high local concentration of Sias is likely to greatly exceed the K_d value of each Siglec, resulting in "masking" of the Sia-binding site. Consequently, the Sia-dependent binding activity of most naturally expressed Siglecs is difficult to demonstrate unless the cells are first treated with sialidase to eliminate the *cis*-interacting sialylated glycans. A notable exception is Sn, which was discovered as a Sia-dependent cell-adhesion molecule on macrophages isolated from various tissues. The "masked" state of most Siglecs is a dynamic equilibrium with multiple ligands. Thus, an external probe or cell surface bearing high-affinity ligands or very high densities of Sia residues can effectively compete even with the binding domains of "masked" Siglecs. In addition, changes in expression of glycosyltransferases or sialidases could influence masking and unmasking of Siglecs at the cell surface, especially during immune and inflammatory responses.

Expression in a Cell-type-restricted Manner

Siglecs show restricted patterns of expression in unique or related cell types. This is most striking for Sn, CD22, and MAG, which are expressed on macrophages, B lymphocytes, and myelin-forming cells, respectively. This theme also extends to some of the CD33rSiglecs (most notably in humans), Siglec-6 on placental trophoblasts, Siglec-7 on NK (natural killer) cells, Siglec-8 on eosinophils, and Siglec-11 on tissue macrophages, including brain microglia. In the mouse, Siglec-H and CD33 are excellent markers of plasmacytoid dendritic cells (DCs) and neutrophils, respectively, and Siglec-F is a useful marker of eosinophils. These cell-type-restricted expression patterns are thought to reflect discrete, cell-specific functions mediated by each of these Siglecs. However, certain key cells of the immune system such as monocytes and conventional DCs express multiple CD33rSiglecs in humans.

Cytoplasmic Tyrosine-based Signaling Motifs

Most Siglecs have one or more tyrosine-based signaling motifs. Exceptions are Sn, Siglec-14, Siglec-15, mCD33, and Siglec-H. The most prevalent motif is the immunoreceptor tyrosine-based inhibitory motif (ITIM) with the consensus sequence (V/I/L)XYXX(L/V), where X is any amino acid. More than 100 ITIM-containing membrane receptors have been identified in the human gemone, and many of these are established inhibitory receptors of the hematopoietic and immune systems. They function by recruiting certain SH2-domain-containing effectors, the best characterized of which are the the protein tyrosine phosphatases SHP-1 and SHP-2 or the inositol 5′ phosphatase SHIP. These counteract activating signals triggered by receptors containing immunoreceptor tyrosine-based activatory motifs (ITAMs). Some Siglecs without a prominent cytoplasmic domain instead have a positively charged residue within the transmembrane region, which can associate with the DAP-12 (DNAX activation protein-12) ITAM-containing adaptor.

STRUCTURAL BASIS OF SIGLEC BINDING TO SIALYLATED GLYCANS

The three-dimensional structures of the mouse (m) Sn and human (h) Siglec-7 V-set domains have been determined by X-ray crystallography, in the presence and absence of Sia ligands (Figure 32.2). These provide a structural template for Sia recognition by Siglecs that is likely to be shared by other family members. In both instances, the "essential" arginine residue is located in the middle of the F β-strand, making a bidentate salt bridge with the carboxylate of Sia. In the absence of bound ligand, the essential arginine of both Sn and Siglec-7 is masked by a basic residue (either arginine or lysine). On binding ligands, this basic residue moves away to allow access of the essential arginine to the Sia carboxylate group. Siglecs also contain a conserved hydrophobic amino acid (either tryptophan or tyrosine) on the G β-strand that interacts with the glycerol side chain of Sia. The C-4 hydroxyl of Sia makes a hydrogen bond either directly or indirectly via a water molecule, and the N-acetyl group interacts with a tryptophan of Sn via hydrophobic interactions and with a similarly positioned tyrosine of Siglec-7 via hydrogen bonding. Although all Siglecs probably share this common template for binding glycosidically linked Sias, their binding preferences for extended glycan chains vary greatly. The peptide loop between the C and C′ β-strands is highly variable among Siglecs and has a key role in determining their fine sugar specificity. For example, molecular grafting of the C-C′ loop between Siglecs-7 and -9 resulted in switched sugar-binding specificities. Structural studies have shown that this loop appears to be highly flexible, being able to make specific and varied interactions with long glycan chains.

EXPRESSION PATTERNS AND FUNCTIONS OF THE CONSERVED SIGLECS

Sialoadhesin (Siglec-1, CD169)

Sn was identified in 1983 as a Sia-dependent sheep erythrocyte receptor (SER) expressed by mouse stromal macrophages isolated from various tissues. Studies with anti-Sn monoclonal antibodies established that expression in humans and mice is highly specific for macrophage subsets, especially those in lymphoid tissues and those recruited to inflammatory sites, as seen in inflamed joints in individuals with rheumatoid arthritis and other autoimmune inflam-

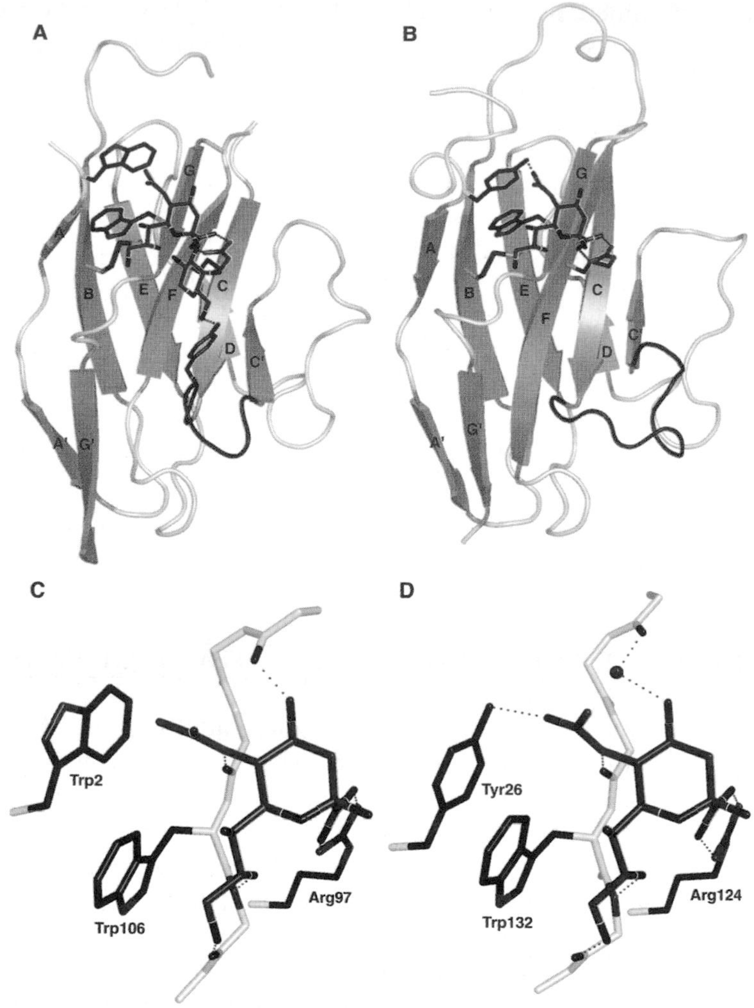

FIGURE 32.2. Structural basis of Siglec binding to ligands. X-ray crystal structures of the V-set domains of sialoadhesin (Sn) (*A*) and Siglec-7 (*B*) are shown complexed with sialic acid. (*C,D*) Molecular details of the interactions of sialic acid with Sn and Siglec-7. (Figure prepared by Dr. Helen Attrill.)

matory disorders. Full-length Sn is predicted to be a transmembrane protein with 17 Ig-like domains. Multiple splice variants encoding truncated and secreted forms of Sn exist, but their biological significance is unknown. Sn can be expressed at very high levels on macrophages, with up to 1 million molecules per cell, generating the potential to mediate cell–cell and cell–matrix interactions. Microscopy studies provided evidence for Sn clustering in vivo, especially in the bone marrow where Sn is localized at the contact sites of resident stromal macrophages and developing granulocytes. Several glycoprotein ligands have been identified from cell lysates using affinity chromatography methods with soluble recombinant forms of Sn. In all cases, the glycoconjugates that bind in a Sia-dependent manner are transmembrane mucins known to display multiple clustered O-linked glycans. These include MUC1 expressed by breast cancer cells and P-selectin glycoprotein ligand-1 (PSGL-1) and CD43 expressed on myeloid cells and T cells. It is unclear if these are indeed specific ligands or if they simply have the highest densities of cognate Sia residues.

The unusually large number of 17 Ig domains in Sn appears to be conserved in mammals and is thought to be important for extending the Sia-binding site away from the plasma membrane to promote intercellular interactions. Electron microscopy shows that Sn is an extended molecule of about 50 nm. Besides sialic acid binding mediated by the V-set

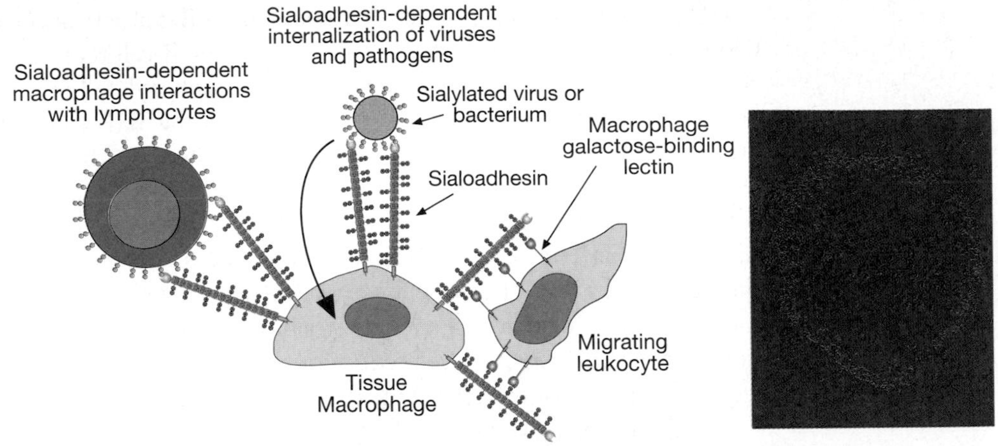

FIGURE 32.3. Biological functions mediated by sialoadhesin: Interactions of sialoadhesin on macrophages with cells and pathogens. (*Right*) Red staining shows a ring of sialoadhesin expressed by macrophages in the marginal zone of the spleen and green staining shows Siglec-H on the plasmacytoid dendritic cells.

domain, the C2-set domains of Sn display N-linked glycans that can act as ligands for other mammalian lectins such as the cysteine-rich domain of the mannose receptor and the macrophage galactose-type C-type lectin-1 (see Chapter 31). Although the biological significance of these interactions is uncertain, it is interesting that both lectins are found on dendritic cell populations that traffic into lymph nodes where Sn is abundantly expressed (Figure 32.3).

A role for Sn in interactions with pathogens has also been suggested. On the one hand, Sn can act as a phagocytic receptor for bacterial and protozoal pathogens such as *Neisseria meningitidis* and *Trypanosoma cruzi*, which coat themselves with Sias in an attempt to evade other forms of immune recognition. On the other hand, Sn expressed on alveolar macrophages can be "hijacked" as an endocytic receptor for the porcine reproductive and respiratory syndrome virus, which derives its envelope surface Sias from the mammalian cells from which it originally buds.

Sn-deficient mice appear essentially normal under specific pathogen-free conditions, with only subtle alterations in numbers of CD8 T cells and some subsets of B cells and reduced levels of circulating IgM. Granulocyte numbers appear normal in the bone marrow, in the blood, and following acute inflammation, suggesting that Sn–granulocyte interactions are not essential for maintenance of myelopoiesis. Interestingly, in mouse models of inherited neuropathy in which Sn-deficient mice were bred onto a P0 heterozygous background or proteolipid protein transgenic mice, reduced infiltration of CD8 T cells and macrophages was observed, accompanied by attenuated demyelination. In a mouse model of autoimmune uveoretinitis, Sn-deficient mice exhibited reduced inflammation, lowered T-cell proliferative responses, and a reduced time of onset to disease. Taken together, these findings suggest that Sn is important for the fine-tuning of adaptive immune responses, but the mechanisms remain to be established.

CD22 (Siglec-2)

CD22 was initially identified in 1985 as a developmentally regulated cell-surface glycoprotein on B cells. It is expressed at approximately the time of Ig gene rearrangement and is lost when mature B cells differentiate into plasma cells. CD22 was cloned in 1990 and was shown to have seven Ig-like domains. The intracellular region of CD22 has six tyrosine-based signaling motifs, four of which function as ITIMs.

CD22 is a well-established negative regulator of B-cell activation, making an important contribution toward the threshold for signaling via the B-cell receptor (BCR) complex. Following BCR cross-linking, CD22 is rapidly tyrosine-phosphorylated on its ITIMs by the Lyn tyrosine kinase. This leads to recruitment and activation of the SHP-1 tyrosine phosphatase and subsequent inhibition of downstream signaling mediated via the BCR. Besides SHP-1, the Ca^{++} pump PMCA4 is also recruited and activated and plays a key role in efflux of intracellular Ca^{++}, which results in dampening of signals that depend on elevated intracellular Ca^{++} concentration. Although additional activatory signaling molecules are recruited to the phosphorylated tyrosine motifs in CD22, the net phenotype of CD22-deficient mice is consistent with a primary role of CD22 in negative regulatory signaling, manifest by up-regulated MHC class II expression, enhanced B-cell turnover, reduced numbers of recirculating B cells in the bone marrow, reduced numbers of marginal zone B cells, and reduced anti-IgM-induced proliferation.

Of all the Siglecs, CD22 has the highest specificity for sialylated ligands, binding primarily to α2-6-linked Sias of the type Neu5Ac(Gc)α2-6Galβ1-4GlcNAc, which are common capping structures of many N-glycans. Additional specificity can be conferred by the nature of the Sia moiety: Neither hCD22 nor mCD22 binds 9-O-acetylated Sias; mCD22 has a strong preference for Neu5Gc over Neu5Ac, whereas hCD22 binds both of the latter forms. Binding of mCD22 to NeuGcα2-6Galβ1-4GlcNAc has a K_d value of 250 μM at 37°C and displays very rapid dissociation kinetics. Recombinant soluble CD22 can precipitate a subset of glycoproteins from cell lysates including CD45, a major sialoprotein of T and B cells carrying up to 18 N-linked glycans. Detailed kinetic studies with a range of native and enzymatically derivatized glycoconjugates have established that binding of CD22 to ligands is dependent on Sia density and linkage and appears not significantly influenced by the nature of the glycan carrier.

In common with many Siglecs, the Sia-binding site of CD22 on B cells is "masked" by *cis*-interactions with α2-6-sialylated ligands. However, when a B cell contacts another cell expressing high levels of α2-6-sialylated ligands, CD22 can redistribute to the points of cell contact, suggesting that *trans*-cellular communication can occur under physiological conditions. This could be important for altering B-cell activation thresholds and may help to ensure that signaling through the B-cell IgM receptor can only occur in lymphoid tissues where CD22 α2-6-sialylated ligands are particularly abundant on both T cells and B cells. In addition, some evidence for "unmasking" of the CD22-binding site has been observed following B-cell activation in vitro, and the specialized subset of B1 marginal zone cells that respond to carbohydrate antigens independently of T cells have constitutively unmasked forms of CD22. This could be important for lowering their activation thresholds, allowing efficient antibody production to foreign carbohydrate antigens.

Mouse mutants have shed light on how the lectin function of CD22 regulates its immunomodulatory role in B cells. ST6Gal-I-deficient mice lack ligands for CD22 and their B cells exhibit an anergic phenotype, essentially the opposite of the phenotype observed with CD22-deficient mice, which have hyperactivated B cells and enhanced signaling via the BCR. Crossing the ST6Gal-I-deficient mice with CD22-deficient mice restores B-cell signaling function, suggesting that CD22 is essential for the reduced BCR signaling seen in CD22-ligand-deficient mice. CD22 is known to associate with the BCR and this still occurs in ST6Gal-I-deficient mice, as well as in B cells transfected with a non-Sia-binding mutant of CD22. This argues strongly for a conserved non-Sia-dependent association of CD22 with BCR. These results, together with biochemical data showing that CD22 is homomultimerized via protein and Sia interactions, suggest that CD22 at the cell surface is the major counterreceptor for itself, possibly to the exclusion of other cell surface sialylated glycoproteins such as CD45. Furthermore, CD22 can associate with BCR in distinct microdomains of the B-cell plasma membrane, and its potential association with clathrin-rich microdomains and recycling at the cell surface are important topographical

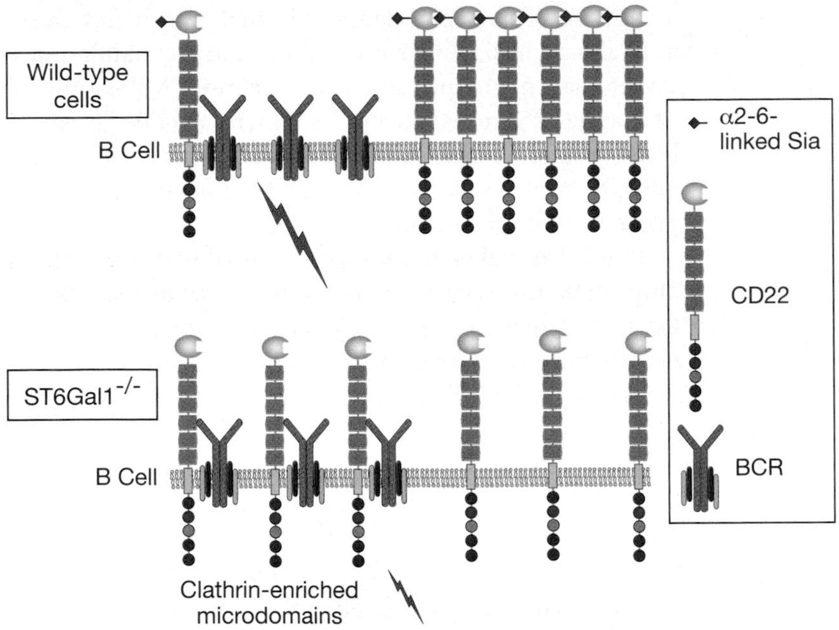

FIGURE 32.4. Proposed biological functions mediated by CD22: CD22 glycan-dependent homotypic interactions in equilibrium with CD22–BCR interactions. The actual situation seems to vary between different cell types and analysis conditions. (BCR) B-cell receptor; (Sia) sialic acid.

issues that require more study in the future. Thus, the anergic phenotype of ST6Gal-I-deficient mice could be due to excessive interactions of BCR with CD22, leading to mislocalization and aberrant signaling. In this model, the role of α2-6-sialylation could be to promote CD22 homomultimerization and sequestration of CD22 away from the BCR (Figure 32.4). Complementary data have arisen from the use of mouse mutants in which the sialic acid–binding capacity of CD22 has been selectively inactivated, but conflicting results were obtained using potent Sia-based inhibitors of CD22. Overall, the somewhat inconsistent data on functions of CD22 Sia recognition may reflect the reality of the biological role of this interaction in "tuning" the B-cell response appropriately to a given circumstance.

Myelin-associated Glycoprotein (Siglec-4)

MAG was identified in 1972 as a minor constituent of central nervous system (CNS) and peripheral nervous system (PNS) myelin. Cloned in 1987, it has five Ig-like domains and is highly conserved in mammals. MAG has a clear-cut ortholog in fish and may represent a primordial Siglec that gave rise to the immune Siglecs via gene duplication and "exon shuffling." Alternative transcripts of MAG (L-MAG and S-MAG) contain long (105 amino acids) or short (74 amino acids) cytoplasmic tails, respectively. The two isoforms are developmentally and anatomically regulated: Equivalent amounts are found in the adult CNS, whereas S-MAG predominates in the adult PNS. MAG is expressed by myelin-forming cells, oligodendrocytes in the CNS, and Schwann cells in the PNS. In mature myelinated axons, it is found on the innermost (periaxonal) myelin wrap but not in the multilayers of compacted myelin. MAG has features that indicate a role in adhesion and signaling in axon–glia and/or glia–glia interactions. L-MAG can recruit the nonreceptor tyrosine kinase Fyn, the calcium-binding protein S100β, and the phospholipase Cγ, whereas S-MAG has been reported to bind to tubulin and microtubules, supporting a role for IT as a cell-adhesion molecule linking the axonal surface and the myelinating glial cell cytoskeleton.

MAG-deficient mice develop normal myelin, but defects in myelin and axons increase as animals age, indicating a role for MAG in the maintenance of myelin and myelinated axons, rather than in the process of myelination. MAG-null mice display late-onset progressive PNS and CNS axonal atrophy and increased "Wallerian" degeneration. Myelinated axons of MAG-null mice fail to show characteristic myelin-induced increases in neurofilament phosphorylation and axon diameter, indicating that MAG signaling is required for appropriate myelin–axon communication.

MAG also makes an important contribution to the inhibitory activity of myelin on axon outgrowth and repair, a major factor in poor recovery from nervous system injury. MAG extracted from myelin, as well as expressed forms of soluble and cell-surface MAG, inhibits neurite outgrowth from a wide variety of neuronal cell types in vitro.

Genetic and biochemical evidence indicates that gangliosides are important physiological ligands for MAG, mediating both myelin-axon stability and inhibition of axon outgrowth. Recombinant forms of MAG bind selectively to the abundant axonal gangliosides GD1a and GT1b, and they bind with higher affinity to the minor ganglioside GQ1bα (see Chapter 10). The phenotype of MAG-deficient mice is similar to that of mice lacking an *N*-acetylgalactosaminyltransferase (GalNAcT) required for synthesis of gangliosides. In addition, the disialyl T antigen (NeuAcα2-3Galβ1-3[NeuAcα2-6]GalNAc-R), a structure found on O-glycans and gangliosides, effectively inhibited the binding of soluble MAG to a target ligand and reversed MAG-dependent inhibition of axon outgrowth. Binding of soluble MAG to certain neurons is Sia-dependent and blocked with antibodies to gangliosides. Neurons from GalNAcT-null mice are less responsive to MAG, whereas MAG still inhibits neurite outgrowth from mice lacking the "b-series" gangliosides (lacking GT1b but expressing GD1a; see Chapter 10) due to a mutation of a sialyltransferase, GD3 synthetase. These findings suggest that gangliosides GD1a or GT1b both act as functional docking sites for MAG on neuronal cells.

MAG also binds specifically and with high affinity to a family of GPI-anchored proteins, the Nogo receptor (NgR) family. Gangliosides and NgRs may act independently or interactively as MAG receptors, linking MAG binding to axonal signaling in different neuronal cell types (Figure 32.5). In one model, a lipid-raft-associated signaling complex on the surface of neurons mediates MAG-dependent inhibition of axon outgrowth. Binding of MAG to NgRs (and perhaps gangliosides) is accompanied by movement of the p75 neurotrophin receptor into glycolipid-enriched membrane microdomains. The p75 is then cleaved by α- and γ-secretase, leading to activation of the small GTP-binding protein Rho A in a protein kinase C–dependent manner. GTP-bound Rho A then activates a Rho-A-dependent kinase leading to changes in actin filaments and microtubules, resulting in inhibition of axon outgrowth. It remains unresolved whether MAG binding to NgRs is Sia-dependent. One possibility is that MAG has two binding sites, one in the amino-terminal V-set domain, which mediates Sia-dependent binding, and another in domain 4 and/or 5, which binds a protein determinant on NgRs. Thus, MAG appears to mediate its effects on neurite outgrowth via protein–glycan and protein–protein interactions with gangliosides and NgR, converging on Rho-A activation and the control of the axon cytoskeleton.

GENOMIC ORGANIZATION AND EVOLUTION OF CD33-RELATED SIGLECS

Genes encoding most of this Siglec subfamily are clustered together on human chromosome 19q13.3-13.4 or the syntenic region of mouse chromosome 7. They include CD33 (Siglec-3) and Siglecs-5, -6, -7, -8, -9, -10, -11, -12, and -14 in humans and CD33, Siglecs-E, -F, -G, and -H in the mouse. Similar clusters are found in other primates and rodents. It is difficult to assign definitive orthologs between Siglecs in primates and rodents, resulting in the current use of different nomenclatures. One reason for this is that most IgSF domains are encoded by

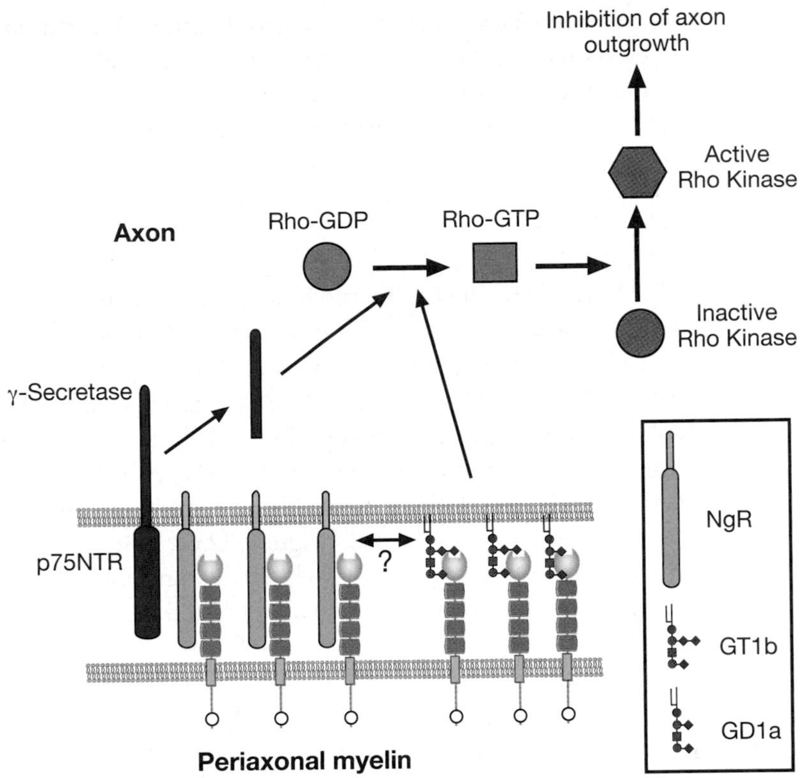

FIGURE 32.5. Proposed biological functions mediated by myelin-associated glycoprotein (MAG): Interactions between MAG and molecules of the axonal membrane lead to inhibition of neurite outgrowth. For a full explanation, see text. (NgR) Nogo receptor.

exons with phase-1 splice junctions. This permits exon shuffling without disrupting the open reading frame, resulting in generation of species-restricted hybrid genes that are difficult to distinguish from similarly organized genes in other species. A second reason is that the Sia-binding sites in the V-set domains of the Siglecs appear to have been rapidly evolving, presumably to change their binding specificity in response to the rapid evolution of the endogenous host sialome (see Chapter 14). There is also evidence for gene conversion events between adjacent genes and pseudogenes within this cluster in a species-specific manner. Of particular interest is the finding that humans show many CD33rSiglec differences compared with their closest evolutionary cousins (the chimpanzees), more than the differences between mice and rats (which shared a common ancestor much earlier). For example, Siglec-13 is specifically deleted in humans, but present in chimpanzees and baboons.

STRUCTURE, EXPRESSION, AND FUNCTIONS OF CD33-RELATED SIGLECS

The CD33rSiglecs share sequence similarity and certain structural features, for example, the presence of a linker region encoded by a separate exon between domains 2 and 3. Most contain a membrane-proximal ITIM and a membrane-distal ITIM-like motif. Although the latter is similar to the ITSM (switch motif) found in other signaling receptors of the immune system, there is no evidence currently that it is essential for phosphatase recruitment or signaling functions. The ITIM recruits and activates both SHP-1 and SHP-2 and is thought to be important for negative inhibitory signaling and modulating CD33rSiglec-dependent adhesion. Most CD33rSiglecs are restricted in expression to the immune system. A shared property may therefore be to regulate leukocyte functions during inflammatory and immune responses, including cell proliferation, differentiation, activation, and survival, perhaps via recognition of sialic acids as "self." In addition, CD33rSiglecs are actively endocytic and could be important in the clearance and/or antigen presentation functions of myeloid cells, especially when involving sialylated pathogens. Brief notes on the properties and putative functions of each of the CD33rSiglecs in humans and mice are provided below.

CD33 (Siglec-3)

CD33 was identified with monoclonal antibodies in 1983 as a marker of early human myeloid progenitors also found on mature monocytes and some macrophages and cloned in 1988. The later cloning of Sn and the recognition of Sia-binding by CD22 prompted experiments to investigate CD33's Siglec activity. CD33 has some preference for α2-6- over α2-3-sialylated glycans. It was the first of the CD33rSiglecs to be characterized as an inhibitory receptor. Cross-linking of CD33 with the activating FcγRI was shown to reduce Ca^{++} signaling and CD33 was able to recruit and activate the inhibitory tyrosine phosphatases SHP-1 and SHP-2 (Figure 32.6). Mutation of the proximal ITIM prevented recruitment of these phosphatases and also resulted in increased Sia-dependent binding of RBC. CD33 is also rapidly phosphorylated on serine residues as a downstream consequence of protein kinase C activation, but the biological significance of this is not clear. Antibodies to CD33 have been reported to inhibit the proliferation of both normal and leukemic cell populations and the differentiation of DCs from bone marrow precursors, suggesting a role of CD33 in regulating hematopoiesis. The endocytic property of CD33 is currently being exploited in the treatment of acute myeloid leukemia using Gemtuzumab, a humanized anti-CD33 monoclonal antibody coupled to the toxic antibiotic calicheamicin. The endocytic pathway is as yet poorly characterized for any member of the CD33-related family, but it requires intact tyrosine motifs and appears to be via a clathrin-independent mechanism, in contrast to endocytosis mediated by CD22.

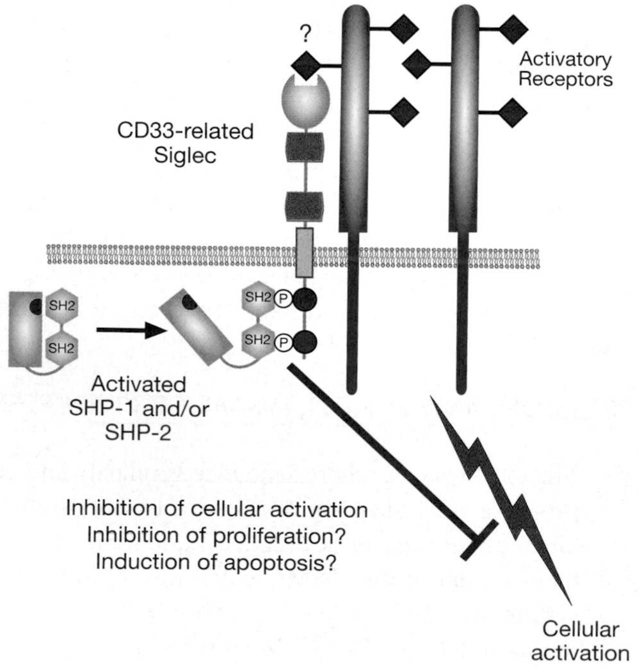

FIGURE 32.6. Proposed biological functions mediated by CD33-related Siglecs: A generic CD33-related Siglec is represented, showing the location of the immunoreceptor tyrosine-based inhibitory motif (ITIM) and the potential for inhibitory signaling.

Prior to the discovery of the CD33rSiglec family, a murine ortholog of CD33 was isolated with two Ig-like domains (61% amino acid identity), yet with a cytoplasmic domain showing considerably less homology. Two alternatively spliced forms of mCD33 that differ in the cytoplasmic region have been described, but neither contains the typical ITIM found in most other CD33rSiglecs. This may explain the lack of a robust phenotype in CD33-null mice. Furthermore, mCD33 has a lysine residue in the transmembrane sequence and may therefore couple to the DAP-12 transmembrane adaptor, as shown recently for mouse Siglec-H and human Siglecs-14 and -15. In contrast to hCD33, mCD33 is expressed mainly on neutrophils rather than monocytes, which also suggests a nonconserved function of this receptor.

Siglec-5 (CD170) and Siglec-14

Siglec-5 contains four Ig domains and mediates Sia-dependent binding to α2-3Gal, α2-6Gal(NAc), and α2-8Sia linkages, indicating a potentially broad binding specificity for protein- and lipid-bound glycoconjugates. Siglec-5 is prominently expressed in the myeloid lineage, but at a later stage in differentiation than CD33. Rather than being lost from neutrophils on exit from marrow into the blood, Siglec-5 is retained and is also expressed on circulating monocytes and subsets of tissue macrophages. It seems to be the only Siglec in humans expressed on plasmacytoid DCs, a specialized cell type responsible for rapid production of type I interferons following viral infections. Furthermore, Siglec-5 can bind and internalize sialylated strains of *N. meningitidis*, raising the possibility that it could play a role in recognition of these pathogens in humans. Siglec-5 has been shown to function as an inhibitory receptor following co-cross-linking with the activating high-affinity IgεR in transfected rat basophilic leukemia cells.

Siglec-14 was recently characterized as a novel human Siglec with three Ig domains and no ITIM-like motifs (Figure 32.7). The first two Ig domains are greater than 99% identical to those of Siglec-5 with only one amino acid difference. In contrast, the third Ig domain of Siglec-14 is much less similar, and its transmembrane region contains an arginine

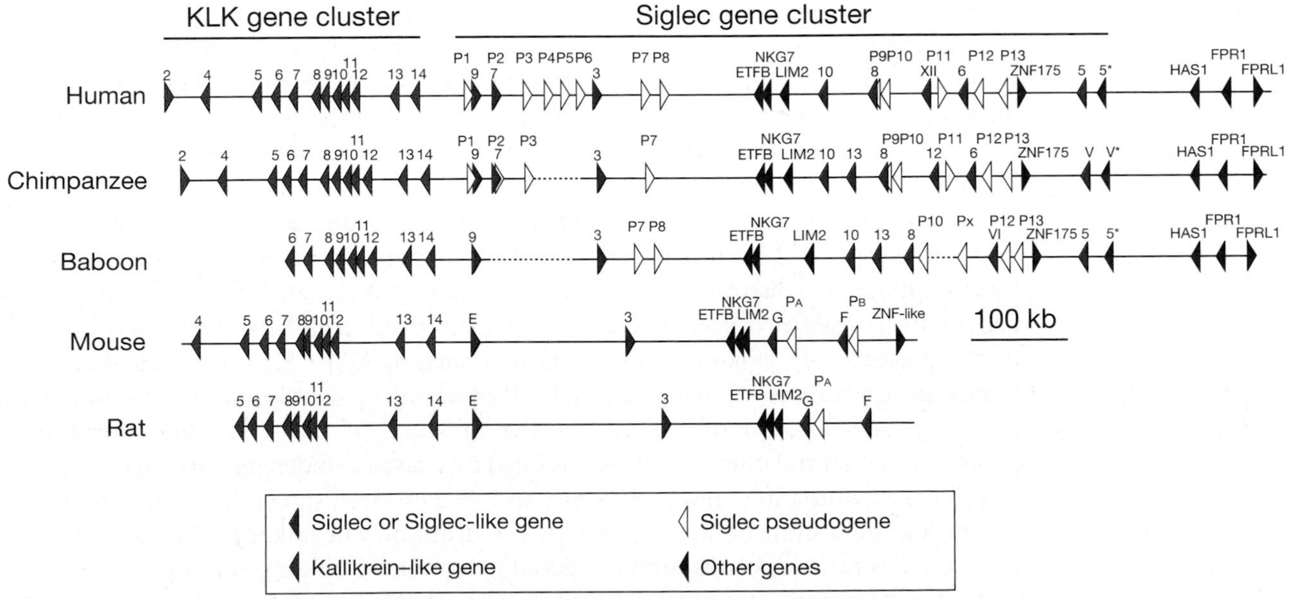

FIGURE 32.7. Chromosomal organization of CD33-related Siglec clusters in some rodents and primates. Note that the locus marked as 5* is now known to encode Siglec-14 in primates. (Reprinted, with permission, from Angata et al. 2004. *Proc. Natl. Acad. Sci.* **101:** 13251–13256.)

residue that can associate with the DAP-12 ITAM-containing adaptor. This curious partial homology is explained by the finding that the Siglec-5 and Siglec-14 genes appear to be undergoing concerted evolution in multiple primate species, suggesting that these receptors may function as paired activating and inhibitory receptors. It should be noted that almost all antibodies currently available against Siglec-5 cross-react with Siglec-14, raising uncertainty about the expression patterns of Siglec-5 described above.

Siglec-6

Siglec-6 was cloned from a human placental cDNA library and also during a screen for proteins that bind leptin, a hormone that regulates body weight. It has three Ig-like domains and the typical arrangement of ITIM and ITIM-like motifs in its cytoplasmic tail. Siglec-6 binds dimeric leptin with a K_d of approximately 90 nM, which reflects an affinity for leptin that is tenfold weaker than the leptin receptor (Ob-R), yet tenfold stronger than the related Siglecs-3 and -5. These findings indicate that although it is unlikely to mediate leptin signaling, Siglec-6 could function as a leptin "sink" and contribute to regulation of leptin plasma concentration. Another unusual feature of Siglec-6 is its high expression in placenta, being localized to cytotrophoblastic and syncytiotrophoblastic cells, where its levels seem to increase during the progress of labor and delivery. Lower levels are expressed on B cells. Siglec-6 does not have an obvious ortholog in mice, but one is present in the chimpanzee and baboon.

Siglecs-7, -8, and -9

These three Siglecs have three Ig domains, share a high degree of sequence similarity, and appear to have evolved by gene duplication from a three-domain ancestral Siglec. Siglec-7 is the major Siglec expressed by NK cells and functions as an inhibitory receptor when expressed naturally on these cells, both in antibody-directed killing assays and in assays using target cells overexpressing GD3, which is a Siglec-7 ligand. Siglec-7 also mediates selective interactions with sialylated lipooligosaccharides of the human pathogen, *Campylobacter jejuni*, suggesting a role in host–pathogen interactions. Siglec-9 is prominently expressed on neutrophils, monocytes, and to a much lesser extent on subsets of NK cells and CD8 T cells. In contrast, Siglec-8 is expressed on eosinophils, with weaker expression on basophils. The original Siglec-8 cDNA clone lacked the typical ITIM and ITIM-like motifs. However, subsequent analyses showed that this represents a minor alternatively spliced form and that a transcript encoding the longer form is more abundantly expressed in eosinophils.

Inhibition of cellular activation by Siglecs-7 and -9 can be demonstrated following co-cross-linking with ITAM-coupled activating FcRs (U937 and RBL cells) or following T-cell receptor engagement (in Jurkat cells). A novel proapoptotic function for Siglec-8 expressed by eosinophils was discovered following antibody-induced cross-linking. This depended on generation of reactive oxygen species and caspase activation and was paradoxically enhanced in the presence of cytokine "survival" factors such as GM-CSF and interleukin-5 (IL-5). Similar observations were made with Siglec-9 expressed on neutrophils. In addition to inducing apoptosis, ligation of Siglec-9 in the presence of GM-CSF (granulocyte-macrophage–colony-stimulating factor) also resulted in a caspase-independent nonapoptotic form of cell death. Autoantibodies to Siglecs-8 and -9 have been shown to be present in human serum, and their induction of granulocyte cell death may be linked to the anti-inflammatory properties of intravenous immunoglobulin, a pooled human serum preparation used to treat certain human autoimmune disorders. The role of the Siglec tyrosine-based motifs and SHP recruitment in triggering cell death are currently unknown.

Analyses of the glycan-binding specificities of these proteins revealed striking differences among Siglecs-7, -8, and -9. Siglec-8 was found to bind best to a single unique ligand, 6′-sulfo-SLe^x, whose natural expression pattern is only partly known. In contrast, Siglec-9 preferred a related structure, 6-sulfo-SLe^x, and Siglec-7 bound well to both forms.

The closest mouse relative to Siglecs-7, -8, and -9 is Siglec-E, which falls into the same evolutionary clade and shares about 70% sequence similarity with all three proteins. Siglec-E appears to exhibit a combination of features found in Siglec-7 and Siglec-9, being expressed on similar cell populations and exhibiting similar sugar-binding activity. Although there is no ortholog of Siglec-8 in mice, the four-Ig domain mouse Siglec-F is expressed in a similar way to Siglec-8 on eosinophils and has a similar glycan-binding preference. It therefore appears to have acquired the same functions through convergent evolution. Siglec-F-null mice show exaggerated eosinophilic responses in a lung allergy model, suggesting that its normal role is to dampen such responses. Interestingly, Siglec-F ligands in the airways and lung parenchyma were also up-regulated during allergic inflammation.

Siglecs-10 and -11

Siglec-10 has five Ig-like domains and displays an additional tyrosine-based motif in its cytoplasmic tail. It is expressed at relatively low levels on several cells of the immune system, including monocytes, eosinophils, and B cells. It is the only CD33-related human Siglec that has a clear-cut ortholog in mice, designated Siglec-G. This, combined with phylogeny analyses, suggests that Siglec-10 represents an ancestral CD33rSiglec that may have given rise to the other members of this subgroup via gene duplication, exon loss, and exon shuffling. Mice deficient in Siglec-G show a tenfold increase in numbers of a specialized subset of B lymphocytes, the B1a cells, that make natural antibodies and fast T cell–independent antibodies to some bacteria. In addition, the B1a cells show exaggerated Ca-fluxing following BCR cross-linking. Siglec-11 also has five Ig domains that are 90% identical to Siglec-10, but with much lower similarity in the transmembrane and cytoplasmic regions. Siglec-11 in humans appears to be a chimeric molecule that underwent gene conversion with an adjacent Siglec pseudogene. Despite the high sequence similarity in the extracellular region, the sugar-binding properties are distinct, with Siglec-10 binding to both α2-3- and α2-6-linked Sias and Siglec-11 binding only weakly to α2-8-linked Sias. Siglec-11 also exhibits different expression patterns, being absent from circulating leukocytes, but expressed widely on populations of tissue macrophages, including resident microglia in the brain, where high levels of α2-8-linked Sias are present on gangliosides. Interestingly, this microglial expression appears unique to humans.

Siglec-15

Siglec-15 has two Ig-like domains: a short cytoplasmic tail and a transmembrane domain containing a lysine residue that allows association with the activating adaptor proteins DAP-12 and DAP-10. Siglec-15 preferentially recognizes the Neu5Acα2-6GalNAcα-(Sialyl-Tn) structure and is expressed on macrophages and/or dendritic cells of human spleen and lymph nodes. While Siglec-15 has the potential to be an activating receptor, it does not have an inhibitory counterpart like Siglec-14. Siglec-15 has been conserved throughout vertebrate evolution, and it presumably plays a conserved, regulatory role in the immune system. As with Sialyl-Tn in a tumor marker, one suggested possibility is that it functions in tumor surveillance.

NATURAL MUTATIONS OF THE "ESSENTIAL" ARGININE RESIDUE

As mentioned earlier, an "essential" arginine residue in all of the known Siglecs is important for binding Sia-containing ligands. Surprisingly, this arginine is frequently mutated in nature, resulting in loss of binding ability. Examples of this include Siglec-12 in humans, Siglecs-5 and -14 in the chimpanzee, gorilla, and orangutan, Siglec-6 in the baboon, and Siglec-H in the rat. The common arginine codon (CGN, where N is any nucleotide) tends to be highly mutable because of the CpG sequence. However, the frequency with which such events occurs is surprising, suggesting that it might be a natural mechanism to eliminate Sia binding of a given Siglec when such activity becomes inappropriate under changing evolutionary pressures without requiring a complete loss of the Siglec.

HUMAN-SPECIFIC CHANGES IN SIGLEC BIOLOGY

The ancestral condition of some hominid Siglecs (e.g., Siglec-7 and Siglec-9) appears to have been preferential binding to Neu5Gc, a Sia that was specifically lost in human evolution about 2–3 million years ago. The loss of Neu5Gc could thus have resulted in extensive Siglec unmasking, possibly leading to a state of heightened innate immune reactivity. Some human Siglecs have undergone an adjustment to allow increased Neu5Ac binding, and the question arises as to whether the adjustment is yet complete. Possibly as a consequence of this event, several Siglecs seem to have undergone human-specific changes in comparison to our great ape evolutionary cousins. For example, Sn seems to be expressed on most human macrophages, whereas only subsets of chimpanzee macrophages are positive for Sn. This may be related to the fact that Sn in humans has a strong binding preference for Neu5Ac over Neu5Gc, similar to that seen in other species. Human Siglec-5 and Siglec-14 appear to have undergone a restoration of the "essential" arginine residue needed for Sia recognition, which is mutated in chimpanzees, gorillas, and orangutans. As mentioned earlier, the gene encoding Siglec-11 has undergone a human-specific gene conversion, resulting in a new protein with altered binding properties and new expression in brain microglia. Siglec-12 has suffered a human-specific inactivation of the essential arginine residue in humans, with subsequent permanent pseudogenization by a frame-shift in some humans. Siglec-13 has undergone a human-specific gene deletion. Expression patterns of some Siglecs also appear to have undergone changes, with the placental expression of Siglec-6 being human-specific and a general suppression of all CD33rSiglecs on human T cells compared with the chimpanzee. The functional implications of these human-specific changes in Siglec biology for physiology and disease deserve further exploration.

FURTHER READING

Powell L.D. and Varki A. 1995. I-type lectins. *J. Biol. Chem.* **270:** 14243–14246.

Crocker P.R. and Feizi T. 1996. Carbohydrate recognition systems: Functional triads in cell–cell interactions. *Curr. Opin. Struct. Biol.* **6:** 679–691.

Kelm S. and Schauer R. 1997. Sialic acids in molecular and cellular interactions. *Int. Rev. Cytol.* **175:** 137–240.

Varki A. 1997. Sialic acids as ligands in recognition phenomena. *FASEB J.* **11:** 248–255.

Crocker P.R. and Varki A. 2001. Siglecs, sialic acids and innate immunity. *Trends Immunol.* **22:** 337–342.

Angata T. and Brinkman-Van der Linden E. 2002. I-type lectins. *Biochim. Biophys. Acta* **1572:** 294–316.

Crocker P.R. 2002. Siglecs: Sialic-acid-binding immunoglobulin-like lectins in cell–cell interactions and signaling. *Curr. Opin. Struct. Biol.* **12:** 609–615.

Crocker P.R. 2005. Siglecs in innate immunity. *Curr. Opin. Pharmacol.* **5:** 431–437.

Nitschke L. 2005. The role of CD22 and other inhibitory co-receptors in B cell activation. *Curr. Opin. Immunol.* **17:** 290–297.

Varki A. and Angata T. 2006. Siglecs—The major subfamily of I-type lectins. *Glycobiology* **16:** 1R–27R.

Crocker P.R., Paulson J.C., and Varki A. 2007. Siglecs and their roles in the immune system. *Nat. Rev. Immunol.* **7:** 255–266.

CHAPTER 33

Galectins

Richard Cummings and Fu-Tong Liu

GALECTINS TYPICALLY BIND β-GALACTOSE-CONTAINING GLYCOCONJUGATES and share primary structural homology in their carbohydrate-recognition domains (CRDs). Previously termed S-type lectins, galectins represent a group of proteins that are the most widely expressed class of lectins in all organisms. This chapter describes the diversity of the galectin family and presents an overview of what is known about their biosynthesis, secretion, and biological activities.

HISTORICAL BACKGROUND

Following the discovery of agglutinins in plants and lectins (discoidin I) in *Dictyostelium discoideum* in the early 1970s, many investigators began to search for lectins in animal tissues. As noted in Chapter 26, the first reported lectin in animal cells was the hepatic asialoglycoprotein receptor, a C-type lectin. The next lectin found in animals was the protein now recognized as the first galectin. It was originally described in 1975 during studies on the possible presence of lectins in the electric organs of the electric eel. The protein, termed electrolectin, had hemagglutinating activity with trypsinized rabbit erythrocytes that was inhibitable by β-galactosides and could be isolated by affinity chromatography on β-galactoside supports. Notably, this protein required the inclusion of β-mercaptoethanol in isolation buffers to maintain its activity, suggesting the presence of one or more free cysteine residues. The galectins were originally referred to as S-type lectins to denote their sulfhydryl dependency, the presence of cysteine residues, their solubility, and their shared primary sequence, in much the same way that the designation C-type denotes Ca^{++} dependency and shared primary structure for that class of lectins. However, electrolectin, unlike most galectins, does not contain cysteine residues, but its key tryptophan residue in the binding site of its CRD can be oxidized, causing loss of activity. Electrolectin is approximately 15 kD in size and occurs as a noncovalently linked homodimer.

The first galectins found in vertebrates were isolated in 1976 from chick muscle and from extracts of calf heart and lung. Lactose was required to isolate the mammalian lectin, now termed galectin-1, from macromolecular glycoconjugates in calf heart/lung extracts. Galectin-1 was purified by affinity chromatography on asialofetuin-Sepharose and it also required reducing conditions to maintain activity. The calf heart/lung galectin-1 is approximately 15 kD in size and occurs as a noncovalent dimer. In the early 1980s, a 35-kD carbohydrate-binding protein (CBP35; now known as galectin-3) that also bound to β-galactosides was identified from mouse fibroblasts. The same protein had been studied by other groups and was known as IgE-binding protein (εBP), L-29, and L-31. All of these proteins demonstrated hemagglutinin activity, but the choice of erythrocytes was crucial. Trypsinized rabbit erythrocytes, which display more terminal galactose residues than human erythrocytes, are readily agglutinated by most galectins, whereas human erythrocytes require treatment with neuraminidase to enhance their agglutinability. The nomenclature for galectins was systematized in 1994. The first galectin found (variously termed electrolectin, β-galactoside-binding lectin, galaptin, L-14, etc., depending on its source) was renamed galectin-1. Its nearest homolog was termed galectin-2. CBP35, εBP, L-29, and L-31 were termed galectin-3, and other members of this family were numbered consecutively by order of discovery.

DIFFERENT MEMBERS OF THE GALECTIN FAMILY

The canonical CRD of galectins has approximately 130 amino acids, although only a small number of residues within the CRD directly contact glycan ligands. A comparison of the sequences of approximately 130 galectins from many different sources reveals that eight residues, which have been shown to be involved in carbohydrate binding by X-ray crystallographic analyses, are invariant. In addition, another dozen residues appear to be highly conserved. Part of the highly conserved sequence motif used to identify galectins is shown in Figure 33.1, along with a comparison of several human galectins. On the basis of sequence homologies, two general subgroups of galectins can be distinguished: the galectin-1 subgroup, which includes galectin-1 and galectin-2, and the galectin-3 subgroup, which includes all others. In comparison to human galectin-1, the mushroom galectin from *Coprinus cinereus* is overall about 20% identical and the 14-kD and 16-kD galectins in chickens are both about 60% identical.

A large number of galectins have now been identified in animals based on the conserved galectin CRD, and although most of them recognize simple β-galactosides, the binding affinity for such structures is relatively weak. A list of the human galectins known to bind β-galactosides is shown in Figure 33.1. A total of 15 galectins have now been found in mammals, but only 12 galectin genes are found in humans, including two for galectin-9. These have been classified into three major groups:

1. The prototypical galectins, which contain a single CRD that may associate to form homodimers.
2. The chimeric galectins, of which galectin-3 is the only known species found in vertebrates. Galectin-3 is characterized by having a single CRD and a large amino-terminal domain,

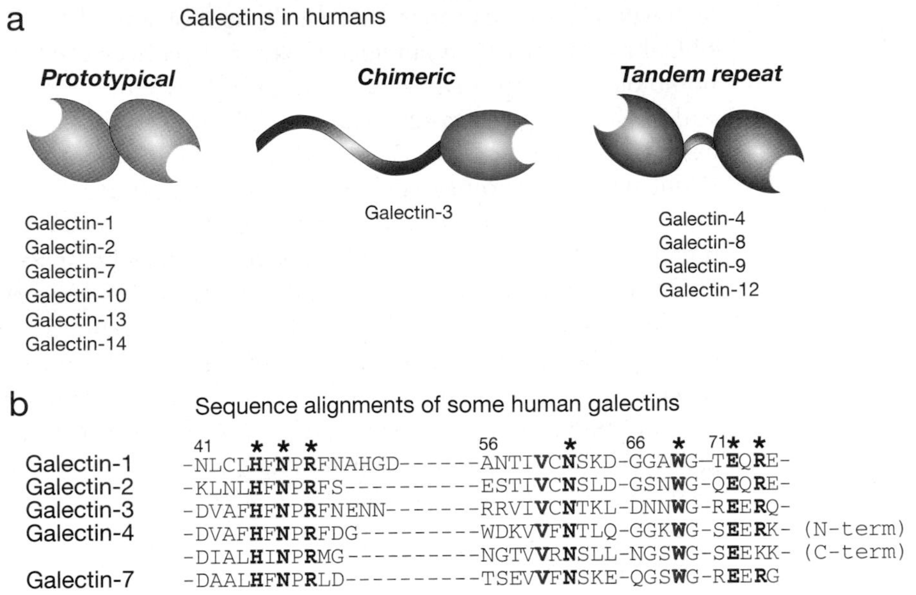

FIGURE 33.1. Different types of galectins in humans. (*a*) Human galectins have been classified into three groups according to their structure: prototypical, chimeric, and tandem repeat. The carbohydrate-recognition domain (CRD) of most galectins is approximately 130 amino acids, and this is indicated by the oval domain. (*b*) Examples of the sequence alignments between several human galectins. The amino acid numbering is shown for galectin-1, which has 135 amino acids in total, but the other galectins are aligned without showing their numbers. Those residues that are highly conserved between galectins and those that are known to make contacts with carbohydrate ligands are indicated by *asterisks*. (*Red*) conserved hydrophilic residues; (*blue*) conserved hydrophobic residues. (N-term) Amino-terminal; (C-term) carboxy-terminal.

which is rich in proline, glycine, and tyrosine residues, similar to those found in some other proteins (such as synexin and synaptophysin), where it may contribute to self-aggregation. In addition, the amino-terminal domain is sensitive to metalloproteinases, such as MMP-2 and MMP-9. Chimeric galectins are more common in invertebrates.

3. The tandem-repeat galectins, in which at least two CRDs occur within a single polypeptide. They are bridged or linked by a small peptide domain. These link domains can range from 5 to more than 50 amino acids in length.

In addition, many of the galectin transcripts may be differentially spliced to generate many different isoforms. For example, at least seven different mRNAs have been identified for human galectin-8, some encoding a tandem-repeat form and others a prototypical form. These isoforms of galectin-8 may be differentially expressed in different tissues. Three isoforms of galectin-9 differing in the length of the linker domains have been identified. Similarly, different splice forms of pig galectin-4 generate long and short forms. Galectin-5 (prototypical) and galectin-6 (tandem-repeat) are found in rodents, but not humans, and galectin-11 (ovagal11; prototypical) has been reported in sheep. Confusingly, the GRIFIN protein (prototypical), which in mammals may not bind to carbohydrates, has also been termed galectin-11 (but see below).

There are also a number of galectin-related proteins that have homology with galectins, but they may not bind carbohydrates, or at least not bind to typical β-galactosides. For example, galectin-10 appears to bind better to β-mannosides than to β-galactosides. Galectin-10 is primarily expressed in eosinophil granules and occurs there in a crystalline form, known as the Charcot–Leyden crystal protein. Several proteins in vertebrates related to galectin-10 do not appear to bind sugars (GRIFIN, PP-13, PPL-13, and the sheep protein ovgal11). These proteins may be thought of as a subfamily of galectin-10. It is also possible that the correct glycan ligands have not yet been identified. Interestingly, the GRIFIN homolog in zebrafish *Danio rerio*, DrGRIFIN, is a functional β-galactoside-binding protein and has a more galectin-like sequence. Like mammalian GRIFIN, DrGRIFIN is also highly expressed in the eye lens. Galectin-10 family members may be generally thought of as crystalline proteins. Another galectin-related protein (GRP) is HSPC159, which is expressed in human hematopoietic stem cell precursors, but it lacks a critical tryptophan residue that is highly conserved in all other galectins; thus, it may be unable to bind β-galactosides.

Galectins are found in virtually all organisms. Birds express 14-kD and 16-kD galectins that are related to galectin-1 and other galectins that are homologs of galectin-3 and -8. Galectins have also been found in the skin (16 kD) and oocytes (15 kD) of the amphibians *Xenopus laevis* and *Bufo arenarum*, respectively, and homologs of galectin-1, -3, -4, -8, and -9, and HSPC159 have been found in *X. laevis*. Fish galectins have also been found and partly characterized, including homologs of galectin-1, -3, -4, and -9, and HSPC159. Zebrafish has three genes that have homology with galectin-1, along with genes for homologs of galectin-4 and galectin-9 and HSPC159. Galectins are also expressed in *Drosophila melanogaster* (six candidate genes) and *Caenorhabditis elegans* (26 candidate genes). At least 14 of these candidate galectin genes in *C. elegans* have been shown to encode proteins and are found as expressed sequence tags (ESTs). Galectins have also been found in sponges (*Geodia cydonium*), and there are galectin-like sequences predicted from the plant genome (*Arabidopsis thaliana*). Interestingly, some of the predicted galectin genes in *C. elegans* and plants encode proteins that might contain a galectin domain in tandem with a glycosyltransferase domain, and some of these proteins may have standard signal sequences. Galectin-like proteins are even expressed in some viruses that infect pigs and fish, including porcine adenovirus and lymphocystis disease virus.

GLYCAN LIGANDS FOR GALECTINS

The interactions of galectins with glycans are complex and several factors contribute to high-affinity binding, including the natural multivalency and oligomeric state of the galectins, the multivalency of their natural glycoconjugate ligands, and the mode of presentation of the glycans. In simple terms, most members of the galectin family tested so far bind simple β-galactosides, such as disaccharides or trisaccharides, but the affinity is relatively weak (i.e., in the high micromolar to low millimolar range). In contrast, galectin binding to natural glycoconjugate ligands expressed on cell surfaces or in the extracellular matrix is usually of much higher affinity (i.e., in the micromolar or submicromolar range). Each galectin CRD recognizes different types of glycan ligands and shows highest affinity binding to different structures. For example, galectin-3 binds tightly to glycans with repeating $[\text{-3Gal}\beta\text{1-4GlcNAc}\beta\text{1-}]_n$ or poly-*N*-acetyllactosamine sequences containing three to four repeating units, regardless of the presence of a terminal β-galactose residue. In contrast, human galectin-1 also binds well to long poly-*N*-acetyllactosamine chains, but it requires a terminal β-galactose residue. Although galectin-3 binds weakly to single *N*-acetyllactosamine units, its binding to these units is enhanced if the penultimate galactose residues are substituted with Galβ1-3, GalNAcα1-3, or Fucα1-2 residues. In contrast, such substitutions dramatically decrease binding by galectin-1. Galectin-8 has two CRDs within its single polypeptide, and the amino- and carboxy-terminal CRDs bind different glycans. For example, the amino-terminal CRD of human galectin-8 binds α2-3-sialylated glycans with high affinity (K_d of ~50 nM), whereas the carboxy-terminal CRD shows lower-affinity binding, primarily to the blood group

A determinant GalNAcα1-3(Fucα1-2)Gal- on either a LacNAc or Galβ1-3GlcNAc core, and does not bind sialylated glycans.

Cocrystallization of galectins with simple β-galactose-containing disaccharides has revealed that many galectins bind to the C-4 and C-6 hydroxyls of galactose and the C-3 hydroxyl of *N*-acetylglucosamine. The binding sites of galectins may be viewed as containing several subsites: one for galactose, another for *N*-acetylglucosamine, and still other subsites that may be filled by other sugars and the aglycone moiety, such as a peptide or lipid.

Galectins appear to bind selectively to some cell-surface and extracellular matrix ligands. However, the precise physiological roles of these interactions with each galectin are not well understood. Potential ligands for galectin-1 and galectin-3 include basement membrane proteins (such as laminin and fibronectin), membrane receptors (such as integrins α7β1 and α1β1, CD43, CD7, and CD45), lysosome-associated membrane proteins (LAMP-1 and LAMP-2), vitronectin, and fibronectin. Galectin-1 interactions with T-cell glycoproteins (such as CD43 and CD45) depend on appropriate glycosylation by specific glycosyltransferases (such as the core-2 β1-6 N-acetylglucosaminyltransferase that forms core-2 O-glycans and the β1-6 N-acetylglucosaminyltransferase V that forms branched N-glycans). However, the precise glycan structures on these macromolecules recognized by galectins are not well defined.

Galectin-8 displays high-affinity binding in the extracellular matrix to integrins and to a variant of CD44, and its binding to integrins (α3β1 and α6β1) modulates their adhesive and signaling properties. Depending on its expression level, galectin-8 may diminish integrin-dependent adhesion or enhance it. Galectin-8 binding to CD44v is high affinity (K_d of 6 nM). Galectin-9 binds to TIM-3, which is a membrane-bound T-cell immunoglobulin-like protein with O-glycans. TIM-3/galectin-9 appears to contribute to T-helper 1 (Th1) cell immunity and tolerance induction.

BIOSYNTHESIS AND EXPORT OF GALECTINS

All members of the galectin family lack a classical signal sequence and membrane-anchoring domains and appear to be synthesized on free polysomes in the cytoplasm and accumulate there prior to secretion. Galectins are probably unique among all types of animal lectins in that they can be found in the nucleus, cytoplasm, outer plasma membrane, and extracellular matrix. One rather common modification of galectins in animal cells is blockage of the amino terminus. Galectin-3 has been shown to also have a serine residue in its amino-terminal region that can be phosphorylated. Newly synthesized galectins isolated directly from the cytoplasm of cells are functional in binding β-galactosides, indicating that they are synthesized and potentially functional in that compartment. Interestingly, there is little evidence for the presence of galectin glycan ligands in the cytoplasm, indicating that they may have other functions there.

As discussed below, however, there is compelling evidence for galectins having noncarbohydrate-binding partners in the cytoplasm. The complexity of galectin biosynthesis, secretion, and oligomerization is illustrated schematically for galectin-1 in Figure 33.2. Curiously, the export of galectins from cells does not involve direct movement through the secretory apparatus. The mysterious process, termed nonclassical export, by which these proteins are exported has been explored in several ways, but the basic mechanism is still unknown. Several other relatively small-sized growth factors and cytokines (e.g., fibroblast growth factors FGF-1 and FGF-2 and interleukin IL-1β) are also secreted by a nonclassical pathway, but whether their secretory pathway converges with that of galectins is not known.

In exploring a possible mechanism for export, recombinant galectin-1 expressed in yeast was found to be exported by a transmembrane protein, but so far this transporter appears

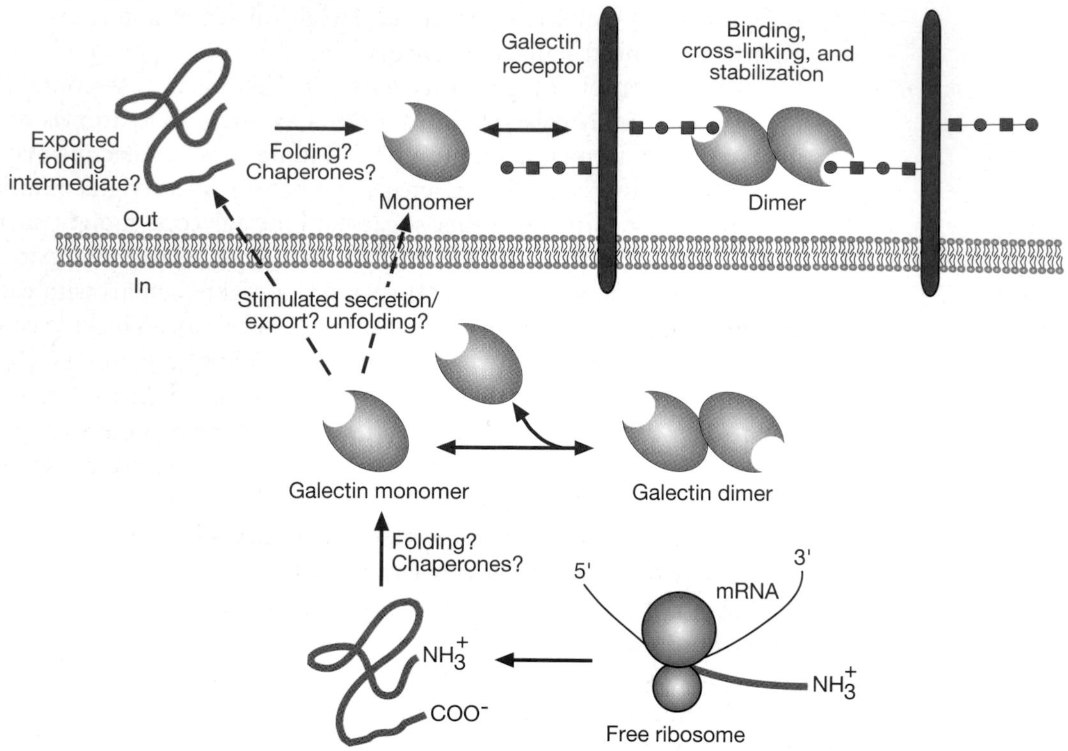

FIGURE 33.2. Possible biosynthetic routes for galectins in animal cells. The mRNA for the protein is translated on free polysomes in the cytoplasm, and the newly synthesized protein is capable of binding carbohydrate ligands or interacting with other proteins within the cell. Secretion or export occurs by an undefined mechanism termed nonclassical export. The newly synthesized galectin appears to be unstable, but its association with carbohydrate ligands can stabilize its structure. The stable monomeric protein may form homodimers and other oligomeric structures and interact with ligands on the cell surface and in the extracellular matrix. It may also interact directly with ligands on other cells.

to be limited to yeast. Whether different transporters occur in animal cells is unknown. It is also possible that export involves membranous structures. For example, galectin-1 appears to exit from myoblasts via evaginations of the plasma membrane, which pinch off to form lectin-enriched vesicles. Similarly, galectin-3 assembles into patches that eventually appear to underlie the plasma membrane as a prelude to deposition in the extracellular space, possibly through vesicular extravasation. Removal of the unique amino-terminal 11 amino acids in galectin-3 prevents its export. Although the mechanism of export is unclear, several galectins, including galectin-1, rapidly lose activity in a nonreducing environment at 37°C, as exists outside cells. In contrast, other galectins, such as galectin-3 and galectin-4, are highly stable in such an environment. Studies on the biosynthesis of galectin-1 found that the newly exported protein is extremely unstable in the absence of glycan ligands, but when high-affinity ligands are available, the protein is stable. Thus, the regulated secretion and availability of glycan ligands may also regulate activity and stability of galectins.

STRUCTURE OF GALECTINS

The crystal structures of several galectins have now been reported, including bovine galectin-1 (complexed with either *N*-acetyllactosamine or diantennary N-glycans), human galectin-2, -3, and -7, and individual domains of galectin-9 (complexed with monosaccharides or small glycans such as lactose). All of the structures show that the CRD of galectin subunits is composed of five- and six-stranded antiparallel β-sheets arranged in a β-sandwich or jelly-roll configuration that completely lacks an α-helix (Figure 33.3). In the dimeric proteins, such as galectin-1, -2, and -7, the subunits are related by a twofold rotational axis perpendicular to the plane of the β-sheets. The glycan-binding sites in the CRD are located at opposite ends of the dimer, although the orientation of the subunits in the galectin-7 dimer is different from that of the other canonical galectin dimers. The compactly arranged structure of the CRD partly explains the protease resistance of the galectin CRD and the high degree of conservation and requirement for 130 amino acids in the CRD. Such folded structures are similar in some ways to the folds in many plant L-type lectins from legumes and those in pentraxins, although the primary sequences of galectins have no similarities to those of other classes of proteins. The complete three-dimensional structures of many other galectins, such as tandem-repeat galectins, have not been determined, because such galectins do not crystallize well, possibly because of disorder introduced by the bridging or linking peptide.

The galectin-1 CRD displays highly specific interactions with galactose and *N*-acetylglucosamine residues. Interactions with carbohydrates generally are through hydrogen bonding, electrostatic interactions, and van der Waals interactions through ring stacking with galactose

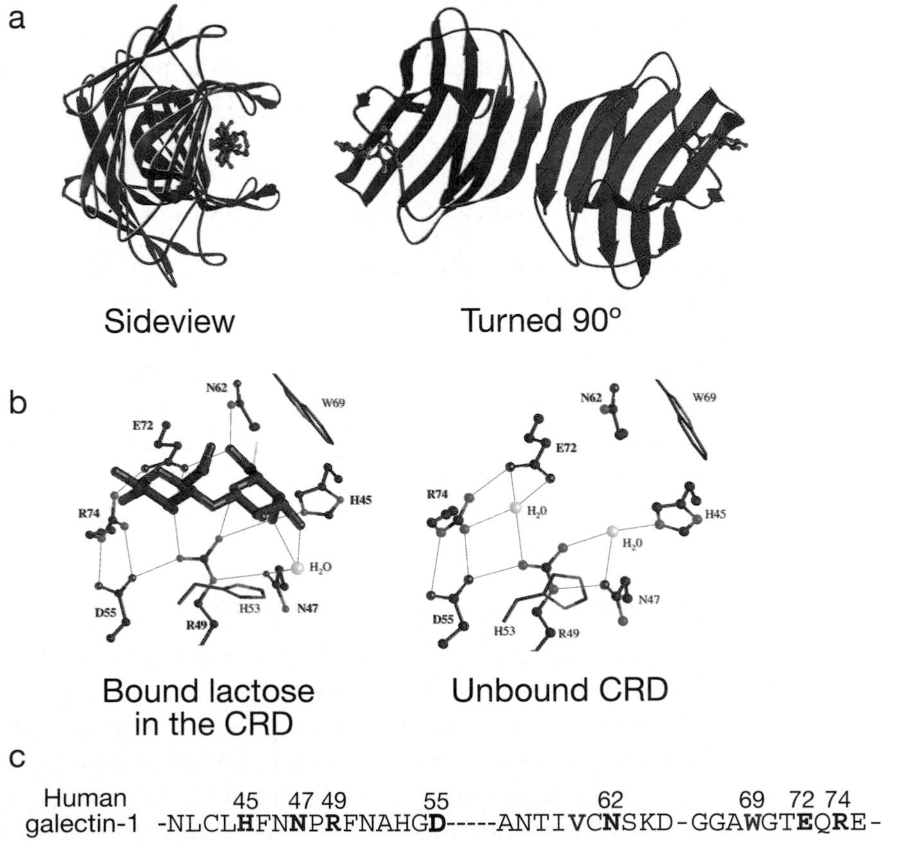

FIGURE 33.3. (*a*) Ribbon diagram of the crystal structure of human galectin-1, based on X-ray crystallographic analyses of the protein complexed with lactose. The homodimer is shown with each monomer colored differently and orthogonal views are presented. The subunit interface is based on interactions between the carboxy- and amino-terminal domains of each subunit. (*b*) Interactions between key amino acid residues within the CRD of galectin-1 when lactose is bound (*left panel*) and when no sugar is bound (*right panel*). (*c*) Primary sequence of human galectin-1 with the numbered residues corresponding to those highlighted in the crystal structure above.

and the highly conserved tryptophan residue (Figure 33.3). In general, the open-ended structure of the carbohydrate-binding site is predicted to allow access to extended galactose-containing glycans, such as the poly-*N*-acetyllactosamines and blood-group-related structures. Several known and predicted subsites on galectins near the carbohydrate-binding site could serve to enhance affinity for more extended glycans. Some of these subsites have been identified by cocrystal structural analysis with glycans and by glycan-binding studies. The crystal structure of bovine galectin-1 was derived for the protein in complex with a biantennary N-glycan containing two terminal β-galactose residues. In this extended crystal structure, the N-glycan is bridged between two galectin dimers, thus effectively creating a crystal latticework. This type of crystal latticework may be unique among galectins in regard to vertebrate galectins and may be critical for their signaling and adhesive functions.

Some galectins rapidly lose activity (within a day) if not kept in reducing buffers, perhaps due to cross-linking and oxidation of cysteine residues or the key tryptophan residue in the CRD. The most labile lectin in this regard is vertebrate galectin-1. The presence of even weak binding ligands, such as lactose, can help stabilize vertebrate galectin-1 in the absence of reducing conditions. However, most others such as galectin-3 or galectin-4 are stable in the absence of reducing conditions or ligands. Interestingly, many of the galectins have free reduced cysteine residues, a situation that is perhaps predictable since they are synthesized in the cytoplasm, which is a highly reducing environment. Interestingly, for galectin-1 there is evidence that, upon oxidation, it can also form intramolecular disulfides, which are coincident with a loss of carbohydrate-binding activity. Curiously, this alternative oxidized galectin-1 has activity in promoting axonal regeneration in adult rat dorsal root ganglia.

FUNCTIONS OF THE GALECTINS

Galectins are probably the most ancient class of glycan-binding proteins, and they are found in all metazoans examined, from sponges and fungi to both invertebrates and vertebrates. Galectins can contribute to cell–cell and cell–matrix interactions, and galectin signaling at the cell surface can also modulate cellular functions. In addition, intracellular galectins may interact with intracellular ligands to regulate cellular activities and may contribute to some fundamental processes such as pre-mRNA splicing (Figure 33.4). Examples of the functions and activities associated with different galectins, with emphasis on their roles in immune regulation and inflammation, are shown in Figure 33.5.

Roles of Galectins in Immune Responses and Inflammation

One of the major functions of galectins is to regulate immune and inflammatory responses. Galectins are expressed by activated T and B cells, regulatory T cells, dendritic cells, mast cells, eosinophils, monocytes/macrophages, and neutrophils. In addition, galectins can promote pro- or anti-inflammatory responses, depending on the inflammatory stimulus, microenvironment, and target cells. Immune cell responses to galectins also depend on the specific glycosylation of surface glycoproteins in those cells to generate galectin ligands.

Galectin-1 function is generally associated with attenuating inflammatory responses. In contrast, galectin-3 has a proinflammatory role. For example, galectin-1 can contribute to the balance between Th1 and Th2 immune responses, which are characterized by the types of cytokines produced. Galectin-1 can induce some anti-inflammatory cytokines, such as IL-5, IL-10, and transforming growth factor-β (TGF-β) in activated T cells, and it can inhibit production of proinflammatory cytokines, such as IL-2, tumor necrosis factor-

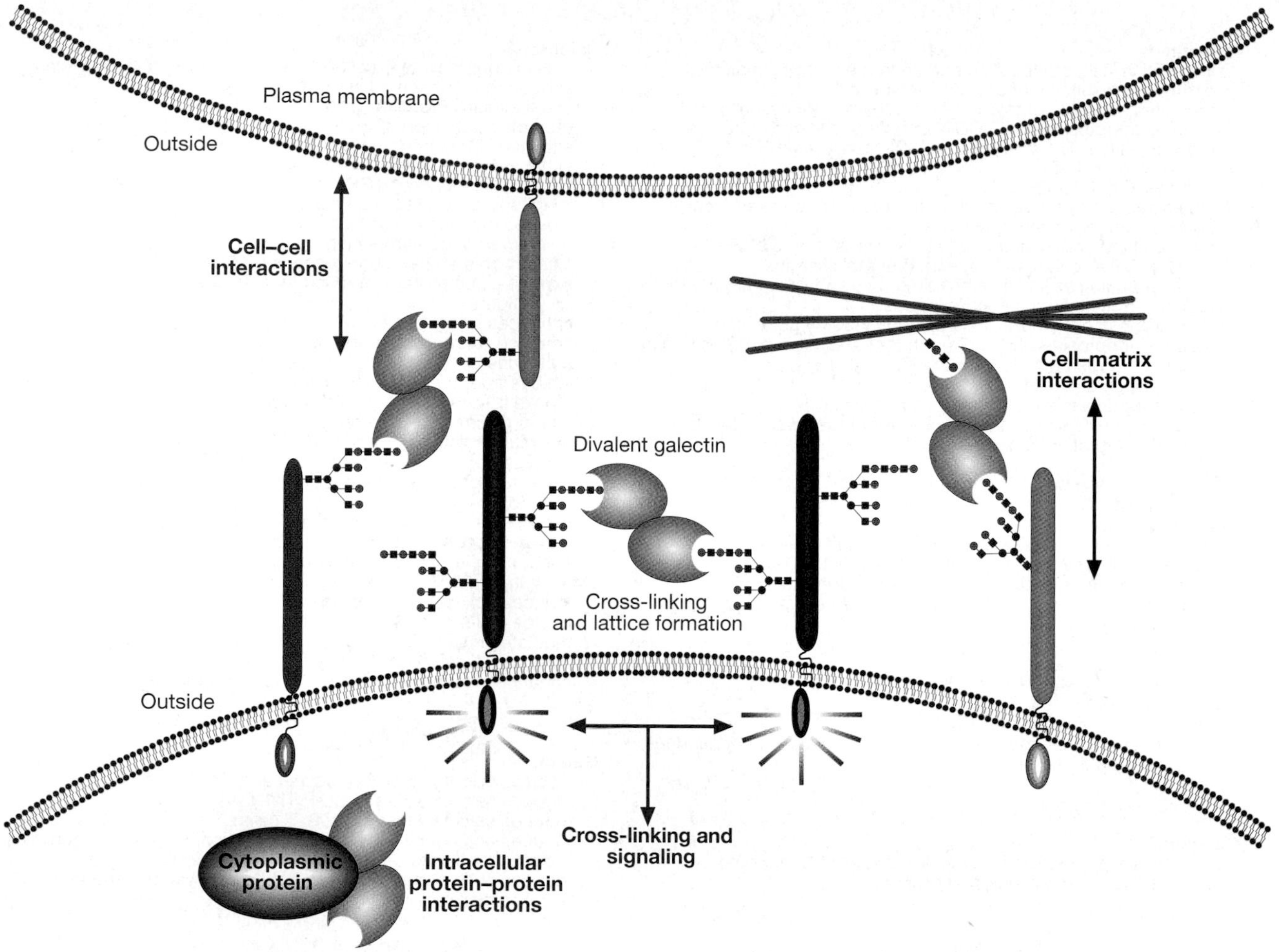

FIGURE 33.4. Functional interactions of galectins with cell-surface glycoconjugates and extracellular glycoconjugates can lead to cell adhesion and cell signaling. Interactions of galectins with intracellular ligands may also contribute to the regulation of intracellular pathways.

α (TNF-α), and interferon-γ (IFN-γ). Th1 and Th17 differentiated cells express the types of nonsialylated glycan ligands that are essential for their binding to galectin-1 and signal responses leading to their down-regulation, whereas Th2 cells may lack these types of ligands and may have more sialylated ligands that reduce galectin-1 binding and signaling. Thus, galectin-1 knockout mice show a "hyper-Th1 and Th17 response" when challenged by antigens in vivo. Overall, it appears that the differential glycosylation of T-helper cells and their differential responses to galectins may determine the overall immune responses. Galectin-1 knockout mice also exhibit heightened sensitivity to experimental autoimmune encephalomyelitis associated with elevated levels of Th1 and Th17 lymphocytes. Interestingly, subcutaneous injections of recombinant human galectin-1 can reduce the severity of several autoimmune diseases in animal models, including experimental autoimmune encephalomyelitis, experimental autoimmune myasthenia gravis, and collagen-induced arthritis.

Galectin-1 interacts with several specific T-cell glycoproteins, including CD45, CD43, and CD7, but these interactions depend on appropriate glycosylation by specific glycosyltransferases, such as the core-2 β1-6 *N*-acetylglucosaminyltransferase that forms core-2 O-glycans and the β1-6 *N*-acetylglucosaminyltransferase V that forms branched N-glycans.

Galectin-1 
involved in Treg cell function and enhances Treg formation
conflicting results on effects on T-cell viability
mediates adhesion of thymocytes to thymic epithelium
induces apoptosis in CD4⁺CD8⁺ double positive thymocytes
induces shift in Th1 response to Th2 (decreases IFNγ; increases IL-5)
reduces TNFα, IL-1β, IL-12, IL-2 and IFNγ
increases IL-10 production in both naive and activated T cells
inhibits mast cell degranulation
reduces pathology-associated graft-versus-host disease, Con A-induced hepatitis, experimental allergic encephalomyelitis, myasthenia gravis and rheumatoid arthritis
reduces acute inflammatory responses
expression in endothelial cells up-regulated by activation
induces apoptosis-independent phosphatidylserine (PS) exposure (Ca^{++}-dependent) in neutrophils
inhibits chemotaxis of neutrophils
inhibits extravasation of neutrophils
activates NADPH-dependent respiratory burst in neutrophils
induces maturation of dendritic cells

Galectin-2 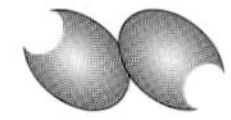
induces T-cell apoptosis under some conditions
decreases IFNγ and TNFα while increasing IL-10 and IL-5
involved in the pathogenesis of atheroma formation
induces apoptosis-independent PS exposure (Ca^{++}-dependent) of neutrophils

Galectin-4 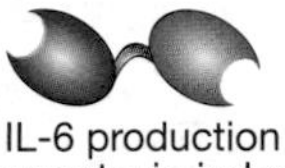
induces IL-6 production in T cells
induces apoptosis-independent PS exposure (Ca^{++}-independent) of neutrophils

Galectin-7 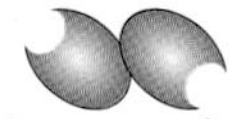
intracellular expression induces apoptosis of tumor cells
extracellularly can inhibit growth of cells

Galectin-8 
activates Rac-1 in T cells
activates NADPH-dependent respiratory burst of neutrophils
modulates integrin-mediated neutrophil adhesion of neutrophils

Galectin-10 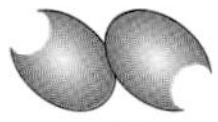
highly expressed in eosinophils (Charcot-Leyden crystal protein)
involved in Treg function

Galectin-3 
blocks apoptosis of T cells when overexpressed intracellularly
endogenously involved in T-cell viability
extracellularly induces apoptosis of T cells
promotes adhesion of thymocytes to thymic epithelium
enhances Th2 immune responses
enhances adhesion of naive T cells to DCs
binds TCR, reducing TCR mediated T cell activation
inhibits IL-5 production in eosinophils
induces mast cell degranulation independent of antigen-mediated IgE stimulation
exacerbates Th2 immune responses (asthma)
expressed on surface of macrophages (also called Mac-2 antigen)
enhances phagocytosis of macrophages
enhances respiratory burst of macrophages
enhances LPS-induced IL-1β secretion of macrophages
inhibits apoptosis (intracellularly)
blocks IL-4-induced survival of activated B cells
favors plasma cell differentiation
exhibits an anti-apoptotic role in B-cell lymphomas
expression induced in dendritic cells by *T. cruzi* infection
enhances pro-inflammatory cytokine release in endothelial cells
expression up-regulated in tumor endothelial cells
induces chemotaxis of neutrophils
enhances extravasation of neutrophils
activates NADPH-dependent respiratory burst of neutrophils
induces activation of neutrophils
induces release of IL-8 of neutrophils
mediates interaction of neutrophils with laminin and fibronectin (both directly and indirectly)
enhances leukocyte adhesion to endothelium

Galectin-9 
induces apoptosis in thymocytes and T cells
induces selective loss of CD4⁺ Th1 cells
induces selective loss of CD8⁺ T cells
induces eosinophil chemotaxis, activation, superoxide generation
induces moderate degranulation of eosinophil
expression in endothelial cells induced by virus infection
induces maturation of dendritic cells

Galectin-12 
intracellular expression induces apoptosis of tumor cells
can cause cell cycle arrest and growth suppression

FIGURE 33.5. A list of known and putative functions and biological activities of galectins toward cells in the immune system.

Galectin-3 is associated with activation of T cells perhaps by interacting with the poly-*N*-acetyllactosamine-containing N-glycans on the T-cell receptor (TCR). Expression of such N-glycans is partly regulated by branching of N-glycans through β1-6 *N*-acetylglucosaminyltransferase V (Mgat5). *Mgat5*-null mice show enhanced clustering of TCRs, indicating that branched N-glycans interacting with galectin-3 may normally restrict TCR clustering and serve as a hindrance to the development of T-cell responses. Galectin-3 can also inhibit IL-5 production in several immune cells, including human eosinophils. On the other hand, galectin-3 can activate mast cells, neutrophils, and monocytes, in terms of mediator release and production of reactive oxygen species. Lack of galectin-3 in knockout mice is associated with reduced mast cell function, reduced accumulation of asthma-associated leukocytes in airway inflammation, and reduced peritoneal inflammatory responses. Endogenous galectin-3 has also been shown to play a role in phagocytosis by macrophages and mediator release/cytokine production by mast cells by functioning intracellularly. Overall, such results suggest a complex set of functions for galectin-3 as a proinflammatory mediator and in regulating many aspects of the inflammatory response.

Roles of Galectins in Apoptosis and Induction of Cell-Surface Phosphatidylserine Exposure

Several galectins (including galectin-1, -2, -3, -7, -8, -9, and -12) have been shown to be able to induce apoptosis in some types of blood cells. For galectin-1, this activity has been studied most in human T cells, where apoptotic pathways may involve cell-surface glycoproteins including CD7, CD29, and CD43, whereas induction of apoptosis in T cells by galectin-3 involves CD71 and CD45. In some cells, apoptotic signaling may function through down-regulation of Bcl-2 and activation of caspases. The interactions of galectin-9 with Tim-3 on Th1 cells may induce their apoptosis. In addition, overexpression of intracellular galectin-3 exhibits antiapoptotic activity, whereas overexpression of galectin-7 and galectin-12 may promote apoptosis in cells. Some potential intracellular binding partners for galectins, especially galectin-3, include several proteins involved in regulating apoptosis, such as Bcl-2, Fas receptor (CD95), synexin (which is a Ca^{++}- and phospholipid-binding protein), and Alg-2. In addition, some galectins, such as galectin-1, -2, and -4, also have the unusual ability to induce exposure of cell-surface phosphatidylserine independently of apoptotic events. In activated human neutrophils, this process induced by galectin-1 requires binding to cell-surface receptors in lipid rafts or microdomains and involves mobilization of intracellular Ca^{++} and signaling through Src kinase and phospholipase Cγ.

Roles of Galectins in Animal Development

Galectins play important, but rather subtle, roles in animal development. Lack of galectin-3 in knockout mice is associated with several phenotypic changes, such as fatty liver disease, reduced mast cell function, reduced liver fibrosis upon induced liver damage, and age-dependent glomerular lesions. In contrast, lack of galectin-1 in mice is associated with a different set of interesting phenotypic changes, including decreased sensitivity to noxious thermal stimuli, altered primary afferent neural anatomy, aberrant topography of olfactory axons, and reduced muscle regeneration ability after injury.

It is possible that the redundancy of galectin family members contributes to survival of these null mutants or that these particular galectins are involved only in postdevelopmental processes, such as immune regulation. Consistent with this possibility, the inflammatory response in galectin-3-null mice is dampened, and there is a decline in infiltrating neutrophils. It will be interesting to observe the phenotypes of mice with deletions of two or more galectins when they are provided with a variety of environmental, pathogenic, and antigenic challenges.

Roles of Galectins in Cancer

Many types of tumors, including melanomas, astrocytomas, and bladder and ovarian tumors overexpress various galectins, and their heightened expression usually correlates with clinical aggressiveness of the tumor and the progression to a metastatic phenotype. Three galectins that have shown importance in cancer progression and metastasis are galectin-1, -3, and -9. The immunosuppressive and apoptotic effects of galectin-1 can contribute to tumor survival, as revealed by knockdown studies, where decreased galectin-1 expression is associated with decreased tumor survival, due to increased survival of IFN-γ-producing Th1 cells and heightened T-cell-mediated tumor rejection. Recent studies using galectin-1 knockout cells have shown that expression of galectin-1 in tumor cell endothelium is essential for tumor angiogenesis. Thus, galectins are likely to play important roles in tumor progression and metastasis through indirect effects in regulating tumor immune responses and direct effects in tumor angiogenesis. Overexpression of galectin-3 correlates well with neoplastic transformation and tumor progression toward metastasis, and expression of galectin-3 may be a histological tumor marker. Some studies even suggest that blocking galectin-3 function may limit tumor metastasis.

Roles of Galectins in Innate Immunity

Although the expression of galectins in animal tissues is tightly regulated, their expression can be induced and this may be especially important in innate immune responses. For example, both galectin-1 and galectin-3 are up-regulated in gastric epithelial cells that are infected by *Helicobacter pylori*, and galectin-9 can be induced upon exposure of periodontal ligament cells to *Porphylomonas gingivalis* lipopolysaccharide. One of the galectins in *Amphioxus* (BbtGal-L) is up-regulated in immune organs by challenge with different pathogens. Galectins, such as galectin-3, have also been shown to bind pathogen-derived glycans directly and may function in stimulating macrophage uptake of pathogen materials for antigen presentation. Finally, galectin-3 can directly induce death of the yeast *Candida albicans* through binding to β1-2 oligomannosyl residues. The roles of galectins in innate defense against microorganisms have been revealed by studying genetically engineered mice deficient in specific galectins. For example, galectin-3-null mice had impaired capacities in clearing late infection of *Mycobacterium tuberculosis* compared to wild-type mice, suggesting involvement of galectin-3 in innate defense against *Mycobacterial* infection.

FURTHER READING

Drickamer K. and Taylor M.E. 1993. Biology of animal lectins. *Annu. Rev. Cell Biol.* **9:** 237–264.

Barondes S.H., Cooper D.N., Gitt M.A., and Leffler H. 1994. Galectins. Structure and function of a large family of animal lectins. *J. Biol. Chem.* **269:** 20807–20810.

Leffler H. 2001. Galectins structure and function—A synopsis. *Results Probl. Cell. Differ.* **33:** 57–83.

Brewer C.F., Miceli M.C., and Baum L.G. 2002. Clusters, bundles, arrays and lattices: Novel mechanisms for lectin-saccharide-mediated cellular interactions. *Curr. Opin. Struct. Biol.* **12:** 616–623.

Cooper D.N. 2002. Galectinomics: Finding themes in complexity. *Biochim. Biophys. Acta* **1572:** 209–231.

Hernandez J.D. and Baum L.G. 2002. Ah, sweet mystery of death! Galectins and control of cell fate. *Glycobiology* **12:** 127R–136R.

Liu F.T., Patterson R.J., and Wang J.L. 2002. Intracellular functions of galectins. *Biochim. Biophys. Acta* **1572:** 263–273.

Lahm H., Andre S., Hoeflich A., Kaltner H., Siebert H.C., Sordat B., von der Lieth C.W., Wolf E., and Gabius H.J. 2004. Tumor galectinology: Insights into the complex network of a family of endogenous lectins. *Glycoconj. J.* **20:** 227–238.

Leffler H., Carlsson S., Hedlund M., Qian Y., and Poirier F. 2004. Introduction to galectins. *Glycoconj. J.* **19:** 433–440.

Liu F.T. and Rabinovich G.A. 2005. Galectins as modulators of tumour progression. *Nat. Rev. Cancer* **5:** 29–41.

Rabinovich G.A. and Gruppi A. 2005. Galectins as immunoregulators during infectious processes: From microbial invasion to the resolution of the disease. *Parasite Immunol.* **27:** 103–114.

Camby I., Le Mercier M., Lefranc F., and Kiss R. 2006. Galectin-1: A small protein with major functions. *Glycobiology* **16:** 137R–157R.

Hsu D.K., Yang R.Y., and Liu F.T. 2006. Galectins in apoptosis. *Methods Enzymol.* **417:** 256–273.

Nakahara S. and Raz A. 2006. On the role of galectins in signal transduction. *Methods Enzymol.* **417:** 273–289.

Rabinovich G.A., Liu F.T., Hirashima M., and Anderson A. 2007. An emerging role for galectins in tuning the immune response: Lessons from experimental models of inflammatory disease, autoimmunity and cancer. *Scand. J. Immunol.* **66:** 143–158.
Thijssen V.L., Poirier F., Baum L.G., and Griffioen A.W. 2007. Galectins in the tumor endothelium; opportunities for combined cancer therapy. *Blood* **110:** 2819–2827.
Stowell S.R. and Cummings R.D. 2008. Interactions of galectins with leukocytes. In *Animal lectins: A functional view* (ed. G.R. Vasta and H. Ahmed), pp. 415–431. CRC Press, Boca Raton, Florida.

CHAPTER 34

Microbial Lectins: Hemagglutinins, Adhesins, and Toxins

Jeffrey D. Esko and Nathan Sharon

MANY MICROORGANISMS EXPLOIT HOST CELL-SURFACE GLYCANS as receptors for cell attachment and tissue colonization, and a large number of pathogenic species depend on these interactions for infection. This chapter describes examples of proteins on the surface of microorganisms (adhesins or hemagglutinins), secreted proteins (heat-labile toxins), their glycan partners on mammalian cell surfaces (receptors), and insights into the molecular interactions that take place.

BACKGROUND

Viruses, bacteria, and protozoa express an enormous number of glycan-binding proteins or lectins. Many of these microbial lectins were originally detected based on their ability to aggregate red blood cells (i.e., to induce hemagglutination). The first microbial hemagglutinin identified was in the influenza virus, and it was shown by Alfred Gottschalk in the early 1950s to bind to erythrocytes and other cells though sialic acid residues of cell-surface glycoconjugates. Don Wiley and his associates crystallized the viral hemagglutinin and determined its structure in 1981, and later they solved the structure of cocrystals prepared with sialyllactose. Since then, a number of viral hemagglutinins have been identified and crystallized. These studies set the stage for analyzing other types of microbial lectins produced by bacteria and protozoa.

Nathan Sharon and colleagues first described bacterial surface lectins in the 1970s. These lectins also have hemagglutinating activity, but their primary function is to facilitate attachment or adherence of bacteria to host cells, a prerequisite for bacterial colonization and infection (see Chapter 39). Thus, bacterial lectins are often called adhesins, and the glycan ligands on the surface of the host cells are called receptors. Note that the term "receptor" in this case is equivalent to "ligand" for animal cell lectins. Bacteria also produce toxins, whose actions often depend on glycan-binding subunits that allow the toxin to combine with membrane glycoconjugates and deliver the functionally active toxic subunit across the plasma membrane. During the last 30 years, many adhesins and toxins have been described, cloned, and characterized. Additionally, adhesins have been discovered on various parasites. The interactions of adhesins with glycan receptors can help determine the tropism of the organism.

Colonization of tissues by microorganisms is not always pathogenic. For example, the normal flora of the lower gastrointestinal tract is determined by appropriate and desirable colonization by beneficial bacteria. The initial events in the formation of nitrogen-fixing nodules in leguminous root tips by species of *Rhizobium* may also involve lectins on the root tip binding to Nod factors generated by the bacterium (see Chapter 37).

Many adhesins contain carbohydrate-recognition domains (CRDs) that bind to the same carbohydrates as endogenous mammalian lectins (see Chapter 26). Like animal cell lectins, some microbial adhesins bind to terminal sugar residues, whereas others bind to internal sequences found in linear or branched oligosaccharide chains. Detailed studies of the specificity of microbial lectins have led to the identification and synthesis of powerful inhibitors of adhesion that may form the basis for therapeutic agents for treating infection (see Chapter 51).

VIRAL GLYCAN-BINDING PROTEINS

By far, the best-studied example of a viral glycan-binding protein is the influenza virus hemagglutinin, which binds to sialic acid–containing glycans. The affinity of this interaction is relatively low, like that of other glycan-binding proteins with their glycan ligands, but the avidity for cell membranes increases because of oligomerization of the hemagglutinin into trimers and the high density of glycan receptors present on the host cell (see Chapter 27). Binding is a prerequisite for fusion of the viral envelope with the plasma membrane and for uptake of the virus into cells. The specificity of the interaction of the hemagglutinin with host glycans varies considerably for different subtypes of the virus. Human strains of influenza-A and -B viruses bind primarily to cells containing Neu5Acα2-6Gal-, whereas chicken influenza viruses bind to Neu5Acα2-3Gal- and porcine strains bind to both Neu5Acα2-3Gal- and Neu5Acα2-6Gal-containing receptors. This linkage preference is due to certain amino acid changes in the hemagglutinin. Influenza-C virus, in contrast, binds exclusively to glycoproteins and glycolipids containing 9-O-acetylated *N*-acetylneuraminic acid (see Chapter 39).

The specificity of the hemagglutinin correlates with the structures of sialylated glycans expressed on target epithelial cells in animal hosts. For example, trachea epithelial cells in humans express glycans with a preponderance of Neu5Acα2-6Gal linkages, whereas other tissues contain many more Neu5Acα2-3Gal-terminated glycans. Thus, the specificity of the hemagglutinin determines the tropism of the virus with respect to species and target cells. The hemagglutinin is also the major antigen against which neutralizing antibodies are produced, and antigenic changes in this protein are in part responsible for new viral outbreaks. In addition, it plays a critical role inside the cell where it facilitates pH-dependent fusion of the viral envelope with endosomal membranes after internalization.

The crystal structure of the hemagglutinin shows two disulfide-bonded subunits, HA1 and HA2, which are derived by cleavage of a precursor protein. The monomer consists of a hydrophilic carboxy-terminal domain located inside the viral envelope and a hydrophobic membrane-spanning domain. An elongated triple-helical coiled stem region extends 135 Å from the membrane topped by a globular domain that contains the carbohydrate-recognition domain (Figure 34.1). The crystal structure also revealed a hydrophobic pocket near the CRD, which explains why sialosides containing a hydrophobic aglycone bind the hemagglutinin with greater affinity than simple glycans.

In addition to the hemagglutinin, influenza-A and -B virions express a sialidase (traditionally and incorrectly called neuraminidase) that cleaves sialic acids from glycoconjugates. Its functions may include (1) prevention of viral aggregation by removal of sialic acid residues from virion envelope glycoproteins, (2) dissociation of newly synthesized virions inside the cell or as they bud from the cell surface, and (3) desialylation of soluble mucins at sites of infection in order to improve access to membrane-bound sialic acids. In influenza-C virus, a single glycoprotein contains both the hemagglutinin activity and the receptor-destroying activity, which in this case is an esterase that cleaves the 9-O-acetyl group from acetylated sialic acid receptors. Powerful inhibitors have been designed based on the crystal structure of the sialidase from influenza-A virus. Some of these inhibit the enzyme activity at nanomolar concentrations and are in clinical use as antiviral agents (see Chapter 51).

Rotaviruses, the major killer of children worldwide, also can bind to sialic acid residues. These viruses only bind to the intestinal epithelium of newborn infants during a period that appears to correlate with the expression of specific types and arrangements of sialic acids on glycoproteins. Many other viruses (e.g., adenovirus, reovirus, Sendai virus, and polyomavirus) also appear to use sialic acids for infection, and crystal structures are now available for several of their sialic acid–binding domains (Table 34.1).

A number of viruses use heparan sulfate proteoglycans as adhesion receptors (e.g., herpes simplex virus [HSV], foot-and-mouth disease virus, and dengue flavivirus [Table 34.1]). In

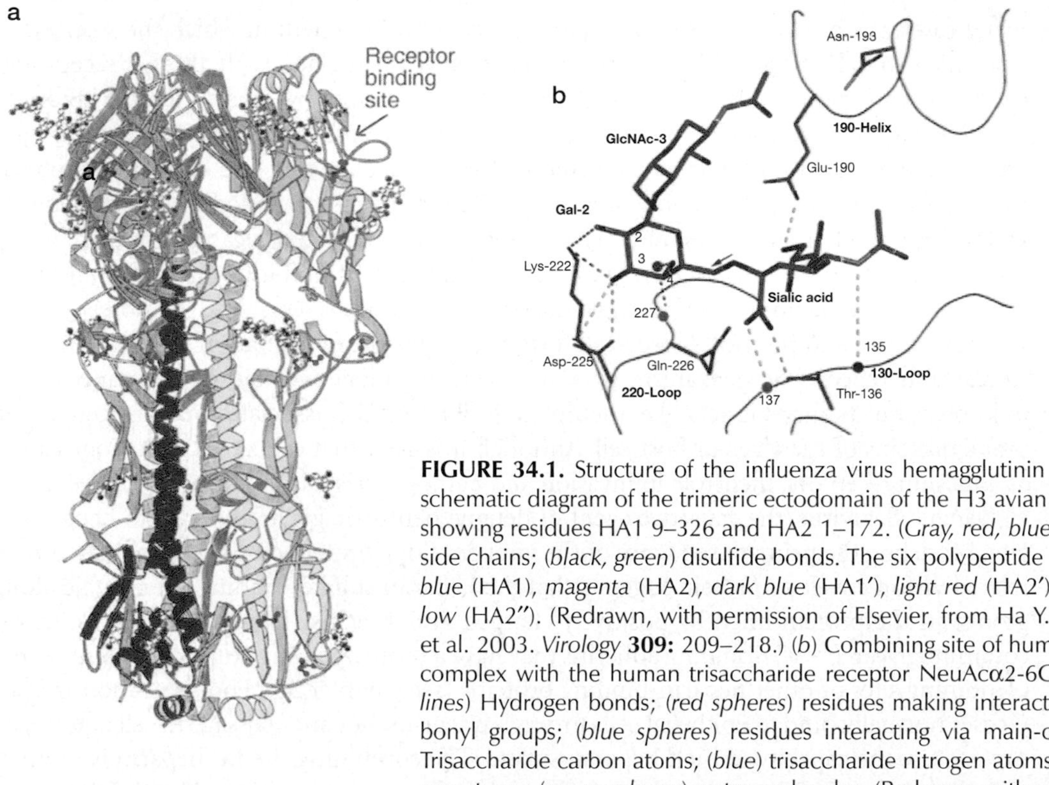

FIGURE 34.1. Structure of the influenza virus hemagglutinin (HA) ectodomain. (*a*) A schematic diagram of the trimeric ectodomain of the H3 avian HA from A/duck/Ukr/63 showing residues HA1 9–326 and HA2 1–172. (*Gray, red, blue*) Modeled carbohydrate side chains; (*black, green*) disulfide bonds. The six polypeptide chains are shown in *light blue* (HA1), *magenta* (HA2), *dark blue* (HA1′), *light red* (HA2′), *green* (HA1″), and *yellow* (HA2″). (Redrawn, with permission of Elsevier, from Ha Y., Stevens D.J., Skehel J.J, et al. 2003. *Virology* **309:** 209–218.) (*b*) Combining site of human influenza virus HA in complex with the human trisaccharide receptor NeuAcα2-6Galβ1-4GlcNAc. (*Dashed lines*) Hydrogen bonds; (*red spheres*) residues making interactions via main-chain carbonyl groups; (*blue spheres*) residues interacting via main-chain nitrogens. (*Yellow*) Trisaccharide carbon atoms; (*blue*) trisaccharide nitrogen atoms; (*red*) trisaccharide oxygen atoms; (*green spheres*) water molecules. (Redrawn, with permission of AAAS, from Gamblin S.J., Haire L.F., and Russell R.J. 2004. *Science* **303:** 1838–1842.)

TABLE 34.1. *Examples of viral lectins and hemagglutinins*

Virus	Lectin	Glycan receptor specificity	Site of infection
Myxoviruses			
Influenza A and B (human)	hemagglutinin	Neu5Acα2-6Gal	upper respiratory tract mucosa
Influenza A and B (avian and porcine)	hemagglutinin	Neu5Acα2-3Gal	intestinal mucosa
Influenza C	hemagglutinin-esterase	9-O-acetyl-Neu5Acα-	
Newcastle disease	hemagglutinin-neuraminidase	Neu5Acα2-3Gal-	
Sendai	hemagglutinin-neuraminidase	Neu5Acα2-8Neu5Ac-	
Papoviruses			
Polyoma		Neu5Acα2-3Gal-, Neu5Acα2-3Galβ1-3 (Neu5Acα2-6)GalNAc- (e.g., gangliosides GD1α, GT1aα, GQ1bα)	
Herpesviruses			
Herpes simplex	glycoproteins gB, gC, gD	3-O-sulfated heparan sulfate	mucosal surfaces of mouth, eyes, genital and respiratory tracts
Picornaviruses			
Foot-and-mouth disease	caspid proteins	heparan sulfate	
Retroviruses			
HIV	gp120 V3 loop	heparan sulfate	CD4 lymphocytes
Flaviviruses			
Dengue	envelope protein	heparan sulfate	macrophages?

many cases, the proteoglycans may be part of a coreceptor system in which the microorganisms make initial contact with a cell-surface proteoglycan and later with another receptor. For example, herpesvirus infection is thought to involve viral glycoproteins gC and/or gB binding to cell-surface heparan sulfate proteoglycans, followed by binding of viral glycoprotein gD to one of several cell-surface receptors, including heparan sulfate and one or more members of the Hve (herpesvirus entry) family of receptors. This step leads to fusion of the viral envelope with the host-cell plasma membrane. The coreceptor role of proteoglycan is reminiscent of the formation of ternary complexes required for antithrombin inhibition of thrombin and for fibroblast growth factor (FGF) cell signaling (see Chapter 35). The HSV glycoprotein gB binds heparan sulfate and promotes virus-cell fusion and syncytium formation (cell–cell fusion) as well as adherence. The mechanism by which heparan sulfate facilitates membrane fusion is unknown, but perhaps it acts like a template facilitating the association of fusogenic membrane proteins of the virus or host cell. Although it is clear that cell-surface proteoglycans act as adhesion receptors, their role in invasion and pathogenesis has not been established.

Dengue flavivirus, the causative agent of dengue hemorrhagic fever, binds to heparan sulfate. Modeling the primary sequence of the viral envelope protein on the crystal structure of a related virion envelope protein suggests that the heparan sulfate–binding site may lie along a groove in the protein lined by positively charged amino acids (Figure 34.2). Thus, the glycosaminoglycan (GAG)-binding adhesins may have a more open structure, consistent with the combining sites of other heparin-binding proteins (see Chapter 35). Foot-and-mouth disease viruses normally bind to epithelial cells through integrins, but after passage in cell culture, they can acquire the capacity to bind heparan sulfate. The combining site for heparin is located in a shallow depression formed by three of the major capsid proteins (Figure 34.3). HIV also can

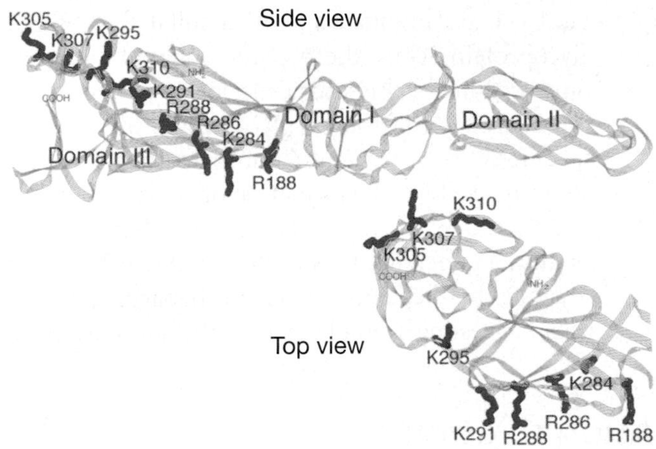

FIGURE 34.2. Two views of a putative heparin-binding site in dengue virus envelope protein. The envelope protein monomer is shown in ribbon form, displayed along its longitudinal axis and as an external side view. (*Top*) Note the alignment of positively charged amino acids along an opening on the face of the protein. (Redrawn, with permission, from Chen Y.P., Maguire T., Hileman R.E., et al. 1997. *Nature Med.* **3:** 866–871, ©Macmillan Publishers Ltd.)

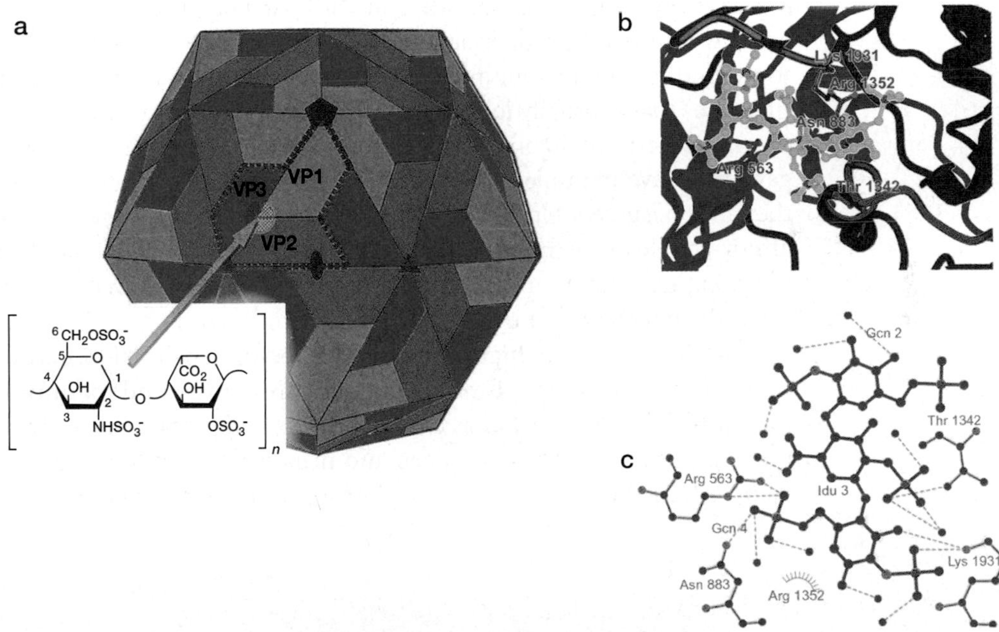

FIGURE 34.3. Model of foot-and-mouth disease virus (FMDV) in complex with heparin trisaccharides. (a) A schematic depiction of the icosahedral capsid of FMDV. VP1-3 contribute to the external features of the capsid: (*blue*) VP1; (*green*) VP2; (*red*) VP3. The heparin-binding site is located in the center of a protomer derived from the uncleaved polyprotein (*bold outline*). (Redrawn, with permission, from Fry E.E., Lea S.M., Jackson T., et al. 1999. *EMBO J.* **18:** 543–554, ©Macmillan Publishers Ltd.) (*b*) Ligand-binding site in serotype $A_{10}61$. Viral proteins are depicted as a *ribbon* diagram with the same color coding as in *a*. The sugar residues of a heparin trisaccharide ligand is drawn as *purple balls* with sulfur atoms in *yellow*. (*c*) Sugar–protein interactions. Only protein side chains that interact directly are shown. Ligand bonds are drawn in *purple* and nonligand bonds in *brown*. Hydrogen bonds are depicted by *olive-green, dashed lines*. Nonligand residues involved in hydrophobic contacts are shown as *red, fringed semicircles*. Water molecules are *red*. Idu, Iduronic acid; Gcn, glucosamine residue. (*b,c*: Redrawn, with permission, from Fry E.E., Newman J.W., Curry S., et al. 2005. *J. Gen. Virol.* **86:** 1909–1920.)

bind to heparan sulfate and other sulfated polysaccharides by way of the V3 loop of the gp120 glycoprotein. Thus, the heparan-sulfate-binding adhesins appear to pick out carbohydrate units within the polysaccharide chains as opposed to binding to terminal sugars. Interestingly, the interaction of the gD glycoprotein of Herpes simplex virus with heparan sulfate shows specificity for a particular substructure in heparan sulfate containing a 3-O-sulfated glucosamine residue, the formation of which is catalyzed by specific isozymes of the glucosaminyl 3-O-sulfotransferase gene family (see Chapter 16). However, it is unclear whether other viruses show a preference for specific heparan sulfate oligosaccharides. A growing body of literature suggests that heparan sulfate and heparin can block the interaction of the viruses with host cells, suggesting a simple therapeutic approach for treating viral infections.

BACTERIAL ADHESION TO GLYCANS

Bacterial lectins occur commonly in the form of elongated, submicroscopic, multisubunit protein appendages, known as fimbriae (hairs) or pili (threads), which interact with glycoprotein and glycolipid receptors on host cells. The best characterized of these are the mannose-specific type-1 fimbriae, the galabiose-specific P fimbriae, and the *N*-acetylglucosamine-binding F-17 fimbriae produced by different strains of *Escherichia coli.* Fimbriated bacteria express 100–400 of these appendages, which typically have a diameter of 5–7 nm and can extend hundreds of nanometers in length (Figure 34.4). Thus, pili extend well beyond the bacterial glycocalyx formed from lipopolysaccharide and capsular polysaccharides (see Chapter 20), which can actually interfere with their activity. The carbohydrate-recognition domain of the fimbriae is found in the minor subunit (FimH of type-1 fimbriae, PapG of P fimbriae, and F17G of F17 fimbriae) that is usually located at the tip of the fimbriae. Some of the other bacterial lectins are monomeric or oligomeric membrane proteins. Most bacteria (and possibly other microorganisms) have multiple adhesins with different carbohydrate specificities, which help define the range of susceptible tissues (i.e., the microbe's ecological niche). Binding is generally of low affinity, but because the adhesins and the receptors often cluster in the plane of the membrane, the resulting avidity can be great. Perhaps an appropriate analogy for adhesin-receptor binding is the interaction of the two faces of Velcro™ strips.

Examination of the high-resolution, three-dimensional structure of the glycan-binding subunit FimH of type-1 fimbriae in complex with bound mannose revealed that although mannose exists in solution as a mixture of α and β anomers, only the former was found in the complex. It is buried at a deep and negatively charged site at the edge of FimH (Figure 34.5). All of the mannose hydroxyl groups, except the anomeric one, interact extensively

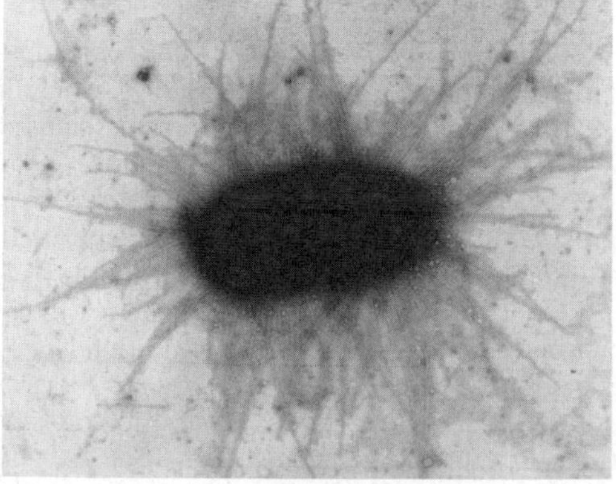

FIGURE 34.4. *E. coli* express multiple pili as indicated by the fine filaments surrounding the cells. (Reprinted, with permission of Elsevier, from Sharon N. 2006. *Biochim. Biophys. Acta* **1760:** 527–537; courtesy of David L. Hasty, University of Tennessee, Memphis, TN.)

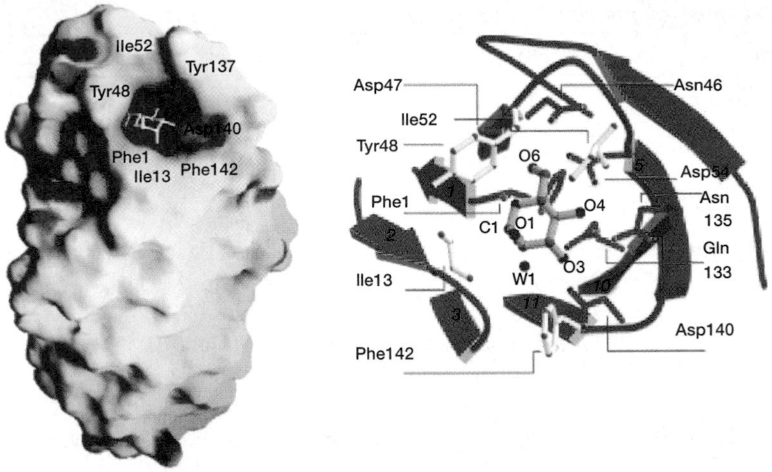

FIGURE 34.5. α-Anomer of mannose in the combining site of FimH. The mannose is buried in a unique site at the tip of the receptor-binding domain (*left*) in a deep and negatively charged pocket (*right*). FimH selects the α configuration around the free reducing anomeric oxygen of D-mannose. The hydroxyl groups of D-mannose interact with residues Phe1, Asn46, Asp47, Asp54, Gln133, Asn135, Asp140, and Phe142 by hydrogen bonding and hydrophobic interactions. Contact residues are shown as a *ball-and-stick* model. W1 stands for water. (Redrawn, with permission, from Hung C.S., Bouckaert J., Hung D., et al. 2002. *Mol. Microbiol.* **44:** 903–915, © Blackwell Publishing Ltd.)

with combining-site residues, almost all of which are situated at the ends of β-strands or in the loops extending from them. Part of the hydrogen-bonding network is identical to that found in mannose complexes of other lectins, such as those of plants.

In the urinary tract, the type-1 fimbriae of *E. coli* mediate binding of the bacteria to a protein called uroplakin Ia. This glycoprotein presents high levels of terminally exposed mannose residues that are capable of specifically interacting with FimH. Anchorage of *E. coli* to the urothelial surface via type-1 fimbriae–uroplakin Ia interactions may play a role in their colonization of the bladder and eventual ascent through the ureters against urine flow to invade the kidneys. Another receptor for these fimbriae is the urinary Tamm–Horsfall glycoprotein, which acts as a soluble inhibitor of the adhesion of the bacteria to the uroplakins and helps to clear the bacteria from the urinary tract. Indeed, mice lacking the Tamm–Horsfall gene are considerably more susceptible to bladder colonization by type-1-fimbriated *E. coli* than normal mice, whereas they are equally susceptible to P-fimbriated *E. coli* that do not bind to the same glycoprotein. Type-1 fimbriae also are instrumental in the attachment of *E. coli* to human polymorphonuclear cells and to human and mouse macrophages. This is often followed by the ingestion and killing of the bacteria, a phenomenon named lectinophagocytosis, which is an early example of innate immunity (see Chapter 39).

Other binding specificities have been described as well (Table 34.2). The specificity of binding can explain the tissue tropism of the organism. The columnar epithelium that lines the large intestine expresses Galα1-4Gal-Cer, whereas the cells lining the small intestine do not. Thus, *Bacterioides, Clostridium, E. coli,* and *Lactobacillus* only colonize the large intestine under normal conditions. P-fimbriated *E. coli* as well as different toxins bind specifically to galabiose (Galα1-4Gal) and galabiose-containing oligosaccharides, most commonly as constituents of glycolipids. Binding occurs either to terminal nonreducing galabiose units or to internal ones (i.e., when the disaccharide is capped by other sugars). P-fimbriated *E. coli* adhere mainly to the upper part of the kidney, where galabiose is abundant. *E. coli* K99 provides a striking illustration of the fine specificity of bacterial surface lectins and their relationship to the animal tropism of the bacteria. This organism binds to glycolipids

TABLE 34.2. *Examples of interactions of bacterial adhesins with glycans*

Microorganism	Adhesin	Carbohydrate specificity	Target tissue
Actinomyces naeslundii	fimbriae	Galβ1-3GalNAcβ-	oral
Bordetella pertussis	filamentous hemagglutinin (FHA)	sulfated glycolipids; heparin	ciliated epithelium in respiratory tract
Borrelia burgdorferi		heparan sulfate	endothelium, epithelium, extracellular matrix
Campylobacter jejuni		Fucα1-2Galβ1-4GlcNAcβ- (H-antigen)	intestine
Escherichia coli	P fimbriae	Galα1-4Galβ-	urinary tract
Escherichia coli	S fimbriae	gangliosides GM3, GM2	neural
Escherichia coli	Type-1 fimbriae	Manα1-3(Manα6)Man	urinary tract
Escherichia coli	K99 fimbriae	gangliosides GM3, Neu5Gcα2-3Galβ1-4Glc	intestinal cells
Haemophilus influenzae		Neu5Acα2-3Galβ1-4GlcNAcβ-; heparan sulfate	respiratory epithelium
Helicobacter pylori	BabA	Le^b	stomach
	SabA	Neu5Acα2-3Galβ1-4GlcNAcβ	
Mycobacterium tuberculosis		Neu5Acα2-3Galβ1-4GlcNAcβ-; heparan sulfate	respiratory epithelium
Neisseria gonorrhoeae	?	LacCer; Neu5Acα2-3Galβ1-4GlcNAcβ-; heparan sulfate	genital tract
Propionobacterium		LacCer	skin, intestine
Pseudomonas aeruginosa	type IV pili	Asialo GM1 and GM2	respiratory tract
Staphylococcus aureus		heparan sulfate	connective tissues, endothelial cells
Streptococcus pneumoniae		GlcNAcβ1-3Gal-	respiratory tract

that contain *N*-glycolylneuraminic acid (Neu5Gc), in the form of Neu5Gcα2-3Galβ1-4Glc, but not to those that contain *N*-acetylneuraminic acid. These two sugars differ in only a single hydroxyl group, present in the acyl substituent on the amino group at C-5 of *N*-glycolylneuraminic acid and absent in that of *N*-acetylneuraminic acid. *N*-glycolylneuraminic acid is found on intestinal cells of newborn piglets, but it disappears when the animals develop and grow, and it is not formed normally by humans. This can explain why *E. coli* K99 can cause often lethal diarrhea in piglets but not in adult pigs or humans.

The relationship between microbes and the host can be quite complex. For example, colonization of germ-free mice with *Bacteroides thetaiotaomicron,* a normal resident microbe of the small intestine, can induce an α1-2 fucosyltransferase in the mucosal epithelial cells. The bacteria bind to L-fucose residues and also use it as a carbon source.

TOXINS THAT BIND GLYCANS

A number of secreted bacterial toxins also bind to glycans (Table 34.3). The best-studied example is the toxin from *Vibrio cholera* (cholera toxin), which consists of A and B subunits in the ratio AB_5. Its crystal structure shows that the B subunits bind to the Galβ1-3GalNAc moiety of GM1 ganglioside receptors through CRDs located on the base of the subunits (Figure 34.6). The A subunit is loosely held above the plane of the B subunits, with a single α-helix penetrating through a central core created by the pentameric B subunits. Upon binding to membrane glycolipids through the B subunits, the A subunit is delivered to the interior of the cell by an unknown mechanism. In cholera toxin, the affinity of the B subunit for its glycan ligand (GM1) is unusually high compared to the bind-

TABLE 34.3. *Examples of receptors for bacterial toxins*

Microorganism	Toxin	Proposed receptor sequence	Target tissue
Bacillus thuringiensis	crystal toxins	Galβ1-3/6Galα/β1-3(±Glcβ1-6)GalNAcβ GlcNAcβ1-3Manβ1-4GlcβCer	intestinal epithelia of insects
Clostridium botulinum	botulinum toxins (A–E)	gangliosides GT1b, GQ1b	nerve membrane
Clostridium difficile	toxin A	GalNAcβ1-3Galβ1-4GlcNAcβ1-3Galβ1-4GlcβCer	large intestine
Clostridium tetani	tetanus toxin	ganglioside GT1b	nerve membrane
Escherichia coli	heat-labile toxin	GM1	intestine
Shigella dysenteriae	Shiga toxin	Galα1-4GalβCer Galα1-4Galβ1-4GlcβCer	large intestine
Vibrio cholerae	cholera toxin	GM1	small intestine

Cer indicates ceramide. Gangliosides are defined using the Svennerholm nomenclature (see Chapter 10).

ing of related AB_5 toxins from *Shigella dysenteria*, *Bordetella pertussis*, and *E. coli* to their glycan receptors (K_d in the nanomolar range for cholera toxin B subunit vs. millimolar range for Shiga toxin B subunit binding to P^k determinants). In both examples, the formation of the pentameric complex greatly increases the avidity of the interaction. This phenomenon is being exploited to make oligovalent glycan ligands as protective agents against the toxin.

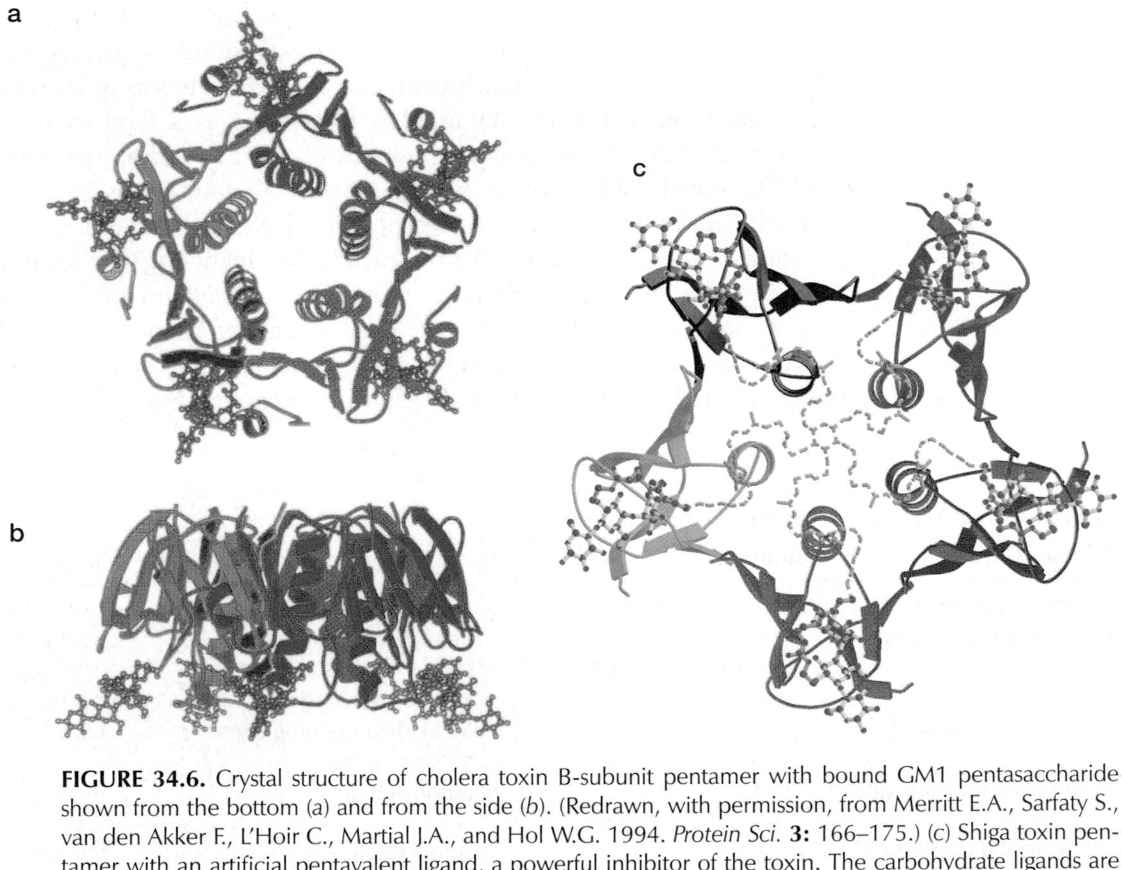

FIGURE 34.6. Crystal structure of cholera toxin B-subunit pentamer with bound GM1 pentasaccharide shown from the bottom (*a*) and from the side (*b*). (Redrawn, with permission, from Merritt E.A., Sarfaty S., van den Akker F., L'Hoir C., Martial J.A., and Hol W.G. 1994. *Protein Sci.* **3:** 166–175.) (*c*) Shiga toxin pentamer with an artificial pentavalent ligand, a powerful inhibitor of the toxin. The carbohydrate ligands are shown in a *ball-and-stick* representation. (*Dashed magenta lines*) Possible conformation for the linker. (Redrawn, with permission of Macmillan Publisher Ltd., from Kitov P.I., Sadowska J.M., Mulvey G., et al. 2000. *Nature* **403:** 669–672.)

Bacillus thuringiensis, which lives in the soil, produces crystal toxins (Bt toxins) that can kill larval stages of plant-pathogenic insects, but they are harmless to most other organisms, including humans. Bt toxins are used in crop protection by spraying plants or by genetically engineering crops to express the toxins. Recent work demonstrates that Bt toxins act by binding to glycolipids that line the gut in nematodes and presumably other invertebrates (see Chapter 23). These receptors belong to the arthroseries of glycolipids and include in their structure the characteristic ceramide-linked, mannose-containing core tetrasaccharide GalNAcβ1-4GlcNAcβ1-3Manβ1-4GlcβCer, which is found only in invertebrates and is conserved between nematodes and insects but is absent in vertebrates (see Chapter 24). Thus, the specificity of Bt toxins determines their tissue and species sensitivity.

Shiga toxin produced by *Shigella dysenteria* and the homologous Shiga-like toxins of *E. coli* (also called verotoxins) will bind to Galα1-4Gal determinants on both glycolipids and glycoproteins, but only the binding to glycolipids results in cell death. The apparent preference of most toxins for glycolipids may be related to the juxtaposition of glycolipid glycans to the membrane surface, compared to the more distal location of glycans attached to glycoproteins and proteoglycans. Binding of a toxin or bacterium to a glycolipid also might increase the likelihood of further interactions with the membrane (e.g., binding to another receptor or membrane intercalation).

PARASITE LECTINS

In addition to viruses and bacteria, a number of parasites also utilize glycans as receptors for adhesion (Table 34.4). *Entamoeba histolytica* expresses a 260-kD heterodimeric lectin that binds to terminal galactose/*N*-acetylgalactosamine residues on glycoproteins and glycolipids. Binding may have a role in attachment, invasion, and cytolysis of intestinal epithelium, and it may function in binding the amoeba to bacteria as a food source. The lectin is heterodimeric, with a transmembrane subunit of 170 kD and a glycosylphosphatidylinositol (GPI)-anchored subunit of 35 kD. The glycan-binding site is located in a cysteine-rich domain. The importance of this receptor in virulence has been established by antisense silencing of the adhesin and by expression of a dominant-negative form of the light subunit. Thus, the adhesin is a potential target to manage *E. histolytica* infection (see Chapter 40).

The interaction of *Plasmodium falciparum* (malaria) merozoites with red blood cells depends on sialic acids present on the host cell, in particular on the major erythrocyte membrane protein glycophorin. In this organism, attachment is mediated by a family of specific

TABLE 34.4. *Examples of glycan receptors for parasites*

Microorganism	Adhesin	Proposed receptor sequence	Target tissue
Entamoeba histolytica	260-kD receptor	terminal Gal/GalNAc residues	small intestinal mucosa
Plasmodium falciparum	EBA175; circumsporozoite (CS) protein	sialic acid–containing glycans (Neu5Acα2-3Galβ-); heparan sulfate	erythrocytes (infected cells bind to placental vasculature); hepatocytes
Trypanosoma cruzi		sialic acid–containing glycans; heparan sulfate	blood
Leishmania amazonensi		heparan sulfate	macrophages; fibroblasts; epithelium
Cryptosporidium parum		terminal Gal-GalNAc	intestinal epithelium
Giardia lamblia		mannose-terminated oligosaccharides	duodenum and small intestine

sialic acid–binding adhesin on merozoites, the most prominent of which is called EBA-175 (erythrocyte-binding antigen-175). The adhesin shows specificity for the type of sialic acid, with preference for Neu5Ac, rather than for 9-*O*-acetyl-Neu5Ac or Neu5Gc. Soluble Neu5Ac and Neu5Acα2-6Gal-containing oligosaccharides do not competitively inhibit the binding of EBA-175 to erythrocytes, but Neu5Acα2-3Gal-containing oligosaccharides are effective inhibitors, indicating that the adhesin is sensitive to the linkage of the sialic acid to the underlying galactose. Binding to erythrocytes leads to invasion and eventual production of additional merozoites. Other organisms expressing sialic acid adhesins can also bind to erythrocytes (e.g., influenza virus), but these interactions do not lead to productive infections in these nonnucleated cells. Thus, under these circumstances, the erythrocyte might be considered to be a clearance mechanism for these agents (see Chapter 3).

Plasmodium-infected erythrocytes also express GAG-binding proteins that are thought to facilitate adherence of infected cells to tissues. The circumsporozoite form of *P. falciparum* binds to heparan sulfate in a tissue-specific manner, with preferred binding to the basal surface of hepatocytes and the basement membrane of kidney tubules. The carboxyl terminus of the protein contains positively charged residues. Clustering of the circumsporozoite protein on the surface of the organism may generate a high concentration of positively charged residues that facilitate binding. Recent studies show that the extent of sulfation of heparan sulfate that the parasite encounters determines whether it migrates or productively invades host cells.

MICROBIAL GLYCAN LIGANDS FOR ANIMAL CELL LECTINS

Like mammalian cells, bacteria, viruses, and parasites are covered by glycans. The enveloped viruses contain virally encoded membrane glycoproteins and may pick up host glycoproteins and glycolipids during the budding process. Bacteria contain a cell wall composed of peptidoglycan, teichoic acids (Gram-positive organisms), and lipopolysaccharide (Gram-negative organisms) and also, in some species, a capsule (see Chapter 20). All of these glycans can potentially interact with host-cell lectins. Thus, it is not surprising that interactions between microbial glycans and host lectins can also aid in infection and colonization.

Trypanosoma cruzi has developed an interesting strategy of molecular camouflage in which a parasite-encoded *trans*-sialidase transfers sialic acid from serum glycoproteins in the host to membrane proteins on its own surface. The primary function of this reaction is most likely to cover surface glycans as a way of preventing host immune reactivity, although the *trans*-sialidase may also act as an adhesin. *Neisseria gonorrhoeae* uses low levels of tissue CMP-sialic acid to cover itself with sialic acid residues, making it resistant to the alternative pathway of complement. It also appears likely that these and other sialylated pathogens interact with host via the Siglec family of sialic acid–binding lectins (see Chapter 32). Schistosomes, which are parasitic filarial worms, contain the Lewis^x antigen that is also found on human leukocytes. Because Lewis^x is recognized by selectins, the presence of these carbohydrates may provide a mechanism for attachment or transcellular migration of the parasite. However, these glycans also generate a massive anti-Lewis^x antibody response in the host.

In a similar way, the capsules and lipopolysaccharide that surround bacteria, as well as yeast cell walls, contain glycans that may be recognized by mammalian cell lectins. For example, yeast mannans are recognized by both soluble and macrophage mannose-binding proteins, which have an important role in innate immune defense during the preimmune phase in infants. Highly virulent forms of pathogenic bacteria such as *Streptococcus* contain a hyaluronan capsule, which may interact with hyaluronan-binding proteins (such as CD44) present on host-cell surfaces and facilitate colonization. The structure and biology of these types of glycans and glycan-binding proteins are discussed in Chapters 20, 21, and 40.

ROLE IN INFECTIOUS DISEASE

As pointed out earlier, the major function of the microbial lectins is to mediate adhesion of the organisms to host cells or tissues, which is a prerequisite for infection to occur. This has been extensively demonstrated both in vitro (in studies with isolated cells and cell cultures) and in vivo (in experimental animals) and is supported in some cases by clinical data. For example, lectin-deficient microbial mutants often lack the ability to initiate infection. In addition, type-1 fimbrial expression is associated with the severity of urinary tract infection in children. Moreover, mono- or oligosaccharides have been shown to protect against infection by lectin-carrying bacteria in experimental models.

Glycans recognized by microbial surface lectins block the adhesion of the organisms to animal cells not only in vitro, but also in vivo, and thus protect animals against infection by such organisms. For example, coadministration of methyl α-mannoside with type-1-fimbriated *E. coli* into the urinary bladder of mice reduces significantly the rate of urinary tract infection, whereas methyl α-glucoside, which is not inhibitory to the fimbriae, has no effect. The protective effect of antiadhesive sugars has been demonstrated in a variety of studies with different pathogenic bacteria and animals, from rabbits to monkeys. Multivalent ligands should prove even more potent.

The ability of exogenous heparin and related polysaccharides to inhibit viral replication suggests that this approach might lead to polysaccharide-based antiviral pharmaceutical agents. As more crystal structures become available, the ability to custom design small-molecule inhibitors that fit into the carbohydrate-recognition domains of adhesins should improve. Already, the structures of influenza hemagglutinin and sialidase have suggested numerous ways to modify sialic acid to fit better into the active sites. Some of these compounds are presently in use to limit the spread of virus (see Chapters 50 and 51).

FURTHER READING

Rostand K.S. and Esko J.D. 1997. Microbial adherence to and invasion through proteoglycans. *Infect. Immun.* **65:** 1–8.

Kitov P.I., Sadowska J.M., Mulvey G., Armstrong G.D., Ling H., et al. 2000. Shiga-like toxins are neutralized by tailored multivalent carbohydrate ligands. *Nature* **403:** 669–672.

Williams S.J. and Davies G.J. 2001. Protein–carbohydrate interactions: Learning lessons from nature. *Trends Biotechnol.* **19:** 356–362.

Hooper L.V., Midtvedt T., and Gordon J.I. 2002. How host-microbial interactions shape the nutrient environment of the mammalian intestine. *Annu. Rev. Nutr.* **22:** 283–307.

Sharon N. and Lis H. 2004. History of lectins: From hemagglutinins to biological recognition molecules. *Glycobiology* **14:** 53R–62R.

Spear P.G. 2004. Herpes simplex virus: Receptors and ligands for cell entry. *Cell Microbiol.* **6:** 401–410.

Griffitts J.S. and Aroian R.V. 2005. Many roads to resistance: How invertebrates adapt to Bt toxins. *BioEssays* **27:** 614–624.

Newburg D.S., Ruiz-Palacios G.M., and Morrow A.L. 2005. Human milk glycans protect infants against enteric pathogens. *Annu. Rev. Nutr.* **25:** 37–58.

Olofsson S. and Bergstrom T. 2005. Glycoconjugate glycans as viral receptors. *Ann. Med.* **37:** 154–172.

Sharon N. 2006. Carbohydrates as future anti-adhesion drugs for infectious diseases. *Biochim. Biophys. Acta* **1760:** 527–537.

Sinnis P. and Coppi A. 2007. A long and winding road: The Plasmodium sporozoite's journey in the mammalian host. *Parasitol Int.* **56:** 171–178.

Mandlik A., Swierczynski A., Das A., and Ton-That H. 2008. Pili in Gram-positive bacteria: Assembly, involvement in colonization and biofilm development. *Trends Microbiol.* **16:** 33–40.

CHAPTER 35

Proteins that Bind Sulfated Glycosaminoglycans

Jeffrey D. Esko and Robert J. Linhardt

SULFATED GLYCOSAMINOGLYCANS (CHAPTER 16) INTERACT with a variety of proteins. This chapter focuses on examples of binding proteins, methods for measuring glycosaminoglycan-protein interaction, and information about three-dimensional structures of the complexes. Hyaluronan, a nonsulfated glycosaminoglycan, also engages in biologically important protein interactions, and this subject is covered in Chapter 15.

GLYCOSAMINOGLYCAN-BINDING PROTEINS ARE COMMON

More than 100 glycosaminoglycan (GAG)-binding proteins have been described in the literature, falling into the broad classes presented in Table 35.1. To a large extent, these studies have focused on protein interactions with heparin, which is a more highly sulfated, iduronic acid (IdoA)-rich form of heparan sulfate (HS; Chapter 16). This bias may reflect the commercial availability of heparin, which is frequently used for fractionation studies and heparin-Sepharose affinity chromatography. The binding of protein ligands to heparin is thought to mimic the physiological interaction of proteins with the HSs present on cell surfaces and in the extracellular matrix. In comparison, relatively few proteins are known to interact with chondroitin sulfate (CS) or keratan sulfate (KS) with comparable avidity and affinity. In some cases, CS and the related GAG dermatan sulfate (DS) may be physiologically relevant binding partners because these GAGs predominate in many tissues. Determining the physiological relevance of these interactions is a major area of research.

TABLE 35.1. *Examples of glycosaminoglycan-binding proteins and their biological activity*

Class	Examples	Physiological/pathophysiological effects of binding
Enzymes	glycosaminoglycan biosynthetic enzymes, thrombin and coagulation factors (proteases), complement proteins (esterases), extracellular superoxide dismutase, topoisomerase	multiple
Enzyme inhibitors	antithrombin III, heparin cofactor II, secretory leukocyte proteinase inhibitor, C1-esterase inhibitor	coagulation, inflammation, complement regulation
Cell adhesion proteins	P-selectin, L-selectin, some integrins	cell adhesion, inflammation, metastasis
Extracellular matrix proteins	laminin, fibronectin, collagens, thrombospondin, vitronectin, tenascin	cell adhesion, matrix organization
Chemokines	platelet factor IV, γ-interferon, interleukins	chemotaxis, signaling, inflammation
Growth factors	fibroblast growth factors, hepatocyte growth factor, vascular endothelial growth factor, insulin-like growth factor–binding proteins, TGF-β-binding proteins	mitogenesis, cell migration
Morphogens	hedgehogs, TGF-β family members	cell specification, tissue differentiation, development
Tyrosine-kinase growth factor receptors	fibroblast growth factor receptors, vascular endothelium growth factor receptor	mitogenesis
Lipid-binding proteins	apolipoproteins E and B, lipoprotein lipase, hepatic lipase, annexins	lipid metabolism, cell membrane functions
Plaque proteins	prion proteins, amyloid protein	plaque formation
Nuclear proteins	histones, transcription factors	unknown
Pathogen surface proteins	malaria cirumsporozoite protein	pathogen infections
Viral envelope proteins	herpes simplex virus, dengue virus, human immunodeficiency virus, hepatitis C virus	viral infections

The interaction between GAGs and proteins can have profound physiological effects on processes such as hemostasis, lipid transport and absorption, cell growth and migration, and development. Binding to GAGs can result in immobilization of proteins at their sites of production and in the matrix for future mobilization, regulation of enzyme activity, binding of ligands to their receptors, and protection of proteins against degradation. In some cases, the interaction may reflect complementarity of charge (e.g., histone-heparin interactions) rather than any specific biologically relevant interaction. In other cases, the interaction has been shown to depend on rare but very specific sequences of modified sugars in the GAG chain (e.g., antithrombin binding).

METHODS FOR MEASURING GLYCOSAMINOGLYCAN-PROTEIN BINDING

Numerous methods are available for analyzing GAG-protein interactions, and some provide direct measurement of K_d values. A common method involves affinity fractionation of proteins on Sepharose columns containing covalently linked GAG chains, usually heparin. The bound proteins are eluted with different concentrations of NaCl, and the concentration required for elution is generally proportional to the K_d. High-affinity interactions require at least 1 M NaCl to displace bound ligand, which translates into K_d values of 10^{-7}–10^{-9} M (determined under physiological salt concentrations by equilibrium binding). Proteins with low affinity (10^{-4}–10^{-6} M) either do not bind under normal conditions (0.15 M NaCl) or require only 0.3–0.5 M NaCl to elute. This method is based on the assumption that GAG-protein interaction is entirely ionic, which may not be entirely correct. Nevertheless, it can provide an assessment of relative affinity compared to other proteins.

A number of more sophisticated methods are now in use that provide detailed thermodynamic data (ΔH [change in enthalpy], ΔS [change in entropy], ΔCp [change in molar specific heat], etc.), kinetic data (on rate, off rate), and high-resolution data on atomic contacts in GAG-protein interactions (Table 35.2). Regardless of the technique one uses, it must be kept in mind that in vitro binding measurements are not likely to be the same as those of binding to proteoglycans on the cell surface or in the extracellular matrix, where the density of ligands, receptors, and other interacting factors varies greatly. To determine the physiological relevance of the interaction, one should consider measuring binding under conditions that can lead to a biological response. For example, one can measure binding to cells with altered GAG composition (Chapter 46) or after treatment with specific lyases to remove GAG chains from the cell surface (Chapter 16) and then determine whether the same response occurs as that in the presence of GAG chains. The interaction can then be studied more intensively using the in vitro assays described above.

TABLE 35.2. *Methods to measure glycosaminoglycan-protein interaction*

Method	Type	Principle
Affinity chromatography	M	immobilized ligand or glycosaminoglycan chains on column matrix
Affinity coelectrophoresis	M/S	gel retardation through protein-impregnated gel
Analytical ultracentrifugation	S	equilibrium sedimentation at different carbohydrate:protein ratios
Circular dichroism	S	change in rotation of plane-polarized light upon binding
Competition ELISA	M	solution- and solid-phase ligands compete for binding
Equilibrium dialysis	S	semipermeable membrane partitions protein but not carbohydrate
Fluorescence spectroscopy	S	conformational change, ligand binding induces change in fluorescence
Fourier transform infrared spectroscopy	S	measures protein and carbohydrate bond energies
Isothermal titration calorimetry	S	measures enthalpy of binding directly and K_d values
Laser light scattering	S	intrinsic scattering intensities of carbohydrate-protein complex used to calculate stoichiometry
Nuclear magnetic resonance	S	chemical shift, coupling constant, and nuclear Overhauser effect to determine contact points, distances, and conformation
Scintillation proximity	M	proximity of radiolabeled carbohydrate to immobilized protein results in emitted photon from scintillant
Surface plasmon resonance	M	mass-induced refractive index change in real time for direct measurement of association and dissociation rate constants
X ray	M	solid state cocrystal structure

(M) Mixed phase; (S) solution phase.

CONFORMATIONAL AND SEQUENCE CONSIDERATIONS

As mentioned above, most GAG-binding proteins interact with HS or heparin. The likely basis for this preference is greater sequence heterogeneity and highly sulfated domains in HS and heparin. The unusual conformational flexibility of iduronic acid (IdoA) found in heparin, HS, and DS also has a role in their ability to bind proteins. GAGs are linear helical structures, consisting of alternating residues of *N*-acetylglucosamine or *N*-acetylgalactosamine with glucuronic acid or iduronic acid (except for keratan sulfates, which consist of alternating *N*-acetylglucosamine and galactose residues; see Chapter 16). Inspection of heparin oligosaccharides containing highly modified domains ([GlcNS6S-IdoA2S]n) shows that the N-sulfo and 6-O-sulfo groups of each disaccharide repeat lie on opposite sides of the helix from the 2-O-sulfo and carboxyl groups (Figure 35.1). Because of rotational restrictions about the anomeric linkages from the N-acetyl and carboxyl groups, the chains are relatively rigid with limited end-to-end bending possible compared to polypeptides. Analysis of the conformation of individual sugars shows that *N*-acetylglucosamine and glucuronic acid residues assume a preferred conformation in solution, designated $^4C_1$ (indicating that carbon 4 is above the plane defined by carbons 2, 3, and 5 and the ring oxygen, and that carbon 1 is below the plane). In contrast, IdoA2S assumes the $^1C_4$ or the $^2S_0$ conformation (Figure 35.1), which reorients the position of the sulfo substituents and therefore creates a different orientation of charged groups. In many cases when a protein binds to an HS chain, it induces a change in conformation of the IdoA2S residue resulting in a better fit and enhanced binding. IdoA2S residues are always found in domains rich in N-sulfo and O-sulfo groups (for biosynthetic reasons; see Chapter 16), which is also where proteins usually bind. Thus, the greater degree of conformational flexibility in these modified regions may explain why so many more proteins bind with high affinity to heparin, HS, and DS than to other GAGs. The presence of an N-acetyl group in an *N*-acetylglucosamine residue changes the preferred conformation of the neighboring IdoA2S residue, showing that even minor modifications can influence conformation and chain flexibility. The

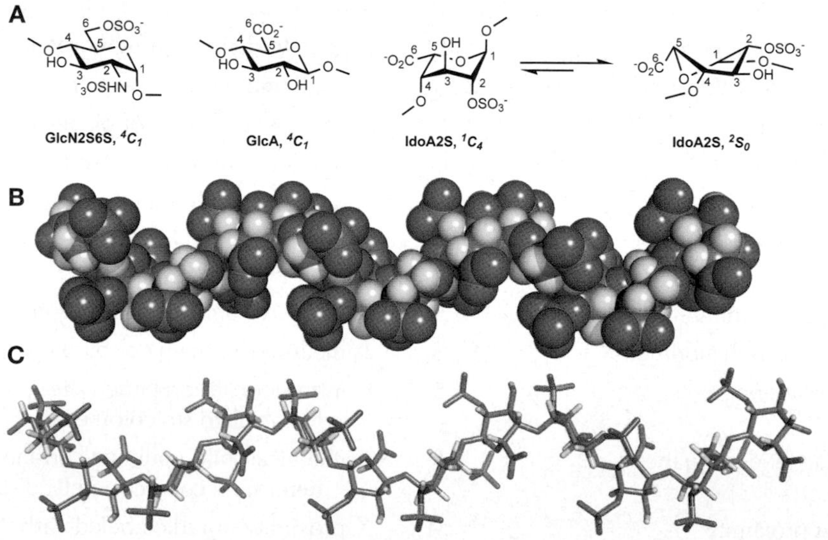

FIGURE 35.1. Conformation of heparin oligosaccharides. (*A*) Glucosamine (GlcN) and glucuronic acid (GlcA) exist in the $^4C_1$ conformation, whereas iduronic acid (IdoA) exists in equally energetic conformations designated $^1C_4$ and $^2S_0$. (*B*) Space-filling model of a heparin oligosaccharide (14 mer) deduced by nuclear magnetic resonance. (*C*) The same structure in stick representation. The renderings in *B* and *C* were made with RASMOL using data from the Molecular Modeling Database (MMDB Id: 3448) at the National Center for Biotechnology Information (NCBI).

evolution of GAG-binding proteins may be directed toward closely approximating the structures of their GAG ligands. Binding to GAGs that have a low degree of sulfation may require larger domains in the protein to interact with longer stretches of oligosaccharide. Molecular dynamic simulations on large heparin oligosaccharides have recently become possible with the availability of supercomputers (see Simulation 35.1 on the accompanying Web site). Such simulations can be used to predict the conformational flexibility of different domains within the chain, providing additional insights into GAG-protein interactions.

HOW SPECIFIC ARE GLYCOSAMINOGLYCAN-PROTEIN INTERACTIONS?

The discovery of multiple GAG-binding proteins led a number of investigators to examine whether a consensus amino acid sequence for GAG binding exists. In retrospect, this strategy was overly simplistic because it assumed that all GAG-binding proteins would recognize the same oligosaccharide sequence within heparin, or at least, sequences that would share many common features. We now know that some GAG-binding proteins interact with different oligosaccharide sequences (Table 35.3). The binding sites in the protein always contain basic amino acids (Lys and Arg) whose positive charges presumably interact with the negatively charged sulfates and carboxylates of the GAG chains. However, the arrangement of these basic amino acids can be quite variable, consistent with the variable positioning of sulfo groups in the GAG partner.

Most proteins are formed from α-helices, β-strands, and loops. Therefore, to engage a linear GAG chain, the positively charged amino acid residues would have to line up along the same side of the protein segment. α-Helices have periodicities of 3.4 residues per turn, which would require the basic residues to occur every third or fourth position along the helix in order to align with an oligosaccharide. In β-strands, the side chains alternate sides every other residue. Thus, positively charged residues should be located very differently if the peptide chain folds into a β-strand.

On the basis of the structure of several heparin-binding proteins that were available in 1991, Alan Cardin and Herschel Weintraub proposed that typical heparin-binding sites had the sequence XBBXBX or XBBBXXBX, where B is lysine or arginine and X is any other

TABLE 35.3. *Examples of oligosaccharides preferentially recognized by glycosaminoglycan-binding proteins*

Protein	Glycosaminoglycan partner	Oligosaccharide
Antithrombin	heparin/heparan sulfate	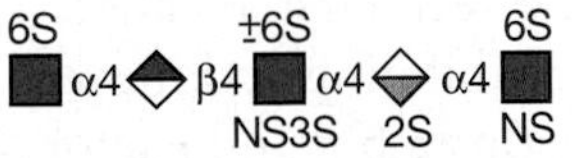
Fibroblast growth factor 2	heparin/heparan sulfate	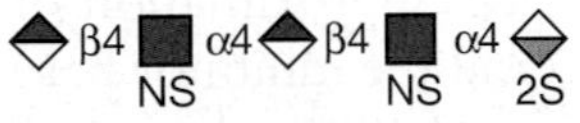
Lipoprotein lipase	heparin/heparan sulfate	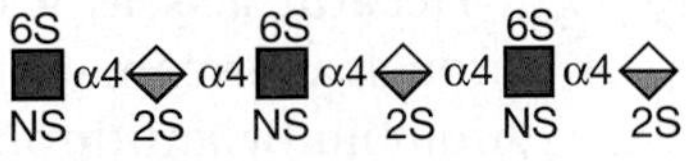
Heparin cofactor II	dermatan sulfate	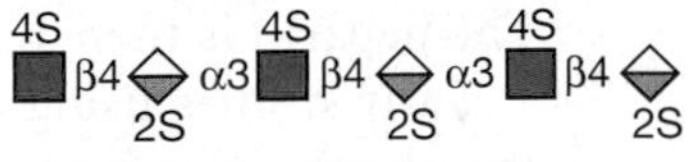
Herpes simplex virus glycoprotein gD	heparin/heparan sulfate	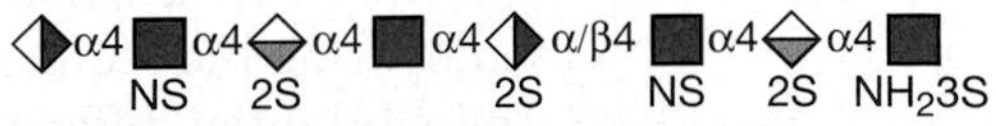

amino acid. From the structural arguments provided above, it should be obvious that only some of the basic residues in these sequences could participate in GAG binding, the actual number being determined by whether the peptide sequence exists as an α-helix or a β-sheet. We now know that the presence of these sequences in a protein merely suggests a possible interaction with heparin (or another GAG chain), but it does not prove that the interaction occurs under physiological conditions. In fact, the predicted binding sites for heparin in fibroblast growth factor 2 (FGF2) turned out to be incorrect once the crystal structure was determined. It is likely that binding involves multiple protein segments that juxtapose positively charged residues into a three-dimensional turn-rich recognition site. The specific arrangement of residues should vary according to the type and fine structure of those oligosaccharides involved in binding.

In plant and animal lectins and antibodies that recognize glycans, the glycan recognition domains are typically shallow pockets that engage the terminal sugars of the oligosaccharide chain (Chapters 27 and 34). In GAG-binding proteins, the protein usually binds to sugar residues that lie within the chain or near the terminus. Therefore, the binding sites in GAG-binding proteins consist of clefts or sets of juxtaposed surface residues rather than pockets. Given that GAG chains generally exist in a helical conformation, only those residues on the face toward the protein interact with amino acid residues; the ones on the other side of the helix might be free to interact with a second ligand. Alternatively, residues in a binding cleft could interact with both sides of the helix. Finally, one should keep in mind that the oligosaccharides that bind represent only a small segment of the GAG chain. Thus, a single glycan chain can bind multiple protein ligands.

Using phage display technology, a library of antiheparin and antiheparan sulfate antibodies was prepared. These antibodies are unreactive with other GAGs, showing specificity toward heparin and HS. Furthermore, they recognize specific structural features, such as heparin's antithrombin-binding site, and show tissue and organ specificity. These antibodies will undoubtedly be useful in future studies aimed at understanding the specificity of GAG-protein interactions.

ANTITHROMBIN-HEPARIN: A PARADIGM FOR STUDYING GLYCOSAMINOGLYCAN-BINDING PROTEINS

The best-studied example of protein-GAG interaction is the binding of antithrombin to heparin and HS. This interaction is of great pharmacological importance in hemostasis because heparin is used clinically as an anticoagulant. Antithrombin is a member of the serpin family of protease inhibitors, many of which bind to heparin. Binding has a twofold effect: First, it causes a conformational change in the protein and activation of the protease inhibiting action, resulting in a 1000-fold enhancement in the rate at which it inactivates thrombin and Factor Xa. Second, the heparin chain acts as a template, enhancing the physical approximation of thrombin and antithrombin. Thus, both the protease (thrombin) and the inhibitor have GAG-binding sites.

Heparin acts as a catalyst in these reactions by enhancing the rate of the reaction through approximation of substrates and conformational change. After the inactivation of thrombin by antithrombin occurs, the complex loses affinity for heparin and dissociates. The heparin is then available to participate in another activation/inactivation cycle.

Early studies using affinity fractionation schemes showed that only about one third of the chains in a heparin preparation actually bind with high affinity to antithrombin. Comparing the sequence of the bound chains with those that did not bind failed to reveal any substantial differences in structure, consistent with the later discovery that the binding site consists of only five sugar residues (Figure 35.2) (the average heparin chain is about 50

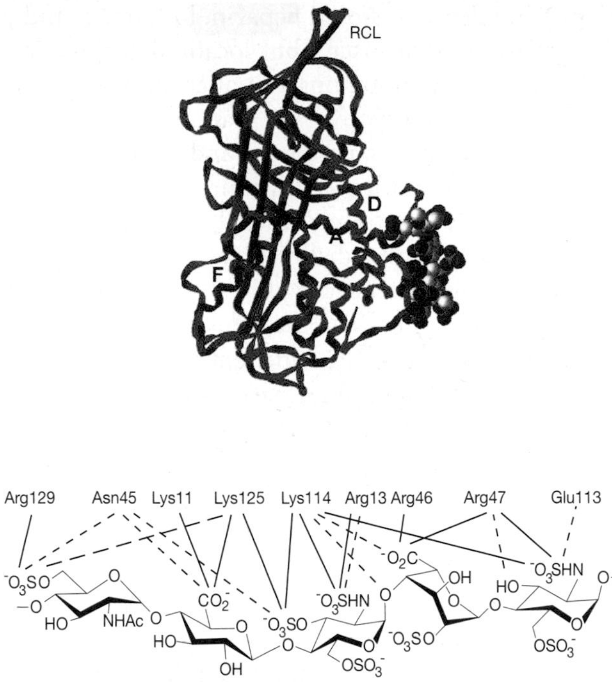

FIGURE 35.2. Crystal structure of the antithrombin-pentasaccharide complex (from Protein Data Bank). (A,D) α-Helices that make contact with heparin; (RCL) the reactive center loop that inactivates thrombin and Factor X; (F) another α-helix in the protein. (*Lower panel*) Interactions between key amino acid residues and individual elements in the pentasaccharide. (*Solid lines*) Electrostatic interactions between positively charged residues and sulfate groups; (*broken lines*) hydrogen bonds; (*alternately broken and solid line*) bridging water molecule.

sugar residues). This observation can be extended to virtually all GAG-binding proteins, inferring that the binding sites represent a very small segment of the chains (Table 35.3).

Crystals of antithrombin were prepared and analyzed by X-ray diffraction to 2.6-Å resolution. The docking site for the heparin pentasaccharide is formed by the apposition of helices A and D, which both contain critical arginine and lysine residues at the interface. The sequence in the D helix ($_{124}$AKLNCRLYRKANKSSKLVSANR$_{145}$) places many of the positively charged residues on one face of the helix, in proximity to the arginine residues in the A helix ($_{41}$PEATNRRVW$_{49}$) (Figure 35.2). The pentasaccharide is sufficient to activate antithrombin binding toward Factor Xa, but it will not facilitate the inactivation of thrombin. For this to occur, a larger oligosaccharide of at least 18 residues is needed. As mentioned above, thrombin also contains a heparin-binding site, and the larger heparin oligosaccharide is thought to act as a template for the formation of a ternary complex with thrombin and antithrombin. In contrast to antithrombin, thrombin exhibits little oligosaccharide specificity. As might be expected, adding high concentrations of heparin actually inhibits the reaction, because the formation of binary complexes of heparin and thrombin or heparin and antithrombin predominate. This important principle of "activation at low concentrations and inhibition at high concentrations" also occurs in other systems where ternary complexes form (Chapter 27).

Heparin is a pharmaceutical formulation produced by partial fractionation of natural GAGs derived primarily from porcine intestines (Chapter 16). Mast cells are known to produce a highly sulfated version of HS resembling heparin, although highly sulfated, iduronic acid–rich heparin oligosaccharides are present in HS isolated from other tissues, especially the liver. Although heparin has proven to be of great therapeutic use, its role in vivo remains unclear. Mast cells degranulate in response to specific antigen stimulation, resulting in release of stored heparin, histamine, and proteases. When this occurs, local anticoagulation might occur, but localized coagulation defects have not been described in animals bearing mutations that alter mast cells or heparin. Antithrombin-binding sequences are also found in ovarian granulosa cell HS, where they may have a role in regulating extravascular coagulation around ovulatory follicles. A small percentage of endothelial cell HS contains antithrombin-binding sequences as well. However, these binding sites appear to be located on the ablumenal side of blood vessels, and mice lacking the central 3-O-sulfated GlcNS unit, a hallmark of the antithrombin-binding sequence (Figure 35.2), do not exhibit any systemic coagulopathy after birth. Nevertheless, antithrombin deficiency causes massive disseminated coagulopathy. Perhaps these findings indicate that lower-affinity binding sequences are sufficient to activate antithrombin. This system illustrates an important caveat: One cannot necessarily ascribe functions to endogenous proteoglycans based on the effects of GAGs added to experimental systems.

Heparin cofactor II (HCII), another thrombin inhibitor, will bind to DS as well as heparin, albeit to different sites on the protein. The structure of HCII and its thrombin complex is very similar to antithrombin and its kinetically competent complex. However, unlike antithrombin, HCII is the only serpin known to associate with DS. As shown in Table 35.3, the DS that binds HCII consists of a repeating structure rich in 2-*O*-sulfoiduronic acid. Furthermore, the kinetic mechanism of inactivating thrombin differs in HCII and antithrombin. HCII function as an anticoagulant is believed to be restricted to damaged tissue in which DS proteoglycans in the matrix become exposed.

FGF-HEPARIN INTERACTIONS ENHANCE STIMULATION OF FGF RECEPTOR SIGNAL TRANSDUCTION

A large number of growth factors can be purified based on their affinity for heparin. The heparin-binding family of fibroblast growth factors has grown to more than 22 members and includes the prototype FGF2, otherwise known as basic fibroblast growth factor. FGF2 has a very high affinity for heparin ($K_d \sim 10^{-9}$ M) and requires 1.5–2 M NaCl to elute from heparin-Sepharose. FGF2 has potent mitogenic activity in cells that express one of the FGF signaling receptors (four FGFR genes are known and multiple splice variants exist). Cell-surface HS binds to both FGF2 and FGFR, facilitating the formation of a ternary complex. Both binding and the mitogenic response are greatly stimulated by heparin or HS, which promote dimerization of the ligand-receptor complex.

The costimulatory role of HS (and heparin) in this system is reminiscent of the heparin/antithrombin/thrombin story. Indeed, the minimal binding sequence for FGF2 also consists of a pentasaccharide (Table 35.3). However, this pentasaccharide is not sufficient to trigger a biological response (mitogenesis). For this to occur, a longer oligosaccharide (10 mer) containing the minimal sequence and additional 6-O-sulfo groups are needed to bind FGFR. The sequence that binds to both FGF2 and FGFR is prevalent in heparin but rare in HS. The requirement for this rare binding sequence reduces the probability of finding this particular arrangement in naturally occurring HSs. Thus, some preparations of HS are inactive in mitogenesis, and those containing only one half of the bipartite binding sequence are actually inhibitory.

The structure of FGF2 cocrystallized with a heparin hexasaccharide has been obtained (Figure 35.3). The heparin fragment ($[GlcNS6S\alpha1\text{-}4IdoA2S\alpha1\text{-}4]_3$) was helical and bound to a turn-rich heparin-binding site on the surface of FGF2. Only one N-sulfo group and the 2-O-sulfo group from the adjacent iduronic acid are bound to the growth factor in the turn-rich binding domain, and the next GlcNS residue is bound to a second site, consistent with the minimal binding sequence determined with oligosaccharide fragments (Table

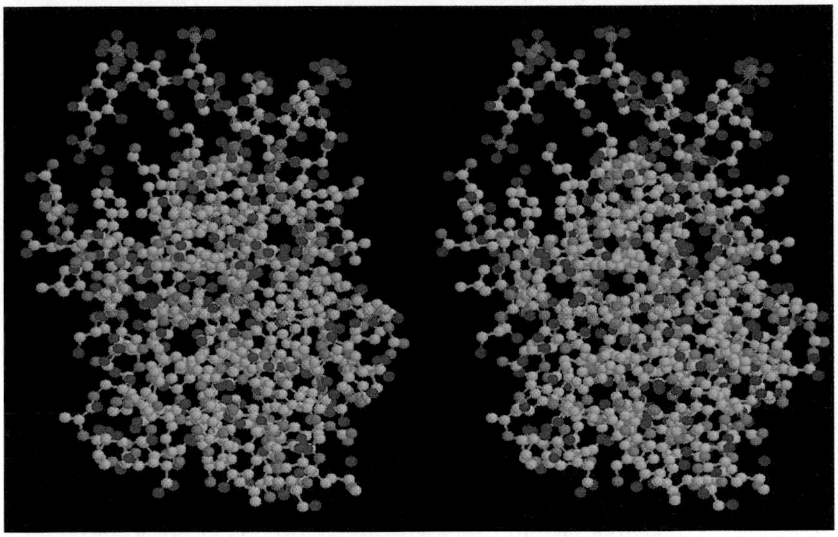

FIGURE 35.3. Stereo view of the crystal structure of FGF2 with a heparin hexasaccharide (shown at the top of the figure; *yellow balls* indicate sulfur atoms). The stereo rendering was made with RASMOL using data from the Molecular Modeling Database (MMDB Id: 4322, http://www.ncbi.nlm.nih.gov/cgi-bin/Structure/mmdbsrv) at the NCBI.

35.3). No significant conformational change in FGF2 occurs upon heparin binding, consistent with the idea that heparin primarily serves to dimerize FGF2 and juxtapose components of the FGF signal-transduction pathway. The crystal structure of acidic FGF (FGF1) has also been solved and shows similar sequences on its surface. However, the oligosaccharide sequence that binds with high affinity to FGF1 contains 6-O-sulfo groups.

The cocrystal structure of the complex of $(FGF2\text{-}FGFR)_2$, first solved in the absence of heparin/HS ligand, showed a canyon of positively charged amino acid residues, suggestive of an unoccupied heparin-binding site. Subsequently, the heparin-oligosaccharide-containing complex was solved. The stoichiometry is still controversial, but recent biochemical evidence supports a 2:2:2 complex of HS: FGF2:FGFR for signaling (Figure 35.4).

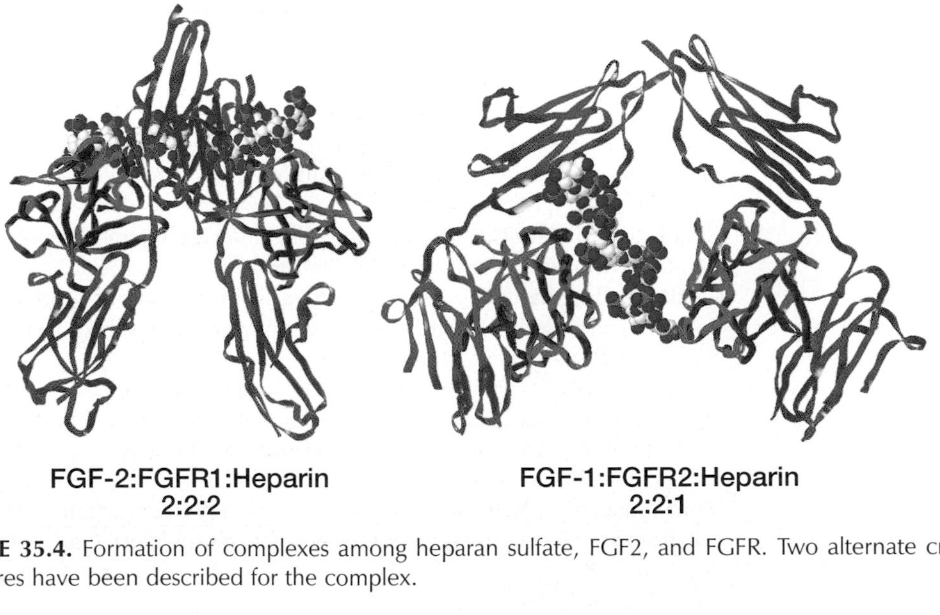

FIGURE 35.4. Formation of complexes among heparan sulfate, FGF2, and FGFR. Two alternate crystal structures have been described for the complex.

OTHER ATTRIBUTES OF GLYCOSAMINOGLYCAN-PROTEIN INTERACTIONS

Chapters 34 and 39 discuss how sulfated GAGs can serve as receptors for microbes, viruses, and parasites. Chapters 38–44 discuss how GAG deficiencies can affect physiology and how aberrations in GAG composition can cause disease.

In some cases, the interaction of GAG chains with proteins may depend on metal cofactors. For example, L- and P-selectins have been shown to bind to a subfraction of HS chains and heparin in a divalent-cation-dependent manner. This observation raises the possibility that other types of cation-dependent receptors for GAG chains may exist. Recent studies show that GAG binding to L-selectin helps in leukocyte rolling. Furthermore, the interaction can be pharmacologically manipulated by exogenous heparin, including chemically modified derivatives that lack anticoagulant activity.

HS proteoglycans are often expressed in a spatially and temporally limited fashion. The temporary placement of an HS proteoglycan at a specific tissue site might or might not coincide with the presence of its appropriate protein ligand. Furthermore, if the binding partner has no access to the HS proteoglycan, it cannot interact—adding an additional level of specificity. Recent studies demonstrate that the fine structure of HS chains also changes during development, thus enabling or disabling specific associations between ligands and receptors.

Gradients of morphogens, factors that determine cell fates based on concentration, also determine the patterns of cell and tissue organization during development (see Chapter 24). The mechanism of gradient formation is controversial, but interestingly, virtually all morphogens can interact with heparin and HS. These interactions can affect transport of ligands, receptor interactions, endocytosis, and degradation, which together may have a role in determining the robustness of the gradient. The GAG chains of proteoglycans also offer a linear domain over which ligand proteins can diffuse. By limiting the space available to these proteins from the three-dimensional space of extracellular fluids and the extracellular matrix to one-dimensional space along the chains, the chance of encounters among heparin-binding proteins, such as FGF and its receptor (FGFR), may be enhanced. Thus, HS proteoglycans may have their most important role in controlling the kinetics of protein–protein interactions rather than the thermodynamics of such encounters.

FURTHER READING

Cardin A.D. and Weintraub H.J. 1989. Molecular modeling of protein-glycosaminoglycan interactions. *Arteriosclerosis* **9:** 21–32.

Maimone M.M. and Tollefsen D.M. 1990. Structure of a dermatan sulfate hexasaccharide that binds to heparin cofactor II with high affinity. *J. Biol. Chem.* **265:** 18263–18271.

Conrad H.E. 1998. *Heparin-binding proteins.* Academic Press, San Diego.

Mulloy B. and Forster M.J. 2000. Conformation and dynamics of heparin and heparan sulfate. *Glycobiology* **10:** 1147–1156.

Casu B. and Lindahl U. 2001. Structure and biological interactions of heparin and heparan sulfate. *Adv. Carbohydr. Chem. Biochem.* **57:** 159–206.

Capila I. and Linhardt R.J. 2002. Heparin–protein interactions. *Angew Chem. Int. Ed. Engl.* **41:** 390–412.

Esko J.D. and Selleck S.B. 2002. Order out of chaos: Assembly of ligand binding sites in heparan sulfate. *Annu. Rev. Biochem.* **71:** 435–471.

Lander A.D., Nie Q., and Wan F.Y. 2002. Do morphogen gradients arise by diffusion? *Dev. Cell* **2:** 785–796.

Li W., Johnson D.J., Esmon C.T., and Huntington J.A. 2004. Structure of the antithrombin-thrombin-heparin ternary complex reveals the antithrombotic mechanism of heparin. *Nat. Struct. Mol. Biol.* **11:** 857–862.

Mohammadi M., Olsen S.K., and Goetz R. 2005. A protein canyon in the FGF–FGF receptor dimer selects from an à la carte menu of heparan sulfate motifs. *Curr. Opin. Struct. Biol.* **15:** 506–516.

Fernandez C., Hattan C.M., and Kerns R.J. 2006. Semi-synthetic heparin derivatives: Chemical modifications of heparin beyond chain length, sulfate substitution pattern and *N*-sulfo/*N*-acetyl groups. *Carbohydr. Res.* **341:** 1253–1265.

Kreuger J., Spillmann D., Li J.P., and Lindahl U. 2006. Interactions between heparan sulfate and proteins: The concept of specificity. *J. Cell Biol.* **174:** 323–327.

Nandini C.D. and Sugahara K. 2006. Role of the sulfation pattern of chondroitin sulfate in its biological activities and in the binding of growth factors. *Adv. Pharmocol.* **53:** 253–279.

Kurup S., Wijnhoven T.J., Jenniskens G.J., Kimata K., Habuchi H., Li J.P., Lindahl U., van Kuppevelt T.H., and Spillmann D. 2007. Characterization of anti-heparan sulfate phage display antibodies AO4B08 and HS4E4. *J. Biol. Chem.* **282:** 21032–21042.

CHAPTER 36

Glycans in Glycoprotein Quality Control

Hudson H. Freeze, Jeffrey D. Esko, and Armando J. Parodi

N-GLYCANS HELP GLYCOPROTEINS FOLD BY CREATING a series of checkpoints that dictate the life or the death of newly made membrane and secreted proteins. This chapter describes the quality control process in the endoplasmic reticulum (ER) and Golgi and what happens to glycoproteins that fail their "final folding exam."

CHAPERONES FACILITATE PROTEIN FOLDING

During protein synthesis, nascent polypeptides begin to assume their final three-dimensional conformation by passing through folding intermediates that are dictated to a large extent by their primary sequence. Proper folding involves formation of secondary structures (α-helices and β-strands), burying of hydrophobic residues in the interior of the protein, formation of disulfide bonds, and quaternary associations via oligomerization or multimerization. Together, these processes prevent unwanted protein aggregation that would interfere with normal protein function and integrated metabolic processes. To make folding more efficient, cells utilize various chaperones. Two major chaperone families present in the cytoplasm consist of heat-shock proteins that are encoded by members of the *hsp60* and *hsp70* gene families. Expression of these chaperones is increased when cells undergo a rapid shift to higher temperature and in other stress conditions that increase the likelihood of protein misfolding. Chaperones bind to hydrophobic patches exposed on misfolded proteins and maintain their solubility as they acquire their final conformation. They also participate in the repair of damaged proteins and proteins that have not properly multimerized. Improperly folded proteins that fail to mature are eventually tagged by ubiquitination, which then leads to degradation by the proteasome.

Membrane proteins and proteins destined for secretion also undergo a quality control process similar to that described above for soluble proteins. Membrane and secretory proteins originate on membrane-bound ribosomes, and translation and translocation into the lumen of the ER occur simultaneously. The lumen is highly specialized for protein folding; its oxidizing environment promotes disulfide bond formation and it is the main cellular reservoir of Ca^{++}. To aid proper protein folding, a battery of molecular chaperones is present, including BiP/Grp78 (a glucose-regulated protein and member of the hsp70 family of chaperones), Grp94, and Grp170. In addition, the ER contains enzymes that promote proline *cis–trans* isomerization and protein disulfide bond formation, such as protein disulfide isomerases (PDI and endoplasmic reticulum proteins ERp59, ERp72, and ERp57). The expression of many of these proteins increases during stress responses.

Unlike cytoplasmic proteins, most membrane and secreted proteins undergo cotranslational modification with N-glycans (see Chapter 8). Although these N-glycans have independent functions after the protein appears on the cell surface or in the extracellular environment, one of the primary functions of the N-glycosylation is to ensure proper protein folding has occurred prior to exit from the ER. To achieve this goal, the N-glycans are recognized by two lectin-like chaperones in the ER, calnexin (CNX) and calreticulin (CRT). These specialized chaperones require Ca^{++} for activity and bind to monoglucosylated forms of the N-glycan, thus retaining the protein in the ER until proper folding occurs. The ER also contains three enzymes central to the process: α-glucosidases I and II (GI and GII), which trim glucose residues from the $Glc_3Man_9GlcNAc_2$ glycan initially transferred to the protein and release the glycoprotein from CNX/CRT; and an enzyme called UDP-Glc:glycoprotein glucosyltransferase (UGGT), which reglucosylates the glycan if proper folding has not occurred. Misfolded proteins that fail to fold or oligomerize properly are eventually retrotranslocated into the cytoplasm and destroyed by N-deglycosylation and proteasomal degradation, a process called ER-associated degradation (ERAD). These processes are described in greater detail in the following sections.

CNX/CRT AND UGGT DETERMINE WHEN GLYCOPROTEINS ARE PROPERLY FOLDED

N-Glycosylation in the ER is considerably different from other types of glycosylation (see Chapter 8). It begins with the "en bloc" transfer of a 14-monosaccharide glycan to the nascent polypeptide as it exits the translocon complex. In vertebrates, addition of the N-glycan occurs on incompletely folded polypeptides. In bacteria, N-glycosylation can occur both on nascent polypeptides and on fully mature proteins (see Chapter 20). N-Glycans modify the physical properties of the glycoproteins by providing noncharged, bulky, hydrophilic groups that keep the protein in solution during folding. They also modulate protein conformation by forcing amino acids near the glycan into a hydrophilic environment.

N-Glycan processing starts immediately after transfer of the precursor sugar chain to the protein in the lumen of the ER. α-Glucosidase I removes the terminal α1-2-linked glucose and glucosidase II then removes both α1-3-linked glucose units (Figure 36.1). Although glucosidase II may remove the second glucose residue from glycoproteins with a single glycan chain (designated m in Figure 36.1), the trimming of the second glucose occurs more efficiently when there is a second N-glycan chain on the same protein. Depending on the cell type, a variable number of mannose residues also may be removed while glycoproteins remain in the ER.

CNX and CRT will bind to glycoproteins containing a single α-linked glucose residue (Figure 36.2). CNX is a type I transmembrane protein and CRT is soluble, but both have

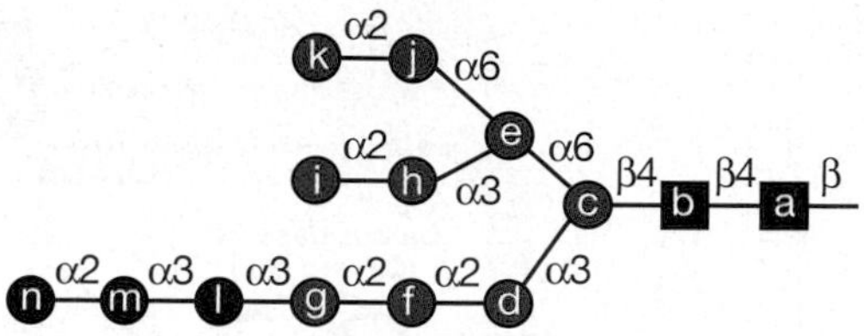

FIGURE 36.1. N-glycan structures. Lettering (a–n) follows the order of addition of monosaccharides in the synthesis of $Glc_3Man_9GlcNAc_2$-P-P-dolichol. α-Glucosidase I removes residue n, and α-glucosidase II removes residues l and m. The glucosyltransferase UGGT adds residue l to residue g. $Man_8GlcNAc_2$ isomer A (M8A) lacks residues l–n and g; M8B, formed by mammalian cell or *Saccharomyces cerevisiae* ER α-mannosidase I, lacks residues i and l–n, and isomer M8C is devoid of residues k and l–n.

ER retrieval signals at their carboxyl termini, which retain them in the ER. They have structurally similar lectin domains and comprise part of a large, weakly associated, heterogeneous protein network that includes BiP/Grp78, ERp57, Grp94, and other ER-resident proteins. Both CNX and CRT are monovalent, low-affinity lectins for monoglucosylated oligomannosyl glycans, but their in vivo specificities are not identical. CNX mainly interacts with N-glycans close to the ER membrane, whereas CRT binds to proteins present in the lumen or to those that have large lumenally oriented domains. Binding of incompletely folded glycoproteins to the CNX/CRT complex prevents their exit from the ER and enhances folding efficiency by preventing aggregation and premature oligomerization/degradation and by facilitating formation of native disulfide bonds.

Removal of the terminal glucose residue from $Glc_1Man_9GlcNAc_2$ glycans by α-glucosidase II prevents binding of the protein to the CNX/CRT complex. If properly folded, these proteins are packaged in COPII-coated vesicles and transferred to the Golgi. However, if they remain unfolded, UGGT reglucosylates the glycan, which then can rebind to CNX/CRT (Figure 36.2). Thus, UGGT has the remarkable property of exclusively glucosylating glycoproteins displaying nonnative conformations. It discriminates between different nonnative conformers and preferentially glucosylates molten globule-like structures, a term given to the partially folded state. UGGT achieves this specificity by recognizing exposed hydrophobic amino acid patches in acceptor conformers and modifies N-glycans in the vicinity of the abnormal structures. UGGT also glucosylates incompletely assembled multimeric complexes because it recognizes hydrophobic surfaces exposed in the absence of the appropriate subunits. UGGT is one of the few soluble glycosyltransferases within the ER. It uses UDP-glucose as a substrate donor and requires millimolar Ca^{++} concentrations typically found within the ER lumen. The amino-terminal domain of UGGT, representing about 80% of the molecule, recognizes the abnormal protein conformer. The carboxy-terminal domain has homology to other glycosyltransferases and is probably responsible for the glucosylation reaction. Together, UGGT and the CNX/CRT complex ensure that only properly folded and multimerized proteins move from the ER to the Golgi.

EXIT FROM THE CNX/CRT/UGGT CYCLE

In spite of the CNX/CRT/UGGT cycle, the process of protein folding is relatively inefficient, with as much as 80% of some newly made proteins never maturing properly. How do cells recognize glycoproteins that are irreparably misfolded, pull them from futile reglucosylation–deglucosylation cycles, and drive them to proteasomal degradation? Despite intensive work in this area, a clear answer is not yet available.

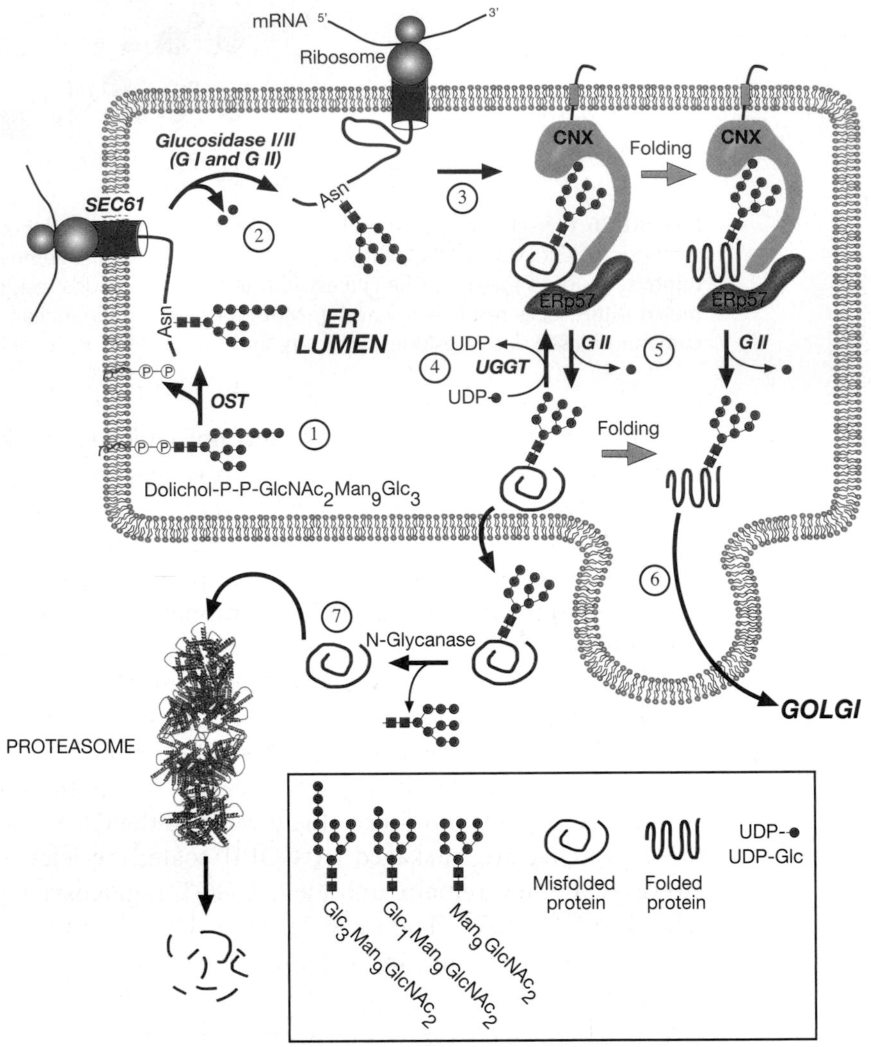

FIGURE 36.2. Model proposed for the quality control of glycoprotein folding. Proteins entering the ER are N-glycosylated by the oligosaccharyltransferase (OST) as they emerge from the translocon (SEC61) (*1*). Two glucose residues are removed by the sequential action of α-glucosidases GI and GII to generate monoglucosylated species (*2*) that are recognized by CNX and/or CRT (only CNX is shown) and are associated with ERp57 (*3*). The complex between the lectins and folding intermediates/misfolded glycoproteins dissociates upon removal of the last glucose by GII, and is reformed by α-glucosyltransferase (UGGT) (*4*). Once glycoproteins have acquired their native conformations, either free or bound to lectins, GII hydrolyzes the remaining glucose residue and releases the glycoproteins from the lectin anchors (*5*). These species are not recognized by UGGT and are transported to the Golgi (*6*). Glycoproteins remaining in misfolded conformations are retrotranslocated to the cytoplasm, where they are deglycosylated and degraded by the proteasome (*7*). One or more mannose residues may be removed during the whole folding process. (Modified, by permission of Oxford University Press, from Winchester B. 2005. *Glycobiology* **15:** R1–R15.)

One idea is that ER α-mannosidase I, a membrane-bound enzyme, converts $Man_9GlcNAc_2$ to $Man_8GlcNAc_2$ isomer B (M8B described in the legend for Figure 36.1). Smaller glycans, such as $Man_5GlcNAc_2$ and $Man_6GlcNAc_2$ also arise, but the enzymatic activities in the ER responsible for their formation are unproven and in fact their appearance might depend on transfer to the Golgi, which contains several other mannosidases. The other group of proteins that can interact with the deglucosylated chain are called endoplasmic reticulum degradation enhancing α-mannosidase-like proteins (EDEMs).

Early work suggested that they were membrane-bound because they retained uncleaved signal sequences, but it is now clear that most mammalian cells convert EDEMs into soluble ER proteins. A major unresolved question is whether EDEMs are α-mannosidases or lectins. Because EDEMs bind to totally deglucosylated misfolded glycoproteins, they were thought to be lectins that physically interacted with the glycans, blocking UGGT-mediated reglucosylation. EDEMs were thought to have broad specificity, because their overexpression enhances misfolded glycoprotein degradation in cells making only $Glc_3Man_5GlcNAc_2$ or $Glc_3Man_9GlcNAc_2$. They also lack a particular disulfide bond conserved in most α-mannosidases.

Other evidence suggests that EDEMs have catalytic activity. Yeast have only one EDEM homolog, but genomic analysis shows that mice and humans have three. They have a 450-residue domain that shares 35% sequence identity with the catalytic domain of ER α-mannosidase I. Overexpression of EDEMs results in more extensive demannosylation of misfolded glycoproteins and enhanced degradation of misfolded proteins, whereas small interfering RNA (siRNA)-mediated reduction of EDEM expression delays degradation. When mutations are inserted to remove amino acids predicted to play a catalytic role, demannosylation of misfolded proteins does not occur. Although these findings suggest that EDEMs have enzymatic activity, no EDEM has been purified to homogeneity and α-mannosidase activity has not yet been detected in recombinant EDEMs. Thus, the issue of whether EDEMs are enzymes or lectins (or both) remains unsettled. In any event, the trimming of mannose residues from N-glycans on misfolded proteins allows their recognition and export from the ER to the cytoplasm.

RETROTRANSLOCATION AND DEGRADATION OF MISFOLDED GLYCOPROTEINS

When permanently misfolded glycoproteins have exhausted their time in the ER, they are transported into the cytoplasm, ubiquitinated, and degraded by the 26S proteasome (Figure 36.2). For this to occur, glycoproteins are disentangled from aggregates, reduced, and unfolded using BiP/Grp78, PDI, and ERp57. The precise mechanism of transport of the misfolded glycoprotein to the cytoplasm as well as the identification of the pore by which misfolded species are retrotranslocated (often referred to as dislocation) have not been defined. Ubiquitination may occur as proteins exit the ER via the membrane E3 ligase, AMFR (autocrine motility factor receptor).

Prior to proteasomal degradation, the glycans are removed. It appears that a cytoplasmic peptide:N-glycanase (PNGase) plays an important role in both removing the glycan and constructing an efficient predegradation complex. PNGase recognizes only misfolded or denatured glycoproteins and the enzyme is bound to the proteasome by subunit S4 and HR23B as a complex with cytoplasmic protein, p97. Degradation of the released glycan occurs in two stages. First, partial cleavage occurs between the chitobiose core via a cytoplasmic endo-β-N-acetylglucosaminidase or possibly a neutral-pH cytoplasmic chitobiase. A cytoplasmic α-mannosidase cleaves four mannose residues to generate Manα1-2Manα1-2Manα1-3[Manα1-6]Manβ1-4GlcNAc. The $Man_5GlcNAc$ is then taken into the lysosome for final degradation to monosaccharides (see Chapter 41) via an ATP-dependent, but as yet unidentified, lysosomal membrane transporter that recognizes the reducing terminal *N*-acetylglucosamine. The reason for this elaborate degradation scheme is not known. Interestingly, more complex free N-glycans have been found in the cytoplasm of plant and fish cells. Having selective degradation of $Man_5GlcNAc$ glycans might prevent clearance of these other more complex chains, which could have other biological activities.

THE UNFOLDED PROTEIN RESPONSE

ER homeostasis is constantly monitored and maintained in the face of stresses including nutrient deprivation, altered redox balance, impaired glycosylation, or simply increased secretory protein synthesis. These demands or insults to the system can activate a coordinated program of ER-associated signaling events collectively called the unfolded protein response (UPR) as a way of dealing with "ER stress." Many agents can induce UPR. For example, aberrant glycosylation can occur in the presence of inhibitors such as tunicamycin, which reduces the synthesis of lipid-linked oligosaccharides, or the alkaloids castanospermine, deoxynojirimycin, deoxymannojirimycin, and kifunensin, which inhibit glucose or mannose processing of protein-bound glycans in the ER (see Chapter 50). Addition of dithiothreitol, which prevents disulfide bond formation, and thapsigargin, a tight-binding inhibitor of the ER Ca^{++} ATPase pump, can alter protein folding. Simply depleting glucose from the culture medium will stress cells by reducing metabolic precursors required for glycosylation. Viral infections can place heavy demands on glycoprotein synthesis and cause ER stress, as well. Accumulation of unfolded (or misfolded) proteins triggers UPR, which results in the general attenuation of translation initiation, increased expression of the resident chaperones and proteins that aid protein folding (e.g., CNX, CRT, BiP/Grp78, UGGT, and PDI), expansion of the ER membrane system, and increase in ERAD.

The mechanism of transcriptional and translational control during the stress response has been worked out in some detail. In the nonstressed state, the master ER lumenal chaperone BiP/Grp78 is associated with inactive sensor proteins, the protein kinases PERK/PEK and IRE1, but when unfolded proteins accumulate, BiP/Grp78 dissociates and instead binds to stretches of hydrophobic amino acids on client proteins. Loss of BiP/Grp78 activates the sensor proteins: PERK (protein kinase–like ER kinase) dimerizes and phosphorylates eukaryotic initiation factor, eIF2a, slowing the rate of protein translation. PERK also activates transcription of ATF4. Releasing BiP/Grp78 from IRE1 leads to its autophosphorylation creating an RNAse activity that splices an intron from an X-box binding protein mRNA (XBP-1) and its translation generates a potent transcription factor, which together with ATF4 up-regulates CNX, CRT, PDI, BiP/Grp78, and Grp94 to manage the overload of misfolded glycoproteins.

LIPID-LINKED OLIGOSACCHARIDES

Under normal conditions, a portion of the lipid-linked oligosaccharide (LLO) glycans in the ER are hydrolyzed in the lumen rather than being transferred to protein. Free LLO-derived glycans are rapidly deglucosylated and exported from the ER into the cytoplasm (Figure 36.3). Although no specific transporter has been found, the process requires both ATP and Ca^{++} and removal of the glucose residues. Free mannose can block their export, suggesting that the carrier may recognize the nonreducing end of the chain. Small N-linked glycopeptides can arise within the ER as well, presumably by proteolysis. These are transported by a different mechanism that does not depend on deglucosylation and exhibits different cation requirements. After entry into the cytoplasm, PGNase cleaves the glycan, as described for the degradation of glycoproteins retrotranslocated into the cytoplasm.

Another source of free glycans in the cytoplasm is cleavage of $Man_5GlcNAc_2$-P-P-dolichol that normally assembles on the cytoplasmic face of the ER but may fail to flip across the membrane during the formation of $Glc_3Man_9GlcNAc_2$-P-P-dolichol. Its origin most likely involves a pyrophosphatase and an endo-β-N-acetylglucosaminidase, which cleaves the chitobiose linkage.

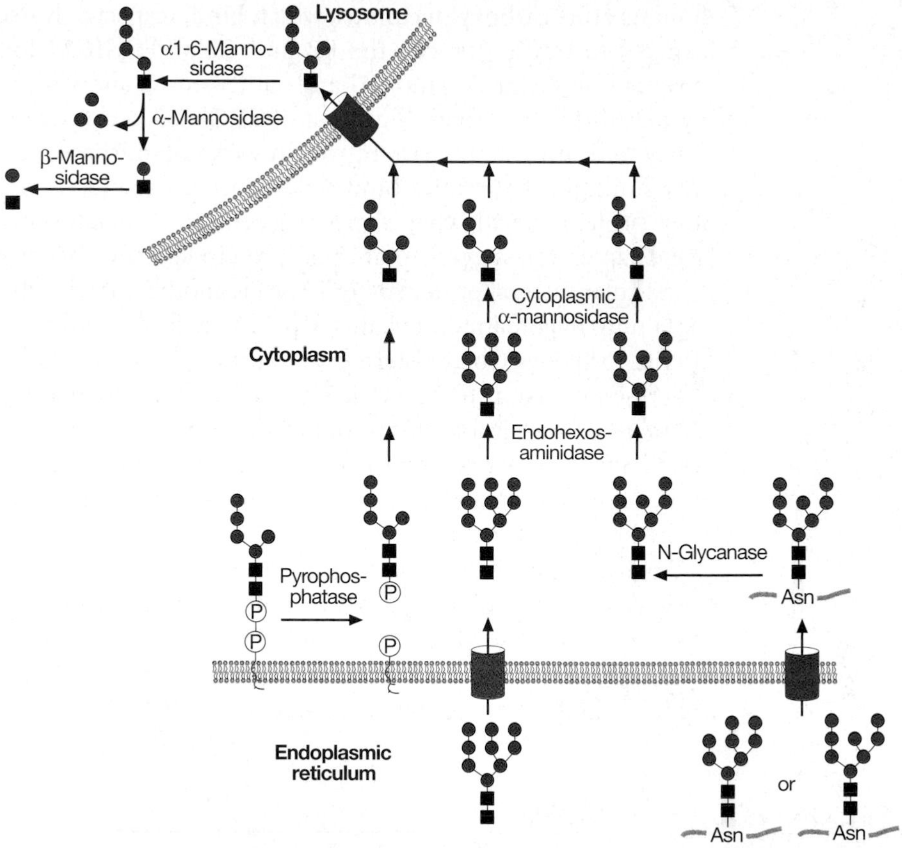

FIGURE 36.3. Degradation of polymannosylglycans in the ER, cytoplasm, and lysosomes. Details are provided in the text. Specific pathways exist for the degradation of free glycans released from misfolded glycoproteins, lipid-linked oligosaccharides (LLOs), and glycopeptides generated within the ER. LLO glycans facing the cytoplasmic side of the ER can also be released.

Mounting evidence points to a fine control between ER stress, the level of LLOs, and protein synthesis. For example, formation and transfer of incomplete glycans from LLO precursors or failure to add a glycan to an N-glycosylation site can generate an ER stress response. Activation of the PERK/PEK pathway then decreases the flux of proteins through the ER, which promotes extension of LLO intermediates to normal size and counteracts abnormal N-glycosylation.

QUALITY CONTROL BEYOND THE ENDOPLASMIC RETICULUM

The ER quality-control process is not foolproof. Some misfolded proteins escape the ER and others may misfold after they exit. Thus, there are additional checkpoints to correct mistakes that occur later in the secretory pathway. Whether lectins recognizing specific glycan structures are involved in post-ER quality control is still an open question, but several candidates may be involved in the process.

Endoplasmic reticulum Golgi intermediate compartment 53 (ERGIC53) is a type I membrane protein that binds to oligomannosyl glycans in a Ca^{++}-dependent manner. Its lumenal, carbohydrate-binding domain is similar to those of soluble lectins from leguminous plants (see Chapter 28). ERGIC53 cycles between the ER and the ERGIC as the mammalian protein displays both ER-targeting (dilysine) and ER-exit (diphenylalanine) determinants at its carboxyl terminus, which bind, respectively, to COPI and COPII coatomers involved in traffic between the ER and Golgi. ERGIC53 loads a subset of glycoproteins (coagulation factors V and VIII and cathepsins C and Z) into COPII-coated vesicles leaving the ER for the Golgi. Mutation of ERGIC53 causes combined factor V and VIII deficiency indicating an important role in secretion of these and perhaps other glycoproteins. Why multiple chaperones bind only selected proteins is unknown, but the discovery of other protein-specific chaperones (e.g., the β1-3 galactosyltransferase chaperone COSMC [see Chapter 9]), suggests that other protein-specific chaperones may exist.

Vesicular integral protein 36 (VIP 36) is another Golgi lectin that also binds glycoproteins containing oligomannosyl chains. VIP36 may facilitate transport of glycoproteins from the ERGIC to the *cis*-Golgi cisternae or retrieve glycoproteins bearing N-glycans that did not undergo conversion to $Man_5GlcNAc_2$ by *cis*-Golgi α-mannosidases IA, IB, and/or IC (see Chapter 8). Conceivably, this would provide an opportunity for new rounds of trimming and eventual formation of complex N-glycans. Both ERGIC53 and VIP36 are up-regulated as part of the UPR. Since many misfolded proteins are transiently diverted to the *cis*-Golgi cisternae before being retrieved to the ER and to ERAD, these lectins may play a role in this process.

Finally, some of the best-characterized lectins of the secretory pathway are the *trans*-Golgi and *trans*-Golgi network cation-dependent and cation-independent mannose-6-phosphate receptors that divert soluble lysosomal proteins to lysosomes (see Chapter 30). Although these lectins are not involved in quality control, they play a pivotal role in directing trafficking of lysosomal enzymes.

QUALITY CONTROL AND DISEASE

In addition to combined factor V and VIII deficiency caused by ERGIC53 mutations, other congenital diseases result from defects in the quality-control system. For instance, the most common mutation in cystic fibrosis patients, CFTR-ΔF508, leads to improper folding of the chloride channel protein and its retention in the ER. A mutated form of α_1-antitrypsin inhibitor that causes emphysema behaves in a similar way. The CNX/CRT system retains both mutant glycoproteins and eventually directs them toward degradation.

Although most glycoproteins transiently associate with CNX/CRT during their maturation in the ER, this lectin-based system is expendable in cultured cells. However, some glycoproteins such as vesicular stomatitis virus G protein, HIV-1 glycoprotein gp120, and M protein from hepatitis B virus absolutely require CNX/CRT for proper folding and exit from the ER. The loss of CNX and CRT in multicellular organisms causes a more severe phenotype. Mice deficient in CNX reach full term, but half die within 2 days of birth, and very few survive beyond 3 months. The runts that survive have obvious motor disorders with loss of large myelinated nerve fibers. CRT-deficient mice and ERp57-null mice also exhibit embryonic lethality.

Another important lectin resident in the ER is calmegin, which is expressed only in the testes during spermatogenesis. This protein shares 60% homology with mouse CNX; the human homolog is 80% identical to mouse calmegin. Mice deficient in calmegin are nearly sterile despite producing normal-looking sperm. The sperm are defective in migration into the oviducts and they do not adhere to the zona pellucida. Sterility occurs because calmegin normally binds to α- and β-fertilin, the sperm-surface disintegrin-metalloproteinases that are key proteins for normal binding and fertilization. Since CNX cannot substitute for calmegin, it explains why both calmegin- and fertilin-knockout male mice are infertile.

ER stress has come to center stage in many diseases because it is clear that it is closely connected to cell survival. This is especially important for professional secretory cells, such as plasma cells, pancreatic β cells, hepatocytes, and osteoblasts. Prolonged or aggravated ER stress can have profound effects on the survival of insulin-producing pancreatic cells. Recently, ER stress has been implicated in the development of amyotrophic lateral sclerosis (ALS), Parkinson's, Alzheimer's, and prion disease. For example, the brains of patients showing sporadic Parkinson's or Alzheimer's disease accumulate an S-nitrosylated version of PDI, which inhibits its activity, leading to UPR and ultimately to enhanced cell death. Mutations in the cytoplasmic protein, α-synuclein, have a dramatic effect on the trafficking of proteins between the ER and Golgi, leading to ER stress and onset of familial Parkinson's disease. Overexpression of the small GTPase Rab1 rescues neuron loss in models of α-synuclein-induced neurodegeneration. This leads to the speculation that Rab1 serves a chaperone function for folding of synaptic trafficking proteins that are important for neurotransmitter release.

Mouse models of the lysosomal storage diseases and GM1 gangliosidosis (see Chapter 41) show up-regulation of BiP/Grp78, and it is likely that many disorders in which mutant proteins misfold and fail to exit the ER can overload the pathway. The term ER storage disorders (ERSDs) has been applied to genetic disorders that overload the ER quality-control process. An estimated 20–30% of normal proteins misfold, so is not surprising that small differences in the kinetics of conformational maturation can tip the balance toward ERAD versus safe passage to the ERGIC and beyond. For example, the substrate specificity of P-glycoprotein (MDR1) can be altered by a "silent single nucleotide polymorphism" that does not change the identity of the amino acid, but changes the codon to a lower abundance tRNA, causing a temporary delay in translation. This alters the kinetics of protein folding, which then, in turn, alters substrate specificity.

Bacteria cleverly exploit the ERAD retrotranslocation pathway to deliver exotoxins such as cholera and shiga toxins into the cytoplasm. These toxins bind to the host cell-surface receptors and are endocytosed, trafficked into the Golgi, to the ER, and ultimately into the cytoplasm, where the toxin expresses its activity.

FURTHER READING

Winchester B.G. 1996. Lysosomal metabolism of glycoconjugates. *Subcell. Biochem.* **27:** 191–238.

Trombetta E.S. and Parodi A.J. 2003. Quality control and protein folding in the secretory pathway. *Annu. Rev. Cell. Dev. Biol.* **19:** 649–676.

Helenius A. and Aebi M. 2004. Roles of N-linked glycans in the endoplasmic reticulum. *Annu. Rev. Biochem.* **73:** 1019–1049.

Hebert D.N., Garman S.C., and Molinari M. 2005. The glycan code of the endoplasmic reticulum: Asparagine-linked carbohydrates as protein maturation and quality-control tags. *Trends Cell Biol.* **15:** 364–370.

Lederkremer G.Z. and Glickman M.H. 2005. A window of opportunity: Timing protein degradation by trimming of sugars and ubiquitins. *Trends Biochem. Sci.* **30:** 297–303.

Chua C.E. and Tang B.L. 2006. α-synuclein and Parkinson's disease: The first roadblock. *J. Cell. Mol. Med.* **10:** 837–846.

Lehrman M.A. 2006. Stimulation of N-linked glycosylation and lipid-linked oligosaccharide synthesis by stress responses in metazoan cells. *Crit. Rev. Biochem. Mol. Biol.* **41:** 51–75.

Moremen K.W. and Molinari M. 2006. N-linked glycan recognition and processing: The molecular basis of endoplasmic reticulum quality control. *Curr. Opin. Struct. Biol.* **16:** 592–599.

Ruddock L.W. and Molinari M. 2006. N-glycan processing in ER quality control. *J. Cell Sci.* **119:** 4373–4380.

Suzuki T. and Funakoshi Y. 2006. Free N-linked oligosaccharide chains: Formation and degradation. *Glycoconj. J.* **23:** 291–302.

Caramelo J.J. and Parodi J. 2008. Getting in and out from calnexin/calreticulin cycles. *J. Biol. Chem.* **283:** 10221–10225.

CHAPTER 37

Free Glycans as Signaling Molecules

Marilynn E. Etzler and Jeffrey D. Esko

IN RECENT YEARS, THERE HAS BEEN A GROWING RECOGNITION that free glycans are used as signals for the initiation of a wide variety of biological processes. Such signaling events are found in development, in defense responses of plants and animals, and in interactions of organisms with one another. This chapter covers the state of current information on this emerging field of study.

NATURE AND SCOPE OF GLYCAN SIGNALING SYSTEMS

Glycan signaling systems are diverse. Simple sugars (such as glucose, fructose, and sucrose) employ various sensing systems often linked with the metabolism of the sugar and form complex webs of signaling events linked to hormones. Glycan signaling systems can also involve various glycoconjugates. For example, the addition of O-GlcNAc to cytoplasmic and nuclear proteins results in changes in the cytoskeleton, gene transcription, and enzyme activation (see Chapter 18). Glycosphingolipids can form lipid rafts, which act as a platform for sequestering signaling receptors, or can associate with receptor tyrosine kinases and modulate their activity (see Chapter 10). Membrane proteoglycans containing sulfated glycosaminoglycans (such as syndecans and phosphacan) may act as signaling molecules by interactions with kinases or phosphatidylinositol-4,5-bisphosphate (see Chapters 16 and 35). Most plasma membrane signaling receptors, including receptor tyrosine kinases and G-protein-coupled receptors, contain N-glycans and O-glycans that can modulate their stability and activity (see Chapters 8 and 9). These types of signaling events are covered in other chapters in this book and are not discussed further here.

In recent years, it has become clear that specific free glycans in picomolar to micromolar concentrations can act as signals in the initiation of a number of biological processes. These signals were first discovered as part of the defense response in plants and are now found to play roles in both plant and animal development, innate immunity, and the initiation of the nitrogen-fixing *Rhizobium*–legume symbiosis. Present knowledge regarding each of these areas is described in the following sections. In most of these processes, the glycans that serve as signals have been identified, but we are still in the beginning stages of identifying and characterizing their receptors and the mechanisms of signal transduction.

GLYCAN SIGNALS TRIGGER THE INITIATION OF THE PLANT DEFENSE RESPONSE

The plant defense response is a multifaceted one that includes the recognition of a pathogen, changes in ion flux across the plasma membrane, formation of reactive oxygen species, the activation of a number of genes that lead to changes in the plant cell wall, the production of glycanases that can degrade the pathogen cell wall, and the production of phytoalexins, which are antimicrobial components that kill the invading pathogen (Figure 37.1). This response leads to the formation of necrotic spots at the site of infection and it limits the spread of the pathogen. Early studies of this defense response showed that a variety of oligosaccharides derived from plant or pathogen cell wall glycans have the ability to elicit this response. This was the first indication that glycans may play signaling roles in nature, and these signals were initially called "oligosaccharins."

Several different oligosaccharides from plant and fungal cell walls have been found to be extremely effective elicitors at low nanomolar concentrations (~10 nM) (Figure 37.2). One of these is a heptaglucoside. This glycan was isolated from a mixture estimated to contain about 300 inactive structural isomers of this compound. Comparisons of the biological activity of this heptaglucoside with the activities of other isolated or chemically synthesized isomers and derivatives showed that both the size and position of the branches were essential for activity (Figure 37.3). Elongating the oligosaccharide at the reducing end had no significant effect on activity. Most activity was also retained by removal of a single glucose from the reducing end. However, the hexaglucoside was found to be the minimal structure that has appreciable activity.

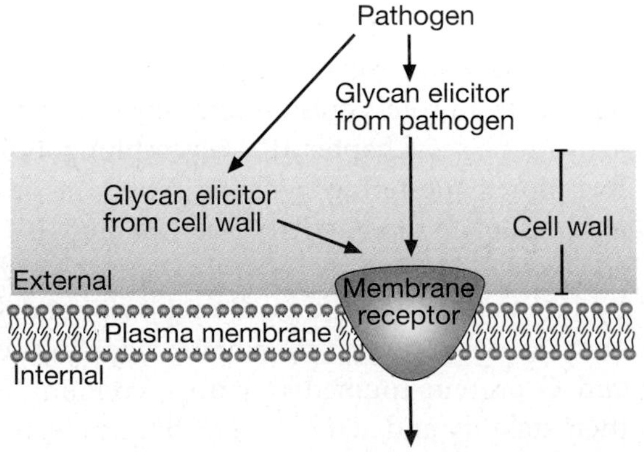

FIGURE 37.1. Plant defense response. This response is initiated by a glycan elicitor produced either by the pathogen or following the interaction of the pathogen with the cell wall. The elicitor interacts with a membrane receptor to initiate a variety of cell responses that culminate in the killing of the pathogen and limit its spread.

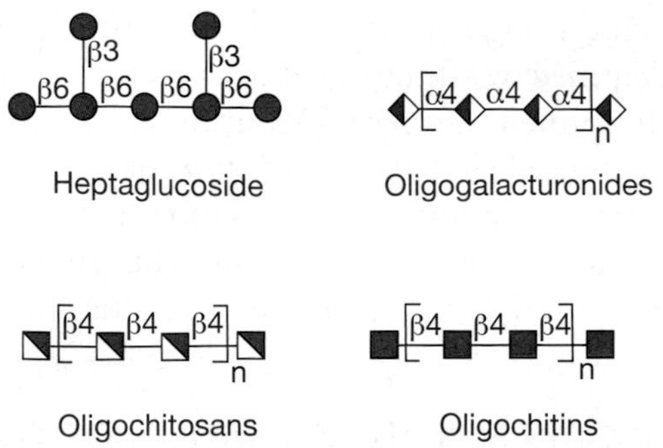

FIGURE 37.2. Oligosaccharide elicitors of the plant defense response. (For the monosaccharide symbol code, see Figure 1.5, which is also reproduced on the inside front cover.)

Other glycan elicitors that have been found are oligogalacturonides, chitosan, and chitin oligosaccharides (Figure 37.2). These glycans are linear homopolymers of various lengths, and studies have focused primarily on the degree of polymerization (dp) necessary for biological activity. Such studies suggest that a dp of 10–14 is necessary to elicit a defense response by oligogalacturonides. In contrast, dps of greater than 7 and greater than 4 are necessary for the oligochitosans and oligochitins, respectively.

The low quantities of glycan signal molecules that are necessary to elicit a defense response and the different types of oligosaccharides that can act as elicitors suggest that specific receptors are available on the plasma membrane that recognize these glycans. Indeed, the plant defense system may have similarities to the innate immune system in animals, in which specific pattern recognition occurs (see below and Chapter 39).

Studies with plant cell cultures and isolated plasma membranes have demonstrated the existence of specific cell-surface or membrane-binding sites that are saturable and have binding specificities similar to those required for biological behavior. Recently, several proteins have been identified that are candidates for such receptors. One such protein is CEBiP,

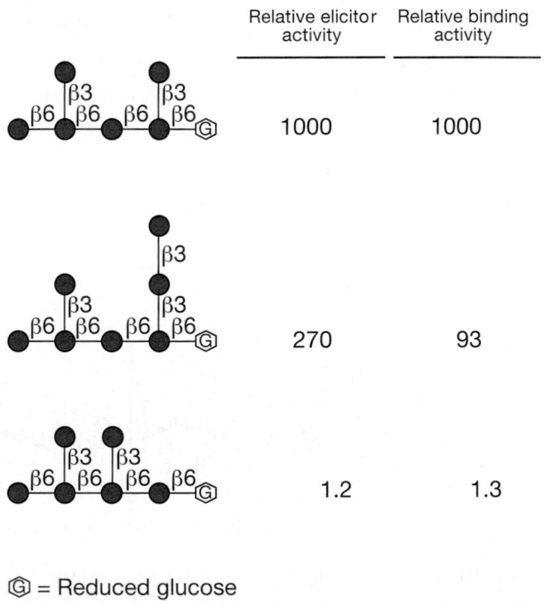

FIGURE 37.3. Relative biological activities of synthetic oligosaccharide isomers and derivatives of heptaglucoside. (Adapted, with permission, from Darvill A., Augur C., Bergmann C., et al. 1992. *Glycobiology* **2:** 181–198, © Oxford University Press.)

a chitin oligosaccharide elicitor binding protein that binds chitin elicitors and was first identified by affinity labeling as a 75-kD plasma membrane protein from cultured rice cells. This protein was recently purified and its gene was cloned. RNA interference (RNAi) studies showed that reducing expression of the gene resulted in suppression of the defense response. Although it is a membrane protein, there is no appreciable portion on the cytoplasmic side of the membrane. This protein may thus be part of an elicitor–receptor complex, but the other components in the system have not been identified.

NOD FACTORS ARE SIGNALS FOR THE INITIATION OF THE NITROGEN-FIXING *RHIZOBIUM*–LEGUME SYMBIOSIS

The interaction between *Rhizobium* and leguminous plants is one of the most agriculturally and economically important symbiotic relationships that occur in nature, because it leads to the fixation of atmospheric nitrogen. A major step in the initiation of this process is the recognition by the plant of a lipooligosaccharide signal, called the Nod factor, which is produced by the bacteria. This signal stimulates a number of changes in the roots of the plant, including the formation of a new organ (the nodule) and changes in the root hairs that enable the bacteria to enter the plant and migrate to the nodule where nitrogen fixation occurs. The importance of the Nod factors in this symbiosis is demonstrated by the finding that picomolar to nanomolar levels of purified Nod factor applied to the roots of an appropriate legume species can initiate the plant responses that lead to nodule formation.

Both the initiation of nodule formation and *Rhizobium* entry are host strain–specific; this specificity is determined by the type of Nod factor produced by a particular Rhizobium strain and the ability of a leguminous species to recognize that signal. All Nod factors consist of a short chitin oligosaccharide backbone, but they differ among *Rhizobium* strains in the types of modification of this backbone. A generic structure of a Nod factor is shown in Figure 37.4 with sites of potential modification. The number of *N*-acetylglucosamine residues in this backbone has been found to vary from three to five. Modifications that have been identified to date include methylation, acylation (usually with a C_{16} or C_{18} fatty acid), acetylation, carbamylation, sulfation, glycosylation, and the addition of glycerol.

Both genetic and biochemical approaches have been useful in identifying potential Nod factor receptors in plant roots. Genetic analyses have identified a few genes from mutants of different legume species that encode proteins that may play a role in this signaling system. It is of interest that these proteins are transmembrane proteins with a serine/threonine receptor kinase motif on the cytoplasmic side of the membrane and lysine motif

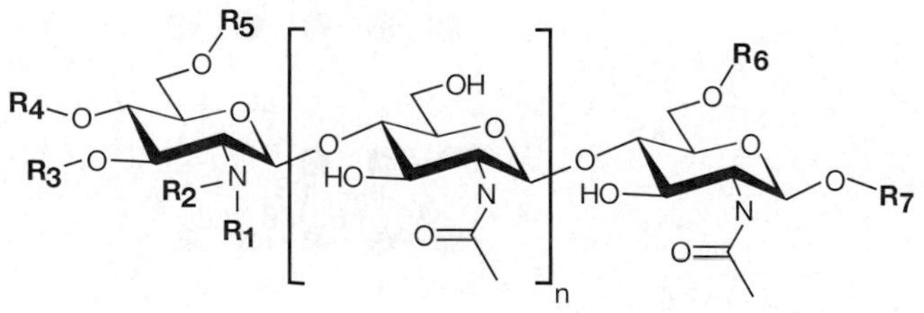

FIGURE 37.4. Generic structure of a Nod factor. Sites on the molecule where species-specific modifications can occur are designated by R_1–R_7. R_1 = H, methyl; R_2 = C16:2, C16:3, C18:1, C18:3, C18:4, C20:3, C20:4; R_3 = H, carbamate; R_4 = H, carbamate; R_5 = H, Ac; R_6 = H, Ac, SO_4, Fuc, AcFuc, MeFuc; R_7 = H, glycerol. *n* = 1–4. (Adapted, with permission, from Dénarié J., Debellé F., and Promé J.C. 1996. *Annu. Rev. Biochem.* **65:** 503–535, ©Annual Reviews.)

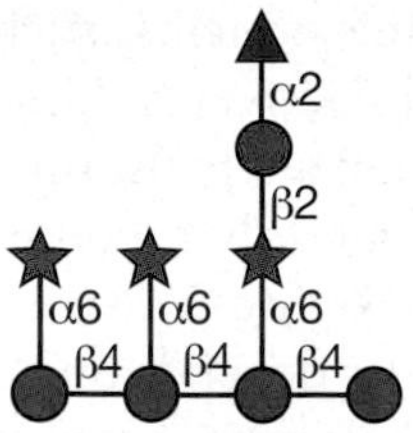

FIGURE 37.5. A nonasaccharide from xyloglucan that exhibits signaling properties.

(LysM) domains that may recognize glycans on the exterior of the membrane. It has been postulated that these LysM domains may bind to Nod factors, although no carbohydrate-binding studies have been conducted. On the other hand, a biochemical study identified a novel lectin nucleotide phosphohydrolase (LNP) from the roots of a legume that binds Nod factors from *Rhizobium* symbionts of the plant species from which it was obtained. LNP is a peripheral membrane protein and it is possible that it may function in a type of receptor complex with one or more of the above proteins.

OLIGOSACCHARIDE SIGNALS IN EARLY PLANT AND ANIMAL DEVELOPMENT

Some of the glycan signals described above were found to have effects on plant growth and organogenesis. For example, oligogalacturonides (see Figure 37.2) with a dp of 12–14 are active in inducing flower formation, and oligogalacturonides with a dp of 11–14 can inhibit root formation. These glycans exert their effect on organogenesis at a concentration of approximately 400 nM and are mainly studied using tissue explants. Pectin fragments (see Chapter 22) have also been shown to affect plant growth and development by enhancing expansion of the cell, and a nonasaccharide-rich fragment of xyloglucan at a concentration of approximately 10^{-8} M inhibited auxin-induced elongation of pea stem segments (Figure 37.5). Some studies have shown that Nod factors can activate developmental pathways in nonlegumes, which has led to the suggestion that plants may utilize endogenous Nod-factor-like signals to regulate growth and organogenesis.

Chitin oligosaccharides may also play a role in animal embryogenesis. A gene called *DG42* has chitin synthase activity and it is expressed briefly in *Xenopus* endoderm cells during the mid-late gastrulation stage. As discussed in Chapter 25, this gene is homologous to the *Rhizobium NodC* gene, which encodes a chitin synthase. Homologs of this gene were found in zebrafish and mice, and transgenic expression of the gene resulted in chitin oligomers that were digestible by chitinase. However, this gene is also homologous to a gene responsible for the synthesis of hyaluronan, and studies have shown that this gene may be able to assemble both chitin and hyaluronan. Although the biochemical function of *DG42* is not certain, injection of chitinases or expression of *NodZ* (which encodes a fucosyltransferase that can modify chitin) also has profound effects on development. Thus, chitin oligosaccharides are examples of free glycans that act as intracellular signaling molecules.

GLYCOSAMINOGLYCANS AND CELL SIGNALING

Glycosaminoglycans (GAGs) can be considered signaling glycans because they interact with receptor tyrosine kinases and/or their ligands and facilitate changes in cell behavior (see Chapters 15, 16, and 35). For example, hyaluronan oligosaccharides can bind to specific membrane proteins, such as CD44. In some cells, binding results in clustering of CD44, which activates kinases such as c-Src and focal adhesion kinase (FAK). Phosphorylation alters the interaction of the cytoplasmic tail of CD44 with regulatory and adaptor molecules that modulate cytoskeletal assembly/disassembly and cell survival and proliferation (Figure 37.6). Like the systems described above, signaling by hyaluronan oligosaccharides depends on the degree of polymerization of the glycans, with low-molecular-weight chains more active than high-molecular-weight chains.

In contrast to hyaluronan-dependent signal transduction, signaling via sulfated GAGs (heparan sulfate [HS] and chondroitin/dermatan sulfate) occurs by a more indirect mechanism. Few membrane receptors have been described where binding actually causes a specific downstream response, such as phosphorylation of the receptor or activation of a kinase. Instead, sulfated GAGs bind to many ligand/receptor pairs, thereby lowering the effective concentration of ligand required to engage the receptor or increasing the duration of the response. An example of this is the ability of exogenous heparin or endogenous HS proteoglycans to activate fibroblast growth factor (FGF) receptors by FGF. No significant conformational change in the ligand occurs upon binding to sulfated GAG, consistent with the idea that the glycan primarily aids in the juxtaposition of components of the signal transduction pathway. Free HS oligosaccharides can be released by the action of secreted heparanase. These glycans could facilitate signaling through the mechanism described above or by the release of growth factors from stored depots in the extracellular matrix. Sulfated GAGs also facilitate the formation of morphogen gradients in tissues during early development. Because the gradient determines cell specification during development, the glycan indirectly affects signaling responses in receptive cells.

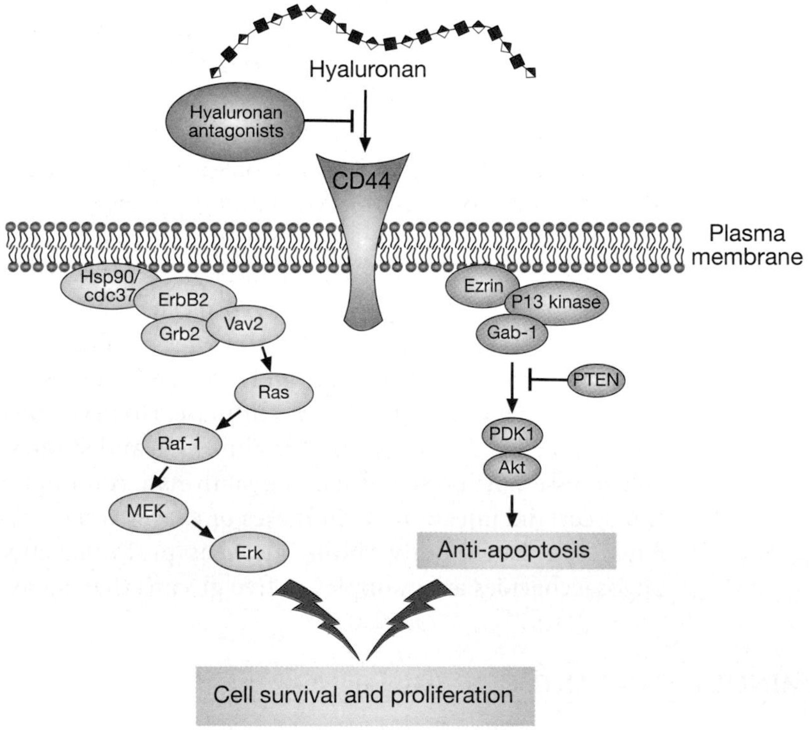

FIGURE 37.6. Schematic diagram of signaling pathways activated by binding of hyaluronan to CD44. In tumor cells, activation results in proliferation and invasion. In embryonic stem cells, it can result in epithelial to mesenchymal transition.

BACTERIAL GLYCANS AND ACTIVATION OF INNATE IMMUNITY

The innate immune system developed early in evolution as the first line of defense of eukaryotes against infection by microorganisms. A key prerequisite of this system is the ability to distinguish self from infectious nonself. In higher eukaryotes, this has been accomplished by the evolution of a range of receptors that recognize conserved molecular patterns on pathogens that are not found in the host. The term pathogen-associated molecular patterns (PAMPs) has been coined to refer to such motifs. The cognate receptors on the host cells are referred to as pattern-recognition receptors (PRRs). Many of these PAMPs are glycans found on the surfaces of bacteria that are not produced by the host. Examples are the lipopolysaccharides (LPS) of Gram-negative bacteria and the peptidoglycans of Gram-positive bacteria (see Chapter 20). A variety of PRRs are present in mammals and they recognize various PAMPs and induce host-defense pathways (see Chapter 39).

One of the best-studied models of innate immunity is the LPS of Gram-negative bacteria, which plays a role in causing septic shock (see Chapter 20). Lipid A (endotoxin), which is the glucosamine-based phospholipid anchor of LPS, is responsible for activating the innate immune system. Lipid A has a highly conserved structure among Gram-negative bacteria, which makes it an excellent PAMP. This glycan is detected at picomolar levels by a PRR, called TLR-4, which is one of the Toll-like receptors (see Figure 39.2). It is first opsonized and complexed with another host cell-surface protein, CD14. The Toll-like receptors are a major class of PRRs found on the surfaces of cells and, upon recognizing PAMPs, they activate various signaling pathways that induce inflammation and antimicrobial effector responses. Some of these receptors are expressed on antigen-presenting cells and help in the activation of the adaptive immune response. A number of other glycan PAMPs have been identified, such as peptidoglycan (which is recognized by TLR-2) and mannans (which are recognized by a soluble PRR that is a mannan-binding lectin).

FURTHER READING

Darvill A., Augur C., Bergmann C., Carlson R.W., Cheong J.J., Eberhard S., Hahn M.G., Lo V.M., Marfa V., Meyer B., Mohnen D., O'Neill M.A., Spiro M.D., van Halbeek H., York W.S., and Albersheim P. 1992. Oligosaccharins—Oligosaccharides that regulate growth, development and defense responses in plants. *Glycobiology* **2:** 181–198.

Denarie J., Debelle F., and Prome J.C. 1996. Rhizobium lipo-chitooligosaccharide nodulation factors: Signaling molecules mediating recognition and morphogenesis. *Annu. Rev. Biochem.* **65:** 503–535.

Ebel J. 1998. Oligoglucoside elicitor-mediated activation of plant defense. *Bioessays* **20:** 569–576.

Bakkers J., Kijne J.W., and Spaink H.P. 1999. Function of chitin oligosaccharides in plant and animal development. *EXS* **87:** 71–83.

Cullimore J.V., Ranjeva R., and Bono J.-J. 2001. Perception of lipo-chitooligosaccharidic Nod factors in legumes. *Trends Plant Sci.* **6:** 24–30.

Rolland F., Winderickx J., and Thevelein J.M. 2001. Glucose-sensing mechanisms in eukaryotic cells. *Trends Biochem. Sci.* **26:** 310–317.

Takeda K., Kaisho T., and Akira S. 2003. Toll-like receptors. *Annu. Rev. Immunol.* **21:** 335–376.

Nurnberger T. and Brunner F. 2005. Innate immunity in plants and animals: Emerging parallels between the recognition of general elicitors and pathogen-associated molecular patterns. *Curr. Opin. Plant Biol.* **5:** 318–324.

Jiang D., Liang J., and Noble P.W. 2007. Hyaluronan in tissue injury and repair. *Annu. Rev. Cell. Dev. Biol.* **23:** 435–461.

Raetz C.R.H., Reynolds C.M., Tent M.S., and Bishop R.E. 2007. Lipid A modification systems in Gram-negative bacteria. *Annu. Rev. Biochem.* **76:** 295–329.

CHAPTER 38

Glycans in Development and Systemic Physiology

Ajit Varki, Hudson H. Freeze, and Victor D. Vacquier

GIVEN THEIR UBIQUITOUS PRESENCE AND VARIED complexity on all cell surfaces, it is not surprising that glycans have many diverse roles in various physiological systems of complex multicellular organisms (see Chapter 6). This brief chapter serves as a guide to the rest of the book from the perspective of a reader interested in specific physiological systems, and it provides examples of the roles of glycans in development and systemic physiology. The focus of this chapter is on vertebrate systems. Pathological processes involving glycans are mainly discussed in other chapters. However, because they often provide clues to normal functions, some examples are mentioned here.

REPRODUCTIVE BIOLOGY

Abundant evidence indicates that both male and female reproductive processes are affected by glycans and glycan-binding proteins. Fertilization begins with contact between swimming sperm and the extracellular matrix of the egg and it ends with the fusion of sperm and egg haploid pronuclei, which restores the diploid genome and creates the zygote. Fertilization has been studied extensively in sea urchins, fish, frogs, and mammals. Evidence for the involvement of glycans in many steps is compelling. Sea urchin and mouse fertilization are the best-characterized systems, and these are covered in more detail in Chapter 25. Glycans also have many roles in other aspects of reproduction. Examples include the significance of glycan recognition in sperm interactions with the lining of the fallopian tube and functions of glycans in the process of implantation of the early embryo. Interactions between oocytes and cumulus cells involving hyaluronan in the ovary are discussed in Chapter 15. Genetic modifications of glycosylation in mice have also revealed examples of male infertility caused by glycan structural perturbations (see Chapter 13).

EMBRYOLOGY AND DEVELOPMENT

As a general rule, genetic modifications that affect early stages in the assembly or initiation of a major glycan pathway result in embryonic lethality. Indeed, with the exception of the mucin O-glycan pathway (where complete elimination is difficult because of multiple polypeptide:O-GalNAc transferase isoforms), gene-knockout experiments in mice have demonstrated a critical role in embryogenesis for all other major classes of glycans, as well as for certain classes of monosaccharides such as sialic acids. As might be expected, the resulting embryonic phenotypes are complex, and no single mechanism can easily explain the causes of lethality. For example, disruption of protein O-fucosylation causes a defect in Notch receptors that results in a severe embryonic phenotype (see Chapter 12). Major modifications of glycosaminoglycans also cause developmental abnormalities, most likely because of their roles in modulating growth factor function and their proposed roles in setting up morphogen gradients (see Chapter 16). However, mutations in proteoglycan core proteins can have variable effects, usually in a tissue-specific manner (see Chapters 16, 23, and 25).

In contrast to the effects of eliminating entire glycan classes, specific modifications of terminal structures on glycans (such as specific sialic acid linkages) usually do not have dramatic effects in development, allowing the production of viable offspring with varying and usually limited abnormalities. Such mutants are more likely to show specific defects in restricted cell types (see below).

MUSCULOSKELETAL BIOLOGY

Glycans appear to have a critical role in the interactions of extracellular matrix molecules like laminin with glycan chains on α-dystroglycan, which is a key component of muscle. Multiple defects in the pathway for assembly of these O-mannose-linked glycans are known to be associated with muscular dystrophies of various kinds, both in humans and in mice (see Chapter 42). Glycan-related interactions may also have important roles in the functioning of the muscle–nerve junctions, including the clustering of acetylcholine receptors. Other evidence suggests that sialic acids may have roles in modulating calcium fluxes in skeletal and cardiac muscle cells. The process of formation and ossification of cartilage into bone intimately involves a variety of glycosaminoglycans, including hyaluronan, heparan and chondroitin sulfate, and keratan sulfate (see Chapters 15 and 16).

CARDIOVASCULAR PHYSIOLOGY

Gene knockouts of hyaluronan synthase indicate that hyaluronan has a critical role in the development of the heart (see Chapter 15). There is considerable evidence that glycosaminoglycans have a role in modulating angiogenesis, partly by virtue of their ability to bind a variety of growth factors (most notably vascular endothelial growth factor and fibroblast growth factor) (see Chapter 16). The high density of sialic acids at the luminal surface of endothelial cells and the presence of glycosaminoglycans within the basement membrane underlying the endothelial cells are thought to contribute to the structural integrity of the vessel wall. As mentioned above with skeletal muscle, evidence suggests that sialic acids have as yet unclear roles in modulating calcium fluxes in cardiac muscle cells.

AIRWAY AND PULMONARY PHYSIOLOGY

As with all organs that contain lumens, the lining epithelia of the upper and lower airways are coated with a dense and complex layer of glycans, both in the form of structural glycoproteins and glycolipids of the luminal border of the epithelial cells and also as secreted soluble mucin molecules. Both membrane-bound and soluble glycoconjugates have roles in the effective functioning of the airways, in hydration of the surfaces, and in protection against external agents, both physical and microbial. Embryonic stem cells that lack complex N-glycans do not form part of the organized layer of bronchial epithelium. Normal N-glycans are also important for healthy lung function, because mice lacking the core α1-6 fucose of N-glycans develop emphysema-like symptoms due to overexpression of matrix metalloproteinases that degrade the lung tissue. This is apparently caused by a misregulation of the transforming growth factor-β1 signaling pathway, most likely through its misglycosylated receptor.

ENDOCRINOLOGY

There is abundant evidence that O-GlcNAc has a role in modulating the actions of insulin and in explaining some of the effects of hyperglycemia on a variety of systems (see Chapter 18). In the thyroid gland, the glycosylation of thyroglobulin is thought to be involved in the targeting of the molecule for its eventual degradation, which is required for the generation of thyroid hormones. One possible mechanism is the recognition of mannose 6-phosphate residues on the thyroglobulin N-glycans by lysosomal receptors (see Chapter 30). Pituitary glycoprotein hormones are known to carry unusual sulfated N-glycans (see Chapters 13 and 28), which appear to be involved in the rapid clearance of these hormones from the blood stream. This pharmacodynamic effect of sulfated N-glycans optimizes the response of the target organs (i.e., the gonads) to these hormones. Mice deficient in their ability to make triantennary N-glycans develop the characteristics of type II diabetes, especially when they are fed a high-fat diet. This appears to result from altered glycosylation of the GLUT2 glucose transporter in the pancreatic islet cells. The improper glycosylation leads to accelerated endocytosis of the transporter, leaving an insufficient amount on the surface to perform its critical role in the ultimate action of insulin.

GASTROENTEROLOGY

The comments above regarding the lining of the airways also apply to the luminal lining of the gastrointestinal system. Given the microbial contents of the gut, the importance of glycans in physical protection against luminal contents is likely even greater in this instance. The glycosphingolipids of gastrointestinal epithelial cells are highly concentrated at the outer leaflet of the apical domain, such that they may even outnumber phospholipids as the dominant component of this leaflet. There is also abundant evidence for the involvement of glycans in the interactions of pathogens and symbionts with the gastrointestinal epithelium, ranging from interaction of *Helicobacter* species with the stomach mucosa to the symbiotic relationships of anaerobic bacteria in the colon, which selectively bind to Galα1-4Gal sequences found in the internal regions of glycosphingolipids (see Chapters 34 and 39). Also of interest is the fact that *Helicobactor pylori* infection is rarely found in the duodenum, where certain unusual α1-4GlcNAc-terminated O-linked mucins are expressed. This glycan apparently acts as a natural antibiotic against *H. pylori* infections by inhibiting the biosynthesis of Glcα-O-cholesterol, a major cell-wall component. There is also evidence for extensive "glycan foraging" by various organisms in the gastrointestinal tract, as part of their complex relationship with the host (see Chapter 34). Heparan sulfate in the basement membrane also serves a critical role as a permeability barrier, preventing protein loss into the gut.

HEPATOLOGY

The great majority of proteins secreted by the liver are heavily glycosylated. Hepatocytes have thus been traditionally an excellent system for studying the organization and function of the Golgi apparatus. Various cell types of the liver also express a variety of receptor systems that mediate clearance, based on recognition of specific glycans on circulating molecules (see Chapters 26, 28, 29, and 31 for examples of liver receptor specificities). These receptor systems appear to cooperate to remove unwanted molecules from the circulation. There is also emerging evidence for a role of glycosaminoglycans in controlling lipoprotein clearance in the liver via sequestration of lipoproteins in the space of Disse, which is located between the fenestrated endothelium and the hepatocytes, and by affecting endocytosis.

NEPHROLOGY

There is extensive evidence that heparan sulfate glycosaminoglycans (see Chapter 15) and sialic acid residues (see Chapter 14) on podocalyxin are involved in assuring the optimal filtering function of the glomerular basement membrane. As with any organ system with hollow cavities, mucin-like molecules and glycosaminoglycans have an important role in providing a barrier function at the luminal surface of the ureter and bladder. Reduced branching of complex N-glycans causes kidney pathology that may result from an autoimmune response.

SKIN BIOLOGY

Glucosylceramide and related glycosphingolipids and adducts appear to have a critical role in maintaining the barrier function of the skin. There is also evidence for a role for dermatan sulfate glycosaminoglycans in maintaining the structure of the dermis and in facilitating wound repair. A lack of O-fucose glycans on Notch receptors results in skin lesions due to changes in hair cell differentiation (see Chapter 12).

ORAL BIOLOGY

Glycosaminoglycans (see Chapters 15 and 16) have a critical role in the development, organization, and structure of both the gums and teeth. Interaction of various oral commensal organisms with the host epithelium can involve recognition of glycans. Mucins produced by the salivary glands may have protective effects in the oral cavity, preventing bacterial biofilm formation on teeth (see Chapters 9 and 39).

HEMATOLOGY

Many aspects of the structure and function of blood components are affected by glycosylation. The trafficking of leukocytes throughout the body is regulated by glycan recognition, particularly with regard to ligands of the selectins (see Chapter 31). Variable glycosylation of red blood cells is responsible for explaining many of the intraspecies blood group differences that affect the practice of blood transfusion (see Chapter 13). There is conflicting evidence about whether or not the half-life of red blood cells in circulation is determined by changes in cell-surface glycans. Nearly all blood proteins are N-glycosylated, which is important for maintaining their stability in the circulation. Patients with impaired N-glycosylation often have insufficient levels of coagulation factors such as antithrombin-III and proteins C and S (see Chapter 42).

IMMUNOLOGY

Terminal components of N- and O-glycans appear to have important roles not only in the trafficking of lymphocytes and other immune cells, but also in their differentiation and/or apoptosis (see Chapter 42). Sialic acid recognition by Siglec adhesion molecules has a role in regulating immune responses (see Chapter 32), and the O-fucose glycans on Notch receptors regulate many cell differentiation processes, including thymic development (see Chapter 12).

NEUROBIOLOGY

The unusual polysialic acid structure attached to the neural cell-adhesion molecule appears to modulate the plasticity of the nervous system with respect to neural changes during embryogenesis in adult life (see Chapter 14). Neural cells are also highly enriched in sialic-acid-containing glycolipids (gangliosides), and alterations in these glycans affect neurological function (see Chapter 10). There are two instances wherein specific glycans appear to inhibit nerve regeneration after injury. In the first, recognition of certain sialylated glycolipids by myelin-associated glycoprotein appears to send a negative signal against neuronal sprouting following injury (see Chapter 32). Similar inhibitory effects appear to be mediated by the glycosaminoglycan chondroitin sulfate (see Chapter 16). In both instances, targeted degradation of the glycan in vivo (by local injection of sialidase or chondroitinase) can stimulate growth and repair, supporting the hypothesis that these glycans normally act to block regeneration. On the basis of genetic defects induced in animals, there is also evidence that complex N-glycans and glycosaminoglycans have critical roles in the development and organization of the nervous system (see Chapters 8 and 16). Indirect evidence indicates a role for fucosylated N-glycans in modulating various aspects of neural development and function (see Chapter 42).

FURTHER READING

Varki A. 1993. Biological roles of oligosaccharides: All of the theories are correct. *Glycobiology* **3:** 97–130.

Mahley R.W. and Ji Z.S. 1999. Remnant lipoprotein metabolism: Key pathways involving cell-surface heparan sulfate proteoglycans and apolipoprotein E. *J. Lipid. Res.* **40:** 1–16.

Mengerink K.J. and Vacquier V.D. 2001. Glycobiology of sperm–egg interactions in deuterostomes. *Glycobiology* **11:** 37R–43R.

Yamaguchi Y. 2002. Glycobiology of the synapse: The role of glycans in the formation, maturation, and modulation of synapses. *Biochim. Biophys. Acta* **1573:** 369–376.

Diekman A.B. 2003. Glycoconjugates in sperm function and gamete interactions: How much sugar does it take to sweet-talk the egg? *Cell Mol. Life Sci.* **60:** 298–308.

Haines N. and Irvine K.D. 2003. Glycosylation regulates Notch signaling. *Natl. Rev. Mol. Cell. Biol.* **4:** 786–797.

Furukawa K., Tokuda N., Okuda T., Tajima O., and Furukawa K. 2004. Glycosphingolipids in engineered mice: Insights into function. *Semin. Cell Dev. Biol.* **15:** 389–396.

Haltiwanger R.S. and Lowe J.B. 2004. Role of glycosylation in development. *Annu. Rev. Biochem.* **73:** 491–537.

Sonnenburg J.L., Xu J., Leip D.D., Chen C.H., Westover B.P., Weatherford J., Buhler J.D., and Gordon J.I. 2005. Glycan foraging in vivo by an intestine-adapted bacterial symbiont. *Science* **307:** 1955–1959.

Stanley P. 2007. Regulation of Notch signaling by glycosylation. *Curr. Opin. Struct. Biol.* **17:** 530–535.

CHAPTER 39

Bacterial and Viral Infections

Victor Nizet and Jeffrey D. Esko

THIS CHAPTER ILLUSTRATES SOME KEY MECHANISMS by which glycans influence the pathogenesis of bacterial and viral infections and describes examples of opportunities for therapeutic intervention.

BACKGROUND

Infectious diseases remain a major cause of death, disability, and social and economic disorder for millions of people throughout the world. Poverty, poor access to health care, human migration, emerging disease agents, and antibiotic resistance all contribute to the expanding impact of infectious diseases. Prevention and treatment strategies for infectious diseases derive from a thorough understanding of the complex interactions between specific viral or bacterial pathogens and the human (or animal) host.

Just as glycans are major components of the outermost surface of all animal and plant cells, so too are oligosaccharides and polysaccharides found on the surface of all bacteria and viruses. Thus, most (if not all) interactions of microbial pathogens with their hosts are influenced to an important degree by the pattern of glycans and glycan-binding receptors that each expresses. This holds true at all stages of infection, from initial colonization of host epithelial surfaces, to tissue spread, to the induction of inflammation or host-cell injury that results in clinical symptoms. The microbial molecules responsible for disease manifestations are known as virulence factors.

In a complex environment with many microbial threats, higher organisms have evolved systems of immunity that can discriminate between potential pathogens and mount appropriate antimicrobial responses to block systemic spread and reduce damage to their cells and tissues. Glycan–receptor interactions play crucial roles in microbial pattern recognition as well as in the regulatory signals that govern the normal activities of immune cells. One important reason why certain microbes cause disease is that they have evolved to display their own sugars and receptors in a fashion that mimics or interferes with host glycan-based immune functions.

BACTERIAL SURFACE GLYCANS AS VIRULENCE FACTORS

Polysaccharide Capsules

A human in good health is colonized by as many as 10^{14} bacteria on their skin and mucosal surfaces (particularly in the gastrointestinal tract), a number that exceeds by an order of magnitude the number of cells in our own body. Despite all these direct encounters, only a small handful of bacterial species are known to spread into the body to produce serious infections. Although not restricted to pathogenic species, one feature that these disease-causing agents share in common is the presence of a polysaccharide capsule that covers the bacterial surface (see Chapter 20). Capsule expression by the bacteria poses a particular challenge to immune clearance.

Effective killing of bacteria by phagocytes such as neutrophils or macrophages requires opsonization, a process in which the bacterial surface is tagged with complement proteins or specific antibodies. Phagocytes express receptors for activated complement or antibody Fc domains, which allow host defense cells to bind, engulf, and kill the bacteria. Anionic sugars such as sialic acid present in bacterial surface capsules (e.g., those of the neonatal pathogens, group B *Streptococcus* [GBS] and *Escherichia coli* K1) can bind the host regulatory protein factor H, thereby attenuating the activity of the alternative complement pathway. Through their surface polysaccharide capsules, bacteria also cloak protein structures on their surfaces to which antibodies might be directed.

Generally, humans can generate good antibody responses against bacterial polysaccharide capsules, but this ability is diminished at extremes of age, so that infants and the elderly are particular prone to invasive infection with encapsulated pathogens. Certain bacteria avoid antibody defenses through molecular mimicry of common host glycan structures, masquerading as "self" to avoid immune recognition. An example is the leading pathogen, group A *Streptococcus* (GAS), which expresses a nonimmunogenic capsule of hyaluronan, identical to the nonsulfated glycosaminoglycan so abundant in host skin and cartilage (see Chapter 15). The contribution of capsule-based host mimicry to bacterial immune evasion is also well illustrated by the homopolymeric sialic acid capsules of *Neisseria meningitidis* (meningococcus), an important cause of sepsis and meningitis. The group C meningococcal capsule is composed of an α2-9-linked sialic acid polymer that is a unique bacterial structure. In contrast, the group B meningococcal capsule is composed of an α2-8-linked sialic acid polymer that resembles a motif present on neural cell adhesion molecules (NCAMs) found in human neural tissues (see Chapter 14). The group C capsule has proven to be a successful vaccine antigen in human populations, whereas the group B capsule is essentially nonimmunogenic.

Another challenge posed to host immunity by certain pathogens is the great diversity of capsular structures, which is reflected in the different compositions and linkages of repeating sugar units that are produced by different strains of the same bacterial species. Often, these structures are immunologically distinct, allowing classification of different capsule "serotype" strains; for example, there are five major capsule serotypes of meningococcus (A, B, C, Y, and W-135), six different capsule serotypes of the respiratory pathogen *Haemophilus influenzae* (A–F), nine capsule serotypes of GBS (Ia, Ib, and II–VIII), and more than 90 different serotypes of *Streptococcus pneumoniae* (pneumococcus), which is a leading cause of bacterial pneumonia, sepsis, and meningitis. Antibodies generated by the host against the capsule of one serotype strain typically do not provide cross-protective immunity. Thus, individuals can be repeatedly infected over their lifetime by different serotype strains of the same bacterial pathogen. Figuratively, the strategy of capsular molecular mimicry (e.g., GAS) provides the pathogen with invisibility to the immune system and the strategy of antigenic diversity of capsule types presents a moving target to the immune system. Genetic exchange of capsule biosynthetic genes among serotype strains of an individual species (e.g., the polysialyltransferase gene of meningococcus) can lead to capsule switching in vivo, which provides another means of pathogen escape from protective immunity.

The key role of the capsule in the virulence of multiple bacterial pathogens has been demonstrated through genetic mutagenesis of essential biosynthetic genes and infectious challenge in small animal models of disease. Compared to the wild-type parent bacterial strains, isogenic capsule-deficient mutants of GAS, GBS, pneumococcus, *H. influenzae*, meningococcus, *Salmonella typhi* (typhoid fever), *Bacillus anthracis* (anthrax), and several other important human pathogens are rapidly cleared from the bloodstream by opsonophagocytosis and they are unable to establish systemic infections. Perhaps the most historically significant bacterial virulence factor of all is the pneumococcal capsule, because in classical experiments by Frederick Griffith in 1928, the disease-causing capacity of virulent, encapsulated (smooth) strains was transferred to nonvirulent, nonencapsulated (rough) strains, thereby providing the framework for the subsequent discoveries of Oswald Avery, Colin MacLeod, and Maclyn McCarty that demonstrated DNA to be the genetic material (Figure 39.1).

Lipopolysaccharide

Lipopolysaccharide (LPS) is a major component of the outer membrane of Gram-negative bacteria (see Chapter 20). LPS contains a lipid A moiety, which is embedded in the outer membrane, and two carbohydrate components that extend outward: a core oligosaccharide containing sugars not found in vertebrates (such as ketodeoxyoctulonate [Kdo] and heptose) and a polysaccharide side chain known as the O-antigen that can vary from strain to strain within an individual species. The biosynthesis of LPS is described in Chapter 20. Many mucosal pathogens such as *H. influenzae* and *Neisseria gonorrhoeae* lack true O-antigens; instead, they produce lipooligosaccharides (LOSs) that contain a recognizable core structure from which one or more monosaccharide or short oligosaccharide branches extend.

LPS is a pathogen-associated molecular pattern (PAMP) that is recognized by the innate immune system and stimulates inflammatory responses to clear bacteria that have breeched the barrier defenses of the skin or mucosal epithelium. Soluble LPS released by invading bacteria, and particularly its lipid A component, interacts with the opsonic receptor CD14 and the membrane protein Toll-like receptor 4 (TLR4) to initiate the immune signaling process (Figure 39.2). TLR4 belongs to an evolutionarily conserved family of receptors (TLRs) that

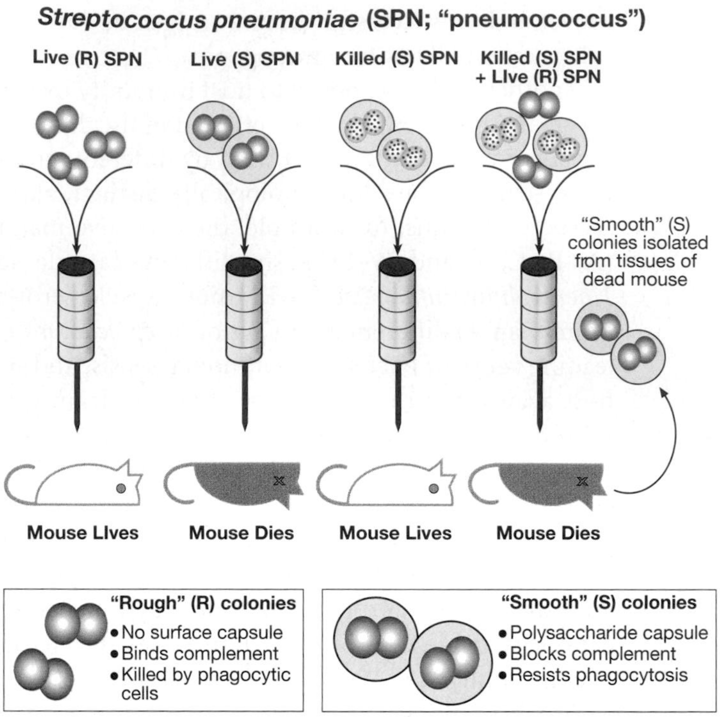

FIGURE 39.1. Classical experiments on the role of the pneumococcal polysaccharide capsule in virulence. *Streptococcus pneumoniae* (SPN) strains can be identified with either a "rough" (R) or a "smooth" (S) phenotype, the latter being due to expression of a thick polysaccharide capsule on their surface. In 1928, Frederick Griffith found that (R) SPN strains were avirulent for mice, whereas (S) strains were highly lethal. Heat-killed (S) strains did not produce disease, but when mixed with live (R) bacteria, the mouse died, and the recovered bacteria expressed the (S) phenotype. Thus, the live (R) strains had been "transformed" to (S) strains by a factor present in the heat-killed preparation of (S) SPN. The factor later proved to be DNA, providing the first evidence that DNA was the basis of the genetic code.

can distinguish closely related microbially derived ligands. For example, although Gram-positive bacteria lack LPS, the TLR2 receptor can recognize peptidoglycan or lipoteichoic acid derived from their cell walls. TLR intracellular signaling is regulated by a group of interleukin-1 (IL-1) receptor-associated kinases (IRAKs), which bind to the TLR intracellular TIR (Toll/interleukin-1 receptor) domain, a process that requires the presence of adapter proteins (e.g., MyD88). A signaling cascade ultimately leads to the activation of the transcription factor NF-κB and its translocation to the nucleus, where it positively regulates the promoters for genes encoding several proinflammatory cytokines (e.g., TNF-α and IL-1).

Although TLR4 detection of LPS is a critical element in triggering host innate immunity, a dangerous syndrome known as sepsis can develop in the setting of overwhelming infection, where patterns of immune activation become dysregulated and spin out of control. Symptoms include fever, low blood pressure, rapid heart rate, abnormal white blood cell counts, and dysfunction of multiple organ systems that may lead to lung or kidney failure and death. Original studies by Richard Pfeiffer in the 1890s revealed that a purified heat-stable component of Gram-negative bacteria was itself sufficient to trigger sepsis in experimental animals; this compound was later discovered to be LPS, which consequently is often referred to as bacterial "endotoxin." In the 1960s, it was discovered that a particular strain of mice (C3H/HeJ) was completely resistant to the sepsis-inducing properties of LPS. In the 1990s, a mutation in these mice was mapped to the *TLR4* gene, thereby identifying TLR4 as the functional receptor for LPS in mammals.

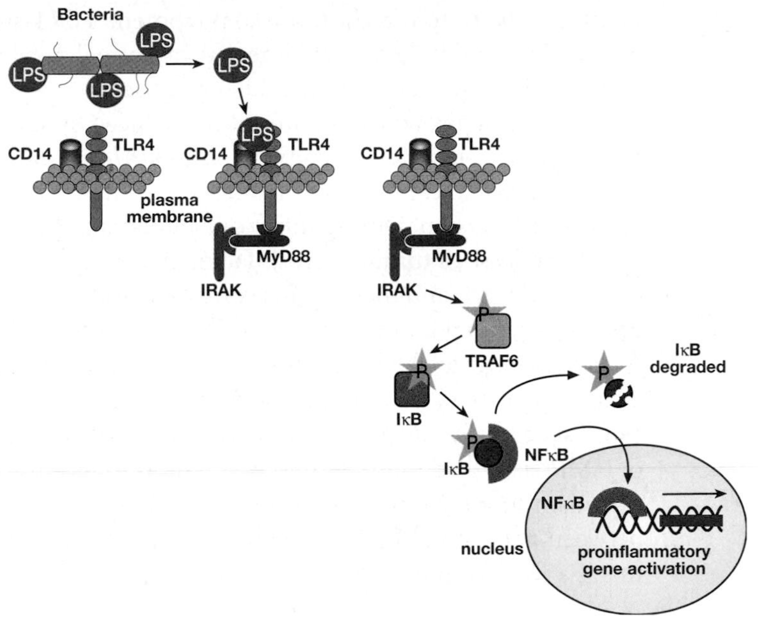

FIGURE 39.2. Activation of immune signaling by bacterial lipopolysaccharide (LPS). LPS from the cell wall of Gram-negative bacteria is bound by the pattern-recognition molecule Toll-like receptor 4 (TLR4) in conjunction with the cell-surface receptor CD14. The binding of LPS leads to recruitment of the adaptor proteins MyD88 and IRAK to the cytoplasmic domain of TLR4. This complex initiates a signaling cascade of phosphorylation events through TRAF 6 and the kinase IκK. Finally, IκK phosphorylates IκB, an inhibitor bound to the transcription factor NF-κB. Phosphorylated IκB is degraded, releasing NF-κB, which migrates to the nucleus where it activates the transcription of proinflammatory genes. Similar signal transduction pathways are activated by Gram-positive cell wall constituents such as peptidoglycan and lipoteichoic acid via TLR2 or TLR6.

Many Gram-negative bacteria vary or modify their LPS to interfere with host immune defense mechanisms. For example, by incorporating modifications that reduce the overall negative charge of LPS, bacteria can repel cationic host antimicrobial peptides (e.g., defensins) away from their cell wall target of action. *Salmonella* species produce gastroenteritis and systemic infections including typhoid fever in humans. The relative resistance of *Salmonella* to antimicrobial peptide killing has been studied extensively in the model organism *S. enterica* serovar Typhimurium. A prominent *S. enterica* LPS modification that confers resistance to antimicrobial peptides is the addition of 4-aminoarabinose to the phosphate group of the lipid A backbone, a genetically controlled phenotype that can be up-regulated 3000-fold during infection. During its adaptation to chronic colonization of the airway of cystic fibrosis patients, *Pseudomonas aeruginosa* has been shown to synthesize a unique hexa-acylated lipid A containing palmitate and 4-aminoarabinose that confers resistance to antimicrobial peptides and stimulates increased cytokine release, thereby perpetuating the chronic pneumonia and lung inflammation that characterize this inherited disorder.

In response to a shift in growth temperature, the plague bacillus *Yersinia pestis* changes the number and type of acyl groups on the lipid A of its LPS. At environmental temperatures (21°C), *Y. pestis* expresses predominantly hexa-acylated lipid A, whereas at the body temperature of the mammalian host (37°C), the pathogen expresses mostly tetra-acylated lipid A. The more complicated hexa-acylated version of the LPS strongly induces cytokine release from host cells, which suggests that the production of a less immunostimulatory form of LPS upon entry into the mammalian host might represent a virulence mechanism to avoid immune detection. Recently, this hypothesis was validated by genetically engineering *Y. pestis* to force it to make the potent TLR4-stimulating (hexa-acetylated) version of its LPS even at 37°C. When tested in animal models, the modified *Y. pestis* was completely avirulent even at high-challenge doses.

LOS plays a pivotal role in infection produced by several mucosal pathogens. For example, the sexually transmitted bacterium *N. gonorrhoeae* can express several antigenically distinct types of LOS and can switch from one type to another by an unknown mechanism as a means of immune evasion. During infection, autolysis of the gonococci releases LOS, which stimulates the release of tumor necrosis factor, proteases, and phospholipases from mucosal cells of the genital tract, and these agents contribute to the pathogenesis of pelvic inflammatory disease, a major cause of female infertility. Gonococcal LOS is also involved in the resistance of *N. gonorrhoeae* to the bactericidal activity of normal human serum. Specific LOS oligosaccharide types are known to be associated with serum-resistant phenotypes of *N. gonorrhoeae.* More specifically, the pathogen can directly and efficiently transfer sialic acid from the traces of CMP-sialic acid found in host body fluids to modify its LOS, converting a serum-sensitive organism to a serum-resistant organism. There are also considerable structural similarities between gonococcal LOS and glycosphingolipid antigens present on human red blood cells; this is a form of molecular mimicry that may hamper immune detection.

MECHANISMS OF COLONIZATION AND INVASION

Adhesins and Receptors

Adherence to skin or mucosal surfaces is a fundamental characteristic of the normal human microflora and also an essential first step in the pathogenesis of many important infectious diseases (see Chapter 34). Most microorganisms express more than one type of adherence factor or "adhesin." A large fraction of microbial adhesins are lectins that bind directly to cell surface glycoproteins, glycosphingolipids, or glycosaminoglycans; adhesion may be mediated through terminal sugars or internal carbohydrate motifs. In other cases, the bacteria express adhesins that bind matrix glycoproteins (e.g., fibronectin, collagen, or laminin) or mucin, providing a form of attachment to the mucosal surface. The specific carbohydrate ligands for bacterial attachment on the animal cell are often referred to as adhesin receptors and they are quite diverse in nature. The tropism of individual bacteria for particular host tissues (e.g., skin vs. respiratory tract vs. gastrointestinal tract) is effectively determined by the array of available adhesin-receptor pairs (see Table 34.2).

In a number of cases, the key adhesive factor is an assembly of protein subunits that project from the bacterial surface in hair-like threads known as pili or fimbriae (Figure 39.3a). Such pili are usually composed of a repeating structural subunit providing extension and a different "tip adhesin" that actually mediates the host-cell interaction. Often the structural genes and enzymes for pilus assembly are encoded in a bacterial operon. Lateral mobility of pili structures in the bacterial membrane provides a Velcro-like binding effect to epithelial surfaces. Certain strains of *E. coli* express pili that bind avidly to P-blood group-related glycosphingolipids in the bladder epithelium, leading to urinary tract infection. Pathogenic strains of *Salmonella* produce pili that facilitate adherence to human intestinal cell mucosa, thereby causing food poisoning and infectious diarrhea. In other cases, a surface-anchored protein (afimbrial adhesin) expressed by the bacteria represents a critical colonization factor (Figure 39.3b). For example, the filamentous hemagglutinin (FHA) of *Bordetella pertussis* promotes strong attachment of the bacteria to the ciliated epithelial cells of the bronchi and trachea, triggering local inflammation and tissue injury that results in the syndrome of "whooping cough." FHA is a component of modern pertussis vaccines given in infancy and early childhood to block infection.

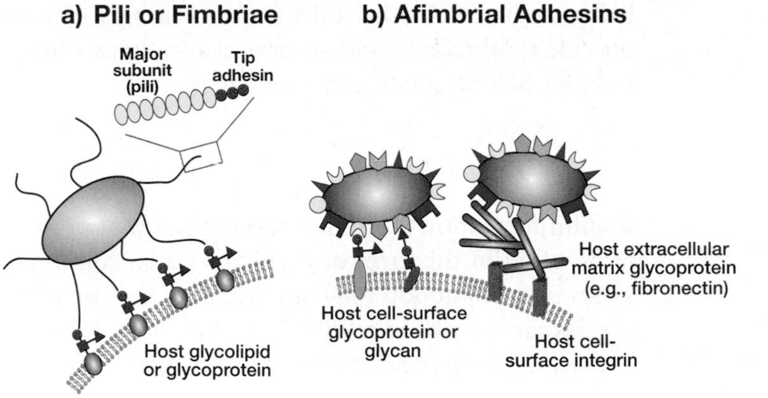

FIGURE 39.3. Examples of mechanisms of bacterial adherence to host-cell surfaces. (*a*) Pili or fimbriae are organelles that project from the cell surface. They are made up of a repeating structural subunit and a protein at their tip that mediates recognition of a specific host-cell glycan motif. (*b*) Afimbrial adhesins are integral bacterial cell wall proteins or glycoproteins that directly engage host-cell receptors to promote colonization.

For certain pathogens, epithelial attachment is a two-step process, in which a microbial glycosidase acts upon a target cell polysaccharide to modify its structure into a novel glycan, which then serves as the adhesin receptor. For example, a secreted *P. aeruginosa* neuraminidase produces increased numbers of asialoglycolipid receptors, which may promote colonization of the cystic fibrosis airway. Likewise, the neuraminidase of *S. pneumoniae* removes sialic acid from respiratory epithelial cells to expose underlying *N*-acetylglucosamine and galactose residues to which the bacterium binds with higher affinity.

Invasion Factors

Glycan–lectin interactions play pivotal roles in enabling certain pathogens to penetrate or invade through epithelial barriers, whereupon they may disseminate through the bloodstream to produce deep-seated infections. *S. enterica* serovar Typhi causes typhoid fever in humans, a process that begins with intracellular invasion of intestinal epithelial cells. The outer core oligosaccharide structure of the serovar Typhi LPS is required for internalization in epithelial cells. Removal of a key terminal sugar residue on the outer core markedly reduces the efficiency of bacterial uptake. *Streptococcus pyogenes*, the common cause of strep throat but also an agent of serious invasive infections, attaches to human pharyngeal and skin epithelial cells through specific recognition of its hyaluronan capsular polysaccharide by the hyaluronan-binding protein CD44 (see Chapter 15). This binding process induces marked cytoskeletal rearrangements manifested by membrane ruffling and opening of intercellular junctions that allow tissue penetration by GAS through a paracellular route.

The human malaria parasite *Plasmodium vivax* is completely dependent on interaction with the Duffy blood group antigen for invasion of human erythrocytes. The Duffy blood group antigen is a 38-kD glycoprotein with seven putative transmembrane segments and 66 extracellular amino acids at the amino terminus. The binding site for *P. vivax* has been mapped to a 35-amino-acid segment of the extracellular region at the amino terminus of the Duffy antigen. Unlike *P. vivax*, *P. falciparum* does not use the Duffy antigen as a receptor for invasion. A 175-kD *P. falciparum* sialic-acid-binding protein (also known as EBA-175 [erythrocyte-binding antigen-175]) binds sialic acid residues on glycophorin A during invasion of the erythrocyte. Some *P. falciparum* laboratory strains use sialic acid residues on alternative sialoglycoproteins such as glycophorin B as invasion receptors, with binding being mediated by other EBA family members. The use of multiple invasion pathways may provide *P. falciparum* with a survival advantage when faced with host immune responses or receptor heterogeneity in host populations.

Biofilms

Biofilm formation is another mechanism that promotes bacterial attachment to host surfaces, often in the form of a polymicrobial community. For example, oral biofilms comprise, in total, about 1000 species, only half of which are culturable and the remaining species can only be identified by nucleic acid detection methods. *Streptococcus* species predominate (60–90%), but *Eikenella*, *Haemophilus*, *Prevotella*, and *Priopionibacterium* species can also be found. Dental plaque represents an oral biofilm in which dense, mushroom-like clumps of bacteria pop up from the surface of the tooth enamel, interspersed with bacteria-free channels filled with extracellular polysaccharide (EPS) produced by the bacteria that can serve as diffusion channels (Figure 39.4). Bacteria within biofilms communicate with one another through soluble signaling molecules in a process known as "quorum sensing" to optimize gene expression for survival. In biofilms, bacteria live under nutrient limitation and in a dormant state in which defense molecules (e.g., antimicrobial peptides) produced by the immune system and pharmacologic antibiotics are less effective. Moreover, the EPS matrix can bind and inactivate these same agents, contributing to the persistence of the biofilm and difficulty in medical treatment of biofilm infections, such as those that arise on catheters and other medical devices.

The EPS synthesized by biofilm bacteria vary greatly in their composition and in their chemical and physical properties. The majority of EPS types are polyanionic because of the presence of either uronic acids (D-glucuronic, D-galacturonic, or D-mannuronic acids) or ketal-linked pyruvate. Inorganic residues, such as phosphate or sulfate, also contribute to the negative charge. In rare cases, EPS is polycationic, as exemplified by the adhesive polymer obtained from *Staphylococcus epidermidis* strains that produce biofilms on catheters. Ordered secondary configuration frequently takes the form of aggregated helices. In some of these polymers, the backbone composition of sequences of β1-4 or β1-3 linkages confers rigidity (as is seen in the cellulosic backbone of xanthan from *Xanthomonas*), whereas the β1-2 or β1-6 linkages found in many dextrans provide more flexible structures. It is thought that the EPS itself can serve as the primary carbon reserve for biofilm microorganisms during substrate deprivation.

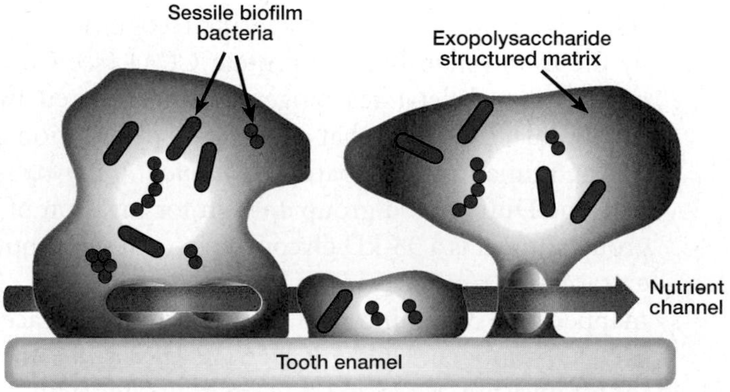

FIGURE 39.4. Structure of a polymicrobial biofilm. Dental plaque is an example of a polymicrobial biofilm in which *Streptococcus* species and other bacteria secrete a thick exopolysaccharide matrix and exist within this matrix in a dormant or sessile state of low metabolic activity. Biofilm bacteria have increased resistance to host immune clearance and antibiotic medicines.

Exploiting Heparan Sulfate Proteoglycans

A number of bacteria utilize heparan sulfate proteoglycans such as syndecans (see Chapter 16) as attachment factors for host epithelia, including the genitourinary pathogens *Chlamydia trachomatis* and *N. gonorrhoeae.* Another mechanism by which pathogens exploit host glycan structures is illustrated by the common bacterial pathogens *Staphylococcus aureus* and *Pseudomonas aeruginosa*, each of which is capable of producing infections in a number of body sites, especially in hospitalized patients. Neither organism binds to syndecans; rather, they release factors that promote the shedding of the syndecan ectodomain from the surface of the host epithelial cell. The negatively charged heparan sulfate side chains of the shed ectodomain bind tightly to cationic molecules such as antimicrobial defensin peptides and lysozyme, neutralizing their antibacterial activities. Thus, these organisms have evolved ways to co-opt the host-cell shedding machinery to release surface proteoglycans, neutralize innate host defenses, and promote their own survival and pathogenicity.

VIRAL INFECTION

The specific binding of a virus particle to a target receptor on the host-cell surface is a prerequisite for viral entry and the subsequent intracellular replication steps in the viral life cycle. As most cell-surface constituents are glycoconjugates, it is not surprising that most viral receptors are mapped to the glycan components of cell-surface glycoproteins, glycolipids, or proteoglycans. These virus–glycan interactions are responsible for species and tissue tropism (see Table 34.1), as is illustrated here with examples provided by the important human infectious agents influenza virus, herpes simplex virus, and human immunodeficiency virus.

Influenza Virus

In humans, influenza viruses are common pathogens of the upper respiratory tract, and seasonal epidemics affect 10–20% of the general population. However, the virus can also be deadly; the influenza pandemic of 1918–1919 killed up to 40 million people worldwide. Influenza virus subtypes are designated by a nomenclature that is based on their surface glycoproteins—namely, hemagglutinin (H) and neuraminidase (N). The first human influenza viruses to be isolated in the 1930s were designated H1N1 based on serological reactions, and this subtype included the pandemic 1918 strain. In 1958, an antigenic shift resulted in the emergence of human H2N2 viruses, and in 1968, a shift to H3N2 viruses occurred; the latter strains have remained the most prevalent in recent years. Emergence of new influenza strains in the human population occurs via transmission from other animal species, especially poultry. Typically, human and avian influenza viruses are different and are not infectious for both species. However, pigs can become infected with both types of viruses, and act as a "mixing vessel" to produce recombinant viruses capable of transmission to humans. Occasionally, direct avian–human transmission can occur, often with enhanced pathogenicity, as demonstrated by the emergence of the recent H5N1 avian influenza in many countries throughout Southeast Asia; human-to-human spread of H5N1 avian influenza has not been conclusively documented.

Interactions between influenza virus hemagglutinin and sialic acid determine the tissue and species tropism of the virus. Human influenza viruses bind to host cell-surface targets that contain *N*-acetylneuraminic acid; however, in the pig, *N*-glycolylneuraminic acids can be used. Binding to sialic acids occurs via a shallow depression near the tip of the hemagglutinin glycoprotein. Viruses that infect humans bind preferentially to terminal sialic acids containing α2-6 linkages, whereas others (such as the bird influenza viruses) favor binding to α2-3-linked sialic acid, a receptor-binding specificity that correlates with a specific amino acid at position 226 of hemagglutinin. Hemagglutinins that have leucine at position 226 selectively bind to α2-6 sialic acid and occur preferentially in human strains; hemagglutinins that have glutamine at position 226 are specific for α2-3 linkages and occur mostly in avian strains of the virus. Pigs can become infected with both avian and human strains because both α2-3- and α2-6-linked sialic acids occur in the trachea of swine. However, a direct binding switch from avian to human specificity can apparently also occur, as happened with the 1918 influenza virus (see cover figure of this book).

The cell-surface receptor(s) for influenza viruses is widely considered to be sialic acid linked to either glycoprotein or glycosphingolipid. However, recent data suggest that the process by which influenza virus enters cells may have further levels of complexity that also depend on protein–glycan interactions. For example, in influenza infection of macrophages, the viruses undergo an additional lectin-like interaction with host mannose receptors after the initial sialic acid binding. In addition, recent experiments in cell lines deficient in terminal N-linked glycosylation (due to a mutation in the N-acetylglucosaminyltransferase I gene) showed deficient cell entry, even though the initial sialic-acid-dependent interactions occurred normally. These results suggest that influenza virus requires one or more N-linked glycoproteins as a cofactor for cell entry after the sialic acid efficiently promotes the initial attachment. In this sense, influenza would share similarities with adenoviruses, another common cause of upper respiratory tract infections, in which host cell-surface integrins function as internalization coreceptors following initial sialic-acid-dependent binding.

Upon completing its cellular replication cycle, the final release of influenza virus from an infected cell surface relies on the action of the viral neuraminidase, which acts to remove sialic acid (the viral receptor) from the surface of the host cells. Without this step, the newly forming virus particles would immediately rebind to their receptor and not be efficiently released into the extracellular space, remaining attached to the cell in large clumps. Thus, the establishment of a productive influenza virus infection is dependent on both neuraminidase and hemagglutinin and a delicate balance between the functions of the two glycoproteins (Figure 39.5a). It is also possible that the neuraminidase assists in evading host mucosal soluble mucin decoys during the process of infection.

Herpes Simplex Virus

Herpes simplex viruses-1 and -2 (HSV-1 and HSV-2) are human pathogens capable of infection and spread in a number of human cell types, with establishment of latent, recurrent infections. The most common infections include recurrent cold sores of the mouth and lips (usually HSV-1) or sexually transmitted genital ulcer disease (usually HSV-2). HSV infection is mediated by a family of ten viral envelope glycoproteins, which have a variety of specific, and sometimes redundant, functional roles in the establishment of cell infection and spread throughout the host. HSV infection first requires virus attachment to the cell-surface membrane before penetration and entry of the nucleocapsid into the cytoplasm can occur. HSV glycoproteins B (gB) and C (gC) have been shown to be involved in the initial attachment phase through the interaction of positively charged residues with negatively charged heparan sulfate (HS) of cell-surface proteoglycans. This HS-dependent attachment facilitates a second high-affinity attachment in which glycoprotein D (gD) binds to a member of the tumor necrosis factor–nerve growth factor (TNF/NGF) receptor family. Following attachment, the virus penetrates the cell by fusion of the virus envelope with the cell plasma membrane (Figure 39.5b).

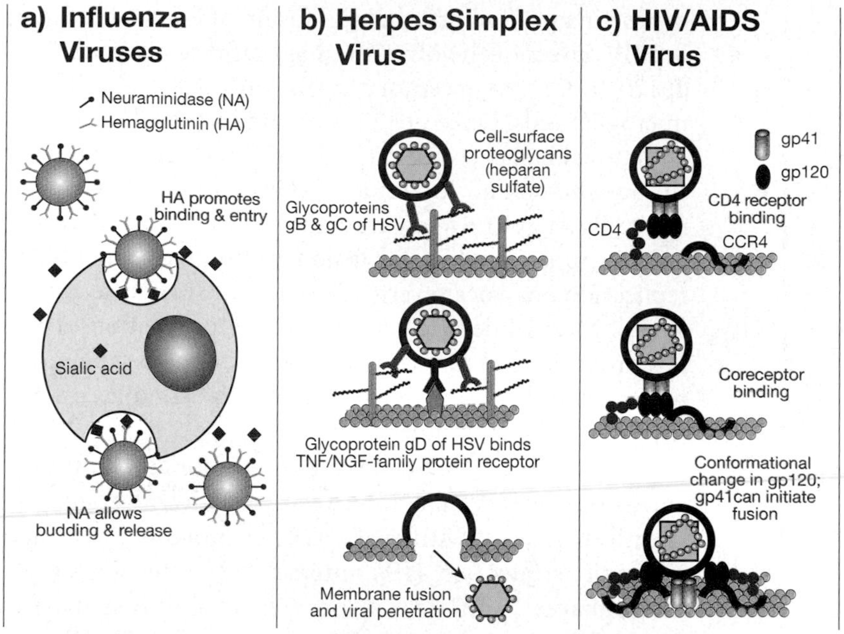

FIGURE 39.5. Mechanisms of viral entry into host cells. (*a*) Influenza virus initiates host cell contact and entry by binding to cell-surface sialic acid receptors through its surface glycoprotein hemagglutinin. After intracellular replication, a cell-surface neuraminidase cleaves sialic acid from the cell membrane allowing viral escape. (*b*) Herpes simplex virus (HSV) engages host cells first through a low-affinity engagement of heparan sulfate proteoglycans via its surface glycoproteins gB and gC. Subsequently, a higher-affinity binding of viral protein gD to a member of the tumor necrosis factor–nerve growth factor (TNF/NGF) receptor family promotes membrane fusion. (*c*) Human immunodeficiency virus (HIV) surface glycoprotein gp120 binds sequentially to the CD4 receptor on T cells and then to a coreceptor such as chemokine receptor CCR4. The latter interaction triggers a conformational change in gp120, which exposes gp41, the HIV factor capable of initiating membrane fusion.

Several lines of evidence identified HS as the critical initial receptor for HSV infection. HS proteoglycans are commonly found on the surface of most vertebrate cell types susceptible to HSV infection, and removal of HS from the cell surface by enzymatic treatment or by selection of mutant cell lines defective in HS expression renders the cells resistant to HSV infection by reducing virus attachment to the cell surface. Heparin, a more heavily modified form of HS, can inhibit viral infection by masking the HS-binding domain on the virus envelope. Immobilized HS columns bind to the principal mediators of virus attachment (gB and gC), and HSV deletion mutants lacking gB and gC glycoproteins exhibit impaired virus binding to the cell surface. The gC glycoprotein also binds to the C3b component of human complement, which aids in immune evasion by protecting against antibody-independent neutralization.

Human Immunodeficiency Virus

Human immunodeficiency virus (HIV) is a retrovirus and the etiologic agent of the acquired immunodeficiency syndrome (AIDS), a pandemic disease now affecting nearly 40 million people worldwide, especially in sub-Saharan Africa. The genome of HIV contains only three major genes (*env*, *gag*, and *pol*) that direct the formation of the basic components of the virus. Glycoprotein products of the *env* gene include an envelope precursor protein gp160 (which undergoes proteolytic cleavage to form the outer envelope glycoprotein gp120 that is responsible for the cellular tropism of the virus) and transmembrane glycoprotein gp41 (which catalyzes fusion of HIV to the target cell's membrane). The first step in HIV infection involves the high-affinity attachment of the CD4-binding domains of gp120 to CD4, a receptor present on certain T cells, macrophages, dendritic cells, and microglial cells. Once gp120 is bound to the CD4 protein, the HIV envelope undergoes a structural change, exposing additional binding domains of gp120 that interact with a cell-surface chemokine coreceptor (CCR5 or CXCR4). This more stable two-pronged attachment allows gp41 to penetrate the cell membrane, bringing the virus and cell membranes in close approximation for fusion and subsequent entry of the viral capsid containing the replication enzymes reverse transcriptase, integrase, and ribonuclease into the cell (Figure 39.5c). Meanwhile, the very heavy N-glycosylation of gp120/41 poses a challenge to the host immune system, as potentially immunogenic peptides are blocked from processing for presentation by major histocompatibility complex (MHC) class II and/or blocked from recognition by antibodies.

The differences in chemokine coreceptors present on cells can also explain how different strains of HIV may infect cells selectively. Strains of HIV known as T-tropic strains selectively interact with the CXCR4 chemokine coreceptor to infect lymphocytes, whereas M-tropic strains of HIV interact with the CCR5 chemokine coreceptor to infect macrophages. Natural resistance to a viral disease may occur in a human subpopulation due to the presence of a CCR5 mutation (CCR5D32) in certain individuals. CCR5 is also posttranslationally modified by O-linked glycans and both CCR5 and CXCR4 contain sulfated amino-terminal tyrosines. Sulfated tyrosine residues contribute to the binding of CCR5 to gp120/CD4 complexes and to the ability of HIV-1 to enter cells.

Syndecans also have a role in HIV pathogenesis. Primary isolates of HIV from $CD4^+$ T cells can bind to all members of the syndecan family via their HS side chains. By binding to syndecans, the fragile HIV-1 virion avoids rapid degradation in the bloodstream. HIV can exploit syndecan binding as an attachment receptor for macrophage entry; in addition, HIV captured by syndecans on nonpermissive cells for replication (e.g., epithelial or endothelial cells) can be transmitted in *trans* to $CD4^+$ T cells. In effect, the syndecan-rich endothelial surface provides a microenvironment that amplifies the kinetics of HIV-1 replication in T cells.

SOME BACTERIAL TOXINS ARE GLYCOSYLTRANSFERASES

The disease symptoms associated with many important bacterial infections can largely be attributed to the action of secreted exotoxins that initiate their action by binding to glycosphingolipid receptors on host cells to alter the function of host cells. These include cholera toxin, botulinum toxin, and tetanus toxin as reviewed in Chapter 34. The family of large clostridial cytotoxins exemplifies another glycan-dependent pathogenic effect of bacterial toxins. This family includes toxin A and toxin B from *Clostridium difficile* (agents that cause diarrhea and colitis, especially in patients whose normal flora has been altered by previous antibiotic therapy) and the hemorrhagic and lethal toxins of *C. sordellii* (agents that cause gas gangrene in wound infections that receive soil contamination). These bacterial products turn out to be glucosyltransferases, and this enzymatic activity is critical to their mechanism of toxic action.

In *C. difficile*, the genes encoding toxin A and toxin B are encoded on a pathogenicity locus, along with negative and positive regulators of their expression. Following release from the bacterium, the two toxins translocate to the cytoplasm of target cells and inactivate a series of small GTP-binding proteins, including Rho, Rac, and Cdc42. Inactivation of these substrates occurs through monoglucosylation of a single reactive threonine with UDP-glucose as a cosubstrate. This threonine lies within the effector-binding loop of the target protein and is responsible for coordinating a divalent cation that is critical to binding GTP. By glucosylating small GTPases, toxin A and toxin B cause actin condensation and cell rounding, activation of the transcription of a number of inflammatory genes such as cytokines, and ultimately the death of the cell through apoptosis. In the intestine, these changes lead to neutrophil infiltration, disruption of tight junctions, fluid loss, and diarrhea.

GLYCAN-BASED INTERACTIONS OF HOST AND GUT MICROFLORA: COMMENSALS AND PATHOGENS

The nature of the relationship between microbes and the human host spans the spectrum from mutually beneficial (symbiotic), to benefiting the microbe without harming the host (commensal), to benefiting the microbe at the expense of the host (pathogenic). One of the key factors determining the placement of any particular microbe–host interaction along this continuum of potential outcomes is the repertoire of glycans expressed on host-cell surfaces.

Bacteroides thetaiotaomicron is an anaerobic bacterium that is one of the most abundant members of the normal colonic microflora in mice and humans. This interesting microbe has evolved a mechanism to establish and maintain a nonpathogenic, mutually beneficial relationship with its mammalian host. A clue to this relationship came from examination of the gut epithelium of mice that are raised germ-free and then infected with specific bacteria. Without bacterial exposure, the intestinal epithelium lacks expression of fucosylated glycoconjugates; when normal colonic bacteria are present, Fucα1-2Gal glycan expression is abundant on the surface of these host cells. *B. thetaiotaomicron* preferentially utilizes fucose both as an energy source and for incorporation into its own surface capsule and glycoproteins, phenotypes that are required for successful colonization. When dietary fucose is low, the bacterium sends a signal to the intestinal cells that increases their expression of mRNA for an α1-2-fucosyltransferase, resulting in incorporation of fucose into surface Fucα1-2Gal glycoconjugates on the epithelial lining. *B. thetaiotaomicron* also expresses multiple fucosidases to cleave these terminal fucose residues and a fucose permease for uptake of the released sugar. Thus, the gut commensal has a system for engineering the production of its own nutrient source from its host; but because the system is regulated for use only in times of need, the host in turn only has to synthesize enough fucosylated glycans to support the maintenance of this important member of its normal microflora. Moreover, the well-adapted *B. thetaiotaomicron* has evolved additional elaborate systems for regulating its expression of particular polysaccharide-binding proteins and glycosidases to forage and consume hexose sugars from the host's dietary intake when abundant, or to switch over to glycans in the host mucus lining when sufficient polysaccharides are missing from the diet.

Helicobacter pylori infects nearly half the world's population, but it triggers chronic gastritis and stomach ulcers (conditions that are known to increase the risk of stomach cancer) in only a small subset of these individuals. Patterns of glycan expression in both host and microbe appear to help determine whether *H. pylori* persists as a benign commensal or triggers disease pathology. *H. pylori* expresses an adhesin (BabA) that can interact directly with gastric epithelium expressing glycans that terminate with the Lewis^b blood group antigen. Expression shows that no binding of *H. pylori* occurs to similar glycans modified to express the Lewis^a antigen. Lewis^b expression in human intestine is limited to mucus-producing pit cells in the gastric epithelium. Transgenic mice engineered to express Lewis^b show enhanced binding of *H. pylori* to their gastric epithelium, which triggers an enhanced cellular immune response and more severe gastritis. This microenvironment of immune activation appears to set the stage for a glycan-based process of molecular mimicry that can promote further host-cell damage. *H. pylori* also expresses Lewis^x-containing structures in the O-antigen of its own LPS, which resemble Lewis^{x}-modified glycans on the surface of parietal cells in the gastric lining. Intimate *H. pylori* binding to the Lewis^{b}-positive pit cells is associated with induction of antibodies against the bacterial LPS that can cross-react with the Lewis^{x}-positive parietal cells, leading to cell death and a condition of chronic gastritis. Parietal cell depletion results in the induction of gastric epithelial progenitor cells that express high levels of terminal NeuAcα2-3Gal glycans, which represents another receptor for the *H. pylori* SabA adhesin, further promoting a persistent infection and altering the balance of the gastric ecosystem. Variation in both organism and host in the expression of Lewis^{x} glycan structures and/or the adhesins may help explain the wide range of potential clinical outcomes following colonization by *H. pylori.*

THERAPEUTIC STRATEGIES BASED ON MICROBIAL PROTEIN–GLYCAN INTERACTIONS

A detailed understanding of key steps in a microorganism's infectious process aids in the identification of molecular targets for rational drug design. An excellent example exploited the crucial role of neuraminidase in the life cycle of influenza virus. Following determination of the crystal structure of the neuraminidase enzyme, a concerted effort was made to identify small-molecule analogs of sialic acid that would bind to and block the highly conserved sialic acid–binding pocket of the enzyme (see Chapter 50). Two such drugs, zanamivir and oseltamivir, were approved for therapy of influenza and are effective at preventing virus release and cell-to-cell spread, reducing the magnitude and durations of symptoms, as well as the incidence of infections in contacts.

Most of the HIV envelope surface is covered by high-mannose-type N-glycans attached to the viral envelope protein, gp120. These glycans are located in a region of the protein known to be antigenically silent and genetically stable. A link exists between glycosylation and viral tropism with respect to CCR5 and CXCR4. Furthermore, the glycans have immune modulating properties, for example, by interacting with CD22 on B cells (which can modulate the immune response) or by preventing natural killer cell function. Thus, drugs that alter N-glycan biosynthesis have potential therapeutic value. *N*-butyl-deoxynojirimycin (NB-DNJ), an inhibitor of glucosidases involved in processing, blocks maturation of the chains and causes structural alterations in gp120 that reduce infectivity (see Chapter 50). NB-DNJ is moderately well tolerated in humans, suggesting the possibility of using it or related N-glycan inhibitors for treating HIV infection.

Cyanovirin-N is a protein derived from blue-green algae that was discovered to bind to the unusual oligomannose-type N-glycans present on gp120. Binding prevents virions from attaching to mucosal cell surfaces. Cyanovirin-N is also active against herpesviruses. Other glycan-binding proteins, such as certain plant lectins and pradimicin A (an antifungal, nonpeptidic, low molecular weight glycan-binding agent), also have antiviral activity in vitro. One advantage of agents that target the N-glycans of viral glycoproteins is that development of drug resistance would require multiple mutations at N-glycosylation sites. However, deletion of N-linked chains in resistant virions would render them more susceptible to the host immune system.

The initial critical interaction of viral pathogens with glycosaminoglycans has been investigated as a target for therapeutic interference. A topical preparation of an acidic molecule that can "trap" the virus before it accesses the host cell can represent an effective microbicide. Cell binding of both HIV particles and papillomavirus capsid proteins are inhibited by preparations of heparin and other polysulfated molecules. Carrageenan, a polysulfated glycosaminoglycan-like extract from red seaweed, blocks attachment of HSV and protects against infection in an animal model. An anionic peptide that binds capsid proteins of the respiratory syncytial virus (RSV; the leading cause of pneumonitis in infants) inhibited viral replication in vivo. Finally, a new therapeutic concept that is being actively explored is the use of lectins and other carbohydrate-binding agents to bind to critical glycoproteins in the envelope of pathogens required for transmission or cell entry; several such molecules that bind the mannose-rich gp120 have been shown to block HIV infection in cultured cells, and analogous targets can be contemplated in other important disease agents such as hepatitis C virus, *H. pylori*, and *Mycobacterium tuberculosis.*

Anti-adhesion therapies based on glycans have also been suggested. For example, urinary tract infections by uropathogenic *E. coli* are mediated by adhesion of FimH on type I pili to mannose-rich glycans on the urinary mucosa, and they can be blocked by administration of α-methylmannoside or free mannose. Although soluble glycans recognized by bacterial lectins can block adhesion of bacteria to animal cells in vitro and have been shown to protect mice, rabbits, calves, and monkeys against experimental infection, they have not yet been proven to be effective in humans. In theory, polyvalent glycans, such as neoglycoproteins or dendrimers, should be more potent, and such molecules have been synthesized. Soluble glycans can also be used to block the action of exotoxins (see Chapter 34). Similarly, analogs of lipid A are under development as endotoxin antagonists.

FURTHER READING

Hooper L.V. and Gordon J.I. 2001. Glycans as legislators of host-microbial interactions: Spanning the spectrum from symbiosis to pathogenicity. *Glycobiology* **11:** 1R–10R.

Spear P.G. 2004. Herpes simplex virus: Receptors and ligands for cell entry. *Cell Microbiol.* **6:** 401–410.

Backhed F., Ley R.E., Sonnenburg J.L., Peterson D.A., and Gordon J.I. 2005. Host-bacterial mutualism in the human intestine. *Science* **307:** 1915–1920.

Olofsson S. and Bergstrom T. 2005. Glycoconjugate glycans as viral receptors. *Ann. Med.* **37:** 154–172.

Voth D.E. and Ballard J.D. 2005. *Clostridium difficile* toxins: Mechanism of action and role in disease. *Clin. Microbiol. Rev.* **18:** 247–263.

Comstock L.E. and Kasper J. 2006. Bacterial glycans: Key mediators of diverse host responses. *Cell* **126:** 847–856.

Ji X., Chen Y., Faro J., Gewurz H., Bremer J., and Spear G.T. 2006. Chen interaction of human immunodeficiency virus (HIV) glycans with lectins of the human immune system. *Curr. Protein Pept. Sci.* **7:** 317–324.

Mazmanian S.K. and Kasper D. 2006. The love–hate relationship between bacterial polysaccharides and the host immune system. *Nat. Rev. Immunol.* **6:** 849–858.

Munford R.S. and Varley A.W. 2006. Shield as signal: Lipopolysaccharides and the evolution of immunity to Gram-negative bacteria. *PLoS Pathog.* **2:** e67.

Balzarini J. 2007. Targeting the glycans of glycoproteins: A novel paradigm for antiviral therapy. *Nat. Rev. Microbiol.* **5:** 583–597.

Balzarini J. 2007. Carbohydrate-binding agents: A potential future cornerstone for the chemotherapy of enveloped viruses? *Antivir. Chem. Chemother.* **18:** 1–11.

Scanlan C.N., Offer J., Zitzmann N., and Dwek R.A. 2007. Exploiting the defensive sugars of HIV-1 for drug and vaccine design. *Nature* **446:** 1038–1045.

CHAPTER 40

Parasitic Infections

Richard Cummings and Salvatore Turco

MANY PARASITIC PROTOZOANS AND HELMINTHS SYNTHESIZE unusual glycan structures and glycan-binding proteins (GBPs) that are often antigenic and involved in host invasion and parasitism. This chapter discusses these parasites and the roles of glycoconjugates in the disease process. Protozoan parasites have evolved unique lifestyles: They shuttle between insect vectors and vertebrate hosts, encountering extremely harsh environments specifically designed to keep microbial invaders at bay. Their survival strategies frequently involve the participation of glycoconjugates that form a protective barrier against hostile forces. The diversity of the glycoconjugate structures and the range of functions that have been ascribed to parasite glycoconjugates, from host-cell invasion to deception of the host's immune system, are remarkable.

BACKGROUND ON PARASITIC INFECTIONS

Parasitism may be defined as a condition in which one organism (the parasite) either harms its host or in some way lives at the expense of the host. Parasites affect millions of people worldwide and cause tremendous suffering and death, especially in less-developed countries. Some sobering statistics are listed in Table 40.1. Worldwide, several million people die each year from parasitic diseases, with the bulk of those deaths coming from malaria and parasitic protozoans. Research into parasite glycobiology and biochemistry is important because of these worldwide human health problems. In addition, we may acquire important new insights into molecular pathology by studying such organisms, which have evolved to deceive and compromise the immune systems of infected animals with great success. However, parasite glycobiology can be frustrating because of the diffi-

TABLE 40.1. *Worldwide distribution of some major parasitic diseases*

Type of disease	Estimated human infections	Estimated deaths per year (World Health Report 2000)
Parasitic helminths		
Roundworm (*Ascaris*)	531 million	3,000
Schistosomiasis	200 million	14,000
Hookworms (*Necator/Ancylostoma*)	194 million	7,000
Whipworm (*Trichuris*)	212 million	2,000
Filarial worms	657 million	20,000–50,000
Pinworms or threadworms (*Enterobius vermicularis*)	200 million	rare
Parasitic protozoans		
Malaria	300–500 million	1–3 million
Leishmaniasis	12 million	57,000
Onchoceriasis (river blindness)	18 million	270,000 (blinded)
African trypanosomiasis (sleeping sickness)	450,000	66,000
South American trypanosomiasis (Chagas' disease)	18 million	50,000

The figures show the total number of infections worldwide (some individuals may have more than one infection) and the number of infection-related deaths per year.

culty in obtaining sufficient amounts of material for study and the difficulty of doing in vitro experimentation. In addition, many parasites have specific primary and intermediate hosts, thus making it difficult to study all stages of the life cycle.

Some of the major parasitic diseases infecting humans and animals are listed here in two categories: those caused by protozoans (single-celled organisms) (Table 40.2) and those caused by helminths (worms/metazoans) (Table 40.3). The major protozoan parasites include *Plasmodium* species (causing malaria), *Entamoeba histolytica* (causing amebiasis),

TABLE 40.2. *Some of the major parasitic protozoans of humans*

Parasite	Comments
Amoeba infecting humans	
Entamoeba histolytica	causes amebic dysentery; can cause liver abscesses
Intestinal and genital flagellates	
Giardia lamblia	causes diarrhea; one of the most common parasites in North America
Trichomonas vaginalis	causes inflammation of reproductive organs; very common
Hemoflagellates	
Leishmania donovani	causes visceral leishmaniasis (kala-azar); hepatosplenomegaly
L. mexicana	causes fulminating, cutaneous ulcers
L. major	causes cutaneous ulcers
Trypanosoma brucei sp.	causes sleeping sickness in humans and nagana in cattle (African trypanosomiasis)
T. cruzi	causes Chagas' disease (South American trypanosomiasis)
Gregarines, coccidia, and related organisms	
Plasmodium falciparum	major cause of human malaria
P. vivax, P. ovale, P. malariae	also cause human malaria
Causes of opportunistic infections in immunodeficiency states	
Toxoplasma gondii	causes muscle pain
Pneumocystis carinii	causes interstitial cell pneumonia
Cryptosporidium parvum	intracellular parasite of intestinal cells that causes diarrhea

TABLE 40.3. *Some of the major parasitic helminths of mammals*

Parasite	Comments
Trematodes	
Blood flukes	
Schistosoma mansoni	causes human schistosomiasis (affects mesenteric veins draining large intestine)
S. haematobium	causes human schistosomiasis (affects urinary bladder plexus)
S. japonicum	causes human schistosomiasis (affects mesentery veins in the small intestine)
Liver flukes	
Fasciola hepatica	primarily infects ruminants and occasionally humans (worms live in biliary tract)
Clonorchis sinensis	most prevalent liver fluke in humans (can be acquired by eating raw fish)
Cestodes	
Taenia solium	long human tapeworm acquired by eating undercooked pork
Echinococcus granulosus	shorter human tapeworm acquired by eating undercooked lamb (parasitic cysts [hydatids] occur in liver and elsewhere)
Taeniaarhynchus saginatus	long human tapeworm acquired by eating undercooked beef
Nematodes	
Ascaris lumbricoides	most common intestinal roundworm in humans
Trichurus trichuiura	intestinal whipworm in humans
Enterobius vermicularis	tiny intestinal roundworm (causes peri-anal night itch in children)
Necator americanus	intestinal hookworm of humans (causes anemia)
Ancylostoma duodenale	intestinal hookworm of humans (causes anemia)
Strongyloides stercoralis	intestinal parasite (causes autoinfection)
Haemonchus contortus	intestinal parasite of sheep and goats
Trichinella spiralis	smallest nematode parasite of humans (trichinosis) residing in muscle fibers (parasite acquired from undercooked pork)
Onchocera volvulus	filarial parasite (causes river blindness)
Wuchereria bancrofti	filaria live in lymph nodes causing elephantiasis
Brugia malayi	filaria live in lymph nodes causing elephantiasis
Dirofilaria immitis	dog heartworm

Leishmania species (causing leishmaniasis), and *Trypanosoma* species (causing sleeping sickness and Chagas' disease). Parasitic worms (helminths) live within tissues outside of cells and have evolved a variety of infective and protective pathways involving complex glycoconjugates and their recognition. Some of the major parasitic helminths include nematodes, such as *Ascaris lumbricoides*, trematodes, such as *Schistosoma mansoni* (causing schistosomiasis), and cestodes or tapeworms, such as *Taenia solium* (causing taeniiasis).

It is often difficult to study parasites because, by definition, most of them require animal hosts to survive and parasites cannot be grown independently. However, exciting new results demonstrate that glycoconjugates are very important in the life cycles and pathology of most major parasites. Some of the parasitic protozoans and helminths rely on GBPs in the host to promote their parasitism, and they have elaborated intriguing strategies to defeat the antiglycan immunity of the host.

PLASMODIA

Malaria in humans is caused by *Plasmodium* species and several major species infect humans, with *P. falciparum* being the most virulent. These parasites lead a complicated life cycle, alternating between a sexual stage within the female *Anopheles* mosquito vector and an asexual stage within mammalian tissues (hepatocytes and erythrocytes) and the bloodstream (Figure 40.1). Cell–cell interactions between the parasite and host are critical for the successful completion of each stage.

Following inoculation into the bloodstream, the sporozoite's major circumsporozoite protein interacts with the liver's heparan sulfate (HS), thereby enabling invasion of hepatocytes, the initial site of replication in the mammalian host. HS from the liver possesses an unusually high degree of sulfation compared to similar glycosaminoglycans from other organs, suggesting the basis for the selective targeting of *Plasmodium* to hepatocytes.

Upon exit from the liver, merozoites can use multiple ligand–receptor interactions to invade host erythrocytes, which vary in their dependency on sialic acid residues on the surface. Multiple proteins on *Plasmodium* merozoites mediate invasion of erythrocytes (Table 40.4). These fall into two broad superfamilies, the erythrocyte-binding-like (EBL; or more appropriately, the Duffy-binding-like [DBL]) and the reticulocyte-binding-like (RBL) proteins. The merozoite protein EBA-175 (erythrocyte-binding antigen-175; a type of EBL) was originally identified based on its ability to bind erythrocytes, and it is now known that the *P. falciparum* genome contains six EBL paralogs. EBA-175 recognizes clusters of sialylated O-glycans attached to erythrocyte glycophorin A, particularly within a 30-amino-acid region that carries 11 O-glycans. Desialylation of erythrocytes precludes interactions of some strains of *P. falciparum*, and individuals lacking glycophorin A or glycophorin B are refractory to invasion. The glycophorins are the major sialic-acid-containing glycoproteins

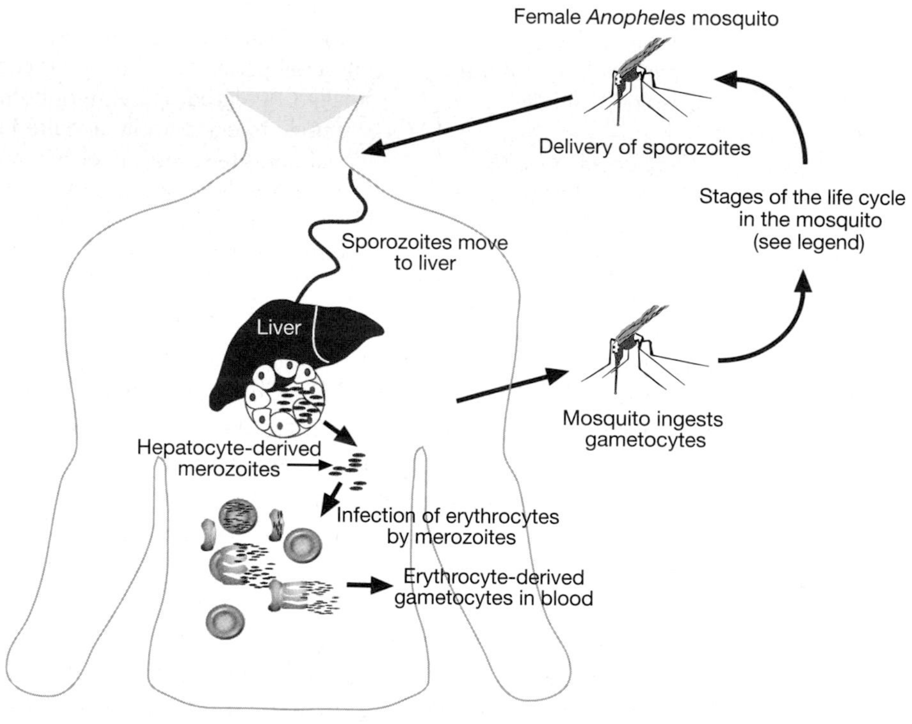

FIGURE 40.1. Life cycle of *Plasmodium falciparum*, the parasitic protozoan that causes malaria in humans. The bite of the female *Anopheles* mosquito introduces sporozoites into the human host and they develop as shown. After a blood meal from an infected person, the malarial gametocytes enter the midgut of the mosquito and continue their life cycle. In the mosquito midgut, the gametocytes transform into male microgametes and female macrogametes. Their union leads to a zygote, which transforms into an ookinete, which penetrates the intestinal wall of the mosquito and is transformed into a circular oocyst. Inside the oocyst, the sporozoites develop from germinal cells known as sporoblasts. The sporozoites emerge from the oocysts and migrate to the salivary gland where they enter the human hosts during the blood meal of the mosquito.

TABLE 40.4. *Some major parasites and their glycan-recognizing proteins*

Parasite	Stage	Protein	Specificity
Plasmodium falciparum	merozoite	EBA-175	Neu5Acα2-3Gal/ glycophorin A
	merozoite	EBA-140	sialic acid/glyco- phorin B?
	merozoite	EBA-180	sialic acid
	sporozoite	circumsporozoite protein	heparan sulfate
Trypanosoma cruzi	trypomastigote	*trans*-sialidase	Neu5Acα2-3Gal
	trypomastigote	penetrin	heparan sulfate
Entamoeba histolytica	trophozoite	Gal/GalNAc lectin	Gal/GalNAc
Entamoeba invadens (a reptilian pathogen)	cyst	cyst wall protein (Jacob lectin)	chitin
Giardia lamblia	trophozoite	taglin (α-1 giardin)	Man-6-phosphate heparan sulfate
Cryptosporidium parvum	sporozoite	Gal/GalNAc lectin	Gal/GalNAc
	sporozoite	Cpa135 protein	?
Acanthamoeba keratitis	trophozoite	136-kD mannose- binding protein	mannose
Toxocara canis	larval	TES-32	?
Haemonchus contortus	gut-localized	galectin	β-galactosides

on erythrocytes. Although EBA-175 plays an important role in invasion, the roles of EBA-140/BAEBL and EBA-181/JSEBL are less clear. Some strains of *P. falciparum* can reversibly switch from sialic-acid-dependent to sialic-acid-independent invasion, which depends on parasite ligand use and involves the expression of a *P. falciparum* RBL-like homolog 4 (PfRh4). The ability to switch receptor usage for erythrocyte invasion from sialic-acid-dependent to sialic-acid–independent pathways has important implications for vaccine design against malaria parasites.

Eruption of the merozoites from infected erythrocytes results in the release of glycosylphosphatidylinositols (GPIs), which are believed to be prominent virulence factors that contribute to malaria pathogenesis (see Chapter 11). These *P. falciparum* GPIs, which are free or arise by proteolytic processing of GPI-anchored proteins, may mimic host-cell GPIs and activate GPI-associated signaling pathways, such as Src-related protein tyrosine kinases. The GPIs can activate the host's macrophages, leading to the production of inflammatory cytokines as well as cell-adhesion molecules, such as intracellular adhesion molecule-1 (ICAM-1), vascular cell-adhesion molecule-1 (VCAM-1), and E-selectin in endothelial cells. Antibodies to these GPIs can neutralize their effects and mitigate the pathology of the disease independently of infection.

The presence of N- and O-linked glycans in *Plasmodia* has been controversial. Several *Plasmodium* proteins (MSP-1, MSP-2, and EBA-175) are believed to be nonglycosylated in vivo, even though potential N-glycosylation sites are present in the primary amino acid sequences. Recent studies show that *Plasmodium* and *Giardia* can synthesize dolichol-P-P-$GlcNAc_2$, which is a normal intermediate in the synthesis of the longer types of dolichol-P-P-oligosaccharides commonly found in vertebrates and that are precursors for N-glycosylation in glycoproteins. Thus, *Plasmodium* and *Giardia* may contain N-glycans with only the disaccharide motif GlcNAcβ1-4GlcNAcβ-Asn. Little is known about O-glycosylation in *Plasmodium*, if it occurs at all. Thus, GPI anchor synthesis emerges as the major form of protein glycosylation, a situation unique in eukaryotic cells.

TRYPANOSOMES

African trypanosomes of the species *brucei* are the etiologic agents of nagana disease in cattle and sleeping sickness in humans. After transmission by blood-sucking tsetse flies, a remarkable feature of these organisms is their ability to survive extracellularly in the bloodstream of the host where they are constantly exposed to the host immune system. Evasion of the host immune response depends on "antigenic variation," a highly evolved strategy of survival that relies heavily on structural variance of the surface GPI-anchored glycoproteins (VSGs) (Figure 40.2). As the main component of the dense glycocalyx, the VSGs are dimeric proteins, consisting of two 55-kD monomers, each of which carries N-linked oligomannose-type oligosaccharides. As parasites multiply in the host bloodstream, the host immune system mounts an immune response that is effective against only a certain population of trypanosomes, those expressing the antigenic VSG. Those that have switched to an alternative VSG coat (encoded among 1000 distinct VSG genes) escape immunological destruction.

Within the gut of the tsetse fly vector, the trypanosome replaces the entire VSG coat with acidic glycoproteins called procyclins (Figure 40.2). These GPI-anchored proteins form a dense glycocalyx and are composed of polyanionic polypeptide repeat domains projecting from the membrane. Unusual features are the presence of a single type of N-glycan ($Man_5GlcNAc_2$) and GPI anchors that are modified with branched poly-*N*-acetyllactosamine

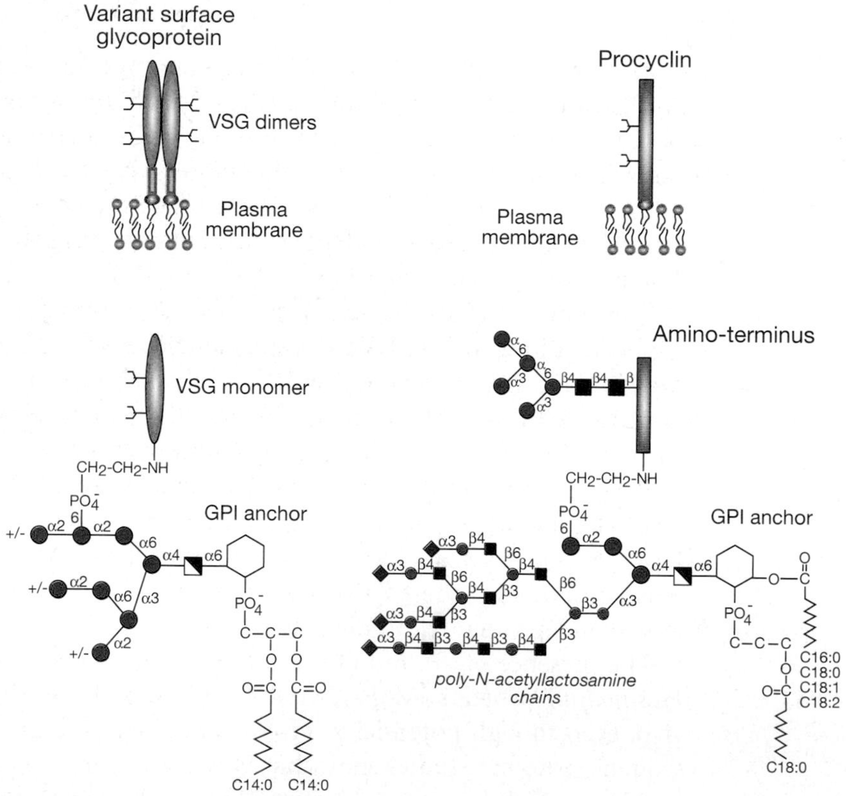

FIGURE 40.2. Schematic representation of the major surface glycoconjugates of procyclic and metacyclic *Trypanosoma brucei*. VSG (variant surface glycoprotein) is the major component of the metacyclic form, and each molecule consists of two GPI-anchored N-glycosylated monomers. (*Shaded ovals*) Protein component. The surface of the procyclic form is densely covered with procyclins. These are GPI-anchored polypeptides with polyanionic repeat domains. The anchor structures are detailed below the schematic.

[Galβ1-4GlcNAc]$_n$ glycans. The terminating galactose residue can be substituted with sialic acids by the action of a *trans*-sialidase. The surface sialic acids appear to protect the parasite from the digestive and trypanocidal environments in the midgut of the tsetse fly.

Trypanosoma cruzi is the etiologic agent of Chagas' disease or South American trypanosomiasis and it is transmitted by reduviid bugs. *T. cruzi* has a dense coat composed of a layer of glycosylinositolphospholipids (GIPLs) and mucins that project above the GIPL layer (Figure 40.3). The GIPLs contain the same basic structure as other GPI anchors, except that they are heavily substituted with galactose, *N*-acetylglucosamine, and sialic acid. The mucins contain large amounts of O-linked glycans composed of serine- or threonine-linked *N*-acetylglucosamine with one to five additional galactose residues. The terminal β-galactose can be further substituted by α2-3-linked sialic acid, which arises from a parasitic *trans*-sialidase that transfers sialic acid from host glycoconjugates. The surface coat has a protective function, providing the parasite with the ability to survive in hydrolytic environments and promoting adherence to macrophages for invasion. Sialylation is also believed to reduce the susceptibility of the parasite to anti-α-Gal antibodies that are normally present in the mammalian bloodstream. These parasites also express a surface heparin-binding protein (~60 kD) called penetrin that interacts with HS of host cells.

Another important glycoconjugate of *T. cruzi* is lipopeptidophosphoglycan (LPPG), which is the major surface glycan of the insect stage of the parasite (Figure 40.3).

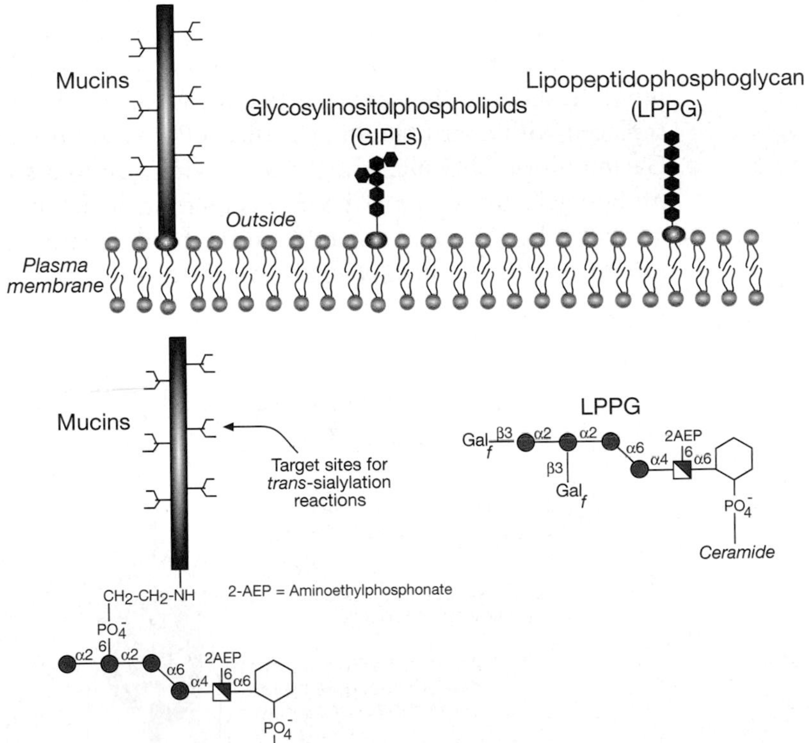

FIGURE 40.3. Schematic representation of the major surface glycoconjugates of *Trypanosoma cruzi*. The cell surface of *T. cruzi* is covered with a dense layer of mucins, glycosylinositolphospholipids (GIPLs), and lipopeptidophosphoglycan (LPPG). The structures of the mucin anchors and the predominant LPPG species are outlined. (2-AEP) Aminoethylphosphonate. (For the monosaccharide symbol code, see Figure 1.5, which is also reproduced on the inside front cover.)

Depending on the life-cycle stage, LPPG is composed of an inositolphosphoceramide-anchored glycan or an alkylacylphosphatidylinositol-anchored glycan that includes nonacetylated glucosamine, mannose, galactofuranose, and 2-aminoethylphosphonate. The lack of ceramide anchors and galactofuranose in mammalian cells suggests potential targets for the development of chemotherapeutic agents.

LEISHMANIA

Leishmania are responsible for a spectrum of human diseases, termed leishmaniasis. Species-specific leishmaniasis manifests clinically in three forms: cutaneous, mucocutaneous, and visceral, the last being fatal if untreated. These parasites have a remarkable capacity to avoid destruction in the hostile environments they encounter during their life cycle, alternating between intracellular macrophage parasitism and extracellular life in the gut of their sandfly vector (Figure 40.4).

Stage-specific adhesion is mediated by structural variation involving the abundant cell-surface glycoconjugate lipophosphoglycan (LPG), which contributes to parasite survival in the hydrolytic midgut environment (Figure 40.5). The basic LPG structure in all *Leishmania* species consists of a 1-*O*-alkyl-2-*lyso*-phosphatidyl(*myo*)inositol lipid anchor, a heptasaccharide glycan core, a long phosphoglycan (PG) polymer composed of (-6Galβ1-4Manα1-PO_4-) repeat units (n ~10–40), and a small oligosaccharide cap. In many species, the PG repeats contain additional substitutions that mediate key roles in stage-specific adhesion. For example, in *L. major*, the LPG phosphoglycan $[\text{Gal-Man-P}]_n$ backbone repeat units bear β1-3-galactosyl side-chain modifications, which form the attachment ligand for the sandfly galectin, PpGalec. As parasites differentiate, the LPG is replaced with a structurally modified LPG where α1-2 arabinosyl residues cap the β1-3 Gal-modified LPG phosphoglycans, giving rise to a structure that does not bind to the midgut galectin receptor, thereby facilitating detachment of the parasite from the sandfly midgut. The abundance of LPG on the parasite surface, which is the site of the primary interface with the host, suggests a central role for the glycoconjugate in the parasite's

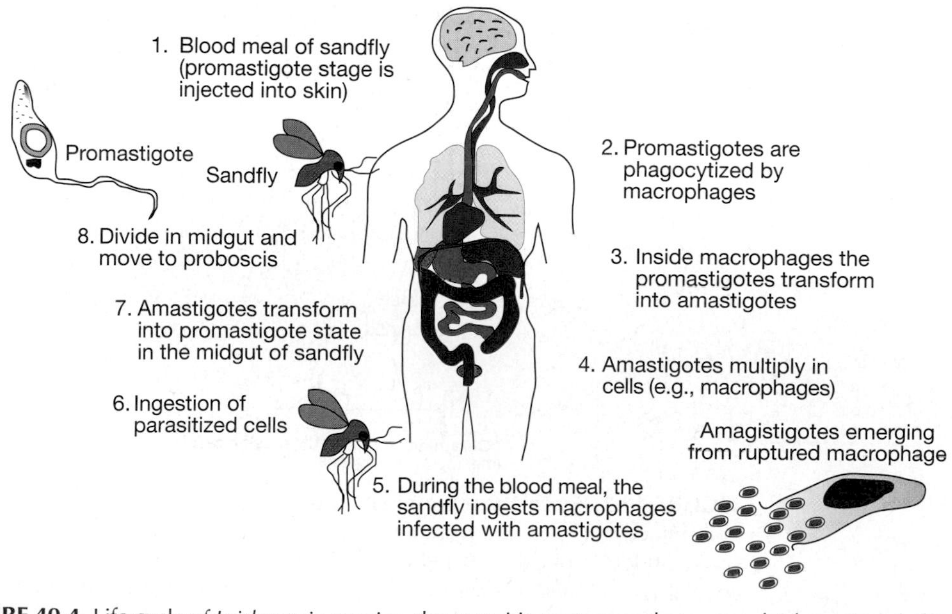

FIGURE 40.4. Life cycle of *Leishmania* species, the parasitic protozoan that causes leishmaniasis in humans.

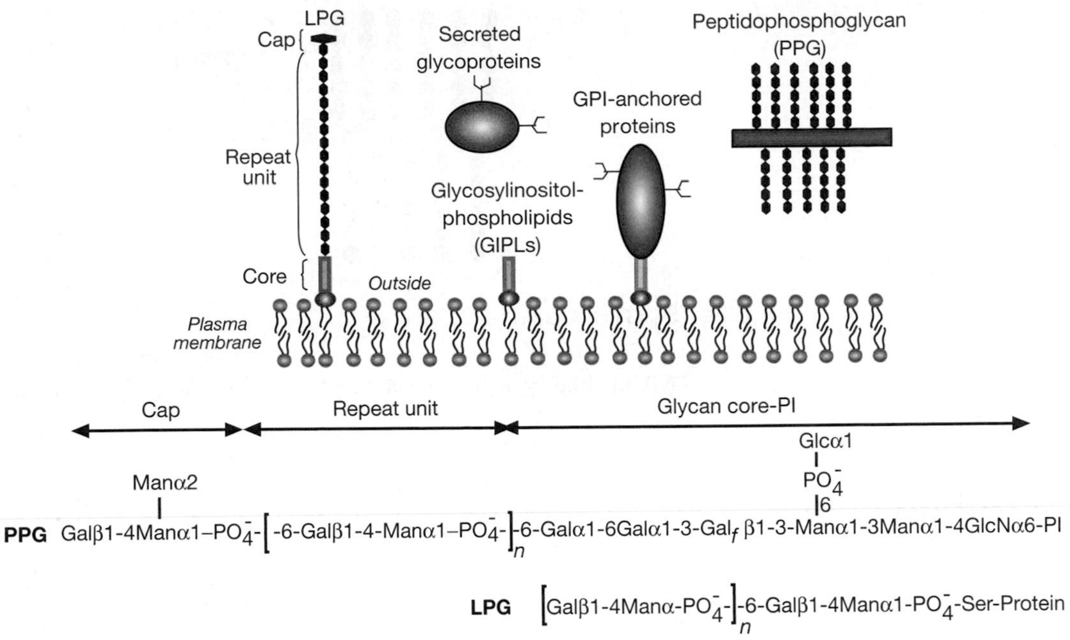

FIGURE 40.5. Schematic representation of the major cell-surface glycoconjugates of *Leishmania*. Lipophosphoglycan (LPG) consists of a Gal-Man-P repeat unit backbone attached to a lipid anchor via a glycan core. Secreted glycoproteins and mucin-like molecules called proteophosphoglycans (PPGs) are composed of serine/threonine-rich peptides that are heavily glycosylated on serine with phosphoglycan chains, similar to those found on LPG. The structures of *L. donovani* LPG and PPG are shown below.

infectious cycle. Besides a role in binding and detachment of the parasite in the midgut of the sandfly, LPG has also been implicated in resistance to complement-mediated lysis upon inoculation into the mammalian host, binding and uptake by macrophages, modulating macrophage signal transduction, resistance to oxidative attack, and, ultimately, allowing the parasite to establish successful infections.

In addition to LPG, *Leishmania* express an abundance of other important glycoconjugates, such as GIPLs, GPI-anchored proteins, and secreted proteophosphoglycans (PPG) (Figure 40.5) (also see Chapter 11). In PPG molecules, most of the serine residues are phosphoglycosylated with Gal-Man-P repeat units via unique phosphodiester linkages. In *L. mexicana*, these highly anionic substances form a gel-like matrix composed of interlocking filaments that enhance parasite development in the sandfly and contribute to the formation of a parasitophorous vacuole in macrophages of the mammalian host, where the parasite replicates.

The early steps in LPG and GPI biosynthesis (up to Man-Man-GlcN-PI) occur in the endoplasmic reticulum (see Chapter 11). The distal mannose of all GPI anchors and some GIPLs is α1-6-linked, whereas it is α1-3-linked to the inner mannose in LPG, and some GIPLs contain the α1-3 mannose. Galactosylation of the LPG glycan core and assembly of the PG repeat unit occur in the Golgi apparatus. Virtually every known glycoconjugate of *Leishmania* shows some intersection with the LPG biosynthetic pathway, especially in the synthesis of the repeat units and GPI anchors. The presence of molecules bearing similar modifications raises the possibility that they share biosynthetic steps. The phosphoglycan portions of LPG and PPG are assembled by the sequential and alternating transfer of mannose-P and galactose from their respective nucleotide-sugar donors, forming the characteristic -Galβ1-4Manα1-P repeat units. Depending on the species of *Leishmania*, additional branching sugars can then be added, creating a remarkable array of side chains that drive *Leishmania*–sandfly vectorial competence.

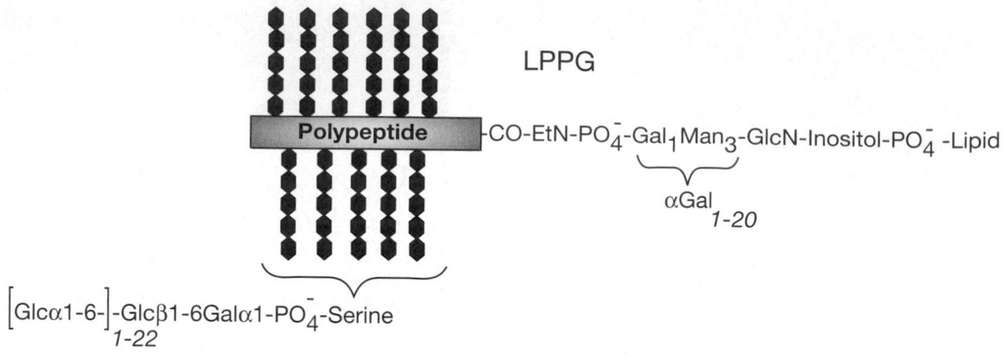

FIGURE 40.6. Structure of *Entamoeba histolytica* lipopeptidophosphoglycan (LPPG).

ENTAMOEBA

Entamoeba histolytica, the etiologic agent of amebic dysentery and hepatic abscesses, has a life cycle that includes two stages, the disease-inducing amebic or trophozoite stage and the infectious cyst stage. Binding to the colonic mucins is mediated by a galactose/*N*-acetylgalactosamine lectin, a GPI-anchored protein with a critical role in parasite viability. This lectin is a 260-kD heterodimeric glycoprotein consisting of heavy subunit (170 kD) and light subunit (either 35 or 31 kD) disulfide bonded together.

Entamoeba trophozoites synthesize a cell-surface lipoglycoconjugate (LPG) and a lipopeptidophosphoglycan (LPPG) (Figure 40.6). The LPG consists of a lipid anchor and a phosphoglycan component that resembles the phosphoglycans of *Leishmania* LPG. These molecules are important as virulence factors because antibodies raised against them inhibit the ability of the parasites to kill target cells, and vaccination retards the development of liver abscess in animal models. The serine residues of LPPG are extensively modified with chains of 2–23 α1-6-linked glucose residues that are presumably attached to the peptide backbone via a Gal-1-P-serine linkage. The LPPG GPI anchors are unique in that they have a novel glycan backbone that contains the sequence Gal_1Man_2GlcN-*(myo)*inositol, the terminal α1-2 mannose residues of other protein anchors being replaced by the α-Gal residue. The anchor is also modified with 1–20 α-Gal residues.

The *Entamoeba* cyst wall is mostly composed of chitin, and there is evidence that the encystation process is accompanied by the appearance of sialylated glycoconjugates on the surface. It has been hypothesized that the negative charge on the cyst surface afforded by the sialylated glycoconjugates may repel the cyst from the intestinal mucosa and thus promote expulsion from the host's intestine.

SCHISTOSOMA

Schistosomiasis is caused by a parasitic trematode, and three major species infect humans worldwide: *Schistosoma japonicum*, *S. mansoni*, and *S. haematobium* (Figure 40.7). This parasite is unique among helminths in that the male/female pair live together in the blood vessels of the human host. It lays eggs that adhere to the endothelium and migrate through vessels into tissues. The eggs may become lodged in the host tissues, where they induce granulomatous inflammatory responses and egg sequestration, and such entrapped eggs in the peripheral circulation can lead to portal hypertension and fibrosis, which are characteristic of chronic schistosomiasis and lead to the associated morbidity and mortality.

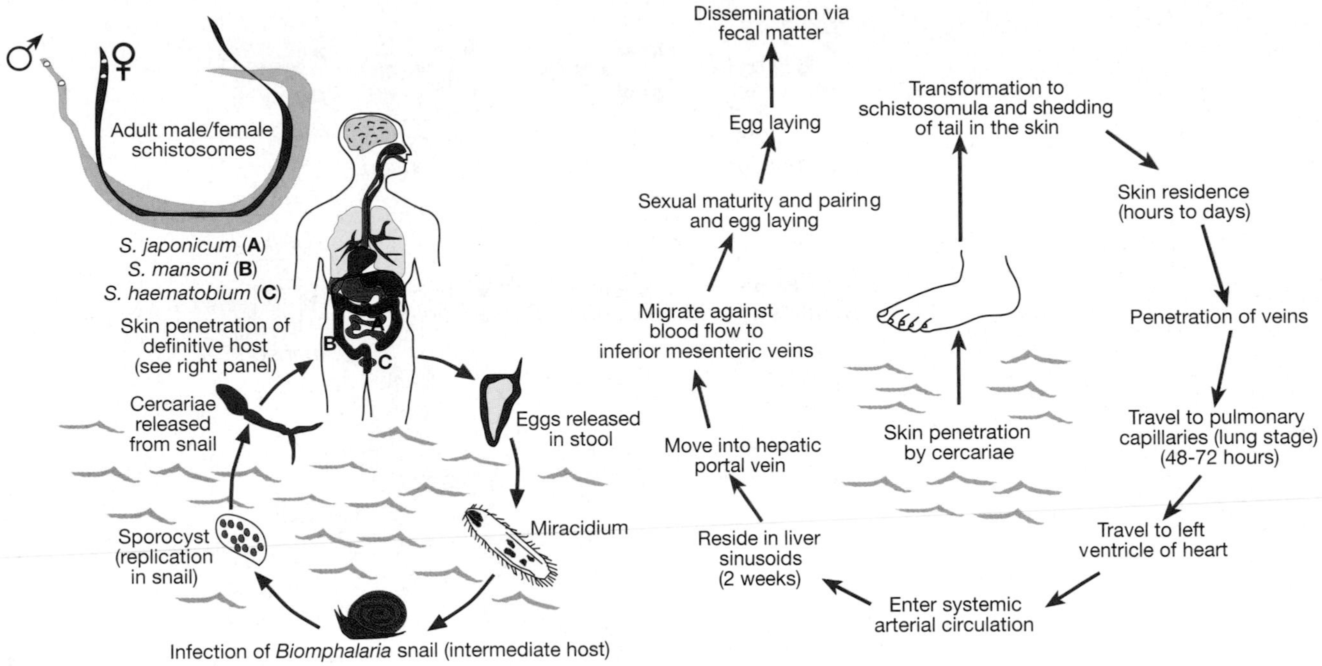

FIGURE 40.7. Life cycle of *Schistosoma* species, the parasitic helminth that causes schistosomiasis in humans. (*Left*) Alternating cycle between the invertebrate snail host and the human definitive host; (*right*) passage of the worms and their maturation to adult sexually active male/female pairs within the human host.

However, many eggs eventually pass into the stool and continue the cycle through intermediate snail hosts, which are unique for each *Schistosoma* species. Thus, the real geographical limitation to the spread of the disease is the range of the intermediate host snails.

Schistosomes, and especially their eggs (which are laid about 4–6 weeks after infection when sexual maturity is attained), generate huge quantities of membrane-bound and circulating glycoproteins containing fucosylated antigens. Some of the notable antigenic glycan structures found in schistosome glycans include Lewis^x (Le^x) antigen, LacdiNAc (LDN), fucosylated LacdiNAc (LDNF), and polyfucose branches (Figure 40.8). Overall, fucosylation is a common theme for most schistosome glycoconjugates. Interestingly, other helminths (such as *Echinococcus granulosus*, *Dirofilaria immitis*, and *Haemonchus contortus*) also synthesize glycoproteins containing LDN and LDNF, in addition to other fucosylated and xylosylated glycans (Figure 40.8). In general, schistosomes are especially rich in complex glycan structures and contain an impressive array of glycosphingolipids and O- and N-linked glycans on a multitude of glycoproteins. It is interesting that a few helminths, including schistosomes and *Dictyocaulus viviparus*, a parasitic nematode in cattle, appear to synthesize the Le^x antigen. Another theme in glycoconjugates derived from schistosomes and other helminths studied so far is the absence of sialic acid and sialyltransferase activities.

Schistosomes also synthesize many other interesting glycoconjugates in their so-called cercarial glycocalyx and in their eggs. Glycoproteins derived from the tegument, gut, and eggs of the parasite are highly antigenic and occur in the circulation of the infected animal. The expression of many of these glycan structures is developmentally regulated and stage specific, but their fundamental roles in parasite development and host pathogenesis are not clear. It is also likely that the different schistosome species differ in several ways in terms of their glycoconjugate structures. For example, the *S. mansoni* glycosphingolipids have extended difucosylated oligosaccharides, but the terminal difucosylated *N*-acetylgalactosamine is absent from glycosphingolipids of *S. japonicum*.

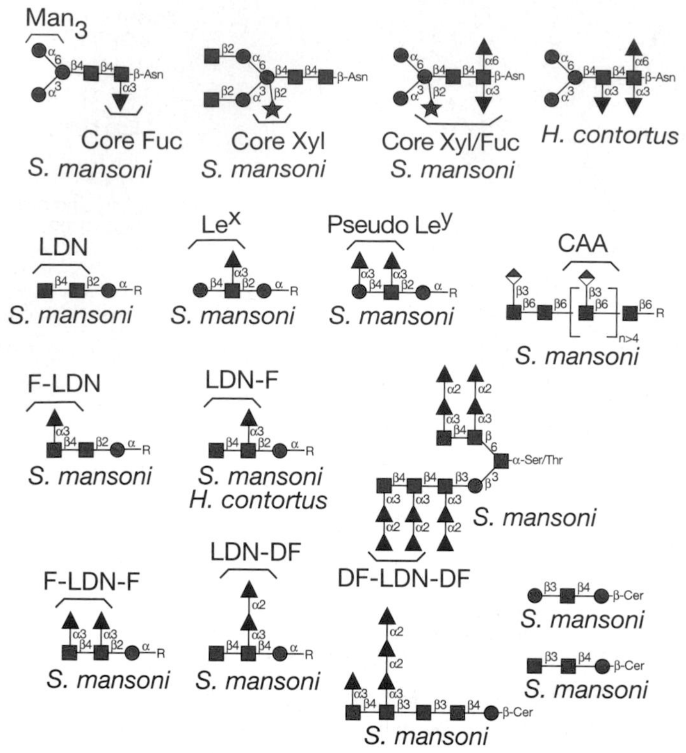

FIGURE 40.8. Structures of glycans found in parasitic helminths, including *Schistosoma mansoni* and *Haemonchus contortus*. The antigens are indicated by the structures within the brackets. (LDN) LacdiNAc; (Le^x) Lewis^x; (CAA) circulating anodic antigen; (F–) fucosylated; (DF–) difucosylated.

Individuals infected with *Schistosoma* species develop a wide variety of antibodies to glycan antigens, and these may be partly protective against further infections, because the adult and mature worms are refractory to immune killing. The adult worms have a rough tegument composed of a syncytial membrane that is highly regenerative, even in response to complement attack. There is mounting evidence that the glycan antigens expressed by schistosomes can influence the immune system and are bioactive in compromising host cellular immunity. During chronic schistosome infection, T-helper 2 (Th2) immune responses (promoting humoral immunity) predominate over Th1 responses (promoting cellular immunity). In response to glycans containing the Le^x antigen, murine B-1 cells secrete large amounts of interleukin-10 (IL-10) in vitro. Because IL-10 can depress Th1 responses in animals, this may partly contribute to Th2 dominance in early stages of schistosomiasis. In addition, schistosome glycans are recognized by C-type lectins, such as DC-SIGN, and other types of GBPs in antigen-presenting cells such as dendritic cells and macrophages. Internalization and processing of parasite glycans by these cells can lead to altered dendritic cell maturation (see Chapter 31).

The identification of many different antigenic glycoconjugates from schistosomes is helping in the design of new diagnostic procedures for schistosomiasis. In addition, the characterization of glycosyltransferases from schistosomes and the enzymes responsible for antigenic glycan biosynthesis, together with enzymes from the intermediate snail hosts, may help in the identification of new drug targets and the development of glycan-based vaccines.

GLYCOBIOLOGY OF OTHER PARASITES

Glycoconjugates and GBPs, as discussed above, play an impressive role in host infection by many different parasites. Several additional examples are shown in Table 40.4. Many protozoan parasites appear to use GBPs as a major mechanism for host-cell attachment and invasion (see Chapter 34). As mentioned earlier, one of the major glycoproteins expressed by *E. histolytica*, a major cause of amebic dysentery, is a lectin that recognizes galactose/*N*-acetylgalatcosamine residues, resulting in adherence of trophozoites to host cells. Adhesion is followed by contact-dependent cytolysis of host cells. *Acanthamoeba keratitis*, which causes severe eye infections involving the corneal epithelium, also adheres via a lectin interaction. This lectin-mediated adhesion of *A. keratitis* to host cells is a prerequisite for ameba-induced cytolysis of target cells. The adhesion occurs via a mannose-binding protein that is highly inhibited by Manα1-3Man disaccharides. Interestingly, mannose and mannose-6-phosphate can inhibit adhesion of *Giardia lamblia* trophozoites. One of the proteins from the parasite that may bind mannose and mannose-6-phosphate is termed taglin. The cDNAs of several galectin family members, identified initially as antigenic proteins, have been cloned from parasitic nematodes, such as *Teladorsagia circumcincta*. Sporozoites from *Cryptosporidium parvum*, an opportunistic protozoan that infects individuals with compromised immunity, have hemagglutinating activity, and a lectin on the parasite surface may play a crucial role in host-cell attachment.

In addition to these ongoing studies on GBPs in parasites, the glycan antigens of many parasites are being characterized in the hope that the information may lead to development of vaccines and new diagnostics for these diseases. For example, the major antigenic glycoconjugates synthesized by larvae of the parasitic nematodes *Toxocara canis* and *T. cati* are O-methylated trisaccharides that contain 2-O-methyl fucosyl and galactosyl residues (see Chapter 23). The intestinal nematode *Trichinella spiralis* synthesizes several highly immunogenic glycoproteins that contain the unusual sugar tyvelose (3,6-dideoxy-D-arabino-hexose). Tyvelose is found in complex N-glycans of larvae from *T. spiralis* and is a critical antigenic determinant recognized by antibodies in infected animals. Strong immunity to these glycans provides protective immunity, which causes expulsion of the invading larvae from the intestine. Protective immunity is also provided by antibodies to uncharacterized glycan antigens from *Haemonchus contortus*, an intestinal parasitic nematode of ruminants.

FURTHER READING

Ferguson M.A. 1997. The surface glycoconjugates of trypanosomatid parasites. *Philos. Trans. R. Soc. Lond. B Biol. Sci.* **352:** 1295–1302.

Cummings R.D. and Nyame A.K. 1999. Schistosome glysoconjugates. *Biochim. Biophys. Acta* **1455:** 363–374.

Dell A., Haslam S.M., Morris H.R., and Khoo K.H. 1999. Immunogenic glycoconjugates implicated in parasitic nematode diseases. *Biochim. Biophys. Acta* **1455:** 353–362.

Ferguson M.A. 1999. The structure, biosynthesis and functions of glycosylphosphatidylinositol anchors, and the contributions of trypanosome research. *J. Cell Sci.* **112:** 2799–2809.

Gowda D.C. and Davidson E.A. 1999. Protein glycosylation in the malaria parasite. *Parasitol. Today* **15:** 147–152.

Loukas A. and Maizels R.M. 2000. Helminth C-type lectins and host–parasite interactions. *Parasitol. Today* **16:** 333–339.

McConville M.J. and Menon A.K. 2000. Recent developments in the cell biology and biochemistry of glycosylphosphatidylinositol lipids. *Mol. Membr. Biol.* **17:** 1–16.

Guha-Niyogi A., Sullivan D.R., and Turco S.J. 2001. Glycoconjugate structures of parasitic protozoa. *Glycobiology* **11:** 45R–59R.

Khoo K.H and Dell A. 2001. Glycoconjugates from parasitic helminths: Structure diversity and immunobiological implications. *Adv. Exp. Med. Biol.* **491:** 185–205.

Petri W.A. Jr., Haque R., and Mann B.J. 2002. The bittersweet interface of parasite and host: Lectin–carbohydrate interactions during human invasion by the parasite *Entamoeba histolytica*. *Annu. Rev. Microbiol.* **56:** 39–64.

Nyame A.K., Lewis F.A., Doughty B.L., Correa-Oliveira R., and Cummings R.D. 2003. Immunity to schistosomiasis: Glycans are potential antigenic targets for immune intervention. *Exp Parasitol* **104:** 1–13.

Naderer T., Vince J.E., and McConville M.J. 2004. Surface determinants of Leishmania parasites and their role in infectivity in the mammalian host. *Curr. Mol. Med.* **4:** 649–665.

Nyame A.K., Kawar Z.S., and Cummings R.D. 2004. Antigenic glycans in parasitic infections: Implications for vaccines and diagnostics. *Arch. Biochem. Biophys.* **426:** 182–200.

Previato J.O., Wait R., Jones C., DosReis G.A., Todeschini A.R., Heise N., and Previato L.M. 2004. Glycoinositolphospholipid from *Trypanosoma cruzi:* Structure, biosynthesis and immunobiology. *Adv. Parasitol.* **56:** 1–41.

Thomas P.G. and Harn D.A. Jr. 2004. Immune biasing by helminth glycans. *Cell. Microbiol.* **6:** 13–22.

Young A.R. and Meeusen E.N. 2004. Galectins in parasite infection and allergic inflammation. *Glycoconj. J.* **19:** 601–606.

Hokke C.H. and Yazdanbakhsh M. 2005. Schistosome glycans and innate immunity. *Parasite Immunol.* **27:** 257–264.

van Die I., van Liempt E., Bank C.M., and Schiphorst W.E. 2005. Interaction of schistosome glycans with the host immune system. *Adv. Exp. Med. Biol.* **564:** 9–19.

Petri W.A. Jr., Chaudhry O., Haque R., and Houpt E. 2006. Adherence-blocking vaccine for amebiasis. *Arch. Med. Res.* **37:** 288–291.

van Die I. and Cummings R.D. 2006. Glycans modulate immune responses in helminth infections and allergy. *Chem. Immunol. Allergy* **90:** 91–112.

Hokke C.H., Fitzpatrick J.M., and Hoffmann K.F. 2007. Integrating transcriptome, proteome and glycome analyses of *Schistosoma* biology. *Trends Parasitol.* **23:** 165–174.

CHAPTER 41

Genetic Disorders of Glycan Degradation

Hudson H. Freeze

THIS CHAPTER EXPLORES THE DEGRADATION AND TURNOVER of glycans in lysosomes, especially with respect to human genetic disorders. Delineation of the dismantling of a few representative glycans illustrates features unique to different pathways. The degradation of oligomannosyl N-glycans removed from misfolded, newly synthesized glycoproteins is covered in Chapter 36.

THE LYSOSOMAL ENZYMES

Most glycans are degraded in lysosomes by highly ordered and specific pathways employing endo- and exoglycosidases, sometimes aided by noncatalytic proteins. Much of the insight for unraveling these complex pathways emerged from studies of rare human genetic disorders called lysosomal storage diseases. In each of these diseases, selected molecules accumulate in the lysosomes. Clever experiments combining enzymology with glycan structural analyses revealed the steps of the pathways and also unlocked the mechanism of lysosomal enzyme targeting as described in Chapter 30.

Lysosomes contain an estimated 50–60 soluble hydrolases that degrade various macromolecules. Most of the glycan-degrading enzymes (endo- and exoglycosidases and sulfatases) have pH optima between 4 and 5.5, but a few lysosomal enzymes have higher pH optima that are closer to neutral. Exoglycosidases cleave the glycosidic linkage of terminal sugars from the nonreducing end of the glycan (i.e., the residue at the outermost end of the molecule at the extreme left of glycan structures as represented in most of the figures in this book). The exoglycosidases recognize only one monosaccharide (rarely two) in a specific anomeric linkage, but they are much less particular about the structure of the molecule beyond the terminal glycosidic linkage. This lack of specificity allows these enzymes to act on a broad range of substrates. However, exoglycosidases do not usually work unless all of the hydroxyl groups of the terminal sugar are unmodified. Substitutions such as acetate, sulfate, or phosphate groups usually have to be removed prior to further degradation. Esterases cleave acetyl groups and specific sulfatases remove the sulfate groups on glycosaminoglycans and N- or O-linked glycans. Endoglycosidases cleave internal glycosidic linkages of larger chains. These enzymes are often more tolerant of modifications of the glycan; in some cases, they require a modified sugar for optimal cleavage.

Even though the lysosomal glycosidases carry out similar reactions, their amino acid sequences are only about 15–20% identical and at most are 40–45% similar to each other. Thus, there are no highly conserved glycosidase catalytic domains. Lysosomal enzymes are all N-glycosylated and most are targeted to the lysosome by the mannose-6-phosphate pathway discussed in Chapter 30. Therefore, they share aspects of the recognition marker for assembly of the mannose-6-phosphate moiety and have variable affinities for mannose-6-phosphate receptors. The concentration of enzymes within the lysosome is difficult to determine, but proteinases such as cathepsins B, D, and L have been estimated to be present at levels of 1 mM. Glycosidases are probably present at much lower concentrations.

GENETIC DEFECTS IN LYSOSOMAL DEGRADATION OF GLYCANS

More than 45 inherited diseases are known that impair the lysosomal degradation of macromolecules. Loss of a single lysosomal hydrolase leads to the accumulation of its substrate as undegraded fragments in tissues and the appearance of related fragments in urine. Many of the human disorders listed in Table 41.1 have animal models.

Table 41.1, Table 41.2, and Table 41.3 also show some of the major clinical symptoms of diseases associated with the degradation of each type of glycoconjugate. Many of the diseases share overlapping symptoms, and yet each disease has unique features that allow it to be specifically diagnosed by experienced clinicians. Many of the diseases also present with a range of severities. Usually, an infantile onset is the most severe and the juvenile or adult onsets have attenuated (milder) symptoms. The later onset forms may even affect organ systems different from those affected by early onset forms. Hundreds of mutations have been mapped in the different disorders. The severity usually depends on the combination of mutated alleles. Predicting the disease severity (prognosis) from the specific mutation is generally difficult, except for a combination of null alleles, which have severe outcomes. Hypomorphic alleles have residual glycosidase activity, but their prognosis is also difficult. Complete absence of a lysosomal hydrolase is uniformly severe.

TABLE 41.1. *Defects in glycoprotein degradation*

Disorder	Defect	Effects on degradation of: Glycoprotein	Effects on degradation of: Glycolipid	Clinical symptoms
α-Mannosidosis (types I and II)	α-mannosidase	major	none	*type I:* infantile onset, progressive mental retardation, hepatomegaly, death between 3 and 12 years *type II:* juvenile/adult onset, milder, slowly progressive
β-Mannosidosis	β-mannosidase	major	none	severe quadriplegia, death by 15 months in most severe cases; mild cases have mental retardation, angiokeratoma, facial dysmorphism
Aspartylglucosaminuria	aspartyl-glucosaminidase	major	none	progressive, coarse facies, mental retardation
Sialidosis (mucolipidosis I)	sialidase	major	minor	progressive, severe mucopolysaccharidosis-like features, mental retardation
Schindler (types I and II)	α-*N*-acetyl-galactosaminidase	yes	?	*type I:* infantile onset, neuroaxonal dystrophy. severe psychomotor and mental retardation, cortical blindness, neurodegeneration *type II:* mild intellectual impairment, angiokeratoma, corpus diffusum
Galactosialidosis	protective protein/ cathepsin A	major	minor	coarse facies, skeletal dysplasia, early death
Fucosidosis	α-fucosidase	major	minor	spectrum of severities includes psychomotor retardation, coarse facies, growth retardation
GM1 gangliosidosis	β-galactosidase	minor	major	progressive neurological disease and skeletal dysplasia in severe infantile form
GM2 gangliosidosis	β-hexosaminidase	minor	major	severe form: neurodegeneration with death by 4 years less severe form: slower onset of symptoms and variable symptoms, all relating to various parts of the central nervous system

It is not clear whether accumulating different types of undegraded material in a lysosome leads to the different symptoms characteristic of each disease. There is no evidence that the stored material causes lysosomes to burst and spew their contents into the cytoplasm. Some leakage may occur or the cell may sense an "engorged" lysosome. The pathology likely depends on the cell type and the cellular balance of synthesis and turnover rates. For instance, dermatan sulfate predominates in connective tissue, which might explain the bone, joint, and skin problems in mucopolysaccharidose (MPS) I, II, VI, and VII. Keratan sulfate is present in cartilage; therefore, MPS IV is largely a skeletal disease. GM2 ganglioside is abundant in neurons but not in other tissues; therefore, gangliosidosis is predominantly a brain disorder. The importance of glycogen for muscle explains the impact of Pompe disease on the heart and diaphragm, leading to rapid lethality in that disease.

In the balance between synthesis and degradation of multiple glycans, reducing synthesis somewhat by the use of an inhibitor helps to retard the accumulation of undegraded material in mouse models and reduces the pathology of disease.

TABLE 41.2. *Defects in glycosaminoglycan degradation—the mucopolysaccharidoses*

Number	Common name	Enzyme deficiency	Glycosamino-glycan affected	Clinical symptoms
MPS I H	Hurler, Hurler/Scheie, Scheie	α-L-iduronidase	DS, HS	*Hurler:* corneal clouding, organomegaly, heart disease, mental retardation, death in childhood *Hurler/Scheie and Scheie:* less severe, individuals survive longer
MPS II	Hunter	iduronate-2-sulfatase	DS, HS	*severe:* organomegaly, no corneal clouding, mental retardation, death before 15 years *less severe:* normal intelligence, short stature, survival age 20–60
MPS III A	Sanfilippo A	heparan N-sulfatase	HS	profound mental deterioration, hyperactivity, relatively mild somatic manifestations
MPS III B	Sanfilippo B	α-*N*-acetylglucos-aminidase	HS	similar to III A
MPS III C	Sanfilippo C	acetyl CoA: α-glucosaminide acetyltransferase	HS	similar to III A
MPS III D	Sanfilippo D	*N*-acetylglucosamine 6-sulfatase	HS	similar to III A
MPS IV A	Morquio A	galactose-6-sulfatase	KS, CS	distinctive skeletal abnormalities, corneal clouding, odontoid hypoplasia, milder forms known to exist
MPS IV B	Morquio B	β-galactosidase	KS	same as IV A
MPS VI	Maroteaux-Lamy	*N*-acetylgalactosamine 4-sulfatase	DS	corneal clouding, normal intelligence, survival to teens in severe form; milder forms known to exist
MPS VII	Sly	β-glucuronidase	DS, HS, CS	wide spectrum of severity, including hydrops fetalis and neonatal form
	multiple sulfatase deficiency	sulfatase modifying factor converts cysteine→formyl glycine	all sulfated glycans	hypotonia, retarded psychomotor development, quadriplegia

(MPS) Mucopolysaccharidoses; (DS) dermatan sulfate; (HS) heparan sulfate; (KS) keratan sulfate; (CS) chondroitin sulfate.

GLYCOPROTEIN DEGRADATION

The great majority of N- and O-glycans reaching the lysosome contain only six sugars linked in one or two anomeric configurations: β-*N*-acetylglucosamine (βGlcNAc), α/β-*N*-acetylgalactosamine (α/βGalNAc), α/β-galactose (α/βGal), α/β-mannose (α/βMan), α-fucose (αFuc), and α-sialic acid (αSia). Each linkage should theoretically require only one anomer-specific glycosidase, assuming that each glycosidase ignores the underlying glycan. This number is actually quite close to the known number of enzymes in the degradation pathways. However, some linkages require a specific enzyme outside of this group. For example, β-*N*-acetylhexosaminidase cleaves both βGlcNAc and βGalNAc residues. Degradation of the GlcNAcβAsn and GalNAcαSer/Thr linkages also requires specific enzymes.

Lysosomal Degradation of Complex N-Glycans

Much of what we know about this pathway comes from analysis of products that accumulate in patients' tissues or urine due to the absence of one of the degradative enzymes (Table

TABLE 41.3. *Defects in glycolipid degradation*

Disease name	Enzyme or protein deficiency	Clinical symptoms
Tay–Sachs	β-hexosaminidase A	*severe:* neurodegeneration, death by 4 years *less severe:* slower onset of symptoms, variable symptoms all relating to parts of the nervous system
Sandhoff	β-hexosaminidase A and B	same as Tay–Sachs
GM1 gangliosidosis	β-galactosidase	see Table 41.1
Sialidosis	sialidase	see Table 41.1
Fabry	α-galactosidase	severe pain, angiokeratoma, corneal opacities, death from renal or cerebrovascular disease
Gaucher's	β-glucoceramidase	*severe:* childhood or infancy onset, hepatosplenomegaly, neurodegeneration *mild:* child/adult onset, no neurodegenerative course
Krabbe	β-galactoceramidase	early onset with progression to severe mental and motor deterioration
Metachromatic leukodystrophy	arylsulfatase A (cerebroside sulfatase)	infantile, juvenile, and adult forms can include mental regression, peripheral neuropathy, seizures, dementia
Saposin deficiency	saposin precursor	similar to Tay–Sachs and Sandhoff

41.1). Structural analysis of monosaccharide-labeled glycoproteins during degradation in perfused rat liver also aided elucidation of the pathway. By conducting the latter studies in the presence of inhibitors of different lysosomal enzymes, a picture of simultaneous and independent bidirectional degradation of the protein and carbohydrate chains emerged (Figure 41.1). The relative degradation rates vary depending on structural and steric factors of the protein and the sugar chains. The accumulation of GlcNAcβ1-4GlcNAcβAsn in cells that cannot cleave the GlcNAcβAsn linkage clearly shows that degradation of the sugar chain does not require prior cleavage of the Asn. Much of the protein is probably degraded before N-glycan catabolism begins. Removal of core fucose (Fucα1-6GlcNAc) and probably any peripheral fucose residues linked to the outer branches of the chain (e.g., Fucα1-3GlcNAc) appears to be the first step in degradation because patients lacking this enzyme still have intact N-glycans bound to asparagine. Glycosylasparaginase (aspartyl-*N*-acetyl-β-D-glucosaminidase) then cleaves the GlcNAcβAsn bond, provided the α-amino group is not in peptide linkage. In rodents and primates, chitobiase (an endo-β-N-acetylglucosaminidase) removes the reducing *N*-acetylglucosamine, leaving the oligosaccharide with only one terminal *N*-acetylglucosamine. In many other species, splitting of the chitobiose linkage (GlcNAcβ1-4GlcNAc) uses the β-N-acetylhexosaminidase mentioned below as the last step in degradation. Either pathway appears to be effective, leaving the presence of chitobiase in some species unexplained. The oligosaccharide chain is then sequentially degraded by sialidases and/or α-galactosidase, followed by β-galactosidase, β-N-acetylhexosaminidase, and α-mannosidases. The remaining Manβ1-4GlcNAc is cleaved by β-mannosidase to mannose and *N*-acetylglucosamine or, in those species that do not have chitobiase, to chitobiose, which is then degraded by β-N-acetylhexosaminidase.

Lysosomal sialidase (neuraminidase), β-galactosidase, and a serine carboxypeptidase called protective protein/cathepsin A form a complex in the lysosome that is required for efficient degradation of sialylated glycoconjugates. Cathepsin A protects β-galactosidase from rapid degradation and also activates the sialidase precursor, but the protection does not depend on the catalytic activity of cathepsin A. Mutations in this protective protein lead to galactosialidosis in which the simultaneous deficiencies in β-galactosidase and sialidase are secondary effects of defective cathepsin A.

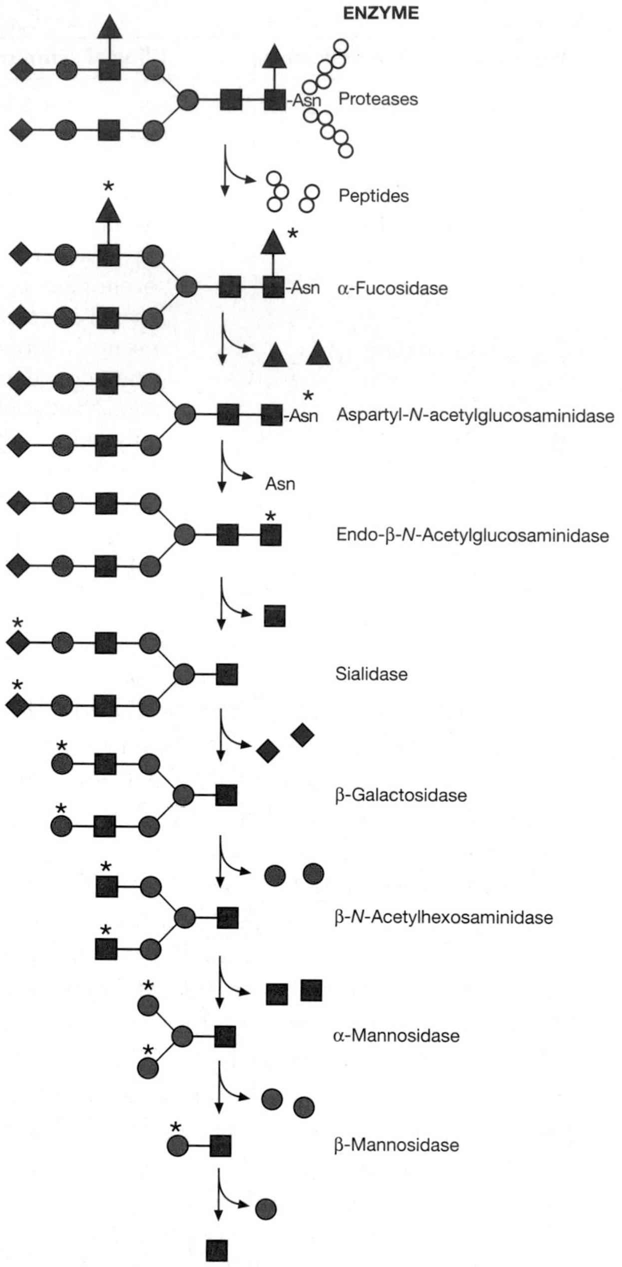

FIGURE 41.1. Degradation of complex N-glycans. The lysosomal degradation pathway of glycoproteins carrying complex-type glycans proceeds simultaneously on both the protein and glycan moieties. The N-glycans are sequentially degraded by the indicated exoglycosidases in a specific order as discussed in the text.

Other glycans that have GalNAcβ1-4GlcNAc, GlcAβ1-3Gal, or Galα1-3Gal on the outer branches must first have these residues removed by β-*N*-acetylhexosaminidase, β-glucuronidase, and α-galactosidase, respectively, prior to any further digestion of the underlying oligosaccharide chain.

Lysosomal Degradation of Oligomannosyl N-Glycans

Oligomannosyl N-glycans that enter the lysosome are hydrolyzed by an α-mannosidase to yield Manα1-6Manβ1-4GlcNAc (a common intermediate derived from hybrid and complex N-glycans). A second α1-6-specific mannosidase can cleave this linkage in humans and rats, but only on molecules that have a single core region *N*-acetylglucosamine (i.e., those generated by chitobiase cleavage). Finally, β-mannosidase completes the degradation. Oligomannosyl N-glycans derived from dolichol-linked precursors or misfolded glycoproteins (primarily $Man_5GlcNAc$) are handled differently as described in Chapter 36.

Degradation of O-Glycans

Degradation of typical αGalNAc-initiated O-glycans has not been systematically studied. Many of the outer structures seen on N-glycans are also found on O-glycans (see Chapter 13); therefore, degradation of these glycans probably uses the same group of exoglycosidases as discussed above. The exception to this, of course, is the linkage region, GalNAcα-O-Ser/Thr. Patients with Schindler disease lack an α-*N*-acetylgalactosaminidase, which is specific for αGalNAc and will not cleave αGlcNAc. This same enzyme probably removes terminal α-GalNAc from blood group A–containing glycans (GalNAcα1-3Gal) and some glycolipids such as the Forsmann antigen (GalNAcα1-3GalNAcβ1-3Galα1-4Galβ1-4GlcβCer). Patients lacking α-*N*-acetylgalactosaminidase accumulate GalNAc-containing glycopeptides in their urine, but curiously they also accumulate more complex, extended glycopeptides containing *N*-acetylglucosamine, galactose, and sialic acid. The structures are the same as those found on some native glycoconjugates. Their production could result from a general slowdown of oligosaccharide degradation, or they may arise by reassembly of glycans on GalNAcα-O-Ser/Thr glycopeptides that accumulate. How could this happen? There are several possibilities. One is that some of the partially degraded glycopeptides enter a compartment (perhaps the Golgi) that contains the appropriate glycosyltransferases and nucleotide sugars. Monosaccharides are then added sequentially before the products exit the cell. In this way, the glycopeptides generated in the lysosome would be behaving like oligosaccharide primers (see Chapter 50). However, another possibility is that the concentration of GalNAcα-O-Ser/Thr is high enough in the lysosome that lysosomal glycosidases use them as acceptors in a series of transglycosylation reactions (see Chapter 49). Because lysosomal glycosidases form complexes to degrade glycoconjugates more efficiently, partially degraded substrates may be in a preferred location as acceptors in transglycosylation reactions.

GLYCOSAMINOGLYCAN DEGRADATION

The glycosaminoglycans (GAGs), including heparan sulfate (HS), chondroitin sulfate (CS), dermatan sulfate (DS), keratan sulfate (KS), and hyaluronan, are degraded in a highly ordered fashion. The first three are O-xylose-linked to core proteins. KS can be both N- and O-linked depending on the tissue source, whereas hyaluronan is made as a free glycan (see Chapter 15). Some proteoglycans are internalized from the cell surface and the protein portion is degraded. The GAG chains are then partially cleaved by enzymes such as endo-β-glucuronidases or endohexosaminidases that clip at a few specific sites. Endoglycosidase cleavage creates multiple terminal residues that can be degraded by unique or overlapping sets of sulfatases and exoglycosidases. Structural analysis of partially degraded fragments in the lysosomes of cells from patients with genetic defects in these pathways cause mucopolysaccharidoses (MPS) and these were critical to dissecting the degradation pathways (see Table 41.2). Note that there is a range of clinical severities and manifestations with mutations in the same gene. For instance, MPS I is clinically subdivided into Hurler, Hurler/Scheie, and Scheie syndromes, although the three disorders represent a continuum of the same disease. Hurler is the most severe form of the disease (Table 41.2). Hurler/Scheie patients progress more slowly and die in early adulthood, whereas Scheie patients can survive to middle or old age. The milder forms of this disease do not cause mental retardation.

Hyaluronan

Hyaluronan (see Chapter 15) is the largest and most abundant GAG: A 70-kg person degrades 5 g of hyaluronan (molecular mass 10^7 daltons) per day. Degradation of hyaluronan involves a series of two endo-β-*N*-acetylhexosaminidases called hyaluronidases (Hyal-1 and Hyal-2), β-glucuronidase, and finally β-*N*-acetylhexosaminidase. Hyal-2 is active at low pH and is GPI-anchored on the cell surface. This enzyme associates both with hyaluronan in lipid rafts and with an Na^+/H^+ exchanger to create an acidic microenvironment. Cleavage generates fragments of approximately 20 kD (about 50 disaccharides), which are internalized, delivered to endosomes, and then finally to lysosomes, where the fragments are degraded by Hyal-1 into tetra- and disaccharides (Figure 41.2). The exoglycosidases are thought to participate in the degradation of the larger fragments as well as that of the di- and tetrasaccharide units.

Heparan Sulfate

HS (see Chapter 16) is first degraded by an endoglucuronidase, followed by a well-ordered sequential degradation. An example is shown in Figure 41.3. A terminal iduronic acid-2-sulfate must be desulfated by a specific iduronic acid-2-sulfatase to make the modified sugar a suitable substrate for α-iduronidase. If a glucuronic acid (GlcA)-2-sulfate were at this position, a GlcA-2-sulfatase would first remove the sulfate, followed by β-glucuronidase cleavage. The new terminal glucosamine sulfate ($GlcNSO_4$) is the next sugar for cleavage, but this process requires two steps. Sulfate is first removed by N-sulfatase forming glucosamine, but this sugar cannot be cleaved by α-*N*-acetylglucosaminidase. The amino group must be N-acetylated by an N-acetyltransferase embedded in the lysosomal membrane in order to convert glucosamine to *N*-acetylglucosamine, which is then a suitable substrate for enzymatic cleavage. In the first step, acetyl CoA donates the acetyl group to a histidine residue in the cytoplasmic domain of the enzyme. The acetyl group then becomes available on the luminal side of the lysosomal membrane and is transferred at low pH to the amino group of glucosamine on the partially degraded HS chain. The terminal αGlcNAc can now be cleaved. Cleavage of the 2-sulfated glucuronic acid residue requires that the sulfate first be removed, followed by β-glucuronidase cleavage. If the next αGlcNAc is 6-O-sulfated, the sulfate is removed by a specific GlcNAc-6-sulfatase. Table 41.2 provides a list of the enzymatic defects in HS degradation and the disorders that result.

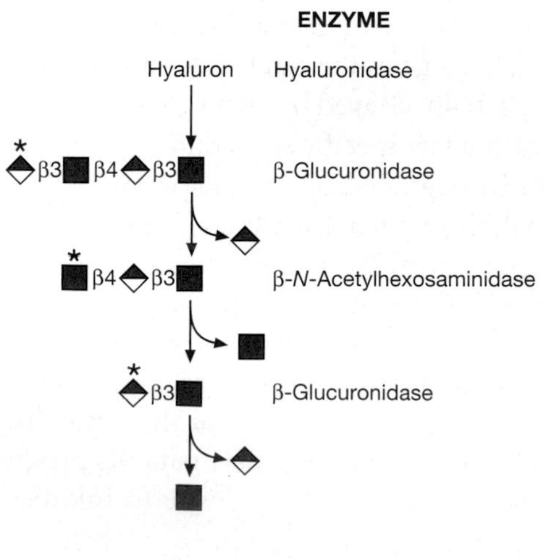

FIGURE 41.2. Degradation of hyaluronan. Hyaluronidase (an endoglycosidase) cleaves large chains into smaller fragments, each of which is then sequentially degraded from the nonreducing end.

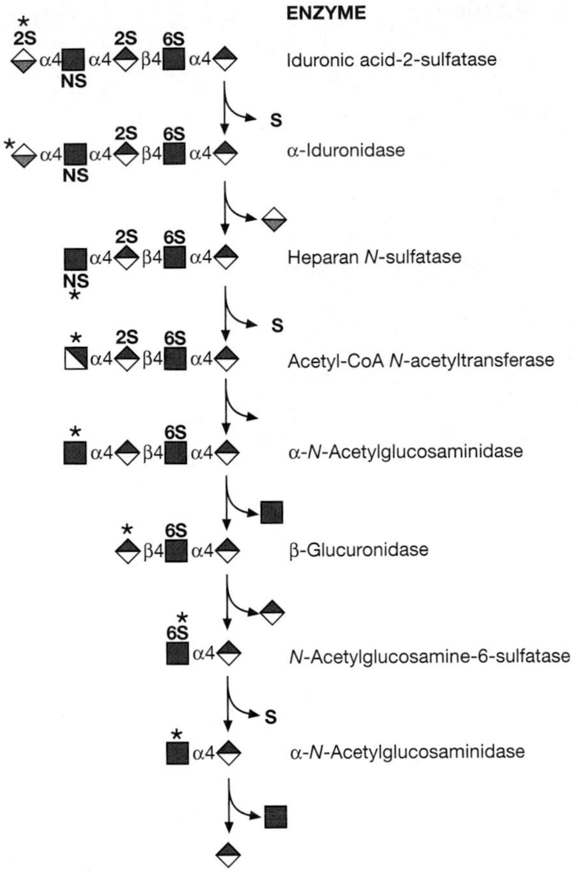

FIGURE 41.3. Degradation of heparan sulfate (HS). An endo-β-glucuronidase first cleaves large chains into smaller fragments and each monosaccharide is then removed from the nonreducing end as described in the text. N- and O-sulfate groups must first be removed before exoglycosidases can act. An unusual feature of HS degradation is that this process also involves a synthetic step. After removal of the N-sulfate residue on $GlcNSO_4$, the nonacetylated glucosamine must first be N-acetylated using acetyl-CoA before α-*N*-acetylglucosaminidase can cleave this residue.

Dermatan Sulfate and Chondroitin Sulfate

DS and CS are related GAG chains based on a repeating polymer of βGlcA and βGalNAc (see Chapter 16). Figure 41.4 shows that a combination of endoglycosidases, sulfatases, and exoglycosidases degrades DS in the lysosome. Iduronate-2-sulfatase is followed by α-iduronidase. The terminal GalNAc-4-SO_4 can be removed by either of two pathways. In the first pathway, a GalNAc-4-SO_4 sulfatase acts followed by β-*N*-acetylhexosaminidase A or B to remove *N*-acetylgalactosamine. In the second, β-*N*-acetylhexosaminidase A removes the entire GalNAc-4-SO_4 unit followed by sulfatase cleavage. Absence of the sulfatase (MPS VI) is unique to the DS pathway. β-Glucuronidase cleaves the β-glucuronic acid residue and the process is repeated on the rest of the molecule.

To degrade CS, GalNAc-6-SO_4 sulfatase and GalNAc-4-SO_4 sulfatase work in combination with β-*N*-acetylhexosaminidase A or B and β-glucuronidase. Hyaluronidases can also degrade CS, but no CS-specific endoglycosidases have been found.

Keratan Sulfate

KS is an N- or O-glycan with a heavily sulfated poly-*N*-acetyllactosamine chain. Mammalian cells do not have an endoglycosidase to break down KS (Figure 41.5). The sequential action of sulfatases and exoglycosidases is needed. Galactose-6-SO_4 sulfatase is the same enzyme that desulfates GalNAc-6-SO_4 in CS degradation. This hydrolysis is followed by β-galactosidase digestion, leaving a terminal GlcNAc-6-SO_4. Desulfation by the sulfatase followed by cleavage with β-*N*-acetylhexosaminidase A or B eliminates GlcNAc-

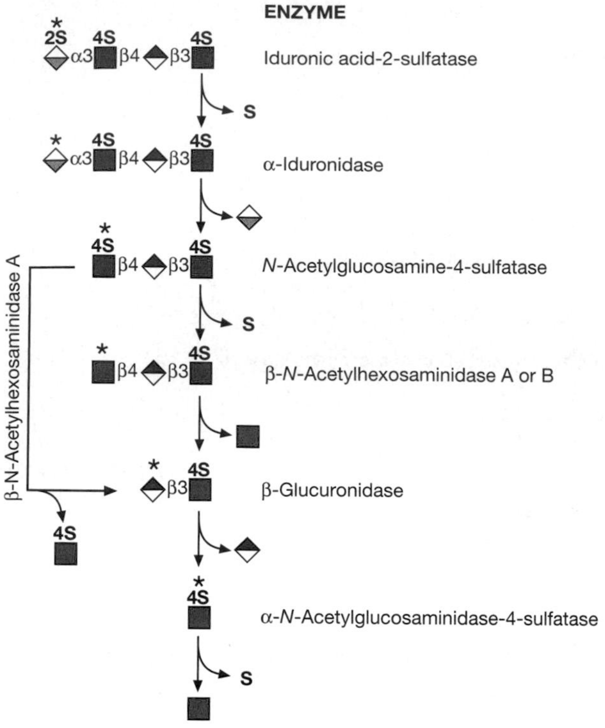

FIGURE 41.4. Degradation of dermatan sulfate and chondroitin sulfate. Sequential degradation can proceed by two different routes. One route uses GalNAc-4-SO_4 sulfatase followed by cleavage with β-*N*-acetylhexosaminidase A or B. The other route uses only β-*N*-acetylhexosaminidase A, which is one of the few exoglycosidases that can cleave sulfated amino sugars at low pH.

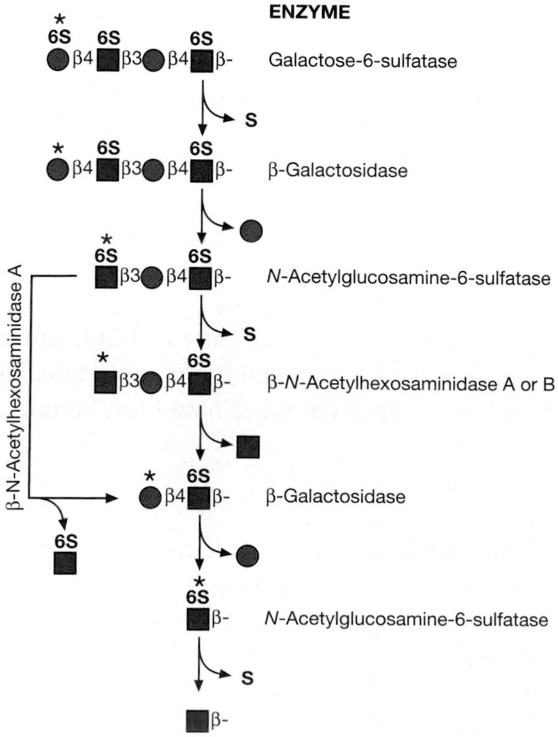

FIGURE 41.5. Degradation of keratan sulfate. Sequential degradation of KS occurs from the nonreducing end like the degradation of DS and CS. The terminal GlcNAc-6-SO_4 can be cleaved sequentially by a sulfatase and then by β-*N*-acetylhexosaminidase A or B or, alternatively, β-*N*-acetylhexosaminidase A can cleave GlcNAc-6-SO_4 directly at low pH.

6-SO_4. Alternatively, β-N-acetylhexosaminidase A can directly release GlcNAc-6-SO_4, followed by desulfation of the monosaccharide.

Linkage Region

Degradation of the core region of O-linked KS (skeletal type; type II) probably occurs by the same route as other O-glycans. The N-glycan corneal-type KS (type I) is probably degraded by the same set of enzymes used for N-glycan degradation. The more typical O-xylose-linked GAG chains (DS, CS, HS) all share a common core tetrasaccharide: GlcAβ1-3Galβ1-3Galβ1-4Xylβ-O-Ser. An endo-β-xylosidase has been detected in rabbit liver. The details of how the linkage regions of GAGs are degraded remain to be established.

Multiple Sulfatase Deficiency

A very rare human disorder is the multiple sulfatase deficiency (MSD). All sulfatases are casualties of this defect since they all undergo a posttranslational conversion of an active site cysteine residue to a 2-amino-3-oxopropionic acid for activity. In essence, the SH group of cysteine is replaced by a double-bonded oxygen atom that probably acts as an acceptor for cleaved sulfate groups. Loss of the enzyme carrying out that reaction leads to inactive sulfatases. This deficit affects GAG degradation and any other sulfated glycan such as sulfatides.

GLYCOSPHINGOLIPID DEGRADATION

Glycosphingolipids (see Chapter 10) are degraded from the nonreducing end by exoglycosidases while they are still bound to the lipid moiety ceramide. Since glycosphingolipids have many of the same outer sugar sequences that are found in N- and O-glycans (see Chapter 13), many of the same glycosidases are used for their degradation (Figure 41.6). However, specialized hydrolases are needed for cleaving the glucose-ceramide and galactose-ceramide bonds and other linkages near the membrane. Besides these specific enzymes, additional noncatalytic sphingolipid activator proteins (SAPs or saposins) help to present the substrate to the enzyme for cleavage. Endoglycoceramidases have been reported in leeches and earthworms. Like PNGase that releases N-glycans from protein, these enzymes release the entire glycan chain from the lipid.

Specialized Issues for Glycolipid Degradation

Some exoglycosidases for glycoprotein and GAG chain degradation also degrade glycolipids, but others are unique to glycolipid degradation. Absence of these unique enzymes causes glycolipid storage diseases, which are listed in Table 41.3. Glucocerebrosidase, also called β-glucoceramidase, is specific for the degradation of the GlcβCer bond. The loss of this enzyme causes Gaucher's disease.

A specialized β-galactosidase, β-galactoceramidase, hydrolyzes the bond between galactose and ceramide and can also cleave the terminal galactose from lactosylceramide. Loss of this enzyme produces Krabbe disease. Galactosylceramide is often found with a 3-sulfate ester (sulfatide) and a specific sulfatase (arylsulfatase A) is needed for its removal prior to β-galactosylceramidase action. Loss of this sulfatase causes metachromatic leukodystrophy and the accumulation of sulfatide. Glycolipids terminated with α-galactose residues are degraded by a specific α-galactosidase, the loss of which causes Fabry disease.

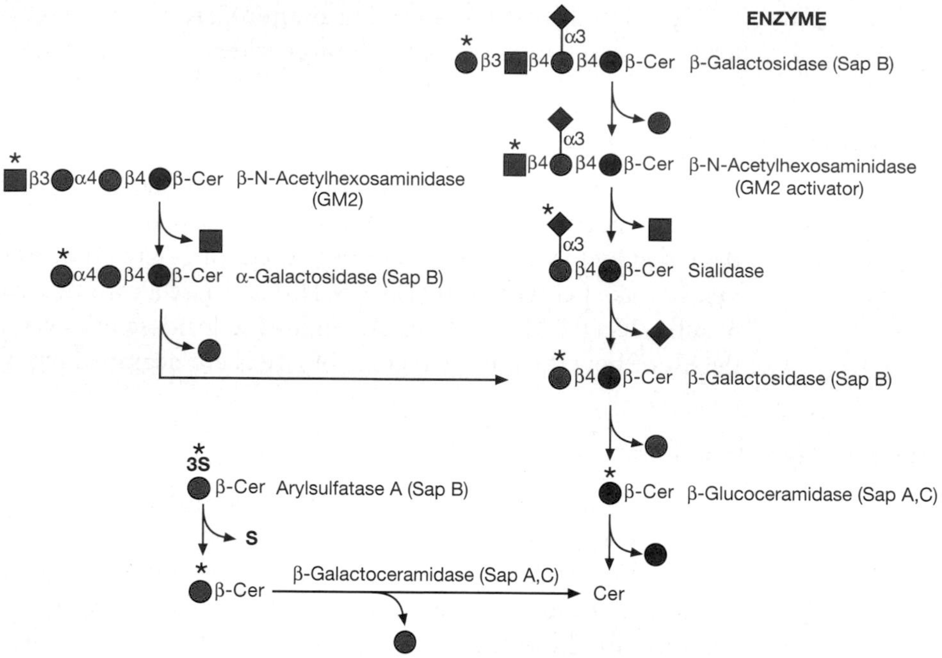

FIGURE 41.6. Degradation of glycosphingolipids. Required activator proteins are shown in parentheses and individual enzymes are as shown in previous figures in this chapter. (Sap) saposin.

Activator Proteins

Sugars that lie too close to the lipid bilayer apparently have limited access to soluble lysosomal enzymes, and additional proteins called saposins are required to present the substrates to the enzymes. Two genes code for all known SAPs. These proteins can also be used to form complexes with more than one degradative enzyme for more efficient hydrolysis of short glycolipids close to the membrane. SAPs are derived from a 524-amino-acid precursor called prosaposin, which is processed into four homologous activator proteins each comprising about 80 amino acids: saposins A, B, C, and D. Despite the homology, each saposin has different properties. Saposin A and saposin C both help β-glucosyl- and β-galactosylceramidase degradation. Saposin B assists arylsulfatase A, α-galactosidase, sialidase, and β-galactosidase. The activation mechanism of SAPs is thought to be similar to that of the GM2 activator, which is described below. Saposins D and B also assist in sphingomyelin degradation by sphingomyelinase. Complete absence of prosaposin is lethal in humans, and deficiencies in saposin B and saposin C lead to defects that clinically resemble aryl sulfatase A deficiency (metachromatic leukodystrophy) and Gaucher's disease. These symptoms might be predicted based on the enzymes that these saposins assist.

GM2 activator protein forms a complex with either GM2 or GA2 and presents them to β-*N*-acetylhexosaminidase A for cleavage of β-GalNAc. The activator protein binds a molecule of glycolipid, forming a soluble complex, which can be cleaved by the hexosaminidase. The resulting product is then inserted back into the membrane, and the activator presents the next GM2 molecule, and so on. Genetic loss of this activator protein causes the accumulation of GM2 and GA2, resulting in the AB variant of GM2 gangliosidosis.

Topology for Degradation and Roles of Lipids

Endocytic vesicles deliver membrane components to the lysosome for degradation and are often seen as intralysosomal multivesicular storage bodies (MVB). They appear like vesicles within the lysosome and are especially prominent in patients with glycolipid storage disorders. How do "vesicles within vesicles" form and why do we have them? MVBs serve an important function. The internal surface of the lysosomal membrane has a thick, degradation-resistant glycocalyx of integral and peripheral membrane proteins decorated with poly-*N*-acetyllactosamines, which protect the lysosomal membrane against destruction. However, these proteins would also shield incoming membranes from degradation if fusion simply occurred. By creating multiple internal membranes seen in typical MVBs, the target molecules are exposed to the soluble lysosomal enzymes, and digestion proceeds efficiently on the membrane surfaces of these internal vesicles.

MVB formation starts with inward budding of the limiting endosomal membrane. Lipids and proteins are sorted to either the internal or the limiting membrane. Ubiquitination of cargo proteins targets them to internal vesicles of MVBs, but ubiquitin-independent routes also exist. Intra-endosomal membranes and limiting endosomal membrane have different lipid and protein compositions. Membrane segregation and lipid sorting prepare the internal membranes for lysosomal degradation. During maturation of the internal membranes, cholesterol is continually stripped away (to <1%) and a negatively charged lipid bis(monoacylglycero)phosphate (BMP) increases up to 45%. This molecule is highly resistant to phospholipases. BMP also stimulates sphingolipid degradation on the inner membranes of acidic compartments. Since BMP is not present in the lysosomal outer membrane, the unique lipid profile ensures that degradation by hydrolases and membrane-disrupting SAPs occurs without digesting the lysosomal outer membrane.

DEGRADATION AND RESYNTHESIS

Degradation of glycans is not always complete. Partially degraded or incomplete glycans on glycoproteins and glycosphingolipids at the cell surface can be internalized and then elongated within a functional Golgi compartment containing the sugar nucleotide and the appropriate glycosyltransferase. In the case of glycosphingolipids, this pathway makes a substantial contribution to total cellular synthesis, but in glycoproteins it probably makes a relatively small contribution. These processes could simply be salvage and repair mechanisms or might play an integral part in an unidentified physiological pathway.

Reglycosylation and Recycling of Glycoproteins

Experiments clearly show that the half-life of specific membrane proteins is longer than the half-life of their sugar chains. Terminal monosaccharides turn over faster than those near the reducing end of the glycan, suggesting that terminal sugars are removed by exoglycosidases. Cleavage may occur at the cell surface or when proteins are endocytosed in the course of normal membrane recycling. Because the mildly acidic late endosomes contain lysosomal enzymes with a fairly broad pH range, terminal sugars such as sialic acid can be cleaved. If the endocytosed proteins are not degraded in lysosomes, the proteins may encounter sialyltransferases in the Golgi, become resialylated, and appear again on the cell surface. A similar situation may occur if a protein has lost both sialic acid and galactose residues from its glycans. Some membrane proteins synthesized in the presence of oligosaccharide processing inhibitors can still reach the cell surface in an unprocessed form.

Subsequent incubation in the absence of inhibitors leads to normal processing over time. The extent of processing and the kinetics depend on the protein and the cell line. Because Golgi enzymes are not distributed identically in all cells, general statements about how much "reprocessing" can occur are difficult to make. Most studies have monitored N-glycans, but membrane proteins with O-glycans probably behave similarly.

Glycosphingolipid Recycling

The majority of the newly synthesized glucosylceramide arrives at the cell surface by a Golgi-independent cytoplasmic pathway. Some of this material, as well as glucosylceramide generated by partial degradation of complex glycosphingolipids, can recycle to the Golgi. There, like glycoproteins, simple glycosphingolipids can serve as acceptors for the synthesis of longer sugar chains. In the lysosome, sphingosine and sphinganine are produced by complete degradation of glycosphingolipids and are reutilized. Together, these pathways, especially the latter, may account for the majority of complex glycolipid synthesis in many cells. Depending on the physiological state of the cell and synthetic demands, the de novo pathway starting from serine and palmitoyl CoA may account for only 20–30% of the total synthesis. Shuttling the recycled components through and among the various organelles appears to involve vimentin intermediate filaments. Mechanisms and details of this process are presently lacking, but these studies offer a physiological function for glycolipid transfer proteins that were purified from brain preparations many years ago (see Chapter 10).

SALVAGE OF MONOSACCHARIDES

Monosaccharide salvage is discussed in Chapter 4. Monosaccharides derived from glycan degradation are transported back into the cytoplasm using transporters specific for neutral sugars, N-acetylated hexoses, anionic sugars, sialic acid, and glucuronic acid. Turnover of glycans is high and the salvage of sugars can be quite substantial. There have been few studies comparing the actual contributions of exogenous monosaccharides, those salvaged from glycan turnover, and those generated from de novo synthesis. Cells probably differ in their preference for each source of monosaccharides. These differences may explain why monosaccharide therapy used to treat some glycosylation disorders is effective in certain cells and not others.

BLOCKING DEGRADATION

There are several ways to prevent lysosomal degradation of glycoconjugates. One is by raising the intralysosomal pH. Another is to inactivate a particular glycosidase genetically or by using a specific inhibitor. Although proteolysis inhibitors may also be an effective treatment, these drugs will not completely prevent degradation,

Because nearly all lysosomal enzymes have acidic pH optima, degradation can be curtailed by increasing the intralysosomal pH. This process can be accomplished with chemicals such as ammonium chloride and chloroquine. These agents lower the rate of degradation, but may not completely block it. Some lysosomal enzymes have relatively broad pH optima, and a few of these have pH optima higher than the lysosome. It has also been suggested that lysosomes may not always remain at low pH. Some "lysosomal" enzymes are actually present and active in early and late endosomes, which have a higher pH than lysosomes.

Glycosylation inhibitors are covered in Chapter 50, but some of these agents also inhibit specific lysosomal enzymes. For example, swainsonine blocks lysosomal α-mannosidase as well as the α-mannosidase II involved in glycoprotein processing. Sheep and cattle become neurologically deranged by eating food rich in swainsonine, which is also called locoweed. These temporary symptoms are likely induced because lysosomal α-mannosidase is inhibited. Undegraded oligosaccharides probably accumulate in affected animals. Although many inhibitors can block various enzymes when tested in enzymatic assays in vitro, the inhibitors may not be effective in cells or whole animals because the drugs may not enter the lysosome, and, even if they do, their concentration may be insufficient for efficacy.

THERAPY FOR LYSOSOMAL ENZYME DEFICIENCIES

As already discussed, genetic defects in lysosomal enzymes provided major insights into catabolic pathways and their importance to human health. Mannose-6-phosphate targeting of lysosomal enzymes was discovered by showing that cocultivation of fibroblasts from patients with different storage diseases led to the disappearance of stored material in both types of cells. Each supplied the corrective factors (i.e., lysosomal enzymes) that the other cell lacked by delivery of secreted enzymes through cell-surface mannose-6-phosphate receptors (see Chapter 30). Cross-correction highlights the fact that only a relatively small amount of enzyme may be needed to prevent accumulation of storage products and the resulting complications caused by these products. Some estimates suggest that less than 5% of normal β-N-acetylhexosaminidase activity may be sufficient to prevent pathological symptoms of Tay–Sachs disease. In fact, the index Scheie patient with α-L-iduronidase deficiency had less than 1% normal activity in fibroblasts but lived until 77 years of age.

These observations have led to two therapeutic approaches: enzyme replacement therapy (ERT) and substrate reduction therapy (SRT). Because accumulation of stored material depends on the relative rates of glycan synthesis and degradation, reducing a glycan's synthetic rate by SRT can offset the effects of low glycosidase activity. This hypothesis was tested in a mouse model of Sandhoff disease by using *N*-butyldeoxynojirimycin, an inhibitor of glucosylceramide synthase, which catalyzes the first step in glycosphingolipid biosynthesis. When wild-type mice were treated with this compound, the amount of glycosphingolipids fell 50–70% in all tissues without obvious pathological effects. When this compound was given to Sandhoff disease mice, the accumulation of GM2 in the brain was blocked and the amount of stored ganglioside was reduced. Thus, reducing the synthesis of the primary precursor reduced the load of GM2 to levels that were degradable by the glycosidase-deficient mice. Because this compound inhibits the first biosynthetic step, theoretically, this drug could reduce the accumulation of storage products in any glycolipid storage disorder.

In clinical trials, *N*-butyldeoxynojirimycin seemed to be effective for patients with non-severe Gaucher's disease, but their improvement was modest and the treatment caused significant side effects. Small preclinical trials of a few infantile Tay–Sachs disease patients did not arrest their neurological deterioration, but the treatment prevented, the development of macrocephaly. Even though SRT seems promising, the expected benefits to date have been minimal.

A recent development called enzyme enhancement therapy (EET) may provide another therapeutic approach. Here, the inhibitors of the enzymes behave as molecular chaperones to stabilize the mutated enzymes in the endoplasmic reticulum to prevent their misfolding and proteasomal degradation. For example, glucocerebrosidase can be stabilized en route to the lysosome, but the inhibitor dissociates at lysosomal pH generating an active enzyme. The overall increase in activity is small, but this small amount can have significant clinical benefits. Chemical chaperones should be able to treat other enzyme deficiencies. This is likely to be a promising approach for other lysosomal storage disorders, because some of the stabilizing molecules may be able to cross the blood-brain barrier.

ERT has been quite successful for treating Gaucher's disease. Injection of glucocerebrosidase carrying mannose-terminated N-glycans targets the enzyme to the macrophage/monocytes, which are the primary sites of substrate accumulation. In use now for many years, the added enzyme consistently improves patients' clinical features with minimal side effects. The cost of treatment is quite substantial, and high doses do not work better than lower ones for improving visceral and hematological features. The proven effectiveness makes glucocerebrosidase the favorite therapy for these patients, but terminal mannose residues cannot be used to target the enzyme to other cells or organs or to enable treatment of other lysosomal storage disorders. Insufficient infused enzyme crosses the blood-brain barrier, making this therapy ineffective for treating neurological symptoms. In addition, patients sometimes develop antibodies against the injected human protein. ERT trials with other recombinant lysosomal enzymes have progressed. Thus, the enzyme (marketed as Cerezyme®) works very well for type I disease without neurological involvement but not for Gaucher's with neurological involvement. In this disorder, the nonneurological form is more common.

Recombinant α-L-iduronidase for MPS I and recombinant α-galactosidase for Fabry disease were approved for clinical use in 2003. In MPS I, the enzyme works very well with the intermediate severity form (known as Hurler/Scheie), which was the focus group of the MPS I clinical trial. It is unclear whether the enzyme will have any effect on the neurological manifestations of the severe form (Hurler syndrome), which unfortunately is the form affecting the majority of MPS I patients. Recombinant N-acetylgalactosamine 4-sulfatase (arylsulfatase B; Naglazyme®) was approved for MPS VI, as was α-glucosidase (Myozyme®) for treating Pompe disease patients. Iduronate sulfatase (Elaprase®) is approved for treating Hunter syndrome. These results clearly validate this approach.

Gene therapy is also an option that is under study. The availability of both naturally occurring and engineered animals deficient in specific lysosomal enzymes offers appropriate systems to test the efficacy of corrective genes. Progression to human therapy may occur when successful gene therapy is more commonly used.

FURTHER READING

Neufeld E.F., Lim T.W., and Shapiro L.J. 1975. Inherited disorders of lysosomal metabolism. *Annu. Rev. Biochem.* **44:** 357–376.

Winchester B.G. 1996. Lysosomal metabolism of glycoconjugates. *Subcell. Biochem.* **27:** 191–238.

Scriver C.R., Beaudet A.L., Valle D., Sly W.S., Childs B., Kinzler K.W., and Vogelstein B., eds. 2001. *The metabolic and molecular basis of inherited diseases*, Vol. 3, pp. 3371–3877. McGraw-Hill, New York.

Meikle P.J. and Hopwood J.J. 2003. Lysosomal storage disorders: Emerging therapeutic options require early diagnosis. *Eur. J. Pediatr.* (suppl. 1) **162:** S34–S37.

Suzuki K., Ezoe T., Tohyama J., Matsuda J., Vanier M.T., and Suzuki K. 2003. Are animal models useful for understanding the pathophysiology of lysosomal storage disease? *Acta Paediatr. Suppl.* **92:** 54–62.

Desnick R.J. 2004. Enzyme replacement and enhancement therapies for lysosomal diseases. *J. Inherit. Metab. Dis.* **27:** 385–410.

Ellinwood N.M., Vite C.H., and Haskins M.E. 2004. Gene therapy for lysosomal storage diseases: The lessons and promise of animal models. *J. Gene. Med.* **6:** 481–506.

Futerman A.H. and van Meer G. 2004. The cell biology of lysosomal storage disorders. *Nat. Rev. Mol. Cell. Biol.* **5:** 554–565.

Stern R. 2004. Hyaluronan catabolism: A new metabolic pathway. *Eur. J. Cell. Biol.* **83:** 317–325.

Tettamanti G. 2004. Ganglioside/glycosphingolipid turnover: New concepts. *Glycoconj. J.* **20:** 301–317.

Kolter T. and Sandhoff K. 2005. Principles of lysosomal membrane digestion: Stimulation of sphingolipid degradation by sphingolipid activator proteins and anionic lysosomal lipids. *Annu. Rev. Cell. Dev. Biol.* **21:** 81–103.

Winchester B. 2005. Lysosomal metabolism of glycoproteins. *Glycobiology* **15:** 1R–15R.

Beutler E. 2006. Lysosomal storage diseases: Natural history and ethical and economic aspects. *Mol. Genet. Metab.* **88:** 208–215.

Brady R.O. 2006. Enzyme replacement for lysosomal diseases. *Annu. Rev. Med.* **57:** 283–296.

Brown J.R, Crawford B.E., and Esko J.D. 2007. Glycan antagonists and inhibitors: A fount for drug discovery. *Crit. Rev. Biochem. Mol. Biol.* **42:** 481–515.

CHAPTER 42

Genetic Disorders of Glycosylation

Hudson H. Freeze and Harry Schachter

THIS CHAPTER DISCUSSES INHERITED HUMAN DISEASES that affect glycan biosynthesis and metabolism. Mutuations affecting each of the major glycan families are described. Disorders affecting the degradation of glycans are described in Chapter 41.

INHERITED PATHOLOGICAL MUTATIONS OCCUR IN ALL MAJOR GLYCAN FAMILIES

Because 1–2% of the genome encodes enzymes involved in glycan formation, it is surprising that inherited disorders in glycan biosynthesis were not discovered until relatively recently. These rare human diseases are biochemically and clinically heterogeneous and usually affect multiple organ systems. Several of the diseases directly or indirectly impact one or more classes of glycans. Defects have been found in the activation, presentation, and transport of sugar precursors, in the glycosidases and glycosyltransferases involved in glycan synthesis and processing, and in proteins that control the traffic of components of the glycosylation machinery within the cell. In rare instances, patients have been successfully treated by oral administration of simple sugars. Increasing awareness of this group of disorders in the clinical community should foster new insights into their pathogenesis and treatment.

Table 42.1 lists the disorders in humans. Most of the human diseases have been discovered recently, although some are well-known diseases whose links to glycosylation are only now appreciated. The disorders are grouped together according to the major pathway affected. The specific nomenclature of each group is still evolving.

DEFECTS IN N-GLYCAN BIOSYNTHESIS

Clinical and Laboratory Features and Diagnosis

The congenital disorders of glycosylation (CDG) were originally called carbohydrate-deficient glycoprotein syndromes (CDGS) and are a subset of genetic defects affecting primarily N-glycan assembly. The broad clinical features involve many organ systems but especially the development of certain regions of the brain and functions of the gastrointestinal, hepatic, visual, and immune systems, indicating the importance of normal glycosylation in their functions. The variability of clinical features makes it difficult for physicians to recognize CDG patients. The first were identified in the early 1980s based on clinical symptoms and deficiencies in multiple plasma glycoproteins. The patients had psychomotor retardation, low muscle tone, incomplete brain development, visual problems, coagulation defects, and endocrine abnormalities. Many of these symptoms are seen in patients with other inherited multisystemic metabolic disorders, such as mitochondrial-based diseases. However, many CDG patients can be identified because they display abnormal glycosylation. Nearly all patients show undersialylation of serum glycoproteins as detected by examination of serum transferrin. This protein has two N-glycosylation sites with a typical disialylated biantennary glycan on each site ("tetrasialo" glycoform). About 10–15% of the protein molecules contain one or two trisialylated triantennary glycans ("pentasialo" or "hexasialo" forms), but only a very small proportion of glycans have just a single sialic acid residue ("disialo" or "monosialo" forms). These glycoforms are identified using isoelectric focusing (IEF), using ion-exchange chromatography, and, more recently, by mass spectrometric analysis. In CDG, many other serum glycoproteins have altered glycosylation, but transferrin is the most reliable, most sensitive, and simplest indicator. Transferrin provides a simple litmus test that has enabled physicians to spot CDG patients without knowing the molecular basis of the diseases.

For convenience, CDG defects are divided into types, I and II. The type I patients lack sialic acid residues on transferrin because they have unoccupied Asn-X-Ser/Thr glycosylation sequons; that is, entire sugar chains are absent. However, the N-glycan chains that remain in CDG type I patients are normal or only slightly altered. In the type II patients, both glycosylation sequons of transferrin are occupied, but the structure of the protein-bound glycan is altered. All defects in the biosynthesis of the lipid-linked oligosaccharide (LLO) and its transfer to proteins in the endoplasmic reticulum (ER) fall into type I, whereas defects in all subsequent processing steps are in type II. Individual type I CDGs (caused by defects in different genes) are given lowercase letters (e.g., CDG-Ia). The CDG nomenclature is in transition and will be replaced soon by the gene name, for example, CDG-PMM2. A list of these disorders is given in Table 42.1, and a few highlights are discussed below. The synthetic pathways and locations of the defects are shown in Figures 42.1 and 42.2.

TABLE 42.1. *Genetic defects of glycan synthesis in humans*

Disorder	Gene	Enzyme	OMIM	Key features
N-Glycan or multiple pathway defects				
CDG-Ia	*PMM2*	phosphomannomutase II	212065	mental retardation (MR), hypotonia, esotropia, lipodystrophy, cerebellar hypoplasia, stroke-like episodes, seizures
CDG-Ib	*MPI*	phosphomannose isomerase	602579	hepatic fibrosis, protein-losing enteropathy, coagulopathy, hypoglycemia
CDG-Ic	*ALG6*	glucosyltransferase I; Dol-P-Glc: $Man_9GlcNAc_2$-PP-Dol glucosyltransferase	603147	moderate MR, hypotonia, esotropia, epilepsy
CDG-Id	*ALG3*	Dol-P-Man: $Man_5GlcNAc_2$-PP-Dol mannosyltransferase	601110	profound psychomotor delay, optic atrophy, acquired microcephaly, iris colobomas, hypsarrhythmia
CDG-Ie	*DPM1*	Dol-P-Man synthase I GDP-Man: Dol-P-mannosyltransferase	603503	severe MR, epilepsy, hypotonia, mild dysmorphism, coagulopathy
CDG-If	*MPDU1*	noncatalytic protein for Dol-P-Man Dol-P-Glc addition	608799	short stature, icthyosis, psychomotor retardation, pigmentary retinopathy
CDG-Ig	*ALG12*	Dol-P-Man: $Man_7GlcNAc_2$PP-Dol mannosyltransferase	607143	hypotonia, facial dysmorphism, psychomotor retardation, acquired microcephaly, frequent infections
CDG-Ih	*ALG8*	glucosyltransferase II Dol-P-Glc: $Glc_1Man_9GlcNAc_2$-PP-Dol glucosyltransferase	608104	hepatomegaly, protein-losing enteropathy, renal failure, hypoalbuminemia, edema, ascites
CDG-Ii	*ALG2*	mannosyltransferase II GDP-Man: Man_1GlcNAc2-PP-Dol mannosyltransferase	607906	normal at birth, mental retardation, hypomyelination, intractable seizures, iris colobomas, hepatomegaly, coagulopathy
CDG-Ij	*DPAGT1*	UDP-GlcNAc: dolichol phosphate *N*-acetylglucosamine-1-phosphate transferase	608093	severe MR, hypotonia, seizures, microcephaly, exotropia
CDG-Ik	*ALG1*	mannosyltransferase I GDP-Man: $GlcNAc_2$-PP-Dol mannosyltransferase	608540	severe psychomotor retardation, hypotonia, acquired microcephaly, intractable seizures, fever, coagulopathy, nephrotic syndrome, early death
CDG-IL	*ALG9*	mannosyltransferase Dol-P-Man: Man_6 and $Man_8GlcNAc_2$-PP-Dol mannosyltransferase	608776	severe microcephaly, hypotonia, seizures, hepatomegaly
CDG-IIa	*MGAT2*	GlcNAc-transferase 2 (GnT II)	212066	MR, dysmorphism, stereotypies, seizures
CDG-IIb	*GLS1*	α1-2 glucosidase I	606056	dysmorphism, hypotonia, seizures, hepatomegaly, hepatic fibrosis (death at 2.5 months)
CDG-IIc	*SLC35C1/ FUCT1*	GDP-fucose transporter	266265	recurrent infections, persistent neutrophilia, MR, microcephaly, hypotonia (normal Tf)
CDG-IId	*B4GALT1*	β1-4 galactosyltransferase	607091	hypotonia (myopathy), spontaneous hemorrhage, Dandy–Walker malformation
CDG-IIe	*COG7*	conserved oligomeric Golgi complex subunit 7	608779	fatal in early infancy, dysmorphism, hypotonia, intractable seizures, hepatomegaly, progressive jaundice, recurrent infections, cardiac failure
CDG-IIf	*SLC35A1*	CMP-sialic acid transporter	605634	thrombocytopenia, no neurologic symptoms, normal Tf, abnormal platelet glycoproteins

Continued

TABLE 42.1. *(Continued)*

Disorder	Gene	Enzyme	OMIM	Key features
COG8 deficiency	*COG8*	member of the COG complex for Golgi trafficking	611182	hypotonia, mental retardation, encephalopathy, lack of muscle
COG1 deficiency	*COG1*	member of the COG complex for Golgi trafficking	606973	feeding problems, failure to thrive, growth retardation, mild mental retardation, enlarged liver and spleen, cerebral and cerebellar atrophy
RFT1 (flippase) deficiency	*RFT1*	thought to be needed to flip LLO donor to ER lumen	612015	developmental deay, hypotonia, seizures, hepatomegaly
Oligosaccharyl transferase subunit	*TUSC3*	subunit of the OST complex	611093	nonsyndromic mental retardation; abnormal glycosylation not proven; does not affect serum protein glycans
Oligosaccharyl transferase subunit	*IAP*	subunit of OST complex found on X-chromosome		X-linked nonsyndromic mental retardation; abnormal glycosylation not proven; does not affect serum protein glycans
Dolichol kinase deficiency	*DK1*	CTP+ dolichol→ dolichol-P +CDP, activation of dolichol	610768	dry, thin skin, ichthyosis, baldness, hypotonia, dialyated cardiomyopathy, hypoketotic hypoglycemia
Vacuolar ATPase	*ATP6V0A2*	regulates pH in Golgi needed for protein trafficking	611716	cutis laxa,wrinkly skin,connective tissue weakness, large fontanelle, +/– mental retardation
Mucolipidoses II and III	*GNPTA*	UDPGlcNAc: lysosomal enzyme: GlcNAc-1 phosphotransferase	252500	coarsening, organomegaly, joint stiffness, dysostosis, median neuropathy at the wrist; MLIII less severe than MLII, which presents in infancy
Muscular dystrophies				
Walker–Warburg syndrome (WWS)	*POMT1/ POMT2*	O-mannosyltransferase 1	236670	type II lissencephaly, cerebellar malformations, ventriculomegaly, anterior chamber malformations, severe delay, death in infancy
Muscle–eye–brain disease (MEB)	*POMGNT1*	O-mannosyl-β1-2 *N*-acetylglucosaminyltransferase 1	253280	type II lissencephaly, progressive myopia, developmental delay, weakness, hypotonia; resembles but is less severe than WWS
Fukuyama muscular dystrophy	*FCMD*	fukutin, a putative glycosyltransferase	253800	cortical dysgenesis, myopia, weakness, and hypotonia; 40% have seizures
Congenital muscular dystrophy type 1C (MDC1C)	*FKRP*	fukutin-related protein; putative glycosyltransferase	606612	hypotonia, impaired motor development, respiratory muscle weakness
Congenital muscular dystrophy type 1D (MDC1D)	*LARGE*	putative glycosyltransferase	608840	muscular dystrophy with profound mental retardation
Hereditary inclusion body myopathy-II (IBM2)	*GNE*	UDP-GlcNAc epimerase/ kinase	600737	adult onset with progressive distal and proximal muscle weakness; spares quadriceps
Glycosaminoglycan defects				
Ehlers–Danlos syndrome	*B4GALT7*	β1-4 galactosyltransferase 7	130070	progeroid Ehlers–Danlos syndrome: macrocephaly, joint hyperextensibility
Hereditary multiple exostosis	*EXT1/EXT2*	glucuronyltransferase/ *N*-acetylglucosaminyltransferase	133700	multiple exostoses (diaphyseal, juxtaepiphyseal)
Chondrodysplasias	*DTDST/ SLC26A2*	sulfate anion transporter	222600 600972 256050	diastrophic dysplasia: scoliosis, talipes equinovarus, "hitchhiker thumb," malformed ears, airway collapse, and early death in severe cases; adult survival reported achondrogenesis Ib: short-limbed dwarfism, thin ribs with fractures and respiratory failure

Continued

TABLE 42.1. *(Continued)*

Disorder	Gene	Enzyme	OMIM	Key features
Spondyloepimetaphyseal dysplasia	*ATPSK2*	PAPS synthase	603005	abnormal skeletal development and linear growth
Molecular corneal dystrophy types I and II	*CHST6*	keratan sulfate 6-0-sulfotransferase	217800	corneal clouding and erosions, painful photophobia
Schneckenbecken dysplasia	*SLC35D1*	UDP-GlcA/UDP-GalNAc Golgi transporter deficiency	610804	severely shortened long bones, bowing of limb bones and unossified vertebral bodies; mouse model has similar features
Glycosphingolipid defects				
Amish infantile epilepsy	*SIAT9*	GM3 synthase	609056	tonic-clonic seizures, arrested development, neurological decline
O-GalNAc defects				
Tn syndrome	*COSMC*	chaperone of β1-3GalT	230430	anemia, leukopenia, thrombocytopenia
Familial tumorous calcinosis	*GALNT3*	GalNAc transferase	211900	massive calcium deposits in skin and tissue
GPI anchor defects				
Paroxysmal nocturnal hemoglobinuria	*PIGA*	PI-GlcNAcT	311770	complement-mediated hemolysis
Autosomal recessive GPI anchor deficiency	*PIGM*	first mannosyltransferase in GPI anchor synthesis	610273	venous thrombosis and seizures caused by loss of SP1 binding site in promotor
Other				
Peters plus syndrome	*B3GALTL*	β1-3 glucosyltransferase specific for O-fucose-linked glycans on thrombospondin type 1 repeats	261540	short stature, brachymorphism, short limbs, prenatal growth retardation; mental retardation, and sometimes cleft palate
Congenital dyserythropoietic anemia (CDA type II, HEMPAS)	?	?	224100	anemia, jaundice spleenomegaly, gall bladder disease

Adapted, with permission, from Freeze H.H. 2006. *Nat. Rev. Genet.* **7:** 537–551, ©Macmillan.
(CDG) Congenital disorder of glycosylation; (OMIM) Online Mendelian Inheritance in Man database.

Type I Congenital Disorders of Glycosylation

A complete absence of N-glycans is lethal. Therefore, the type I CDG mutations create hypomorphic alleles, not complete knockouts. A deficiency in any of the steps required for the assembly of LLO in the ER (e.g., nucleotide sugar synthesis or sugar addition catalyzed by a glycosyltransferase) produces a structurally incomplete LLO. Because the oligosaccharyltransferase prefers full-sized LLO glycans, this results in hypoglycosylation of multiple glycoproteins. Importantly, many deficiencies in LLO synthesis produce incomplete intermediates. Most of the LLO assembly steps are not easy to assay by standard biochemical methods, but functional assays for LLO assembly have been developed. Because LLO assembly is conserved from yeast to humans, the intermediates that accumulate in CDG patients often correspond to the intermediates that accumulate in mutant *Saccharomyces cerevisiae* strains with known defects in LLO assembly. Some mutant mam-

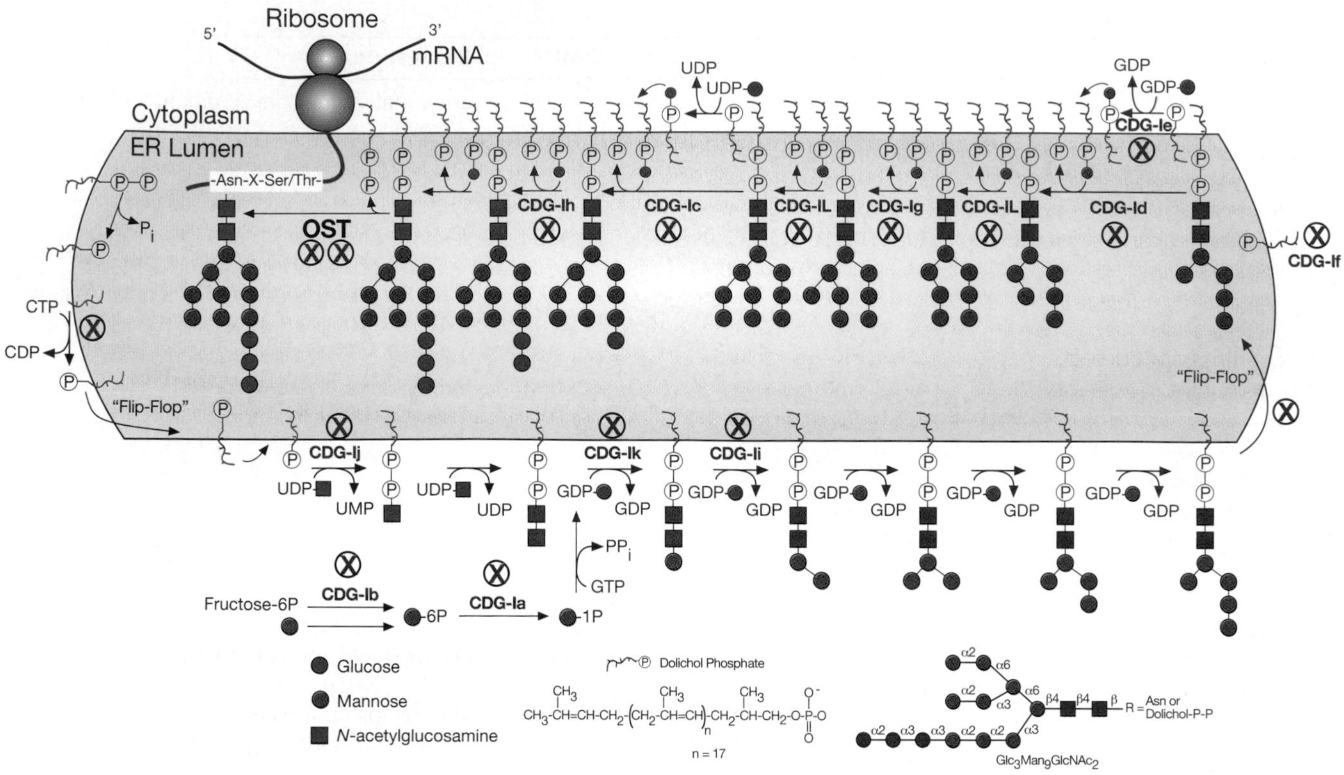

FIGURE 42.1. Location of defects in type I congenital disorders of glycosylation (CDG-I). The figure shows the biosynthesis of the lipid-linked oligosaccharide (LLO) precursor and the defective steps leading to type I CDG. The monosaccharides must first be activated: mannose to mannose-6-phosphate (Man-6-P), which can occur either by direct phosphorylation of mannose or by conversion of fructose-6-P. Man-6-P is converted to Man-1-P and then on to GDP-Man and dolichol-P-mannose (Dol-P-Man). *N*-Acetylglucosamine (GlcNAc) is activated to UDP-GlcNAc (not shown). Glucose (Glc) is activated to UDP-Glc (not shown) and then on to Dol-P-Glc. The assembly of the LLO donor begins with dolichol kinase activating dolichol to Dol-P. Then separate transferases use UDP-GlcNAc to add GlcNAc-1-P to Dol-P and GlcNAc to GlcNAc-P-P-Dol, respectively. Three separate mannosyltransferases catalyze the transfer of mannose from GDP-Man on the cytoplasmic side of the ER (see Chapter 8). The resulting $Man_5GlcNAc_2$-P-P-Dol structure flips into the ER lumen where it is completed with four mannose and three glucose residues using donors Dol-P-Man and Dol-P-Glc. Reactions using these donors employ a protein (MPDU1) that helps localize them within the ER for efficient transfer. Once the LLO is completed, the oligosaccharyltransferase complex transfers glycan to asparagine residues on the peptide emerging from the translocation complex. (*Red crosses*) Steps defective in type I CDG. Additional defects have been found in genes encoding dolichol kinase, "flippase," and two genes (*TUSC3* and *IAP*) that encode subunits of the OST complex. (Adapted, with permission, from Freeze H.H. and Aebi M. 2005. *Curr. Opin. Struct. Biol.* **15:** 1–9, © Elsevier.)

malian cells (e.g., Chinese hamster ovary cells) have been shown to have similar defects (see Chapter 46). The close homology between yeast and human genes enabled the normal human ortholog to rescue defective glycosylation in mutant yeast strains, whereas the cDNA from patients with mutations do not. This provided substantial clues to the likely human defect, and sequencing of the patients' genes can reveal mutations. This approach was invaluable for tracking the human defects that cause CDG-Ic, -Id, -Ig, -Ih, -Ii, -Ik, and -IL. In a few types of CDG where yeast and mammalian pathways diverged, mammalian cell lines with known glycosylation defects are available for complementation by a similar strategy. This is especially important for identifying the defects in CDG-Ie and -If. Established enzyme assays help identify the defects in CDG-Ia, -Ib, -Ie, and -Ij. One critical clue for solving CDG-Ia, the most common type, is that patient cells synthesize a series

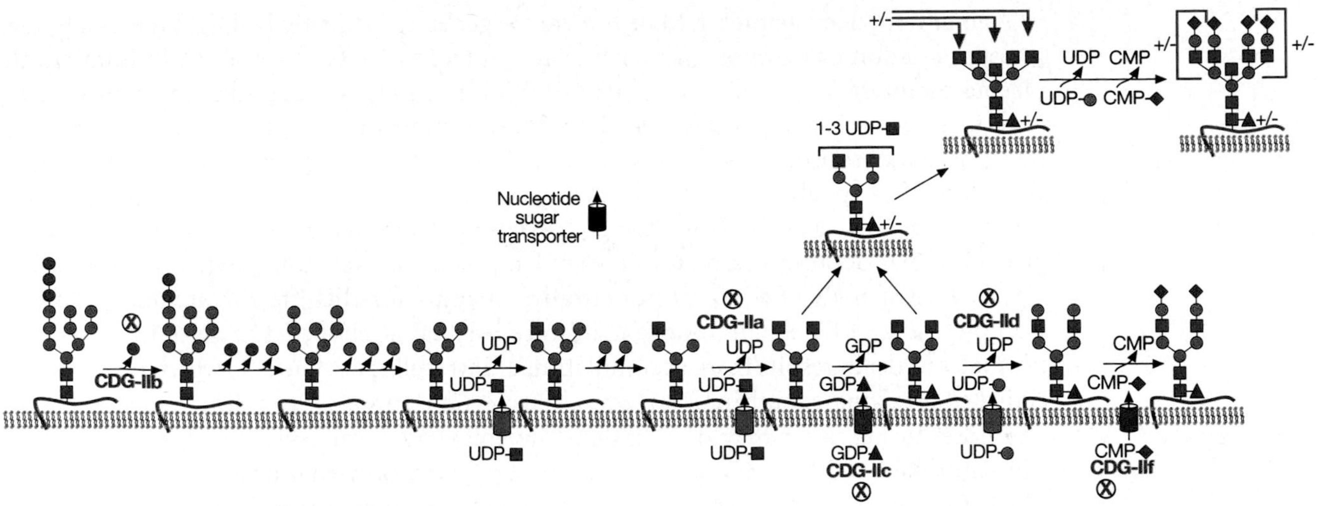

FIGURE 42.2. Location of defects in type II congenital disorders of glycosylation (CDG-II). After the glycan is added to the protein, the trimming/processing reactions generate multiantennary complex-type glycans. Three glucose and four mannose residues are removed and an *N*-acetylglucosamine (GlcNAc) is added to $Man_5GlcNAc_2$-protein by GlcNAcT-I. This is followed by the removal of two mannose residues and the addition of a second (or more) *N*-acetylglucosamine residue. Fucose is added to the core region of the molecule and the antennae can be capped with galactose and sialic acid. (*Red crosses*) Defects in glycosidases, glycosyltransferases, and nucleotide sugar transporters that cause type II CDGs. The shuttling of the glycosylation machinery within the Golgi is organized and regulated by a variety of cytoplasmic complexes including the conserved oligomeric Golgi (COG) complex (not shown). Defects in different subunits of this complex cause a series of type II CDG. (Adapted, with permission, from Patterson M.C. 2006. Disorders of glycosylation. In *Pediatric neurology*, 4th ed. [ed. K. Swaiman et al.], p. 648, © Elsevier.)

of truncated LLO species when incubated in medium containing reduced glucose, whereas normal cells do not. This points to a general limitation in early precursors that originally led to the identification of phosphomannomutase as the defective enzyme in CDG-Ia, resulting in deficiency of GDP-mannose.

CDG-Ia is the most common type of CDG, with more than 600 cases identified worldwide. The patients have moderate to severe psychomotor retardation, hypotonia, dysmorphic features, failure to thrive, liver dysfunction, coagulopathy, abnormal endocrine functions, and a pronounced susceptibility to infection. Scores of mutations have been found in *phosphomannomutase 2* (*PMM2*), the defective gene in CDG-Ia. *PMM2* encodes an enzyme that catalyzes the conversion of Man-6-P to Man-1-P, which is a precursor required for the synthesis of GDP-mannose (GDP-Man) and dolichol-P-mannose (Dol-P-Man). Both donors are substrates for the mannosyltransferases involved in the synthesis of $Glc_3Man_9GlcNAc_2$-PP-Dol and their levels are decreased in CDG-Ia patients. Patients have hypomorphic alleles and complete loss of activity is lethal. In fact, mouse embryos lacking *Pmm2* die 2–4 days after fertilization, whereas those with homozygous hypomorphic alleles survive. There are currently no therapeutic options for CDG-Ia patients. In vitro studies suggested that supplements of mannose might improve glycosylation, but mannose therapy for CDG-Ia patients is ineffective.

CDG-Ib is caused by a deficiency in phosphomannose isomerase activity encoded by the gene *MPI.* The enzyme interconverts fructose-6-phosphate and Man-6-P, and deficiency leads to the synthesis of glycoproteins with unoccupied glycosylation sequons. In contrast to CDG-Ia, patients with CDG-Ib do not have psychomotor or developmental abnormalities. Instead, they show hypoglycemia, coagulopathy, severe vomiting and diarrhea, protein-losing enteropathy, and hepatic fibrosis. Several patients died of severe bleeding before the basis of this CDG had been described.

As mentioned in Chapter 4, Man-6-P can be generated directly by hexokinase-catalyzed phosphorylation of mannose. This pathway is intact in CDG-Ib patients. In humans, the plasma mannose level is about 50 μM, resulting from glycan degradation or processing. Mannose can be taken up into cells by a hexose transporter. Oral mannose therapy for phosphomannose-isomerase-deficient CDG-Ib patients was quite successful and corrected coagulopathy, hypoglycemia, protein-losing enteropathy, and intermittent gastrointestinal problems, as well as normalizing the glycosylation of plasma transferrin and other serum glycoproteins. Because orally administered mannose is well tolerated, this approach is clearly a satisfyingly effective, if not curative, therapy for this life-threatening condition. Complete loss of the single *Mpi* gene in mice is lethal at about E11.5. N-Glycosylation is normal, but death results from accumulation of intracellular Man-6-P, which depletes ATP, activates hexokinase, and inhibits several glycolytic enzymes. Providing dams with extra mannose during pregnancy only hastens the embryo's demise via the "honeybee effect." This unusual condition happens when bees are given only mannose instead of glucose. They continue flying for a short time and then literally drop dead. The reason is that they have low phosphomannose isomerase activity compared to hexokinase and therefore accumulate Man-6-P, which they degrade, only to phosphorylate it again using an ever-decreasing amount of ATP. Entry of Man-6-P into glycolysis is very slow and thus the bees become energy starved and die within a few minutes. CDG-Ib patients have sufficient residual phosphomannose isomerase activity that they do not accumulate intracellular Man-6-P when given mannose, but it is sufficient to correct impaired glycosylation.

Other types of CDG-I have a broad range of clinical phenotypes. Some have characteristic phenotypes: for example, low LDL, low IgG, kidney failure, genital hypoplasia, and cerebellar hypoplasia. The reasons for these effects are unknown, but accumulated LLO intermediates may be toxic or there may be specific requirements for polymannosyl glycans that a truncated LLO cannot provide. It is likely that patients will be found with mutations in all the remaining steps of LLO biogenesis. See Table 42.1 and Figure 42.1 for known other types of CDG-I. Defects were recently found in genes encoding dolichol kinase, a putative LLO flippase, and two oligosaccharyltransferase subunits.

Type II Congenital Disorders of Glycosylation

Type II CDG disorders (Figure 42.2) are more diverse than type I disorders because mutations occur in glycosyltransferases, nucleotide sugar transporters, and cytoplasmic proteins that traffic the glycosylation machinery into and within the Golgi.

Structural analysis of serum protein glycans helped to pinpoint the defect in CDG-IIa. It is caused by mutations in *MGAT2*, which encodes N-acetylglucosaminyltransferase-II (GlcNAcT-II), the enzyme that adds the second *N*-acetylglucosamine of biantennary complex-type chains, the most common glycan in serum glycoproteins. In CDG-IIa patients, these glycans are replaced with monosialylated hybrid chains. The patients have dysmorphic features and psychomotor retardation, but do not have peripheral neuropathy or abnormal cerebellar development. Very rare (1%) survivors of a mouse line ablated in *Mgat2* provide a near phenocopy of the patients' pathology, including some of the dysmorphic features (see Chapter 8).

In CDG-IId, serum protein glycan structural analysis showed the loss of both galactose and sialic acid from transferrin as a result of the loss of nearly all β1-4 galactosyltransferase activity. Determining the cause of CDG-IIc, also called leukocyte adhesion deficiency type II (LAD-II), was more complex. Here transferrin sialylation was normal, so this defect was not detected by the usual test, but the N-glycans of IgM and O-linked glycans on leukocyte surface proteins are deficient in fucose. This group of proteins includes the selectin ligand carrying the glycan, sialyl Lewis^x that mediates leukocyte rolling prior to extravasation of leukocytes from the capillary lumen into the tissues (see Chapter 31). This defect greatly elevates circulating leukocytes and leads to frequent infections. The defect is due to mutations in the GDP-fucose transporter, which explains why it affects fucosylation of both N- and O-linked glycans. The success of mannose therapy for CDG-Ib prompted the use of orally administered fucose as a therapy for CDG-IIc. Several patients responded well to daily supplements of fucose in their diet. Sialyl Lewis^x reappeared on the leucocytes, and elevated circulating neutrophils promptly returned to normal levels. As mentioned in Chapter 4, fucose is converted into Fuc-1-P by fucose kinase and Fuc-1-P is converted to GDP-Fuc by GDP-Fuc pyrophosphorylase. A mouse model of fucose deficiency has been described that lacks de novo biosynthesis of GDP-Fuc from GDP-Man (see Chapter 4) because of the loss of the FX protein (GDP-4-keto-6-deoxymannose 3,5-epimerase-4-reductase). The mice die without fucose supplements, but providing fucose in the drinking water rapidly and reversibly normalizes their elevated neutrophils. The treatment also corrects abnormal hematopoeisis resulting from disrupted Notch signaling, emphasizing the crucial importance of its O-fucose glycans for signaling. CDG-IIf is caused by a defect in the CMP-sialic acid transporter (see Chapter 4). The patient completely lacks sialyl Lewis^x on leukocytes because of a deficiency of sialic acid, leading to severe neutropenia, formation of giant platelets, and abnormal platelet glycoproteins.

CDG-IIe is a new type of CDG-II defect caused by mutations in the *COG7* gene that disrupt multiple glycosylation pathways by impeding the normal trafficking of multiple glycosyltransferases and nucleotide sugar transporters. This mutation affects the synthesis of both N- and O-linked glycans including glycosaminoglycan (GAG) chains, so the phenotypic effects are far reaching. Cog7 is part of the eight-subunit COG (conserved oligomeric Golgi) complex, which is thought to have multiple roles in trafficking within the Golgi. Mammalian cells deficient in Cog1, Cog2, Cog3, and Cog5 also show various degrees of altered glycosylation. Many unclassified CDG-II patients who show defects in multiple pathways may have alterations in trafficking proteins that will likely impact synthesis of GAG chains and the normal assembly of the extracellular matrix. Several patients have defects in Cog8, and one has a defect in Cog1, but the phenotypes are surprisingly mild compared to the Cog7 deficiency. It is likely that mutations will be found in other Cog subunits as well as in other proteins that control protein trafficking or lumenal pH homeostasis in the Golgi. A series of mutations in a vacuolar H^+-ATPase subunit causes a CDG-II that affects multiple glycosylation pathways.

Other genetic defects in N-glycan synthesis are known. I-cell disease, which results from the lack of Man-6-P on lysosomal enzyme N-glycans, is covered in Chapter 30, but for historical reasons it was not classified as a CDG.

Is it possible to have diseases caused by "excessive" glycosylation? Not all potential N-glycosylation sites are occupied, and site occupancy for some proteins varies by tissue. The location of N-glycan sites tends to be conserved in most proteins, and point mutations that randomly create novel glycosylation sites could cause protein misfolding and rapid degradation. Perhaps the survival of these abnormally glycosylated proteins would have even worse consequences because new glycans might impair an important function or the ability to form multimeric complexes. For example, Marfan syndrome is caused by known mutations in fibrillin1 (FBN1). One of these creates an N-glycosylation site that disrupts multimeric assembly.

A good example of hyperglycosylation is seen in patients with heightened susceptibility to mycobacterial infections because they have an N-glycan site mutation in interferon γ receptor 2 (IFNγR2). Protein folding and surface localization are normal, but the function of the hyperglycosylated protein is dramatically decreased. Enzymatic removal of the additional glycan at the cell surface restores IFNγR2 activity. Thus, inserting the N-glycan at that site destroyed the receptor's function. This may not be an isolated case. A survey of 577 missense mutations of proteins traveling the ER-Golgi pathway showed that 13% of them potentially create inappropriate glycosylation sites. This is far greater than predicted for the effects of random mutations and may mean that hyperglycosylation has a greater impact on function than loss of glycosylation. It is therefore not surprising that random insertion of bulky glycans into a finely tuned receptor has negative effects on formation of signaling complexes.

Congenital Dyserythropoieic Anemia Type II

More than 200 cases of autosomal recessive congenital dyserythropoieic anemia type II (CDA-II) are known. This disorder is also called HEMPAS (hereditary erythroblastic multinuclearity with a positive acidified serum lysis test). Patients generally have a normal life span, although complications may develop with age, including an enlarged liver, jaundice, gallstones, or diabetes. Ineffective erythropoiesis causes anemia and morphological abnormalities of the majority of erythroblasts in the bone marrow. In erythroid cells, the presence of poly-*N*-acetyllactosamines on glycosphingolipids is greatly increased along with a loss of protein-bound complex N-glycans and an increase in hybrid forms. This suggested a deficiency of N-glycan processing enzymes α-mannosidase II or GlcNAcT-II (encoded by the *MGAT2* gene). Reduced expression of Golgi α-mannosidase II and/or GlcNAcT-II activities has been observed in some, but not all, cases of HEMPAS. The α-mannosidase II knockout mouse also develops dyserythropoietic anemia. However, no CDA-II patients have proven mutations in these genes and *MGAT2* mutations cause CDG-IIa. Linkage analysis located the defective locus in CDA type II to 20q11.2, which eliminates both of these candidates. Sequencing candidate genes in this region in a large CDA-II population also failed to find mutations in other known genes. The cause(s) of CDA-II therefore remains unknown; however, alteration of a transcription factor or other regulatory molecule that affects several enzymes has been suggested. Alternatively, appreciation of the function of glycosylation-promoting chaperone proteins in the ER (e.g., COSMC) and cytoplasm (e.g., COG complex) offers new possibilities to consider.

GALACTOSEMIA

Galactosemia refers to a group of diseases caused by inherited defects in the genes encoding three enzymes involved in galactose metabolism. One of these disorders, termed "classical galactosemia", is caused by a deficiency of galactose-1-phosphate uridyltransferase (GALT; Figure 42.3). This disease increases intracellular galactose-1-phosphate and decreases syn-

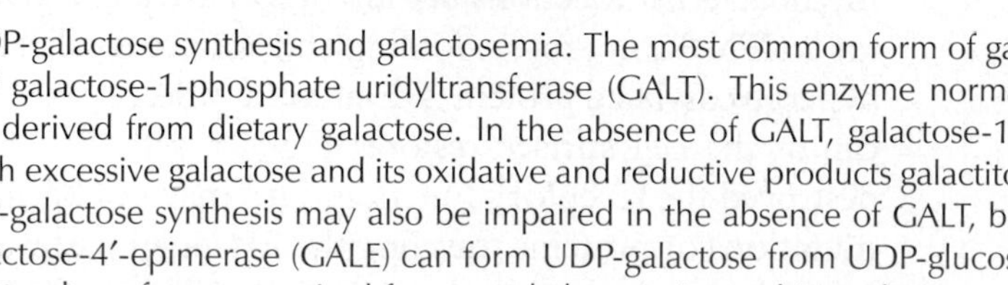

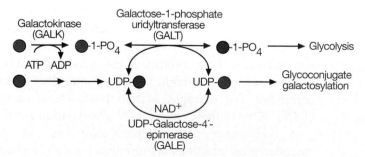

FIGURE 42.3. UDP-galactose synthesis and galactosemia. The most common form of galactosemia is due to a deficiency of galactose-1-phosphate uridyltransferase (GALT). This enzyme normally utilizes galactose-1-phosphate derived from dietary galactose. In the absence of GALT, galactose-1-phosphate accumulates, along with excessive galactose and its oxidative and reductive products galactitol and galactonate (not shown). UDP-galactose synthesis may also be impaired in the absence of GALT, but not completely because UDP-galactose-4′-epimerase (GALE) can form UDP-galactose from UDP-glucose and can supply the donor to galactosyltransferases required for normal glycoconjugate biosynthesis.

thesis and availability of UDP-galactose. Defects in UDP-galactose-4′-epimerase (GALE; Figure 42.3) or galactokinase (GALK; Figure 42.3) also cause the disease, but they are rare.

As infants, GALT-deficient patients fail to thrive and have enlarged liver, jaundice, and cataracts. Lactose-free diet ameliorates most of the acute symptoms because it reduces the amount of galactose entering the metabolic pathway and diminishes the accumulation of galactose and galactose-1-phosphate that is thought to contribute to the symptoms of the disease. Reducing galactose accumulation also helps to inhibit the formation of galactitol and galactonate, which are produced via reductive or oxidative metabolism of galactose, respectively. Galactitol is not metabolized further and has osmotic properties that contribute to cataract formation. Unfortunately, a galactose-free diet apparently does not prevent the appearance of cognitive disability, ataxia, growth retardation, and ovarian dysfunction that are characteristic of this disease. The long-term complications in treated GALT-deficient individuals may be due to small amounts of toxic metabolites that continue to accumulate in these patients. Another possibility is that the complications are long-term outcomes of dysfunctions that originated during fetal life. GALT deficiency may decrease UDP-galactose and galactosylated glycans. Hypogalactosylation of glycoproteins and glycolipids has been observed in some GALT-deficient individuals. Some patients who mistakenly receive galactose develop a hypoglycosylated transferrin that is missing both entire N-glycans and individual monosaccharides on the remaining chains. The basis of this combined loss is not understood, but the pattern returns to normal when the patients are placed on a galactose-free diet. It is possible that accumulation of hypogalactosylated glycans is secondary to a general metabolic abnormality in these patients, but how abnormal structures lead to the complications in these patients remains unsolved.

MUSCULAR DYSTROPHIES

Congenital Muscular Dystrophies

Mutations altering glycans on α-dystroglycan (αDG) cause at least five types of congenital muscular dystrophies (CMDs) (Table 42.1 and Figure 42.4) and other CMDs with unknown etiologies will likely involve glycosylation. The O-mannosylation pathway is presented in Chapter 12. The initial step occurs in the ER lumen using a Dol-P-Man donor and a het-

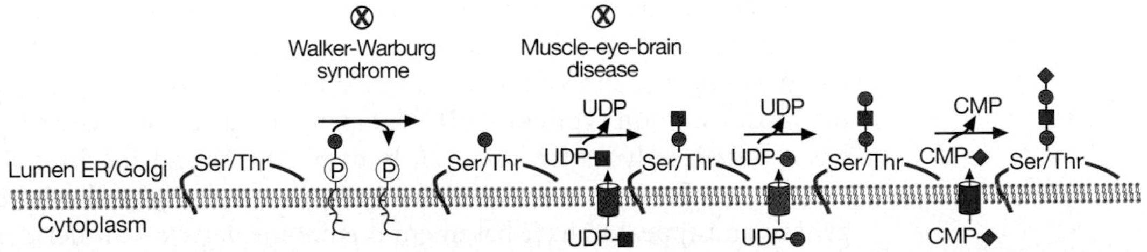

FIGURE 42.4. Location of glycosylation defects in the O-mannosyl glycosylation pathway leading to congenital muscular dystrophies. The biosynthesis of O-mannosyl glycans primarily on α-dystroglycan first involves the addition of mannose to serine or threonine residues catalyzed by the *POMT1/POMT2*-encoded mannosyltransferase complex, which uses dolichol-P-mannose as donor. A deficiency of this enzyme activity causes some cases of Walker–Warburg syndrome (WWS). β1-2GlcNAc is then added to mannose using the enzyme encoded by *POMGNT1*. Disruption of this step causes muscle–eye–brain (MEB) disease. The O-mannosyl glycan is extended with galactose and sialic acid residues using unknown transferases. The genes *Fukutin*, *FKRP*, and *LARGE* have glycosyltransferase motifs, but have not been shown to catalyze any specific reaction. (Adapted, with permission, Patterson M.C. 2006. Disorders of glycosylation. In *Pediatric neurology*, 4th ed. [ed. K. Swaiman et al.], p. 653, © Elsevier.)

erodimeric enzyme (protein O-mannosyltransferase; POMT1/POMT2) to add mannose in α1-O linkage to serine/threonine. In the Golgi, a pathway-specific protein O-mannoside β1-2GlcNAc-transferase (POMGNT1) creates a disaccharide and then galactose and sialic acid are added by unspecified transferases. αDG is the only identified carrier of these glycans. It is a vital component of the dystrophin glycoprotein complex (DGC) on the sarcolemma of skeletal muscle cells. Dystroglycan is a major component of the DGC that connects the ECM to the cytoskeleton in many tissues. Both components (αDG and βDG) are derived from a single gene, *DAG1*. In muscle, actin in the cytoskeleton is linked to βDG, which spans the cell membrane. The extracellular domain of βDG binds to αDG, which in turn binds laminin-2 in the ECM via its glycan-containing extracellular domain. Traditional naming of CMDs based on clinical symptoms is being replaced by designation of the specific defective genes.

The degree and type of αDG glycosylation vary in different tissues. The presence of sialic acid may be required for laminin-2 binding. Monoclonal antibodies against the glycans show very low or undetectable binding to αDG in some patients and provide the key tool to identify glycosylation-related defects, similar to the use of transferrin for N-glycosylation defects in the CDGs. Normal αDG runs as a diffuse band of 130–190 kD as detected by an antibody that recognizes the protein core, but in CMD patients anti-αDG recognizes a less diffuse band of about 90 kD, regardless of which step is defective. The size and amount of βDG are unaffected. Although no pathological mutations have been found in αDG itself, the name α-dystroglycanopathy has been suggested to describe these glycosylation disorders.

Walker–Warburg syndrome (WWS) is the most severe form of the CMD caused by defective αDG glycosylation. Patients have a short life span (less than 1 year on average), multiple brain abnormalities, and severe muscular dystrophy. About 20% of patients have mutations in *POMT1* and a few have mutations in *POMT2*. As with *DAG1*-null mice, *POMT1* ablation is an early embryonic lethal; both mutant strains of mice fail to synthesize Reichert's membrane, a basement membrane that surrounds the blastocyst, supporting a functional connection between the two genes. A few clinically defined WWS patients have defects in the *fukutin* and *FKRP* (fukutin-related protein) genes, which are thought to function in this glycosylation pathway. Mutations in *fukutin* and *FKRP* also cause other, milder forms of CMD (see below). Linkage studies of consanguineous WWS families suggest that mutations in at least two other genes can cause disease.

The *POMGnT1* gene is mutated in muscle–eye–brain disease (MEB), which is characterized by symptoms similar to, but milder than, WWS. The most severely affected patients die during the first years of life, but the majority of mild cases survive to adulthood.

Fukuyama muscular dystrophy (FCMD) is caused by a single 3-kb 3′-retrotransposon insertional event into the *fukutin* gene, which occurred 2000–2500 years ago. This partially reduces the stability of the mRNA, making it a relatively mild mutation. FCMD is one of the most common types of CMD in Japan with a carrier frequency of 1/188. The protein has a putative glycosyltransferase domain (DXD) and localizes to the *cis*-Golgi, but no transferase activity has yet been demonstrated. *Fukutin*-null mice die by E9.5 in embryogenesis and appear to have basement membrane defects. Chimeric mice with less than 50% contribution of cells that are heterozygous for loss of *fukutin* show typical muscular dystrophy, compromised survival, and disorganized laminar structures in the brain, along with ocular and retinal abnormalities.

Studies of two other CMDs provide further examples of novel, as yet uncharacterized, proteins (putative enzymes) that play a role in αDG glycosylation. MDCIC is a relatively mild disorder that is caused by mutations in *FKRP*. The protein has a glycosyltransferase signature (DXD) but no known transferase activity. It is Golgi localized, but a portion may also reside in the ER. Patients with MDC1D, a limb-girdle muscular dystrophy, contain mutations in *LARGE*, originally described in myodystrophic mice (*myd*, now called

Large^{myd}). The protein has two glycosyltransferase signatures (DXD motifs) in different domains, suggesting the possibility of a bifunctional enzyme. No enzymatic activity has been demonstrated, but mutations of the DXD motifs prevent LARGE from rescuing defective glycosylation. The protein resides in the Golgi and appears to recognize an amino-terminal region of αDG defining an enzyme-substrate recognition motif necessary to initiate functional glycosylation. Molecular recognition of αDG by the product of *LARGE* is required for the biosynthesis of a functional dystroglycan and to prevent muscle degeneration.

Overexpression of *LARGE* results in hyperglycosylation of αDG by an as yet unknown compensatory mechanism and phenotypically rescues the *Large*^{myd} mouse. The glycan-enriched αDG shows a high affinity for extracellular ligands. LARGE circumvents the αDG glycosylation defect in cells from individuals with genetically distinct types of CMD without increasing the expression of the other dystroglycan complex proteins. Transfection of the *LARGE* gene into cultured cells from CMD patients restores αDG receptor function and rescues defective FCMD myoblasts, MEB fibroblasts, and WWS myoblasts. These results are significant because they may mean that *LARGE* could be a broad-based gene therapy for CMD. LARGE is not the only protein with this effect. Overexpression of a specific cytotoxic T cell α-N-acetylgalactosaminyltransferase can also create a novel glycan in αDG and prevent the onset of pathology in the *mdx* mouse; this mouse is deficient in dystrophin, the cause of Duchenne's muscular dystrophy, showing that formation of an effective extracellular complex can override and substitute for the loss of the intracellular link to actin. The mechanism appears to involve increased binding to utrophin, a dystrophin homolog, which is not normally present at the muscle membrane in sufficient levels.

Inclusion Body Myopathy 2

Autosomal recessive, adult-onset inclusion body myopathy type 2 (IBM2) occurs worldwide, but is especially common among Persian Jews (1:1500). It is allelic with distal myopathy with rimmed vacuoles (Nonaka myopathy) and patients accumulate vacuoles containing β-amyloid, tau, and presenilin. They are caused by mutations in the *GNE* gene, which encodes a bifunctional, two-domain enzyme UDP-GlcNAc 2-epimerase/N-acetylmannosamine kinase that catalyzes successive steps in the de novo pathway for sialic acid biosynthesis (see Chapter 12). Dominant mutations in *GNE* also cause sialuria, resulting in continuous secretion of sialic acid, but IBM2 patients do not show this phenotype. Sialuria patients have mutations that inactivate an allosteric binding site for CMP-sialic acid. In IBM2, mutations occur in various combinations in both domains, with the most common (M712T) occurring in the kinase domain. In vitro assays show only moderately reduced enzyme activity (20–60%) in both domains and it is unproven whether such modest decreases in measured activity cause defective sialylation, especially on αDG. It is unknown whether GNE has additional roles besides sialic acid biosynthesis. Since sialic acid is efficiently salvaged from degraded glycoproteins, some cells may be less reliant on de novo synthesis, but there is little information on the cell-type preference or age-dependent contributions of the de novo versus salvage pathways. *GNE*-null mice have an embryonic lethal outcome. Most of the mice homozygous for the M712T mutation die a few days after birth; however, they do not develop a myopathy. Instead, they have severe hematuria and proteinuria as a result of abnormalities in the glomerular basement membrane. The major sialoprotein in foot podocytes (podocalexin) is undersialylated. Providing *N*-acetylmannosamine to the pups rescues a portion of the pups and increases sialylation of podocalexin. This may provide a therapy for IBM2 patients, although it is not yet proven that the muscle defect is due to reduced sialyation. Another mouse model, carrying a *GNE* mutation common in the Japanese population, develops a pathological muscle phenotype involving β-amyloid deposition that precedes the accumulation of inclusion bodies. No therapeutic studies have been done on this model.

DEFECTS IN O-GalNAc GLYCANS

A defect in O-linked glycosylation causes familial tumoral calcinosis. This severe autosomal recessive metabolic disorder shows phosphatemia and massive calcium deposits in the skin and subcutaneous tissues. The disorder is due to mutations in *GALNT3*, one of the O-GalNAc transferases in mucin O-glycosylation. Mutations in the O-glycosylated fibroblast growth factor 23 (FGF23) also cause phosphatemia, suggesting that *GALNT3* modifies FGF23. The rare autoimmune disease, Tn syndrome, is caused by somatic mutations in the X-linked gene *COSMC*, which encodes a highly specific chaperone required for the proper folding and normal activity of β1-3galactosyltransferase needed for synthesis of core 1 O-glycans (see Chapter 43 for additional details).

DEFECTS IN PROTEOGLYCAN SYNTHESIS

Proteoglycans and their GAG chains are critical components in extracellular matrices. For a discussion of their biosynthesis, core proteins, and function, see Chapter 16.

Ehlers–Danlos Syndrome (Progeroid Type)

Ehlers–Danlos syndrome (progeroid type) is a connective tissue disorder characterized by failure to thrive, loose skin, skeletal abnormalities, hypotonia, and hypermobile joints, together with delayed motor development and delayed speech. The molecular basis of the disorder is in the synthesis of the core region of xylose-based GAG chains. Decorin, a dermatan sulfate proteoglycan that binds to collagen fibrils, is partially deficient, and some molecules are made without an extended GAG chain. Galactosyltransferase I, the enzyme that adds galactose to xylosyl-serine has only 5% of normal activity. The activity of galactosyltransferase II, the enzyme responsible for adding the second galactose residue to the GAG chain core, has only 20% of normal activity. One possible explanation for the dual effect is that the primary mutation affects the formation or stability of a biosynthetic complex involving several GAG-chain biosynthetic enzymes. Because all proteoglycans contain this common linkage region, it is unclear why the defect has a selective effect on decorin.

Congenital Exostosis

Defects in the formation of heparan sulfate (HS) cause hereditary multiple exostosis (HME), an autosomal dominant disease with a prevalence of about 1:50,000. It is caused by mutations in two genes *EXT1* and *EXT2*, which are involved in HS synthesis. HME patients have bony outgrowths, usually at the growth plates of the long bones. Normally, the growth plate contains chondrocytes in various stages of development, which are enmeshed in an ordered matrix composed of collagen-chondroitin sulfate. In HME, however, the outgrowths are often capped by disorganized cartilagenous masses with chrondrocytes in different stages of development. About 1–2% of patients also develop osteosarcoma.

HME mutations occur in *EXT1* (60–70%) and *EXT2* (30–40%). The proteins encoded by these genes are thought to exist as a complex in the Golgi and both are required for polymerizing GlcNAcα1-4 and GlcAβ1-3 into HS. However, the partial loss of one allele of either gene appears sufficient to cause HME. This means that haploinsufficiency decreases the amount of HS and that EXT activity is rate limiting for HS biosynthesis. This is unusual because most glycan biosynthetic enzymes are in substantial excess.

The mechanism of HME pathology is likely rooted in a disruption of the normal distribution of HS-binding growth factors, which include FGF and morphogens such as hedgehog, Wnt, and members of the TGF-β family. The loss of HS disrupts these pathways in *Drosophila*. Mice that are null for either *Ext* gene are embryonic lethal and fail to gastrulate; however, *Ext* heterozygous animals are viable and about one third develop a visible exostoses on the ribs. No exostoses develop on the long bones of these animals (in contrast to patients with HME), but subtle chondrocyte growth abnormalities were seen in the growth plates of these bones. Further studies are needed to understand how truncation of the HS chains leads to ectopic growth plate formation and the phenotype abnormalities.

Achondrogenesis, Diastrophic Dystrophy, and Atelosteogenesis

Three autosomal recessive disorders, diastrophic dystrophy (DTD), atelosteogenesis type II (AOII), and achondrogenesis type IB (ACG-IB), all result from defective cartilage proteoglycan sulfation. These forms of osteochondrodysplasia have various outcomes. AOII and ACG-IB are perinatally lethal because of respiratory insufficiency, whereas DTD patients develop symptoms only in cartilage and bone, including cleft palate, club feet, and other skeletal abnormalities. Those DTD patients surviving infancy often live a nearly normal life span. All of these disorders result from different mutations in the DTD gene that encodes a plasma membrane sulfate transporter. Unlike monosaccharides, sulfate released from degraded macromolecules in the lysosome is not salvaged well. The heavy demand for sulfate in bone and cartilage proteoglycan synthesis probably explains why the symptoms are most evident in these locations. Defects in the UDP-GlcA/UDP-GalNAc Golgi transporter cause Schneckenbecken dysplasia. Patients have bone abnormalities similar to those seen in other chrondrodysplasias, and a mouse model of the disease shows similar features.

Macular Corneal Dystrophy

Keratan sulfate (KS) in the cornea is an N-linked oligosaccharide with poly-*N*-acetyllactosamine repeats (Galβ1-4GlcNAcβ1-3) variably sulfated at the 6-positions. Macular corneal dystrophy (MCD), an autosomal recessive disease, causes the cornea to become opaque and corneal lesions to develop. Two types of MCD have been described. MCD I appears to be due to a deficiency in sulfating the repeating units. Both galactose and *N*-acetylglucosamine are sulfated in KS; sulfation of galactose and *N*-acetylgalactosamine in chondroitin sulfate are also affected in MCD patients.

DEFECTS IN GLYCOSPHINGOLIPID SYNTHESIS

Positional cloning identified *SIAT9* as the cause of autosomal recessive Amish infantile epilepsy syndrome. This gene encodes a sialyltransferase required for the synthesis of the ganglioside GM3 (Siaα2-3Galβ1-4Glc-ceramide) from lactosylceramide (Galβ1-4Glc-ceramide). GM3 is also a precursor for some more complex gangliosides. The nonsense mutation in these families truncates the protein and abolishes GM3 synthase activity. Analysis of plasma glycosphingolipids showed the accumulation of multiple nonsialylated glycolipids. In contrast to the human form of the disease, mice that lack GM3 do not have seizures or a shortened life span. However, mouse strains that are null for the sialyltransferase and an N-acetylgalactosaminyltransferase that is required for making other complex gangliosides do develop seizures, suggesting that it is the absence of these more complex gangliosides that may be the underlying problem (see Chapter 10).

PHENOTYPES, MULTIPLE ALLELES, AND GENETIC BACKGROUND

One puzzling feature of the genetic disorders of glycosylation is that the phenotypic expression of the same mutation can have widely variable impact, even among affected siblings. Explanations based on the level of residual enzymatic activity for these "simple Mendelian disorders" are neither simple nor generally satisfying. Genotype–phenotype correlations are often difficult to establish. The most likely explanation is differences in the "genetic background," a term meaning all the individual's other genes. For instance, a very frequent single-nucleotide polymorphism (SNP) in *ALG6*, the cause of CDG-Ic, has a barely discernible effect on glycosylation of a model protein in yeast and yet when examined in CDG-Ia patients (PMM2 deficiency), the SNP is twice as frequent in severe cases relative to mild cases. In yeast, *Alg6* deletion decreases glycosylation, but it is not lethal; however, when it is combined with another nonlethal mutation in the oligosaccharyltransferase gene (*Wbp1*), lethality occurs. The effects of mutations are context dependent. As mentioned above, a knockout mutation may be lethal in one highly inbred mouse strain, but not in another because compensatory pathways may exist. Dietary and environmental impacts are substantial as seen in CDG-Ib patients with and without oral mannose therapy. As discussed in Chapter 43, the genetic deficiency in these patients makes them more susceptible to infections and to the impact of pro-inflammatory cytokines on their intestinal protein leakage, probably as a result of loss of epithelial heparan sulfate. The synergism of multiple simultaneous or sequential environmental insults on genetic insufficiencies may create a cascade leading to overt disease.

FURTHER READING

Zak B.M., Crawford B.E., and Esko J.D. 2002. Hereditary multiple exostoses and heparan sulfate polymerization. *Biochim. Biophys. Acta* **1573:** 346–355.

Martin P.T. 2005. The dystroglycanopathies: The new disorders of O-linked glycosylation. *Semin. Pediatr. Neurol.* **12:** 152–158.

Freeze H.H. 2006. Genetic defects in the human glycome. *Nat. Rev. Genet.* **7:** 537–551.

Fridovich-Keil J.L. 2006. Galactosemia: The good, the bad, and the unknown. *J. Cell. Physiol.* **209:** 701–705.

Kanagawa M. and Toda T. 2006. The genetic and molecular basis of muscular dystrophy: Roles of cell-matrix linkage in the pathogenesis. *J. Hum. Genet.* **51:** 915–926.

Ungar D., Oka T., Krieger M., and Hughson F.M. 2006. Retrograde transport on the COG railway. *Trends Cell Biol.* **16:** 113–120.

Wopereis S., Lefeber D.J., Morava E., and Wevers R.A. 2006. Mechanisms in protein O-glycan biosynthesis and clinical and molecular aspects of protein O-glycan biosynthesis defects: A review. *Clin. Chem.* **52:** 574–600.

Eklund E.A., Bode L., and Freeze H.H. 2007. Diseases associated with carbohydrates/glycoconjugates, In *Comprehensive glycoscience* (ed. J. Kamerling, G.-J. Boons, Y. Lee, et al.), Vol. 4, pp. 339–372. Elsevier, Amsterdam.

Jaeken J. and Matthijs G. 2007. Congenital disorders of glycosylation: A rapidly expanding disease family. *Annu. Rev. Genomics Hum. Genet.* **8:** 261–278.

CHAPTER 43

Glycans in Acquired Human Diseases

Ajit Varki and Hudson H. Freeze

GIVEN THE DIVERSE AND UBIQUITOUS PRESENCE OF GLYCANS on all cell surfaces, it is not surprising that several human disease conditions involve acquired (noninherited) changes in glycosylation and/or in the recognition of glycans. This chapter discusses some examples of these situations and considers the mechanisms of the changes seen, as well as the pathophysiological roles of glycans. Wherever relevant, the potential therapeutic significance of the information is mentioned. Details regarding some of these situations are covered elsewhere in the text. Glycosylation changes in cancer and pathologies resulting from inherited human genetic disorders are discussed separately in Chapters 44 and 42, respectively. It will be evident from the examples presented in this chapter that acquired changes in glycans and/or in their recognition have a significant role in a variety of human diseases. In some cases, the evidence remains circumstantial, and further work is needed to define whether the glycan changes have a primary role. In many of the situations, detailed knowledge of the nature of glycan–receptor interactions could result in improved diagnostic or therapeutic approaches.

CARDIOVASCULAR MEDICINE

Role of Selectins in Reperfusion Injury

A variety of common cardiovascular disorders (e.g., stroke, myocardial infarction, and hypovolemic shock) are characterized by a period of decreased or absent blood flow followed by a state of reperfusion, which occurs either by natural mechanisms or because blood flow has been restored by medical intervention. Despite rescue of the tissue from permanent anoxemic necrosis, the entry of leukocytes into the reperfused area can initiate a cascade of events that ultimately results in tissue damage (called reperfusion injury). P-Selectin on the activated endothelium in the reperfused area and/or L-selectin on leukocytes have vital roles in mediating the initial steps of this cascade (see Chapter 31). Substantial data in animal model systems indicate that blockade of this initial selectin-based recognition can ameliorate the subsequent tissue damage. A major goal of some pharmaceutical and biotechnology companies has been to make small-molecule inhibitors that can be used to achieve this blockade in human patients (see Chapters 49 and 50 regarding the synthesis of small glycan molecules designed to be selectin inhibitors). Interestingly, some forms of heparin currently used to effect anticoagulation under some of these conditions also have the ability to block P- and L-selectin at clinically tolerable doses.

Roles of Selectins, Glycosaminoglycans, and Sialic Acids in Atherosclerosis

High levels of low-density lipoprotein (LDL) cholesterol and decreased high-density lipoprotein (HDL) cholesterol are associated with an increased risk of atherosclerotic lesions in the large arteries, which are the major cause of heart attacks, strokes, and other serious diseases. The very earliest phase of the development of atherosclerotic lesions (the fatty streak) involves the entry of monocytes into the subendothelial regions of the blood vessels. There is evidence that this process involves the expression of P- and/or E-selectin on the endothelium, which recognizes P-selectin glycoprotein ligand-1 (PSGL-1) or sialyl-Lewis^x on circulating monocytes. Indeed, atherosclerotic lesions in atherosclerosis-prone mice showed delayed progression in a P-selectin-deficient background, and even slower progression occurs in a combined P- and E-selectin-deficient state. The induction of endothelial P-selectin expression may result from oxidized lipids that are present in LDL particles and/or the inflammatory process that occurs in the early atheromatous plaque. It remains to be seen whether it is feasible to intervene in this process, because early lesions probably develop very slowly and relatively early in life. The subsequent subendothelial retention of LDLs in the early plaque is thought to occur at least partly via their interactions with proteoglycans. The interaction is thought to cause irreversible structural alterations of LDL, potentiating oxidation and uptake by macrophages and smooth-muscle cells. At the molecular level, clusters of basic amino acids present in apolipoprotein B (the protein moiety of LDL) appear to bind the negatively charged glycosaminoglycans of proteoglycans. Meanwhile, heparan sulfate (HS) found in the liver may regulate the turnover of lipoprotein particles. Several reports also indicate a lowered overall sialylation of LDL in patients with coronary artery disease. The pathophysiological significance of this finding and the mechanism(s) involved remain unclear. One hypothesis is that the desialylated LDL is more prone to be taken up and incorporated into atheromatous plaques.

DERMATOLOGY

Role of Selectins in Inflammatory Skin Diseases

Several inflammatory skin diseases (e.g., atopic dermatitis and contact dermatitis) are characterized by the entry of leukocytes into the dermis, where they have a pathogenic role in recruiting other types of cells and in mediating tissue damage. These types of skin lesions are sometimes associated with the chronic persistent expression of E-selectin on the endothelial cells. Independent evidence indicates that E-selectin can recruit circulating lymphocytes carrying the cutaneous lymphocyte antigen (detected by the antibody HECA452), which appears to be a specific E-selectin ligand epitope carried on a subset of PSGL-1 molecules (see Chapter 31). There is also evidence that some T-helper-1 (Th1) lymphocytes can be recruited into the skin by virtue of their expression of the PSGL-1 ligand for P-selectin. Many of these observations have been made only in experimental models. The potential for therapeutic intervention in these selectin-mediated processes has not been fully pursued.

ENDOCRINOLOGY AND METABOLISM

Pathogenesis and Complications of Diabetes Mellitus

Diabetes mellitus is a disease of dysregulated glucose metabolism, resulting from relative or absolute lack of insulin action. It is accompanied by characteristic long-term vascular and neurologic complications. One mechanism appears related to the high levels of free glucose in body fluids, which cause acceleration of a well-known nonenzymatic process in which the open-chain (aldehyde) form of the glucose reacts randomly with lysine residues on various proteins, resulting in reversible Schiff bases. With time, some of these adducts undergo the irreversible Amadori rearrangement. These then undergo a series of "browning" (Maillard) reactions, which eventually progress to advanced glycation end products (AGE). The resulting protein cross-links can damage cellular functions, and such adducts can also be recognized by receptors, for example, the receptor for advanced glycation end products (RAGE) and the macrophage scavenger receptor, perhaps participating in the process of atherogenesis. A current view is that this is a normal process of aging, which is accelerated in the setting of the chronic persistent hyperglycemia of uncontrolled diabetes mellitus. It is important to differentiate mechanistically and semantically between this nonenzymatic glycation (or "glucosylation") process and enzymatic glycosylation that takes place in the endoplasmic reticulum (ER), Golgi apparatus, and cytoplasm, utilizing glycosyltransferases and nucleotide sugar donors.

Another metabolic change of particular interest in diabetes mellitus is the increased production of UDP-GlcNAc caused by the conversion of excess glucose via the glucosamine:fructose aminotransferase (GFAT pathway). A current hypothesis is that this increase in cytoplasmic UDP-GlcNAc gives a secondary increase of O-GlcNAc levels on nuclear and cytoplasmic glycoproteins, secondarily altering the phosphorylation of the same proteins and their functions (see Chapter 18). Specific molecular mechanisms involving such altered O-GlcNAcylation have been defined in animal models for complications such as diabetic cardiomyopathy (increased O-GlcNAcylation of various nuclear proteins) and erectile dysfunction (O-GlcNAcylation of endothelial nitric oxide synthase). Interestingly, several of the cytoplasmic proteins involved in insulin receptor signaling and resulting nuclear transcription changes are themselves O-GlcNAcylated and are functionally altered in diabetes.

Nephropathy is a diabetic complication associated with high mortality. It begins with low levels of urinary albumin excretion (microalbuminuria), which progresses to frank macroalbuminuria. Ultimately, nephrotic syndrome and a concomitant decrease in glomerular filtration rate progress to end-stage renal disease. The proteinuria has been correlated with a reduction in the HS proteoglycan content of the glomerular basement membrane. The underlying mechanism may involve a reduction in HS synthesis by glomerular epithelial cells that may, in turn, be caused by the high glucose in the environment. One theory is that the resulting decrease in anionic change and loss of HS proteoglycan are thought to affect the porosity of the glomerular basement membrane. However, recent genetic evidence in mice questions this hypothesis. Interestingly, high glucose also mediates increased plasminogen activator inhibitor-1 (PAI-1) gene expression in renal glomerular mesangial cells, via O-GlcNAc-mediated alterations in Sp1 transcriptional activity.

GASTROENTEROLOGY

Role of Gut Epithelial Glycans in Gastrointestinal Infections

Many gastrointestinal pathogens interact with the gut mucosa via recognition of glycan structures (see Chapter 34). Prominent examples include cholera toxin (which binds GM1 ganglioside) and *Helicobacter pylori*, the causative agent of peptic ulcer disease and gastritis (which binds Lewis type glycans in the stomach mucosa). Consideration is now being given to using orally administered soluble glycan inhibitors to impede the attachment of such pathogens in the gut. In this regard, it is interesting that a time-honored treatment for peptic ulcer disease was a combination of antacids and milk (which contains large amounts of free sialyloligosaccharides). In addition, the variety and high concentrations of free glycans found in human milk (especially in the early days after birth of the baby) are thought to impede the ability of gut pathogens to bind to the mucosa and initiate infections.

Autosomal Dominant Polycystic Liver Disease

Autosomal dominant polycystic liver disease is thought to arise by somatic mutations in individuals who already have a mutated allele in one of two different genes. One is *SEC63*, which facilitates recognition of proteins by ER chaperones. The other is *PRKCSH* (also known as hepatocystin or the β subunit of α-glucosidase II), which was previously described as a protein kinase C substrate (80K-H) and is localized on the cell surface, in intracellular vesicles, and in the ER. The products of both these genes are involved in the translocation, folding, and quality control of newly synthesized proteins. The cysts in the liver usually develop later in life, and outgrowth is restricted to the biliary epithelium, suggesting a specific impact on a regulator of their proliferation rather than a gross effect on all glycoproteins. The specific proteins affected have not been identified, but recent studies of genetically altered mice suggest that HS proteoglycans play a role.

Heparan Sulfate Proteoglycans in the Pathogenesis of Protein-losing Enteropathy

Protein-losing enteropathy (PLE) is defined as the enteric loss of plasma proteins, which become life-threatening. The cellular and molecular mechanisms of this disease are not well understood, but it develops in some patients with congenital disorder of glycosylation type Ib (CDG-Ib) and CDG-Ic (see Chapter 42) or as a complication months to years following Fontan surgery to correct congenital heart malformations in patients with normal N-glycosylation. PLE appears to result from a collision of genetic insufficiencies and environmental stress. Impaired N-glycosylation, increased proinflammatory cytokines, and increased venous pressure all synergize to create PLE. In each of these pathologies, HS is specifically absent from the basolateral surface of intestinal epithelial cells during the episodes, and it returns when PLE subsides. The inflammatory cytokines tumor necrosis factor-α (TNF-α) and interferon-γ (IFN-γ) both bind to HS, which may thus serve as a buffer. In vitro studies show that removal of HS from the basolateral surface increases cytokine signaling through their receptors, thereby loosening tight junctions that normally prevent protein leakage. Bacterial and viral infections increase these cytokines and often trigger PLE in patients. Some CDG-Ib and post-Fontan patients have increased venous pressure, which can down-regulate some of the genes involved in extracellular matrix biosynthesis. Pressure synergizes with the effects of cytokines and the localized loss of HS to create a downward spiral of disease. Traditional therapy for PLE includes treatment of the underlying conditions if possible, maintenance of nutritional state, and sometimes albumin infusions and steroid hormones or other anti-inflammatory drugs. The mother of a post-Fontan patient made the astute observation that her son's PLE disappeared when he was given heparin injections as an anticoagulant prior to surgery. Later, in vitro studies showed that a few micrograms per milliliters of heparin or HS completely reverse the synergistic effects of epithelial cell HS loss and cytokine-induced breakdown of tight junctions. Nonanticoagulant heparin therapy may hold new promise for a variety of PLE patients.

Changes in Sialic Acid O-Acetylation in Ulcerative Colitis

Ulcerative colitis is an inflammatory disease typically affecting the superficial epithelial layer of the rectum and the distal colon. Although the primary cause of the disease is unknown, a large body of evidence suggests both genetic and environmental factors, and remissions and exacerbations are common. The sialic acids of the colonic mucosa, which are normally heavily O-acetylated, lose this modification in ulcerative colitis. Whether or not this is of pathogenic significance is uncertain, but these modifications normally do render the sialic acids more resistant to bacterial sialidases that are found in the gut. There have been conflicting claims about the efficacy of heparin treatments in improving the symptoms of this disease.

HEMATOLOGY

Clinical Use of Heparin as an Anticoagulant

Preparations of heparin (a highly sulfated form of HS; see Chapter 16) are routinely purified from animal tissues (particularly porcine intestines) and are used as a fast-acting and potent anticoagulant in a wide variety of diseases that involve thrombosis, in medical procedures such as dialysis, and in surgical procedures such as open heart surgery. As described in Chapter 16, the mechanism of anticoagulation is precise, involving a specifically sulfated heparin pentasaccharide that interacts with circulating antithrombin and markedly enhances its ability to inactivate coagulation factors Xa and IIa (thrombin). The use of "unfractionated heparin" is being partially supplanted by various forms of low-molecular-weight heparins that seem to be easier to use and are associated with fewer complications. One explanation is that the unfractionated heparin effects on factor IIa require a long chain that interacts both with the antithrombin and with the IIa itself, in a tripartite complex. In contrast, the shorter chains found in low-molecular-weight heparins only facilitate antithrombin inactivation of factor Xa. Thus, low-molecular-weight heparins affect Xa but not IIa levels. Most recently, a synthetic pentasaccharide that binds and facilitates antithrombin inactivation of factor Xa has been introduced as an alternative to heparin. Although these improvements have been valuable, it must be kept in mind that the original unfractionated heparin has a variety of other biological effects besides anticoagulation. Thus, other beneficial effects of heparin, such as the blockade of P- and L-selectin, are being reduced or even eliminated during the course of the switch to the low-molecular-weight heparins and the synthetic pentasaccharide.

A rare but feared complication of heparin treatment is heparin-induced thrombocytopenia. The pathogenesis appears to involve the formation of complexes between heparin and platelet factor-4 and the generation of antibodies against the complexes. These antibodies in turn then deposit on platelets, causing their aggregation and loss from circulation. Somewhat paradoxically, this process results in exaggerated thromboses, rather than bleeding. The incidence of this complication appears to be lower with the use of low-molecular-weight heparins, and it may be absent with the use of the pure pentasaccharide.

Hemolytic Transfusion Reactions

The invention of blood transfusion uncovered the existence of the ABO blood group system, which is dictated by different alleles of an α-Gal(NAc) transferase (for details, see Chapter 13). These and other less prominent glycan antigens are responsible for many of the hemolytic transfusion reactions that can occur when errors are made in blood typing. Active attempts have recently been under way to generate "universal donor" red blood cells, via enzymatic conversion of blood group A and B antigens to the O state.

Acquired Anticoagulation Due to Circulating Heparan Sulfate

Occasionally, patients with diseases such as cirrhosis and hepatocellular carcinoma spontaneously secrete a circulating anticoagulant and have an unusual coagulation test profile that makes it appear as if the patient has been treated with heparin. The anticoagulant activity can be purified from the plasma and has been identified as an HS glycosaminoglycan. The precise source of secretion has not been defined, and therapy is often difficult unless the underlying disease can be corrected or the liver transplanted.

Abnormal Glycosylation of Plasma Fibrinogen in Hepatoma and Liver Disorders

Plasma fibrinogen is heavily sialylated and the sialic acids are involved in binding calcium. Certain genetic disorders of fibrinogen are known to be associated with altered glycosylation of its N-glycans, which causes altered function in clotting. Patients with hepatomas and other liver disorders can also sometimes manifest increased branching and/or number of N-glycans, resulting in an overall increase in sialic acid content. This can present clinically as a bleeding disorder associated with a prolonged thrombin time. Patients with congenital genetic disorders affecting N-glycan biosynthesis (see Chapter 42) can also have thrombotic or bleeding disorders that may be partly explained by altered glycosylation of plasma proteins and/or platelets involved in blood coagulation.

Paroxysmal Nocturnal Hemoglobinuria

Paroxysmal nocturnal hemoglobinuria (PNH) is an unusual form of acquired hemolytic anemia (excessive destruction of red blood cells) that usually appears in adults. The defect arises through a somatic mutation in bone marrow stem cells that causes the production of one or more abnormal clones. The defect is an inactivation of the single active copy of the *PIGA* gene, an X-linked locus involved in the first stage of biosynthesis of glycosylphosphatidylinositol (GPI) anchors (for details on GPI anchor biosynthesis, see Chapter 11). Although several blood cell types show abnormalities, the red cell defect is the most prominent, being characterized by an abnormal susceptibility to the action of complement. This is now known to be due to the lack of expression of certain GPI-anchored proteins, such as decay-accelerating factor, that normally down-regulate complement activation on "self" surfaces. However, hypercoagulability also occurs, presumably due to loss of GPI-anchored proteins on other cells, such as monocytes. Interestingly, many of these patients later develop either bone marrow failure (aplastic anemia) or acute leukemia. It is now known that most normal humans already have a tiny fraction of circulating cells with the PNH defect. These presumably represent the products of one or more bone marrow stem cells that develop this acquired defect because of a single hit on the active X chromosome but then did not become prominent contributors to the total pool of circulating red blood cells. In this scenario, the independent occurrence of a process damaging other stem cells allows the "unmasking" of the PNH defect.

Paroxysmal Cold Hemoglobinuria

Patients with this rare disorder have a cold-induced intravascular destruction of red cells (hemolysis), which appears to be caused by a circulating IgG antibody directed against the red cell P blood group system. The pathogenesis of this disorder is unknown, but it tends to occur in the setting of some viral infections and in syphilis. The IgG antibody is demonstrated by the so-called "Donath–Landsteiner test," where the patient's serum is mixed either with the patient's own red cells or with those from a normal person and chilled to 4°C. Hemolysis occurs after warming back to 37°C.

Cold Agglutinin Disease

This disease is caused by autoimmune IgM antibodies directed against glycan epitopes on erythrocytes. High titers of IgM agglutinins are present in serum and are maximally active at 4°C. This IgM is presumed to bind to erythrocytes that are circulating in the cooled blood of peripheral regions of the body. The antibody fixes complement, which then destroys the cells when they reach warmer areas of the body. There are several variants of the syndrome. One affects young adults and follows infection with *Mycoplasma pneumoniae* or Epstein-Barr virus (infectious mononucleosis). This antibody is typically directed against the so-called "i" antigen (poly-*N*-acetyllactosamine), is polyclonal, and is generally short-lived, disappearing when the infection subsides. Because *M. pneumoniae* is itself known to have a receptor that recognizes sialylated poly-*N*-acetyllactosamine, it is hypothesized that the autoimmune antibody results from a mirror-image, anti-idiotypic reaction to the initial antibody directed against the mycoplasma's binding site. An idiopathic variant of cold agglutinin disease affects older individuals, involves a monoclonal IgM, and can be a precursor or an accompaniment to a lymphoproliferative disease such as Waldenström's macroglobulinemia, chronic lymphocytic leukemia, or other lymphomas. These antibodies are typically directed against the "I" antigen (β1-6-branched poly-*N*-acetyllactosamine) present on erythrocytes. Some less common variants of cold agglutinin disease involve antibodies directed against sialylated *N*-acetyllactosamines. In some patients on chronic hemodialysis, the syndrome occurs due to the formation of antibody directed against the sialylated blood group antigen N.

Tn Polyagglutinability Syndrome

Tn polyagglutinability syndrome is an acquired condition in which the blood cells made by the bone marrow express the Tn antigen (O-linked *N*-acetylgalactosamine, GalNAcα-O-Ser/Thr) and sialyl-Tn (Siaα2-6GalNAcα-O-Ser/Thr), thus becoming susceptible to hemagglutination by the naturally occurring anti-Tn antibodies present in most normal human sera. The defect tends to be incomplete, with some circulating cells expressing the more complete sialylated tri- and tetrasaccharide O-glycans as well. These observations are best explained as an acquired stem-cell-based loss of expression of the O-glycan core-1 β1-3 galactosyltransferase activity (also called the T synthase). This in turn has now been explained by the acquired inactivation of Cosmc, a chaperone required for the biosynthesis of the T synthase. As with PNH, the existence of the *COSMC* gene on the X chromosome allows a single hit on the active X chromosome to cause a glycosylation defect in a single bone marrow stem cell. Patients with this syndrome show a wide range of symptoms. Some are picked up simply because the polyagglutinability of their red blood cells is detected when blood typing is done for a possible transfusion. Others have varying degrees of hemolytic anemia and/or decreases in other blood cell types. Some of these patients can subsequently progress into frank leukemia. It is unclear how the primary syndrome predisposes to the development of the malignancy. As with PNH, the possibility exists that an underlying bone marrow disorder simply allows the "unmasking" of preexisting minor stem cell clones with the defect. In keeping with this, the leukemic clones that arise later need not necessarily have the same defect.

IMMUNOLOGY AND RHEUMATOLOGY

Changes in IgG Glycosylation in Rheumatoid Arthritis

The IgG class of circulating immunoglobulins carry N-glycans, and those in the constant (CH2 or Fc) region of human IgG are reported to have several unusual properties. First, the glycans are buried between the folds of the two constant regions. Second, they are often sufficiently immobilized by carbohydrate–protein interactions that can be seen in the crystal structure of the protein (most glycans are not visible in crystal structures). Third, although processed into biantennary complex type-N glycans, they are hardly ever completed into fully sialylated molecules. Instead, most of the molecules remain with one or two terminal β-linked galactose residues (so-called G1 and G2 molecules, respectively). It was previously noted that in patients with a chronic systemic disease called rheumatoid arthritis, a major fraction of the serum IgG molecules have decreased galactosylation of N-glycans, some carrying no galactose at all (so-called G0 molecules). The severity of the disease tends to correlate with the extent of the glycosylation change, and the spontaneous improvement that occurs during pregnancy is correlated with a restoration in galactosylation. One function attributed to the Fc N-glycans is to maintain the conformation of the Fc domains as well as the hinge regions. These structural features are necessary for effector functions such as complement binding and Fc-dependent cytotoxicity. Nuclear magnetic resonance (NMR) studies have shown that the G0 N-glycans have an increased mobility resulting from the loss of interactions between the glycan and the Fc protein surface. Thus, it is thought that regions of the protein surface that are normally covered by the glycan are exposed in rheumatoid arthritis. In addition, some studies suggest that the more mobile G0 N-glycan may be recognized by the circulating mannose-binding protein, which can activate complement directly. Rheumatoid arthritis is also characterized by circulating immune complexes consisting of antibody molecules (called rheumatoid factor) that seem to be directed against the Fc region of other IgG molecules. However, the epitopes involved here do not seem to be glycan-related. Another possibility being considered is that the altered glycosylation changes interactions with Fc receptors. With regard to the mechanism for the underglycosylation, some groups have reported lowered activities of β-galactosyltransferase activity in lymphocytes from patients with rheumatoid arthritis. It remains an open question whether the altered glycosylation of IgG has a primary pathogenic role in rheumatoid arthritis, because the appearance of G0 molecules is a general feature of other unrelated chronic granulomatous diseases, for example, Crohn's disease and tuberculosis. Furthermore, the glycan change is also seen to a lesser extent in osteoarthritis, a form of chronic degenerative arthritis with a completely different pathogenesis. Overall, the change in IgG glycosylation in rheumatoid arthritis remains an interesting phenomenon whose precise significance and pathogenic role need further study.

Secondary Changes in the O-Glycans of CD43 in Wiskott–Aldrich Syndrome

This inherited genetic disease is characterized by skin eczema, altered cellular immune responses, and low platelet counts, symptoms that appear in childhood. Early studies suggested that the disease was associated with an absence of CD43 (also called leukosialin or sialophorin), the major O-glycosylated protein of lymphocytes. However, in retrospect, it is clear that this polypeptide is still expressed normally, but it has changed gel mobility because of markedly increased branching of O-glycans. Recent data indicate that the primary defect in this syndrome is not in glycosylation but in a transcription factor. However, the glycan changes seen in resting T cells of these patients are exactly the same as those that can be induced upon activation of normal T cells. Thus, it remains possible that some aspects of the immune disorders in this disease are due to a secondary change in glycosylation.

INFECTIOUS DISEASE

Recognition of Glycans by Bacterial Adhesins, Toxins, and Viral Hemagglutinins

A wide variety of pathogens initiate infection by specifically recognizing cell-surface glycans (see Chapter 34). In some instances, the differences in infection rates between individuals can be attributed to variations in the expression of the glycan target. For example, adhesion of certain pathogenic strains of *Escherichia coli* to cells in the urinary tract can be mediated by P fimbriae, involving a specific glycan receptor on the P blood group antigens. Infections do not occur in individuals who are P negative. P fimbriae also appear to be important in determining the propensity for bacterial bloodstream invasion from the kidney.

Desialylation of Blood Cells by Circulating Microbial Sialidases during Infections

Several microorganisms produce sialidases (classically called neuraminidases) that are involved in the pathogenesis of the diseases that they cause. In most instances, this enzyme remains localized to the site of infection. However, in some severe cases, for example, *Clostridium perfringens*-mediated gas gangrene, a sufficient amount of the sialidase is produced so that it can appear in the plasma. In this situation, the surface of circulating blood cells can become desialylated, resulting in enhanced clearance and anemia. The detection of the circulating sialidase has been proposed to have diagnostic and prognostic significance. Some cases of hemolytic-uremic syndrome are also associated with sialidase-producing *Streptococcus pneumoniae* infections. It is possible that blocking the sialidase with appropriate inhibitors could have therapeutic value in these situations.

NEPHROLOGY

Loss of Glomerular Sialic Acids in Nephrotic Syndrome

Nephrotic syndrome occurs when the kidney glomerulus fails to retain serum proteins during the initial filtration of plasma and these proteins then leak into the urine. The epithelial/endothelial mucin molecule called podocalyxin, which is present on the foot processes (pedicles) of glomerular podocytes, is thought to have a role in maintaining pore integrity and in excluding large molecules, such as proteins, from the glomerular filtrate. The sialic acid residues of podocalyxin molecules are believed to be critical in this process. Loss of glomerular sialic acid is seen in spontaneous minimal-change renal disease in children and in the nephrotic syndrome that follows some bacterial infections. Several animal models seem to mimic this situation. Proteinuria and renal failure develop in a dose-dependent manner after a single inoculation of *Vibrio cholerae* sialidase, and this correlates with loss of sialic acids from the glomerulus. This was also accompanied by the effacement of foot processes and the apparent formation of tight junctions between podocytes. The anionic charge returned to endothelial and epithelial sites within two days of sialidase inoculation, but the foot process loss remained. Another model is termed aminonucleoside nephrosis, and it is induced in rats by injection of puromycin. Again, defective sialylation of a podocalyxin and glomerular glycosphingolipids has been detected in this model.

Changes in the O-Glycans in IgA Nephropathy

In humans, only IgA1 and IgD contain O-glycans in the hinge regions, whereas all immunoglobulin classes contain N-glycans in the Fc domain. Aggregation of the IgA1 molecule is thought to be involved in a form of nephrotic syndrome called IgA nephropathy. Studies of the O-glycans on serum IgA1 showed glycan truncations in the IgA nephropathy group compared with a negative control group. One of the functions of the IgA1 O-glycan chains is thought to be to stabilize the three-dimensional structure of the molecule. Studies of heat-induced aggregation support the notion that the altered glycosylation on the hinge region of IgA1 results in a loss of conformational stiffness, perhaps explaining the aggregation phenomenon. Removal of glycans from the IgA1 molecule also results in noncovalent self-aggregation and a significant increase in adhesion to the extracellular matrix proteins. It is therefore suggested that the underglycosylation of the IgA1 molecule found in IgA nephropathy is involved in the nonimmunologic glomerular accumulation of IgA1. The primary mechanism of underglycosylation remains unknown. A likely scenario is a defect in the *COSMC* gene, similar to that found in the Tn polyagglutinability syndrome (see above). The difference is that instead of affecting a bone marrow stem cell, the defect would involve a clone of B cells that specifically expresses IgA.

Heparan Sulfate in Systemic Lupus Erythematosus

Systemic lupus erythematosus (SLE) is an autoimmune disorder characterized by deposition of antigen-antibody complexes in various organs, especially the skin and the kidney. The precise initiating mechanisms of SLE are unknown, but it appears that the resulting pathology may involve both cytokines and HS. Amounts of HS are reduced on the glomerular basement membrane, and this was thought to result from masking of HS by complexes of nucleosomes and antinuclear antibodies, but the actual situation is likely to be more complex. Even though anti-double-stranded DNA antibodies are the hallmark of SLE, circulating antibodies to HS strongly correlate with disease activity. In some studies, HS injections into dogs induce SLE symptoms within several weeks. Elevated HS is found in the urine of SLE patients, especially in severe cases. SLE patients also have systemic elevated TNF-α. Anticytokine therapy against systemic and locally expressed cytokines suggests that blocking the proinflammatory cascade may be of significant value. Glomeruli have locally increased TNF-α, which acts to induce interleukin-1β (IL-1β) and IL-6. IFN-γ is thought to induce TNF-α, precipitating glomerulonephritis due to local inflammation. Some SLE patients also develop PLE, perhaps as a consequence of misplaced or degraded HS and elevated cytokines, creating the appropriate environment for PLE (see above).

NEUROLOGY AND PSYCHIATRY

Pathogenic Autoimmune Antibodies Directed against Neuronal Glycans

A variety of diseases are associated with circulating antibodies directed against specific glycan molecules that are enriched in the nervous system. These patients suffer from symptoms related to autoimmune neural damage. Such antibodies can arise via at least three distinct pathogenic mechanisms. In the first situation, patients with benign or malignant B-cell neoplasms (e.g., benign monoclonal gammopathy of unknown significance [MGUS], Waldenstrom's macroglobulinemia, and plasma cell myeloma) secrete monoclonal IgM or IgA antibodies that are highly specific for either ganglio-series gangliosides or, more commonly, for sulfated glucuronosyl glycans (the so-called HNK-1 epitope). These antibodies react with the glycolipid bearing this epitope 3-O-SO_3-GlcAβ1-4Galβ1-4GlcNAcβ1-3Galβ1-4Glc-Cer (3′sulfoglucuronosylparagloboside) and against the N-glycans on a variety of CNS glycoproteins (MAG, P0, L1, N-CAM) that bear the same terminal sequence (3-O-SO_3-GlcAβ1-4Galβ1-4GlcNAcβ1-). The resulting peripheral demyelinating neuropathy can sometimes be more damaging than the primary disease itself. Therapy consists of attempts to treat the primary disease with chemotherapy or to remove the immunoglobulin by plasmapheresis. Both approaches are usually unsuccessful at lowering the immunoglobulin to a level sufficient to diminish the symptoms. The second situation is an immune reaction to the molecular mimicry of neural ganglioside structures by the lipooligosaccharides of bacteria such as *Campylobacter jejuni.* Following an intestinal infection with such organisms, circulating cross-reacting antibodies against gangliosides such as GM1 and GQ1b appear in the plasma. These are typically associated with the onset of symptoms of a demyelinating neuropathy involving the peripheral and central nervous systems (the Guillain–Barré and Miller–Fisher syndromes, respectively). The third situation is a human-induced disease arising from recent attempts to treat patients with disease such as stroke using intravenous injections of mixed bovine brain gangliosides. Although some evidence exists that this treatment may be beneficial for the primary disease, several cases of Guillain–Barré syndrome have been reported as a likely side effect. One suggested scenario is that the presence of small amounts of gangliosides with the nonhuman sialic acid *N*-glycolylneuraminic acid facilitates the formation of antibodies that cross-react with ganglioside containing human sialic acid *N*-acetylneuraminic acid.

Role of Glycans in the Histopathology of Alzheimer's Disease

Alzheimer's disease is a common primary degenerative dementia of humans, with an insidious onset and a progressive course. The ultimate diagnosis is made by postmortem histological examination of brain tissue, which shows characteristic amyloid plaques with neurofibrillary tangles that are associated with neuronal death. Two types of glycans have been implicated in the histopathogenesis of the lesions: O-GlcNAc and HS glycosaminoglycans. Paired helical filaments are the major component of the neurofibrillary tangle. These are primarily composed of the microtubule-associated protein Tau, which is present in a hyperphosphorylated state. This abnormally hyperphosphorylated Tau no longer binds microtubules and self-assembles to form the paired helical filaments that may contribute to neuronal death. Normal brain Tau is known to be multiply modified by Ser(Thr)-linked O-GlcNAc, the dynamic and abundant posttranslational modification that is often reciprocal to Ser(Thr) phosphorylation (see Chapter 18). The hypothesis currently being investigated is that site-specific or stoichiometric changes in O-GlcNAc addition may modulate Tau function and may also play a part in the formation of paired helical filaments by allowing excessive phosphorylation. The hyperphosphorylated Tau in Alzheimer's disease brain is found in association with HS proteoglycans. Nonphosphorylated Tau isoforms with three microtubule-binding repeats form paired helical-like filaments under physiological conditions in vitro when incubated with HS. Heparin prevents Tau from binding to microtubules and promotes microtubule disassembly. These findings, together with previous evidence that heparin stimulates Tau phosphorylation by protein kinases, have been used to argue that sulfated glycosaminoglycans may be a critical factor in the formation of the neurofibrillary tangles. However, no significant difference was noted between the detailed structure of HS obtained from control brains and that from Alzheimer's disease brains. Furthermore, the topological separation of Tau (which occurs in the cytoplasm) from glycosaminoglycans (which are extracellular) indicates that this physical association can only occur after cell death. On the other hand, HS proteoglycans may also have an important role in amyloid plaque deposition. Investigators have demonstrated high-affinity binding between HS proteoglycans and the amyloid precursor, as well as with the A4 peptide derived from the precursor. In addition, a specific vascular HS proteoglycan found in senile plaques bound with high affinity to two amyloid protein precursors. Overall, the data indicate that HS chains may have a significant role in the pathogenesis of the histological lesions.

ONCOLOGY: ALTERED GLYCOSYLATION IN CANCER

Altered glycosylation is a universal feature of cancer cells, but only certain specific glycan changes are frequently associated with tumors. These include (1) increased β1-6GlcNAc branching of N-glycans; (2) changes in the amount, linkage, and acetylation of sialic acids; (3) truncation of O-glycans, leading to expression of Tn and sialyl Tn antigens; (4) expression of the nonhuman sialic acid *N*-glycolylneuraminic acid, likely incorporated from dietary sources;

(5) expression of sialylated Lewis structures and selectin ligands; (6) altered expression and enhanced shedding of glycosphingolipids; (7) increased expression of galectins and poly-*N*-acetyllactosamines; (8) altered expression of ABH(O) blood-group-related structures; (9) alterations in sulfation of glycosaminoglycans; (10) increased expression of hyaluronan; and (11) loss of expression of GPI lipid anchors. Some of these changes, for example, increased β1-6GlcNAc branching of N-glycans and expression of selectin ligands, have been shown to have pathophysiological significance in model tumor systems, and some are also targets for diagnostic and therapeutic approaches to cancer. For details regarding these topics, see Chapter 44.

PULMONARY MEDICINE

Role of Selectins, Siglecs, and Mucins in Bronchial Asthma

Asthma is a disease that is characterized by a hyperresponsiveness of the tracheobronchial tree to various stimuli, resulting in widespread narrowing of the airways, and it changes in severity, either spontaneously or as a result of therapy. The two dominant pathological features of asthma are airway wall inflammation and luminal obstruction of the airways by inflammatory exudates, consisting predominantly of mucins. Most cases are due to the presence of antigen-specific IgE antibodies, which then bind to mast cells as well as to basophils and certain other cell types. Subsequently, antigen can cross-link adjacent IgE molecules, triggering an explosive release of vasoactive, bronchoactive, and chemotactic agents from mast cell granules into the extracellular milieu. Eosinophils also contribute to the pathogenesis of asthma in several ways, by synthesizing leukotrienes, stimulating histamine release from mast cells and basophils, providing a positive feedback loop, and releasing major basic protein, a granule-derived protein that has toxic effects on the respiratory epithelium. Underlying all this, it appears that $CD4^+$ Th2 cells are responsible for orchestrating the responses of other cell types. Recent evidence indicates that the selectins are intimately involved in the recruitment of eosinophils and basophils (and possibly T lymphocytes) into the lung, raising the hope that small-molecule inhibitors of selectin function and/or heparin can be used to treat the early stages of an asthmatic attack. Likewise, chemokine interactions with HS are important in leukocyte trafficking. Recent evidence from Siglec-F knockout mice also suggests that the functionally equivalent human paralog Siglec-8 is a good target for reducing the contributions of eosinophils to the pathology (see Chapter 32). Finally, the large increase in mucus production is at least partly mediated by an up-regulation of synthesis of mucin polypeptides, under the influence of various cytokines that stimulate the goblet cells of the airway epithelium.

Role of Selectins in Acute Respiratory Distress Syndrome

Acute respiratory distress syndrome is a serious pathophysiological process that is the final common pathway of lung injury arising from a variety of events, such as shock, trauma, or sepsis. It is characterized by diffuse pulmonary endothelial injury, progressing to pulmonary edema, which results from a marked increase in capillary permeability. Selectins and integrins help circulating neutrophils to adhere to the endothelium and release injurious oxidants, proteolytic enzymes, and arachidonic acid metabolites, resulting in endothelial cell dysfunction and destruction. The presence of many neutrophils and secretory products in bronchoalveolar lavage liquid emphasizes the critical role of the underlying inflammatory response. Again, the hope is that small molecule selectin inhibitors and/or the right kind of heparin can be used in the early stages of this syndrome, before it progresses to extensive lung damage and respiratory failure.

Altered Glycosylation of Epithelial Glycoproteins in Cystic Fibrosis

Cystic fibrosis is a very common genetic disorder caused by a mutation in the cystic fibrosis transmembrane conductance regulator (*CFTR*). This causes defective chloride conduction across the apical membrane of involved epithelial cells. Cystic fibrosis is associated with increased accumulation of viscous mucins in the pancreas, gut, and lungs, which leads to many of the symptoms of the disease. In the lung airways, there are known to be widespread increases in sialylation of secreted proteins and increases in the sulfation and fucosylation of mucus glycoproteins. One possible explanation is that the primary CFTR defect allows a higher Golgi pH, resulting in the abnormalities in glycosylation: however, there is currently some controversy about this conclusion. Curiously, the CFTR is mainly expressed within nonciliated epithelial cells, duct cells, and serous cells of the tubular glands, but it is not highly expressed in the goblet cells and mucous glands of the acinar cells, which are the cells that synthesize respiratory mucins. Thus, the *CFTR* mutation may indirectly affect mucin glycosylation through the generation of inflammatory responses. Another major cause of morbidity in the disease is the colonization of respiratory epithelium by an alginate-producing form of *Pseudomonas aeroginosa.* Certain glycolipids and mucin glycans have been suggested to be the *Pseudomonas* receptors that help to maintain the colonization. The changes in glycolipid and mucin glycosylation could enhance the production of potential binding targets for organ colonization. The presence of bacterial products is also a proinflammatory condition, since the bacterial capsular polysaccharides may activate Toll receptors and lead eventually to neutrophil accumulation and organ damage.

FURTHER READING

(Because of the wide range of topics covered in this chapter, it is not feasible to provide literature citations for all of them. Some examples are provided but the reader should consult references at the end of the other cited chapters.)

Berger E.G. 1999. Tn-syndrome. *Biochim. Biophys. Acta* **1455:** 255–268.

Karlsson K.A. 2000. The human gastric colonizer *Helicobacter pylori*: A challenge for host–parasite glycobiology. *Glycobiology* **10:** 761–771.

Lefer D.J. 2000. Pharmacology of selectin inhibitors in ischemia/reperfusion states. *Annu. Rev. Pharmacol. Toxicol.* **40:** 283–294.

Raats C.J., Van Den Born J., and Berden J.H. 2000. Glomerular heparan sulfate alterations: Mechanisms and relevance for proteinuria. *Kidney Int.* **57:** 385–400.

Petz L.D. 2001. Treatment of autoimmune hemolytic anemias. *Curr. Opin. Hematol.* **8:** 411–416.

Shriver Z., Liu D., and Sasisekharan R. 2002. Emerging views of heparan sulfate glycosaminoglycan structure/activity relationships modulating dynamic biological functions. *Trends Cardiovasc. Med.* **12:** 71–77.

Ley K. 2003. The role of selectins in inflammation and disease. *Trends Mol. Med.* **9:** 263–268.

Coppo R. and Amore A. 2004. Aberrant glycosylation in IgA nephropathy (IgAN). *Kidney Int.* **65:** 1544–1547.

Rose M.J. and Page C. 2004. Glycosaminoglycans and the regulation of allergic inflammation. *Curr. Drug Targets Inflamm. Allergy* **3:** 221–225.

Drenth J.P., Martina J.A., van de Kerkhof R., Bonifacino J.S., and Jansen J.B. 2005. Polycystic liver disease is a disorder of cotranslational protein processing. *Trends Mol. Med.* **11:** 37–42.

Gertz M.A. 2005. Cold agglutinin disease and cryoglobulinemia. *Clin. Lymphoma* **5:** 290–293.

Ju T. and Cummings R.D. 2005. Protein glycosylation: Chaperone mutation in Tn syndrome. *Nature* **437:** 1252.

Romano S.J. 2005. Selectin antagonists: Therapeutic potential in asthma and COPD. *Treat. Respir. Med.* **4:** 85–94.

Wickramasinghe S.N. and Wood W.G. 2005. Advances in the understanding of the congenital dyserythropoietic anaemias. *Br. J. Haematol.* **131:** 431–446.

Yuki N. and Odaka M. 2005. Ganglioside mimicry as a cause of Guillain–Barré syndrome. *Curr. Opin. Neurol.* **18:** 557–561.

Bode L. and Freeze H.H. 2006. Applied glycoproteomics—Approaches to study genetic-environmental collisions causing protein-losing enteropathy. *Biochim. Biophys. Acta* **1760:** 547–559.

Girolami B. and Girolami A. 2006. Heparin-induced thrombocytopenia: A review. *Semin. Thromb. Hemost.* **32:** 803–809.

Goldin A., Beckman J.A., Schmidt A.M., and Creager M.A. 2006. Advanced glycation end products: Sparking the development of diabetic vascular injury. *Circulation* **114:** 597–605.

Kneuer C., Ehrhardt C., Radomski M.W., and Bakowsky U. 2006. Selectins—Potential pharmacological targets? *Drug Discov. Today* **11:** 1034–1040.

Luzzatto L. 2006. Paroxysmal nocturnal hemoglobinuria: An acquired X-linked genetic disease with somatic-cell mosaicism. *Curr. Opin. Genet. Dev.* **16:** 317–322.

Patey S.J. 2006. The role of heparan sulfate in the generation of Abeta. *Drug News Perspect.* **19:** 411–416.

Rose M.C. and Voynow J.A. 2006. Respiratory tract mucin genes and mucin glycoproteins in health and disease. *Physiol. Rev.* **86:** 245–278.

Soldatos G. and Cooper M.E. 2006. Advanced glycation end products and vascular structure and function. *Curr. Hypertens. Rep.* **8:** 472–478.

Ulloa L. and Messmer D. 2006. High-mobility group box 1 (HMGB1) protein: Friend and foe. *Cytokine Growth Factor Rev.* **17:** 189–201.

Yan S.F., Naka Y., Hudson B.I., Herold K., Yan S.D., Ramasamy R., and Schmidt A.M. 2006. The ligand/RAGE axis: Lighting the fuse and igniting vascular stress. *Curr. Atheroscler. Rep.* **8:** 232–239.

Arnold J.N., Wormald M.R., Sim R.B., Rudd P.M., and Dwek R.A. 2007. The impact of glycosylation on the biological function and structure of human immunoglobulins. *Annu. Rev. Immunol.* **25:** 21–50.

Kuwabara S. 2007. Guillain–Barré syndrome. *Curr. Neurol. Neurosci. Rep.* **7:** 57–62.

Varki N.M. and Varki A. 2007. Diversity in cell surface sialic acid presentations: Implications for biology and disease. *Lab. Invest.* **87:** 851–857.

CHAPTER 44

Glycosylation Changes in Cancer

Ajit Varki, Reiji Kannagi, and Bryan P. Toole

ALTERED GLYCOSYLATION IS A UNIVERSAL FEATURE OF CANCER CELLS, and certain glycan structures are well-known markers for tumor progression. This chapter discusses some glycan biosynthetic pathways that are frequently altered in cancer cells, the correlation between altered glycosylation and clinical prognosis, the genetic basis of some of these changes, and in vitro and in vivo studies that indicate the pathological importance of these pathways.

HISTORICAL BACKGROUND

Like normal cells during embryogenesis, tumor cells undergo activation and rapid growth, adhere to a variety of other cell types and cell matrices, and invade tissues. Embryonic development and cellular activation in vertebrates are typically accompanied by changes in cellular glycosylation profiles. Thus, it is not surprising that glycosylation changes are also a universal feature of malignant transformation and tumor progression. The earliest evidence came from observing that plant lectins (e.g., wheat germ agglutinin) showed enhanced binding to and agglutination of tumor cells. Next, it was found that in vitro transformation was frequently accompanied by a general increase in the size of metabolically labeled glycopeptides produced by trypsinization of surface molecules from cancer cells. With the advent of monoclonal antibody technology in the late 1970s, investigators in search of a "magic bullet" against cancer cells found that many of their "tumor-specific" antibodies were directed against glycan epitopes, especially those borne on glycosphingolipids. In most cases, further studies showed that these epitopes were "onco-fetal antigens"; that is, they were also expressed in embryonic tissues and, in a few cell types, in the normal adult. Significant correlations between certain types of altered glycosylation and the prognosis of tumor-bearing animals or patients increased interest in these changes. In several instances, in vitro cellular assays and in vivo animal studies have further supported the view that these changes are critical to aspects of tumor cell behavior. This chapter outlines the most common changes that have been described and their potential relevance.

GLYCOSYLATION CAN BE ALTERED IN VARIOUS WAYS IN MALIGNANCY

Glycan changes in malignant cells take a variety of forms. Examples have been found of loss of expression or excessive expression of certain structures, the persistence of incomplete or truncated structures, the accumulation of precursors, and, less commonly, the appearance of novel structures. Alterations in early branch points in the normal pathways of biosynthesis can markedly affect the relative amount of one class of structure while allowing the dominance of another. However, this is not simply the random consequence of disordered biology in tumor cells. It is striking that of all the possible glycan biosynthetic changes, only a limited subset of changes are frequently correlated with malignant transformation and tumor progression. Given that cancer is a "microevolutionary" process in which only the fittest cells in a genetically heterogeneous population survive, it is reasonable to suggest that these specific glycan changes are selected for during tumor progression. The commonest of these changes are discussed below, including consideration of the likely biosynthetic mechanisms and the possible biological consequences.

ALTERED BRANCHING OF N-GLYCANS

Classic reports of increased size of tumor cell–derived glycopeptides have now been convincingly explained by an increase in β1-6 branching of N-glycans (Figure 44.1), which results from enhanced expression of UDP-GlcNAc:N-glycan GlcNAc transferase V (GlcNAcT-V; see Chapter 8). The change in expression of this enzyme seems to result from increased transcription of its gene (also called *MGAT5*) and can be induced by various mechanisms, including viral and chemical carcinogenesis. Such responses have been linked to specific features of the 5′-promoter region of *MGAT5*. Cell lines with increased GlcNAcT-V expression show an increased frequency of metastasis in animal models, and spontaneous revertants for loss of enzyme activity lose this metastatic phenotype. Clinical specimens of some human tumors show increased staining with the plant lectin L-phytohemagglutinin (L-PHA), which preferentially recognizes branched N-glycans bearing the β1-6 branched GlcNAcT-V product (see Chapter 45). Transfection of *MGAT5* cDNA into cultured cells causes a visually obvious transformed phenotype associated with colony formation in soft agar, increased cell spreading, enhanced invasiveness through membranes, and tumorigenic behavior by previously nontumorigenic cells. Most convincingly,

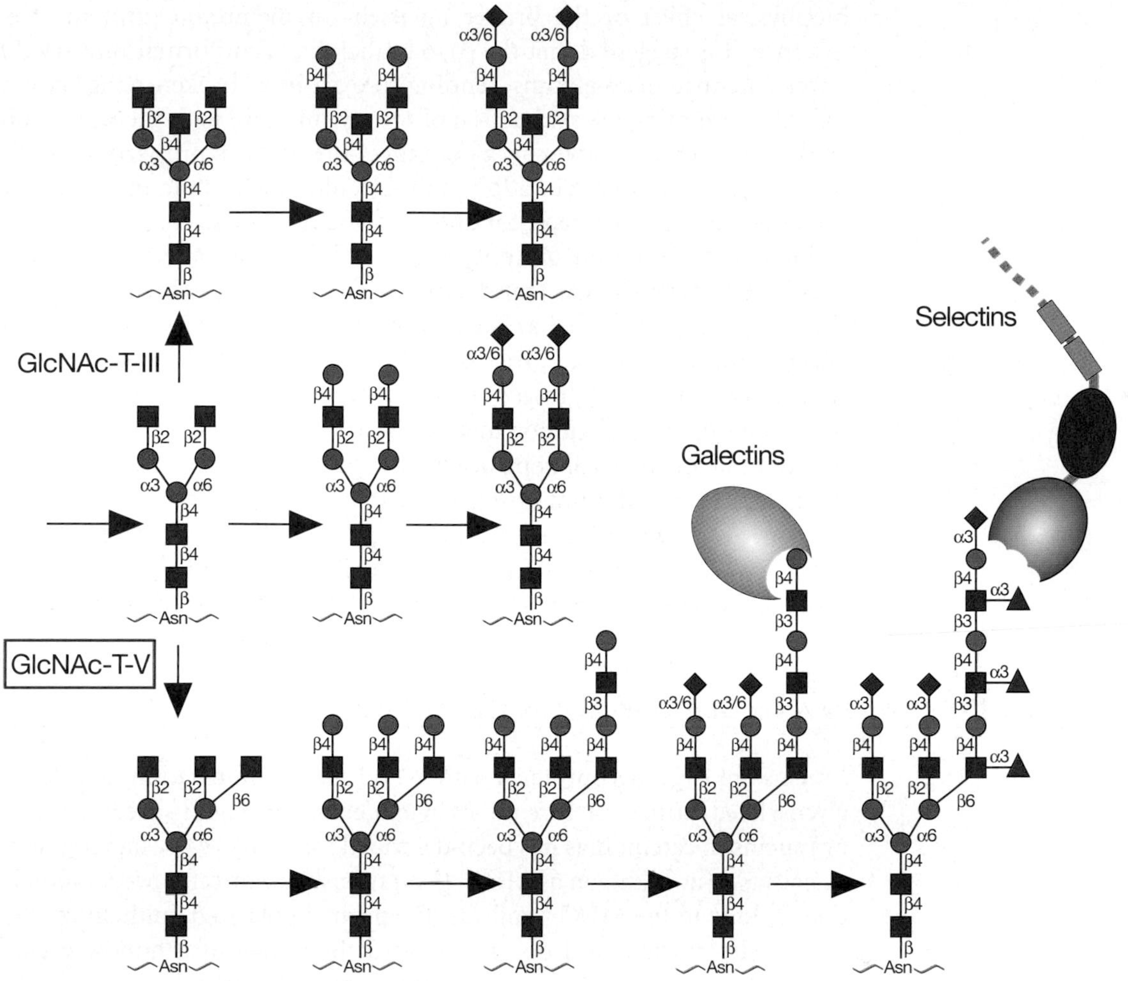

FIGURE 44.1. The increased size of N-glycans that occurs upon transformation can be explained by an elevation in GlcNAc transferase-V (GNT-V) activity, which catalyzes the β1-6 branching of N-glycans. This, in turn, may lead to enhanced expression of poly-*N*-acetyllactosamines, which can also be sialylated and fucosylated. These structures are potentially recognized by galectins and selectins, respectively. The structural consequences of increased expression of GlcNAc transferase-III (GNT-III) are also shown.

MGAT5-deficient mice show a striking reduction in the growth and metastasis of breast tumors induced by a viral oncogene. Conversely, metabolic inhibition of N-glycan processing by the plant alkaloid swainsonine (which blocks α-mannosidase II, thereby preventing complete processing of N-glycans and abrogating addition of the β1-6 branch) gives some reversal of tumorigenic behavior.

Taken together, all of these data indicate that GlcNAcT-V plays an important part in the biology of cancer. Indeed, by conventional criteria, such a suite of characteristics is typical of a true "oncogene." What remains to be clarified are the precise mechanism(s) by which these biochemical and structural changes result in the biological outcomes observed. Some possibilities (Figure 44.1) include (1) an increase in poly-*N*-acetyllactosamine-containing glycans (potentially recognized by galectins) that are preferentially found on this β1-6 branch; (2) alterations in the cell-surface half-life of growth factor receptors caused by changes in galectin-mediated lattice formation; (3) increased outer-chain polyfucosylation and sialyl Lewis^x production (potentially recognized by the selectins); and (4) a general biophysical effect of the branching itself on membrane protein structure. In the last instance, it is suggested that the β1-6 branch has a conformation very different from other outer antennae of N-glycans, tending to exist in a "broken wing" conformation, perhaps directly promoting the association of the glycan chain with the nearby polypeptide surface. In this manner, β1-6 branching may affect the physical properties and functional behavior of glycosylated cell-surface adhesion molecules, such as the integrins, and signaling receptors, such as the T-cell receptor and cytokine receptors.

Enhanced expression of another glycosyltransferase affecting N-glycan structure UDP-GlcNAc:N-glycan GlcNAc transferase III (GlcNAcT-III), which catalyzes the addition of the bisecting GlcNAc branch, has been reported in certain tumors, such as rat hepatomas (Figure 44.1). However, deliberate overexpression of GlcNAcT-III in other tumor cell types caused suppressed function of growth factor receptors and altered cellular morphology, and actually reduced the rate of tumor metastasis. Mice genetically deficient in GlcNAcT-III show a reduced rate of chemical hepatocellular carcinogenesis, despite the fact that (unlike the case in rats) GlcNAcT-III is not up-regulated in mouse hepatomas. The data suggest that an independent glycoprotein factor with bisecting GlcNAc residues facilitates tumor progression in the mouse liver. Overall, although GlcNAcT-III expression appears to affect the biology of tumors, the results are not as clear-cut and consistent as those seen with GlcNAcT-V.

ALTERED EXPRESSION AND GLYCOSYLATION OF MUCINS

Mucins are large glycoproteins with a "rod-like" conformation, which carry many clustered glycosylated serines and threonines in tandem repeat regions (see Chapter 9). Overexpression of mucins in carcinomas has been described for many years, starting with the classic studies of episialin (now known as MUC-1) on mouse tumor cells. Most epithelial mucin polypeptides belong to the MUC family. In the normal polarized epithelium, mucins are expressed exclusively on the apical domain, toward the lumen of a hollow organ. Likewise, soluble mucins are secreted exclusively into the lumen. However, loss of correct topology in malignant epithelial cells (Figure 44.2) allows mucins to be expressed on all aspects of the cells, and soluble mucins can then enter the extracellular space and body fluids such as the blood plasma. The simultaneous expression of both membrane-bound and secreted forms of mucins by many carcinoma cells confounds discussion of their pathophysiological roles, because the

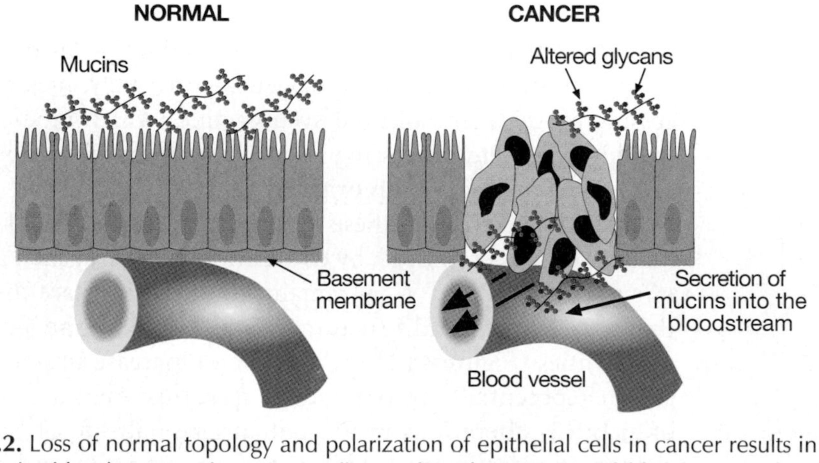

FIGURE 44.2. Loss of normal topology and polarization of epithelial cells in cancer results in secretion of mucins into the bloodstream. The tumor cells invading the tissues and bloodstream also present such mucins on their cell surfaces.

two forms of mucins could have opposing effects. Regardless, the secreted mucins often appear in the bloodstream of patients with cancer and can be detected by monoclonal antibodies. In cancers of epithelial origin (carcinomas) in particular, mucins appear to be the major carriers of altered glycosylation. A critical pathophysiological role of mucins in malignancy is suggested by the inhibitory effects of preincubation with GalNAcα-benzyl or acetylated GlcNAcβ1-3Galβ-naphthalenemethanol, which blocks O-glycosylation or the assembly of outer structures on mucins (see Chapters 9 and 50). Apart from specific interactions of the O-glycans with endogenous lectins (see below), the rod-like structures of the mucins and their negative charge are thought to repel intercellular interactions and sterically prevent other adhesion molecules such as cadherins and integrins from carrying out their functions. Thus, in some instances, mucins may act as "anti-adhesins" that can also promote displacement of a cell from the primary tumor during the initiation of metastasis. Evidence suggests that they might also physically block interactions between blood-borne carcinoma cells and the host cytolytic cells such as natural killer cells. In addition, mucins may mask presentation of antigenic peptides by major histocompatibility complex (MHC) molecules.

Another abnormal feature of carcinoma mucins is incomplete glycosylation. One common consequence with the O-glycans is the expression of Tn and T antigens (Figure 44.3; see also Chapter 9). Because such structures occur infrequently, in normal tissues, it is thought that they may provoke immune responses in the patient. Indeed, a correlation exists between the expression of the T (Galβ1-3GalNAc-α1-*O*-Ser/Thr) and Tn (GalNAc-α1-*O*-Ser/Thr) antigens, the spontaneous expression of antibodies directed against them, and the prognosis of patients with carcinomas. The most extreme form of underglycosylation results in expression of "naked" mucin polypeptides. Clinical trials are under way to deliberately provoke or enhance these immune responses by injecting patients with synthetic peptide antigens, sometimes bearing Tn or sialyl-Tn (Siaα2-6GalNAc-α1-*O*-Ser/Thr) structures. It is of note that the best immune responses to the glycosylated antigens are seen when they are presented in arrays, exactly as seen on the mucins. This is possibly because of multivalent binding of the antigen to the surface immunoglobulin "antigen receptor" of B cells, giving maximum cellular activation.

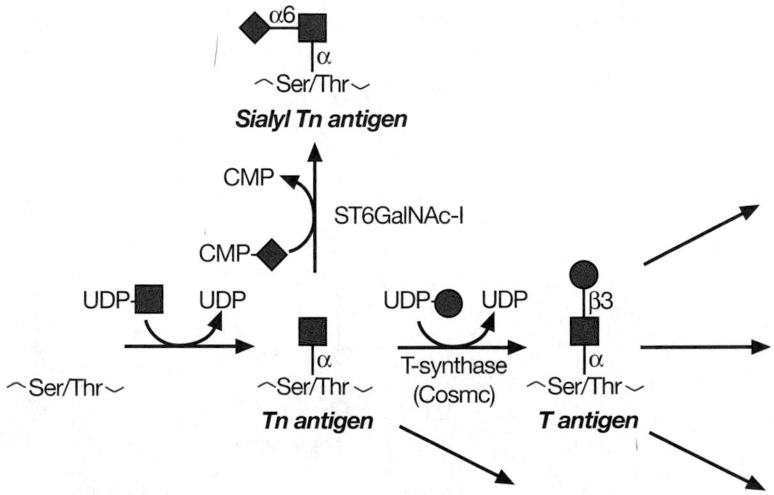

FIGURE 44.3. Incomplete glycosylation in the O-linked pathway results in expression of the Tn antigen, the sialylated Tn antigen (a "dead-end" structure), or the T antigen (Thomsen–Friedenreich antigen or unmodified core 1 structure). Multiple copies of these structures may occur in a closely spaced array on the polypeptide. Some regions of the mucin backbone may also be "naked" (i.e., completely unglycosylated). *Arrows* indicate biosynthetic pathways.

As for the mechanism causing excessive sialyl-Tn to appear on tumor cells, current evidence does not support simple overexpression of an ST6GalNAc Golgi enzyme that can produce this structure. The more likely explanation appears to be a mutation of the gene encoding Cosmc, a chaperone required for the expression of the β1-3 galactosyltransferase that synthesizes the Galβ1-3GalNAc-α1-O-Ser/Thr core-1 unit of O-glycans. As the gene for Cosmc is on the X chromosome (Xq24), a single mutation would be sufficient to eliminate expression. In this scenario, sialyl-Tn accumulation would then result as a side effect of the loss of ability to make the core-1 O-glycan and its extensions along with expression of the specific ST6GalNAc sialyltransferase (Figure 44.3). Given the frequency of sialyl-Tn accumulation by cancer cells, it is likely that its expression confers some as yet unknown advantage to the tumor cells. Conversely, an increase of UDP-galactose transporter has been implicated in the induction of T-antigen expression.

CHANGES IN SIALIC ACID EXPRESSION IN MALIGNANCY

The classic observation of increased wheat germ agglutinin binding to animal tumor cells is likely explained by an overall increase in cell-surface sialic acid content, which in turn reduces attachment of metastatic tumor cells to the matrix, and may help protect them from recognition by the alternative pathway of complement activation (by recruiting the factor H controlling protein). The increase in sialylation is often manifested as specific increases in α2-6-linked sialic acids attached to outer *N*-acetyllactosamine (Galβ1-4GlcNAc units) or to inner GalNAc-α1-O-Ser/Thr units on O-glycans. There is some evidence that the overexpression of Siaα2-6Galβ1-4GlcNAc units on N-glycans may enhance β1-integrin action. As discussed above, sialyl-Tn expression may well be a "side effect" of decreased O-glycan extension, and it is currently a target for immunotherapy.

Apart from the amount and linkage of sialic acids, there can also be significant changes in their modifications. Sialic acid 9-O-acetylation either can be up-regulated (e.g., the ganglioside epitope 9-O-acetylated GD3 [Figure 44.4] is increased in melanoma cells in species

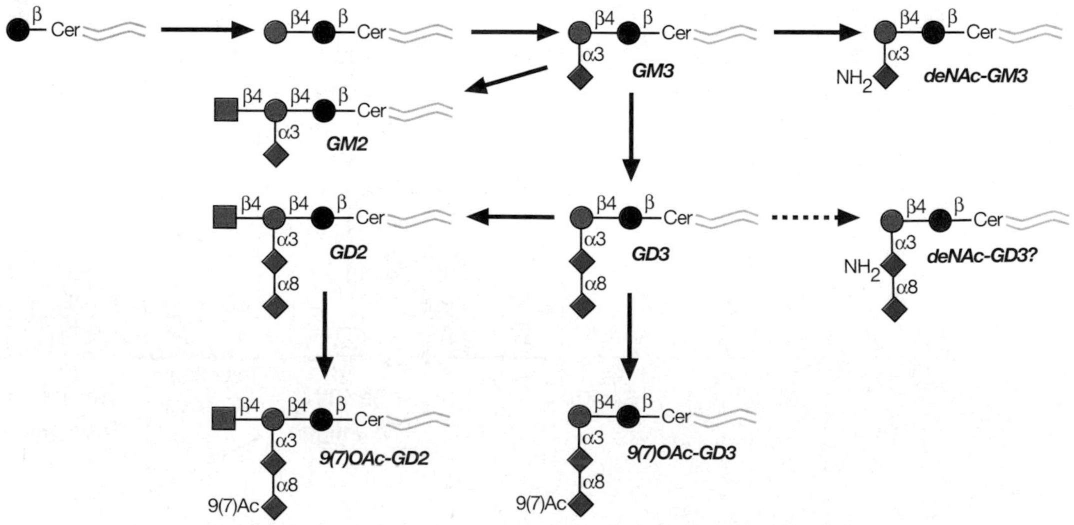

FIGURE 44.4. Some pathways for expression of gangliosides in human neuroectodermal tumors. The *heavy arrows* indicate pathways that are up-regulated. The *dashed arrows* indicate pathways that have not yet been formally proven. O-Acetylation of the terminal sialic acid can occur at the 7- or 9-position, with the former migrating to the latter position under physiological conditions.

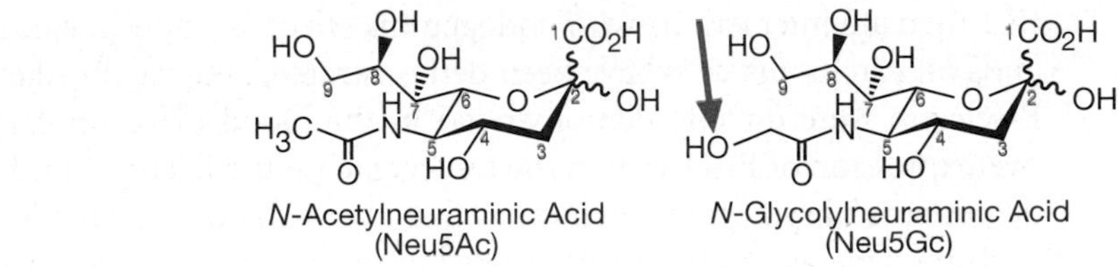

FIGURE 44.5. *N*-Acetylneuraminic acid (Neu5Ac) and *N*-glycolylneuraminic acid (Neu5Gc). The latter is a nonhuman sialic acid that accumulates in tumor cells, apparently from dietary origins. A negatively charged carboxylate occurs at the C-1 position of Neu5Gc and linkages to the underlying sugar chain occur from the C-2 position. The *arrow* points to the additional oxygen atom that distinguishes Neu5Gc from its precursor, Neu5Ac.

ranging from humans to fish) or may be decreased (e.g., on the O-glycans of colon carcinomas). Some tumor cell types have also been reported to express small amounts of de-N-acetyl gangliosides (Figure 44.4), wherein the N-acetyl group typical of common sialic acids is removed, exposing a free amino group. Again, the pathological significance of these molecules is uncertain. Some studies have suggested that O-acetylated gangliosides may protect cells from apoptosis, and that de-N-acetylgangliosides act by stimulating tyrosine phosphorylation of the epidermal growth factor (EGF) receptor. As described in Chapter 32, a subgroup of I-type lectins called Siglecs specifically recognize many structural features of sialic acids and are found on innate immune cells. It is possible that altered sialylation of tumor cells affects interactions with some Siglecs. Such interactions could be potentially beneficial to the tumor cell by sending an inhibitory signal to innate immune cells.

Another interesting phenomenon is the aberrant expression of *N*-glycolylneuraminic acid (Neu5Gc) in human tumor cells. This sialic acid differs from the usual *N*-acetylneuraminic acid (Neu5Ac) by the addition of a single oxygen atom (Figure 44.5). Adult humans do not express significant levels of Neu5Gc on their normal cells, and they mount an immune response to this epitope when infused with Neu5Gc-containing animal serum. This is because the hydroxylase enzyme responsible for creating Neu5Gc is mutated in humans (see Chapter 14). As there is no definitive alternate pathway for Neu5Gc production, reports of aberrant Neu5Gc expression in human tumors are best explained by uptake of Neu5Gc from dietary sources (or from fetal bovine serum in the case of cultured human cancer cells). Regardless of the mechanism, this re-expression of Neu5Gc may explain why some patients with cancer spontaneously develop so-called "Hanganutziu–Deicher" antibodies that are directed against gangliosides that contain Neu5Gc. The presumed selective advantage to tumor cells of accumulating Neu5Gc in the face of an immune response needs explanation. One possibility under study is that the resulting weak immune response is actually beneficial to tumor growth, by enhancing chronic inflammation and angiogenesis.

SIALYLATED LEWIS STRUCTURES AND SELECTIN LIGANDS ON CANCER CELLS

Immunohistochemical studies on tumor specimens have shown that Lewis^{x} and Lewis^{a} structures (see Chapters 13 and 45) are frequently overexpressed in carcinomas, being carried on O-glycans as well as on N-glycans and glycosphingolipids. Indeed, sialyl Lewis^{x} and sialyl Lewis^{a} were first identified as tumor antigens. The expression of these antigens by epithelial carcinomas correlates with tumor progression, metastatic spread, poor prognosis in humans, and metastatic potential in mice. These sialylated fucosylated structures also form critical components of most natural ligands for the selectins (see Chapter 31). Thus, it was reasonable to postulate that tumor cells gain a selective advantage by presenting pathological selectin ligands that mediate interactions with endogenous selectins. Indeed, calcium-dependent selectin ligands on carcinoma cells have been demonstrated, and mucin-like tumor antigens binding to E-selectin were directly demonstrated in the blood of colon carcinoma patients. Likewise, overexpression of E-selectin in the transgenic mouse liver induced redirection of the metastatic patterns of syngeneic carcinomas that normally colonize the lung.

Tumor metastasis is also attenuated in mice with P-selectin or L-selectin deficiency. These and other studies indicate that interactions between tumor-derived mucins and selectin molecules play a part in the metastatic cascade of some carcinoma cells (Figure 44.6). This ties in with the classic observation that cancer cells entering the bloodstream form complex thromboemboli with platelets and leukocytes, which are thought to facilitate arrest at ectopic sites, assist interactions with the endothelium, and help in evasion of the immune system. Current data suggest that this phenomenon can be explained by interactions between platelet and/or endothelial P-selectin and carcinoma mucins. Thus, carcinoma cells show a reduced metastatic rate in P-selectin-deficient mice, which can be explained at least in part by a lack of P-selectin-dependent rosetting of platelets on the tumor cells. With regard to L-selectin, one of the mechanisms appears to involve leukocyte interactions with fucosyltransferase-7 (FucT-7)-dependent endothelial ligands, which are induced at the site of tumor embolization in the vasculature. Of practical relevance is the finding that the glycosaminoglycan heparin is a potent inhibitor of P- and L-selectin interactions. Although the physiological relevance of selectin interactions with heparan sulfate is still being explored (see Chapter 31), the phar-

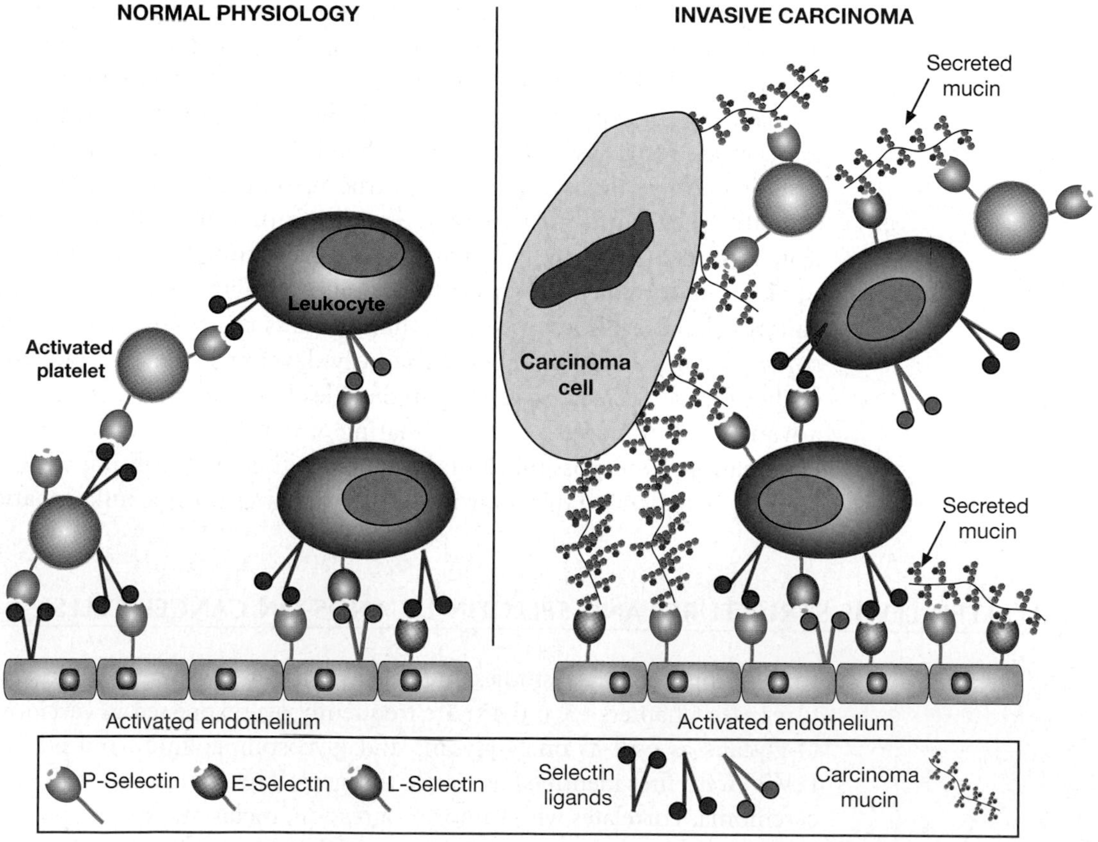

FIGURE 44.6. Potential interactions that could occur between tumor cells and selectins. Several of these have been formally proven to be important in tumor biology. See text for discussion. (Modified from Stevenson J.L., et al. 2005. *Clin. Cancer Res.* **11:** 7003–7011.)

macological effects of heparin on selectin interactions are thought to explain some prior reports of benefits of heparin therapy in cancer. Another strategy under development involves small-molecule inhibitors of selectin ligand formation (Chapter 49).

A related issue is that cancer patients sometimes develop thromboemboli and hypercoagulable states (Trousseau's syndrome), which are frequently associated with mucin-producing carcinomas. In this regard, secreted mucins from tumor cells have been shown to initiate a similar form of platelet-rich microthrombosis in mice, in a P- and L-selectin-dependent manner. Notably, as with tumor metastasis, this process can be blocked by heparin but not by other mechanisms that directly interfere with the fluid-phase coagulation pathway.

"INCOMPLETE SYNTHESIS" AND ALTERED EXPRESSION OF BLOOD GROUP–RELATED STRUCTURES

In addition to changes in Lewis^{x}- and Lewis^{a}-related structures mentioned above, loss of normal AB blood group expression (accompanied by increased expression of the underlying H and Lewis^{y} structures) (see Chapter 13) is associated with a poorer clinical prognosis of carcinomas in several studies. There are some other instances where glycan determinants strongly expressed on normal cells are lost upon transformation. For example, the Sd^{a} antigen (also called Cad), a blood group glycan structure abundantly expressed in the left half of the colon, is lost in colon carcinoma. Sulfation of the C-3 position of terminal galactose residues is also reduced in cancers. Sialyl-6-sulfo Lewis^{x} and disialyl Lewis^{a}, which are more complex structures derived from the same synthetic pathway as sialyl Lewis$^{x/a}$, are also expressed preferentially on normal colonic epithelial cells but are reduced in colon cancer cells. All of these changes can be considered as biosynthetically related to the enhanced production of sialyl Lewis$^{x/a}$ in cancers. DNA methylation and histone deacetylation, the epigenetic mechanisms for suppression of normal gene transcription commonly observed in cancers, are proposed to lie behind these glycan alterations.

As mentioned earlier, synthesis of a variety of complex glycans in normal cells can be impaired in malignant cells, leading to the accumulation of less complex precursor structures in cancers, a concept summarized as "incomplete synthesis." The loss of sialyl-6-sulfo Lewis^{x}, disialyl Lewis^{a}, 3′-sulfation, Sd^{a} antigen, and normal AB blood group determinants in cancers are good examples supporting this concept. Another example mentioned earlier is the accumulation of T, Tn, and sialyl-Tn and reduction of more complex structures in cancer cell O-glycans.

There are also rare instances in which a tumor may present a "forbidden" blood group structure (i.e., expression of a B blood group antigen in an A-positive patient; see Chapter 13 for details of the structure of ABO blood group antigens). The genetic basis for such a change remains unexplained, because the "B" transferase would theoretically require four independent amino acid changes to convert it into an "A" transferase. Regardless of the underlying mechanism, tumor regression has been noted in a few such cases, presumably mediated by the naturally occurring endogenous antibodies directed against the illegal structure. The appearance of P antigen in cancers of rare P blood type individuals is another example.

TRANSCRIPTIONAL REGULATION OF ALTERED GLYCAN EXPRESSION IN MALIGNANT CELLS

Unlike proteins, which are generally encoded by a single gene, a glycan determinant is produced by the concerted action of several related genes, making it difficult to elucidate the genetic regulatory mechanism for expression of some cell-surface glycans. With the availability of the total human genome sequence, application of DNA microarray and other molecular biological techniques to glycosyltransferases and related genes has facilitated accumulation of knowledge on the transcriptional background of cancer-associated glycan alteration. Several examples have been elucidated that indicate a direct association between glycan alteration and the genetic mechanism for malignant transformation of cells. One example mentioned earlier is transcriptional induction of *MGAT5* (the gene for GlcNAc transferase-V) by v-*src*, H-*ras*, and v-*fps*. A binding site for a transcription factor *ets-1* was shown to be present in the 5′-regulatory region of the gene. Another example is enhanced expression of sialyl Lewis^x on adult T-cell leukemia cells, which are known to exhibit extremely strong tissue infiltrative activity, likely mediated by selectins. The etiologic agent for this leukemia is a retrovirus called HTLV-1. The transcriptional activator protein, Tax, encoded by the virus, binds to the 5′-regulatory region of the gene for fucosyltranserase-7, the rate-limiting enzyme of sialyl Lewis^x synthesis in leukocytes, and constitutively activates its transcription (Figure 44.7). This is another good example of the direct association of malignant transformation and glycan alteration.

Under the poorly oxygenated conditions found in locally advanced tumors, hypoxia-resistant cancer cells survive by the principle of natural selection, acquiring hypoxia tolerability through a transcription factor HIF (hypoxia inducible factor), the nuclear translocation of which is facilitated by inactivation of tumor suppressors such as *VHL* and *p53*. Recently, HIF was shown to induce transcription of several genes for glycan synthesis, leading to the significant alteration of glycan profiles, including enhanced sialyl Lewis$^{x/a}$ expression in cancer cells. Tumor hypoxia may also affect the expression of Neu5Gc in human tumor cells mentioned earlier by influencing transcription of genes for the lysosomal sialic acid transporter.

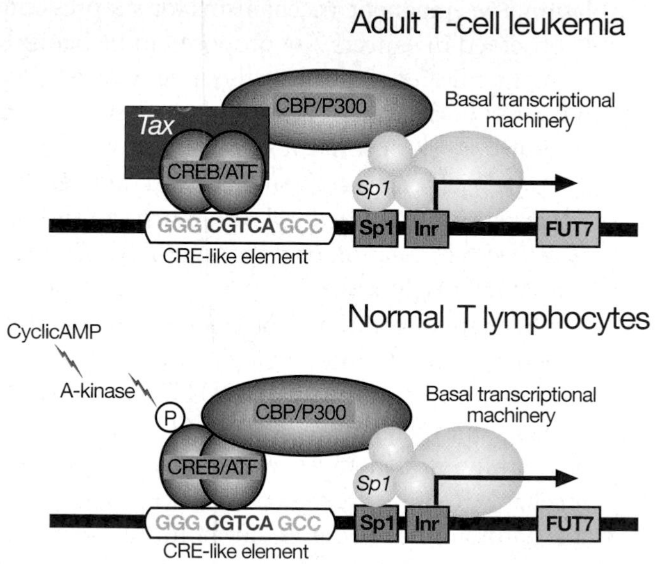

FIGURE 44.7. An example of a direct association of glycan alteration and malignant transformation is shown. The oncogenic protein Tax (encoded by HTLV-1, the etiologic virus for adult T-cell leukemia) induces transcription of the gene for fucosyltransferase VII (*FUT7*), which encodes the rate-limiting enzyme for sialyl Lewis^x synthesis (*upper panel*). Transcription of *FUT7* is regulated at the CRE element in the promoter region by the phosphorylation of CREB/ATF family transcription factors catalyzed by cyclic-nucleotide-dependent kinase (A-kinase) in normal lymphocytes (*lower panel*). The Tax protein bypasses the action of A-kinase and causes constitutive transcription of *FUT7*, which results in the strong expression of sialyl Lewis^x and vigorous tissue infiltration of leukemic cells. (Modified from Kannagi R., et al. 2004. *Cancer Sci.* **95:** 377–384.)

ALTERED EXPRESSION AND SHEDDING OF GLYCOSPHINGOLIPIDS

Many of the "tumor-specific" monoclonal antibodies raised against cancer cells are reactive with the glycan portion of glycosphingolipids. Some of these structures are highly enriched in specific types of tumors (e.g., Gb3/CD77 in Burkitt's lymphoma and GD3 in melanomas). Several types of tumors (particularly those of neuroectodermal origin such as melanoma and neuroblastoma) are characterized by the synthesis of very high levels of sialylated glycosphingolipids (gangliosides; see Chapter 10). Some of these (e.g., GD2) are not normally found at high levels in extraneural gangliosides and they are therefore considered targets for both passive immunotherapy (monoclonal antibody infusion) and active immunotherapy (immunization with purified glycolipid preparations). In some cases, gangliosides are also the major carriers of modified sialic acids (see above). Many in vitro studies suggest that some gangliosides have effects upon growth control, and it is suggested (but not proven) that this also may be the case in vivo. Some investigators have noted strong immunosuppressive effects that correlate with the large quantities of gangliosides "shed" from the cell surface by some tumors. It is not known whether this type of immunosuppression is a purely pathological phenomenon with no natural counterpart. Regardless, the levels of free gangliosides found in the body fluids of some patients with these tumors are high enough that such immunosuppression is likely to be medically relevant.

LOSS OF GLYCOPHOSPHOLIPID ANCHOR EXPRESSION

A complete loss of expression of glycosylphosphatidylinositol (GPI)-anchored proteins is seen in some cases of malignant and premalignant states involving the hematopoietic system. This results from acquired somatic mutations of hematopoietic stem cells in the *PIGA* gene (required for an early step in the biosynthesis of the GPI anchor precursor; see Chapter 11). Like the *Cosmc* gene discussed above, the *PIGA* gene is also on the X chromosome (Xp22.1). The consequence is a marked or complete loss of cell-surface expression of several GPI-anchored proteins on the progeny of a single hematopoietic stem cell clone. Because this mutation arises from a single stem cell, one would imagine that its progeny would not be easily detected. However, this clone often gradually replaces all of the others, giving rise to a syndrome called paroxysmal nocturnal hemoglobinuria or PNH (see Chapter 11). Although the mutation itself does not confer a malignant phenotype, the affected clone appears to be prone to become malignant (i.e., give rise to a leukemia). The mechanism(s) predisposing to the subsequent transforming event are unknown. It is also possible that the preexisting PNH cells are simply unmasked by the loss of normal cells due to an unrelated insult to the bone marrow.

CHANGES IN POLY-*N*-ACETYLLACTOSAMINE EXPRESSION AND GALECTIN FUNCTION

Increased expression of galectins (especially galectin-3) has also been associated with tumor progression. A molecular significance of this correlation is proposed to be the interactions of galectins with poly-*N*-acetyllactosamines on matrix proteins such as laminin, which aids cellular invasion. Because poly-*N*-acetyllactosamines are also expressed on cancer mucins and are enriched on the β1-6-branched glycans of tumor N-glycans (see above and Figure 44.1), this molecular interaction could mediate homotypic adhesion of carcinoma cells as well. Galectin recognition may also explain how adding cell-surface galactose to tumor mutants lacking the Golgi UDP-Gal transporter enhances metastasis. Another possibility is the formation of cell-surface lattices involving multiple galectin and poly-*N*-

acetyllactosamine molecules. Overall, it remains to be elucidated exactly how galectin–poly-*N*-acetyllactosamines interactions alter the biology of cancer.

CHANGES IN HYALURONAN

Hyaluronan is a very large negatively charged polysaccharide composed of the repeating disaccharide [GlcAβ1-3GlcNAcβ1-4]$_n$. It differs from other glycosaminoglycans in that it is nonsulfated and exists as a free polymer, rather than being covalently linked to a protein. Furthermore, it is synthesized and extruded from the cell directly at the plasma membrane, rather than being processed through the ER-Golgi pathway (see Chapter 15).

Many classes of malignant tumors express high levels of hyaluronan. In carcinomas, hyaluronan is usually enriched in the tumor-associated stroma (i.e., connective tissue elements and blood vessels). This stroma is more or less prominent depending on tumor type; for example, it is usually prominent in breast cancer. However, hyaluronan is also often localized immediately around the tumor cell surface. In normal tissues, hyaluronan serves at least three functions, which may also contribute to tumor progression. First, it increases levels of tissue hydration, which can facilitate movement of cells through tissues. Second, it is intrinsic to the assembly of extracellular matrices through specific interactions with other macromolecules, and thus it participates in tumor cell–matrix interactions that facilitate or inhibit tumor cell survival and invasion. Finally, hyaluronan interacts with several types of cell-surface receptors, especially CD44 and the receptor for hyaluronan-mediated motility (RHAMM/CD168). Hyaluronan–CD44 interactions are often crucial to tumor malignancy and are a current target for novel therapies.

Various alternatively spliced isoforms of CD44 are often elevated in cancer cells. However, this is not universal among different tumor types, leading to some confusion regarding the role of CD44 in cancer. Regardless, it is now clear that activation of hyaluronan–CD44 signaling is much more important in cancer progression than actual levels of CD44. In normal adult tissues, hyaluronan appears to be relatively inert with respect to cell signaling and behavior. However, during embryonic development, during tissue healing and regeneration, and in various pathological situations, hyaluronan-induced signaling via interaction with CD44 becomes activated. The consequences of this signaling are dramatic because they are essential to or promote cell behaviors such as proliferation, survival, migration, and invasion, which are also key elements of the malignant phenotype. Hyaluronan–CD44 interaction at the tumor cell surface is required for the constitutive activation of some well-known oncogenes, especially the receptor tyrosine kinase, *ErbB2*, which is amplified or mutated in a large number of carcinomas. Accordingly, hyaluronan–CD44 interaction promotes downstream intracellular pathways that are also hallmarks of cancer, such as the phosphatidylinositol-3-kinase/AKT and mitogen-activated protein (MAP) kinase pathways. It has been demonstrated that antagonists of hyaluronan–CD44 interaction cause inactivation of these pathways in malignant cells in culture and can inhibit tumor growth and metastasis in animal models. An exciting new development is the finding that hyaluronan stimulates multidrug resistance and that its antagonists sensitize resistant cancer cells to chemotherapeutic drugs. Evidently, stimulation of receptor tyrosine kinase activities by hyaluronan leads to increased phosphatidylinositol-3-kinase levels, which in turn stimulate new hyaluronan synthesis, thus setting up a positive feedback loop that amplifies anti-apoptotic pathways and expression of ABC-type multidrug transporters (Figure 44.8).

Despite convincing evidence for the involvement of hyaluronan-induced signaling in malignant cell behavior, there are several unsolved conundrums. For example, the mecha-

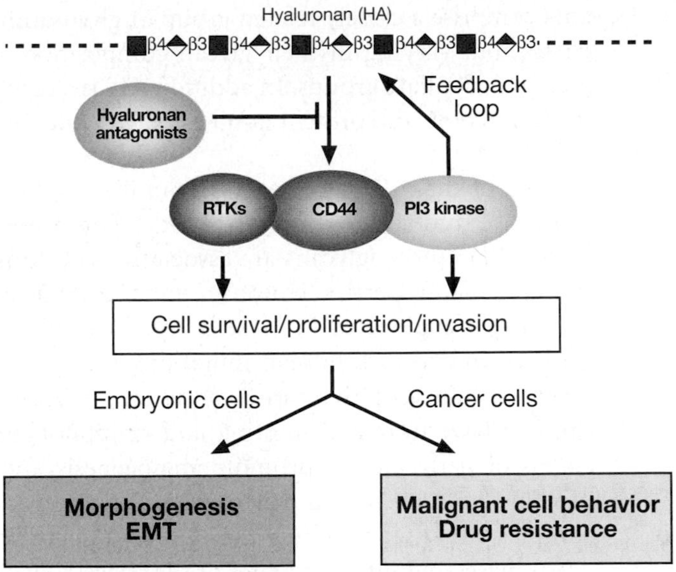

FIGURE 44.8. Effects of hyaluronan (HA) on oncogenic signaling. Endogenous HA–CD44 interaction is required for the formation of constitutive signaling complexes containing CD44, activated receptor tyrosine kinases (RTKs), phosphatidylinositol-3-kinase (PI3 kinase), and several other signaling molecules involved in cell survival, proliferation, and invasiveness. In addition, PI3 kinase stimulates HA production, thus setting up a positive feedback loop that amplifies these pathways. Antagonists of HA–CD44 interaction (e.g., HA oligomers, soluble HA-binding proteins, or siRNA against CD44) cause disassembly of constitutive complexes present in tumor cells with high levels of RTK activity and consequently inhibit their activity. On the other hand, increased HA production induces assembly of these complexes in cells with low levels of RTK activity and consequently stimulates RTK activity. HA-stimulated RTK activity is essential for correct embryonic morphogenesis and epithelial-mesenchymal transitions (EMT), whereas increased, deregulated RTK activity promotes the progression of malignant properties such as invasiveness, increased cell survival, and drug resistance in cancer cells.

nism of signaling activation is not understood. Two possibilities under consideration are alterations in CD44 status and effects of size on the signaling properties of hyaluronan. A current theme in pathophysiology is that emergent activities arise from breakdown products of extracellular matrix macromolecules, for example, the derivation of pro- and anti-angiogenic fragments from physiological proteolysis of collagens and proteoglycans. Some studies suggest a similar phenomenon for the mechanism of hyaluronan action and that hyaluronan activity may depend on its partial degradation by hyaluronidases. In agreement with this, both hyaluronan and hyaluronidase levels are elevated in several cancers and hyaluronan oligosaccharides are angiogenic. However, the mechanisms by which these breakdown products act are not understood. Despite all these remaining questions, it is clear that hyaluronan is an important therapeutic target in cancer.

CHANGES IN SULFATED GLYCOSAMINOGLYCANS

Numerous sulfated proteoglycans contribute to tissue structure and function during development and adult homeostasis. These proteoglycans contain core proteins that are characteristically decorated with covalently attached negatively charged glycosaminoglycan side chains—namely, chondroitin sulfate, dermatan sulfate, keratan sulfate, and heparan sulfate (HS) (see Chapter 16). Although the content and distribution of many proteoglycans are altered during tumorigenesis, only HS proteoglycans have been implicated in tumor pathogenesis in a widespread and convincing manner. A large number of HS and heparin variants comprise a closely related group of glycosaminoglycans derived from a common precursor, but varying in their glycan composition, especially with respect to number and location of sulfate groups. In addition, HS is a component of several proteoglycans derived from different core protein families (e.g., syndecans, glypicans, and perlecan) (see Chapter 16).

Each of these HS proteoglycan families is closely associated with positive or negative aspects of tumor progression (or both). For example, early work suggested that decreased levels of HS proteoglycans are associated with tumor progression and that HS suppresses tumorigenic properties. However, mutant CHO cells that do not produce HS and human tumor cells engineered to make less perlecan or glypican proteoglycans lose their tumorigenicity. In contrast, human mutations affecting HS polymerization result in osteochondromas and loss of glypican-3 causes an overgrowth syndrome. Even a specific proteoglycan can have tumor-promoting and -inhibitory properties. For example, while HS side chains of perlecan can promote angiogenesis, the carboxy-terminal domain of the core protein (endorepellin) is potently anti-angiogenic. More recent studies illustrate a rather complex situation in which the type and degree of processing of the HS side chains by heparanases and other enzymes or the processing of the proteoglycan core proteins by proteases can cause the various HS proteoglycans to have positive or negative effects on tumorigenesis. Despite this complexity, recent data have opened up the possibility of novel therapeutic interventions, based on the nature of HS processing and the activities of different heparanase products.

The major functions of HS proteoglycans relevant to tumor formation are (1) to promote cell–cell and cell–matrix interactions that are important in tissue assembly and (2) to bind a wide array of bioactive factors such as FGF-2 (fibroblast growth factor-2), VEGF (vascular endothelial growth factor), and numerous members of the interleukin, Wnt, and chemokine and growth factor families. Various studies have shown that the former activity can serve to build either inhibitory barriers or permissive pathways for cell invasion and that the latter activity can either sequester factors away from their receptors or present them efficaciously to these receptors. Thus, it is easy to visualize how HS proteoglycans could have opposite effects on cellular functions under varying cell and tissue arrangements. Syndecan-1 is particularly illustrative with respect to the multiple important roles of HS, especially in promoting metastasis and angiogenesis. Early work suggested that syndecan-1 is important in maintaining the normal differentiated state of epithelia and that low levels of tumor-associated syndecan-1 correlate with malignancy. However, recent work also highlights the importance of enzymatic processing by proteases and heparanase. First, a high level of proteolytically shed HS-rich ectodomain of syndecan-1 is present in the sera of patients with myeloma and some types of carcinoma and is a predictor of poor prognosis. Second, high levels of syndecan-1 ectodomain also accumulate in the tumor microenvironment, where it plays an important role in activation of chemokine and growth factor signaling and consequently tumor cell behavior. This conclusion is strongly supported by studies in which cells that are genetically engineered to produce large amounts of syndecan-1 ectodomain generate increased tumor growth, angiogenesis, and metastasis in animal models. Especially significant is the finding that this ectodomain promotes metastasis to bone tissue, which is characteristic of myeloma and certain carcinomas (e.g., of breast or prostate) in human patients. The mechanism of ectodomain shedding appears to involve members of the matrix metalloproteinase and ADAM/ADAM-TS protease families.

HS side chains of several HS proteoglycans are cleaved to fragments containing 10–20 sugar moieties by vertebrate heparanase (as opposed to smaller HS saccharides produced by bacterial eliminases, i.e., heparinases and heparitinases). Heparanase is elevated in numerous types of cancer and increased heparanase activity can lead to induction of angiogenesis and metastasis. Among the protumorigenic molecular and cellular consequences of this type of HS cleavage are (1) disassembly of basement membranes allowing penetration by tumor cells; (2) blood vessel remodeling required for angiogenesis; (3) release of HS-bound angiogenic factors, growth factors, and chemokines; and (4) increased bioactivity of the heparanase products compared to intact HS. Another mechanism of HS modification, the removal of sulfate moieties by endosulfatases, may alter HS activities and thus affect tumor growth. These studies have led to increased interest in HS as a therapeutic target in cancers (see below).

CLINICAL SIGNIFICANCE

Diagnostics

Although many tumor-specific monoclonal antibodies have been generated by immunization with tumor-derived glycosphingolipids, the appearance of corresponding glycan epitopes in the bloodstream is probably caused by the spillover of mucin glycoproteins (see Figure 44.2) bearing similar terminal glycan structures. The latter are of well-established value for detecting and monitoring the growth status of tumors (e.g., assays for CA19-9 in pancreatic cancers, CA125 and sialyl-Tn in ovarian carcinoma, and sialyl Lewis^x-related glycans in lung and breast cancers allow a physician to monitor the amount of tumor remaining in a patient after surgery or chemotherapy). However, these circulating mucins also pose a potential problem for the therapeutic use of monoclonal antibodies directed against these antigens, because they represent a "sink" that would absorb any infused antibody, preventing it from reaching the tumor cell surface. In recent times, the failure of proteomics to deliver reliable cancer biomarkers has rekindled interest in glycomic profiling of serum and other body fluid glycoproteins for this purpose. Also being reinvestigated are classic findings of glycan-specific antibodies associated with cancer, such as the heterophile antibodies against the nonhuman sialic acid, Neu5Gc. Another potential advance lies in improving the specificity of known cancer biomarkers by defining glycoforms uniquely expressed by cancer cells.

Therapy

Although altered glycosylation is of clear value in diagnosis, its role in specific therapies is less clear. As mentioned above, attempts are being made at specific passive and active immunotherapy against certain tumor-associated gangliosides and against incompletely glycosylated mucins. Some trials using humanized antiganglioside antibodies have shown promising results. With regard to the increased β1-6 branching of N-glycans, clinical studies of the metabolic inhibitor swainsonine have not yet yielded definitive results.

Cimetidine, a histamine H2 antagonist, suppresses E-selectin expression on vascular beds and reportedly improves patient prognosis after surgical therapy. Recent studies have also suggested that the potent effects of heparin in attenuating tumor metastasis might be due not to its anticoagulant properties, but rather to its ability to block P- and L-selectin binding to tumor and/or host ligands (see Chapter 31). In this regard, it is disconcerting that the recent move to supplant the traditional unfractionated heparins with very low-molecular-weight heparins and heparinoids may be associated with a loss of selectin-inhibitory properties. Several compounds have also been shown to antagonize the potentially protumorigenic activities of HS. For example, low-molecular-weight heparins may inhibit tumor progression not only by blocking selectins, but also by blocking heparanase activity or by interfering with constitutive HS activities. Other inhibitors of heparanase, such as suramin, laminarin sulfate, and the sulfated phosphomannopentaose PI-88, also inhibit angiogenesis and metastasis, but some of these compounds may also potentially work by blocking selectins. Some of these compounds may act by competitively inhibiting HS binding and function. Recent advances in our understanding of HS structure and function are likely to promote further development of efficient HS-based anticancer therapies. Low-molecular-weight oligosaccharides of hyaluronan could also be useful therapeutically, because they inhibit the prooncogenic influences of constitutive polymeric hyaluronan, especially drug resistance and signaling events induced by hyaluronan–CD44 interaction. Disaccharides that can enter the cell and act as decoys to divert glycosylation pathways are also showing promise (see Chapter 50). Further studies of all these possibilities are needed.

FURTHER READING

Hakomori S. and Kannagi R. 1983. Glycosphingolipids as tumor-associated and differentiation markers. *J. Natl. Cancer Inst.* **71:** 231–251.

Feizi T. 1985. Demonstration by monoclonal antibodies that carbohydrate structures of glycoproteins and glycolipids are onco-developmental antigens. *Nature* **314:** 53–57.

Hakomori S. 1986. Tumor associated glycolipid antigens, their metabolism and organization. *Chem. Phys. Lipids* **42:** 209–233.

Fukuda M. 1996. Possible roles of tumor-associated carbohydrate antigens. *Cancer Res.* **56:** 2237–2244.

Kim Y.S., Gum J., and Brockhausen I. 1996. Mucin glycoproteins in neoplasia. *Glycoconj. J.* **13:** 693–707.

Kim Y.J. and Varki A. 1997. Perspectives on the significance of altered glycosylation of glycoproteins in cancer. *Glycoconj. J.* **14:** 569–576.

Dennis J.W., Granovsky M., and Warren C.E. 1999. Glycoprotein glycosylation and cancer progression. *Biochim. Biophys. Acta* **1473:** 21–34.

Hakomori S. 1999. Antigen structure and genetic basis of histo-blood groups A, B and O: Their changes associated with human cancer. *Biochim. Biophys. Acta* **1473:** 247–266.

Varki N.M. and Varki A. 2002. Heparin inhibition of selectin-mediated mechanisms in the hematogenous phase of carcinoma metastasis: Rationale for clinical trials in humans. *Semin. Thromb. Hemost.* **28:** 53–66.

Kannagi R. 2004. Molecular mechanism for cancer-associated induction of sialyl Lewis X and sialyl Lewis A expression—The Warburg effect revisited. *Glycoconj. J.* **20:** 353–364.

Kannagi R., Izawa M., Koike T., Miyazaki K., and Kimura N. 2004. Carbohydrate-mediated cell adhesion in cancer metastasis and angiogenesis. *Cancer Sci.* **95:** 377–384.

Sanderson R.D., Yang Y., Suva L.J., and Kelly T. 2004. Heparan sulfate proteoglycans and heparanase—Partners in osteolytic tumor growth and metastasis. *Matrix Biol.* **23:** 341–352.

Toole B.P. 2004. Hyaluronan: From extracellular glue to pericellular cue. *Nat. Rev. Cancer* **4:** 528–539.

Fuster M.M. and Esko J.D. 2005. The sweet and sour of cancer: Glycans as novel therapeutic targets. *Nat. Rev. Cancer* **5:** 526–542.

Kannagi R., Yin J., Miyazaki K., and Izawa M. 2008. Current relevance of incomplete synthesis and neo-synthesis for cancer-associated alteration of carbohydrate determinants. *Biochim. Biophys. Acta* **1780:** 525–531.

CHAPTER 45

Antibodies and Lectins in Glycan Analysis

Richard D. Cummings and Marilynn E. Etzler

ANTIGLYCAN ANTIBODIES AND LECTINS ARE WIDELY USED in glycan analysis because their specificities enable them to discriminate among a variety of glycan structures and their multivalency ensures high-affinity binding to the glycans and cell surfaces containing those glycans. This chapter describes the variety of commonly used antibodies and lectins and the types of analyses to which they may be applied.

BACKGROUND

The first evidence that carbohydrates were antigenic arose from the discovery that the human blood group ABO antigens were glycans (see Chapter 13). A key tool in these studies was the use of plant lectins, and by the mid-1940s they had found widespread use in typing blood.

Since this early beginning, hundreds of different plant and animal lectins have been identified and characterized. Because of their abundance in many plant seeds and tissues, lectins from these sources have been analyzed in detail and used for practical research on glycans. Thus, although monoclonal antibodies might be more specific for glycan determinants, plant and animal lectins have useful specificities, are usually less expensive, are better characterized with respect to binding specificity, and are more stable. The availability of the plant lectins and some animal lectins and their exquisite specificity for complex glycans helped to catapult the field of glycobiology into the modern era.

LECTINS MOST COMMONLY USED IN ANALYSIS OF GLYCANS

Many of the lectins currently used as tools in glycobiology come from plants and are commercially available. Most of these lectins were characterized initially by inhibition assays, in which monosaccharides, monosaccharide derivatives, or small oligosaccharides are used to block lectin binding to cells or some other glycan-coated target. Small-sized molecules that compete with binding of a lectin or antibody to a larger-sized ligand are termed haptens. These lectins are grouped by specificity depending on the monosaccharide(s) for which they show the highest affinity and their distinct preference for α- or β-anomers of the sugar. However, lectins within a particular specificity group also may differ in their affinities for different glycans. The common method for grouping lectins according to monosaccharide specificity should thus be used with caution because it does not reflect the complex specific determinants a given lectin may recognize with high affinity. The binding affinity (K_d) of lectins for complex glycans is often in the range of 1 to 10 μM. For complex glycoconjugates with multiple determinants or multivalency, the binding affinity of lectins may approach nanomolar values. In contrast, the affinity of most lectins for monosaccharides is in the millimolar range. The specificity of concanavalin A (ConA), perhaps the most widely used lectin, demonstrates this point. This lectin (which is an α-mannose/α-glucose-binding lectin) binds to N-glycans and is not known to bind O-glycans on animal cell glycoproteins. However, it binds oligomannose-type N-glycans with much higher affinity than it binds complex-type biantennary N-glycans, and it fails to bind more highly branched complex-type N-glycans (Figure 45.1).

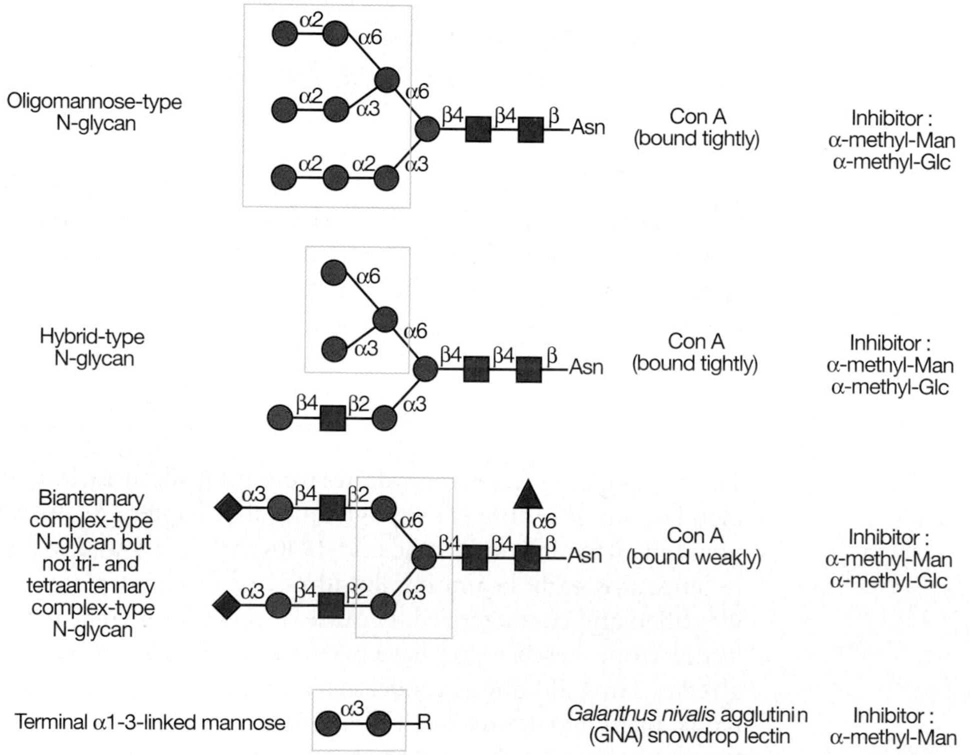

FIGURE 45.1. Examples of N-glycans recognized by concanavalin A (ConA) from *Canavalia ensiformis* and *Galanthus nivalis* agglutinin (GNA). The determinants required for binding are indicated in the *boxed areas*. Hapten sugars that can competitively inhibit binding of the lectin to the indicated glycans are shown on the *right*.

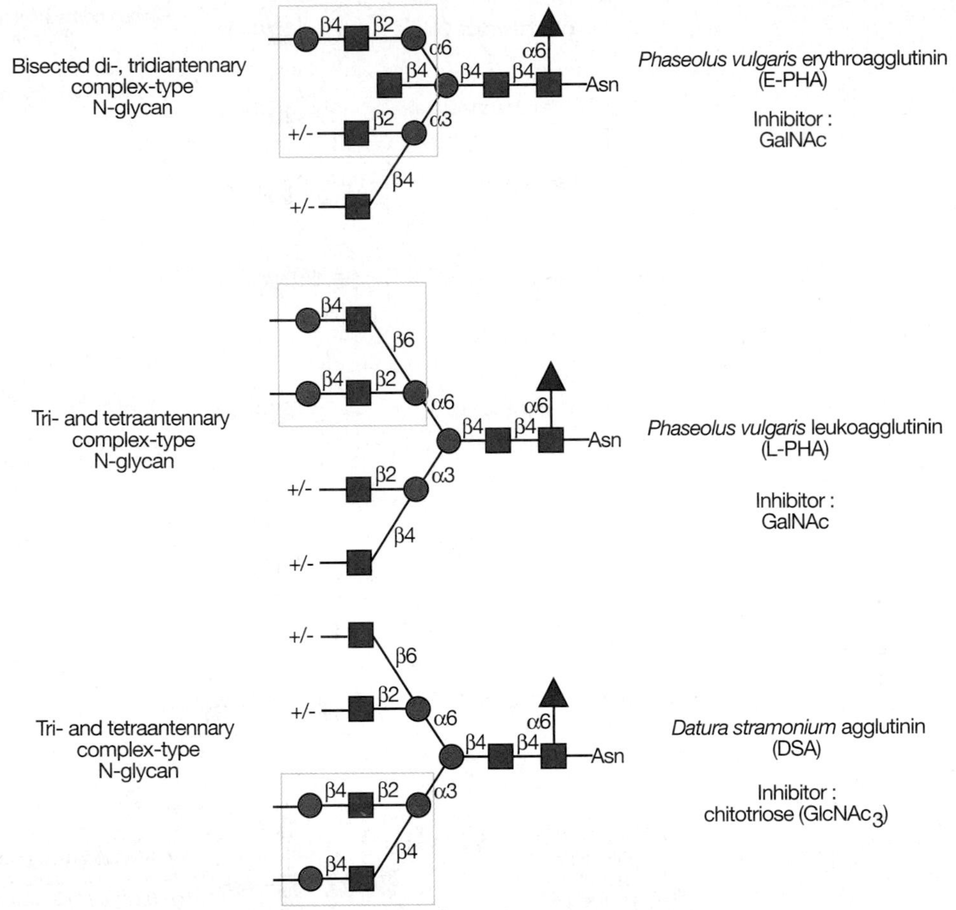

FIGURE 45.2. Examples of types of N-glycans recognized by L-PHA, E-PHA, and DSA. The determinants required for binding are indicated in the *boxed areas*. Hapten sugars that can competitively inhibit binding of the lectin to the indicated glycans are shown on the *right*.

Other lectins, such as L-phytohemagglutinin (L-PHA) and E-PHA from *Phaseolus vulgaris* and lentil lectin (LCA) from *Lens culinaris*, also recognize specific aspects of N-glycan structures. Many lectins, such as *Ulex europaeus* agglutinin I (UEA-I), *Sambucus nigra* agglutinin (SNA), *Maackia amurensis* leukoagglutinin (MAL), and *Griffonia simplicifolia* I-B4 agglutinin (GSI-B4), bind terminal structures with specific determinants (Figures 45.2–45.4) (see also Chapters 28 and 29).

Some of the animal lectins that are widely used in glycobiology include *Helix pomatia* agglutinin (HPA) from the snail, *Limulus polyphemus* agglutinin (LPA) from the hemolymph of the horseshoe crab, *Limax flavus* agglutinin (LFA) from the garden slug, and *Anguilla anguilla* agglutinin (AAA) from the freshwater eel.

GENERATION OF MONOCLONAL ANTIBODIES TO GLYCAN ANTIGENS

There are several approaches for generating antibodies to glycan antigens.

1. Whole cells have been used to immunize mice to generate specific monoclonal antibodies to various glycoprotein and glycolipid antigens, including the Tn antigen (GalNAcα1-Ser/Thr) and the stage-specific embryonic antigen-1 (SSEA-1), now

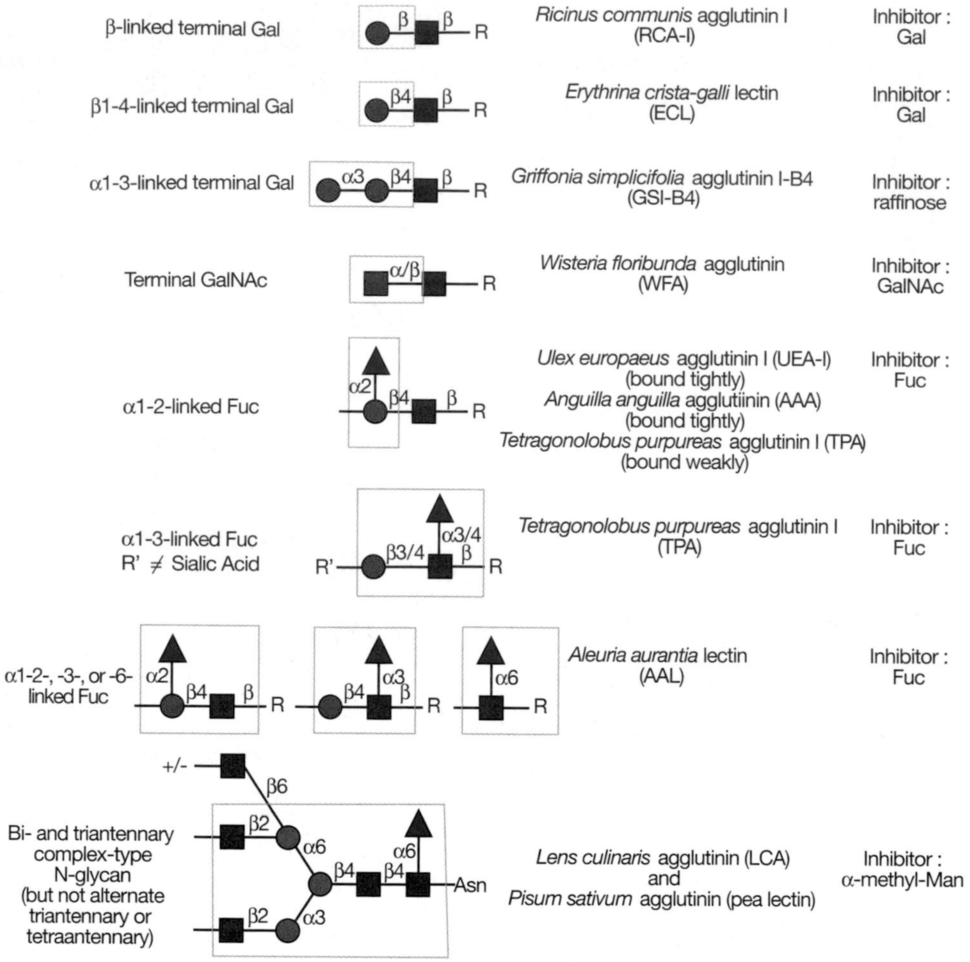

FIGURE 45.3. Examples of types of glycan determinants bound with high affinity by different plant and animal lectins. The determinants required for binding are indicated in the *boxed areas*. Hapten sugars that can competitively inhibit binding of the lectin to the indicated glycans are shown on the *right*.

known as the Lewis^x (Le^x) antigen (Figure 45.5). In this approach, hybridomas are screened for monoclonal antibodies that recognize the immunizing cell but not other types of cells. A large number of antibodies to glycan determinants were generated by immunizing mice with different types of cancer cells.

2. Glycan-protein conjugates (glycans coupled to carrier proteins such as bovine serum albumin [BSA] or keyhole limpet hemocyanin [KLH]) have been used to generate monoclonal antibodies to specific structures. A common variation of this approach is to immunize mice directly with a preparation of glycoproteins (such as a membrane fraction) or even a purified glycoprotein, glycolipid, or glycosaminoglycan. For example, antibodies to the plant glycoprotein horseradish peroxidase detect the presence of the unusual modifications Xylβ1-2Man-R and Fucα1-3GlcNAc-R in the core regions of complex-type N-glycans. Immunization with glycan conjugates has also been used to generate polyclonal antisera in rabbits and chickens. Using such antisera, it is possible to purify antibodies to specific glycan determinants by affinity chromatography on immobilized glycans. Knockout mice lacking specific glycoconjugates are also useful

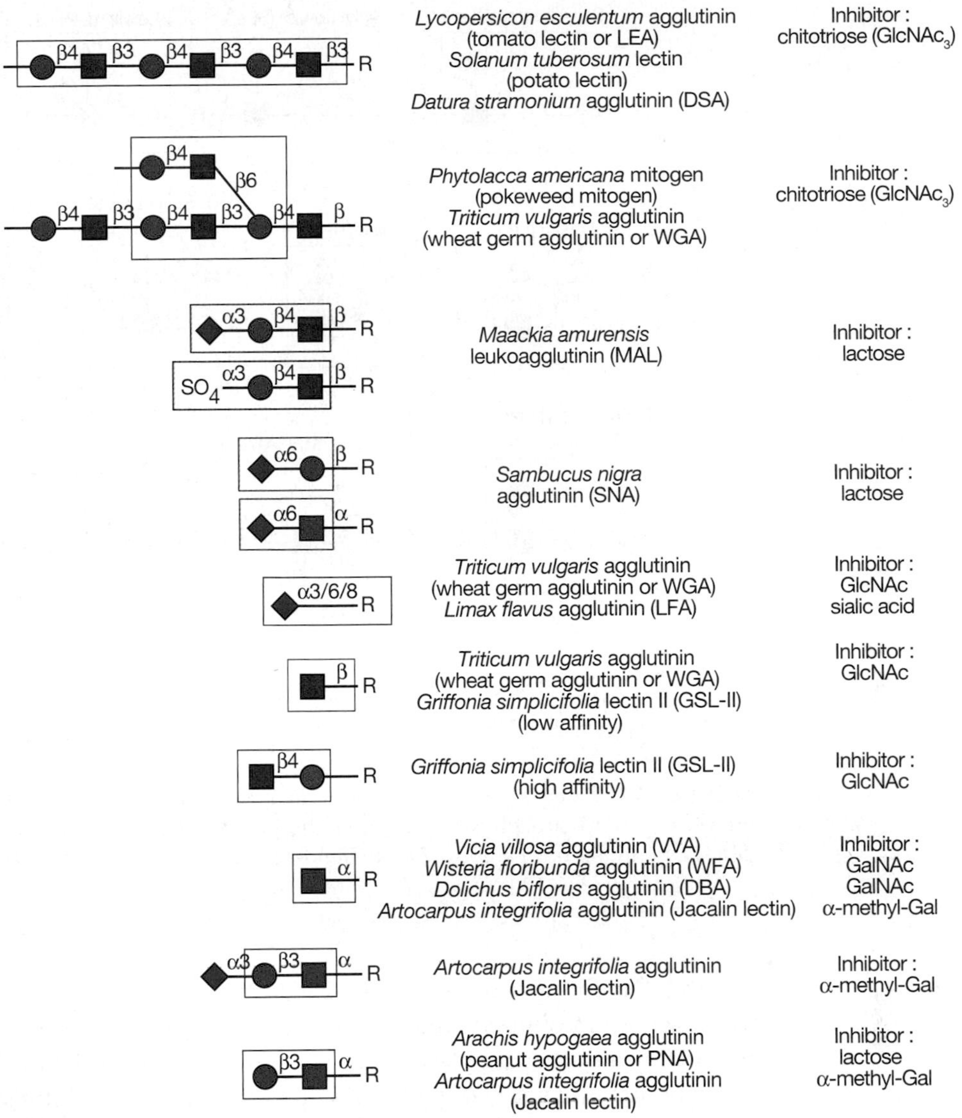

FIGURE 45.4. Examples of types of glycan determinants bound with high affinity by different plant lectins. The determinants required for binding are indicated in the *boxed areas*. Hapten sugars that can competitively inhibit binding of the lectin to the indicated glycans are shown on the *right*.

for generating antibodies to common structures. For example, antibodies to the common glycolipid sulfatide (3-O-sulfate-Galβ1-ceramide) were generated by taking mice that lacked the cerebroside sulfotransferase and immunizing them with sulfatide. If desired, it is possible to produce recombinant single-chain antibodies to glycan determinants after cloning the V_H and V_L domains of antibodies from hybridomas expressing specific monoclonal antibodies.

3. A novel approach for obtaining monoclonal antibodies to specific glycan antigens has been to use mice that were infected with specific parasites or bacteria and then preparing hybridomas from the splenocytes of the infected animals. This approach has been used to generate many specific monoclonal antibodies to pathogen-specific glycan antigens.

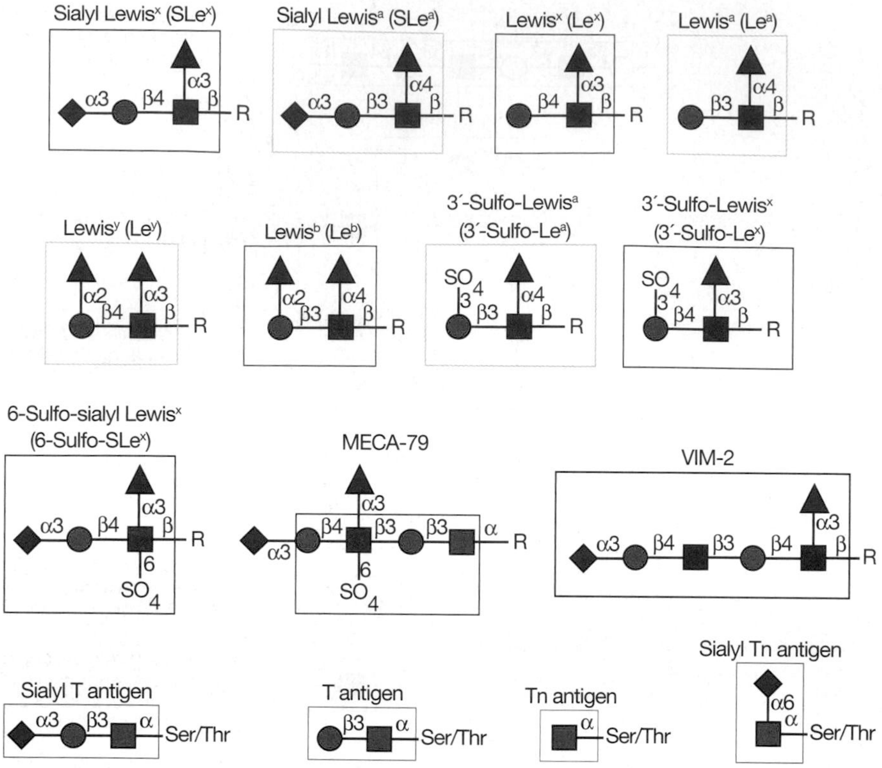

FIGURE 45.5. Examples of different glycan antigens recognized by specific monoclonal antibodies. The antigens have the structures shown within the *boxed area* and are named as indicated. Usually, the antigen shown in the box can be linked to almost any glycan and antibodies will still recognize the antigen.

Screening for appropriate antibodies to the desired antigen usually involves ELISA-type assays with immobilized target glycans. These targets may be specific glycolipids, which can be directly adsorbed to plastic wells; neoglycoproteins, such as BSA derivatives with covalently bound glycans; glycoproteins with known glycan structures; glycosaminoglycans, such as keratan sulfate, chondroitin sulfate, dermatan sulfate, and heparin; or biotinylated glycans captured on streptavidin-coated plates. However, in many cases, it is very difficult to define the specific epitopes recognized by these antibodies, because a wide variety of related compounds are not readily available for comparison. Thus, it is always advisable to be cautious in interpreting results based on monoclonal antibodies to glycan antigens, unless the antibody specificities have been well defined. In the future it will be important to define the specificities of each monoclonal antibody precisely using glycan microarrays and related approaches, where binding to a large number of glycans can be compared (see Chapter 27). Antiglycan antibodies are widely used in glycobiology and some of the more common antigens they recognize are shown in Figures 45.5 and 45.6. Most of the murine monoclonal antibodies to these antigens are of the IgM variety, but some are IgGs. The higher valency of IgM antibodies can affect the binding specificity. Some antibodies to glycan antigens are commercially available, whereas others are often obtained directly from individual laboratories.

Most of the antibodies against the antigens shown in Figures 45.5 and 45.6 recognize terminal glycan determinants, although subterminal sequences may also be required for binding. In addition, the context of expression (glycoprotein vs. glycolipid) or the class of glycan (N- vs. O-linked) carrying the antigen may play a significant role in determining specificity. For example, antibodies to the 6-sulfo-SLe^{x} antigen require the fucose, sialic acid, and

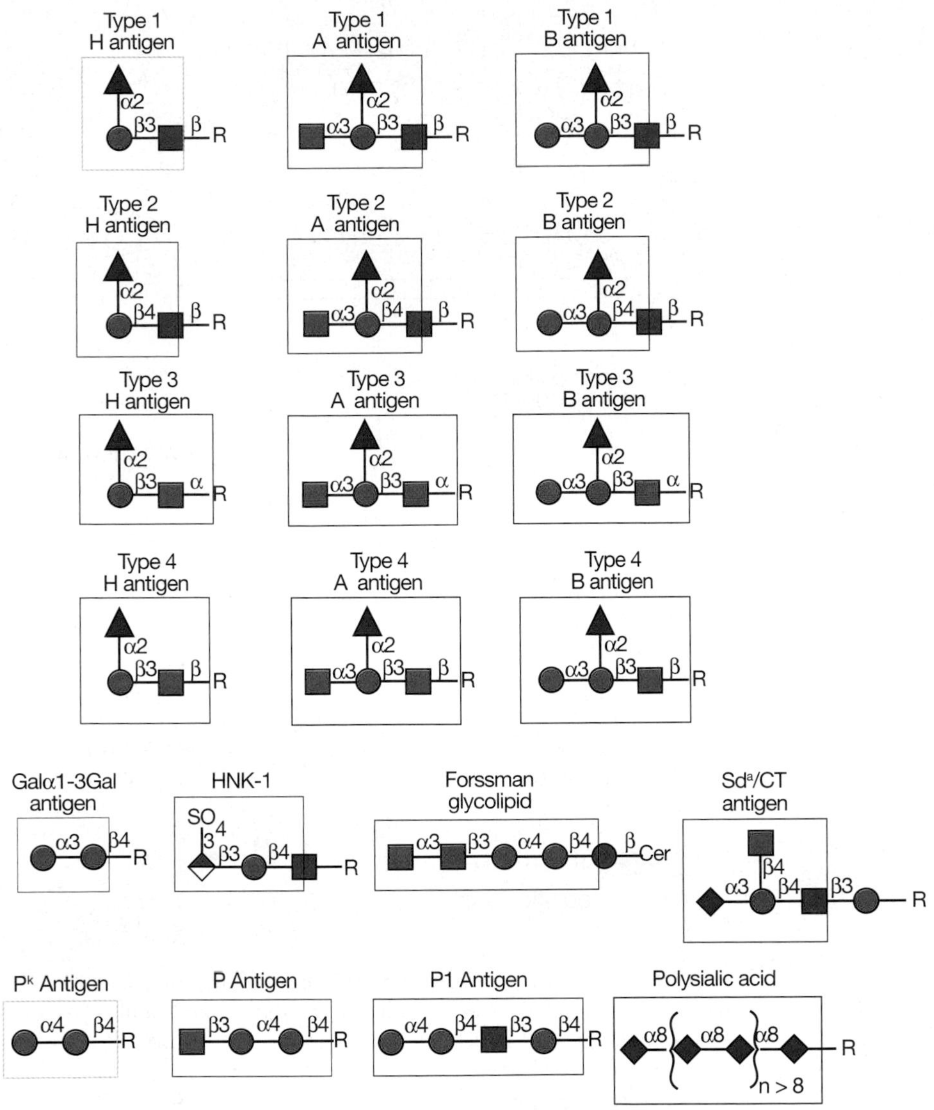

FIGURE 45.6. Additional examples of different glycan antigens recognized by specific monoclonal antibodies. The antigens have the structures shown within the *boxed area* and are named as indicated. Usually, the antigen shown in the box can be linked to almost any glycan and antibodies will still recognize the antigen. Antibodies are also available to glycosaminoglycans and to many glycolipid antigens, including globoside, GD3, GD2, GM2, GM1, asialo-GM1, GD1a, GD1b, GD3, and GQ1b.

GlcNAc-6-O-sulfate residues. In contrast, the MECA-79 antibody recognizes extended core-1 O-glycans that contain internal GlcNAc-6-O-sulfate residues; it does not require either the fucose or sialic acid for recognition, but it does require the core-1 O-glycan. There are also many monoclonal antibodies that recognize different glycolipids and glycosaminoglycans.

USES OF ANTIBODIES AND LECTINS IN GLYCAN IDENTIFICATION

Lectins and antibodies are useful reagents for aiding in glycan identification. Some of the important uses are illustrated in Figure 45.7. They include agglutination of cells and blood typing, cell separation and analysis, bacterial typing, identification and selection of mutat-

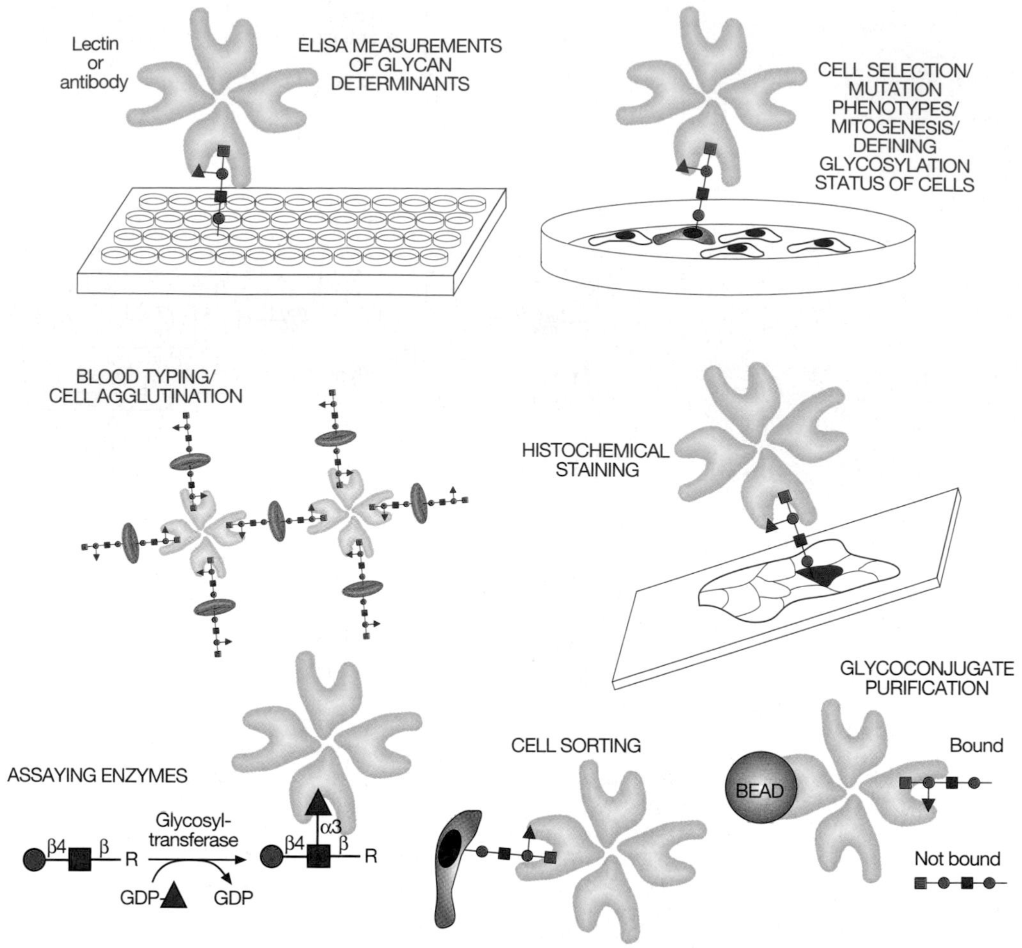

FIGURE 45.7. Examples of different uses of plant and animal lectins and antibodies in glycobiology. Many plant and animal lectins are multivalent as shown, and antibodies are always multivalent. They can be used to detect glycan structures in all of the formats shown.

ed cells with altered glycosylation, toxic conjugates for tumor cell killing, cytochemical characterization/staining of cells and tissues, inducing mitogenesis of cells, acting as growth inhibitors, mapping neuronal pathways, purification and characterization of glycoconjugates, assays of glycosyltransferases and glycosidases, and defining glycosylation status of target glycoconjugates and cells.

Antibodies and lectins each have distinct advantages. For example, plant and animal lectins are less expensive than antibodies, a major consideration in affinity chromatography approaches where milligram amounts of the glycan-binding protein might be needed for purification. However, antibodies may be needed to bind specifically to certain determinants. For example, no lectin specific for the SLe^a antigen has been found. Conversely, no monoclonal antibodies have been identified that bind general determinants, such as α2-6-linked sialic acid and α1-2-linked fucose, whereas lectins have this specificity (e.g., SNA and *Aleuria aurantia* lectin [AAL], respectively). Of course antibodies that are specific to some restricted determinants, such as Le^x and SLe^a antigens, still cannot distinguish their presentation on O-glycans, N-glycans, or glycolipids. In addition, although antibodies cannot distinguish N-glycan structural motifs, such features are well recognized by some plant lectins. For example, the plant lectin ConA does not bind mucin-type O-glycans in animal cells, but binds only to some specific classes of N-glycans (Figure 45.1). Additionally, E-PHA binds "bisected" complex-type N-glycans (Figure 45.2) and does not bind any known glycolipid or O-glycan. In all studies, care should always be taken to use lectins and antibodies at appropriate concentrations where their specificity can be exploited. Thus, it can be seen that antibodies and lectins each offer distinct advantages in defining glycan structures.

The glycan determinants bound with highest affinity by each of these probes have been identified by a combination of approaches, including affinity chromatography, glycan synthesis, and binding to specific glycoconjugates and cells. A good example of this is the phytohemagglutinin L-PHA, which is often used by immunologists as a mitogen to stimulate quiescent T cells to divide. L-PHA originates from the red kidney bean *Phaseolus vulgaris*, which also contains isolectins to L-PHA, notably E-PHA. L-PHA binds to certain branched, complex-type N-glycans containing the pentasaccharide sequence Galβ1-4GlcNAcβ1-2(Galβ1-4GlcNAcβ1-6)Manα1-R (the so-called "2,6-branch"), as shown in the boxed portion of the glycan in Figure 45.2. Curiously, the only monosaccharide that effectively inhibits either L-PHA or E-PHA is *N*-acetylgalactosamine, although this monosaccharide is not part of the determinants in N-glycans recognized by these lectins (Fig. 45.2). This finding arose from a combination of studies. First, L-PHA does not bind to any known glycolipid or O-glycan. Interestingly, L-PHA does not bind to human erythrocytes; thus, it is not a hemagglutinating lectin. It can, however, agglutinate white blood cells, and thus it is called a leukoagglutinin. Affinity chromatography approaches using immobilized L-PHA showed that it binds well to highly branched complex-type N-glycans containing the 2,6-branched structure, but it does not bind to alternate branched N-glycans lacking the 2,6-branch. In addition, L-PHA is highly inhibited by synthetic glycans containing the pentasaccharide sequence Galβ1-4GlcNAcβ1-2(Galβ1-4GlcNAcβ1-6)Man. Finally, L-PHA binds to the murine thymic leukemia cell line BW5147, but not to a mutated derivative cell line $PHA^R2.1$. This mutant lacks the β1-6 *N*-acetylglucosaminyltransferase that creates the 2,6-branched complex-type N-glycan containing the pentasaccharide sequence. Thus, it has become clear that differential binding to erythrocytes versus leukocytes results because 2,6-branched N-glycans are not expressed on glycoproteins in human erythrocytes. Indeed, the binding of L-PHA has been used by glycobiologists to identify these types of branched N-glycans in cells. Interestingly, such L-PHA-binding glycoproteins are increased in many tumor cells (see Chapter 44), whereas they are dramatically decreased in mice genetically null for the branching β1-6 *N*-acetylglucosaminyltransferase (GNT-V).

Another good example of lectin specificity is *G. simplicifolia* I-B4 agglutinin (GSI-B4). This isolectin (from the seeds of a shrub native to West and Central Africa) agglutinates human blood group B and AB erythrocytes, poorly agglutinates A erythrocytes, and does not agglutinate H(O) erythrocytes. These seeds also contain other isolectins, such as GSI-A4, which binds terminal α-linked *N*-acetylgalactosamine residues. Glycoconjugates with terminal α-linked galactose, but not β-linked galactose, inhibited agglutination by GSI-B4. Affinity chromatography of glycoproteins from murine tumor cells on immobilized GSI-B4 led to purification of a subset of glycoproteins strongly enriched in terminal α-linked galactose determinants. The lectin has also been used to expression clone the cDNA encoding the α1-3-galactosyltransferase that synthesizes the terminal sequence Galα1-3Galβ1-4GlcNAc-R, as discussed below. Thus, GSI-B4 shows a strong preference for glycoconjugates containing the terminal sequence Galα1-3Galβ1-R.

Using a variety of lectins and antibodies to probe glycoproteins or cells, it is possible to deduce many aspects of glycan structure. This approach is especially sensitive in regard to defining whether two samples differ in glycosylation. For example, such approaches have been adapted to study differential glycosylation of prion glycoproteins using a panel of biotinylated lectins in ELISA-type formats.

USES OF ANTIBODIES AND LECTINS IN GLYCAN PURIFICATION

There are several approaches to using antibodies and lectins in glycan purification, including affinity chromatography or affinity binding and immunoprecipitation or lectin-induced precipitation. The proteins may be covalently coupled to a carrier such as Sepharose or biotinylated and captured on streptavidin-Sepharose. In addition, antibodies may be noncovalently captured on protein A (or G)-Sepharose. These bound antibodies and lectins can then be used in affinity chromatography or affinity capture to isolate glycoconjugates expressing specific glycan determinants. ConA-Sepharose is commonly used to isolate glycoproteins as it has little binding to nonglycosylated proteins. However, it does not bind all glycoproteins because it recognizes specific N-glycan structures. Using free glycans, ConA-Sepharose has been used to isolate oligomannose-, hybrid-, and complex-type biantennary N-glycans.

When combined in a serial format, multiple lectins can be used in affinity chromatography to isolate most of the major glycan structures present in animal cells, with glycans being separated as classes that share common determinants. An example of serial lectin affinity chromatography is shown in Figure 45.8. A mixture of cell glycans, containing the structures shown, may be applied to columns of immobilized ConA, LCA, and MAL. On the basis of their binding to these lectins, the glycans are separated into bound and unbound fractions, which are serially analyzed as shown. In this example, all glycans are bound by at least one lectin. The binding, or lack of binding, allows structural features of the glycans to be predicted. For example, the biantennary glycans bind weakly to ConA and are eluted with 10 mM α-methylglucoside, whereas the oligomannose- and hybrid-type N-glycans bind strongly and elution requires 100 mM α-methylmannoside. The biantennary N-glycans are fractionated into glycans that bind LCA (which are predicted to have a core α1-6-fucose residue) and those that are unbound (which are predicted to lack this fucose linkage). The unbound biantennary N-glycans from LCA are then bound by MAL and they are predicted to contain terminal α2-3-sialic acid linked to Galβ1-4GlcNAc-R residues. When serial lectin affinity chromatography is coupled with ion-exchange chromatography and HPLC, it is possible to obtain highly purified glycans with predicted structures, which can then be confirmed by mass spectrometry of native and permethylated derivatives (see Chapter 47).

It should also be noted that mixed-bed lectin chromatography, where a number of different immobilized lectins are combined in a single column or tube, is useful for generally isolating all types of glycoconjugates simultaneously from nonglycosylated material. For example, this can be used to separate glycopeptides from peptides. Thus, the ability of glycans to be recognized by lectins dependent on specific structural features in the glycans is a powerful tool for glycan identification and isolation. In such approaches, the glycans may be tagged at the reducing end by fluorophores and radioisotopes or may be obtained by metabolic radiolabeling from cells or tissues grown in the presence of radiolabeled sugar precursors, such as [2-^{3}H]mannose or [6-^{3}H]glucosamine. Interestingly, glycan fractionation shown on immobilized lectins in Figure 45.8 is currently not possible with antibodies because no antibodies are known that can distinguish such core structural features in glycans.

When intact glycoproteins are analyzed for their interactions with plant lectins, the interpretation of data may be complicated by the multivalency of the glycoprotein and the density of the immobilized lectin. For example, glycoproteins containing multiple high mannose-type N-glycans bind so tightly to immobilized ConA that it is difficult to elute the bound glycoprotein, even with extremely high concentrations of hapten and under

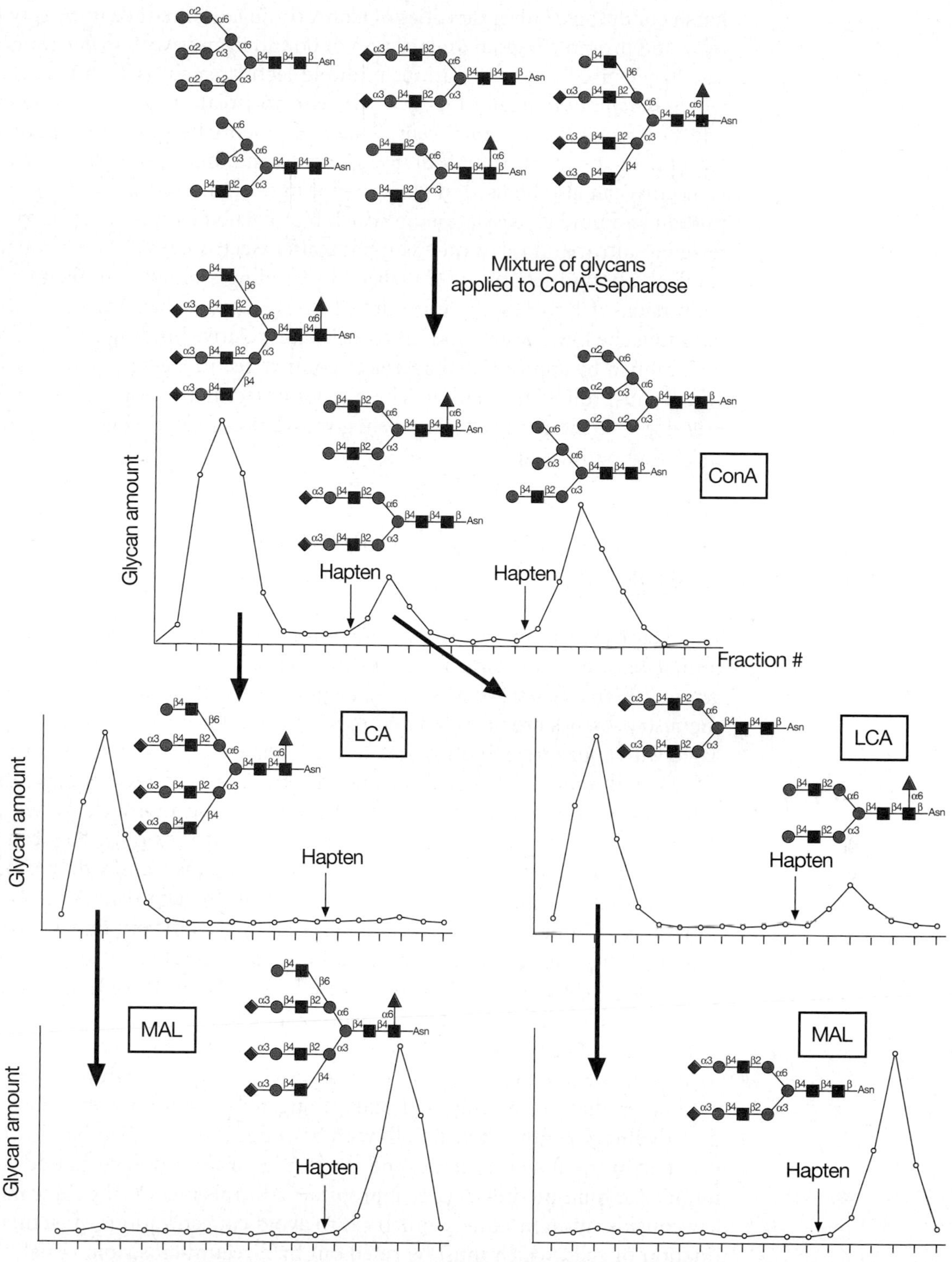

FIGURE 45.8. An example of the use of different immobilized plant lectins in serial lectin affinity chromatography of complex mixtures of glycopeptides. In this example, a mixture of glycopeptides is applied to a column of immobilized ConA, and the bound glycans are eluted with increasing concentrations of hapten sugars (α-methylglucose and then α-methylmannose) (*arrows*). The recovered glycans are then applied to a second set of columns containing immobilized LCA and bound glycans are eluted with the hapten α-methylmannose. This process is repeated over other immobilized lectins, such as MAL, and bound glycans are eluted with the hapten lactose. The glycans bound at each step are shown in the *panels* above the chromatographic peaks.

harsh conditions. Lower densities of ConA conjugation reduce its avidity for the glycoproteins and promote hapten dissociation of bound ligands with lower concentrations of sugars. When used in combination, multiple lectins, such as ConA, AAL, LCA, and RCA (*Ricinus communis* agglutinin), can be used to isolate most glycoproteins containing N- and O-glycans from animal cells. This is a potentially powerful approach for glycoproteomics, or the identification of glycoproteins and their glycosylation status.

Lectins can also be used in western blotting (lectin blotting) approaches to characterize protein and lipid glycosylation, in which biotinylated lectins are applied to material transferred to nitrocellulose or other supports after electrophoresis or chromatography. In such cases, the bound lectins are visualized by binding streptavidin-alkaline phosphatase and conversion of luminescent substrates. In these approaches, the concentrations of lectins used must be low enough both to reduce false-positive binding and to allow the binding to be inhibited by appropriate haptens to confirm lectin-sugar binding, rather than nonspecific lectin-protein association. These types of lectin blotting studies have been especially useful in characterizing alterations in glycosylation following genetic manipulation of glycosylation pathways.

USES OF ANTIBODIES AND LECTINS IN CHARACTERIZING CELL-SURFACE GLYCOCONJUGATES

The major approaches for using antibodies and lectins to characterize cell-surface glycoconjugates are histochemistry for lectins and immunohistochemistry for antibodies, flow cytometry and cell sorting, and cell agglutination. In histochemistry and immunohistochemistry, tissues are prepared and fixed as usual for histological staining, and then incubated with appropriate biotinylated or peroxidase-labeled lectins or antibodies. The bound lectins or antibodies are visualized after the binding of secondary reagents, such as streptavidin-peroxidase or labeled secondary antibody. These approaches often yield information that is difficult to obtain by any other approach. For example, they can reveal the spatial orientation of different glycans, their relative abundance, and whether they are intracellular and/or extracellular. Three important controls in such studies are (1) use of lectins or antibodies at limiting concentrations to avoid nonspecific binding; (2) confirmation of the specificity of binding by appropriate inhibition by haptens or by destruction of the predicted target glycans with glycosidases; and (3) use of multiple lectins or antibodies to provide further confirmation of the conclusions.

Lectins and antibodies to glycans have also been widely used in flow cytometry and cell sorting. In such studies, cells are incubated with low, nonagglutinating levels of lectins or antibodies that are biotinylated and conjugated to fluorescently labeled streptavidin or directly fluorescently labeled. Cells with bound lectins can then be identified by their fluorescence in the flow cytometer, and the degree of fluorescence can be correlated with the number of binding sites. Again, appropriate controls include the three mentioned above. A key consideration in flow cytometry is to avoid concentrations of lectins that cause agglutination of cells, which must be ruled out by direct microscopic visualization. Lectins and antibodies can also be used to identify specific membrane localization of different glycans, using confocal microscopy and electron microscopy.

Lectins and antibodies are also especially useful for characterizing cell-surface glycans when limited numbers of cells are available. For example, studies on glycosylation of embryonic stem cells have been greatly aided by using panels of specific lectins to identify unique glycan determinants and changes in their expression during cellular differentiation. A recent variation of this approach is to use a microarray of immobilized lectins, which are probed with fluorescently labeled extracts of cells. Such assays can reveal minor differences in protein glycosylation between different samples and give insight into the glycan structures that are present.

Finally, one of the oldest uses of lectins is in the agglutination and precipitation of glycoconjugates, cells, and membrane vesicle preparations. The easiest assay for many soluble lectins that are multivalent is agglutination of target cells, such as erythrocytes, leukocytes, or even bacteria or fungi. Agglutination can often be easily observed without a microscope, but it is also measurable in instruments such as aggregometers. In these assays, a lectin solution is serially diluted and the reciprocal of the final dilution that gives measurable cell agglutination is taken to define the activity. Bacterial agglutination by plant and animal lectins is often used to explore the surface glycocalyx and changes in glycocalyx upon culture conditions and to define phenotypes of different serotypes. Lectin precipitation and aggregation can be used to define the glycan composition and overall architecture of polysaccharides, as has been done for bacterial, algal, plant, and animal polysaccharides. Some lectins also have antiviral activity, as a result of aggregation of the viruses and blocking of viral adhesion and infection. These antiviral activities are being explored as new treatments for human diseases.

USES OF ANTIBODIES AND LECTINS FOR GENERATING ANIMAL CELL GLYCOSYLATION MUTANTS

An important use of lectins and antibodies has been in the selection of cell lines that express altered cell-surface glycans. A good example of this approach involves studies using the Chinese hamster ovary cell line (CHO) as a model system. The common lectins that have been used are ConA, WGA (wheat germ agglutinin), L-PHA, LCA, PSA (*Pisum sativum* agglutinin), E-PHA, ricin, modeccin, and abrin. The latter three lectins are heterodimeric, disulfide-bonded proteins that are classified as type II ribosome-inactivating proteins; they contain an A subunit that encodes an enzyme called RNA-N-glycosidase, which inactivates the 28S ribosome, and a B subunit that is a galactose-binding lectin. Many plant lectins such as ConA, WGA, or LCA (which lack an enzymatic or toxic A subunit) are toxic to animal cells via poorly understood mechanisms, whereas other lectins, such as soybean agglutinin (SBA) and peanut agglutinin (PNA), are not toxic to animal cells in culture. One mechanism of toxicity of plant lectins is the induction of apoptosis, perhaps by blocking receptor or transport functions in cells. Noncytotoxic lectins and antibodies to specific glycans can be rendered toxic by conjugating them with ricin A subunit or other toxic proteins. Using these agents, both loss-of-function (e.g., loss of a glycosyltransferase or glycosidase) and gain-of-function (e.g., activation of a latent transferase gene) mutants have been obtained. More details about glycosylation mutants of cultured cells are presented in Chapter 46.

USES OF ANTIBODIES AND LECTINS FOR CLONING GLYCOSYLTRANSFERASE GENES BY EXPRESSION

The specificity of lectins and antibodies to particular glycans has made them especially useful in cloning genes encoding glycosyltransferases or other proteins required for proper glycosylation, such as nucleotide sugar transporters. To illustrate the approaches that are usually taken, several examples will be discussed. CHO cells and the African green monkey kidney cell line COS lack glycans with terminal α-galactose residues. Consequently, the cells do not bind the plant lectin GSI-B4 (Figure 45.3). When the cells are transfected with a cDNA library prepared from cells that do express terminal α-galactose residues (such as murine teratocarcinoma cells F9), a few of the transfected cells will have taken up a plasmid encoding the cognate α1-3 galactosyltransferase and consequently express terminal α-galactose residues. Those cells will bind to GSI-B4 and can be identified by binding to plates coated with the lectin. Isolation of the transfected plasmids from the bound cells and repeated recloning and reexpression by this technique (called expression cloning) led to the identification of a specific gene encoding the murine α1-3 galactosyltransferase.

Related approaches using antibodies and lectins to other glycan structures and wild-type CHO cells and CHO mutants led to the identification of the genes encoding many other glycosyltransferases, including some involved in extending glycosphingolipids and nucleotide sugar transporters, such as the transporter for CMP-NeuAc. Such approaches based on lectin selection are also useful with yeast. For example, the gene encoding a yeast *N*-acetylglucosaminyltransferase (GlcNAcT) was identified by expression cloning in yeast using a GlcNAcT-deficient yeast. The mannan chains of the yeast *Kluyveromyces lactis* normally contain some terminal *N*-acetylglucosamine residues and are bound by the plant lectin GSL-II, which binds terminal *N*-acetylglucosamine residues (Figure 45.4). A mutant lacking the GlcNAcT and lacking terminal *N*-acetylglucosamine residues on mannoproteins was identified. Transformation of yeast with DNA containing the gene encoding the GlcNAcT led to production of some yeast that were bound by the fluorescently labeled GSL-II. This strategy was used to identify the gene encoding the GlcNAcT. The transporters for UDP-Gal in *Leishmania* parasites and in the plant *Arabidopsis* were also identified by expression cloning in Lec8 CHO cells; these cells have a mutation in their endogenous UDP-Gal transporter, and consequently lack galactose-containing glycans on their surface. For the UDP-Gal transporter from *Arabidopsis*, Lec8 cells were co-transfected with a cDNA library encoding the putative *Arabidopsis* UDP-Gal transporter, along with a glucuronosyltransferase that can lead to synthesis of the unsulfated version of the HNK epitope (Figure 45.6), GlcAβ1-3Galβ-R that is bound by a specific antibody. This identification strategy led to the expression cloning of several UDP-Gal transporters from *Arabidopsis*.

CHO and COS cell lines have also been useful in characterizing the activities of novel glycosyltransferase genes that were identified by other approaches. For example, candidate fucosyltransferase genes encoding α1-2 or α1-3 fucosyltransferases have been expressed in CHO and COS cells, which lack these enzymes. Expression of these enzymes then leads to expression on the cell surface of antigens recognized by specific antibodies and lectins, such as the H-antigen (for the α1-2 FucT) or Le^x and SLe^x (for the α1-3 FucT).

USES OF ANTIBODIES AND LECTINS IN ASSAYING GLYCOSYLTRANSFERASES AND GLYCOSIDASES

Lectins and antibodies to glycan antigens have been very useful in assaying specific glycosyltransferases and glycosidases in a variety of formats (Figure 45.7). Immobilized lectins have been used to isolate products of glycosyltransferases assays, such as chitin polysaccharides on WGA or glycosylated peptides on mixed-bed lectin columns. Assays for several specific glycosyltransferases have relied on lectin and antibody binding to the products. For example, α1-3 fucosyltransferases that synthesize the Le^x and SLe^x antigens have been assayed based on capture of the product on immobilized antibodies to these antigens or binding of antibody to the immobilized fucosylated product in an ELISA-type format. Likewise, α2-3 sialyltransferases and α2-6 sialyltransferases have been assayed using immobilized acceptors in ELISA-type format and in BIAcore formats (see Chapter 27), and their products have been measured with MAL (which binds to the α2-3-sialylated product) or SNA (which binds to the α2-6-sialylated product). Similarly, α1-3 galactosyltransferases have been assayed using immobilized acceptors in an ELISA-type format, and the product has been measured with GSI-B4 or *Viscum album* agglutinin, which binds to the α1-3 galactosylated product. The glycoprotein-specific β1-4 *N*-acetylgalactosaminyltransferase has been assayed using glycoprotein acceptors in solution, capture by a specific monoclonal antibody in a microtiter plate, and measurement of product by ELISA-type assay with a *Wisteria floribunda* lectin (WFA) that binds to terminal β1-4-linked GalNAc residues generated by the enzyme. These are just some of the examples of ELISA-type, relatively high-throughput, and nonradioactive assays that have been developed for various glycosyltransferases using plant lectins and antibodies to detect the specific products. Conversely, glycosidases can be assayed by measuring the gain in binding of lectins. For example, the lectin PNA, which binds to nonsialylated Galβ1-3GalNAcα1-Ser/Thr in O-glycans, can be used to measure bacterial sialidases by agglutination of treated erythrocytes. It is easy to envision how lectins and antibodies can be used to probe the products of other specific glycosidases and glycosyltransferases, given the specificities of the lectins described here.

FURTHER READING

Hakomori S. 1984. Tumor-associated carbohydrate antigens. *Annu. Rev. Immunol.* **2:** 103–126.

Merkle R.K. and Cummings R.D. 1987. Lectin affinity chromatography of glycopeptides. *Methods Enzymol.* **138:** 232–259.

Osawa T. and Tsuji T. 1987. Fractionation and structural assessment of oligosaccharides and glycopeptides by use of immobilized lectins. *Annu. Rev. Biochem.* **56:** 21–42.

Osawa T. 1988. The separation of immunocyte subpopulations by use of various lectins. *Adv. Exp. Med. Biol.* **228:** 83–104.

Esko J.D. 1992. Animal cell mutants defective in heparan sulfate polymerization. *Adv. Exp. Med. Biol.* **313:** 97–106.

Kobata A. and Endo T. 1992. Immobilized lectin columns: Useful tools for the fractionation and structural analysis of oligosaccharides. *J. Chromatogr.* **597:** 111–122.

Cummings R.D. 1994. Use of lectins in analysis of glycoconjugates. *Methods Enzymol.* **230:** 66–86.

Lis H. and Sharon N. 1998. Lectins: Carbohydrate-specific proteins that mediate cellular recognition. *Chem. Rev.* **98:** 637–674.

Bush C.A., Martin-Pastor M., and Imberty A. 1999. Structure and conformation of complex carbohydrates of glycoproteins, glycolipids, and bacterial polysaccharides. *Annu. Rev. Biophys. Biomol. Struct.* **28:** 269–293.

Morgan W.T. and Watkins W.M. 2000. Unravelling the biochemical basis of blood group ABO and Lewis antigenic specificity. *Glycoconj. J.* **17:** 501–530.

Goldstein I.J. 2002. Lectin structure-activity: The story is never over. *J. Agric. Food Chem.* **50:** 6583–6585.

Paschinger K., Fabini G., Schuster D., Rendic D., and Wilson I.B. 2005. Definition of immunogenic carbohydrate epitopes. *Acta Biochim. Pol.* **52:** 629–632.

Akama T.O. and Fukuda M.N. 2006. N-Glycan structure analysis using lectins and an α-mannosidase activity assay. *Methods Enzymol.* **416:** 304–314.

Lehmann F., Tiralongo E., and Tiralongo J. 2006. Sialic acid-specific lectins: Occurrence, specificity and function. *Cell. Mol. Life Sci.* **63:** 1331–1354.

Maeda Y., Ashida H., and Kinoshita T. 2006. CHO glycosylation mutants: GPI anchor. *Methods Enzymol.* **416:** 182–205.

Patnaik S.K. and Stanley P. 2006. Lectin-resistant CHO glycosylation mutants. *Methods Enzymol.* **416:** 159–182.

Varki N.M. and Varki A. 2007. Diversity in cell surface sialic acid presentations: Implications for biology and disease. *Lab. Invest.* **87:** 851–857.

CHAPTER 46

Glycosylation Mutants of Cultured Cells

Jeffrey D. Esko and Pamela Stanley

Rapid progress in our understanding of glycosylation pathways in eukaryotes came with the application of genetic strategies to isolate mutants of mammalian cells and yeast with defects in glycan synthesis. This chapter reviews general methods used to isolate glycosylation mutants, the diversity of mutants that may be obtained, and the application of these mutants to address questions in glycobiology research. Many of the cell lines described in this chapter are available through the American Type Culture Collection. Glycosylation mutants of yeast are discussed in Chapter 21.

HISTORY

The success of bacterial genetics as an approach to defining genetic and biochemical pathways, together with the desire to understand the molecular basis of human genetic disorders and general metabolism, led to the development of somatic cell genetics in mammalian cells. The field began in the late 1950s with pioneering studies of cultured fibroblasts and the derivation of cell lines from Chinese hamster ovary tissue (CHO cells). Stable cell lines from animal tissues were obtained readily and propagated in vitro under partially defined growth conditions. Furthermore, mutants were isolated at a respectable frequency and their phenotypes remained stable over many generations. Studying various processes in cultured cell mutants circumvented the long generation times inherent in genetic studies of whole organisms and allowed systematic control of environmental factors, such as nutrients. Techniques already established for microbial organisms could now be applied to somatic cells.

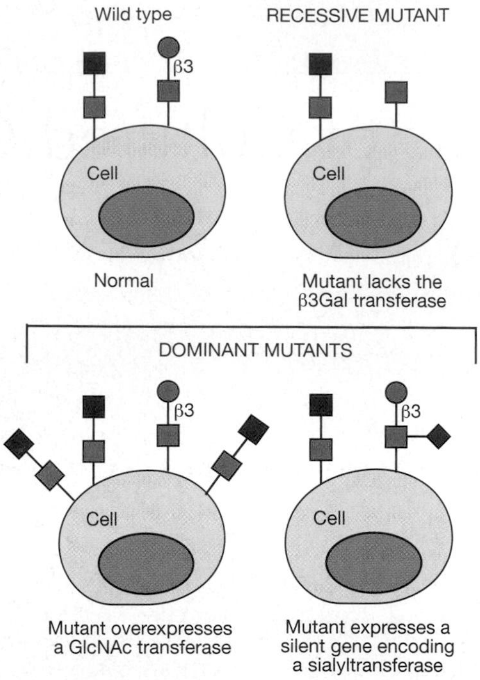

FIGURE 46.1. Alteration of cell-surface glycans by recessive and dominant glycosylation mutations.

Somatic cell genetic techniques were applied early on to glycobiology, yielding numerous mutants in glycoprotein biosynthesis and later in proteoglycan, glycosylphosphatidylinositol (GPI) anchor, and glycolipid biosynthesis. Similar strategies were used to obtain yeast glycosylation mutants (see Chapter 21). N-glycan synthesis in yeast and mammals is very similar in the early part of the pathway, including the formation of the mature 14-sugar dolichol-glycan, transfer of the glycan to protein, and the trimming of three glucose residues and one mannose residue in the endoplasmic reticulum. The ability to isolate glycosylation mutants in culture made it possible to unravel pathways of glycan synthesis and degradation and to identify, isolate, and map structural and regulatory genes. Mutants often accumulate intermediates upstream of the block in a pathway and thereby reveal the chemical structure of substrates and the nature of reactions that constitute a metabolic pathway. Sequencing of mutant alleles reveals changes in specific amino acids that affect a glycosylating activity. In most cases, glycosylation mutations are loss-of-function mutations and they depress the activity of an enzyme in a pathway; but there are also gain-of-function mutations that activate a silent glycosylation gene, elevate the expression of an existing activity, or inactivate a negative regulatory factor (Figure 46.1). In nearly all situations, the mutations lead to the presence of altered glycans on cell-surface glycoconjugates and changes in biological responses that correlate glycan structure to function.

INDUCTION AND ISOLATION OF MUTANTS

In cell culture, mutations occur randomly at a low rate ($<10^{-6}$ mutations per generation), and thus the likelihood of finding mutants is low. To increase this probability, mutations may be induced by treating cells with chemical (e.g., alkylating agents), physical (e.g., ionizing radiation), or biological (e.g., a virus) mutagens, thereby increasing the number of mutants in a population by several orders of magnitude. However, mammalian cells are diploid and cultured cells are often hyperdiploid or tetraploid. Because the ploidy state determines the number of gene copies per cell, it will affect the frequency of obtaining a mutant phenotype, particularly if the mutation is recessive, as expected for the majority of loss-of-function mutations. Surprisingly, however, the frequency of finding recessive mutants is often much higher than predicted. In CHO cells, many loci are functionally hemizygous (single copy), which means that a single hit generates a recessive mutant phenotype. The frequency of dominant mutations, which in most cases induce a gain-of-function phenotype, is usually independent of ploidy state.

Even with mutagenesis, the incidence of mutants with defects specifically in glycosylation genes is low, consistent with the observation that these genes represent only about 1% of the genome (see Chapter 7). Thus, selection or enrichment is needed in order to find rare mutants bearing a desired glycosylation phenotype (Table 46.1). Direct selection schemes based on resistance to cytotoxic plant lectins that bind to cell-surface glycans are especially useful for identifying mutants altered in N-glycans. Dozens of lectins are available, each with a different specificity for sugars in various arrangements (see Chapters 28 and 29). Mutants with a glycosylation defect become resistant to cytotoxic lectins by reducing the expression of a single sugar or a group of sugars on cell-surface glycans (Figures 46.1 and 46.2). Importantly, many mutants that are resistant to one or more lectins because of the loss of specific sugars become supersensitive to killing by a different group of lectins that recognize sugar residues exposed by the mutation. The latter group of lectins may be used to select for revertants in the original mutant population. Nontoxic lectins are also useful for selecting lectin-binding mutants. For example, lectins may interfere with cell adhesion, and mutant cells lacking a particular glycan may be selected because they continue to adhere in the presence of these lectins. Mutations that affect all stages of glycosylation reactions, including the generation and transport of nucleotide sugars, have been detected using lectins as selective agents (Table 46.2).

Virtually any agent that recognizes cell-surface glycans or a specific surface glycoprotein, GPI-anchored protein, or glycolipid can be used to isolate mutants with a glycosylation defect (Figure 46.2). Thus, glycosylation defects in GPI-anchor biosynthesis reduce expression of GPI-anchored proteins at the cell surface (see Chapter 11). Conjugation of glycan-binding proteins to a cytotoxin that has no receptor provides another agent to select mutants in systems where lectins are not available. For example, basic fibroblast growth factor (FGF-2)/saporin complexes have been used for the selection of mutants deficient in heparan sulfate (HS). Lectins, antibodies, or ligands that are fluorescently tagged also may be used to enrich for mutants that are either deficient in binding or have acquired a novel binding ability due to altered glycosylation or reduced expression of an antigen at the cell surface. Panning is a related technique that is based on lack of adhesion of cells to surfaces

TABLE 46.1. *Classes of glycosylation mutants obtained from different selections or screens*

Selection/enrichment scheme	Types of mutants isolated
Lectin resistance	N-linked glycosylation dolichol-P-P-oligosaccharide assembly nucleotide sugar formation or transport gain-of-function mutants
Radiation suicide	N-linked glycosylation nucleotide sugar conversion
Cell sorting Antibody/complement-mediated lysis	GPI-anchor biosynthesis
Replica plating	GAG biosynthesis N-linked glycosylation nucleotide sugar formation

(GP1) Glycosylphosphatidylinositol; (GAG) glycosaminoglycan.

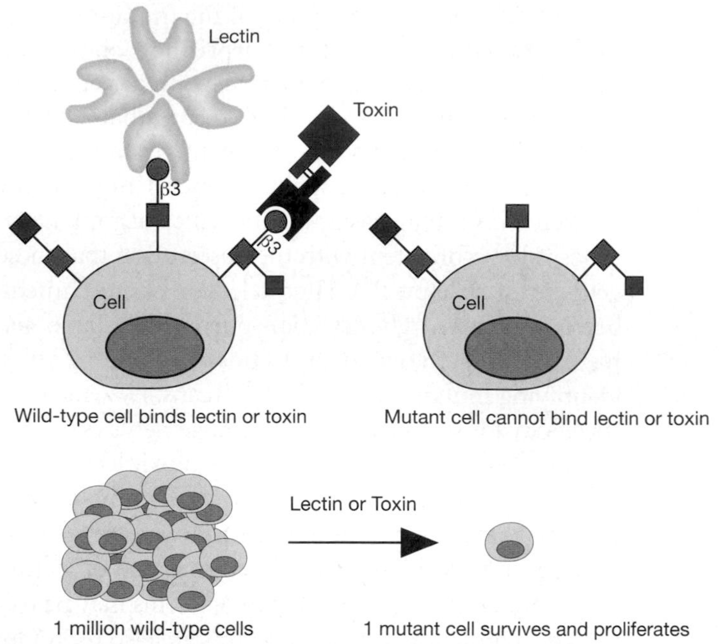

FIGURE 46.2. Selection of mutants with lectins or cytotoxic agents that bind to specific sugar residues.

coated with a glycan-binding agent. For example, coating a plate with FGF-2 allows selection of mutant cells that fail to produce HS proteoglycans, because those mutants fail to adhere to a FGF-2-coated plate.

Radiation suicide is another direct selection method for obtaining glycosylation mutants. Incubation of cells with a radioactive sugar, sulfate, or other precursor of high radiospecific activity leads to labeled glycoproteins, glycolipids, or proteoglycans that, upon prolonged storage, will cause radiation damage and death to wild-type cells, whereas mutants with reduced incorporation of the label survive. Animal cells can also be replica-plated, much like

TABLE 46.2. *Examples of mutants with defects in nucleotide sugar formation or transport*

Mutant	Biochemical defect	Glycosylation phenotype
Lec32 (CHO)	CMP-NeuAc synthetase	reduced CMP-Neu5Ac; glycans lack terminal sialic acid; terminate in galactose
Lec2 (CHO)	CMP-NeuAc transporter	reduced CMP-Neu5Ac in Golgi; glycans lack terminal sialic acid; terminate in galactose
Lec8 (CHO)	UDP-Gal transporter	reduced UDP-Gal in Golgi; N-glycans terminate in *N*-acetylglucosamine; O-glycans terminate in *N*-acetylgalactosamine
Lec13 (CHO)	GDP-Man-4,6-dehydratase	reduced GDP-Fuc; glycans lack fucose
ldlD (CHO)	UDP-Gal/ UDP-Glc-4-epimerase	reduced UDP-Gal, UDP-Glc, UDP-GalNAc and UDP-GlcNAc; N-glycans lack galactose; O-glycans and chondroitin sulfate not synthesized in the absence of UDP-GalNAc
D33W25-1 (MDAY-D2) SAP (CHO)	activation of CMP-Neu5Ac hydroxylase	CMP-Neu5Gc synthesized glycans terminate in Neu5Gc instead of Neu5Ac

(Neu5Ac) *N*-Acetylneuraminic acid; (Neu5Gc) *N*-glycolylneuraminic acid.

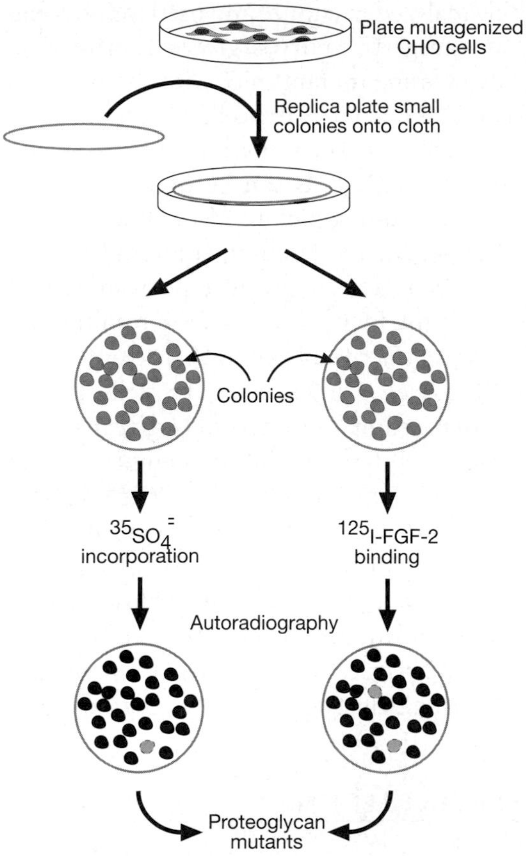

FIGURE 46.3. Screening for mutants using animal cell replica plating. Animal cell colonies transferred to discs can be screened for defects in incorporation of radioactive precursors, binding to lectins and antibodies, or direct enzymatic assay. Mutants are depicted as colonies lacking a strong signal in the lower discs.

microbial colonies, using porous cloth made of polyester or nylon as the replica (Figure 46.3). Colonies of cells on the disc can be used to measure incorporation of macromolecular precursors (e.g., radioactive sugars or sulfate) or to identify mutants that fail to bind to a lectin, an antibody, or a growth factor. An adaptation of this technique allows detection of mutants affecting a specific enzyme by direct assay for activity in colony lysates generated on the discs. Although this technique has great specificity, its limited capacity makes detection of rare mutants difficult, and mutagenesis prior to screening is usually a requirement.

Regardless of the technique used to isolate mutants, the resulting strains must be cloned and carefully characterized for stability and the molecular basis of mutation. Additional genetic analyses include somatic cell hybridization for dominance/recessive testing and assigning mutants to different genetic complementation groups. Biochemical analysis involves the characterization of glycan structures produced by mutant cells (see Chapter 47), the quantitation and analysis of intermediates, and assays for activities thought to be missing or acquired based on the properties of the mutant. Identifying the molecular basis of mutation requires isolation of a complementing cDNA that reverts the mutant phenotype and determining whether the mutation arose from defective transcription, translation, or stability of the gene product or from a missense or nonsense mutation in the coding region of the gene.

DERIVATION OF CELL LINES FROM MICE OR HUMANS WITH A GLYCOSYLATION MUTATION

Transgenic mice that overexpress a glycosylation gene or mutant mice that lack a glycosylation activity due, in most cases, to targeted gene inactivation (see Chapter 38) are a source of mutant cells that may be analyzed in culture and used for glycobiology research. Cells may be derived from mutant or transgenic embryos, grown as primary cultures, or immortalized by viral transformation. By crossing mutant mice with the Immortomouse, which carries a temperature-sensitive SV40 T antigen in every cell, immortalized mutant cell strains can be derived from essentially any cell type. For mutations that cause embryos to die during gestation, mutant embryonic stem (ES) cells can be derived from blastocysts, provided the mutation is not cell-lethal. The resulting mutant ES cell lines can be used to investigate functions for specific glycans during differentiation in embryoid cell culture or in vivo in mouse chimeras. A chimera is obtained by injecting wild-type or mutant ES cells into the inner cell mass of a mouse blastocyst. If the ES cells survive, the resulting mouse is a mosaic of cells derived from the ES cells and cells derived from the blastocyst. Mutant ES cells may not contribute equally well to all tissues. For example, ES cells lacking GlcNAc transferase I (GlcNAcT-I) are unable to make complex or hybrid N-glycans (see Chapter 8), but they differentiate normally into many cell types in cultured embryoid bodies. However, following introduction into blastocysts, mutant ES cells lacking GlcNAcT-I do not contribute to the organized layer of bronchial epithelium in chimeric embryos.

Similarly, cell lines can be obtained from other organisms, although with greater difficulty compared to mice. For example, human fibroblasts and lymphocytes are easy to obtain but difficult to convert to immortal cell lines. Studies of fibroblasts from patients with defects in glycosylation have led to the elucidation of the underlying defect (see Chapter 42). Many lines bearing defects in lysosomal degradation are available as well.

MUTANTS SELECTED FOR RESISTANCE TO PLANT LECTINS

Selection schemes based on isolating rare mutants resistant to cytotoxic plant lectins have yielded a large number of glycosylation mutants affected in diverse aspects of glycan synthesis. Table 46.2 lists examples of mutant strains altered in nucleotide sugar formation or transport into the Golgi. As might be expected, some defects are pleiotropic. For example, the UDP-Gal transporter defect in the Lec8 mutant affects transfer of galactose to O-linked and N-linked structures on glycoproteins as well as to glycosaminoglycans (GAGs) and glycolipids. The ldlD mutant is particularly interesting in this regard, because it lacks the epimerase responsible for converting UDP-Glc to UDP-Gal and UDP-GlcNAc to UDP-GalNAc (Figure 46.4). Because there are salvage pathways for

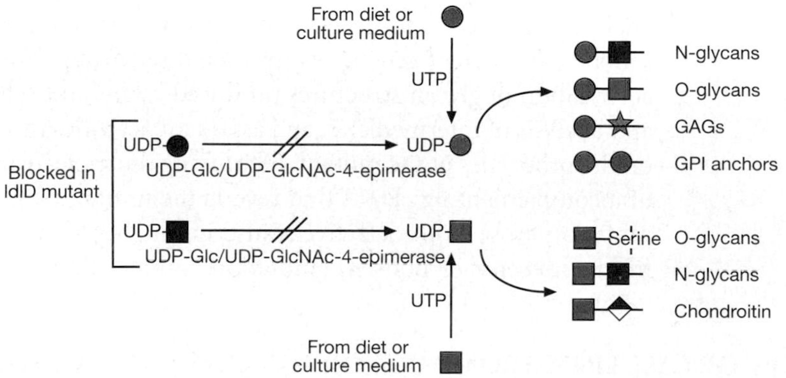

FIGURE 46.4. Pleiotropic effects of mutations in UDP-Glc/UDP-GlcNAc-4-epimerase. ldlD mutant CHO cells may be rescued by salvage reactions that generate the UDP-Gal and UDP-GalNAc necessary for the synthesis of many classes of glycans.

importing galactose and *N*-acetylgalactosamine into cells (see Chapter 4), the composition of different classes of glycans can be controlled in ldlD cells by nutritional supplementation with either galactose or *N*-acetylgalactosamine. Alternatively, supplementing cells with increased serum glycoproteins such as fetuin will also bypass the defect because galactose and *N*-acetylgalactosamine are salvaged following lysosomal degradation. Most of the mutants in Table 46.2 lack a glycosylation activity or fail to make a precursor. Two CHO cell mutants, ldlB and ldlC (not shown in the table) carry mutations in the conserved oligomeric Golgi (COG) complex used for trafficking glycan biosynthetic enzymes between the endoplasmic reticulum and Golgi. SAP mutants of CHO and D33W25-1 cells have dominant mutations that activate a latent enzyme, in this case, a hydroxylase that converts CMP-Neu5Ac to CMP-Neu5Gc. Such gain-of-function mutants provide access to gene products that may normally be expressed only in a few, very specialized cells in the body. Therefore, dominant mutants are important in glycosylation gene discovery, in identifying mechanisms of glycosylation gene regulation, and for defining pathways of glycan biosynthesis.

Some lectin-resistant mutants are defective in the formation of dolichol-P-oligosaccharides or in the processing reactions that remove glucose and mannose residues after transfer of the glycan chain from the dolichol intermediate to glycoproteins (see Chapter 8). The latter mutants revealed the identity and importance of α-mannosidases in the formation of N-linked glycans in cultured cells. However, when the α-mannosidase II gene was ablated in mice, no effect was seen in certain tissues because another previously unknown α-mannosidase allowed N-glycans to be synthesized. This finding emphasizes a limitation of somatic cell mutants—the investigation is restricted to the cell line in which the gene is mutated. Because many glycosyltransferases appear to be developmentally regulated in a tissue-specific manner, studying mutants of a single cell type might preclude the discovery of alternate pathways.

Other examples of defects in N-linked glycan synthesis are given in Table 46.3. Note that some mutations affect the kinetic properties of an enzyme (e.g., Lec1A) or its subcellular localization (e.g., Lec4A). Sequencing mutant alleles provides leads for further site-directed mutagenesis of the gene in order to define important functional domains of the protein required for catalysis or compartmentalization.

Another class of lectin-resistant mutants consists of strains with a gain-of-function dominant phenotype due to the increased expression of a glycosyltransferase that is normally silent or expressed at very low levels. The activation of these glycosyltransferase genes may reflect a mutation in a regulatory region of the gene or in a *trans*-acting factor. The generation of gain-of-function mutants provides the opportunity to detect and analyze the

TABLE 46.3. *Examples of mutants altered in late N-glycan synthesis*

Mutant	Biochemical defect	Glycosylation phenotype
Lec1 (CHO) Clone 15B (CHO)	GlcNAcT-I	$Man_5GlcNAc_2Asn$ at N-glycan sites normally carrying complex or hybrid N-glycans
Lec1A (CHO)	GlcNAcT-I (Km defect)	$Man_5GlcNAc_2Asn$ at some N-glycan sites but complex and hybrid N-glycans present
RicR21 (BHK)	GlcNAcT-II	complex N-glycans only mono- or biantennary
Lec4A (CHO)	GlcNAcT-V mislocalized	complex N-glycans lack the β6GlcNAc branch on the α6Man of the N-glycan core
Sil (KB)	α2-3 sialyltransferase	reduced Neu5Ac on a subset of glycans

(GlcNAcT) *N*-acetylglucosaminyltransferase.

TABLE 46.4. *Dominant gain-of-function mutants expressing a new activity*

Mutant	Biochemical change	Glycosylation phenotype
LEC10 (CHO)	GlcNAcT-III expressed	complex N-glycans have the bisecting *N*-acetylglucosamine residue
LEC11 (CHO)	α3FucT-VI expressed	fucose on polylactosamine generates Le^X SLe^X and VIM-2 determinants
LEC12 (CHO)	α3FucT-IX expressed	fucose on polylactosamine generates Le^X and VIM-2 determinants
LEC14 (CHO)	GlcNAcT-VII expressed	N-glycan core has an extra *N*-acetylglucosamine on β1-4-linked Man
LEC18 (CHO)	GlcNAcT-VIII expressed	N-glycan core has an extra *N*-acetylglucosamine on β1-4GlcNAc

Note on nomenclature: Uppercase is used for gain-of-function mutants (e.g., LEC10); lowercase is used for loss-of-function mutants (e.g., Lec1).

effects of novel genes, which in many cases, were not previously known to exist. For example, *LEC14* and *LEC18* mutations activate previously unknown transferases that add branching *N*-acetylglucosamine residues to the core of N-glycans (Table 46.4).

MUTANTS IN GPI-ANCHOR BIOSYNTHESIS

Lectins that selectively bind to GPI anchors have not been described, but there are bacterial toxins that bind these glycans and they may be used to select GPI-anchor mutants. Originally, however, many GPI mutants were isolated by strategies that took advantage of antibodies to a GPI-anchored protein (e.g., Thy-1 on T-cell lymphoma cells). Cells expressing Thy-1 on their surface were incubated with an antibody to Thy-1 and serum-containing complement components, which lysed cells expressing the Thy-1 antigen. Loss of GPI-anchor biosynthesis reduced the expression of Thy-1 on the surface and conferred resistance to the cytolytic effect. Other mutants have been obtained by sorting cells that do not bind to a fluorescent antibody. The mutants obtained to date fall into more than 20 genetic complementation groups, each having a different lesion in GPI-anchor biosynthesis (Table 46.5; see Chapter 11). These mutants reveal the complexity of GPI-anchor biosynthesis: Multiple gene products are involved in forming the *N*-acetylglucosamine linkage to PI, the first committed intermediate in the pathway; Dol-P-Man is utilized as the donor of mannose; at least three enzymes are involved in the attachment of ethanolamine phosphate residues; and five genes are required for the transfer of the GPI anchor to protein. The available strains demonstrate the importance of genetic approaches for identifying genes that might not be obvious from measuring biosynthetic reactions in vitro.

TABLE 46.5. *Mutants defective in GPI-anchor biosynthesis*

Mutant	Biochemical defect	Glycosylation phenotype
Pig-A, Pig-C, Pig-H	GlcNAc to PI transferase	formation of GlcNAc-PI inhibited
Pig-J	GlcNAc N-deacetylase	accumulates GlcNAc-PI
Pig-E	Dol-P-Man synthase	accumulates GlcN-PI
Pig-B	α1-2 mannosyltransferase	accumulates Man_2GlcN-PI
Pig-F, Pig-K	ethanolamine transferases	accumulates Man_3(EtNP)1-2GlcN-PI

TABLE 46.6. *Mutants defective in proteoglycan assembly*

Strain	Biochemical defect	Phenotype
pgsA (CHO)	XT-1: xylosyltransferase	lack of HS and CS
pgsB (CHO)	GalT-I: galactosyltransferase I	lack of HS and CS
pgsG (CHO)	GlcAT-I: glucuronyltransferase I	lack of HS and CS
pgsD (CHO)	EXT-1: GlcA and GlcNAc transferase	HS deficient and accumulates CS
ldlD (CHO)	UDP-Glc/UDP-GlcNAc-4-epimerase	lack of CS when starved for *N*-acetylgalactosamine and fed galactose lack of all GAG chains when starved for galactose
Lec8	UDP-Gal transporter	reduced KS
pgsC (CHO)	sulfate transporter	normal GAG biosynthesis due to salvage of sulfate from oxidation of sulfur containing amino acids
pgsE (CHO)	NDST1: GlcNAc N-deacetylase/N-sulfotransferase	undersulfated HS
pgsF (CHO)	heparan sulfate 2-O-sulfotransferase	defective 2-O-sulfation of uronic acids in HS; defective FGF-2 binding
pgsH (CHO)	6OST-1: HS 6-O-sulfotransferase	reduced 6-O-sulfation of glucosamine residues
Mouse LTA cells	3OST-1: GlcNS 3-O-sulfotransferase	defective 3-O-sulfation of GlcNS units; defective antithrombin binding

(HS) Heparan sulfate; (CS) chondroitin sulfate; (KS) keratan sulfate.

MUTANTS IN PROTEOGLYCAN ASSEMBLY

A large collection of mutants defective in glycosaminoglycan (GAG)/proteoglycan biosynthesis has been isolated (Table 46.6). Many of these mutants were obtained by replica plating methods using sulfate incorporation to monitor GAG production in colonies (see Figure 46.3). Mutants in the early steps of GAG biosynthesis (complementation groups A, B, and G) lack both chondroitin sulfate (CS) and heparan sulfate (HS) chains, and enzymatic assays showed that they lack enzymes responsible for the assembly of the core protein linkage tetrasaccharide shared by both these GAG species (see Chapter 16). Another class of mutants (group D) is defective only in HS biosynthesis. This mutation defines a bifunctional enzyme (EXT1) that catalyzes the alternating addition of *N*-acetylglucosamine and glucuronic acid residues to growing HS chains. Some of the mutant alleles depress both enzyme activities, whereas others only affect the glucuronic acid transfer activity. Thus, the mutants define different functional domains of the protein, which have been mapped by sequencing various mutant alleles. Mutants in the GlcNAc N-deacetylase/N-sulfotransferase (Ndst1) (another bifunctional enzyme) have only a partial deficiency in N-sulfation of HS chains. Further analysis of the mutant showed that more than one isozyme is present in CHO cells and that the defect affects only one locus. Thus, the mutants revealed early on that the assembly of HS is much more complex than had been appreciated on the basis of known structures, enzymatic reactions measured in cell extracts, or intermediates observed in pulse-labeling experiments.

MUTANTS DEFECTIVE IN GLYCOLIPID OR O-GLYCAN SYNTHESIS

Glycolipid and O-glycan structures are often relatively simple in cultured cells. For example, CHO cells synthesize mainly ganglioside GM3 and lactosylceramide with a small amount of glucosylceramide. O-glycans initiated with *N*-acetylgalactosamine contain up to only four sugars in glycoproteins from CHO cells. All of these structures are affected in the mutants described in Table 46.2 in which CMP-sialic acid, UDP-Gal, UDP-GalNAc, or GDP-fucose syntheses are reduced or altered. Similarly, a defective sialyltransferase or galactosyltransferase may cause these structures to be truncated, depending on its acceptor specificity. A mutant of B16 melanoma cells that is defective in cerebroside glucosyltransferase lacks all glycolipids because this enzyme catalyzes the first committed step in the synthesis pathway (see Chapter 10). However, cultured cell mutants defective in protein O-GalNAc transferases or protein O-fucosyltransferase have not been isolated. This is most likely due to lack of application of the selection technologies described above, in part because of the paucity of cytotoxic lectins or toxins that bind to O-glycans and glycolipids or because of redundancy of enzymes in the system (see Chapters 9 and 10). Mice lacking specific glycolipid synthetic enzymes and glycosyltransferases that transfer *N*-acetylgalactosamine or fucose to protein have been generated and provide a source of mutant cells that may be studied in culture. Interestingly, cells lacking the O-GlcNAc transferase that acts in the cytoplasm to transfer *N*-acetylglucosamine to protein have not been obtained, and mouse mutants defective in this transferase become arrested in development at the two-cell-stage embryo, demonstrating that O-GlcNAc addition is essential for cell viability.

USES OF SOMATIC CELL GLYCOSYLATION MUTANTS

Glycosylation mutants of cultured cells have been used to address many questions in glycobiology and for glycosylation engineering of recombinant glycoproteins. Because mutant selections are broad and often not intentionally biased, they generate mutants defective in both known and novel reactions. Thus, glycosylation mutants play an important role in research to define the pathways and regulation of glycosylation in mammals. In this regard, they are more useful tools than mutant mice because cells in culture are viable in the absence of glycolipids, GPI anchors, proteoglycans, O-GalNAc and O-fucose glycans, and complex or hybrid N-glycans.

Glycosylation mutants make glycans with truncated or altered structures and thus provide an opportunity to study functional roles for cell-surface glycans in the context of a living cell. Important insights have been gained into specific sugars required for viral, bacterial, or parasite adhesion and infection and for leukocyte cell adhesion and motility. In addition, functional roles for glycans in the intracellular sorting and secretion of glycoproteins, in growth factor binding and activation, and in receptor functions have been identified using glycosylation mutants. For example, a panel of CHO glycosylation mutants was used in a coculture assay to show that ligand-induced Notch signaling is reduced when GDP-fucose levels are low, but this is not affected by reductions in sialic acid or Gal. Similarly, one of the first demonstrations for coreceptor functions for HS employed mutant CHO cells defective in HS synthesis and engineered to express the FGF receptor.

Although glycosylation is in many cases dispensable for survival of isolated cells in a culture dish, it is often crucial in vivo. Gene ablation studies in mice have identified several instances in which an intact glycosylation pathway is essential for embryogenesis. Examples include mutants that lack complex and hybrid N-glycans and proteoglycan mutants defective in HS, whereas the corresponding mutants in CHO cells do not cause an obvious phenotype. Thus, one theme that emerges from the study of mutants is that glycosylation is critical in the context of a multicellular organism but dispensable in isolated cells. This conclusion has been driven home in recent years by the discovery of human genetic diseases termed congenital disorders of glycosylation, which arise from mutations in genes involved in glycosylation (see Chapter 42).

CHO cells have become the cells of choice for the biotechnology industry in the production of recombinant therapeutic glycoproteins and in glycosylation engineering (see Chapter 51). For example, CHO cells with a *Lec1* mutation have been used to produce the lysosomal enzyme glucocerebrosidase for the treatment of patients with Gaucher's disease, who lack this enzyme. Glucocerebrosidase from cells lacking GlcNAcT-I have only oligomannosyl N-glycans and are thereby efficiently targeted to the mannose receptor on reticuloendothelial cells and ultimately to lysosomes. LEC11 CHO cells have been used to generate recombinant soluble complement receptor carrying sialyl Lewis^x, which targets the molecule to damaged endothelium where it is most effective. In a third example, CHO cells with multiple mutations that simplify N- and O-glycans are being used by X-ray crystallographers to produce homogeneous preparations of membrane glycoproteins with highly truncated N- and O-glycans that do not inhibit their crystallization.

Somatic cell genetics arose from the desire to manipulate the genome of cultured cells in vitro. Today, the availability of genomic sequences from multiple organisms has shifted the emphasis in genetics toward the generation of mutant organisms using the techniques of transgenesis, homologous recombination for gene replacement, or conditional gene inactivation. However, the study of somatic cell mutants still plays an important role in glycobiology research because it provides a less-expensive and faster method for studying the effects of deleting or newly expressing particular glycosylation gene products in a cell. Gain-of-function mutants may of course be generated by transfection of cDNAs encoding glycosylation genes, and reduced expression of any gene can be achieved by the use of RNA interference (RNAi) or antisense cDNA strategies. Although extremely valuable, the latter approaches generally target only known genes, whereas cell-based genetics makes it possible to discover new genes by screening for phenotypic changes directly related to glycosylation changes. Additionally, cells and mutants with well-characterized glycosylation pathways are ideal hosts for investigating the activity encoded by a putative glycosylation gene identified in genome sequence databases. These mutant cells also provide a platform to test the severity of human mutations in a complementation test: The normal human gene rescues defective glycosylation when transfected into the mutant cell, but the same gene with pathological mutations does not. Thus, somatic cell mutants provide access to novel genes involved in glycosylation, which in turn guide strategies for sophisticated gene-manipulation experiments in animals. By combining the two approaches, the biological function of a particular glycosyltransferase, sugar residue, or lectin can be defined. Coupled with powerful new mass spectrometry techniques for determining glycan structures from small samples of tissue or cells, glycosylation mutants of cells and animals provide complementary material for structure/function analyses and identifying mechanistic bases of glycan functions in mammals.

FURTHER READING

Puck T.T. and Kao F.-T. 1982. Somatic cell genetics and its application to medicine. *Annu. Rev. Genet.* **16:** 225–271.

Stanley P. 1984. Glycosylation mutants of animal cells. *Annu. Rev. Genet.* **18:** 525–552.

Esko J.D. 1989. Replica plating of animal cells. *Methods Cell Biol.* **32:** 387–422.

Esko J.D. 1991. Genetic analysis of proteoglycan structure, function and metabolism. *Curr. Opin. Cell Biol.* **3:** 805–816.

Stanley P. 1992. Glycosylation engineering. *Glycobiology* **2:** 99–107.

Kinoshita T., Inoue N., and Takeda J. 1995. Defective glycosyl phosphatidylinositol anchor synthesis and paroxysmal nocturnal hemoglobinuria. *Adv. Immunol.* **60:** 57–103.

Stanley P., Raju T.S., and Bhaumik M. 1996. CHO cells provide access to novel N-glycans and developmentally regulated glycosyltransferases. *Glycobiology* **6:** 695–699.

Esko J.D. and Selleck S.B. 2002. Order out of chaos: Assembly of ligand binding sites in heparan sulfate. *Annu. Rev. Biochem.* **71:** 435–471.

Patnaik S.K. and Stanley P. 2006. Lectin-resistant CHO glycosylation mutants. *Methods Enzymol.* **416:** 159–182.

Zhang L., Lawrence R., Frazier B.A., and Esko J.D. 2006. CHO glycosylation mutants: Proteoglycans. *Methods Enzymol.* **416:** 205–221.

CHAPTER 47

Structural Analysis of Glycans

Barbara Mulloy, Gerald W. Hart, and Pamela Stanley

THIS CHAPTER SURVEYS VARIOUS APPROACHES for determining the structures of complex glycans. Specific chemical, enzymatic, and other analytical strategies as well as mass spectrometric (MS) and nuclear magnetic resonance (NMR) spectroscopic methods that lead to complete glycan sequence determination are described. Finally, methods for solving the three-dimensional structures of glycans are considered.

BACKGROUND

The primary structure of a glycan is defined not only by the nature and order of constituent monosaccharides, but also by the configuration and position of glycosidic linkages and the nature and location of nonglycan substituents (see Chapter 2). For a typical mammalian glycoprotein, the aim is often to identify the correct structure from a range of known or predictable candidate structures. For glycans from bacteria or less well-characterized organisms, it is hard to make predictions, and structural determinations are performed without any assumptions. Choice of methodology is often dictated by the amount and purity of material available and its source (e.g., from tissue samples or cultured cell lines). If quantities are not limiting, the complete primary structure may be determined. In most situations where purity and/or amounts are not optimal, partial characterization will usually be possible. The sensitivity of methods for glycan structural analysis continues to increase with technological advances. Many of these techniques are referred to by acronyms (Table 47.1).

DETECTION OF GLYCANS

Methods for initial glycan detection in glycoconjugates include direct chemical reactions with constituent monosaccharides, metabolic labeling with either radioactive or chemically reactive monosaccharides, and detection with specific lectins or antibodies. A general method for detecting the presence of glycans on proteins involves periodate oxidation of vicinal hydroxyl groups followed by Schiff base formation with amine- or hydrazide-based probes (see Chapter 2 for an example of this reaction). This chemical modification procedure, also referred to as the periodic acid–Schiff (PAS) reaction, can be used to identify glycoproteins in gels. Commercially available kits allow detection of 5–10 ng of a glycoprotein using the periodate reaction with subsequent amplification by means of biotin hydrazide/streptavidin-alkaline phosphatase.

Lectin overlays of blots of SDS-PAGE gels can also be used to detect the presence of specific glycans with comparable sensitivity and greater specificity. For example, the agglutinins from *Sambucus nigra* (SNA) bind to glycans that terminate in α2-6-linked sialic acid. Lectins specific for terminal fucose, galactose, *N*-acetylgalactosamine, and *N*-acetylglucosamine are also commercially available (see Chapter 45).

Metabolic labeling of glycoconjugates with radioactive sugars is another powerful tool for detecting glycans and determining the composition of their attached glycans (Figure 47.1). Cells incubated in medium containing ^{3}H- or ^{14}C-labeled monosaccharides will incorporate the label into the glycan chains of glycoconjugates. The radiolabeled molecules can be detected following gel electrophoresis (SDS-PAGE) or thin-layer chromatography (TLC) by autoradiography or fluorography. Proteins with a glycosylphosphatidylinositol (GPI) anchor may also be specifically labeled with radioactive precursors such as *myo*-inositol or ethanolamine (see Chapter 11). Glycosaminoglycan (GAG) chains on proteoglycans (see Chapter 16) can be metabolically labeled with $^{35}SO_4$- or [^{3}H]-glucosamine and separated from other glycoproteins by ion-exchange chromatography or cetylpyridinium chloride/ethanol precipitation.

Another approach is to transfer a radioactive label in vitro from a radioactive nucleotide sugar to glycans on glycoproteins or cells using a purified glycosyltransferase. For example, the O-GlcNAc modification (see Chapter 18) may be detected by the transfer of labeled galactose using β4-galactosyltransferase and UDP-[^{3}H]-galactose. Purified glycosyltransferases may also be used to label cells by modifying terminal sugars exposed on the glycans of cell-surface glycoproteins.

Radiolabels can also be introduced by chemical reactions that are selective for the structural features of specific glycan types. For example, terminal sialic acids have a unique arrangement of hydroxyl groups on their glycerol side chain that distinguishes this monosaccharide from the others found in mammalian glycans. Mild periodate oxidation of such terminal sialic acid side chains creates aldehyde groups that can be reduced by subsequent treatment with sodium [^{3}H]-borohydride, thereby labeling the sialic acids (Figure 47.1).

The use of radiolabeling strategies for structural analysis also has the advantage of being easy to perform and monitor. Labeled glycans can be isolated by the same methods used for unlabeled sugars with the advantage that radiometric purity is much easier to achieve, and sufficient incorporation is usually obtained for further analysis. However, the infor-

TABLE 47.1. *Separation techniques and their acronyms*

Acronym	Technique	Description	Use
FACE	fluorophore-assisted carbohydrate electrophoresis	gel-electrophoresis-based chromatographic technique for separating samples derivatized with an anionic fluorophore	separation, identification, and quantification of labeled mono- and oligosaccharides
GLC or GC	gas-liquid chromatography or gas chromatography	gas-phase chromatographic technique for separating volatile derivatized samples	sugar composition and linkage analysis; usually interfaced with MS
HPAEC-PAD	high-pH anion-exchange chromatography–pulsed amperometric detection	ion-exchange liquid chromatographic separation technique carried out at high pH	separation, identification, and quantification of mono- and oligosaccharides without derivatization
HPCE	high-performance capillary electrophoresis	chromatographic technique for separating charged molecules	separation, identification, and quantification of charged glycans; sometimes interfaced with MS
HPLC	high-pressure liquid chromatography	chromatographic technique for analytical and preparative separations	separation of all classes of glycans and glycoconjugates; may be interfaced with MS
HPTLC	high-performance thin-layer chromatography	chromatographic technique for analytical separations	glycolipid characterization
SDS-PAGE	sodium dodecyl sulfate–polyacrylamide gel electrophoresis	gel electrophoresis technique for separation of proteins according to molecular weight	glycoprotein characterization
PAS	periodic acid–Schiff reaction	colorimetric determination of sugars	detection of glycans
NMR	nuclear magnetic resonance	1D NMR spectroscopy	number and anomeric configuration of monosaccharides in a glycan
COSY	correlation spectroscopy	2D NMR spectra; cross-peaks indicate protons joined by few bonds	identity and anomeric configuration of monosaccharides in a glycan
TOCSY	total correlation spectroscopy	2D NMR spectra; cross-peaks define whole spin system (e.g., one monosaccharide residue)	identity and anomeric configuration of monosaccharides in a glycan
NOESY	nuclear Overhauser effect spectroscopy	2D NMR spectra; cross-peaks indicate protons close in space	sequence analysis, conformational analysis
ROESY	rotating-frame NOESY	2D NMR spectra; cross-peaks indicate protons close in space; better than NOESY for oligosaccharides	sequence analysis, conformational analysis
HMBC	heteronuclear multiple bond spectroscopy	2D NMR spectra; cross-peaks indicate proton and C, N, or P atom linked by few bonds	assignment of NMR signals to atoms in structure; sequence and substitution analysis
HSQC	heteronuclear single-quantum coherence spectroscopy	2D NMR spectra; cross-peaks indicate proton and C, N, or P atom linked by one bond	assignment of NMR signals to atoms in structure
MS	mass spectrometry	technique for mass measurement of gas-phase ions	primary structure analysis of biopolymers
FAB	fast atom bombardment	MS ionization technique	mass mapping and sequence analysis of glycans and glycolipids
MALDI	matrix-assisted laser desorption ionization	MS ionization technique	mass mapping of glycans and glycoconjugates; important for glycomics
ESI	electrospray ionization	MS ionization technique	molecular weight and sequence analysis of glycans and glycoconjugates; important for glycomics and glycoproteomics
CAD-MS/MS	collisionally activated decomposition mass spectrometry/mass spectrometry	tandem MS technique in which fragment ions are produced from a selected parent ion via collisions with an inert gas	sequence analysis of glycans and glycoconjugates

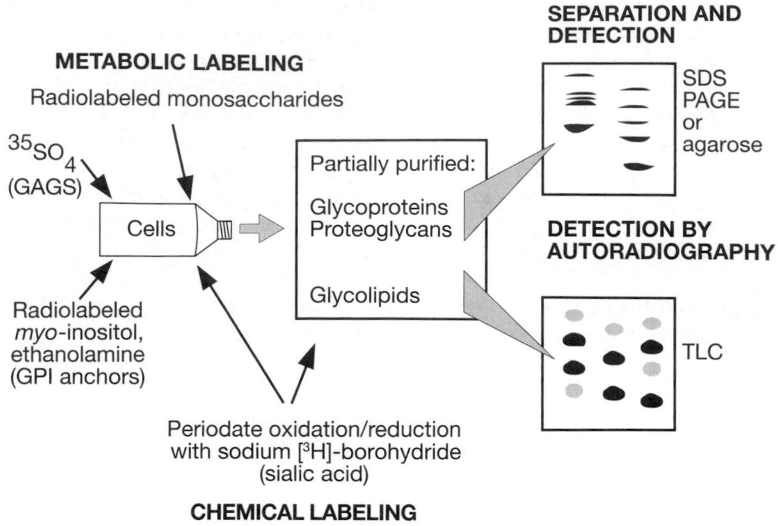

FIGURE 47.1. Radiolabeling strategies for the detection of glycans. Metabolic labeling with [^{3}H]- or [^{14}C]-monosaccharides allows detection of any glycoconjugate into which the monosaccharide is incorporated. $^{35}SO_4$ can be used to trace glycosaminoglycans (GAGs), and labeled *myo*-inositol or ethanolamine can be used to follow GPI anchors. An example of the chemical introduction of a label using sodium [^{3}H]-borohydride is also shown, and this technique can be applied to cell-surface glycoconjugates or to partially purified mixtures. Radiolabeled glycoconjugates can then be separated by gel electrophoresis (for glycoproteins or proteoglycans) or TLC (for glycolipids), the labeled compounds being detected by autoradiography. Radiolabeling is an efficient and sensitive method, but it is decreasing in use due to safety considerations.

mation obtained from such analyses is limited, and a complete structural identification generally requires the isolation of unlabeled material.

Metabolic labeling can also be performed with synthetic monosaccharides that are modified with chemically reactive groups. For example, the azido monosaccharide *N*-azidoacetylmannosamine is converted by cells to *N*-azidoacetyl sialic acid, which is incorporated into sialylated glycans in place of the natural sialic acid residue. The azide group can then be selectively reacted with phosphine or alkyne reagents (see Chapter 49) that introduce a fluorescent dye or an affinity probe such as biotin, thereby enabling detection of the sialic acid residue. Commercially available azido analogs of *N*-acetylgalactosamine and *N*-acetylglucosamine can be used to label O-GalNAc glycans (see Chapter 9) and O-GlcNAc-modified cytoplasmic and nuclear proteins (see Chapter 18), respectively.

Glycoproteins and Mucins

During gel electrophoresis, a glycosylated protein typically presents as one or more diffuse bands, which result from heterogeneity in the glycan. Visualized by protein-staining reagents, this phenomenon is often the first indication of the presence of glycans. Some mucins of very high molecular weight do not enter ordinary gels or, if they do, they migrate as heterogeneous smears. Agarose gels or combination polyacrylamide-agarose gels may be useful in this situation. Several analytical options are available to investigate the presence of glycans further, for example, the classical PAS stain. Treatment of glycoproteins with endoglycosidases is another option (e.g., peptide-N-glycosidase F, endoglycosidase F2, and endoglycosidase H; see Table 47.2), and if it results in a mobility change for one or more of the bands on the gel, the presence of N-glycans is indicated. O-Sialoglycoprotease can be used for the identification of glycoproteins or mucins containing clustered sialylated O-

TABLE 47.2. *Glycan-degrading enzymes*

Enzyme name	EC number	Specificity	Uses and notes
Endoglycosidases			
Endoglycosidase H	3.2.1.96	cleaves between the two *N*-acetylglucosamine residues in the core of oligomannose N-glycans	detection of N-glycosylation; release of glycans
Endoglycosidase F2	3.2.1.96	cleaves between the two *N*-acetylglucosamine residues in the core of oligomannose or biantennary N-glycans	detection of N-glycosylation; release of glycans
Peptide-N-glycosidase F (PNGase F) or glycopeptidase F	3.5.1.52	cleaves between Asn of oligomannose, complex, or hybrid N-glycans; requires at least one amino acid at both the amino terminal and carboxyl terminal of Asn	may not work if core is α1-3-fucosylated
Peptide-N-glycosidase A (PNGase A) or glycopeptidase A	3.5.1.52	cleaves between Asn of oligomannose, complex, or hybrid N-glycans; requires at least one amino acid at both the amino terminal and carboxyl terminal of Asn	will work if core is α1-3-fucosylated; efficacy for use on mammalian glycoproteins poorly defined
Endo-β-galactosidases	3.2.1.102	cleave between galactose and *N*-acetylglucosamine of poly-*N*-acetyllactosmines	detection of poly-*N*-acetyllactosamines and some keratan sulfates
Endoneuraminidases (endo-α-sialidase)	3.2.1.129	cleave between sialic acid units of polysialic acids	detection of polysialic acids
Exoglycosidases			
Sialidases, neuraminidases	3.2.1.18	nonreducing terminal sialic acids	detailed sequence analysis of glycans
Fucosidase	3.2.1.51	nonreducing terminal fucose	some types and linkages can be resistant
α1-2 Fucosidase	3.2.1.63	nonreducing terminal fucose linked α1-2	
β-galactosidase	3.2.1.23	nonreducing terminal β-galactose	
α-mannosidase	3.2.1.24	nonreducing terminal α-mannose	
Glycosaminoglycan lyases			
Chondroitinase ABC	4.2.2.4	degrades galactosaminoglycans (chondroitin sulfates A and C, dermatan sulfate) to disaccharides	detection and characterization of galactosaminoglycans; purification of other glycosaminoglycans
Chondroitinase AC	4.2.2.5	degrades chondroitin sulfates A and C to disaccharides	detection and characterization of chondroitins A and C
Chondroitinase B	4.2.2.5	degrades dermatan sulfate (chondroitin sulfate B) at iduronic acids	detection and characterization of dermatan sulfate
Heparinase, heparin lyase I	4.2.2.7	cleaves between N-sulfated glucosamine and 2-O-sulfated iduronate residues	detection and characterization of heparin and heparan sulfate
Heparitinase I, heparin lyase III	4.2.2.8	cleaves between N-acetylated glucosamine and glucuronic acid residues	detection and characterization of heparan sulfate
Keratanase	3.2.1.103	cleaves between galactose and *N*-acetylglucosamine	detection and characterization of keratan sulfate
Keratanase II	3.2.1.103	cleaves between 6-O-sulfated *N*-acetylglucosamine and galactose (+/– 6-O-sulfate)	detection and characterization of keratan sulfate

Many other enzymes are available, with varying specificities; it is wise to consult the supplier's technical information and the literature for the exact specificity of any product.

glycans, as such glycoconjugates are specifically degraded by this protease. Removal of individual sugars by exoglycosidases such as sialidase or β-galactosidase may also result in a mobility change, depending on the number of residues removed. However, not all glycans may be detected by these treatments due to resistance to the enzymes used. Such resistance can result from modifications to glycan hydroxyl groups (e.g., sulfation, acetylation, or phosphorylation; see Chapter 2), glycosidic linkages that are not recognized by the enzymes, or steric inaccessibility of the glycan. Complete removal of N- and O-glycans can be achieved by chemical treatments (e.g., hydrazinolysis, β-elimination, or hydrogen fluoride treatment), but peptide damage usually precludes further analysis by gel electrophoresis. Aspects of the glycan may also be modified (e.g., O-acetyl groups may be lost).

Proteoglycans

Proteoglycans typically contain more glycan than protein (see Chapter 16). They may be separated by agarose gel electrophoresis and by ion-exchange chromatography, which separates on the basis of the charge conferred by sulfate groups. Treatment of proteoglycans with GAG lyases (Table 47.2) will produce a shift in mobility on a gel, condensing the proteoglycan smear into discrete bands. After the removal of much of the glycan portion, antibodies that recognize the remaining structures ("stubs") may be used in western analysis. The lyases cleave a 4,5 unsaturated uronic acid at the nonreducing end. Anti-"stub" antibodies recognize the sulfation of the penultimate *N*-acetylglucosamine or *N*-acetylgalactosamine residue.

Glycolipids

Typically, the analysis of glycolipid glycans by NMR or mass spectrometry is preceded by their purification using chromatographic methods. Mixtures of glycolipids can be fractionated by TLC, and staining of TLC plates with glycan-reactive reagents may allow detection of individual glycolipids. Using different reagents, it is possible to recognize gangliosides (e.g., resorcinol-HCl detects sialic acids) or bands that contain only neutral monosaccharides (e.g., orcinol-sulfuric acid detects all monosaccharides). Reagents are also available for the detection of sulfate and phosphate groups on glycolipids. Some prepurification of the crude extract is usually preferred (e.g., Folch partitioning and ion-exchange chromatography). These procedures separate nonpolar or nonionic lipids from polar lipids (e.g., glycosphingolipids) and those that contain charged groups (e.g., gangliosides, phospholipids, and sulfatides). Following TLC or HPLC separation of the enriched mixture, target glycolipids may be detected more easily by glycan-reactive reagents. It is also common practice to deduce the presence of specific sugars by evaluating the shifts produced in the migration position of a band following a chemical or enzymatic treatment. Glycolipids on TLC plates can also be detected by reagents that recognize specific glycan features including monoclonal antibodies, lectins, or even intact microorganisms expressing glycan-binding receptors (see Chapter 45). More detailed structural features may be identified by running the TLC in a second dimension following a specific treatment. On a larger scale, glycolipids are separated using column chromatography or by HPTLC on silica plates.

GPI Anchors

GPI-anchored proteins (see Chapter 11), with their lipid, protein, and glycan moieties, have unique physicochemical properties that can be exploited for detection purposes. The nonionic detergent Triton X-114 at low temperature (4°C) extracts soluble and integral membrane proteins as well as GPI-anchored proteins. When the solution is warmed, two phases separate, and GPI-anchored and other amphiphilic proteins remain associated with the detergent-enriched phase. GPI-specific phospholipases can be used to cleave GPI anchors for further characterization. Successful cleavage by GPI-specific phospholipases can be assessed by subsequently analyzing samples by SDS-PAGE, because removal of the GPI anchor causes a shift in molecular mass. This is a common diagnostic method for identifying the presence of a GPI anchor on a protein of interest. Another method is to treat the GPI-anchored protein with nitrous acid, which cleaves the unsubstituted glucosamine residue that links the glycan to the phosphatidylinositol (PI).

Plant and Bacterial Polysaccharides

This family of glycans contains many structures, including homo- and heteropolysaccharides, neutral and ionic polysaccharides, and linear and branched structures, with widespread molecular sizes ranging from a few monosaccharide units to thousands (see Chapters 20 and 22). These polysaccharides are typically extracted with water, salts, chaotropic agents, or detergents and are isolated by precipitation with alcohols. Detection is based on refractive index or colorimetric reactions, because sample quantity is not usually a limitation.

RELEASE, PROFILING, AND FRACTIONATION OF GLYCANS

Once the presence and general type of glycan has been established, the next challenge is to determine how many structurally different glycans are present in a glycoconjugate. The approach varies, but the answer is generally attained by some type of chromatographic or mass spectrometric profiling. When glycans are released prior to chromatographic profiling, it is necessary first to consider the need for a quantitative release procedure that neither destroys nor structurally alters the glycan. Ideally, information regarding the nature of the linkage between the glycan and its liberated protein or lipid should be retained, although this is not always possible.

Release of Glycans from Glycoconjugates

The glycan moiety of glycosphingolipids can also be removed enzymatically using endoglycoceramidase or chemically by ozonolysis. The profiles obtained often provide glycan structural information based on their similarity to known standards. Chemical approaches suitable for the release of glycans from a protein include hydrazinolysis, which releases both N-glycans and O-glycans or, under controlled conditions, cleaves only the N-glycans. Alkaline borohydride treatment (termed β-elimination) is a procedure that under carefully controlled conditions releases only O-glycans. Complex, hybrid, and oligomannose N-glycans can also be released by the peptide-N-glycosidases PNGase F or PNGase A (often termed N-glycanases) (Figure 47.2). An endoglycosidase termed Endo H may be used for the selective release of oligomannose and hybrid N-glycans, but complex N-glycans are resistant (Figure 47.2). N-Glycans and O-glycans can be obtained nonselectively by degradation of the protein by proteases to generate glycopeptides. GPI anchors may be cleaved from protein by phospholipase treatment or obtained following proteolysis of the protein. Free glycans obtained by all of these methods are subsequently analyzed by HPLC, HPAEC, HPCE, or FACE (Table 47.1). The profiles obtained often provide glycan structural information based on their similarity to known standards.

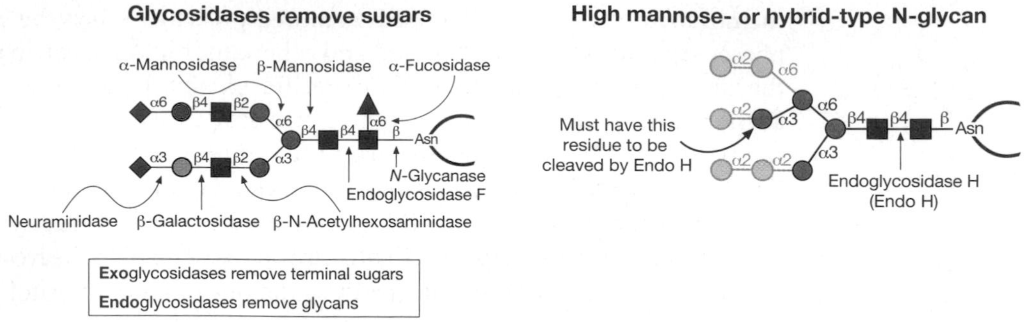

FIGURE 47.2. Glycosidases used for structural analysis. (*Left*) A biantennary N-glycan is shown with exoglycosidases that can be used to remove each monosaccharide sequentially. Exoglycosidases act only on terminal sugars. Also shown are endoglycosidases that remove the intact N-glycan. N-Glycanase cleaves the GlcNAc-Asn bond, releasing the N-glycan and converting asparagine into aspartate. Endoglycosidase F cleaves between the core *N*-acetylglucosamine residues and therefore leaves an *N*-acetylglucosamine attached to the protein. (*Right*) Endoglycosidase H (Endo H) cleaves between the core *N*-acetylglucosamine residues of oligomannose or hybrid N-glycans that have at least four mannose residues as shown. Endo H does not act on complex N-glycans.

Profiling of Glycoprotein Glycans

Glycans released from a glycoprotein are usually a complex mixture. Even when only one glycosylation site in the protein is occupied, it can bear many different glycans resulting in many glycoforms of the glycoprotein. Chromatographic profiles are used for comparative studies and to obtain a preliminary indication of the number, relative quantities, and types of glycans present in a glycoprotein.

Profiling strategies are chosen based on the quantity of sample available. For large amounts (>5 mg), HPAEC-PAD or other HPLC-based profiling is feasible, provided the mixture contains fewer than about 50 different glycans. Individual fractions can be analyzed by MS or NMR. Radiolabeling using the chemical or metabolic methods described above is used to enhance the sensitivity of glycan detection. Indeed, scintillation detectors can be directly linked to HPLC equipment to monitor the purification of radiolabeled glycans. Once liberated from their glycoconjugates, glycans with free reducing termini (see Chapter 2) can be chemically labeled with fluorescent tags such as 2-aminopyridine (2-AP), 2-aminobenzamide (2-AB), 2,6-diaminopyridine (DAP), or biotinylated 2,6-diaminopyridine (BAP), providing detection sensitivity that rivals the level achieved with radiolabels. Advantages of this method include more facile purification of the labeled glycans and a wider variety of options for chromatographic separations and analytical techniques.

If a label is introduced at the reducing end, structural information may be obtained by sequential exoglycosidase treatments (Figure 47.2 and Table 47.2) and chromatography to detect shifts in glycan elution or migration (e.g., by paper chromatography, HPLC, or TLC) that indicate susceptibility to the enzyme. Comparison with known standards treated in the same manner allows tentative glycan identification. However, well-characterized standards are difficult to obtain in pure form, and there are nearly always species in a chromatogram that appear at unusual elution times. It is very important to note that separation profiling should not be confused with actual structural analysis, because coelution with a standard does not necessarily connote a structure identical to that standard.

Profiling of Glycosaminoglycans

Structural analysis of GAGs is an area in which methodologies are rapidly improving (see Chapter 16). Molecular size profiles of GAGs can be determined by chromatographic or electrophoretic methods. Various hydrolases and chemical degradation methods (such as nitrous acid deamination) are available to define the class and/or structures of GAG chains further (Table 47.2). Characterization of a heterogeneous sample might be achieved by fingerprinting techniques (such as chromatography or electrophoresis of enzyme-generated oligosaccharides) and analysis of the disaccharide products of exhaustive depolymerization. Where an oligosaccharide of homogeneous sequence is available, various strategies for precise sequencing can be used, including end-labeling, specific enzyme digestions, separation techniques, and mass spectrometry. For example, sequencing of heparan sulfate can be achieved by treatment with heparanase followed by MS or NMR spectroscopy.

MONOSACCHARIDE COMPOSITION ANALYSIS

Some qualitative information concerning the monosaccharide composition of a glycan may be derived from the procedures described above for the detection, release, and profiling of glycans. Conversely, it is often convenient and informative to determine the monosaccharide composition of a glycoconjugate without prior release of glycans. After total hydrolysis of a glycan into its monosaccharide constituents, colorimetric reactions can be used to determine the total amount of hexose, hexuronic acid, or hexosamine in the sample. These approaches only require common reagents and a spectrophotometer, but determination of total glycan content may not always be accurate because of variations in the sensitivities of different linkages to hydrolysis, variations in the degradation of individual saccharides, or a lack of specificity and/or sensitivity in the assays.

Quantitative monosaccharide analysis provides estimated molar ratios of individual sugars and may suggest the presence of specific oligosaccharide classes (e.g., N-glycans vs. O-glycans). The analysis involves the following steps: cleavage of all glycosidic linkages (typically by acid hydrolysis), fractionation of the resulting monosaccharides, detection, and quantification. Since the early 1960s, a variety of gas-liquid chromatography (GLC) methods have been developed to quantify monosaccharides. The most useful involve coupling of GLC and MS for linkage and composition information. These methods are most successful when the monosaccharides are first chemically modified at their hydroxyl and aldehyde groups. Reduction of the aldehyde of a free monosaccharide followed by acetylation of its hydroxyl groups provides a derivative termed the "peracetylated alditol acetate." These modified monosaccharides can be readily analyzed by GLC and MS and compared with authentic standards. The hydroxyl groups of free monosaccharides generated by glycan hydrolysis can also be converted to trimethylsilyl ethers. These per-*O*-trimethylsilyl derivatives are widely used for monosaccharide compositional analysis by GLC-MS. Incorporation of an optically pure chiral aglycone (e.g., a [–]-2-butyl group), in combination with trimethylsilylation, allows the GLC separation of the D and L pair of isomers and thus determination of the absolute configuration of each monosaccharide.

Chemical derivatization of monosaccharides was once required for HPLC or GLC separation and analysis. However, in recent years, these classical methods have been supplanted by HPAEC-PAD, which does not require monosaccharide derivatization. Fluorescent derivatives produced by reductive amination (e.g., with 2-AB, 2-AP, or 8-amino-1,3,6-naphthalene trisulfonic acid [ANTS]) became popular for detection by reversed-phase HPLC with online fluorophore-assisted carbohydrate electrophoresis (FACE), or HPCE. For example, tagging sialic acids with a fluorescent compound (1,2-diamino-4,5-methylene-dioxybenzene [DMB]) has allowed an increase in detection sensitivity to the femtomole range.

Monosaccharide compositional analysis can be performed on glycoproteins separated by SDS-PAGE and blotted onto polyvinylidene difluoride (but not nitrocellulose) membranes. The membrane is hydrolyzed, the hydrolysate is easily recovered, and monosaccharides are measured as described above. Depending on conditions, peptide or protein may remain bound to the membrane. Sequential analyses are also possible. For example, sialic acids can be released selectively with mild acid. Strong acid can then be added to release the remaining monosaccharides.

LINKAGE ANALYSIS

Determination of Linkage Positions

Methylation analysis is a well-established and ingenious approach for determining linkage positions. The principle of this method is to introduce a stable substituent (an ether-linked methyl group) onto each free hydroxyl group of the native glycan. The glycosidic linkages, which are much more labile than the ether-linked methyl groups, are then cleaved by acid hydrolysis, producing individual methylated monosaccharides with free hydroxyl groups at the positions that were previously involved in a linkage. The partially methylated monosaccharides are derivatized to produce volatile molecules amenable to GLC-MS analysis. The most common strategy involves reduction of the monosaccharides to produce alcohols at C-1 (eliminating the formation of ring structures), followed by derivatization (usually acetylation) of free hydroxyl groups. Individual components of the mixture of partially methylated (methyl groups mark the hydroxyl groups that were originally free), partially acetylated (acetyl groups mark hydroxyl groups originally at substituted, linked, or ring-closure positions) monosaccharide alditols can be identified by GLC-MS (Figure 47.3).

Partially methylated alditol acetates are identified by a combination of GLC retention time and electron impact (EI)-MS fragmentation pattern. The fragmentation patterns of similarly substituted isomeric monosaccharides (e.g., aldohexoses) are the same. Thus, definitive identification requires, in addition to the analysis of the MS pattern, the comparison of GLC retention times with those of known standards (e.g., all 2,3,4-tri-*O*-methyl-hexoses produce the same EI-MS spectrum, but peracetylated 2,3,4-tri-*O*-methylgalactitol elutes later than peracetylated 2,3,4-tri-*O*-methylglucitol). This type of analysis identifies terminal residues (they are methylated at every position except the hydroxyl group at C-1 and C-5), indicates how each monosaccharide is substituted including the occurrence of branching points, and allows the determination of the ring size (pyranose *p* or furanose *f*) for each monosaccharide. However, methylation analysis gives no sequence information and cannot determine whether a particular linkage is of the α or β anomeric configuration.

Determination of Anomericity

The anomeric configuration of linkages is often determined by NMR spectroscopy (see below) and can also be obtained from sequential exoglycosidase digestions (Table 47.2 and Figure 47.2). Cleavage by α- or β-exoglycosidases indicates the anomericity of specific terminal sugar residues. Cleavage by specific endoglycosidases can give added information regarding internal regions of the glycan. Many glycosidases are specific for both monosaccharide residue and linkage type, allowing detailed structural conclusions, although the number of such enzymes available is limited.

NMR Spectroscopy

When enough sample is available (typically a milligram or more but see below), the anomericity of a particular monosaccharide residue in a glycan can usually be determined by ^{1}H-NMR spectroscopy. The anomeric resonances (H-1 signals) appear in a well-resolved

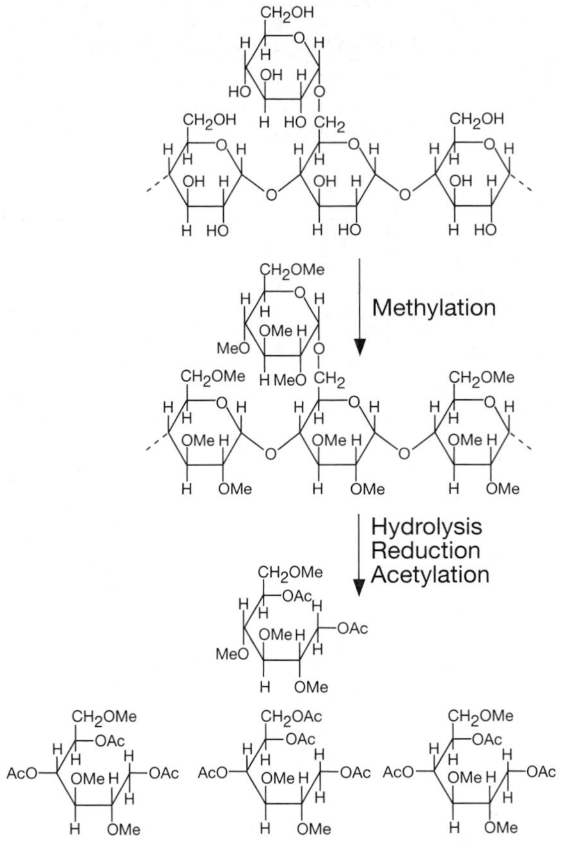

FIGURE 47.3. A simple example of methylation analysis, showing a structural motif that may be found in the polysaccharide glycogen. An α1-4-linked glucose chain has an α1-6-linked glucose branch. Successive steps of methylation, hydrolysis, reduction, and acetylation result in a set of compounds in which the linkage positions can be identified by their acetylation. For the sequence illustrated, the terminal α1-6 glucose gives rise to glucitol acetylated at C-1 and C-5; the two α1-4 glucose units on either side of the branch point are acetylated at C-1, C-5, and C-4, and the branch-point α-glucose is acetylated at C-5, C-1, C-4, and C-6. Methylation analysis cannot give details of sequence or anomericity, but when combined with NMR spectroscopy, it has been used to elucidate the structures of many complex and unusual glycans.

region of the spectrum and show characteristic doublets with a splitting that is significantly larger for β anomers than for α anomers. Thus, a first glance at the ^{1}H-NMR spectrum typically indicates how many residues there are (by counting anomeric signals) and how many of them belong to each anomeric type. A simple ^{1}H-NMR spectrum can provide the entire primary structure of a glycan if ^{1}H-NMR spectra of well-characterized glycans of related structures are available for comparison. As an example, the ^{1}H-NMR spectrum of a mixture of two triantennary N-glycans obtained from bovine fetuin is shown in Figure 47.4.

Limitations on the use of NMR spectroscopy are the cost of spectrometers and the level of expertise required for interpreting NMR spectra. However, access to high-field (i.e., 500 MHz and above) NMR spectrometers fitted with very sensitive probes (e.g., nano-NMR probes) allows ^{1}H-NMR profiling of individual HPLC fractions using minute quantities of sample (2–5 nmoles of glycan).

NMR spectroscopy is a powerful tool for de novo full structural characterization of a glycan. Because this method is nondestructive, the same sample can later be used for other, destructive approaches (e.g., MS and methylation analysis). Complete structural elucidation requires full assignment of both the ^{1}H- and ^{13}C-NMR spectra of a glycan. This is accomplished by a combination of two-dimensional NMR techniques such as correlation

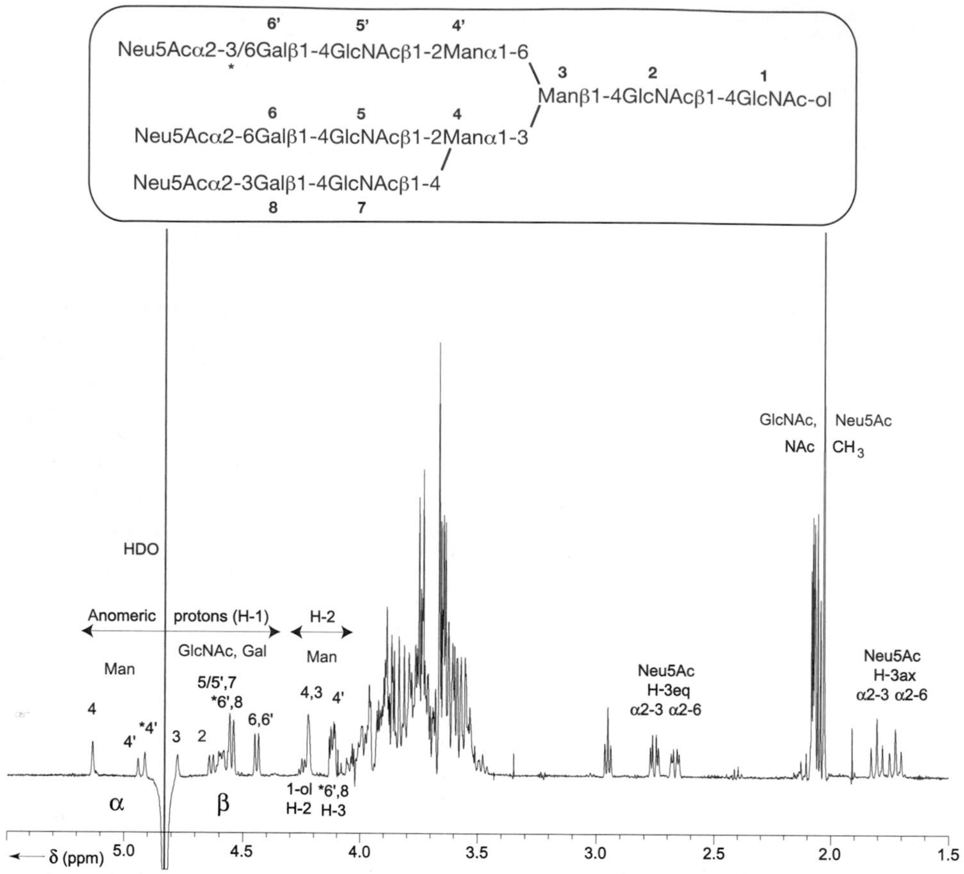

FIGURE 47.4. ^{1}H-NMR spectrum of a mixture of two trisialyl triantennary N-glycans obtained as alditols from bovine fetuin by hydrazinolysis, followed by purification on HPAEC (Table 47.1) and subsequent reduction with sodium borohydride. The spectrum was recorded at 500 MHz, using a solution of 100 µg of the glycan mixture in 0.7 ml of D_2O in a 5-mm NMR tube at pH 6.5 and 23°C. (*Inset*) Structures of the two glycans. Note that they differ only in the linkage position of sialic acid to Gal on the Manα1-6 branch (marked with an *asterisk*). The numbers in the spectrum refer to the corresponding residues in the structures. Assignments are given for the signals of structural reporter groups, including those of the anomeric protons (H-1 signals), the Man H-2, Gal H-3, Neu5Ac H-3eq (equatorial) and H-3ax (axial) signals, and the N-acetyl amino sugar methyl signals. Signals marked by an *asterisk* (for residues 4′ and 6′) refer to the component with Neu5Ac in α2-3 linkage to Gal-6′. The H-1 signals to the left of the residual water (HDO) peak are indicative of the presence of α-linked monosaccharides in the glycan(s), whereas those to the right of HDO originate from β-linked monosaccharide residues in the structures.

spectroscopy (COSY) and total correlation spectroscopy (TOCSY) for ^{1}H, which allows assignment of the ^{1}H signals of individual monosaccharide residues. After this, the heteronuclear single-quantum coherence (HSQC) experiment can be used to extend the assignment to the ^{13}C spectrum. The key experiment for sequencing is the two-dimensional heteronuclear multiple-bond correlation (HMBC) experiment, which detects a coupling between the anomeric proton and the carbon atom on the opposite side of the glycosidic linkage. However, in instances where there is not enough sample for these two-dimensional NMR experiments (HMBC is not a very sensitive technique), other data are required to complete the structural picture. A less rigorous NMR approach for glycan sequencing relies exclusively on two-dimensional ^{1}H-NMR spectroscopy, using through-space effects (nuclear Overhauser effects [NOEs]) as the sole source of evidence for linking, position, and sequence. Use of a 900-MHz NMR spectrometer and a nanoprobe increases the sensitivity so that microgram amounts of a glycan can be analyzed.

Polysaccharides from bacteria (see Chapter 20) give remarkably good NMR spectra (despite their high molecular weight) due to their internal mobility, and it is often possible to determine the structure of the repeat unit by NMR without need for depolymerization. Figure 47.5 illustrates NMR and MS data for a bacterial polysaccharide that is one of the components of a vaccine. The combination of NMR and MS analyses gives a thorough structural assignment.

Mass Spectrometry

The use of EI-MS in monosaccharide composition and linkage analyses is covered above. In this section, three other types of mass spectrometry—fast atom bombardment (FAB), matrix-assisted laser desorption ionization (MALDI), and electrospray ionization (ESI)—are described. All three technologies permit the direct ionization of nonvolatile substances and are applicable to intact glycoconjugates, as well as fragments thereof. Historically, FAB-MS has played an important role in the structural analysis of glycans. However, because of the expense and level of specialized expertise required to operate FAB-MS, this method has been largely supplanted by ESI-MS and MALDI-MS. Among the structural features that can be defined by MS methods are (1) degree of heterogeneity and type of glycosylation (e.g., N-glycan vs. O-glycan; high mannose, hybrid, or complex types, etc.); (2) sites of glycosylation; (3) glycan-branching patterns; (4) the number and lengths of antennae, their building blocks, and the patterns of substitution with fucose, sialic acids, or other capping groups such as sulfate, phosphate, or acetyl esters; (5) complete sequences of individual glycans; and (6) structures of glycolipids, glycopeptides, (lipo)polysaccharides, and GAG-derived glycans.

In the FAB-MS experiment, samples are dissolved in a liquid matrix and ionization/desorption is effected by a high-energy beam of particles fired from an atom or ion gun. High field magnets are the most powerful analyzers for this type of mass spectrometry. In MALDI-MS experiments, the sample is dried on a metal target in the presence of a chromophoric matrix until matrix crystals containing trapped sample molecules are formed. Ionization of the sample is effected by energy transfer from matrix molecules that have absorbed energy from laser pulses. MALDI sources are usually attached to time of flight (TOF) analyzers that can analyze very high-molecular-mass ions (in excess of 200 kD). For ESI-MS, a stream of liquid containing the sample enters the source through a capillary interface, where the sample molecules are stripped of solvent, leaving them as multiply charged species. Electrospray experiments are often performed using instruments with quadrupole analyzers. ESI-MS can be coupled to micro- or nano-bore liquid chromatography (LC) permitting on-line LC/ESI-MS analysis. This method is particularly useful when complex mixtures of peptides and glycopeptides are being examined (e.g., after proteolytic digestion of a glycoprotein).

In principle, MS provides two types of structural information–the masses of intact molecules (the molecular ions) and the masses of fragment ions. MALDI-MS is arguably the most sensitive of the three ionization technologies; hence, it is the preeminent technique for screening for molecular ions ("mass mapping"), especially when high throughput and sensitivity are demanded.

Of the three techniques, FAB-MS is the only one that reliably yields fragment ions. The internal energy acquired during molecular ion formation in MALDI and ESI-MS is usually insufficient for fragmentation to occur. To overcome this, most ESI and some MALDI mass spectrometers have two analyzers in tandem, which allows the detection (using the second analyzer) of fragment ions produced after molecular ions selected by the first analyzer undergo collisions with an inert gas in a chamber placed between the two analyzers. These are referred to as collisionally activated MS/MS experiments. One of the most pow-

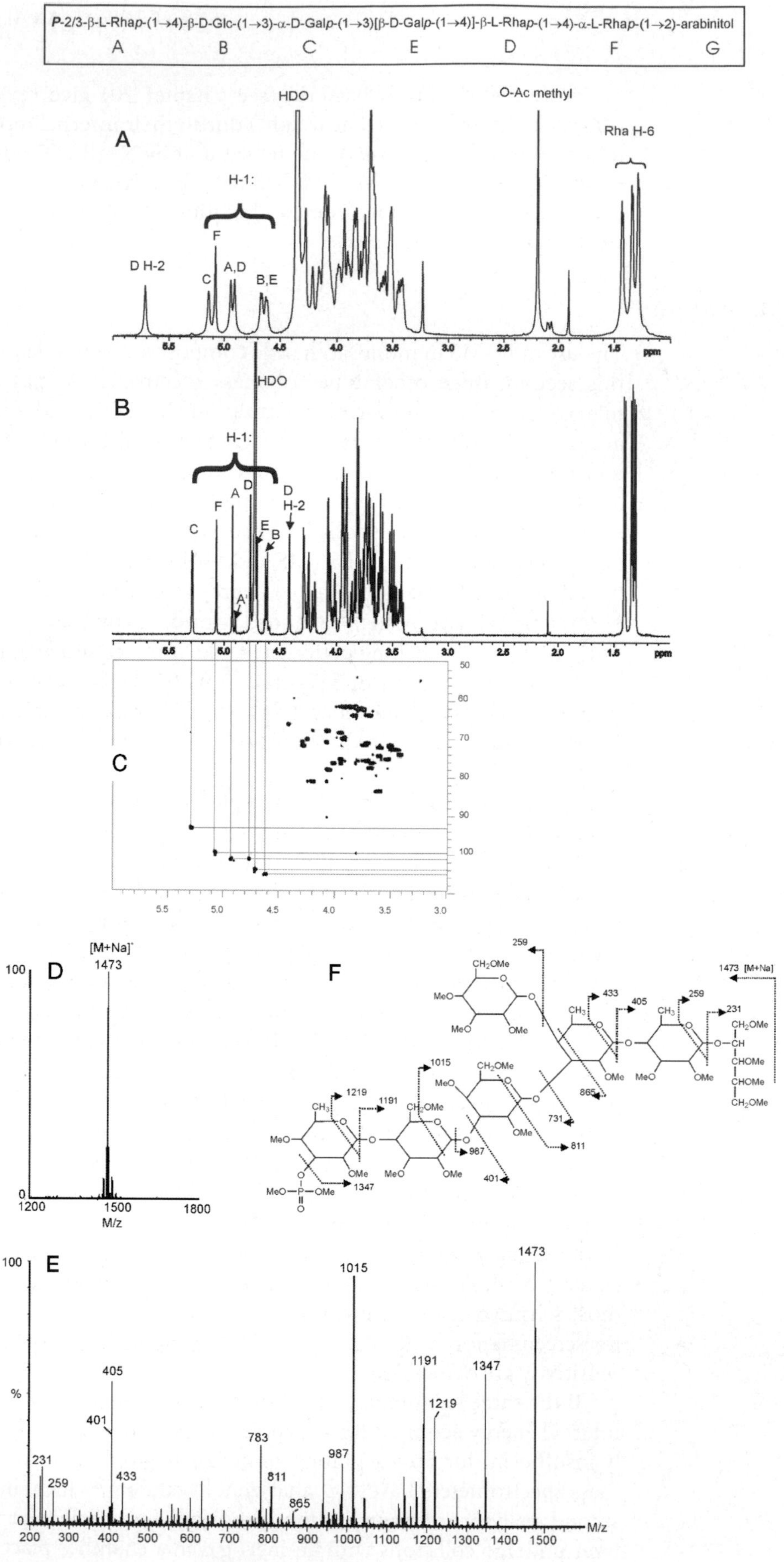

FIGURE 47.5. *(See facing page for legend.)*

erful current technologies for MS/MS is the Q-TOF mass spectrometer, which has a quadrupole as the first analyzer and an orthogonal TOF as the second analyzer.

Fourier transform mass spectrometry (FTMS) with electron capture dissociation (ECD) and lower-cost ion traps with electron transfer dissociation (ETD) are cutting-edge technologies that considerably improve MS analyses of complex posttranslational modifications, including glycans. MS instrumentation and methods continue to improve at an astonishing rate.

Although underivatized glycans can be analyzed by FAB-MS and ESI- or MALDI-MS/MS, far superior data are normally obtained if the glycans are derivatized prior to MS analysis. Derivatization methods can be broadly divided into two categories: (1) "tagging" of reducing ends and (2) protection of most or all of the hydroxyl groups. Commonly used tagging reagents include *p*-aminobenzoic acid ethyl ester (ABEE), 2-AP, 2-AB, and aminolipids. This type of derivatization facilitates chromatographic purification as explained above and enhances the formation of reducing-end fragment ions in MS and MS/MS experiments. Protection of hydroxyl groups by permethylation is by far the most important type of full derivatization employed in glycan MS (although with accompanying destruction of acetyl esters, some sulfate esters, and glycolyl groups during the derivatization process). In FAB-MS experiments, permethylated derivatives form abundant fragment ions arising from cleavage on the reducing side of each HexNAc residue (usually referred to as A-type ions) whose masses define important structural features of N- and O-glycans, including the types of capping sugars and the presence or absence of polylactosamine sequences. In MS/MS experiments, additional fragment ions are produced by cleavage on either side of susceptible glycosidic linkages.

Broadly speaking, the unique strengths of MS can be exploited in two general ways in glycobiology. The first way is to obtain detailed characterization of purified individual glycans or mixtures of glycans. In this type of study, it is essential to acquire sufficient rigorous data to define structure unambiguously; many different MS-based experiments will be required, often complemented by NMR, linkage analysis, and profiling of enzyme digests. An example of this type of application is illustrated in Figure 47.5. The second way is for glycomics investigations in situations where it may not be essential to define structures fully and when high-throughput glycomic profiling or mass-mapping procedures are exploited (see Chapter 48).

FIGURE 47.5. NMR and MS data used in the determination of the structure of the complex pneumococcal capsular polysaccharide 17F. Mild base treatment of the polysaccharide specifically breaks the phosphodiester linkage between rhamnose (Rha) and arabinitol and removes O-acetylation to give an oligosaccharide with the structure shown in the box. *p* denotes a pyranose ring. (*A*) ^{1}H-NMR spectrum of the intact polysaccharide (at 70°C). Prominent, well-resolved signals come from the anomeric protons (H-1), the H-6 of rhamnose, and the methyl protons of O-acetyl groups. The signal from rhamnose D H-2 is shifted downfield (to the left) because of an *O*-acetyl group at D C-2. Residue G is an alditol with no anomeric center; all of its signals are in the crowded central region and are hard to assign. (*B*) ^{1}H-NMR spectrum of the oligosaccharide (at 30°C). The signals from the oligosaccharide are sharper: The O-acetyl methyl signal has disappeared and the D H-2 signal is no longer shifted. Residue A is usually phosphorylated at C-3; the good resolution of the oligosaccharide spectrum allows identification of a small signal (A′) from H-1 of the small proportion of A phosphorylated at C-2. (*C*) Part of the HSQC spectrum used to assign ^{13}C signals of the glycan (rhamnose H-6/C-6 signals omitted). This two-dimensional spectrum has ^{1}H and ^{13}C chemical shifts as X and Y axes. The guidelines (for the anomeric signals) show how ^{13}C shifts may be determined from this spectrum once the ^{1}H spectrum is assigned. These and other NMR spectra, taken together, can provide information on the number and type of residues in the repeating unit and their anomeric configuration, ring size (pyranose [*p*] or furanose [*f*]), linkage positions, substitution positions, and sequence. (*D*) The MALDI spectrum showing a molecular ion at m/z 1473 $[M + Na]^+$ corresponding to a permethylated glycan of the composition *P*deoxyHex$_3$Hex$_3$pent-ol. (*E*) The ESI-MS/MS spectrum of m/z 748 $[M + 2Na]^{2+}$ showing fragment ions that define the sequence shown in the cartoon in panel *F*.

Mass Spectrometry Profiling Underpins Many Glycomics Investigations

As discussed in Chapter 48, the term "glycome" is used to denote the complement of glycans in a cell or organism. Thus, strictly speaking, the term "glycomics" should mean the study of the full complement of glycans from a defined source. In practice, because glycomics investigations are still in their infancy, they are usually confined to studies of subsets of glycans, for example, the most abundant N- or O-glycans present in a particular cell type or tissue. MS strategies have been devised to screen for the types of N- and O-glycans present in a diverse range of biological material, including body fluids, secretions, organs, and cultured cell lines. These methods are based on the analysis of permethylated derivatives, which yield molecular ions at high sensitivity. Putative structures are assigned to each molecular ion based on the usually unique glycan composition for a given mass and prior knowledge of N- and O-glycan biosynthesis. This is called "glycomic profiling" and is most conveniently carried out using MALDI because of its high sensitivity. Assignments can be confirmed in a second experiment employing ESI-MS/MS instrumentation by selecting each molecular ion for collisional activation and recording its fragment ion spectrum. If necessary, additional information can be provided by MS experiments on chemical and enzymatic digests, the choice of which is guided by the sequence information provided by mass mapping and MS/MS experiments. These methodologies are illustrated by data from a glycomics analysis of the mouse kidney in Figure 47.6. Of course, none of these approaches to glycomics address the actual localization of a glycan within a cell type of the tissue being extracted for analysis.

Three-dimensional Glycan Structure

Because of the inherent flexibility of glycosidic linkages, most complex glycans do not have a single, well-defined three-dimensional structure in solution. Crystal structures are available for many mono- and oligosaccharides (in the Cambridge Structural Database; http://www.ccdc.cam.ac.uk). Anyone with suitable molecular modeling software can generate approximate models of more complex glycans from these simple sugar structures. Such models can be useful as an aid to thinking about the overall sizes and shapes of glycans, as long as they are not taken too seriously. Full characterization of glycan conformation and dynamics in solution remains an area of active research and is usually based on experimental data from NMR spectroscopy. For example, H-H coupling constant values around the pyranose ring depend on the ring geometry, and quantitative interpretation of NOEs between adjacent monosaccharides can give clues as to the conformational equilibrium around the glycosidic linkage between them. Recently introduced methods, such as conformational restraints derived from residual dipolar couplings in partially ordered media, can also be applied to glycans. A full treatment of modeling glycan structures is beyond the scope of this chapter (see "Further Reading"). Coordinates for three-dimensional structures of glycoproteins in the Protein Data Bank (PDB: http://www.rcsb.org/pdb) usually define only the stub of any attached glycan. Of more direct interest are complexes between proteins and glycans, of which the PDB contains many, including enzymes, lectins, and heparin-binding proteins among others. In this context, the glycan conformations are both well-defined and biologically relevant.

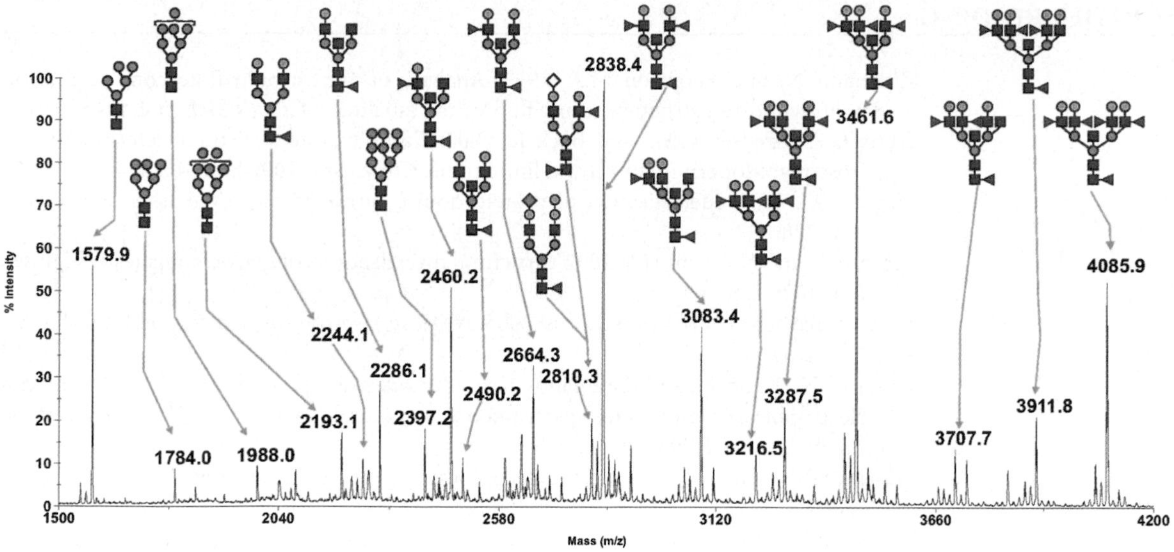

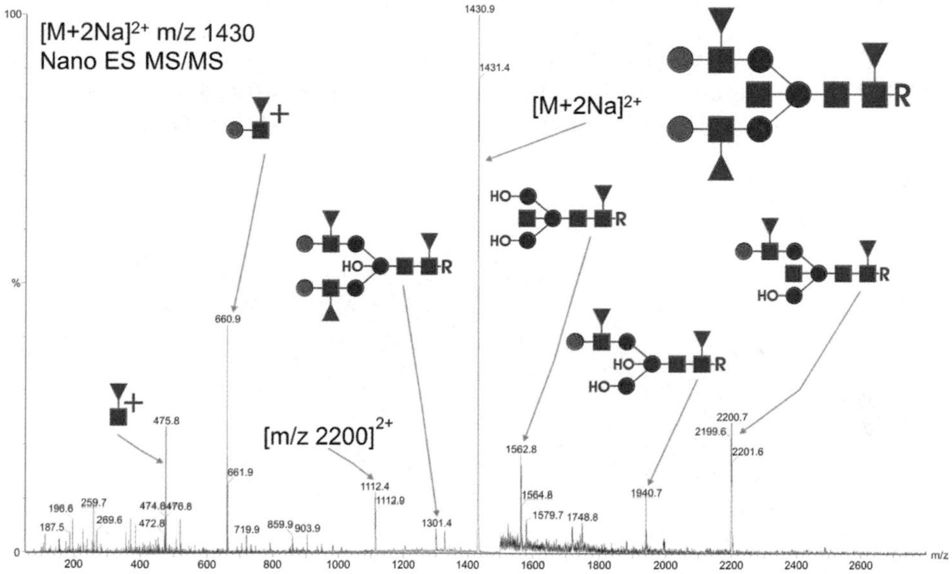

FIGURE 47.6. Data from a glycomics study of N-glycans from mouse kidney (courtesy of Anne Dell, Imperial College London). (*Top*) MALDI-TOF profile of neutral N-glycans released from kidney extracts by peptide N-glycosidase F and permethylated. The glycan structures indicate the most probable sequences attributable to each major molecular ion. (*Bottom*) Part of the evidence for these assignments is provided by fragment ion data derived from ESI-MS/MS experiments. In this experiment, the doubly charged molecular ion at m/z 1430 $[M + 2Na]^{2+}$, which corresponds to the singly charged MALDI-TOF signal at m/z 2837 $[M + Na]^{2+}$, was selected for collision-induced decomposition. Sequence-informative fragment ions arising from the antennae and core are shown in the cartoon. Glycosidic bonds that have been cleaved during the MS/MS process are "tagged" by the presence of a hydroxyl group (HO^-). All other hydroxyl groups on the glycan carry methyl groups because the sample was permethylated prior to analysis.

FURTHER READING

Zitzmann N. and Ferguson M.A. 1999. Analysis of the carbohydrate components of glycosylphosphatidylinositol structures using fluorescent labeling. *Methods Mol. Biol.* **116:** 73–89.

Duus J., Gotfredsen C.H., and Bock K. 2000. Carbohydrate structural determination by NMR spectroscopy: Modern methods and limitations. *Chem. Rev.* **100:** 4589–4614.

Harvey D.J. 2001. Identification of protein-bound carbohydrates by mass spectrometry. *Proteomics* **1:** 311–328.

Mechref Y. and Novotny M.V. 2002. Structural investigations of glycoconjugates at high sensitivity. *Chem. Rev.* **102:** 321–369.

Jiménez-Barbero J. and Peters T., eds. 2003. *NMR spectroscopy of glycoconjugates.* Wiley-VCH, Weinheim, Germany.

Lamari F.N., Kuhn R., and Karamanos N.K. 2003. Derivatization of carbohydrates for chromatographic, electrophoretic and mass spectrometric structural analysis. *J. Chromatogr. B. Analyt. Technol. Biomed. Life Sci.* **793:** 15–36.

Haslam S.M., North S.J., and Dell A. 2006. Mass spectrometric analysis of N- and O-glycosylation of tissues and cells. *Curr. Opin. Struct. Biol.* **16:** 584–591.

Vliegenthart J.F.G. and Woods R.J. 2006. *NMR spectroscopy and computer modeling of carbohydrates: Recent advances.* ACS Symposium, Washington, D.C.

Tissot B., Gasiunas N., Powell A.K., Ahmed Y., Zhi Z.L., Haslam S.M., Morris H.R., Turnbull J.E., Gallagher J.T., and Dell A. 2007. Towards GAG glycomics: Analysis of highly sulfated heparins by MALDI-TOF mass spectrometry. *Glycobiology* **17:** 972–982.

Jiménez-Barbero J., Díaz M.D., and Nieto P.M. 2008. NMR structural studies of oligosaccharides related to cancer processes. *Anticancer Agents Med. Chem.* **8:** 52–63.

CHAPTER 48

Glycomics

Carolyn R. Bertozzi and Ram Sasisekharan

THE TERM "GLYCOME" DESCRIBES THE COMPLETE REPERTOIRE of glycans and glycoconjugates that cells produce under specified conditions of time, space, and environment. "Glycomics," therefore, refers to studies that profile the glycome and is the topic of this chapter.

HISTORICAL PERSPECTIVE OF "OMICS" SCIENCE: GENOMICS, TRANSCRIPTOMICS, AND PROTEOMICS

The field of genomics arose from the availability of complete genome sequence data as well as computational methods for their analysis. One of the surprising findings from analysis of the human genome was the presence of fewer protein-encoding genes (a mere 25,000) than had been predicted earlier. Furthermore, the protein-encoding (i.e., "expressed") genes comprise a small fraction, less than 2%, of the human genome. These genes are transcribed into mRNAs that are often referred to collectively as the "transcriptome." The ability to analyze transcripts in a high-throughput parallel format using a DNA microarray, or "gene chip," has enabled researchers to probe global differences in gene expression, for instance, between healthy and diseased cells, between neurons and muscle cells, and between drug-sensitive and drug-resistant cancer cells. Such "transcriptomic" comparisons have revealed networks of genes whose expression is linked to disease.

Although many scientific discoveries have emerged from genomic and transcriptomic approaches, this information still does not provide a complete picture of the physiology of a cell or organism. The proteins expressed by the cell, collectively termed the "proteome," perform many of the cell's functions. Most eukaryotic proteins are posttranslationally modified (e.g., by phosphorylation, oxidation, ubiquitination, lipidation, or glycosylation). These modifications, combined with alternative splicing in eukaryotes, render the proteome considerably more complex than the transcriptome. Although it is not known how many discrete proteins a particular human cell expresses, estimates between 50,000 and 120,000 have been suggested. Direct characterization of the proteome is required to understand both its complexity and its global functions. The global systems-level analysis of all proteins expressed by cells, tissues, or organisms is referred to as "proteomics."

Unlike the genome, which is fixed for most cells, the proteome is dynamic. The repertoire of proteins expressed by a cell is highly dependent on its tissue type, microenvironment, and stage within its life cycle. As cells receive cues in the form of growth factors, hormones, metabolites, or other agents, various genes are turned on or off. Thus, proteomes vary during cell differentiation, activation, trafficking, and during malignant transformation. Also, many proteins are secreted from cells and circulate in the blood or lymphatic fluid or are excreted in the saliva, mucus, tear fluid, or urine. These bodily fluids also have distinct proteomes.

WHAT IS "GLYCOMICS"?

Glycomic analyses seek to understand how a collection of glycans relates to a particular biological event. As described throughout this book, glycans participate in almost every biological process, from intracellular signaling to organ development to tumor growth. Understanding how the totality of glycans governs these processes is a central goal of glycobiology.

The glycomes of life-forms include all of the glycan and glycoconjugate types that have been described in this book. For example, vertebrates possess protein-associated N- and O-glycans, glycosaminoglycans, and GPI anchors, as well as lipid-associated glycans and free glycans such as hyaluronan (see Chapters 8–18). Other organisms possess their own distinct glycomes, with those of plants (see Chapter 22) and prokaryotes (see Chapter 20) differing greatly in composition from the vertebrate and invertebrate glycomes (see Chapter 25). And as with the proteome, each cell type has its own distinct glycome that is governed by local cues and the cell's internal state. The size of any particular glycome has not yet been established, but we know that glycomes can far exceed proteomes and transcriptomes with respect to complexity. For example, some estimates have placed the vertebrate glycome at more than one million discrete structures. Furthermore, it appears that the glycome is considerably more dynamic than the proteome or transcriptome.

The notion that glycans should be studied as a totality, as well as simply one at a time, is not a radical concept among glycobiologists. Indeed, researchers in the field have long known that glycans form patterns on cells that change during development (see Chapter 38) and cancer progression (see Chapter 44). Also, many glycan-binding proteins are oligomerized on cells and interact with multivalent arrays of glycans on opposing cells (see Chapter 27). In some cases, multiple discrete glycan epitopes work in concert to engage two cells or deliver a signal from one cell to the other. Thus, before "glycomics" was coined, scientists had already concluded that many aspects of glycobiology can be understood only with a systems-level analysis. Conversely, no systems-level analysis of a biological process is complete without interrogating the glycome in addition to the genome, transcriptome, and proteome.

RELATIONSHIP OF THE GLYCOME TO THE GENOME AND PROTEOME

Clues regarding the composition and complexity of the glycome can be found in the cell's genome, transcriptome, and proteome. As discussed in Chapter 7, genome "mining" using known sequences can identify many genes involved in glycan biosynthesis and processing. By such an analysis, more than 250 glycosyltransferases have been found encoded in the human genome as well as many nucleotide sugar biosynthetic enzymes and Golgi transporters (see Chapter 5). Some of the corresponding enzymes have been studied biochemically and their glycosyl donor and acceptor specificities have been defined, whereas others have been assigned predicted functions based on sequence relationships. Furthermore, expression patterns of many glycosyltransferases have been determined in human and mouse tissues using northern blots, quantitative PCR (polymerase chain reaction), and transcriptomic analyses.

In principle, one might use all of this information to construct "virtual glycomes." However, this exercise is of limited value because the combinatorial action of glycosyltransferases in many competing biosynthetic pathways renders the complete glycome very difficult to predict with any accuracy. As an example, the reduced expression of a single glycosyltransferase can perturb the biosynthesis of dozens of glycan structures, some negatively and some positively. The direct glycan products of the glycosyltransferase will be reduced in expression, whereas glycans made by other enzymes that compete for common intermediates might increase in levels. Furthermore, unlike the genome and, to our knowledge, the proteome, the glycome can be sensitive to exogenous nutrient levels. Thus, variations in dietary monosaccharides, such as glucose, galactose, glucosamine, fucose, mannose (see Chapter 18), and *N*-glycolylneuraminic acid (see Chapter 14), can change the composition of the glycome. Because of these complexities, transcriptomic and proteomic data can at best guide hypotheses regarding the presence or absence of specific classes of structures. In contrast, the absence of particular genes (e.g., the sialic acid biosynthesis machinery in *Caenorhabditis elegans*; see Chapter 14) has been useful in assessing the relative compositions of various glycomes.

The numerous factors that influence the glycome (the genome, the proteome, and environmental nutrients, as well as the secretory machinery, pH, and many other determinants) create a system that is highly diverse and dynamic. Thus, the glycome can change dramatically in response to a subtle change in the cellular system. This feature makes glycomics research both exciting and also daunting. Since neither the proteome nor the transcriptome can accurately predict such a moving target, the glycome must be analyzed directly. Techniques that have been employed to characterize the glycome are summarized below.

TOOLS FOR CHARACTERIZING THE GLYCOME

The glycome can be described at many hierarchical levels of complexity (Figure 48.1). First, the glycome can be deconstructed into an inventory of glycan structures separated from their protein or lipid scaffolds and independent of their location in the cell, organ, or organism. This first hierarchical level is essentially a catalog of structures. It is an important starting point for any comprehensive glycome analysis. But how the parts in the catalog assemble to form the intact system is also important for understanding function. Thus, a second hierarchical level of analysis involves defining which glycans are associated with individual proteins or lipids. Analysis of the complete repertoire of a cell's glycoproteins, including

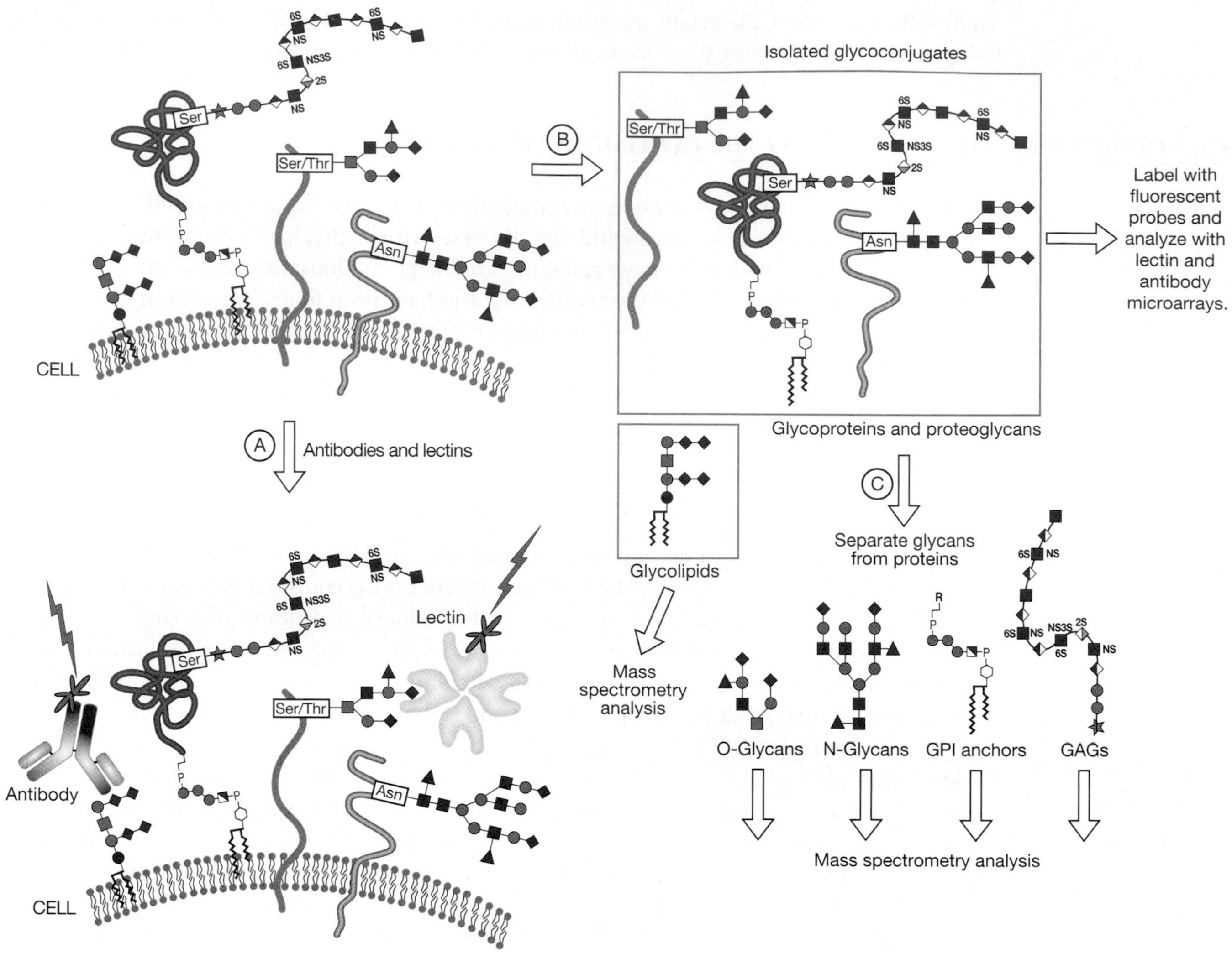

FIGURE 48.1. Multiple approaches for profiling a cell's glycome at various hierarchical levels of complexity. (Step A) Cells can be directly probed for glycan expression using labeled lectins and glycan-specific antibodies. This top–down experiment provides a global view of the distribution of certain glycan epitopes on cells and tissues but does not afford detailed structural information. (Step B) Glycoproteins and glycolipids can be isolated from cell lysates and then analyzed using lectin and antibody microarrays and mass spectrometry methods. Glycolipids can be sequenced directly, whereas glycoproteins are often further deconstructed into separate glycans and proteins before structural analysis. (Step C) Isolated glycans are separated based on type (i.e., N-glycans, O-glycans, glycosaminoglycans, etc.) and their sequences determined by mass spectrometry. Alternatively, intact glycoproteins can be digested with trypsin and the glycopeptides characterized by mass spectrometry. This approach retains the glycan–peptide linkage and allows assignment of sites of protein glycosylation. Collectively, steps B and C comprise a bottom–up glycomic analysis.

their glycan structures and sites of attachment, lies at the intersection of glycomics and proteomics and is often referred to with the term "glycoproteomics." A third level of complexity involves determining which glycans or glycoconjugates are expressed on specific cells or tissues. This level of glycomic profiling is essential if the goal is to reveal new functions in cell–cell communication or to correlate particular glycomes with disease tissue. A final level that has yet to be investigated involves visualizing how glycoconjugates are actually organized relative to each other within the cell, at the cell surface, and in the extracellular matrix.

As described below, numerous techniques have been developed for interrogating the glycome at these various hierarchical levels. No single technique can define all aspects of the glycome. Thus, several approaches are typically employed in parallel, allowing one to assemble a picture of the glycome both from the "bottom up" (i.e., from the individual glycan repertoire) and from the "top down" (i.e., from a global tissue expression analysis). A significant challenge in analyzing the glycome derives from its enormous structural diversity. Different approaches and techniques are required to characterize the structures of glycoproteins versus glycolipids, N-glycans versus O-glycans, and sulfated glycosaminoglycans versus neutral glycans (see Chapter 47). By contrast, a single technique, the DNA microarray, can be used to interrogate all RNA transcripts at once. Thus, at present, the techniques for glycomic analysis remain relatively low-throughput and specific for a particular glycan type, although considerable effort is being directed toward methods that encompass all glycan classes.

Mass Spectrometry

High-resolution mass spectrometry (see Chapter 47) is the primary technique for characterizing the structures of individual glycans when only small quantities are available, as is the case in most glycomic studies. In a typical experiment, a glycoprotein- or glycolipid-enriched sample is prepared from cell lysates and analyzed by multiple rounds of mass spectrometry. In the case of glycoproteins, the N-glycans can be selectively released enzymatically or chemically, separated by HPLC (high-pressure liquid chromatography) methods, and actually sequenced. Separately, the O-glycans are released chemically and sequenced as well. Glycolipids can often be directly sequenced without separation of the lipid component. Glycosaminoglycans are more problematic because of their large size, but small fragments can be sequenced by mass spectrometry in conjunction with enzymatic digestion (see Chapter 16). An advantage of mass spectrometric glycan profiling is that multiple glycans of any given subtype can be profiled at once, increasing the throughput of the glycomic analysis. Still, there is no method at present by which highly complex samples possessing many glycan subtypes can be analyzed in one mass spectrometry experiment. Furthermore, many of the techniques routinely used tend to partially or completely destroy the sample or miss potentially important modifications such as sulfation and O-acetylation.

Mass spectrometry can also be employed to define sites of attachment of glycans to the underlying protein scaffold (i.e., for glycoproteomic analysis). Typically, the glycoprotein-associated glycans are first trimmed to remove peripheral epitopes such as sialic acid and fucose residues. This procedure simplifies the diversity of the pool and therefore sacrifices some of the information in the glycome. Then, the glycoprotein is subjected to tryptic digest and the peptides are analyzed by mass spectrometry using a technique that leaves the glycans attached to the peptide during the analysis (termed electron transfer dissociation [ETD] mass spectrometry). In parallel, the protein can be stripped of its glycans before the tryptic digest and the masses of those naked peptides can be compared to those of the glycopeptides. The differential allows prediction of the attached glycan structure. Furthermore, if an endoglycosidase such as PNGase F is used to release N-glycans (see Chapter 47), the resulting change from asparagine to aspartic acid can mark the site of the original glycosylation. Similarly, β-elimination of O-glycans changes the amino acid at the site of elimination, and can also be followed by Michael addition of nucleophiles such as dithiothreitol to mark the original O-linked sites.

A common problem encountered in proteomic studies is a lack of sensitivity for low-abundance species. The range of protein levels in cells and bodily fluids is thought to span more than eight orders of magnitude. Mass spectrometry detection often suffers from saturation by the most abundant species, leaving those with lower abundance impossible to detect. For glycoproteomic analyses, therefore, it is often beneficial to enrich the sample in glycoproteins and to discard those proteins that do not bear glycans. Lectins have been artfully employed for this purpose. As discussed in Chapter 45, a variety of glycan-binding proteins (e.g., the plant lectins) are commercially available. These proteins can be immobilized on agarose beads and used for affinity purification of glycoproteins from cell lysates or body fluids. Once enriched, the glycoproteins can be analyzed in the absence of abundant unglycosylated protein contaminants.

Lectin and Antibody Arrays

A major benefit of mass spectrometry is the detailed information it provides regarding the structure of a glycan. A drawback, however, is its relatively low throughput and the need for different experimental protocols for each glycan subtype. Lectin and antibody arrays can be employed to interrogate the glycome with much higher throughput, although in considerably less structural detail. As described in Chapter 45, nature has provided a considerable collection of lectins and many have been biochemically characterized. These lectins possess a range of specificities; some recognize a particular monosaccharide in virtually any context, whereas others are very specific for higher-order glycan epitopes or single residues within a defined context. For those epitopes that lack a naturally occurring lectin, one can generate monoclonal antibodies (see Chapter 45). This can be accomplished by immunization of a rodent with a synthetic or isolated version of the glycan attached to an antigenic carrier protein such as keyhole limpet hemocyanin. The monoclonal antibodies generated in this fashion can serve as "artificial lectins." Single-chain-fragment antibodies generated in bacteria have also proven useful for analysis of glycosaminoglycans. However, recent studies have noted that some antibodies once thought to be highly specific for certain glycans can cross-react with others.

The lectin array employs the same architecture as the DNA microarray or the glycan microarray described in Chapter 27. Lectins (or glycan-specific antibodies) are spatially arrayed on a glass chip by covalent attachment (Figure 48.2). Glycoproteins from the cell lysate or fluid sample of interest are nonspecifically labeled with a fluorescent dye. The sample is then incubated with the array and the fluorescence associated with each pixel is quantified. The pattern of bright spots reflects the glycome of the particular sample. For comparative purposes, two samples can be analyzed in parallel, one labeled with a green dye and one with a red dye. Combining the two samples onto one array allows direct analysis of changes in the glycome. In principle, samples that are even more complex than glycoproteins can be probed using lectin arrays, and indeed intact bacterial cells have been probed using this technique.

The lectin array provides global information about the types of glycan epitopes that are present in the sample but does not give any detailed structural information, nor does the experiment provide information regarding which proteins the glycans are attached to. However, the high-throughput platform allows for rapid comparison of many glycomes in search of global changes that might motivate further mass spectrometry studies.

Cell and Tissue Analysis Using Lectins and Antibodies

Glycobiologists routinely use lectins and glycan-specific antibodies as histological probes of glycan expression. This approach still holds an important place in any comprehensive glycomic analysis, as the cellular and tissue distribution of glycans is an important element of the glycome. In the modern era, tissue-expression patterns observed using lectins and antibodies can be correlated with lectin array data and mass spectrometry profiling data as well as genomic and proteomic data to create a more complete picture of the glycome. In the future, such in situ labeling followed by laser-capture microdissection of specific regions from tissue sections could potentially allow all the techniques to come together.

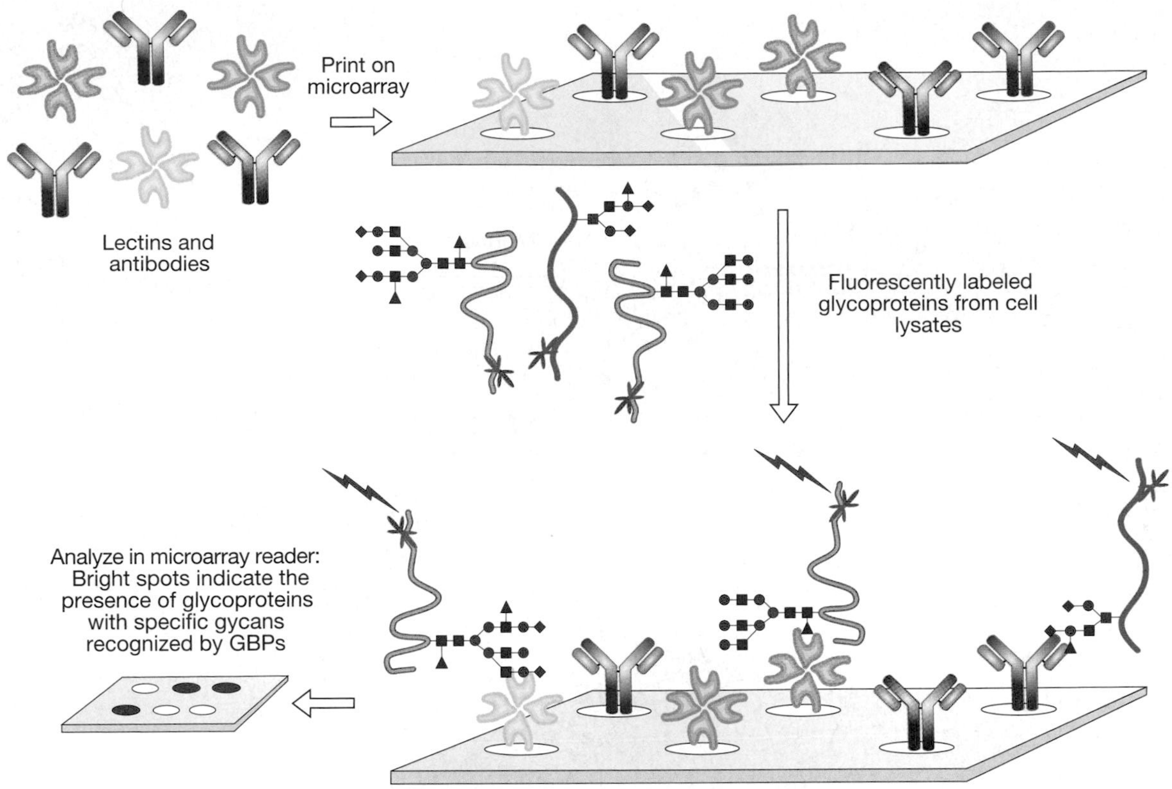

FIGURE 48.2. Analysis of cellular glycomes using lectin and antibody arrays. The arrays are generated by immobilization of lectins and glycan-specific antibodies on a chip. Glycoproteins from cell or tissue samples are labeled with a fluorescent dye and then incubated with the array. The fluorescent spots reflect the presence of glycoproteins bearing glycans recognized by the corresponding lectin or antibody. The technique provides minimal structural detail but permits rapid high-throughput analysis of many samples. Intact cells or virus particles can also be interrogated on lectin micrarrays.

Imaging the Glycome by Metabolic and Covalent Labeling

A recent addition to the arsenal of tools for glycome analysis is the use of metabolic labels that allow covalent tagging of glycans with imaging probes. As shown in Figure 48.3, cells or organisms can be treated with monosaccharide substrates bearing azido groups, and the downstream metabolic products are incorporated into cellular glycans. The azido groups can be covalently reacted with azide-specific imaging reagents. Once labeled, the glycans can be visualized on cells or tissues by fluorescence microscopy. This procedure has been used to image changes in the glycome during zebrafish embryonic development. The technique provides little structural information regarding the elements of the glycome that are labeled but has the advantage that global changes in the glycome can be monitored in vivo and in real time. By contrast, lectin and antibody reagents are largely restricted to ex vivo analysis of glycomes in cultured cells or tissues.

COMPARATIVE GLYCOMICS

Because the glycome is influenced by both genetic and environmental factors, the information contained therein might shed light on intraspecies and interspecies variations as well as on changes that have occurred over evolutionary history. From the immediate clin-

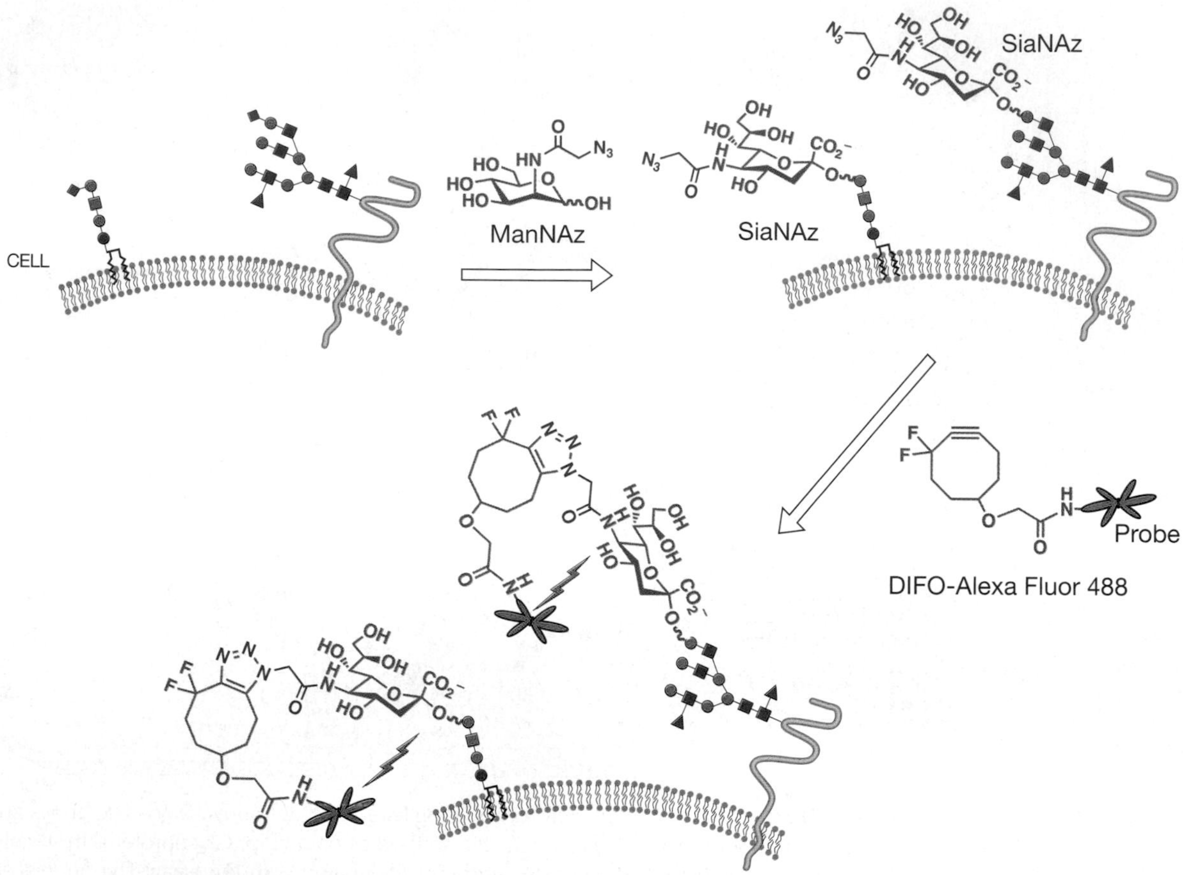

FIGURE 48.3. Metabolic and covalent labeling of glycans for in vivo imaging of the glycome. Cells metabolize azido sugars such as *N*-azidoacetylmannosamine (ManNAz), which is converted into the corresponding azido sialic acid (SiaNAz) and incorporated into cellular glycans. The azides are selectively reacted with a fluorescent probe conjugated to a cyclooctyne reagent termed "DIFO." The fluorescent dye-labeled glycans are then visualized. Using this two-step method, changes in the glycome can be probed within the physiologically authentic environment of a live organism.

ical perspective, the glycome might provide indicators of disease that can be used for diagnosis and for monitoring the efficacy of drugs. Comparative glycomics is therefore an exciting frontier in biology and medicine.

As discussed in detail in Chapter 44, numerous changes in the glycome have been associated with malignancy and metastasis, including altered N- and O-glycosylation, up-regulation of sialylated and fucosylated antigens, and altered heparan sulfates. In some cases, the tumor-associated glycans have been shown to be functionally relevant, whereas in other cases, the association of the glycan with cancer remains correlative. No matter its functional consequence, however, a change in the glycome that is highly correlated with malignancy (or any disease) can serve as a diagnostic biomarker. Given the dearth of such biomarkers for cancer screening, it is not surprising that considerable effort is being directed toward analysis of cancer-associated glycomes.

Already, mass spectrometry has been artfully employed in several glycomic studies of serum samples from healthy and cancer patients. In these studies, O-glycans were released from serum glycoproteins and analyzed by mass spectrometry. Although many structures were similar in the samples from healthy donors and cancer patients, a handful were markedly elevated in the latter. This observation suggests that changes in the serum glycome accompany cancer and that such changes might be used for diagnostic purposes. Notably, we do not need to understand *how* the disease of interest, in this case cancer, causes changes in the serum glycome in order to use that information for clinical benefit. Indeed, glycans that are altered in the disease sample might not be directly related to the disease at all. Rather, they may reflect downstream consequences of the disease on remote organs or they may reflect changes in the patient's immune system.

A fundamental question that remains unanswered is what is the extent of natural variation among individual human glycomes? Since the glycome can respond, in principle, to dietary and environmental changes, an equally interesting question is how glycomes around the world vary as a function of local dietary habits and/or medicine use, and further, how do glycome variations relate to acquired disease susceptibility (see Chapter 43)?

Studies of evolutionary biology have also much to gain from comparative glycomics. Evolution of the vertebrate immune system, for example, was accompanied by the acquisition of new glycan-binding proteins including the Siglec (see Chapter 32) and selectin (see Chapter 31) family members. The process by which the glycome evolved to accommodate these developments is of considerable interest. Likewise, the glycomes of microbes and their vertebrate hosts may have coevolved in some instances. The human blood group antigens, for example, are thought to reflect selective pressure induced by bacterial pathogens bearing similar glycan epitopes (see Chapter 39). This dramatic observation may be one of many examples of glycome coevolution across species. Comparative glycomic analysis of humans and their resident microbes, both pathogenic and symbiotic, may be highly revealing.

FUNCTIONAL GLYCOMICS USING GLYCAN MICROARRAYS

Taking inventory of the glycome provides the basis for hypotheses regarding biological function. Studying the functions of glycans is, of course, the central goal of the field of glycobiology. When this goal is pursued using high-throughput techniques, the term "functional glycomics" is often applied. As an example, the relative binding activity of glycans to glycan-binding proteins can be probed in high-throughput parallel fashion using glycan microarrays. These arrays seek to represent a fraction of the glycome and are often generated using glycans isolated from cells or tissue sources. The glycans are typically immobilized on a chip and then exposed to the protein of interest labeled with a fluorescent dye. Binding of the protein to the various glycans is detected by fluorescence imaging.

Arrays have been used for the analysis of a number of glycan-binding proteins, such as plant and microbial lectins, glycan-binding proteins involved in the innate and adaptive immune system, glycan-specific antibodies, viral glycan-binding proteins, and whole cells (see Chapter 27). Two examples of such applications are briefly described below.

DC-SIGN and DC-SIGNR

DC-SIGN and DC-SIGNR belong to the type II transmembrane receptor subfamily of C-type lectins (see Chapter 31). DC-SIGN is abundantly expressed on dendritic cells and plays a key role in adhesion of T cells as well as in the recognition of pathogens such as HIV. Indeed, binding of HIV to DC-SIGN on dendritic cells enhances T-cell infection. The related protein DC-SIGNR shares 77% sequence identity with DC-SIGN, but screening of these two similar proteins using glycan arrays revealed distinct ligand specificities. In addition to the high-mannose structures bound by both receptors, DC-SIGN recognized certain fucosylated ligands that were not bound by DC-SIGNR. This finding may shed light on the functional distinction between the two related proteins.

Influenza Virus Hemagglutinin

Influenza A virus subtypes are avian viruses that are named according to their surface antigens: hemagglutinin (HA) and neuraminidase (NA) (see Chapter 39). These viruses bind to sialylated glycans on host epithelial cells to initiate infection. The HA glycoprotein mediates host-cell recognition and is therefore an important determinant of species tropism. Human viral HA preferentially recognizes glycans terminated by NeuAcα2-6Gal, whereas avian HA preferentially recognizes glycans containing NeuAcα2-3Gal. Likewise, the upper airway epithelial cells in humans contain mainly NeuAcα2-6Gal, whereas in birds both the airways and intestine contain mainly NeuAcα2-3Gal linkages (see Chapters 13 and 14, and the cover figure).

In view of the rapid geographic spread of avian influenza A subtypes such as H5N1 and the increasing numbers of confirmed human cases, it is critical to survey potential influenza outbreaks and monitor human adaptation, which is a key step in the emergence of a pandemic virus. Glycan array technologies have proven to be powerful tools for this purpose. Arrays patterned with sialylated glycans of various linkages and topologies have been generated and screened with viral HA proteins as well as intact viruses. The microarray studies revealed striking glycan binding preferences that were governed not only by the sialic acid linkage and glycan modifications such as fucosylation or sulfation but also by underlying glycan structures. Once integrated into a portable format, glycan arrays may be employed in the field for influenza surveillance, which is a very practical application of functional glycomics techniques.

THE INFORMATICS CHALLENGES OF DIVERSE GLYCOMIC DATA

As mentioned above, a complete picture of the glycome can only be assembled using both top–down and bottom–up approaches. The structures of the individual glycans, their assembly on proteins and lipids, their distribution on cells and tissues, and their relation to each other on cells and within the extracellular matrix all warrant interrogation at the systems level. However, each corresponding experimental platform produces very different types of data. Mass spectral data, which can guide the assignment of a glycan's primary sequence, have a different form than lectin microarray data. The integration of disparate forms of data to generate a comprehensive picture of the glycome is a major frontier in informatics associated with glycomics research.

A comprehensive systems-level analysis would correlate data that define the glycome at various hierarchical levels with data derived from transcriptomic and proteomic experiments. One would like to know, for example, how the relative expression levels of genes that encode glycan biosynthetic enzymes compare to the glycome observed in a cell or tissue type. The effects of pharmacological agents or gene knockdowns on the glycome as compared to the transcriptome and proteome are also of interest. The functional consequences of perturbing a cell's glycome on its interactions with other cells might also be cataloged and correlated with additional systems-level data.

At present, we do not have a clear picture of how expression levels of glycan biosynthesis and processing genes relate, at the systems level, to the composition of the glycome. Efforts to correlate large data sets obtained from glycomic, transcriptomic, genomic, and proteomic studies have met with several challenges. Representation of glycan chemical structures is difficult because of their complexity and branching patterns. The use of single alphabet codes, as employed to describe nucleic acid and amino acid sequences, is not applicable to glycans. Rather than the conventional character-based codes used for sequence information in transcriptomic and proteomic data sets, numerical or object-based codes are better suited to link the complex glycan structure information in various glycomic data sets to each other and, ultimately, to the transcriptomic and proteomic data sets. The field is in need of a comprehensive bioinformatics platform that stores, integrates, and processes data from glycomic and other "omic" studies and disseminates them in a meaningful fashion via the Internet to the scientific community.

In recent years, academic and commercial organizations have made a significant effort toward building new databases and bioinformatics platforms that fulfill this goal (GlycoSuiteDB, Sweet, KEGG GLYCAN). International organizations have formed to develop community resources. The Consortium for Functional Glycomics (CFG), EuroCarb, and the Japanese Glycomics Consortia are collaborating to develop technologies for advancing glycomics. These collaborative efforts have resulted in the development of novel experimental resources as well as online searchable databases.

With growing effort directed toward new technology and bioinformatics platforms, the future of comprehensive "omics" science is bright and potentially exciting. The reformatting of existing genomic and proteomic data sets for compatibility with emerging glycomic data sets is under way and stand-alone glycan structure databases are already in place. Once achieved, the integration of large data sets that link the glycome, genome, transcriptome, and proteome will generate a wealth of hypotheses for pursuit by future generations of scientists.

FURTHER READING

Hirabayashi J. 2004. Lectin-based structural glycomics: Glycoproteomics and glycan profiling. *Glycoconj. J.* **21:** 35–40.

Campbell C.T. and Yarema K.J. 2005. Large-scale approaches for glycobiology. *Genome Biol.* **6:** 236.

Paulson J.C., Blixt O., and Collins B.E. 2006. Sweet spots in functional glycomics. *Nat. Chem. Biol.* **2:** 238–248.

Prescher J.A. and Bertozzi C.R. 2006. Chemical technologies for probing glycans. *Cell* **126:** 851–854.

Raman R., Venkataraman M., Ramakrishnan S., Lang W., Raguram S., and Sasisekharan R. 2006. Advancing glycomics: Implementation strategies at the consortium for functional glycomics. *Glycobiology* **16:** 82R–90R.

Sasisekharan R., Raman R., and Prabhakar V. 2006. Glycomics approach to structure–function relationships of glycosaminoglycans. *Annu. Rev. Biomed. Eng.* **8:** 181–231.

Pilobello K.T. and Mahal L.K. 2007. Deciphering the glycocode: The complexity and analytical challenge of glycomics. *Curr. Opin. Chem. Biol.* **11:** 300–305.

Timmer M.S., Stocker B.L., and Seeberger P.H. 2007. Probing glycomics. *Curr. Opin. Chem. Biol.* **11:** 59–65.

Turnbull J.E. and Field R.A. 2007. Emerging glycomics technologies. *Nat. Chem. Biol.* **3:** 74–77.

Mahal L.K. 2008. Glycomics: Towards bioinformatic approaches to understanding glycosylation. *Anticancer Agents Med. Chem.* **8:** 37–51.

CHAPTER 49

Chemical and Enzymatic Synthesis of Glycans and Glycoconjugates

Peter H. Seeberger, Nathaniel Finney, David Rabuka, and Carolyn R. Bertozzi

PURE GLYCANS OF DEFINED STRUCTURE ARE ESSENTIAL research tools in glycobiology. Unlike proteins and nucleic acids, which can be obtained in homogeneous forms using biological methods such as recombinant expression and polymerase chain reaction (PCR), glycans produced in biological systems are heterogeneous. Furthermore, the quantities that can be obtained from biological systems are rather small. Chemical synthesis can be used to obtain homogeneous glycans in larger quantities than are available from most cellular production systems. Enzymes can be used together with chemical methods to further simplify the process of glycan synthesis. Chemical synthesis can be further employed to incorporate glycans into homogeneous glycoproteins. This chapter summarizes the chemical and enzymatic synthetic methods to produce glycans and glycoproteins.

CHEMICAL SYNTHESIS OF GLYCANS

Controlling Regiochemistry

The inherent chemical properties of oligosaccharides make their synthesis more complex than the other major classes of biomolecules (i.e., oligonucleotides and oligopeptides). The fundamental challenge of glycan synthesis is the requirement for modifying one specific hydroxyl group in the presence of many others. Therefore, glycan synthesis is characterized by the manipulation of various protecting groups, chemical moieties that mask the hydroxyl groups and prevent them from reacting with other chemical reagents. Hydroxyl-protecting groups are selectively added and removed from glycan structures, allowing for chemical alteration of the exposed hydroxyl groups. Subsequently, in a typical glycan synthetic scheme, the exposed hydroxyl group serves as a point for further elaboration. The selective exposure of one hydroxyl group allows for regioselective addition of another monosaccharide unit. Synthetic schemes of this type are commonly applied to the generation of O- and N-linked glycans (see Chapters 8 and 9) as well as proteoglycans (see Chapter 16) and glycosphingolipids (see Chapter 10).

The choice of protecting groups and the sequence of protecting group installation are essential for a synthetic route to be successful. The most common protecting groups include persistent protecting groups, such as benzyl ethers that stay in place through many synthetic steps, and temporary protecting groups, such as esters that are removed during intermediate steps in a synthesis. A wealth of information exists on the chemical generation of glycans and various protecting group manipulations used in the context of glycan synthesis. The reviews listed at the end of the chapter provide a comprehensive overview of this subject.

Controlling Stereochemistry

A glycosidic linkage is formed through the activation of a glycosyl donor to create a reactive electrophilic species that couples with the glycosyl acceptor's hydroxyl group. This coupling reaction results in either the formation of an α- or β-linkage (see Chapter 2). A synthetic challenge in glycan synthesis is the stereospecific formation of glycosidic bonds (Figure 49.1A; Chapter 2). There are a variety of methods available to generate stereospecific glycosidic linkages. The yield and the stereochemical outcome of these reactions depend on the steric and electronic nature of the glycosylating agent (termed the glycosyl donor) and the nucleophilic hydroxyl group on the glycosyl acceptor. Good reaction yields are considered to range from 60% to 95%, and the ratios of the linkages at the anomeric position (α:β) vary from complete selectivity (all α or all β) to equal mixtures. A current method used to control the stereochemistry at the anomeric center involves the use of certain protecting groups, such as ester moieties, on the 2-hydroxyl group (Figure 49.1B). Under glycosylation reaction conditions, these "neighboring protecting groups" form a cyclic oxonium ion intermediate that shields one face of the molecule, leading exclusively to the formation of a "*trans*" glycosidic linkage (i.e., where the anomeric substituent and the C-2 hydroxyl group are on opposite sides of the ring, as in β-glucosides). The opposite anomeric stereochemistry, termed a *cis* glycosidic bond, is more difficult to construct with high specificity because neighboring group participation is not possible. In the absence of such protecting group participation, mixtures of anomers are often formed.

A Representative Chemical Glycan Synthesis

The synthesis of the tumor-associated glycan antigen Le^x-Le^y is one example of the construction of a complex glycan. Figure 49.2 illustrates the synthetic scheme followed for the prepara-

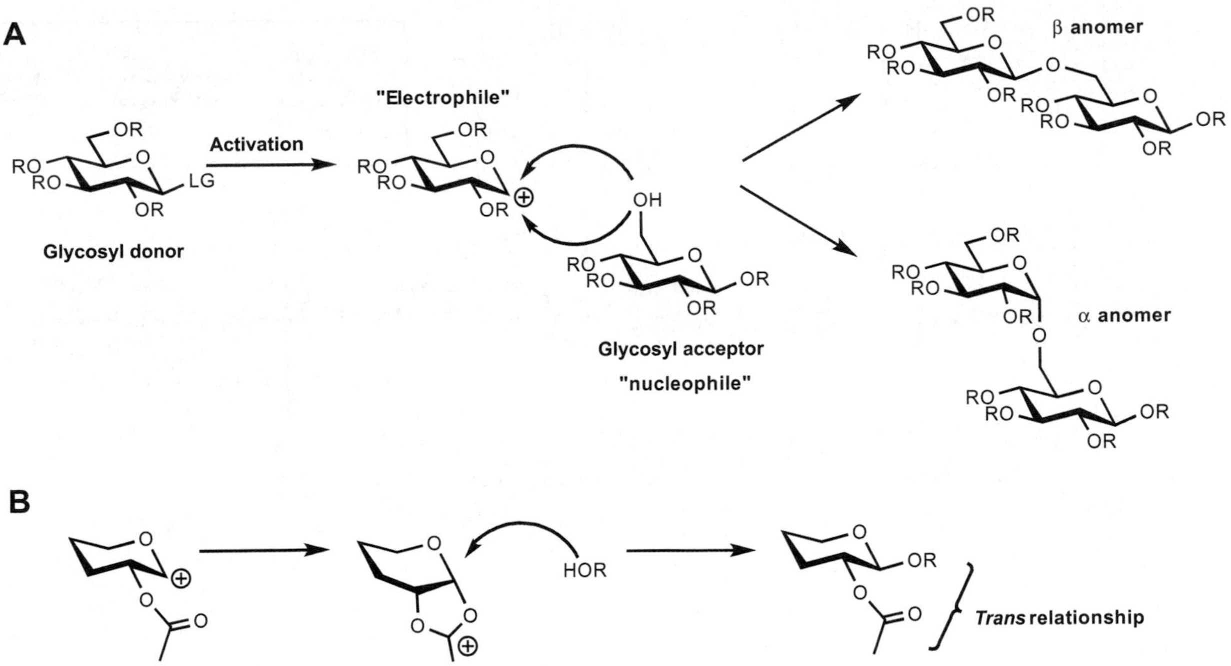

FIGURE 49.1. (*A*) Stereospecific formation of glycosidic bonds as either an α or β linkage. (LG) Leaving group. (*B*) Formation of a cyclic oxonium ion intermediate leading to the formation of a β-glycosidic linkage.

tion of this nonasaccharide and depicts the transformations required in a typical glycan synthetic procedure. These synthetic routes are often time consuming and technically challenging. Indeed, the synthesis of Le^x-Le^y took many months and the overall yield was less than 1%. This synthetic scheme also highlights a major difficulty with glycan synthesis—that is, isolating glycans via chromatography after individual coupling steps. Because of the low yields and complex purifications, traditional methods are not amenable to rapid and efficient synthesis of glycans.

AUTOMATED METHODOLOGY FOR GLYCAN SYNTHESIS

Unlike proteins and nucleic acids, for which automated synthesizers are commercially available, the glycans cannot yet be synthesized using a general commercial system. However, a number of methodologies are currently being developed by synthetic chemists to address the inherent challenges in glycan chemistry that have prohibited automation. Two different approaches to the automation of glycan assembly are presented in this chapter: (1) the one-pot solution-phase synthesis and (2) the solid-phase synthesis method.

One-pot Glycan Synthesis

The one-pot approach uses a computer program to automate glycan synthesis. This strategy exploits differences in the anomeric reactivity of a large set of diverse thioglycoside building blocks. The thioglycoside possesses an –SR group at the anomeric center, rather than the natural –OR group. These derivatives are stable under ambient conditions, but they can be converted to reactive leaving groups using reagents that interact with the thioether functionality (i.e., *N*-iodosuccinimide and dimethylthiosulfonium triflate).

More than 400 thioglycoside analogs with a wide spectrum of protecting groups on the monosaccharide hydroxyl groups have been prepared and characterized with respect to

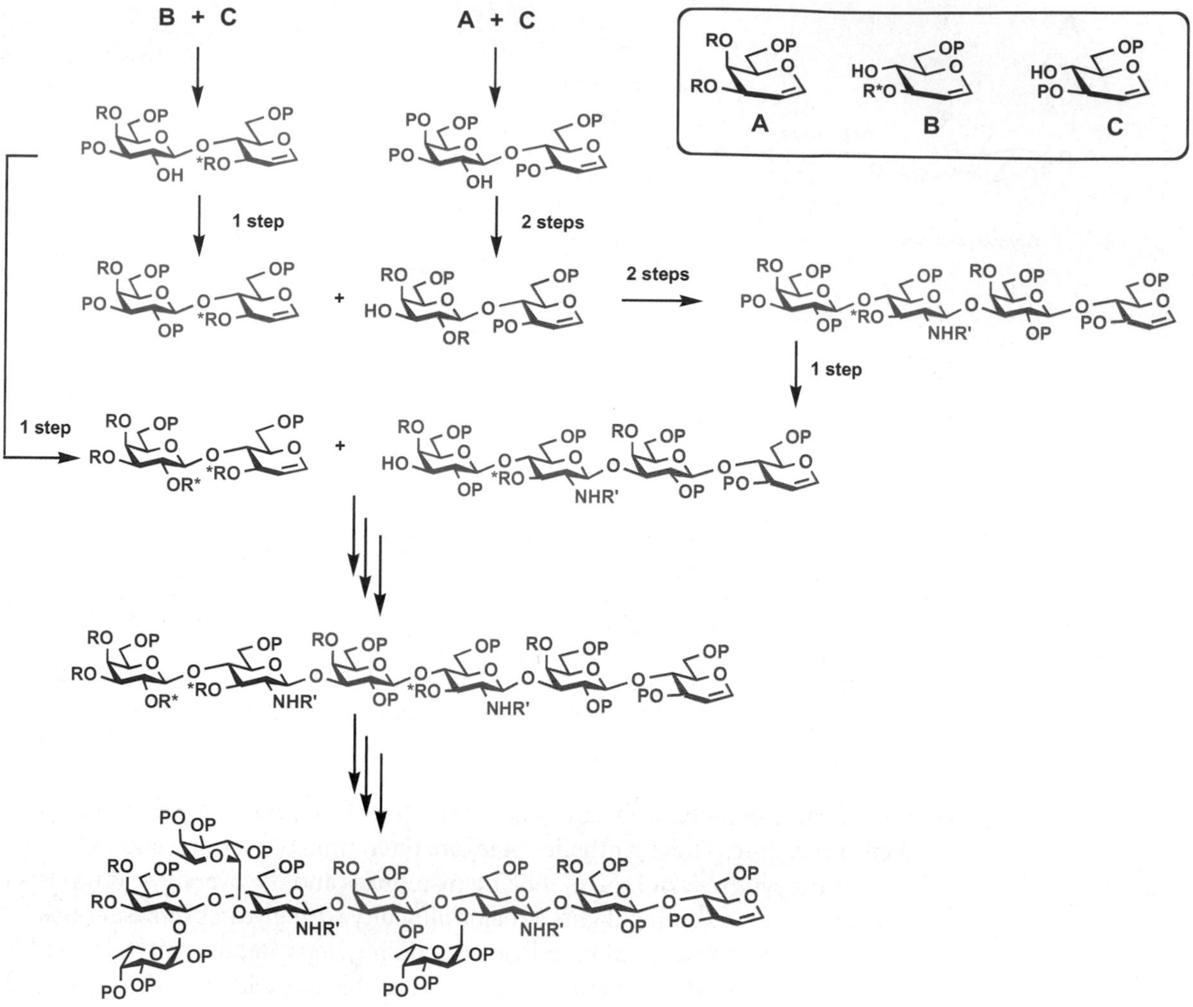

FIGURE 49.2. Various transformations in a typical complex glycan synthesis. (P) Protecting group; (R) P or H; (R′) nitrogen-protecting group; (R*) unique hydroxyl-protecting group and fucosylation site.

their relative reactivity. This study produced a table of relative reactivity values (RRVs) that could be programmed into a computer. A computer program, Optimer, allows the user to input the desired glycan structure, and, in return, the program displays an appropriate combination of building blocks needed to generate that glycan. The program also dictates the stereochemistry as α- or β-directing. Once Optimer provides the synthetic blueprint, the chemist proceeds to follow the procedure by sequential addition of the building blocks to a single reaction vessel. Typically, the building blocks are added starting from the most reactive and ending with the least reactive. The final glycan is globally deprotected and purified.

The assembly of the cancer antigen Globo H, a hexasaccharide, illustrates the power of this streamlined approach (Figure 49.3). The key advantage to using a one-pot approach is the rapid planning and ordering of the synthetic steps. In addition, a programmable one-pot synthesis eliminates intermediate purification steps utilized in traditional synthetic schemes. Depending on the glycan, the entire process of one-pot synthesis can take from minutes to days. However, the solution one-pot approach requires the availability of a large number of monosaccharide building blocks. These building blocks are not yet commercially available, and expertise in chemical synthesis is required to generate them.

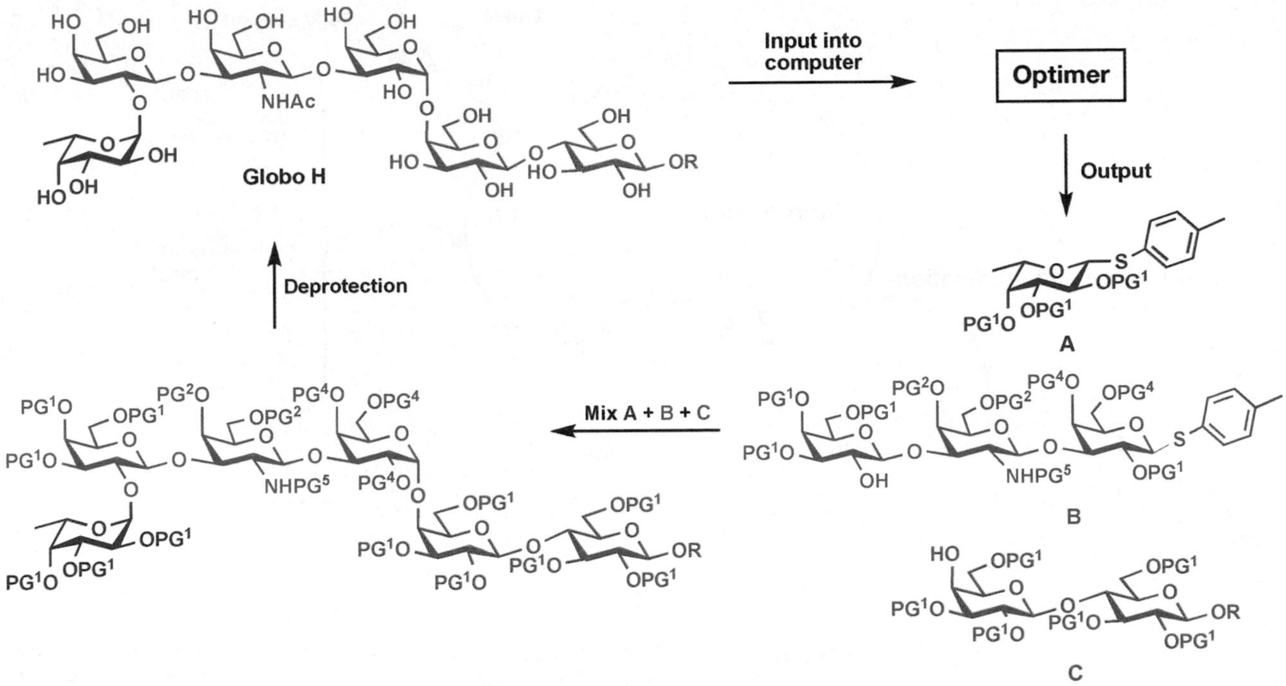

FIGURE 49.3. Synthesis of complex glycan structures using one-pot methodology. (PG) Protecting group; (LG) leaving group.

Solid-phase Glycan Synthesis

A major breakthrough in peptide synthesis was the development of solid-phase synthetic methodology. This development in synthetic chemistry made peptides available to biologists and facilitated studies of their structures and functions. Likewise, the availability of a general solid-phase method for glycan synthesis could vastly accelerate access to homogeneous material for biological studies. Progress toward this end is well under way.

The first automated glycan synthesizer was constructed by modifying a peptide synthesizer, and the synthetic process closely resembles protocols established for the generation of polypetides (Figure 49.4). The first monosaccharide is attached through its reducing end via a cleavable linker to a polystyrene bead. Upon removal of a temporary protecting group, the hydroxyl group to be modified in the coupling reaction is exposed. Addition of a glycosyl donor together with an activator, typically a reagent that generates a good leaving group at the anomeric position, produces the new glycosidic linkage. An excess amount of the glycosyl donor can be used to drive the reaction to completion. Side products, reagents, and unreacted starting material are washed away while the desired glycan remains covalently attached to the polymer support. Next, another protecting group is selectively removed to provide another site for subsequent monosaccharide addition.

The assembly of the tumor-associated antigen Le^x-Le^y using a peptide synthesizer and glycosyl phosphate building blocks illustrates the utility of the solid-phase approach (Figure 49.4). In contrast to the solution-phase synthesis of Le^x-Le^y, which took months (see above), solid-phase synthesis of the nonasaccharide took less than 2 days, including cleavage of fully protected material from the solid support. Removal of the protecting groups required an additional 3 days.

The major advantage of solid-phase synthesis is that excess reagents can be used to drive each reaction to completion without the worry of purifying these reagents from the biopolymer, which is attached to solid support. Similar to the one-pot synthetic approach, only one

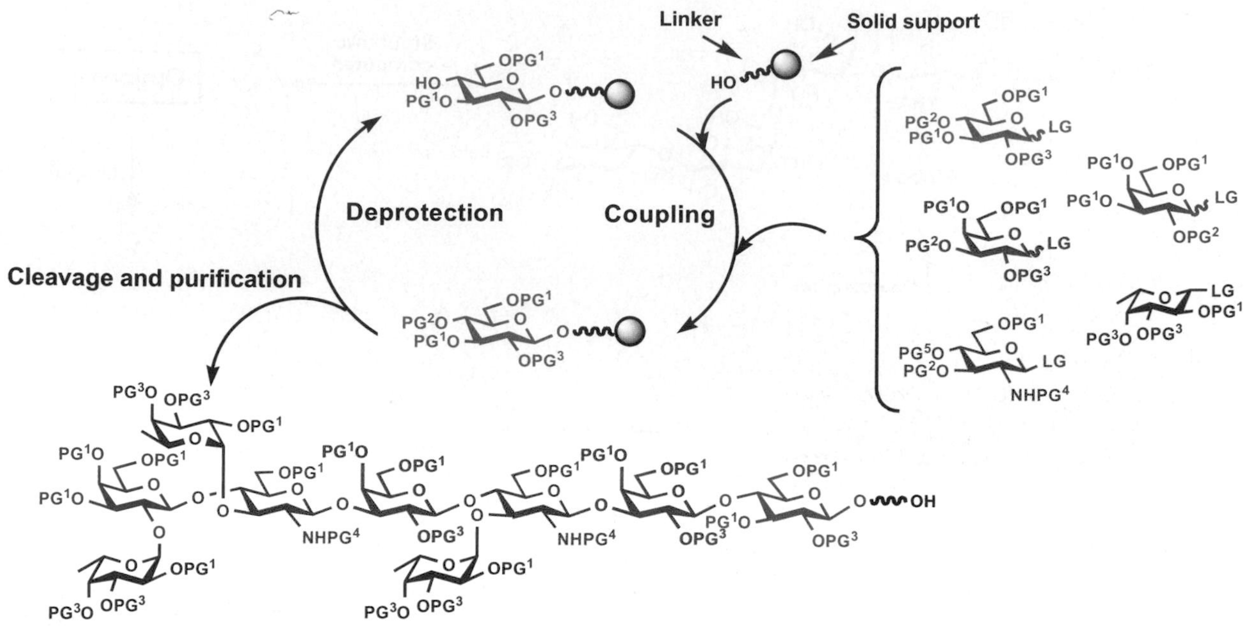

FIGURE 49.4. Solid-phase glycan synthesis using a peptide synthesizer to generate the carbohydrate antigen Le^x-Le^y. (PG) Protecting group; (LG) leaving group.

purification step is required (after the polymer is liberated from the solid phase), which is often the most time-consuming aspect of glycan assembly. The bottleneck of automated solid-phase synthesis is the need for protected monosaccharide building blocks that are not yet commercially available. This issue will need to be addressed in order for solid-phase glycan synthesis to achieve the generality and accessibility of solid-phase peptide synthesis.

ENZYMATIC SYNTHESIS OF GLYCANS

Nature has addressed the challenges of glycan synthesis by evolving enzymes that couple monosaccharides with exquisite regio- and stereospecificity. These enzymes can be harnessed by chemists to catalyze the formation of glycosidic bonds in a synthetic context (see Chapter 5 for an introduction to glycosyltransferases). In addition to glycosyltransferases, glycosidases whose natural function is the cleavage of glycosidic bonds (see Chapters 5 and 41) may be manipulated to "run in reverse" and function as glycosylating enzymes.

Employing glycosyltransferases and glycosidases in a synthetic scheme has strengths and weaknesses that are in many ways complementary to those of chemical synthesis. Chemical synthesis allows for the preparation of diverse natural and unnatural structures but requires the extensive use of protecting groups and the preparation of specialized precursor compounds. In contrast, enzymes are typically much less flexible and/or available but do not require protecting groups or elaborate precursors, and they construct glycosidic linkages with perfect regio- and stereochemical control. Many successful applications of enzymes in glycan production have utilized "chemoenzymatic" syntheses, relying on a hybrid of chemical and enzymatic steps that typically begins with chemical synthesis and ends with enzymatic extension.

Synthesis of Glycans Using Glycosyltransferases

The majority of glycosyltransferases used in glycan synthesis catalyze the transfer of a glycosyl donor to a sugar or amino acid acceptor. These transferases generally utilize nucleotide

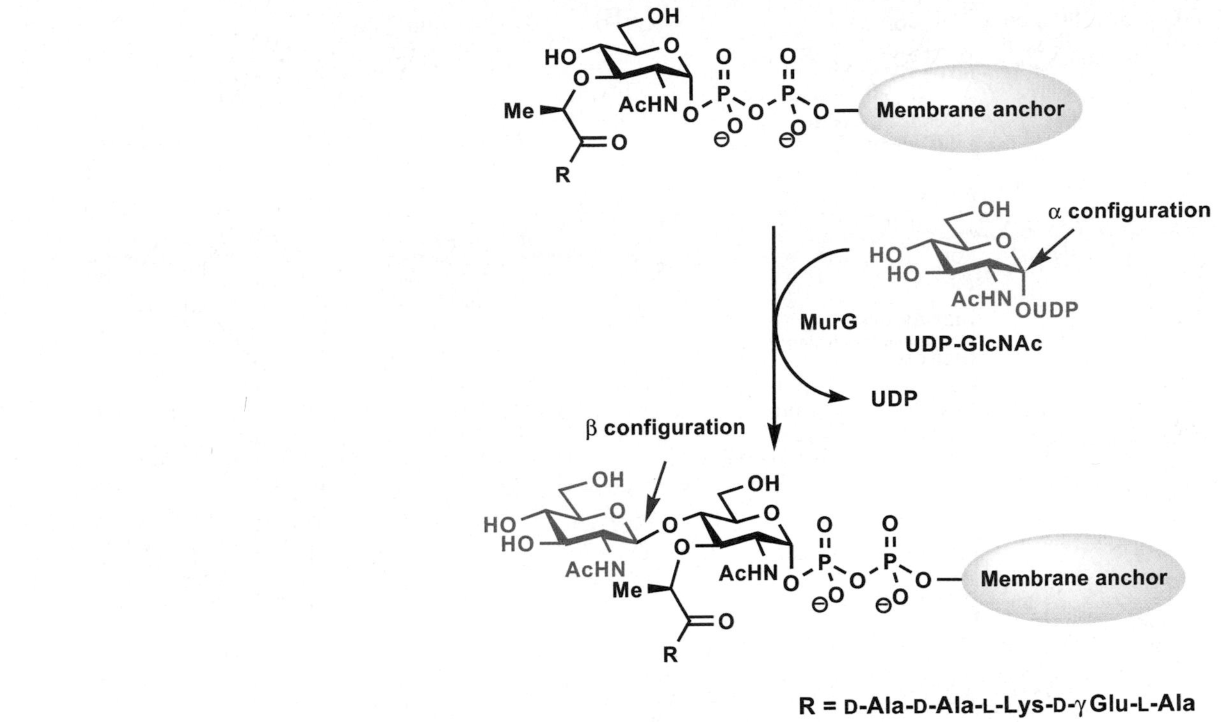

FIGURE 49.5. Example of an inverting GlcNAc transferase.

diphosphate sugar donors (e.g., UDP-GlcNAc), with occasional exceptions such as sialyltransferases, which require CMP-sialic acids. The glycosyltransferases are subdivided into two classes termed retaining and inverting (see Chapter 5). Retaining glycosyltransferases transfer the donor with retention of stereochemistry at the anomeric position, whereas inverting glycosyltransferases transfer the donor with inversion at the anomeric position. A representative example is MurG, the inverting GlcNAc transferase responsible for the final step in bacterial cell wall biosynthesis (Figure 49.5; see Chapter 20). Notable aspects of this glycosylation event are the regiospecificity (only a single hydroxyl group is glycosylated) and complete control of anomeric stereochemistry (only the β-anomer is produced).

Figure 49.6 shows a comparison of the enzymatic and chemical synthesis of tetra- and pentasaccharide cell-surface epitopes from *Neisseria meningitidis* (a causative agent of meningitis). The enzymatic synthesis of the pentasaccharide (Figure 49.6A) relies on the use of three recombinant enzymes: a GlcNAc transferase, a fusion protein combining UDP-Glc epimerase and Gal transferase activities, and a fusion protein combining CMP-Neu5Ac synthase and transferase activities. A lactose derivative was carried through three sequential reactions, all giving remarkable yield. The GlcNAc transfer (96% yield) was followed by in situ epimerization of UDP-Glc to UDP-Gal, Gal transfer (96% yield), in situ synthesis of CMP-Neu5Ac from CTP and Neu5Ac, and subsequent sialylation producing the final pentasaccharide (97% yield).

For comparison, a chemical synthesis of the tetrasaccharide began with the same lactose derivative (Figure 49.6B). Conversion to a suitably protected lactose-based glycosyl acceptor in three steps gave reasonable (55%) yield. Subsequent coupling with a lactose-based glycosyl donor (available in nine steps from lactose) generated the protected tetrasaccharide in modest (40%) yield. The fully deprotected tetrasaccharide was obtained after two further steps (66% yield). In this example, the tetrasaccharide was not sialylated because of difficulties with the final chemical transformations. This is a significant limitation because sialic acid appears as a terminal sugar of many mammalian glycans.

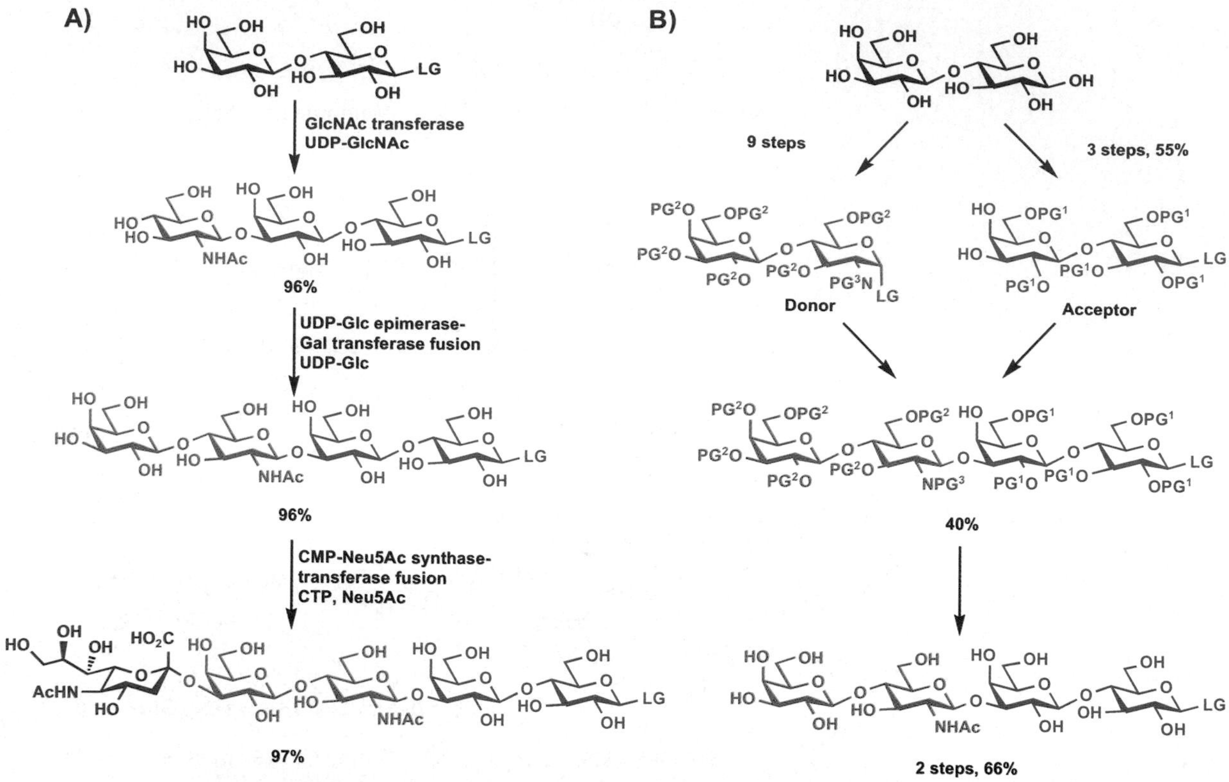

FIGURE 49.6. A comparison of the enzymatic (*A*) and chemical (*B*) syntheses of tetra- and pentasaccharide glycans. (PG) Protecting group; (LG) leaving group.

However, the synthesis of complex glycans with terminal sialyl residues can be readily achieved using sialyltransferases. Overall, the enzymatic procedure was far superior to the chemical counterpart.

The potential advantages of glycan synthesis using glycosyltransferases are immediately obvious. The reactions give high yields, with complete regio- and stereospecificity, without the need for installation and removal of numerous protecting groups. However, glycosyltransferases have a number of significant limitations. First is the issue of availability. All of the enzymes used in the above synthesis were discovered, cloned, overexpressed, and purified in the laboratory and would not otherwise be readily available. However, although obtaining enzymes can be problematic, many glycosyltransferases are currently available from commercial sources and a growing number of academic labs. A second limitation is the regio- and stereospecificity of the enzymes. Glycosyltransferases generally carry out one reaction nearly perfectly, but they are unable to tolerate even minor variations in donor or acceptor (one exception will be noted below). This means that every novel synthetic transformation requires a new enzyme to be isolated, cloned, and purified. In contrast, although chemical methods are seldom as selective or high yielding, they may be applied to the preparation of a wide range of glycans. A third limitation of glycosyltransferases is that their nucleotide sugar substrates are often unstable and/or very expensive (e.g., CMP-Neu5Ac). Although this problem can be addressed by in situ nucleotide sugar generation, such reactions increase the complexity of the enzymatic process. Finally, many glycosyltransferases are strongly inhibited by the nucleotide generated following sugar transfer to the acceptor. This problem can be addressed by using excess nucleotide sugar, enzymatic degradation of the nucleotide by an added phosphatase, or nucleotide recycling in which the nucleotide is enzymatically converted back into the nucleotide sugar.

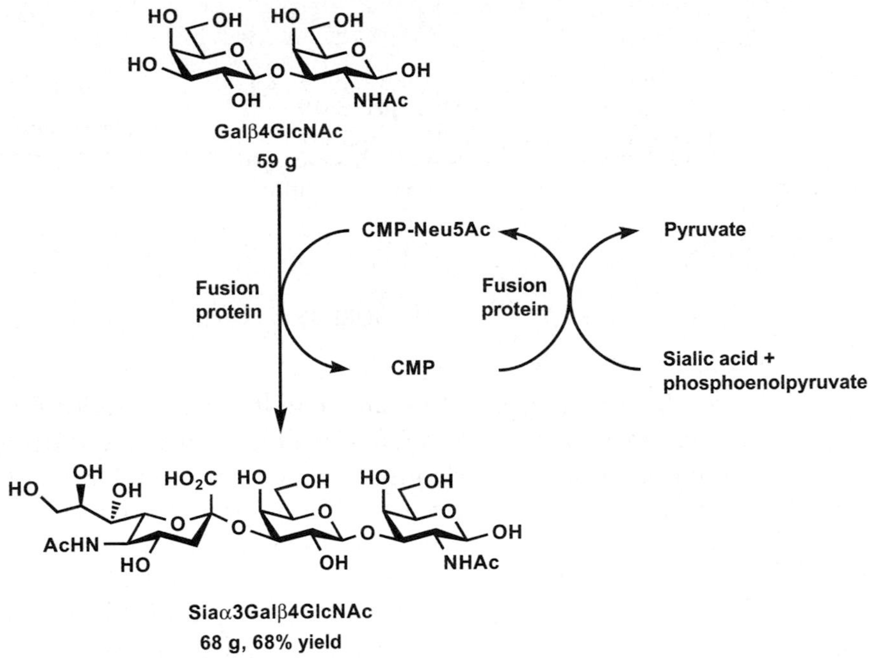

FIGURE 49.7. Example of multigram enzymatic synthesis.

Generating large amounts of a desired glycan with enzymatic reactions is also problematic. Although chemical processes can be carried out on gram or even multi-ton scales, enzymes are scarce and/or too expensive to be feasible alternatives. However, the issue is ameliorated to some extent because most research applications require only small amounts (10–100 mg) of the glycan of interest, and once the challenges associated with enzyme identification and overexpression have been addressed, enzymatic syntheses can be viable on a large scale. This is illustrated by the preparation of a trisaccharide related to the *N. meningitidis* epitopes described above, which was carried out on an approximately 100-g scale in a manner that allowed reisolation and reuse of the enzyme (Figure 49.7). The cost and stability of the nucleotide sugar were addressed by in situ regeneration of the requisite CMP-Neu5Ac, and the complexity of the system was reduced by use of a fusion protein containing two enzyme activities. Smaller-scale experiments with the same fusion enzyme demonstrated some tolerance for variation of both the donor and acceptor components of the reaction. Additionally, Galβ1-4GlcNAcβ-(Fl), Galβ1-4Glcβ-(Fl), and Galα1-4Galβ1-4Glcβ-(Fl) (where Fl is a fluorescent tag) were all found to be effective substrates for glycosylation. In addition, the enzyme was found to transfer Neu5Propyl and Neu5Gc as effectively as Neu5Ac. So far, enzymatic synthesis has proved to be the only viable and cost-effective approach for the commercial-scale preparation of sialyl Lewis^{x} (Chapter 6).

Synthesis of Glycans Using Glycosidases

Glycosidases provide an alternative to the use of glycosyltransferases for the enzymatic synthesis of glycans. Glycosidases usually cleave glycans into smaller fragments or monosaccharides (see Chapter 5). They are separated into two categories: endoglycosidases, which cleave internal glycosidic bonds of glycans, and exoglycosidases, which cleave of the nonreducing terminal sugar of a glycan. A representative example of an exoglycosidase is the β-*N*-acetylhexosaminidase involved in a late stage of hyaluronan degradation (Figure 49.8).

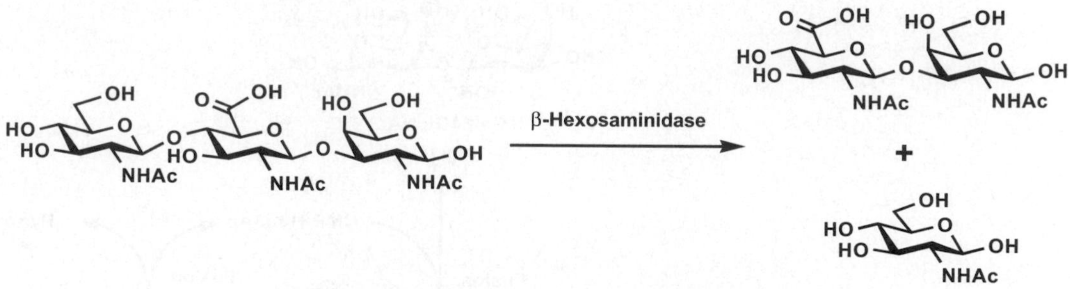

FIGURE 49.8. Example of an exoglycosidase.

In general, glycosidases are more stable and readily available than glycosyltransferases and are also more tolerant to variations in substrate structure. To use glycosidases for synthetic purposes, their normal function must be reversed. All enzymatic processes are (formally) equilibrium processes. It is, in principle, possible to force a glycosidase to run in reverse by exposing the enzyme to a large excess of the reaction products and allowing the system to reach equilibrium. However, because the reverse reaction is invariably endothermic, the equilibrium will always favor the cleavage products. Replacing the anomeric hydroxyl group of the glycosyl donor fragment with a good leaving group, such as *para*-nitrophenol (PNP), shifts the equilibrium further toward the glycosylation product. Figure 49.9 illustrates the glycosidase-mediated synthesis of two sialyl-Tn antigen epitopes. Beginning with the methyl ester of GalNAcα-*N*-Ac-threonine, treatment with Gal-β-PNP in the presence of the β-galactosidase from bovine testes provides the corresponding disaccharide in modest yield (22%) with complete regio- and stereoselectivity. Subsequent treatment with the PNP derivative of sialic acid (Neu5Ac α-PNP) in the presence of *Vibrio cholerae* sialidase provided the α2-6-linked sialylated product in low yield but with excel-

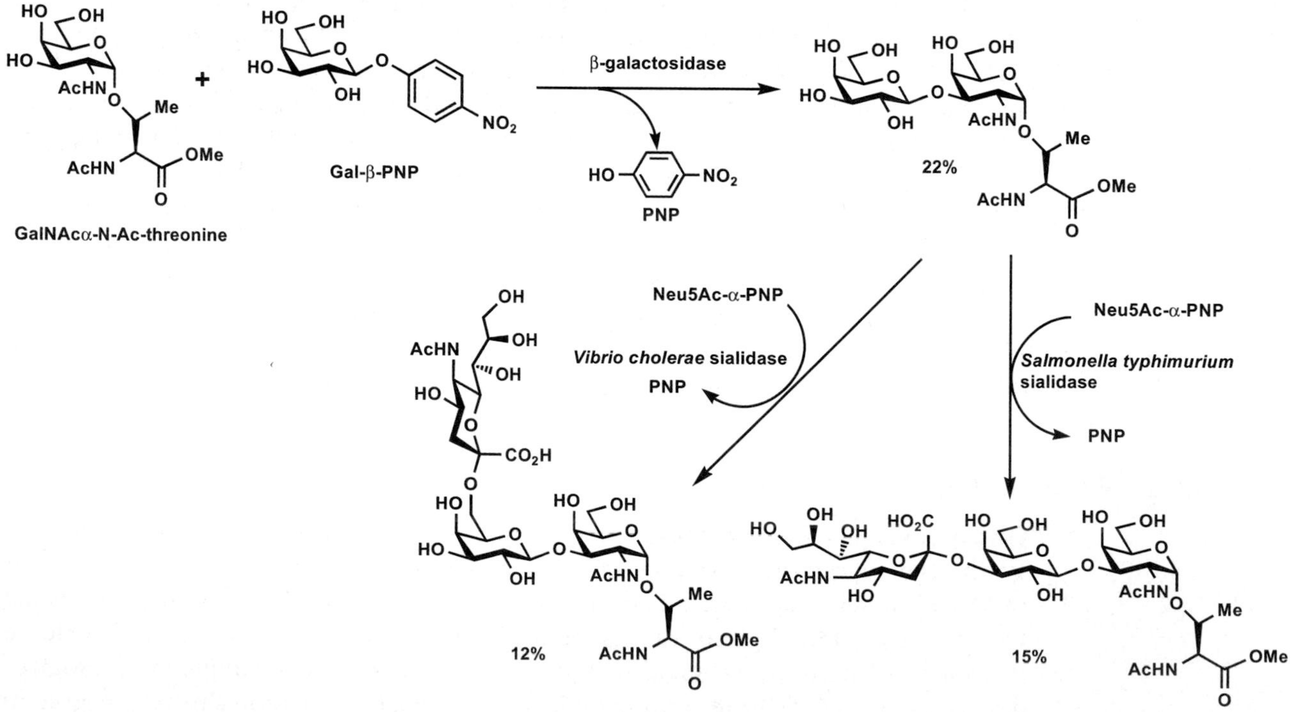

FIGURE 49.9. The glycosidase-mediated synthesis of two sialyl-Tn antigen epitopes.

lent regio- and stereoselectivity (only trace amounts of isomeric products could be detected). Likewise, *Salmonella typhimurium* sialidase provided the α2-3-linked trisaccharide, again in low yield, but with high regio- and stereoselectivity.

Oligosaccharide synthesis with glycosidases retains three of the advantages associated with glycosyltransferases—that is, good-to-excellent regio- and stereoselectivities without extensive protecting group manipulation. Additional benefits include greater stability and accessibility of glycosidases and greater tolerance to variations in substrate structure. Although the yields of glycosidase-based reactions are lower than glycosyltransferase-catalyzed processes, there are transformations for which the requisite glycosyltransferase is not available. In such cases, the use of glycosidases can complement transferase-based methods for synthesis.

SYNTHESIS OF GLYCOCONJUGATES

In nature, glycans are often found in the context of glycoproteins. The heterogeneity of these protein-associated glycans has interfered with protein structure elucidation and complicated attempts to determine the effects of the glycan on protein function. Therapeutic glycoproteins, the fastest growing class of pharmaceutical reagents, are reliant on their glycans for pharmacokinetic properties and stability. The biological and biophysical study of glycoproteins could benefit tremendously from the availability of homogeneously glycosylated material. With the development of improved methods for glycan synthesis, chemists have recently taken on the challenge of synthesizing whole glycoproteins using combinations of chemical and enzymatic methods.

Solid-phase Glycopeptide Synthesis

Short polypeptides (<50 residues) can be readily prepared by solid-phase peptide synthesis (SPPS). Glycans can be incorporated through the glycosylation of amino acid building blocks prior to synthesis of the polypeptide biopolymer. The glycosylated amino acid is often referred to as a "cassette." The hydroxyl groups of the glycan component are typically protected as acetyl or benzoyl esters, which are easily removed with base after the completion of peptide assembly. Cassette strategies have been widely used in the synthesis of glycopeptides from mucin-type glycoproteins (Chapter 9) and can be complemented by both the one-pot and solid-phase glycan synthesis methodologies.

As mentioned earlier in the chapter, glycosyltransferases are powerful tools for the construction of defined carbohydrate structures. The enzymatic transfer of individual monosaccharides to preformed glycopeptides is an excellent alternative to the synthesis of large glycosyl–amino acids. After glycopeptide synthesis using a glycosyl–amino acid cassette, the glycan is deprotected and then elaborated with glycosyltransferases. The synthesis of a PSGL-1 fragment (see Chapter 31) is an elegant example of using both chemical and enzymatic methodologies to generate a complex glycopeptide (Figure 49.10). The peptide backbone was generated following solid phase peptide synthetic procedures with the N-acetylgalactosidase residue incorporated using a cassette strategy. Following the cleavage and deprotection of the glycopeptide, the appropriate glycosyltransferases were used to elaborate the glycan to generate the desired hexasaccharide. Sulfation was also carried out enzymatically to sulfate the tyrosine amino acid residues, generating the final desired compound. Thus, by employing a synthetic cassette strategy in tandem with enzymatic elaboration, a complex glycoconjugate was synthesized that would otherwise be prohibitively difficult to generate.

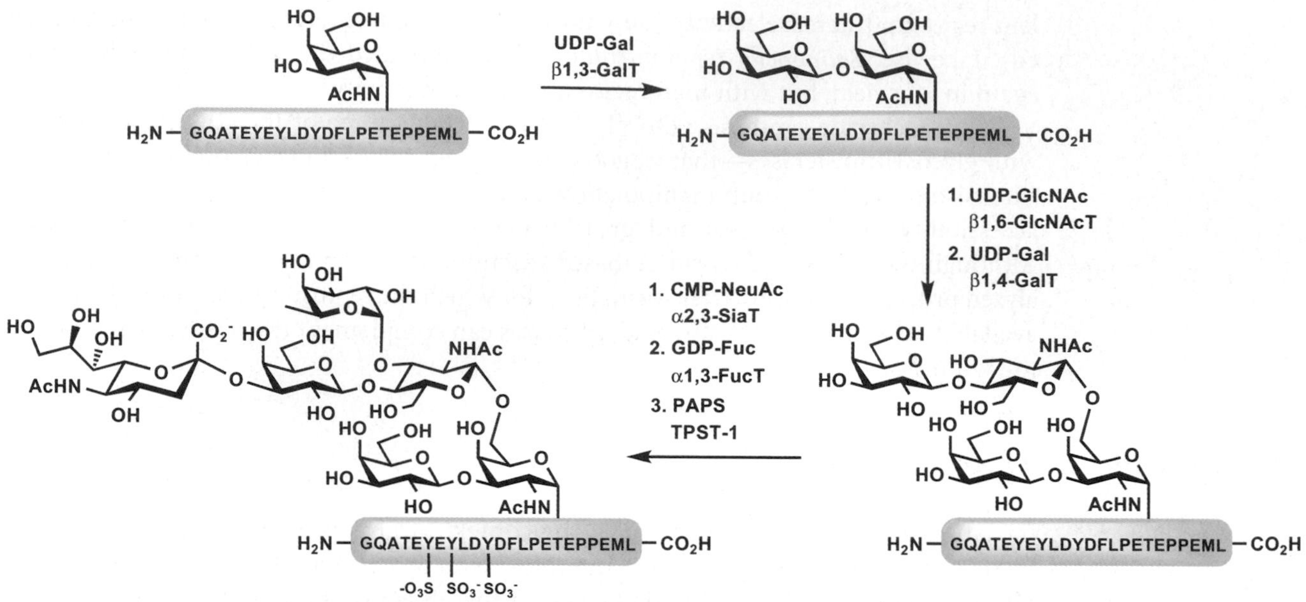

FIGURE 49.10. The chemoenzymatic synthesis of a PSGL-1 glycopeptide fragment. PAPS, 3′-phosphoadenosine 5′-phosphosulfate; TPST-1; tyrosylprotein sulfotransferase 1.

Synthesis of Glycopeptides by Native Chemical Ligation

The synthesis of larger glycoproteins (i.e., >50 amino acid residues) cannot be achieved using this linear synthetic strategy. The field of protein chemistry has provided a convergent approach to the synthesis of larger proteins, termed native chemical ligation (NCL) (Figure 49.11). This technique involves the condensation of two unprotected peptide fragments, one bearing a carboxy-terminal thioester and the other an amino-terminal cysteine residue. The reaction is mild, selective, and compatible with the presence of glycans on the fragments. Thus, SPPS with glycosyl–amino acid cassettes can be used to generate glycopeptide fragments, which are then assembled by NCL into larger glycoproteins.

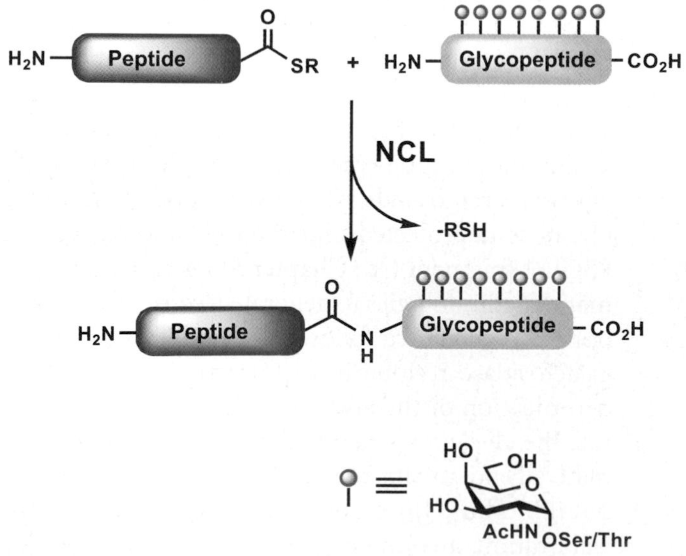

FIGURE 49.11. Native chemical ligation (NCL).

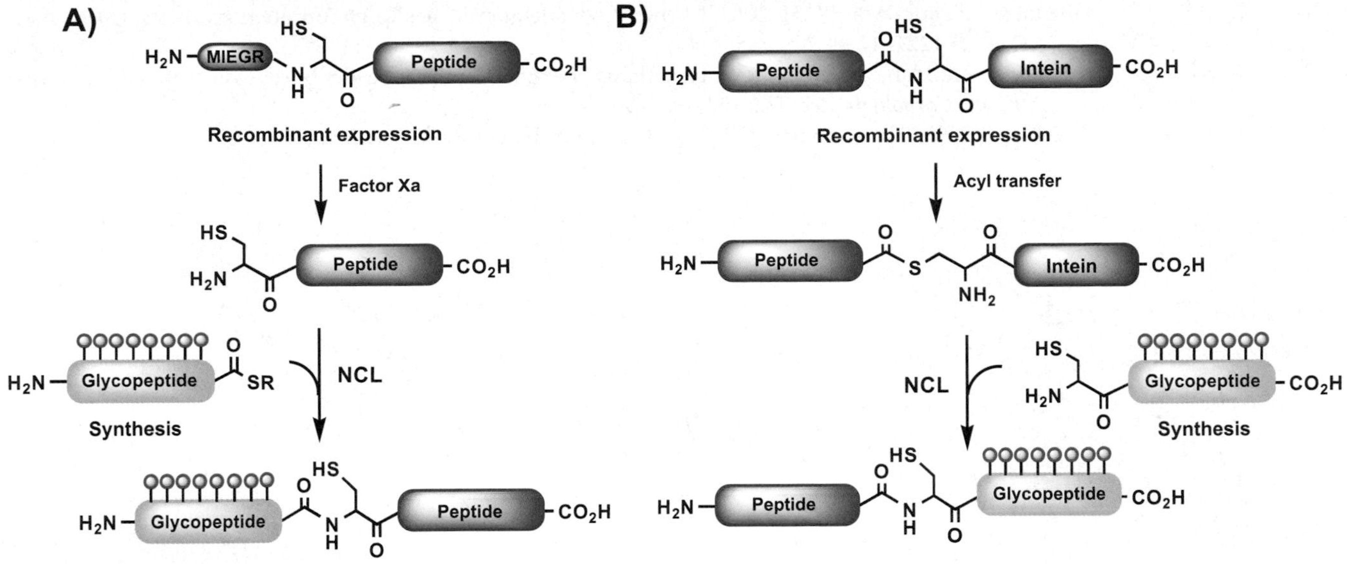

FIGURE 49.12. General approach to glycoprotein synthesis using expressed protein ligation (EPL). (*A*) Assembly of glycoproteins bearing glycans near the amino terminus. (*B*) Assembly of glycoproteins bearing glycans near the carboxyl terminus.

The Semi-synthesis of Glycopeptides by Expressed Protein Ligation

The synthesis of proteins larger than 20 kD in size using native chemical ligation is extremely difficult given the limitations on the size of the fragments that are made by SPPS. To access glycoproteins beyond this size, expressed protein ligation (EPL) has been artfully employed (Figure 49.12). This method generates recombinant proteins with carboxy-terminal thioesters using the natural phenomenon of intein-mediated protein splicing. Thus, a recombinant protein bearing a carboxy-terminal thioester can be coupled by the NCL reaction to a synthetic glycopeptide adorned with an amino-terminal cysteine residue. Alternatively, a recombinant protein encoding an amino-terminal cysteine residue can be ligated to a synthetic glycopeptide functionalized with a carboxy-terminal thioester. This modular combination of recombinant technology and chemical synthesis has produced full-length glycoproteins with homogeneous glycans.

FURTHER READING

Hanessian S. 1997. *Preparative carbohydrate chemistry.* Marcel Dekker, New York.

Wellings D.A. and Atherton E. 1997. Standard Fmoc protocols. *Methods Enzymol.* **289:** 44–67.

Koeller K.M. and Wong C.-H. 2000. Synthesis of complex carbohydrates and glycoconjugates: Enzyme-based and programmable one-pot strategies. *Chem. Rev.* **100:** 4465–4493.

Vocadlo D.J. and Withers S.G. 2000. Glycosidase-catalyzed oligosaccharide synthesis. *Carbohydrates Chem. Biol.* **2:** 723–844.

Marcaurelle L.A. and Bertozzi C.R. 2002. Recent advances in the chemical synthesis of mucin-like glycoproteins. *Glycobiology* **12:** 69R–77R.

Seeberger P.H. 2003. Automated carbohydrate synthesis to drive chemical glycomics. *Chem. Commun.* **2003:** 1115–1121.

Pratt M.R. and Bertozzi C.R. 2005. Synthetic glycopeptides and glycoproteins as tools for biology. *Chem. Soc. Rev.* **34:** 58–68.

Werz D.B. and Seeberger P.H. 2005. Carbohydrates as the next frontier in pharmaceutical research. *Chem. Eur. J.* **11:** 3194–3206.

Liu L., Bennett C.S., and Wong C.-H. 2006. Advances in glycoprotein synthesis. *Chem. Commun.* **2006:** 21–33.

Bennett C.S. and Wong C.H. 2007. Chemoenzymatic approaches to glycoprotein synthesis. *Chem. Soc. Rev.* **36:** 1227–1238.

Faijes M. and Planas A. 2007. In vitro synthesis of artificial polysaccarides by glycosidases and glycosynthases. *Carbohydr. Res.* **342:** 1581–1594.

Seeberger P.H. 2008. Automated oligosaccharide synthesis. *Chem. Soc. Rev.* **37:** 19–28.

CHAPTER 50

Chemical Tools for Inhibiting Glycosylation

Jeffrey D. Esko and Carolyn R. Bertozzi

THIS CHAPTER DISCUSSES VARIOUS TYPES OF INHIBITORS, including natural products, substrate-based tight-binding inhibitors, glycoside primers, monosaccharide analogs for chemically tagging glycans, inhibitors found through screening chemical libraries, and examples of rationally designed inhibitors based on three-dimensional structures of enzymes (Table 50.1).

ADVANTAGES OF INHIBITORS

Chapters 41, 42, and 46 describe various natural and induced mutants with defects in glycosylation. These mutants have helped to define genes that encode various transferases and glycosidases, and in some cases alternate biosynthetic pathways have been uncovered. Mutants also provide insights into the functions of glycosylation in cells and tissues as well as models for human inborn errors in metabolism and disease. However, one limitation of studying mutants is that the analyses are usually restricted to the cell or organism from which the mutant strain was isolated. Additionally, many mutations are lethal in animals, which makes the study of the gene in adult animals more difficult.

TABLE 50.1. *Classes of inhibitors*

Class of inhibitor/modulator	Target
Metabolic inhibitors	steps involved in formation of common intermediates such as PAPS or nucleotide sugars
Tunicamycin	N-linked glycosylation through inhibition of dolichol-PP-GlcNAc formation; peptidoglycan biosynthesis through inhibition of undecaprenyl-PP-GlcNAc assembly
Plant alkaloids	N-linked glycosylation through inhibition of processing glycosidases
Substrate analogs	specific glycosyltransferases or glycosidases
Glycoside primers	glycosylation pathways by diverting the assembly of glycans from endogenous acceptors to exogenous primers
Tagged monosaccharides	different glycosylation pathways

Inhibitors of glycosyltransferases and glycosidases provide another approach for studying glycosylation in cells, tissues, and whole organisms that avoids some of the problems associated with studying mutants. Many of these compounds are small molecules that are taken up readily by a variety of cell types and can be absorbed through the gut, which provides an opportunity for designing drugs to treat human diseases and disorders correlated with altered glycosylation (see Chapter 51). Because the field is quite large, only those inhibitors that act on specific enzymes or metabolic pathways and that illustrate certain basic concepts are discussed here (Table 50.1). Agents that block protein/carbohydrate interactions are surveyed in Chapter 27, and the aminoglycoside antibiotics are discussed briefly in Chapter 51.

INDIRECT INHIBITORS AND METABOLIC POISONS

A number of inhibitors have been described that block glycosylation by interfering with the metabolism of common precursors or intracellular transport activities. Some of these compounds act indirectly by impeding the transit of proteins between the endoplasmic reticulum (ER), Golgi, and *trans*-Golgi network. For example, the fungal metabolite brefeldin A causes retrograde transport of Golgi components located proximal to the *trans*-Golgi network back to the ER. Thus, treating cells with brefeldin A separates enzymes located in the *trans*-Golgi network from those found in the ER and Golgi and uncouples the assembly of the core structures of some glycans from later reactions, such as sialylation or sulfation. The drug can be used to examine if two pathways reside in the same compartment or share the same enzymes. Because the localization and array of the enzymes vary considerably in different cell types, extrapolating the effects of brefeldin A from one system to another is often difficult.

Some inhibitors act at key steps in intermediary metabolism where precursors involved in glycosylation are formed. For example, a glutamine analog, 6-diazo-5-oxo-L-norleucine, blocks glutamine:fructose-6-phosphate amidotransferase, the enzyme that forms glucosamine from fructose and glutamine (see Chapter 4). Depressing glucosamine production in this way has a pleiotropic effect on glycan assembly because all of the major families contain *N*-acetylglucosamine or *N*-acetylgalactosamine. Chlorate is another type of general inhibitor that blocks sulfation. The chlorate anion (ClO_4^{2-}) is an analog of sulfate (SO_4^{2-}) and it forms an abortive complex with the sulfurylase involved in the formation of phosphoadenosine-5′-phosphosulfate (PAPS), the active sulfate donor for all known sulfation reactions. Thus, treating cells with chlorate (usually 10–30 mM) inhibits sulfation by more than 90%, but the effect is not specific for any particular class of glycan or sulfation reaction (e.g., tyrosine sulfation is also affected).

A number of sugar analogs have been made with the hope that they might exhibit selective inhibition of glycosylation. 2-Deoxyglucose and fluorinated analogs of sugars (3-deoxy-3-fluoroglucosamine, 4-deoxy-4-fluoroglucosamine, 2-deoxy-2-fluoroglucose, and 2-deoxy-2-fluoromannose) inhibit glycoprotein biosynthesis as measured by decreased incorporation of radiolabeled glucosamine or fucose, but the mechanism underlying the inhibitory effect is unclear in many cases. Early studies of 2-deoxyglucose showed that the analog was converted to UDP-2-deoxyglucose as well as to GDP-2-deoxyglucose and dolichol-P-2-deoxyglucose. Inhibition of glycoprotein formation apparently occurs as a result of accumulation of various dolichol oligosaccharides containing 2-deoxyglucose, which cannot be elongated or transferred to glycoproteins normally. Care must be taken in interpreting the results of experiments employing these compounds because they may have effects due to metabolic conversion into other reactants.

TUNICAMYCIN: INHIBITION OF DOLICHOL-PP-GlcNAc ASSEMBLY

A number of natural products have been found to alter glycosylation. Tunicamycin belongs to a class of nucleoside antibiotics composed of uridine, an 11-carbon disaccharide called aminodeoxydialdose (tunicamine), and a fatty acid of variable length (13–17 carbons), branching, and unsaturation (Figure 50.1). Tunicamycin was first identified in *Streptomyces lysosuperificus*, and related compounds were found later in other microorganisms. It derives its name from its antiviral activity, which occurs by inhibiting viral coat (or "tunica") formation.

Tunicamycin inhibits N-glycosylation in eukaryotes by blocking the transfer of *N*-acetylglucosamine-1-phosphate (GlcNAc-1-P) from UDP-GlcNAc to dolichol-P (catalyzed by GlcNAc phosphotransferase; GPT), thereby decreasing the formation of dolichol-PP-GlcNAc (see Chapter 8). Other GlcNAc transferase reactions are not inhibited (e.g., GlcNAcTI–V), but the transfer of GlcNAc-1-P to undecaprenyl-P and the formation of undecaprenyl-PP-MurNAc pentapeptide (which is involved in bacterial peptidoglycan biosynthesis) are sensitive to tunicamycin (see Chapter 20). Tunicamycin acts as a tight-binding competitive inhibitor, presumably because it resembles the donor nucleotide sugar. The K_i value for tunicamycin is about 5×10^{-8} M, whereas the K_m value for UDP-GlcNAc is approximately 3×10^{-6} M. The actual amount of tunicamycin needed to inhibit glycosylation varies in different cells (0.1–10 µg/ml), possibly because of variable uptake and culture conditions or differences in the level of expression of the phosphotransferase. Tunicamycin is cytotoxic to cells, and resistant mutants overproduce GPT. Similarly, transfection of cells with the cloned GPT confers resistance, suggesting that the variable dose of inhibitor required in different cells may reflect variation in enzyme levels.

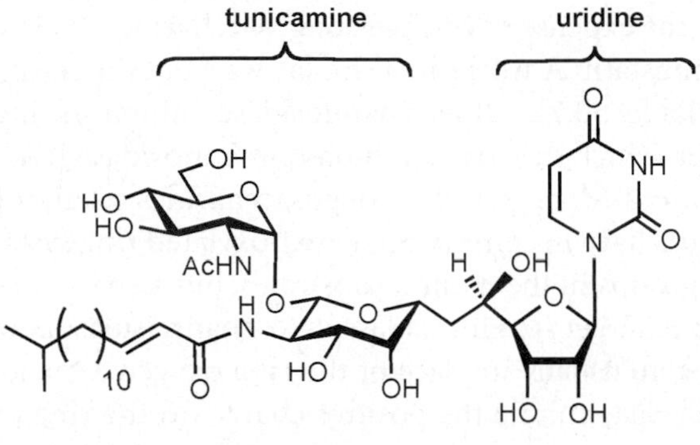

FIGURE 50.1. Structure of tunicamycin, which consists of uridine conjugated to the disaccharide, tunicamine.

Tunicamycin has been used extensively for studying the role of N-glycans in glycoprotein maturation, secretion, and function, as documented by thousands of papers citing its use since its first discovery in 1973. The drug was shown to induce apoptosis preferentially in cancer cells, presumably because of alterations in glycosylation of various cell-surface receptors and signaling molecules. Although the mechanism underlying its apoptosis-inducing activity may be complex, the observation suggests the possibility of a chemotherapeutic treatment based on inhibiting glycosylation. Inhibition of N-glycan formation could be useful for treating patients with naturally occurring mutations that create N-glycosylation sites in cell-surface receptors (gain-of-glycosylation mutants; see Chapter 42).

Another compound called amphomycin inhibits dolichol-P-mannose synthesis by forming a complex with dolichol-P. This compound is a lipopeptide and apparently forms complexes with the carrier lipid. Other lipophilic compounds have been described that bind lipid intermediates in bacterial cell wall synthesis (see Chapter 20).

PLANT ALKALOIDS: NATURAL INHIBITORS OF GLYCOSIDASES

Plant alkaloids block N-linked glycosylation by inhibiting the processing glycosidases involved in trimming nascent chains (Table 50.2). Unlike tunicamycin, which blocks glycosylation of glycoproteins entirely, the alkaloids inhibit the trimming reactions that occur after the $Glc_3Man_9GlcNAc_2$ oligosaccharide is attached to a glycoprotein (see Chapter 8). Thus, the inhibitory alkaloids generally cause the appearance of glycoproteins on the cell surface lacking the characteristic termini found on mature N-glycans (see Chapter 13). One class of alkaloids inhibits the α-glucosidases involved in the initial processing of the N-glycans and in quality control of protein folding (see Chapters 29 and 36). This class includes castanospermine (from the seed of the Australian chestnut tree, *Castanosperum australe*), which inhibits both α-glucosidases I and II; australine (also from *C. australe*), which preferentially inhibits α-glucosidase I; and deoxynojirimycin (from *Streptomyces* species), which preferentially inhibits α-glucosidase II (Table 50.2). As expected, castanospermine and australine cause accumulation of fully glucosylated chains, whereas deoxynojirimycin results in chains containing one to two glucose residues. Treating cells with these inhibitors revealed that some trimming of the mannose residues could occur independently of removal of the glucose residues (see Chapter 8).

Swainsonine was first discovered in plants from the western United States (*Astragalus* species; also known as locoweed) and Australia (*Swainsona canescens*), and it was later found in the fungus *Rhizoctonia leguminocola* that infects red clover. Consumption of these plants causes a severe abnormality called locoism and accumulation of glycoproteins in the lymph nodes. Thus, swainsonine is part of a chemical defense strategy used by plants against grazing animals (and probably insects). Swainsonine inhibits α-mannosidase II, causing the accumulation of high-mannose oligosaccharides ($Man_4GlcNAc_2$ and $Man_5GlcNAc_2$) and hybrid-type chains at the expense of complex oligosaccharides. It also inhibits the lysosomal α-mannosidase. Mannostatin A works in a similar way, but differs significantly in structure from swainsonine (Table 50.2). Other mannosidase inhibitors include deoxymannojirimycin and kifunensin, which selectively inhibit α-mannosidase I. As expected, these agents cause the accumulation of $Man_{7–9}GlcNAc_2$ oligosaccharides on glycoproteins.

All of these inhibitors have in common polyhydroxylated ring systems that mimic the orientation of hydroxyl groups in the natural substrates, but a strict correlation between stereochemistry and enzyme target (α-glucosidase vs. α-mannosidase) does not exist. The compounds contain nitrogen, usually in place of the ring oxygen. One idea is that the nitrogen in the protonated state may mimic the positive charge on the ring oxygen that arises from

TABLE 50.2. *Examples of alkaloids that inhibit glycosidases involved in N-linked glycan biosynthesis*

Alkaloid	Source	Target
Australine	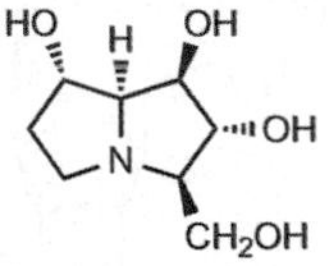	α-glucosidase I
Castanospermine	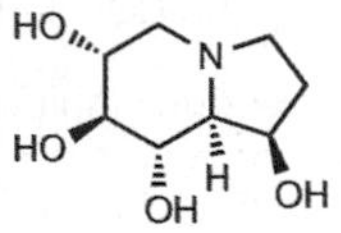	α-glucosidase I and II
Deoxynojirimycin	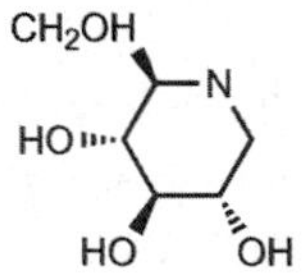	α-glucosidase II (and I)
Deoxymannojirimycin	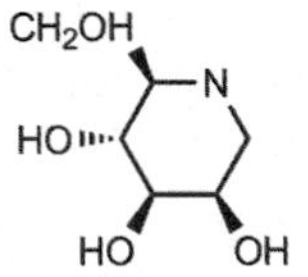	α-mannosidase I
Kifunensin	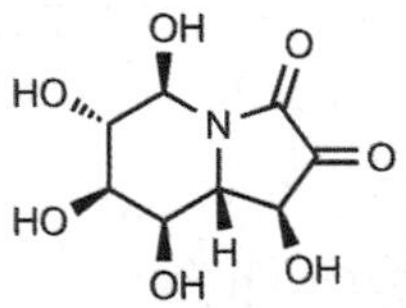	α-mannosidase I
Swainsonine	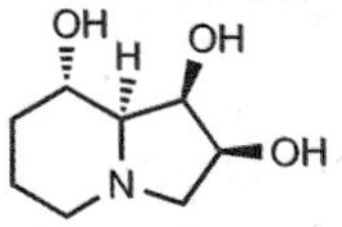	α-mannosidase II
Mannostatin A	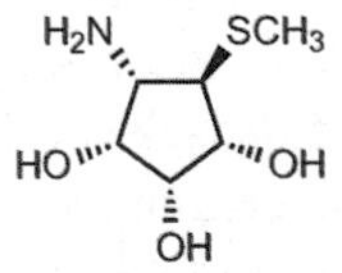	α-mannosidase II

delocalization of charge from the tentative carbocation at C-1 generated during the hydrolysis reaction. Crystal structures for the α-mannosidase are available with bound inhibitors.

Alkylated and acylated analogs of the alkaloids have been made and shown to have interesting and useful properties. N-Butylation of deoxynojirimycin actually converts the glucosidase inhibitor into an inhibitor of glycolipid biosynthesis. Alkylation of the amino group or acylation of the hydroxyl groups can raise the potency of the compound, presumably by facilitating uptake across the plasma and Golgi membranes. Some of these compounds have shown positive effects as drugs for treating diabetes, lysosomal storage diseases, cancer, and HIV infection and for inducing male sterility (see Chapter 51).

INHIBITION OF O-GalNAc INITIATION OF MUCIN-TYPE GLYCANS

In contrast to the biosynthesis of N-linked glycans, comparatively few inhibitors are available that block O-linked glycans. Mucin-type O-linked glycan biosynthesis is initiated by

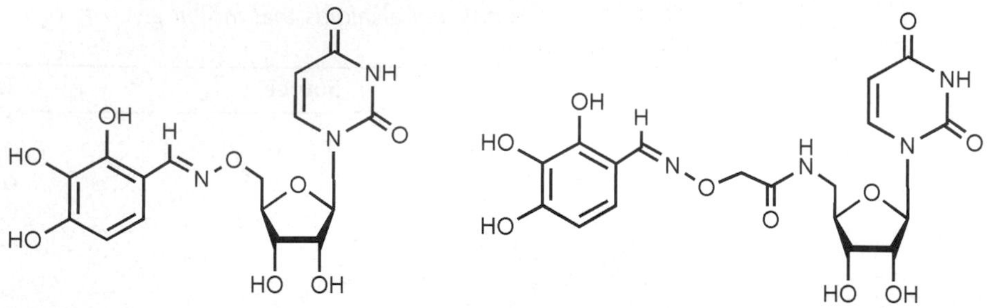

FIGURE 50.2. Broad-spectrum inhibitors of the ppGalNAcTs identified from screening a uridine-based library.

polypeptidyl N-acetylgalactosaminyltransferases (ppGalNAcTs), a large family of enzymes that use UDP-GalNAc as a common donor and various glycoprotein acceptors (see Chapter 9). A synthetic library of uridine analogs was screened against members of the ppGalNAcT family, which led to the identification of two compounds that disrupt O-GalNAc addition (Figure 50.2). These compounds have K_i values of approximately 8 μM with respect to UDP-GalNAc. Like tunicamycin, these inhibitors suppress glycosylation without selectivity for different glycoprotein targets. These compounds represent the first generation of enzyme inhibitors that work on O-linked glycans, and suggest the possibility of obtaining compounds that inhibit specific ppGalNAcT isoforms as well as other types of O-linked glycans, such as O-xylose (see Chapter 16), O-glucose (see Chapter 17), and O-GlcNAc (see Chapter 18).

INHIBITION OF O-GLCNAC MODIFICATION

With the realization of the importance of O-GlcNAc addition to many cytoplasmic and nuclear proteins (see Chapter 18), great interest exists in developing agents to inhibit its addition by O-GlcNAc transferase (OGT) or its removal by O-GlcNAc-specific β-hexosaminidase (O-GlcNAcase). Although alloxan and streptozotocin affect O-GlcNAc addition, these compounds lack specificity. The first potentially useful inhibitors of OGT have been obtained by screening chemical libraries for compounds that can displace a fluorescent derivative of the donor sugar, UDP-GlcNAc. The active compounds do not block other *N*-acetylglucosamine addition reactions, for example one involved in formation of the polysaccharide backbone of bacterial peptidoglycan (see Chapter 20).

Several inhibitors of the O-GlcNAcase have been devised based on the structure of *N*-acetylglucosamine. The first compound in this class, PUGNAc (*O*-[2-acetamido-2-deoxy-D-glucopyranosylidene]amino-*N*-phenylcarbamate; Figure 50.3) inhibits O-GlcNAcase at nanomolar concentrations, but also inhibits lysosomal β-hexosaminidases (HexA and HexB; see Chapter 41). *N*-Acetylglucosamine-thiazoline (NAG-thiazoline) is more specific and inhibits at even lower concentrations than PUGNAc. A new rationally designed glucoimidazole, GlcNAcstatin, inhibits O-GlcNAcase with a K_i of 4.6 pM and exhibits 10^5-fold selectivity over HexA and HexB. These compounds inhibit the enzyme in cultured cells, providing new tools to study the function of O-GlcNAc, and are potential candidates for drug therapy.

SUBSTRATE ANALOGS: DIRECTED SYNTHESIS OF INHIBITORS

A number of inhibitors of specific transferases have been developed based on the concept that substrate analogs might act as tight-binding inhibitors. The general strategy is to modi-

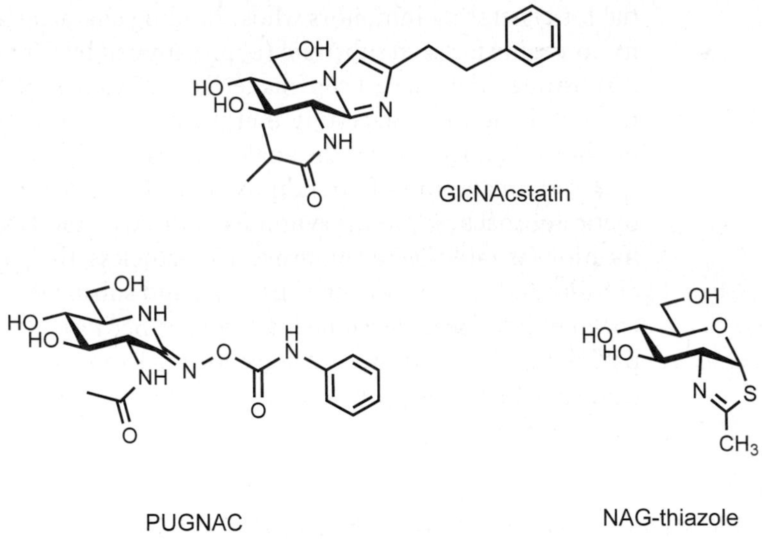

FIGURE 50.3. Inhibitors of O-GlcNAc-specific β-hexosaminidase (O-GlcNAcase).

fy the hydroxyl group that acts as the nucleophile during formation of the glycosidic bond or groups in its immediate vicinity (Table 50.3). About half of the compounds that have been made lack inhibitory activity, presumably because modification of the targeted hydroxyl group prevents binding of the analog to the enzyme by interfering with hydrogen bonding networks that hold the substrate in place. In other cases, the analogs exhibit K_i values in the approximate range of the K_m values for the unmodified substrate. As one might expect, the analogs usually act competitively with respect to the unmodified substrate, but in a few cases the inhibition pattern is more complex, suggesting possible binding outside the active site.

Analogs of acceptors have the advantage that they should be selective for specific enzymes. However, analogs of nucleotide sugars provide opportunities for blocking classes of enzymes that utilize a common donor (e.g., all fucosyltransferases utilize GDP-fucose). A large number of nucleotide sugar derivatives have been made (e.g., N- and O-substituted analogs of UDP-GalNAc) and several exhibit inhibitory activity in vitro. "Bisubstrate" analogs consist of the nucleoside sugar donor or PAPS covalently linked to the acceptor substrate by way of a neutral bridging group. This arrangement has the poten-

TABLE 50.3. *Synthetic substrate-based inhibitors of glycosyltransferases*

Enzyme	Substrate	Inhibitor	Substrate K_m (μM)	Inhibitor K_i (μM)
α2FucT	β3GlcNAcβ-O-R	2-deoxyGalβ3GlcNAcβ-O-R	200	800
β4GalT	GlcNAcβ3Galβ-O-R	6-thioGlcNAcβ3Galβ-O-Me	1000	1000
α3GalT	Galβ4GlcNAcβ-O-R	3-aminoGalβ4GlcNAcβ-O-R	190	104[a]
β6GlcNAcT	Galβ3GalNAcα-O-R	Galβ3(6-deoxy)GalNAcα-O-R	80	560
β6GlcNAcT-V	GlcNAcβ2Manα6Glcβ-O-R	GlcNAcβ2(6-deoxy)Manα6Glc β-O-R	23	30
β6GlcNAcT-V	GlcNAcβ2Manα6Glcβ-O-R	GlcNAcβ2(4-O-methyl)Manα6 Glcβ-O-R	23	14
β6GlcNAcT-V	GlcNAcβ2Manα6Glcβ-O-R	GlcNAcβ2(6-deoxy,4-O-Me)Manα6Glcβ-O-R	23	3
α6SialylT	Galβ4GlcNAcβ-O-R	6-deoxyGalβ2GlcNAcβ-O-R	900	760[a]
α3GalNAcT-A	Fucα2Galβ-O-R	Fucα2(3-deoxy)Galβ-O-R	2	68
α3GalNAcT-A	Fucα2Galβ-O-R	Fucα2(3-amino)Galβ-O-R	2	0.2[a]

The aglycone (R) varies in the different compounds.
[a] Inhibition mixed or noncompetitive.

tial for generating inhibitors whose binding characteristics reflect the product of the affinity constants for each substrate (approximated by the product of the individual K_m values). Bisubstrates that have been made have K_i values in the range of the K_m values for the nucleotide donors, suggesting that the correct geometry for the bridging group may have not been attained or that the analog binds in ways that differ from the natural substrates.

A certain amount of serendipity is needed to obtain active compounds using these synthetic approaches, and the synthesis of di-, tri-, and tetrasaccharides with the desired modifications is rather labor intensive. Nevertheless, the approach has yielded insights into the binding and reactivity of the enzymes, and substrate analogs with selectivity for particular enzymes have been developed in this way. Because many of the transferases have now been purified and cloned, we can look forward to more detailed kinetic and crystallographic studies, which will provide clues for deriving mechanism-based inhibitors in the future (see Chapter 5).

GLYCOSIDE PRIMERS: MIMICKING WHAT ALREADY WORKS

The utility of any glycosyltransferase inhibitor ultimately depends on its ability to cross the plasma membrane and enter the Golgi where the glycosyltransferases reside. Unfortunately, many of the compounds described above lack activity in live cells, presumably because their polarity and charge prevents their uptake. More than 35 years ago, Okayama and colleagues found that D-xylose in β-linkage to a hydrophobic aglycone (the noncarbohydrate portion of a glycoside; e.g., *p*-nitrophenol) was taken up rather efficiently and inhibited the assembly of glycosaminoglycans on proteoglycans. Xylosides mimic the natural substrate, xylosylated serine residues in proteoglycan core proteins, and thus act as a substrate. "Priming" of chains occurs on the added xyloside, which diverts the assembly process from the endogenous core proteins and causes inhibition of proteoglycan formation. In general, cells incubated with xylosides secrete large amounts of individual glycosaminoglycan chains and accumulate proteoglycans containing truncated chains. The enormous success of β-D-xylosides in altering proteoglycan biosynthesis suggested that other glycosides might act as "primers" as well (Table 50.4). Subsequently, studies have shown that β-*N*-acetylgalactosaminides prime oligosaccharides found on mucins and inhibit O-glycosylation of glycoproteins. Other active glycosides include β-glucosides, β-galactosides, β-*N*-acetylglucosaminides, and even disaccharides and trisaccharides. These compounds require conjugation to appropriate aglycones and acetylation to neutralize the polar hydroxyl groups on the sugars. Cells contain several carboxyesterases that remove the acetyl groups. Apparently, this can occur in a way that makes the compounds available to the transferases in the Golgi.

TABLE 50.4. *Examples of glycoside primers*

Glycoside	Pathway affected
Xylββ-O-R	glycosaminoglycans glycolipids
Galβ-O-R	glycosaminoglycans
GalNAcα-O-R	O-linked chains found on glycoproteins and mucins
GlcNAcβ-O-R	polylactosaminoglycans
Peracetylated Galβ4GlcNAcβ-O-R	sialyl Lewis^x
Peracetylated GlcNAcβ3Galβ-O-R	sialyl Lewis^x

Priming by glycosides occurs in a concentration-dependent manner, but the efficiency varies widely among different compounds and cell types. These variations may relate to the relative abundance of endogenous substrates, enzyme concentration and composition, the solubility of different glycosides, their susceptibility to hydrolysis, their uptake across the plasma membrane and into the Golgi, and their relative affinity for the glycosyltransferases. The type of chain made on a given primer also depends on concentration and aglycone structure, which may reflect selective partitioning of primers into different intracellular compartments or into different branches of biosynthetic pathways. Like priming, inhibition of glycoprotein, glycolipid, or proteoglycan formation occurs in a dose-dependent fashion, but the blockade is rarely complete, probably because of the inability of glycosides to mimic completely the endogenous substrates.

Primers represent starting points for making analogs that might have the properties of tight-binding inhibitors described in the previous section. Many of the compounds described in Table 50.3 could be converted to permeable acylated glycosides and tested in live cells for inhibitory activity. Active compounds might have the potential to become carbohydrate-based drugs for treating diseases and disorders dependent on glycosylation. Priming of oligosaccharides may have beneficial effects as well. Xylosides, for example, can be absorbed through the gut, and when consumed at sufficient concentration, they exhibit antithrombotic activity. Many glycosides occur naturally, since various organisms (especially plants) produce hydrophobic compounds as part of chemical defense and conjugate them to sugars in order to render them soluble. Thus, the human diet may contain various types of glycosides with interesting (and unknown) biological activities.

MODULATING GLYCAN STRUCTURES USING ENZYMES AND MONOSACCHARIDES ANALOGS

A general assumption in glycobiology is that each glycan linkage is catalyzed by a highly selective transferase, with specificity both for substrate and donor nucleotide sugar. However, many transferases will catalyze formation of other linkages or utilize sugar analogs with low efficiency, but with sufficient capacity to alter composition under specific conditions. Some transferases will tolerate even large adducts appended to the donor nucleotide sugar. For example, one of the α1-3 fucosyltransferases will transfer a GDP-fucose analog containing bulky substituents at the C-6 position of the sugar (such as the H-blood group antigen). A mutant form of β1-4 galactosyltransferase has been described that can use UDP-2-acetonyl-2-deoxy-D-galactose as donor. Treatment of O-GlcNAc-containing glycoproteins with the nucleotide analog and the mutant enzyme introduces a ketone-tagged sugar that can then be reacted with a biotinylating reagent, thus allowing selection and detection of the modified glycoproteins. Microbial enzymes have also been engineered to tolerate unusual monosaccharides and nucleotide sugars, making it possible to synthesize glycoconjugates with unusually modified glycans.

Naturally occurring transferases will also tolerate modified sugars. For example, cells will metabolically incorporate mannosamine analogs with an acyl, alkyl, or ketone group covalently linked to the amino group instead of acetate (Figure 50.4). Thus, the various kinases and nucleotidyltransferases will apparently accept the analogs as substrates. When ketone-containing derivatives are incorporated into a cell, cell-surface glycoconjugates become sensitive to reagents that react with carbonyl groups. Other highly chemoselective derivatives and reagents have been developed, including azides that can be modified selectively with phosphines (Staudinger ligation) or activated alkynes. The highly selective reactivity of these functionalities allows tagging of endogenous glycoproteins

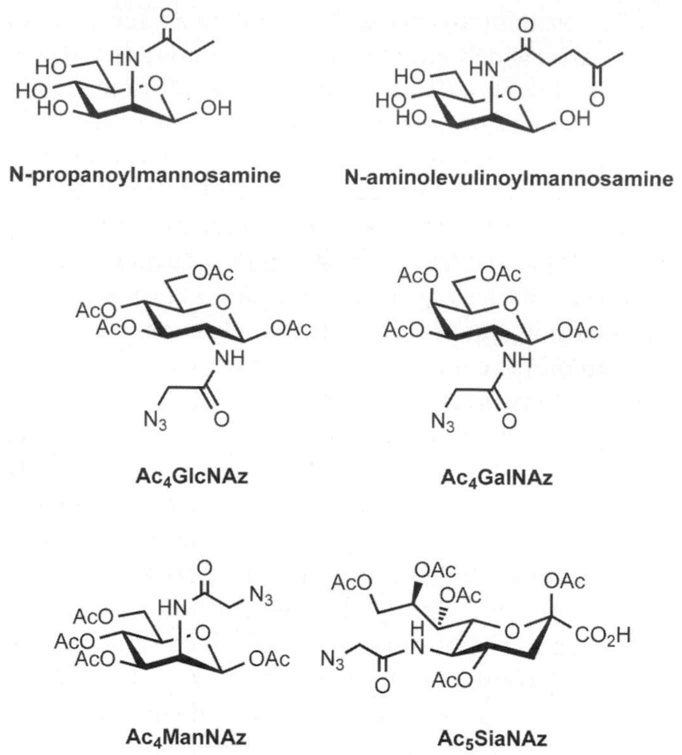

FIGURE 50.4. Unnatural monosaccharides can be transformed into activated nucleotide sugars that are transported into the Golgi compartment and then transferred to glycoconjugates.

for proteomic analysis and imaging studies in vivo when injected systemically in mice. A number of endogenous pathways will use azide-containing monosaccharides, including pathways for incorporation of sialic acid (ManAz), GlcNAc (GlcNAz), and GalNAc (GalNAz) (see Figure 50.4). Metabolic glycan engineering might be used for the inventory of altered glycosylation in normal and diseased tissue or for modulating protein–glycan interactions.

INHIBITORS OF GLYCOLIPIDS AND GPI ANCHORS

Reagents have been described that can alter the assembly of glycolipids in cells. Xylosides have a mild effect on glycolipid formation, possibly because of the similarity between xylose and glucose and the assembly of a GM_3-like compound (Neu5Acα2-3Galβ1-4Xylβ-O-R) on the primer. Because cells will take up intermediates in glycolipid biosynthesis, they behave like synthetic glycoside primers. For example, glucosylceramide will give rise to complex glycolipids when fed to cells. On the basis of this observation, an analog containing a reactive exocyclic epoxide group was prepared. The compound inhibits glycolipid formation (IC_{50} ~ 8 μM), presumably by reaction of the epoxide with a nucleophile in the active site of lactosylceramide synthase.

As mentioned above, the α-glucosidase inhibitor, *N*-butyldeoxynojirimycin, has been used to inhibit glycosphingolipid formation because it blocks glucosylceramide synthesis. This compound is now in clinical use for treating type I Gaucher's disease, a lysosomal storage disorder in which glucocerebrosidase is missing (see Chapter 41). Its beneficial activity occurs through "substrate deprivation" by blocking synthesis of glycosphingolipids, thereby "depriving" the lysosome of substrate. The sphingolipid analogs shown in Figure

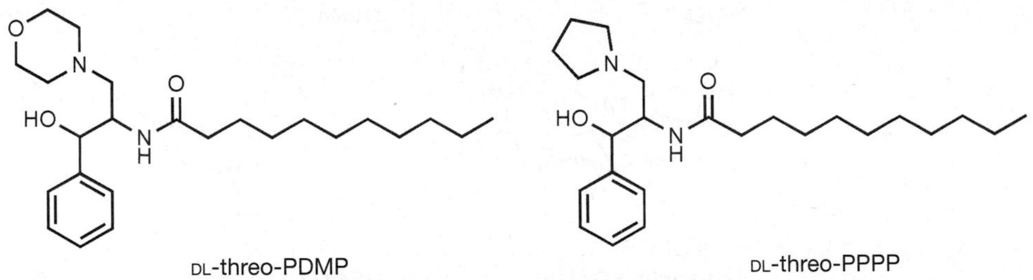

FIGURE 50.5. Inhibitors of glycosphingolipid formation. These commercially available analogs are potent inhibitors of the glucosyltransferase that initiates glycosphingolipid formation.

50.5 are more potent inhibitors. Lengthening the hydrocarbon chain from 10 to 16 carbons further enhances efficacy. These compounds can almost completely deplete glycolipids in cells, and they require only micromolar amounts for activity.

Inhibitors of GPI anchor formation have also been described. Mannose analogs (2-deoxy-2-fluoro-D-glucose and 2-deoxy-D-glucose) inhibit the formation of dolichol-P-mannose in vivo and thus inhibit GPI biosynthesis, but they lack specificity because these agents also can affect other pathways dependent on dolichol-P-mannose. Mannosamine inhibits GPI anchor formation both in *Trypanosoma brucei* and in mammalian cells by formation of $ManNH_2$-Man-GlcN-PI. Apparently, mannosamine in its activated form (GDP-$ManNH_2$) is used as a substrate in the second mannosyltransferase reaction, but the $ManNH_2$-Man-GlcN-PI intermediate will not act as a substrate for the next α2-mannosyltransferase (see Chapter 11). GlcNR-phosphatidylinositols with different substituents (R) act as substrate analogs and some act as suicide inhibitors in vitro. Another class of inhibitors is based on fatty acid analogs that only trypanosomes incorporate into GPI anchors. Trypanosomes, unlike their mammalian hosts, incorporate myristic acid into GPI anchors by exchanging myristic acid for other fatty acids in the phosphatidylinositol moiety. By making a series of analogs, one was found that is highly toxic to trypanosomes in culture and nontoxic to mammalian cells (10-[propoxy]decanoic acid). Reagents like these have the potential to become drugs for treating trypanosomiasis, which is endemic in sub-Saharan regions of Africa.

RATIONAL DESIGN USING CRYSTAL STRUCTURES

Neuraminidase (Sialidase) Inhibitors

Studies of influenza neuraminidase exemplify the power of rationally designed drugs. The crystal structure for influenza neuraminidase was obtained in 1983, and since then many other enzymes have been characterized from other sources. Even before the crystal structure had been obtained, an inhibitor for neuraminidase was deduced by assuming that the hydrolysis reaction probably involved a transition state with a carbocation intermediate at C-2. This would result in C-2 and C-3 adopting a trigonal planar (sp2) configuration, and therefore compounds that mimicked this geometry were hoped to have inhibitory activity. Indeed, Neu5Ac-2-ene (DANA; Figure 50.6) has a micromolar K_i value. Interestingly, this compound works on most sialidases, but not on the trypanosome *trans*-sialidase and only weakly on bacterial sialidases.

Visual inspection of the influenza enzyme with the inhibitor bound showed that two glutamate residues lined a pocket near O4 of the sialic acid residue and formed hydrogen bonds with O4 (Figure 50.7). The pocket was fairly open, suggesting that a bulkier sub-

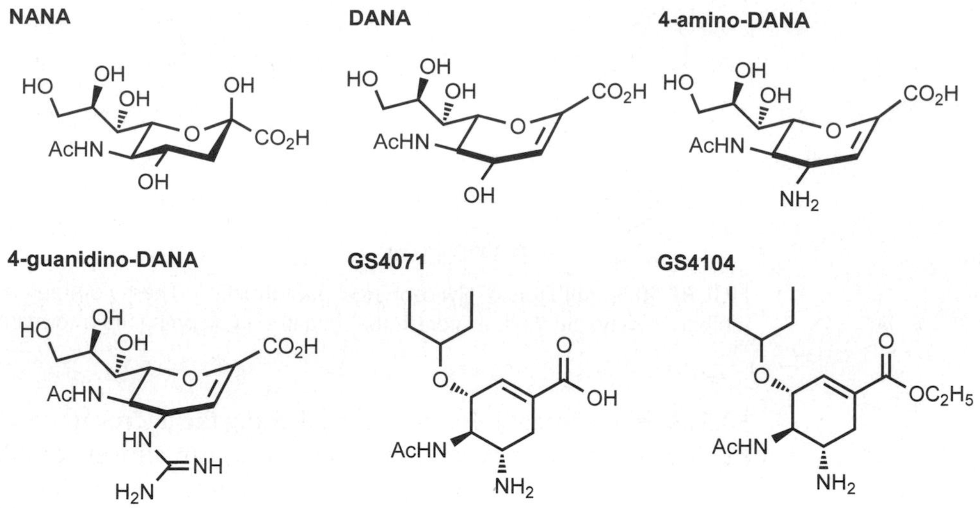

FIGURE 50.6. Structure of influenza neuraminidase inhibitors. Chemical structure of Neu5Ac, NANA; 2-deoxy-2,3-dehydro-*N*-acetyl neuraminic acid, DANA; 4-amino-DANA; 4-guanidino-DANA (Relenza, zanamivir); (3R, 4R, 5S)-4-acetamido-5-amino-3-(1-ethylpropoxyl)-1-cyclohexane-1-carboxylic acid (GS4071); ethyl ester derivative of GS4071 (GS4104, Tamiflu). DANA is thought to resemble the transition state in hydrolysis. The addition of the guanidinium group in Relenza permits higher affinity binding to the active site.

stituent at this position might be tolerated, at least sterically. A substrate analog was produced containing a positively charged guanidinium group instead of O4, and binding studies showed that the analog had a K_i value of 10^{-11} M (4-guanidino-DANA; Figure 50.6). The higher affinity was presumably due to an additional salt bridge formed between the charged guanidinium group and the carboxylates lining the pocket. Interestingly, the analog does not work on bacterial sialidases because the pocket is filled with an arginine group and only one aspartate residue lines the pocket.

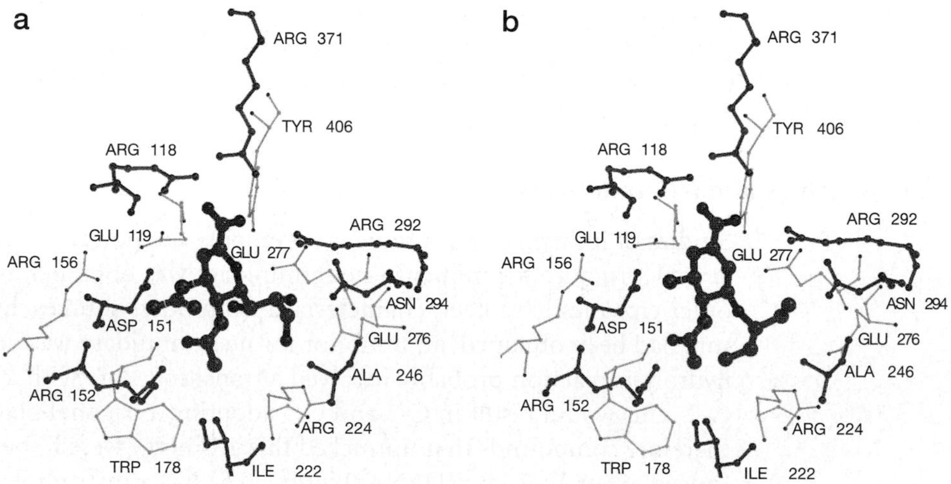

FIGURE 50.7. Crystallographic structures of influenza virus neuraminidase (N9 subtype) with two different rationally designed inhibitors bound in the active site. The inhibitors are shown as *colored ball and stick* models, where *yellow* is carbon, *blue* is nitrogen, and *red* is oxygen. The catalytic site of the enzyme is shown with the closer carbon atoms in *dark green* and those further away in *light green*. (*a*) Relenza (4-guanidino-Neu5Ac2en). (*b*) De-esterified Tamiflu. (Redrawn, with permission, from Garman E. and Laver G. 2004. *Curr. Drug Targets* **5:** 119–136.)

Subsequent studies have attempted to use noncarbohydrate analogs based on the similarity of benzoic acid to the partially planar ring of the proposed intermediate in hydrolysis. Reasonably good inhibitors were found in this way, as long as hydrogen bond donors and acceptors were added to the ring (Figure 50.6). The best inhibitor contained a guanidinium group in the same position relative to the carboxylate anion (as in 4-guanido DANA), but the crystal data showed that the analog actually bound to the enzyme in an orientation very different from that of sialic acid. As the crystal structures for other sialidases are solved, the design of species-specific analogs may be possible. This rational approach for making inhibitors has great potential, not only for neuraminidase inhibitors, but also for the design of compounds that might block the activity of various glycosyltransferases.

Sulfotransferase Inhibitors

A large family of Golgi sulfotransferases installs sulfate esters on a variety of glycans using PAPS as the active sulfate donor. High-resolution crystal structures are now available for both glycan-modifying sulfotransferases as well as soluble drug detoxifying sulfotransferases. These enzymes have in common a conserved fold that binds PAPS (see Chapter 5). A series of small aromatic compounds (such as 2,6-dichloro-4-nitrophenol and pentachlorophenol) and several disaccharide analogs of GlcNAc-6-sulfotransferase substrates have inhibitory activity. Screening targeted libraries of purine derivatives yielded compounds with high selectivity towards individual sulfotransferases, suggesting that subtle differences in the PAPS-binding sites can be exploited.

The ability to screen large libraries against sulfotransferases has been facilitated by the development of high-throughput assays. Most low-throughput assays measure the transfer of [^{35}S] from [^{35}S]PAPS or measure the radiolabel transfer to a carbohydrate substrate bearing a hydrophobic tail that can easily be isolated using a reverse-phase cartridge. Some sulfotransferases will catalyze the reverse reaction in the presence of high concentrations of a sulfated donor. For example, β-arylsulfotransferase IV (β-AST-IV) will catalyze the reverse transfer of the sulfuryl group from *p*-nitrophenol sulfate to PAPS, generating PAPS and *p*–nitrophenolate ion. When coupled to another sulfotransferase of interest, β-AST-IV regenerates PAPS and stoichiometric amounts of the ion, which can be monitored by UV absorbance. The enzyme also will transfer sulfate to water. A library of 35,000 compounds with purine and pyrimidine scaffolds was screened using β-AST-IV and a fluorescence assay that measured desulfation of 4-methylumbelliferone sulfate. Multiple hits were obtained with moderate inhibition, and subsequent structure elaboration of the library resulted in the generation of a very tight binding small molecule inhibitor with a K_m value five orders of magnitude lower then the natural substrate. In theory, this approach can be exploited for other enzymes for which high-throughput assays can be developed.

FURTHER READING

Tifft C.J. and Proia R.L. 2000. Stemming the tide: Glycosphingolipid synthesis inhibitors as therapy for storage diseases. *Glycobiology* **10:** 1249–1258.

Asano N. 2003. Glycosidase inhibitors: Update and perspectives on practical use. *Glycobiology* **13:** 93R–104R.

Brown J.R., Fuster M.M., and Esko J.D. 2003. Glycoside primers and inhibitors of glycosylation. In *Carbohydrate-based drug discovery* (ed. C.-H. Wong), Vol. 2, pp. 883–889. Wiley-VCH, Weinheim.

de Macedo C.S., Shams-Eldin H., Smith T.K., Schwarz R.T., and Azzouz N. 2003. Inhibitors of glycosylphosphatidylinositol anchor biosynthesis. *Biochimie* **85:** 465–472.

Chapman E., Best M.D., Hanson S.R., and Wong C.H. 2004. Sulfotransferases: Structure, mechanism, biological activity, inhibition, and synthetic utility. *Angew Chem. Int. Ed. Engl.* **43:** 3526–3548.

Garman E. and Laver G. 2004. Controlling influenza by inhibiting the virus's neuraminidase. *Curr. Drug Targets* **5:** 119–136.

Butters T.D., Dwek R.A., and Platt F.M. 2005. Imino sugar inhibitors for treating the lysosomal glycosphingolipidoses. *Glycobiology* **15:** 43R–52R.

Gross B.J., Kraybill B.C., and Walker S. 2005. Discovery of O-GlcNAc transferase inhibitors. *J. Am. Chem. Soc.* **127:** 14588–14589.

Hang H.C. and Bertozzi C.R. 2005. The chemistry and biology of mucin-type O-linked glycosylation. *Bioorg. Med. Chem.* **13:** 5021–5034.

Ostash B. and Walker S. 2005. Bacterial transglycosylase inhibitors. *Curr. Opin. Chem. Biol.* **9:** 459–466.

Prescher J.A. and Bertozzi C.R. 2005. Chemistry in living systems. *Nat. Chem. Biol.* **1:** 13–21.

Blanchard S. and Thorson J.S. 2006. Enzymatic tools for engineering natural product glycosylation. *Curr. Opin. Chem. Biol.* **10:** 263–271.

Dorfmueller H.C., Borodkin V.S., Schimpl M., Shepherd S.M., Shapiro N.A., and van Aalten D.M. 2006. GlcNAcstatin: A picomolar, selective O-GlcNAcase inhibitor that modulates intracellular O-GlcNAcylation levels. *J. Am. Chem. Soc.* **128:** 16484–16485.

Zhang C., Griffith B.R., Fu Q., Albermann C., Fu X., Lee I.K., Li L., and Thorson J.S. 2006. Exploiting the reversibility of natural product glycosyltransferase-catalyzed reactions. *Science* **313:** 1291–1294.

Brown J.R., Crawford B.E., and Esko J.D. 2007. Glycan antagonists and inhibitors: A fount for drug discovery. *Crit. Rev. Biochem. Mol. Biol.* **42:** 481–515.

Rexach J.E., Clark P.M., and Hsieh-Wilson L.C. 2008. Chemical approaches to understanding O-GlcNAc glycosylation in the brain. *Nat. Chem. Biol.* **4:** 97–106.

CHAPTER 51

Glycans in Biotechnology and the Pharmaceutical Industry

Carolyn R. Bertozzi, Hudson H. Freeze, Ajit Varki, and Jeffrey D. Esko

GLYCANS ARE COMPONENTS OF MANY BIOTHERAPEUTIC AGENTS, ranging from natural products to molecules based on rational design to recombinant glycoproteins. The glycan components of these agents can be important determinants of their biological activity and therapeutic efficacy. This chapter provides a brief overview of some of the issues pertinent to glycan functions in therapeutics and a few examples of successful glycan or glycoconjugate-based drugs (Table 51.1).

TABLE 51.1. *Examples of glycan-based drugs, their target diseases, and modes of action*

Drug	Company	Disease/disorder	Mode of action
Targeting sialic acids			
Zanamivir (Relenza®)	Biota/GlaxoSmithKline	influenza type A and B	inhibits neuraminidase
Oseltamivir (GS 4104, Tamiflu®)	Gilead/Roche	chemoprophylaxis	inhibits neuraminidase
Targeting glycosaminoglycans			
Heparin	multiple brands	anticoagulant; possible value in cancer metastasis prevention	activates antithrombin; inhibits heparanase and selectins and blocks interactions between growth factors and heparan sulfate
Hyaluronan (HA)	multiple brands	ocular surgery; osteoarthritis; plastic surgery	tissue space filler; anti-inflammatory agent
Laronidase (Aldurazyme®)	Genzyme	mucopolysaccharidosis type I (MPSI); α-iduronidase deficiency	enzyme replacement therapy (ERT)
Galsulfase (Naglazyme®)	Biomarin	mucopolysaccharidosis type VI; arylsulfatase B deficiency	ERT
Hyaluronidase (Cumulase®)	Halozyme	in vitro fertilization; in development as an adjuvant for cancer chemotherapy	degrades HA around oocytes improving fertilization; degrades HA in tumors to decrease intratumor pressure
Tramiprosate (Alzhemed)	phase III trials (Neurochem)	amyloid diseases, Alzheimer's disease, and possibly other amyloidoses	binds to amyloid plaque, blocks its formation
Eprodisate (Kiacta®)	phase II/III trials (Neurochem)	amyloid A amyloidosis	interferes with glycosaminoglycan–amyloid interactions
Targeting glycosphingolipids			
N-butyl-deoxynojirimycin (DNJ) (Miglustat, Zavesca®)	phase III trials (Acetelion)	type 1 Gaucher's disease; Niemann–Pick's disease type C; late-onset Tay–Sach's disease; type 3 Gaucher's disease	substrate reduction therapy; inhibits glucosylceramide synthase
Imiglucerase (Cerezyme®)	Genzyme	type 1 Gaucher's disease	ERT
β-Agalsidase (Fabrazyme®)	Genzyme	Fabry disease; α-galactosidase A deficiency	ERT
Others			
Acarbose (Glucobay®)	Bayer	type 2 diabetes	blocks intestinal α-glucosidases involved in digestion of dietary glycans
Alglucosidase alfa (Myozyme®)	Genzyme	Pompe's disease (glycogen storage disease); α-glucosidase A deficiency	ERT
Allosamidin	Industrial Research	insecticide	chitinase inhibitor

Modified from Brown J.R., Crawford B.E., and Esko J.D. 2007. *Crit. Rev. Biochem. Mol. Biol.* **24:** 481–515.

Compounds targeted at microbial glycans, such as the aminoglycoside antibiotics or other inhibitors of cell wall assembly, have not been included.

GLYCANS AS COMPONENTS OF SMALL-MOLECULE DRUGS

Many well-known small-molecule drugs, such as antibiotics and anticancer therapeutic agents, are natural products that contain glycans as part of their essential structure and/or as a sugar side chain (i.e., a glycoside). Some examples of natural products that bear glycan side chains are shown in Figure 51.1. This is a well-established area of natural product and synthetic/biosynthetic chemistry and will not be reviewed in detail here. Glycans have also been modified to generate synthetic drugs. The most recent examples are the small-molecule inhibitors of influenza virus neuraminidase briefly discussed below (see also Chapter 50).

SMALL-MOLECULE INHIBITORS OF INFLUENZA VIRUS NEURAMINIDASE

Influenza virus has two major surface proteins, termed hemagglutinin and neuraminidase (see Chapter 34). The hemagglutinin initiates infection by binding to cell-surface sialic acids. The neuraminidase assists virus release by cleaving sialic acids to prevent unwanted retention of newly synthesized virus on the cell surface. Neuraminidase may also function during the invasion phase by removing sialic acids on soluble mucins that would otherwise inhibit cell-surface binding. On the basis of the essential requirement for the neuraminidase during the viral life cycle, the crystal structure of the purified enzyme was studied and a rational drug design program was initiated. This approach resulted in the development of Zanamivir (Relenza™), in which the addition of a bulky guanidino side chain at the C-4 position of a previously known neuraminidase inhibitor, 2-deoxy-2,3-dehydro-*N*-acetylneuraminic acid, markedly increased the affinity for influenza neuraminidase, without having similar affects on host-cell neuraminidases (Figure 51.2; see also Chapter 50). Relenza blocks the life cycle of influenza virus by preventing infection and by interrupting the spread of the virus during the early phase of an infection. However, because it is not orally absorbable, the drug must be given as an inhaled medication to work at the mucosal sites of infection in the upper airway. An orally absorbable drug called Oseltamivir (Tamiflu™) was

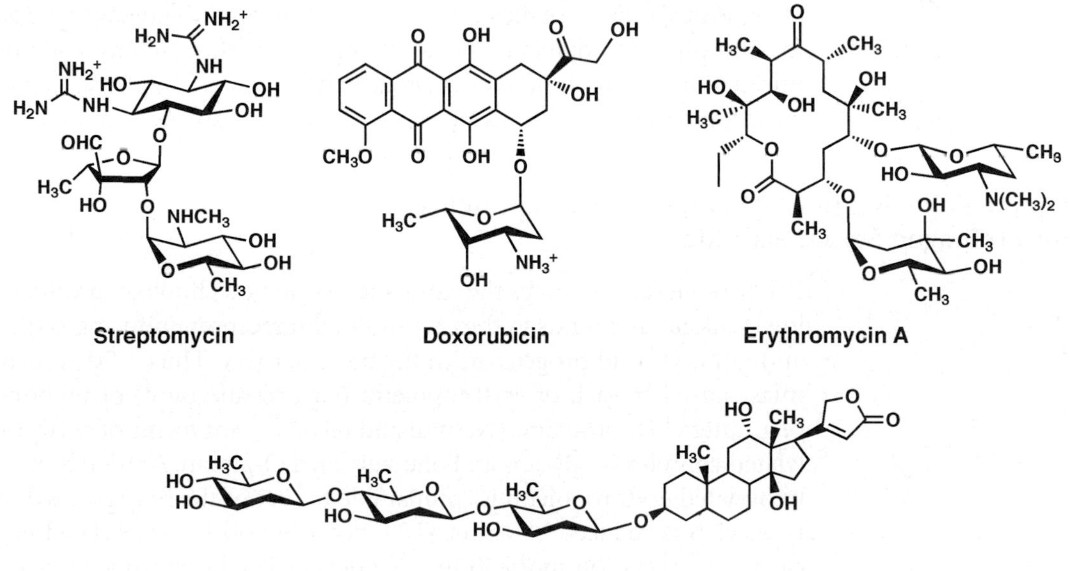

Digoxin

FIGURE 51.1. Examples of natural products that possess glycan components. Streptomycin and erythromycin A are antibiotics, doxorubicin is chemotherapeutic drug, and digoxin is used to treat cardiovascular disease.

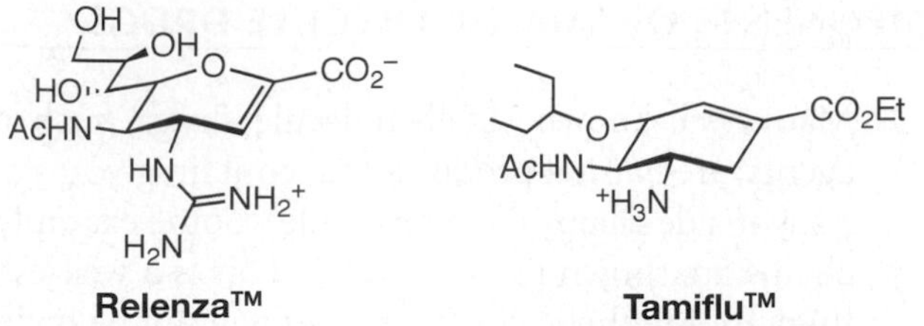

FIGURE 51.2. The synthetic influenza neuraminidase inhibitors Relenza™ and Tamiflu™.

subsequently developed, and it can achieve the same effects. Both drugs are now in clinical use, either for treatment of influenza or for preventing spread of the disease.

The fear that the avian influenza virus ("bird flu") might spread into the human population has resulted in stockpiling of Tamiflu. Although there is a public health rationale for this approach, it is difficult to know when and where the drug will be needed. In addition, drug-resistant strains of influenza are beginning to emerge, wherein the viral neuraminidase is no longer substantially inhibited. Thus, indiscriminate use of the drug might eventually, as in the case of antibiotics, negate its value when it is really needed. Moreover, widespread use may reveal off-target side effects, which may be more harmful to individuals for whom the risk of infection by influenza is low. Nevertheless, this is a classic example of how the study of the basic structural biology of a glycosidase and its glycan substrate resulted in one of the first effective drugs for a common and dangerous disease.

THERAPEUTIC GLYCOPROTEINS

A major proportion of biotherapeutic products are glycoproteins. These include erythropoietin and various other cytokines, antibodies, glycosyltransferases, and glycosidases, which together generate sales of billion of dollars worldwide. Therapeutic glycoproteins are typically produced as recombinant products in cell culture systems or, less commonly, in the milk of transgenic animals. Control of glycosylation is of major importance during the development of these drugs, because their glycan chains have marked effects on stability, activity, antigenicity, and pharmacodynamics in intact organisms. In most cases, glycosylation must be optimized to ensure prolonged circulatory half-life in the blood. Manipulation of glycans to promote targeting to specific tissues and cell types has also been a useful element of drug design.

Optimizing Glycans of Therapeutic Glycoproteins for Prolonged Serum Half-life

Erythropoietin is perhaps the most successful biotechnology product to date. It is a circulating cytokine that binds to the erythropoietin receptor, inducing proliferation and differentiation of erythroid progenitors in the bone marrow. Thus, it has much value in treating anemias caused by lack of erythropoietin (e.g., renal failure) or by bone marrow suppression (e.g., after chemotherapy). Natural and recombinant forms of erythropoietin carry three sialylated complex N-glycans and one sialylated O-glycan. Although in vitro the activity of deglycosylated erythropoietin is comparable to that of the fully glycosylated molecule, its activity in vivo is reduced by about 90%, because poorly glycosylated erythropoietin is rapidly cleared by filtration in the kidney. Undersialylated erythropoietin is also rapidly cleared by galactose receptors in hepatocytes and macrophages (see Chapter 31). These problems can be reduced by having fully sialylated chains and by increasing the amount of tetra-antennary branching. This, in turn, increases its activity in vivo nearly tenfold. Another approach is to deliberately add an N-glycosylation site to increase half-life and activity in vivo. Covalently linking polyethylene glycol to the protein also reduces clearance by the kidney.

Erythropoietin is somewhat unusual because it is small enough to be cleared by the kidney if it is underglycosylated. For most glycoprotein therapeutics, a more important consideration is minimizing clearance by galactose-binding hepatic receptors by ensuring full sialylation of glycans. Because glycans can have such dramatic effects on the properties of these drugs, it is important to ensure that glycosylation is controlled during their production, as well as to satisfy regulatory requirements for batch-to-batch product consistency. Changes in culture pH, the availability of precursors and nutrients, and the presence or absence of various growth factors and hormones can each affect the extent of glycosylation, the degree of branching, and the completeness of sialylation. Sialidases and other glycosidases that are either secreted or released by dead cells can also cause degradation of the previously intact product in the culture medium. Thus, during the development of a glycoprotein drug, considerable effort is devoted to defining appropriate conditions for reproducibly producing relatively homogeneous and complete glycosylation.

Impact of Glycosylation on Licensing and Patentability of Biotherapeutic Agents

Modern patenting of new therapeutics typically requires definition of the composition of matter in the claimed molecule. This is relatively straightforward for small molecules of defined structure and for nonglycosylated proteins. However, with regard to glycoproteins, especially those with multiple glycosylation sites, it is virtually impossible to obtain a single preparation that contains only a single glycoform. Thus, most biotherapeutic agents that are glycoproteins consist of a mixture of glycoforms. Licensing bodies such as the U.S. Food and Drug Administration have come to recognize this and will allow for a certain range of variation in glycoforms and the complexity of the mixture. However, the manufacturer and the agency must agree on the extent such variation is acceptable for a given drug formulation. Biopharmaceutical companies therefore spend considerable effort in assuring that their products fall within these defined ranges, once these are approved by licensing bodies. The inherent difficulty in reproducing complex glycoform mixtures also complicates efforts to make generic forms of recombinant glycoprotein drugs. Given the complexities of producing glycotherapeutic agents in mammalian cells, even the smallest changes in growth conditions (temperature, pH, media, etc.) can have significant effects on the range of glycoforms found in any given batch of the product. In fact, the licensing agencies also use consistency in glycoform composition as an indirect measure of the quality of process control in production. Differences in glycosylation can even have implications for the patentability of agents in which the underlying polypeptide remains constant. Marked differences in glycosylation have been used to define agents as being uniquely different. However, it is usually necessary to show that the differences in glycosylation being claimed also have a significant effect in changing the functionality of the drug in question. The associated pharmaceutical licensing and legal issues are rapidly evolving to keep pace with scientific advances in this area.

Significance of Nonhuman Glycosylation on Biotherapeutic Products

Certain types of nonhuman glycosylation, such as Galα1-3Gal ("α-Gal") units and/or the sialic acid *N*-glycolylneuraminic acid (Neu5Gc), are found in terminal positions of glycans on some biotherapeutic glycoproteins. The addition of α-Gal residues can only occur if the therapeutic product is produced in a nonhuman cell line that expresses an α1-3Gal transferase not present in humans (see Chapter 13). The addition of Neu5Gc occurs when the nonhuman cell line used produces this sialic acid and/or because the animal-derived products used in the media provide a metabolic source of Neu5Gc, which can be taken up, processed, and made available for sialylation reactions (see Chapter 14). Other glycans unique to nonhuman expression systems may exist as well. Since humans have circulating antibodies against α-Gal termini, Neu5Gc, and presumably other nonhuman glycan determinants, a potential for antigen–antibody responses exists that could be deleterious and/or affect therapeutic efficiency. An optimal solution is to use cell lines that do not express α-Gal or Neu5Gc (e.g., CHO cells, which express no α-Gal and only low levels of Neu5Gc), to eliminate Neu5Gc from materials used in tissue culture, and to carefully analyze the glycans present in recombinant protein therapeutics. Further studies of the effects of these antigenic modifications on half-life and pharmacokinetics are needed.

GLYCOSYLATION ENGINEERING

There are limits as to how much of a biotherapeutic glycoprotein an animal cell line can produce. This becomes an issue especially when there is a need to produce very large amounts (e.g., if the glycoprotein needs to be given in high milligram quantities per day). In this situation, it would be beneficial to produce the glycoproteins in plants or yeast that are capable of much higher levels of production. However, it is necessary to eliminate risks arising from the nonhuman glycans of plant and fungal cells, which could cause excessively rapid clearance and/or antigenic reactions. Many plant and yeast glycans are immunogenic and elicit glycan-specific IgE and IgG antibodies in humans when delivered parenterally. To this end, there have been several efforts to add back a variety of mammalian genes into yeast and/or to eliminate genes that are producing nonhuman glycosylation. This has now been achieved using extensively engineered yeast strains that are capable of producing biantennary N-linked glycans with the human sialic acid *N*-acetylneuraminic acid (Neu5Ac). However, the productivity of such yeast strains has yet to be reported. Similar efforts to engineer yeast to make human-like O-glycans are under way, but other glycans such as glycosaminoglycans have not yet been addressed.

Plants and algae have also been used to engineer recombinant glycoproteins, but, as in yeast, the glycans produced by plants differ from those found in vertebrates. The antigenic differences that arise in recombinant glycoproteins produced in plants become less problematic if used for topical or oral administration, since humans are normally exposed to plant glycans in the diet. Another advantage is that the cost of production is much lower than in animal cell culture systems and animal sera are not needed. As in yeast, "humanizing plants" with respect to glycosylation may allow the production of nonimmunogenic glycoproteins. Chemical methods for synthesizing glycoproteins from scratch are also being explored as a means to obtain therapeutic agents with well-defined tailored glycans (see Chapter 49).

GLYCAN THERAPEUTIC APPROACHES TO METABOLIC DISEASES

Salvage versus De Novo Synthesis

All monosaccharides needed for cellular glycan synthesis can be obtained from glucose through metabolic interconversions (see Chapter 4). Alternatively, monosaccharides can be derived from the diet or salvaged from degraded glycans. The relative contributions of different sources can vary with the cell type. For instance, even though all mammalian cells use sialic acid, only some contain high amounts of UDP-GlcNAc epimerase/*N*-acetylmannosamine kinase (GNE), which is required for the de novo synthesis of CMP-sialic acid. But sialic acid salvage from degraded glycans is quite efficient, decreasing the demand on the de novo pathway. Similarly, galactose, fucose, mannose, *N*-acetylglucosamine, and *N*-acetylgalactosamine can come from the diet or be salvaged for glycan synthesis, whereas glucuronic acid, iduronic acid, and xylose cannot. All monosaccharides derived from the diet or degraded glycans can be catabolized for energy, and again, cells vary in their reliance on the different pathways.

The variable contributions of these pathways are important for therapy of some diseases. For instance, patients with congenital disorder of glycosylation type Ib (CDG-Ib), who are deficient in phosphomannose isomerase, benefit greatly from oral mannose supplementation to bypass the insufficient supply of glucose-derived mannose-6-phosphate. A few CDG-IIc patients have been treated with fucose to restore synthesis of sialyl Lewis^x on leukocytes (see Chapter 42). Some patients with Crohn's disease show clinical improvement with oral *N*-acetylglucosamine supplementation, but the mechanism is unknown. Mice deficient in GNE activity have kidney failure, but providing *N*-acetylmannosamine in the diet prevents this outcome. Clinical trials using *N*-acetylmannosamine to treat GNE-deficient patients with hereditary inclusion body myopathy type II (HIBM-II) are proposed, although it is unclear whether sialic acid deficiency is responsible for the human muscle pathology in HIBM-II.

Special Diets

Some monosaccharides and disaccharides can be toxic to humans who lack specific enzymes. For example, people who lack fructoaldolase (aldolase B) accumulate fructose-1-phosphate, which ultimately causes ATP depletion and disrupts glycogen metabolism. Prolonged fructose exposure in these people can be fatal, and fructose-limited diets are critical. Deficiencies in the ability to metabolize galactose (see Chapter 4) are mostly due to a severe reduction in galactose-1-phosphate uridyl transferase activity and cause galactosemia. Although these patients are asymptomatic at birth, ingesting milk leads to vomiting and diarrhea, cataracts, hepatomegaly, and even neonatal death. Low-galactose or galactose-free diets can prevent these life-threatening symptoms. However, even these diets do not prevent unexplained long-term complications, which include speech and learning disabilities and ovarian failure in almost 85% of females with galactosemia.

Infants hydrolyze lactose (Galβ1-4Glc) quite well, but the level of intestinal lactase can be much lower or absent in adults because of down-regulation of lactase gene expression. About two thirds of the human population has lactase nonpersistence, making milk products a dietary annoyance. This is because unabsorbed lactose provides an osmotic load and is metabolized by colonic bacteria, causing diarrhea, abdominal bloating and pain, flatulence, and nausea. Lactase persistence has evolved in certain pastoral populations from northwestern Europe, India, and Africa, allowing milk consumption in adult life. However, many adults either avoid lactose-containing foods or use lactase tablets to improve lactose digestion.

Substrate Reduction Therapy

The failure to turn over glycans by lysosomal degradation causes serious problems for patients with lysosomal storage disorders. Deficiencies in individual lysosomal enzymes lead to pathological accumulation of their substrates in inclusion bodies inside the cells (see Chapter 41). One approach to treating these disorders is to inhibit initial glycan synthesis, a strategy termed substrate reduction therapy (SRT). Reduced synthesis of the initial compound decreases the load on the impaired enzyme, and some patients show significant clinical improvement. One example of a drug used for SRT is *N*-butyldeoxynojirimycin (or *N*-

butyl-DNJ) (Miglustat, Zavesca®), which has shown some efficacy for treating Gaucher's disease (glucocerebrosidase deficiency).

Lysosomal Enzyme Replacement Therapy

Another approach for treating lysosomal storage disorders is enzyme replacement therapy. Unlike most therapeutic glycoproteins that interact with target receptors on the surface of cells, lysosomal enzymes developed for replacement therapy must be delivered intracellularly to lysosomes, their site of action. During the normal biosynthesis of lysosomal enzymes, their N-glycans become modified with mannose-6-phosphate (Man-6-P) residues, which target them to lysosomes using Man-6-P receptors (see Chapter 30). The challenge for enzyme replacement therapy is to get the enzymes targeted properly to lysosomes, where they can degrade accumulated substrate. A special case is enzyme replacement therapy for Gaucher's disease. In this case, the enzyme has been successfully targeted to lysosomes of macrophages through their cell-surface mannose receptor (see Chapter 31). This approach required modification of the complex N-glycans of the enzyme isolated from placenta to expose terminal mannose residues, a process that involved digestion with various glycosidases (sialidase, galactosidase, and hexosaminidase). Nowadays, recombinant enzyme is produced in Lec1 mutant CHO cells (see Chapter 46), in which N-glycans with terminal mannose residues are present at all N-glycan sites.

The success of glucocerebrosidase treatment stimulated the development of lysosomal enzymes for treatment of other lysosomal storage diseases. Clinical trials have demonstrated clinical benefit in Fabry's disease, mucopolysaccharidoses type I, II, and VI, and Pompe's disease. However, the usefulness of replacement therapy is limited by the fact that injected enzymes do not have beneficial effects on all aspects of these diseases, and existing tissue damage is usually not reversible. Moreover, as Man-6-P receptors do not function in all cell types, N-glycan termini are needed to target all the affected cell types. In all instances, an additional challenge is to deliver enzymes efficiently to cells of the nervous system past the blood-brain barrier.

Chaperone Therapy

A third approach for treating lysosomal storage disorders takes advantage of the fact that some genetic defects lead to misfolding of the encoded enzyme in the endoplasmic reticulum (ER). There is evidence that low-molecular-weight molecules that are competitive inhibitors of some of these enzymes can act as "chaperones," stabilizing the folded enzyme in the ER and effectively rescuing the mutation. The result is a higher steady-state concentration of active enzyme in the lysosome. Of course, the dose of such an inhibitor must be carefully adjusted to ensure that the inhibitory effects on enzyme function do not overshadow beneficial effects on folding. Fortunately, only a low level of enzyme restoration is needed to significantly reduce the accumulation of undigested glycan substrates, indicating that lysosomal hydrolases are normally present in large catalytic excess.

THERAPEUTIC APPLICATIONS OF GLYCOSAMINOGLYCANS

The use of purified glycans as therapeutics has received less attention than the development of glycoprotein-based treatments. Difficulties in establishing structure–activity relationships due to the large number of chiral centers and functional groups, undesirable pharmacokinetics of available formulations, poor oral absorption of the compounds, and low-affinity interactions with drug targets have limited their development. Some successful glycan drugs, such as the anticoagulant heparin, are given by injection, although efforts are under way to convert heparin into an orally absorbable form by complexing it with positively charged molecules. It may be possible to deliver other hydrophilic and/or negatively charged glycan drugs in this way to allow penetration of the intestinal barrier. Glycans are also sometimes attached to hydrophobic drugs to improve their solubility and alter their pharmacokinetics.

As discussed in Chapters 16 and 43, heparin is one of the most widely prescribed drugs today because of its anticoagulant activity, which is based on the binding and activation of antithrombin, a protease inhibitor of the coagulation cascade. Activation of antithrombin leads to rapid inhibition of thrombin and factor Xa, shutting down the production of fibrin clots. Heparin is produced by autodigestion of pig intestines, followed by graded fractionation of the products. Annual production is measured in metric tons, translating into a billion doses per year. Unfractionated heparin also binds to several plasma, platelet, and endothelial proteins, producing a variable anticoagulant response. Therefore, there has been much interest in producing heparin preparations with more predictable outcomes. Low-molecular-weight (LMW) heparins are derived by chemical or enzymatic cleavage of heparin to form smaller fragments. The pharmacological properties and the relative efficacy of the various LMW heparins appear generally superior to those of unfractionated heparin and there are fewer secondary complications. As the enzymes of heparin biosynthesis have now been cloned, it may be possible to produce recombinant heparins soon. There is also a synthetic heparin pentasaccharide that functions to bind antithrombin in exactly the same manner as heparin. This compound, known commercially as the drug Arixtra, is used to prevent deep-vein thrombosis and pulmonary embolism.

It is important to be able to neutralize heparin rapidly under some circumstances. The conventional approach uses the basic protein protamine, which binds to heparin, neutralizes its activity, and results in clearance of the complex by the kidney and liver. Recombinant heparinases are also under development as alternative neutralization agents. Heparin's additional effects on blocking P- and L-selectin binding are discussed in Chapter 31. Heparin is also used to treat protein-losing enteropathy (PLE), likely working by competing for proinflammatory heparin-binding cytokines that trigger PLE in susceptible patients (see Chapter 43). It should be noted that the synthetic pentasaccharide Arixtra may not possess all of the nonanticoagulant activities of the natural products, such as selectin inhibition.

Hyaluronan (see Chapter 15) is a naturally occurring glycosaminoglycan that is extensively used in surgical applications. Because of its viscoelastic properties, hyaluronan has lubricating and cushioning properties that have made it useful for protecting the corneal endothelium during ocular surgery. Hyaluronan has antiadhesive properties and therefore might be useful in postsurgical wound healing as well. The mechanism of action is not well understood, but it may involve hyaluronan-binding proteins that mediate cell adhesion (see Chapter 15). Intra-articular injections of hyaluronan are used to treat knee and hip osteoarthritis. Several studies have shown modest improvement in patients treated with hyaluronan, but it is unclear if it is acting in a mechanical (as a viscosupplement) and/or a biological (via signaling pathways) manner. Hyaluronan is also injected as a tissue filler in cosmetic medicine.

"GLYCONUTRIENTS"

"Glyconutrient" is a term used by the nutritional supplement industry to describe some of their products. Numerous preparations are sold as nutritional supplements, often with wide-ranging claims concerning potential benefits. In most cases, these claims have not been substantiated through placebo-controlled, double-blind trials with defined, quantifiable outcomes. Much work is needed in this area to obtain insight into the potential role of dietary glycans on human health and to help consumers make wise decisions regarding their use.

As an example, mixtures of plant polysaccharides such as larchbark arabinogalactan, glucomannan, and others are often referred to "glyconutrients" that are claimed to contain "eight essential monosaccharides" needed for "cell communication." Because all monosaccharides can be made from glucose (except in patients with rare genetic deficiencies; see Chapter 42), none of the other monosaccharides are actually known to be "essential." Moreover, these polysaccharides are not degraded to available monosaccharides in the stomach or small intestine. Instead, anaerobic bacteria in the colon metabolize them and produce short-chain fatty acids. There are no peer-reviewed clinical studies supporting the efficacy of such "glyconutrients" for any disease or condition. Nevertheless, the following examples demonstrate how dietary glycans might have beneficial effects.

Glucosamine and Chondroitin Sulfate

Glucosamine (often mixed with chondroitin sulfate) has been promoted to relieve symptoms of osteoarthritis, which involves the age-dependent erosion of articular cartilage. Cartilage provides a cushion between the bones to minimize mechanical damage, and a net loss of cartilage occurs when the degradation rate exceeds the synthetic rate. A number of clinical trials report that glucosamine improves osteoarthritis symptoms, and some claim to restore partially the structure of the eroded cushion, in particular in the knees. Superficially, this would seem to make sense, because primary glycans of cartilage include hyaluronan (see Chapter 15) and chondroitin sulfate, both of which contain hexosamines within their structure (see Chapter 16). However, there are conflicting reports, and the outcome may depend on study design and the type and source of material (e.g., chloride vs. sulfate salts and the overall level of purity). Nevertheless, veterinarians have treated animals with glucosamine for over two decades with apparently positive results. Furthermore, double-blind, placebo-controlled studies in humans have shown a decreased rate of joint space narrowing. It should be noted that glucosamine might also alter UDP-GlcNAc and potentially UDP-GalNAc levels, thus affecting cellular responses involving any major class of glycans.

Positive effects of chondroitin sulfate on osteoarthritis are less well-documented. Indeed, how the acidic chondroitin sulfate polymer can be absorbed and delivered to its proposed site of action remains an open question. Further studies are needed to determine how dosage affects the circulating levels of these supplements, whether they are absorbed by the target tissue, and if they actually lead to changes in cartilage metabolism.

Xylitol and Sorbitol in Chewing Gum

Many studies suggest that chewing gum containing sugar alditols, specifically xylitol and sorbitol, can help control the development of dental caries. Mothers who chew xylitol-sweetened gum may even block transmission of caries-causing bacteria to their children. The benefit of these reduced sugars seems to be based on stimulation of salivary flow, but an antimicrobial effect is also possible. Xylitol also inhibits the expression and secretion of proinflammatory cytokines from macrophages and inhibits the growth of *Porphyromonas gingivalis,* one of the suspected causes of periodontal disease. Children who drank xylitol solutions also had a lower occurrence of otitis media.

Milk Oligosaccharides

Human milk contains about 70 g/liter of lactose and 5–10 g/liter of free oligosaccharides. More than 130 different glycan species have been identified with lactose at the reducing end, including poly-*N*-acetyllactosamine units. Some glycans are α2-3- and/or α2-6-sialylated and/or fucosylated in α1-2, α1-3, and/or α1-4 linkages. In contrast, bovine milk, the typical mainstay in human infant formulas, contains much smaller amounts of these glycans. These differences may account for some of the physiological advantages seen for breast-fed versus formula-fed infants. The glycans may also favor growth of a nonpathogenic bifidogenic microflora and/or block pathogen adhesion that causes infections and diarrhea. Surprisingly, a substantial number of human milk oligosaccharides remain almost undigested in the infant's intestine and are excreted intact into the urine. Whether supplementing infant formula with specific, biologically active free glycans enhances infant health is unknown.

Chitosan

Chitosan (partially de-N-acetylated chitin, a polymer of β-linked *N*-acetylglucosamine residues) is a mixture of different-sized water-soluble polymers and has been reported to have hypocholesterolemic, antimicrobial, and weight-reduction effects. Combined analyses from various studies show a slight benefit in body weight reduction, but this benefit is less pronounced in more rigorous evaluations. The effects are likely to be of limited clinical significance.

GLYCANS AS VACCINE COMPONENTS

Microbial Vaccines

Vaccines consisting solely of glycan components typically elicit poor immunity, especially in infants. The primary limitation is that glycans are T-cell-independent antigens and therefore do not effectively stimulate T-helper-dependent activation and class switching of B-cell-mediated immunity. Conjugate vaccines with oligosaccharides coupled to carrier proteins have proven to be highly effective. For example, *Haemophilus influenzae* type b (Hib) causes an acute lower respiratory infection among young children. A conjugated form of an Hib-derived oligosaccharide coupled to a protein carrier is now routinely given to infants in developed countries and has resulted in a more than 95% decrease in incidence of infections in vaccinated populations. New vaccines are being developed using conjugated components of other bacterial capsular polysaccharides, and the studies appear promising.

Cancer Vaccines

Several carbohydrate-based vaccines are under development to treat cancer. Some of these vaccines are based on ganglioside immunogens present on certain types of cancer cells (e.g., gangliosides GM2 and GD2 in melanomas and globo H in breast cancer). In some cases, vaccines are composed of glycan-based haptens conjugated to a protein carrier. In one case, a breast cancer antigen sialyl-Tn (sialylα2-6GalNAcα-) is synthesized chemically and conjugated to a protein carrier (keyhole limpet hemocyanin). Sialyl-Tn is found on cancer mucins, and its expression correlates with progression to metastatic disease (see Chapter 44). A more direct strategy thus consists of attaching the sialyl-Tn units in their natural linkages to serine or threonine residues of the Muc-1 mucin polypeptide repeat. Although several cancer vaccine candidates are presently under investigation, they have yet to reach clinical utility.

BLOCKING GLYCAN RECOGNITION IN DISEASES

Blocking Infection

As discussed in Chapter 34, many microbes and toxins bind to mammalian tissues by recognizing specific glycan ligands. Thus, small soluble glycans or glycan mimetics can be used to block the initial attachment of microbes and toxins to cell surfaces (or block their release), and thus prevent or suppress infection. Because many of these organisms naturally gain access through the airways or gut, the glycan-based drugs can be delivered directly without the requirement of being distributed systemically. Examples of such applications currently under study include milk oligosaccharides that are believed to be natural antagonists of intestinal infection in infants (see above) and polymers that will block the binding of viruses such as influenza. Although backed by a strong scientific rationale and robust in vitro studies, such "antiadhesive" therapies have not yet found much practical application.

Inhibition of Selectin-mediated Leukocyte Trafficking

If specific glycan-protein interactions in vivo are responsible for selective cell–cell interactions and a resulting pathology, then infusion of small-molecule analogs or mimetics of the natural ligand could theoretically be useful. The best-studied example is the selectin-mediated recruitment of neutrophils (and other leukocytes) into sites of inflammation or ischemia/reperfusion injury, which involves specific selectin–glycan interactions occurring in the vascular system (see Chapter 31). Initial approaches focused on the use of sialyl Lewis^{x} derivatives. However, the tetrasaccharide is a difficult synthetic target and has poor oral availability and a short serum half-life. To improve these shortcomings, attempts were made to design "glycomimetics," which are compounds that preserve the essential functionality of the parent tetrasaccharide but eliminate unwanted polar functional groups and synthetically cumbersome glycan components. An example of the design of a monosaccharide glycomimetic starting from sialyl Lewis^{x} is shown in Figure 51.3. First, the sialic acid residue was replaced with a charged glycolic acid group, the *N*-acetylglucosamine residue was then replaced with an ethylene glycol linker, and finally the galactose residue was replaced with a linker moiety. The resulting glycomimetic had E-selectin binding affinity comparable to sialyl Lewis^{x} but with a simpler structure.

Simple monovalent compounds such as sialyl Lewis^{x} and its glycomimetics have proven effective in various animal inflammatory models. This finding is somewhat surprising, because the binding constants for some of these monovalent ligands (with K_d values in the millimolar range) are much poorer than that for the physiological ligand. In fact, similar studies in humans did not show a clear benefit. Studies of the PSGL-1 (P-selectin glycoprotein ligand-1) glycosulfopeptide that has high affinity for P- and L-selectin are presently being pursued, as well as dendrimers consisting of multivalent forms of Lewis antigens. These conjugates have considerably longer serum half-lives than the monovalent glycans and glycomimetics. In addition, the inhibitory effects of heparin on P- and L-selectin are being considered. Finally, an alternate approach consists of finding specific inhibitors of the enzymes involved in forming endogenous ligand determinants in vivo. For example, the fucosyltransferase involved in sialyl Lewis^{x} biosynthesis (FUT7) is an attractive target.

TRANSFUSION AND TRANSPLANTATION REJECTION BY ANTIGLYCAN ANTIBODIES

As discussed in Chapter 13, a variety of glycans, including the classical A and B blood group determinants, can act as barriers to blood transfusion and transplantation of organs.

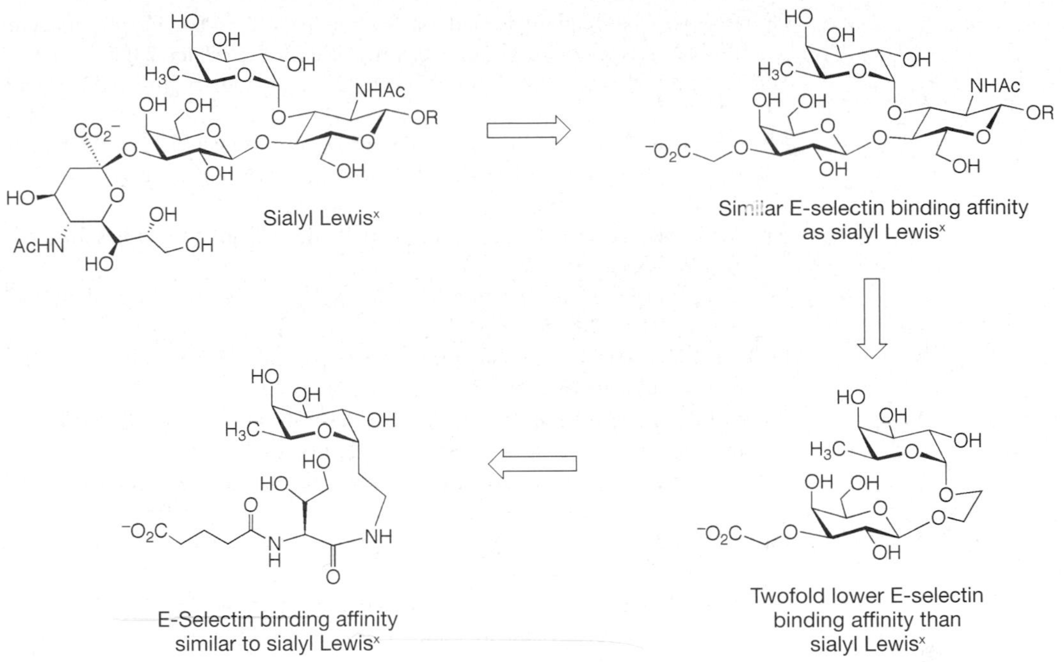

FIGURE 51.3. Glycomimetic E-selectin inhibitors based on sialyl Lewis^x.

Rejection of mismatched blood or organs occurs because hosts have a high titer of preexisting antibodies against the glycan epitopes, presumably as a prior reaction to related structures found on bacteria or other microbes. In the case of the ABO blood groups, incompatibility is routinely managed by blood and tissue typing and finding an appropriate donor for the recipient. A successful strategy recently reported is to use bacterial enzymes in vitro to remove the A and B blood group determinants from A and B red cells, converting them into "universal donor" O red cells.

A related problem is found in xenotransplantation (i.e., the transplantation of organs between species) which is actively being pursued as a solution for the shortage of human organs for patients. The animal donors of preference are pigs, because many porcine organs resemble those of humans in size, physiology, and structure. However, unlike humans and certain other primates, pigs and most other mammals produce the terminal "α-Gal" epitope on glycoproteins and glycolipids. Because humans have naturally occurring high-titer antibodies in blood directed toward this epitope, this results in hyperacute rejection of porcine organ transplants, via reaction of the antibodies with endothelial cells of blood vessels. Attempts have been made to prevent this reaction, including blood filtration over glycan affinity columns to remove xenoreactive antibodies and blockade of the interaction by infusing soluble competing oligosaccharides. Transgenic pigs lacking the reactive epitope have also been produced, as have animals with an excess of complement-controlling proteins on their cell surfaces. Pig organs also have high levels of the nonhuman sialic acid (Neu5Gc), against which most humans have antibodies. Even if this problem is solved, there are other glycan and protein structural differences between humans and pigs that cause later stages of graft rejection, thus necessitating immunosuppression.

FURTHER READING

Kunz C., Rudloff S., Baier W., Klein N., and Strobel S. 2000. Oligosaccharides in human milk: Structural, functional, and metabolic aspects. *Annu. Rev. Nutr.* **20:** 699–722.

Gomord V., Chamberlain P., Jefferis R., and Faye L. 2005. Biopharmaceutical production in plants: Problems, solutions and opportunities. *Trends Biotechnol.* **23:** 559–565.

Joshi L. and Lopez L.C. 2005. Bioprospecting in plants for engineered proteins. *Curr. Opin. Plant. Biol.* **8:** 223–226,

Mhurchu C.N., Dunshea-Mooij C., Bennett D., and Rodgers A. 2005. Effect of chitosan on weight loss in overweight and obese individuals: A systematic review of randomized controlled trials. *Obes. Rev.* **6:** 35–42.

Pastores G.M. and Barnett N.L. 2005. Current and emerging therapies for the lysosomal storage disorders. *Expert Opin. Emerg. Drugs* **10:** 891–902.

Ly K.A., Milgrom P., and Rothen M. 2006. Xylitol, sweeteners, and dental caries [discussion 192–198]. *Pediatr. Dent.* **28:** 154–163.

Beck M. 2007. New therapeutic options for lysosomal storage disorders: Enzyme replacement, small molecules and gene therapy. *Hum. Genet.* **121:** 1–22.

Brown J.R., Crawford B.E., and Esko J.D. 2007. Glycan antagonists and inhibitors: A fount for drug discovery. *Crit. Rev. Biochem. Mol. Biol.* **42:** 481–515.

Butters T.D. 2007. Pharmacotherapeutic strategies using small molecules for the treatment of glycolipid lysosomal storage disorders. *Expert Opin. Pharmacother.* **8:** 427–435.

Divine J.G., Zazulak B.T., and Hewett T.E. 2007. Viscosupplementation for knee osteoarthritis: A systematic review. *Clin. Orthop. Relat. Res.* **455:** 113–122.

Eklund E.A., Bode L., and Freeze H.H. 2007. Diseases associated with carbohydrates/glycoconjugates. In *Comprehensive glycoscience* (ed. J.P. Kamerling, G.J. Boone, and Y.C. Lee), Vol. 4, pp. 339–372. Elsevier, New York.

Hamilton S.R. and Gerngross T.U. 2007. Glycosylation engineering in yeast: The advent of fully humanized yeast. *Curr. Opin. Biotechnol.* **18:** 387–392.

Schultz B.L., Laroy W., and Callewaert N. 2007. Clinical laboratory testing in human medicine based on the detection of glycoconjugates. *Curr. Mol. Med.* **7:** 397–416.

von Itzstein M. 2007. The war against influenza: Discovery and development of sialidase inhibitors. *Nat. Rev. Drug Discov.* **6:** 967–974.

Schnaar R.L. and Freeze H.H. 2008. A "glyconutrient sham." *Glycobiology* **18:** 652–657.

Glossary

ABO Gene locus comprising three major allelic glycosyltransferases that generate the A, B, and O blood group antigens.

Acetal An organic compound derived from a hemiacetal by reaction with an alcohol. If the hemiacetal is a sugar, then the acetal is a glycoside.

Adhesin A protein on the surface of bacteria, viruses, or parasites that binds to a ligand present on the surface of a host cell.

Agglutination The clumping of cells in the presence of a protein (e.g., antibody or lectin). The related term "hemagglutination" denotes the specific case wherein the cells are red blood cells.

Aglycone Noncarbohydrate portion of a glycoconjugate or glycoside that is glycosidally linked to the glycan through the reducing terminal sugar.

Aldose A monosaccharide with an aldehyde group or potential aldehydic carbonyl group (by definition, this is the C-1 position).

Amino sugar A monosaccharide in which an alcoholic hydroxyl group is replaced by an amino group.

Anomeric carbon The carbon atom of a monosaccharide that bears the hemiacetal functionality (C-1 for most sugars; C-2 for sialic acids).

Anomers Stereoisomers of a monosaccharide that differ only in configuration at the anomeric carbon of the ring structure.

Antenna A branch of an oligosaccharide emanating from a "core" structure.

Asparagine-linked oligosaccharide *See* **N-Glycan**.

Azide A functional group comprising three nitrogen atoms bound in a linear arrangement (N_3).

Azido sugar A monosaccharide to which an azido group has been introduced synthetically.

Bactoprenol *See* **Undecaprenol**.

β-elimination The cleavage of a C-O or C-N bond positioned on the β-carbon with respect to a carbonyl group. The process is used to cleave O-glycans from Ser or Thr residues.

Biofilm Community of bacteria that adheres to a moist surface (e.g., surface of ponds or teeth).

Calnexin Membrane-bound protein chaperone that mediates quality control of protein folding in the endoplasmic reticulum.

Calreticulin Soluble protein chaperone that mediates quality control of protein folding in the endoplasmic reticulum.

Capsule A protective extracellular polysaccharide coat surrounding certain bacteria. Presence of a capsular polysaccharide is often associated with virulence.

Carbohydrate A generic term used interchangeably in this book with sugar, saccharide, or glycan. This term includes monosaccharides, oligosaccharides, and polysaccharides as well as derivatives of these compounds.

Carbohydrate-recognition domain (CRD) The domain of a polypeptide that is specifically involved in binding to carbohydrate. In lectins, it is often a highly evolutionarily conserved region of the polypeptide.

Cassette A glycosylated amino acid used in solid-phase peptide synthesis to generate glycopeptides.

CAZy database Denoting "*c*arbohydrate *a*ctive en*zy*mes," this database describes the families of structurally related catalytic and carbohydrate-binding modules (or functional domains) of enzymes that degrade, modify, or create glycosidic bonds.

Cellulose A repeating homopolymer of β1-4-linked glucose residues.

Ceramide The common lipid component of glycosphingolipids, composed of a long-chain amino alcohol (sphingosine) and an amide-linked fatty acid.

Cerebroside A glycolipid composed of ceramide with an attached galactose (galactosylceramide) or glucose (glucosylceramide) residue.

Chemoenzymatic synthesis Glycan synthesis that uses both chemical and enzymatic transformations to obtain the desired product.

Chitin A repeating homopolymer of β1-4-linked *N*-acetylglucosamine residues. It is the main component of the cell walls of fungi and the exoskeletons of arthopods, among other functions.

Chondroitin sulfate A type of glycosaminoglycan defined by the disaccharide unit $(GalNAc\beta1\text{-}4GlcA\beta1\text{-}3)_n$, modified with ester-linked sulfate at certain positions and typically found covalently linked to a proteoglycan core protein.

Complex glycan A glycan containing more then one type of monosaccharide.

C-type lectins A class of Ca^{++}-dependent lectins recognizable by a characteristic sequence that constitutes their carbohydrate-recognition domain.

Deoxy sugar A monosaccharide in which a hydroxyl group is replaced by a hydrogen atom.

Dermatan sulfate A modified form of chondroitin sulfate in which a portion of the D-glucuronate residues are epimerized to L-iduronates.

Dolichol A polyisoprenoid lipid carrier utilized during the assembly of N-glycans and GPI anchors.

Electron transfer dissociation (ETD) mass spectrometry A technique used to determine sites of glycosylation on peptides and proteins.

Endoglycosidase An enzyme that catalyzes the cleavage of an internal glycosidic linkage in an oligosaccharide or polysaccharide.

Endotoxin *See* **Lipid A.**

Epidermal growth factor (EGF)-like repeats Small protein motifs (~40 amino acids) with six conserved cysteine residues that form three disulfide bonds. These often serve as sites for glycan modification.

Epimerase An enzyme that catalyzes racemization of a chiral center in a sugar.

Epimers Two isomeric monosaccharides differing only in the configuration of a single chiral carbon. For example, mannose is the C-2 epimer of glucose.

Epitope The part of a molecule that is recognized by a specific antibody or receptor.

Exoglycosidase An enzyme that cleaves a monosaccharide from the outer (nonreducing) end of an oligosaccharide, polysaccharide, or glycoconjugate.

Exotoxins Heat-labile, proteinaceous toxins secreted by bacteria that cause illness.

Expressed protein ligation (EPL) A method for generating semisynthetic proteins by the condensation of a synthetic peptide and a recombinant protein. Glycoproteins can be generated by condensation of a synthetic glycopeptide and a recombinant protein.

Extracellular matrix A complex array of secreted molecules including glycoproteins, proteoglycans, and/or polysaccharides and structural proteins. In plants, the extracellular matrix is also referred to as the cell wall.

Extrinsic glycan-binding proteins Receptors that recognize glycans from a different organism and consist mostly of pathogenic microbial adhesins, agglutinins, or toxins.

Fimbriae Proteinaceous fiber-like appendages found in many Gram-negative bacteria.

Fischer projection A two-dimensional representation of a three-dimensional organic molecule devised by Hermann Emil Fischer.

Fringe Family of proteins that modify Notch activity by catalyzing the transfer of *N*-acetylglucosamine from UDP-GlcNAc to fucose on an EGF-like repeat.

Furanose Five-membered (four carbons and one oxygen, i.e., an oxygen heterocycle) ring form of a monosaccharide named after the structural similarity to the compound furan.

Galectins S-type (sulfhydryl-dependent) β-galactoside-binding lectins, usually occurring in a soluble form, expressed by a wide variety of animal cell types and distinguishable by the amino acid sequence of their carbohydrate-recognition domains.

Ganglioside Anionic glycosphingolipid containing one or more residues of sialic acid.

Gene chip A DNA microarray used to quantify transcript levels in high-throughput format.

Genome The complete genetic sequence of one set of chromosomes.

Glycan A generic term for any sugar or assembly of sugars, in free form or attached to another molecule, used interchangeably in this book with saccharide or carbohydrate.

Glycan array A collection of glycans attached to a surface in a spatially addressed manner.

Glycan-binding proteins Proteins that recognize and bind to specific glycans and mediate their biological function. *See* **Lectin** and **Glycosaminoglycan-binding protein**.

Glycation The nonenzymatic, chemical modification of proteins by addition of carbohydrate, usually through a Schiff-base reaction with the amino group of the side chain of lysine and subsequent Amadori rearrangement to give a stable conjugate. Not to be confused with (enzymatic) glycosylation.

Glycobiology Study of the structure, chemistry, biosynthesis, and biological functions of glycans and their derivatives.

Glycocalyx The cell coat consisting of glycans and glycoconjugates surrounding animal cells that is seen as an electron-dense layer by electron microscopy.

Glycoconjugate A molecule in which one or more glycan units are covalently linked to a noncarbohydrate entity.

Glycoforms Different molecular forms of a glycoprotein, resulting from variable glycan structure and/or glycan attachment site occupancy.

Glycogen A polysaccharide comprising α1-4 and α1-6-linked glucose residues that functions in short-term energy storage in animals; sometimes referred to as animal starch.

Glycogenin A protein that acts as a primer for glycogen synthesis.

Glycolipid General term denoting a molecule containing a glycan linked to a lipid aglycone. In higher organisms, most glycolipids are glycosphingolipids, but glycoglycerolipids and other types exist.

Glycome The total collection of glycans synthesized by a cell, tissue, or organism under specified conditions of time, space, and environment.

Glycomics Systematic analysis of the glycome.

Glycomimetics Noncarbohydrate compounds that mimic the properties of glycans.

Glycone Carbohydrate component of a glycoconjugate.

Glycopeptide A peptide having one or more covalently attached glycans.

Glycoprotein A protein with one or more covalently bound glycans.

Glycoproteomics The systems-level analysis of glycoproteins, including their protein identities, sites of glycosylation, and glycan structures.

Glycosaminoglycans Polysaccharide side chains of proteoglycans or free complex polysaccharides composed of linear disaccharide repeating units, each composed of a hexosamine and a hexose or a hexuronic acid. *See* **Heparin**, **Heparan sulfate**, **Chondroitin sulfate**, **Dermatan sulfate**, and **Hyaluronan**.

Glycosaminoglycan-binding proteins Proteins that recognize and bind to specific glycosaminoglycans.

Glycosidase An enzyme that catalyzes the hydrolysis of glycosidic bonds in a glycan. *See* **Exoglycosidase** and **Endoglycosidase**.

Glycoside A glycan containing at least one glycosidic linkage to another glycan or an aglycone.

Glycosidic linkage Linkage of a monosaccharide to another residue via the anomeric hydroxyl group. The linkage generally results from the reaction of a hemiacetal with an alcohol (e.g., a hydroxyl group on another monosaccharide or amino acid) to form an acetal. Glycosidic linkages between two monosaccharides have defined regiochemistry and stereochemistry.

Glycosphingolipid Glycolipid comprising a glycan glycosidically attached to the primary hydroxyl group of ceramide.

Glycosyl acceptor The nucleophile in a glycosylation reaction, usually containing a free hydroxyl group.

Glycosylation The enzyme-catalyzed covalent attachment of a carbohydrate to a polypeptide, lipid, polynucleotide, carbohydrate, or other organic compound, generally catalyzed by glycosyltransferases, utilizing specific sugar nucleotide donor substrates.

Glycosyl donor The electrophile in a glycosylation reaction; the nucleotide sugar in an enzymatic glycosylation reaction.

Glycosylphosphatidylinositol (GPI) anchor A membrane anchor that consists of a glycan bridge between phosphatidylinositol and ethanolamine in an amide linkage to the carboxyl terminus of a protein.

Glycosyltransferase Enzyme that catalyzes transfer of a sugar from a sugar nucleotide donor to a substrate.

Hapten A small glycan that competes with a more complex ligand for binding to a lectin. More generally, any small molecule that interacts with a receptor or antibody.

Haworth projection A representation of monosaccharides wherein the cyclic structures are depicted as planar rings with the hydroxyl groups orientated above or below the plane of the ring.

Hemagglutination The clumping of red blood cells in the presence of a protein (e.g., antibody or lectin).

Hemagglutinin A lectin that recognizes carbohydrates on the surface of red blood cells and causes hemagglutination.

Hemiacetal A compound formed by reaction of an aldehyde with an alcohol group, as in ring closure of an aldose.

Hemiketal A compound formed by reaction of a ketone with an alcohol group, as in ring closure of a ketose.

Heparan sulfate A glycosaminoglycan defined by the disaccharide unit (GlcNAcα1-4GlcAβ1-4IdoAα1-4)$_n$, containing N- and O-sulfate esters at various positions, and typically found covalently linked to a proteoglycan core protein.

Heparin A type of heparan sulfate made by mast cells that has the highest amount of iduronic acid and of N- and O-sulfate residues. Pharmaceutical heparin binds and activates antithrombin.

Heteropolysaccharide A polysaccharide containing more than one type of monosaccharide.

Hexosamine Hexose with an amino group in place of the hydroxyl group at the C-2 position. Common examples found in vertebrate glycans are the N-acetylated sugars, *N*-acetylglucosamine and *N*-acetylgalactosamine.

Hexose A six-carbon monosaccharide typically with an aldehyde (or potential aldehyde) at the C-1 position (aldohexose) and hydroxyl groups at all other positions. Common examples in vertebrate glycans are mannose, glucose, and galactose.

Homopolysaccharide A polysaccharide composed of only one type of monosaccharide.

Hyaluronan A glycosaminoglycan defined by the disaccharide unit (GlcNAcβ1-4GlcAβ1-3)$_n$ that is neither sulfated nor covalently linked to protein. It is referred to in older literature as hyaluronic acid.

Hydrazinolysis A chemical method that uses hydrazine to cleave amide bonds (e.g., the glycosylamine linkage between a sugar residue and asparagine or the acetamide bond in *N*-acetylhexosamines).

Intrinsic glycan-binding proteins Receptors that recognize glycans from the same organism. Typically they mediate cell–cell interactions or recognize extracellular molecules, but they can also recognize glycans on the same cell.

I-type lectins A class of lectins belonging to the immunoglobulin superfamily.

Jelly-roll fold Description of tertiary structure common to L-type lectins.

Keratan sulfate A polylactosamine [Galβ1-4GlcNAcβ1-3]$_n$ with sulfate esters at C-6 of *N*-acetylglucosamine and galactose residues, found as a side chain of a keratan sulfate proteoglycan.

Ketal An organic compound derived from a hemiketal by reaction with an alcohol. If the hemiketal is a sugar, then the ketal is a glycoside.

Ketose A monosaccharide with a ketone group or a potential ketonic carbonyl group (typically at the C-2 position in natural monosaccharides).

Lactose The disaccharide Galβ1-4Glc, an abundant milk sugar.

Lectin A protein (other than a glycan-specific antibody) that specifically recognizes and binds to glycans without catalyzing a modification of the glycan.

Lewis blood group antigens (e.g., Le^x, Le^y, and Le^a) A related set of glycans that carry α1-3/α1-4 fucose residues covalently linked to galactose or *N*-acetylglucosamine.

Ligand A molecule that is recognized by a specific receptor. In the case of lectins, the ligands are partly or completely glycan-based and are sometimes called counterreceptors.

Link module A protein fold that interacts specifically with hyaluronan.

Lipid A (also known as endotoxin) Lipid that contains fatty acids linked to glucosamine with a variable number of phosphate groups and 1-4 units of ketodeoxyoctulosonic acid (Kdo). *See* **Lipopolysaccharide**.

Lipid-linked oligosaccharide (LLO) An oligosaccharide linked to dolichol.

Lipid rafts Small lateral microdomains of self-associating lipids in a membrane.

Lipooligosaccharide (LOS) Similar to lipopolysaccharide but lacking the O-antigen polysaccharide side chain repeats.

Lipopolysaccharide (LPS) A bacterial polysaccharide linked to a lipid moiety containing glucosamine rather than glycerol (lipid A) that makes up the major portion of the outer leaflet of the outer membrane of Gram-negative bacteria. A major determinant of antigenic specificity, also known as heat-stable toxin or endotoxin.

L-type lectins Superfamily of glycan-binding proteins with a common feature of tertiary structure called a "jelly-roll fold."

Lysosomal storage disorders Human genetic disorders in which defects in lysosomal enzymes result in the accumulation of various glycans in the lysosomes (e.g., Tay–Sachs disease).

Lysozyme An endo-β-*N*-acetylhexosaminidase that cleaves the polysaccharide backbone of bacterial peptidoglycan.

Mannan Mannose-rich polysaccharide found in certain bacteria, fungi, and plants.

Mannose-6-phosphate receptors *See* **P-type lectins**.

Membrane-derived oligosaccharides (MDOs) Highly charged β-glucans that create an osmotic buffer in the periplasmic space of Gram-negative bacteria.

Methylation analysis A method for carbohydrate structure analysis based on the acid stability of methyl ethers and the acid lability of glycosidic linkages; used to determine the linkage positions of monosaccharide residues in an oligosaccharide chain.

Michael addition The chemical reaction in which a nucleophile attacks the β-carbon of an α,β-unsaturated carbonyl compound. The reaction is used after O-glycan β-elimination in order to attach probes to those sites.

Microarray A collection of molecules (e.g., DNA, proteins, or glycans) spatially addressed on a surface within features that have micrometer dimensions.

Microheterogeneity Structural variations in the glycan at any given glycosylation site on a protein (one source of glycoforms).

Molecular mimicry Strategy some microbial pathogens use to evade immune recognition by decorating themselves with glycans similar to those of their hosts.

Monosaccharide Carbohydrate that cannot be hydrolyzed into a simpler carbohydrate. It is the building block of oligosaccharides and polysaccharides. Simple monosaccharides are polyhydroxyaldehydes or polyhydroxyketones with three or more carbon atoms.

Mucin Large glycoprotein with a high content of serine, threonine, and proline residues and numerous O-GalNAc-linked saccharides, often occurring in clusters on the polypeptide.

Mucopolysaccharide An out-of-date term replaced by the term "glycosaminoglycan." It is still used as a group name for human disorders ("mucopolysaccharidoses") involving glycosaminoglycan accumulation due to genetic deficiency of certain lysosomal enzymes.

Mutarotation The interconversion of stereoisomers at the anomeric center of a monosaccharide.

***N*-acetyllactosamine** A disaccharide with the sequence Galβ1-4GlcNAc.

Native chemical ligation (NCL) A technique used to generate large polypeptides by condensation of smaller peptide fragments.

Neuraminidase *See* **Sialidase.**

N-Glycan (N-linked oligosaccharide, N-linked glycan) Glycan covalently linked to an asparagine residue of a polypeptide chain in the consensus sequence: -Asn-X-Ser/Thr. Unless otherwise stated, the term N-glycan is used generically in this book to denote the most common linkage region, Manβ1-4GlcNAcβ1-4GlcNAcβ1-N-Asn.

Nod factor Lipooligosaccharide produced by *Rhizobium* bacteria that stimulates nodule formation and initiates nitrogen fixation in leguminous plants.

Nonreducing terminus (nonreducing end) Outermost end of an oligosaccharide or polysaccharide chain, which is opposite to that of the reducing end.

Notch Family of cell-surface receptors that are glycosylated on EGF-like repeats. Ligands include Delta and Serrate/Jagged.

Nucleotide sugars Activated forms of monosaccharides, such as UDP-Gal, GDP-Fuc, and CMP-Sia, typically used as donor substrates by glycosyltransferases.

Nucleotide sugar transporters Membrane-bound proteins that specifically transport nucleotide sugars from the cytosol into the lumen of intracellular organelles (e.g., the Golgi).

O-GalNAc glycan *See* **O-Glycan.**

O-GlcNAcylation Dynamic modification of proteins by β-linked *N*-acetylglucosamine (a posttranslational modification similar to protein phosphorylation).

O-Glycan (O-linked oligosaccharide, O-linked glycan) A glycan glycosidically linked to the hydroxyl group of the amino acids serine, threonine, tyrosine, or hydroxylysine. Unless otherwise stated, the term O-glycan is used in this book to denote the common linkage GalNAcα1-O-Ser/Thr.

Oligosaccharide Linear or branched chain of monosaccharides attached to one another via glycosidic linkages. The number of monosaccharide units can vary; the term polysaccharide is usually reserved for large glycans with repeating units.

Optimer A computer program used to determine the appropriate glycan building blocks in one-pot automated glycan synthesis.

Peptidoglycan A bacterial polysaccharide consisting of MurNAcβ1-4GlcNAcβ1-4 repeat units, covalently cross-linked to short peptides. Also known as murein, peptidoglycan represents the major structural component of the periplasm.

Periodate oxidation A reagent that cleaves C-C bonds with vicinal hydroxyl groups (e.g., within carbohydrates) to form the two corresponding aldehydes.

Pili (fimbriae) Hair-like appendages on the surface of some bacteria that often contain adhesins.

Polyisoprenoid A lipid polymer composed of repeating units of the unsaturated five-carbon isoprene unit. *See* **Dolichol** and **Undecaprenol.**

Polymerase chain reaction (PCR) The process used to amplify DNA starting from a template DNA strand and complementary oligonucleotide primers.

Poly-*N*-acetyllactosamine Repeating units of *N*-acetyllactosamines [Galβ1-4GlcNAcβ1-3]$_n$ of variable length (sometimes called PolyLacNAc).

Polysaccharide Glycan composed of repeating monosaccharides, generally greater than ten monosaccharide units in length.

Polysialic acid A homopolymer of sialic acids abundant in the brain and fish eggs and found on certain pathogenic bacteria.

Protecting group A chemical moiety commonly used in glycan synthesis that masks hydroxyl groups in order to prevent them from reacting with other chemical reagents.

Proteoglycan Any protein with one or more covalently attached glycosaminoglycan chains.

Proteome The total collection of proteins in a cell, tissue, or organism, under specific conditions of time, space, and environment.

P-type lectins Class of lectins that recognize mannose-6-phosphate (also called M6P receptors).

Pyranose Six-membered (five carbons and one oxygen, i.e., an oxygen heterocycle) ring form of a monosaccharide; the most common form found for hexoses and pentoses. The name is based on the structural similarity to the compound "pyran."

Receptor A protein that binds to a ligand and initiates signal transmission or other cellular activity. In this book, most receptors are lectins (i.e., they recognize glycans). In microbiology, the terminology is reversed: Adhesins or agglutinins on the microbes bind to receptors, which are glycans on the host cell.

Reducing terminus (reducing end) End of a glycan that has reducing power because it is unattached to an aglycone and is thus a hemiacetal or a hemiketal. In a glycoconjugate, reducing terminus is also used as a synonym for a potential reducing terminus, referring to the end of a glycan covalently attached to the aglycone by a glycosidic bond (i.e., it would have reducing power if it were released).

Regiochemistry The region among many possible regions of a molecule that is involved in a chemical reaction. For glycosidic linkages, regiochemistry denotes the hydroxyl group of one monsaccharide that is bound to the anomeric position of the other (i.e., 1-3 vs. 1-4 linkage).

R-type lectins Superfamily of glycan-binding proteins that contain a carbohydrate-recognition domain similar to that in ricin.

Saccharide A generic term for any carbohydrate or assembly of carbohydrates, in free form or attached to another molecule, used interchangeably in this book with carbohydrate and glycan.

Saccharolipid A glycoconjugate comprising fatty acyl chains covalently attached directly to a sugar backbone (e.g., lipid A).

Selectin A C-type (Ca^{++}-dependent) lectin expressed by cells in the vasculature and bloodstream. The three known selectins are L-selectin/CD62L (expressed by most leukocytes), E-selectin/CD62E (expressed by cytokine-activated endothelial cells), and P-selectin/CD62P (expressed by activated endothelial cells and platelets).

Ser/Thr-linked oligosaccharide *See* **O-glycan**.

Sialic acids Family of acidic sugars with a nine-carbon backbone, of which the most common is *N*-acetylneuraminic acid, in vertebrates.

Sialidase Enzyme that releases sialic acid residues from a glycoconjugate. Its older name was neuraminidase, which is now used only to refer to the influenza sialidase.

Sialome Total array of sialic acid types and linkages expressed by a particular cell tissue or organism, under specified conditions of time, space, and environment.

Siglecs Sialic acid–binding proteins that are members of the I-type lectin family and have an amino-terminal V-set domain with typical conserved residues.

S layer (surface layer) A protein monolayer coating often containing covalently linked glycans and found in the cell envelope of many bacteria and Archaea.

Sphingolipids Lipids with ceramide as their core structure.

Sphingosine Long-chain amino alcohol (forms ceramide when in amide linkage with a fatty acid).

Sugar A generic term often used to refer to any carbohydrate, but most frequently to low molecular weight carbohydrates that are sweet in taste. Table sugar, sucrose, is a nonreducing disaccharide (Fruβ2-1αGlc). Oligosaccharides are sometimes called "sugar chains" and individual monosaccharides in a sugar chain are sometimes referred to as "sugar residues."

Teichoic acid A complex polymer consisting of either phosphoglycerol- or phosphoribitol-modified carbohydrates or amino acids, found on the surface of Gram-positive bacteria.

Thrombospondin repeats (TSRs) Small protein motifs (50–60 amino acids) with six conserved cysteine residues that form three disulfide bonds. They serve as sites for glycan modification.

Transcriptome The total collection of RNA transcripts in a cell, tissue, or organism, under specific conditions of time, space, and environment.

Undecaprenol (bactoprenol, C55 isoprenoid) A polyisoprenoid lipid carrier for glycan synthesis in bacteria.

Study Guide

THE FIRST EDITION OF *ESSENTIALS OF GLYCOBIOLOGY* WAS based on a course with the same name, which has been taught for many years at the University of California, San Diego. The course was aimed at providing an overview of fundamental facts, concepts, and methods in the field. This study guide was stimulated by teaching sessions in the spring of 2008, which were structured around major themes in the field, supported by selected readings from the second edition of the textbook and the original literature. The instructors (Jeff Esko, Ajit Varki, Hudson Freeze, and Pascal Gagneux) provided background information through brief lectures based on the book, with supplementary material from classic or current papers. To stimulate discussion, a series of "study questions" were devised to challenge the participants, to integrate material across the course, and to develop critical thinking skills. Some of the questions derived from material presented in this book, but most arose from our attempts to provide thought-provoking ideas that could invoke active discussion. I would like to thank my co-instructors and all of the students and fellows who participated in the course for their input into the creation of this study guide. I hope that other instructors who teach courses in glycobiology will also find it useful.

Jeffrey D. Esko
June 2008

Coauthors: Carolyn R. Bertozzi, Richard D. Cummings, Tamara L. Doering, Alan D. Elbein, Hudson H. Freeze, Pascal Gagneux, Gerald W. Hart, Robert S. Haltiwanger, Vincent Hascall, Bernard Henrissat, Ulf Lindahl, Robert J. Linhardt, Fu-Tong Liu, Rodger P. McEver, Barbara Mulloy, Victor D. Nizet, James Paulson, James M. Rini, Harry Schachter, Ronald L. Schnaar, Nathan Sharon, Pamela Stanley, Akemi Suzuki, Sam Turco, Victor D. Vacquier, Ajit Varki, and Christopher M. West.

Course participants: Heather Buschman, Adam Cadwallader, Andrew Dix, Erin Foley, Darius Ghaderi, Chris Gregg, Vered Karavani, Roger Lawrence, Max Nieuwdorp, Diego Nino, Jamie Phelps, Jessica Ricaldi, Cory Rillahan, Manuela Schuksz, Kristin Stanford, Liangwu Sun, and Xiaoxia Wang.

STUDY QUESTIONS

Chapter 1: Historical Background and Overview

1. What factors have deterred the integration of studies of the biology of glycans ("glycobiology") into conventional molecular and cellular biology?
2. Why has evolution repeatedly selected for glycans to be the dominant molecules on all cell surfaces?
3. Why are extracellular and nuclear/cytosolic glycans so different from one another?
4. What are the various factors that can affect glycan composition and structure on cell-surface and secreted molecules?

Chapter 2: Structural Basis of Glycan Diversity

1. If there are 21 amino acids and only 10 major monosaccharides in eukaryotes, why are there so many more possible combinations of monosaccharides in a hexasaccharide than amino acids in a hexapeptide?
2. Define the following terms: D- and L-stereochemistry, epimer and anomer, axial and equatorial, reducing end and nonreducing end, α- and β-linkages.
3. α-Glycosides of glucose position the aglycone group in the axial orientation, whereas α-glycosides of sialic acid position this group in the equatorial orientation. Explain this apparent discrepancy by applying the definitions of α and β anomeric stereochemistry to these two monosaccharides.
4. In nature, D-galactose can be converted to L-galactose in just two enzymatic steps. Using Fischer projections, show the chemical transformations that can accomplish this two-step interconversion.
5. Based on the atoms and functional groups within monosaccharides, describe the ways in which they might interact with proteins (e.g., electrostatic interactions, hydrogen bonding, van der Waals forces, hydrophobic interactions).

Chapter 3: Cellular Organization of Glycosylation

1. Consider the advantages and disadvantages of topologically restraining glycosylation to the ER/Golgi compartments.
2. What are the differences between physical and functional localization of glycan-modifying enzymes?
3. Describe mechanisms that determine Golgi localization of transferases.
4. Explain how localization of transferases can affect glycan composition of cell-surface and secreted molecules.
5. Propose functions for secreted soluble glycosyltransferases or sulfotransferases generated from membrane-bound enzymes.

Chapter 4: Glycosylation Precursors

1. "Essential" monosaccharides would be those that an organism cannot make de novo. Do essential monosaccharides exist for mammals?
2. Why do animal cells not require mannose, fucose, or galactose in the diet? Are there situations where an individual would have a requirement for dietary regulation of any of these sugars?
3. Why can humans not metabolize cellulose as a source of energy? How do cows and other ruminants metabolize cellulose?
4. Explain why the inactivation of genes that encode enzymes needed to generate activated monosaccharides is lethal during early development in the mouse, but expendable in cultured cells.
5. Enzyme complexes may exist in the Golgi and in the cytoplasm. Provide examples of how enzyme complexes would affect glycan synthesis.

Chapter 5: Glycosyltransferases and Glycan-processing Enzymes

1. What molecular mechanisms might determine the fidelity of a glycosyltransferase?
2. Explain what is meant by a conserved sequence motif and explain why they are often observed within a given glycosyltransferase family.
3. What features of a glycosyltransferase determine its catalytic mechanism?
4. Give an example of a peptide sequence–dependent glycosyltransferase and explain the evolutionary advantage of such specificity.
5. Why is the K_m of a glycosyltransferase for its substrates an important parameter in establishing the glycan structures produced by a cell?

Chapter 6: Biological Roles of Glycans

1. What are the different ways in which glycans can mediate or modulate biological functions?
2. Explain the difference between intrinsic and extrinsic functions of glycans.
3. What are the possible benefits for pathogens that are able to mimic host glycans?
4. Why are the biological consequences of altering glycosylation in cells and intact animals so variable?
5. Given intra- and interspecies variations in glycosylation, how can one narrow down critical functions?
6. Why does it appear that some glycans may not have specific functions when their assembly is genetically altered?

Chapter 7: A Genomic View of Glycobiology

1. What is a sequence-based classification of glycosyltransferases?

2. Describe the ways in which gene sequences predict or fail to predict functionality in transferases, hydrolases, and glycan-binding proteins.
3. Give examples of bifunctional enzymes involved in glycosylation. Suggest the driving force for the evolution of bifunctional transferases.
4. What can you learn about the way of life of an organism ("ecology") based on the relative number of glycosidase and glycosyltransferase genes in its genome?
5. How could an organism effectively augment the number of glycosidases and glycosyltransferases at its disposal?

Chapter 8: N-Glycans

1. What are some advantages to a glycoprotein in having a large number of N-glycosylation sites?
2. Consider the topology of N-glycosylation and provide possible explanations for segregating the formation of $Man_5GlcNAc_2$-Dol from the formation of $Glc_3Man_9GlcNAc_2$-Dol.
3. How is N-glycan biosynthesis different in yeast, invertebrates, plants, and mammals?
4. What is N-glycan microheterogeneity? What might be some advantages of N-glycan microheterogeneity?
5. Describe how the branching of N-glycans can regulate growth factor signaling.

Chapter 9: O-GalNAc Glycans

1. What are the factors that determine the O-GalNAc glycan composition of a cell?
2. What characteristics make a polypeptide a good acceptor for O-GalNAc glycosylation? Given the available information, can you predict sites of O-GalNAc glycosylation based on these characteristics?
3. How does the assembly of O-GalNAc glycans differ from the assembly of N-glycans?
4. Explain the most important functional features of a typical secreted mucin.
5. What are the advantages of having so many polypeptide-N-acetylgalactosaminyltransferases?

Chapter 10: Glycosphingolipids

1. What are the factors that control the composition of glycosphingolipids (GSLs) in tissues?
2. Explain the function of activator proteins in GSL degradation. What other classes of glycoconjugates would you predict to require activators?
3. How do GSLs function in the plasma membrane?
4. Compare the biosynthesis of GSLs to those of N- and O-glycans.
5. N-Glycans, O-glycans, and GSLs can have common terminal structures, but some glycan-binding proteins recognize only one class of glycans. Explain a basis for this observation.

Chapter 11: Glycosylphosphatidylinositol Anchors

1. What do GSLs and GPI anchors have in common? How do they differ?
2. Describe differences in the behavior of proteins that have transmembrane domains from those with GPI anchors.
3. Explain how GPI-anchored proteins might facilitate signal transduction across the plasma membrane.
4. Devise an assay to measure the distribution of GPI anchor intermediates across the ER membrane and the mechanism for flipping intermediates across the ER.

Chapter 12: Other Classes of ER/Golgi-derived Glycans

1. Propose a mechanism that could explain how altering the glycosylation of Notch affects the selection of different Notch ligands.
2. A number of years ago, N-linked glycosaminoglycans (GAGs) were discovered based on the susceptibility of glycans liberated by PNGase to GAG-degrading enzymes. Propose a reasonable core structure based on what you know about N-glycans and GAG biosynthesis.
3. What changes in the overall glycosylation machinery would be needed for a "normal" cell to become a factory making free glycans, such as those found in milk?
4. What approach would you take to find new glycan structures in an organism?

Chapter 13: Structures Common to Different Glycan Classes

1. Propose a function for the allelic variation observed in the ABO blood group system. If nonprimates do not express the ABO locus as a result of evolutionary loss of the gene, how would this affect your answer?
2. Hyperacute (graft) rejection (HAR) occurs after transplantation of organs from non-human donors into humans and results from an immediate reaction of circulating antiGalα1-3Gal antibodies with the transplanted tissue. Suggest ways to modify the donor or acceptor to prevent HAR.
3. Compare and contrast "LacNAc" and "LacdiNAc" units. How does the presence of these terminal disaccharides affect the addition of sialic acid and fucose?
4. Based on what you know about terminal structures on follicle-stimulating hormone and lutropin, propose several glycan-based mechanisms that could account for infertility in humans.
5. Certain strains of *Escherichia coli* bind to P blood group antigens and cause urinary tract infections. What evolutionary advantage might exist for retaining the transferases for a deleterious glycan?

Chapter 14: Sialic Acids

1. Compare and contrast the structure of sialic acids with other vertebrate monosaccharides.
2. What advantages does sialic acid diversity provide in vertebrate systems?

3. What are the unique features of the sialic acid biosynthetic pathways in comparison to those of other vertebrate monosaccharides?
4. How would you determine if a previously unstudied organism contains sialic acids?
5. Contrast the addition of α2-6-linked sialic acids to O-GalNAc glycans and N-glycans and their recognition by sialic acid–binding lectins.

Chapter 15: Hyaluronan

1. Why do small molecules diffuse readily through a high-molecular-weight hyaluronan (HA) solution such as the vitreous of the eye, whereas larger proteins do not?
2. HA solutions have unusual viscoelastic properties; for example, HA acts like a gel, yet it can function as a lubricant. How do you explain these properties in terms of the molecular structure of the chains?
3. Why are HA-binding proteins considered lectins, but proteins that bind to sulfated glycosaminoglycans are not? How do these two classes of glycan-binding proteins differ?
4. How could a cell-surface HA receptor (e.g., CD44) respond differently to HA oligosaccharides with 6-10 sugar units than to high-molecular-weight HA?
5. How would you demonstrate whether an HA chain assembles from the reducing end versus the nonreducing end?

Chapter 16: Proteoglycans and Sulfated Glycosaminoglycans

1. What are factors that can affect the fine structure of heparan sulfate in cells?
2. Overexpression of Ext2 (which is part of the heparan sulfate copolymerase complex) increases the extent of sulfation of the chain. Provide an explanation for this finding.
3. Compare and contrast the biological functions of GPI-anchored proteoglycans from those that contain transmembrane domains.
4. Proteins that bind to sulfated glycosaminoglycans are not considered lectins. Why?
5. Interactions between proteins and sulfated glycosaminoglycans are important in various physiological and pathophysiological settings. Are they specific?

Chapter 17: Unique Forms of Nucleocytoplasmic Glycosylation

1. What biochemical criteria would you require to demonstrate the attachment of a glycan to a specific nuclear or cytoplasmic protein?
2. What conventional glycosylation pathways have steps that occur on the cytoplasmic side of membranes that could be a source of nucleocytoplasmic glycans?
3. Compare and contrast the initiating glycosylation reactions on mucins, proteoglycans, Notch, glycogenin, and Skp1.
4. How would you demonstrate the presence of glycosaminoglycans in the nucleus?
5. Give examples of glycoconjugates that are initially formed in the cytoplasm but later transit to and function at the cell surface or in the extracellular space.

Chapter 18: The O-GlcNAc Modification

1. O-GlcNAc is now known to be the most common form of glycosylation in the cell. Why did it take so long for this fact to be appreciated? What was the serendipity involved in its discovery?
2. O-GlcNAc is thought to compete with phosphorylation for the same or similar sites on nuclear or cytoplasmic glycoproteins. What are the similarities and differences between O-GlcNAcylation and phosphorylation?
3. What are the mechanistic differences between O-GlcNAc glycosylation and cell-surface glycosylation?
4. How does O-GlcNAc act as a "metabolic sensor"?
5. Speculate as to how O-GlcNAc might contribute to "glucose toxicity" in diabetes.

Chapter 19: Evolution of Glycan Diversity

1. What processes could maintain glycan gene polymorphisms (i.e., structural heterogeneity) within populations?
2. What changes in sialic acid biology occurred during human evolution?
3. Can you provide examples of evolutionary trends in glycosylation?
4. Is it possible to predict glycan function by examining glycan composition across phylogeny?
5. What are the problems in using "comparative glycobiology" for determining evolutionary relationships (phylogeny)?

Chapter 20: Eubacteria and Archaea

1. Compare and contrast the pathways of glycoprotein N-glycosylation in Archaea, Bacteria, and eukaryotes.
2. All cells produce acidic glycans, but the source of the negative charge varies. What are the acidic groups on the glycans present in *E. coli*, Archaea, yeast, and animal cells?
3. Plants, bacteria, and yeast all have cell walls that provide resistance to osmotic pressure. Compare the composition and architecture of these barriers.
4. Both bacteria and animal cells utilize polyisoprenoids for the assembly of glycans. Compare and contrast these lipid intermediates.
5. Compare the structure of lipopolysaccharide to glycerolipids and gangliosides.

Chapter 21: Fungi

1. Compare the composition and structure of yeast cell walls and the envelope of Gram-negative bacteria.
2. What changes in the yeast cell wall might occur in a mutant that produces less β-glucan? What effects might an abnormal cell wall have on the shape, growth, or viability of this mutant?

3. Compare and contrast N-glycan synthesis in yeast and mammals. What is the functional significance of the differences?
4. Describe a unique feature of GPI-linked proteins in fungi. How does this process change protein localization in these organisms?
5. A pharmaceutical company has hired you to assess glycan synthesis as a target for drug development to combat a newly described and highly virulent pathogenic fungus. Describe a set of reasonable targets and some important issues you need to consider.

Chapter 22: Viridiplantae

1. Why do plants that do not express sugars present in animal cells (e.g., sialic acids) have lectins that bind to glycans containing these sugars?
2. Pectins in plants are sometimes compared to glycosaminoglycans in animals. How do they differ? How are they similar?
3. Why are recombinant mammalian glycoproteins generated in plants immunogenic?
4. Compare the structures of glycoglycerolipids in plants, lipid A in bacteria, and glycosphingolipids in animals.
5. Elicitors and Nod factors are active at very low concentration and therefore one might predict that their affinity for their signal-transducing receptors would be very high (in the pM range). Based on what you know about other glycan-binding proteins, how would such high affinity be achieved?

Chapter 23: Nematoda

1. Propose some evolutionary forces driving the large expansion of some glycosyltransferase families in *Caenorhabditis elegans* (e.g., fucosyltransferases) compared with others (e.g., mannosyltransferases).
2. Compare and contrast chondroitin proteoglycan synthesis in *C. elegans* and in vertebrates.
3. How would you go about selecting mutants of *C. elegans* defective in N-glycan formation?
4. In contrast to vertebrate systems, O-GlcNAc addition to nuclear and cytoplasmic proteins is dispensable in *C. elegans.* How do you explain this finding?
5. Given the absence of sialic acids in *C. elegans*, what might you predict about the types and specificity of glycan-binding proteins in *C. elegans*?

Chapter 24: Arthropoda

1. Compare and contrast what happens to the first *N*-acetylglucosamine residue attached to the mannosyl core of an N-glycan in *Drosophila, C. elegans*, and vertebrates.
2. *Drosophila*, like most invertebrates, has little terminal galactose on N-glycans, but instead has *N*-acetylgalactosamine in the form of LacdiNAc. What evolutionary change took place to bring about this difference?

3. Compare the core structure of glycosphingolipids in *Drosophila* with those present in *C. elegans* and vertebrates. How do the outer chains differ?
4. Transgenic expression of a β1-4galactosyltransferase substitutes for Egghead (*egh*), which is a mannosyltransferase. What does this tell you about the function of the glycans present in *Drosophila* glycosphingolipids?
5. Explain how either the overexpression or deletion of *dally*, a glypican homolog, can reduce the diffusion of a morphogen, such as dpp.

Chapter 25: Deuterostomes

1. In studying the glycoproteins that mediate sperm–egg interaction during fertilization, why is it important to use several model animals?
2. If you were an enzymologist, how would you go about studying the synthesis of fucose sulfate polymers?
3. Sulfated fucans are also extremely potent inhibitors of coagulation and inflammation in mammalian systems. Propose a mechanism for this action on the basis of the similarity of their structure to other bioactive glycans.
4. Why do some glycan-related gene knockouts in laboratory mice exhibit no obvious phenotype?
5. Compare the advantages and disadvantages of studying different glycan classes in different model organisms. If you were to discover a new glycan in humans, which model organisms would you pick for further studies?

Chapter 26: Discovery and Classification of Glycan-binding Proteins

1. If every lectin has a carbohydrate-recognition domain (CRD), is every protein with a CRD a lectin? Explain your answer with examples.
2. Why are sulfated glycosaminoglycan-binding proteins distinguished from lectins?
3. Suppose you discovered a new glycan-binding protein (GBP). How would you determine its classification?
4. Compare and contrast glycan recognition by a transferase from glycan recognition by a GBP.
5. What are the circumstances in which a transferase might be considered a GBP?

Chapter 27: Principles of Glycan Recognition

1. What determines the affinity of a glycan for a GBP?
2. Most glycan–protein interactions are low affinity, but high avidity is achieved by clustering receptors and ligands. What are the advantages and disadvantages of achieving high-affinity interactions through multivalency?
3. How does the density of glycan ligands affect binding of a GBP? Is this relevant in vivo?

4. Cholera toxin binds to the ganglioside GM1 with high affinity (K_d ~ 0.1 nM) relative to the binding of many other GBPs to their ligands (which exhibit K_ds in the range of 0.1 μM to 0.1 mM). How do you explain this observation?

5. Provide examples of GBPs that bind with relatively low affinity to highly abundant glycans and other GBPs that bind with relatively high affinity to glycans that are scarce.

Chapter 28: R-type Lectins

1. Describe the differences and similarities between *Ricinus communis* agglutinin-I and ricin.

2. For ricin and other ribosome-inactivating toxins to kill cells, they must first gain access to the cytoplasm. How does this occur? How would you exploit this mechanism to deliver cargo to different sites in a cell?

3. Explain how a cell that becomes resistant to one type of toxic lectin could become sensitive to another.

4. What are the functions of R-type lectin domains found in enzymes such as glycosyltransferases and glycosidases?

5. Describe examples of animal lectins that engage glycan ligands in both *cis* and *trans* configurations.

Chapter 29: L-type Lectins

1. Describe possible functions for L-type plant lectins present in the seeds of leguminous plants.

2. If L-type lectins are involved in defense, why does each plant produce only a very limited number of lectins?

3. Why are both plant seed lectins and glycan-binding proteins involved in protein quality control classified as L-type lectins?

4. Compare and contrast the "jelly-roll" fold in L-type lectins, the C-type lectin fold, and the link module.

5. Plant lectins are typically glycoproteins and therefore mature through the ER/Golgi secretory pathway. Propose a mechanism to prevent their interaction with other Golgi glycoproteins during their assembly and secretion.

Chapter 30: P-type Lectins

1. Why was it important to use a double-labeled substrate donor [β-^{32}P]UDP[^{3}H]GlcNAc in studies of Man-6-P recognition marker biosynthesis?

2. Compare and contrast the process of assembling the Man-6-P recognition marker on lysosomal enzymes via formation of GlcNAc-P-Man and subsequent removal of the *N*-acetylglucosamine moiety versus a mannose-specific ATP-dependent kinase.

3. The Man-6-P recognition marker assembles mainly on lysosomal enzymes by selective recognition of peptide determinants in the substrate proteins by GlcNAc-P-transferase. Describe other examples of selective modification of glycans on subsets of glycoproteins. How do the recognition determinants differ?

4. How would the number of N-glycans on a lysosomal enzyme affect its affinity for one of the Man-6-P receptors?
5. Provide an alternate route for enzyme replacement therapy in cells carrying a mutation in the cation-independent Man-6-P receptor.

Chapter 31: C-type Lectins

1. Many proteins that contain a C-type lectin domain do not bind glycans, and the ones that do are called C-type lectins. What is the difference in structure that distinguishes these two classes of proteins?
2. Why is it difficult to predict the type of glycan to which a C-type lectin will bind?
3. Some C-type lectins can form oligomers, which greatly increase the avidity of interactions with glycan ligands. Explain how oligomerization can also affect the specificity of the interaction.
4. Some C-type lectins, notably the selectins, bind with higher affinity to some glycoproteins than to others on the same cell, even though several glycoproteins may display similar glycan structures. Consider mechanisms that might confer such preferential binding.
5. Compare the interaction of P-selectin with PSGL-1 to the binding of a plant lectin to PSGL-1.

Chapter 32: I-type Lectins

1. There are now more than a dozen human Siglecs known. Why were these and other sialic-acid-binding proteins not discovered until very recently?
2. Compare the potential function of Siglecs with inhibitory motifs in their cytosolic tails with those that can recruit activatory motifs
3. Why are Siglec homologs found primarily in "higher" animals?
4. Explain the likely mechanism and driving forces for the rapid evolution of some Siglecs.
5. Why do plants and invertebrates that do not express sialic acids have sialic acid–binding proteins?

Chapter 33: Galectins

1. How do you explain the finding that galectins are not routinely found in large amounts in body fluids, even though most of them are soluble proteins and are often found extracellularly?
2. Why do changes in glycan branching pathways and sialylation have the potential to impact galectin function?
3. How do galectins achieve high-affinity binding to cell-surface glycans? How do galectins form lattices with cell-surface glycans?
4. Explain how a galectin, as an innate immune effector, might act as a receptor to fight microbial infection.

5. Galectins bind to a variety of cells and trigger various responses in different cell types. How do galectins send signals through cell-surface receptors?

Chapter 34: Microbial Lectins: Hemagglutinins, Adhesins, and Toxins

1. What kinds of cytoplasmic glycosylation events are associated with infection and pathology?
2. Compare the carbohydrate-recognition domains of bacterial and viral adhesins to those of animals and plant lectins.
3. What agents other than simple sugars could be used for anti-adhesion therapy of microbial diseases?
4. A serious problem limiting the use of antibiotics is the rapid emergence of resistant bacteria. To what extent could this also become a problem with anti-adhesion therapy?
5. Multivalent and polyvalent sugars are more powerful inhibitors of microbial lectins than simple monomeric ones. Explain the reasons for this phenomenon and discuss its applications.

Chapter 35: Proteins that Bind Sulfated Glycosaminoglycans

1. Proteins that bind to sulfated glycosaminoglycans (GAGs) are not considered lectins. Why?
2. In the 2:2:2 complex of FGF, FGF receptor, and heparan sulfate, two different heparin oligosaccharides are present. Would it be possible to form a similar complex with a single chain?
3. The extent of modification of heparin is much greater than that of heparan sulfate. How would this affect conformation and the interaction of GAG-binding proteins?
4. In GAG–protein interactions, what are the advantages and disadvantages of the shallow GAG-binding sites in the proteins and the conformational flexibility and linear structure of the GAGs?
5. In the hundreds of known GAG-binding proteins, why are there so few examples known of specific GAG and protein sequences involved in these interactions?
6. Why did evolution use sulfate groups instead of phosphate groups in GAGs?

Chapter 36: Glycans in Glycoprotein Quality Control

1. Why is N-glycosylation more likely to play a role than O-glycosylation in protein folding?
2. Describe the types of chaperones present in the ER.
3. The addition and removal of glucose residues constitutes part of the quality control system for monitoring protein folding. What is the role of mannose trimming?
4. How is the ER stress response (ERAD) coordinated with N-glycan synthesis?
5. Compare the processing of N-glycans in the ER and Golgi with degradative pathways for N-glycans in lysosomes and in the cytoplasm.
6. How might diseases of protein misfolding be managed therapeutically?

Chapter 37: Free Glycans as Signaling Molecules

1. Provide an explanation for the size dependence of signaling by hyaluronan and oligoglucan elicitors.
2. What are the advantages of using glycans derived from host organisms as signals of danger?
3. How can glycans mediate the interaction between nonglycan signals and their receptors?
4. Can you think of disadvantages of the use of glycans as pathogen-associated molecular patterns (PAMPs) by host immune systems?
5. How do signaling mechanisms dependent on heparan sulfate differ from signaling mediated by hyaluronan?

Chapter 38: Glycans in Development and Systemic Physiology

1. If glycans have roles in almost every aspect of systemic physiology, why can one sometimes alter glycan structure without observing any obvious effect?
2. Explain the evidence supporting the idea that glycans and glycan-binding proteins are involved in reproductive biology.
3. Why are genetic modifications of core glycan structures frequently lethal, whereas those of terminal glycans are frequently not?
4. Consider common features of the roles of glycans on mucins on different epithelial surfaces.

Chapter 39: Bacterial and Viral Infections

1. How can bacteria benefit by coating their surface with a polysaccharide capsule?
2. How do pathogenic bacteria initially colonize tissues?
3. Would a mouse lacking Toll-like receptor 4 be more or less susceptible to bacterial infection? What about to lipopolysaccharide-induced sepsis?
4. How do influenza and herpes simplex virus engage the host cell surface to initiate infection?
5. Can one manipulate glycans to prevent or treat microbial infection?

Chapter 40: Parasitic Infections

1. Explain the role of glycoconjugates in the high fever typically associated with the pathogenesis of malaria.
2. How do African trypanosomes avoid destruction by the immune system after inoculation by the bite of the tsetse fly?
3. What is the mechanism by which the protozoan parasite *Leishmania* attaches and eventually detaches from its sandfly vector midgut during transmission?

4. Many glycans made by the parasitic worm *Schistosoma mansoni* are highly antigenic in the infected hosts. What property of these glycans makes them so antigenic and would this offer a possibility to make a vaccine?
5. What glycosyltransferases and sugar/nucleotide sugar transporters may be unique to parasites and therefore potential sites for chemotherapeutic intervention?

Chapter 41: Genetic Disorders of Glycan Degradation

1. Predict which glycans and tissues/organs would be affected most if β-galactosidase was altered.
2. In lysosomal storage disorders, undegraded or partially degraded glycans and glycopeptides are often excreted in the urine. Propose a mechanism for how these partial degradation products escape from lysosomes and cells.
3. Provide possible explanations for the accumulation of glycopeptides with O-glycans in the urine of patients deficient in α-N-acetylgalactosaminidase.
4. How do multivesicular bodies arise and what purpose do they serve?
5. It would seem counterintuitive to use an enzyme inhibitor as a molecular chaperone to restore enzyme activity in a lysosomal storage disorder. Explain the rationale behind this therapeutic approach.

Chapter 42: Genetic Disorders of Glycosylation

1. How do you define a "glycosylation" disorder? Describe the methods used today to identify a glycosylation disorder.
2. Serum transferrin has two N-glycosylation sites and each glycan consists of biantennary sugar chains with sialic acid. What kinds of glycan patterns would you expect in patients with congenital disorders of glycosylation (CDGs)?
3. What types of cells might be especially susceptible to loss of heterozygosity or spontaneous mutations that cause glycosylation disorders?
4. Explain how "gain-of-function" mutations can cause a glycosylation disorder.
5. How would you assess the genetic and environmental contributions to a glycosylation disorder?

Chapter 43: Glycans in Acquired Human Diseases

1. What are the common underlying mechanisms for the roles of selectins in various diseases?
2. Although heparin is primarily used as an anticoagulant, its use has been proposed in connection with several other diseases. How can one drug have relevance to so many different mechanisms?
3. Give two examples where altered glycosylation has resulted in acquired blood cell diseases involving the hematopoietic stem cell. Why is it possible for somatic mutations to give rise to a phenotype?

4. Describe the common underlying molecular mechanism that causes changes in O-glycans in blood cell diseases, in IgA nephropathy, and in the altered glycosylation of cancer.

Chapter 44: Glycosylation Changes in Cancer

1. Explain why many cancer-specific markers detected by monoclonal antibodies turn out to be directed against glycan epitopes.
2. Many cancer cell types exhibit altered branching of N-glycans, excessive expression of mucins, changes in hyaluronan production and turnover, and decreased expression and sulfation of heparan sulfate. Discuss how these changes come about and how they would affect cancer growth and metastasis.
3. Sialyl-Tn expression is a prominent feature of many carcinomas. What explains the high frequency of this expression despite the fact that the enzyme responsible for its synthesis is not always upregulated?
4. Consider the potential roles of selectins and selectin ligands in cancer progression and metastasis.
5. What are the potential ways in which alterations in glycan structure could be used advantageously for diagnosing or treating cancer?

Chapter 45: Antibodies and Lectins in Glycan Analysis

1. What are the advantages and disadvantages of using monoclonal antibodies versus plant lectins for determining the presence or absence of glycans in a preparation?
2. What are important controls when using lectins or antiglycan antibodies to determine the presence or absence of a glycan in a tissue, on a cell, or in a mixture of glycans?
3. Select from the large number of available lectins a subset that would allow you to determine the relative amounts of oligomannosyl, hybrid, and complex type N-glycans in a preparation.
4. Propose methods for using a monoclonal antibody to a glycan determinant for the isolation of a mutant cell line deficient in the expression of the glycan.
5. By observing gene homology, you suspect that insects produce a novel β-glucuronidase that acts on terminal glucuronic acid residues present in insect glycans. Propose a nonradioactive method to measure the activity of this enzyme in cell extracts.

Chapter 46: Glycosylation Mutants of Cultured Cells

1. What are the advantages and disadvantages of isolating mutants in cultured cell lines compared to deriving cell lines from mutant animals or humans afflicted with glycosylation disorders?
2. Discuss the advantages and disadvantages of different schemes used to isolate mutants (i.e., selection with lectins or toxins, selection by complement-mediated lysis, screening by replica plating, and sorting by flow cytometry).
3. How might you use ldlD cells, in the presence and absence of galactose and *N*-acetylgalactosamine, to test the role(s) of glycans in biological processes?

4. Describe various types of gain-of-function glycosylation mutations. Consider mutations that create protein glycosylation sites as well as those that change the expression of glycosylation genes.
5. Propose a method to identify animal cell mutants blocked in the synthesis of O-mannose glycans.

Chapter 47: Structural Analysis of Glycans

1. Explain how the structure of a glycan may be determined by a combination of mass spectrometry, linkage analysis, and NMR spectroscopy.
2. An oligosaccharide has a molecular weight of 972, and yet its NMR spectrum is that of a single monosaccharide of α-glucose. Methylation analysis yields a single product, methylated at the C-2, C-3, and C-6 positions. What is the glycan structure?
3. What is the difference (for saccharides) between anomeric configuration and absolute configuration? What methods are best for determination of these configurations?
4. Design a protocol for isolating the five major classes of glycans from a tissue (N-linked and O-linked glycans, glycosaminoglycans, glycosphingolipids, and GPI anchors).
5. What tools might be useful to find new types of glycosylation or glycan structures in an organism?

Chapter 48: Glycomics

1. What is the "glycome" of an organism? Does it differ for individual cells in that organism?
2. What information from the genome and the proteome might be useful in predicting a cell's glycome?
3. What are some of the limitations of using glycan microarrays for determining the specificity of a glycan-binding protein?
4. What information from the genome and the proteome might be useful in predicting a cell's glycome?
5. Propose an experimental strategy to characterize different glycan subtypes that comprise the glycome. For example, how might glycolipid-associated glycans and protein-associated N- and O-glycans be physically separated and structurally characterized?

Chapter 49: Chemical and Enzymatic Synthesis of Glycans and Glycoconjugates

1. β-Glucosides are readily synthesized by exploiting protecting groups at C-2 capable of neighboring group participation. Without such protecting groups, the preferred product in most chemical glycosylation reactions is the β-glycoside. Explain this finding.
2. Why are β-mannosides so difficult to generate chemically?
3. In solid phase synthesis of glycans, glycosidic bonds are most often constructed with the glycosyl acceptor bound to the solid support and the activated glycosyl donor in solution. Why is this situation preferred to the alternative approach in which the glycosyl donor is bound to the solid support?

4. We think of glycosidases as enzymes that cleave rather than synthesize glycosidic bonds. How are the substrates and reaction conditions of glycosidases manipulated in order to convert them from degrading enzymes to synthetic enzymes?
5. Enzymatic synthesis of glycans can be far more efficient than chemical synthesis of the same structures, but production of large quantities of a glycan requires significant amounts of the required glycosyltransferases or glycosidases. Pick a source of enzymes and explain why you think it is more promising with respect to production of specific glycan products in large quantity.

Chapter 50: Chemical Tools for Inhibiting Glycosylation

1. Explain how an inhibitor of glutamine:fructose aminotransferase (GFAT) would affect glycosylation?
2. From a mechanistic point of view, how can an alkaloid that inhibits a glycosidase also block a glycosyltransferase?
3. How would you go about obtaining an inhibitor of glycans that are initiated by the addition of O-fucose to EGF repeats in Notch?
4. Identify at least two enzymes that might be targets for designing inhibitors of selectin-mediated cell adhesion and propose a strategy for obtaining selective inhibitors.
5. Propose chemical modifications to make to galactose to create an inhibitor of sialyltransferases.
6. How can an enzyme inhibitor also act as a chemical chaperone?

Chapter 51: Glycans In Biotechnology and the Pharmaceutical Industry

1. Explain the mechanism of action of influenza neuraminidase inhibitors.
2. Design a glycan-based therapeutic that acts by blocking the interaction of a naturally occurring glycan with glycan-binding proteins on an intact (or live) microbe.
3. A portion of erythropoietin (EPO) produced by CHO cells is not fully sialylated (i.e., some glycoforms have exposed galactose residues on their N-glycans). What sugars might be added to the cell culture media to increase the overall level of EPO sialylation?
4. Explain how increasing the extent of glycosylation of recombinant glycoproteins can increase their half-life in vivo.
5. Describe the potential deleterious effects of producing recombinant therapeutic proteins in cultured animal cells of nonhuman origin.

Index

Page references followed by f denote figures; page references followed by t denote tables.

D

F

G

H

I

M

N

O

P

Q

T

U